Evolutionary Cell Biology

Evolutionary Cell Biology

The Origins of Cellular Architecture

Michael Lynch

Professor and Director of the Center for Mechanisms of Evolution
Biodesign Institute, Arizona State University, Tempe, USA

OXFORD
UNIVERSITY PRESS

Great Clarendon Street, Oxford, OX2 6DP,
United Kingdom

Oxford University Press is a department of the University of Oxford.
It furthers the University's objective of excellence in research, scholarship,
and education by publishing worldwide. Oxford is a registered trade mark of
Oxford University Press in the UK and in certain other countries

Published in the United States of America by Oxford University Press
198 Madison Avenue, New York, NY 10016, United States of America

British Library Cataloguing in Publication Data

Data available

Library of Congress Control Number: 2023940185

ISBN 9780192847287

DOI: 10.1093/oso/9780192847287.001.0001

Printed in the UK by
Bell & Bain Ltd., Glasgow

Links to third-party websites are provided by Oxford in good faith and
for information only. Oxford disclaims any responsibility for the materials
contained in any third party website referenced in this work.

"Now naturally, like many of us, I have a reluctance to change too much of the old ways. But there is no virtue at all in clinging as some do to tradition merely for its own sake." From: Kazuo Ishiguro. 1988. Remains of the Day. Vintage Books.

Preface

Having released *The Origins of Genome Architecture* in 2007, I was floundering around, trying to figure out what to do next. The emerging field of evolutionary genomics was filling up with young upstarts on a bioinformatics feeding frenzy, and I had contributed about as much as I could (which may not have been much in an absolute sense).

In 2009, Mukund Thattai (Bangalore) and Michael Brenner (Harvard) invited me to join them in organizing a three-month long program on cellular evolution at the Kavli Institute of Theoretical Physics (UC, Santa Barbara). I was skeptical at first, but I had a sabbatical coming up, and so was able to spend three months at the KITP, effectively becoming the token evolutionary biologist in the program. Each week, there was a focus on a different area of cell biology, with the key all-stars giving hours-long talks. Tolerance to seemingly stupid questions, many of which I had, was encouraged and appreciated. I was struck by the similarities between the challenges that organisms face from the myriad internal constituents of cells and those resulting from communities of interacting species outside of cells (the typical focus of most evolutionary biologists). This had a significant impact on me, and I'll forever be grateful to the KITP for providing me the opportunities to participate in this and numerous subsequent programs.

I gradually realized that the logical next step in my effort to understand evolutionary mechanisms was to move up the ladder of biological complexity, from genomes into the world of cells and their multifaceted component parts. All living things are composed of cells, and every aspect of every phenotype is ultimately a product of cellular features, so there is little question that the field of evolutionary biology could profit from an integration with cell biology. However, cell biology is a huge field, expanding at a breath-taking pace, and oftentimes focused on seemingly arcane details down to the level of hydrogen-bond positions. What are the key issues that we need to understand to facilitate progress in evolutionary biology, and what details are better left behind to those working on rare human genetic disorders? One doesn't really understand a field until one gets a feel for the key unknowns, and getting to that point takes time.

Merging the fields of evolution and cell biology is made difficult because most major universities are intentionally constructed to isolate ecologists and evolutionary biologists in one subsection, and molecular, cellular, and developmental biologists in another (with microbiologists who work on prokaryotes oddly being sequestered in still another subsection). This has not been good for biology. Often a twisted sense of mistrust develops between these different camps, sometimes because of limited resources but also owing to a level of disrespect for the types of science being done. Earlier in my career, I experienced these kinds of things at three major institutions, which was not a great situation for someone wanting to be in both camps. In 2016, Arizona State University provided me with an unexpected opportunity to start a new Center for Mechanisms of Evolution, focused on the topic of evolutionary cell biology, where we've been able to establish an outstanding faculty focused on interdisciplinary research: Kerry Geiler-Samerotte, Ke Hu, Pengyao Jiang, John McCutcheon, Navish Wadwha, and Jeremy Wideman, whose company I enjoy greatly.

Evolutionary biology is at an odd crossroads. On the one hand, an enormously fruitful body of quantitative theory has been developed, providing the foundations upon which all evolutionary thought can be built. On the other hand, most of this theory is disconnected from the actual cellular features upon which all evolutionary change is built. Closing this gap will not be easy, as every aspect of cell biology has its unique nuances. However, most of the necessary tools are now available, thanks to the enormous advances in molecular and cellular biology. The main problem at the cell-biological end is the near complete lack of understanding of evolutionary mechanisms and appreciation of why this is even important. Many biologists are content to live in a pure Dawkinsian world in which every aspect of biological diversification was set down for adaptive reasons, but details at the molecular and cellular levels suggest that this vision is far from the truth. There is still a lot to be learned, and it is safe to say that evolutionary biology has not yet matured to the point at which we can shut down the field.

The main issue is how the major advances in evolutionary theory and cell biology can be integrated into a unifying field of evolutionary biology. In the following pages, my goal has been to provide a setting that bridges this gap for the pure evolutionary biologist with little cell biological background, and vice versa, with the hopes of demonstrating why a fusion of ideas from both fields is essential for a mechanistic understanding of evolutionary diversification. Some readers will disagree with what I have to say in the following pages, and I strongly encourage that, subscribing to the motto of the first organized scientific club, England's Royal Society founded in 1660, "Nullius in verba" (take no one's word for it).

In short, this book is an attempt at a new way of thinking about evolution. In doing so, I have tried to distill things down to their essence, in many cases skipping over a lot of technical details, while providing a full compendium of references into which the dedicated reader can dig more deeply. I did not intend this tome to be a textbook, but Oxford University Press would like it to become one, which explains the stylistic layouts of the chapters. Quantitative thinking is central to nearly all of today's life-science subdisciplines, and the details do matter. Yet, many biologists are wary enough about mathematics that they will fake incompetence rather than confront the details. To minimize the pain, I have sequestered a great deal of mathematical theory to foundation sections at the close of each chapter, minimizing the presentation of algebraic expressions in the main text to a level at which the reader will hopefully gain some appreciation for the utility of this way of thinking.

Writing these things down took ~13 years, and although I am neither a great evolutionary biologist nor a great cell biologist, I hope this excursion has been successful in highlighting some of the key questions residing between the two disciplines and why these are important. In "A Mathematician's Apology" (1940, Cambridge University Press), G. H. Hardy noted that "It is sometimes suggested by lawyers or politicians or businessmen, that an academic career is one sought mainly by cautious and unambitious persons who care primarily for comfort and security." That's not me. He also said, "A mathematician may still be competent enough at sixty, but it useless to expect him to have original ideas." I hope that's not me.

Acknowledgment

It will be clear from the preface that I am enormously grateful to Mukund Thattai, Michael Brenner, and the Kavli Institute of Theoretical Physics for providing me with the opportunity to learn why a merging of the fields of evolution and cell biology is essential to our understanding of where biology comes from. Throughout my career, generous support from the National Science Foundation, the National Institutes of Health, the Department of Defense, and the Moore and Simons Foundations has enabled me to maintain a lab of exceptional graduate students, technicians, and postdoctoral fellows, focused on issues at both the theoretical and empirical ends of things, in areas ranging from ecology to the genomic, molecular and cellular/developmental levels. I have learned an enormous amount from all of them and have greatly enjoyed seeing them establish their own careers. Every day, when entering my lab, I still marvel at the opportunities that I've been provided with. My colleagues in the Center for Mechanisms of Evolution have greatly expanded my thinking, and I am especially grateful to Joshua LaBaer, Director of ASU's Biodesign Institute, for providing me with the opportunity to start the CME.

Numerous people provided comments on various chapters. As the project has been going on for well over a decade, I will not remember them all, but they include: Aneil Agrawal, Charlie Baer, Lin Chao, Shelley Copley, Mark Field, Wayne Frasch, Kerry Geiler-Samerotte, Paul Higgs, Georg Hochberg, Christian Landry, John McCutcheon, Sergio Mũnoz-Gómez, Csaba Pál, Balázs Papp, Rob Phillips, Will Ratcliff, Paul Schavemaker, Noah Spencer, Boguljub Trickovic, Alex Varshavsky, Navish Wadwha, Gunter Wagner, Jeremy Wideman, and Xingbo Yang. To all of them, the many additional scientists with whom I have enjoyed interacting with, and the myriad authors whose work I have drawn from, I am very grateful for the windows that you've opened. There have been a few thorns in my side over the years (names unmentioned), but I greatly cherish those as well.

Emília Martins (life sciences professor) has kept me grounded for a long time, as have my children Erin Lynch (food and wine blogger), Adam Lynch (accountant), and Gabriel Martins (concert cellist).

Contents

List of Figures

List of Tables

List of Foundations

PART 1

Introduction

CHAPTER 1

Evolutionary Cell Biology

Evolutionary biology encompasses all aspects of life, living and dead, from the molecular level to emergent phenotypes. Like its subject matter, however, evolutionary research has followed a pattern of descent with modification. Four historical contingencies bias and jade our general understanding of evolutionary mechanisms. First, most evolutionary study focuses on aspects of the environment extrinsic to the organism – resource availability, competitors, predators, pathogens, and potential mates. As a consequence, in academic institutions, evolutionary biologists are invariably housed with ecologists and behavioral biologists to the exclusion of molecular, cell, and developmental biologists. Indeed, all too often, there is an unhealthy level of mistrust between these two different camps.

Without question, the community of organisms with which a species interacts is a major driver of evolution, and ecology is central to this field. However, the molecules and structures internal to cells also comprise a sort of community of interacting partners that channel the possible routes of evolutionary descent with modification. Would this kind of disciplinary bias have existed had Darwin spent his life staring down the barrel of a microscope, or had molecular biology existed at the dawn of evolutionary thinking?

A second pervasive problem in biology is the religious adherence to the idea that natural selection is solely responsible for every aspect of biological diversity. For example, much of the field of evolutionary ecology seeks simply to determine why particular life-history and/or behavioral strategies are optimized to particular environments, leaving no room for alternative interpretations of phenotypic variation. For traits strongly related to fitness in animals and vascular plants, such models are often quite successful (Charnov 1982, 1993; Roff 1993; Krebs and Davies 1997), leading to conceptual models invoking trade-offs such as (trait A) × (trait B) = a constant. This leaves unaddressed deeper questions as to source of the quantitative value of the constant or why such a constant even exists.

Inspired by this way of thinking and digging no deeper, many molecular biologists start with the dubious assumption that natural selection is also the only mechanism of evolution at the cellular level, often asserting that even the most blatantly deleterious features of organisms must actually have hidden favorable effects. Under this view, increased rates of mutation, translation error, and phenotypic aberrations in stressful environments (Galhardo et al. 2007; Jarosz and Lindquist 2010; Schwartz and Pan 2017), aneuploidy in gametes (Wang et al. 2017), and gene location in prokaryotes (Martincorena et al. 2012; Merrikh 2017) are all products of natural selection, maintained to somehow preserve future potential for evolvability. Some have gone so far as to proclaim that virtually any nucleotide that is at least occasionally transcribed or bound to a protein must be maintained by selection (ENCODE Project Consortium 2012).

Such arguments are inconsistent with substantial theory and empirical work suggesting that many aspects of gene and genome evolution are consequences of the limitations of natural selection (Kimura 1983; Lynch 2007). Remaining, however, is the key question as to the level of biological organization above which selection can be safely assumed to be the only driving force of evolution. Does effectively neutral evolution somehow cease to occur at the level of cellular features or at a higher level of emergent properties in multicellular species?

Evolutionary Cell Biology. Michael Lynch, Oxford University Press. © Michael Lynch (2024). DOI: 10.1093/oso/9780192847287.003.0001

Third, although evolution is a process of genetic change, and evolutionary biology has long been endowed with a powerful theoretical framework grounded in genetics, a large fraction of what passes as evolutionary research is completely removed from genetics. For example, the optimization hypotheses in evolutionary ecology noted above focus almost exclusively on verbal or semi-quantitative arguments devoid of genetic details. The field of evolutionary developmental biology is often proudly defiant of any association with conventional genetic understanding.

Finally, the vast majority of research in evolutionary biology is focused on multicellular animals and land plants. It is easy to become enamoured of biodiversity that is readily visualized on a day-to-day basis. It is also easier to work with organisms that can be seen without the aid of a microscope. Nonetheless, animals and vascular plants are the odd-balls of evolutionary biology – interesting in their own right, and containing the only species capable of writing and rejecting a manuscript, but also constituting only a tiny fraction of the phylogenetic Tree of Life and of the planetary census of individuals.

We now have well-established fields of molecular and genome evolution, and some aspects of evolutionary developmental biology are being integrated with modern evolutionary theory. Yet, despite the extraordinary accomplishments in the field of cell

biology, there is as yet no comprehensive field of evolutionary cell biology. Attempts to decipher the Tree of Life, most of which is unicellular, are common, and many aspects of molecular evolution are focused on cell biological issues. However, a general evolutionary framework for explaining the diversity of cell biological structures and processes remains to be developed.

It need not have been this way. Early in the past century, for example, there was an enormous amount of comparative work done on diverse protists. But the late 1900s witnessed a near cessation of this kind of work, as cell biology became increasingly inward looking and medically oriented, focusing on a few model laboratory systems devoid of variation, e.g., the bacterium *Escherichia coli*, the yeast *Saccharomyces cerevisiae*, and mammalian cell cultures. Embracing the many impressive results from these systems, but going well beyond them, the goal here is to plant the seeds for a science of evolutionary cell biology.

The Dominance of Unicellular Life

Taking a phylogenetic perspective, it can be seen that the major foci of life-science research – animals and land plants – comprise only a small fraction of the Tree of Life (Figure 1.1). Most of global diversity at the DNA level resides in prokaryotes, and this is even more true if one further considers the

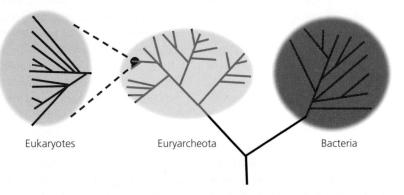

Eukaryotes　　　　Euryarcheota　　　　Bacteria

Figure 1.1 A broad overview of the Tree of Life. The overall structure is presented in an idealized fashion, some details of which are covered in Chapter 3. The main points are that: 1) the overall tree is primarily prokaryotic, with eukaryotes being derived from one small lineage (highlighted by the small blue ball) within the major domain called the Euryarcheota; and 2) the land-plant (green) and animal (red) lineages comprise only a small fraction of the diversity within eukaryotes, as shown by the expansion of the total eukaryotic lineage to the left (dashed lines).

variation of gene functions, as prokaryotes harbor much more diversity in metabolic pathways than do eukaryotes. Even restricting attention to eukaryotes, the vast majority of phylogenetic diversity resides within lineages consisting entirely of unicellular species.

The conclusion that the vast majority of life on Earth is in the provenance of unicellular organisms is retained if the focus is shifted to total numbers of individuals. Achieving accurate census counts in various groups of organisms is made difficult by the uneven sampling of different global ecosystems, the absence of surveys for many phylogenetic groups, and seasonal fluctuations of population sizes in microbes. However, crude order-of-magnitude estimates are possible. For example, the number of viral particles in the open oceans is estimated to be $\simeq 10^{30}$ (Suttle 2005), and even if there were twice as many viruses on land and in freshwater (unlikely), this would not increase the global estimate beyond $\simeq 10^{31}$. The estimated global number of prokaryotic cells is also $\simeq 10^{30}$ (Flemming and Wuertz 2019), and this sums to a total amount of global biomass that exceeds that of all animals by a factor of ~ 40 (Whitman et al. 1998; Kallmeyer et al. 2012; Bar-On et al. 2018). There may be as many as 10^{12} species of prokaryotes (Locey and Lennon 2016), although an alternative upper-bound estimate is $\sim 10^6$ (Amann and Rosselló-Móra 2016). Taking the logarithmic mean of these two estimates, 10^9, implies an average of $\sim 10^{21}$ individuals per prokaryotic species (although substantial variation in this number must exist among taxa).

The total number of unicellular heterotrophic eukaryotic cells is $\sim 0.1\%$ of that for bacteria in the marine environment (Pernice et al. 2015), and drawing from average estimates in Whitman et al. (1998) and Landenmark et al. (2015), the ratio in terrestrial soils is $\simeq 0.5\%$. This suggests that the total number of unicellular eukaryotic cells on Earth exceeds 10^{27} (Bar-On et al. 2018), as the previous estimates exclude fungi and photosynthetic species. Thus, assuming the average volume of a eukaryotic cell is $\simeq 1000\times$ that of a prokaryote (Chapter 8), the global biomass of unicellular eukaryotes likely exceeds that of prokaryotes. Of the estimated 10^7 eukaryotic species on earth (potentially just 1% of

the number for prokaryotes), it has been suggested that $\sim 90\%$ are animals, 6% fungi, 3% plants, and the small remainder protists (Mora et al. 2011). The latter could, however, be vastly underestimated, given the relative lack of attention to the systematics of such groups (Wideman et al. 2020). Assuming 10^6 unicellular eukaryotic species would imply an average $\simeq 10^{21}$ individuals per species, the same order of magnitude as in prokaryotes.

Although crude, these estimates for unicellular organisms dwarf the numbers of individual land plants and metazoans. For example, there are $\sim 10^{13}$ trees on Earth, and $\sim 75,000$ tree species (Crowther et al. 2015; Beech et al. 2017; Gatti et al. 2022), implying an average of $\sim 10^8$ individuals per tree species. Two of the most abundant groups of invertebrates on Earth are the ants, estimated to comprise $\sim 10^{16}$ individuals distributed over $\sim 10^4$ species (Hölldobler and Wilson 1990) and Antarctic krill with $\sim 10^{15}$ individuals in a single species (Atkinson et al. 2008). Nematodes, perhaps the most numerically abundant animal phylum, comprise $\sim 10^{20}$ individuals globally, distributed over some 10^6 species (Kiontke and Fitch 2013; van den Hoogen et al. 2019). As these observations imply that most animals have global population sizes $\ll 10^{15}$ (with a suggested mode of $\sim 10^{10}$; Buffalo 2021), assuming 10^7 animal species suggests that the total number of animals on Earth is $< 10^{20}$, several orders of magnitude below the numbers for both prokaryotes and unicellular eukaryotes.

An upper-bound estimate to the total number of vertebrate individuals on earth is $\sim 10^{16}$, distributed over $\sim 10^5$ species (mostly fish), implying an average of $\sim 10^{11}$ individuals/species (Bar-On et al. 2018). For birds, the average number is $\sim 10^7$ individuals (Callaghan et al. 2021). Notably, the average human harbors a microbiome of $\sim 10^{13}$ bacterial cells, which exceeds the total number of humans that have ever lived (Sender et al. 2016).

What is Evolutionary Cell Biology?

As all organismal features derive from cell-level processes, an ultimate understanding of the mechanisms of evolution cannot be complete without an appreciation for how cellular features emerge on an evolutionary time scale. As nicely summed up in the timeless quote of E. B. Wilson (1925), 'The key to every biological problem must finally be sought in the cell, for every living organism is, or at some time has been, a cell.'

Evolutionary cell biology is the fusion of cell biology with evolutionary thinking, informed by the integration of the great engines of theoretical and quantitative biology – biochemistry, biophysics, and population genetics (Lynch et al. 2014). Despite its centrality, especially for the multitude of species for which the individual cell is also the organism, this intrinsically interdisciplinary field is embryonic in almost every way. For example, evolutionary biologists have only rarely incorporated the concepts of biochemistry and biophysics into their thinking, despite some striking similarities between the underlying theoretical frameworks of statistical physics and population genetics (Sella and Hirsh 2005; Lässig 2007; Barton and de Vladar 2009; Zhang et al. 2012). Likewise, although cell biologists commonly remark on the exquisite design of the traits being studied, they almost never consider the feasibility of the evolutionary paths by which such features are imagined to have emerged.

Understanding evolution at the cellular level requires consideration of three major aspects of the environment, each of which subdivides into at least three other domains (Figure 1.2). First, as mentioned above, the classical intellectual domain of evolutionary biology is ecology, where the usual focus is on challenges imposed by factors outside of the organism. The central issues here include the procurement of resources, the avoidance of predators and pathogens, the acquisition of mates, and various aspects of mutualism and cooperation.

Second, we must consider the cellular environment, which imposes historical contingencies, biophysical constraints, and molecular stochasticity. All cells are endowed with an array of features fundamentally unmodified since the last universal common ancestor of life. These include: the use of double-stranded DNA as genomic material; the expression of genes through intermediate transcriptional products (made of RNA), and in the case of proteins followed by translation at ribosomes; the use of lipid membranes; and the deployment of highly conserved mechanisms for ATP production.

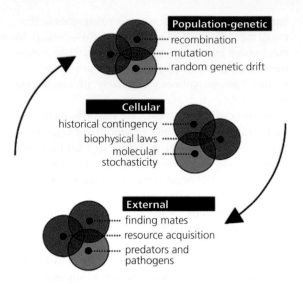

Figure 1.2 Summary of the major environmental components influencing the tempo and mode of evolution.

The strengths of bonds associated with C, H, N, O, and P atoms, life's favored elements, dictate the stability of intermolecular interactions. Small cell size imposes an upper limit on the number of molecules that can be housed. One can imagine other possible forms of cellular organization, but on planet Earth these are the indelible backgrounds upon which all other cellular modifications must develop.

Finally, there is the population-genetic environment. Owing to the imperfections in all molecular interactions, DNA replication is naturally error-prone. Although this ensures the recurrent input of the genetic variation upon which all evolutionary change ultimately depends, the privilege of evolutionary potential comes at a cost – most mutations are deleterious. Recombination assorts variation within and among chromosomes, further generating genetic diversity, which promotes some kinds of evolutionary change while inhibiting others. Random genetic drift, a consequence of finite numbers of individuals within populations and genes being linked on chromosomes, creates noise in all evolutionary processes, more so in smaller populations.

The joint operation of these three dimensions of the population-genetic environment defines the limits to what natural selection can and cannot accomplish in various phylogenetic lineages (Chapters 4, 5, and 6), thereby dictating the mechanisms and directions by which evolution proceeds at the cellular level.

Although cell biology has not been the traditional domain of evolutionary biology, it offers powerful opportunities for identifying the explicit biological connections between genotypes, phenotypes, and fitness, essential to the development of a mature field of evolutionary biology. With these matters in mind, the focus here is on the degree to which selection, effectively neutral processes, historical contingencies, and/or constraints at the biochemical and biophysical levels jointly influence patterns of evolutionary diversification. This way of thinking may ultimately find use in the applied fields of agriculture, medicine, environmental science, and synthetic biology.

The Completeness of Evolutionary Theory

Before proceeding, some comments on the use of theory in biology are in order. Without an explanatory framework, science is reduced to a fact-collecting enterprise. Of course, the emergence of facts from consistent observations is central to science, but theory provides a mechanistic explanation of the facts. A theoretical framework can motivate the development of predictions in areas where observations have not been made previously. Ideally, such a reach is not simply based on statistical extrapolation, but on arguments from first principles. In particular, mathematical theory allows the construction of logical arguments from well-defined assumptions, whereas verbal theorizing can easily go awry in the analysis of complex systems.

Fortunately, evolutionary biology has a well-established framework of quantitative principles from which to draw. In one of the most important scientific papers ever written, Fisher (1918) convincingly explained a simple connection between the Mendelian inheritance of segregating genetic factors and the near-continuous range of phenotypic variation for complex traits within populations, closing a long-standing controversy about the material basis of evolution (Provine 1971). In this same paper, Fisher established one of the primary pillars upon which modern statistics relies, the

analysis of variance. Emanating from these roots, the century-old field of population genetics now forms the foundation for all of evolutionary theory (Walsh and Lynch 2018). Most of the principles for integrating selection, mutation, and random genetic drift were laid down in the first half of the twentieth century, while key findings with respect to recombination were generated over the latter half. With the further integration of diffusion theory and statistical aspects of gene genealogies now commonplace, the field of theoretical population genetics is as well grounded as any other area of quantitative biology.

Notably, the establishment of the primary roots of evolutionary theory substantially preceded any knowledge of the details of the genetic material. Starting in the 1950s, dramatic findings emerged in the field of molecular genetics, including the discovery of DNA as the ultimate genetic material; the basic structure of genes and their component parts; the processes of transcription and translation; and the molecular mechanisms of recombination. Yet, none of these discoveries led to any alteration in the basic structure of evolutionary theory. The discovery of mitochondrial DNA did not alter our fundamental understanding of maternal effects, and the discovery of transposable elements did not alter our appreciation of the mutational process. Such observations simply provided a deeper molecular explanation of modes of production of phenotypic variation. This robustness of evolutionary theory in the face of revolutionary changes in our understanding of genetics at the molecular level speaks volumes. Important specific applications may remain to be developed, but the theoretical foundations of evolutionary genetics provide a solid framework for defining the conditions under which various evolutionary scenarios are and are not possible.

This optimistic viewpoint is periodically confronted with claims that evolutionary biology is in a phase of turmoil. However, the bearers of such messages seldom offer a solution to the field's imagined shortcomings, and without exception, these episodes have gone badly. Most notable are Goldschmidt's (1940) argument that large changes in evolution are products of macromutations with coordinated developmental effects and Lysenko's rejection of Mendelian genetics in favor of the inheritance of acquired characteristics.

Unlike the laws of physics, biology is subject to historical contingencies, and for virtually every set of general observations one can find some kind of exception. Discoverers of such exceptions sometimes claim that their observations are sufficient to dismantle previous theoretical frameworks for broadscale patterns. However, a deeper look almost always reveals underlying explanations for oddities that are fully compatible with the rules of life.

One of the more recent promotional exercises involves a clamour for an 'extended evolutionary synthesis' or EES (Gerhart and Kirschner 1997; Pigliucci and Müller 2010; Goldenfeld and Woese 2011; Shapiro 2011; Laland et al. 2014, 2015). Asserting that population genetics provides an antiquated and inadequate framework for evolution, the claim is that 'the number of biologists calling for change in how evolution is conceptualized is growing rapidly', and that there is a current 'struggle for the very soul of the discipline'. The nature of this discourse is reminiscent of the distant 'bean-bag genetics' diatribe of Mayr (1959, 1963), which was promptly disembowelled by Haldane (1964). No glaring errors in contemporary evolutionary theory have been correctly pointed out by the EESers, there exists little evidence of familiarity with current theory, and no novel predictions have been offered (Stoltzfus 2017; Welch 2017). There is just a warning that once qualified theoreticians come on board, the revolution will begin.

A particularly extreme claim is that the discovery of various epigenetic effects amounts to a game-changer in evolutionary biology, imposing the need to revamp our general understanding of inheritance and its evolutionary implications (Jablonka and Lamb 2005; Caporale 2006; Danchin et al. 2011; Shapiro 2011). Highlighted phenomena include base modifications on DNA, histone modifications on nucleosomes, and mechanisms of gene regulation by small RNAs, all of which can in principle have transient transgenerational effects without imposing permanent changes at the level of genomic DNA. Advocates of epigenetic inheritance as an enhancer of evolvability commonly argue that such phenomena promote beneficial phenotypic responses to environmental induction, which then allows for an acceleration in the rate of adaptive phenotypic evolution, in effect resurrecting the concept of the inheritance of acquired characteristics.

The logic underlying the entire subject has been masterfully dismantled by Charlesworth et al. (2017), and just two points are made here. First, to appreciate the implausibility of a long-term contribution of non-genetic effects to phenotypic evolution, one need only recall the repeated failure of inbred (totally homozygous) lines to respond to persistent strong selection. Many such experiments dating back to the beginning of the twentieth century provide formal support for the necessity of genetic variation for evolutionary progress (Lynch and Walsh 1998). Second, countering the claim that evolutionary theory is incapable of addressing the matter of epigenetic inheritance, one need only point to models for the inheritance of environmental maternal effects developed well before the discovery of the molecular basis of any epigenetic effects (Table 1.1). Existing theory readily demonstrates that variance in maternal effects can contribute to the response to selection, but unless such effects reside at the DNA level, the response is bounded, owing to the fact that transgenerational effects are progressively diluted out. Moreover, if epigenetic effects are sufficiently stochastic, they will reduce, rather than enhance, the response to selection, owing to the reduction in the correspondence between genotype and phenotype.

Although a persistent claim of the EESers is that the environmental induction of a trait in a novel situation can enhance the exposure of the trait to selection, thereby magnifying the response to selection, this is by no means a novel insight. Such effects are central to the concept of genotype × environment interaction, the theory of which dates back decades (Lynch and Walsh 1998). Indeed, breeders have long exploited this concept to determine the optimum environmental setting in which to select for particular phenotypes (Walsh and Lynch 2018). Thus, the idea that evolutionary theory needs to be remodelled to account for phenotypic plasticity is without merit.

The most remarkable EESer claim is that the key flaw of contemporary evolutionary theory is the assumption that change in allele frequencies is a necessary component of the response to selection (Laland et al. 2014). Their counter view is that 'the direction of evolution does not depend on selection alone, and need not start with mutation'. Whereas it has long been appreciated that evolution can (and sometimes does) occur in the absence of selection (for example, by random genetic drift of neutral traits), we await an explanation as to how any form of evolution (aside from cultural) can occur in the absence of genetic variation. Technically speaking, evolution can occur in the absence of allele-frequency change, but only via changes in the form of allelic associations across loci (e.g., via linkage disequilibrium, which necessarily implies transient genotype-frequency change).

Far from providing a weak and/or incomplete caricature of evolving genetic systems, population- and quantitative-genetic theory has generated powerful, general, and sometimes unexpected mechanistic explanations for trait variation and phenotypic evolution, several of which are noted in Table 1.1. Few of these issues would have ever been resolved with simplistic verbal arguments. Indeed, it was Fisher's paper (1918) that rescued the previously verbal debate over evolutionary mechanisms from the high seas of obfuscation. Inspired by quantitative thinking derived from first principles in genetics, most subfields in evolutionary biology were rapidly transformed by the emergence of population-genetic theory. Developmental biology is somewhat of an exception, remaining in many respects in a pre-population-genetics mode of confusion, a condition that evolutionary cell biology need not emulate.

Although the preceding railing on the EES movement may be offensive to some and/or pandering to trivia to others, the implication that a century's worth of theoreticians has been woefully misled

Table 1.1 A few key areas in evolutionary biology where theory has enhanced our understanding of the mechanistic basis of trait variation, and in doing so has provided novel predictions. Many of these issues are covered in depth in subsequent chapters, although this list is by no means complete. LW denotes Lynch and Walsh (1998).

Topic	References
Quantitative-trait variation:	
Phenotypic resemblance between relatives, and its scaling with the degree of relationship.	Fisher 1918; Kempthorne 1954; LW Chapter 7
Inbreeding depression, and how this scales with parental relatedness.	Crow 1948; LW Chapter 10
Quasi-inheritance of familial (including maternal) effects, and transient selection response.	Willham 1963; Falconer 1965; LW Chapter 23
Expression of all-or-none traits as a function of underlying determinants.	Wright 1934a,b; LW Chapter 25
Pleiotropy and genetic correlation between traits.	Mode and Robinson 1959; LW Chapter 21
Long-term patterns of evolution:	
Sudden (saltational) transitions from one discrete character state to another.	Lande 1978
Rates/patterns of evolution in the fossil record.	Charlesworth et al. 1982; Charlesworth 1984a,b
Rapid evolution across adaptive valleys by stochastic tunnelling.	Lynch 2010; Weissman et al. 2010
Mutation bias and the inability of a mean phenotype to attain an optimal state.	Lynch 2013; Lynch and Hagner 2014
Spatial variation in genotypic values in the absence of underlying ecological variation.	Higgins and Lynch 2001
Genome evolution:	
The fate of duplicate genes.	Force et al. 1999; Lynch and Force 2000
Conditions for the spread of mobile elements.	Charlesworth and Charlesworth 1983; Charlesworth and Langley 1986
Evolution of codon bias.	Bulmer 1991
Evolution of transcription-factor binding sites.	Lynch and Hagner 2014
The illusion of evolutionary robustness.	Frank 2007; Lynch 2012
Evolution of the genetic machinery:	
Evolution of the mutation rate.	Lynch 2011; Lynch et al. 2023
Evolutionary consequences of sexual reproduction.	Kondrashov 1988; Charlesworth 1990; Otto and Barton 2001
Evolutionary deterioration of sex chromosomes.	Charlesworth and Charlesworth 2000

is a misrepresentation of the facts, and as Darwin (1871) pointed out, a reliance on false facts is 'highly injurious to the progress of science'. As outlined in Table 1.1 and expanded upon in Chapters 4–6, evolutionary theory developed over the twentieth century has made predictions that are consistent with a wide range of empirical observations. This being said, because evolution is a stochastic process, no theoretical framework can ever be expected to predict the exact trajectories of evolution at the molecular, cellular, or developmental levels in any specific lineage. As Haldane (1964) pointed out, if population genetics could make such specific predictions, it would not be a branch of biology – it would be the entirety of biology.

Evolution Via Nonadaptive Pathways

Darwin's (1859) and Wallace's (1870) grand views about selection as a natural force for the emergence of adaptive change marked a watershed moment in the history of biology. Their narrative has been so convincing that most who now think about evolution simply view all aspects of biology, current and historical, as necessary products of natural selection. However, whereas natural selection is one of the most powerful forces in the biological world, it is not all powerful. As the following pages repeatedly show, the genetic paths open to exploitation by selection are strongly influenced by another pervasive force – the noise in the evolutionary process imposed by random genetic drift.

Evolutionary stochasticity is an inevitable consequence of finite numbers of individuals within populations and the physical linkage of different nucleotide sites on chromosomes. If the power of selection is weak relative to that of drift, as is often the case at the molecular level, evolution will proceed in an effectively neutral manner (Chapter 4). Biased mutation pressure can also modify evolutionary trajectories, but even non-biased mutation can strongly influence the distribution of mean phenotypes if mutation is sufficiently strong relative to the efficiency of selection (Chapter 5).

To understand the degree to which natural selection molds the features of populations, it is essential to know what to expect in the absence of selection. For this reason, neutral models have been repeatedly exploited in evolutionary analyses. The three nonadaptive mechanisms of evolution – random genetic drift, mutation, and recombination – are the sole evolutionary mechanisms under such models. The resultant formulations then provide null hypotheses for testing for natural selection. Neutral models are relatively easy to develop for DNA-level features, as mutation can be explicitly defined in terms of the twelve possible nucleotide substitutions, and such constructs are fundamental to most studies in molecular evolution (Kimura 1983; Jensen et al. 2019). Although the construction of neutral models becomes more challenging in the case of complex cellular/organismal traits, where the phenotypic features of mutations can be more difficult to define (Lynch and Walsh 1998), this is not justification for ignoring the matter. Indeed, neutral models have been particularly useful in attempts to understand long-term phenotypic divergence recorded in the fossil record, where dramatic changes that might seem only achievable by selection are found to be not so impressive when evaluated in the proper context of drift and mutation (Lande 1976; Charlesworth 1984a; Lynch 1990).

Some have suggested that so much evidence for selection has emerged that we should abandon the use of neutral theory (Pigliucci and Kaplan 2000; Hahn 2008; Kern and Hahn 2018), in one case going so far as to argue that 'the implications of our continued use of neutral models are dire', and 'can positively mislead researchers and skew our understanding of nature'. No one argues that selection is unimportant, but the proposition of a selection theory as a null model for hypothesis testing presents a logical challenge. One can concoct a selection argument for essentially any observed pattern, rendering such a restrictive view unfalsifiable. If one form of selection does not adequately fit the data, then one can invoke another, and failing that, still another, never abandoning the pan-selection view. In contrast, when properly constructed, neutral models make very explicit predictions, rescuing arguments for the role of natural selection from an endless loop of qualitative hand-waving. By offering a formal means for testing for the influence of selection, the measurement of deviations between observations and neutral expectations yields a deeper and more defensible understanding of evolutionary processes.

An additional problem with criticisms of neutral theory is their frequent reliance on incorrect biological assumptions. For example, Lewontin (1974) invoked the fact that standing variation in natural populations is only weakly associated with effective population size (N_e) as a dramatic violation of the neutral theory, as standing levels of variation at silent sites in populations should scale with $N_e u$, where u is the mutation rate per nucleotide site (Chapter 4). Although this argument continues to be made (Hahn 2008; Buffalo 2021), the postulated pattern ignores the fact (unknown at Lewontin's time) that mutation rates evolve to be inversely correlated with N_e, rendering the product $N_e u$ relatively constant (Lynch et al. 2023; Chapter 4). Thus, Lewontin's observation is not so paradoxical after all.

Like the call for an extended evolutionary synthesis, the call for a selection theory of evolution has not resulted in any theoretical upheaval. No offering of a novel theory of selection has been presented, and none is likely to emerge for the very simple reason that we already have such a theory. From the very beginning, population- and quantitative-genetic theory has fully embraced selection as a central force in evolution, with the understanding that what selection can accomplish is modulated by the relative power of the nonadaptive forces of evolution (Walsh and Lynch 2018).

The preceding comments have been offered primarily for the benefit of outsiders with only a peripheral understanding of what might appear to be substantive controversies. Conflict and cooperation are the engines that keep science running. Conflict engineered and incessantly repeated with no evidence sometimes has other motivations (Gupta et al. 2017).

The Grand Challenges

Comparative biology has made substantial contributions to evolutionary biology, telling us what has evolved, and hence, revealing the facts that evolutionary theory needs to explain. In some cases, where there is a decent fossil record, comparative biology has also provided insight into rates of phenotypic evolution. Where there is compelling phylogenetic information, ancestral phenotypic states can sometimes be predicted and, in the case of simple molecular features, even resurrected and evaluated (Hochberg and Thornton 2017). Good comparative biology can be done in the complete absence of knowledge of evolution. However, when disconnected from the mechanisms driving genetic change, comparative biology is a far cry from evolutionary biology.

The challenges for evolutionary cell biology are substantial. Owing to cell biology's focus on just a few model organisms, there is no expansive field of comparative cell biology. As a consequence, the range of existing variation for cellular traits is often unclear, leaving even the question of what needs to be explained unsettled. Even so, evolution is still part of the mindset of many cell biologists who focus on a single species throughout their careers. This can be seen from the final paragraphs of numerous papers in cell biological journals where adaptive hypotheses are commonly offered for the phenomenon observed.

The ultimate goal of any area of science is to provide compelling, mechanism-based answers to all of the central questions in the field, bringing things to the point at which all future observations have a pre-existing explanation. No scientific area has yet reached that point, and evolutionary biology might not be the first. What follows is a brief list of some of the major challenges that need to be overcome for evolutionary cell biology to achieve a reasonably mature state. Ways in which their solution might be achieved are explored in detail in the following chapters.

The origin of life

More than three billion years ago, cellular biochemistry became established in such a way as to provide all of the necessities for evolution: metabolism, growth, replication, and variation. Deciphering the ways in which this happened would go a long way towards explaining the seemingly idiosyncratic features shared by all of life. Unfortunately, owing to the absence of fossils for the simplest of cells, we will probably never attain a precise understanding of the first steps by which the ancestor of all life emerged. We may never know whether competing forms of life initially coexisted and/or fused to form the most distant ancestor of us all. However, hypotheses focused on potentially plausible scenarios, combined with research in biochemistry, can help narrow down the alternative possibilities, in turn yielding useful predictions as to where life might have originated independently elsewhere in the universe (Chapter 2), and offering up possibilities for designing synthetic life forms. Fortunately, a lack of clarity on these matters does not bear in any significant way on our ability to understand the mechanisms of evolution in contemporary organisms.

The roots of organismal complexity

Although it is commonly asserted that added layers of cellular complexity make for more robust and evolutionary successful organisms, evidence for this is entirely lacking. If complexity is entirely driven by natural selection, then why has only one lineage (eukaryotes) evolved complex internal cell structure? And why has the apex of biological complexity, multicellularity at the level found in land plants and animals, evolved only twice (or three or four times if one wishes to include kelps and fungi)? One might argue that there is something fundamentally lacking in prokaryotes that prevents such evolution, yet as mentioned above, despite any imagined deficiencies, microbes comprise much of the Earth's biomass. Indeed, from the standpoint of metabolism, prokaryotes are the cradles of diversity, whereas eukaryotes are relatively bland. Such disparities reveal the intrinsic biases that arise if evolutionary thinking is confined to visually perceived morphological differences.

As noted earlier, there is substantial support for the idea that much of evolution at the genomic level has proceeded by effectively neutral processes, guided largely by the forces of mutation and random genetic drift. Moving to higher and higher levels of organization, e.g., protein structure, protein-complex architecture, cellular features, and the emergent properties of multicellular organisms, one might expect that the likelihood of neutral evolution would be dramatically diminished

(Zhang 2018). The following chapters show, however, that because there are often many ways to achieve the same phenotype, the paths open to neutral evolution at the cellular level may often be more plentiful than at lower levels of organization. This means that, rather than being products of adaptive promotion, various aspects of cellular complexity may have arisen via nonadaptive pathways of evolution.

The types of mutations that arise in any particular interval are a matter of chance, not summoned by selective demand, and the molecular spectrum of mutations strongly depends on organismal background. Thus, owing to the combined forces of mutation pressure and genetic drift, biological structures and functions need not evolve in directions that would be most economical from an engineering perspective, and as we further discuss below, some are quite arcane. As a modern analogy to the passive emergence of complexity, consider how software companies modify their computer code over time – not by full-scale rewriting of the code, but by inserting patches for old problems. This slow accrual can lead to a complexity ratchet, whereby a general function is retained despite an irreversible series of cumulative changes at the component level.

Molecular stochasticity

Messenger RNAs are often present in fewer than ten copies per cell, sometimes with a mean less than one for specific genes, especially in small-celled species. Proteins have longer half-lives and tend to be more abundant, but still are often present as only hundreds of copies per gene per cell. Transcription factors are among the rarest of proteins, leading to questions as to how they reliably find their DNA targets. Collectively, these features and more (including asymmetries in cell division; Chapters 7, 9, and 21) lead to substantial stochasticity in cellular composition, even among cells with identical genotypes inhabiting homogenous environments.

Natural selection operates on phenotypic variance, and is more efficient when most of the variance is due to genetic differences. Thus, stochastic cellular noise must impose a speed limit on the rate of evolution, as it blurs the reliability of the phenotype as an indicator of the genotype. How does this problem with phenotypic variation vary across cellular life forms? On the one hand, small cells might be expected to exhibit more phenotypic variation associated with internal and external environmental factors (Chapters 7, 9, and 21). On the other hand, large populations of small cells may harbor more genetic variation that, combined with short cell-division times, might enhance the rate of evolution (Chapters 4–6).

Molecular complexes

The rules of life are such that each messenger RNA almost always encodes for one amino-acid chain. However, the majority of proteins organize into higher-order consortiums, e.g., dimers and tetramers. Such complexes are frequently comprised of subunits derived from the same genetic locus (homomers), often with the multimer having no different function than the subunit components. Heteromers consisting of non-identical components also exist, many of which are higher-order complexes that are more than the sum of their parts, e.g., the ribosome and the nuclear-pore complex. The number of subunits underlying the same protein can vary across species, but not always in ways that reflect organismal complexity. This weak connection is very unlike the situation in genome evolution, where genome architecture becomes enormously complex in large multicellular species (Lynch 2007). Observations on the phylogenetic diversity of molecular complexes raise the suspicion that natural selection is unlikely to be the sole driving force (Chapter 13).

Cellular networks

Very few of the molecular constituents of cells operate alone. Examples of molecular networks include the cell cycle, transport systems, circadian clocks, and pathways in metabolism, transcription regulation, and signal transduction (Chapters 10, 18, 15, 19, 21, 22). The structural features of cellular pathways often border on the baroque, e.g., larger numbers of steps than seemingly necessary and linear chains of enhancing vs. suppressing steps. Given that each component added to a pathway imposes an energetic cost of production on the cell, how do such architectures emerge? Do significantly profitable kinetic and/or dynamical properties emerge with some structures, or do they again just represent evolutionary sojourns along effectively neutral paths? Sometimes different lineages have similar network topologies, but with entirely different underlying protein participants or orders of steps. How does rewiring of an underlying structure evolve without leading to catastrophic intermediate consequences?

Intracellular and extracellular communication systems consist of at least one signalling molecule and one receptor. How does the language of such systems coevolve so as to avoid crosstalk between parallel pathways? When are there sufficient degrees of freedom to allow cellular communication systems to drift over time in an effectively neutral fashion, much like the human languages have diversified across the planet?

Cellular surveillance systems

The internal cellular environment introduces a variety of challenges associated with the accuracy of a wide array of cellular functions: errors introduced at the levels of replication, transcription, and translation; enzyme promiscuity with respect to substrate utilization; protein-folding problems, etc. There are often multiple layers of surveillance and correction for intracellular errors, suggesting highly refined and robust systems. Yet, the error rates for some of these functions can vary by at least a thousandfold among organisms, and some layers of surveillance are lost in some phylogenetic lineages (Chapter 20). Such observations raise numerous questions. High rates of surveillance are costly, but low fidelity can be catastrophic, so what are the limits to the burden of manageable intracellular error proliferation? Owing to the power of random genetic drift, there are limits to the level of molecular perfection that can evolve for any particular cellular function. Does this encourage the expansion of complexity by the evolutionary layering of surveillance mechanisms (e.g., proofreading) and if so, are there any long-term advantages to such embellishments or does the overall performance regress to its original state?

Growth regulation

So-called growth laws have been invoked for years by microbial physiologists, and substantial theoretical work has been devoted to explain these. However, the empirical work has been largely confined to a single species (*E. coli*), leaving open many questions about generality (Chapter 9). Moreover, the models that have been developed are largely phenomenological, leaving mechanistic issues unresolved. When species evolve under different resource conditions, does the evolved 'growth-law' pattern recapitulate the more transient (plastic) pattern found within a genotype in response to varying nutrient availability? That is, do patterns of evolutionary response reflect patterns of physiological response? Are the rules for eukaryotes the same as those for prokaryotes?

Biological scaling laws

Cell biologists have identified a number of 'scaling laws' that transcend species boundaries (Chapter 8), whereby specific cellular features can be approximated as power functions of cell size. The traits involved range from cell division rates to total lifetime energy budgets to internal organelle sizes to swimming speeds. Such patterns provide convincing statistical descriptions of the rules of life. But what are the underlying mechanisms leading to the observed slopes and intercepts of such functions, and why do they often appear to be universal across the Tree of Life?

Summary

- The fact that all evolutionary change begins at the cellular level motivates the need for eliminating the intellectual disconnect between cell biology (including microbiology) and evolutionary theory. Together, these subdisciplines provide the key connections between genotype, phenotype, and fitness that are essential to understanding all evolutionary processes.

- The need for a field of evolutionary cell biology is further justified by the composition of the biosphere. The total number of prokaryotic cells on earth outnumbers that of unicellular eukaryotes by several orders of magnitude, and the latter exceeds the number of animals and land plants by a similar degree.

- A mature field of evolutionary cell biology will ultimately need to integrate the three big engines of quantitative biology (population genetics, biophysics, and biochemistry) with comparative and experimental analyses across the Tree of Life.

- Evolutionary theory, grounded in principles of Mendelian genetics and stochastic transmission of gene frequencies, is as well established as any area of quantitative biology. Thus, an essential platform is in place for developing a mechanistic understanding of the origin and diversification of cellular features by the progressive fixation of new mutations.

- Although natural selection is the most powerful force in the biological world, it is not all powerful. Rather, the efficiency of selection is dictated by the population-genetic environment – defined by the magnitudes of mutation, recombination, and random genetic drift, all of which vary by orders of magnitude among phylogenetic lineages. Many aspects of molecular and genome evolution reflect the inability of natural selection to act, as opposed to being reflections of adaptive refinement. The following chapters demonstrate that this is also commonly true at the cell biological level.

Literature Cited

Amann, R., and R. Rosselló-Móra. 2016. After all, only millions? mBio 7: e00999–16.

Atkinson, A., V. Siegel, E. A. Pakhomov, M. J. Jessopp, and V. Loeb. 2009. A re-appraisal of the total biomass and annual production of Antarctic krill. Deep Sea Research Part I: Oceanogr. Res. Papers 56: 727–740.

Bar-On, Y. M., R. Phillips, and R. Milo. 2018. The biomass distribution on Earth. Proc. Natl. Acad. Sci. USA 115: 6506–6511.

Barton, N. H., and H. P. de Vladar. 2009. Statistical mechanics and the evolution of polygenic quantitative traits. Genetics 181: 997–1011.

Beech, E., M. Rivers, S. Oldfield, and P. P. Smith. 2017. GlobalTreeSearch: The first complete global database of tree species and country distributions. J. Sustainable Forestry 36: 454–489.

Buffalo, V. 2021. Quantifying the relationship between genetic diversity and population size suggests natural selection cannot explain Lewontin's Paradox. eLife 10: e67509.

Bulmer, M. 1991. The selection-mutation-drift theory of synonymous codon usage. Genetics 129: 897–907.

Callaghan, C. T., S. Nakagawa, and W. K. Cornwell. 2021. Global abundance estimates for 9,700 bird species. Proc. Natl. Acad. Sci. USA 118: e2023170118.

Caporale, L. H. (ed.) 2006. The Implicit Genome. Oxford University Press, Oxford, UK.

Charlesworth, B. 1984a. Some quantitative methods for studying evolutionary patterns in single characters. 10: 308–318.

Charlesworth, B. 1984b. The cost of phenotypic evolution. Paleobiol. 10: 319–327.

Charlesworth, B. 1990. Mutation-selection balance and the evolutionary advantage of sex and recombination. Genet. Res. 55: 199–221.

Charlesworth, B., and D. Charlesworth. 1983. The population dynamics of transposable elements. Genet. Res. 42: 1–27.

Charlesworth, B., and D. Charlesworth. 2000. The degeneration of Y chromosomes. Phil. Trans. Roy. Soc. Lond. B Biol. Sci. 355: 1563–1572.

Charlesworth, B., R. Lande, and M. Slatkin. 1982. A neo-Darwinian commentary on macroevolution. Evolution 36: 474–498.

Charlesworth, B., and C. H. Langley. 1986. The evolution of self-regulated transposition of transposable elements. Genetics 112: 359–383.

Charlesworth, D., N. H. Barton, and B. Charlesworth. 2017. The sources of adaptive variation. Proc. Biol. Sci. 284: 20162864.

Charnov, E. L. 1982. The Theory of Sex Allocation. Princeton University Press, Princeton, NJ

Charnov, E. L. 1993. Life History Invariants. Oxford University Press, Oxford, UK.

Crow, J. F. 1948. Alternative hypotheses of hybrid vigor. Genetics 33: 477–487.

Crowther, T. W., H. B. Glick, K. R. Covey, C. Bettigole, D. S. Maynard, S. M. Thomas, J. R. Smith, G. Hintler, M. C. Duguid, G. Amatulli, et al. 2015. Mapping tree density at a global scale. Nature 525: 201–205.

Danchin, É., A. Charmantier, F. A. Champagne, A. Mesoudi, B. Pujol, and S. Blanchet. 2011. Beyond DNA: Integrating inclusive inheritance into an extended theory of evolution. Nat. Rev. Genet. 12: 475–486.

Darwin, C. 1859. On the Origin of Species by Means of Natural Selection. John Murray, London, UK.

Darwin, C. 1871. The Descent of Man, and Selection in Relation to Sex. John Murray, London, UK.

ENCODE Project Consortium. 2012. An integrated encyclopedia of DNA elements in the human genome. Nature 489: 57–74.

Falconer, D. S. 1965. Maternal effects and selection response. Proc. XIth Internat. Cong. Genetics 3: 763–774.

Fisher, R. A. 1918. The correlation between relatives on the supposition of Mendelian inheritance. Trans. Royal Soc. Edinburgh 52: 399–433.

Flemming, H. C., and S. Wuertz. 2019. Bacteria and archaea on Earth and their abundance in biofilms. Nat. Rev. Microbiol. 17: 247–260.

Force, A., M. Lynch, B. Pickett, A. Amores, Y.-L. Yan, and J. Postlethwait. 1999. Preservation of duplicate genes by complementary, degenerate mutations. Genetics 151: 1531–1545.

Frank, S. A. 2007. Maladaptation and the paradox of robustness in evolution. PLoS One 2: e1021.

Galhardo, R. S., P. J. Hastings, and S. M. Rosenberg. 2007. Mutation as a stress response and the regulation of evolvability. Crit. Rev. Biochem. Mol. Biol. 42: 399–435.

Gatti, R. C., P. B. Reich, J. G. P. Gamarra, T. Crowther, C. Hui, A. Morera, J. F. Bastin, S. de-Miguel, G. J. Nabuurs, J. C. Svenning, et al. 2022. The number of tree species on Earth. Proc. Natl. Acad. Sci. USA 119: e2115329119.

Gerhart, J., and M. Kirschner. 1997. Cells, Embryos and Evolution. Blackwell Science, Malden, MA.

Goldenfeld, N., and C. Woese. 2011. Life is physics: Evolution as a collective phenomenon far from equilibrium. Annu. Rev. Condensed Matter Physics 2: 375–399.

Goldschmidt, R. 1940. The Material Basis of Evolution. Yale University Press, New Haven, CT.

Gupta, M., N. G. Prasad, S. Dey, A. Joshi, and T. N. C. Vidya. 2017. Niche construction in evolutionary theory: The construction of an academic niche? J. Genet. 96: 491–504.

Hahn, M. W. 2008. Toward a selection theory of molecular evolution. Evolution 62: 255–265.

Haldane, J. B. S. 1964. A defense of beanbag genetics. Perspect. Biol. Med. 7: 343–359.

Higgins, K., and M. Lynch. 2001. Metapopulation extinction due to mutation accumulation. Proc. Natl. Acad. Sci. USA 98: 2928–2933.

Hochberg, G. K. A., and J. W. Thornton. 2017. Reconstructing ancient proteins to understand the causes of structure and function. Annu. Rev. Biophys. 46: 247–269.

Hölldobler, B., and E. O. Wilson. 1990. The Ants. Harvard University Press, Cambridge, MA.

Jablonka, E., and M. J. Lamb. 2005. Evolution in Four Dimensions: Genetic, Epigenetic, Behavioral, and Symbolic Variation in the History of Life. MIT Press, Cambridge, MA.

Jarosz, D. F., and S. Lindquist. 2010. Hsp90 and environmental stress transform the adaptive value of natural genetic variation. Science 330: 1820–1824.

Jensen, J. D., B. A. Payseur, W. Stephan, C. F. Aquadro, M. Lynch, D. Charlesworth, and B. Charlesworth. 2019. The importance of the neutral theory in 1968 and 50 years on: A response to Kern and Hahn 2018. Evolution 73: 111–114.

Kallmeyer, J., R. Pockalny, R. R. Adhikari, D. C. Smith, and S. D'Hondt. 2012. Global distribution of microbial abundance and biomass in subseafloor sediment. Proc. Natl. Acad. Sci. USA 109: 16213–16216.

Kempthorne, O. 1954. The correlation between relatives in a random mating population. Proc. Royal Soc. Lond. B 143: 103–113.

Kern, A. D., and M. W. Hahn. 2018. The neutral theory in light of natural selection. Mol. Biol. Evol. 35: 1366–1371.

Kimura, M. 1983. The Neutral Theory of Molecular Evolution. Cambridge University Press, Cambridge, UK.

Kiontke, K., and D. H. Fitch. 2013. Nematodes. Curr. Biol. 23: R862–R864.

Kondrashov, A. S. 1988. Deleterious mutations and the evolution of sexual reproduction. Nature 336: 435–440.

Krebs, J. R., and N. B Davies. 1997. Behavioural Ecology: An Evolutionary Approach. Wiley, New York, NY.

Laland, K. N., T. Uller, M. W. Feldman, K. Sterelny, G. B. Müller, A. Moczek, E. Jablonka, and J. Odling-Smee. 2015. The extended evolutionary synthesis: Its structure, assumptions and predictions. Proc. Biol. Sci. 282: 20151019.

Laland, K., T. Uller, M. Feldman, K. Sterelny, G. B. Müller, A. Moczek, E. Jablonka, J. Odling-Smee, G. A. Wray, H. E. Hoekstra, et al. 2014. Does evolutionary theory need a rethink? Nature 514: 161–164.

Lande, R. 1976. Natural selection and random genetic drift in phenotypic evolution. Evolution 30: 314–334.

Lande, R. 1978. Evolutionary mechanisms of limb loss in tetrapods. Evolution 32: 73–92.

Landenmark, H. K., D. H. Forgan, and C. S. Cockell. 2015. An estimate of the total DNA in the biosphere. PLoS Biol. 13: e1002168.

Lässig, M. 2007. From biophysics to evolutionary genetics: Statistical aspects of gene regulation. BMC Bioinformatics 8 Suppl. 6: S7.

Lewontin, R. C. 1974. The Genetic Basis of Evolutionary Change. Columbia University Press, New York, NY.

Locey, K. J., and J. T. Lennon. 2016. Scaling laws predict global microbial diversity. Proc. Natl. Acad. Sci. USA 113: 5970–5975.

Lynch, M. 1990. The rate of morphological evolution in mammals from the standpoint of the neutral expectation. Amer. Natur. 136: 727–741.

Lynch, M. 2007. The Origins of Genome Architecture. Sinauer Associates, Inc., Sunderland, MA.

Lynch, M. 2010. Scaling expectations for the time to establishment of complex adaptations. Proc. Natl. Acad. Sci. USA 107: 16577–16582.

Lynch, M. 2011. The lower bound to the evolution of mutation rates. Genome Biol. Evol. 3: 1107–1118.

Lynch, M. 2012. Evolutionary layering and the limits to cellular perfection. Proc. Natl. Acad. Sci. USA 109: 18851–18856.

Lynch, M. 2013. Evolutionary diversification of the multimeric states of proteins. Proc. Natl. Acad. Sci. USA 110: E2821–E2828.

Lynch, M., F. Ali, T. Lin, Y. Wang, J. Ni, and H. Long. 2023. The divergence of mutation rates and spectra across the Tree of Life. EMBO Rep. e57561.

Lynch, M., M. C. Field, H. Goodson, H. S. Malik, J. B. Pereira-Leal, D. S. Roos, A. Turkewitz, and S. Sazer. 2014. Evolutionary cell biology: Two origins, one objective. Proc. Natl. Acad. Sci. USA 111: 16990–16994.

Lynch, M., and A. Force. 2000. The probability of duplicate-gene preservation by subfunctionalization. Genetics 154: 459–473.

Lynch, M., and K. Hagner. 2014. Evolutionary meandering of intermolecular interactions along the drift barrier. Proc. Natl. Acad. Sci. USA 112: E30–E38.

Lynch, M., and J. B. Walsh. 1998. Genetics and Analysis of Quantitative Traits. Sinauer Associates, Inc., Sunderland, MA.

Martincorena, I., A. S. Seshasayee, and N. M. Luscombe. 2012. Evidence of non-random mutation rates suggests an evolutionary risk management strategy. Nature 485: 95–98.

Mayr, E. 1959. Where are we? Cold Spring Harbor Symp. Quant. Biol. 24: 1–14.

Mayr, E. 1963. Animal Species and Evolution. Belknap Press, Cambridge, MA.

Merrikh, H. 2017. Spatial and temporal control of evolution through replication-transcription conflicts. Trends Microbiol. 25: 515–521.

Mode, C. G., and H. F. Robinson. 1959. Pleiotropism and the genetic variance and covariance. Biometrics 15: 518–537.

Mora, C., D. P. Tittensor, S. Adl, A. G. Simpson, and B. Worm. 2011. How many species are there on Earth and in the ocean? PLoS Biol. 9: e1001127.

Otto, S. P., and N. H. Barton. 2001. Selection for recombination in small populations. Evolution 55: 1921–1931.

Pernice, M. C., I. Forn, A. Gomes, E. Lara, L. Alonso-Sáez, J. M. Arrieta, F. del Carmen Garcia, V. Hernando-Morales, R. MacKenzie, M. Mestre, et al. 2015. Global abundance of planktonic heterotrophic protists in the deep ocean. ISME J. 9: 782–792.

Pigliucci, M., and J. Kaplan. 2000. The fall and rise of Dr. Pangloss: Adaptationism and the spandrels paper 20 years later. Trends Ecol. Evol. 15: 66–70.

Pigliucci, M., and G. B. Müller. 2010. Evolution: the Extended Synthesis. MIT Press, Cambridge, MA.

Provine, W. B. 1971. The Origins of Theoretical Population Genetics. University of Chicago Press, Chicago, IL.

Roff, D. E. 1993. The Evolution of Life Histories. Springer-Verlag, New York, NY.

Schwartz, M. H., and T. Pan. 2017. Function and origin of mistranslation in distinct cellular contexts. Crit. Rev. Biochem. Mol. Biol. 52: 205–219.

Sella, G., and A. E. Hirsh. 2005. The application of statistical physics to evolutionary biology. Proc. Natl. Acad. Sci. USA 102: 9541–9546.

Sender, R., S. Fuchs, and R. Milo. 2016. Revised estimates for the number of human and bacteria cells in the body. PLoS Biol. 14: e1002533.

Shapiro, J. A. 2011. Evolution: A View from the 21st Century. FT Press Science, Upper Saddle River, NJ.

Stoltzfus, A. 2017. Why we don't want another 'Synthesis'. Biol. Direct 12: 23.

Suttle, C. A. 2005. Viruses in the sea. Nature 437: 356–361.

van den Hoogen, J., S. Geisen, D. Routh, H. Ferris, W. Traunspurger, D. A. Wardle, R. G. M. de Goede, B. J. Adams, W. Ahmad, W. S. Andriuzzi, et al. 2019. Soil nematode abundance and functional group composition at a global scale. Nature 572: 194–198.

Wallace, A. R. 1870. Contributions to the Theory of Natural Selection. Macmillan and Co., New York, NY.

Walsh, J. B., and M. Lynch. 2018. Evolution and Selection of Quantitative Traits. Oxford University Press, Oxford, UK.

Wang, S., T. Hassold, P. Hunt, M. A. White, D. Zickler, N. Kleckner, and L. Zhang. 2017. Inefficient crossover maturation underlies elevated aneuploidy in human female meiosis. Cell 168: 977–989.

Weissman, D. B., M. W. Feldman, and D. S. Fisher. 2010. The rate of fitness-valley crossing in sexual populations. Genetics 186: 1389–1410.

Welch, J. J. 2017. What's wrong with evolutionary biology? Biol. Philos. 32: 263–279.

Whitman, W. B., D. C. Coleman, and W. J. Wiebe. 1998. Prokaryotes: The unseen majority. Proc. Natl. Acad. Sci. USA 95: 6578–6583.

Wideman, J. G., A. Monier, R. Rodríguez-Martínez, G. Leonard, E. Cook, C. Poirier, F. Maguire, D. S. Milner, N. A. T. Irwin, K. Moore, et al. 2020. Unexpected mitochondrial genome diversity revealed by targeted single-cell genomics of heterotrophic flagellated protists. Nat. Microbiol. 5: 154–165.

Willham, R. L. 1963. The covariance between relatives for characters composed of components contributed by related individuals. Biometrics 19: 18–27.

Wilson, E. B. 1925. The Cell in Development and Inheritance. Macmillan and Co., London, UK.

Wright, S. 1934a. An analysis of variability in number of digits in an inbred strain of guinea pigs. Genetics 19: 506–536.

Wright, S. 1934b. The results of crosses between inbred strains of guinea pigs, differing in number of digits. Genetics 19: 537–551.

Zhang, F., L. Xu, K. Zhang, E. Wang, and J. Wang. 2012. The potential and flux landscape theory of evolution. J. Chem. Phys. 137: 065102.

Zhang, J. 2018. Neutral theory and phenotypic evolution. Mol. Biol. Evol. 35: 1327–1331.

CHAPTER 2

The Origin of Cells

Ideally, a treatise that claims to be focused on cellular evolution would give substantial coverage to the earliest stages of life, so here is a bit of a let-down. Unfortunately, the cumulative effects of nearly four billion years of chemistry, physics, and geology have erased all traces of pre-cellular life. As a consequence, we will probably never be able to determine with certainty the earliest steps in the emergence of life from an inorganic world. However, this need not dampen our enthusiasm for understanding how life might have evolved. One of the goals of the active field of 'origin-of-life' research is to combine our knowledge of the physical sciences and biochemistry to identify the most plausible scenarios for launching planet Earth into the age of biology.

Now thousands of cells thick in some places and diversified into millions of species, the Earth's biological skin has been molded from the beginning by historical contingencies, most notably the unique mix of elemental resources that make up the planet. The laws of chemistry and physics further dictated how these elements could be organized into biology's structures and functions. Evolution is opportunistic, with all changes reflecting processes involving 'descent with modification', and only a tiny fraction of imaginable evolutionary changes has occurred. The nature of the genetic machinery that happened to evolve at an early stage, combined with the basic rules of population genetics, dictates what pathways remain open to evolutionary exploitation today.

Life on Earth is a peculiar mix of ingredients – nucleotides, amino acids, carbohydrates, and lipids, and even then, relying on just a small subset of the possible types of these building blocks. Phosphorus, a rare element, is essential in energy and information transmission. Transition metals like iron, zinc, manganese, nickel, copper, and molybdenum are widely used in the catalytic cores of proteins. One of life's oddest features is the reliance on proton pumping to generate return gradients for ATP formation, and the use of ATP itself as an energy storage molecule remains an enigma. Are these universal features of cellular biochemistry inevitable necessities of life, or might they simply reflect the specific conditions under which life first arose, i.e., the frozen legacy of the singular successful lineage that gave rise to all other species on the planet?

Living systems distinguish themselves in several key ways: 1) an ability to acquire and convert energy and material resources into new organic compounds – metabolism and growth; 2) a reliable mechanism for storing information and converting it into a phenotype – genetics and individuality; and 3) a means for transmitting information and biotic materials from one generation to the next – reproduction and inheritance. The temporal order by which these three features emerged remains unclear. However, once they were simultaneously present in enough individuals to avoid extermination by vagaries in the environment, a permanent platform was in place for the most powerful force in the natural world – evolution by natural selection, biology's intrinsic mechanism for designing and refining its own features. Given a population with some level of variation among individuals, and a mechanism of heritable transmission of phenotypic differences across generations, natural selection is an inevitable property of life. Thus, if we desire a singular time point for the origin of life, a logical defining event is the origin of evolution by natural selection.

But herein lies the problem. How did life get to the point where an ability to evolve by natural

Evolutionary Cell Biology. Michael Lynch, Oxford University Press. © Michael Lynch (2024). DOI: 10.1093/oso/9780192847287.003.0002

selection (indeed, an inability to avoid such a process) was locked in forever? Did metabolism arise before genetics, providing the fuel for the emergence of the genetic machinery (necessary for heritable variation), or vice versa? Once a genetic system was established, how was it faithfully maintained across generations so that useful variants could be retained? And at what point did membranes (necessary for individuality) come in?

The Earliest Stages

An understanding of the early features of the geosphere, combined with an appreciation of the peculiar shared molecular attributes of today's organisms, provides a logical basis for narrowing down the broad range of possible first steps toward the origin of life. However, although most of those who think about such matters implicitly assume that life initiated on Earth, there is no formal basis for rejecting the hypothesis that the seeds of the biosphere were derived from another planet. This caveat should be kept in mind as the following discussion embarks on an Earth-first view, but we are still confronted with the same fundamental questions about life's origin.

We know that the Earth originated ~4.6 billion years ago (BYA), and was then sporadically bombarded with substantial interstellar debris for one to two billion years (Sleep et al. 1989; Nisbet and Sleep 2001; Bottke et al. 2012; Johnson and Melosh 2012). Because some of the more massive impacts generated enough heat to sterilize the entire planet, it is generally thought that the roots of life must be younger than 3.8 BYA. This does not rule out the possibility that biology experienced a number of false starts prior to this point, and small graphite inclusions (with carbon isotope ratios compatible with a biological origin) have been found in rocks dating to 4.1 BYA (Bell et al. 2015).

Further refinement of this key point in time will not be easy. Rock formations dating earlier than 3.5 BYA are extremely rare, and the first universal common ancestor of life most likely was so simple that no fossils were produced. However, organic signatures pointing to biological activity have been found in rocks from 3.8 to 3.4 BYA (Rosing 1999;

Furnes et al. 2004; Tice and Lowe 2004), and the oldest known fossils, some from filamentous organisms and potentially eukaryotic in nature, date to 3.5 to 3.2 BYA (Schopf 1993; Rasmussen 2000; Schopf et al. 2002; Knoll 2004; Wacey et al. 2011; Brasier et al. 2015).

Our understanding of biology's reliance on a limited set of molecular building blocks (e.g., amino acids, nucleotides, lipids) would be advanced if their production could be linked to contexts in which life might have first emerged. Identifying plausible scenarios requires knowledge of bioenergetic, geochemical, and physical opportunities and constraints in the prebiotic world. For example, given that carbon in the early Earth's atmosphere was dominated by oxidized forms (CO_2 and CO resulting from volcanic out-gassing), some source of sustained external energy would have been required for the construction of reduced-carbon compounds (containing C-H bonds) upon which all life is built. In addition, given the absence of atmospheric oxygen, there would have been no ozone shield, and hence, the damaging effects of UV light at the Earth's surface would have been tens to thousands of times greater than today (Cnossen et al. 2007). This suggests that life likely arose in an energy-rich, photo-protective setting. Some sort of structured environment would also have been essential as a means for co-localizing the interacting molecules necessary for some semblance of individuality.

The most celebrated experiments showing that simple forms of organic matter can be generated in abiotic environments are those of Miller (1953) and Miller and Urey (1959). Based on the assumption that the earliest atmosphere harbored methane, ammonia, hydrogen, and water, these four compounds were sealed into a sterilized glass apparatus and subjected to cycles of electrical discharges followed by cooling periods to condense the resultant products. A few days of such treatment revealed the synthesis of abundant quantities of urea, sugars, formaldehyde, hydrogen cyanide, among other things. Reanalysis of the generated residues decades later revealed all of the amino acids used in today's organisms (Johnson et al. 2008).

Simple biochemical reactions underlying the results of these experiments are known. For example, starting with the one-carbon compound formaldehyde (CH_2O), through a series of steps involving water, hydrogen cyanide, and ammonia, the Strecker reaction yields the spontaneous production of glycine, the simplest amino acid (Figure 2.1). Similar reactions starting with more complex aldehydes (having side residues to the -CHO group other than the H in formaldehyde; Figure 2.1) led to other amino acids (Benner et al. 2010). Carbohydrates are much less stable than amino acids, but plausible scenarios for their accumulation have also been suggested. For example, when in complex with borate, carbohydrates are stabilized and differentially channelled towards pentoses such as ribose (Ricardo et al. 2004; Kim et al. 2011), a potentially important insight given that ribose is a key component of RNA.

Although the Miller–Urey experiments reinforced the popular idea that life emerged spontaneously out of a so-called primordial soup (Haldane 1929; Oparin 1938), a number of doubts exist regarding this hypothesis (Maden 1995). The most prominent problem is that the early atmosphere was likely much more oxidative than the one imposed by Miller and Urey, with CO_2 (rather than methane) being the primary carbon source

and N_2 (rather than ammonia) being the primary nitrogen source (Zahnle et al. 2010). A second issue is that open water is counterproductive to the maintenance of the organic aggregates essential to the nucleation of life. Finally, there is the matter of how long life could have relied on abiotically generated carbon sources before experiencing a resource-limitation crisis. Carbonaceous meteorites are known to harbor numerous organic compounds, including most of the metabolites within the citric-acid cycle deployed by most of today's organisms (Cooper et al. 2011). However, it is unclear that such sporadic delivery could provide the sustenance for emerging life forms relying on inefficient biochemical pathways (not yet refined by natural selection).

These and other doubts about a 'heterotrophy-first' origin of life have inspired several alternative hypotheses focused on settings conducive to the sustained geological production of organic molecules. Under these 'autotrophy-first' hypotheses, the synthesis of simple organic compounds is fuelled by continuous sources of energy, rather than by sporadic bursts of atmospheric electricity, and water is viewed as an aid, not a hindrance. Although there are still many unresolved issues, under this view, there was a prebiotic phase of autotrophic metabolism, with a genetic system somehow arising secondarily and gaining regulatory control of energy harvesting.

These are not simple issues, but Wächtershäuser (1988, 1997, 2007) suggested a plausible autotrophy-first scenario for the emergence of metabolism. He envisioned a particular kind of chemoautotrophy naturally generated in a high-pressure, high-temperature environment associated with underwater volcanic activity, an idea first broached by Baross and Hoffman (1985). Assuming the presence of one-carbon compounds such as CO, CO_2, COS, HCN, and CH_3SH, and with iron and nickel sulfides acting as catalysts, a sort of primitive form of spontaneous carbon fixation was postulated. A specific reaction suggested by Wächtershäuser is the oxidation of FeS with H_2S to produce FeS_2 (pyrite), hydrogen ions (protons), and electrons. The products of this reaction were proposed to drive the reduction of CO_2 to formate (HCO_2), and then to more complex chemicals such as acetate (CH_3CO_2) and pyruvate (CH_3COCO_2).

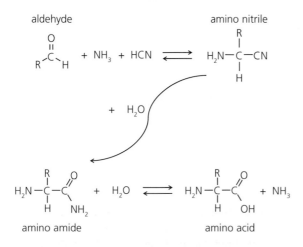

Figure 2.1 The series of chemical steps in amino-acid synthesis via the Strecker reaction, which requires just an aldehyde in the presence of ammonia (NH_3), hydrogen cyanide (HCN), and water. R denotes an arbitrary side chain (the simplest for an amino acid being R = H for glycine).

Two attractive features of this model are that pyruvate is a primary participant in the major metabolic cycles of today's organisms (e.g., Figure 2.2), and that metal-sulfur clusters serve as the catalytic centres of numerous metabolic enzymes. Indeed, Wächtershäuser went so far as to suggest the possibility of the fixation of carbon by the entry of CO_2 into a reverse form of the citric-acid cycle (whose forward reaction is used in today's organisms to break down sugar; Figure 2.2). Under his model, primordial energy transduction was accomplished by thioesters, rather than by the phosphate-bearing molecules essential to today's organisms, i.e., the hypothesized intermediate metabolites are thiol analogues (containing -SH groups) of the components of today's citric-acid cycle, with H_2S (rather than the usual H_2O) entering at various steps. Based on biochemical considerations, the very high CO_2 concentration in hydrothermal-vent environments provides a

kinetically favorable setting for running the citric-acid cycle in reverse (Steffens et al. 2021). Thus, it may be no coincidence that a core of today's metabolism operates by using enzymes enriched in iron-sulfur complexes (Goldford et al. 2017).

The existence of modern hydrothermal-vent microbes dependent solely on a continuous flow of chemistry and energy demonstrates the ability of such environments to support life. Although this need not imply that the current inhabitants of such environments have been derived in a linear line of descent from the time of life's origin, some aspects of Wächtershäuser's model have been validated (Cody 2004). Laboratory experiments imposing hydrothermal vent-like conditions do yield pyruvate in the presence of transition-metal sulfides (Cody et al. 2000), and reactions involving pyruvate can lead to more complex organic molecules, some of which have properties related to the lipids essential to building membranes (Hazen and

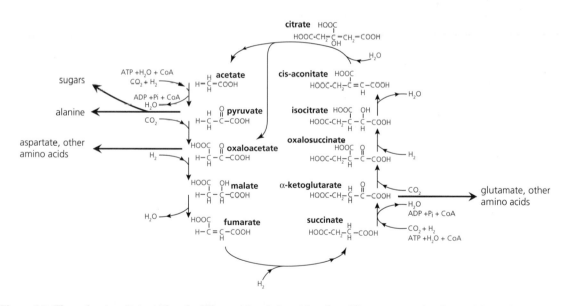

Figure 2.2 The reductive citric-acid cycle. Whereas the citric-acid cycle oxidizes sugar molecules to CO_2 and water, the reverse (reductive) cycle uses CO_2 and water to make sugars, i.e., to fix carbon. In principle, the metabolic intermediates can serve as the precursors for the synthesis of a number of amino acids (as shown in red), as in the normal citric-acid cycle (Chapter 19). This pathway, which occurs in some chemoautotrophic bacteria, was envisioned by Wächtershäuser (1988, 1997, 2007) as a prebiotic form of carbon fixation initiated by metal catalysts and with H_2S taking the place of H_2O as a hydrogen donor (see also DeDuve 1991). The presence of the full set of enzymes in the citric-acid cycle in all major domains of life today implies the existence of the citric-acid cycle in the last universal common ancestor (LUCA). However, the oxidative direction common to today's organisms (and opposite to the direction in the figure) would not have been possible in the earliest stages of life in anoxic environments. Thus, today's citric-acid cycle may have initially run in the reductive (reverse) direction. Modified from Smith and Morowitz (2004).

Deamer 2007). At temperatures compatible with life, pyruvate reacts with H_2S, H_2, and NH_4 to yield products such as aldols, lactate, alanine, proprionic acid, and sulfur-containing organics, with the specific blends of products being strongly dependent upon temperature and the mineral substrates (Novikov and Copley 2013). Starting with the metabolites of glycolysis and the pentose-phosphate pathway (Chapter 19), almost the full set of reactions of these pathways can be generated in a completely abiotic setting (Keller et al. 2014), and a mixture of pyruvate and ferrous iron yields nine of the eleven intermediate metabolites of the citric-acid cycle (Muchowska et al. 2019).

The key point here is that simple metabolites generated by abiotic processes are themselves subject to downstream chemical conversion to a large number of alternative compounds, with specific catalysts channelling reactions down specific pathways. This raises the intriguing possibility that the basic structure of many of today's metabolic pathways are reflections of primordial sets of abiotic chemical reactions that fueled life's origin. Under this view, modern-day enzymes would have then emerged secondarily as catalytic enhancers of pre-existing abiotic reactions.

Regardless of what one thinks of the details, Wächtershäuser's attempt to draw a connection between inorganic chemistry, geology, and the roots of biology inspired a new generation of hypotheses for the catalytic origin of life by geothermal forces, as reviewed in Cody (2004) and Stüeken et al. (2013). Although a ringing endorsement cannot be given to any one of these ideas, the focus is now on narrowing down the rich pool of candidate settings to those most conducive to the origin of life. Notably, almost all current hypotheses for the origin of life view the colonization of open marine waters occurring secondarily, only after the establishment of ion-tight membranes.

Two of the more plausible scenarios are outlined in the following section (Figure 2.3). Each is based on a geological setting in which a freely available, energy-capturing reaction shares key features with metabolic mechanisms in today's organisms. The most notable aspect of such models is the implication that the origin of life is not just a chance event with an infinitesimally small probability, but an essentially unavoidable consequence of the early Earth's geochemical and atmospheric properties. If this view is correct, then life has very likely emerged on other Earth-like planets, perhaps even incorporating similar metabolic processes.

The alkaline hydrothermal-vent hypothesis

The famous 'black smokers' emerging from oceanic hydrothermal vents provide a potential source of geothermal energy and chemistry envisioned by Wächtershäuser (Corliss et al. 1981). However, their brevity, extremely low pH, and lack of compartmentalization prompts questions over their suitability as cradle-of-life candidates (Martin and Russell 2003; Lane et al. 2010; Russell et al. 2014).

Less extreme variants include lower-temperature, alkaline hydrothermal-vent systems (Russell and Hall 1997; Martin et al. 2008; Russell et al. 2010). In such settings, ocean fluids percolate deep into the Earth's crust, where they interact with iron compounds to release hydrogen, a strong electron donor and hence, a source of energy in the presence of suitable electron acceptors. Dissolved CO_2 plays the latter role in such environments, entering from above and then being reduced to simple hydrocarbons such as methane, formate, and acetate. Dinitrogen (N_2) is also reduced to NH_3, and sulfates to H_2S, providing potential paths for the downstream integration of nitrogen and sulfur into organic compounds. Alkaline-hydrothermal vents also support the growth of porous towers of calcium carbonate up to tens of meters in height, providing potential sites of compartmentalization necessary for the origin of individuality. Based on these observations, like Wächtershäuser, proponents of the alkaline hydrothermal-vent hypothesis postulate an initially purely geological mechanism of carbon fixation that was somehow eventually supplanted by evolved biotic mechanisms (Lane and Martin 2012).

There are other attractive aspects of alkaline hydrothermal vents as potential locales for the origin of life. First, they present a ten-thousand fold proton gradient, as the pH for the ocean water influent $\simeq 6.0$ while that for the effluent $\simeq 10.0$. Such a setting is analogous (and some would argue a direct antecedent) to the peculiar mechanism by which almost all of today's cells produce energy – a proton gradient across biological membranes

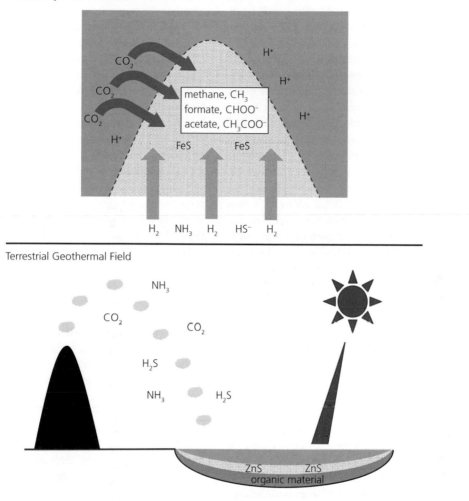

Figure 2.3 Simplified depiction of two proposed settings for the origin of life. **Above:** The alkaline hydrothermal-vent hypothesis. Water enters the Earth's crust, where geological activity leads to the production of hydrogen, ammonia, and hydrogen sulfide ions, which are then subject to chemical reactions as they enter the overlying mound. The latter contains pores harboring metal sulfides, which catalyse the production of simple organic compounds from CO_2 diffusing in from the overlying water. Hydrogen ions (protons) would also be subject to diffusion from the more acidic overlying water into the more alkaline environment of the mound. **Below:** The geothermal-field hypothesis. Volcanic activity releases carbon dioxide, ammonia, and hydrogen sulfide gas, among other things, which accumulate in nearby ponds. The latter are postulated to contain layers of zinc and manganese sulfides, which provide both photoprotection and a mechanism for photocatalysis of simple organic compounds from CO_2.

used to drive ATP production by chemiosmosis (Foundations 2.1). Second, the iron and nickel sulfide groups envisioned as operating in early abiotic metabolism also comprise the catalytic sites for energy transfer in the electron-transport chain deployed in today's organisms (Hall et al. 1974;

Cammack et al. 1981). Third, abiogenic production of amino acids has been detected within such systems (Ménez et al. 2018).

Notably, one of the known biological mechanisms of carbon fixation, the acetyl-CoA (or Wood-Ljungdahl) pathway (Chapter 19), has similarities to the pathway invoked in the alkaline hydrothermal-vent hypothesis (Martin and Russell 2003). The acetyl-CoA pathway is the only known carbon-fixation mechanism that yields energy in the process of reducing carbon. It is deployed by two distantly related groups of anaerobic prokaryotes – the archaeal methanogens and the bacterial acetogens, both of which obtain all of their C, N, and S resources from simple gases, CO_2, CO, N_2, and H_2S. These two lineages utilize apparently unrelated enzymes in the acetyl–CoA pathway and produce different final products (methane vs. acetate), suggesting an ancient episode of parallel evolution (Lane and Martin 2012). Although two independent origins of a similar mechanism for extracting energy and organic material from inorganic gases may appear highly implausible, the following chapters reveal a number of other ways in which prokaryotes carry out similar functions with apparently unrelated molecules.

The terrestrial geothermal-field hypothesis

As enticing as the hydrothermal-vent hypothesis may seem, a marine setting need not have been essential for the origin of life. Some have argued that terrestrial hydrothermal processes (hot springs and geysers) associated with volcanic activity harbor most of the advantages envisioned by the hydrothermal-vent hypothesis as well as others (Mulkidjanian 2009; Mulkidjanian and Galperin 2009; Mulkidjanian et al. 2012a; Damer and Deamer 2015). Under this alternative view, life would have originated in shallow ponds, with solar irradiation being the primary energy source. The condensates found in such ponds likely would have been enriched with potassium, metal sulfides, zinc, and boron, with phosphorus concentrations possibly as much as $100\times$ greater than those at hydrothermal vents. Assuming regularly occurring wet/dry cycles, concentrations of many other reagents would have been elevated as well.

The geothermal-field hypothesis is inspired, in part, by one of the great mysteries in cell biology – the elevated use of various cations relative to their environmental abundances. Intracellular concentrations of potassium, zinc, manganese, and phosphate (and most other elements; Chapter 7) are generally

higher than those found in modern sea water, whereas the opposite is true for sodium (Mulkidjanian et al. 2012b). To achieve intracellular disparities in ion concentrations, modern cells invest considerable energy in the operation of ion pumps (Chapter 18), raising the question as to whether the differential cellular use of potassium and sodium is a historical relic. Because sodium concentrations are low in geothermal fields relative to the situation in marine environments, the geothermal-field hypothesis provides a potential explanation for this conundrum. The earliest membranes almost certainly would have been permeable, and ions with the greatest environmental availability would arguably have been more subject to exploitation in the earliest stages of cell-physiological evolution.

There are other attractive features of the geothermal-field hypothesis. First, although today's terrestrial geothermal fields have extraordinarily low pH levels, this likely would not have been the case on the surface of an early Earth devoid of oxygen, as the H_2S associated with geothermal activity would not have been oxidized to sulfuric acid. Second, metal sulfides likely would have precipitated in shallow waters. Notably, ZnS crystals are powerful photocatalysts, raising the possibility that diverse hydrocarbons might have been produced by a sort of abiotic photosynthesis in shallow, light-rich environments. Third, as they are highly efficient at scavenging UV light, ZnS and MnS crystals also have photo-protective capacity. Thus, Mulkidjanian and colleagues (the primary architects of this hypothesis) envisioned a layered system, with production of organics occurring at the surface, and the lower layer harboring protocells for harvesting such molecules. In accordance with the geothermal-field hypothesis, phylogenetic analyses of various enzymes (which attempt to pinpoint their first appearance in the Tree of Life) suggest a very early origin of those utilizing Zn, Mn, H_2S, and K (but not Na) (Dupont et al. 2010; David and Alm 2011).

Deamer and Georgiou (2015) provide a synopsis of the empirical evidence supporting the hydrothermal-vent and the geothermal-field hypotheses, and propose key tests for further discrimination between the two. It should be emphasized, however, that the scenarios painted by these two hypotheses are by no means the only possible routes to biotic evolution. Notably, although both hypotheses are focused on environments with fairly high temperatures, a number of experiments with RNA have shown that the assembly and maintenance of polymers is actually facilitated at low temperatures (reviewed in Higgs and Lehman 2015). This is an obvious matter of concern with respect to the origin of an information-bearing genome, and the well-known instability of complex molecules at high temperatures. The geothermal-field hypothesis is also confronted with an additional issue – the possibility that the entire Earth was under water at the time of life's origin (Dong et al. 2021).

An Early RNA World?

In all of the above scenarios, an autocatalytic mode of metabolism emerges before the appearance of any genetic machinery. Without a mode of inheritance or replication, such starting points do not meet our definition of life, although they do offer potential explanations for a number of cellular features. Nonetheless, regardless of which (if any) of these suggested links to the past is correct, abiotically induced chemical reactions must have been taken over eventually by protein catalysts, and therein lies the problem. The spontaneous assembly of complex proteins from environmental sources of amino acids is extraordinarily unlikely, and the refinement of catalytic properties in the absence of a replicating genome, required for the operation of natural selection, is even less likely. Because all modern cells are incapable of genome replication in the absence of a substantial set of helper proteins, this is the 'chicken-and-egg' problem for the origin of life.

A potential solution to this problem would be a set of molecules that served simultaneously as catalysts and information-bearing polymers. Proteins carry out a bewildering diversity of tasks, but self-replication is not one of them. DNA provides a superb substrate for information storage but is generally catalytically inert. RNA is the only biomolecule in today's cells for which some variants can both specify a genotype and express a phenotype. This then leads to the suggestion that RNA is the only reasonable candidate for a starting point in evolution. Invoking this default state, the RNA-world hypothesis postulates that at some point early in the evolution of life, genetic continuity was assured by the replication of an RNA-based genome, with all underlying catalysis also being carried out by RNAs, and no involvement of proteins or DNA (Woese 1967; Crick 1968; Orgel 1968; Gilbert 1986; Robertson and Joyce 2012; Higgs and Lehman 2015). Under this view, the complex protein repertoire now employed by all organisms arose secondarily, layered on top of the more fundamental RNA scaffold.

A number of observations provide indirect support for this hypothesis. First, RNA's use of four nucleotides constitutes a language, and the potential for double-strandedness allows for a template-based mechanism for replication. Second, with its ability to fold into complicated stem-loop structures, RNA is structurally diverse and capable of a wide variety of catalytic properties, including binding to proteins and facilitating amino-acid chain formation. Third, across the Tree of Life, all of the major players in today's protein synthesis are derived from RNA – transfer RNAs, messenger RNAs, the catalytic cores of the ribosome and of the eukaryotic spliceosome, and numerous small RNAs involved in transcript silencing and/or proliferation. Fourth, many of the central players in metabolism are nucleotide derivatives, e.g., ATP, coenzyme A, NAD (nicotinamide adenine dinucleotide), and FAD (flavin adenine dinucleotide). Fifth, *in vitro* experiments on populations of RNA molecules demonstrate the evolvability of RNA in simple systems under selection for a wide variety of catalytic activities (e.g., Wilson and Szostak 1999; Joyce 2004).

If there was an early RNA world, it may have coexisted with and even exploited one of the physical energy-generating scenarios outlined previously. Successful members of the population of RNA molecules might then have gradually evolved a coding mechanism for producing proteins for enhancing the efficiency of energy harvesting, and eventually displacing the abiotic pathway entirely. Provided that a population of such proto-genomes inhabited an environment spatially structured enough to ensure the association of metabolites with their source (e.g., a proto-membrane), such a setting might have initiated an auto-catalytic process of self-improvement (i.e., natural selection) on the path to what we now call life. Under this view, just as we see the reliance on a proton-motive force and on particular elements as being evolutionary relics, the diverse roles played by RNA in today's cells, especially in transcript processing and translation, can be viewed as molecular descendants of this early era of biochemistry.

There remains the fundamental question of how an information-bearing system can get started before the arrival of a platform for exploiting information. One simple scenario is outlined in Foundations 2.2, which shows that, provided there is a mechanism of polymerization and recurrent input of alternative monomeric building blocks, an equilibrium population of polymers with variable lengths and sequences will naturally evolve. Such a condition is expected to emerge even in the absence of a mechanism of self-replication, as the alternative states simply grow and decay out of a series of stochastic chemical reactions. Thus, it is not too far-fetched to imagine that a geochemical setting that provided a source of alternative ribonucleotides and a means for concatenating them would be primed towards developing a system of molecules carrying a potentially exploitable language. What remains, however, is the need for a genetic and cellular mechanism for the production and maintenance of heritable variation essential to the operation of natural selection.

Although these observations support the plausibility of an RNA-world episode in the origin of life, it should be kept in mind that the RNA world is a hypothesis, not a confirmed fact. One central caveat is that it is now known that, like RNA, DNA can take on a number of catalytic functions (when maintained in single-stranded form, but allowed to fold into stems and loops); these include RNA cleavage, RNA and DNA ligation, and conjugation of amino acids to nucleotides (Silverman 2016).

Other unresolved issues leave room for doubt about the extreme model in which RNA is the only replicator and the only catalyst (Shapiro 2006; McCollom 2013). First, there is the difficult question of how the basic building blocks of RNA, the ribonucleosides, were sustained and replenished. Mechanisms for the abiotic production of purine and pyrimidines have been demonstrated

(Robertson and Joyce 2012; Becker et al. 2016), and ribose might also have arisen in the environments envisioned in the hydrothermal-vent and geothermal-field hypotheses, especially if boron was present (Kim et al. 2011; McCollom 2013; Neveu et al. 2013; Pearce et al. 2017). Under some conditions, most notably wet–dry cycles that may have occurred in geothermal fields, purines and pyrimidines can even be coupled with ribose to make nucleosides and further polymerized to small RNAs (Powner et al. 2009; Neveu et al. 2013; Becker et al. 2019). The speed of such reactions is not impressive, but early life had the luxury of time and perhaps lack of competition. Still, there remains the central problem that in today's organisms, nucleobases are synthesized from amino-acid precursors, the building blocks of protein, not *de novo* from inorganic materials.

Second, jump-starting an RNA world would not only require a pool of ribonucleosides, but a means for activating them with pyrophosphates to promote chain growth. A mechanism to polymerize these basic building blocks would also be required. Moreover, to maintain continuity across generations, self-replicators would need to avoid accidental replication of other competing genomes, which would stifle the process of natural selection.

Third, although key steps towards self-replication have been accomplished (e.g., Lincoln and Joyce 2009; Mizuuchi et al. 2022), a fully self-replicating ribozyme has not yet been developed, despite considerable effort. One could argue that this is a minor concern, as the emergence of life had eons of time in which to discover a few exceptional molecules. This dwarfs the three decades of research performed in a handful of laboratories in 10-ml test tubes, $\sim 10^{23}$ of which would be required to match the mass of oceanic water. Indeed, recent laboratory experiments involving thermal gradients (Mast et al. 2013) or ice substrates (Attwater et al. 2013) have succeeded in developing ribozymes capable of polymerizing RNA sequences up to 200 bp in length, the approximate size generally necessary for catalytic activity.

Fourth, RNA is known to be maximally stable at slightly acidic pH (4.0–6.0) and unstable in alkaline water. This reduces the appeal of the alkaline hydrothermal-vent hypothesis unless there was a spatial mechanism for decoupling metabolism and replication/information storage, but may be more compatible with the scenario postulated by the geothermal-field hypothesis (Bernhardt and Tate 2012).

Confronted with these numerous requirements for the assembly and maintenance of ribonucleotide polymers, some have suggested a pre-RNA world dominated by some other polymer capable of genetic and catalytic functions (Joyce and Orgel 1999; Orgel 2000; Shapiro 2006). There is no shortage of candidate molecules. For example, numerous nucleotide analogues substitute various moieties, including amino acids, for ribose (Robertson and Joyce 2012), and some of these are capable of non-enzymatic template copying (Pinheiro et al. 2012; Zhang et al. 2013). The abiogenic production of hybrid molecules consisting of RNA pyrimidines and DNA purines has even been demonstrated (Xu et al. 2020). Why oligonucleotides at all? One argument is that such structures may have been chemically selected over other compounds, as they are powerful deactivators of UV light and exceptionally photostable (Mulkidjanian et al. 2003; Serrano-Andrés and Merchán 2009; Dibrova et al. 2012).

Two strong arguments favor DNA arising after both RNA and proteins. First, an early RNA-protein world requires the existence of a genetic code prior to the arrival of DNA, which is consistent with the ubiquitous use of transfer, messenger, and ribosomal RNAs in translation. Second, modern cells derive their DNA building blocks (deoxyribonucleotides) via chemical modifications of ribonucleotides. Ribonucleotide reductases are used in the production of dAMPs, dCMPs, and dGMPs, while thymidylate synthase produces dTMPs by methylating dUMPs (Figure 2.4). Remarkably, the two primary prokaryotic lineages, the bacteria and the archaea (Chapter 3), utilize seemingly unrelated thymidylate synthases (Myllykallio et al. 2002) as well as two apparently unrelated sets of DNA replication proteins (Edgell and Doolittle 1997; Olsen and Woese 1997; Leipe et al. 1999). Such observations raise the intriguing possibility that the shift to a DNA world may have occurred more than once.

Given the assumed early success of a pre-DNA world, why would the transition to a DNA world be so complete as to eradicate all RNA-based genomes (other than RNA viruses) from the cellular domains of life? One attractive answer invokes two chemical features that enhance the stability of DNA-based genomes. First, the additional -OH group on ribose (Figure 2.4) renders RNA much less structurally stable than DNA. Second, one of the most common sources of mutation is spontaneous cytosine deamination, which produces uracil. In thymine-bearing DNA, uracil can be recognized as aberrant and corrected prior to replication (once a mechanism for such recognition is established), but such a distinction is impossible in RNA. Thus, an organism that discovered a way to store its genome as DNA while retaining RNA for non-heritable phenotypic functions would have had a substantial advantage in terms of reliable genome propagation.

Membranes and the Emergence of Individuality

Left unclear in the previous discussion is how a metabolism-first scenario, an RNA world, or a collaboration between the two might have led to an eventual transition to an autonomous membrane-bound cell. Such encapsulation was almost certainly required before life could occupy the vast open space of marine environments. For without discrete individuals, the efficiency of natural selection would have been greatly diminished as any key metabolites produced by a local entity would become public goods, eliminating the genotype-phenotype loop. On the negative side, cellularization reduces access to resources, although the earliest membranes might have been quite permeable.

The most compelling reason to think that the most-recent common ancestor to all of today's life had a cell envelope is the universal use of membrane-bound ATP synthase (noted earlier) as well as several other membrane-associated proteins (Jékely 2006; Mulkidjanian et al. 2009). All membranes in today's organisms consist of some form of lipid, and there are good reasons to think that this was the case in the earliest cells. First, although the concentrations are unknown, we can be virtually certain that lipids were present before the emergence of biology. Fatty acids, from which lipids are built, have been found in extra-terrestrial rocks, such as the Murchison meteorite (Deamer

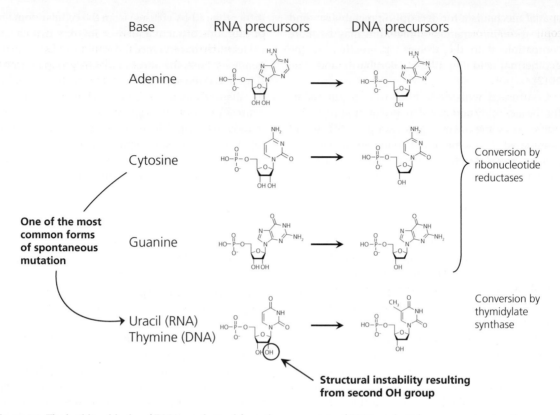

Figure 2.4 The building blocks of DNA are derived from the conversion of RNA nucleotide precursors using the enzymes denoted to the right. Owing to the presence of a second hydroxyl group (circled in red for uracil), RNA nucleotides are more unstable than DNA nucleotides. In addition, cytosine deamination results in the production of uracil.

1985), and can be synthesized under plausible prebiotic conditions. For example, starting with CO, CO_2, and H_2 gases in the presence of metal catalysts at high temperature, a reaction known as Fischer–Tropsch synthesis can lead to long-chain fatty acids (McCollom et al. 1999). Second, lipids spontaneously assemble into organized bi-layered vesicles, with their hydrophobic tails and hydrophilic heads pointing to the inside and outside of the membrane, respectively (Chapter 15). Thus, it is quite plausible that lipids provided a natural starting point for membrane development even prior to the evolution of a genomically encoded mechanism for lipid biosynthesis.

Of course, the emergence of an autonomous cell would be much more likely if a genome, metabolism, and a membrane did not have to

evolve independently in a stepwise fashion, but instead somehow facilitated each other's development. Some intriguing experiments demonstrate such possibilities. For example, using a simple system containing protocells made out of lipid bilayers, some empty and some containing RNA, Chen et al. (2004) showed that vesicles with high RNA concentrations experience osmotic stress that is relieved by the recruitment of lipid molecules from empty vesicles. This neatly demonstrates the intrinsic capacity of a genome-containing protocell to grow in the absence of any encoded mechanism for growth (and indeed, in the absence of any initial genomic function at all). That is, if RNAs and lipids were co-localized in the same environment, the system would not only have been naturally biased towards spontaneous growth, but also may have set up a sort of competition, with the acquisition of membrane components being dominated by genome-containing vesicles. Further work has shown that, in a setting containing DNA templates and DNA primers, polymers can grow within a membrane-bound vesicle permeable to charged nucleotides, again causing osmotic pressure and vesicle growth (Walde et al. 1994; Mansy and Szostak 2009).

Once such a system like this was in place, any cell that contained a faster replicator would experience still more rapid growth, thereby initiating a process of natural selection. There remains, however, the question of how membranes might come to be associated with nucleotides at all. Black et al. (2013) found that when co-localized, nucleotide bases and ribose not only associate with lipids, but also stabilize the resultant aggregates in saline water. The fact that all of these features are a simple consequence of chemistry and physics again leads to the conclusion that, rather than arising as a series of unimaginably low-probability events, molecular liaisons that constitute key aspects of life may have emerged semi-deterministically via natural abiotic processes.

Despite this progress, a number of gaps remain in our understanding of how membrane-bound life might have arisen. The unknowns include the mechanisms by which: (1) repeated rounds of nucleotide polymerization might have been achieved; (2) competition between strand rean-nealing and new chain growth might have been avoided; and (3) an organized mode of cell division might have been established. Prior to the emergence of a precise cell-division mechanism, proto-cell fission might have been governed simply by physical forces associated with the environment. For example, Zhu and Szostak (2009) found that as spherical vesicles grow, they eventually elongate into filamentous forms that are then subject to subdivision into daughter vesicles when the surrounding fluid is agitated. When associated with certain photochemically active compounds, filamentous vesicles can also differentiate into strings of ellipsoid subcompartments that eventually fragment into individual vesicles (Zhu et al. 2012).

From observations on these simple kinds of systems, one can imagine how a natural environment subject to redox cycles (perhaps driven by the diurnal light cycle) might have provided a purely physical mechanism for regulating protocell division. However, the short-chain fatty-acid membranes employed in these studies are highly permeable to a wide variety of nutrients, including amino acids and nucleotides. In contrast, modern cells are generally bounded by phospholipid bilayers, which impose a much stronger barrier to charged ions and polar molecules. Assuming the earliest lipids were simple in form, how might a transition to phospholipids have come about? Budin and Szostak (2011) found that two-chain phospholipids can compete for incorporation into lipid membranes, raising the point that any cell that encoded a mechanism for producing such molecules from single-chain lipids might have gravitated toward the use of phospholipid membranes for purely physical reasons.

This, however, raises still another issue – by conferring reduced permeability, the progressive establishment of a phospholipid membrane would diminish access to the external environment. In principle, the gradual emergence of the phospholipid membrane may have fostered the evolution of internal cellular biosynthetic pathways to compensate for the reduction of external resource availability. However, such an argument is not particularly compelling from an evolutionary perspective, as any realistic scenario for this sort of transition would have required a series of steps in which cellular fitness was never diminished. Natural selection allows populations to increase fitness in response to selective challenges, but does not encourage the continuous exposure to harmful conditions.

Finally, note that all of the preceding views on the origin of membranes are based on the assumption that the cytoplasm of protocells resided inside cell membranes. An alternative view considers the opposite topology, with the cytoplasm initially nucleating on the outside of vesicles called obcells, perhaps being held in place by actin-like filaments (Blobel 1980; Cavalier-Smith 2001; Griffiths 2007). In principle, such entities could have been cup shaped, with the open side attached to a substrate (like a suction cup). Cellularization might then have evolved as the membrane of the liposome somehow completely invaginated (as in embryonic gastrulation) or as pairs of cups fused together, engulfing the previously external protoplasm. Although perhaps not impossible, these kinds of scenarios for the start of life ignore the central requirement for individuality to enable the efficient operation of natural selection.

Genomic Constraints on the Establishment of Life

Most origin-of-life researchers can be subdivided into two opposing camps. Those with a metabolism-first affinity find it unfathomable that an even moderately complex genome could ever be assembled prior to the establishment of a reliable source of energy for biocatalysis. In contrast, the genome-first school argues that complex catalytic pathways could never be assembled without the guiding hand of information-bearing molecules. Both camps put a priority on framing hypotheses based on presumed geochemical/biophysical constraints, an entirely reasonable and desirable enterprise. However, remarkably absent from this debate is the equally important matter of the maintenance of a population with heritable features capable of progressive adaptation. Indeed, most origin-of-life research has proceeded with almost no consideration of fundamental evolutionary processes.

To ensure a sustainable and productive path forward by natural selection, an information-bearing molecule must be capable of generating accurate copies of itself. However, at the dawn of life, replication fidelity would have been much less accurate than in today's refined organisms, imposing a significant restriction on genome size. Given its relative chemical stability, the arrival of DNA might have provided a more permissive environment for genomic expansion and hence, the emergence of more complex biological functions, but there still must have been substantial limitations.

In the first attempt to grapple with this issue, Eigen (1971) and Eigen and Schuster (1977) proposed the concept of a molecular quasi-species, with the master sequence being the genome with maximum fitness. Under this model, an error threshold is reached when the mutation rate is high enough that the master sequence cannot be maintained, i.e., the rate of promotion by selection is offset by mutational degradation to adjacent states.

Loss of the master sequence need not imply extinction of the entire species, as suboptimal molecules may still have sufficient fitness to ensure numerical replacement of the population across generations. However, there exists a still higher mutation rate beyond which the population can no longer even be sustained (Bull et al. 2005). Theory from population genetics (Foundations 2.3) shows that, for a given genome size, extinction avoidance requires a sufficiently high replication fidelity that enough mutation-free offspring genomes are produced each generation to avoid progressive fitness loss by stochastic sampling.

A key parameter dictating the fraction of deleterious mutation-free individuals in an asexual population is the ratio $\phi = U_d/s$, where U_d is the genome-wide deleterious mutation rate, and s is the fractional selective disadvantage of a deleterious mutation. The expected proportion of mutation-free individuals in a very large population in selection-mutation balance is $e^{-\phi}$ (Foundations 2.3), so if ϕ is much larger than 1.0 (i.e., if the rate of introduction of deleterious mutations greatly exceeds the power of selection to remove them), mutation pressure alone will ensure the progressive loss of the highest fitness classes in all but enormous populations, eventually leading to population extinction by mutational meltdown. Indeed, this general principle underlies the application of lethal-mutagenesis strategies for eradicating pathogens (Bull et al. 2005, 2007; Bull and Wilkie 2008; Chen and Shakhnovich 2009; Jensen et al. 2020). Using laboratory populations of RNA molecules with an error rate $> 10^{-5}$ per nucleotide site, Soll et al. (2007) showed that the time to extinction is positively related with population size, as theory predicts.

The concept of mutational meltdown is based on the inevitable consequences of sampling of propagules from one generation to the next (Lynch and Gabriel 1990; Lynch et al. 1993). Imagine that, following a generation of selection, the fraction of individuals in the best class is p and that N random progeny are then derived from all surviving classes. The probability that a single draw does not contain a member of the best class is $(1 - p)$, and that of not drawing any members of this class at all is $(1 - p)^N$, where N is the current population size. If there is just a single individual in the best class ($p = 1/N$), then the probability of not drawing any progeny from this class is 0.367, and if there are two or three such individuals, the probability of loss is still substantial, 0.135 and 0.050, respectively. The main point is that if there are very few individuals in the best class, which will inevitably be the case with high recurrent mutation pressure, there is an appreciable probability that the best class will be lost in any particular generation, and virtually certain probability that it will be lost over multiple generations. Once the best class is lost, the previously second-best class will be advanced to premier status, but it too will eventually suffer the same fate. With the mean fitness of individuals declining, a critical point will ultimately be reached when the average individual cannot replace itself, which causes a reduction in population size. This sets in motion a downward spiral towards extinction, as population-size reduction progressively increases the probability of loss of the best-class individuals, with each such loss leading to still further loss in fitness.

Mutational-meltdown theory (Foundations 2.3) is sufficiently well-developed that one can obtain mathematical approximations for the expected time to extinction given U_d and s. However, a more satisfying theory would start from first principles, generating an expected value of U_d resulting from selection for replication fidelity, rather than simply assuming an arbitrary value. As outlined in Foundations 2.3, the strength of selection operating on the mutation rate in an asexual population is primarily a function of the difference in U_d among different genotypes, independent of the effects of individual mutations. Once selection has driven the mutation rate down to a sufficiently low level that the next increment of possible improvement in replication fidelity is smaller than the role of chance fluctuations in the population (the magnitude of which is related to the reciprocal of population size; Chapter 4), no further improvement is possible. As a first-order approximation, for example, for a small population with 10^4 individuals, U_d cannot evolve to a level much lower than 10^{-4}. Thus, because U_d is approximately equal to the product of the mutation rate per nucleotide site and the number of sites in the genome with functional effects, for a population of any particular size, this puts an upper limit on the genome size consistent with maintaining sufficient numbers of mutation-free individuals to avoid the meltdown. Back and/or compensatory mutations may relieve the scenario outlined above somewhat (Wagner and Gabriel 1990; Poon and Otto 2000; Silander et al. 2007; Goyal et al. 2012), but sufficiently high mutation pressure will still eventually lead to a mutational meltdown (Zeldovich et al. 2007; Chen and Shakhnovich 2009).

The relevance of these results to understanding the origin of life is that levels of replication fidelity were likely quite low in the earliest stages of life. For example, polymerization off an RNA template in simple laboratory experiments typically yields error rates on the order of 0.01 to 0.1 per base incorporated, even under optimal conditions (Johnston et al. 2001; Attwater et al. 2013; Zhang et al. 2013), many orders of magnitude higher than in any of today's cells (Chapter 4). Because a genome with a length much greater than the inverse of the mutation rate per nucleotide site has essentially no chance of spawning an intact offspring molecule, this makes clear the challenge to early life – the need to encode for a high level of replication fidelity in an appropriately small genome. One possible way around this size-limitation problem is the joint operation of a set of suitably small cooperative molecules in a closed cycle with each member in the loop being responsible for the next member's replication (Attwater and Holliger 2012; Vaidya et al. 2012; Higgs and Lehman 2015). However, the salient point remains – to understand the origin of life, a resolution of population-genetic issues is just as critical as a focus on promising geological settings.

Summary

- The basic properties of life were established on Earth ~3.5 to 4.0 BYA. However, the order in which the three major requirements for life – metabolism for resource acquisition, a genome for the heritable transmission of genetic information, and external membranes necessary for individuality – remains a matter of speculation.

- Plausible hypotheses identify certain environmental settings on an early Earth that might have been conducive to the spontaneous emergence of the ancestor of all of today's organisms. However, the commonly invoked open-water primordial soup is not one of them. The peculiar sets of reactions, metal cofactors, and metabolic building blocks that life came to depend on may be a reflection of the ancestral setting.

- Given the core roles played by RNA in several key functions in today's organisms, the RNA-world hypothesis promotes the view that a single type of molecule (RNA) simultaneously provided the means for catalysis and information storage in the earliest stages of biotic evolution.

- One conceptual problem with origin-of-life narratives is that they are just that. Demonstrations of the plausibility of single steps towards life in restricted chemical/physical environments too often lead to increased adherence to specific views on the nature and order of events leading to the origin of life. This weaving of entire series of low-probability events into a convincing, comprehensive scenario should be interpreted with caution.

- Major downstream challenges in the origin of life would have involved the establishment of reliable means of genome transmission, which requires membranes for individuality, accurate RNA/DNA polymerases for replication fidelity, and a platform for the production of catalytic machinery consisting of proteins.

- Putting aside the shortcomings in our understanding of the specific steps towards the establishment of life, considerable experimental evidence suggests that, rather than being improbable, the origin of life may be a nearly inevitable consequence of the geochemical environments on early Earth-like planets.

Foundations 2.1 The proton-motive force and the evolution of ATP synthase

Life requires energy-capturing mechanisms to sustain the work necessary for cell growth, survival, and reproduction. Although numerous sources of external energy are available to today's organisms, these must ultimately be converted to ATP, the universal currency of cellular energy storage and transport. Given that even a small bacterium requires the equivalent of about 30 billion ATP hydrolyses per cell division (Phillips and Milo 2009; Lynch and Marinov 2015; Chapter 8), the convoluted path by which ATP is produced and recycled is all the more remarkable.

Much like a dam generating electricity via water passing through a turbine, a process called 'chemiosmosis' drives the cell's production of ATP by directing a gradient of hydrogen ions through channels in otherwise impermeable membranes (Figure 2.5). The protein complex involved, ATP synthase, sits in the cell membranes of prokaryotes and in the inner mitochondrial and chloroplast membranes of eukaryotes. However, unlike the situation with hydroelectric power, proton flow is not free. Instead, the proton gradient essential to the process is set up by the cell itself – using the electron-transport system, protons derived from the oxidation of food are translocated to the exterior of the membrane. They are then reimported through ATP synthase, where the energy associated with the proton-motive force is used to convert ADP and inorganic phosphate (usually denoted as Pi) into ATP. Although cases of substrate-level phosphorylation of ADP are known, almost all organisms rely on chemiosmosis to regenerate ATP from ADP. Thus, we can be fairly certain that a membrane-embedded ATP synthase was used by the last (universal) common ancestor of all of life (generally denoted as LUCA).

Given the centrality of ATP synthase to bioenergetics, the establishment of this complex can be viewed as one of the key events in the history of cellular evolution. But why did life adopt such an arcane mechanism of energy harvesting? One possibility is that the reliance on a proton-motive force is a historical relic of the exploitable energy sources present at the time of life's foundation. Under this hypothesis, early life would have relied on an environmentally derived proton gradient (such as passive vent-associated energy) until establishing its own membrane-based mechanism for self-generating such a gradient and converting the mechanical energy from the returning proton movement to chemical energy in the form of ATP (Lane et al. 2010). If this view is correct, life could not have inhabited open-water environments (which do not provide strong, small-scale energy gradients) without first acquiring bioenergetic cell membranes and a sophisticated genome encoding them. An additional implication is that the use of ATP as an energy carrier emerged prior to the evolution of ATP synthase.

ATP synthase is a complex molecular machine, generally consisting of at least two dozen protein subunits (Figure 2.5), assembled into a membrane-bound pore (F_0), which in turn is connected to a central stalk that rotates within a large internal ring (F_1) kept stationary by a membrane-attached stator. Pushed by the pH gradient, protons flow through the complex, causing the stalk to rotate ~100 to 150 times/second (about 10 × more rapidly than the rotation of the wheels of a car moving at 60 miles/hour). Synthesis of ATP from ADP occurs as the rotating stalk interacts with the stationary F_1 ring carrying the catalytic subunits (Walker 2013).

Two types of ATP synthases are known: the so-called F-type found in bacteria and organelles, and the V-type found in archaea, some bacteria, and eukaryotic vesicle membranes (Cross and Müller 2004; Mulkidjanian et al. 2007, 2008). Owing to the complex structure of ATP synthase, the mechanisms of its origin are far from clear. The probability of the sudden *de novo* evolutionary emergence of such a machine is minuscule, so the origin of ATP synthase most likely involved the exploitation of pre-existing modules engaged in other functions. The membrane subunit (F_0) could plausibly have been derived from a membrane pore (Walker 1998); a relationship to the membrane-bound motor of the bacterial flagellum has also

been suggested (Mulkidjanian et al. 2007; Chapter 16). Similarities also exist between the internal catalytic subunit (F_1) and the ring-like helicases that use energy from ATP hydrolysis to separate the strands of DNA (Walker 1998; Patel and Picha 2000).

How these changes became established without the loss of ancestral gene functions remains unclear, but some involvement of gene duplication and reassignment seems almost certain (Chapter 5). Indeed, the catalytic component (F_1) of ATP synthase consists of a hexameric ring of two alternating subunits (α and β) derived from an ancient gene duplication (with just one of the subunit types carrying out catalysis).

Despite the centrality of ATP synthase to energy harvesting, the enzyme exhibits significant structural variation with apparent functional implications. Bioenergetic efficiency is directly related to the structure of the membrane subunit (Soga et al. 2017). Each full rotation of the F_0 ring leads to the production of three ATP molecules (one for each of the F_1 subunits with catalytic properties), and the number of protons required per rotation is equal to the number of subunits in the rotating ring. Thus, in the yeast *Saccharomyces cerevisiae*, where there are ten subunits in the membrane ring, the bioenergetic cost is 10/3 = 3.33 protons/ATP molecule produced. In bovine ATP synthase, however, there are only eight ring subunits, and sequence comparisons suggest that a similar structure may exist in

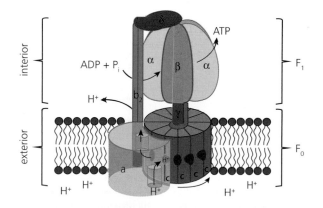

Figure 2.5 Idealized structure of ATP synthase. As a result of the export of hydrogen ions (protons) across a membrane (not shown), a proton concentration differential is maintained by cells. The resultant chemiosmotic gradient encourages a focused re-entry of protons through the ATP synthase complex, as the rest of the membrane is impermeable. The membrane-embedded rotor (containing the c subunits, and called F_0) rotates as protons pass through. This, in turn, causes the attached central rod (denoted by γ) to rotate, which activates the catalytic sites on the stationary, internal ring (with subunits α and β, and called F_1), converting ADP to ATP. This internal ring is kept stable by the stator apparatus (denoted by subunits b_2 and δ), which is anchored to the membrane with an additional structure (shown in light blue). After passing through the rotor, the protons exit to the internal side of the membrane and are eventually re-exported, maintaining the gradient. For prokaryotes, the interior refers to the cytoplasm; for eukaryotes, it refers to the inside of the mitochondrion. From Boyer (2002).

all metazoans (Watt et al. 2010). For these species, the cost of each ATP is just $8/3 = 2.67$ protons. Notably, the stoichiometry of the (F_0) ring is conferred by the sequence of the monomeric subunit (c), so that gene transfer from one species to another maintains the donor structure. This enables the experimental analysis of shifts in gear structure (Meier et al. 2005; Matthies et al. 2009).

Across the Tree of Life, the number of ring subunits among characterized species ranges from 8 to 15, with known prokaryotic structures covering nearly the full range of variation (Walker 2013). Thus, in terms of proton utilization, there is a nearly twofold range of variation in the cost of ATP production among species. (A slight correction is that, in mitochondria, there is an additional cost of one proton per ATP associated with the import (Walker 2013), so the total cost of ATP production for eukaryotes is the above plus 1.0; this cost of membrane transport is not incurred in prokaryotes). Also not included in this analysis is the cost of producing the hydrogen-ion gradient itself.

These observations raise the question as to why all species don't utilize a system with the efficiency of that in the mammalian mitochondrion. But this is not the only mystery posed by ATP synthase, as still other structural variants are known. For example, two key protein components normally present in the stator appear to be absent from parasitic apicomplexans and free-living ciliates, whereas the complex in the ciliate *Tetrahymena* contains at least thirteen novel proteins not found in other organisms, including the use of two stators, rather than the one found in other characterized species (Balabaskaran Nina et al. 2010). Members of the Chlamydomonadales (a group of green algae) have nine unique stator proteins (Lapaille et al. 2010), and *Euglena* ATP synthase has eight unique subunits (Mühleip et al. 2019). Some of these modifications have effects on the higher-order assembly

of ATP synthases into multimeric clusters in mitochondria (Flygaard et al. 2020; Mühleip et al. 2021). Many other differences are known in the bacteria and eukaryotes (Müller et al. 2005; Walker 2013).

If the preceding observations are not complicated enough, note that ATP synthases are actually reversible molecular machines that in certain contexts (e.g., vesicle acidification in eukaryotes) act as ATPases, with hydrogen ions being pumped with energy derived from the conversion of ATP to ADP. Moreover, a number of F- and V-type ATP synthases couple the synthesis/hydrolysis of ATP with the transport of sodium ions rather than protons, and some are capable of using both. The interwoven phylogenetic relationships of the Na- and H-utilizing enzymes (Mulkidjanian et al. 2008) leave the ancestral state ambiguous.

Thus, ATP synthase, a central requirement for all of life, has undergone substantial structural modifications despite the retention of a highly conserved function. There is, as yet, no evidence that any of these changes has been driven by adaptive processes, and many questions remain unanswered. Why the use of a membrane-bound machine, and why the use of a rotary motor? Why the reliance on ATP and not CTP or some other nucleoside triphosphate (or tetra- or higher-order phosphate)? Does the use of Na as a driving ion imply anything about the context in which ATP synthase originated?

Here, we have only considered the ultimate molecular machine involved in membrane-based bioenergetics, ATP synthase. An excellent overview of the knowns and unknowns of the complexes involved in the proton pumping upon which the system depends, the electron-transport chain and its many variants, is provided by Goldman et al. (2023).

Foundations 2.2 Evolution prior to self-replication

Although ample evidence exists that most of the basic building blocks of life can emerge via abiotic processes, life requires the polymerization of monomeric subunits into linear arrays. How might populations of such polymers have initially come about in the absence of any mechanisms to specify their sequence or to exploit the information carried? And to what extent is the population of such molecules capable of evolution before a mechanism of replication has been acquired?

To examine these questions, Nowak and Ohtsuki (2008) considered a hypothetical prebiotic situation involving just two types of activated monomers, *0 and *1, each capable of joining a pre-existing polymer on one side

(like the joining of nucleotides only at the 3′ end of a nucleic-acid chain). This type of model readily extends to situations with more than two types of monomers, but the general principles are most easily seen with just two alternative states at each site.

The possible polymeric states under this system consist of binary strings of various lengths: 0 and 1 for monomers; 00, 01, 10, and 11 for dimers; 000, 001, 010, 011, 100, 101, 110, and 111 for trimers, etc., so there are 2^L possible sequences of length L. Denoting an arbitrary polymer as i, there are two possible elongation reactions:

$$i + *0 \rightarrow i0$$

$$i + *1 \rightarrow i1$$

occurring at rates a_{i0} and a_{i1} (the absence of an ∗ indicates a recipient molecule). The population of possible molecules can then be viewed as two nested trees (starting with either 0 or 1) (Figure 2.6). Each sequence i has a single possible precursor (i') and two possible descendants ($i0$ and $i1$).

Letting the rate of conversion of i' to i be a_i, assuming a death rate of i equal to d_i, and assuming a very large population size (so that stochastic fluctuations of frequencies can be ignored), a general description of the dynamics of the system is given by

$$\frac{dn_i}{dt} = a_i n_{i'} - (d_i + a_{i0} + a_{i1})n_i, \qquad (2.2.1)$$

where n_i is the abundance of string i. The first term on the right denotes the net flux from class i' into i, while the second term denotes the flux out of i resulting from either death (rate d_i) or the production of the next higher-order polymeric states (rate $a_{i0} + a_{i1}$). Assuming the monomeric precursors (0 and 1) are kept at a steady state arbitrarily scaled to $n_{0'} = n_{1'} = 1.0$, and letting $b_i = a_i/(d_i + a_{i0} + a_{i1})$, provided all $b_i > 0$, this system will eventually evolve to an equilibrium composition from any starting point,

$$n_i = b_i b_{i'} b_{i''} \cdots, \qquad (2.2.2)$$

where the string of b coefficients goes back to the base of the tree (0 or 1). In words, the expected abundance of sequence i is simply equal to the product of all coefficients leading from its starting monomer. At this steady-state condition, for every unique string the total influx from the precursor is equal to total efflux into the two descendent string classes plus the death rate.

Although the general solution may be difficult to visualize, now imagine that both monomers behave identically, so that: 1) the baseline conversion rates from the activated monomers ∗0 and ∗1 to states 0 and 1,

respectively, are both equal to $\lambda/2$ (so the total rate of birth of polymerizable strings is equal to λ); 2) $a_i = a$ for all other classes (so all members of a particular length class grow at identical rates); and 3) there is a constant death rate per chain. At equilibrium, each of the 2^L possible sequences of length L then have abundance

$$n_L = \left(\frac{\lambda}{2a}\right)\left(\frac{a}{2a + d}\right)^L. \qquad (2.2.3)$$

Because the fraction on the right is smaller than one, this shows that the abundance of sequences declines exponentially with the length, so that even though there is no physical upper limit imposed on L, long sequences become diminishingly rare. The mean sequence length is $1 + (2a/d)$, which is quite small unless $a \gg d$. Note, however, that long sequences are not rare because of any intrinsic fitness disadvantage, but simply because of the cumulative mortality of the $L - 1$ precursor sequences leading up to them. Scaled to the abundance of activated monomers (arbitrarily set equal to 1.0), the summed abundance over all strings (the total population size) is λ/d, i.e., the ratio of the rates of input and output for the population.

A number of key points arise from these results. First, although there is no self-replication in the system, a steady-state abundance distribution is maintained by the recurrent introduction of unit length strings (0 and 1). Second, if the rate constants take on different values, quite different distributions of string types will be obtained, i.e., the underlying conversion rates determine into which classes the overall system is channelled. In effect, with unequal transition rates, some classes will grow more rapidly than others, so that the system undergoes a kind of natural selection. Third, because the products of this birth-death process are polymers with alternative states at

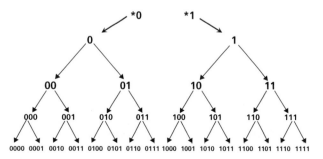

Figure 2.6 A simple tree of polymeric strings involving two alternative initial monomers, denoted by 0 and 1. Each parental molecule of length L can give rise to two alternative molecules of length $L + 1$ by adding one of the activated monomers.

each site, the evolved system contains sequence diversity within each length class and hence, potential information. Although this information is not actually utilized, such a system would be poised for exploitation once mechanisms of inheritance and replication were in place. Fourth, if some sequences do acquire the capacity for self-replication, there will be a critical rate of self-replication above which the behavior of the system can be radically altered, e.g., if only the largest molecules, which are normally kept rare by attrition, are capable of self-replication (Wu and Higgs 2009). More details on all of these matters and others can be found in Nowak and Ohtsuki (2008), Manapat et al. (2009, 2010), Ohtsuki and Nowak (2009), and Bianconi et al. (2013).

Foundations 2.3 The limits to genome replication fidelity

To understand the ultimate limits to any evolutionary phenomenon, we require theory to describe the average state of a population in the face of mutation, selection, and random genetic drift (where the latter is a result of stochasticity in inheritance in finite populations, as is discussed in Chapter 4). Here, we apply some basic results from population-genetic theory to gain insight into the fundamental features of genomic stability essential for the establishment of life. Some of the results are given without detailed explanations, which are postponed until subsequent chapters.

We assume an asexual population (with no exchange of genetic information among individuals) such that the genome of each offspring is a direct copy of that of its parent, barring mutation. The population is assumed to be of constant size (with N individuals), set by the level of resource availability and other ecological features. All mutations are assumed to be deleterious, and these arise at rate U_d per genome per replication. Further assuming that each deleterious mutation independently reduces individual fitness by a fraction s (the selection coefficient), the fitness of an individual with k mutations is

$$W_k = (1 - s)^k. \tag{2.3.1}$$

Under this model, the fitness of mutation-free ($k = 0$) individuals is equal to 1.0. This arbitrary baseline setting has no influence on the following results because selection operates via fitness differences relative to the population mean.

The relentless pressure from mutation will always result in a population with a breadth of fitness classes, with the relative abundances of the various classes being functions of the joint pressures of mutation towards higher k and selection favoring smaller k. A central question is whether the number of individuals in the $k = 0$ class is sufficiently large to avoid loss by stochastic sampling across generations (Haigh 1978).

Muller (1964) first pointed out that, unless s is large enough that mutations are eliminated rapidly, only a small minority of individuals (if any) typically occupy the $k = 0$ class. Unless this number, N_0, is sufficiently large, the best class will eventually be lost, as there will ultimately be a chance generation in which no member of the best class leaves mutation-free progeny. At that point, the second-best class will be elevated to superior status, but it too will eventually suffer the fate of being replaced by the third best class, and so on. This phenomenon, which leads to a progressive loss of fitness, was called Muller's ratchet by Felsenstein (1974), and numerous authors have attempted to solve the difficult problem of the rate at which the ratchet clicks (e.g., Gabriel et al. 1993; Stephan et al. 1993; Gessler 1995; Gordo and Charlesworth 2000). Once initiated, Muller's ratchet can eventually lead to the point at which a population can no longer replace itself. This then leads to an accelerating approach to extinction via a mutational meltdown, whereby progressive declines in population size encourage still faster stochastic clicks of the ratchet (Lynch and Gabriel 1990; Lynch et al. 1993).

The key determinant of whether a population can avoid the ratchet is the expected number of individuals in the $k = 0$ class. For very large N, the frequency distribution of the numbers of individuals in the various classes is Poisson, with U_d/s being the average number of deleterious mutations per individual (Haigh 1978). The Poisson distribution is a simple function of the mean, with the expected number of individuals in the lowest ($k = 0$) class being

$$N_0 = Ne^{-U_d/s}, \tag{2.3.2}$$

where N is the total population size. This number is critical to a population's ability to withstand mutation pressure. The remaining $N - N_0$ mutation-carrying individuals are effectively the 'living dead', as all future descendants of the population must ultimately trace back to the $k = 0$ class if the population is to avoid descent down the path of mutational degradation.

We now consider two fundamental issues: 1) the critical genome size above which U_d is so high that Muller's ratchet will rapidly proceed; and 2) the degree to which natural selection to reduce the mutation rate can ameliorate this process. If the ratchet is to be stopped, the power of selection against new mutations arising in the $k = 0$ subclass must substantially exceed the random fluctuations in allele frequencies caused by drift, the variance of which is proportional to the inverse of the population size (Chapter 5). In this case, we are concerned with fluctuations in the best class, so avoidance of the ratchet requires $s \gg 1/N_0$, or equivalently, $sN_0 \gg 1$. Substituting for N_0 from Equation 2.3.2 and rearranging, the critical genomic mutation rate below which the ratchet effectively stops is found to be

$$U_d \ll s \ln(sN). \qquad (2.3.3a)$$

Now, note that for a genome size of n nucleotides, U_d can be expressed as nu_d, where u_d is the deleterious mutation rate per nucleotide site. Thus, Equation 2.3.3a implies an upper limit to a sustainable genome size of

$$n_s \simeq [s \ln(sN)]/u_d. \qquad (2.3.3b)$$

Strictly speaking, n_s refers only to genomic sites at which a nucleotide substitution has fitness consequences. For example, RNA molecules with catalytic properties typically assemble into complex structures containing stems (consisting of complementary base pairs) and loops, with the loop sequences often being of negligible importance so long as the loop is retained. Thus, the total sustainable genome size in an RNA-world organism could have exceeded n_s to the extent that effectively neutral sites were present in the genome (Kun et al. 2005).

The preceding derivation makes clear that, with a lower mutation rate, there is more room for genome-size expansion, but leaves unexplained the evolution of the mutation rate itself. To achieve such an understanding, we require an expression for the selective advantage of high replication fidelity. In an asexual species, such selection operates through the deleterious mutation loads that become trapped in lineages with different mutation rates. Letting ΔU_d denote the difference in deleterious mutation rates between two lineages, and recalling that the selection coefficient s is equivalent to the rate of removal of individuals with an excess mutation, the excess equilibrium mean number of mutations in the lineage with the higher mutation rate is the ratio of the elevated rate of input to the rate of removal by selection, $\Delta U_d/s$. The fitness difference among genotypes is the product of this excess number and the reduction in fitness per mutation,

$(\Delta U_d/s) \times s = \Delta U_d$. Thus, the long-term selective disadvantage of a genotype with an elevated mutation rate is simply equal to the increase in the genome-wide deleterious mutation rate, independent of the effects of individual mutations (Kimura 1967; Johnson 1999a; Lynch 2008).

This result allows a simple statement on the degree to which selection can reduce U_d. As just noted, selection is ineffective unless its magnitude exceeds the power of genetic drift. Because the maximum possible selective disadvantage of a hypothetical genotype is obtained by contrasting with the expectations for a genotype with perfect replication fidelity ($U_d = 0$), the absolute lower limit to the evolvable genome-wide deleterious mutation rate is on the order of

$$U_d^* = 1/N. \qquad (2.3.4)$$

This follows because the power of drift can be no smaller than $1/N$, and from the above-point that selection is ineffective if its magnitude is lower than the diffusive power of drift (Lynch 2010, 2011).

Thus, Equation 2.3.4 tells us that selection is unable to drive U_d below $1/N$, whereas Equation 2.3.3a tells us that $U_d > s \ln(sN)$ is inconsistent with sustainable life in the absence of recombination. It then follows that there must be a critical population size (N^*) below which selection is incapable of driving U_d to a low enough level to avoid eventual extermination by a mutational meltdown. The solution, obtained by equating Equations 2.3.3a and 2.3.4, is

$$(sN^*) \ln(sN^*) = 1,$$

which simplifies to

$$N^* = 1.76/s. \qquad (2.3.5)$$

Notice that this critical population size is independent of the mutation rate, depending only on the average effect of deleterious mutations. It is difficult to say what the magnitude of s might have been at the early stages of life, although in modern-day species the average value of s may be on the order of 0.001 to 0.01 for fitness-altering mutations (Lynch et al. 1999). Thus, Equation 2.3.5 implies that the absolute minimum population size critical to the establishment of a stable primordial life form is a few hundred to a few thousand individual genomes.

Aside from the potentially serious problem of simple population loss by physical accidents, there are a number of reasons why this number is certainly an underestimate (perhaps by orders of magnitude). First, the drift barrier to mutation-rate evolution of $1/N$ is strictly valid only if mutations influencing the mutation rate are equally distributed in the upward and downward directions. The critical population size would be elevated if mutator alleles arise more frequently than antimutators (Lynch et al. 2016). Second, Equation 2.3.4 is derived by drawing a contrast with the extreme case of a genome with perfect replication fidelity. In reality, the contrast should be made between adjacent possible changes on the scale of replication fidelity, i.e., between the current and next best rate, which will be much smaller. Third, virtually all populations behave genetically as though they are much smaller than their absolute sizes, owing to the selective interference that operates when multiple mutations are simultaneously competing for promotion by natural selection (Chapter 4).

The previous results rely on the assumption that essentially all mutations are deleterious. Attempts have been made to define the optimal mutation rate for maximizing the long-term rate of adaptive evolution when unconditionally beneficial and deleterious mutations are occurring simultaneously (Leigh 1970, 1973; Orr 2000; Bull 2008; Gerrish et al. 2013), but because selection operates on the immediate time scale (rather than looking to the future), it is unclear whether such rates are ever achievable (Sturtevant 1937; Johnson 1999b; Clune et al. 2008; Desai and Fisher 2011). Occasionally, a mutator allele may be brought to high-frequency by hitchhiking with a tightly linked beneficial mutation, as in cases of mismatch-repair deficient pathogens acquiring antibiotic resistance (LeClerc et al. 1996; Denamur and Matic 2006; Giraud et al. 2001). However, such events are generally transient, as they are quickly followed by reversion of the mutation rate (André and Godelle 2006; Gerrish et al. 2007; Raynes et al. 2012). Thus, taken together, theory and empirical observations (Chapter 4) lead to the conclusion that selection primarily drives mutation rates in a downward direction, with an ultimate barrier to what can be achieved being set by the size of the population (Lynch 2011; Lynch et al. 2016).

Literature Cited

André, J. B., and B. Godelle. 2006. The evolution of mutation rate in finite asexual populations. Genetics 172: 611–626.

Attwater, J., and P. Holliger. 2012. Origins of life: The cooperative gene. Nature 491: 48–49.

Attwater, J., A. Wochner, and P. Holliger. 2013. In-ice evolution of RNA polymerase ribozyme activity. Nat. Chem. 5: 1011–1018.

Balabaskaran Nina, P., N. V. Dudkina, L. A. Kane, J. E. van Eyk, E. J. Boekema, M. W. Mather, and A. B. Vaidya. 2010. Highly divergent mitochondrial ATP synthase complexes in *Tetrahymena thermophila*. PLoS Biol. 8: e1000418.

Baross, J. A. and S. E. Hoffman. 1985. Submarine hydrothermal vents and associated gradient environments as sites for the origin and evolution of life. Origins of Life 15: 327–345.

Becker, S., I. Thoma, A. Deutsch, T. Gehrke, P. Mayer, H. Zipse, and T. Carell. 2016. A high-yielding, strictly regioselective prebiotic purine nucleoside formation pathway. Science 352: 833–836.

Becker, S., J. Feldmann, S. Wiedemann, H. Okamura, C. Schneider, K. Iwan, A. Crisp, M. Rossa, T. Amatov, and T. Carell. 2019. Unified prebiotically plausible synthesis of pyrimidine and purine RNA ribonucleotides. Science 366: 76–82.

Bell, E. A., P. Boehnke, T. M. Harrison, and W. L. Mao. 2015. Potentially biogenic carbon preserved in a 4.1 billion-year-old zircon. Proc. Natl. Acad. Sci. USA 112: 14518–14521.

Benner, S. A., H. J. Kim, M. J. Kim, and A. Ricardo. 2010. Planetary organic chemistry and the origins of biomolecules. Cold Spring Harbor Perspect. Biol. 2: a003467.

Bernhardt, H. S., and W. P. Tate. 2012. Primordial soup or vinaigrette: Did the RNA world evolve at acidic pH? Biol. Direct 7: 4.

Bianconi, G., K. Zhao, I. A. Chen, and M. A. Nowak. 2013. Selection for replicases in protocells. PLoS Comput. Biol. 9: e1003051.

Black, R. A., M. C. Blosser, B. L. Stottrup, R. Tavakley, D. W. Deamer, and S. L. Keller. 2013. Nucleobases bind to and stabilize aggregates of a prebiotic amphiphile, providing a viable mechanism for the emergence of protocells. Proc. Natl. Acad. Sci. USA 110: 13272–13276.

Blobel, G. 1980. Intracellular protein topogenesis. Proc. Natl. Acad. Sci. USA 77: 1496–1500.

Bottke, W. F., D. Vokrouhlický, D. Minton, D. Nesvorný, A. Morbidelli, R. Brasser, B. Simonson, and H. F. Levison. 2012. An Archaean heavy bombardment from a destabilized extension of the asteroid belt. Nature 485: 78–81.

Boyer, P. D. 2002. A research journey with ATP synthase. J. Biol. Chem. 277: 39045–39061.

Brasier, M. D., J. Antcliffe, M. Saunders, and D. Wacey. 2015. Changing the picture of Earth's earliest fossils (3.5–1.9 Ga) with new approaches and new discoveries. Proc. Natl. Acad. Sci. USA 112: 4859–4864.

Budin, I., and J. W. Szostak. 2011. Physical effects underlying the transition from primitive to modern cell membranes. Proc. Natl. Acad. Sci. USA 108: 5249–5254.

Bull J. J. 2008. The optimal burst of mutation to create a phenotype. J. Theor. Biol. 254: 667–673.

Bull J. J., L. A. Meyers, and M. Lachmann. 2005. Quasispecies made simple. PLoS Comput. Biol. 1: e61.

Bull, J. J., R. Sanjuán, and C. O. Wilke. 2007. Theory of lethal mutagenesis for viruses. J. Virol. 81: 2930–2939.

Bull, J. J., and C. O. Wilke. 2008. Lethal mutagenesis of bacteria. Genetics 180: 1061–1070.

Cammack, R., K. K. Rao, and D. O. Hall. 1981. Metalloproteins in the evolution of photosynthesis. Biosystems 14: 57–80.

Cavalier-Smith, T. 2001. Obcells as proto-organisms: Membrane heredity, lithophosphorylation, and the origins of the genetic code, the first cells, and photosynthesis. J. Mol. Evol. 53: 555–595.

Chen, I. A., R. W. Roberts, and J. W. Szostak. 2004. The emergence of competition between model protocells. Science 305: 1474–1476.

Chen, P., and E. I. Shakhnovich. 2009. Lethal mutagenesis in viruses and bacteria. Genetics 183: 639–650.

Clune, J., D. Misevic, C. Ofria, R. E. Lenski, S. F. Elena, and R. Sanjuán. 2008. Natural selection fails to optimize mutation rates for long-term adaptation on rugged fitness landscapes. PLoS Comput. Biol. 4: e1000187.

Cnossen, I., J. Sanz-Forcada, F. Favata, O. Witasse, T. Zegers, and N. F. Arnold. 2007. Habitat of early life: Solar X-ray and UV radiation at Earth's surface 4–3.5 billion years ago. J. Geophys. Res. 112: E02008.

Cody, G. D. 2004. Transition metal sulfides and the origins of metabolism. Annu. Rev. Earth Planet. Sci. 32: 569–599.

Cody, G. D., N. Z. Boctor, T. R. Filley, R. M. Hazen, J. H. Scott, A. Sharma, and H. S. Yoder, Jr. 2000. Primordial carbonylated iron–sulfur compounds and the synthesis of pyruvate. Science 289: 1337–1340.

Cooper, G., C. Reed, D. Nguyen, M. Carter, and Y. Wang. 2011. Detection and formation scenario of citric acid, pyruvic acid, and other possible metabolism precursors in carbonaceous meteorites. Proc. Natl. Acad. Sci. USA 108: 14015–14020.

Corliss, J. B., J. A. Baross, and S. E. Hoffman. 1981. An hypothesis concerning the relationships between submarine hot springs and the origin of life on Earth. Oceanologica Acta 4: 59–69.

Crick, F. H. 1968. The origin of the genetic code. J. Mol. Biol. 38: 367–379.

Cross, R. L., and V. Müller. 2004. The evolution of A-, F-, and V-type ATP synthases and ATPases: Reversals in function and changes in the H+/ATP coupling ratio. FEBS Lett. 576: 1–4.

Damer, B., and D. Deamer. 2015. Coupled phases and combinatorial selection in fluctuating hydrothermal pools: A scenario to guide experimental approaches to the origin of cellular life. Life (Basel) 5: 872–887.

David, L. A., and E. J. Alm. 2011. Rapid evolutionary innovation during an Archaean genetic expansion. Nature 469: 93–96.

Deamer, D. W. 1985. Boundary structures are formed by organic components of the Murchison carbonaceous chondrites. Nature 317: 792–794.

Deamer, D. W., and C. D. Georgiou. 2015. Hydrothermal conditions and the origin of cellular life. Astrobiology 15: 1091–1095.

DeDuve, C. 1991. Blueprint for a Cell. Neil Patterson Publishers, Burlington, NC.

Denamur, E., and I. Matic. 2006. Evolution of mutation rates in bacteria. Mol. Microbiol. 60: 820–827.

Desai, M. M., and D. S. Fisher. 2011. The balance between mutators and nonmutators in asexual populations. Genetics 188: 997–1014.

Dibrova, D. V., M. Y. Chudetsky, M. Y. Galperin, E. V. Koonin, and A. Y. Mulkidjanian. 2012. The role of energy in the emergence of biology from chemistry. Orig. Life Evol. Biosph. 42: 459–468.

Dong, J., R. A. Fischer, L. P. Stixrude, and C. R. Lithgow-Bertelloni. 2021. Constraining the volume of earth's early oceans with a temperature-dependent mantle water storage capacity model. AGU Advances 2: e2020AV000323.

Dupont, C. L., A. Butcher, R. E. Valas, P. E. Bourne, and G. Caetano-Anollés G. 2010. History of biological metal utilization inferred through phylogenomic analysis of protein structures. Proc. Natl. Acad. Sci. USA 107: 10567–10572.

Edgell, D. R., and W. F. Doolittle. 1997. Archaea and the origin(s) of DNA replication proteins. Cell 89: 995–998.

Eigen, M. 1971. Selforganization of matter and the evolution of biological macromolecules. Naturwissenschaften 58: 465–523.

Eigen, M., and P. Schuster. 1977. The hypercycle. A principle of natural self-organization. Part A: Emergence of the hypercycle. Naturwissenschaften 64: 541–565.

Felsenstein, J. 1974. The evolutionary advantage of recombination. Genetics 78: 737–756.

Flygaard, R. K., A. Mühleip, V. Tobiasson, and A. Amunts. 2020. Type III ATP synthase is a symmetry-deviated dimer that induces membrane curvature through tetramerization. Nat. Commun. 11: 5342.

Furnes, H., N. R. Banerjee, K. Muehlenbachs, H. Staudigel, and M. de Wit. 2004. Early life recorded in Archean pillow lavas. Science 304: 578–581.

Gabriel, W., M. Lynch, and R. Bürger. 1993. Muller's ratchet and mutational meltdowns. Evolution 47: 1744–1757.

Gerrish, P. J., A. Colato, A. S. Perelson, and P. D. Sniegowski. 2007. Complete genetic linkage can subvert natural selection. Proc. Natl. Acad. Sci. USA 104: 6266–6271.

Gerrish, P. J., A. Colato, and P. D. Sniegowski. 2013. Genomic mutation rates that neutralize adaptive evolution and natural selection. J. R. Soc. Interface 10: 20130329.

Gessler, D. D. 1995. The constraints of finite size in asexual populations and the rate of the ratchet. Genet. Res. 66: 241–253.

Gilbert, W. 1986. The RNA world. Nature 319: 618.

Giraud, A., M. Radman, I. Matic, and F. Taddei. 2001. The rise and fall of mutator bacteria. Curr. Opin. Microbiol. 4: 582–585.

Goldford, J. E., H. Hartman, T. F. Smith, and D. Segré. 2017. Remnants of an ancient metabolism without phosphate. Cell 168: 1126–1134.

Goldman, A. R., J. M. Weber, D. E. LaRowe, and L. M. Barge. 2023. Electron transport chains as a window into the earliest stages of evolution. Proc. Natl. Acad. Sci. USA 120: e2210924120.

Gordo, I., and B. Charlesworth. 2000. The degeneration of asexual haploid populations and the speed of Muller's ratchet. Genetics 154: 1379–1387.

Goyal, S., D. J. Balick, E. R. Jerison, R. A. Neher, B. I. Shraiman, and M. M. Desai. 2012. Dynamic mutation–selection balance as an evolutionary attractor. Genetics 191: 1309–1319.

Griffiths, G. 2007. Cell evolution and the problem of membrane topology. Nat. Rev. Mol. Cell Biol. 8: 1018–1024.

Haigh, J. 1978. The accumulation of deleterious genes in a population – Muller's ratchet. Theor. Popul. Biol. 14: 251–267.

Haldane, J. B. S. 1929. The origin of life. Rationalist Annual 3: 148–169.

Hall, D. O., R. Cammack, and K. K. Rao. 1974. The iron–sulphur proteins: Evolution of a ubiquitous protein from model systems to higher organisms. Orig. Life 5: 363–386.

Hazen, R. M., and D. W. Deamer. 2007. Hydrothermal reactions of pyruvic acid: Synthesis, selection, and self-assembly of amphiphilic molecules. Orig. Life Evol. Biosph. 37: 143–152.

Higgs, P. G., and N. Lehman. 2015. The RNA world: Molecular cooperation at the origins of life. Nat. Rev. Genet. 16: 7–17.

Jékely, G. 2006. Did the last common ancestor have a biological membrane? Biol. Direct 1: 35.

Jensen, J. D., R. A. Stikeleather, T. F. Kowalik, and M. Lynch. 2020. Imposed mutational meltdown as an antiviral strategy. Evolution 12: 2549–2559.

Johnson, A. P., H. J. Cleaves, J. P. Dworkin, D. P. Glavin, A. Lazcano, and J. L. Bada. 2008. The Miller volcanic spark discharge experiment. Science 322: 404.

Johnson, B. C., and H. J. Melosh. 2012. Impact spherules as a record of an ancient heavy bombardment of Earth. Nature 485: 75–77.

Johnson, T. 1999a. The approach to mutation–selection balance in an infinite asexual population, and the evolution of mutation rates. Proc. R. Soc. Lond. B 266: 2389–2397.

Johnson, T. 1999b. Beneficial mutations, hitchhiking and the evolution of mutation rates in sexual populations. Genetics 151: 1621–1631.

Johnston, W. K., P. J. Unrau, M. S. Lawrence, M. E. Glasner, and D. P. Bartel. 2001. RNA-catalyzed RNA polymerization: Accurate and general RNA-templated primer extension. Science 292: 1319–1325.

Joyce, G. F. 2004. Directed evolution of nucleic acid enzymes. Ann. Rev. Biochem. 73: 791–836.

Joyce, G. F., and L. E. Orgel. 1999. Prospects for understanding the origin of the RNA World, pp. 49–77. In R. F. Gesteland, T. R. Cech, and J. F. Atkins (eds.) The RNA World, 2nd edn. Cold Spring Harbor Laboratory Press, Cold Spring Harbor, NY.

Keller, M. A., A. V. Turchyn, and M. Ralser. 2014. Non-enzymatic glycolysis and pentose phosphate pathway-like reactions in a plausible Archean ocean. Mol. Syst. Biol. 10: 725.

Kim, H. J., A. Ricardo, H. I. Illangkoon, M. J. Kim, M. A. Carrigan, F. Frye, and S. A. Benner. 2011. Synthesis of carbohydrates in mineral-guided prebiotic cycles. J. Am. Chem. Soc. 133: 9457–9468.

Kimura, M. 1967. On the evolutionary adjustment of spontaneous mutation rates. Genet. Res. 9: 23–34.

Knoll, A. H. 2004. Life on a Young Planet: The First Three Billion Years of Evolution on Earth. Princeton University Press, Princeton, NJ.

Kun, A., M. Santos, and E. Szathmáry. 2005. Real ribozymes suggest a relaxed error threshold. Nat. Genet. 37: 1008–1011.

Lane, N., J. F. Allen, and W. Martin. 2010. How did LUCA make a living? Chemiosmosis in the origin of life. Bioessays 32: 271–280.

Lane, N., and W. F. Martin. 2012. The origin of membrane bioenergetics. Cell 151: 1406–1416.

Lapaille, M., A. Escobar-Ramírez, H. Degand, D. Baurain, E. Rodríguez-Salinas, N. Coosemans, M. Boutry, D. Gonzalez-Halphen, C. Remacle, and P. Cardol. 2010. Atypical subunit composition of the chlorophycean mitochondrial F_1F_0-ATP synthase and role of Asa7 protein in stability and oligomycin resistance of the enzyme. Mol. Biol. Evol. 27: 1630–1644.

LeClerc, J. E., B. Li, W. L. Payne, and T. A. Cebula. 1996. High mutation frequencies among *Escherichia coli* and *Salmonella* pathogens. Science 274: 1208–1211.

Leigh, E. G., Jr. 1970. Natural selection and mutability. Amer. Natur. 104: 301–305.

Leigh, E. G., Jr. 1973. The evolution of mutation rates. Genetics (Suppl.) 73: 1–18.

Leipe, D. D., L. Aravind, and E. V. Koonin. 1999. Did DNA replication evolve twice independently? Nucleic Acids Res. 27: 3389–3401.

Lincoln, T. A., and G. F. Joyce. 2009. Self-sustained replication of an RNA enzyme. Science 323: 1229–1232.

Lynch, M. 2008. The cellular, developmental, and population-genetic determinants of mutation-rate evolution. Genetics 180: 933–943.

Lynch, M. 2010. Evolution of the mutation rate. Trends Genet. 26: 345–352.

Lynch, M. 2011. The lower bound to the evolution of mutation rates. Genome Biol. Evol. 3: 1107–1118.

Lynch, M., M. Ackerman, J.-F. Gout, H. Long, W. Sung, W. K. Thomas, and P. L. Foster. 2016. Genetic drift, selection, and evolution of the mutation rate. Nat. Rev. Genetics 17: 704–714.

Lynch, M., J. Blanchard, D. Houle, T. Kibota, S. Schultz, L. Vassilieva, and J. Willis. 1999. Spontaneous deleterious mutation. Evolution 53: 645–663.

Lynch, R. Bürger, D. Butcher, and W. Gabriel. 1993. Mutational meltdowns in asexual populations. J. Heredity 84: 339–344.

Lynch, M., and W. Gabriel. 1990. Mutation load and the survival of small populations. Evolution 44: 1725–1737.

Lynch, M., and G. K. Marinov. 2015. The bioenergetic costs of a gene. Proc. Natl. Acad. Sci. USA 112: 15690–15695.

Maden, B. E. 1995. No soup for starters? Autotrophy and the origins of metabolism. Trends Biochem. Sci. 20: 337–341.

Manapat, M. L., I. A. Chen, and M. A. Nowak. 2010. The basic reproductive ratio of life. J. Theor. Biol. 263: 317–327.

Manapat, M., H. Ohtsuki, R. Bürger, and M. A. Nowak. 2009. Originator dynamics. J. Theor. Biol. 256: 586–595.

Mansy, S. S., and J. W. Szostak. 2009. Reconstructing the emergence of cellular life through the synthesis of model protocells. Cold Spring Harb. Symp. Quant. Biol. 74: 47–54.

Martin, W., J. Baross, D. Kelley, and M. J. Russell. 2008. Hydrothermal vents and the origin of life. Nat. Rev. Microbiol. 6: 805–814.

Martin, W., and M. J. Russell. 2003. On the origins of cells: A hypothesis for the evolutionary transitions from abiotic geochemistry to chemoautotrophic prokaryotes, and from prokaryotes to nucleated cells. Philos. Trans. R. Soc. Lond. B Biol. Sci. 358: 59–83.

Mast, C. B., S. Schink, U. Gerland, and D. Braun. 2013. Escalation of polymerization in a thermal gradient. Proc. Natl. Acad. Sci. USA 110: 8030–8035.

Matthies, D., L. Preiss, A. L. Klyszejko, D. J. Muller, G. M. Cook, J. Vonck, and T. Meier. 2009. The c13 ring from a thermoalkaliphilic ATP synthase reveals an extended diameter due to a special structural region. J. Mol. Biol. 388: 611–618.

McCollom, T. M. 2013. Miller-Urey and beyond: What have we learned about prebiotic organic synthesis reactions in the past 60 years? Ann. Rev. Earth Planet. Sci. 41: 207–229.

McCollom, T. M., G. Ritter, and B. R. T. Simoneit. 1999. Lipid synthesis under hydrothermal conditions by Fischer–Tropsch–Type reactions. Origins Life Evol. Biosphere 29: 153–166.

Meier, T., J. Yu, T. Raschle, F. Henzen, P. Dimroth, and D. J. Muller. 2005. Structural evidence for a constant c11 ring stoichiometry in the sodium F-ATP synthase. FEBS J. 272: 5474–5483.

Ménez, B., C. Pisapia, M. Andreani, F. Jamme, Q. P. Vanbellingen, A. Brunelle, L. Richard, P. Dumas, and M. Réfrégiers. 2018. Abiotic synthesis of amino acids in the recesses of the oceanic lithosphere. Nature 564: 59–63.

Miller, S. L. 1953. Production of amino acids under possible primitive earth conditions. Science 117: 528–529.

Miller, S. L., and H. C. Urey. 1959. Organic compound synthesis on the primitive Earth. Science 130: 245–251.

Mizuuchi, R., T. Furubayashi, and N. Ichihashi. 2022. Evolutionary transition from a single RNA replicator to a multiple replicator network. Nat. Commun. 13: 1460.

Muchowska, K. B., S. J. Varma, and J. Moran. 2019. Synthesis and breakdown of universal metabolic precursors promoted by iron. Nature 569: 104–107.

Mühleip, A., R. K. Flygaard, J. Ovciarikova, A. Lacombe, P. Fernandes, L. Sheiner, and A. Amunts. 2021. ATP synthase hexamer assemblies shape cristae of *Toxoplasma* mitochondria. Nat. Commun. 12: 120.

Mühleip, A., S. E. McComas, and A. Amunts. 2019. Structure of a mitochondrial ATP synthase with bound native cardiolipin. eLife 8: e51179.

Mulkidjanian, A. Y. 2009. On the origin of life in the zinc world: 1. Photosynthesizing, porous edifices built of hydrothermally precipitated zinc sulfide as cradles of life on Earth. Biol. Direct 4: 26.

Mulkidjanian, A. Y., A. Y. Bychkov, D. V. Dibrova, M. Y. Galperin, and E. V. Koonin. 2012a. Open questions on the origin of life at anoxic geothermal fields. Orig. Life Evol. Biosph. 42: 507–516.

Mulkidjanian, A. Y., A. Y. Bychkov, D. V. Dibrova, M. Y. Galperin, and E. V. Koonin. 2012b. Origin of first cells at terrestrial, anoxic geothermal fields. Proc. Natl. Acad. Sci. USA 109: E821–E830.

Mulkidjanian, A. Y., D. A. Cherepanov, and M. Y. Galperin. 2003. Survival of the fittest before the beginning of life: Selection of the first oligonucleotide-like polymers by UV light. BMC Evol. Biol. 3: 12.

Mulkidjanian, A. Y., and M. Y. Galperin. 2009. On the origin of life in the zinc world. 2. Validation of the hypothesis on the photosynthesizing zinc sulfide edifices as cradles of life on Earth. Biol. Direct 4: 27.

Mulkidjanian, A. Y., M. Y. Galperin, K. S. Makarova, Y. I. Wolf, and E. V. Koonin. 2008. Evolutionary primacy of sodium bioenergetics. Biol. Direct 3: 13.

Mulkidjanian, A. Y., M. Y. Galperin, and E. V. Koonin. 2009. Co-evolution of primordial membranes and membrane proteins. Trends Biochem. Sci. 34: 206–215.

Mulkidjanian, A. Y., K. S. Makarova, M. Y. Galperin, and E. V. Koonin. 2007. Inventing the dynamo machine: The evolution of the F-type and V-type ATPases. Nat. Rev. Microbiol. 5: 892–899.

Muller, H. J. 1964. The relation of recombination to mutational advance. Mutation Res. 1: 2–9.

Müller, V., A. Lingl, K. Lewalter, and M. Fritz. 2005. ATP synthases with novel rotor subunits: New insights into structure, function and evolution of ATPases. J. Bioenerg. Biomembr. 37: 455–460.

Myllykallio, H., G. Lipowski, D. Leduc, J. Filée, P. Forterre, and U. Liebl. 2002. An alternative flavin-dependent mechanism for thymidylate synthesis. Science 297: 105–107.

Neveu, M., H. J. Kim, and S. A. Benner. 2013. The 'strong' RNA world hypothesis: Fifty years old. Astrobiol. 13: 391–403.

Nisbet, E. G., and N. H. Sleep. 2001. The habitat and nature of early life. Nature 409: 1083–1091.

Novikov, Y., and S. D. Copley. 2013. Reactivity landscape of pyruvate under simulated hydrothermal vent conditions. Proc. Natl. Acad. Sci. USA 110: 13283–13288.

Nowak, M. A., and H. Ohtsuki. 2008. Prevolutionary dynamics and the origin of evolution. Proc. Natl. Acad. Sci. USA 105: 14924–14927.

Ohtsuki, H., and M. A. Nowak. 2009. Prelife catalysts and replicators. Proc. R. Soc. B 276: 3783–3790.

Olsen, G. J., and C. R. Woese. 1997. Archaeal genomics: An overview. Cell 89: 991–994.

Oparin, A. I. 1938. Origin of Life. Macmillan and Co., New York, NY.

Orgel, L. E. 1968. Evolution of the genetic apparatus. J. Mol. Biol. 38: 381–393.

Orgel, L. E. 2000. Self-organizing biochemical cycles. Proc. Natl. Acad. Sci. USA 97: 12503–12507.

Orr, H. A. 2000. The rate of adaptation in asexuals. Genetics 155: 961–968.

Patel, S. S., and K. M. Picha. 2000. Structure and function of hexameric helicases. Annu. Rev. Biochem. 69: 651–697.

Pearce, B. K. D., R. E. Pudritz, D. A. Semenov, and T. K. Henning. 2017. Origin of the RNA world: The fate of nucleobases in warm little ponds. Proc. Natl. Acad. Sci. USA 114: 11327–11332.

Phillips, R., and R. Milo. 2009. A feeling for the numbers in biology. Proc. Natl. Acad. Sci. USA 106: 21465–21471.

Pinheiro, V. B., A. I. Taylor, C. Cozens, M. Abramov, M. Renders, S. Zhang, J. C. Chaput, J. Wengel, S. Y. Peak-Chew, S. H. McLaughlin, et al. 2012. Synthetic genetic polymers capable of heredity and evolution. Science 336: 341–344.

Poon, A., and S. P. Otto. 2000. Compensating for our load of mutations: Freezing the meltdown of small populations. Evolution 54: 1467–1479.

Powner, M. W., B. Gerland, and J. D. Sutherland. 2009. Synthesis of activated pyrimidine ribonucleotides in prebiotically plausible conditions. Nature 459: 239–242.

Rasmussen, B. 2000. Filamentous microfossils in a 3,235-million-year-old volcanogenic massive sulphide deposit. Nature 405: 676–679.

Raynes, Y., M. R. Gazzara, and P. D. Sniegowski. 2012. Contrasting dynamics of a mutator allele in asexual populations of differing size. Evolution 66: 2329–2334.

Ricardo, A., M. A. Carrigan, A. N. Olcott, and S. A. Benner. 2004. Borate minerals stabilize ribose. Science 303: 196.

Robertson, M. P., and G. F. Joyce. 2012. The origins of the RNA world. Cold Spring Harbor Perspect. Biol. 1: 4(5).

Rosing, M. T. 1999. ^{13}C-depleted carbon microparticles in > 3700-Ma sea-floor sedimentary rocks from west Greenland. Science 283: 674–676.

Russell, M. J., L. M. Barge, R. Bhartia, D. Bocanegra, P. J. Bracher, E. Branscomb, R. Kidd, S. McGlynn, D. H. Meier, W. Nitschke, et al. 2014. The drive to life on wet and icy worlds. Astrobiology 14: 308–343.

Russell, M. J., and A. J. Hall. 1997. The emergence of life from iron monosulphide bubbles at a submarine hydrothermal redox and pH front. J. Geol. Soc., London 154: 377—402.

Russell, M. J., A. J. Hall, and W. Martin. 2010. Serpentinization as a source of energy at the origin of life. Geobiology 8: 355–371.

Schopf, J. W. 1993. Microfossils of the early Archean Apex chert: New evidence of the antiquity of life. Science 260: 640–646.

Schopf, J. W., A. B. Kudryavtsev, D. G. Agresti, T. J. Wdowiak, and A. D. Czaja. 2002. Laser-Raman imagery of Earth's earliest fossils. Nature 416: 73–76.

Serrano-Andrés, L., and M. Merchán. 2009. Are the five natural DNA/RNA base monomers a good choice from natural selection? A photochemical perspective. J. Photochem. Photobiol. 10: 21–32.

Shapiro, R. 2006. Small molecule interactions were central to the origin of life. Quart. Rev. Biol. 81: 105–125.

Silander, O. K., O. Tenaillon, and L. Chao. 2007. Understanding the evolutionary fate of finite populations: The dynamics of mutational effects. PLoS Biol. 5: e94.

Silverman, S. K. 2016. Catalytic DNA: Scope, applications, and biochemistry of deoxyribozymes. Trends Biochem. Sci. 41: 595–609.

Sleep, N. H., K. J. Zahnle, J. F. Kasting, and H. J. Morowitz. 1989. Annihilation of ecosystems by large asteroid impacts on the early Earth. Nature 342: 139–142.

Smith, E., and H. J. Morowitz. 2004. Universality in intermediary metabolism. Proc. Natl. Acad. Sci. USA 101: 13168–13173.

Soga, N., K. Kimura, K. Kinosita, Jr., M. Yoshida, and T. Suzuki. 2017. Perfect chemomechanical coupling of F_oF_1-ATP synthase. Proc. Natl. Acad. Sci. USA 114: 4960–4965.

Soll, S. J., C. Díaz Arenas, and N. Lehman. 2007. Accumulation of deleterious mutations in small abiotic populations of RNA. Genetics 175: 267–275.

Steffens, L., E. Pettinato, T. M. Steiner, A. Mall, S. König, W. Eisenreich, and I. A. Berg. 2021. High CO_2 levels drive the TCA cycle backwards towards autotrophy. Nature 592: 784–788.

Stephan, W., L. Chao, and J. G. Smale. 1993. The advance of Muller's ratchet in a haploid asexual population: Approximate solutions based on diffusion theory. Genet. Res. 61: 225–231.

Stüeken, E. E., R. E. Anderson, J. S. Bowman, W. J. Brazelton, J. Colangelo-Lillis, A. D. Goldman, S. M. Som, and J. A. Baross. 2013. Did life originate from a global chemical reactor? Geobiol. 11: 101–126.

Sturtevant, A. H. 1937. Essays on evolution. I. On the effects of selection on mutation rate. Quart. Rev. Biol. 12: 464–476.

Tice, M. M., and D. R. Lowe. 2004. Photosynthetic microbial mats in the 3,416-myr-old ocean. Nature 431: 549–552.

Vaidya, N., M. L. Manapat, I. A. Chen, R. Xulvi-Brunet, E. J. Hayden, and N. Lehman. 2012. Spontaneous network formation among cooperative RNA replicators. Nature 491: 72–77.

Wacey, D., M. R. Kilburn, M. Saunders, J. Cliff, and M. D. Brasier. 2011. Microfossils of sulphur-metabolizing cells in 3.4-billion-year-old rocks of Western Australia. Nat. Geosci. 4: 698–702.

Wächtershäuser, G. 1988. Before enzymes and templates: Theory of surface metabolism. Microbiol. Rev. 52: 452–484.

Wächtershäuser, G. 1997. The origin of life and its methodological challenge. J. Theor. Biol. 487: 483–494.

Wächtershäuser, G. 2007. On the chemistry and evolution of the pioneer organism. Chem. Biodivers. 4: 584–602.

Wagner, G. P., and W. Gabriel. 1990. Quantitative variation in finite parthenogenetic populations: What stops Muller's ratchet in the absence of recombination? Evolution 44: 715–731.

Walde, P., A. Goto, P.-A. Monnard, M. Wessicken, and P. L. Luisi. 1994. Oparin's reactions revisited: Enzymic synthesis of poly(adenylic acid) in micelles and self-reproducing vesicles. J. Am. Chem. Soc. 116: 7541–7547.

Walker, J. E. 1998. ATP synthesis by rotary catalysis. Angew. Chem. Int. Ed. Engl. 37: 2309–2319.

Walker, J. E. 2013. The ATP synthase: The understood, the uncertain and the unknown. Biochem. Soc. Trans. 41: 1–16.

Watt, I. N., M. G. Montgomery, M. J. Runswick, A. G. Leslie, and J. E. Walker. 2010. Bioenergetic cost of making an adenosine triphosphate molecule in animal mitochondria. Proc. Natl. Acad. Sci. USA 107: 16823–16827.

Wilson, D. S., and J. W. Szostak. 1999. *In vitro* selection of functional nucleic acids. Annu. Rev. Biochem. 68: 611–647.

Woese, C. R. 1967. The Genetic Code: The Molecular Basis for Genetic Expression. Harper & Row, New York, NY.

Wu, M., and P. G. Higgs. 2009. Origin of self-replicating biopolymers: Autocatalytic feedback can jump-start the RNA world. J. Mol. Evol. 69: 541–554.

Xu, J., V. Chmela, N. J. Green, D. A. Russell, M. J. Janicki, R. W. Góra, R. Szabla, A. D. Bond, and J. D. Sutherland. 2020. Selective prebiotic formation of RNA pyrimidine and DNA purine nucleosides. Nature 582: 60–66.

Zahnle, K., L. Schaefer, and B. Fegley. 2010. Earth's earliest atmospheres. Cold Spring Harbor Perspect. Biol. 2: a004895.

Zeldovich, K. B., P. Chen, and E. I. Shakhnovich. 2007. Protein stability imposes limits on organism complexity and speed of molecular evolution. Proc. Natl. Acad. Sci. USA 104: 16152–16157.

Zhang, S., J. C. Blain, D. Zielinska, S. M. Gryaznov, and J. W. Szostak. 2013. Fast and accurate nonenzymatic copying of an RNA-like synthetic genetic polymer. Proc. Natl. Acad. Sci. USA 110: 17732–17737.

Zhu, T. F., K. Adamala, N. Zhang, and J. W. Szostak. 2012. Photochemically driven redox chemistry induces proto-cell membrane pearling and division. Proc. Natl. Acad. Sci. USA 109: 9828–9832.

Zhu, T. F., and J. W. Szostak. 2009. Preparation of large monodisperse vesicles. PLoS One 4: e5009.

CHAPTER 3

The Major Lines of Descent

Once the foundation for cellular life was established, various paths towards diversification were set in motion. Populations that are physically isolated from each other for sufficiently long periods naturally accumulate independent mutations, in some cases promoted by natural selection for optimal phenotypes in local environments, and in others simply by random genetic drift. When the genomes of isolated lineages diverge to a sufficient degree, the internally co-adapted gene complexes will be mutually incompatible, preventing the production of viable downstream hybrids. Such genetic isolation constitutes the speciation process, ensuring the survival and evolution of independent lineages on their own merits, as reflected in the millions of species now inhabiting the planet.

Understanding the genealogical relationships (phylogeny) of existing lineages is critical to biology, as it provides a historical overview of what evolution has been able to accomplish, thereby facilitating the development of hypotheses for how evolution occurs. When related species share the same trait, we can often be fairly certain that their common ancestor (at the basal node of the clade) also carried the trait. Because sister taxa evolve from common ancestors, their differences also provide insight into the kinds of changes that are possible from a shared beginning. For example, gains and losses of traits can be inferred when single lineages deviate from their surrounding relatives. Information on many pairs of taxa can then begin to reveal commonalities among traits and parallel paths of evolution.

To avoid being circular, analyses like these require well-resolved phylogenies, built from observations on traits other than those under investigation. The modern age of whole-genome sequencing has brought us to the limits of such information. Nevertheless, despite the millions of informative nucleotide sites now known in thousands of species, many of the earliest branching patterns in the Tree of Life remain ambiguous. By outlining what we do know about the relationships among the major lineages of life, this chapter sets the stage for more in-depth comparative analyses in subsequent chapters.

First, we examine the degree of phylogenetic affinity between the two major organizational grades of cellular life: the prokaryotes and eukaryotes. Although the two groups are generally distinguished by the absence/presence of a nuclear envelope and other membrane-bound organelles, this morphological distinction turns out to be misleading with respect to genealogical relationships. Not only are there two distantly related groups of prokaryotes, the bacteria and the archaea, but the eukaryotic lineage emerged from the latter, rather than being genealogically distinct from both.

Second, we briefly consider the eukaryotes, the lineage with the greatest morphological diversification. The main point here is that a very large suite of intracellular embellishments became established prior to the divergence of the major eukaryotic lineages, leaving no intermediate-state traces (at least as so far discovered). The establishment of the last eukaryotic common ancestor (LECA) was then followed fairly quickly by the emergence of major subclades with their own unique features.

Third, analogues of many of the shared (so-called universal) eukaryotic traits can be found in one or more prokaryotic lineages. Although these are not always orthologous in origin, this does indicate that their emergence was not strictly dependent on eukaryogenesis. However, the stem eukaryote distinguished itself in assembling a unique mixture of

Evolutionary Cell Biology. Michael Lynch, Oxford University Press. © Michael Lynch (2024). DOI: 10.1093/oso/9780192847287.003.0003

features into a single lineage. The conditions that drove the subsequent Big Bang of eukaryotic diversity remain unclear, and may not have been ecological in nature. Instead, the radical shift in the genetic system of the eukaryotic cell may have been the primary enabler of species diversification.

The Primary Domains of Life

Just as pairs of individuals within a population are related to various degrees in a pedigree sense, species relationships can be described in the form of a phylogenetic tree. Sibling species reside on adjacent branches separated by a single node (branch point), with pairs of species with lower affinities residing on more distant branches. Historically, the field of taxonomy sought to classify organisms by their physical appearances, but owing to the possibility of convergent phenotypic evolution, such an approach is fraught with interpretative problems. Information at the nucleotide level has an explicit genetic interpretation and is less ambiguous. Thus, almost all attempts to infer phylogenetic relationships are now based on observations on DNA-sequence divergence among extant species (Felsenstein 2004).

Although this is a highly technical field involving computationally demanding algorithms to obtain genealogical relationships that are most compatible with the data, the conceptual basis for these analyses is straightforward. Because DNA naturally acquires nucleotide substitutions and rearrangements by mutation, some of which are nearly neutral (Chapter 4), related species experience DNA-sequence divergence over evolutionary time. The simple fact that species with higher levels of sequence similarity tend to be more closely related forms the logical basis for virtually all statistical methods for estimating phylogenetic trees and dating evolutionary events.

With the advent of molecular-genetic methods, insights into the broad form of the Tree of Life began to emerge in the 1970s. Up to this point, based on the obvious morphological void between prokaryotes and eukaryotes, the former had been viewed as one large, monophyletic group, ill-defined internally but assumed to be deeply separated from the eukaryotes (Sapp 2005). However, noting that the genomes of all organisms encode for ribosomal RNAs (which comprise the catalytic hearts of ribosomes), Woese and Fox (1977) reasoned that a higher degree of resolution could be obtained by comparative analysis of such sequences. They quickly discovered a deep phylogenetic furrow within the prokaryotes, implying the existence of two major lineages, seemingly as distinct from each other as they are from eukaryotes. These two prokaryotic groups came to be known as the archaea (often called archaebacteria) and the bacteria (sometimes called eubacteria) (Woese et al. 1990).

This division of life into three major groups raised several questions about the base of the Tree of Life. Are the bacterial, archaeal, and eukaryotic groups all monophyletic, each with independent common ancestors (a three-domains of life model), or are one or more clades embedded within another (a one- or two-domains model)? Assuming the groups are monophyletic, are eukaryotes more closely related to archaea or bacteria, or do they have affinities with both? Can we formally rule out the possibility that eukaryotes are ancestral to prokaryotes?

The key to answering these questions is a correctly rooted phylogenetic tree denoting the location of the most recent common ancestor from which all species in the tree ultimately descend. This hypothetical taxon is often referred to as LUCA (for last universal common ancestor) (Figure 3.1), whereas the last common ancestors for bacteria, archaea, and eukaryotes are designated LBCA, LACA, and LECA. The first common ancestor for a lineage (e.g., FUCA, FBCA, FACA, and FECA) denotes the most remote point on the branch leading to the last common ancestor not containing any other major clade. Traits that are shared by all members of a clade were almost certainly present in the last common ancestor of the clade, but one cannot rule out an earlier origin on the branch extending from the first common ancestor. For example, a feature shared by all bacteria may have arisen anywhere along the FBCA–LBCA branch.

Two problems conspire to make the ascertainment of the relationships between the three major groups a difficult enterprise. First, although placing a root on a phylogeny is usually a simple matter of including in the analysis a compelling outgroup (i.e., a bird for a mammalian phylogeny), this is

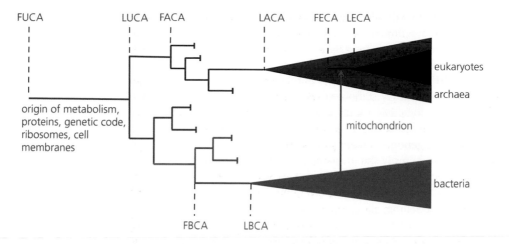

Figure 3.1 An idealized view of the two-domains view of the Tree of Life. The last universal common ancestor (LUCA) diverged into the bacterial and archaeal lineages, with eukaryotes then emerging out of the archaeal clade. Horizontal lines with blunt ends denote extinct lineages, and the relative temporal positions of lineage origins are not meant to be taken literally. The vertical blue line denotes the origin of the mitochondrion via endosymbiosis of a colonizing bacterium. Abbreviations used in acronyms for ancestors: F = first; L = last (or most recent); U = universal; B = bacterial; A = archaeal; E = eukaryote; CA = common ancestor.

not an option when the entire Tree of Life is being considered. Second, the amount of molecular divergence among the few hundred genes shared by all three ancient groups is so vast that the signal of genealogical relationships has been greatly diluted by the accumulation of multiple nucleotide substitutions per site.

This being said, a consensus seems to have emerged on the deepest branches of the Tree of Life. Virtually all analyses indicate that the bacterial lineage is monophyletic and separate from the lineage containing the archaea and eukaryotes (e.g., Raymann et al. 2015; Coleman et al. 2021). Phylogenetic analysis implies that LBCA was a sophisticated cell, with a cell wall sandwiched between two membranes, a capacity for flagellar swimming and chemotaxis, and a CRISPR-Cas system for warding off invasive DNA elements. As the archaeal lineage also appears to be monophyletic (Williams et al. 2017), this leaves the positioning of eukaryotes as the main issue. In principle, eukaryotes could simply join as a separate monophyletic clade at a single node outside of bacteria and archaea. Alternatively, one of the groups might emerge as a sublineage within the other. The first pattern would be consistent with the three-domain model postulated by Woese and colleagues. The second condition would

imply a two-domain scenario in which eukaryotes are simply a derived lineage within one of the prokaryotic groups, or vice versa.

Resolving this issue has been challenging, owing to complications beyond the statistical problems outlined above. Most notable are: 1) the occurrence of substantial horizontal gene transfer among lineages early in the history of life (Doolittle et al. 2003); and 2) the additional massive transfer of bacterial genes to their eukaryotic host cells following the endosymbiotic establishment of the mitochondrion (derived from a bacterium) in the basal eukaryote (Chapter 23). Gene relocations blur the deep branches on the Tree of Life, as different genes have different phylogenetic histories. Nonetheless, most large-scale analyses now seem to support the eocyte hypothesis of Lake et al. (1984), which postulates eukaryotes as being most closely related to one particular archaeal group (Cox et al. 2008; Guy and Ettema 2011; Kelly et al. 2011; Thiergart et al. 2012; Williams et al. 2012, 2013, 2020; Raymann et al. 2015). This hypothesis essentially eliminates the possibility that eukaryotes are the primordial cellular lineage, rejects the three-domains view, and implicates a member of the Archaea as the ultimate source of the eukaryotic nuclear genome.

It has been argued that the closest living relatives to eukaryotes reside within a lineage called the Asgard archaea, known mainly from the sequencing of environmental samples from deep-sea sediments (Spang et al. 2015; Hug et al. 2016; Zaremba-Niedzwiedzka et al. 2017; Tahon et al. 2021), although greater phylogenetic affinity with an alternative archaeal lineage cannot be ruled out entirely (Liu et al. 2021). The genome contents for members of the Asgard archaea imply the presence of actin- and tubulin-related (cytoskeletal) proteins as well as components associated with vesicle trafficking and membrane remodelling, all of which are classical attributes of eukaryotic cells (Ettema et al. 2011; Yutin and Koonin 2012; Akil and Robinson 2018; Liu et al. 2021). The only members of the Asgard archaea that are cultivatable in the lab so far produce long protuberances and may depend on a symbiotic relationship with another member of the Archaea, but do not exhibit complex internal cell structure (Imachi et al. 2020; Rodrigues-Oliveira et al. 2023).

Notably, eukaryotic proteins involved in information processing (e.g., transcription and translation) tend to be more similar to those in archaea than bacteria, as expected if the nuclear genome is derived from a member of the archaea. In contrast, proteins involved in housekeeping functions (e.g., metabolism) tend to most closely resemble those in bacteria (Brown and Doolittle 1997; Rivera et al. 1998; Leipe et al. 1999; Brown et al. 2001; Horiike et al. 2001), many of which may be derived from the colonizing bacterium that became the mitochondrion.

One concern with the two-domains hypothesis relates to the types of phospholipids deployed in the cell membranes of the different major lineages (Figure 3.2). All cells are enveloped by phospholipid bilayers, with the individual molecules comprised of a glycerol-phosphate sandwiched between a head group and two hydrocarbon chains (Chapter 15). However, whereas glycerol 1-phosphate (G1P) is bound to methyl-branched isoprenoid chains by ether linkages in archaea, glycerol 3-phosphate (G3P) is bound to straight fatty-acid chains by ester linkages in bacteria and eukaryotes (Boucher et al. 2004; Peretó et al. 2004). Likewise, the ability to produce membrane steroids appears to be restricted to bacteria and eukaryotes, and absent from archaea (Hoshino and Gaucher 2021). This affiliation of membrane composition in bacteria and eukaryotes is clearly inconsistent with the topology of the Tree of Life suggested earlier, unless LACA and its early descendants had membranes containing a mixture of both types of lipids (Lombard et al. 2012).

The latter idea has some support. The dehydrogenase enzymes that make G1P and G3P are found in all major lineages, raising the possibility of a non-specific glycerol-phosphate dehydrogenase in LUCA. In addition, some bacteria and eukaryotes have phospholipids with ether linkers; some archaea have fatty acids; and isoprenoids are universally distributed, although they are synthesized by different pathways in the three major groups (Lange et al. 2000; Lombard and Moreira 2011; Villanueva et al. 2021). It has been argued that a mixed population of lipid molecules will reduce membrane stability, but there are doubts about this idea (Shimada and Yamagishi 2011), and indeed, *E. coli* has been engineered to contain up to 30% archaeal lipids with little negative effects on growth rate (Caforio et al. 2018). Thus, it is plausible that LUCA had a membrane consisting of a mixture of the molecules found in modern-day prokaryotic lineages, with alternative mechanisms for catalysing

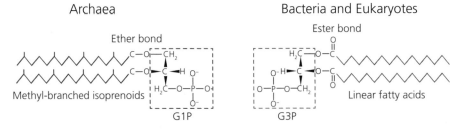

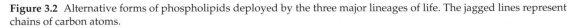

Figure 3.2 Alternative forms of phospholipids deployed by the three major lineages of life. The jagged lines represent chains of carbon atoms.

pure populations of G1P or G3P molecules evolving independently in isolated lineages (Koga et al. 1998; Martin and Russell 2003; Wächtershäuser 2003).

Finally, note that the two-domains model makes the implicit assumption that the root of the entire Tree of Life falls between the bacterial and archaeal domains. A more formal way of evaluating the problem uses genes that duplicated prior to the divergence of the main domains, as each member of such a gene pair can serve to root the phylogeny of the other. In the ideal scenario, both trees resulting from such reciprocal rooting would yield the same topology. To exploit this strategy, Gogarten et al. (1989) used anciently duplicated subunits of ATP synthase (Foundations 2.1) to show that archaea and eukaryotes consistently group together to the exclusion of bacteria. The same result has been obtained with several other pairs of ancient duplicate genes (Iwabe et al. 1989; Brown and Doolittle 1995; Baldauf et al. 1996; Lawson et al. 1996; Gribaldo and Cammarano 1998; Zhaxybayeva et al. 2005).

Although the emerging view is that the Tree of Life is rooted as illustrated in Figure 3.1, there is still some dissent on the matter (Philippe and Forterre 1999; Cavalier-Smith and Chao 2020). Devos (2021) goes so far as to advocate a one-domain model in which bacteria are ancestral to all of life, with one bacterial lineage (putatively related to the planctomycetes, which have internal membranes and in some cases are capable of phagocytosis) spawning a common ancestor to archaea and eukaryotes (LAECA). Under this hypothesis, Eukarya and the Archaea are each monophyletic sister taxa derived from a common ancestor, with the former embarking down a pathway of increasing complexity and the latter retaining greater simplicity. Although this one-domain model does not enjoy phylogenetic support based on gene-sequence data, given the limited power to confidently reveal relationships at the base of the Tree of Life, there are reasons to still be cautious in embracing the two-domain model as an established fact.

Times of Origin

The preceding description of the basic topology of the main trunks of the Tree of Life leaves unresolved the times of origin of various lineages, i.e., the temporal positions of the first and last common ancestors of the key clades. The gold standard for such estimates is a fossil record. However, only a small fraction of species leaves such traces, and even in the best of circumstances, the vagaries of geological activity generally result in substantial gaps and uncertain time horizons in the fossil record. Although there is a well-established fossil record for many groups of land plants and animals, few unicellular organisms are fossilizable, and a wide range of abiotic events can leave traces that can be nearly indistinguishable from those induced by real cells (Javaux 2019). Today's smallest bacteria have diameters $< 1\,\mu m$, and the earliest cells were likely even smaller, reducing the likelihood of detection. Rock formations older than 3.5 BY (billion years) are extremely rare, further restricting the opportunities of directly inferring the earliest stages of evolution.

Methods for detecting organic material in ancient rocks expand the potential for inferring life's presence (Brasier et al. 2015), although as noted in Chapter 2, numerous geological mechanisms can yield organic molecules in the absence of any biology. Given the inference that the complex processes of photosynthesis and methanogenesis were present by 3.4 BYA (billion years ago) (Ueno et al. 2006; Javaux 2019), this further implies the establishment by this time of many of the metabolic/molecular processes from which all subsequent cellular lineages were built. Thus, it is not far-fetched to suggest that cells were present as early as 4 BYA, and some indirect evidence for biological activity as early as 4.1 BYA has been suggested (Bell et al. 2015).

The first evidence of eukaryotic cells appears in shale deposits containing putative molecular biomarkers of membrane components from ~2.7 BYA (Brocks et al. 1999), with the first presumptive algal fossils dating to ~2.1 BYA (Han and Runnegar 1992). Many other fossils of unicellular eukaryotes with complex surface ornamentations date to 1.5 to 1.7 BYA (Shixing and Huineng 1995; Javaux et al. 2001; Knoll 2004). However, complex multicellularity remained absent for at least another billion years. A dramatic shift occurred ~550 MYA (million years ago), when all of the major groups of multicellular animals appear suddenly in the fossil record in what is popularly known as the Cambrian explosion (keeping in mind that 'sudden' from a paleontological perspective can exceed tens of millions of years). The most visible biota on today's Earth, the jawed vertebrates and land plants, emerged only ~440 and ~400 MYA, respectively.

Of course, the time of first appearance of a group in the fossil record must postdate the actual time of origin. To work around this problem, attempts have been made to estimate key early divergence points in the Tree of Life using molecular clocks for protein-coding sequences calibrated with more recent fossils from well-understood taxonomic groups. Although numerous assumptions underlie these analyses, the current prognosis is an initial point of divergence of the eukaryotic branch from its archaeal ancestor ~1.9 BYA, demarcating the position of the first eukaryotic common ancestor (FECA), with the last eukaryotic common ancestor (LECA), at the base of the tree of diverging eukaryotic lineages) dating to ~1.0 to 1.7 BYA (Parfrey et al. 2011; Shih and Matzke 2013; Eme et al. 2014). These dates are roughly compatible with the fossil-record data noted above.

If this interpretation is correct, the first two billion years or so of life's timeline was written entirely by prokaryotes, with > 80% of biological history involving a world containing only single-celled organisms. Significant surprises may still be in store, as genome sequences from environmental samples continue to reveal new microbial lineages (Hug et al. 2016; Zaremba-Niedzwiedzka et al. 2017; Liu et al. 2021; Tahon et al. 2021).

The Emergence of Eukaryotes

Evolutionary cell biology is equally concerned with prokaryotes and eukaryotes. However, considering the disproportionate attention given to yeast, plant, and animal cells, which represent only a sliver of the massive expansion of morphological complexity in eukaryotes, a brief excursion on the unity and diversity of the main lineages of the latter group is warranted.

Based on evidence from comparative genomics, there is little question that FECA was a chimera between members of the archaea and bacteria, but how such a liaison came about is less clear. As discussed above, key features of the nuclear genome (genes involved in replication and translation, in particular) were derived from an archaeum, but this leaves open a number of possible scenarios (López-García and Moreira 2020; Martin et al. 2015). In one view, FECA was an archaeal cell that acquired bacterial genes from an endosymbiotic bacterium, which became the mitochondrion. Although the latter point is well-established (Chapter 23), an alternative view is that the original host cell was a bacterium (which became the source of internal membranes) harboring an endosymbiotic archaeum (which became the nucleus), with the mitochondrion joining secondarily (López-García and Moreira 1999, 2020). Whether these alternatives, or any other suggested models, can ever be definitively resolved remains unclear, but numerous features of LECA are more certain.

The stem eukaryote

Provided that a group of species is monophyletic, as seems to be the case for eukaryotes, we can generally be confident that any feature shared across all members of the clade must have been present in its most recent common ancestor (in this case, LECA). Based on the logic that highly complex cellular traits are unlikely to have arisen independently in multiple lineages, comparative biology tells us that LECA was a flagellated heterotroph, capable of phagocytosis, with quite complex internal structure, and distinguished from prokaryotes in dozens of other ways at the level of cell structure, intracellular processes, gene structure, and genome organization (Cavalier-Smith 2009; Koumandou et al. 2013). The order in which these features emerged on the path from FECA to LECA is less clear, and will likely remain so unless basal lineages lacking subsets of such traits are discovered. This raises significant challenges for determining the key innovations that might have precipitated the evolutionary cascade of events known as eukaryogenesis. The following provides just a brief overview of the primary changes, with fuller details appearing in subsequent chapters.

The most celebrated eukaryotic attributes are physical. Most notably, a nuclear envelope allows a spatial separation between gene transcription within the nucleus and translation of messenger RNAs in the cytoplasm. Unique cytoskeletal structures based on actin and tubulin provide physical support for a variety of cellular functions. These include: platforms for membrane bending essential for vesicle formation, food engulfment by phagocytosis, and osmotic regulation by contractile vacuoles; scaffolds for the ordered transport of chromosomes during cell division; and highways for molecular motors engaged in transporting vesicles and powering flagella. Molecular motors are eukaryotic inventions, and the eukaryotic flagellum is completely different from that deployed in bacteria. Finally, internal membrane-based structures, such as the endoplasmic reticulum and the Golgi, provide sites for molecular processing unique to eukaryotes.

A key eukaryotic organelle is the mitochondrion, which became established at some point between FECA and LECA and is one of the only eukaryotic features whose origin is known. Unlike other organelles, mitochondria contain genomes whose sequences reveal alphaproteobacterial ancestry, with the original colonist eventually becoming an obligate endosymbiont now known as the powerhouse of eukaryotic cells (Chapter 23). Prior to the establishment of the mitochondrion, ATP synthase (Chapter 2), resided on the cell membrane (as it does in all of today's prokaryotes), but in eukaryotes ATP synthase is sequestered to internal mitochondrial membranes. Some have argued that this relocation provided a solution to the reduced surface:volume ratio in larger cells, essentially generating a bioenergetics revolution necessary for the establishment of all other things eukaryotic (Lane 2002, 2015; Lane and Martin 2010). Under this view, colonization of the mitochondrion would have been the causal event in eukaryogenesis, and therefore the first key innovation to appear on the branch from FECA to LECA.

However, despite its superficial attractiveness, an association alone does not indicate the direction of causality, and the idea that the establishment of the mitochondrion spawned a quantum leap in bioenergetic capacity is inconsistent with numerous lines of evidence outlined in subsequent chapters. Nonetheless, once established, the mitochondrion generated

numerous secondary effects to accommodate its use. For example, substantial transfer of mitochondrial genes to the nuclear genome occurred prior to LECA. Many of these transferred genes generate products that must be sent back to the mitochondrion, in some cases providing components to protein complexes that also contain mitochondrially encoded subunits. This necessitates reliable mechanisms for coordinating the activities of organelle and nuclear genomes and targeting the transport of proteins to their appropriate destinations.

The transition to eukaryotes was also accompanied by major alterations in the mode of genome replication and transmission (Lynch 2007; Chapter 10). Almost all bacterial genomes consist of single circular chromosomes that replicate bidirectionally in two continuous streams from a single origin of replication, with the daughter genomes moving to opposite ends of the parental cell by fairly simple mechanisms. In contrast, the nuclear genomes of eukaryotes consist of multiple linear chromosomes, spooled around protein complexes called histones, with multiple origins of replication and ends capped by repetitive arrays of short motifs called telomeres.

Eukaryotic cell division requires an organized set of events, known as mitosis, by which multiple chromosomes duplicate simultaneously, with complete offspring sets then being dragged to opposite poles along a microtubule-based spindle apparatus. Moreover, eukaryotes have another specialized form of genome replication called meiosis, which has no counterpart in prokaryotes. During this process, homologous pairs of chromosomes (one haploid set from each parent) line up in parallel in a diploid cell, where they reciprocally exchange material by recombination, ultimately producing four haploid daughter cells (with single copies of each chromosome). The fusion of two such haploid cells reconstitutes the diploid form, completing the sexual life cycle.

The mode of transcript processing also underwent considerable modification in the stem eukaryote (Lynch 2007). Most, if not all, prokaryotic genomes contain operons (cassettes of co-transcribed and often functionally related genes). Such multigene transcripts constitute a significant challenge for the membrane-bound genomes of eukaryotes, as the entire units have to be either exported from the nucleus in their entirety or pre-processed into single-gene messages prior to export. The few known cases of eukaryotic operons (e.g., in nematodes and euglenoids) do, in fact, involve such processing, along with the *trans*-splicing of a small leader sequence to the front end of each individual transcript, a process that is unknown in prokaryotes.

Finally, the nuclear envelope provided a genomic environment that promoted the emergence of more complex gene structure, most notably the colonization of genes by intragenic spacers called introns. Because introns are transcribed along with their surrounding exons, this genes-in-pieces architecture imposes another significant challenge for information processing – introns must be precisely excised and exons spliced back together (*cis*-splicing) prior to the export of mature mRNAs through a nuclear pore to the cytoplasm. Splicing is carried out by a complex molecular machine unique to eukaryotes, the spliceosome, consisting of five small RNA subunits and more than 100 proteins. In striking contrast, nearly all prokaryotic genes consist of a single uninterrupted coding region, and in the very few instances where this is not the case, the introns are self-splicing.

These are just a few of the many features unique to the eukaryotic lineage, the main point being that an enormous remodelling of cell biology occurred on the lineage from FECA to LECA. Notably, however, parallels of many 'eukaryotic-specific' attributes can be found in isolated prokaryotic lineages, so one need not invoke *de novo* invention. For example, as already noted, many proteins previously thought to be restricted to eukaryotes are now known to have orthologous relatives in the Asgard archaea. In addition, organelles of a wide variety of types bounded by lipid or protein membranes are known for several members of the bacteria and archaea (Grant et al. 2018; Greening and Lithgow 2020), the planctomycetes in particular, with at least one such lineage being capable of phagocytosis (the engulfment and digestion of other cells) (Boedeker et al. 2017; Shiratori et al. 2019). These types of observations, along with other indirect inferences (Pittis and Gabaldón 2016), clearly indicate that many of the embellishments of eukaryotic cells did not have to await the origin of the mitochondrion as an energy support system.

What remains unclear is how so many odd features of prokaryotic cells came to be co-localized in the same FECA–LECA lineage. Although one might argue that FECA was a highly polymorphic species, with different individuals harboring subsets of traits (O'Malley et al. 2019), it is difficult to conceive of individuals with different constellations of complex traits still being reproductively compatible. Any such exchange would have had to occur prior to the emergence of meiosis, which requires sequence homology between pairing chromosomes.

Thus, the early steps of eukaryogenesis remain a mystery. We do not know the events that triggered eukaryogenesis, nor do we know the extent to which the peculiar features that arose did so via the encouragement of natural selection. Some modifications, such as intron colonization, may have emerged in population settings that enabled mildly deleterious mutations to accumulate passively by mutation pressure alone. Once established, however, the vast set of changes bestowed upon LECA provided the substrate for the evolutionary explosion in cell architectural diversity that is the hallmark of eukaryotes.

The eukaryotic radiation

As with investigations of the prokaryote–eukaryote divide, progress on revealing phylogenetic relationships among the major eukaryotic groups has largely relied on comparative gene-sequence analysis. However, even with whole-genome analyses, multiple issues still conspire to cloud our understanding of the phylogeny of eukaryotes. These include idiosyncratic changes in rates of evolution, divergent nucleotide compositions across lineages, possibilities of early horizontal gene transfer, gene duplications, and inadequate taxon sampling. Two things are agreed upon. First, the primary eukaryotic lineages are deeply branching in time, with the major groups upon which most biological research is performed (metazoans, fungi, and plants) constituting only a small fraction of eukaryotic phylogenetic diversity. Second, although these three favored sets of study organisms are sometimes viewed as members of a 'crown group' of eukaryotes or 'higher forms' of life, they do not even comprise a monophyletic lineage.

An attempt to summarize what is known about eukaryotic phylogeny is presented in Figure 3.3, with two caveats. First, this description is by no means complete, as it contains only the groups that are encountered in the following chapters. Even if all of the major known groups of eukaryotes were included, the story would be an abstract at best, as agnostic searches for molecular sequences from environmental samples suggest that many novel lineages of microbial eukaryotes, never before visualized, reside in our midst (Dawson and Pace 2002). Second, the phylogenetic relationships of many of the main eukaryotic groups remain unresolved. Depending on the authors, between five and eight monophyletic supergroups are recognized (e.g., Baldauf et al. 2000; Richards and Cavalier-Smith 2005; He et al. 2014; Derelle et al. 2015; Katz and Grant 2015; Ren et al. 2016; Burki et al. 2020; Wideman et al. 2020), and these will likely change to some degree as further data emerge (e.g., Tikhonenkov et al. 2022).

In one view, the vast majority of eukaryotes fall into two major morphological groups based on the ancestral number of flagella being one or two (Cavalier-Smith 1998). The first of these, the unikonts, are united by the general presence of cells with a single flagellum at some stage of the life cycle (Cavalier-Smith 1998; Steenkamp et al. 2006; Paps et al. 2013). The unikonts contain the opisthokont group, an assemblage of metazoans, choanoflagellates, and fungi (top of Figure 3.3), as well as the amoebozoan group, comprised of the lobose amoeba and the slime molds (Bapteste et al. 2002). Along with a few biflagellate lineages, the unikonts appear to be separated from the remaining supergroups (all of which are biflagellate, and referred to as bikonts) at the root of the eukaryotic tree (Derelle et al. 2015). This motivates the suggestion that LECA was a biflagellate.

The large bikont assemblage contains the remaining supergroups, whose interrelationships remain unresolved. One of these groups, the Archaeplastida, encompasses the chloroplast-bearing green plants (including the green algae), red algae (rhodophytes), and glaucophyte algae. The excavate supergroup contains the Euglenozoa, which unites the euglenoids (e.g., *Euglena*) with the parasitic kinetoplastids (e.g., the trypanosomes *Trypanosoma* and *Leishmania*), as well as several other groups of flagellates.

Another large supergroup is dubbed the SAR clade, based on its primary component lineages, the stramenopiles, alveolates, and rhizarians. The diverse stramenopile subclade contains the diatoms, brown algae, and oomycetes, whereas the alveolates (united by the presence of alveoli, a system of sacs underlying the cell surface) encompass the ciliates (e.g., *Paramecium* and *Tetrahymena*), the dinoflagellates (a group of aquatic flagellates), and the obligately parasitic apicomplexans (including the malarial parasite *Plasmodium*) (Fast et al. 2002). The rhizaria consist of cercozoans, foraminiferans, and radiolarians, most of which are amoeboid and produce external skeletons (Nikolaev et al. 2004).

Monophyly of the entire bikont group has drawn support from a unique fusion between two key genes (dihydrofolate reductase and thymidylate synthase), which are encoded separately in all unikonts and prokaryotes (Stechmann and Cavalier-Smith 2002). However, some exceptions have been found within the bikonts (Burki 2014), which might represent secondary reversions. In addition, the amitochondriate diplomonad (including *Giardia*) and trichomonad lineages appear not to contain either gene and so cannot be assigned phylogenetic positions on this basis, although they may be members of the excavate supergroup.

A eukaryotic Big Bang?

Given that the deep lines of descent between bacteria, archaea, and eukaryotes have been resolved with less data, the inability to fully decipher the more recent relationships among the main lines of eukaryotes is unlikely to be a matter of a shortage of genomic information. The relatively short internal branches of the eukaryotic tree, which imply a rapid early radiation of such groups, is the major issue – shorter branches between related groups lead to lower discriminating power. Thus, the bushy form of the eukaryotic tree has inspired a 'Big Bang' hypothesis suggesting that most of the major lineages became established in a period of 10–100 million years (Philippe et al. 2000; Cavalier-Smith 2002;

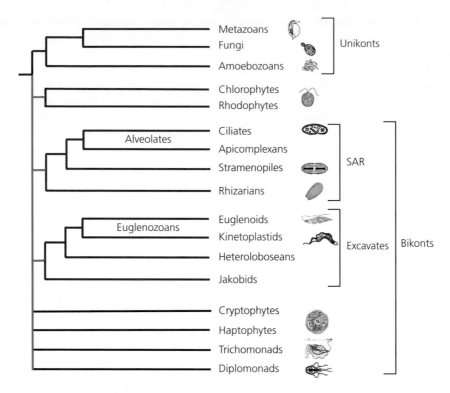

Figure 3.3 An approximate phylogenetic tree for some of the major eukaryotic 'supergroups', generalized from the references in the text. The branch lengths are not proportional to time, although all external branches are expected to be in excess of 700 million years in length. Grey lines at the base of the tree denote areas of uncertainty.

Koonin 2007). If this idea is correct, the arguments presented above, along with other molecular estimates of the age of LECA, would suggest a radiation set down in a window roughly between 1.7 and 2.0 BYA (Wang et al. 1999; Yoon et al. 2004; Parfrey et al. 2011; Eme et al. 2014).

What might have precipitated such an active phase of lineage isolation? Most attempts at explaining evolutionary radiations resort to ecological arguments, either invoking a dramatic change in the environment or the chance appearance of an evolutionary novelty allowing the exploitation of new ecological niches, e.g., predation as a new way of living (Knoll 2014). However, a species radiation requires more than ecological opportunity. There must also be genetic isolating mechanisms to keep lineages distinct. Ultimately, opportunities for speciation require that populations be kept separate for long enough periods to allow the accumulation of sufficient mutational changes that viability and/or fertility will be compromised by parental-genome incompatibilities that arise within hybrids.

Post-reproductive isolating barriers can arise by many different mechanisms (Coyne and Orr 2004), but microchromosomal rearrangements in which genes relocate from one chromosome to another are of particular relevance to the early eukaryotic radiation, which experienced two novel forms of genomic upheaval. Consider first the primordial mitochondrion. Most prokaryotic genomes contain a few thousand genes, while mitochondrial genomes contain no more than a few dozen. Thus, it is clear that hundreds of organelle-to-nuclear gene transfers occurred early in the establishment of mitochondria, although many were probably simply lost (Chapter 23). Because mitochondrial genomes are haploid and generally inherited uniparentally, a relocation of an essential mitochondrial gene to the nuclear genome would create an imbalance in hybrid progeny resulting from a

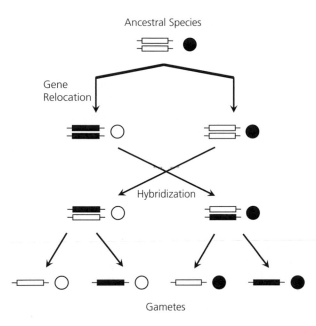

Figure 3.4 The development of reproductive incompatibility following the relocation of a mitochondrial gene. A diploid phase of the nuclear genome is shown. Rectangles and circles denote autosomal and organelle gene copies, respectively, with open symbols indicating gene absence. Following a geographic isolating event, the incipient species on the left experiences an organelle-gene transfer to the nucleus. Subsequent hybridization yields presence/absence heterozygotes at the autosomal locus, with the status of the uniparentally inherited mitochondrial genome depending on maternal identity. As a consequence of Mendelian segregation of the diploid autosomal locus, gene imbalance arises following meiosis: half of the gametes of the individual in the lower left will lack the gene entirely, and half of those on the right will have a double dose of the gene. After Lynch and Force (2000).

cross with any lineage having the ancestral (non-rearranged) type (Figure 3.4). Regardless of the direction of the cross, both types of diploid hybrids would be presence/absence heterozygotes for the nuclear gene. In addition, one would also harbor a mitochondrial genome devoid of the gene. As a consequence, half of the gametes produced by the latter individual would lack the gene entirely (and half of those produced by the other hybrid type would acquire a double dose of the gene).

Although a single genomic transfer of this sort does not produce complete reproductive isolation, just a few independent transfers have a powerful effect. Imagine an incipient pair of species experiencing n independent organelle-to-nuclear gene transfers in each lineage. Assuming independent assortment of the nuclear genes during meiosis, then the fraction of F_1 gametes entirely lacking in a functional gene at one or more loci is $1 - 0.5^n$, which is 0.969 for $n = 5$. Thus, when one considers

the hundreds of organelle-to-nuclear gene transfers that may have occurred soon after the colonization of the primordial mitochondrion, and probably extended over several million years, such gene traffic would have played a significant role in the passive development of isolating barriers among the earliest eukaryotes. Such microchromosomal rearrangements only yield reproductive isolating barriers in species with multiple chromosomes and sexual reproduction, as both are necessary for the independent segregation of unlinked loci. Notably, however, both speciation-facilitating features were among the novelties that emerged on the branch from FECA to LECA.

A second mechanism of gene relocation relevant to the eukaryotic radiation involves nuclear gene-duplication events, which can passively lead to rearrangements when the original copy is silenced and a descendant copy is preserved on a separate chromosome (Chapter 6). Such events are of interest here because, as discussed in subsequent chapters, there was a massive amount of gene duplication at the base of eukaryotes, possibly owing to one or two complete genome duplications (Chapter 24). Such activities left their imprint on a wide variety of cellular features, including mitosis and meiosis (Ramesh et al. 2005; Malik et al. 2007; Onesti and MacNeill 2013; Liu et al. 2015), the cytoskeleton (McKean et al. 2001; Goodson and Hawse 2002; Dutcher 2003) and the flagellum (van Dam et al. 2013), proteasomes (Bouzat et al. 2000) and chaperones (Fares and Wolfe 2003), the nuclear-pore complex (Alber et al. 2007), and other organelles (Schledzewski et al. 1999; Hirst et al. 2011; Mast et al. 2014).

Thus, the indirect consequences of two of the defining cytological attributes of the stem eukaryote, a genome-bearing mitochondrion and meiotic recombination, along with rampant duplication in the nuclear genome, may have played a central role in the passive and relatively rapid emergence of the basal eukaryotic lineages. Although ecological divergence need not have played any initiating role in such processes, the resultant reproductive isolation would have allowed such lineages to descend down independent evolutionary pathways driven by adaptation to local environmental settings.

Summary

- Although life has classically been divided into eukaryotes and prokaryotes, molecular analyses indicate that these are not meaningful phylogenetic labels. Instead, there appear to be two domains of life, with the two prokaryotic groups (bacteria and archaea) residing on opposite sides of the root of the Tree of Life, and eukaryotes being the most recent newcomer, emerging from a member of the archaea.

- Prokaryotes were established on Earth ~4 BYA, with eukaryotes appearing ~3 BYA. Although many eukaryotic lineages may have coexisted during this early period, only one (called LECA) eventually gave rise to today's eukaryotes, forming the base of the tree of extant lineages ~2 BYA. Each of these time points has a level of uncertainty of a few hundred million years.

- From the standpoint of morphological diversification, the emergence of eukaryotes marked a dramatic phase in Earth's history. With dozens of eukaryote-specific changes having become established prior to LECA, this keystone species was extraordinarily unique in terms of cellular and genomic architecture. However, the order in which these features arose remains unknown, and many of them are difficult to explain with adaptive arguments.

- Once established, LECA gave rise to an explosive radiation of the major eukaryotic groups on a relatively short time scale. This rapid episode of lineage isolation may have had little to do with ecological factors, instead being an inevitable consequence of two pre-LECA genomic upheavals – the origin of the mitochondrion and a period of rampant nuclear gene duplication. Combined with the evolution of sex and independently segregating chromosomes, these changes would have led to the passive accumulation of microchromosomal rearrangements and reproductive isolation in ways that are inoperable in prokaryotic lineages.

Literature Cited

Akil, C., and R. C. Robinson. 2018. Genomes of Asgard archaea encode profilins that regulate actin. Nature 562: 439–443.

Alber, F., S. Dokudovskaya, L. M. Veenhoff, W. Zhang, J. Kipper, D. Devos, A. Suprapto, O. Karni-Schmidt, R., B. T. Chait, et al. 2007. The molecular architecture of the nuclear pore complex. Nature 450: 695–701.

Baldauf, S. L., J. D. Palmer, and W. F. Doolittle. 1996. The root of the universal tree and the origin of eukaryotes based on elongation factor phylogeny. Proc. Natl. Acad. Sci. USA 93: 7749–7754.

Baldauf, S. L., A. J. Roger, I. Wenk-Siefert, and W. F. Doolittle. 2000. A kingdom-level phylogeny of eukaryotes based on combined protein data. Science 290: 972–977.

Bapteste, E., H. Brinkmann, J. A. Lee, D. V. Moore, C. W. Sensen, P. Gordon, L. Durufle, T. Gaasterland, P. Lopez, M. Muller, and H. Philippe. 2002. The analysis of 100 genes supports the grouping of three highly divergent amoebae: *Dictyostelium*, *Entamoeba*, and *Mastigamoeba*. Proc. Natl. Acad. Sci. USA 99: 1414–1419.

Bell, E. A., P. Boehnke, T. M. Harrison, and W. L. Mao. 2015. Potentially biogenic carbon preserved in a 4.1-billion-year-old zircon. Proc. Natl. Acad. Sci. USA 112: 14518–14521.

Boedeker, C., M. Schüler, G. Reintjes, O. Jeske, M. C. van Teeseling, M. Jogler, P. Rast, D. Borchert, D. P. Devos,

M. Kucklick, et al. 2017. Determining the bacterial cell biology of Planctomycetes. Nat. Commun. 8: 14853.

Boucher, Y., M. Kamekura, and W. F. Doolittle. 2004. Origins and evolution of isoprenoid lipid biosynthesis in archaea. Mol. Microbiol. 52: 515–527.

Bouzat, J. L., L. K. McNeil, H. M. Robertson, L. F. Solter, J. E. Nixon, J.E. Beever, H. R. Gaskins, G. Olsen, S. Subramaniam, M. L. Sogin, et al. 2000. Phylogenomic analysis of the alpha proteasome gene family from early diverging eukaryotes. J. Mol. Evol. 51: 532–543.

Brasier, M. D., J. Antcliffe, M. Saunders, and D. Wacey. 2015. Changing the picture of Earth's earliest fossils (3.5–1.9 Ga) with new approaches and new discoveries. Proc. Natl. Acad. Sci. USA 112: 4859–4864.

Brocks, J. J., G. A. Logan, R. Buick, and R. E. Summons. 1999. Archean molecular fossils and the early rise of eukaryotes. Science 285: 1033–1036.

Brown, J. R., and W. F. Doolittle. 1995. Root of the universal tree of life based on ancient aminoacyl-tRNA synthetase gene duplications. Proc. Natl. Acad. Sci. USA 92: 2441–2445.

Brown, J. R., and W. F. Doolittle. 1997. Archaea and the prokaryote-to-eukaryote transition. Microbiol. Mol. Biol. Rev. 61: 456–502.

Brown, J. R., C. J. Douady, M. J. Italia, W. E. Marshall, and M. J. Stanhope. 2001. Universal trees based on large combined protein sequence data sets. Nat. Genet. 28: 281–285.

Burki, F. 2014. The eukaryotic tree of life from a global phylogenomic perspective. Cold Spring Harb. Perspect. Biol. 6: a016147.

Burki, F., A. J. Roger, M. W. Brown, and A. G. B. Simpson. 2020. The new tree of eukaryotes. Trends Ecol. Evol. 35: 43–55.

Caforio, A., M. F. Siliakus, M. Exterkate, S. Jain, V. R. Jumde, R. L. H. Andringa, S. W. M. Kengen, A. J. Minnaard, A. J. M. Driessen, and J. van der Oost. 2018. Converting *Escherichia coli* into an archaebacterium with a hybrid heterochiral membrane. Proc. Natl. Acad. Sci. USA 115: 3704–3709.

Cavalier-Smith, T. 1998. A revised six-kingdom system of life. Biol. Rev. 73: 203–266.

Cavalier-Smith, T. 2002. The phagotrophic origin of eukaryotes and phylogenetic classification of protozoa. Internat. J. Syst. Evol. Microbiol. 52: 297–354.

Cavalier-Smith, T. 2009. Predation and eukaryote cell origins: A coevolutionary perspective. Int. J. Biochem. Cell Biol. 41: 307–322.

Cavalier-Smith, T., and E. E. Chao. 2020. Multidomain ribosomal protein trees and the planctobacterial origin of neomura (eukaryotes, archaebacteria). Protoplasma 257: 621–753.

Coleman, G. A., A. A. Davin, T. A. Mahendrarajah, L. L. Szánthó, A. Spang, P. Hugenholtz, G. J. Szöllosi, and T. A. Williams. 2021. A rooted phylogeny resolves early bacterial evolution. Science 372: eabe0511.

Cox, C. J., P. G. Foster, R. P. Hirt, S. R. Harris, and T. M. Embley. 2008. The archaebacterial origin of eukaryotes. Proc. Natl. Acad. Sci. USA 105: 20356–20361.

Coyne, J. A., and H. A. Orr. 2004. Speciation. Sinauer Associates, Inc., Sunderland, MA.

Dawson, S. C., and N. R. Pace. 2002. Novel kingdom-level eukaryotic diversity in anoxic environments. Proc. Natl. Acad. Sci. USA 99: 8324–8329.

Derelle, R., G. Torruella, V. Klimeš, H. Brinkmann, E. Kim, Č. Vlček, B. F. Lang, and M. Eliáš. 2015. Bacterial proteins pinpoint a single eukaryotic root. Proc. Natl. Acad. Sci. USA 112: E693-E699.

Devos, D. P. 2021. Reconciling Asgardarchaeota phylogenetic proximity to eukaryotes and planctomycetes cellular features in the evolution of life. Mol. Bio. Evol. 38: 3531–3542.

Doolittle, W. F., Y. Boucher, C. L. Nesbo, C. J. Douady, J. O. Andersson, and A. J. Roger. 2003. How big is the iceberg of which organellar genes in nuclear genomes are but the tip? Phil. Trans. Roy. Soc. Lond. B Biol. Sci. 358: 39–57.

Dutcher, S. K. 2003. Long-lost relatives reappear: Identification of new members of the tubulin superfamily. Curr. Opin. Microbiol. 6: 634–640.

Eme, L., S. C. Sharpe, M. W. Brown, and A. J. Roger. 2014. On the age of eukaryotes: Evaluating evidence from fossils and molecular clocks. Cold Spring Harb. Perspect. Biol. 6: a016139.

Ettema, T. J., A. C. Lindå s, and R. Bernander. 2011. An actin-based cytoskeleton in archaea. Mol. Microbiol. 80: 1052–1061.

Fares, M. A., and K. H. Wolfe. 2003. Positive selection and subfunctionalization of duplicated CCT chaperonin subunits. Mol. Biol. Evol. 20: 1588–1597.

Fast, N. M., L. Xue, S. Bingham, and P. J. Keeling. 2002. Re-examining alveolate evolution using multiple protein molecular phylogenies. J. Eukaryot. Microbiol. 49: 30–37.

Felsenstein, J. 2004. Inferring Phylogenies. Sinauer Associates, Inc., Sunderland, MA.

Gogarten, J. P., H. Kibak, P. Dittrich, L. Taiz, E. J. Bowman, B. J. Bowman, M. F. Manolson, R. J. Poole, T. Date, T. Oshima, et al. 1989. Evolution of the vacuolar H+-ATPase: Implications for the origin of eukaryotes. Proc. Natl. Acad. Sci. USA 86: 6661–6665.

Goodson, H. V., and W. F. Hawse. 2002. Molecular evolution of the actin family. J. Cell Sci. 115: 2619–2622.

Grant, C. R., J. Wan, and A. Komeili. 2018. Organelle formation in bacteria and archaea. Annu. Rev. Cell Dev. Biol. 34: 217–238.

Greening, C., and T. Lithgow. 2020. Formation and function of bacterial organelles. Nat. Rev. Microbiol. 18: 677–689.

Gribaldo, S., and P. Cammarano. 1998. The root of the universal tree of life inferred from anciently duplicated genes encoding components of the protein-targeting machinery. J. Mol. Evol. 47: 508–516.

Guy, L., and T. J. Ettema. 2011. The archaeal 'TACK' superphylum and the origin of eukaryotes. Trends Microbiol. 19: 580–587.

Han, T. M., and B. Runnegar. 1992. Megascopic eukaryotic algae from the 2.1-billion-year-old Negaunee iron-formation, Michigan. Science 257: 232–235.

He, D., O. Fiz-Palacios, C. J. Fu, J. Fehling, C. C. Tsai, and S. L. Baldauf. 2014. An alternative root for the eukaryote tree of life. Curr. Biol. 24: 465–470.

Hirst, J., L. D. Barlow, G. C. Francisco, D. A. Sahlender, M. N. Seaman, J. B. Dacks, and M. S. Robinson. 2011. The fifth adaptor protein complex. PLoS Biol. 9: e1001170.

Horiike, T., K. Hamada, S. Kanaya, and T. Shinozawa. 2001. Origin of eukaryotic cell nuclei by symbiosis of Archaea in Bacteria is revealed by homology-hit analysis. Nat. Cell Biol. 3: 210–214.

Hoshino, Y., and E. A. Gaucher. 2021. Evolution of bacterial steroid biosynthesis and its impact on eukaryogenesis. Proc. Natl. Acad. Sci. USA 118: e2101276118.

Hug, L. A., B. J. Baker, K. Anantharaman, C. T. Brown, A. J. Probst, C. J. Castelle, C. N. Butterfield, A. W. Hernsdorf, Y. Amano, K. Ise, et al. 2016. A new view of the tree of life. Nat. Microbiol. 1: 16048.

Imachi, H., M. K. Nobu, N. Nakahara, Y. Morono, M. Ogawara, Y. Takaki, Y. Takano, K. Uematsu, T. Ikuta, M. Ito, et al. 2020. Isolation of an archaeon at the prokaryote–eukaryote interface. Nature 577: 519–525.

Iwabe, N., K. Kuma, M. Hasegawa, S. Osawa, and T. Miyata. 1989. Evolutionary relationship of archaebacteria, eubacteria, and eukaryotes inferred from phylogenetic trees of duplicated genes. Proc. Natl. Acad. Sci. USA 86: 9355–9359.

Javaux, E. J. 2019. Challenges in evidencing the earliest traces of life. Nature 572: 451–460.

Javaux, E. J., A. H. Knoll, and M. R. Walter. 2001. Morphological and ecological complexity in early eukaryotic ecosystems. Nature 412: 66–69.

Katz, L. A., and J. R. Grant. 2015. Taxon-rich phylogenomic analyses resolve the eukaryotic tree of life and reveal the power of subsampling by sites. Syst. Biol. 64: 406–415.

Kelly, S., B. Wickstead, and K. Gull. 2011. Archaeal phylogenomics provides evidence in support of a methanogenic origin of the Archaea and a thaumarchaeal origin for the eukaryotes. Proc. Biol. Sci. 278: 1009–1018.

Knoll, A. H. 2004. Life on a Young Planet: The First Three Billion Years of Evolution on Earth. Princeton University Press, Princeton, NJ.

Knoll, A. H. 2014. Paleobiological perspectives on early eukaryotic evolution. Cold Spring Harb. Perspect. Biol. 6: a016121.

Koga, Y., T. Kyuragi, M. Nishihara, and N. Sone. 1998. Did archaeal and bacterial cells arise independently from noncellular precursors? A hypothesis stating that the advent of membrane phospholipid with enantiomeric glycerophosphate backbones caused the separation of the two lines of descent. J. Mol. Evol. 46: 54–63.

Koonin, E. V. 2007. The biological Big Bang model for the major transitions in evolution. Biol. Direct 2: 21.

Koumandou, V. L., B. Wickstead, M. L. Ginger, M. van der Giezen, J. B. Dacks, and M. C. Field. 2013. Molecular paleontology and complexity in the last eukaryotic common ancestor. Crit. Rev. Biochem. Mol. Biol. 48: 373–396.

Lake, J. A., E. Henderson, M. Oakes, and M. W. Clark. 1984. Eocytes: A new ribosome structure indicates a kingdom with a close relationship to eukaryotes. Proc. Natl. Acad. Sci. USA 81: 3786–3790.

Lane, N. 2002. Power, Sex, Suicide: Mitochondria and the Meaning of Life. Oxford University Press, Oxford, UK.

Lane, N. 2015. The Vital Question. W. W. Norton & Co., Inc., New York, NY.

Lane, N., and W. F. Martin. 2010. The energetics of genome complexity. Nature 467: 929–934.

Lange, B. M., T. Rujan, W. Martin, and R. Croteau. 2000. Isoprenoid biosynthesis: The evolution of two ancient and distinct pathways across genomes. Proc. Natl. Acad. Sci. USA 97: 13172–13177.

Lawson, F. S., R. L. Charlebois, and J. A. Dillon. 1996. Phylogenetic analysis of carbamoylphosphate synthetase genes: Complex evolutionary history includes an internal duplication within a gene which can root the tree of life. Mol. Biol. Evol. 13: 970–977.

Leipe, D. D., L. Aravind, and E. V. Koonin. 1999. Did DNA replication evolve twice independently? Nucleic Acids Res. 27: 3389–3401.

Liu, Y., K. S. Makarova, W. C. Huang, Y. I. Wolf, A. N. Nikolskaya, X. Zhang, M. Cai, C. J. Zhang, W. Xu, Z. Luo, et al. 2021. Expanded diversity of Asgard archaea and their relationships with eukaryotes. Nature 593: 553–557.

Liu, Y., J. Pei, N. Grishin, and W. J. Snell. 2015. The cytoplasmic domain of the gamete membrane fusion protein HAP2 targets the protein to the fusion site in *Chlamydomonas* and regulates the fusion reaction. Development 142: 962–971.

THE MAJOR LINES OF DESCENT

Lombard, J., P. López-García, and D. Moreira. 2012. The early evolution of lipid membranes and the three domains of life. Nat. Rev. Microbiol. 10: 507–515.

Lombard, J., and D. Moreira. 2011. Origins and early evolution of the mevalonate pathway of isoprenoid biosynthesis in the three domains of life. Mol. Biol. Evol. 28: 87–99.

López-García, P., and D. Moreira. 1999. Metabolic symbiosis at the origin of eukaryotes. Trends Biochem. Sci. 24: 88–93.

López-García, P., and D. Moreira. 2020. The syntrophy hypothesis for the origin of eukaryotes revisited. Nat. Microbiol. 5: 655–667.

Lynch, M. 2007. The Origins of Genome Architecture. Sinauer Assocsiates, Inc., Sunderland, MA.

Lynch, M., and A. Force. 2000. Gene duplication and the origin of interspecific genomic incompatibility. Amer. Natur. 156: 590–605.

Malik, S. B., M. A. Ramesh, A. M. Hulstrand, and J. M. Logsdon, Jr. 2007. Protist homologs of the meiotic *Spo11* gene and topoisomerase VI reveal an evolutionary history of gene duplication and lineage-specific loss. Mol. Biol. Evol. 24: 2827–2841.

Martin, W. F., S. Garg, and V. Zimorski. 2015. Endosymbiotic theories for eukaryote origin. Philos. Trans. R. Soc. Lond. B Biol. Sci. 370: 20140330.

Martin, W., and M. J. Russell. 2003. On the origins of cells: A hypothesis for the evolutionary transitions from abiotic geochemistry to chemoautotrophic prokaryotes, and from prokaryotes to nucleated cells. Philos. Trans. R. Soc. Lond. B Biol. Sci. 358: 59–83.

Mast, F. D., L. D. Barlow, R. A. Rachubinski, and J. B. Dacks. 2014. Evolutionary mechanisms for establishing eukaryotic cellular complexity. Trends Cell Biol. 24: 435–442.

McKean, P. G., S. Vaughan, and K. Gull. 2001. The extended tubulin superfamily. J. Cell Sci. 114: 2723–2733.

Nikolaev, S. I., C. Berney, J. F. Fahrni, I. Bolivar, S. Polet, V. V. Aleshin, and N. B. Petrov. 2004. Molecular phylogenetic analysis places *Percolomonas cosmopolitus* within Heterolobosea: Evolutionary implications. J. Eukaryot. Microbiol. 51: 575–581.

O'Malley, M. A., M. M. Leger, J. G. Wideman, and I. Ruiz-Trillo. 2019. Concepts of the last eukaryotic common ancestor. Nat. Ecol. Evol. 3: 338–344.

Onesti, S., and S. A. MacNeill. 2013. Structure and evolutionary origins of the CMG complex. Chromosoma 122: 47–53.

Paps, J., L. A. Medina-Chacón, W. Marshall, H. Suga, and I. Ruiz-Trillo. 2013. Molecular phylogeny of unikonts: New insights into the position of apusomonads

and ancyromonads and the internal relationships of opisthokonts. Protist 164: 2–12.

Parfrey, L. W., D. J. Lahr, A. H. Knoll, and L. A. Katz. 2011. Estimating the timing of early eukaryotic diversification with multigene molecular clocks. Proc. Natl. Acad. Sci. USA 108: 13624–13629.

Peretó, J., P. López-Garcia, and D. Moreira. 2004. Ancestral lipid biosynthesis and early membrane evolution. Trends Biochem. Sci. 29: 469–477.

Philippe, H., and P. Forterre. 1999. The rooting of the universal tree of life is not reliable. J. Mol. Evol. 49: 509–523.

Philippe, H., P. Lopez, H. Brinkmann, K. Budin, A. Germot, J. Laurent, D. Moreira, M. Müller, and H. Le Guyader. 2000. Early-branching or fast-evolving eukaryotes? An answer based on slowly evolving positions. Proc. Roy. Soc. Lond. B 267: 1213–1221.

Pittis, A. A., and T. Gabaldón. 2016. Late acquisition of mitochondria by a host with chimaeric prokaryotic ancestry. Nature 531: 101–104.

Ramesh, M. A., S. B. Malik, and J. M. Logsdon Jr. 2005. A phylogenomic inventory of meiotic genes: Evidence for sex in *Giardia* and an early eukaryotic origin of meiosis. Curr. Biol. 15: 185–191.

Raymann, K., C. Brochier-Armanet, and S. Gribaldo. 2015. The two-domain tree of life is linked to a new root for the Archaea. Proc. Natl. Acad. Sci. USA 112: 6670–6675.

Ren, R., Y. Sun, Y. Zhao, D. Geiser, H. Ma, and X. Zhou. 2016. Phylogenetic resolution of deep eukaryotic and fungal relationships using highly conserved low-copy nuclear genes. Genome Biol. Evol. 8: 2683–2701.

Richards, T. A., and T. Cavalier-Smith. 2005. Myosin domain evolution and the primary divergence of eukaryotes. Nature 436: 1113–1118.

Rivera, M. C., R. Jain, J. E. Moore, and J. A. Lake. 1998. Genomic evidence for two functionally distinct gene classes. Proc. Natl. Acad. Sci. USA 95: 6239–6244.

Rodrigues-Oliveira T., F. Wollweber, R. I. Ponce-Toledo, J. Xu, S. K.-M. R. Rittmann, A. Klingl, M. Pilhofer, and C. Schleper. 2023. Actin cytoskeleton and complex cell architecture in an Asgard archaeon. Nature 613: 332–339

Sapp, J. 2005. The prokaryote–eukaryote dichotomy: Meanings and mythology. Microbiol. Mol. Biol. Rev. 69: 292–305.

Schledzewski, K., H. Brinkmann, and R. R. Mendel. 1999. Phylogenetic analysis of components of the eukaryotic vesicle transport system reveals a common origin of adaptor protein complexes 1, 2, and 3 and the F subcomplex of the coatomer COPI. J. Mol. Evol. 48: 770–778.

Shih, P. M., and N. J. Matzke. 2013. Primary endosymbiosis events date to the later Proterozoic with cross-calibrated phylogenetic dating of duplicated ATPase proteins. Proc. Natl. Acad. Sci. USA 110: 12355–1260.

Shimada, H., and A. Yamagishi. 2011. Stability of heterochiral hybrid membrane made of bacterial *sn*-G3P lipids and archaeal *sn*-G1P lipids. Biochemistry 50: 4114–4120.

Shiratori, T., S. Suzuki, Y. Kakizawa, and K. I. Ishida. 2019. Phagocytosis-like cell engulfment by a planctomycete bacterium. Nat. Commun. 10: 5529.

Shixing, Z., and C. Huineng. 1995. Megascopic multicellular organisms form the 1700-million-year-old Tuanshanzi formation in the Jixian area, north China. Science 270: 620–622.

Spang, A., J. H. Saw, S. L. Jørgensen, K. Zaremba-Niedzwiedzka, J. Martijn, A. E. Lind, R. van Eijk, C. Schleper, L. Guy, and T. J. G. Ettema. 2015. Complex archaea that bridge the gap between prokaryotes and eukaryotes. Nature 521: 173–179.

Stechmann, A., and T. Cavalier-Smith. 2002. Rooting the eukaryote tree by using a derived gene fusion. Science 297: 89–91.

Steenkamp, E. T., J. Wright, and S. L. Baldauf. 2006. The protistan origins of animals and fungi. Mol. Biol. Evol. 23: 93–106.

Tahon, G., P. Geesink, and T. J. G. Ettema. 2021. Expanding archaeal diversity and phylogeny: Past, present, and future. Annu. Rev. Microbiol. 75: 359–381.

Thiergart, T., G. Landan, M. Schenk, T. Dagan, and W. F. Martin. 2012. An evolutionary network of genes present in the eukaryote common ancestor polls genomes on eukaryotic and mitochondrial origin. Genome Biol. Evol. 4: 466–485.

Tikhonenkov, D. V., K. V. Mikhailov, R. M. R. Gawryluk, A. O. Belyaev, V. Mathur, S. A. Karpov, D. G. Zagumyonnyi, A. S. Borodina, K. I. Prakina, A. P. Mylnikov, et al. 2022. Microbial predators from a new supergroup of prokaryotes. Nature 612: 714–719.

Ueno, Y., K. Yamada, N. Yoshida, S. Maruyama, and Y. Isozaki. 2006. Evidence from fluid inclusions for microbial methanogenesis in the early Archaean era. Nature 440: 516–519.

van Dam, T. J., M. J. Townsend, M. Turk, A. Schlessinger, A. Sali, M. C. Field, and M. A. Huynen. 2013. Evolution of modular intraflagellar transport from a coatomer-like progenitor. Proc. Natl. Acad. Sci. USA 110: 6943–6948.

Villanueva, L., F. A. B. von Meijenfeldt, A. B. Westbye, S. Yadav, E. C. Hopmans, B. E. Dutilh, and J. S. S. Damsté. 2021. Bridging the divide: Bacteria synthesizing archaeal membrane lipids. ISME J. 15: 168–182.

Wächtershäuser, G. 2003. From pre-cells to Eukarya – a tale of two lipids. Mol. Microbiol. 47: 13–22.

Wang, D. Y., S. Kumar, and S. B. Hedges. 1999. Divergence time estimates for the early history of animal phyla and the origin of plants, animals and fungi. Proc. Roy. Soc. Lond. B 266: 163–171.

Wideman, J. G., A. Monier, R. Rodríguez-Martínez, G. Leonard, E. Cook, C. Poirier, F. Maguire, D. S. Milner, N. A. T. Irwin, K. Moore, et al. 2020. Unexpected mitochondrial genome diversity revealed by targeted single-cell genomics of heterotrophic flagellated protists. Nat. Microbiol. 5: 154–165.

Williams, T. A., C. J. Cox, P. G. Foster, G. J. Szöllösi, and T. M. Embley. 2020. Phylogenomics provides robust support for a two-domains tree of life. Nat. Ecol. Evol. 4: 138–147.

Williams, T. A., P. G. Foster, C. J. Cox, and T. M. Embley. 2013. An archaeal origin of eukaryotes supports only two primary domains of life. Nature 504: 231–236.

Williams, T. A., P. G. Foster, T. M. Nye, C. J. Cox, and T. M. Embley. 2012. A congruent phylogenomic signal places eukaryotes within the Archaea. Proc. Biol. Sci. 279: 4870–4879.

Williams, T. A., G. J. Szöllösi, A. Spang, P. G. Foster, S. E. Heaps, B. Boussau, T. J. G. Ettema, and T. M. Embley. 2017. Integrative modeling of gene and genome evolution roots the archaeal tree of life. Proc. Natl. Acad. Sci. USA 114: E4602–E4611.

Woese, C. R., and G. E. Fox. 1977. Phylogenetic structure of the prokaryotic domain: The primary kingdoms. Proc. Natl. Acad. Sci. USA 74: 5088–5090.

Woese, C. R., O. Kandler, and M. L. Wheelis. 1990. Towards a natural system of organisms: Proposal for the domains Archaea, Bacteria, and Eucarya. Proc. Natl. Acad. Sci. USA 87: 4576–4579.

Yoon, H. S., J. D. Hackett, C. Ciniglia, G. Pinto, and D. Bhattacharya. 2004. A molecular timeline for the origin of photosynthetic eukaryotes. Mol. Biol. Evol. 21: 809–818.

Yutin, N., and E. V. Koonin. 2012. Archaeal origin of tubulin. Biol. Direct 7: 10.

Zaremba-Niedzwiedzka, K., E. F. Caceres, J. H. Saw, D. Bäckström, L. Juzokaite, E. Vancaester, K. W. Seitz, K. Anantharaman, P. Starnawski, K. U. Kjeldsen, et al. 2017. Asgard archaea illuminate the origin of eukaryotic cellular complexity. Nature 541: 353–358.

Zhaxybayeva, O., P. Lapierre, and J. P. Gogarten. 2005. Ancient gene duplications and the root(s) of the tree of life. Protoplasma 227: 53–64.

PART 2

The Genetic Mechanisms of Evolution

CHAPTER 4

The Population-Genetic Environment

All evolutionary processes operate through change in extant genotypic variation. But to understand the kinds of modifications that are evolutionarily possible, one must start with an appreciation of the three dimensions of the population-genetic environment that ultimately dictate what natural selection can and cannot accomplish. Mutation creates the variation upon which evolution depends. Recombination reassorts variation between nucleotide sites among chromosomes in ways that can either accelerate or impede evolutionary progress. Finally, random genetic drift serves as a lens on the evolutionary process, modulating the level of noise in allele transmission across generations, and hence, dictating the efficiency of selection. Although it is tempting to stare at biodiversity and spin adaptive stories as to why things are so, doing this in the absence of an understanding of evolutionary-genetic processes is not much more reliable than trying to understand biochemistry without acknowledging the existence of hydrogen bonds.

Chapters 4–6 outline in a non-technical way the minimum set of principles required to construct a logical evolutionary argument. In addition to introducing some of the most elementary aspects of evolutionary theory, this chapter summarizes the state of knowledge on the three dimensions of the population-genetic environment noted above. It shows that all three factors vary by several orders of magnitude across the Tree of Life, albeit in non-independent ways. In particular, mutation and recombination rates are strongly associated with the power of random genetic drift, and therefore with each other.

Quantitative information on how these forces vary across species is critical to understanding inter-species divergence. For example, if the power of genetic drift exceeds the strength of selection operating on a particular variant, the latter will be essentially immune to selection and will evolve in the direction dictated by any prevailing mutation bias. In species experiencing low levels of random genetic drift (small microbes), natural selection can take advantage of mutations of small effects that are unavailable to adaptive exploitation in species experiencing higher levels of noise.

We start with a more formal presentation of the preceding ideas. First, we show that random genetic drift is influenced by both population size and chromosomal architecture, leading to the concept of a genetic effective population size (N_e), which can be orders of magnitude smaller than the actual census size. Second, we show how drift competes with the selection process, with $1/N_e$ defining an approximate benchmark above and below which selection operating on a mutation is effective versus ineffective. Third, we show how patterns of variation in rates of mutation and recombination across the Tree of Life are consistent with some relatively simple models based on random genetic drift. Further technical details on these matters can be found in Walsh and Lynch (2018, Chapters 2–7). Drawing from the information presented here, Chapters 5 and 6 take things a step further by considering more specifically how the population-genetic environment modulates various processes involved in cellular evolution.

Demystifying Random Genetic Drift

Critical to understanding all aspects of evolution is the large role of chance in determining the fates

Evolutionary Cell Biology. Michael Lynch, Oxford University Press. © Michael Lynch (2024). DOI: 10.1093/oso/9780192847287.003.0004

of mutant alleles. Consider a newly arisen neutral mutation in a haploid population of N individuals, each of which produces a large number of potential offspring. To produce the next generation, N newborns must be drawn from this pool. Because a new mutant allele is present in a single copy, it has an initial frequency of $1/N$, and the probability that a randomly drawn offspring is not of this type is $[1 - (1/N)]$. There is then a $[1 - (1/N)]^N \simeq e^{-1} \simeq 0.368$ chance that none of the newborns contains the mutation, in which case the new mutation is lost in the first generation.

Although this result assumes a neutral mutation, the probability of immediate loss is not much different for an allele with fractional benefit s, as the previous expression generalizes to $\{1 - [(1 + s)/N]\}^N \simeq e^{-(1+s)}$, which is 0.333 when $s = 0.1$ (an enormous 10% selective advantage, far beyond what is typically observed in nature). If the population is a sexual diploid, a $2N$ is substituted for N in the preceding expressions, but the numerical results are the same. Thus, natural selection is a wasteful process in that the vast majority of new mutations, no matter how beneficial nor how large the population size, are lost by chance.

If the mutant allele is fortunate enough to survive the first generation, the same process of stochastic sampling occurs again. In each generation, the allele frequency can wander up or down, but due to the cumulative effects of this sorting process over many generations, all mutant alleles eventually suffer the fate of either loss (returning to a frequency of 0.0) or fixation (progressing to a frequency of 1.0) (Figure 4.1). For a neutral mutation, there is no directional pressure on the frequency change across generations, and the probability of fixation is simply equal to the initial frequency ($1/N$ for a new mutation in a haploid population, and $1/(2N)$ in a sexual diploid).

If over a series of generations, a beneficial mutation does wander to a sufficiently high density, so that the probability of chance loss in any particular generation becomes small relative to the strength of selection, then natural selection can propel it to eventual fixation in a nearly deterministic fashion. The critical frequency depends on a key feature

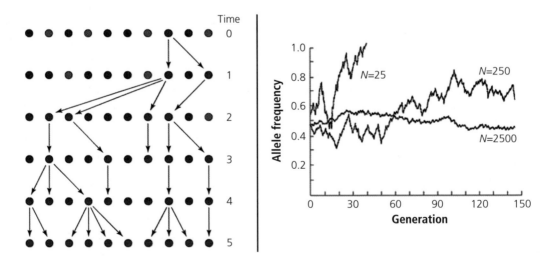

Figure 4.1 **Left:** Random genetic drift of a single mutation to fixation (denoted in red), with the population initially containing two additional alleles (denoted in yellow and purple). Each ball can be viewed as a single haploid individual. In each generation, population replacement is achieved by randomly drawing offspring from the members of the parental population (with replacement). After three generations, the yellow allele has been lost by chance, and the purple allele is lost soon thereafter. The arrows trace the full gene genealogy of the population at generation 5 back to the single red mutation at time 0. Different numbers of descendants of each gene, in this case between 0 and 4 per generation, are purely a matter of chance. Ultimately, the red allele would be replaced by a subsequently arising, novel mutation. **Right:** Long-term sample trajectories for three populations with different sizes (N), all starting at allele frequency 0.5. Each trajectory is unique, and the magnitude of fluctuations increases with decreasing population size.

known as the *effective* population size, generally abbreviated as N_e. Only in an ideal population, where each adult has an equal probability of contributing to each offspring and family sizes are completely random, does N_e equal the *absolute* population size N.

Nearly every conceivable property of natural populations conspires to cause a population to behave genetically as though it is much smaller than its actual size, i.e., $N_e \ll N$. For example, owing to some individuals acquiring more resources than others or attracting more predators or pathogens, non-random variation in family size causes gene transmission to the next generation to be dominated by the most successful individuals. In addition, because each individual in a sexual population has two parents, if the sex ratio is uneven, the genetic effective population size will more closely resemble the number of the rarer sex. Population subdivision can further reduce the species-wide N_e, as individual demes become increasingly inbred and reflect the states of smaller numbers of gametes than expected under panmixia. Finally, fluctuations in population size have a major effect because reductions in numbers of individuals have a much more substantial influence on sampling noise than do increases.

All of these effects can be viewed as ecological/demographic factors imposed on species without regard to their genetic constitution. However, an even more important determinant of gene-transmission stochasticity, particularly in large populations, is the structure of the genetic machinery itself. Because genes are physically connected on chromosomes, the fate of a new mutation is determined by the states of the nucleotide sites to which it is initially chromosomally linked (Figure 4.2). For example, a beneficial mutation that arises adjacent to other segregating alleles with sufficiently deleterious collective effects will be removed from the population unless it can be rapidly freed from such a background by recombination. Because the vast majority of mutations have deleterious effects, this kind of background-selection process is expected to be on-going in all populations (Charlesworth 2012), but will be more significant in large populations, which generally harbor more variation. In addition, a sufficiently beneficial mutation that is propelled forward by natural selection will drag all other linked mutations to fixation as well (including those that are mildly deleterious). Such selective sweeps have the same effect as a bottleneck in population size, although the effects of each sweep are confined to specific chromosomal regions.

To more fully understand the impact of drift on the efficiency of natural selection, we need to know the expected magnitude of generation-to-generation allele-frequency fluctuations. We start with the simple situation in which the noise in the evolutionary process is entirely due to random sampling of parental gametes across generations, as this is identical to a coin-flipping problem. Imagine two alleles, A and a, with respective frequencies in the population equal to p and $(1-p)$. If N alleles are randomly drawn to produce the next generation, as in an ideal haploid population, the frequency of A in the next generation will almost certainly be slightly different than p, just as the fraction of heads drawn from throws of an unbiased coin will deviate slightly from 0.5. The variance in allele frequency among independent sets of N draws is $p(1-p)/N$, showing that the smaller the number of draws (N), the larger the change in p across generations. (This becomes $p(1-p)/(2N)$ in a diploid population).

This simple result implies that the consequences of drift unfold on longer time scales in larger populations. Consider the early generations of the drift process for a neutral allele with intermediate frequency. As the process has no memory, the variance in allele-frequency change is cumulative across generations, i.e., $p(1-p)/N$ after the first generation, and approximately $2p(1-p)/N$ after the second generation, and $tp(1-p)/N$ after the tth generation (provided $t \ll N$). Thus, if the population size is doubled, it will take twice the number of generations to achieve the same level of allele-frequency change as in a population of size N. The key issue is that the time scale of random genetic drift is inversely proportional to the population size (Figure 4.1). (As discussed further below, this linear scaling eventually breaks down as all alleles become fixed or lost).

The deeper problem here is that the absolute number of individuals in the population, N, is generally not sufficient to define the stochasticity of allele-frequency change. To accommodate the substantial complexity of the problem, population geneticists rely on the concept of an effective

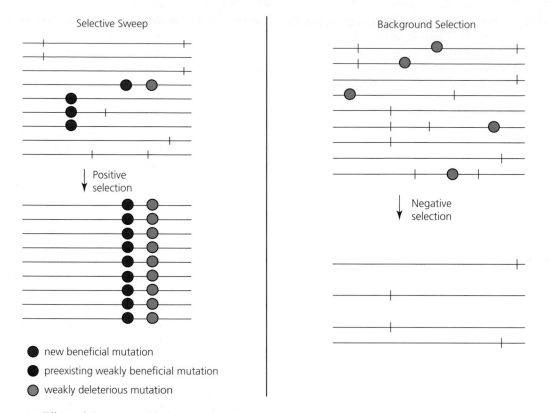

Figure 4.2 Effects of chromosomal linkage on the efficiency of natural selection. **Left:** A selective sweep involving a strongly beneficial mutation (orange circle) drags to fixation the chromosomal segment upon which this mutation arose, and in doing so jointly fixes a weakly deleterious mutation (blue circle) to which it is linked, while also causing the loss of another more weakly beneficial mutation (red circles). The tick marks denote additional nucleotide polymorphisms at other sites in the ancestral population, which are also lost in the region of the sweep. With free recombination, both the orange and red mutations would eventually become fixed with high probability, and the deleterious mutations would be purged. **Right:** Background selection against recurrent deleterious mutations (blue) causes their associated chromosomal segments to be purged from the population by negative selection, removing variation.

population size. The goal is to determine the size of an ideal population (like that envisioned in the preceding paragraph) that most closely mimics the between-generation drift of allele frequencies experienced by the actual population. In other words, given the myriad of usually undefined ecological and genome-architectural issues, we desire a composite measure of the effective population size (N_e) that yields a dispersion of allele frequencies across adjacent generations approximately equal to $p(1-p)/N_e$.

This concept of an effective measure of an assumed underlying parameter may be viewed as an oversimplification by some readers. However, although usually unstated, the use of surrogate measures underlies a multitude of model-fitting exercises in science – we attempt to reason out a theoretical framework to explain an expected set of observations based on assumed underlying mechanisms, and then obtain model-parameter estimates that best fit the data. The limitations of the human mind demand such simplification, and in biology, we are often not bothered if an approximation leads us astray by no more than a few per cent.

Numerous mathematical formulations have been developed to explicitly link N_e to N under various conditions associated with demography and chromosomal architecture (Charlesworth 2009; Chapter 3 in Walsh and Lynch 2018). Although these can be difficult to implement in a practical sense, the general view is that, owing to the totality of genetic-interference effects, for populations with absolute sizes in the range of unicellular species, N_e grows only logarithmically with N, increasing $\sim 2\times$ with each tenfold increase in the latter (Neher 2013; Lynch 2020). The following section shows that estimates of N_e indirectly derived from empirical data are in rough accord with this weak scaling.

The Genetic Effective Sizes of Populations

In Chapter 1, the case was made that the average number of individuals per species is on the order of 10^{21} for both bacteria and unicellular eukaryotes, and at least nine to ten orders of magnitude lower for multicellular eukaryotes (with a very large range of variation within each group). Absolute numbers like these are of relevance to the field of ecology, but they may also leave the false impression that there is little room for the role of chance in long-term evolutionary processes.

There are numerous ways to estimate genetic effective population sizes, the most logically compelling being an evaluation of the fluctuations in allele frequencies across generations. As noted previously, the variance in allele-frequency change across generations for a neutral nucleotide site is simply $p(1-p)/N_e$ or $p(1-p)/(2N_e)$, for haploid and diploid populations, respectively, where p is the initial allele frequency. However, observations of allele-frequency changes are only reliable with very small populations and very large sample sizes or very long periods of elapsed generations, as otherwise the bulk of such change is simply due to sampling error on the part of the investigator.

The most powerful alternative approach is to evaluate the standing level of variation in a population under the assumptions that the nucleotide sites being observed are neutral and have reached levels of variation expected under the balance between recurrent input by mutation and loss by drift. As noted in Foundations 4.1, the expected average level of nucleotide variation (defined as hypothetical heterozygosity under random mating, or equivalently the probability that two randomly chosen alleles from a population are different in state) at neutral sites is $\theta \simeq 2N_e u$ or $4N_e u$, again for

haploid versus diploid populations, where u is the rate of base-substitution mutation per nucleotide site per generation. Note that the composite parameter θ has a simple interpretation. It is equivalent to the ratio of the power of mutation ($2u$ for two sequences being compared) and that of drift ($1/N_e$ or $1/2N_e$).

Standing levels of heterozygosity at neutral nucleotide sites are generally estimated by confining attention to synonymous positions within codons of protein-coding genes (e.g., third positions in codons that specify the same amino acid whether they are occupied by A, C, G, or T) or deep within introns or intergenic regions (where no functional sites are thought to reside). In practice, θ is estimated by obtaining average estimates of neutral heterozygosity over a large number of sites in a sample of individuals. Such estimates integrate information over approximately the past N_e generations, which is equivalent to the average number of generations separating two random alleles in a haploid population (Kimura and Ohta 1969; Gale 1990; Ewens 2004). This point can be seen by noting that two alleles will accumulate mutational differences at a rate $2u$ per site, which, after an average of N_e generations, sums to $\theta = 2N_e u$ (for a diploid population, the average separation time is $2N_e$ generations, yielding $\theta = 4N_e u$).

Estimates of θ derived with this approach have been summarized for a wide range of species across the Tree of Life by Lynch (2007) and Leffler et al. (2012), and more specifically for metazoans and land plants by Romiguier et al. (2014), Corbett-Detig et al. (2015), and Chen et al. (2017). For this diverse assemblage of eukaryotic and prokaryotic species, there is a negative association between organism size and θ, with estimates for prokaryotes averaging ~ 0.10, those for unicellular eukaryotes averaging ~ 0.05, invertebrates ~ 0.03, and land plants and vertebrates generally being < 0.01. These are very approximate averages, and there is considerable variation around the mean. The main point is that, for diverse sets of organisms, the range in estimated θ (in part, a reflection of N_e) is only ~ 10-fold. In contrast, as noted previously, the absolute numbers of individuals per species differ by many orders of magnitude.

There are a number of sources of potential bias in these estimates. For example, for both bacteria and unicellular eukaryotes, most silent-site heterozygosity measures are derived from surveys of pathogens, whose N_e may be abnormally low because of the restricted distributions of their multicellular host species. In addition, observed levels of silent-site diversity will deviate from the neutral expectation if such sites experience some form of selection. The direction of bias depends on whether selection opposes or reinforces any prevailing mutation bias. If there is a conflict between selection and mutation bias, expected levels of heterozygosity can exceed the neutral expectation. Most results are consistent with this type of conflict, but the resultant levels of heterozygosity are inflated by no more than three- to fourfold (Long et al. 2017). With these caveats in mind, the existing data lead to a compelling conclusion with respect to the relative power of mutation and random genetic drift – in essentially no species does the former exceed the latter (as this would cause $\theta = 4N_e u > 1$).

Taken at face value, the preceding results might suggest that average N_e varies by no more than an order of magnitude between organisms as diverse as bacteria and microbes. However, the problem with this sort of interpretation is that θ is a function of the product of N_e and the mutation rate. The conclusion that N_e is relatively constant only follows if mutation rates are also relatively constant, which the next section shows to be far from the case. Factoring out known estimates of u from θ yields estimates of N_e ranging from 10^4 in some vertebrates to $> 10^8$ in many bacteria (Figure 4.3). Over a nearly 10^{20} range of variation in adult size, N_e exhibits a negative power-law relationship with organism size.

Why the structure of life has led to this particular range and scaling of N_e remains unclear. The fact that N_e estimates fall many orders of magnitude below the actual numbers of individuals per species, especially in the case of microbes, is consistent with the view that the predominant source of drift is the stochasticity in gene transmission resulting from jointly segregating polymorphisms linked on chromosomes. Recall the theoretical expectation that N_e increases with the natural logarithm of N (Neher 2013; Lynch 2020). If we assume that species with the lowest N_e ($\simeq 10^4$) have $N \simeq 10^6$, then an average bacterial species with $N \simeq 10^{21}$ (from above) would be expected to have an $N_e \simeq e^{15} \times 10^4 \simeq 10^{10}$. Considering the crudeness of this estimate, it is remarkably close to the upper limit of observed $N_e \simeq 10^9$ (Figure 4.3; and Bobay and Ochman 2018).

The next section discusses in detail how the fact that no species appears to have $N_e > 10^9$ means that beneficial mutations with selective advantages smaller than 10^{-9} cannot be exploited by natural selection in any lineage. For bacterial species with $N_e \simeq 10^8$, selection is capable of operating on mutations with $|s|$ as small as 10^{-8}. In contrast, mutations with advantages smaller than about 10^{-5} are unavailable to selection in multicellular species with $N_e \simeq 10^5$. Thus, the range of fitness effects of mutations susceptible to selection expands by about four orders of magnitude with decreasing organism size, enabling selection to operate in a more fine-grained manner in small organisms.

Probability of Fixation of a Mutant Allele

We now consider in more quantitative detail the specific issue of the probability of fixation of a newly arisen mutant allele, i.e., of rising to frequency 1.0. This will clarify the different roles played by both absolute and effective population sizes, while also demonstrating more formally the way in which the efficiency of natural selection is dampened by the magnitude of random genetic drift. As noted above, fixation probabilities are virtually always $\ll 1.0$. Their magnitude depends on: 1) the initial allele frequency p_0, which for a new mutation is a function of the actual population size, with $p_0 = 1/N$ or $1/(2N)$ for haploid and diploid populations; 2) the strength of selection s, which is a function of the mutational effect; and 3) the effective population size N_e.

Here, we focus on a diploid population, with the mutant allele having additive fitness effects (such that each copy of the allele changes fitness by an amount s, yielding genotypic fitnesses of 1, $1 + s$, and $1 + 2s$, with heterozygotes being intermediate to the two homozygotes); see Walsh and Lynch (2018, Chapter 7) for more complex situations. Taking into consideration the stochastic effects noted, Malécot (1952) and Kimura (1957) found that the probability

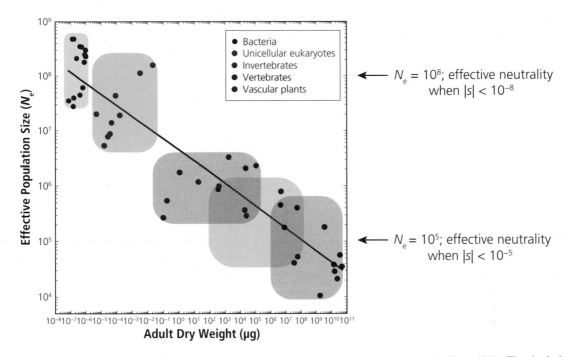

Figure 4.3 The negative scaling of effective population size (N_e) with organism size across the Tree of Life. The shaded squares envelope the full range of estimated variation for each major phylogenetic group. The line is the least-squares fit to the overall data $N_e = (4.5 \times 10^6)B^{-0.20}$, where B is adult dry weight (in μg). As indicated on the right, mutations with selective effects much smaller than $1/N_e$ fall in the range of effective neutrality and are impervious to natural selection. From Lynch and Trickovic (2020).

of fixation of an allele starting at initial frequency p_0 is

$$\phi_f(p_0) \simeq \frac{1 - e^{-4N_esp_0}}{1 - e^{-4N_es}}. \qquad (4.1a)$$

(The same formula applies to haploid populations if a 2 is substituted for each 4). For a newly arisen mutation, $p_0 = 1/(2N)$ for diploids, and Equation 4.1a reduces to

$$\phi_f(1/2N) \simeq \frac{1 - e^{-2(N_e/N)s}}{1 - e^{-4N_es}}. \qquad (4.1b)$$

As discussed above, in almost all natural settings $N_e/N \ll 1$, and most mutations have only minor effects on overall fitness (Chapter 5), so that $|s| \ll 1$. Thus, noting that $e^{-x} \simeq 1 - x$ for $x \ll 1$, the numerator is closely approximated by $2s(N_e/N)$, simplifying things further to

$$\phi_f(1/2N) \simeq \frac{2s(N_e/N)}{1 - e^{-4N_es}}. \qquad (4.1c)$$

It is useful to note that $2N_es = s/[1/(2N_e)]$ is equivalent to the ratio of the power of selection to that of drift.

Four limiting conditions are clear from Equations 4.1a–c. First, for strong selection relative to drift, $2N_es \gg 1$, $e^{-4N_es} \simeq 0$, and the denominator is essentially equal to 1.0, showing that the probability of fixation of a new beneficial mutation is just $2s(N_e/N)$. Thus, even in populations with very large N_e, the probability of fixation of a beneficial mutation is less than twice the selective advantage. These results formalize the point made earlier – owing to the high probability of stochastic loss in the earliest generations, even strongly beneficial mutations only rarely proceed to fixation.

Second, only after the frequency of a beneficial allele becomes sufficiently high does the probability of fixation become almost certain. For example, from Equation 4.1a, if $N_esp_0 > 0.5$, the probability of fixation exceeds 0.70, and if $N_esp_0 > 1$, $\phi_f(p_0) > 0.93$.

Thus, as a matter of convention, it is often argued that to be assured of a high probability of fixation, a beneficial mutation requires a starting frequency of $p_0 \gg 1/(2N_e s)$.

Third, as $|4N_e s| \to 0$, the probability of fixation converges to the initial frequency (p_0). This should be intuitive, as under these conditions drift is so dominant that there is effectively no directional pressure on allele-frequency change. For this reason, the domain in which $|s| < 1/(4N_e)$ is known as the realm of effective neutrality. The salient point is that natural selection is unable to purge deleterious mutations or promote beneficial mutations with absolute effects $< 1/(4N_e)$. Thus, the range of variation recognizable by natural selection is expanded in large populations.

Finally, it follows that the probability of fixation of a newly arising neutral mutation ($s = 0$) is always equal to its initial frequency, $1/(2N)$ in a diploid population, regardless of N_e. This has interesting implications for some forms of molecular evolution. Letting the mutation rate per nucleotide site equal u, then $2Nu$ mutations arise in the population each generation, so that the long-term rate of evolution at a neutral site is simply equal to the product $2Nu \cdot [1/(2N)] = u$. Thus, the rate of molecular evolution at neutral sites is equal to the mutation rate, independent of the mode of gene action, rates of recombination, and population size (Kimura 1983).

Evolution of the Mutation Rate

The fact that no organism has evolved to have 100% replication fidelity is consistent with basic thermodynamic principles (Chapter 20), but the issues are much deeper than this. There is a thousand fold range of variation in the mutation rate among species, raising the question as to why particular rates are associated with particular lineages. Kimura (1967) first pointed out that selection operates on the mutation rate indirectly, via the effects of mutations linked to alleles associated with their production. Under this view, a newly arisen mutator allele progressively acquires an excess linked mutation load (Figure 4.4). Thus, given that the vast majority of mutations are deleterious (Lynch et al. 1999; Baer et al. 2007; Eyre-Walker and Keightley 2007; Katju and Bergthorsson 2019),

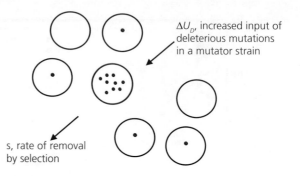

Figure 4.4 The excess equilibrium mutation load in a mutator strain in an asexual population. Open circles denote cells, and red dots represent deleterious mutations. The ancestral population of cells (black circles) has a low background mutation rate, with each cell containing zero or one deleterious mutation. Each mutation has a deleterious fitness effect s, which is equivalent to the fractional reduction in survivorship associated with the addition of each mutation. The mutator strain (red circle) has a genome-wide excess rate of input of deleterious mutations equal to ΔU_D, which leads to an excess equilibrium mutation load within this sublineage of cells.

it follows that natural selection generally strives to minimize the mutation rate.

The ability of natural selection to eradicate a mutator allele is a function of the magnitude of this associated mutation load, which is equal to the product of three terms: 1) the excess genome-wide rate of production of deleterious mutations relative to the pre-existing population mean, ΔU_D; 2) the reduction in fitness per mutation, with individuals harboring an additional mutation leaving a fraction s fewer progeny, and those with n mutations having fitness $(1 - s)^n$; and 3) the average number of generations that a mutation remains associated with the mutator, $\bar{t}$. The persistence time $\bar{t}$, in turn, is determined by two factors: 1) the selective disadvantage of mutations (s); and 2) the rate of recombination (r), which physically dissociates the mutator from the load that it creates.

The strength of selection against a mutator is greatest in the case of asexual reproduction, as the mutator is never separated from its mutation load by recombination. Provided $s \gg 1/N_e$, s is equivalent to the rate of removal of a deleterious mutation from the population by selection, and the mean persistence time is simply equal to the reciprocal of the average rate of removal of a mutation (i.e., $\bar{t} = 1/s$). The selective disadvantage of a mutator allele is then the product of the three terms noted above: $s_m = \Delta U_D \cdot s \cdot (1/s) = \Delta U_D$, and hence, is simply equal to the increased per-generation rate of production of deleterious mutations. This shows that, under asexuality, the selective disadvantage is completely independent of the effects of mutations, as mutations with larger effects are selectively removed from a population at higher rates (along with the linked mutator allele).

Recombination weakens the selective disadvantage of a mutator allele by exporting the initially linked mutations to other members of the population. With free recombination ($r = 0.5$), mutant alleles are statistically uncoupled from their source in an average of just $\bar{t} = 1/r = 2$ sexual generations. The mutator-allele disadvantage then becomes $s_m = \Delta U_D \cdot s \cdot (1/2) = 2s\Delta U_D$. Not every mutation will be freely recombining with respect to a mutator allele, but because most eukaryotic genomes contain multiple chromosomes, most new mutations will arise unlinked to the mutator, and the load is unlikely to be greater than twice the preceding value (Lynch 2008). The key point is that sexual reproduction reduces the magnitude of selection against a mutator allele (relative to the case of complete linkage) by a factor of $\sim 2s$. Chapter 5 outlines how the average fitness effects of new mutations are generally <0.1, implying that recombination reduces the strength of selection operating on the mutation rate by at least 80% relative to the case under asexuality.

Impeding the ability of selection to reduce the mutation rate is random genetic drift, which begins to prevail once the level of replication fidelity becomes so highly refined that the next incremental improvement is effectively neutral, i.e., $s_m \ll 1/N_e$. The general principle can be understood by reference to the logic outlined in the preceding section. Letting $\eta < 1$ be the fraction by which the mutation rate is reduced by an antimutator, so that $\Delta U_D = \eta U_D$, the drift-barrier hypothesis postulates that once the genome-wide deleterious mutation rate U_D in an asexual (haploid) population is reduced to the point that $\eta U_D < 1/N_e$, selection for further reduction in the mutation rate will be overwhelmed by stochastic noise.

To reiterate, owing to the stochastic consequences of finite population size, the degree of refinement that natural selection can achieve increases with N_e, which allows alleles with smaller incremental benefits to be promoted (Lynch 2011, 2020; James and Jain 2016; Lynch et al. 2016). This drift-barrier hypothesis leads to the prediction that the mutation rate will evolve to be negatively associated with N_e (Figure 4.5). If this hypothesis is correct, the very

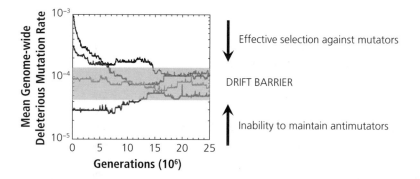

Figure 4.5 The evolution of quasi-equilibrium distributions of mutation rates under the drift-barrier hypothesis. Mutator and antimutator alleles are recurrently introduced, and these acquire associated loads of linked deleterious mutations, imposing indirect selection on the mutation rate. Regardless of the starting point, the mutation rate eventually evolves to reside within the confines of the grey area (the upper and lower bounds to the drift barrier). Above the drift barrier, the mutation rate is sufficiently high that antimutators have a large enough fitness advantage to be advanced by selection, reducing the mutation rate; below the drift barrier, the selective disadvantage of mutators is sufficiently small that they are able to drift to fixation, increasing the mutation rate. Within the grey domain, the mutation rate fluctuates stochastically through time owing to the sporadic introduction of mutator and antimutator alleles. From Lynch (2011).

process necessary for producing adaptive mutations is selected against, with the fuel for evolution (the small fraction of beneficial mutations) being largely an inadvertent by-product of an imperfect process.

How do the data accord with the drift-barrier hypothesis? Prior to this century, almost all estimates of the mutation rate were derived indirectly using reporter constructs in microbes (Drake 1991; Drake et al. 1998), leaving considerable uncertainties with respect to the accuracy of inferred values. With the advent of whole-genome sequencing, it became possible to precisely evaluate the genome-wide appearance of mutations in replicate lines maintained by single-progeny (or single full-sib mating) descent (i.e., $N_e \simeq 1$ or 2) for large numbers of generations (Lynch et al. 2016; Katju and Bergthorsson 2019). By maximizing the power of drift relative to selection, such treatment ensures that essentially all mutations (other than the small fraction causing lethality or sterility) will accumulate in an effectively neutral fashion.

The numerous results from such work demonstrate a thousandfold range of interspecific variation in the per-generation mutation rate per nucleotide site, from a low of $\sim 10^{-11}$ in some unicellular eukaryotes to a high of $\sim 10^{-8}$ in some mammals (including humans) (Lynch et al. 2016, 2023; Long et al. 2017; Lynch and Trickovic 2020). Consistent with expectations under the drift-barrier hypothesis, the rate of base-substitution mutation per nucleotide site (u) scales negatively with N_e (Figure 4.6, left). Although the rates for three major groups (bacteria, unicellular eukaryotes, and multicellular eukaryotes) appear as fairly discrete clusters, recall that the theory outlined above implies that selection operates on the *genome-wide* deleterious rate, which is the product of the mutation rate per nucleotide site (per generation) and the number of nucleotides under selection (the effective genome size, P_e). Multiplying u by P_e (approximated by the amount protein-coding DNA) unifies the overall set of results, leading to a nearly inverse scaling between uP_e (the approximate genome-wide

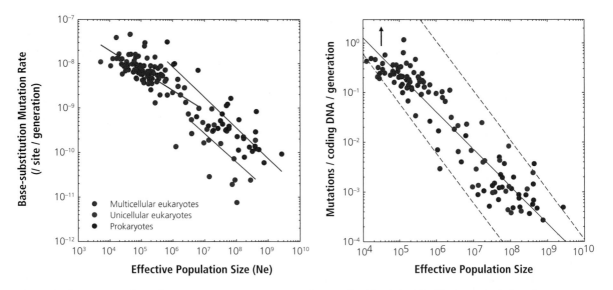

Figure 4.6 Negative scaling of per-generation mutation rates (per nucleotide site) with effective population sizes across the Tree of Life. **Left:** Regressions are given for three major phylogenetic groups: bacteria, $0.00011 N_e^{-0.76}$; unicellular eukaryotes, $0.0000080 N_e^{-0.64}$; and multicellular eukaryotes, $0.0000013 N_e^{-0.45}$. **Right:** Rates are given as the total haploid genomic rate for sites likely to be under selection (the sum over all nucleotide sites in protein-coding genes), with the overall regression being $1250 N_e^{-0.75}$. Coding nucleotides comprise almost the entire genomes of bacteria, so there is little room for underestimation here; the upper left arrow denotes the likely magnitude of the elevation of the slope due to additional regulatory sites in eukaryotes not accounted for in the plot (see Lynch et al. (2016) for further details). Dashed lines denote slopes of −1. The plotted data are from Lynch et al. (2016), Long et al. (2018), and Lynch et al. (2023).

deleterious mutation rate) and N_e (Figure 4.6, right).

Thus, mutation rates are lower in unicellular eukaryotes than in bacteria with similar effective population sizes because the genomes of the former contain many more genes than those of the latter, and hence, are larger targets for deleterious mutations. The scaling patterns in Figure 4.6 also hold for insertion-deletion mutations, which are about 10% as common as base-substitution mutation rates across the Tree of Life (Sung et al. 2016; Lynch et al. 2023).

High mutability of mutation rates

The preceding results make clear that the mutation rate is evolutionarily malleable, but what is the timescale of such change? The mutational target size for the mutation rate is likely to be very large, as it includes multiple DNA polymerases, DNA-repair proteins, and essentially all genes whose products alter the mutagenicity of the intracellular environment (including those influencing the production of free oxygen radicals via metabolic activity and those modulating the relative abundances of free nucleotides). Thus, with the expectation that both mutators and antimutators are recurrently introduced into all populations (Denamur and Matic 2006; Lynch 2008; Raynes and Sniegowski 2014; Sasani et al. 2022), virtually all natural populations are expected to harbor polymorphisms for the mutation rate.

The idea that the mutation rate is capable of rapid change is supported by a diversity of observations. For example, Boe et al. (2000) estimate that E. coli cells with mutation rates elevated 20–80× arise at rates of 5×10^{-6} per cell division, and one can imagine even higher rates of origin of milder (and less easily detected) mutators (as well as antimutators). Indeed, in an E. coli mutation-accumulation experiment initiated with a mutator strain that allowed accumulation of diversity over a period of 1250 generations, numerous lines evolved mutation rates < 10% of the baseline rate (antimutators), while a small fraction of them experienced up to tenfold increases in the mutation rate (Singh et al. 2017).

Given this potential for rapid change in the level of replication fidelity, it is not surprising that microbes commonly evolve mutator genotypes when confronted with strong selective challenges (such as antibiotic treatment). Swings et al. (2017) found that in lethally stressful environments, laboratory E. coli populations rapidly evolve a mutator phenotype (on a time scale of ~100 generations), and then revert to background mutation-rate levels once adaptation has been achieved. In contrast, in constant environments, bacterial populations founded with a mutator genotype frequently evolve lower mutation rates on relatively short time scales via compensatory molecular changes at genomic sites not involved in the initial mutator construct (McDonald et al. 2012; Turrientes et al. 2013; Wielgoss et al. 2013; Williams et al. 2013; Wei et al. 2022). Taken together, these observations indicate that the mutation rate is among the most rapidly evolving traits known. Moreover, the common appearance of antimutators implies the presence of substantial unexploited potential for improvement in replication fidelity.

As further evidence that, despite their extremely low values, evolved mutation rates are not constrained by biophysical limitations, consider the fact that although mammals harbor the highest known eukaryotic mutation rates per generation, the rate per germline cell division rivals the very low per-generation rates for unicellular species (Lynch 2010). The human germline mutation rate per nucleotide site is $\sim 6 \times 10^{-11}$ per cell division, approaching the lowest rates observed in unicellular eukaryotes, and 10–100× lower than rates in various human somatic tissues (Lynch 2010; Behjati et al. 2014; Milholland et al. 2017; Cagan et al. 2022). The key point here is that although selection operates on the per-generation mutation rate, this is accommodated by changes in replication fidelity at the cell-division level – an increased number of germline cell divisions is balanced by enhanced replication fidelity per cell division.

Error-prone polymerases

In all known organisms, almost all DNA replication is carried out by one or two major polymerases, each of which has a high baseline level of accuracy, with a substantial fraction of the few errors arising at this step being removed secondarily via a proofreading step. However, the genomes of nearly all organisms also encode for one or more error-prone polymerases, whose usage is restricted mostly to times of stress or to dealing with bulky lesions in DNA. Stress-induced mutagenesis (SIM) has been found in virtually all organisms that have been examined, e.g., *E. coli* and many other bacteria (Kang et al. 2006; Foster 2007; Kivisaar 2010); yeast (Heidenreich 2007); *Chlamydomonas* (Goho and Bell 2000); *Caenorhabditis* (Matsuba et al. 2013); and *Drosophila* (Sharp and Agrawal 2012). The distinction between SIM and normal mutagenesis is often blurry, in that a variety of environmental stresses (e.g., nature of the limiting nutrient) alter the molecular spectrum of mutations without affecting the mutation rate itself (Maharjan and Ferenci 2017; Shewaramani et al. 2017).

An elevation in error rates under extreme environmental situations should not be too surprising, as physiological breakdown can be expected for virtually all traits. Nonetheless, some have argued that high mutation rates associated with error-prone polymerases have been promoted by selection as a means for generating adaptive responses to changing environments (Radman et al. 2000; Rosenberg 2001; Tenaillon et al. 2001; Earl and Deem 2004; Foster 2007; Galhardo et al. 2007; Rosenberg et al. 2012). Direct empirical support for such an argument is lacking, although special scenarios have been shown in theory to encourage selection for SIM, e.g., situations in which two mutations are required for an adaptation, with the first conferring reduced fitness and an elevated mutation rate when alone (Ram and Hadany 2014), or when stresses are sufficiently diverse and persistently fluctuating (Lukačišinová et al. 2017). However, establishing that an evolutionary outcome is theoretically possible is quite different than demonstrating a high likelihood of it actually occurring.

A simpler and more compelling explanation for the error-prone nature of some polymerases follows directly from the drift-barrier hypothesis – the net selection pressure to improve accuracy is expected to be proportional to the average number of nucleotide transactions that a DNA polymerase engages in per generation. Because error-prone polymerases generally replicate only small patches of DNA and do so quite infrequently, the strength of selection on accuracy will be correspondingly reduced (Lynch 2008, 2011; MacLean et al. 2013).

This 'use it or lose it' hypothesis is also consistent with the high error rates for polymerases deployed in the replacement of small RNA primers used in replication initiation (Lynch 2011). In addition, the secondary and tertiary fidelity mechanisms associated with replication (proofreading and mismatch repair), which necessarily involve far fewer nucleotide transactions than the earlier polymerization step, have greatly elevated error rates (Lynch 2008). None of these latter observations can be explained as specific adaptations to stress, as all of the factors are fundamental to normal replication cycles.

This view does not deny the critical importance of error-prone polymerases as mechanisms for dealing with bulky lesions or other forms of DNA damage, nor does it deny that induced mutagenesis can play a role in generating an appropriate adaptation in extreme times, sometimes being the only means for survival. It does, however, eliminate the need for an adaptive explanation for high error rates, implying instead that there is no way to avoid such an outcome.

Optimizing the mutation rate

Mutational processes generate a large fraction of detrimental variants that must be removed by natural selection, layered over a small fraction of beneficial mutations essential for adaptation in changing environments. Thus, considerable attention has been given to the idea that natural selection might fine-tune the mutation rate so as to maximize the long-term rate of adaptive evolution in the face of an onslaught of deleterious mutations. In contrast to the drift-barrier hypothesis, under this view, selection does not constantly push the mutation rate to the lowest achievable level, but instead promotes specific levels of mutation via indirect effects associated with the small pool of beneficial mutations. This is a difficult area for theory development as the relative merits of increasing versus decreasing the mutation rate depend on the distribution of mutational effects, the population size, the recombination rate, and the pattern of environmental change.

It is especially unclear how natural selection operating at the individual level in a sexual population can promote an elevated mutation rate by associated beneficial effects. The primary problem is that, in a sexual population, hitchhiking of a mutator allele with a linked beneficial mutation will generally be thwarted by their dissociation by recombination (on average in just two generations when the two loci are on different chromosome arms). Continuous reinforcement necessary for the promotion of a mutator allele requires a substantial rate of input of closely linked beneficial mutations. However, as noted above, because the vast majority of mutations are deleterious, there will be a steady-state background deleterious load associated with all mutator alleles, regardless of whether the mutator is involved in a transient (and most likely incomplete) beneficial sweep. An additional limitation in multicellular organisms is the direct negative effects that mutators impose via the production of somatic mutations, e.g., cancer, with immediate detrimental effects on fitness (Lynch 2008, 2010).

Several attempts have been made to estimate the theoretically optimal mutation rates for maximizing long-term rates of adaptive evolution in non-recombining, asexual populations, but the resultant models do not explain the most prominent pattern in the data – the inverse relationship between u and N_e (Figure 4.6). Nor do they explain why, if optimized, mutation rates are nearly 1000 × higher in multicellular sexual species than in most microbes. Indeed, most models concerned with optimal mutation rates in persistently changing environments imply a positive association between N_e and u (Kimura 1967; Leigh 1970; Orr 2000; Johnson and Barton 2002; Desai and Fisher 2007; Good and Desai 2016). Thus, one could argue that the utility of these models is not that they explain the data, but that they formally highlight the inconsistencies of optimization arguments with observed mutation rates.

In summary, both data and theory are incompatible with the idea that mutation-rate evolution is guided by a population-level goal of maximizing the long-term rate of incorporation of beneficial mutations. Nor do hypotheses based on numbers of cell-divisions per generation explain the data in Figure 4.6, as nearly the full range of variation in mutation rates per generation is encompassed by unicellular species alone. Lastly, generation length fails to explain the patterns, as unicellular eukaryotes have longer cell-division times but lower mutation rates than prokaryotes.

One lingering concern may be that, after accounting for effective proteome and population sizes, bacteria have no lower mutation rates than unicellular eukaryotes. As noted above, theory predicts that the efficiency of selection on the mutation rate is increased in the absence of recombination, and bacteria are commonly viewed as being clonal in nature. However, as discussed below, although bacteria lack meiotic recombination, they nonetheless experience roughly the same amount of recombination per nucleotide site (via other mechanisms) at the population level as do eukaryotes.

The non-random nature of mutation

Decades of observations are consistent with the postulate that mutations arise randomly with respect to the forces imposed by natural selection. However, some have argued that selection is capable of modulating mutation rates on a gene-by-gene basis by, for example, locating genes in regions with or without potentially mutagenic collisions between DNA and RNA polymerase or by somehow providing protection against mutagenic aspects of high rates of transcription (Martincorena et al. 2012; Paul et al. 2013; Monroe et al. 2022). These claims have not held up to close scrutiny, and the theory outlined above explains why. The differences in mutation rates among genes associated with chromosomal locations and/or transcriptional activities are simply too small to be promoted by selection (Chen and Zhang 2013; Lynch et al. 2016; Liu and Zhang 2022).

This being said, although mutations are random with respect to desirable gene targets, they are nonetheless non-random in essentially every physical way. For example, in some bacteria there is a symmetrical wave-like pattern of the mutation rate around the circular chromosome (Foster et al. 2013; Long et al. 2015), although the amplitude of differences does not exceed 2.5 × and the pattern differs among species. Up to twofold differences have also been found among locations on eukaryotic chromosomes on spatial scales ranging from 200 bp to

100 kb (Stamatoyannopoulos et al. 2009; Lang and Murray 2011; Chen et al. 2012). The molecular mechanisms driving these large-scale patterns remain unclear, but may be associated with variation in the nucleotide pool composition during the cell cycle, regional variation in transcription rates and their influence on replication, protection by nucleosomes in eukaryotes, and/or alterations in the rates of processivity of DNA polymerase across different chromosomal regions.

On a more local scale, every genome that has been assayed reveals uneven frequencies of the twelve types of base-substitution mutations (Long et al. 2017). In most prokaryotes, and all eukaryotes so far observed, there is mutation bias in the direction of $G + C \rightarrow A + T$. In addition, the mutabilities of the individual nucleotides are context dependent, influenced by the nature of the neighboring nucleotides (Sung et al. 2015).

The appearance of mutations can also be temporally correlated within the same genomes. The naive view is that if mutations arise at an average rate u per nucleotide site, the rate of simultaneous origin of mutations at two specific sites would be u^2, at three sites would be u^3, etc. Given that average u is on the order of 10^{-9}, this would imply that double mutants would rarely ever occur except in large microbial populations. However, data from mutation-accumulation experiments suggest that on spatial scales of 100 bp or so, multinucleotide mutations commonly comprise 1 to 3% of mutational events in diverse lineages (Drake 2007; Schrider et al. 2011; Terekhanova et al. 2013; Harris and Nielsen 2014; Uphoff et al. 2016). Potential reasons for mutational clusters include local patches of DNA damage, the occasional use of a defective DNA polymerase molecule, accidental deployment of an innately error-prone polymerase, and mutagenic repair of double-strand breaks (Drake 2007; Hicks et al. 2010; Malkova and Haber 2012; Chan and Gordenin 2015; Seplyarskiy et al. 2015).

The key point here is that transient, localized hypermutation is common enough that rates of occurrence of double mutations are often many orders of magnitude above the u^2 expectation under independent origin, and more commonly on the order of $u/1000$ to $u/100$. Triple-mutation rates may be only a few orders of magnitude lower.

The occurrence of mutation clusters has major implications for the evolution of complex features, as modifications requiring multiple nucleotide changes on small spatial scales (e.g., within genes) need not await the sequential fixation of individual mutations, but can arise *de novo* and be promoted together. This change in view becomes particularly important with respect to complex adaptations in which first-step single-nucleotide variants are deleterious (Chapters 5 and 6).

Recombination

Recombination is a double-edged sword in evolution. On the one hand, by eliminating peculiarities associated with individual genetic backgrounds, the physical scrambling of linked loci increases the ability of natural selection to perceive mutations on the basis of their individual average effects. In addition, recombination can create favorable genetic interactions by bringing together mutations that have arisen on independent backgrounds. On the other hand, high rates of recombination can inhibit the permanent establishment of pairs of mutations with favorable interactive effects if they are separated more rapidly than they are advanced as a unit by selection.

Before proceeding, it may be useful to review the mechanics of recombination, as this will help clarify how the recombination rate scales with distance between nucleotide sites (Foundations 4.2; Figure 4.7). Contrary to common belief, the recombination rate between sites is not equal to the crossover rate, except in the case of distantly located sites (typically > 10 kb). This is because much of recombination involves localized patches of gene conversion that do not, in themselves, cause exchange of flanking chromosomal regions.

Two general approaches provide insight into the level of recombination per physical distance along chromosomes. The first of these involves the construction of genetic maps, usually from observations on the frequency of meiotic crossovers between molecular markers in controlled crosses. Such maps have the power to yield accurate estimates of average recombination rates over fairly long physical distances (usually with markers being separated by millions of nucleotide sites). However,

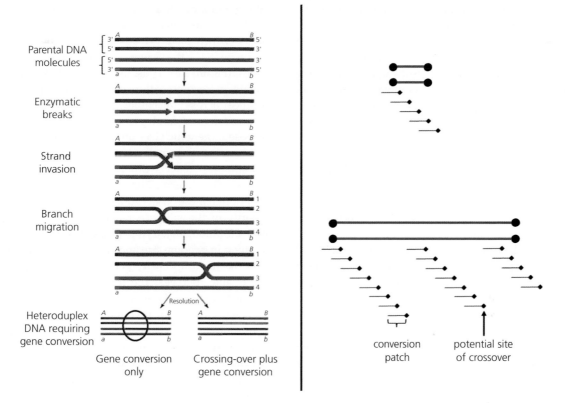

Figure 4.7 Left: The physical mechanics of meiotic recombination. After the first stage of meiosis in a diploid organism, chromosomes exist as sister chromatids, and these undergo pairing with their homologous partners (one from the paternal and the other from the maternal chromosome that formed the individual). Here, double-stranded sisters (each a DNA molecule) are shown for each of the homologous parental chromosomes (one yellow, and one blue). Single-strand breaks are created enzymatically during meiosis, and then resected, with strand invasion resulting in the two homologous chromosomes becoming intercalated at the breakpoint. This creates patches of heteroduplex DNA consisting of complementary molecules derived from each parental chromosome. To complete meiosis, this complex must be restored to the two-chromosome state, and depending on how the intermediates are separated, this may lead to a complete exchange of material distal to the breakpoint, a process known as crossing-over. However, regardless of whether crossing-over occurs, there is potential for a gene conversion in the heteroduplex (yellow–blue) patches – if such patches contain base mismatches (which will be the case if the individual is heterozygous in the region), these must be restored to Watson:Crick (A:T or G:C) states by the mismatch-repair pathway. If, for example, one parental chromosome had an A:T double-strand state, and the other C:G, invasion would produce an A:G or a C:T mismatch, which would then be restored to one of the parental states. Depending on the directionality of change, this can lead to a short patch of exchange between chromosomes, independent of whether crossing-over occurred. **Right:** The relative importance of gene conversion and crossing over in recombination between sites separated by different distances. The small black diamonds represent nucleotide sites at which a crossover can occur, and horizontal black lines denote conversion tracts. Recombination between sites represented by the black balls occurs if a conversion tract covers a single site or a crossover event falls between sites. At the top, the sites are close together, and relative to the number of relevant gene-conversion patches, only a few recombination events can lead to potential crossovers. At the bottom, sites are farther apart, and whereas the potential for crossing-over increases linearly with distance, the potential for single-site gene conversion does not, as most conversion tracts reside between the markers of interest.

without enormous numbers of evaluated progeny, such exercises cannot reveal recombination rates at small spatial scales simply because of the absence of observed recombination events over short intervals. We do not elaborate on the details of constructing genetic maps here, and only note that they rely on mapping functions that convert observed recombination frequencies into the expected numbers of crossovers between pairs of markers (Chapter 14 in Lynch and Walsh 1998).

Despite the limitations, results from genetic-map construction allow a compelling general statement about average genome-wide levels of crossing-over. Although eukaryotic genome sizes (G, the total number of nucleotides per haploid genome) vary by four orders of magnitude, the range of variation in genetic-map lengths (in units of the total number of crossovers per genome per meiosis) is only about tenfold among species (Lynch 2007; Lynch et al. 2011; Stapley et al. 2017). This behavior can be explained by a very simple physical constraint, which appears to be nearly invariant across the eukaryotic phylogeny. During meiosis, there are typically only one to two crossover events per chromosome arm, regardless of chromosome size. Because phylogenetic increases in genome size are generally associated with increases in average chromosome length rather than chromosome number (Lynch 2007), the little variation in the total number of meiotic crossover events per genome that exists among eukaryotes is due to variation in chromosome number.

These observations lead to a simple structural model for the average crossover rate per physical distance across a genome, which is technically equal to the product of the rate of recombination initiation per nucleotide site (c_0), the distance between sites, and the fraction of such events that lead to a crossover (x) (Foundations 4.2). Letting M be the haploid number of chromosomes per genome, G/M is the average physical length of chromosomes. Letting κ be the average number of crossovers per chromosome per meiosis, then the average amount of recombination per nucleotide site associated with crossing over is $\bar{c}_0 \simeq \kappa M/G$. If this model is correct, a regression of $\bar{c}_0$ on G on a log scale should have a slope of -1.0, with the vertical distribution (residual deviations) around the regression line being defined largely by variation in M (with species with the same genome size but more chromosomes having proportionally more crossing-over per nucleotide site).

The data closely adhere to this predicted pattern, with the smallest genomes of microbial eukaryotes

having recombination rates per physical distance $\sim 1000\times$ greater than those for the largest land plants (which have $\sim 1000\times$ larger genomes, but approximately the same numbers of chromosomes) (Figure 4.8). Thus, the smooth, overlapping decline in recombination intensity (per physical distance) across unicellular species, invertebrates, vertebrates, and land plants reflects the general increase in genome sizes among the latter eukaryotic domains. The smaller level of vertical variation in Figure 4.8 reflects differences in chromosome numbers (Lynch et al. 2011).

Although these observations suggest that the vast majority of the variance in the average crossing-over rate among eukaryotic species is simply due to variation in genome size and chromosome number, even the highest density genetic maps are generally unable to reveal the features of short chromosomal regions. Finer-scale molecular analyses have shown that up to hundred-fold differences in recombination rates can exist on spatial scales of a few kb, although the locations and molecular mechanisms dictating their distributions are highly variable among species (Petes 2001; de Massy 2003, 2013; Jeffreys et al. 2004; Myers et al. 2005; Arnheim et al. 2007; Coop et al. 2008; Mancera et al. 2008; Kohl and Sekelsky 2013; Lam and Keeney 2015; Haenel et al. 2018). For example, recombination hotspots in budding yeast are enriched in the transcriptional promoters of genes, whereas in fission yeast, they are enriched in intergenic regions. Although in neither yeast species is the guiding mechanism known, in numerous mammals, a zinc-finger protein (PRDM9) marks chromosomes in the vicinity of specific DNA-sequence motifs that guide the recruitment of Spo11, an enzyme involved in double-strand breaks (Baudat et al. 2013; Capilla et al. 2016; Wells et al. 2020).

In the genomes of mice and great apes, > 20,000 hotspots are defined in this way, although the localization of sites varies dramatically even among closely related species. Such shifts are associated with the rapid sequence evolution of the zinc-finger used in DNA motif recognition, leading to recombination-hotspot variation among subspecies and even among individuals within species (Brick et al. 2012). A key issue with respect to the PRDM9 system is that, if the positions of double-strand

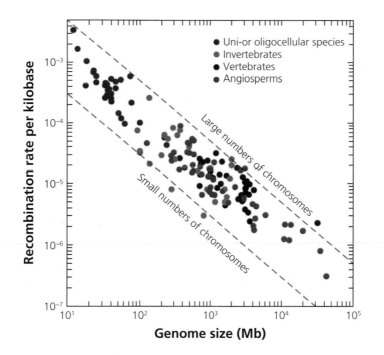

Figure 4.8 Average rates of crossover recombination per physical distance for four major groupings of eukaryotes, determined from information on total physical and genetic map sizes. Each data point is based on the genetic map of a different species. The two dashed reference lines have slopes of −1.0 in accordance with the theory discussed in the text. Letting κ be the average number of crossovers per chromosome, and M be the number of haploid chromosomes, the upper line assumes $\kappa M = 50$, i.e., 50 chromosomes with an average of one crossover, 25 with averages of two crossovers, etc. The lower line assumes $\kappa M = 3$. For the plotted species, κ is in the range of 0.3 to 3.1 (with one exception) and M is in the range of 3 to 44. From Lynch et al. (2011).

breaks are closely associated with the recognition motif, the latter will eventually be lost by gene conversion to the allele on the recipient chromosome, thereby leading to the eventual loss of the hotspot. A shift in the zinc-finger recognition motif in PRDM9 could then lead to altered hotspot localization, unfolding a new series of events. Notably, some mammals (e.g., dogs) do not even have such a system. Moreover, despite their potential origin by drive-like processes, there is no evidence that recombination hotspots are promoted by natural selection, and there is no evidence of their existence in unicellular species.

A second approach to estimating recombination rates uses measures of linkage disequilibrium (LD, a measure of the covariance of allelic status at paired sites) in natural populations to quantify the statistical degree of association between allelic states at linked chromosomal sites. Just as theory predicts

an equilibrium average level of molecular heterozygosity (variance) within neutral nucleotide sites at mutation-drift equilibrium (Foundations 4.1), the average level of LD between sites is expected to reach a balance between the forces of recombination, mutation, and drift. The equilibrium LD between two sites (i and j) is a function of the composite parameter $4N_e c_{ij}$, which is proportional to the ratio of the rate of recombination between sites (c_{ij}) and the power of random genetic drift ($1/2N_e$ for diploids) (Chapter 4 in Walsh and Lynch 2018).

The attractiveness of the LD approach is that the observations reflect the historical outcome of many thousands of generations (and equivalently, thousands of meioses). This provides the power to obtain much more refined (kilobase scale) views of the recombinational landscape than is possible with short-term breeding experiments. To understand this benefit, note that for a mapping cross involving n gametes with recombination frequency c_{ij} between sites i and j, the expected number of recombinants is nc_{ij}, so for sufficiently close sites ($c_{ij} \ll 1/n$), the typical outcome of a cross will be a complete absence of recombinants. On the other hand, if n random diploid individuals are sampled from a natural population, because the mean time to a common ancestor between random neutral alleles is $\sim 2N_e$ generations, the expected number of recombination events is $4N_e nc_{ij}$.

Numerous population-level surveys have been made to estimate the recombination parameter $4N_e c_0$, where c_0 is the rate of recombination between adjacent nucleotide sites. This is usually done by first estimating the population-level parameter at various distances between sites (L_{ij}), and then dividing by L_{ij} under the assumption that $c_{ij} = c_0 L_{ij}$, i.e., a linear relationship between the recombination rate and physical distance between sites. As noted in Foundations 4.2, this is a reasonable approximation provided the distance between sites is less than the average length of a gene-conversion tract, but will lead to an underestimate of $4N_e c_0$ when greater distances are relied upon (by a factor up to $20\times$; the discrepancy being due to the fact that most recombination events do not lead to crossovers between distant sites). Using this procedure, all estimates of the per-site parameter $4N_e c_0$ are smaller than 0.1, with many falling below 0.01 (Chapter 4 in Walsh and Lynch 2018). These observations provide support for the idea that, as with mutation, random genetic drift is generally a more powerful force than recombination at the level of individual nucleotide sites, with the caveat that most existing analyses involve animals and land plants.

By dividing estimates of $4N_e c_0$ by parallel estimates of $\theta = 4N_e u$, the effective population size cancels out, yielding an estimate of the ratio of recombination and mutation rates at the nucleotide level (c_0/u). All such estimates for eukaryotes are smaller than 5.0, and nearly half are smaller than 1.0 (Lynch 2007; Walsh and Lynch 2018). For example, the average estimate of c_0/u for *Drosophila* species is ~ 2.7, whereas that for humans is ~ 0.8, and the average for land plants is 1.1. These observations imply that the power of recombination

between adjacent sites is often of the same order of magnitude as the power of mutation, or perhaps somewhat larger owing to the downward bias in c_0 estimates noted above.

Contrary to common belief, relative to the background rate of mutation, recombination at the nucleotide level is not exceptionally low in bacteria (Shapiro 2016; Bobay and Ochman 2017; Garud and Pollard 2020; Sakoparnig et al. 2021). Although bacteria do not engage in the kinds of organized meiotic activities of eukaryotes, they have several other pathways that can lead to homologous recombination, including transduction of sequences by bacteriophage, physical conjugation and DNA exchange between conspecifics, and even consumption and integration of free DNA from dead cells. Consistent with high levels of recombination induced by these alternative mechanisms, estimates of c_0/u for bacterial species are often of the same order of magnitude as those for eukaryotes (Lynch 2007; Vos and Didelot 2009; Rosen et al. 2015).

Finally, as suggested above, not all recombination events involve crossovers. For example, direct empirical observations suggest that the fraction of recombination events accompanied by crossing over is $x \simeq 0.30$ in the budding yeast *S. cerevisiae* (Malkova et al. 2004; Mancera et al. 2008), and $\simeq 0.15$ in the fly *D. melanogaster* (Hilliker et al. 1994). Indirect LD-based estimates suggest $x \simeq 0.14$ in humans (Frisse et al. 2001; Padhukasahasram and Rannala 2013), 0.09 on average in other vertebrates (Lynch et al. 2014), $\simeq 0.08$ in *D. melanogaster* (Langley et al. 2000; Yin et al. 2009), and $\simeq 0.10$ in plants (Morrell et al. 2006; Yang et al. 2012). Thus, the data universally point to $\sim 70-90\%$ of recombination events being simple local gene conversions unaccompanied by crossovers. This implies that recombination rates at short distances are typically 3–10 times greater than expected based on crossing-over alone. Average conversion-tract lengths tend to be several hundred to a few thousand bp in diverse eukaryotes (Lynch et al. 2014; Liu et al. 2018).

Evolution of the recombination rate

Considerable attention has been devoted to understanding how selection might favor recombination-rate modifiers in various contexts (e.g., Feldman et al. 1996; Barton and Otto 2005; Keightley and Otto 2006; Barton 2010; Hartfield et al. 2010). As in the case of mutation-rate evolution, selection on modifiers of the recombination rate is expected to involve second-order effects, operating via fitness-altering recombination effects elsewhere in the genome. To this end, virtually all theoretical work on the evolution of the recombination rate is motivated by the idea that natural selection inadvertently encourages the build-up of linkage disequilibrium in ways that inhibit full evolutionary potential unless unleashed by recombination.

Two general aspects of genetic systems can encourage the development of hidden genetic variance. First, synergistic epistasis (with fitness declining at an increasing rate with increasing numbers of deleterious alleles) tends to encourage the maintenance of intermediate genotypes, thereby providing a selective advantage for recombinational production of the extreme genotypes and their more efficient promotion/elimination by selection (Eshel and Feldman 1970; Kondrashov 1988; Charlesworth 1990; Barton 1995). In contrast, diminishing-returns epistasis (with fitness declining at a decreasing rate with increasing numbers of deleterious alleles) has the opposite effect, thereby potentially encouraging reduced recombination rates. As the evidence on the relative incidences of these two forms of epistasis is mixed (Chapter 12), the role of epistasis in the evolution of recombination rates remains unclear.

Second, as noted previously, linkage reduces the efficiency of selection on multilocus systems by inducing background selection and selective sweeps. These general effects are expected to be more pronounced in larger populations, which generally harbor larger numbers of co-segregating polymorphic loci. Plausible arguments have been made that the power of this second effect in selecting for modifiers for increased recombination rates may substantially outweigh that resulting from synergistic epistasis (Felsenstein and Yokoyama 1976; Otto and Barton 2001; Pálsson 2002; Barton and Otto 2005; Keightley and Otto 2006; Roze and Barton 2006).

Despite all of the theory, the extent to which recombination-rate modifiers ever arise with substantial enough fitness consequences to be promoted by these kinds of associative effects remains unclear. Most modelling attempts have focused on rather extreme situations in which selection coefficients and/or the magnitude of the modifier's effect on the recombination rate are quite large, and yet even under these conditions the selective advantage of the modifier can be quite small (Barton and Otto 2005), perhaps too small to overcome the likelihood of being lost by drift in most cases.

This is not to dispute the adaptive utility of sexual reproduction, which promotes both independent segregation of chromosomes and recombination within chromosomes. Laboratory experiments with yeast populations support the idea that fitness increases more rapidly in response to a selective challenge in the presence of outcrossing than under clonal propagation (Goddard et al. 2005; McDonald et al. 2016). The latter study is most notable in that it used genome-wide sequencing to follow the fates of newly arising mutations, showing that many more mutations go to fixation in a hitchhiking fashion in asexual populations, with mildly deleterious mutations commonly being dragged along by linked beneficials (Figure 4.9). In contrast, in sexual populations, fewer mutations arise to high frequency, but most that do are beneficial. In effect, sexual reproduction reduces the effects of selective interference (between beneficial mutations at different linked loci competing with each other for fixation), while also reducing the 'ruby in the rubbish' effect (beneficial mutations being permanently linked to a background containing deleterious mutations; Chapters 5 and 6).

The key issue here is whether, conditional on sexual reproduction, natural selection further fine-tunes the level of within-chromosome recombination. As noted above (Figure 4.8), the fact remains that nearly all interspecific variation in the recombination rate per physical distance can be explained by a simple, phylogenetically invariant physical model of meiosis, leaving very little residual variation to be potentially assigned to adaptive fine-tuning. Variation in the recombination rate does exist among individuals and between closely related species (Dapper and Payseur 2017; Ritz et al. 2017), so evolutionary modification is certainly possible. However, the level of variation is generally

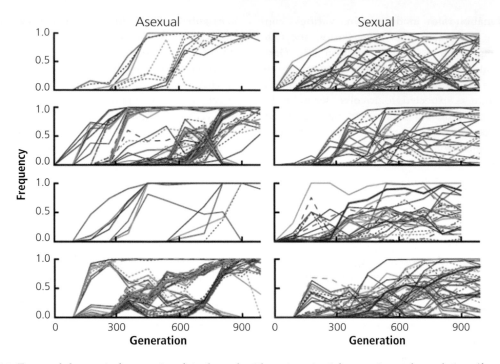

Figure 4.9 Temporal changes in frequencies of single-nucleotide variants in eight experimental populations (four asexual and four sexual) of yeast (*S. cerevisiae*) over a period of 1000 generations. All populations were initiated with the same single clone, and the frequencies of spontaneously arising mutations were estimated by pooling entire populations and sequencing the entire genome at high coverage. Solid lines denote amino acid-altering mutations; dashed lines denote silent changes in coding regions that leave the encoded amino acid unaltered; and dotted lines denote mutations in intergenic regions. Note that in the asexual populations, bundles of trajectories of individual mutations follow the same pattern of frequency change; these represent the history of individual clones within which all relevant variants are permanently linked and therefore dragged together to fixation or loss. In sexual populations, segregation and/or recombination places individual mutations on multiple genetic backgrounds, rendering the trajectories of individual variants more independent. From McDonald et al. (2016).

less than twofold, and may simply reflect the recurrent introduction of minor recombination-rate variants by mutation.

The most widely cited evidence in favor of the idea of adaptive modification of recombination rates involves examples of moderate increases in the crossover rate in metazoan populations exposed to strong directional selection (e.g., domestication, and insecticide resistance) (Ritz et al. 2017). In principle, such results might be examples of hitchhiking of recombination-rate enhancers with strongly favored gene combinations, much like the situation with transient increases in mutator-allele frequencies. However, enough counter-examples have been presented (Muñoz-Fuentes et al. 2015; Stapley et al. 2017) that one must be concerned with

reporting bias towards positive results. Notably, Dumont and Payseur (2007) find that variation in recombination rates across mammalian species evolves in a manner that cannot be discriminated from the expectations of a neutral model.

An alternative view, consistent with the pattern in Figure 4.8, is that natural selection generally operates to minimize the amount of meiotic recombination (one crossover per arm being a minimum requirement for proper chromosome segregation), with phylogenetic divergence in recombination rates being largely an indirect and passive response to changes in the population-genetic environment. With reduced N_e in organisms with increasing cell/body size (Figure 4.3), genome sizes passively expand as selection becomes less capable of resisting the accumulation of intronic and mobile-element associated DNAs (Lynch 2007). With an increase in chromosome length, but not in the number of crossovers per chromosome, the recombination rate per physical distance then naturally declines. Thus, as with the mutation rate, the bulk of the variation in the recombination rate among species may be largely governed by differences in the cumulative effects of random genetic drift.

Summary

- Evolution is a population-genetic process governed by the joint forces of mutation, recombination, and random genetic drift, all of which vary by more than four orders of magnitude across the Tree of Life. As these three features define the playing field upon which evolution operates, such quantitative knowledge is essential for understanding the limits to all adaptive and nonadaptive evolutionary pathways.

- In their early stages, all newly arisen mutations experience stochastic fluctuations in allele frequencies, with only a small fraction of them being harvested by natural selection, even if highly beneficial. The magnitude of noise in the evolutionary process (random genetic drift) is dictated by the genetic effective population size (N_e), which is influenced by the non-independence of simultaneously interfering mutations at linked chromosomal sites, by the absolute numbers of individuals in the population (N), and by various ecological and behavioral factors. N_e is generally orders of magnitude smaller than N, scaling negatively with organism size, with no known species having $N_e > 10^9$ individuals.

- The mutation rate is generally under persistent selection in the downward direction, with the rate per nucleotide site per generation declining from $\sim10^{-8}$ to $\sim10^{-11}$ with increasing N_e. This gradient can be explained by the drift-barrier hypothesis, with the efficiency of selection on replication fidelity and DNA repair becoming stalled by random genetic drift as the room for improvement declines. This hypothesis also explains why microbial eukaryotes (with more functional DNA) have lower mutation rates than prokaryotes with the same N_e and why specialized polymerases that engage in relatively small numbers of nucleotide transactions have elevated error rates.

- Mutations arise randomly with respect to the selective demands operating on target genes. Nonetheless, they are non-random in almost all other respects, including chromosomal location and nucleotide identity. Mutations are also commonly clustered, so that the incidence of double and triple mutants can be orders of magnitude greater than expected by chance.

- Average rates of recombination per physical distance decline in larger organisms with decreased N_e because eukaryotic meiosis almost always involves just one or two crossovers per chromosome arm. In lineages with decreasing N_e, average chromosome lengths increase via the passive accumulation of noncoding DNA, whereas the number of crossovers per chromosome remains relatively constant.

- A unifying view of these observations is that ecological and behavioral factors, combined with the influence of chromosomal linkage, conspire to define the effective sizes of populations, which in turn indirectly modulate the evolution of mutation and recombination rates. Small organisms with high N_e tend to have relatively low mutation rates but high recombination rates per nucleotide site. In contrast, larger organisms have lower N_e, and owing to higher levels of random genetic drift, passively evolve higher mutation rates but lower recombination rates. These co-varying aspects of the population-genetic environment modify the ways in which evolution by natural selection can proceed in different phylogenetic lineages.

- Because mutations with selective effects $\ll 1/N_e$ are overwhelmed by drift, small organisms with higher N_e are capable of utilizing a wider range of mutational effects in adaptive evolution. Larger organisms, with correspondingly smaller N_e, have a reduced capacity for evolutionary fine-tuning and hence, are constrained to more coarse-grained evolution.

Foundations 4.1 The amount of neutral nucleotide variation maintained at mutation-drift equilibrium

Consider a population with a base-substitution mutation rate of u per genomic site per generation, with a long-term average effective population size of N_e. Here, we assume that each nucleotide base mutates to each of the three other types at rate $u/3$. In each generation, new variation (defined as heterozygosity, which is the probability that randomly paired chromosomes differ at a nucleotide site) will arise by mutation. In addition, a fraction of pre-existing variation will be lost by drift. If u and N_e are kept approximately constant, an expected steady-state level of heterozygosity per nucleotide site will eventually be reached, at which point the rates of gain and loss of heterozygosity will be equal. Such heterozygosity can be measured as either the fraction of all neutral sites that are heterozygous within single, random diploid individuals (assuming a randomly mating population) or as the average level of heterozygosity over a large number of neutral sites at the population level.

To obtain the expected equilibrium, we start with a formulation for the dynamics of neutral-site heterozygosity, and then seek the point at which the loss and gain rates are equal. As the focus is on the neutral situation, we will assume that A, C, G, and T have equivalent fitness effects, as might be the case for a fourfold redundant site in protein-coding sequence (e.g., in the third positions of a number of amino-acid codons).

Letting H_t denote the expected level of heterozygosity in generation t, and assuming a diploid population with k allelic types per site, we wish to determine the dynamics of change in H_t and its eventual equilibrium value. It follows from basic theory (Chapter 2 in Walsh and Lynch 2018) that drift causes a fractional loss of heterozygosity equal to $1/(2N_e)$ per generation. Letting $\lambda = 1 - (1/2N_e)$, the expected frequency of heterozygotes in generation $t+1$ in the absence of mutation is then λH_t, whereas the expected frequency of homozygotes is $1 - \lambda H_t$. Following mutation, the heterozygous state will be retained if: 1) neither allele mutates, the probability of which is $(1 - 2u)$, ignoring the very small

probability of double mutations to the same state; or 2) one of the alleles mutates to a different state than the other, the probability of which is $[2u(k-2)/(k-1)]$ assuming that all allelic types are equally mutationally exchangeable. For nucleotide sites, there are $k=4$ alternative states, and the preceding expression reduces to $4u/3$, that is, each of two sites mutates to two other possible nucleotides at rate $2u/3$. On the other hand, homozygotes will be mutationally converted to heterozygotes at rate $2u$.

Summing up, the expected dynamics of neutral-site heterozygosity under random mating can be expressed as

$$H_{t+1} = \lambda H_t \left((1 - 2u) + \frac{4u}{3} \right) + 2u(1 - \lambda H_t). \quad (4.1.1)$$

Setting $H_{t+1} = H_t = \widetilde{H}$, the expected level of heterozygosity under drift-mutation balance is found to be

$$\widetilde{H} = \frac{\theta}{1 + (4\theta/3)}, \quad (4.1.2a)$$

where $\theta = 4N_e u$ (Malécot 1948; Kimura 1968). The same expression applies to haploidy by setting $\theta = 2N_e u$.

There are two key points to note about Equation 4.1.2a. First, the final result is a function of one composite parameter, θ, which is equivalent to the ratio of the rates of mutational production of heterozygotes from homozygotes ($2u$) and the rate of loss of heterozygosity by drift ($1/2N_e$). Second, if θ is $\ll 1$, as is almost always the case in natural populations,

$$\widetilde{H} \simeq \theta = 4N_e u. \quad (4.1.2b)$$

Although Equation 4.1.1 needs to be modified if the different nucleotides mutate at different rates (Kimura 1983; Cockerham 1984), provided $4N_e u \ll 1$, the equilibrium approximation given by Equation 4.1.2b still holds.

Foundations 4.2 Relationship of the recombination rate to physical distance between sites

Meiotic recombination events involve heteroduplex formations between paired homologous chromosomes in diploid cells, i.e., the invasion of one double-stranded DNA by a single strand from another. Temporary physical

annealing of homologous regions occurs as a single strand from one chromosome invades the double-stranded recipient homolog (Figure 4.7). Upon separation of recombining chromosomes, the heteroduplex DNA (containing one strand from each of the contributing chromosomes) remains. If heterozygous sites are contained within such a patch, the non-matching pairs have to be resolved by the mismatch-repair pathway. This leads to a process of gene conversion, as each mismatched pair of sites is restored to a Watson-Crick state, yielding either the recipient- or donor-strand state. Depending on how the heteroduplex is resolved, gene conversion may be accompanied by a crossover, which leads to a complete swapping of chromosomal material to one side of the conversion event. Gene conversion involves unidirectional exchange of information, whereas a crossover generates reciprocal exchange.

It is often assumed that the recombination rate is equivalent to the crossover rate between sites, but this is generally not true. Although all recombination events involve gene conversion, only a fraction lead to crossovers. If a gene-conversion tract unaccompanied by a crossover occurs within the span between two distantly located sites, there will be no recombination between the pair of sites, as they will retain their original status. Thus, when the sites under consideration are far apart, most recombination events involve crossing-over because most conversion events are irrelevant. On the other hand, when sites are close together, recombination mostly results from the conversion of single sites (Figure 4.7).

To understand this behavior in a more quantitative way, let c_0 be the total rate of initiation of recombination events per nucleotide site (with or without crossing-over), L be the number of sites separating the two focal positions (with $L = 1$ for adjacent sites), and x be the fraction of recombination events accompanied by crossing over. Using Haldane's (1919) mapping function, which assumes equal probabilities of recombination at all sites, the Poisson probability of no crossover between two homologous chromosomes during a meiotic event is $e^{-2c_0 xL}$. The crossover rate can then be represented as $0.5(1 - e^{-2c_0 xL})$, which $\simeq c_0 xL$ for $c_0 xL \ll 1$, and asymptotically approaches 0.5 for large $c_0 xL$. This asymptotic value follows from the fact that, as the number of crossovers between markers increases, even and odd numbers of events become equally likely, with even numbers restoring the parental state.

How does gene conversion alter the recombination rate between sites? For ease of presentation, we assume distances between sites that are small enough that the crossover rate $\simeq c_0 xL$. As noted by Andolfatto and Nordborg (1998), from the perspective of two sites, a gene-conversion event causes recombination between a pair of sites if the conversion tract encompasses just one of the sites. Under the assumption of an exponential distribution of tract lengths with mean length T (in bp), the total conversion rate per site is $(1 - x)c_0 T(1 - e^{-L/T})$ (Langley et al. 2000; Frisse et al. 2001; Lynch et al. 2014). The total recombination rate between sites separated by distance L is then

$$c_L \simeq c_0[xL + (1 - x)T(1 - e^{-L/T})]. \qquad (4.2.1a)$$

For sites that are much more closely spaced than the average conversion-tract length, $L \ll T$,

$$c_L \simeq c_0 L, \qquad (4.2.1b)$$

whereas for $L \gg T$,

$$c_L \simeq c_0 Lx. \qquad (4.2.1c)$$

These results show that unless all recombination events are accompanied by crossovers ($x = 1$), the use of recombination rates between distantly related sites to extrapolate to closely spaced sites will underestimate the true rate by a factor of $1/x$.

Literature Cited

Andolfatto, P., and M. Nordborg. 1998. The effect of gene conversion on intralocus associations. Genetics 148: 1397–1399.

Arnheim, N., P. Calabrese, and I. Tiemann-Boege. 2007. Mammalian meiotic recombination hot spots. Annu. Rev. Genet. 41: 369–399.

Baer, C. F., M. M. Miyamoto, and D. R. Denver. 2007. Mutation rate variation in multicellular eukaryotes: Causes and consequences. Nat. Rev. Genet. 8: 619–631.

Barton, N. H. 1995. A general model for the evolution of recombination. Genet. Res. 65: 123–145.

Barton, N. H. 2010. Mutation and the evolution of recombination. Philos. Trans. R. Soc. Lond. B Biol. Sci. 365: 1281–1294.

Barton, N. H., and S. P. Otto. 2005. Evolution of recombination due to random drift. Genetics 169: 2353–2370.

Baudat, F., Y. Imai, and B. de Massy. 2013. Meiotic recombination in mammals: Localization and regulation. Nat. Rev. Genet. 14: 794–806.

Behjati, S., M. Huch, R. van Boxtel, W. Karthaus, D. C. Wedge, A. U. Tamuri, I. Martincorena, M. Petljak, L. B. Alexandrov, G. Gundem, et al. 2014. Genome sequencing of normal cells reveals developmental lineages and mutational processes. Nature 513: 422–425.

Bobay, L. M., and H. Ochman. 2017. Biological species are universal across Life's domains. Genome Biol. Evol. 9: 491–501.

Bobay, L. M., and H. Ochman. 2018. Factors driving effective population size and pan-genome evolution in bacteria. BMC Evol. Biol. 18: 153.

Boe, L., M. Danielsen, S. Knudsen, J. B. Petersen, J. May-mann, and P. R. Jensen. 2000. The frequency of muta-tors in populations of *Escherichia coli*. Mutat. Res. 448: 47–55.

Brick, K., F. Smagulova, P. Khil, R. D. Camerini-Otero, and G. V. Petukhova. 2012. Genetic recombination is directed away from functional genomic elements in mice. Nature 485: 642–645.

Cagan, A., A. Baez-Ortega, N. Brzozowska, F. Abascal, T. H. H. Coorens, M. A. Sanders, A. R. J. Lawson, L. M. R. Harvey, S. Bhosle, D. Jones, et al. 2022. Somatic mutation rates scale with lifespan across mammals. Nature 604: 517–524.

Capilla, L., M. Garcia Caldés, and A. Ruiz-Herrera. 2016. Mammalian meiotic recombination: A toolbox for genome evolution. Cytogenet. Genome Res. 150: 1–16.

Chan, K., and D. A. Gordenin. 2015. Clusters of multi-ple mutations: Incidence and molecular mechanisms. Annu. Rev. Genet. 49: 243–267.

Charlesworth, B. 1990. Mutation-selection balance and the evolutionary advantage of sex and recombination. Genet. Res. 55: 199–221.

Charlesworth, B. 2009. Fundamental concepts in genet-ics: Effective population size and patterns of molec-ular evolution and variation. Nat. Rev. Genet. 10: 195–205.

Charlesworth, B. 2012. The effects of deleterious mutations on evolution at linked sites. Genetics 190: 5–22.

Chen, J., S. Glémin, and M. Lascoux. 2017. Genetic diver-sity and the efficacy of purifying selection across plant and animal species. Mol. Biol. Evol. 34: 1417–1428.

Chen, X., and J. Zhang. 2013. No gene-specific optimiza-tion of mutation rate in *Escherichia coli*. Mol. Biol. Evol. 30: 1559–1562.

Chen, X., Z. Chen, H. Chen, Z. Su, J. Yang, F. Lin, S. Shi, and X. He. 2012. Nucleosomes suppress spontaneous mutations base-specifically in eukaryotes. Science 335: 1235–1238.

Cockerham, C. C. 1984. Drift and mutation with a finite number of allelic states. Proc. Natl. Acad. Sci. USA 81: 530–534.

Coop, G., X. Wen, C. Ober, J. K. Pritchard, and M. Prze-worski. 2008. High-resolution mapping of crossovers reveals extensive variation in fine-scale recombination patterns among humans. Science 319: 1395–1398.

Corbett-Detig, R. B., D. L. Hartl, and T. B. Sackton. 2015. Natural selection constrains neutral diversity across a wide range of species. PLoS Biol. 13: e1002112.

Dapper, A. L., and B. A. Payseur. 2017. Connecting theory and data to understand recombination rate evolution. Philos. Trans. R. Soc. Lond. B Biol. Sci. 372: 20160469.

de Massy, B. 2003. Distribution of meiotic recombination sites. Trends Genet. 19: 514–522.

de Massy, B. 2013. Initiation of meiotic recombination: How and where? Conservation and specificities among eukaryotes. Annu. Rev. Genet. 47: 563–599.

Denamur, E., and I. Matic. 2006. Evolution of mutation rates in bacteria. Mol. Microbiol. 60: 820–827.

Desai, M. M., and D. S. Fisher. 2007. Beneficial mutation selection balance and the effect of linkage on positive selection. Genetics 176: 1759–1798.

Drake, J. W. 1991. A constant rate of spontaneous mutation in DNA-based microbes. Proc. Natl. Acad. Sci. USA 88: 7160–7164.

Drake, J. W. 2007. Too many mutants with multiple muta-tions. Crit. Rev. Biochem. Mol. Biol. 42: 247–258.

Drake, J. W., B. Charlesworth, D. Charlesworth, and J. F. Crow. 1998. Rates of spontaneous mutation. Genetics 148: 1667–1686.

Dumont, B. L., and B. A. Payseur. 2008. Evolution of the genomic rate of recombination in mammals. Evolution 62: 276–294.

Earl, D. J., and M. W. Deem. 2004. Evolvability is a selectable trait. Proc. Natl. Acad. Sci. USA 101: 11531–11536.

Eshel, I., and M. W. Feldman. 1970. On the evolutionary effect of recombination. Theor. Popul. Biol. 1: 88–100.

Ewens, W. J. 2004. Mathematical Population Genetics, 2nd edn. Springer-Verlag, Berlin.

Eyre-Walker, A., and P. D. Keightley. 2007. The distribution of fitness effects of new mutations. Nat. Rev. Genet. 8: 610–618.

Feldman, M. W., S. P. Otto, and F. B. Christiansen. 1996. Population genetic perspectives on the evolution of recombination. Annu. Rev. Genet. 30: 261–295.

Felsenstein, J., and S. Yokoyama. 1976. The evolutionary advantage of recombination. II. Individual selection for recombination. Genetics 83: 845–859.

Foster, P. L. 2007. Stress-induced mutagenesis in bacteria. Crit. Rev. Biochem. Mol. Biol. 42: 373–397.

Foster, P. L., A. J. Hanson, H. Lee, E. M. Popodi, and H. Tang. 2013. On the mutational topology of the bacterial genome. G3 (Bethesda) 3: 399–407.

Frisse, L., R. R. Hudson, A. Bartoszewicz, J. D. Wall, J. Donfack, and A. Di Rienzo. 2001. Gene conversion and different population histories may explain the con-trast between polymorphism and linkage disequilib-rium levels. Amer. J. Hum. Genet. 69: 831–843.

Gale, J. S. 1990. Theoretical Population Genetics. Unwin Hyman Ltd., London, UK.

Galhardo, R. S., P. J. Hastings, and S. M. Rosenberg. 2007. Mutation as a stress response and the regula-tion of evolvability. Crit. Rev. Biochem. Mol. Biol. 42: 399–435.

Garud, N. R., and K. S. Pollard. 2020. Population genetics in the human microbiome. Trends Genet. 36: 53–67.

Goddard, M. R., H. C. Godfray, and A. Burt. 2005. Sex increases the efficacy of natural selection in experimental yeast populations. Nature 434: 636–640.

Goho, S., and G. Bell. 2000. Mild environmental stress elicits mutations affecting fitness in *Chlamydomonas*. Proc. Biol. Sci. 267: 123–129.

Good, B. H., and M. M. Desai. 2016. Evolution of mutation rates in rapidly adapting asexual populations. Genetics 204: 1249–1266.

Haenel, Q., T. G. Laurentino, M. Roesti, and D. Berner. 2018. Meta-analysis of chromosome-scale crossover rate variation in eukaryotes and its significance to evolutionary genomics. Mol. Ecol. 27: 2477–2497.

Haldane, J. B. S. 1919. The combination of linkage values, and the calculation of distance between the loci of linked factors. J. Genetics 8: 299–309.

Harris, K., and R. Nielsen. 2014. Error-prone polymerase activity causes multinucleotide mutations in humans. Genome Res. 24: 1445–1454.

Hartfield, M., S. P. Otto, and P. D. Keightley. 2010. The role of advantageous mutations in enhancing the evolution of a recombination modifier. Genetics 184: 1153–1164.

Heidenreich, E. 2007. Adaptive mutation in *Saccharomyces cerevisiae*. Crit. Rev. Biochem. Mol. Biol. 42: 285–311.

Hicks, W. M., M. Kim, and J. E. Haber. 2010. Increased mutagenesis and unique mutation signature associated with mitotic gene conversion. Science 329: 82–85.

Hilliker, A. J., G. Harauz, A. G. Reaume, M. Gray, S. H. Clark, and A. Chovnick. 1994. Meiotic gene conversion tract length distribution within the *rosy* locus of *Drosophila melanogaster*. Genetics 137: 1019–1026.

James, A., and K. Jain. 2016. Fixation probability of rare nonmutator and evolution of mutation rates. Ecol. Evol. 6: 755–764.

Jeffreys, A. J., J. K. Holloway, L. Kauppi, C. A. May, R. Neumann, M. T. Slingsby, and A. J. Webb. 2004. Meiotic recombination hot spots and human DNA diversity. Phil. Trans. Roy. Soc. Lond. B Biol. Sci. 359: 141–152.

Johnson, T., and N. H. Barton. 2002. The effect of deleterious alleles on adaptation in asexual populations. Genetics 162: 395–411.

Kang, J. M., N. M. Iovine, and M. J. Blaser. 2006. A paradigm for direct stress-induced mutation in prokaryotes. FASEB J. 20: 2476–2485.

Katju, V., and U. Bergthorsson. 2019. Old trade, new tricks: Insights into the spontaneous mutation process from the partnering of classical mutation accumulation experiments with high-throughput genomic approaches. Genome Biol. Evol. 11: 136–165.

Keightley, P. D., and S. P. Otto. 2006. Interference among deleterious mutations favours sex and recombination in finite populations. Nature 443: 89–92.

Kimura, M. 1957. Some problems of stochastic processes in genetics. Ann. Math. Stat. 28: 882–901.

Kimura, M. 1967. On the evolutionary adjustment of spontaneous mutation rates. Genet. Res. 9: 23–34.

Kimura, M. 1968. Genetic variability maintained in a finite population due to mutational production of neutral and nearly neutral isoalleles. Genet. Res. 11: 247–269.

Kimura, M. 1983. The Neutral Theory of Molecular Evolution. Cambridge University Press, Cambridge, UK.

Kimura, M., and T. Ohta. 1969. The average number of generations until fixation of a mutant gene in a finite population. Genetics 61: 763–771.

Kivisaar, M. 2010. Mechanisms of stationary-phase mutagenesis in bacteria: Mutational processes in pseudomonads. FEMS Microbiol. Lett. 312: 1–14.

Kohl, K. P., and J. Sekelsky. 2013. Meiotic and mitotic recombination in meiosis. Genetics 194: 327–334.

Kondrashov, A. S. 1988. Deleterious mutations and the evolution of sexual reproduction. Nature 336: 435–440.

Lam, I., and S. Keeney. 2015. Nonparadoxical evolutionary stability of the recombination initiation landscape in yeast. Science 350: 932–937.

Lang, G. I., and A. W. Murray. 2011. Mutation rates across budding yeast chromosome VI are correlated with replication timing. Genome Biol. Evol. 3: 799–811.

Langley, C. H., B. P. Lazzaro, W. Phillips, E. Heikkinen, and J. M. Braverman. 2000. Linkage disequilibria and the site frequency spectra in the $su(s)$ and $su(w^a)$ regions of the *Drosophila melanogaster* X chromosome. Genetics 156: 1837–1852.

Leffler, E. M., K. Bullaughey, D. R. Matute, W. K. Meyer, L. Ségurel, A. Venkat, P. Andolfatto, and M. Przeworski. 2012. Revisiting an old riddle: What determines genetic diversity levels within species? PLoS Biol. 10: e1001388.

Leigh, E. G., Jr. 1970. Natural selection and mutability. Amer. Natur. 104: 301–305.

Liu, H., J. Huang, X. Sun, J. Li, Y. Hu, L. Yu, G. Liti, D. Tian, L. D. Hurst, and S. Yang. 2018. Tetrad analysis in plants and fungi finds large differences in gene conversion rates but no GC bias. Nat. Ecol. Evol. 2: 164–173.

Liu, H., and J. Zhang. 2022. Is the mutation rate lower in genomic regions of stronger selective constraints? Mol. Biol. Evol. 39: msac169.

Long, H., W. Sung, S. F. Miller, M. S. Ackerman, T. G. Doak, and M. Lynch. 2015. Mutation rate, spectrum, topology, and context-dependency in the DNA mismatch-repair deficient *Pseudomonas fluorescens* ATCC948. Genome Biol. Evol. 7: 262–271.

Long, H., W. Sung, S. Kucukyildirim, E. Williams, S. W. Guo, C. Patterson, C. Gregory, C. Strauss, C. Stone, C. Berne, et al. 2017. Evolutionary determinants of genome-wide nucleotide composition. Nat. Ecol. Evol. 2: 237–240.

Lukačišinová, M., S. Novak, and T. Paixão. 2017. Stress-induced mutagenesis: Stress diversity facilitates the persistence of mutator genes. PLoS Comput. Biol. 13: e1005609.

Lynch, M. 2007. The Origins of Genome Complexity. Sinauer Associates, Inc., Sunderland, MA.

Lynch, M. 2008. The cellular, developmental, and population-genetic determinants of mutation-rate evolution. Genetics 180: 933–943.

Lynch, M. 2010. Evolution of the mutation rate. Trends Genet. 26: 345–352.

Lynch, M. 2011. The lower bound to the evolution of mutation rates. Genome Biol. Evol. 3: 1107–1118.

Lynch, M. 2020. The evolutionary scaling of cellular traits imposed by the drift barrier. Proc. Natl. Acad. Sci. USA 117: 10435–10444.

Lynch, M., M. Ackerman, J.-F. Gout, H. Long, W. Sung, W. K. Thomas, and P. L. Foster. 2016. Genetic drift, selection, and evolution of the mutation rate. Nat. Rev. Genet. 17: 704–714.

Lynch, M., F. Ali, T. Lin, Y. Wang, J. Ni, and H. Long. 2023. The divergence of mutation rates and spectra across the Tree of Life. EMBO Rep. e57561.

Lynch, M., J. Blanchard, D. Houle, T. Kibota, S. Schultz, L. Vassilieva, and J. Willis. 1999. Spontaneous deleterious mutation. Evolution 53: 645–663.

Lynch, M., L. M. Bobay, F. Catania, J.-F. Gout, and M. Rho. 2011. The repatterning of eukaryotic genomes by random genetic drift. Annu. Rev. Genomics Hum. Genet. 12: 347–366.

Lynch, M., and B. Trickovic. 2020. A theoretical framework for evolutionary cell biology. J. Mol. Biol. 432: 1861–1879.

Lynch, M., and J. B. Walsh. 1998. Genetics and Analysis of Quantitative Traits. Sinauer Associates, Inc., Sunderland, MA.

Lynch, M., S. Xu, T. Maruki, P. Pfaffelhuber, and B. Haubold. 2014. Genome-wide linkage-disequilibrium profiles from single individuals. Genetics 198: 269–281.

MacLean, R. C., C. Torres-Barceló, and R. Moxon. 2013. Evaluating evolutionary models of stress-induced mutagenesis in bacteria. Nat. Rev. Genet. 14: 221–227.

Maharjan, R. P., and T. Ferenci. 2017. A shifting mutational landscape in 6 nutritional states: Stress-induced mutagenesis as a series of distinct stress input-mutation output relationships. PLoS Biol. 15: e2001477.

Malécot, G. 1948. Les Mathématiques de l'Hérédité. Masson, Paris.

Malécot, G. 1952. Les processus stochastiques et la méthode des fonctions génératrices ou caracteréstiques. Publ. Inst. Stat. Univ. Paris 1: Fasc. 3: 1–16.

Malkova, A., and J. E. Haber. 2012. Mutations arising during repair of chromosome breaks. Annu. Rev. Genet. 46: 455–473.

Malkova, A., J. Swanson, M. German, J. H. McCusker, E. A. Housworth, F. W. Stahl, and J. E. Haber. 2004. Gene conversion and crossing over along the 405-kb left arm of *Saccharomyces cerevisiae* chromosome VII. Genetics 168: 49–63.

Mancera, E., R. Bourgon, A. Brozzi, W. Huber, and L. M. Steinmetz. 2008. High-resolution mapping of meiotic crossovers and non-crossovers in yeast. Nature 454: 479–485.

Martincorena, I., A. S. Seshasayee, and N. M. Luscombe. 2012. Evidence of non-random mutation rates suggests an evolutionary risk management strategy. Nature 485: 95–98.

Matsuba, C., D. G. Ostrow, M. P. Salomon, A. Tolani, and C. F. Baer. 2013. Temperature, stress and spontaneous mutation in *Caenorhabditis briggsae* and *Caenorhabditis elegans*. Biol. Lett. 9: 20120334.

McDonald, M. J., Y. Y. Hsieh, Y. H. Yu, S. L. Chang, and J. Y. Leu. 2012. The evolution of low mutation rates in experimental mutator populations of *Saccharomyces cerevisiae*. Curr. Biol. 22: 1235–1240.

McDonald, M. J., D. P. Rice, and M. M. Desai. 2016. Sex speeds adaptation by altering the dynamics of molecular evolution. Nature 531: 233–236.

Milholland, B., X. Dong, L. Zhang, X. Hao, Y. Suh, and J. Vijg. 2017. Differences between germline and somatic mutation rates in humans and mice. Nat. Commun. 8: 15183.

Monroe, J. G., T. Srikant, P. Carbonell-Bejerano, C. Becker, M. Lensink, M. Exposito-Alonso, M. Klein, J. Hildebrandt, M. Neumann, D. Kliebenstein, et al. 2022. Mutation bias reflects natural selection in *Arabidopsis thaliana*. Nature 602: 101–105.

Morrell, P. L., D. M. Toleno, K. E. Lundy, and M. T. Clegg. 2006. Estimating the contribution of mutation, recombination and gene conversion in the generation of haplotypic diversity. Genetics 173: 1705–1723.

Muñoz-Fuentes, V., M. Marcet-Ortega, G. Alkorta-Aranburu, C. Linde Forsberg, J. M. Morrell, E. Manzano-Piedras, A. Söderberg, K. Daniel, A. Villalba, A. Toth, et al. 2015. Strong artificial selection in domestic mammals did not result in an increased recombination rate. Mol. Biol. Evol. 32: 510–523.

Myers, S., L. Bottolo, C. Freeman, G. McVean, and P. Donnelly. 2005. A fine-scale map of recombination rates and hotspots across the human genome. Science 310: 321–324.

Neher, R. A. 2013. Genetic draft, selective interference, and population genetics of rapid adaptation. Ann. Rev. Ecol. Evol. Syst. 44: 195–215.

Orr, H. A. 2000. The rate of adaptation in asexuals. Genetics 155: 961–968.

Otto, S. P., and N. H. Barton. 2001. Selection for recombination in small populations. Evolution 55: 1921–1931.

Padhukasahasram, B., and B. Rannala. 2013. Meiotic gene-conversion rate and tract length variation in the human genome. Eur. J. Hum. Genet. 2013: 1–8.

Pálsson, S. 2002. Selection on a modifier of recombination rate due to linked deleterious mutations. J. Hered. 93: 22–26.

Paul, S., S. Million-Weaver, S. Chattopadhyay, E. Sokurenko, and H. Merrikh. 2013. Accelerated gene evolution through replication-transcription conflicts. Nature 495: 512–515.

Petes, T. D. 2001. Meiotic recombination hot spots and cold spots. Nat. Rev. Genet. 2: 360–369.

Radman, M., F. Taddei, and I. Matic. 2000. Evolution-driving genes. Res. Microbiol. 151: 91–95.

Ram, Y., and L. Hadany. 2014. Stress-induced mutagenesis and complex adaptation. Proc. Biol. Sci. 281: 20141025.

Raynes, Y., and P. D. Sniegowski. 2014. Experimental evolution and the dynamics of genomic mutation rate modifiers. Heredity 113: 375–380.

Ritz, K. R., M. A. F. Noor, and N. D. Singh. 2017. Variation in recombination rate: Adaptive or not? Trends Genet. 33: 364–374.

Romiguier, J., P. Gayral, M. Ballenghien, A. Bernard, V. Cahais, A. Chenuil, Y. Chiari, R. Dernat, L. Duret, N. Faivre, et al. 2014. Comparative population genomics in animals uncovers the determinants of genetic diversity. Nature 515: 261–263.

Rosen, M. J., M. Davison, D. Bhaya, and D. S. Fisher. 2015. Fine-scale diversity and extensive recombination in a quasisexual bacterial population occupying a broad niche. Science 348: 1019–1023.

Rosenberg, S. M. 2001. Evolving responsively: Adaptive mutation. Nat. Rev. Genet. 2: 504–515.

Rosenberg, S. M., C. Shee, R. L. Frisch, and P. J. Hastings. 2012. Stress-induced mutation via DNA breaks in Escherichia coli: A molecular mechanism with implications for evolution and medicine. Bioessays 34: 885–892.

Roze, D., and N. H. Barton. 2006. The Hill–Robertson effect and the evolution of recombination. Genetics 173: 1793–1811.

Sakoparnig, T., C. Field, and E. van Nimwegen. 2021. Whole genome phylogenies reflect the distributions of recombination rates for many bacterial species. eLife 10: e65366.

Sasani, T. A., D. G. Ashbrook, A. C. Beichman, L. Lu, A. A. Palmer, R. W. Williams, J. K. Pritchard, and K. Harris. 2022. A natural mutator allele shapes mutation spectrum variation in mice. Nature 605: 497–502.

Schrider, D. R., J. N. Hourmozdi, and M. W. Hahn. 2011. Pervasive multinucleotide mutational events in eukaryotes. Curr. Biol. 21: 1051–1054.

Seplyarskiy, V. B., G. A. Bazykin, and R. A. Soldatov. 2015. Polymerase ζ activity is linked to replication timing in humans: Evidence from mutational signatures. Mol. Biol. Evol. 32: 3158–3172.

Shapiro, B. J. 2016. How clonal are bacteria over time? Curr. Opin. Microbiol. 31: 116–123.

Sharp, N. P., and A. F. Agrawal. 2012. Evidence for elevated mutation rates in low-quality genotypes. Proc. Natl. Acad. Sci. USA 109: 6142–6146.

Shewaramani, S., T. J. Finn, S. C. Leahy, R. Kassen, P. B. Rainey, and C. D. Moon. 2017. Anaerobically grown Escherichia coli has an enhanced mutation rate and distinct mutational spectra. PLoS Genet. 13: e1006570.

Singh, T., M. Hyun, and P. Sniegowski. 2017. Evolution of mutation rates in hypermutable populations of Escherichia coli propagated at very small effective population size. Biol. Lett. 13: 20160849.

Stamatoyannopoulos, J. A., I. Adzhubei, R. E. Thurman, G. V. Kryukov, S. M. Mirkin, and S. R. Sunyaev. 2009. Human mutation rate associated with DNA replication timing. Nat. Genet. 41: 393–395.

Stapley, J., P. G. D. Feulner, S. E. Johnston, A. W. Santure, and C. M. Smadja. 2017. Variation in recombination frequency and distribution across eukaryotes: Patterns and processes. Philos. Trans. R. Soc. Lond. B Biol. Sci. 372: 20160455.

Sung, W., M. S. Ackerman, M. M. Dillon, T. G. Platt, C. Fuqua, V. S. Cooper, and M. Lynch. 2016. Evolution of the insertion-deletion mutation rate across the Tree of Life. G3 (Bethesda) 6: 2583–2591.

Sung, W., M. S. Ackerman, J. F. Gout, S. F. Miller, P. Foster, and M. Lynch. 2015. Asymmetric context-dependent mutation patterns revealed through mutation-accumulation experiments. Mol. Biol. Evol. 32: 1672–1683.

Swings, T., B. Weytjens, T. Schalck, C. Bonte, N. Verstraeten, J. Michiels, and K. Marchal. 2017. Network-based identification of adaptive pathways in evolved ethanol-tolerant bacterial populations. Mol. Biol. Evol. 34: 2927–2943.

Tenaillon, O., F. Taddei, M. Radman, and I. Matic. 2001. Second-order selection in bacterial evolution: Selection acting on mutation and recombination rates in the course of adaptation. Res. Microbiol. 152: 11–16.

Terekhanova, N. V., G. A. Bazykin, A. Neverov, A. S. Kondrashov, and V. B. Seplyarskiy. 2013. Prevalence of multinucleotide replacements in evolution of primates and Drosophila. Mol. Biol. Evol. 30: 1315–1325.

Turrientes, M. C., F. Baquero, B. R Levin, J.-L. Martínez, A. Ripoll, J.-M. González-Alba, R. Tobes, M. Manrique,

M.-R. Baquero, M.-J. Rodríguez-Domínguez, et al. 2013. Normal mutation rate variants arise in a mutator (Mut S) *Escherichia coli* population. PLoS One 8: e72963.

Uphoff, S., N. D. Lord, B. Okumus, L. Potvin-Trottier, D. J. Sherratt, and J. Paulsson. 2016. Stochastic activation of a DNA damage response causes cell-to-cell mutation rate variation. Science 351: 1094–1097.

Vos, M., and X. Didelot. 2009. A comparison of homologous recombination rates in bacteria and archaea. ISME J. 3: 199–208.

Walsh, J. B., and M. Lynch. 2018. Evolution of Quantitative Traits. Sinauer Associates, Inc., Sunderland, MA.

Wei, W., W.-C. Ho, M. Behringer, and M. Lynch. 2022. Rapid evolution of mutation rate and spectrum in response to environmental and population-genetic challenges. Nat. Comm. 13: 4752.

Wells, D., E. Bitoun, D. Moralli, G. Zhang, A. Hinch, J. Jankowska, P. Donnelly, C. Green, and S. R. Myers. 2020. ZCWPW1 is recruited to recombination hotspots by PRDM9 and is essential for meiotic double strand break repair. eLife 9: e53392.

Wielgoss, S., J. E. Barrick, O. Tenaillon, M. J. Wiser, W. J. Dittmar, S. Cruveiller, B. Chane-Woon-Ming, C. Médigue, R. E. Lenski, and D. Schneider. 2013. Mutation rate dynamics in a bacterial population reflect tension between adaptation and genetic load. Proc. Natl. Acad. Sci. USA 110: 222–227.

Williams, L. N., A. J. Herr, and B. D. Preston. 2013. Emergence of DNA polymerase ε antimutators that escape error-induced extinction in yeast. Genetics 193: 751–770.

Yang, S., Y. Yuan, L. Wang, J. Li, W. Wang, H. Liu, J. Q. Chen, L. D. Hurst, and D. Tian. 2012. Great majority of recombination events in *Arabidopsis* are gene conversion events. Proc. Natl. Acad. Sci. USA 109: 20992–20997.

Yin, J., M. I. Jordan, and Y. S. Song. 2009. Joint estimation of gene conversion rates and mean conversion tract lengths from population SNP data. Bioinformatics 25: i231–i239.

CHAPTER 5

Evolution as a Population-Genetic Process

With knowledge on rates of mutation, recombination, and random genetic drift in hand, we now consider how the magnitudes of these population-genetic features dictate the paths that are open versus closed to evolutionary exploitation in various phylogenetic lineages. Because historical contingencies exist throughout the Tree of Life, we cannot expect to derive from first principles the evolutionary source of every detail of cellular diversification. We can, however, use established theory to address more general issues, such as the degree of attainable molecular refinement, rates of transition from one state to another, and the degree to which nonadaptive processes (mutation and random genetic drift) contribute to phylogenetic diversification.

Much of the field of evolutionary theory is concerned with the mechanisms maintaining genetic variation within populations, as this ultimately dictates various aspects of the short-term response to selection (Charlesworth and Charlesworth 2010; Walsh and Lynch 2018). Here, however, we are primarily concerned with long-term patterns of phylogenetic diversification, with a focus on mean phenotypes. This still requires knowledge of the principles of population genetics, as evolutionary divergence is ultimately a manifestation of the accrual of genetic modifications at the population level. All evolutionary change initiates as a transient phase of genetic polymorphism, during which mutant alleles navigate the rough sea of random genetic drift, often being evaluated on diverse genetic backgrounds, with some paths being more accessible to natural selection than others.

The primary goals of this most technical of chapters are to summarize some of the more general challenges to understanding how evolutionary change is accomplished and to endow the reader with an appreciation for why the population-genetic details matter. With a specific focus on the ways in which selection acts to promote novel adaptive changes, emphasis is placed on how the efficiency of selection is compromised or enhanced in different population-genetic environments, sometimes in counter-intuitive ways. Special attention is paid to the ways in which evolutionary rates and outcomes are expected to vary with the effective sizes of populations (N_e).

Most of the theory presented here is discussed in a generic way, focusing for example, on a mutation with selective advantage or disadvantage s, with no connection to the actual underlying trait(s). Such an approach is a necessary prelude to more explicit exploration of particular traits where genotypes can be directly connected to phenotypes and then to fitness. Specific examples to be presented in later chapters include the evolution of protein-protein interfaces, the coevolution of transcription factors and their binding sites, and the evolution of maximum growth-rate potential. Although the rudimentary level of presentation here may be disappointing to high-level theoreticians, the goal is not to overwhelm the reader with a litany of equations and formal derivations (some of which can be found in the Foundations sections), but to facilitate understanding as to how population-genetic theory can help transform comparative cell biology into evolutionary cell biology.

Evolutionary Cell Biology. Michael Lynch, Oxford University Press. © Michael Lynch (2024). DOI: 10.1093/oso/9780192847287.003.0005

The Perils of the Adaptive Paradigm

Ever since Darwin, most discussions with any connection to evolutionary thought start with the implicit assumption that all organismal traits are products of natural selection. Under this extreme view, the genetic details are irrelevant, as the belief is that natural selection is capable of finding the optimal solution to any environmental challenge, extinction being the alternative. Such logic underlies virtually every study in the field of evolutionary ecology. Closer to the subject material herein, a massive number of papers in cell biology end with a speculative paragraph on why the trait being studied (and its sometimes arcane structure) must have been refined to its current state by selective forces, almost always in the absence of any direct evidence or even an awareness that such evidence ought to be sought.

An appreciation for the power of natural selection is one of the great advances of the life sciences over the past century. However, problems arise when the wand of selection is deemed to be the only mechanism relevant to evolutionary change, as this eliminates any hope for broader understanding of evolutionary processes, and often leads to false narratives. Starting with the conclusion that the phenotype under investigation is a necessary product of natural selection, the only remaining challenge is to identify the actual agent of selection. If one hypothesis fails, one moves on to another possibility, but always with unwavering certainty that selection must somehow be involved. Many biologists have spent entire careers wandering down such paths in search of an adaptive explanation for a particular biological feature, and sometimes never finding it.

This is not to say that optimization thinking has completely mislead us with respect to the evolution of alternative behavioral and/or life-history strategies, and theory certainly explains how natural selection is often able to bring most phenotypes within the vicinity of adaptive peaks (Fisher 1930; Rice et al. 2015). However, as mentioned in Chapter 4, the evolutionary outcomes achievable by natural selection depend critically on levels of mutation, drift, and recombination. Moreover, owing to the stochastic nature of these processes, even

under constant selective pressures, the phenotypic states of populations are expected to wander over time. Finally, depending on the bias and granularity of mutational effects, the most common phenotype need not even be the optimum for a given environment.

In support of these ideas, this chapter closes with an overview of the concept of long-term steady-state distributions of mean phenotypes. This entrée also provides a more formal analysis of the drift-barrier hypothesis introduced in Chapter 4 in the context of mutation-rate evolution, highlighting the conditions under which traits are expected to exhibit gradients in performance scaling with N_e. The points made here are not just arcane technical nuances. Subsequent chapters on particular cellular traits show that, owing to the population-genetic and molecular features of biology, many aspects of evolution at the cellular level are best understood not by invoking the all-powerful guiding hand of natural selection, but by appreciating the factors that limit the reach of selection.

The Fitness Effects of New Mutations

Before proceeding with the theory, an overview of the fitness effects of mutations is necessary, as this defines the landscape and potential granularity of evolutionary change accessible to natural selection. As pointed out in Chapter 4, mutations with an absolute selective advantage/disadvantage (s) much smaller than the reciprocal of the effective population size ($1/N_e$) are essentially invisible to the eyes of natural selection. Thus, for random genetic drift to impose significantly different barriers to the evolution of a trait in different lineages, there must be a substantial pool of mutations with small enough deleterious effects that they can drift to fixation in species with small but not large N_e. Such differentiation is further facilitated if mutations are biased in the negative direction. As we discuss next, numerous lines of evidence are consistent with both conditions.

First, studies of serially bottlenecked mutation-accumulation (MA) lines across diverse species consistently reveal a slow decline in growth rate and other fitness traits (Keightley and Eyre-Walker

1999; Lynch et al. 1999; Baer et al. 2007; Katju and Bergthorsson 2019). Such experiments start with a set of isogenic lines, which are then maintained for large numbers of generations by propagation of just one (clone or selfer) or two (full-sib) individuals per generation. With an N_e this small, natural selection is incapable of promoting or removing individual mutations with fitness effects < 25%, so the data from these experiments are in full accord with a strong bias of mutations towards deleterious effects. Furthermore, statistical inferences based on the distribution of MA-line performances imply highly skewed distributions of fitness effects. The modes for such distributions are often indistinguishable from zero, with the bulk of mutations having absolute effects < 1% (almost all negative), although mean deleterious effects can sometimes be as high as 1–10% owing to the presence of rare mutations with large negative effects (Keightley 1994; Robert et al. 2018; Böndel et al. 2019).

Second, indirect inferences derived from allele-frequency distributions in natural populations of diverse multicellular species commonly suggest that 10–50% of mutations have deleterious effects smaller than 10^{-5}, with the inferred distribution sometimes being bimodal, but always with one mode being near 0.0 (Keightley and Eyre-Walker 2007; Bataillon and Bailey 2014; Huber et al. 2015; Kim et al. 2017; Lynch et al. 2017; Booker and Keightley 2018; Johri et al. 2020). Because many of these studies focus only on amino-acid substitutions in protein-coding genes, the full distributions of effects (which would include the often much more numerous sites in introns, intergenic DNA, and third positions in codons) can be expected to be even more skewed towards near-zero values. Substantial evidence also supports the idea that even silent (synonymous) sites in protein-coding genes are subject to weak selection for usage of particular nucleotides, such that the scaled strength of selection (ratio of the power of selection relative to drift) favoring G/C content is typically in the range of $N_e s = 0.1$ to 1.0, i.e., on the edge of the domain of effective neutrality (Long et al. 2017).

Third, the costs of some kinds of mutations can be derived from first principles. For example, from a knowledge of the total energy budget of a cell

and the biosynthetic costs of its building blocks, it is possible to estimate the fractional reduction in cell growth rates resulting from various kinds of mutations (Chapter 17; Foundations 5.1). Bioenergetic considerations of the costs of small nucleotide insertions, which typically comprise ~10% of *de novo* mutations (Sung et al. 2016), imply fractional reductions in fitness far below 10^{-5} (Lynch and Marinov 2015). Likewise, from both the standpoint of biosynthetic expenditures and elemental (e.g., C, N, or S) composition (Chapter 18), the costs of using alternative amino acids or nucleotides imply that s for such substitutions is generally $\ll 10^{-5}$ unless there are additional functional consequences. Broad experimental surveys of single amino-acid substitutions in a range of proteins generally imply that most such changes influence protein performance by < 1%, with a large class indistinguishable from zero, and a secondary peak with very substantial effects (Chapter 12). Single-residue changes in protein-protein interfaces or DNA binding sites are also expected to have small effects (Chapters 13 and 21). This diverse set of observations makes clear that there are large fractions of deleterious mutations with fitness effects $< 10^{-5}$ in all species.

Drawing from these observations, as well as other considerations regarding genome content outlined in Foundations 5.1, the genome-wide distribution of fitness effects (DFE) of new mutations is expected to take on different forms in large- and small-N_e species (Figure 5.1). In the former, prokaryotes in particular, > 95% of the genome consists of coding DNA, ~75% of which constitutes amino-acid substitution sites. A small fraction of silent-site mutations that retain identical nucleotides but alter their strand associations (e.g., A:T vs. T:A) may be absolutely neutral, but otherwise the lower bound of the fitness effect of a silent-site mutation is based on the differential cost of (C/G) versus (A/T) nucleotide pairs relative to the total cost of building a cell (in ATP equivalents). An expected central peak in the idealized prokaryotic DFE in Figure 5.1 is associated with the substitution of amino acids with biosynthetic costs at sites that are otherwise insensitive to amino-acid identity, whereas the potential lower peak to the right is associated with amino-acid altering mutations with additional functional effects on protein performance.

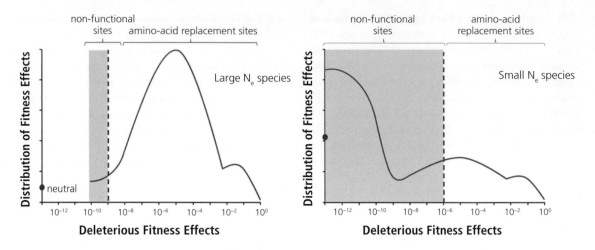

Figure 5.1 Idealized representations of the distributions of fitness effects (DFEs) of new mutations in large-N_e species (with genomes consisting primarily of coding DNA) versus small-N_e species (with genomes consisting primarily of non-coding DNA), based on arguments presented in Foundations 5.1. Although it is clear that most mutations have relatively small effects, there remain substantial uncertainties about the precise form of the DFE, and the absolute placements of the peaks are somewhat arbitrary. The small peak to the right is meant to represent a class of mutations of relatively large functional effects, as expected with amino-acid substitutions in catalytic sites of proteins, premature stop codons, and gene deletions. Other amino-acid-altering mutations may have minor effects on protein function, while nonetheless imposing small biosynthetic costs. The distribution of beneficial effects is not shown, but in total this is expected to be < 1% of that for deleterious mutations, again with the mode being near zero. The large blue dots on the left denote the fraction of absolutely neutral mutations, whereas a deleterious effect of $10^0 = 1$ denotes a lethal mutation. Note that the curves in both examples are normalized to have constant areas, as they are relative frequency distributions; if these were scaled to the absolute numbers of mutations, the total heights of the abundance distributions would depend on the mutation rate per nucleotide site, which tends to be lowest in unicellular eukaryotes, somewhat higher in prokaryotes, and highest in multicellular eukaryotes (Chapter 4). Hypothetical positions of the drift barrier, $\simeq 1/N_e$, denoted by the vertical red lines, are given for $N_e = 10^9$ and 10^6 (left and right panels, respectively); mutations to the left of this line cannot be perceived by natural selection.

An altered form of the DFE is expected in eukaryotic species, owing to the dramatic expansion of genome content. In many multicellular species, > 95% of nucleotide sites are non-coding and non-functional (with mutational effects limited to small differential biosynthetic costs of nucleotides) with the genome structures of unicellular eukaryotes typically being intermediate between this extreme and that of prokaryotes (Lynch 2007). Combined with a several order-of-magnitude increase in the cost of construction of large eukaryotic cells relative to prokaryotes, which dilutes the relative biosynthetic costs of individual mutations (Foundations 5.1), this shifts the DFE to the left and creates a more L-shaped form, owing to the elevated number of non-functional sites dominating the overall distribution (Figure 5.1).

Taking all of these observations into consideration, the existence of large pools of mutations with deleterious effects small enough to allow fixation in some lineages but large enough to ensure removal by selection in others is not in doubt. This supports the contention that traits under identical selection pressures in species with different N_e will evolve different mean phenotypes, as discussed in the context of mutation-rate evolution in Chapter 4. Nonetheless, significant caveats remain.

First, statistical and empirical limitations prevent us from knowing with certainty the full form of the distribution of mutations of small effects, e.g., the fractions of mutations with effects $< 10^{-5}$, $< 10^{-6}$, $< 10^{-7}$, $< 10^{-8}$, etc. Thus, the exact form of these regions of the hypothetical DFEs in Figure 5.1 should not be taken too seriously. This is a significant concern. As we know that N_e ranges from $\sim 10^4$ to 10^9 (Chapter 4), these are the mutations that can be utilized/purged in some lineages but not in others. Second, and most importantly, although most of the observations noted above address the general fitness properties of random mutations, they are generally disconnected from the actual cellular traits that we wish to explore. Further genetic dissection is essential to inform us as to the precise molecular targets and phenotypic effects of fitness-altering mutations.

The Classical Model of Sequential Fixation

A common starting point for thinking about the temporal dynamics of adaptive evolution invokes the case in which a trait is under persistent directional selection, with the pace of evolution being slow enough that each consecutive adaptive mutation is fixed before the next beneficial mutation destined to fix arises. In principle, such a scenario can exist if the supply of adaptive mutations is quite limited owing to either a relatively small population size, a low mutation rate to beneficial variants, or both. In this limiting situation, recombination is irrelevant because no two loci are ever simultaneously segregating polymorphisms at meaningful frequencies.

This sequential model of molecular evolution (sometimes also called the strong selection/weak mutation model or the origin-fixation model; Gillespie 1983; McCandlish and Stoltzfus 2014) may only rarely represent reality as it assumes a constant march towards higher fitness. Nonetheless, it serves as a useful heuristic for addressing issues concerning the limits to rates of adaptive evolution and how they might scale with population size.

Under this model, the long-term rate of adaptation is equal to the product of three terms: the rate of introduction of beneficial mutations, the fixation probability of such mutations, and their fitness effects.

1) The population-wide rate of origin of new beneficial mutations is NU_b for haploids and $2NU_b$ for diploids, where U_b is the mutation rate (per generation) to beneficial alleles. Depending on the focus, U_b can represent a single nucleotide site, a single gene, or an entire haploid genome.

2) The probability of fixation of a new beneficial mutation is $\simeq 2s(N_e/N)$, where N and N_e are the actual and effective population sizes, and s is the selective advantage relative to the ancestral allele (Chapter 4). As N_e/N is almost always less than one, this ratio can be thought of as the efficiency with which selection promotes new beneficial mutations, with the maximum fixation rate being $2s$.

3) The increase in fitness per fixation is s for haploids, but $2s$ for diploids. (Note that for diploids, it is assumed here that mutational effects are additive, with heterozygotes having a fitness advantage $1 + s$ intermediate to that for the two homozygotes, 1 and $1 + 2s$).

The expected rate of evolution in terms of fixation numbers is the product of the first two terms,

$$r_e = NU_b \cdot 2s(N_e/N) = 2N_eU_bs, \qquad (5.1a)$$

for haploids, and twice this for diploids. The increase in fitness is then equal to

$$\Delta W = r_e \cdot s = 2N_eU_bs^2, \qquad (5.1b)$$

for haploids, and quadruple this for diploids. (Note that this difference between haploids and diploids is merely a function of how the fitness effects are annotated in diploids. If s is redefined so that mutant heterozygotes and homozygotes have fitnesses $1 + (s/2)$ and $1 + s$, respectively, then Equations 5.1a,b hold for diploids. To keep this simple, the remaining discussion will rely on these expressions.)

This model for the speed limit to the rate of adaptation is idealized in many ways, as it assumes long-term persistent selection in one direction, constant beneficial mutation rates, and constant effect sizes of fixed mutations. Nonetheless, this simple approach highlights the key roles that the individual population-genetic parameters play in dictating the potential for evolutionary change. For example, Equations 5.1a,b suggest that, all other things being equal, the rate of adaptive evolution should scale linearly with the effective population size (which need not be equivalent to the absolute population size; Chapter 4) and with the genome-wide beneficial mutation rate.

However, there is room for caution in interpreting expressions like Equations 5.1a,b. First, the conditions under which mutations are likely to fix in a stepwise manner are limited. Sequential fixation requires that the average time between fixations (the inverse of $r_e = 2N_eU_bs$ for haploids) be greater than the mean time required for each mutation to fix, which $\simeq (2/s)\ln(N)$ generations for a haploid population (Walsh and Lynch 2018, Equation 8.4c). It then follows that for sequential fixation to be the rule, $2N_eU_b$ must be smaller than $1/[2\ln(N)]$. Because $\ln(N)$ falls in the range of 9 to 46 for $N = 10^4$ to 10^{20}, as a first-order approximation, the sequential model will only hold if the effective number of beneficial mutations arising per generation is $N_eU_b < 0.01$, i.e., if no more than one beneficial mutation for the trait arises per 100 generations at the effective-population size level.

How likely is this condition to be met? Recall from Chapter 4 that the product of $2N_e$ and the mutation rate per nucleotide site per generation (u) generally falls in the range of $2N_eu = 10^{-3}$ to 10^{-2}. Multiplying $2N_eu$ by the number of selected sites in a chromosomal region of interest and the fraction of mutations that are beneficial converts this quantity to $2N_eU_b$. With a moderately sized region of 10^5 fitness-relevant sites and just 0.01% of mutations being beneficial, then $2N_eU_b$ for such a region would be in the range of 10^{-2} to 10^{-1}. As this is on the edge of the strict cut-off for the sequential model, it follows that the latter cannot be assumed to be generally valid unless the fraction of beneficial mutations is very small. These issues are evaluated in great detail in Weissman and Barton (2012), Weissman and Hallatschek (2014), and Lynch (2020).

The primary reason for concern with violations of the sequential model is that simultaneously segregating mutations (both beneficial and deleterious) at different nucleotide sites interfere with each other in the selection process (Chapter 4), diminishing the probabilities of fixation of the good and purging of the bad (Campos and Wahl 2009, 2010; Frenkel et al. 2014; Pénisson et al. 2017). If tightly linked to a segregating deleterious mutation, a beneficial mutation will not experience its full intrinsic advantage, and in some cases will be completely overshadowed by the linked background load (Good and Desai 2014).

The central point here is that, although the effects of background variation do not alter the expectation that the rate of adaptive evolution will scale positively with N_e, the gradient of scaling is expected to decline substantially with increasing N_e. Theoretical work suggests that selective interference reduces the scaling of the rate of adaptation from linear with N_e in the sequential domain to as weakly as logarithmic in N_e beyond that point (Weissman and Barton 2012; Neher 2013; Weissman and Hallatschek 2014).

Selective interference is commonly observed in long-term laboratory evolution experiments involving microbes. For example, Figure 5.2 illustrates the trajectories of genome-wide mutant-allele frequencies in three replicate populations of E. coli grown in just 10 ml of medium. Over a period of 60,000 generations, these populations experienced an average 30% increase in fitness in the culture conditions, albeit at a diminishing rate. The norm in these kinds of experiments is for many mutations to be simultaneously polymorphic, and more often than not, groups of mutations increase (and sometimes decline) in a coordinated manner. As noted earlier for asexual populations of yeast (Figure 4.9), this is a simple consequence of the clonal nature of the experimental populations, as a single, positively selected mutation driving to fixation sweeps along all linked 'passenger' mutations (some of which themselves have beneficial or deleterious fitness effects).

Figure 5.2 also shows examples of mutations reaching very high frequencies in a short period of time, followed by a subsequent decline to 0.0 as other, fitter mutant clones take over. In one population, two major clones, each containing multiple mutations, appear to reach equilibrium frequencies, with neither going to fixation (middle panel); this may be a result of some form of frequency-dependent selection, with each clone providing a metabolic product beneficial to the other (Behringer et al. 2018). In another case, there is a massive accumulation of mutations near the midpoint of the experiment (lower panel), owing to

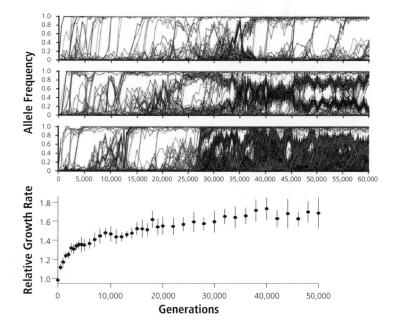

Figure 5.2 Upper panels: Allele-frequency trajectories in three replicate cultures of *E. coli* grown in 10-mL cultures serially diluted 100-fold on a daily basis. To estimate allele frequencies, the complete genomes of each mixed culture were subject to pooled-population sequencing, to an average 50 × depth of coverage, every 500 generations over a 60,000-generation period. Each individual line in the plots denotes a mutation that arose to frequency 0.1 on at least one occasion. Results are shown for three replicate populations. All cultures were genetically identical and monomorphic at time zero. From Good et al. (2017). **Lower panel:** Evolutionary trajectory of mean fitness (cell-division rate relative to a time-zero control) averaged over multiple replicate populations. From Wiser et al. (2013).

the appearance of a mutator strain, which may have hitchhiked to fixation in linkage with a beneficial mutation that it promoted.

Compensating for these constraining effects from selective interference is the fact that a larger fraction of beneficial mutations is exploitable in larger populations. Owing to the fact that efficient selection requires $|N_e s| > 1$ (Chapter 4), populations with larger N_e have access to mutations with smaller s. However, although this expansion in the pool of available beneficial mutations will further tip the balance in favor of higher rates of evolution in larger populations, from Equation 5.1 we see that the contribution of beneficial mutations to increases in fitness scales with the square of the selective advantage, s^2. Thus, because average s is typically $\ll 1$, broadening the window of mutational availability will have a less than linear effect on the rate of adaptation unless the pool of beneficial mutations

is very strongly skewed towards those with small fitness effects.

Surveys of the clonal dynamics in long-term evolution experiments illustrate these principles in a more general way (Nguyen Ba et al. 2019). Clonal competition can lead to a 'rich get richer' scenario, whereby currently successful clones ride a wave to high abundance, and in doing so can more efficiently utilize additional beneficial mutations arising during their clonal expansion. On the other hand, currently low-fit clones can occasionally vault their way to success via a fortuitous secondary mutation, but only if the latter has a very high fitness effect.

With respect to the scaling of the rate of adaptation with population size, there are still additional issues to consider. Recall from Chapter 4 that, at the interspecies level, there is a nearly inverse relationship between N_e and the mutation rate (u). From

Equation 5.1b, such a compensatory effect would yield near-independence of the rate of adaptation and N_e, although the potential expansion of the window of exploitable mutations can modify this result.

On the other hand, there is the matter of timescale. The previous derivations consider the rate of adaptive evolution on a per-generation basis. However, smaller organisms (with large N_e) typically have shorter generation times, which will elevate the rate of evolution on an absolute timescale. If both mutation rates and generation lengths were to scale inversely with N_e, the two effects would cancel out, and the scaling of rates of evolution with N_e on an absolute timescale would follow the expectations of Equations 5.1a,b.

Taken together, for simple adaptations involving mutations with additive effects, the above observations point toward the potential rate of evolutionary change (per absolute time unit) being greater for organisms with small size, short generation times, and large N_e. However, whether the scaling of adaptive evolutionary potential with N_e is sublinear, linear, or superlinear remains uncertain. As we discuss in the following section, for changes involving interactions among loci, the expected scaling of evolutionary rates with N_e can deviate from the above expectations.

Before proceeding, one final point merits discussion. Although the quantity $r_e = 2N_e U_b s$ is a measure of the expected rate of long-term fixations under the ideal model, because mutation and fixation events are stochastic, considerable variation is expected around this expectation. For a Poisson process, where each rare event is independent of the others, the variance in the amount of long-term change among replicate populations is equal to the expectation. If, for example, the time interval is such that one beneficial mutation is expected per lineage on average, the probability that one event actually accrues is just 0.368, but the probability of no fixations is also 0.368, of two is 0.182, of three is 0.061, and of four or more is 0.021. The central point is that considerable variation in observed rates of evolution is expected among lineages exposed to identical selection pressures. Thus, it is risky to assume that such dispersion is evidence of adaptive differentiation, of the adaptive emergence of mutators/antimutators, or of intrinsic differences in evolutionary potential. Foundations 5.2 provides an even more dramatic example of how the magnitude of divergence among populations exposed to identical selection pressures can exceed that expected under neutral drift.

Vaulting Barriers to More Complex Adaptations

To this point, we have been assuming that the fitness effect of an allele is independent of the genetic background on which it resides. Under this view, Equations 5.1a,b provide the simplest possible model for the rate of adaptation by new mutations, as prior fixations have no bearing on subsequent events. However, this assumption can be violated for at least two reasons. First, for the case of stabilizing selection for an intermediate optimum phenotype or directional selection up the edge of a fitness plateau, each fixed mutation will alter the selection coefficients of future mutations by moving the mean phenotype closer to the optimal state, reducing the capacity for further improvement.

Second, when mutations have epistatic effects (i.e., interact in a non-additive fashion), the possibility exists for neutral or even deleterious mutations to become beneficial in certain contexts (Figure 5.3). Multilocus traits exhibiting the latter types of genetic behavior are referred to here as complex adaptations, as the paths for their evolution and the rapidity with which they are acquired are much less straightforward than under conditions of additive fitness effects. Mutations that pave the way for an increase in fitness via secondary mutations that otherwise would be deleterious are sometimes referred to as enabling mutations (Tóth-Petróczy and Tawfik 2013).

One broad category of complex-trait evolution involves compensatory mutations, wherein single mutations at either of two loci cause a fitness reduction, while their joint appearance can restore or even elevate fitness beyond the ancestral state. Such epistatic interactions play a prominent role in Wright's (1931, 1932) shifting-balance theory of evolution, which postulates that adaptive valleys between fitness peaks are typically traversed in small subpopulations that facilitate drift through a deleterious intermediate state in an effectively neutral fashion. Compensatory mutations play important roles in protein-sequence evolution (Chapter 12), in the composition of nucleotides in the stems of RNA molecules (Stephan and Kirby 1993; Kondrashov et al. 2002; Kulathinal et al. 2004; Azevedo et al. 2006; DePristo et al. 2007; Breen et al. 2012; Wu et al. 2016), and in the coevolution of complexes consisting of components jointly encoded in nuclear and organelle genomes (Chapter 23).

Deciphering the molecular paths of establishment of complex adaptations from observations on evolutionary endpoints is challenging. However, clear examples do exist. For example, in a long-term (> 40,000 generations) evolution experiment with *E. coli* selected for growth in flasks on a defined medium, the novel ability to utilize citrate as a carbon source emerged in one of twelve cultures (Blount et al. 2008; Quandt et al. 2014). Drawing

from the historical record of evolution by resurrecting frozen samples, it was found that a weak variant for citrate utilization arising from a promoter-region mutation provided a potential mutational target for further refinement of the trait. While this initial mutation was still infrequent in the population (and possibly effectively neutral), a linked mutation appearing at a second locus conveyed a much greater ability to take up citrate, conferring a substantial increase in fitness that drove the double mutant to fixation.

Sequential fixation versus stochastic tunnelling

Even when only two loci are involved, ascertaining the population-genetic conditions under which complex adaptations are likely to occur brings in challenges not encountered with the ideal single-site model. This is because, unlike the situation in

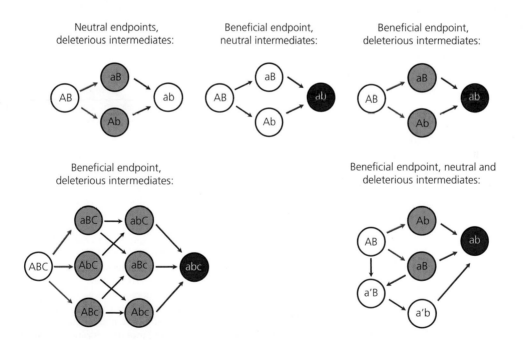

Figure 5.3 Some possible routes to the establishment of adaptations involving two or more mutations at different genetic loci (which might be two nucleotide sites within single loci). In each case, the starting genotype is on the left. Mutations that are deleterious with respect to their ancestral allele are denoted in blue, whereas neutral changes are in white, and beneficial changes are denoted in red. Intermediate states can also be beneficial, although such instances are not shown. The lower left denotes a case involving three loci, with all intermediate states being deleterious, and the final three-mutation genotype being beneficial. The lower right illustrates a situation in which the first site has three alternative allelic states, with one derived allele (a) being deleterious and the other (a') providing a three-step neutral path to the final adaptation.

which a single mutation fixes at a rate depending only on its own initial frequency, the success of a mutation involved in an interlocus interaction depends on the frequency of alleles at the interacting locus, on the fitnesses associated with all possible multi-locus genotypes, and on the recombination rate between the two loci.

The focus here is on the rate of establishment of a complex adaptation, defined to be the inverse of the expected arrival time of the ultimate multi-mutation configuration destined to be fixed in the population. Although this excludes the additional time required for fixation, the latter is generally considerably smaller than the time to establishment, and ignoring it does not influence the following conclusions.

Population size alone can dictate the kinds of evolutionary pathways that are open to the establishment of complex traits (Figure 5.4). For populations of sufficiently small size, the path toward adaptation almost always involves sequential fixations of the contributing mutations, owing to the extreme rarity of occasions in which multiple mutations are simultaneously segregating at key sites. This issue was explored in the previous section for the situation in which mutational effects are independent, but we now consider an adaptation requiring two genetic changes.

The conditions under which a population is constrained to acquiring the two-site adaptation in two sequential steps is a function of three factors. Assuming a first-step (neutral) mutation is destined to fixation, its mean time to fixation is $4N_e$ generations in a diploid population (Kimura 1983). Because the frequency increases from near 0.0 to 1.0, the average frequency of this mutation during its sojourn to fixation is 0.5, and this implies the presence of an average $0.5 \cdot 2N = N$ copies during the polymorphic phase. Letting μ_2 be the rate of second-step mutations linked to the first-step background, from the theory outlined earlier the rate of appearance of second-step mutations destined to fix is then $\mu_2 \cdot 2s(N_e/N)$ per mutational target per generation. The product of these three quantities gives the approximate probability of a secondary mutation arising on a segregating first-step mutation background (and also being destined to fix) of $4N_e \cdot N \cdot \mu_2 \cdot 2s(N_e/N) = 8N_e^2\mu_2 s$. This shows that

there is a negligible chance of arrival of a successful secondary mutation before fixation of the first if $N_e \ll (8\mu_2 s)^{-1/2}$. With $\mu_2 = 10^{-9}$ and $s = 10^{-4}$, for example, the critical effective population size is $\simeq 10^6$. For N_e below this threshold value, selection is restricted to exploring the fitness landscape by sequential mutational steps.

In contrast, in large populations, key secondary (and even tertiary) mutations can arise prior to the fixation of earlier-step contributors (Figure 5.4). This raises the possibility of the joint, simultaneous fixation of combinations of mutations as single haplotypes without any enabling mutation having been previously common in isolation as heterozygotes. For example, a conditionally beneficial secondary

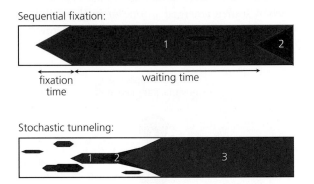

Figure 5.4 Origin of a complex adaptation involving three mutations (blue (1), red (2), and green (3)) in small and large populations. The vertical axis denotes the frequency in the population, with fixation implied when a vertical line through the temporal profile is all of one colour. **Top:** Here, the population size is small enough and mutation sufficiently weak that each contributing mutation goes to fixation prior to the appearance of the next mutation destined to fix. This leads to sequential fixation of single-step changes, with potentially long waiting times between fixation events. As illustrated, a few secondary mutations may arise and go extinct (closed red polygons) before one reaches a high enough frequency to be susceptible to fixation. In the illustrated case, the final (green) mutation has not arisen yet. **Bottom:** In large populations, allelic variants will typically be segregating at low frequencies at all times, even if deleterious, owing to the recurrent input by mutation. Secondary and tertiary mutations can then arise on such backgrounds, occasionally generating a beneficial combination that leads to fixation of the entire linked haplotype by positive selection.

mutation may arise in linkage with a low-frequency deleterious first-step mutation, with the joint fixation of the double-mutation haplotype in effect rescuing the first-step mutation otherwise destined to be lost. Such a process, often referred to as stochastic tunnelling (Komarova et al. 2003; Iwasa et al. 2004), provides a smooth route for the establishment of complex adaptations, allowing large populations to explore the fitness surface more broadly than possible by single-step mutations. Most notably, stochastic tunnelling allows progression through intermediate deleterious alleles without the population ever experiencing the transient decline in fitness that would necessarily occur with sequential fixation (Gillespie 1984; Weinreich and Chao 2005; Gokhale et al. 2009; Weissman et al. 2009, 2010; Lynch and Abegg 2010; Lynch 2010). This shows how deleterious mutations with conditional effects can play a central role in evolutionary diversification.

The following analyses focus on the domain in which stochastic tunnelling dominates, i.e., populations of moderate to very large size, as will often be the case in single-celled organisms. However, before proceeding, note that the theory explored below assumes that each mutation contributing to a final adaptation arises independently of all others. Recall from Chapter 4 that mutations sometimes arise in clusters, which means that adaptations involving two or three, and perhaps even more, site-specific mutations will occasionally arise spontaneously in a single individual on realistic timescales. In such cases, assuming negligible recombination between these sites, the rate of fixation of the mutant haplotype follows directly from the single-mutation theory noted in Chapter 4, Equations 4.1a-c. That is, we simply inquire as to the probability of fixation of a newly arisen beneficial haplotype, $\simeq 2s(N_e/N)$.

Two-locus transitions

We start with a simple selection scenario, first explored by Kimura (1985), in which haplotypes Ab and aB have reduced but equivalent fitness $(1 - s_d)$ relative to AB and ab, both of which have fitnesses of 1.0 (Figure 5.3, upper left). In this case, two-step transitions between pure population states of AB and ab render no gain in fitness, but do involve

an intermediate deleterious genotype. Initially, the two sites will be assumed to be completely linked, and μ_d and μ_b will denote the rate of mutation to first (potentially deleterious) and second (potentially conditionally beneficial) step variants. (If $s_d = s_b = 0$, the intermediate states are neutral).

As an explicit example, AB and ab might represent beneficial pairs of amino acids involved in protein folding or stability, with Ab and aB representing non-matching combinations. The Watson–Crick pairs in the stems of ribosomal RNAs provide another compelling example. Although the overall stem/loop structure of rRNAs is highly conserved across species, complementary nucleotide pairs in stems often have different states in different species. Barring a rare double mutation, such shifts require passage through an intermediate deleterious state, e.g., A:T $\rightarrow$ A:C $\rightarrow$ G:C. Provided the overall secondary structure is maintained, which is presumably essential for proper ribosome assembly, rRNAs from different bacterial species with up to 20% divergence can substitute for each other with only small effects on fitness (Kitahara et al. 2012).

Starting with a population in state AB, we wish to determine the mean time for the population to reach an alternative state of fixation at both loci, i.e., with haplotype ab. Mutation will recurrently introduce deleterious Ab and aB haplotypes, and provided the strength of selection exceeds that of mutation, both the aB and Ab haplotypes will then be expected to have steady-state frequencies of μ_d/s_d. This condition results from the balance between the rates of mutational input μ_d (from the ancestral AB haplotype) and selective removal s_d (Walsh and Lynch 2018).

In a diploid population, there will be a steady-state total of $2N \cdot 2 \cdot \mu_d/s_d$ of these low-frequency aB and Ab haplotypes, as the $2N$ chromosomes each contain two loci with expected deleterious-allele frequency μ_d/s_d. These intermediate types then serve as reliable substrates for secondary mutations to the ab type, which arise at rate μ_b from each intermediate background. However, even though the ab type has an advantage over its deleterious parental Ab or aB haplotype, ab mutations fix in an essentially neutral fashion with probability $1/(2N)$. This follows under the assumption that $\mu_d/s_d \ll 1$, which ensures that almost all resident haplotypes are of type *AB*, which have equivalent fitness to *ab*. Thus, the rate of establishment of the *ab* type by stochastic tunnelling from *AB* is

$$r_e \simeq \frac{4N\mu_d}{s_d} \cdot \mu_b \cdot \frac{1}{2N} = \frac{2\mu_d\mu_b}{s_d} \qquad (5.2)$$

(Gillespie 1984; Stephan 1996). This rate is independent of population size (provided the conditions for selection-mutation balance, $4N_e s_d \gg 1$, are met, and ignoring for the moment any population-size dependence of the mutation rate).

Now suppose that the secondary mutation is advantageous, such that the fitness of the *AB* and *ab* haplotypes are 1 and $1 + s_b$, respectively (Figure 5.3, upper right). Modifying Equation 5.2 to account for the fact that the fixation probability of the double mutant is $\simeq 2s_b(N_e/N)$ leads to

$$r_e \simeq \frac{4N\mu_d}{s_d} \cdot \mu_b \cdot 2s_b(N_e/N) = \frac{8N_e\mu_d\mu_b s_b}{s_d}. \qquad (5.3)$$

As in the case of selectively equivalent end states, the rate of establishment scales with the square of the mutation rate, but now also with the strength of positive selection scaled relative to that of drift, $4N_e s_b$.

Finally, consider the special situation in which first-step mutations are effectively neutral, so that again there should be a non-zero probability of their being present at some low frequency at the outset. Denoting this initial frequency as p_0, and simply substituting this for μ_d/s_d in Equation 5.3 yields

$$r_e = 4Np_0 \cdot \mu_b \cdot 2s_b(N_e/N) = 8N_e p_0 \mu_b s_b. \qquad (5.4)$$

This is a potentially much higher rate of tunnelling than implied by Equations 5.2 and 5.3, owing to the expectation that mutations at neutral sites will have much higher expected frequencies than deleterious mutations. If, for example, the genetic substrate here is one particular nucleotide at a genomic site, assuming no mutational bias, the long-term average frequency of each of the four nucleotide types is 0.25 (even a small population harboring little polymorphism will be fixed for the enabling mutation with probability 0.25). This then yields $r_e = 2N_e\mu_b s_b$, which is just half the expectation for the single-site model, Equation 5.1a, because a potential starting nucleotide is present at the outset at each of the

two sites with probability 0.25. This kind of scenario would apply to the situation in which a codon requires two changes for a transition to a more beneficial amino acid.

What can be inferred about the likely scaling of two-site adaptations from these results? As noted above, a key issue is that the algebraic scaling implied by the preceding expressions is confounded by the non-independent behavior of the biological components. Again, suppose that there is an inverse scaling between the mutation rate per nucleotide site per generation and N_e across the Tree of Life, and consider Equation 5.2 for the rate of transition between two equivalent fitness states via deleterious intermediates. Although this expression suggests that the evolutionary rate scales with $\mu_d\mu_b$, independent of N_e, if both mutation rates scale inversely with N_e, the expectation is that r_e will actually scale inversely with N_e^2, i.e., $r_e \propto 2/(N_e^2 s_d)$. Extending this logic, Equation 5.3 for the rate of transition to a higher fitness state through deleterious intermediates would imply an inverse scaling of r_e with N_e. On the other hand, for the rate of transition to a higher fitness state through neutral intermediates, Equation 5.4 would be expected to be independent of N_e, as $N_e\mu_b$ is approximately constant.

As noted above for the single locus case, there is the added issue of generation length, which the reader can pursue further. The key point is that the rate of exploitation of various kinds of evolutionary pathways can depend critically on the nature of the adaptive change, on the effective population size, and on mutation rates.

More complex scenarios

While the previous analyses assume an evolutionary path to a final adaptation through just a single intermediate step, the routes to many molecular/cellular modifications may involve pathways through any number of mutations, e.g., Figure 5.3 (bottom). The rates of establishment under these alternative scenarios have been examined in some detail, again with the primary focus on situations in which the intermediate states are neutral or deleterious (Gokhale et al. 2009; Weissman et al. 2009; Lynch and Abegg 2010; Santiago 2015). The mathematics necessarily becomes more complex, and simple analytical approximations have been found in only a few cases.

For the case of deleterious intermediates, suppose that all haplotypes involving 1 to $d-1$ mutations are equally deleterious (with fitness $1 - s_d$), with the final mutation conferring an advantage s_b ($d = 2$ represents the two-step case noted earlier). First-step mutations then arise from mutation-free individuals at rate $2Nd\mu_d$, but owing to selection have an expected survivorship time of $1/s_d$ generations, during which period $d - 2$ additional intermediate-step mutations must be acquired, followed by the appearance of a final-step mutation destined to fixation. This leads to a rate of establishment via tunnelling of

$$r_e \simeq 4N_e d! (\mu_d/s_d)^{d-1} \mu_b s_b, \qquad (5.5)$$

which reduces to Equation 5.3 when $d = 2$. Here, we see that the rate of establishment scales with the dth power of the mutation rate, owing to the limited opportunities for secondary mutations during the short sojourn times of deleterious mutations. Thus, the acquisition of a novel adaptation involving multiple, deleterious intermediate steps is a very low probability event, potentially diminishingly so for populations with large N_e, as assuming an inverse relationship between the mutation rate and N_e, the expected scaling is now with N_e^{1-d} (on a per-generation timescale).

For the case of neutral intermediates with d mutations required for the final adaptation (and the order of events assumed to be irrelevant), there is again the conceptual issue of the starting conditions. The worst-case scenario is the one in which all contributing mutations are absent at time zero, with establishment then requiring a series of nested tunnelling events. Consider first the special situation in which $d = 2$ (the two-step model introduced earlier) and enabling neutral mutations arise at rate μ_n per site (Figure 5.3, upper middle). With two sites in a diploid population, $4N\mu_n$ neutral first-step mutations arise per generation. To obtain the expected rate of tunnelling in this case, we also require the probability that tunnelling occurs within a descendant lineage of a first-step mutation before the latter becomes lost from the population. Assuming complete linkage, this probability is

approximately $\sqrt{2\mu_b s_b (N_e/N)}$ in large populations (Komarova et al. 2003; Iwasa et al. 2004; Weissman et al. 2009, 2010; Lynch and Abegg 2010). Thus, the rate of establishment via tunnelling becomes

$$r_{e2} \simeq 4N\mu_n \sqrt{2\mu_b s_b (N_e/N)} = 4\mu_n \sqrt{2\mu_b s_b N_e N}. \qquad (5.6a)$$

The key observation here is that, when the intermediate steps are neutral but the first-step mutations are initially absent from the population, the probability of tunnelling scales positively with the square root of the product of absolute and effective population sizes.

Now consider the case of $d = 3$ (with two neutral enabling mutations required before the final adaptation is assembled with a third mutation). In this situation, a secondary (conditionally neutral) mutation must arise on a haplotype lineage containing the first such mutation, and before being lost by drift, the still smaller two-mutation lineage must acquire a third mutation destined to fixation. This nested set of events expands Equation 5.6a to

$$r_{e3} = 3N \left(2\mu_n \sqrt{r_{e2}/(2N)} \right)$$
$$= 6N\mu_n \sqrt{2\mu_n \sqrt{2\mu_b s_b (N_e/N)}}. \qquad (5.6b)$$

Note that the first term is now $6N\mu_n$ because first-step mutations can arise at three diploid sites. The next step then initiates at either of the two remaining sites, bringing in the additional $2\mu_n$ term, with the final stage being initiated at the one remaining site and fixing at the usual rate for a single beneficial mutation. With $d = 4$, this expression expands one step further to

$$r_{e4} = 4N \left(2\mu_n \sqrt{r_{e3}/(2N)} \right)$$
$$= 8N\mu_n \sqrt{3\mu_n \sqrt{2\mu_n \sqrt{2\mu_b s_b (N_e/N)}}}, \qquad (5.6c)$$

and so on.

These results show that with neutral intermediates, the rate of establishment of complex adaptations can be much more rapid than expected under the naive assumption that independently arising mutations would lead to a scaling with the dth power of the mutation rate. Regardless of the number of sites involved with neutral intermediates, the rate of establishment by tunnelling scales with no more than the square of the mutation rate. Again, adhering to the empirical observation of an approximately inverse relationship between mutation rates and population size, this implies that the expected rate of establishment of beneficial features dependent on a constellation of new enabling mutations is nearly inversely proportional to N_e on a per-generation timescale; and if the generation length also scales inversely with N_e, the rate of establishment is N_e-independent.

These examples are just a small sample of the kinds of evolutionary pathways that can exist between two complex genotypes. In principle, multiple pathway types may connect two presumed endpoints, including those with mixtures of neutral and deleterious intermediates, different numbers of links (Figure 5.3, lower right), and so on. The kind of theory just outlined can be used to evaluate the relative probabilities of such alternative routes, as well as the possibility of becoming transiently trapped at points with suboptimal fitness, and subsequently back-tracking and exploring alternative paths (McCandlish 2018). Experimental-evolution studies with microbes are being increasingly exploited to evaluate these issues (e.g., Lind et al. 2019; Rodrigues and Shakhnovich 2019; Zheng et al. 2019). Although much work remains to be done, the key point is that complex adaptations can arise in large asexual populations (e.g., unicellular organisms) much more rapidly than one might imagine under the assumption of sequential fixation of interactive mutations.

Effects of recombination

All of the above analyses assume an absence of recombination. In the sequential-fixation regime, recombination can be ignored simply because multiple polymorphic sites are never present simultaneously. However, in the stochastic-tunnelling domain, opportunities may exist for both the recombinational creation and breakdown of optimal haplotypes. Examination of this problem with a broad class of models leads to the conclusion that recombination is likely to have either a minor

advantageous or a strong inhibitory effect on the *de novo* establishment of a complex adaptation (Higgs 1998; Lynch 2010; Weissman et al. 2010; Santiago 2015).

Consider, for example, the case of a two-site adaptation involving a deleterious intermediate, starting with a population fixed for the *ab* haplotype. The overall influence of recombination on the rate of establishment of the *AB* haplotype will then be a function of two opposing effects. On the one hand, the rate of origin of *AB* haplotypes by recombination between the two single-mutation haplotypes (*aB* and *Ab*) will be proportional to the rate of recombination between the sites (*c*). On the other hand, recombinational breakdown discounts the net selective advantage of resultant *AB* haplotypes from s_b to $s_b - c$. This occurs because in the early stages of *AB* establishment, *ab* haplotypes still predominate, and hence, are the primary partners in recombination events with *AB*, generating the maladaptive *Ab* and *aB* products. Thus, because the product $c(s_b - c)$ is maximized at $c = s_b/2$, two-site adaptations are expected to emerge most rapidly in chromosomal settings where the recombination rate is half the selective advantage of the final adaptation. This is a rather specific requirement, as the optimal recombination rate depends on the advantage of the final adaptation. Moreover, even in the case of neutral intermediates, the rate of establishment at the optimal recombination rate is generally enhanced by much less than an order of magnitude relative to the situation with complete linkage (Lynch 2010).

The situation is especially bleak when first-step mutations are deleterious. In this case, if the rate of recombination exceeds the selective advantage of the *AB* haplotype, recombination presents an extremely strong barrier to establishment of the *AB* haplotype, as most recombinant (*Ab* and *aB*) intermediates are removed by selection more rapidly than the *AB* type can be promoted. Thus, because the role that recombination plays in the origin of specific adaptations depends on both the selective advantage of the final product, the selective disadvantages of the intermediate states, and the physical distance between the genomic sites of the underlying mutations, recombination is far from universally advantageous.

The Phylogenetic Dispersion of Mean Phenotypes

The theory discussed above provides insight into the rapidity with which populations can respond to novel and/or persistent directional selective challenges. Such scenarios might be encountered in a continuous coevolutionary arms race between a host cell and a pathogen, or in situations involving a sudden environmental shift. However, numerous cellular traits may have been under very similar selective pressures across phylogenetic lineages since their origin. This is likely to be true, for example, for enzymes whose sole function has always been to convert a specific substrate into a specific product, membrane channels specialized to admitting and/or excluding specific ions, or polymerases responsible for generating complementary base-pair matches.

In such cases, the dynamical response to changing selection pressures is no longer the key issue. The more relevant evolutionary focus is the long-term steady-state probability distribution of alternative genotypes. Although natural selection relentlessly strives to improve trait performance, there are numerous reasons why perfection will seldom, if ever, be achievable. First, absolute limits to the refinements of chemical and physical processes are dictated by factors such as diffusion rates and effectively discrete processes such as the energy associated with individual hydrogen bonds. Second, the stochastic processes of mutation and drift can result in the dispersion of mean phenotypes around an expected value to a degree that depends on the range of effectively neutral parameter space. Third, mutation pressure will almost never be perfectly aligned with the goals of selection, and this will cause the average phenotype to deviate from an optimum, even in the absence of mutation bias.

The latter two points have significant implications for interpreting patterns of phenotypic divergence. Most notably, populations under identical selection pressures will not necessarily have identical mean phenotypes, but instead will exhibit a dispersion of such measures. Moreover, the most common phenotype need not be the optimum phenotype. Finally, gradients of mean phenotypes with respect to N_e are likely to be molded by differences in the power of random genetic drift across the Tree of Life. To reduce the likelihood of evolutionary cell biology succumbing to the common practice of interpreting all phylogenetic variation in phenotypes as necessary reflections of differences in selective environments, these basic principles are first sketched out for the case of a very simple two-state trait, and then further explored for traits encoded by multiple genetic loci.

Two-state traits

Consider a single biallelic locus with one allele (denoted as state 1) having a fractional selective advantage s over the other allele (denoted as state 0). Although allele 1 has the highest achievable fitness, this does not ensure that once fixed, it will be immune to replacement by the suboptimal type. Denoting the mutation rate from allele 0 to 1 as μ_{01}, with μ_{10} being the reciprocal mutation rate, and assuming a haploid population, we wish to determine the long-term mean frequency of allele 1.

The simplest situation involves a population with a small enough effective size that the waiting times between mutations destined to fixation are large enough that the population is nearly always fixed for one allele or the other, with probabilities $\tilde{p}_0$ and $\tilde{p}_1$. The domain of N_e and mutation rates necessary for such a situation is equivalent to that noted earlier for the sequential-fixation model, with the more general model (allowing for any N_e) outlined in Foundations 5.3.

Under these weak-mutation, strong-selection conditions, a lineage spends a long period of time in one particular monomorphic state before making a stochastic shift to another. The intervening intervals (waiting times for transitions) are functions of the relative strengths of selection, mutation, and random genetic drift, but over a very long time period, the rate of movement from state 0 to 1 must equal that in the opposite direction (a principle known as detailed balance in the statistical physics literature). This is because, whether favorable or encouraged by mutation pressure, the most abundant state provides more opportunities for transitions, which are individually less likely to proceed to fixation; in contrast, the rarer state provides fewer transition opportunities, but when these arise, they are more likely to fix. It then follows that

$$\widetilde{p}_0 \cdot (N\mu_{01}) \cdot \phi_{01} = \widetilde{p}_1 \cdot (N\mu_{10}) \cdot \phi_{10}, \quad (5.7)$$

where ϕ_{01} and ϕ_{10} denote the probabilities of fixation of newly arising beneficial and deleterious alleles, defined by Equation 4.1b. Each side of this equation is the product of the expected frequency of a state, the rate of origin of mutations to the opposite state, and the probability of fixation of the mutant allele. Using the useful identity $\phi_{01}/\phi_{10} = e^S$, where $S = 2N_e s$, and the fact that $\widetilde{p}_1 = 1 - \widetilde{p}_0$, Equation 5.7 rearranges to

$$\widetilde{p}_0 = \frac{1}{1 + \beta e^S}, \quad (5.8)$$

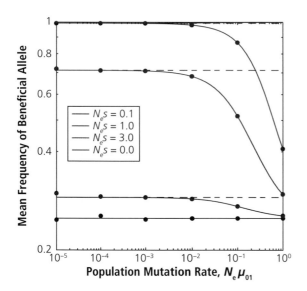

Figure 5.5 Expected frequency of the beneficial allele at a biallelic locus under the joint forces of drift, mutation, and selection. The mutation rate to deleterious alleles (μ_{10}) is assumed to be 3× that to beneficial alleles (μ_{01}), as would be approximately the case for complementary nucleotides in an RNA stem. Results are given for four intensities of selection (with s denoting the selective advantage of the beneficial allele) relative to drift. Solid lines give the exact results from Foundations 5.3, whereas the dashed lines are the results under the sequential model, Equation 5.8, which assumes the population to be nearly always in a monomorphic state and becomes increasingly unreliable as $N_e\mu_{01}$ exceeds 0.01. Note that when $s = 0$ (neutrality), the expected frequency (0.25 in this case) is independent of the population mutation rate.

where $\beta = \mu_{01}/\mu_{10}$ is the ratio of the mutation rates in both directions (mutation bias being implied when $\beta \neq 1$).

This simple model illustrates three key points. First, unless completely lethal, the low-fitness state has a non-zero probability of occurrence. Thus, despite constant selection pressure, a lineage is not expected to remain in a stable fixed state forever. Second, the two alleles approach equal probabilities as $\beta e^S \to 1$. This composite parameter is simply equal to the product of the mutation and fixation biases in favor of state 1, so that even if state 1 is selectively favored ($S > 0$), state 0 will be more common if mutation bias in the opposite direction is of sufficient strength that $\beta e^S < 1$. The relevance of this point is that maximum divergence occurs when $\beta e^S = 1$, again demonstrating the potential for substantial variation in the face of uniform selection. Third, if the effective population size and/or strength of selection is sufficiently small that $S \ll 1$ (the domain of effective neutrality), the equilibrium frequency of the disfavored allele will be entirely driven by mutation pressure. In this case, because $e^S \simeq 1$, $\widetilde{p}_0 \simeq 1/(1 + \beta)$, which is a function of the relative (but not absolute) mutation rates.

These expectations are altered when the population-level mutation rate exceeds the limits of the domain of the sequential model (Foundations 5.3). Most notably, Equation 5.8 defines the lower bound to the expected frequency of the low-fitness allele. The expected frequency of the beneficial allele declines once the population-level mutation rate ($N_e\mu_{01}$) exceeds ~0.01 (i.e., a new mutation enters the population at least every 100 generations), asymptotically approaching the neutral expectation $\widetilde{p}_1 \simeq \beta/(1 + \beta)$ as $N_e\mu_{01}$ exceeds 1.0 (Figure 5.5). The latter condition arises when mutation brings in allelic variants faster than natural selection can promote beneficial over detrimental alleles. In this case, the population is also almost always represented by a polymorphic collection of both alleles, rather than by a state of fixation.

Multistate-traits and the drift-barrier hypothesis

Extension of the preceding single-locus model to an arbitrary number of L linked sites (loci) yields additional insights into the limits to what natural selection can accomplish. To appreciate the fundamental points in a relatively simple manner, we assume that all genetic loci are equivalent, with two alternative (+ and −) allelic states contributing positively and negatively to the trait. Depending on the context, the loci may be viewed as single nucleotides, amino-acid codons, or entire genes.

For all but the two most extreme genotypes (containing all + or all − alleles), a multiplicity of functionally equivalent classes exists with respect to the number of + alleles, i. These are defined by binomial coefficients. As an example, for the case of $L = 4$, there are $2^4 = 16$ possible genotypes, but just five genotypic classes (having 0, 1, 2, 3, 4, and 5 + alleles, with respective multiplicities 1, 4, 6, 4, and 1) (Figure 5.6). With equivalent fitness for all members (haplotypes) within a particular class, this variation in multiplicity of states plays a significant role in determining the long-term evolutionary distribution of alternative classes, as classes with higher multiplicities are more mutationally accessible. This type of biallelic model has been widely exploited in theoretical studies of the genetic structure of quantitative (multilocus) traits (Walsh and Lynch 2018), and is encountered in a number of different contexts in subsequent chapters, including the evolution of protein–protein interfaces, transcription-factor binding sites, and growth rate. Here, we assume a haploid, non-recombining population of N individuals. The site-specific per-generation mutation rates

from the − to the + state, and vice versa, are again defined as μ_{01} and μ_{10}, respectively.

As with the single-factor model, the multiple-factor model has a long-term steady-state probability distribution of population residence in the $L+1$ alternative states. Again, starting with the assumption that population-level mutation rates are low enough that transitions only occur between adjacent classes (satisfying the conditions for a sequential-fixation scenario), the relative flux rates between classes are equal to the expressions on the arrows in Figure 5.6. These rates are proportional to the products of rates of mutational production and probabilities of subsequent fixation, with the numerical coefficients being defined by the numbers of − and + sites within each class. The absolute population size N defines the number of mutational targets per generation, but because N influences all mutational flux rates in the same way, it is omitted as a prefactor, although both N and N_e still influence the equilibrium solution via the fixation probabilities (Equation 4.1b).

This linear sequential model has a relatively simple solution (Berg et al. 2004; Sella and Hirsh 2005; Lynch and Hagner 2014; Lynch 2020). As a reference point, consider first the extreme case of effective neutrality. Here, $\eta = \mu_{01}/(\mu_{01} + \mu_{10}) = \beta/(1 + \beta)$ is the expected equilibrium frequency of + alleles at each site, which arises when the net flux of + → − and − → + mutations is balanced. With no

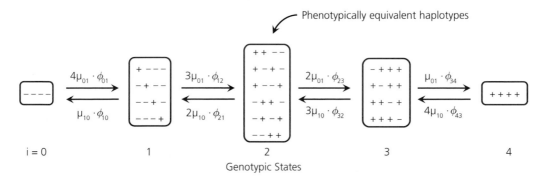

Figure 5.6 Schematic for the transition rates between adjacent genotypic classes under the sequential-fixation model for the case of $L = 4$ sites. This layout readily generalizes to any value of L. μ_{01} and μ_{10} are the mutation rates from − to + allelic states, and vice versa, and ϕ_{ij} is the probability of fixation of a newly arisen mutation to allele j from a background of i, defined by Equation 4.1b. The number of + alleles in a class is denoted by i, and except for the two extreme classes ($i = 0$ and $i = 4$), there are multiple equivalent genotypic states within each class of genotypic values.

selection for particular combinations of alleles, each site evolves independently, so the steady-state distribution of phenotypes under neutrality is simply equal to

$$\widetilde{p}_{n,i} = \binom{L}{i} \eta^i (1 - \eta)^{L-i}$$

$$= C \cdot \binom{L}{i} \beta^i \qquad (5.9)$$

which is the binomial probability distribution, where n denotes the neutral condition, and i denotes the number of + alleles in a genotypic class. The term in large parentheses is the binomial coefficient $L!/[i!(L-i)!]$ (which is the multiplicity of alternative orderings of + and − positions within a particular class i), and $C = (1 + \beta)^{-L}$. Equation 5.9 defines the long-term probability of a population residing in each of the $L+1$ possible genotypic classes, i.e., the fractional time wandering over the evolutionary landscape spent in each class. Note that the neutral steady-state distribution depends only on the ratio of mutation rates (β), not on their absolute values. Denoting the overall genotypic state as the sum of + alleles, the long-term mean and variance of the trait are $L\eta$ and $L\eta(1 - \eta)$, respectively.

Under the sequential model, selection transforms the neutral distribution in a remarkably simple way, with each class being weighted by the exponential function of the scaled strength of selection $S_i = 2N_e s_i$,

$$\widetilde{p}_i = C \cdot \widetilde{p}_{n,i} \cdot e^{S_i}, \qquad (5.10)$$

where C is a new normalization constant that ensures that the $\widetilde{p}_i$ sum to 1.0. The selection coefficients associated with each class are generally defined as deviations from some benchmark in the population (say the optimum type), but this does not matter, as the reference is a constant that simply enters the normalization constant. The utility of this approach is that, provided there are mutational connections between all adjacent states, Equation 5.10 can be applied to any fitness function describing the relationship between s and i.

Taken together, Equations 5.9 and 5.10 show that the equilibrium frequencies of the genotypic classes are functions of three factors: 1) the multiplicity of

configurations, as defined by the binomial coefficients; 2) the ratio of mutation rates; and 3) the strength of selection scaled by the power of random genetic drift. All other things being equal, the within-class multiplicity magnifies the likelihood of residing in such a state. This demonstrates that mutation need not be directionally biased to have an impact on the overall distribution. All that is required is that neutral distribution deviates from the expectations under selection alone, which will almost always be the case.

We now explore two examples to illustrate how these three factors jointly define the distribution of phenotypes within and among alternative population-genetic environments. First, consider the simple case in which a trait determined by just $L=2$ factors is under stabilizing selection, with an optimum phenotype θ, and fitness (W) dropping off at a rate determined by the width of the fitness function ω (analogous to the standard deviation of a normal distribution). This Gaussian (bell-shaped) function is defined by

$$W_i = e^{-(i-\theta)^2/(2\omega^2)}. \qquad (5.11)$$

With two sites, there are three possible genotypic classes, $i = 0, 1$, and 2, with the phenotypically equivalent +− and −+ states being lumped into the $i = 1$ class. Selection is purely directional if the optimum is at or beyond an end state, i.e., $\theta \leq 0$ or $\geq L$, and neutrality is approached as the fitness function becomes flatter, i.e., $\omega \to \infty$. Although i is confined to integer values, θ need not be, and if θ has a value other than 0, 1, or 2, the optimum phenotype is unattainable. The selection coefficients can be simply defined as deviations of fitness from the maximum value of 1, $s_i = 1 - W_i$. The mean phenotype (in this case, the average genotypic value of i) is $\widetilde{p}_1 + 2\widetilde{p}_2$, which reduces to 2η in the case of neutrality.

This expansion to a second site introduces complexities not encountered with the one-site model (Figure 5.7a). For example, for the case of $\theta = 1.5$, where the optimum is straddled by the class 1 and 2 genotypes, assuming mutation bias towards − alleles ($\beta < 1$), the long-term genotypic mean never reaches the optimum, even at very large N_e, and instead remains much closer to 1. This bias results because, although the class 1 and 2 genotypes have

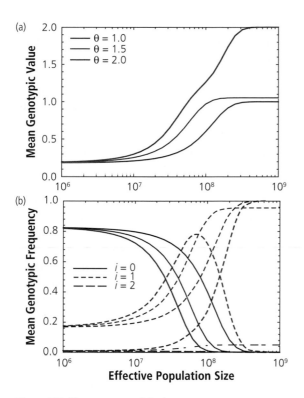

Figure 5.7 The response of the long-term mean genotypic state (number of + alleles; panel **a**) and the underlying mean class frequencies (for i = 0, 1, and 2 + alleles; panel **b**) over a gradient of effective population sizes for a two-locus, two-allele, sequential model. The mutation rate to beneficial alleles is 10% of that to deleterious alleles ($\beta = 0.1$). Results are given for three phenotypic optima, with the width of the fitness function $\omega = 5000$. For each color-coded optimum, the mean frequencies of the three genotypic classes (i = 0, 1, 2) are given as solid, short-dashed, and long-dashed lines, which for each N_e sum to 1.0. Results are derived from Equations 5.10 and 5.11. From Lynch (2020).

equivalent fitness, mutation pressure towards – alleles weights the frequency of class 1 by a factor of 2β (the two being the multiplicity of this class), but class 2 by the smaller factor of β^2 (from Equation 5.10)

For the case in which $\theta = 2$ (pure directional selection), there is a progressive succession of the prevailing genotype classes with increasing N_e (Figure 5.7b). When N_e is sufficiently low to impose effective neutrality, class 0 predominates owing to the mutation bias towards – alleles. With increasing

N_e, selection becomes more effective at promoting class 1, but there remains effective mutation pressure against class 2. Finally, with very large N_e, selection becomes efficient enough to drive class 2 to near fixation, thereby decreasing the incidence of class 1. These results show that in the face of a constant pattern and strength of selection, the genotypic mean can exhibit a considerable gradient with N_e owing to changes in the power of drift. They also show that appreciable incidences of all three genotypic classes can be expected over time in lineages with intermediate N_e, i.e., the population mean phenotype is expected to wander temporally between alternative states despite the imposition of constant selection.

As a second example, consider the case in which fitness declines with the number of – alleles ($L - i$) in a multiplicative fashion,

$$W_i = (1 - s)^{(L-i)}, \qquad (5.12)$$

where i is the number of beneficial (+) alleles in the genome. Three examples are shown in Figure 5.8 for different numbers of completely linked loci, $L = 10^4$, 10^5, and 10^6. In each case, the overall performance is equally subdivided across all L factors, such that $s = 1/L$. With such large numbers of loci, the analytical solution noted earlier is not reliable at large N_e, as the number of mutations arising per generation may exceed the limit of the domain of the sequential model, although results have been obtained by computer simulations (Lynch 2020).

Three general features are clear in Figure 5.8. First, as already noted, when N_e is sufficiently small that the mutational effects are rendered effectively neutral, the mean fraction of + alleles is defined by the neutral expectation, with the probability of a + allele being η ($\simeq 0.01$ in the given example) at each locus. Second, with increasing N_e, the mean fraction of + alleles progressively increases, converging on 1.0 as N_e becomes much greater than $1/(2s)$. This gradient in trait means with N_e is a result of the drift barrier, which increasingly compromises the ability of natural selection to alter the frequencies of mutations as the population size declines. The exact location of the drift barrier is defined by the relative power of drift and selection, becoming lower with smaller s, but also by the mutation bias and by the multiplicity effect. Third, the observed gradients are much shallower than the expectations under free recombination. This illustrates the point made in Chapter 4 that, owing to selective interference among linked loci, populations with large absolute sizes behave genetically as though they are much smaller than implied by their absolute size. For example, in the absence of selective interference, sites with selection coefficients equal to 10^{-5} are expected to be nearly fixed for + alleles once the absolute population size exceeds 500,000, but with linked loci at the same absolute population size, the vast majority of alleles are of the − type owing to the combination of mutation pressure and random genetic drift (Figure 5.8).

These results highlight the riskiness of an evolutionary biology that assumes that all phenotypes simply reflect optimal outcomes dictated by natural selection. In addition to the pervasive influence of drift, mutation can cause mean phenotypes to deviate from the optimum in substantial and often unexpected ways that are not simply functions of the magnitude of mutation bias. Rather, when alternative, functionally equivalent underlying genotypes exist for a trait, the multiplicity of certain intermediate combinations can result in a mutational pull of the mean phenotype away from the optimum. This effect becomes particularly significant when the phenotypic optimum is far from the expected mean under neutrality alone, especially if the level of multiplicity for the optimum is small relative to that for other phenotypic states. Cases may even exist in which the opposing pressures of selection and mutation are sufficiently strong that the equilibrium mean-phenotype distribution can have two peaks, one driven by selection and the other by mutation (Lynch and Hagner 2014; Tuğrul et al. 2015; Lynch 2018).

Summary

- Information on the distribution of fitness and phenotypic effects of new mutations is critical to understanding the evolutionary potential and limitations of phylogenetic lineages. Based on bioenergetic considerations alone, few mutations can have zero fitness effects. Multiple lines of evidence indicate that the vast majority of mutations are deleterious, with most being very mildly so and the mode being near zero, but many details remain unknown.

- The frequency of mutations with very small effects almost certainly increases in organisms of larger size, and there is strong evidence for a substantial pool of mutations that are only sensitive to selection in species with large effective population sizes (N_e). These mutations play a key role in defining the limits to adaptation in different phylogenetic lineages.

- Despite the common view that populations under identical selection pressures will tend to be highly similarly phenotypically, many plausible situations exist in which uniform selection combined with random genetic drift and/or mutation bias can lead to substantial interspecies

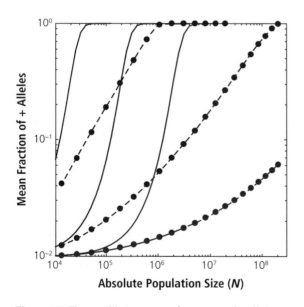

Figure 5.8 The equilibrium mean frequency of + alleles as a function of the absolute population size for the case in which the mutation bias to + alleles is 0.01. Results are given for three numbers of loci (L) with multiplicative effects on fitness. Throughout, the selective disadvantage of a − allele is $s = 1/L$, the inverse of the number of factors contributing to the trait: red, $L = 10^4, s = 10^{-4}$; green, $L = 10^5, s = 10^{-5}$; blue, $L = 10^6, s = 10^{-6}$, so that the total fitness effects of mutations is the same in all cases. Analytical results (solid lines) are given for the case of free recombination, Equation 5.3.3, whereas simulation results (dashed lines and data points) are given for the case of complete linkage. From Lynch (2020).

divergence, sometimes more than expected under drift alone.

- Many molecular adaptations require the co-occurrence of two or more mutations to elicit a phenotype with elevated fitness. Theory suggests that the rapidity (on an absolute timescale) of acquiring such adaptations is roughly independent of N_e if the intermediate states are neutral, but scales negatively with N_e if the intermediate states are deleterious, and more rapidly so with increasing numbers of intermediate steps. Thus, the likelihood of alternative paths of adaptive evolution can be strongly modulated by changes in N_e. However, sufficiently high rates of multi-nucleotide mutation can instantaneously embark a complex genotype on a more rapid path to establishment, the more so in larger populations.

- Many cellular traits have retained the same function for hundreds of millions of years, and may have been under nearly invariant selection pressures over this same timescale. This shifts the evolutionary focus away from dynamical changes in allele frequency under directional selection to the long-term steady-state probability distribution of alternative phenotypic states. The drift-barrier hypothesis predicts that, despite the operation of persistent directional or stabilizing selection, the mean phenotypes of such traits will commonly exhibit gradients with respect to N_e, with the level of functional refinement increasing with the latter.

- Aside from being the source of variation upon which natural selection operates, mutation impacts the expected distribution of mean phenotypes because genotypic states differ in the multiplicity of ways in which they can be constructed from the underlying set of genetic loci. Mutation bias further influences the evolutionary attraction towards a particular region of phenotypic space, in ways that may conflict with or reinforce prevailing selection pressures.

- Taken together, these results from evolutionary theory call into question the common practice of assuming that observed mean phenotypes provide a perfect reflection of prevailing selection pressures. Even under constant environmental conditions, the most common phenotype can be an unreliable indicator of the optimum defined by natural selection.

Foundations 5.1 The distribution of fitness effects (DFEs) of newly arising mutations

The success of natural selection depends critically on the presence of mutations with large enough effects to overcome the stochastic effects of random genetic drift. Yet, the DFE of *de novo* mutations is one of the most poorly known aspects of genetics relevant to evolution. The central problem is that biology is structured in such a way that a large fraction of mutations has effects that are too small to be perceived as allele-frequency changes in laboratory experiments on reasonable timescales. Relying on results from subsequent chapters to make indirect inferences, here the case is made that the genome-wide DFE is highly skewed towards deleterious mutations with tiny effects, although many details remain to be worked out.

First, because the costs of biosynthesis of the four nucleotides differ, even among unexpressed mutations with no functional effects, only a very small fraction is likely to be absolutely neutral. From Foundations 17.2, the total bioenergetic cost of synthesizing the A and T in an A:T (or T:A) bond exceeds that for the C and G in a C:G (or G:C) bond by $\sim$2 ATP equivalents. From Equation 8.2b, it is known that the cost of building a cell (in 10^9 ATPs) is a function of its volume (V, in μm^3), $\sim$27$V^{0.96}$. Thus, as a first-order approximation a transversion from A/T to C/G (or vice versa) alters the cell's energy budget (and the cell-division time) by a fraction 10^{-10} in a bacterial-sized cell of 1 μm^3 and by 10^{-12} in a medium-sized eukaryotic cell of 100 μm^3. The only truly neutral mutations may be those that alter A:T to T:A or C:G to G:C in nucleotide sites that are fully non-functional (i.e., are not transcribed and do not serve regulatory purposes).

Second, there are no absolutely neutral insertions and deletions. Because a complementary nucleotide pair in DNA has a total bioenergetic cost of $\sim$100 ATP equivalents, a single base-pair insertion will impose a fractional reduction in fitness $\simeq 100/[(27 \times 10^9)V] \simeq 10^{-9}/V$ (again, rounding off to an order of magnitude), with an insertion x base pairs in length having a x-fold higher cost. These are the minimal costs of insertions, as would be expected in non-functional DNA. The total magnitude of effect would be substantially larger in functional regions owing to frameshifts in coding DNA and/or alterations in regulatory regions.

Third, owing to the additional energetic costs of gene expression (Lynch and Marinov 2015), nucleotide sites in protein-coding genes have average costs that are $\sim100\times$ those for unexpressed nucleotides, and owing to the different costs of synthesizing alternative amino acids, an amino-acid altering mutation can further alter the overall cost per site by a factor of 0.1 to 6.0 (Foundations 17.2). This means that, based on energetic costs alone, depending on the level of gene expression, amino-acid-altering mutations will have average fitness effects on the order of 10–$600\times$ those for the most innocuous silent-site substitutions noted above. The dispersion around these expectations following roughly a log-normal distribution with a range of at least an order of magnitude in both directions. For example, in a bacterial-sized cell, the fitness effects of amino-acid substitutions based on biosynthetic costs alone will be distributed over a range of 10^{-10} to 10^{-6}. In addition to these energetic consequences, amino-acid substitutions can have effects associated with protein function ranging from near zero to essentially lethal ($s = 1$), with most having $s \ll 0.1$ and the average being of order $s = 0.01$ (Chapter 12). Summing both types of effects, we expect the distribution of fitness effects of amino-acid-altering mutations to have a very wide range, likely with a mode of at least 10^{-8} in bacteria, and perhaps several orders of magnitude higher.

These biological considerations allow a crude genome-wide view of the DFE for *de novo* mutations. In most bacteria, $\sim95\%$ of the genome is coding DNA, and perhaps half of the intergenic sites and silent sites will have functional consequences associated with gene regulation. However, in eukaryotes, with increasing organism size, genomes become progressively more bloated with non-functional intergenic and intronic DNA as well as with more complex regulatory regions (Lynch 2007). As a consequence, in many vertebrates and land plants, $<2\%$ of the genome is coding DNA and $>50\%$ of the remaining nucleotide sites are likely non-functional and so experience fitness-altering mutations only via small energetic effects. Owing to the nature of the genetic code, $\sim25\%$ of coding-region sites are silent and the remaining 75% are amino-acid replacement sites (Lynch 2007), and assuming no mutation bias, one in three nucleotide substitutions at silent sites will have no energetic consequences (e.g., A→T vs. A→C or G). In addition, across the Tree of Life, $\sim10\%$ of mutations involve insertions and deletions of one to several nucleotides, with the remainder being base substitutions (Chapter 4). Taking all of these observations into consideration leads to DFEs with the provisional forms outlined in Figure 5.1, as further discussed in the main text.

Foundations 5.2 Divergence under uniform selection

Although it is generally thought that selection will increase the evolutionary determinism of a system, causing pairs of populations under identical selection pressures to be more similar than expected on the basis of random assortment of variation, this is not necessarily the case (Cohan 1984; Lynch 1986). Consider a pair of populations exposed to identical conditions and starting with two alleles, A and a, with identical frequencies of p and $(1 - p)$, respectively. Letting $\phi(p)$ be the probability of fixation of allele A, the probability that a pair of populations will ultimately experience fixation for different alleles is $\Delta = 2\phi(p)[1 - \phi(p)]$, which reaches a maximum value when $\phi(p) = 0.5$. Under the naive view that the beneficial allele always fixes, one expects $\phi(p) = 1$ and $\Delta = 0$, but this is incorrect because the probability of fixation of beneficial alleles is <1.

That populations can sometimes diverge to a greater extent under uniform selection than under pure neutral drift can be seen as follows. In the absence of selection, the probability of fixation of allele A is simply p, and the probability of alternative outcomes is $\Delta = 2p(1 - p)$, which is maximized when $p = 0.5$. For the same reason, the probability of divergence is increased by selection if $\phi(p)$ under selection is closer to 0.5 than the initial frequency p. Because $\phi(p) > p$ for a selectively favored allele, it follows that a minimum requirement for increased divergence under pan-selection is that the starting frequency of the advantageous allele be <0.5.

The conditions for excess divergence under drift plus selection to exceed that under drift alone are not very restrictive. Consider two replicate populations with identical initial frequencies of the A allele, $p = 0.25$. Under pure drift, the probability that one replicate becomes fixed for A and the other for a is $2 \cdot 0.25 \cdot (1 - 0.25) = 0.375$. Now suppose that A is weakly favored by selection, with $N_e s = 0.5$. Again assuming $p_0 = 0.25$, Equation 4.1a gives the fixation probability of A as 0.46, implying a probability of fixing alternative alleles of $2 \cdot 0.46 \cdot 0.54 = 0.496$, close to the maximum level of divergence expected under neutrality. Thus, even when experiencing identical directional selection pressures, populations that initiate with low-frequency, advantageous alleles can exhibit levels of divergence conventionally interpreted as

being associated with diversifying selection. Of course, this analysis ignores the recurrent downstream introduction of mutations, which could ultimately lead to convergence.

Foundations 5.3 Mean probabilities of alternative alleles at steady state

A concern with the sequential model outlined in the text is that large populations are expected to reside in polymorphic states for significant amounts of time. This would not be a problem if the average frequencies of alleles when monomorphic were the same as those while polymorphic, but such a condition is only met in the special case of neutrality. This disparity arises because deleterious mutations that are strongly inhibited from going to fixation can nonetheless maintain measurable within-population frequencies owing to recurrent mutational input. As population sizes increase, the likelihood of residing in a polymorphic state necessarily increases, owing to the greater total influx of variation per generation. Under such conditions, one can still inquire as to the average frequency of a sampled allele over a long-term steady-state equilibrium, but this average must also factor in all possible polymorphic states, ranging from allele frequency $(1/N)$ to $[1 - (1/N)]$ for haploids.

Let P_1, P_0, and P_p denote the steady-state probabilities of a population being monomorphic for the optimal allele (1), monomorphic for the suboptimal allele (0), or polymorphic (p). Under the sequential model, $P_1 + P_0 \simeq 1$. Here, we make use of a result from diffusion theory that describes the steady-state probability distribution of frequency x for the deleterious allele 0 (which is equivalent to the beneficial allele 1 being present at frequency $1 - x$), described more fully by Kimura et al. (1963), Wright (1969), and Charlesworth and Jain (2014). Although actual allele-frequency distributions are discrete, with large N, the probability that a population has allele frequency x can be accurately approximated by the continuous distribution

$$\phi(x) = Cx^{U_{10}-1}(1 - x)^{U_{01}-1}e^{-Sx}, \quad (5.3.1a)$$

where $U_{01} = 2N_e\mu_{01}$, $U_{10} = 2N_e\mu_{10}$, $S = 2N_es$ and the normalization constant

$$C = \frac{\Gamma(U_{01} + U_{10})}{\Gamma(U_{01}) \cdot \Gamma(U_{10}) \cdot {_1}F_1(U_{10}; (U_{01} + U_{10}); -S)}, \quad (5.3.1b)$$

ensures that integration of the distribution over the full range of allele frequencies sums to 1.0. Γ denotes the gamma function, and ${_1}F_1$ is the confluent hypergeometric function. These two functions can be calculated numerically using series expansions defined respectively as Equations 6.1.2 and 13.1.2 in Abramowitz and Stegun (1964).

The probability of being monomorphic for state 1 can be approximated by integration of the end class, implying an absence of the deleterious allele,

$$P_1 = \int_0^{1/N} \phi(x) \cdot dx.$$

Because x is very small in this region, both $(1 - x)$ and e^{-Sx} can be approximated as 1, leading to

$$P_1 \simeq \left(\frac{C}{U_{10}}\right)\left(\frac{1}{N}\right)^{U_{10}}. \quad (5.3.2a)$$

At the opposite end of the spectrum, using $x \simeq 1$ and $e^{-Sx} \simeq e^{-S}$ yields the probability of being monomorphic for state 0,

$$P_0 = \int_{1-(1/N)}^{1} \phi(x) \cdot dx \simeq \left(\frac{C}{U_{01}}\right)\left(\frac{1}{N}\right)^{U_{01}} e^{-S}. \quad (5.3.2b)$$

Here, it can be seen that the ratio P_1/P_0 obtained with this approach deviates from the prediction of the sequential model, $\tilde{p}_1/\tilde{p}_0 = (\mu_{01}/\mu_{10})e^S$ (inferred from Equation 5.8), unless $\mu_{01} = \mu_{10}$. Although the details are not covered here, it can be shown that the probability of polymorphism, $P_p = 1 - P_0 - P_1$, is only weakly dependent on the magnitude of selection, and generally does not exceed 0.1 until $N_e\mu_{01} > 0.01$.

The average frequencies of the two alleles over the stationary distribution can be obtained by weighting the frequency classes by their densities, Equation 5.3.1a, and integrating over $(0,1)$, which yields

$$\bar{p}_0 = \frac{\mu_{10} \cdot {_1}F_1[(U_{10} + 1); (U_{01} + U_{10} + 1); -S]}{(\mu_{01} + \mu_{10}) \cdot {_1}F_1[U_{10}; (U_{01} + U_{10}); -S]}. \quad (5.3.3)$$

Foundations 5.4 The detailed-balance solution for the evolutionary distribution of alternative molecular states

Here we assume a linear array of alternative molecular states, with population-level transitions only occurring between adjacent states (Figure 5.6). For the latter condition to be met, each transition rate must be sufficiently small that a population generally resides in one state for an extended period of time before fixation of a subsequent mutation leads to a switch between states. Under these conditions, a relatively simple model defines the probability of residing in each class after a sufficient amount of time has elapsed to ensure potential occupancy over the entire distribution of states. At this equilibrium, for any particular state, the rates of entry and exit must be equal, a condition known as detailed balance. The overall form of the steady-state distribution, which depends on the full set of transition rates, is reached regardless of the starting conditions.

Letting $m_{i,j}$ denote the rate of evolutionary transition from state i to state j, we have a system of $L+1$ simultaneous equations (where L denotes the final state in the series, which starts with index 0),

$$p_0(t+1) = (1 - m_{0,1})p_0(t) + m_{1,0}p_1(t),$$

$$p_i(t+1) = m_{i-1,i}p_{i-1}(t) + (1 - m_{i,i-1} - m_{i,i+1})p_i(t)$$
$$+ m_{i+1,i}p_{i+1}(t),$$

$$p_L(t+1) = m_{L-1,L}p_{L-1}(t) + (1 - m_{L,L-1})p_L(t).$$

Assuming non-zero transition rates between all adjacent classes in this linear array, the equilibrium solution (the steady-state probability of being in state i) takes on a simple form (Lynch 2013),

$$\widetilde{p}_i = \frac{\left(\prod_{j=0}^{i-1} m_{j,j+1}\right)\left(\prod_{k=i+1}^{L} m_{k,k-1}\right)}{C} \tag{5.4.1}$$

where the first term in the numerator is equal to the product of all transition rates pointing up toward the class, the second term is the product of all transition rates pointing down toward the class, and C is simply a normalization constant that ensures that all of the $\widetilde{p}_i$ sum to one (it is equal to the sum of numerators for all i, and is generally called the partition function).

As an example, with four alternative states (indexed 0, 1, 2, 3), the equilibrium probabilities become

$$\widetilde{p}_0 = m_{1,0}m_{2,1}m_{3,2}/C,$$

$$\widetilde{p}_1 = m_{0,1}m_{2,1}m_{3,2}/C,$$

$$\widetilde{p}_2 = m_{0,1}m_{1,2}m_{3,2}/C,$$

$$\widetilde{p}_3 = m_{0,1}m_{1,2}m_{2,3}/C,$$

where C is the sum of the numerators in all four expressions. The steady-state probabilities, $\widetilde{p}_i$, can be

equivalently viewed as the proportion of time a specific lineage spends in state i over a long evolutionary period, or as the fraction of independent populations experiencing identical population-genetic environments that are expected to reside in class i at any specific time.

In the context of the model introduced in the text, each of the m coefficients can be viewed as the product of the number of mutations arising in the population per generation and the probability of fixation. Letting N be the population size, $\mu_{i,j}$ be the mutation rate from allelic state i to j (where j can be only $i-1$ or $i+1$), and $\phi_{i,j}$ be the probability of fixation of a newly arisen j allele in a population currently occupied by allele i, the transition rates are equal to the products of the relevant numbers of new mutant alleles arising per generation and their probabilities of fixation, i.e., $m_{i,j} = 2N\mu_{i,j}\phi_{i,j}$ assuming diploidy (or one half that assuming haploidy). Because every coefficient has the same prefactor, $2N$ or N, this can be ignored, reducing the coefficients to $m_{i,j} = \mu_{i,j}\phi_{i,j}$. (In the text, the $\mu_{i,j}$ are functions of per-site mutation rates and the number of sites relevant to the particular transition).

A second key simplification arises from the behavior of the probability of fixation in opposite directions between adjacent states. Letting s_i denote the selective disadvantage of allele i, measured relative to a perfect fitness of 1.0, then $s_{i+1} < s_i$ implies that allele $i+1$ is beneficial compared to allele i. Assuming mutations with additive effects on fitness, application of the formula for the fixation probability of new mutations, Equation 4.1b, yields the convenient result that $\phi_{i,i+1}/\phi_{i+1,i} = e^{4N_e(s_i - s_{i+1})}$ for diploids (with a 2 substituted for the 4 in haploids) (Berg et al. 2004; Sella and Hirsh 2005; Lynch 2013; Lynch and Hagner 2014).

As a simple application of the preceding methods, consider the situation in which there are just two alternative states, A and a, with the mutation rate from A to a being u, from a to A being v, and s being the selective advantage of a (negative if a is disadvantageous). The combined mutation/selection pressure towards a is then $u\phi_{A,a}$, while that towards A is $v\phi_{a,A}$, implying that

$$\widetilde{p}_a = \frac{u\phi_{A,a}}{u\phi_{A,a} + v\phi_{a,A}}, \tag{5.4.2a}$$

where the denominator is the partition function. Dividing all terms by $v\phi_{a,A}$, and using the relationship just noted for ratios of opposite fixation probabilities for diploids, leads to the simplification

$$\widetilde{p}_a = \frac{(u/v)e^{4N_e s}}{1 + (u/v)e^{4N_e s}}. \tag{5.4.2b}$$

Literature Cited

Abramowitz, M., and I. A. Stegun (eds.) 1964. Handbook of Mathematical Functions. Dover Publications, Inc., New York, NY.

Azevedo, L., G. Suriano, B. van Asch, R. M. Harding, and A. Amorim. 2006. Epistatic interactions: How strong in disease and evolution? Trends Genet. 22: 581–585.

Baer, C. F., M. M. Miyamoto, and D. R. Denver. 2007. Mutation rate variation in multicellular eukaryotes: Causes and consequences. Nat. Rev. Genet. 8: 619–631.

Bataillon, T., and S. F. Bailey. 2014. Effects of new mutations on fitness: Insights from models and data. Ann. N. Y. Acad. Sci. 1320: 76–92.

Behringer, M. G., B. I. Choi, S. F. Miller, T. G. Doak, J. A. Karty, W. Guo, and M. Lynch. 2018. *Escherichia coli* cultures maintain stable subpopulation structure during long-term evolution. Proc. Natl. Acad. Sci. USA 115: E4642–E4650.

Berg, J., S. Willmann, and M. Lässig. 2004. Adaptive evolution of transcription factor binding sites. BMC Evol. Biol. 4: 42.

Blount, Z. D., C. Z. Borland, and R. E. Lenski. 2008. Historical contingency and the evolution of a key innovation in an experimental population of *Escherichia coli*. Proc. Natl. Acad. Sci. USA 105: 7899–7906.

Böndel, K. B., S. A. Kraemer, T. Samuels, D. McClean, J. Lachapelle, R. W. Ness, N. Colegrave, and P. D. Keightley. 2019. Inferring the distribution of fitness effects of spontaneous mutations in *Chlamydomonas reinhardtii*. PLoS Biol. 17: e3000192.

Booker, T. R., and P. D. Keightley. 2018. Understanding the factors that shape patterns of nucleotide diversity in the house mouse genome. Mol. Biol. Evol. 35: 2971–2988.

Breen, M. S., C. Kemena, P. K. Vlasov, C. Notredame, and F. A. Kondrashov. 2012. Epistasis as the primary factor in molecular evolution. Nature 490: 535–538.

Campos, P. R. A., and Wahl, L. M. 2009. The effects of population bottlenecks on clonal interference, and the adaptation effective population size. Evolution 63: 950–958.

Campos, P. R. A., and Wahl, L. M. 2010. The adaptation rate of asexuals: Deleterious mutations, clonal interference and population bottlenecks. Evolution 64: 1973–1983.

Charlesworth, B., and D. Charlesworth. 2010. Elements of Evolutionary Genetics. Roberts and Co. Publishers, Greenwood Village, CO.

Charlesworth, B., and K. Jain. 2014. Purifying selection, drift, and reversible mutation with arbitrarily high mutation rates. Genetics 198: 1587–1602.

Cohan, F. M. 1984. Can uniform selection retard random genetic divergence between isolated populations? Evolution 38: 495–504.

DePristo, M. A., D. L. Hartl, and D. M. Weinreich. 2007. Mutational reversions during adaptive protein evolution. Mol. Biol. Evol. 24: 1608–1610.

Fisher, R. A. 1930. The Genetical Theory of Natural Selection. Oxford University Press, Oxford, UK.

Frenkel, E. M., B. H. Good, and M. M. Desai. 2014. The fates of mutant lineages and the distribution of fitness effects of beneficial mutations in laboratory budding yeast populations. Genetics 196: 1217–1226.

Gillespie, J. H. 1983. Some properties of finite populations experiencing strong selection and weak mutation. Amer. Natur. 121: 691–708.

Gillespie, J. H. 1984. Molecular evolution over the mutational landscape. Evolution 38: 1116–1129.

Gokhale, C. S., Y. Iwasa, M. A. Nowak, and A. Traulsen. 2009. The pace of evolution across fitness valleys. J. Theor. Biol. 259: 613–620.

Good, B. H., and M. M. Desai. 2014. Deleterious passengers in adapting populations. Genetics 198: 1183–1208.

Good, B. H., M. J. McDonald, J. E. Barrick, R. E. Lenski, and M. M. Desai. 2017. The dynamics of molecular evolution over 60,000 generations. Nature 551: 45–50.

Higgs, P. G. 1998. Compensatory neutral mutations and the evolution of RNA. Genetica 102/103: 91–101.

Huber, C. D., B. Y. Kim, C. D. Marsden, and K. E. Lohmueller. 2015. Determining the factors driving selective effects of new nonsynonymous mutations. Proc. Natl. Acad. Sci. USA 114: 4465–4470.

Iwasa, Y., F. Michor, and M. A. Nowak. 2004. Stochastic tunnels in evolutionary dynamics. Genetics 166: 1571–1579.

Johri, P., B. Charlesworth, and J. D. Jensen. 2020. Towards an evolutionarily appropriate null model: Jointly inferring demography and purifying selection. Genetics 215: 173–192.

Katju, V., and U. Bergthorsson. 2019. Old trade, new tricks: Insights into the spontaneous mutation process from the partnering of classical mutation accumulation experiments with high-throughput genomic approaches. Genome Biol. Evol. 11: 136–165.

Keightley, P. D. 1994. The distribution of mutation effects on viability in *Drosophila melanogaster*. Genetics 138: 1315–1322.

Keightley, P. D., and A. Eyre-Walker. 1999. Terumi Mukai and the riddle of deleterious mutation rates. Genetics 153: 515–523.

Keightley, P. D., and A. Eyre-Walker. 2007. Joint inference of the distribution of fitness effects of deleterious mutations and population demography based on nucleotide polymorphism frequencies. Genetics 177: 2251–2261.

Kim, B. Y., C. D. Huber, and K. E. Lohmueller. 2017. Inference of the distribution of selection coefficients for new nonsynonymous mutations using large samples. Genetics 206: 345–361.

Kimura, M. 1983. The Neutral Theory of Molecular Evolution. Cambridge University Press, Cambridge, UK.

Kimura, M. 1985. The role of compensatory neutral mutations in molecular evolution. J. Genetics 64: 7–19.

Kimura, M., T. Maruyama, and J. F. Crow. 1963. The mutation load in small populations. Genetics 48: 1303–1312.

Kitahara, K., Y. Yasutake, and K. Miyazaki. 2012. Mutational robustness of 16S ribosomal RNA, shown by experimental horizontal gene transfer in *Escherichia coli*. Proc. Natl. Acad. Sci. USA 109: 19220–19225.

Komarova, N. L., A. Sengupta, and M. A. Nowak. 2003. Mutation-selection networks of cancer initiation: Tumor suppresser genes and chromosomal instability. J. Theor. Biol. 223: 433–450.

Kondrashov, A. S., S. Sunyaev, and F. A. Kondrashov. 2002. Dobzhansky–Muller incompatibilities in protein evolution. Proc. Natl. Acad. Sci. USA 99: 14878–14883.

Kulathinal, R. J., B. R. Bettencourt, and D. L. Hartl. 2004. Compensated deleterious mutations in insect genomes. Science 306: 1553–1554.

Lind, P. A., E. Libby, J. Herzog, and P. B. Rainey. 2019. Predicting mutational routes to new adaptive phenotypes. eLife 8: e38822.

Long, H., W. Sung, S. Kucukyildirim, E. Williams, S. W. Guo, C. Patterson, C. Gregory, C. Strauss, C. Stone, C. Berne, et al. 2017. Evolutionary determinants of genome-wide nucleotide composition. Nat. Ecol. Evol. 2: 237–240.

Lynch, M. 1986. Random drift, uniform selection, and the degree of population differentiation. Evolution 40: 640–643.

Lynch, M. 2007. The Origins of Genome Architecture. Sinauer Associates, Inc., Sunderland, MA.

Lynch, M. 2010. Scaling expectations for the time to establishment of complex adaptations. Proc. Natl. Acad. Sci. USA 107: 16577–16582.

Lynch, M. 2013. Evolutionary diversification of the multimeric states of proteins. Proc. Natl. Acad. Sci. USA 110: E2821–E2828.

Lynch, M. 2018. Phylogenetic diversification of cell biological features. eLife 7: e34820.

Lynch, M. 2020. The evolutionary scaling of cellular traits imposed by the drift barrier. Proc. Natl. Acad. Sci. USA 117: 10435–10444.

Lynch, M., and A. Abegg. 2010. The rate of establishment of complex adaptations. Mol. Biol. Evol. 27: 1404–1414.

Lynch, M., M. Ackerman, K. Spitze, Z. Ye, and T. Maruki. 2017. Population genomics of *Daphnia pulex*. Genetics 206: 315–332.

Lynch, M., J. Blanchard, D. Houle, T. Kibota, S. Schultz, L. Vassilieva, and J. Willis. 1999. Spontaneous deleterious mutation. Evolution 53: 645–663.

Lynch, M., and K. Hagner. 2014. Evolutionary meandering of intermolecular interactions along the drift barrier. Proc. Natl. Acad. Sci. USA 112: E30–E38.

Lynch, M., and G. K. Marinov. 2015. The bioenergetic costs of a gene. Proc. Natl. Acad. Sci. USA 112: 15690–15695.

McCandlish, D. M. 2018. Long-term evolution on complex fitness landscapes when mutation is weak. Heredity 121: 449–465.

McCandlish, D. M., and A. Stoltzfus. 2014. Modeling evolution using the probability of fixation: History and implications. Quart. Rev. Biol. 89: 225–252.

Neher, R. A. 2013. Genetic draft, selective interference, and population genetics of rapid adaptation. Ann. Rev. Ecol. Evol. Syst. 44: 195–215.

Nguyen Ba, A. N., I. Cvijović, J. I. Rojas Echenique, K. R. Lawrence, A. Rego-Costa, X. Liu, S. F. Levy, and M. M. Desai. 2019. High-resolution lineage tracking reveals travelling wave of adaptation in laboratory yeast. Nature 575: 494–499.

Pénisson S., T. Singh, P. Sniegowski, and P. Gerrish. 2017. Dynamics and fate of beneficial mutations under lineage contamination by linked deleterious mutations. Genetics 205: 1305–1318.

Quandt, E. M., D. E. Deatherage, A. D. Ellington, G. Georgiou, and J. E. Barrick. 2014. Recursive genomewide recombination and sequencing reveals a key refinement step in the evolution of a metabolic innovation in *Escherichia coli*. Proc. Natl. Acad. Sci. USA 111: 2217–2222.

Rice, D. P., B. H. Good, and M. M. Desai. 2015. The evolutionarily stable distribution of fitness effects. Genetics 200: 321–329.

Robert, L., J. Ollion, J. Robert, X. Song, I. Matic, and M. Elez. 2018. Mutation dynamics and fitness effects followed in single cells. Science 359: 1283–1286.

Rodrigues, J. V., and E. I. Shakhnovich. 2019. Adaptation to mutational inactivation of an essential gene converges to an accessible suboptimal fitness peak. eLife 8: e50509.

Santiago, E. 2015. Probability and time to fixation of an evolving sequence. Theor. Popul. Biol. 104: 78–85.

Sella, G., and A. E. Hirsh. 2005. The application of statistical physics to evolutionary biology. Proc. Natl. Acad. Sci. USA 102: 9541–9546.

Stephan, W. 1996. The rate of compensatory evolution. Genetics 144: 419–426.

Stephan, W., and D. A. Kirby. 1993. RNA folding in *Drosophila* shows a distance effect for compensatory fitness interactions. Genetics 135: 97–103.

Sung, W., M. S. Ackerman, M. Dillon, T. Platt, C. Fuqua, V. Cooper, and M. Lynch. 2016. Evolution of the insertion-deletion mutation rate across the tree of life. G3 (Bethesda) 6: 2583–2591.

Tóth-Petróczy, A., and D. S. Tawfik. 2013. Protein insertions and deletions enabled by neutral roaming in sequence space. Mol. Biol. Evol. 30: 761–771.

Tuğrul, M., T. Paixão, N. H. Barton and G. Tkačik. 2015. Dynamics of transcription factor binding site evolution. PLoS Genet. 11: e1005639.

Walsh, J. B. 1995. How often do duplicated genes evolve new functions? Genetics 139: 421–428.

Walsh, J. B., and M. Lynch. 2018. Evolution of Quantitative Traits. Sinauer Associates, Inc., Sunderland, MA.

Weinreich, D. W., and L. Chao. 2005. Rapid evolutionary escape by large populations from local fitness peaks is likely in nature. Evolution 59: 1175–1182.

Weissman, D. B., and N. H. Barton. 2012. Limits to the rate of adaptive substitution in sexual populations. PLoS Genet. 8: e1002740.

Weissman, D. B., M. M. Desai, D. S. Fisher, and M. W. Feldman. 2009. The rate at which asexual populations cross fitness valleys. Theor. Pop. Biol. 75: 286–300.

Weissman, D. B., D. S. Fisher, and M. W. Feldman. 2010. The rate of fitness-valley crossing in sexual populations. Genetics 186: 1389–1410.

Weissman, D. B., and O. Hallatschek. 2014. The rate of adaptation in large sexual populations with linear chromosomes. Genetics 196: 1167–1183.

Wiser, M. J., N. Ribeck, and R. E. Lenski. 2013. Long-term dynamics of adaptation in asexual populations. Science 342: 1364–1367.

Wright, S. 1931. Evolution in Mendelian populations. Genetics 16: 97–159.

Wright, S. 1932. The roles of mutation, inbreeding, crossbreeding and selection in evolution. Proc. 6th Internat. Cong. Genet. 1: 356–366.

Wright, S. 1969. Evolution and Genetics of Populations. The Theory of Gene Frequencies. University of Chicago Press, Chicago, IL.

Wu, N. C., L. Dai, C. A. Olson, J. O. Lloyd-Smith, and R. Sun. 2016. Adaptation in protein fitness landscapes is facilitated by indirect paths. eLife 5: e16965.

Zheng, J., J. L. Payne, and A. Wagner. 2019. Cryptic genetic variation accelerates evolution by opening access to diverse adaptive peaks. Science 365: 347–353.

CHAPTER 6

Evolution of Cellular Complexity

Having gained an appreciation for how various population-genetic forces interact to define the accessibility of alternative evolutionary pathways, we now turn to more specific issues relevant to the diversification of cellular features. That natural selection provides a powerful mechanism for advancing adaptive mutations is well established, so there is no need to belabor that issue further. Likely less familiar and/or less fathomable is the idea that the nonadaptive forces of mutation and drift can often dictate the paths down which phenotypic evolution is most likely to travel. In certain settings, the net result can be a gradual, passive increase in organismal complexity, with little (if any) increase in fitness throughout the process.

The goal now is to instil an appreciation for the shallowness of the assumption that natural selection is a process in relentless pursuit of biological complexity. The initial focus is on general issues regarding the evolution of complex features, with details specific to particular cellular structures and functions unfolding in subsequent chapters. To maximize the accessibility of the key points, a distinctly non-mathematical sojourn is taken, which is not to say that the mathematical details outlined in Chapter 5 are irrelevant.

Before proceeding, a brief recap of the population-genetic principles relevant to phenotypic divergence is in order. First, the classes of mutations available to selection depend on the effective population size (N_e), the inverse of which defines the power of random genetic drift. Selection will be ineffective if the randomizing potential of genetic drift is sufficiently strong. As a consequences, small populations can only advance beneficial mutations with relatively large effects and cannot prevent the accumulation of deleterious mutations with small effects. Large populations are more capable of evolutionary fine tuning.

Second, owing to the granularity and directional biases of mutations, phenotypic optima may only occasionally, if at all, be attainable for cellular traits. Large-N_e species are expected to evolve higher levels of efficiency and functionality of molecular attributes. However, small N_e enables populations to move into domains that can dramatically shift the course of evolution by natural selection, with mutation playing a powerful role in directing the paths open for exploration. As these fundamental evolutionary principles are unavoidable consequences of the nature of life's genetic material, they must be kept in mind in any attempt to explain cellular diversification.

Illusions of Grandeur

A common view is that biological complexity represents the crown jewel of the awesome power of natural selection (e.g., Lane 2020), with metazoans (humans in particular) representing the pinnacle of what can be achieved. This is a peculiar assumption, as there is no evidence that increases in complexity are intrinsically advantageous. Nor is there any evidence that biology's metabolic, morphological, and behavioral features have reached a maximum level of refinement or ever will. To think that a mammal is superior to a bacterium is as meaningful as proclaiming that an Olympic athlete is superior to an award-winning cellist. In the evolutionary arena, ecological context is paramount, and the currency of natural selection (relative fitness) is only exchangeable for members of the same gene pool. Bacteria can outperform vertebrates in myriad ways with respect to metabolism and environmental sensing.

Evolutionary Cell Biology. Michael Lynch, Oxford University Press. © Michael Lynch (2024). DOI: 10.1093/oso/9780192847287.003.0006

Vertebrates can harvest different food types and have complex visual and auditory systems. However, whereas a brain can be useful in certain settings, is there any objective basis for concluding that the streamlined signal-transduction systems of prokaryotes are fundamentally inferior to the baroque and error-prone nervous systems of animals? Consider the frequency with which humans make errors.

Although there are mathematical indices for quantifying complexity in physical systems, things are not so straightforward in living systems, and the term is used loosely here to simply reflect differences in the numbers of unique parts and interactions within organisms. Even these measures are not always easily enumerated, rendering comparisons among closely related organisms difficult. However, aspects of cellular complexity that most pique the interest of biologists are features such as large protein complexes, the emergence of the eukaryotic cell plan from a prokaryotic ancestor, and the transition from unicellularity to multicellularity. In these cases, there is no disagreement on where things lie on the complexity gradient.

In contrast to eukaryotes, most prokaryotes have not evolved internal cell structure or complex multicellularity. Is this a sign of evolutionary inferiority, that is of an innate inability to generate increased morphological complexity despite the potential benefits? Given their enormous population sizes, their ability to recombine, and their presence on the planet for ~4 billion years, the supply of variation is hardly limiting for microbes, and as noted in Chapters 2, 3, and 24, aspects of intracellular complexity and even multicellularity have in fact emerged in some prokaryotes. Thus, the unavoidable conclusion is that morphological complexity is actively selected against in the prokaryotic world. And if that is the case, what is the evidence that increased complexity is universally advantageous in eukaryotes?

The evolution of root systems and support tissues enabled land plants to occupy ecological niches unavailable to microbes, and the evolution of predatory capacity in animals opened up new ways of living. Surely, such transitions were promoted by natural selection. However, with such transitions, other modes of living were left behind, new survivorship

challenges were encountered, and rapid rates of reproduction were relinquished. Moreover, the question remains as to whether all of the underlying genetic and cellular changes in such organisms were necessary antecedents to such adaptation, as opposed to being inadvertent by-products of such changes. For example, relative to their unicellular ancestors, in just a few tens of millions of years of evolution, the genomes of metazoans and land plants independently became bloated with nonfunctional, energetically costly, and mutationally hazardous DNAs such as mobile-genetic elements and large introns (Lynch 2007). Were all such embellishments essential tickets to the evolution of organismal complexity, somehow maintained in anticipation of future benefits? No credible mechanisms exist for such evolutionary prescience. More likely, many aspects of increased genome complexity simply reflect the reduced efficiency of natural selection against genomic insertions in larger organisms with reduced effective population sizes.

There are at least three reasons why cellular/organismal complexity can be suppressed in certain lineages, while passively increasing in others. First, more complex features inevitably impose greater bioenergetic costs for construction and maintenance. For small cells with relatively low total energy budgets and large effective population sizes, even minor additions to the cellular repertoire can be efficiently opposed by selection unless there are immediate benefits. In contrast, for larger cells with higher total energy budgets, a given genomic addition comprises a smaller fraction of the total energy budget. Combined with a higher power of random genetic drift resulting from populations with smaller effective population sizes (Chapter 4), moderately sized cellular additions will then be less visible to the eyes of natural selection, and subject to fixation in an effectively neutral fashion. These issues are addressed more formally in Chapter 17. The main point here is that cell size alone can dictate the degree to which initially unnecessary (and sometimes weakly harmful) embellishments can become established in a population.

Second, virtually all gene-structural embellishments increase the vulnerability of genes to inactivating mutations (Lynch 2007). Typically, the increased mutational susceptibility is relatively small (on the order of the product of the mutation rate per nucleotide site, u, and the number of key nucleotide sites for proper gene function imposed by the embellishment, n). As a consequence, weakly mutationally hazardous genomic alterations will only be effectively selected against in populations with very large effective sizes. As an example, n is on the order of 25 for proper intron splicing, and u is in the range of 10^{-10} to 10^{-8} (Chapter 4). If nu is smaller than the power of drift ($1/N_e$ for a haploid), the mutational excess associated with such a gene addition cannot be countered by purifying selection.

Finally, all other things being equal, the drift-barrier hypothesis implies that organisms with lower N_e will also evolve to have less refined structural and functional features. The negative correlation between the mutation rate and N_e (Chapter 4) provides a case in point, and other examples appear in subsequent chapters. In some cases, the reduced functionality of a system can open up opportunities for the establishment of additional layers of complexity, which can in turn lead to further relaxation of selection on previously established mechanisms, leading to the false impression that robust systems represent adaptive improvement (Chapter 20). This is discussed in further detail in the following sections.

Taken together, these arguments highlight the fact that N_e limitations, driven by fundamental constraints associated with ecology and the genetic machinery, play a central role in encouraging particular lineages to ascend up the hierarchy of complexity by nonadaptive mechanisms. That is, certain population-genetic environments are conducive to the passive operation of a complexity ratchet, with small incremental changes accruing on short timescales cumulatively leading a lineage to a new location in phenotypic space. One might still expect that, in moving up the ladder of biological organization – from nucleotide sequences to translated protein products to higher-order structural and biochemical features of cells, there will be a diminishing probability of effectively neutral evolution. However, as discussed next, the very nature of genome and cellular organization facilitates the emergence of neutral evolutionary pathways at higher levels. Just as the third positions of codons

for amino acids with fourfold redundancy in the genetic code renders some nucleotide substitutions effectively neutral, many aspects of cellular architecture are structured in ways that provide multiple degrees of freedom for making molecular shifts with minor fitness consequences. Thus, the evolution of increased complexity need not imply increased superiority in any sense of the word, and evolution driven by nonadaptive mechanisms (mutation, recombination, and random genetic drift) need not imply a descent towards overtly maladaptive change.

Constructive Neutral Evolution

A verbal model presented by Stoltzfus (1999) and colleagues (Gray et al. 2010; Lukeš et al. 2011; Brunet and Doolittle 2018) suggests ways in which seemingly gratuitous cellular complexity might grow in the absence of direct selection for such features. The process they call constructive neutral evolution (CNE) has some antecedents in earlier verbal models of Woese (1971) and Zuckerkandl (1997).

Consider an ancestral cellular function carried out by the product of a single gene (A) (Figure 6.1). Suppose a fortuitous physical interaction then develops with another protein B, with such binding having negligible effects on both A and B's functionality. By hiding part of A's surface from the cellular environment, B may suppress the effects of future mutations arising at the A–B interface that would be destabilizing to A if exposed (Chapter 13). Over time, this permissive interfacial environment could then lead to enough mutational build-up that A would no longer be functional without B. In principle, this evolved functional dependence of A on B could be followed by a similar scenario involving a third protein, C, and so on.

Under this scenario, the intricate interdependencies of the components of molecular complexes need not always have been advanced by positive selection for functional improvement. Rather, they may simply be the result of a series of effectively neutral coevolutionary steps accompanied by relaxed selection against previously forbidden mutations.

Although this verbal model provides a plausible argument for the passive origin of complexity, three key assumptions underlie the CNE

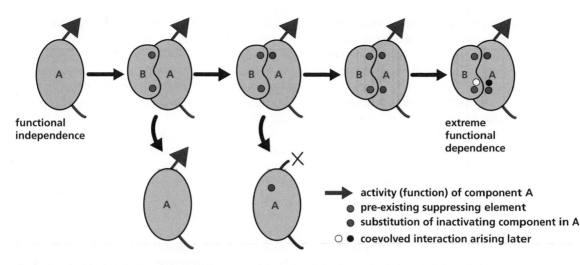

Figure 6.1 An idealized scenario by which increased complexity might arise by constructive neutral evolution – a transition from an independently functioning molecule A to an obligatory A:B interaction. Initially, a fortuitous interaction with B suppresses subsequent deleterious mutational effects in A (denoted by the red dot, which would otherwise be eliminated by selection), rendering A dependent on B. The complex then becomes further entrenched evolutionarily, as A acquires an additional conditionally silent mutation (red) that if exposed by elimination of B leads to loss of function. In the final stage, two additional mutations (white and black), potentially refining A:B function beyond its initial state, have become established. From Lukeš et al. (2011).

hypothesis. Foremost is the idea that biological systems often harbor excess capacity. In particular, the process requires that the evolutionary diversion of B molecules to A has negligible effects on any preexisting benefits of B, at least to the extent that could be opposed by natural selection. As excess capacity implies a superfluous energetic drain on the cell, why would such conditions exist? As discussed in the following section, although redundancy is unlikely to be promoted on its own merits, recurrent gene duplication may lead to a sort of quasi-equilibrium level of redundancy at the population level. The specific genes in duplicate form at any particular time will vary, but some such genes will nearly always be present. In addition, transient conditions may exist in which a change in environment may render the prior function of B obsolete such that its diversion has no fitness consequences.

The second issue is that the evolution of A's dependency on B requires that the fortuitous A–B interaction survives for a long enough period for A to acquire the conditionally harmful mutations essential to the development of dependency on B. This returns us to the kinds of scenarios outlined in Chapter 5 whereby a small number of mutations are required for a transition to an alternative semi-stable state.

But this also brings us to the third assumption of the CNE model. Unlike the situation in which populations can shift in both directions, the transition to complexity under CNE is viewed as being a one-way street. The assumption here is that once the complex is established, the accumulation of conditionally lethal mutations becomes extreme enough to essentially eliminate the possibility of an evolutionary reversion to the simpler condition.

Unfortunately, the population-genetic requirements for the operation of CNE have not been formally worked out except in the case of evolution by gene duplication (covered in the following section). However, based on the theory outlined in the previous chapters, one can at least envision scenarios under which the process is most likely to proceed. All of these involve a relaxation in the efficiency of selection, in particular an initial A:B state that is no worse than very weakly deleterious, combined with a sufficiently small effective population size to render the initial transition effectively neutral.

In potential support of the CNE model, numerous examples exist in which molecular complexes with universally conserved functions have larger numbers of subunits in eukaryotes than in prokaryotes. Consider, for example, oxidative phosphorylation. Carried out in the mitochondria of eukaryotes, and on the plasma membranes of prokaryotes, this energy-generating mechanism involves multiple complexes with conserved functions throughout the Tree of Life. Well over 100 subunits encoded by different genes are distributed among the multiple electron-transport chain (ETC) complexes in eukaryotes, more than double the number found in bacteria (Hirst 2011; Huynen et al. 2013). Although most of these additions occurred prior to LECA, there have been numerous subsequent lineage-specific accruals. Nearly all of the accessory proteins are encoded in the nuclear genome. Although a common explanation for their existence is their essential role in maintaining structural stability of the complexes, the larger eukaryotic complexes are no more stable than those in bacteria. It has been argued that the subunit additions evolved as structural compensations for defects in the mitochondrially encoded components (resulting from deleterious-mutation accumulation in organelle genomes; Chapter 23) (Angerer et al. 2011; Hirst 2011; van der Sluis et al. 2015). However, a CNE scenario in which structural dependency arose as a consequence, rather than a cause, of subunit recruitment has not been ruled out.

A second example of the apparently gratuitous evolution of complexity involves the ribozyme RNase P, a complex of proteins surrounding a single catalytic RNA molecule that processes precursor transfer RNAs to their mature form. Although the RNA subunit is similar in all organisms, bacterial RNase P deploys just a single protein, whereas the archaeal and eukaryotic complexes contain five to ten proteins. This is a substantial investment in complexity for an enzyme whose sole role is to cleave a single phosphodiester bond. Again, the primary function of the additional proteins appears to be in stabilizing the overall complex, although there is no evidence that the eukaryotic RNase P is exceptionally stable (Lan et al. 2018). Thus, such dependencies likely arose secondarily as initially fortuitous interactions became entrenched by the accumulation of otherwise harmful mutations for interfacial residues. Indeed, whereas the RNA core of the bacterial complex is internally stabilized by tertiary RNA–RNA interactions, these structural RNA features are reduced in archaeal and eukaryotic RNAs (Gopalan et al. 2018), as expected under the CNE model.

It has been argued that the evolution of higher-order RNase P complexes is a by-product of their having evolved additional cellular functions (Gopalan et al. 2018), but the possibility that any such functions could also be fulfilled by less-elaborate structures has not been ruled out. In fact, a few bacteria and eukaryotes have lost the RNA component of RNase P and carry out the usual function solely with an enzymatic protein complex, showing that a simpler structure can indeed suffice. Complementation studies have shown that these RNA-free proteins function with no apparent harmful effects when they are expressed in species that normally utilize RNA-containing RNase P (Weber et al. 2014; Lechner et al. 2015; Nickel et al. 2017).

Although a number of open questions remain, the simplest explanation for these observations on ETC complexes and RNase P is that excess complexity arose within eukaryotes by effectively neutral processes, the result being the conservation of ancestral functions but with increased bioenergetic cost to the organism. Other examples of apparent overdesign of eukaryotic features include the circadian clock, which typically is based on products of no more than three genes in prokaryotes (Chapter 18) but involves a complex web of many more genes in eukaryotes (Sancar 2008), and the spliceosome, a complex of five RNAs and dozens of proteins involved in intron splicing, which evolved from a single-component self-splicing intron in prokaryotes (Lynch 2007). Elaborating on earlier ideas of Stoltzfus (1999) and Lukeš et al. (2011), we now consider in more depth another potential example of CNE involving an even larger ribonucleoprotein complex.

Ribosomes

In all cells in all organisms, the ribosome has a single, conserved role – the translation of messenger RNAs. The catalytic core of the ribosome consists of three to four ribosomal RNAs (rRNAs), which collectively operate as a complex ribozyme. However, no ribosome can operate unless the rRNAs are co-assembled with dozens of structural proteins. The question of why a molecular machine of this sort would require such a large endowment of protein components is further motivated by the substantial variation in the set of ribosomal proteins utilized in different phylogenetic lineages.

Thirty-four ribosomal proteins are universally deployed in all eukaryotes and prokaryotes and often referred to as the common core. However, at least 34 other ribosomal proteins are also shared by eukaryotes and archaea but absent from bacteria, whereas bacteria share no ribosomal protein just with eukaryotes or just with archaea (Lecompte et al. 2002; Hartman et al. 2006). This phylogenetic distribution is entirely consistent with the hypothesis that bacteria form an outgroup to archaea/eukaryotes (Chapter 3).

Not only do the protein constituents of ribosomes vary among the major domains of life, but the numbers of distinct proteins deployed vary as well. Each domain harbors unique ribosomal proteins not found in either of the other groups. In bacteria, ∼21 and 33 proteins are contained within

the small and large ribosomal subunits (denoted SSUs and LSUs, and respectively responsible for decoding mRNA information and forming peptide bonds). In eukaryotes, these numbers expand to 33 and 46, respectively (Melnikov et al. 2012), with most of the additional proteins joining the external surfaces of the ribosome, like rings on an onion (Hsiao et al. 2009).

The two major rRNAs, occupying the small and large subunits, also vary in size among organisms, with an average 50% expansion of both in eukaryotes relative to prokaryotes, and with weak coordination in size changes between the two subunits (Figure 6.2). Most rRNA enlargements occur by the addition of expansion segments that leave the common core structure undisturbed (Petrov et al. 2014).

In eukaryotes, separate ribosomes are deployed in the cytosol and in mitochondria. The rRNAs are encoded in their respective genomes, but the proteins of both are almost always nuclear-encoded. The rRNAs deployed within mitochondria are often reduced in size relative to those in bacteria. For example, the mammalian mitochondrial LSU rRNA contains less than a third of the number of nucleotides as its counterpart in the cytosolic ribosome (1559 vs. 5347) and only half that in typical

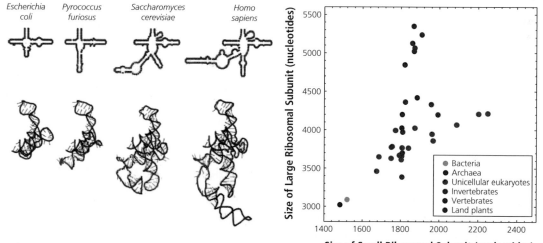

Figure 6.2 Ribosomal RNA evolution. **Left:** Adventitious growth of one particular set of helices associated with the SSU ribosomal RNA. The nearly invariant functional core is represented by the dark blue subsections. Upper and lower diagrams depict two- and three-dimensional structures. From Petrov et al. (2015). **Right:** Sizes of the small and large rRNA subunits in various taxa (in terms of sequence length). From Petrov et al. (2014).

bacteria, although those in yeast and other protists can be comparable in size to those in bacteria.

Substantial modifications in the protein contribution to ribosomal structure have also evolved in mitochondria. Despite having to typically translate just a dozen or so mitochondrial genes, the protein repertoire of mitochondrial ribosomes is typically quite large. Overall, mitochondrial ribosomes contain 10–20 proteins not found in their alphaproteobacterial ancestors, with these again largely being distributed over the ribosome surface (Desmond et al. 2011). For example, the human mitochondrial LSU contains 48 proteins, all of which are encoded in the nuclear genome and 21 of which are mitochondrial-specific (Brown et al. 2014). Eleven of these 21 are not found in the yeast mitochondrial ribosome LSU, which nevertheless contains 39 proteins (Amunts et al. 2014).

The overall picture that one gets here is that ribosome expansion likely followed the emergence of eukaryotes, with further gains and losses of components then occurring on individual lineages, and with all such changes leaving the internal catalytic core intact. However, not just the structure of the ribosome, but also the pathways involved in ribosome biogenesis became more elaborate in eukaryotes (Strunk and Karbstein 2009). In bacteria, ribosome assembly involves just a few additional proteins, whereas on the order of 200 accessory proteins are essential for the development of mature eukaryotic ribosomes. The operation of many of these ribosome-biogenesis proteins requires hydrolysis of nucleotide triphosphates (ATP or GTP) and hence, is energetically demanding. Thus, considering the expanded number of nucleotides and amino acids in eukaryotic ribosomal RNAs and proteins, it is clear that the overall energetic cost of the translational machinery in eukaryotes is substantially greater than that in prokaryotes.

Given their association with organismal complexity, it has been argued that ribosome expansions and elaborations reflect a long-term pattern of adaptive divergence of ribosome architecture (Petrov et al. 2014, 2015). However, such a view is confronted with two fundamental problems: 1) the apparent inability of prokaryotes to achieve such changes despite having existed for longer periods of time and in much larger populations than eukaryotes; and 2) the absence of evidence that either the expansion segments of rRNAs or the additional ribosomal proteins confer any intrinsic benefits or novel functions.

The maximum rate of translation per ribosome (codons per second) in eukaryotes $\simeq 17$ in *Neurospora crassa* (Alberghina et al. 1975), 10 in *Saccharomyces cerevisiae* (Boehlke and Friesen 1975; Waldron and Lacroute 1975; Bonven and Gulløv 1979), and 6 in mouse embryonic stem cells (Ingolia et al. 2011). Estimates in bacteria are 20 in *E. coli* (Forchhammer and Lindahl 1971; Dennis and Bremer 1974; Young and Bremer 1976), 16 in *Staphylococcus aureus* (Martin and Iandolo 1975), and 3 in *Streptomyces coelicolor* (Cox 2004). Thus, although some of these estimates are likely more reliable than others, there is no indication of an elevated processing rate in larger eukaryotic ribosomes. Nor is there any indication that translation accuracy is improved in eukaryotes (Chapter 20).

These kinds of observations have not inhibited some authors from claiming that ribosomes are optimally designed. Focusing on *E. coli*, Reuveni et al. (2017) have argued that ribosomes consist of large numbers of similarly sized but unusually small proteins because such features maximize cellular efficiency. However, this conclusion seems to be another example of the perils of the adaptive-paradigm syndrome – the inevitable arrival at some kind of optimization argument if one searches hard enough. In fact, the proposed hypothesis is readily rejected upon a closer look at the data (Wei and Zhang 2018). Likewise, although Kostinski and Reuveni (2020) argue that the 2:1 mass ratio for rRNA: protein in bacterial ribosomes maximizes growth rate, their analysis is conditional on other ribosomal features and fails to address why proteins are required at all.

Evolution by Gene Duplication

We now turn to a major route to the evolutionary origin of novelty and complexity with an ample body of empirical support. Although much of the theory reviewed in Chapter 5 focused on small incremental changes to individual genes, such as single-nucleotide substitutions, larger-scale changes are also common. Duplications of entire genes or fragments thereof are of special interest because they generally contain fully functional domains tested under a prior history of selection. In this sense, novel gene functions do not have to be built from scratch, but more often than not can arise as elaborations of pre-existing functions. The potential contribution of gene duplication to evolutionary innovation is substantial, as individual genes duplicate at rates that are comparable to the rates at which base-substitution mutations arise at individual nucleotide sites (Lynch and Conery 2000; Konrad et al. 2018).

The fates of duplicate genes depend on the mechanisms by which they arise and the population-genetic environments within which they reside. Owing to the random breakpoints of duplicated DNA spans, duplication events will not necessarily encompass the full regulatory and/or coding regions of parental genes, and hence, may have divergent features at birth (Katju and Lynch 2006). At the other extreme, exceptional cases involve whole-genome duplications in which all genes are simultaneously duplicated in entirety. Such events have occurred in the ancestry of numerous eukaryotic lineages, including yeast (Wolfe and Shields 1997), ciliates (Aury et al. 2006; McGrath et al. 2014; Gout et al. 2023), vertebrates (Jaillon et al. 2004; Chain and Evans 2006; Putnam et al. 2008), arthropods (Kenny et al. 2016; Li et al. 2018), and land plants (Soltis and Soltis 2016).

Like all mutations, gene duplicates are initially present in just a single copy in a single individual. This will also be true for genes arising by other mechanisms, such as fortuitous *de novo* origin from pre-existing non-coding sequence (Wissler et al. 2013; Bornberg-Bauer et al. 2015; McLysaght and Hurst 2016; Neme and Tautz 2016; Vakirlis et al. 2020) or via horizontal transfer from exogenous sources (Keeling and Palmer 2008; Vos et al. 2015). Thus, all of the population-genetic issues fundamental to the establishment of point mutations (Chapter 5), and more, apply to gene duplication. To be successful in the long term, a new gene must first drift towards fixation, and having arisen to high frequency, must then be preserved by sufficiently strong selective forces to prevent rapid loss by degenerative mutation.

The vast majority of duplicates arising by single-gene duplications are lost from populations on time scales of less than a few million generations (Lynch and Conery 2000), most never even

proceeding to fixation. Basic population-genetic principles (Chapter 5) indicate why. Letting N be the population size and assuming diploidy, in the absence of immediate positive (or negative) selection, a fraction $[1 - (1/2N)]$ of newly arisen gene duplicates will be lost by random genetic drift in an average of just $\sim 2\ln(2N_e)$ generations (Kimura and Ohta 1969), a flash on the evolutionary time scale, as $2\ln(2N_e) \simeq 43$ with N_e at the upper limit of 10^9. Moreover, the small remaining fraction that manages to drift to fixation, $1/(2N)$, is also expected to fall victim to silencing mutations relatively quickly unless a preservational mechanism is acquired. Letting μ denote the rate of appearance of gene-silencing mutations, the average time to gene inactivation is on the order of the mean waiting time for the appearance of a null mutation at one of the two loci, $\simeq 1/(2\mu)$ generations, which will generally be on the order of 10^6 or so generations (Watterson 1983; Lynch et al. 2001).

Although it is often argued that an increase in gene number is a sign of evolutionary success and superiority (e.g., Lane and Martin 2010), there is little support for this point of view. Indeed, the number of genes per genome is nearly decoupled from organismal complexity (Chapter 24). For example, the genomes of the most behaviorally sophisticated animals contain fewer genes than found in many protists and only a few times more than in most bacteria. Only a few hundred genes are conserved across the entire Tree of Life (Tatusov et al. 2003; Koonin et al. 2004), and there can even be substantial differences in the numbers of genes among individuals within a species. This being said, the evidence is overwhelming that the repatterning of gene functions and gene locations by duplication events plays a central role in organismal diversification, although the connections often have little to do with adaptive processes.

The goals here are to summarize the ways in which gene duplication opens up novel pathways for evolutionary elaboration, provide insight into how the likelihoods of such processes are influenced by the population-genetic environment, and address some of the concerns with the more general model of constructive neutral evolution. More thorough reviews on the rates of origin, fates, and consequences of duplicate genes can be found Lynch (2007), Conant and Wolfe (2008),

Innan and Kondrashov (2010), and Katju (2012). The small minority of duplicates that are retained for long periods of time are thought to owe their preservation to one of four mechanisms, one of which will first be dispensed with.

The masking effect

All populations harbor low-frequency, suboptimal alleles resulting from the recurrent introduction of deleterious mutations, and this has encouraged the common view that duplicate genes have an intrinsic selective advantage associated with their ability to mask the effects of deleterious mutations at the ancestral locus. However, the frequency at which a back-up is useful is proportional to the incidence of deleterious genotypes at the opposite locus, which is on the order of the mutation rate to degenerative alleles. Thus, the selective advantage of a back-up gene is approximately equal to the rate of its own silencing by deleterious mutations. This leads to a miniscule selective advantage of the masking effect, generally smaller than the power of random genetic drift (Fisher 1935; Clark 1994; Lynch et al. 2001; Proulx and Phillips 2005).

The most serious and obvious challenge to the masking hypothesis for duplicate-gene retention is the general paucity of duplicate genes in haploid microbes despite their exceptionally high effective population sizes (which would maximize the efficiency of selection for weakly favorable redundancy). As discussed in Chapter 17, the energetic cost of a gene in bacterial species (relative to the total cellular energy budget) is generally sufficiently large for selection to efficiently remove redundant gene duplicates on this basis alone.

Neofunctionalization

Historically, the origin of a new gene function was thought to be the only preservational mechanism for the long-term survival of gene duplicates, with the much more common fate being the mutational silencing of one copy by degenerative mutations (Haldane 1933; Muller 1940; Ohno 1970). The idea here is that gene duplication can free one copy for evolutionary exploration and eventual acquisition of a new adaptive function. If the modifications underlying this new function are acquired at the expense of essential ancestral gene functions, the joint maintenance of both members of the pair will be enforced. A key issue here, of course, is that duplicate-gene preservation by neofunctionalization requires a setting in which there is indeed utility for a new gene function.

Neofunctionalization is expected to be more common in large populations for at least three reasons (Lynch et al. 2001; Walsh 2003). First, the larger the population size, the greater the population-level rate of origin of a rare neofunctionalizing mutation, and hence, the higher the probability of fixation of such a mutation prior to one locus being silenced by a degenerative mutation. Second, in a sufficiently large population, even a duplicate gene initially destined to be lost by random genetic drift has a non-trivial chance of being rescued and propelled forward by a neofunctionalizing mutation. Third, in very large populations, the process need not depend on new neofunctionalizing mutations at all, as the requisite alleles may be maintained at low frequency by selection-mutation balance in the base population (but incapable of spreading to fixation prior to duplication because individuals lacking the essential ancestral allele are inviable). As discussed further in Chapter 13, cases of balancing selection (e.g., heterozygote superiority) may sometimes maintain two alleles at moderate frequencies at single genetic loci (Spofford 1969).

Subfunctionalization

With the emergence of genome-sequence data in a wide variety of lineages, it became clear that the levels of retention of duplicate genes following whole-genome duplications are far too high to be consistent with a model in which most are preserved by the evolution of novel functions. Given the mutation rate to degenerative mutations, duplicate genes are also far too common to be fortuitous avoidances of gene silencing. Thus, something other than neofunctionalization must often be responsible for duplicate-gene retention. The fact that the vast majority of newly arising mutations are deleterious, combined with the emerging understanding of gene-structural complexity, suggested a resolution to this dilemma – a mechanism by which duplicate-gene preservation can be completely driven by degenerative mutations. Under the DDC (duplication-degeneration-complementation) model, both members of a gene pair acquire complementary negative changes that necessitate joint preservation (Force et al. 1999; Lynch and Force 2000; Lynch et al. 2001).

In the case of multifunctional genes, subfunctionalization can involve the partitioning of independently mutable, essential gene functions, leading to specialized copies with nonoverlapping features (qualitative subfunctionalization; Figure 6.3). Subfunctionalization can also be instigated by partial reduction in the efficiencies of the same functions in both members of a pair down to the total level required in the single-copy state (quantitative subfunctionalization) (Lynch and Force 2000; Duarte et al. 2006; Gout and Lynch 2015; Thompson et al. 2016). In both cases, subfunctionalization eliminates the need for beneficial mutations in the gene

preservational process, although this need not rule out the emergence of secondary, adaptive modifications, as noted in the following section.

Contrary to the situation with neofunctionalization, the probability of subfunctionalization is expected to diminish with increasing effective population sizes, for at least three reasons (Lynch et al. 2001; Walsh 2003). First, there are prices to be paid for a pair of subfunctionalized genes. With respect to the coding region, the system will be roughly twice as mutationally vulnerable as a single-copy gene, thereby imposing a selective disadvantage equivalent to the null mutation rate per gene; and as just noted, there will also be an energetic cost of duplicate-gene maintenance and operation. As both of these costs are relatively small, they will only be opposed by selection in large populations. Second, a subfunctionalized allele en route to fixation is vulnerable to acquiring secondary silencing mutations, and the likelihood of such an effect is magnified in a large-N_e setting owing to the longer time to drift

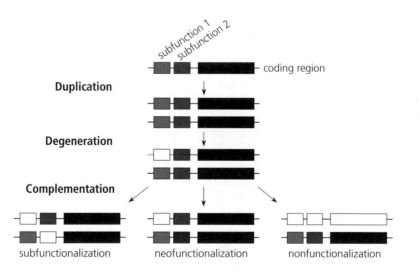

Figure 6.3 The DDC model for the alternative fates of duplicate genes. The ancestral gene is depicted as having two independently mutable subfunctions (blue and green), e.g., two regulatory elements, each driving expression in a particular tissue, subcellular location, or environmental condition. Solid boxes denote fully functional regulatory and coding regions, whereas open boxes denote loss of function, and an orange box denotes the gain of a new beneficial function. Each pair of genes reflects the fixed state of the population. In this example, following the duplication event, the first degenerative mutation eliminates a subfunction from one of the copies. The second mutational event then dictates the final fate of the pair: subfunctionalization, with the second copy acquiring a complementary loss-of-subfunction mutation; neofunctionalization, with the second copy acquiring a novel, beneficial expression pattern at the expense of an ancestral subfunction; or nonfunctionalization, with the first copy losing all functionality. From Force et al. (1999).

to fixation. Finally, qualitative subfunctionalization requires the presence of independently mutable regulatory mechanisms or protein domains, and as discussed below, the evolution of such modularity is reduced in large-N_e settings.

Adaptive-conflict resolution

The action of subfunctionalization and/or neofunctionalization may lead to gene copies that are not only largely distinct from each other but also have improved functionalities relative to what is possible with a single-copy ancestral gene (Piatigorsky and Wistow 1991; Hughes 1994; Stoltzfus 1999). Consider, for example, a single-copy locus subject to a 'jack-of-all-trades is a master-of-none' syndrome, that is with an adaptive conflict between its subfunctions. In such a situation, following duplication, complementary loss-of-subfunction mutations may alter the selective landscape experienced by the two pair members, enabling each copy to become more refined to a specific subset of tasks, potentially even opening up previously unavailable pathways to neofunctionalization. By this means, two of the most common forms of genomic upheaval, gene duplication and degenerative mutation, can provide a unique mechanism for the creation of novel evolutionary opportunities through the elimination of pleiotropic constraints. Again, however, whether such an adaptive-conflict resolution leads to a net selective advantage will depend on the degree to which the improvement(s) in gene functions exceed the cost of maintaining two genes.

A variant on the adaptive-conflict model is the IAD (innovation-amplification-divergence) model of Bergthorsson et al. (2007), which postulates that a common path to the origin of a new function starts with the duplication of a gene with a promiscuous secondary function, which in times of extreme need might suffice to provide enough functional rescue to buy time for further evolutionary refinement. Additional duplications would increase the number of mutational targets for such improvements, with deletions of the excess copies after establishment of the neofunctionalized gene eliminating the cost of gene amplification. Näsvall et al. (2012) and Newton et al. (2017) demonstrated the operation of this mechanism in the bacterium *Salmonella*, focusing

on a gene involved in histidine biosynthesis with weak promiscuous involvement in tryptophan biosynthesis. When placed on a genetic background lacking the primary tryptophan-synthesis pathway, evolutionary rescue was accomplished as duplicates of the histidine gene arose and in some cases became specialized to alternative pathways leading to tryptophan. Other examples of this sort are discussed in Chapter 19, which shows that adaptive-conflict resolution and gene duplication play a major role in the evolutionary remodelling of metabolic pathways.

The Case for Subfunctionalization

Prior to the development of the DDC model, circumstantial evidence for duplicate-gene preservation via subfunctionalization was suggested by studies of polyploid fishes, which repeatedly revealed tissue-specific expression of duplicated enzyme loci (Ferris and Whitt 1977, 1979). Such observations have now been supplemented by a wide array of investigations in other ray-finned fish lineages, zebrafish in particular, all of which arose following a whole-genome duplication event (e.g., Pasquier et al. 2017). Without an outgroup, it is difficult to determine whether duplicate-gene specialization is an outcome of neofunctionalization versus subfunctionalization. However, the evolutionary interpretations of divergent-expression-patterns of duplicate genes in fishes have been greatly facilitated by observations of orthologous single-copy genes in tetrapods (usually mouse or chicken). These lineages, which branched off prior to the ray-finned fish-specific polyploidization event, generally reveal the presence of both gene subfunctions in their single-copy gene. Similar observations have been made in the tetraploid frog *Xenopus laevis* in comparison to its diploid relatives (Morin et al. 2006; Sémon and Wolfe 2008), as well as in numerous land plants (Rutter et al. 2012; Freeling et al. 2015). Indeed, there are now hundreds of examples of qualitative subfunctionalization of duplicated genes via the partitioning of tissue-specific expression in multicellular organisms.

Although this particular mechanism of duplicate-gene preservation (i.e., tissue-specific expression divergence) is unavailable to unicellular species, there many other potential paths to subfunction partitioning in such organisms. For example, gene products may become specialized for use in different subcellular locations. Genes can also be regulated in modular ways with respect to timing of expression during the cell cycle or in response to different environmental conditions. In addition, proteins that assemble as homomeric multimers can acquire complementary interfacial changes after duplication, enforcing assembly as heteromers between the duplicate-gene products (Diss et al. 2017; Chapter 13).

Thus, although unicellular species often have large effective population sizes, which might be expected to reduce the incidence of subfunctionalization, the process is by no means restricted to multicellular species. Indeed, as outlined in subsequent chapters, key episodes of the process may have been facilitated during small-N_e phases in early eukaryotic history.

A striking example of subfunctionalization deep in the eukaryotic phylogeny involves the dynamin family of proteins, which are used to pinch membranes. Phylogenetic analysis suggests the presence in LECA of a bifunctional dynamin with dual roles in vesicle scission from cell membranes and in mitochondrial division (Leger et al. 2015; Purkanti and Thattai 2015). Although this single dual-function gene is retained in numerous eukaryotic lineages, in three independent lineages duplicate copies became specialized to the two alternative ancestral functions.

Given the enormous amount of cell biological work done on yeast, and the whole-genome duplication that preceded the emergence of *Saccharomyces cerevisiae* (Wolfe and Shields 1997), much has been learned about the mechanisms preserving duplicate genes in this species. In particular, empirical studies in which *S. cerevisiae* duplicates have been replaced with the single-copy gene from a closely-related outgroup species have provided some of the most compelling evidence for subfunctionalization (van Hoof 2005). For example, *Orc1* and *Sir3* are sister genes in *S. cerevisiae*, with the former playing a role in chromosomal replication and the latter being part of a nucleosome-binding complex involved in chromosome-silencing functions. In *Kluyveromyces lactis*, a related taxon that branched off prior to

whole-genome duplication, both functions are carried out by a single-copy gene (Hickman and Rusche 2010).

An example of subfunctionalization's role in adaptive-conflict resolution has also been revealed by molecular dissection in *S. cerevisiae*, where two sister genes are involved in galactose utilization, one (*Gal3*) playing a regulatory role in pathway induction and the other (*Gal1*) serving as a galactokinase (Hittinger and Carroll 2007). Again by reference to *K. lactis*, it was determined that the ancestral single-copy gene served both functions. Gene duplication then allowed the refinement of binding-site configurations that had previously been constrained in the ancestral gene, thereby enabling the emergence of a more tightly regulated system (Figure 6.4).

A striking example of subfunctionalization based on structural alterations is provided by the hexameric membrane ring for the vacuolar ATP synthase pump in yeast (Figure 6.5). In most eukaryotes, the ring consists of five copies of one protein (Vma16) and one of another (Vma3), both of which arose from an ancient gene duplication. In fungi, a third duplicate (Vma11) that arose by duplication of Vma3 replaces one subunit of Vma16, specifically residing between Vma16 and Vma3. Experimental modifications of the subunit interfaces revealed that one side of Vma3 lost the ability to bind to one side of Vma16, whereas the other side of Vma11 lost the ability to bind to Vma3 (Finnigan et al. 2012). There is no evidence that this increase in the complexity of vacuolar ATP synthase has endowed yeast with increased fitness.

As noted above, it is unlikely that duplicate genes are selectively preserved on the basis of having back-up features. Nonetheless, observations from *S. cerevisiae* show that such properties can arise fortuitously as an indirect consequence of overlapping gene functions retained after partial subfunctionalization. For example, two ancient yeast paralogs, *Sir2* and *Hst1*, operate as histone deacetylases with rather different functions in the cell (Hickman and Rusche 2007). However, when one gene is absent, the other can partially compensate by engaging in the non-cognate function. Comparison with a pre-duplication outgroup species makes clear that this is a case of quantitative subfunctionalization,

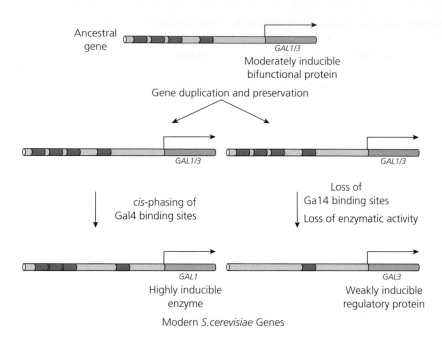

Figure 6.4 Evolution of the galactose-utilization pathway following duplication and subfunctionalization of the ancestral gene (*Gal1/3*) in the yeast *S. cerevisiae*, giving rise to the duplicates *Gal1* and *Gal3*. Blue represents the coding regions, and orange the binding sites of the transcription factor *Gal4*. The loss of binding sites in *Gal3* and the rearrangement of sites in *Gal1* removed an adaptive conflict involving the regulatory efficiency of the single-copy gene. From Hittinger and Carroll (2007).

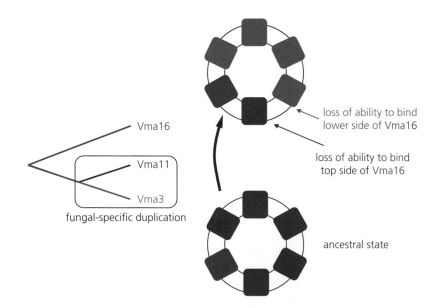

Figure 6.5 Duplication and subfunctionalization of components of the internal-membrane bound ring of vacuolar ATP synthase in the ancestral fungal genome. After their origin by gene duplication from a protein denoted by green, subunits Vma3 (blue) and Vma11 (olive) acquired complementary changes in interface residues, preventing each from binding one side of Vma16 (pink), as demonstrated by experimental manipulation of protein sequences by Finnigan et al. (2011).

illustrating the risks of assuming that because duplicate genes have redundant functions, they must have been preserved on the basis of their backup capacities.

Finally, a potentially common mode of duplicate-gene preservation in eukaryotes involves the partitioning of gene functions via the modification of transit signals for localization of mRNAs and/or proteins to particular subcellular regions (Kumar et al. 2002; Silva-Filho 2003; Krogan et al. 2006). For example, immediately after transcription, eukaryotic mRNAs are typically decorated with one or more RNA-binding proteins, many of which attach to specific motor proteins for delivery to a specific subcellular location prior to translation (Besse and Ephrussi 2008; Holt and Bullock 2009; Buxbaum et al. 2015). There are numerous cases in which modifications of transit signals in post-duplication genes have led to sub- or neolocalization. As a case in point, following duplication of one of the subunits of cytochrome c oxidase (a terminal complex in the electron transport chain) in an ancestral vertebrate, one member came to specialize on localization to the mitochondrion, whereas the other is delivered to the Golgi (Schmidt et al. 2003). Likewise, NADP-dependent isocitrate dehydrogenase has been duplicated independently in both yeast and mammals, and in both cases the descendant copies partitioned their localizations to either the nucleus or the cytoplasm (Nekrutenko et al. 1998; Szewczyk et al. 2001).

Marques et al. (2008) suggest that about a third of duplicated genes surviving the whole-genome duplication in the ancestry of *S. cerevisiae* exhibit spatial subcellular partitioning, and similar estimates have been given for other taxa. For example, up to 25% of gene duplicates in the plant *Arabidopsis* (another descendant of a whole-genome duplication event) have experienced relocalization or sublocalization of their gene products (Byun and Singh 2013; Liu et al. 2014). There is, however, some uncertainty as to whether such partitioning is typically a cause or consequence of duplicate-gene preservation, as singleton genes in *S. cerevisiae* also appear to frequently acquire novel relocalization patterns (Qian and Zhang 2009). Although bacteria also exhibit spatial organization of translation (Montero Llopis et al. 2010; Nevo-Dinur et al. 2011),

this is dictated primarily by the cellular locations of genes on the chromosome. In this case, there appears to be less opportunity for partitioning subcellular localization following gene duplication, and indeed duplicate genes are rare in prokaryotes.

The Emergence of Modular Gene Subfunctions

Taken together, these results (along with many others to appear in subsequent chapters) make clear that duplicate-gene subfunctionalization has played a major role in the evolution of structural and enzymatic features of eukaryotic cells. One simple reason for the rarity of subfunctionalization in prokaryotes is the population-size constraint associated with the mutational and energetic costs of duplicate genes (Adler et al. 2014), but another is the general absence of independently mutable regulatory elements and localization zip codes necessary for subfunction partitioning. This raises the broader question as to how modular gene architectural features essential to subfunctionalization evolve in the first place.

Resolution of this matter resides in the fact that the same types of duplication and degeneration processes that lead to the subfunctionalization of duplicate genes promote the emergence of the subfunctions themselves (Force et al. 2005). To simplify discussion, we will assume that subfunctions are defined by transcription-factor binding sites (TFBSs) or integrated regions of such sites (simply referred to here as promoters) that are separable from other such sites, both mutationally and functionally (as further elaborated on in Chapter 21). However, the same principles apply to subfunctions defined by functional motifs in coding regions, binding interfaces in multimers, or any other gene features that can be mutationally separated.

The goal here is to understand how a gene that is initially ubiquitously controlled in the same manner under all conditions comes to be regulated by more specialized mechanisms while retaining the same overall expression pattern. The process envisioned, subfunction fission, involves the progressive reconfiguration of a general-purpose enhancer via consecutive processes of partial duplication and loss of regulatory information, with each step proceeding in a nearly neutral fashion (Figure 6.6).

The first phase involves the accretion of new regulatory elements, followed by the degeneration of one or more ancestral sites to yield two semi-independent promoters. The second phase involves tandem duplication of the regulatory region, followed by the formation of two entirely independent regulatory subfunctions by complementary degenerative mutations. Other than the fact that smaller DNA elements are involved, the events during the second phase are conceptually identical to those noted above for the subfunctionalization of entire genes.

Under this model, there is not necessarily a permanent allelic state, as the alternative classes of shared and semi-independently regulated alleles are free to mutate back and forth (hence, the two-way arrows in the top left of Figure 6.6). Thus, it is necessary to consider the circumstances under which semi-independently regulated alleles are likely to rise to high frequency, as this is a requirement for completing the transition to an allele with two entirely independent subfunctions.

There are three reasons why gene structure is more likely to gravitate to the modular state in small populations. First, the stochastic gain of specific regulatory elements can occur either by *de novo* mutation to an appropriate motif within existing sequence or by the insertion of a pre-existing element via duplication from alternative genomic sites. The rate of *de novo* origin of an appropriate TFBS motif by mutation will depend on the mutation rate per nucleotide site and the mutational

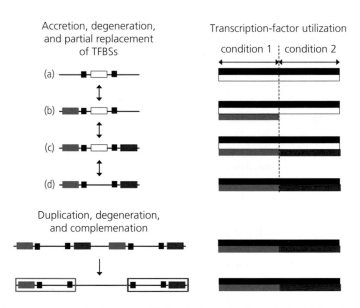

Figure 6.6 A hypothetical scenario in which a gene with two independently mutable subfunctions can arise from an ancestral state with a single generalized expression mechanism. Regulatory regions are depicted on the left, with each regulatory element colour-coded according to the transcription factor that binds to it. On the right, the patterns of allele-specific utilization of transcription factors are colour-coded. Transcription factors denoted by black and white are ubiquitously expressed, whereas those denoted by blue and red are expressed in single, non-overlapping conditions. The original gene has a promoter that requires occupancy of three transcription-factor binding sites (one white and two black) for expression. In the first phase of gene evolution, the regulatory region undergoes the sequential accretion of blue and red elements (a-c), which together are redundant with respect to the white element, which is then lost in a neutral fashion by degenerative mutation (d). At this point, the evolved allele has a semi-independent mode of expression, as the two black elements are still essential for expression in both tissues. In the second phase, the entire enhancer region is tandemly duplicated, with each component then losing a complementary (red/blue) element. The resultant gene now has two independent subfunctions denoted by the blue and red open boxes, as a mutation in either region has effects that are confined to a single condition. Note that throughout all of these evolutionary steps, there has been no change in the pattern of expression of the gene; only the mechanism of expression has been altered. From Force et al. (2005).

target size (amount of intergenic spacer DNA), both of which scale approximately inversely with N_e (Chapter 4; Lynch 2007). Second, the large, more gene-laden genomes of species with small N_e (e.g., eukaryotes vs. prokaryotes) have more potential sources of TFBSs for duplicative transpositions. Third, although alleles with more complex regulatory regions have a higher mutational vulnerability and impose an excess energetic cost at the DNA level, both of these are small effects that will only be efficiently opposed by selection in populations with large N_e.

The salient point here is that the same population-genetic environments that favor the subdivision of gene functions following gene duplication are expected to favor the emergence of gene-structural architectures necessary to fuel subfunctionalization. Such reinforcement provides further support for the contention that reductions in N_e, which naturally occurred as eukaryotes arose (and was further exacerbated in the metazoan and land-plant lineages), promoted a setting for the passive evolution of complexity with essentially no involvement of positive selection. Consistent with such a march towards complexity is the observation that whereas duplicate genes gradually lose their shared expression patterns over evolutionary time, the total numbers of regulatory motifs and interacting protein partners remain roughly constant for each member of the pair, suggesting an approximate balance between gains and losses of such elements. Such patterns have been observed in yeast (Papp et al. 2003; He and Zhang 2005), mammals (Huminiecki and Wolfe 2004), and *Arabidopsis* (Arsovski et al. 2015).

Taken together, these observations raise significant questions about the frequently assumed necessity and sufficiency of natural selection as a determining force in the emergence of complex patterns of gene regulation and protein deployment. In sufficiently small populations, modular forms of gene structure are expected to emerge in the absence of any direct selection for such architectural features. In sufficiently large populations, such changes are opposed by selection (unless immediately accompanied by phenotypic advantages that substantially offset the mutational and energetic disadvantages).

The Passive Origin of Species via Gene Duplication

In addition to playing a central role in the evolutionary divergence of cellular traits within lineages, gene duplication also has a powerful indirect role in the second major engine of evolution – the process of speciation, that is the emergence of new phylogenetic lineages (Lynch and Force 2000). Genetic theories of speciation have traditionally focused on two competing hypotheses (reviewed in Orr 1996; Rieseberg 2001; Coyne and Orr 2004). The Bateman–Dobzhansky–Muller (BDM) model postulates the accumulation of lineage-specific gene-sequence changes that are mutually incompatible when brought together in a hybrid genome. The chromosomal model invokes the accumulation of genomic rearrangements that result in gene loss in hybrid backgrounds.

Both models are based on rather stringent assumptions. For example, the BDM model invokes the evolution of mutually incompatible co-adaptive complexes of epistatically interacting factors, few of which have been convincingly identified as instigating the speciation process (as opposed to being downstream consequences). Chromosomal models generally focus on major rearrangements, for which within-population fixation can be greatly inhibited by the reduction in fitness in chromosomal heterozygotes owing to problems during meiosis. Notably, the gene-duplication model for speciation is consistent with both the BDM and the chromosomal models, while requiring fewer assumptions than either of them.

The passive reassignment of gene (sub)functions to novel locations following gene duplication is central to the gene-duplication model of speciation. To see this, consider a diploid ancestral species with an unlinked pair of duplicate autosomal genes, which then experience divergent non- or subfunctionalization in two descendent species. This results in different chromosomal locations of the active genes (Figure 6.7). Because the F_1 hybrids of such species will be 'presence–absence' heterozygotes at the two independently segregating loci, $1/4$ of the F_1 gametes will contain null (absentee) alleles at both loci. In a predominantly haploid species, this single divergently resolved duplication would result in an

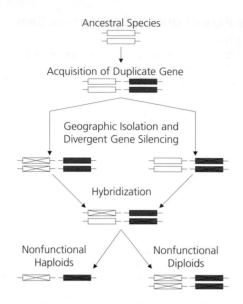

Figure 6.7 Divergent silencing of an ancestral duplicate gene in two geographically isolated lineages. One gene copy is denoted by a white box, the other by a green box, with × denoting an inactivated gene. Gene pairs represent alleles at diploid loci. Progeny genotypes other than those indicated in the bottom row might have compromised fitness, for example, individuals with just a single active gene.

expected 25% reduction in functional progeny. In a predominantly diploid species, $(1/4)^2 = 1/16$ of the F_2 offspring from the interspecific cross would lack functional alleles at both loci, and another $1/4$ would carry only a single functional allele. Thus, if the gene is haploinsufficient, $5/16$ of the F_2 zygotes of such a cross would be inviable (and/or sterile). With n divergently resolved duplicates, the expected fitness of hybrid progeny is $W = (1 - \delta)^n$, with δ denoting the reduction in hybrid fitness per map change. For example, with $\delta = 5/16$, as in a zygotically acting haploinsufficient viability gene, $W = 0.024$ in the F_2 generation when $n = 10$, and 5×10^{-17} when $n = 100$.

Observed rates of gene duplication indicate that this type of process is sufficiently powerful to yield nearly complete genomic incompatibility within a few million generations of cessation of gene flow (Lynch and Force 2000; Shpak 2005). This is also the approximate timescale over which postzygotic isolation generally occurs in animals (Parker et al. 1985; Coyne and Orr 1997; Sasa et al. 1998;

Presgraves 2002; Price and Bouvier 2002). Unfortunately, knowledge on the timescale of speciation in unicellular organisms is scant. However, genomic comparisons of the yeasts *S. cerevisiae* and *Candida albicans* imply an overall rate of microchromosomal rearrangement of ∼2.3/lineage/MY (Seoighe et al. 2003), likely driven in large part by divergent resolution of duplicate genes, as further discussed below.

The gene-duplication model for speciation is effectively a chromosomal model, but because the rearrangements are microchromosomal, they are unlikely to cause significant pairing problems during meiosis. Such changes can then accumulate passively without any alteration in within-species fitness, only being revealed after crossing to a lineage with a deviant gene location. The gene-duplication model also masquerades as a BDM model, in that reassignments of genes to new locations operate like epistatic interactions because the loss-of-function phenotype is determined by the total number of active alleles across the two duplicate loci in hybrid progeny.

A key feature of the gene-duplication model is that speciation can occur without any molecular evolution at the gene-sequence level. All that is required is the reciprocal silencing of ancestral-gene (sub)functions in sister taxa following ancestral gene duplications. Nonetheless, this process can also proceed via paths of neofunctionalization provided the latter occurs in the ancestral gene copy in one lineage (Lynch and Force 2005), and this can lead to misinterpretations regarding the underlying genetic mechanism of postzygotic isolation. Often it is assumed that speciation is a by-product of local adaptation generating physiologically incompatible alleles. However, incompatibilities resulting from the neofunctionalization of a duplicate gene need not be a direct function of adaptive changes at the neofunctionalized locus, but simply an indirect consequence of relocation of the ancestral-gene function.

Of course, the divergent resolution of duplicate genes is by no means the only possible route to the origin of post-zygotic species isolating barriers. However, given the frequency of gene duplication, it is difficult to escape the conclusion that it is a common and pervasive mechanism for speciation.

As an example, a duplicate pair of a genes involved in histidine biosynthesis was present in the ancestor of the plant *Arabidopsis thaliana*, with different copies becoming silent in different *A. thaliana* sublineages. When plants containing the reciprocally silenced genes are crossed, the hybrids (presence/absence heterozygotes at both loci) segregate out different haplotypes in the next round of gametes, with progeny lacking both copies being inviable (Bikard et al. 2009; Blevins et al. 2017). A similar scenario, involving a different gene duplication, has been found in the genus *Mimulus* (Zuellig and Sweigart 2018).

The fruit fly *Drosophila* has been one of the major workhorses for research on the genetics of speciation, providing well-documented examples of the involvement of duplicate genes in reproductive isolation. In two cases, a strong phase of positive selection operating on single duplicate copies has been implicated (Ting et al. 2004; Greenberg et al. 2006), suggesting the possibility of neofunctionalization. But in some *D. melanogaster* × *D. simulans* hybrids, sterility appears to be a simple consequence of the movement of an essential gene to a new chromosomal location via an intermediate phase of gene duplication (and without a change in function) (Masly et al. 2006).

Finally, it bears emphasizing that under the gene-duplication model, certain groups of organisms are expected to be more prone to speciation than others. For example, for lineages experiencing a doubling in genome size (polyploidization), the process noted previously will be essentially unavoidable, owing to the very large number of gene targets. Moreover, following the first map changes induced by reciprocal silencing in sister polyploid taxa, the thousands of duplicate pairs still remaining will be free to become divergently resolved in subsequently isolated lineages, potentially yielding a large number of nested speciation events, i.e., a species radiation.

A particularly striking example of reproductive isolation by this form of divergent resolution is provided by the *Paramecium aurelia* complex, consisting of at least 14 cryptic species of ciliates. All of these emerged after two ancestral whole-genome duplication events led to hundreds of map changes as ancestral single-copy genes came to be represented by one, two, or three copies located on different chromosomes (McGrath et al. 2014; Gout et al. 2023). Remarkably, although the members of the *P. aurelia* complex have evolved unique pairs of mating types, despite $> 10^8$ years of isolation, there has been no discernible morphological divergence among lineages.

Another observation that appears to be compatible with the gene-duplication model for the origin of isolating barriers involves the yeast *S. cerevisiae* and its close relatives, which exhibit hundreds of differences in gene-order changes resulting from divergently resolved pairs of gene duplicates following a whole-genome duplication (Scannell et al. 2006). Although the haploid offspring of crosses between such species are almost always sterile, engineering of the chromosomes to partially restore large-scale co-linearity increases fertility to levels of ~25% (Delneri et al. 2003). Restoration to complete co-linearity might have even a greater effect. Notably, Selmecki et al. (2015) demonstrated that whole-genome duplication in yeast can facilitate adaptation by providing more opportunities for modifying gene balance by large deletions and/or chromosome loss, all of which will lead to the chromosomal repatterning essential to the gene-duplication model of speciation.

The key point here is that, as in the case of phenotypic change within lineages, ample mechanisms exist for the passive origin of new species via nonadaptive processes. One potential example of such a key event, touched upon in Chapter 3, involves the base of the eukaryotic lineage – the colonization of LECA by the mitochondrion. Considering the very large number of organelle-to-nucleus gene transfers that apparently occurred soon after the establishment of the mitochondrial progenitor (Martin et al. 1998), divergent resolution of duplicated organelle genes may have provoked the passive development of isolating barriers among basal eukaryotic lineages (Chapters 3 and 24).

Summary

- To minimize energetic costs and mutational vulnerability, natural selection is expected to favor simplicity over complexity. Yet, many aspects of cell biology are demonstrably over-designed, particularly in eukaryotes, and most notably in multicellular species.

- Constructive neutral evolution provides a vision for how organismal complexity can emerge by nonadaptive mechanisms. The key idea is that the fortuitous development of initially neutral inter-actions between different gene products can alter the selective environment in ways that enable the fixation of previously forbidden mutations, thereby leading to permanent mutual depen-dence. Although the formalities of the theory remain to be worked out, the model provides a plausible explanation for the origin of a wide variety of cellular features, including the large number of protein subunits associated with com-plexes such as the electron transport chain and the ribosome.

- Gene duplication is one of the primary mechanisms for the origin of organismal complexity. Although neofunctionalization of one member of a pair provides a facile route to the origin of novel gene features, duplicate genes are more commonly preserved by other nonadaptive mechanisms. Most notably, sub-functionalization occurs when complementary degenerate mutations result in the partitioning of ancestral gene functions in duplicated descen-dants. The probability of this outcome is elevated in populations with small effective sizes.

- The same processes that lead to subfunctionaliza-tion of duplicate genes promote the evolution of modular forms of gene structure upon which the process of subfunctionalization depends. Thus, by facilitating the recurrent emergence and par-titioning of gene subfunctions, reduced effec-tive population sizes can lead to the passive increase in organismal complexity without any direct selection for such changes.

- Gene duplication also provides a powerful mech-anism for the passive origin of reproductively isolated species, particularly in lineages that have experienced whole-genome duplications, as has happened repeatedly throughout the eukaryotic phylogeny.

Literature Cited

Adler, M., M. Anjum, O. G. Berg, D. I. Andersson, and L. Sandegren. 2014. High fitness costs and instability of gene duplications reduce rates of evolution of new genes by duplication-divergence mechanisms. Mol. Biol. Evol. 31: 1526–1535.

Alberghina, F. A., E. Sturani, and J. R. Gohlke. 1975. Levels and rates of synthesis of ribosomal ribonucleic acid, transfer ribonucleic acid, and protein in *Neurospora crassa* in different steady states of growth. J. Biol. Chem. 250: 4381–4388.

Amunts, A., A. Brown, X. C. Bai, J. L. Llácer, T. Hussain, P. Emsley, F. Long, G. Murshudov, S. H. Scheres, and V. Ramakrishnan. 2014. Structure of the yeast mitochon-drial large ribosomal subunit. Science 343: 1485–1489.

Angerer, H., K. Zwicker, Z. Wumaier, L. Sokolova, H. Heide, M. Steger, S. Kaiser, E. Nübel, B. Brutschy, M. Radermacher, et al. 2011. A scaffold of accessory sub-units links the peripheral arm and the distal proton-pumping module of mitochondrial complex I. Biochem. J. 437: 279–288.

Arsovski, A. A., J. Pradinuk, X. Q. Guo, S. Wang, and K. L. Adams. 2015. Evolution of *cis*-regulatory elements and regulatory networks in duplicated genes of *Arabidopsis*. Plant Physiol. 169: 2982–2991.

Aury, J. M., O. Jaillon, L. Duret, B. Noel, C. Jubin, B. M. Porcel, B. Ségurens, V. Daubin, V. Anthouard, N. Aiach, et al. 2006. Global trends of whole-genome duplications revealed by the ciliate *Paramecium tetraurelia*. Nature 444: 171–178.

Bergthorsson, U., D. I. Andersson, and J. R. Roth. 2007. Ohno's dilemma: Evolution of new genes under con-tinuous selection. Proc. Natl. Acad. Sci. USA 104: 17004–17009.

Besse, F., and A. Ephrussi. 2008. Translational control of localized mRNAs: Restricting protein synthesis in space and time. Nat. Rev. Mol. Cell Biol. 9: 971–980.

Bikard, D., D. Patel, C. Le Metté, V. Giorgi, C. Camilleri, M. J. Bennett, and O. Loudet. 2009. Divergent evolution of duplicate genes leads to genetic incompatibilities within *A. thaliana*. Science 323: 623–626.

Blevins, T., J. Wang, D. Pflieger, F. Pontvianne, and C. S. Pikaard. 2017. Hybrid incompatibility caused by an epiallele. Proc. Natl. Acad. Sci. USA 114: 3702–3707.

Boehlke, K. W., and J. D. Friesen. 1975. Cellular content of ribonucleic acid and protein in *Saccharomyces cerevisiae* as a function of exponential growth rate: Calculation of the apparent peptide chain elongation rate. J. Bacteriol. 121: 429–433.

Bonven, B., and K. Gulløv. 1979. Peptide chain elongation rate and ribosomal activity in *Saccharomyces cerevisiae* as a function of the growth rate. Mol. Gen. Genet. 170: 225–230.

Bornberg-Bauer, E., J. Schmitz, and M. Heberlein. 2015. Emergence of *de novo* proteins from 'dark genomic matter' by 'grow slow and moult'. Biochem. Soc. Trans. 43: 867–873.

Brown, A., A. Amunts, X. C. Bai, Y. Sugimoto, P. C. Edwards, G. Murshudov, S. H. Scheres, and V. Ramakrishnan. 2014. Structure of the large ribosomal subunit from human mitochondria. Science 346: 718–722.

Brunet, T. D. P., and W. F. Doolittle. 2018. The generality of constructive neutral evolution. Biol. Philos. 33: 2.

Buxbaum, A. R., G. Haimovich, and R. H. Singer. 2015. In the right place at the right time: Visualizing and understanding mRNA localization. Nat. Rev. Mol. Cell Biol. 16: 95–109.

Byun, S. A., and S. Singh. 2013. Protein subcellular relocalization increases the retention of eukaryotic duplicate genes. Genome. Biol. Evol. 5: 2402–2409.

Chain, F. J., and B. J. Evans. 2006. Multiple mechanisms promote the retained expression of gene duplicates in the tetraploid frog *Xenopus laevis*. PLoS Genet. 2: e56.

Clark, A. G. 1994. Invasion and maintenance of a gene duplication. Proc. Natl. Acad. Sci. USA 91: 2950–2954.

Conant, G. C., and K. H. Wolfe. 2008. Turning a hobby into a job: How duplicated genes find new functions. Nat. Rev. Genet. 9: 938–950.

Cox, R. A. 2004. Quantitative relationships for specific growth rates and macromolecular compositions of *Mycobacterium tuberculosis*, *Streptomyces coelicolor* A3(2) and *Escherichia coli* B/r: An integrative theoretical approach. Microbiology 150: 1413–1426.

Coyne, J. A., and H. A. Orr. 1997. 'Patterns of speciation in *Drosophila*' revisited. Evolution 51: 295–303.

Coyne, J. A., and H. A. Orr. 2004. Speciation. Sinauer Associates, Inc., Sunderland, MA.

Delneri, D., I. Colson, S. Grammenoudi, I. N. Roberts, E. J. Louis, and S. G. Oliver. 2003. Engineering evolution to study speciation in yeasts. Nature 422: 68–72.

Dennis, P. P., and H. Bremer. 1974. Macromolecular composition during steady-state growth of *Escherichia coli* B-r. J. Bacteriol. 119: 270–281.

Desmond, E., C. Brochier-Armanet, P. Forterre, and S. Gribaldo. 2011. On the last common ancestor and early evolution of eukaryotes: Reconstructing the history of mitochondrial ribosomes. Res. Microbiol. 162: 53–70.

Diss, G., I. Gagnon-Arsenault, A. M. Dion-Coté, H. Vignaud, D. I. Ascencio, C. M. Berger, and C. R. Landry. 2017. Gene duplication can impart fragility, not robustness, in the yeast protein interaction network. Science 355: 630–634.

Duarte, J. M., L. Cui, P. K. Wall, Q. Zhang, X. Zhang, J. Leebens-Mack, H. Ma, N. Altman, and C. W. dePamphilis. 2006. Expression pattern shifts following duplication indicative of subfunctionalization and neofunctionalization in regulatory genes of *Arabidopsis*. Mol. Biol. Evol. 23: 469–478.

Ferris, S. D., and G. S. Whitt. 1977. Duplicate gene expression in diploid and tetraploid loaches (Cypriniformes, Cobitidae). Biochem. Genet. 15: 1097–1112.

Ferris, S. D., and G. S. Whitt. 1979. Evolution of the differential regulation of duplicate genes after polyploidization. J. Mol. Evol. 12: 267–317.

Finnigan, G. C., V. Hanson-Smith, T. H. Stevens, and J. W. Thornton. 2012. Evolution of increased complexity in a molecular machine. Nature 481: 360–364.

Fisher, R. A. 1935. The sheltering of lethals. Amer. Natur. 69: 446–455.

Force, A., W. A. Cresko, F. B. Pickett, S. Proulx, C. Amemiya, and M. Lynch. 2005. The origin of subfunctions and modular gene regulation. Genetics 170: 433–446.

Force, A., M. Lynch, B. Pickett, A. Amores, Y.-L. Yan, and J. Postlethwait. 1999. Preservation of duplicate genes by complementary, degenerate mutations. Genetics 151: 1531–1545.

Forchhammer, J., and L. Lindahl. 1971. Growth rate of polypeptide chains as a function of the cell growth rate in a mutant of *Escherichia coli* 15. J. Mol. Biol. 55: 563–568.

Freeling, M., M. J. Scanlon, and J. E. Fowler. 2015. Fractionation and subfunctionalization following genome duplications: Mechanisms that drive gene content and their consequences. Curr. Opin. Genet. Dev. 35: 110–118.

Gopalan, V., N. Jarrous, and A. S. Krasilnikov. 2018. Chance and necessity in the evolution of RNase P. RNA 24: 1–5.

Gout, J. F., Y. Hao, P. Johri, O. Arnaiz, T. G. Doak, S. Bhullar, A. Couloux, F. Guérin, S. Malinsky, A. Potekhin, et al. 2023. Dynamics of gene loss following ancient whole-genome duplication in the cryptic *Paramecium* complex. Mol. Bio. Evol. 40: msad107.

Gout, J. F., and M. Lynch. 2015. Maintenance and loss of duplicated genes by dosage subfunctionalization. Mol. Bio. Evol. 32: 2141–2148.

Gray, M. W., J. Lukes, J. M. Archibald, P. J. Keeling, and W. F. Doolittle. 2010. Cell biology. Irremediable complexity? Science 330: 920–921.

Greenberg, A. J., J. R. Moran, S. Fang, and C.-I. Wu. 2006. Adaptive loss of an old duplicated gene during incipient speciation. Mol. Biol. Evol. 23: 401–410.

Haldane, J. B. S. 1933. The part played by recurrent mutation in evolution. Amer. Natur. 67: 5–9.

Hartman, H., P. Favaretto, and T. F. Smith. 2006. The archaeal origins of the eukaryotic translational system. Archaea 2: 1–9.

He, X., and J. Zhang. 2005. Rapid subfunctionalization accompanied by prolonged and substantial neofunctionalization in duplicate gene evolution. Genetics 169: 1157–1164.

Hickman, M. A., and L. N. Rusche. 2007. Substitution as a mechanism for genetic robustness: The duplicated deacetylases Hst1p and Sir2p in *Saccharomyces cerevisiae*. PLoS Genet. 3: e126.

Hickman, M. A., and L. N. Rusche. 2010. Transcriptional silencing functions of the yeast protein Orc1/Sir3 subfunctionalized after gene duplication. Proc. Natl. Acad. Sci. USA 107: 19384–19389.

Hirst, J. 2011. Why does mitochondrial complex I have so many subunits? Biochem. J. 437: e1–e3.

Hittinger, C. T., and S. B. Carroll. 2007. Gene duplication and the adaptive evolution of a classic genetic switch. Nature 449: 677–681.

Holt, C. E., and S. L. Bullock. 2009. Subcellular mRNA localization in animal cells and why it matters. Science 326: 1212–1216.

Hsiao, C., S. Mohan, B. K. Kalahar, and L. D. Williams. 2009. Peeling the onion: Ribosomes are ancient molecular fossils. Mol. Biol. Evol. 26: 2415–2425.

Hughes, A. L. 1994. The evolution of functionally novel proteins after gene duplication. Proc. Roy. Soc. Lond. B 256: 119–124.

Huminiecki, L., and K. H. Wolfe. 2004. Divergence of spatial gene expression profiles following species-specific gene duplications in human and mouse. Genome Res. 14: 1870–1879.

Huynen, M. A., I. Duarte, and R. Szklarczyk. 2013. Loss, replacement and gain of proteins at the origin of the mitochondria. Biochim. Biophys. Acta 1827: 224–231.

Ingolia, N. T., L. F. Lareau, and J. S. Weissman. 2011. Ribosome profiling of mouse embryonic stem cells reveals the complexity and dynamics of mammalian proteomes. Cell 147: 789–802.

Innan, H., and F. Kondrashov. 2010. The evolution of gene duplications: Classifying and distinguishing between models. Nat. Rev. Genet. 11: 97–108.

Jaillon, O., J. M. Aury, F. Brunet, J. L. Petit, N. Stange-Thomann, E. Mauceli, L. Bouneau, C. Fischer, C. Ozouf-Costaz, A. Bernot, et al. 2004. Genome duplication in the teleost fish *Tetraodon nigroviridis* reveals the early vertebrate proto-karyotype. Nature 431: 946–957.

Katju, V. 2012. In with the old, in with the new: The promiscuity of the duplication process engenders diverse pathways for novel gene creation. Int. J. Evol. Biol. 2012: 341932.

Katju, V., and M. Lynch. 2006. On the formation of novel genes by duplication in the *Caenorhabditis elegans* genome. Mol. Biol. Evol. 23: 1056–1067.

Keeling, P. J., and J. D. Palmer. 2008. Horizontal gene transfer in eukaryotic evolution. Nat. Rev. Genet. 9: 605–618.

Kenny, N. J., K. W. Chan, W. Nong, Z. Qu, I. Maeso, H. Y. Yip, T. F. Chan, H. S. Kwan, P. W. H. Holland, K. H. Chu, and J. H. L. Hui. 2016. Ancestral whole-genome duplication in the marine chelicerate horseshoe crabs. Heredity 116: 190–199.

Kimura, M., and T. Ohta. 1969. The average number of generations until fixation of a mutant gene in a finite population. Genetics 61: 763–771.

Konrad, A., S. Flibotte, J. Taylor, R. H. Waterston, D. G. Moerman, U. Bergthorsson, and V. Katju. 2018. Mutational and transcriptional landscape of spontaneous gene duplications and deletions in *Caenorhabditis elegans*. Proc. Natl. Acad. Sci. USA 115: 7386–7391.

Koonin, E. V., N. D. Fedorova, J. D. Jackson, A. R. Jacobs, D. M. Krylov, K. S. Makarova, R. Mazumder, S. L. Mekhedov, A. N. Nikolskaya, B. S. Rao, et al. 2004. A comprehensive evolutionary classification of proteins encoded in complete eukaryotic genomes. Genome Biol. 5: R7.

Kostinski, S., and S. Reuveni. 2020. Ribosome composition maximizes cellular growth rates in *E. coli*. Phys. Rev. Lett. 125: 028103.

Krogan, N. J., G. Cagney, H. Yu, G. Zhong, X. Guo, A. Ignatchenko, J. Li, S. Pu, N. Datta, A. P. Tikuisis, et al. 2006. Global landscape of protein complexes in the yeast *Saccharomyces cerevisiae*. Nature 440: 637–643.

Kumar, A., S. Agarwal, J. A. Heyman, S. Matson, M. Heidtman, S. Piccirillo, L. Umansky, A. Drawid, R. Jansen, Y. Liu, et al. 2002. Subcellular localization of the yeast proteome. Genes Dev. 16: 707–719.

Lan, P., M. Tan, Y. Zhang, S. Niu, J. Chen, S. Shi, S. Qiu, X. Wang, X. Peng, G. Cai, et al. 2018. Structural insight into precursor tRNA processing by yeast ribonuclease P. Science 362: eaat6678.

Lane, N. 2020. How energy flow shapes cell evolution. Curr. Biol. 30: R471–R476.

Lane, N., and W. Martin. 2010. The energetics of genome complexity. Nature 467: 929–934.

Lechner, M., W. Rossmanith, R. K. Hartmann, C. Thölken, B. Gutmann, P. Giegé, and A. Gobert. 2015. Distribution of ribonucleoprotein and protein-only RNase P in Eukarya. Mol. Biol. Evol. 32: 3186–3193.

Lecompte, O., R. Ripp, J. C. Thierry, D. Moras, and O. Poch. 2002. Comparative analysis of ribosomal proteins in complete genomes: An example of reductive evolution at the domain scale. Nucleic Acids Res. 30: 5382–5390.

Leger, M. M., M. Petrů, V. Zárský, L. Eme, Č. Vlček, T. Harding, B. F. Lang, M. Eliáš, P. Doležal, and A. J. Roger. 2015. An ancestral bacterial division system is widespread in eukaryotic mitochondria. Proc. Natl. Acad. Sci. USA 112: 10239–10246.

Li, Z., G. P. Tiley, S. R. Galuska, C. R. Reardon, T. I. Kidder, R. J. Rundell, and M. S. Barker. 2018. Multiple large-scale gene and genome duplications during the evolution of hexapods. Proc. Natl. Acad. Sci. USA 115: 4713–4718.

Liu, S. L., A. Q. Pan, and K. L. Adams. 2014. Protein subcellular relocalization of duplicated genes in *Arabidopsis*. Genome Biol. Evol. 6: 2501–2515.

Lukeš, J., J. M. Archibald, P. J. Keeling, W. F. Doolittle, and M. W. Gray. 2011. How a neutral evolutionary ratchet can build cellular complexity. IUBMB Life 63: 528–537.

Lynch, M. 2007. The Origins of Genome Architecture. Sinauer Associates, Inc., Sunderland, MA.

Lynch, M., and J. S. Conery. 2000. The evolutionary fate and consequences of duplicate genes. Science 290: 1151–1154.

Lynch, M., and A. Force. 2000. The probability of duplicate-gene preservation by subfunctionalization. Genetics 154: 459–473.

Lynch, M., M. O'Hely, B. Walsh, and A. Force. 2001. The probability of fixation of a newly arisen gene duplicate. Genetics 159: 1789–1804.

Marques, A. C., N. Vinckenbosch, D. Brawand, and H. Kaessmann. 2008. Functional diversification of duplicate genes through subcellular adaptation of encoded proteins. Genome Biol. 9: R54.

Martin, S. E., and J. J. Iandolo. 1975. Translational control of protein synthesis in *Staphylococcus aureus*. J. Bacteriol. 122: 1136–1143.

Martin, W., B. Stoebe, V. Goremykin, S. Hansmann, M. Hasegawa, and K. V. Kowallik. 1998. Gene transfer to the nucleus and the evolution of chloroplasts. Nature 393: 162–165.

Masly, J. P., C. D. Jones, M. A. Noor, J. Locke, and H. A. Orr. 2006. Gene transposition as a cause of hybrid sterility in *Drosophila*. Science 313: 1448–1450.

McGrath, C. L., J. F. Gout, P. Johri, T. G. Doak, and M. Lynch. 2014. Differential retention and divergent resolution of duplicate genes following whole-genome duplication. Genome Res. 24: 1665–1675.

McLysaght, A., and L. D. Hurst. 2016. Open questions in the study of *de novo* genes: What, how and why. Nat. Rev. Genet. 17: 567–578.

Melnikov, S., A. Ben-Shem, N. Garreau de Loubresse, L. Jenner, G. Yusupova, and M. Yusupov. 2012. One core, two shells: Bacterial and eukaryotic ribosomes. Nat. Struct. Mol. Biol. 19: 560–567.

Montero Llopis, P., A. F. Jackson, O. Sliusarenko, I. Surovtsev, J. Heinritz, T. Emonet, and C. Jacobs-Wagner. 2010. Spatial organization of the flow of genetic information in bacteria. Nature 466: 77–81.

Morin, R. D., E. Chang, A. Petrescu, N. Liao, M. Griffith, W. Chow, R. Kirkpatrick, Y. S. Butterfield, A. C. Young, J. Stott, et al. 2006. Sequencing and analysis of 10,967 full-length cDNA clones from *Xenopus laevis* and *Xenopus tropicalis* reveals post-tetraploidization transcriptome remodeling. Genome Res. 16: 796–803.

Muller, H. J. 1940. Bearing of the *Drosophila* work on systematics, pp. 185-268. In J. S. Huxley (ed.) The New Systematics. Clarendon Press, Oxford, UK.

Näsvall, J., L. Sun, J. R. Roth, and D. I. Andersson. 2012. Real-time evolution of new genes by innovation, amplification, and divergence. Science 338: 384–387.

Nekrutenko, A., D. M. Hillis, J. C. Patton, R. D. Bradley, and R. J. Baker. 1998. Cytosolic isocitrate dehydrogenase in humans, mice, and voles and phylogenetic analysis of the enzyme family. Mol. Biol. Evol. 15: 1674–1684.

Neme, R., and D. Tautz. 2016. Fast turnover of genome transcription across evolutionary time exposes entire non-coding DNA to *de novo* gene emergence. eLife 5: e09977.

Nevo-Dinur, K., A. Nussbaum-Shochat, S. Ben-Yehuda, and O. Amster-Choder. 2011. Translation-independent localization of mRNA in *E. coli*. Science 331: 1081–1084.

Newton, M. S., X. Guo, A. Söderholm, J. Näsvall, P. Lundström, D. I. Andersson, M. Selmer, and W. M. Patrick. 2017. Structural and functional innovations in the real-time evolution of new ($\beta\alpha$) (8) barrel enzymes. Proc. Natl. Acad. Sci. USA 114: 4727–4732.

Nickel, A. I., N. B. Wäber, M. Gößringer, M. Lechner, U. Linne, U. Toth, W. Rossmanith, and R. K. Hartmann. 2017. Minimal and RNA-free RNase P in *Aquifex aeolicus*. Proc. Natl. Acad. Sci. USA 114: 11121–11126.

Ohno, S. 1970. Evolution by Gene Duplication. Springer-Verlag, Berlin, Germany.

Orr, H. A. 1996. Dobzhansky, Bateson, and the genetics of speciation. Genetics 144: 1331–1335.

Papp, B., C. Pal, and L. D. Hurst. 2003. Evolution of *cis*-regulatory elements in duplicated genes of yeast. Trends Genet. 19: 417–422.

Parker, H. R., D. P. Philipp, and G. S. Whitt. 1985. Relative developmental success of interspecific *Lepomis* hybrids as an estimate of gene regulatory divergence between species. J. Exp. Zool. 233: 451–466.

Pasquier, J., I. Braasch, P. Batzel, C. Cabau, J. Montfort, T. Nguyen, E. Jouanno, C. Berthelot, C. Klopp, L. Journot, et al. 2017. Evolution of gene expression after whole-genome duplication: New insights from the spotted gar genome. J. Exp. Zool. B Mol. Dev. Evol. 328: 709–721.

Petrov, A. S., C. R. Bernier, C. Hsiao, A. M. Norris, N. A. Kovacs, C. C. Waterbury, V. G. Stepanov, S. C. Harvey, G. E. Fox, R. M. Wartell, et al. 2014. Evolution of the ribosome at atomic resolution. Proc. Natl. Acad. Sci. USA 111: 10251–10256.

Petrov, A. S., B. Gulen, A. M. Norris, N. A. Kovacs, C. R. Bernier, K. A. Lanier, G. E. Fox, S. C. Harvey, R. M. Wartell, N. V. Hud, et al. 2015. History of the ribosome and the origin of translation. Proc. Natl. Acad. Sci. USA 112: 15396–15401.

Piatigorsky, J., and G. Wistow. 1991. The recruitment of crystallins: New functions precede gene duplication. Science 252: 1078–1079.

Presgraves, D. C. 2002. Patterns of postzygotic isolation in Lepidoptera. Evolution 56: 1168–1183.

Price, T. D., and M. M. Bouvier. 2002. The evolution of F_1 postzygotic incompatibilities in birds. Evolution 56: 2083–2089.

Proulx, S. R., and P. C. Phillips. 2005. The opportunity for canalization and the evolution of genetic networks. Amer. Natur. 165: 147–162.

Purkanti, R., and M. Thattai. 2015. Ancient dynamin segments capture early stages of host-mitochondrial integration. Proc. Natl. Acad. Sci. USA 112: 2800–2805.

Putnam, N. H., T. Butts, D. E. Ferrier, R. F. Furlong, U. Hellsten, T. Kawashima, M. Robinson-Rechavi, E. Shoguchi, A. Terry, J. K. Yu, et al. 2008. The amphioxus genome and the evolution of the chordate karyotype. Nature 453: 1064–1071.

Qian, W., and J. Zhang. 2009. Protein subcellular relocalization in the evolution of yeast singleton and duplicate genes. Genome Biol. Evol. 1: 198–204.

Reuveni, S., M. Ehrenberg, and J. Paulsson. 2017. Ribosomes are optimized for autocatalytic production. Nature 547: 293–297.

Rieseberg, L. H. 2001. Chromosomal rearrangements and speciation. Trends Ecol. Evol. 16: 351–358.

Rutter, M. T., K. V. Cross, and P. A. Van Woert. 2012. Birth, death and subfunctionalization in the *Arabidopsis* genome. Trends Plant Sci. 17: 204–212.

Sancar, A. 2008. The intelligent clock and the Rube Goldberg clock. Nat. Struct. Mol. Biol. 15: 23–24.

Sasa, M. M., P. T. Chippindale, and N. A. Johnson. 1998. Patterns of postzygotic isolation in frogs. Evolution 52: 1811–1820.

Scannell, D. R., K. P. Byrne, J. L. Gordon, S. Wong, and K. H. Wolfe. 2006. Rapid speciation associated with reciprocal gene loss in polyploid yeasts. Nature 440: 341–345.

Schmidt, T. R., J. W. Doan, M. Goodman, and L. I. Grossman. 2003. Retention of a duplicate gene through changes in subcellular targeting: An electron transport protein homologue localizes to the golgi. J. Mol. Evol. 57: 222–228.

Selmecki, A. M., Y. E. Maruvka, P. A. Richmond, M. Guillet, N. Shoresh, A. L. Sorenson, S. De, R. Kishony, F. Michor, R. Dowell, and D. Pellman. 2015. Polyploidy can drive rapid adaptation in yeast. Nature 519: 349–352.

Sémon, M., and K. H. Wolfe. 2008. Preferential subfunctionalization of slow-evolving genes after allopolyploidization in *Xenopus laevis*. Proc. Natl. Acad. Sci. USA 105: 8333–8338.

Seoighe, C., C. R. Johnston, and D. C. Shields. 2003. Significantly different patterns of amino acid replacement after gene duplication as compared to after speciation. Mol. Biol. Evol. 20: 484–490.

Shpak, M. 2005. The role of deleterious mutations in allopatric speciation. Evolution 59: 1389–1399.

Silva-Filho, M. C. 2003. One ticket for multiple destinations: Dual targeting of proteins to distinct subcellular locations. Curr. Opin. Plant Biol. 6: 589–595.

Soltis, P. S., and D. E. Soltis. 2016. Ancient WGD events as drivers of key innovations in angiosperms. Curr. Opin. Plant Biol. 30: 159–165.

Spofford, J. B. 1969. Heterosis and the evolution of duplications. Amer. Natur. 103: 407–432.

Stoltzfus, A. 1999. On the possibility of constructive neutral evolution. J. Mol. Evol. 49: 169–181.

Strunk, B. S., and K. Karbstein. 2009. Powering through ribosome assembly. RNA 15: 2083–2104.

Szewczyk, E., A. Andrianopoulos, M. A. Davis, and M. J. Hynes. 2001. A single gene produces mitochondrial, cytoplasmic, and peroxisomal NADP-dependent isocitrate dehydrogenase in *Aspergillus nidulans*. J. Biol. Chem. 276: 37722–37729.

Tatusov, R. L., N. D. Fedorova, J. D. Jackson, A. R. Jacobs, B. Kiryutin, E. V. Koonin, D. M. Krylov, R. Mazumder, S. L. Mekhedov, A. N. Nikolskaya, et al. 2003. The COG database: An updated version includes eukaryotes. BMC Bioinform. 4: 41.

Thompson, A., H. H. Zakon, and M. Kirkpatrick. 2016. Compensatory drift and the evolutionary dynamics of dosage-sensitive duplicate genes. Genetics 202: 765–774.

Ting, C. T., S. C. Tsaur, S. Sun, W. E. Browne, Y. C. Chen, N. H. Patel, and C.-I. Wu. 2004. Gene duplication and speciation in *Drosophila*: Evidence from the *Odysseus* locus. Proc. Natl. Acad. Sci. USA 101: 12232–12235.

Vakirlis, N., O. Acar, B. Hsu, N. Castilho Coelho, S. B. Van Oss, A. Wacholder, K. Medetgul-Ernar, R. W. Bowman 2nd, C. P. Hines, J. Iannotta, et al. 2020. *De novo* emergence of adaptive membrane proteins from thymine-rich genomic sequences. Nat. Commun. 11: 781.

van der Sluis, E. O., H. Bauerschmitt, T. Becker, T. Mielke, J. Frauenfeld, O. Berninghausen, W. Neupert, J. M. Herrmann, and R. Beckmann. 2015. Parallel structural evolution of mitochondrial ribosomes and OXPHOS complexes. Genome Biol. Evol. 7: 1235–1251.

van Hoof, A. 2005. Conserved functions of yeast genes support the duplication, degeneration and complementation model for gene duplication. Genetics 171: 1455–1461.

Vos, M., M. C. Hesselman, T. A. Te Beek, M. W. J. van Passel, and A. Eyre-Walker. 2015. Rates of lateral gene transfer in prokaryotes: High but why? Trends Microbiol. 23: 598–605.

Waldron, C., and F. Lacroute. 1975. Effect of growth rate on the amounts of ribosomal and transfer ribonucleic acids in yeast. J. Bacteriol. 122: 855–865.

Walsh, B. 2003. Population-genetic models of the fates of duplicate genes. Genetica 118: 279–294.

Watterson, G. A. 1983. On the time for gene silencing at duplicate loci. Genetics 105: 745–766.

Weber, C., A. Hartig, R. K. Hartmann, and W. Rossmanith. 2014. Playing RNase P evolution: Swapping the RNA catalyst for a protein reveals functional uniformity of highly divergent enzyme forms. PLoS Genet. 10: e1004506.

Wei, X., and J. Zhang. 2018. On the origin of compositional features of ribosomes. Genome Biol. Evol. 10: 2010–2016.

Wissler, L., J. Gadau, D. F. Simola, M. Helmkampf, and E. Bornberg-Bauer. 2013. Mechanisms and dynamics of orphan gene emergence in insect genomes. Genome Biol. Evol. 5: 439–455.

Woese, C. R. 1971. Evolution of macromolecular complexity. J. Theor. Biol. 33: 29–34.

Wolfe, K. H., and D. C. Shields. 1997. Molecular evidence for an ancient duplication of the entire yeast genome. Nature 387: 708–713.

Young, R., and H. Bremer. 1976. Polypeptide-chain-elongation rate in *Escherichia coli* B/r as a function of growth rate. Biochem. J. 160: 185–194.

Zuckerkandl, E. 1997. Neutral and nonneutral mutations: The creative mix – evolution of complexity in gene interaction systems. J. Mol. Evol. 44 Suppl 1: S2–S8.

Zuellig, M. P., and A. L. Sweigart. 2018. Gene duplicates cause hybrid lethality between sympatric species of *Mimulus*. PLoS Genet. 14: e1007130.

Basic Cellular Features

CHAPTER 7

The Cellular Environment

Armed with an appreciation for the variation in the population-genetic environment experienced by different lineages and the principle factors governing evolutionary change, we now consider a few of the basic chemical and physical constraints dictating the properties of cells. Unlike the population-genetic environment, several aspects of the cellular environment are largely invariant across the Tree of Life. These include the elemental make-ups of cells, the diffusion properties of molecules, the effects of temperature on biological processes, and the amounts of energy accessible from various food types. Some lineages have evolved special attributes to cope with such challenges, e.g., increased protein stability in thermophiles, and the use of motors for molecular transport in eukaryotes. Nonetheless, immutable laws of physics and chemistry ultimately dictate what natural selection can and cannot do.

The cellular environment is in large part defined by the ancient foundational features of biology. For example, the earliest stages of evolution set the elemental requirements of the biochemical building blocks from which all of today's cell bodies are constructed. Life depends on $< 20\%$ of the 119 elements, but aside from carbon, hydrogen, and oxygen, most of these have environmental concentrations thousands to millions of times lower than those found in cellular biomass, highlighting the power of cells to sequester nutrients. Of the myriad forms of organic compounds, life has come to rely on just a handful of fundamental types – amino acids, nucleotides, lipids, carbohydrates, and a few others.

Here, we consider some of the quantitative consequences of biophysical and chemical constraints on cell biology. With an overview of what cells are made of, how many molecules are present per cell, and how much carbon and energy are required for cellular reproduction, the stage is then set for understanding the breadth of issues covered in subsequent chapters. As introduced here and further elaborated on in Chapter 8, numerous cellular features scale with cell size in predictable ways that transcend phylogenetic boundaries. Finite numbers of molecules per cell, combined with the physical constraints associated with molecular diffusion and temperature, dictate the possible rates of intracellular biochemical reactions. The energy content of resources constrains the rate at which new biomass can be constructed. These and many other 'rules of life' define the ultimate limits of the evolutionary playing field. An excellent overview of many of the points discussed here can be found in Milo and Phillips (2016).

The Molecular Composition of Cells

Given that life evolved in an aqueous environment, it is not surprising that the primary component of all of today's cells is water, albeit with a much higher solute load than in the surrounding environment. Cell dry weights scale with cell volume in what appears to be a near-universal relationship across all phylogenetic groups. Over a range of eleven orders of magnitude in cell volume, there is a smooth power-law relationship of

$$W \simeq 0.57 V^{0.92}, \qquad (7.1)$$

where cell dry weight W has units of pg (picograms, or 10^{-12} grams) and cell volume has units of μm^3 (cubic microns, or 10^{-12} ml) (Figure 7.1). The exponent is significantly less than 1.0, indicating that cell density (W/V) decreases with the ~ 0.08 power of cell volume. Because 1 μm^3 of water weighs

Evolutionary Cell Biology. Michael Lynch, Oxford University Press. © Michael Lynch (2024). DOI: 10.1093/oso/9780192847287.003.0007

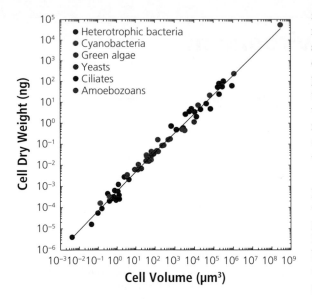

Figure 7.1 Relationship between dry weights and volumes of individual cells. The regression line is applied to all groups simultaneously; $\log_{10}(W) = -3.244(0.040) + 0.920(0.013)\log_{10}(V)$; standard errors of the parameters are in parentheses; $r^2 = 0.99$, $n = 68$. Data taken from various sources in the literature are recorded in Supplemental Table 7.1.

1 pg, these results imply that between one-fifth (large eukaryotic cells) and one-third (small bacterial-sized cells) of total cell weights are comprised of biomolecules and ions. Exceptions occur in diatoms, haptophytes, and foraminiferans, whose cells have hard outer coverings.

Water

Because of life's association with water from the start, many of the features of biology have been permanently molded by the unique properties of this simple molecule. Consisting of a bent complex of two hydrogen atoms and one oxygen atom, H_2O molecules have polarity, with a slight negative charge on the oxygen side and a slight positive charge on the hydrogen side. As a result, liquid water naturally forms a three-dimensional network with each molecule being connected to three to four others via hydrogen bonds in a sort of tetrahedral arrangement (Figure 7.2).

These unique organizational features enable water to operate as a highly effective solvent for other polar molecules. Solubility is an essential feature of most biomolecules involved in chemical reactions requiring diffusive encounters with dissolved reactants. On the other hand, the exclusion of non-polar molecules from the water network provides a pathway for the spontaneous construction of certain cellular features. For example, in water, the hydrophobic tails of lipid molecules naturally aggregate in a highly coordinated fashion (Chapter 15), generating the membranes upon which cells rely.

The hydrogen-bonding ability of water can also present a problem. First, the inner hydrophobic cores that maintain protein structure can be compromised by the intrusion of water molecules. Hydrophobic surface residues also cause proteins to be promiscuously sticky. This imposes strong selective pressure for soluble proteins to achieve their globular structures by populating their outer surfaces with hydrophilic amino acids (Chapter 12). Second, the cohesion within networks of water molecules imposes a drag on large molecules moving through the cytoplasm and on cells moving through aqueous environments, limiting rates of intracellular reactions, and extracellular nutrient uptake and swimming speeds of mobile species (Chapters 16, 18, and 19).

Finally, the thermal properties of water are unique. The viscosity of water declines by nearly 50% from 4 to 40°C, so warm water imposes less resistance to the directed movements of cells but also provides less buoyancy (e.g., imposing higher sinking velocities in aquatic settings). At normal atmospheric pressure, pure water freezes at 0°C, imposing a lower temperature barrier to single-celled organisms incapable of thermoregulation. However, the fact that water has a maximum density at 4°C provides a buffer against such an extreme, as aquatic environments freeze from the top down, with bottom waters rarely colder than 4°C. Ball (2008, 2017) provides a comprehensive overview of many additional knowns and unknowns regarding the biological consequences of water.

Elemental composition

Of the many dozens of chemical elements found in the natural world, only about 20 are essential to life. Ignoring hydrogen and oxygen, carbon is

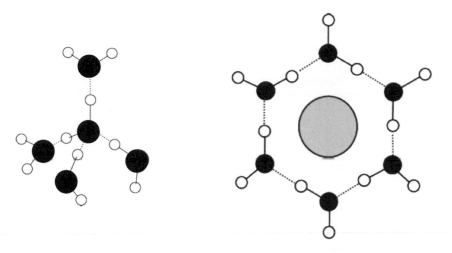

Figure 7.2 The organizational features of water molecules. **Left:** Free water, H_2O, consists of a fluid network of hydrogen bonds wherein each oxygen atom (black ball) is conjoined with an average of ~3.5 hydrogens (white balls). **Right:** The general view is that water builds a cage around soluble particles (interior grey ball). Further details can be found in Ball (2008).

always predominant in terms of molar composition, followed by nitrogen (Table 7.1). The bulk of the remaining biomass is associated with two other elements, phosphorus and sulfur, incorporated into the basic building blocks of cells (e.g., nucleic acids, amino acids, and lipids), along with five other major ions – sodium, calcium, magnesium, potassium, and chloride. All of these elements generally have intracellular concentrations >1 mM. Essential trace metals (e.g., iron, manganese, cobalt) that serve as cofactors of individual enzymes are present at 10–100× lower concentrations. Redfield (1934) first proposed that the ratio of C, N, and P atoms in cells is typically on the order of 106:16:1, and the average of the exemplars in Table 7.1, 100:13:1, is close to this expectation.

A comparison of cellular and environmental elemental concentrations reveals the extent to which cells go to sequester nutrients. There can be considerable variation in the biogeochemistry of different environments, but reliable average estimates exist for the dissolved content of ocean water. As many of the species in Table 7.1 derive from marine environments, molar concentrations in seawater will be used as a reference point. This shows that the degree of cellular enrichment averages ~5000× for carbon, and 50,000 to 60,000× for nitrogen and phosphorus. The remaining major ions range from being nearly isotonic with sea water to enriched by no more than 25×. On the other hand, several essential trace metals (iron, manganese, and cobalt) are enriched by factors >10^6.

To appreciate the challenges imposed by such nutrient acquisition, consider as an example phosphorus, which has an average cellular enrichment of ~60,000×. Living in an average marine environment, in order to produce an offspring, a bacterial cell with volume $1\,\mu m^3$ would need to accomplish the equivalent of fully clearing a surrounding volume of ~$60,000\,\mu m^3$ of P, and for the trace metals noted above, the equivalent of ~10^6 cell volumes would need to be scrubbed clean. For a moderately sized eukaryotic cell, $100\,\mu m^3$ in volume, the necessary volumes of environmental clearance are 100 × higher.

When viewed in the context of cell-division times, the impressive rate at which cells harvest nutrients becomes clear. Again, consider a cell with volume $1\,\mu m^3$ (equivalent to 10^{-15} litres) at birth. With an average internal concentration of 115 mM for phosphorus (Table 7.1), such a cell would contain ~7×10^7 P atoms. Cells of this size have a minimum doubling time of ~0.4 days at 20°C (Chapter 8), implying an incorporation rate of ~2000 P atoms/sec at maximum growth rate. Similar calculations for cells of volume 10, 100, and 1000 μm^3, growing at maximum rates, indicate incorporation rates of ~1×10^4, 9×10^4, and 6×10^5

Table 7.1 Contents of the major elemental constituents (other than hydrogen and oxygen) in a variety of unicellular species. Concentrations in the top half of the table are in units of mM, whereas as those in the bottom half are in units of μM. Species are listed in order of increasing cell volume (μm^3). The means for Ca and Sr exclude the haptophytes *E. huxleyi* and *Gephyrocapsa oceanica*, which have hard outer shells consisting of these elements. *Prochlorococcus* and *Synechococcus* are cyanobacteria; *Vibrio* and *Escherichia* are heterotrophic bacteria; *Pycnococcus*, *Nannochloris*, *Pyramimonas*, and *Dunaliella* are green algae; *Saccharomyces* is budding yeast; *Nitzschia*, *Amphidinium*, and *Thalassiosira* are diatoms; and *Prorocentrum* and *Thoracosphaera* are dinoflagellates. Seawater concentrations are taken from Nozaki (1997). References: cyanobacteria (Heldal et al. 2003); heterotrophic bacteria (Fagerbakke et al. 1996, 1999); yeast (Lange and Heijnen 2001); and all others (Ho et al. 2003).

Species	Size	C	N	P	S	K	Na	Mg	Ca	Cl
Prochlorococcus sp.	0.16	15323	1682	87	82	49	410	371	25	173
Synechococcus sp.	1.00	14906	1755	122	72	78	248	104	49	120
Vibrio natriegen	3.50	8333	1837	157	116	320	400	73	8	1320
Escherichia coli	3.80	7675	1880	263	74	62	210	61	10	104
Pycnococcus provasoli	10	14000	1900	72	77	89		19	4	
Nannochloris atomus	14	14000	2000	81	29	78		19	2	
Saccharomyces cerevisiae	67	15809	2218	131	27	39	7	26	0	
Nitzschia brevirostris	119	11000	1700	250	290	610		150	67	
Emiliania huxleyi	142	10000	1200	130	100	110		18	19000	
Gephyrocapsa oceanica	142	8900	1000	140	140	130		18	18000	
Dunaliella tertiolecta	227	11000	1900	49	14	18		18	1	
Amphidinium carterae	514	1200	160	9	12	1		5	3	
Pyramimonas parkeae	587	6800	570	32	47	27			55	
Prorocentrum minimum	833	22000	1800	16	350	210		160	61	
Thoracosphaera heimii	1353	5100	400	63	82	63		30	2800	
Thalassiosira eccentrica	6627	18000	1900	240	470	790		520	160	
Means		11503	1494	115	124	167	255	106	232	429
Seawater		2.25	0.03	0.002	28	10.2	469	52.7	10.3	546
Cellular enrichment		5,100	50,000	57557	4.4	16.4	0.5	2.0	22.6	0.8

	Sr	Fe	Mn	Zn	Cu	Co
Pycnococcus provasoli	8	910	150	66	38	7
Nannochloris atomus	4	1100	93	140	19	7
Saccharomyces cerevisiae		354	31	642	46	
Nitzschia brevirostris	330	790	590	69	46	14
Emiliania huxleyi	44000	460	940	50	9	39
Gephyrocapsa oceanica	39000	560	990	57	16	50
Dunaliella tertiolecta	4	560	93	74	33	1
Amphidinium carterae	11	120	47	12	5	3
Pyramimonas parkeae	390	500	250	48	20	8
Prorocentrum minimum	470	1100	980	140	440	73
Thoracosphaera heimii	5000	110	79	7	4	6
Thalassiosira eccentrica	950	1600	500	240	68	59
Means	796	680	395	129	62	24
Seawater	89	0.00054	0.00036	0.0054	0.0024	0.000020
Cellular enrichment	8.9	1,260,000	1,086,000	24,000	26,000	1,182,000

P atoms/sec. Given the average 100:13:1 ratio for C:N:P, these incorporation requirements would be 100 and 13 × higher for C and N atoms, respectively. Thus, depending on their size, when growing at maximum rates, cells incorporate on the order of 10^6 to 10^{10} atoms per minute.

Table 7.1 shows the variation among species in elemental composition, some of which may relate to cell size. Menden-Deuer and Lessard (2000) summarized the scaling of carbon content with cell volume in a wide variety of unicellular marine eukaryotes. Aside from chrysophytes, which have inexplicably low carbon estimates, the average exponent on the power-law relationship across groups is 0.91 (SE = 0.03), so there is a decline in carbon content per cell volume in larger cells. For cells of volume 1, 10, 100, and 1000 μm^3, mean carbon contents are 0.30, 0.23, 0.18, and 0.14 pg/μm^3, implying a reduction in cell density with increasing cell size, consistent with the results in Figure 7.1. Using Equation 7.1, the average fractional contributions of carbon to dry weight for cells of these sizes are $\simeq$0.53, 0.49, 0.46, and 0.44, respectively. Thus, a rough rule of thumb from these and other studies (Ho and Payne 1979; Roels 1980; Finlay and Uhlig 1981; Williams et al. 1987; de Queiroz et al. 1993; von Stockar and Marison 2005) is that ~50% of average dry weight in both prokaryotic and eukaryotic cells consists of carbon.

Biomolecules

The organic fraction of cells consists primarily of macromolecules such as proteins, nucleic acids, lipids, and carbohydrates (as well as their precursor building blocks). Most information on this fundamental issue is confined to quite old literature, sometimes based on methods that are not terribly reliable, and variation is also associated with growth conditions during assays (Chapter 9). The most reliable statement that can be made is that proteins comprise the largest fraction of the organic component of cellular biomass (on a dry weight basis), typically in the range of 40 to 60%, but somewhat lower in eukaryotes than in prokaryotes (Figure 7.3). The other primary contributors are RNA (including messenger, ribosomal, and transfer RNAs), carbohydrates (especially in species with cell walls – most bacteria, and some eukaryotes such as fungi and plants), and lipids (which are more enriched in eukaryotic cells, owing to the presence of internal membranes).

Although the fractional contributions to biomass from protein, RNA, lipids, and carbohydrates do not obviously scale with cell volume, the data are scant and noisy enough that such patterns cannot be ruled out. However, the matter is readily accessible for DNA, as genomes have been sequenced for a substantial number of species, and 10^9 bp of DNA is equivalent to ~1 pg dry weight. Here, there is a very strong negative scaling of proportional contribution with cell volume (Figure 7.3). Despite its centrality to all of life, DNA almost never constitutes >10% of the biomass of any cell, and this fraction declines to 0.001% in relatively large eukaryotic cells. Thus, although larger cells tend to have larger genomes (Lynch 2007), scaling as $\sim V^{0.25}$, the proportional investment in DNA is progressively diminished.

Numbers of Biomolecules per Cell

The preceding results provide a generic view of cellular contents per unit biomass. Finer details (e.g., numbers of molecules per cell volume) are required to understand issues related to the properties of specific gene products, such as reaction rates among colliding particles, cellular stochasticity, random variation in inheritance, etc. High-throughput methods for characterizing and quantifying individual mRNA and protein molecules provide insight into these matters. Although data are only available for a few species, over a range of five orders of magnitude in cell size (including both prokaryotes and eukaryotes), the total number of protein molecules/cell scales nearly isometrically with cell volume (V, in units of μm^3),

$$N_{tot,p} = (2.0 \times 10^6)V^{0.95}, \qquad (7.2a)$$

(Figure 7.4). The smallest known bacterial cells harbor <10^5 total protein molecules, whereas larger eukaryotic cells (like those in metazoans) contain >10^9.

To resolve the degree of gene-expression stochasticity, a view at the gene-specific level is necessary. Owing to the fact that large cells often harbor more genes, the average number of proteins within a cell per active gene,

$$\overline{N}_p = 1820V^{0.68}, \qquad (7.2b)$$

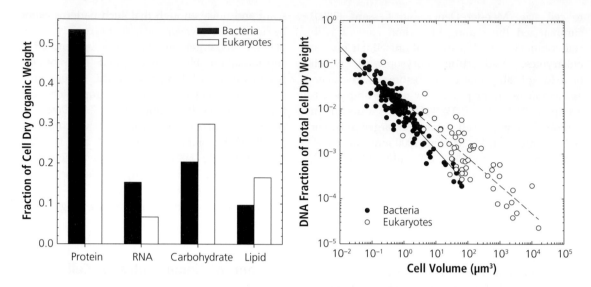

Figure 7.3 Left: The fractional contributions of major biomass components to the total dry weights of cells, determined by averaging over 18 species of bacteria and 15 species of eukaryotes (nine of which are photosynthesizers). The plotted fractions for each group sum to 1.0, and exclude some biomass such as chlorophyll in plants. Data are provided in Supplemental Table 7.2. **Right:** The negative scaling between the fractional contribution of DNA to cellular dry weight and cell volume. DNA dry weight per cell was obtained from the number of nucleotides in the total genome sequences of species, and assumes a single haploid genome per cell; cell dry weights were obtained by applying the function in Figure 7.1 to species with known cell volumes. The fitted power functions for the fractional dry weight of DNA are $0.0072V^{-0.77}$ for bacteria and $0.014V^{-0.62}$ for eukaryotes, where V is the cell volume in units of μm^3. Too much credence should not be attached to the apparent scaling differences between bacteria and eukaryotes, as both sets of data assume cells with haploid genomes, whereas a number of the eukaryotes may be diploid, and bacterial cells in active growth phases often contain several genomes. As a consequence, the plotted estimates may be somewhat downwardly biased (although by no more than a factor of two in eukaryotes). Data are provided in Supplemental Table 7.3.

scales with cell volume more weakly than the total number of proteins per cell.

Moreover, there is substantial variation in the amount of protein product associated with different genes within a cell around the overall mean $\overline{N}_p$. Distributions of the numbers of proteins for individual genes are approximately log-normal (a normal 'bell-shaped' distribution on a logarithmic scale), with the mean being considerably larger than the median, owing to the long tail to the right. With such distributions, the smallest known cells are on the edge of having just one (or fewer) proteins per cell for some genes. For a cell the size of *E. coli*, ~ 1 μm^3, a substantial number of genes are represented by fewer than 100 protein molecules per cell (Figure 7.4). This means that genetically identical offspring resulting from binary fission can vary substantially in their protein contents. If each of

the n copies of a protein in a parental cell is randomly partitioned to daughters, the coefficient of variation (ratio of standard deviation to the mean) among sisters will equal $\sqrt{1/n}$. Further complications arise when proteins are aggregated in vesicles (Chapter 9).

What do the preceding numbers mean in terms of cellular concentrations? Focusing on the number of proteins representing an average gene, Equation 7.2b, the concentration on a per μm^3 basis becomes $1820V^{-0.32}$. Multiplying this by 10^{15} μm^3/litre, and dividing by the number of molecules per mol (Avogadro's number, 6.023×10^{23}), yields an average concentration of $3.0V^{-0.33}$ μM (μmol/litre, where 1 $\mu mol = 10^{-6}$ mol). With protein numbers 10× above and 10× below the average, this concentration would be multiplied by 10 and 0.1, respectively. Thus, cellular concentrations

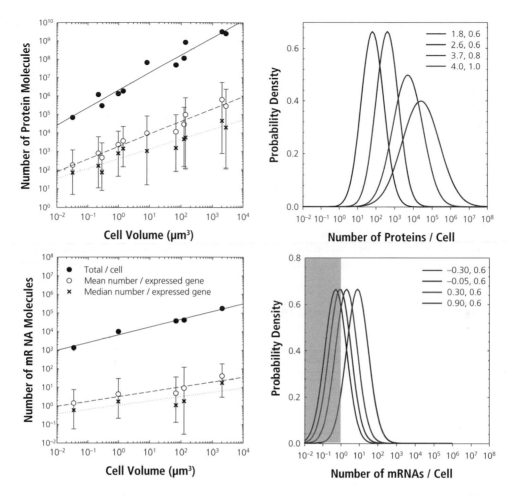

Figure 7.4 Numbers of proteins and messenger RNAs per cell. **Left panels:** Upper solid lines denote the total number of molecules per cell, summed over all gene products. The brackets for numbers of molecules per gene denote the lower 2.5% and upper 97.5% cut-offs in the overall distributions; and the dashed and dotted lines are the regressions involving the means and medians. **Right panels:** Approximate distributions are given for the numbers per cell for different proteins in four differently sized cells, ranging from (left to right) the bacteria *Mycoplasma* (0.05 μm^3) and *E. coli* (1 μm^3) to an approximate yeast (100 μm^3) to a mammalian cell (2400 μm^3). These distributions are based on the empirical values given in the left panels; means and standard deviations on the log$_{10}$ scale are given in the insets. The grey region for the mRNA plot denotes the fraction of cells expected to harbor zero transcripts at any point in time. Data are from the previous survey in Lynch and Marinov (2015), a more recent quantitative transcriptomic study on an unclassified marine picobacterium OM43 (Huggett et al. 2012; Gifford et al. 2016), and proteomics studies of the unicellular eukaryotes *Trichomonas* (Dias-Lopes et al. 2018) and *Leishmania* (Pinho et al. 2020).

of proteins are typically in the nM (nanomolar, or 0.001 μM) to μM range, with concentrations tending to decline with increasing cell volume.

The situation is much more extreme for messenger RNAs, as even large, well-nourished cells typically harbor orders of magnitude fewer mRNAs than the total number of proteins (Figure 7.4). The total number of transcripts per cell scales more weakly with cell volume than for the case of proteins,

$$N_{tot,t} = 6760V^{0.42}, \tag{7.3a}$$

and the mean number of transcripts per active gene is just

$$\overline{N}_t = 3.2V^{0.26} \tag{7.3b}$$

(Figure 7.4). Typically, there are hundreds to thousands of more copies of proteins than transcripts per gene within cells, and the average gene is represented by fewer than ten mRNAs at any particular time. As a consequence, substantial numbers of genes are at least transiently devoid of transcripts in small cells, and this is even true for a small subset of genes in species with the largest of cells.

As discussed in Chapter 8, the numbers of ribosomes per cell also scale across the Tree of Life with cell volume in a predictable manner. Here, however, cells are much more guarded against stochastic loss, as the average number of ribosomes per cell is generally > 100 even in the smallest cells, ranging up to 10^8 in the largest cells. This should not be too surprising, as complete loss of ribosomes is equivalent to a death sentence.

Passive Transport of Particles through the Cytoplasm

To carry out their key functions, biomolecules often have to travel to particular destinations to encounter specific substrate molecules. Except for large complexes and cargoes within vesicles in eukaryotic cells, most molecules spend the majority of their time moving by passive diffusion. Thus, to understand the ultimate biophysical constraints on cellular functions, we require information on how rapidly molecules can diffuse from one location to another. Due to background thermal motion, each molecule within a cell is continuously jostled in random ways (often referred to as Brownian motion), and until encountering an impervious barrier, such as the cell membrane, will diffuse at a roughly constant average rate, depending on the nature of the medium. The average distance moved after t time units is a function of the diffusion coefficient D, defined as the average squared distance of molecular movement per unit time (Foundations 7.1).

The reason for focusing on the squared distance is most easily understood in the context of a random one-dimensional diffusion process. In this case, at each time point a particle has an equal probability of moving to the left versus the right, so the average directional movement of particles is zero. Nonetheless, when molecules move randomly, such that there is no memory in the process, the noise of each incremental move is cumulative. Thus, although the mean location remains constant, with increasing time a diminishing fraction of molecules will remain in the vicinity of their initial location. With respect to the starting position, the probability distribution of locations of individual molecules becomes wider and wider with time (t). Taking the square root of the mean-squared distance, $\sqrt{2Dt}$, provides a measure of average absolute dispersion on the original scale for a one-dimensional process.

Of course, not all diffusion processes in biology are one-dimensional. Diffusion of individual molecules within a fluid lipid membrane is a two-dimensional process, whereas diffusion through the cytoplasm is three-dimensional. There is, however, a simple algebraic relationship between the expected magnitude of diffusion and the dimensionality of the process. As just noted, under one-dimensional diffusion, a particle can move in only two directions, right versus left. Adding a dimension increases the magnitude of dispersion, owing to the reduction in the degree of backtracking (Figure 7.5). For example, considering a two-dimensional grid, a particle can move in four

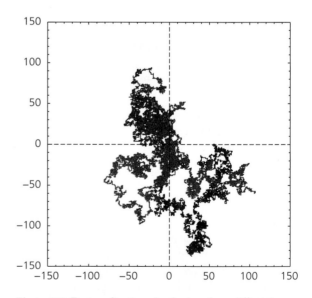

Figure 7.5 Four realizations (each given by a different colour) of a two-dimensional random walk, all starting at the same point (the intersection of the vertical and horizontal lines). In each case, 10,000 steps of unit length were recorded with the movement (right, left, up, down) being directionally random.

directions (e.g., north, south, east, west), and on a three-dimensional lattice, there are six possible routes of movement. In these higher-dimensional cases, the dispersion distance is the radial (straight-line) distance from the initial point, and with two- and three-dimensional diffusion, the root mean-squared distances after t time units become $\sqrt{4Dt}$ and $\sqrt{6Dt}$, respectively. Thus, the rate of diffusion relative to an initial location increases with dimensionality, but the scaling with the square root of time is retained. From these expressions, it can be seen that for an n-dimensional process, the expected time required for a particle to move an absolute distance of d units is $d^2/(2nD)$.

To understand the implications of diffusion limitation for cellular processes, we require information on how the diffusion coefficient depends on the features of a particle and the medium through which it moves. In its most elementary form, a diffusion coefficient is defined as

$$D = \frac{k_B T}{\gamma}, \tag{7.4}$$

where k_B is the Boltzmann constant (1.38×10^{-16} cm$^2\cdot$ g $\cdot$ sec$^{-2}\cdot$ K^{-1}), which relates energy at the particle level to temperature T in degrees Kelvin, and γ is the friction coefficient, which is a net measure of the resistance imposed on particle movement by the medium (with units of g $\cdot$ sec^{-1}). The form of this expression is reasonably intuitive – the numerator is a measure of the jostling due to thermal noise, and the denominator is a measure of resistance to such jostling. Because most of life (other than thermophiles) exists in the range of $T \simeq 280$ to 315 K, T can be approximated as 300 K with only slight loss of accuracy. A sampling of diffusion coefficients for small molecules in an aqueous environment is provided in Figure 7.6, which shows that large proteins diffuse up to 100× more slowly than small ions.

The friction coefficient depends on the medium as well as on the shape and form of the particle, and many expressions have been developed to accommodate such effects (He and Niemeyer 2003; Dill and Bromberg 2011; Soh et al. 2013). For a perfectly spherical particle with radius r (in

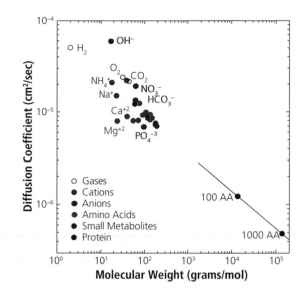

Figure 7.6 Some diffusion coefficients for simple substances in water at 20°C, as a function of molecular weight; a few of these are specifically annotated. The solid line is obtained with Equation 7.8 by multiplying the numbers of amino acids in the chain by the average molecular weight of an amino acid, 137 g/mol. The units may be changed by noting that 1 cm = 10^4 μm = 10^7 nm.

units of cm), the Stokes–Einstein equation tells us that

$$\gamma = 6\pi\eta r, \tag{7.5}$$

where $\pi \simeq 3.142$ is the universal constant (equal to the ratio of a circle's circumference to diameter), and η is the viscosity of the medium (with units g $\cdot$ cm$^{-1}\cdot$ sec^{-1}). For water, η is temperature dependent, taking on values of 0.013, 0.011, 0.010, 0.0089, and 0.0080 g $\cdot$ cm$^{-1}\cdot$ sec^{-1} at 10, 15, 20, 25, and 30°C. For simplicity, the 20°C value will be assumed in the following calculations. The diffusion coefficient of a sphere in a typical aqueous environment then becomes

$$D \simeq \frac{22 \times 10^{-6}}{r}, \tag{7.6}$$

where the numerator has units μm^3/sec, and r has units of μm. If r is in units of nm, $D \simeq 22/r$ nm^2/sec.

Biology introduces numerous complications. For example, most biomolecules depart from a perfectly spherical geometry, and cytoplasm is substantially more viscous than water. Here we focus on proteins, which typically fold into globular structures. The problem of particle shape can then be dealt with by considering the effective particle radius. For the ideal case of perfectly packed spherical proteins composed of N_{AA} amino acids, the radius would scale as $N_{AA}^{1/3}$ (because the volume of a sphere is proportional to the cube of the radius). However, empirical study implies that the average distance of a protein molecule's parts to a central point (often called the radius of gyration) scales as

$$r_g = (2.2 \times 10^{-8}) N_{AA}^{0.4} \qquad (7.7)$$

(in units of cm; Hong and Lei 2009). Subdivision of proteins into domains, less than perfect packing, and various elastic features contribute to this elevated scaling relative to the ideal situation. Tyn and Gusek (1990) find that a protein with radius of gyration r_g behaves hydrodynamically on average as though the effective radius is $r \simeq 1.3 r_g$. Applying this correction factor and Equation 7.7 to Equation 7.6, we obtain an expected diffusion coefficient for a protein in an aqueous environment of

$$D \simeq 770 N_{AA}^{-0.4}, \qquad (7.8)$$

with units of $\mu m^2/sec$. Proteins diffuse at rates that are typically < 10% of the rates for individual amino acids (Figure 7.6).

Intracellular crowding imposes an additional impediment to molecular diffusion. The internal milieu of a cell is hardly the open-water environment assumed in most diffusion theory. Rather, 20–40% of the cytoplasmic volume of a typical cell is occupied by proteins and other macromolecules (Zimmerman and Trach 1991; Luby-Phelps 2000; Ellis 2001). As a consequence, the average distance between proteins is on the order of the width of the proteins themselves. This then raises questions as to how the basic composition of cells alters the freedom of movement of the very molecules upon which life depends. On the one hand, molecular crowding reduces the aqueous volume that must be searched to locate a small solute. On the other hand, transient molecular confinement, aggregation with molecules of opposite charge, and sieving effects can inhibit the free diffusion of proteins. Although

the net consequences of these added complications are minor for small metabolites, the diffusion coefficients for proteins are reduced by ten to fifty-fold in *E. coli* (Elowitz et al. 1999; Konopka et al. 2006; Nenninger et al. 2010), and perhaps somewhat less in eukaryotic cells (Luby-Phelps 2000; Dix and Verkman 2008).

For example, in an aqueous environment, green fluorescent protein (GFP), with a chain length of 238 amino acids, has a diffusion coefficient of 87 $\mu m^2/sec$, almost exactly as predicted by Equation 7.8. In contrast, empirical estimates of GFP diffusion coefficients within the cytoplasm of multiple bacteria (*Caulobacter crescentus*, *E. coli*, *Lactococcus lactis*, and *Pseudomonas aeruginosa*) are in the range of 5 to 15 $\mu m^2/sec$ (Konopka et al. 2009; Nenninger et al. 2010; Montero Llopis et al. 2012; Guillon et al. 2013; Mika et al. 2014), and on the order of 25 to 30 $\mu m^2/sec$ in the slime mold *Dictyostelium* and mammalian cells (Swaminathan et al. 1997; Potma et al. 2001). Examination of a diversity of proteins in *E. coli* demonstrates that, despite the crowdedness of bacterial cytoplasm, diffusion is still well-described as a Brownian process once the effective viscosity of the medium is accounted for (Bellotto et al. 2022). Large complexes diffuse much more slowly. For example, the estimated rate for a ribosome is 0.04 $\mu m^2/sec$ in *E. coli* (Bakshi et al. 2012). Membrane proteins undergoing two-dimensional diffusion through a much more densely packed lipid milieu have diffusion coefficients in the range of 0.02 to 0.03 $\mu m^2/sec$ in bacteria, with the rate declining with the number of transmembrane domains in the protein (Kumar et al. 2010; Mika et al. 2014).

The preceding observations suggest that general diffusion processes may speed up in eukaryotes. On the one hand, the average protein chain length for eukaryotes, $N_{AA} = 532$, is 45–60% larger than the means in bacteria (365) and archaea (329) (Wang et al. 2011). On this basis, assuming similar folding architectures, all other things being equal, Equation 7.8 implies that a ~1.5× increase in total chain length should yield a 15% reduction in the average diffusion coefficient for proteins in eukaryotes. However, given that the density of eukaryotic cytoplasm is lower than in prokaryotes, reduced crowding effects may essentially cancel this particle-size effect. A third effect that merits further consideration is that active processes in eukaryotic cells, such as those created by molecular motors, generate as by-products random diffusion-like forces, thereby enhancing rates of molecular movement throughout the cytoplasm even by non-carrier molecules (Guo et al. 2014).

Whether these cytoplasmic features of eukaryotic cells have evolved to facilitate long-distance diffusion and/or result in relaxed selection against protein stickiness remains a matter of speculation (Soh et al. 2013). There is, however, some evidence that the diffusive properties of proteins coevolve with their proteomic environment. For example, Mu et al. (2017) found that when placed in the cytoplasm of *E. coli*, human proteins tend to stick to their foreign environment, but that modification of a few surface amino-acid residues can yield diffusion rates equivalent to the native *E. coli* proteins.

Finally, to appreciate the timescale of passive molecular diffusion, consider a protein of moderate length with a diffusion coefficient of $D \simeq 20$ $\mu m^2/sec$. In a three-dimensional setting (e.g., cytoplasmic diffusion), the root mean-squared distance travelled after t seconds will be $d = \sqrt{6Dt} \simeq 11\sqrt{t}$ μm. The expected time to travel d μm is then $(d/11)^2$ sec. A spheroid bacterial cell with a 1 μm^3 volume has a diameter of 1.2 μm, so it would take ~ 0.01 seconds for the protein to travel the width of the cell. For a moderately sized eukaryotic cell with volume 100 μm^3, traversing the 5.8-μm width requires ~ 0.28 seconds. For a large spherical cell with volume 10^5 μm^3 (which is attained in some marine diatoms and dinoflagellates), travelling the cell width of 58 μm requires ~ 28 seconds. Thus, molecular delivery across a cell based on diffusion alone is effectively instantaneous in bacteria, and comes with no cost, as it is entirely fuelled by background thermal noise. In contrast, diffusion becomes dramatically less efficient in large eukaryotic cells, which often transport material by use of molecular motors, which run on ATP (Chapter 16). One final caveat with respect to all of these results is that the viscosity of cytoplasm appears to vary significantly with the level of cell nutritional state, increasing in starved cells (Joyner et al. 2016). All of these issues, and more, are reviewed in Schavemaker et al. (2018).

Intermolecular Encounter Rates

Proteins do not operate in isolation. More often than not, they aggregate into multimeric complexes, and most engage with particular substrate molecules. Diffusion theory explains the rates of dispersion of individual particles, but the rate of encounter of interacting particles depends on particle sizes and concentrations. As an entrée into this area, we consider the simple situation in which the two interacting particle types are products of the same genetic locus, as in the case of two monomeric subunits coalescing to form a homodimer, a very common situation for proteins. (The more general case of two different particles is derived in Foundations 7.2).

To move forward, we require a measure of the encounter rate per unit concentration, k_e, which is a function of the particle diffusion rate (Foundations 7.2) and has units of events $\cdot$ cm$^3 \cdot$ sec^{-1} (or some other combination of distance and time units). This must be multiplied by the product of the concentrations of the particles to be joined to account for the fact that both interacting partners are randomly diffusing; in this particular example, each particle has the same concentration [C]. The resultant rate of encounter per unit volume, which has units of events $\cdot$ cm$^{-3} \cdot$ sec^{-1}, must then be multiplied by the cell volume V (in units of cm^3 per cell) to give the total rate of encounter events within the cell. A small modification arises because a particle cannot interact with itself, necessitating a correction factor of $1 - (1/n)$, where $n = [C] \cdot V$ is the expected number of particles per cell. (For $n > 100$, this modification can be ignored). The final expression for the total rate of encounter then becomes

$$R_E = k_e \cdot [C]^2 \cdot V[1 - (1/n)]$$
$$= (11 \times 10^{-12}) \cdot [C]^2 V[1 - (1/n)] \qquad (7.9)$$

with units of events/cell/sec, and the constant substituted for k_e applying to the specific case of two spherical particles of the same size (Foundations 7.2).

To gain some appreciation for the constraints on such encounters, and hence the viability of a strategy to dimerize, consider a cell with a 1 μm^3 volume (bacterial sized) and a molecule with a concentration of $1\ \mu M$, which as noted previously is within the range typically seen for proteins. Using the conversions $10^6 \mu M/M$, $1000\ cm^3/litre$, and 6.02×10^{23} molecules/mol, a $1\ \mu M$ concentration transforms to $[C] = 6.02 \times 10^{14}$ molecules/cm^3. Thus, because there are $10^{12}\ \mu m^3$ in $1\ cm^3$, the $1\ \mu m^3$ cell is expected to contain $n \simeq 602$ molecules. Application of Equation 7.9 then leads to an encounter rate of 4×10^6 events/cell/sec. Increasing the concentration by a factor of x will increase the encounter rate by a factor of x^2.

In the preceding example, n is sufficiently large that the correction factor has essentially no effect. However, decreasing $[C]$ and/or V begins to have a non-linear effect at sufficiently low values. For example, if the concentration is reduced to $0.01\ \mu M$, the expected number of molecules/cell is reduced to $n \simeq 6$, and the encounter rate is reduced by a factor of $(0.01)^2 (5/6)$ to ~ 333 events/cell/sec. Further reducing the cell volume to $0.1\ \mu m^3$, then $n < 1$, and a protein would almost always be without partners in a cell. These results demonstrate that constraints on the number of molecules contained within small cells (Figure 7.4) must ultimately limit the reaction rates that can be carried out (Klumpp et al. 2013).

Finally, it should be noted that all of the above considers only the physical encounter rate between particles, assuming an ideal homogeneous setting with no attractive or repulsive forces between particles. Should the surface of each particle contain a restricted reactive patch, this will reduce the effective encounter rate by a factor related to the effective patch size per particle, after also taking into consideration the process of rotational diffusion (which refers to random movement of a particle on its axes, apart from movement across space). These issues are taken in up in Chapters 13 and 18, focused on protein multimerization and nutrient uptake.

Temperature-Dependence of Biological Processes

Through its effects on rates of molecular motion, temperature influences virtually all biological processes. For most biochemical interactions, elevated temperature increases the reaction rate, at least up to the point beyond which the stability of the reactants is compromised. Chemical reaction rates depend on the frequency of successful encounters between participating molecules, and most reactions require some amount of energy to go forward. The energetic barrier to a reaction is called the activation energy (E_a), with a higher value of E_a implying a slower response to temperature. A powerful result from statistical mechanics, the Boltzmann distribution, relates the distribution of energy states of molecules to ambient temperature (Foundations 7.3).

This distribution has the useful property of being exponential in form, with the mean energy state of molecules being the familiar $k_B T$. For a system in thermodynamic equilibrium, the fraction of molecules with an energy state above the activation energy is simply

$$f_e = e^{-E_a/(k_B T)}. \qquad (7.10)$$

As temperature increases, more molecules have high enough energy to overcome the activation barrier, and $f_e \to 1$ at a rate that depends on E_a. The overall reaction rate is the product of the encounter rate between reactants (R_E) and the fraction of successful encounters,

$$R_{\text{tot}} = R_E \cdot f_e = R_E \cdot e^{-E_a/(k_B T)}. \qquad (7.11)$$

Taking the log of this expression demonstrates that a plot of the log of a reaction rate against the inverse of temperature ($1/T$) is expected to yield a straight line

$$\ln(R_{\text{tot}}) = a - b(1/T) \qquad (7.12)$$

with the slope (b) estimating $-E_a/k_B$, and the intercept (a) estimating the log of the encounter rate, which is a function of the properties of reactants and their concentrations (Foundations 7.2). Such an inverse relationship between the rate of a molecular reaction and $1/T$ is known as Arrhenius-rate behavior, after its early advocate (Arrhenius 1889), who derived the expression in a different way than pursued in Foundations 7.3. Because k_B is a constant, Equation 7.12 provides a simple means for estimating the activation energy of a reaction. (It may be noticed from Equation 7.2.2 that temperature appears in the expression for R_E, which is inversely related to the viscosity of the fluid, but this is generally ignored under the assumption that the exponential dependence of f_e on temperature dominates the overall behavior).

Although the Arrhenius equation often provides an excellent description of the temperature-dependence of simple chemical reactions, organisms consist of mixtures of hundreds to thousands of biomolecules. Each biochemical reaction will have its own activation energy, with the concentrations and stabilities of the interacting partners changing with environmental conditions, including temperature (e.g., Hunter and Rose 1972; Alroy and Tannenbaum 1973; Herendeen et al. 1979). Many of these reactions will operate in parallel (as, for example, independent pathways for uptake of different nutrients), whereas others will operate in series (as in consecutive steps in metabolic pathways). Thus, although there may be one rate-limiting step at any particular temperature, the nature of this step (and its associated activation energy) is likely to change among temperatures. Further complicating matters is the fact that complex biomolecules tend to become increasingly unstable at high temperatures and can have altered properties at low temperatures (Dill et al. 2011).

All of these issues motivate the question as to whether rates of higher-order biological functions scale in accordance with Equation 7.13, and if they do, whether there is any simple mechanistic interpretation of the fitted slopes and intercepts. At best, any estimate of E_a for a cell-biological process would seem to be a composite 'effective' barrier to activation of the process. Nonetheless, it is often suggested that processes such as metabolic and developmental rates, and the 'rate of living' (inverse of life span), scale in close accordance with the Arrhenius equation, at least below temperatures at which key molecular/cellular processes begin to break down (Gillooly et al. 2001; Savage and West 2006). It has been suggested that in *E. coli*, most biochemical reaction rates have similar responses to temperature under non-extreme conditions, leading to an overall adherence to Arrhenius-rate behavior (Mairet et al. 2021).

Herein lies the problem. Although the range of temperatures consistent with Arrhenius-rate behavior are often referred to as being 'biologically relevant', this is usually little more than a matter of convenience and/or conjecture, with the edges of such regions often being quite arbitrary. When taken to even moderately extreme temperatures,

the responses of cellular growth rates to temperature are virtually always curvilinear, in contrast to the expectations from Equation 7.12, with the optimal temperature and the form of the response curve often varying substantially among species (Figure 7.7). Even within the range of 'meaningful' temperatures, not all biological rates scale exponentially with temperature, with the response of growth rate to $1/T$ approaching linearity in various unicellular eukaryotes (Montagnes et al. 2003). Nor do all biological features respond in a positive way to a thermal increase. For example, the cell sizes of unicellular eukaryotes often decline by $\sim$2–4% for each 1°C increase in temperature (Montagnes and Franklin 2001; Atkinson et al. 2003).

Some have suggested that these kinds of variations in temperature response curves can be accommodated by relatively simple modifications of the Arrhenius equation, such as by subtracting or dividing one exponential expression by another to account for contrasting responses of cell features to temperature (Mohr and Krawiec 1980; Ratkowsky et al. 1983, 2005; Corkrey et al. 2014; Arroyo et al. 2022). For example, Dill et al. (2011) show how deviations from ideal Arrhenius behavior can be accommodated by multiplying Equation 7.10 by a function that accounts for increasing protein denaturation with temperature. Although the fits of such mathematical relationships to biological features are often quite good over a substantial temperature range, and the underlying models are frequently viewed as first-principles sorts of derivations, caution is warranted in attaching too much biological meaning to them. With four or more parameters, a wide variety of nonlinear functions can yield essentially identical fits to the same data. Indeed, more than 24 alternative mathematical functions have been proposed for the relationship between reaction rates and temperature (Noll et al. 2020). Such general statistical fits remain of great interest, as they suggest the operation of universal scaling features of biological traits, begging the question as to the underlying mechanisms that apply across the Tree of Life.

One of the alternative approaches is a common rule-of-thumb in biology, the so-called Q_{10} rule, which states that biological rates typically increase by a factor of 2–3 with a 10°C increase in

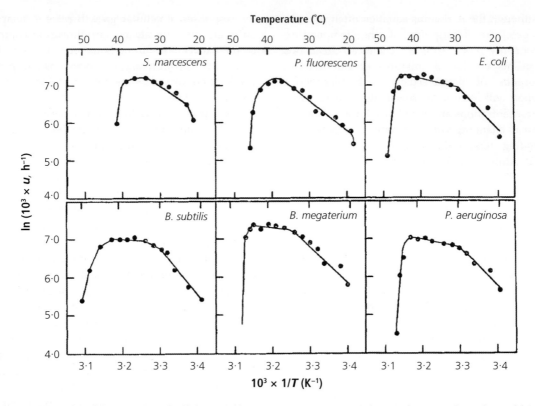

Figure 7.7 Examples of the response of cellular growth rate to temperature in six bacterial species. Growth rate (u) is plotted on a log scale as a function of the inverse of temperature in degrees Kelvin ($1/T$). Temperatures in Celsius are given along the top axis. Only in the central regions of the plots do the growth rate responses follow the linear decline with $1/T$ expected under Arrhenius rate behavior, and even then this is often just an approximation to a broader curvilinear pattern. From Mohr and Krawiec (1980).

temperature (Raven and Geider 1988; Hansen et al. 1997), again with a presumed focus on a 'biologically relevant' temperature range. The idea was first raised by Arrhenius' PhD advisor, Van't Hoff, and can be crudely related to the Arrhenius equation. For example, considering two commonly used temperatures, 12 and 22°C (i.e., $T = 285$ and 295), then from Equation 7.10 the ratio of Arrhenius rates at the high versus low temperature is $\sim e^{0.00012(E_a/k_B)}$. If $E_a/k_B \simeq 8333$, then $e^1 \simeq 2.72$, which is within the range of commonly observed Q_{10} estimates. This implies that E_a/k_B must typically be on the order of 8333. However, somewhat different results will be obtained with different 10°C limits on the temperature range. For example, applying temperatures of 22 and 32°C yields a ratio of $e^{0.00011(E_a/k_B)}$. Assuming $E_a/k_B = 8333$ still holds, this would imply $Q_{10} = 2.52$. Thus, the Q_{10} approach is an approximation, albeit a fairly good one, if the system behaves in

accordance with the Arrhenius equation. However, there is little justification for claiming the superiority of one approach over the other, as both are phenomenological descriptions of biological functions.

Energy, Carbon Skeletons, and Cell Yield

Heterotrophic organisms, incapable of fixing CO_2, are reliant on the uptake and assimilation of organic compounds for the production of new cellular biomass. The key resources consist of reduced carbon compounds containing hydrogen, usually with some oxygen, nitrogen, phosphorus, and/or sulfur atoms also present. In today's organisms, these substances are almost always ultimately derived from cellular materials or excretory products of photoautotrophs, with many undergoing secondary modification in herbivores and detritivores before again being ingested by carnivores. Food materials provide both the carbon skeletons necessary for biosynthesis of the monomeric building blocks of the cell, e.g., amino acids, nucleotides, and lipids. They are also the source of energy for subsequent transformation into the cell's energetic currency, ATP.

The specific organic composition of food ultimately dictates the rate at which a heterotroph can invest biomass and energy into self-maintenance, growth, and reproduction. In the organism, as in the furnace, the oxidation of organic substrates releases energy, and some organic substances have higher energy contents than others. The maximum amount of extractable energy of a substance is equivalent to its heat of combustion, ΔH_C, with the absolute limit to biological energetics being set by the product of the latter and the consumption rate (ignoring the costs of building and maintaining the metabolic machinery itself).

A deeper understanding of the biological relevance of heats of combustion can be achieved by considering the chemical composition of a substrate and the fates of carbon-associated electrons upon combustion. Kharasch and Sher (1925) classified organic compounds on the basis of the number of electrons that experience a transition from a methane-type bond (C–H) to a carbon dioxide-type bond (C=O) upon combustion,

$$N_E = 4N_C + N_H - 2N_O \qquad (7.13)$$

for a molecule containing N_C carbon, N_H hydrogen, and N_O oxygen atoms. The structure of this expression follows from the fact that each carbon atom has four outer-shell (valence) electrons, sharing one additional electron with each bonded hydrogen atom and two with each bonded oxygen. The electrons shared with each hydrogen atom are free to move upon combustion, whereas the two associated with each oxygen are already in the position expected after oxidation. Complete combustion reconfigures hydrogen atoms into water, and oxygen atoms into CO_2, which from Equation 7.13 has $N_E = 0$. For glucose, $C_6H_{12}O_6$, $N_E = 24$.

For carbon substrates commonly employed in laboratory growth experiments, this composite measure of the degree of electron movement upon transformation to CO_2 and water is nearly perfectly correlated with known heats of combustion

determined in chemistry labs (Figure 7.8), with ΔH_C (in units of kcal/mol) being closely approximated by $27N_E$. Extensions to organic substrates containing nitrogen and/or sulfur can be found in Kharasch and Sher (1925) and Williams et al. (1987).

These purely physico-chemical descriptors of substrate molecules are informative with respect to growth rates of pure cultures of unicellular organisms raised in chemostats (Chapters 8 and 17). For situations in which a single substrate is the sole source of carbon and energy (and all other nutrients are in excess supply), a compilation of data from studies involving alternative carbon sources indicates that the growth yield (g cell dry weight/g carbon consumed) increases with the heat of combustion per carbon atom in the substrate, with no obvious differences between bacteria and eukaryotes (Figure 7.8). However, beyond the point at which the caloric content of the substrate exceeds 10 kcal/g carbon, the cell yield levels off at ~1.3 g dry weight/g carbon consumed. Because the values in Figure 7.8 are derived from cultures growing at maximum rates, only a small fraction of the cell's energy budget is allocated to maintenance (Chapter 8), so the estimates provided are close to the maximum biomass yields associated with the substrate.

This overall pattern, first suggested by Linton and Stephenson (1978), with many fewer data than contained in Figure 7.8, implies that for low-energy substrates (heats of combustion < 10 kcal/g carbon), heterotrophic cells are intrinsically energy limited, i.e., they are incapable of experiencing the maximum possible yield of ~1.3 g dry weight/g carbon consumed. Above a substrate heat of combustion of 10 kcal/g carbon, the constant cell growth yield per unit carbon implies a progressive decline in the efficiency of energy extraction with increasing energetic content of the substrate. Thus, an energy content of ~10 kcal/g carbon appears to separate a lower domain in which the substrate provides insufficient energy to assimilate the available carbon from an upper domain where energy is in excess of the requirements for carbon assimilation. Foundations 2.3 takes these kinds of observations a step further to estimate the amount of energy in units of ATP hydrolyses needed to produce a unit of biomass.

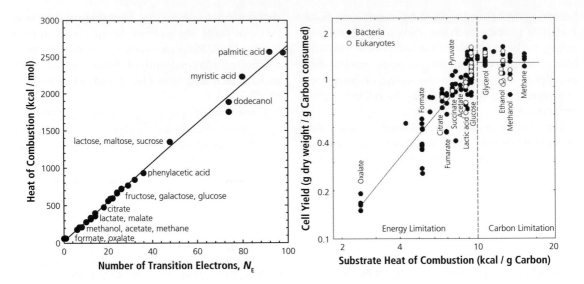

Figure 7.8 Left: The heats of combustion of organic substrates commonly applied in growth experiments of microbes. The fitted regression is $\Delta H_C = 26.2 N_E + 25.8$. **Right:** The yield of cell dry weight per gram of carbon consumed as a function of the heat of combustion of the substrate in units of carbon follows a power-law relationship, $Y = 0.042 H_C^{1.47}$ ($r^2 = 0.81$), where H_C is the normalized heat of combustion, up to a carbon-specific heat of combustion $\simeq 10$, and thereafter levels off with constant value $\simeq 1.3$ (units are defined in the axis labels). Note that the heats of combustion plotted here are equivalent to the plotted values on the y axis in the left panel divided by the number grams of carbon/mol of substrate. From Lynch and Trickovic (2020).

Finally, recalling from above that the average fractional carbon mass per cell dry weight is ~0.5, the cell yields in Figure 7.8 can be rescaled to units of g cellular carbon/g substrate carbon, providing a measure of assimilation efficiency for carbon. With dry-weight cell yields/g substrate carbon being in the range of 0.8 to 1.6 for nearly all common substrates (Figure 7.8), this implies typical carbon assimilation efficiencies in the range of 0.4 to 0.8. After nearly four billion years, this is the best that natural selection has been able to achieve. Complete conversion of substrate carbon into biomass is unobtainable, as energy must be extracted from some of the substrate to carry out cellular functions, and some carbon is lost as CO_2.

Summary

- Between 65 and 80% of the wet weight of cells consists of H_2O, eukaryotic cells being more watery than those of bacteria. Across the Tree of Life ~50% of cell dry weight is comprised of carbon atoms, and one- to two-thirds of the dry weights of most cells consist of protein.

- The unique physical properties of water govern almost every aspect of biology, as they dictate the folding stability of proteins, the ability of lipid molecules to aggregate into membranes, the diffusion rates of molecules, and the challenges to motility.

- Of the 20 chemical elements essential to life, many have intracellular concentrations enriched by factors of 10^3 to 10^6 relative to environmental levels. Such factors are equivalent to the volume of the environment relative to that of the cell that needs to be fully harvested to produce an offspring cell.

- Despite its centrality to life, the fractional contribution of genomic DNA to cellular biomass scales negatively with cell volume, declining from ~10% in the smallest bacterial cells to < 0.001% in the largest eukaryotic cells.

- The total number of protein molecules per cell and the average number per gene increase sublinearly with cell volume, consistent with larger cells being less dense with biomaterials. Messenger RNAs are typically 100- to 10,000-fold less abundant per cell than their cognate proteins, with the mean number per gene often being in the range of 1 to 10. With the distributions of both mRNAs and protein molecules per gene per cell being approximately log-normal in form, there can be significant stochastic variation in gene expression among genetically uniform cells. Thus, there must be a lower bound to cell size below which adequate numbers of molecules cannot be harbored to reliably sustain key biochemical reaction rates.

- Many molecules travel through cells by passive diffusion processes. Fuelled by background thermal noise, such transport imposes no costs to the host cell. For small bacterial-sized cells, an average protein can diffuse across a cell diameter in several milliseconds, whereas such a sojourn can require up to half a minute in some of the larger eukaryotic cells. Thus, diffusion limits to intracellular transactions can ultimately constrain the rates of biological processes in eukaryotic cells.

- Through its influence on the motion of all molecules, temperature plays a governing role in all reaction rates. A number of mathematical expressions have been proposed as summary descriptors for the response of biological processes to temperature, although the mechanistic interpretation of the fitted parameters is open to debate.

- Life ultimately depends on the acquisition of energy. For aerobic heterotrophs (most organisms other than photosynthesizers), food comes in the form of reduced carbon compounds, which provide both carbon skeletons for constructing biomass and energy for carrying out cellular functions. The heats of combustion of substrates provide reliable measures of the energy that organisms can extract from such compounds. More reduced carbon compounds provide more energy, but there is an intermediate level of substrate reduction (approximately equivalent to that in glucose) above which carbon starts to be limiting.

- The upper limit of evolved assimilation efficiency of carbon compounds (fraction of ingested carbon atoms incorporated into biomass) is ~ 0.8, and the construction of 10 grams dry weight of cellular biomass requires the energy released from the hydrolysis of ~ 1 mol of ATP.

Foundations 7.1 Intracellular diffusion

In a homogeneous medium, small particles are subject to random walks as a consequence of background thermal perturbations. This leads to diffusive particle movement from a starting location in a symmetric fashion. To minimize the mathematical details, the focus here is on a one-dimensional diffusion process, with a summary of the general results for two and three dimensions following the initial details.

Consider a particle moving randomly to the right and left with equal probabilities of 0.5 and fixed jump lengths, independent of prior motion at each time unit. Let t be the total number of jumps, with t_+ being the number to the right and t_- the number to the left, so that

$$t = t_+ + t_-.$$

The net displacement relative to a starting point at position 0 is then

$$x = t_+ - t_-.$$

Given t jostling episodes, the probability of t_+ draws in the positive direction is given by the binomial distribution

$$P(x) = \frac{t!}{t_+! t_-!} \left(\frac{1}{2}\right)^{t_+} \left(\frac{1}{2}\right)^{t - t_+} = \frac{t!}{t_+! t_-!} \left(\frac{1}{2}\right)^{t},$$

$$(7.1.1a)$$

where $y! = y \cdot (y - 1) \cdot (y - 2) \cdots 1$ is the factorial function.

For large t, this discrete-state formula can be simplified to a continuous distribution by first noting that $t_+ = (t + x)/2$ and $t_- = (t - x)/2$, substituting into Equation 7.1.1a, and then logarithmically transforming to obtain

$$\ln[P(x)] = \ln(t!) - \ln\left\{\left[\frac{t}{2}\left(1 + \frac{x}{t}\right)\right]!\right\}$$

$$- \ln\left\{\left[\frac{t}{2}\left(1 - \frac{x}{t}\right)\right]!\right\} - t \ln 2. \quad (7.1.1b)$$

Factorial functions can be unwieldy, but large t allows the use of Stirling's approximation for the logarithm of large factorials,

$$\ln(y!) \simeq \frac{\ln(2\pi y)}{2} + y\ln(y) - y, \qquad (7.1.2)$$

application of which simplifies Equation 7.1.1b to

$$\ln[P(x)] = \ln[(2/\pi t)^{0.5}] - \left(\frac{t+x+1}{2}\right)\ln\left(1 + \frac{x}{t}\right)$$
$$- \left(\frac{t-x+1}{2}\right)\ln\left(1 - \frac{x}{t}\right).$$

Further simplification is accomplished by noting that for $y < 0.5$,

$$\ln(1+y) \simeq y - (y^2/2), \qquad (7.1.3a)$$

$$\ln(1-y) \simeq -y - (y^2/2). \qquad (7.1.3b)$$

Applying these approximations to the preceding expression, followed by exponentiation to return to the original scale eventually leads to

$$P(x) \simeq \left(\frac{2}{\pi t}\right)^{1/2} \exp\left(-\frac{x^2}{2t}\right). \qquad (7.1.4a)$$

We now modify Equation 7.1.4a to a more familiar and general form. First, we note that the variance in the number of jumps to the right follows from the properties of the binomial distribution. When the probability of each type of event is 0.5, the binomial variance associated with each event is $0.5 \cdot 0.5$, and summing over t independent events leads to variance $\sigma^2 = t/4$. Second, the disparity between

numbers of right and left jumps, x, can be rewritten as $(t_+ - t_-) = (2t_+ - t)$, and because the expected number of jumps to the right (the mean) can be written as $\mu = t/2$, this further reduces to $x = 2(t_+ - \mu)$. Substituting the latter expression and $t = 4\sigma^2$ into Equation (7.1.4a), we obtain

$$P(t_+) = \left(\frac{1}{2\pi\sigma^2}\right)^{1/2} \exp\left(-\frac{(t_+ - \mu)^2}{2\sigma^2}\right). \qquad (7.1.4b)$$

This is the widely used normal (or Gaussian) distribution of a variable (in this case t_+) with mean μ and variance σ^2.

In the current case, diffusion results in movement from the initial point, but with no net bias, so we can rescale to a mean of zero. The variance σ^2 can also be written as the mean-squared deviation $2Dt$, where D is the diffusion coefficient, with units of length^2/time (see main text). The one-dimensional diffusion distance d then has probability distribution

$$P(d) = \left(\frac{1}{4\pi Dt}\right)^{1/2} \exp\left(-\frac{d^2}{4Dt}\right). \qquad (7.1.4c)$$

Note that in the one-dimensional case, the diffusion variance is proportional to $2D$, and increases linearly with time. In two dimensions, $2Dt$ becomes $4Dt$, and with three dimensions, it becomes $6Dt$. The standard deviation is the root mean-squared distance that a particle is expected to have travelled (with equal probability in all directions) after t time units. Thus, regardless of the dimensionality, the expected distance travelled increases with the square root of time. Berg (1993) provides a useful compendium of results and biological applications of diffusion theory.

Foundations 7.2 Rates of encounter by molecular diffusion

A purely physical limit to the encounter rate between two molecules can be derived from diffusion theory developed by Smoluchowski (1915), who independently of Einstein outlined a number of the general principles of Brownian motion. We start by considering the diffusion of two spherical molecules, with respective radii r_a and r_b, moving randomly through an otherwise homogeneous environment. A collision between these two molecules will occur whenever their centers come within a distance $r_c = r_a + r_b$ from each other. To simplify the overall analysis, one may then consider an imaginary sphere around the center of either particle, with radius r_c, whose overall surface area $4\pi r_c^2$ represents the entire boundary across which a flux of one particle or the other constitutes a collision (Figure 7.9).

To proceed further, we require the total rate of particle movement, which is determined by the sum of the diffusion coefficients associated with each particle type. From Equations 7.4 and 7.5,

$$D = D_a + D_b = \frac{k_B T(r_a + r_b)}{6\pi\eta(r_a r_b)}, \qquad (7.2.1)$$

with units of cm^2/sec, where k_B is Boltzmann's constant, T is the temperature (in Kelvins), and η is the viscosity of the medium (see main text for the values of these parameters).

To complete the derivation of the encounter rate, we require an expression for the rate of diffusion across a surface. This is given by Fick's first law, which states that

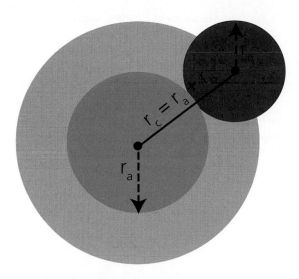

Figure 7.9 Two spheres with respective radii r_a and r_b will contact each other whenever the distance between their centers is $r_c < r_a + r_b$.

the flux rate of a diffusing substrate across a point is equal to the product of the concentration gradient at that point and the diffusion coefficient. Here, the concentration gradient can be approximated by treating the concentration inside the sphere of radius r_c as zero and denoting the bulk concentration (outside the sphere) as [C], implying a concentration gradient of $([C] - 0)/r_c$ and flux rate $[C]D/r_c$. After multiplying by the total surface area ($4\pi r_c^2$) and dividing by the concentration, this scales up to a flux rate per unit concentration of $(4\pi r_c^2)(D/r_c) = 4\pi r_c D$. Substituting Equation 7.2.1 for D, we then obtain an expression for the encounter-rate coefficient,

$$k_e = 4\pi r_c D = \left(\frac{2k_B T}{3\eta} \right) \left(\frac{(r_a + r_b)^2}{r_a r_b} \right). \quad (7.2.2)$$

After substituting for the average temperature of life (see main text) and the viscosity of water at 20°C, this reduces further to

$$k_e = (2.8 \times 10^{-12}) \left(\frac{(r_a + r_b)^2}{r_a r_b} \right), \quad (7.2.3)$$

with units $cm^3 \cdot sec^{-1}$. (Note that if [C] is expressed in molar-concentration units, then k_e needs to be divided by 1000 to convert to liters and multiplied by Avogadro's number to convert to molecules, making the prefix 1.7×10^9). The product of this encounter-rate coefficient and the concentrations of both particle types (each in units of molecules/cm^3) yields the expected number of collisions between the two particle types in a 1 cm^3 volume per second,

$$R_e = k_e [C_a][C_b]. \quad (7.2.4)$$

For two spherical particles identical in size ($r_a = r_b$), as in the case of two monomeric subunits forming a homodimeric protein, Equation 7.2.3 reduces to

$$k_e \simeq 11 \times 10^{-12}. \quad (7.2.5a)$$

In this case, the rate coefficient is independent of the particle size because any increase in target size is perfectly balanced by a reduction in the rate of diffusion. On the other hand, if one particle type is much larger than the other, $r_b \ll r_a$,

$$k_e \simeq (2.8 \times 10^{-12}) \left(\frac{r_a}{r_b} \right), \quad (7.2.5b)$$

showing that the encounter rate depends only on the ratio of particle sizes, not on their absolute sizes.

The encounter rates denoted by these expressions are sometimes called the Smoluchowski limits. They denote the encounter rate in the ideal situation in which there are no attractive or repulsive forces between colliding particles, and otherwise no barriers for diffusion through a homogeneous medium, any of which can become important in various biological contexts.

Foundations 7.3 The Boltzmann probability distribution for alternative molecular states

Numerous situations are encountered in cell biology where it is necessary to know the distribution of alternative states of the individual members of a population of molecules, as these often determine the average rates and stochasticities of cellular processes. Theoretical results in this area are generally derived from the field of statistical mechanics, which takes a microscopic view of particle

states within a closed system assumed to be in thermodynamic equilibrium. There are numerous ways to achieve the final result (e.g., Schroeder 2000; Phillips et al. 2012). The route taken here uses the properties of combinatorics, along with a few mathematical approximations.

The starting assumption is a system containing n molecules, which together harbor a fixed amount of energy, S. We assume discrete energy states, taking on values of $0, \epsilon, 2\epsilon, \ldots, k\epsilon$, so $N = S/\epsilon$ represents the total number of discrete energy packets available to the system. Individual particles are free to change energy states, but the overall probability distribution of alternative states remains constant under the assumption of equilibrium. It is this steady-state probability distribution that we wish to determine, i.e., the probability that a random particle is in energy states $i = 0, 1, \ldots, k$. To accomplish this, we must account for the full distribution of the alternative states that a set of n molecules can take on, conditional on their sum equalling N. Given the large number of particles typically involved, this can be a dauntingly complex problem, but a few mathematical tricks simplify the overall derivation.

We first note that the total number of ways that N packets of energy can be partitioned among n molecules is given by

$$T(N, n) = \frac{(N + n - 1)!}{N!(n - 1)!}, \qquad (7.3.1a)$$

where ! again denotes the factorial function. To obtain this general result, note that there are n bins within which N energy packets must be partitioned. The numerator is the total number of ordered ways that N distinct packets can be randomly assigned to n bins. But because the energy packets are all identical in content, the ordering in which they are assigned is irrelevant, and the two terms in the denominator discount the numerator to account for the redundancy associated with ordering of packets and bins.

Now consider the situation where one specific molecule has energy $i\epsilon$, so there are a remaining $(N - i)$ packets to partition among $(n-1)$ molecules. Modification of the previous expression then leads to

$$T(N - i, n - 1) = \frac{(N + n - 2 - i)!}{(N - i)!(n - 2)!}. \qquad (7.3.1b)$$

Thus, the probability of a particle having energy content $i\epsilon$ is

$$p(i) = \frac{T(N - i, n - 1)}{T(N, n)} = (n - 1) \cdot \frac{N!(N + n - 2 - i)!}{(N - i)!(N + n - 1)!}. \qquad (7.3.2)$$

Further simplification is possible if it is assumed that the energy in the system is substantial enough that $N \gg n$, which makes reasonable the approximations $N!/(N-x)! \simeq N^x$, and $n/(N + n) \simeq n/N$. Noting as well that the number of molecules is large, so that $n - 1 \simeq n$,

$$p(i) \simeq nN^i \cdot (N + n)^{-(i+1)} \simeq \frac{n}{N} \cdot [1 + (n/N)]^{-i}, \quad (7.3.3a)$$

which further reduces to

$$p(i) \simeq \frac{n}{N} \cdot e^{-in/N}, \qquad (7.3.3b)$$

using $e^x \simeq (1 + x)$ for $x \ll 1$. Thus, what started as a complex problem reduces to a relatively simple expression (a negative exponential distribution) under the assumption of large numbers.

Letting $\overline{E} = N\epsilon/n$ denote the average energy per particle, the preceding expression implies that the probability of a particle having energy state $E_i = \epsilon i$ is

$$p(E_i) = C \cdot e^{-E_i/\overline{E}}, \qquad (7.3.4)$$

where C is a normalization constant that ensures that the total probability distribution sums to 1.0, satisfied in this case by $C = 1/\overline{E}$. Letting $\overline{E} = k_B T$ be the average energy per particle yields the Boltzmann distribution,

$$p(E_i) = C \cdot e^{-E_i/(k_B T)}. \qquad (7.3.5)$$

Note that the cumulative function for this exponential distribution, which defines the probability of being in a state below E_i is $1 - e^{-E_i/\overline{E}}$.

Foundations 7.4 The yield of cellular biomass per ATP usage

Observations in the final section of the main text allow for a crude estimate of the amount of energy required to build new cellular material (in terms of ATP $\rightarrow$ ADP hydrolyses), an issue that will be addressed in more detail in Chapter 17. Here we assume a relatively high-energy carbon substrate with a heat of combustion of 9.3 kcal/g C (carbon), the approximate value for most six-carbon sugars (including glucose). From Figure 7.8, such a substrate leads to an expected 1.3 g DW (dry weight) produced/g C consumed, or inversely $(1/1.3) = 0.77$ g C consumed/g DW produced. Multiplying by 9.3 kcal/g C leads to an

estimated cellular energy-intake requirement of 7.2 kcal/g DW produced.

How much of this required consumption is diverted to energy production for cell functions? Surveys of multiple bacterial and eukaryotic species suggest average caloric contents of 5.41 (0.05) and 5.13 (0.04) kcal/g cell DW (Supplemental Table 8.1), respectively. Assuming an average value of 5.3, this implies that of the 7.2 kcal consumed/g DW produced, $7.2 - 5.3 = 1.9$ kcal (26%) must be used in cellular processes required to produce new cellular material (with the rest of the substrate providing carbon skeletons used in the construction of the monomeric building blocks of the cell). Thus, $\sim 0.77 \times 0.26 = 0.20$ g C of substrate must be converted to energy in order to produce 1 g of cell DW.

What does this energetic investment mean in units of ATP, the cellular currency of bioenergetics? One mole of glucose contains 72 g C, and assuming complete aerobic metabolism, observations from biochemistry tell us that each mol of metabolized glucose generates ~ 32 mol of ATP (Chapter 17). This suggests that, in units of ATP, the energetic requirement for the production of 1 g DW of cells is ~ 0.20 g C consumption $\times$ (1 mole glucose/72 g C) $\times$ (32 mol ATP/mol glucose) = 0.089 mole ATP. It then follows that the yield of cells is $\sim 1/0.089 = 11.2$ g DW/mol ATP. This rough estimate is quite close to the average value of 10.5 for more direct estimates found in a wide variety of organisms raised on a diversity of carbon substrates (Payne 1970).

Literature Cited

Alroy, Y., and S. R. Tannenbaum. 1973. The influence of environmental conditions on the macromolecular composition of *Candida utilis*. Biotechnol. Bioeng. 15: 239–256.

Arrhenius, S. 1889. Über die Reaktionsgeschwindigkeit bei der Inversion von Rohrzucker durch Saeuren. Z. Physik. Chem. 4: 226–248.

Arroyo, J. I., B. Díez, C. P. Kempes, G. B. West, and P. A. Marquet. 2022. A general theory for temperature dependence in biology. Proc. Natl. Acad. Sci. USA 119: e2119872119.

Atkinson, D., B. J. Ciotti, and D. J. Montagnes. 2003. Protists decrease in size linearly with temperature: Ca. 2.5% $^\circ C^{-1}$. Proc. Biol. Sci. 270: 2605–2611.

Bakshi, S., A. Siryaporn, M. Goulian, and J. C. Weisshaar. 2012. Superresolution imaging of ribosomes and RNA polymerase in live *Escherichia coli* cells. Mol. Microbiol. 85: 21–38.

Ball, P. 2008. Water as an active constituent in cell biology. Chem. Rev. 108: 74–108.

Ball, P. 2017. Water is an active matrix of life for cell and molecular biology. Proc. Natl. Acad. Sci. USA 114: 13327–13335.

Bellotto, N., J. Agudo-Canalejo, R. Colin, R. Golestanian, G. Malengo, and V. Sourjik. 2022. Dependence of diffusion in *Escherichia coli* cytoplasm on protein size, environmental conditions, and cell growth. eLife 11: e82654.

Berg, H. C. 1993. Random Walks in Biology. Princeton University Press, Princeton, NJ.

Corkrey, R., T. A. McMeekin, J. P. Bowman, D. A. Ratkowsky, J. Olley, and T. Ross. 2014. Protein thermodynamics can be predicted directly from biological growth rates. PLoS One 9: e96100.

de Queiroz, J. H., J.-L. Uribelarrea, and A. Pareilleux. 1993. Estimation of the energetic biomass yield and efficiency of oxidative phosphorylation in cell-recycle cultures of *Schizosaccharomyces pombe*. Appl. Microbiol. Biotech. 39: 609–614.

Dias-Lopes, G., J. R. Wiśniewski, N. P. de Souza, V. E. Vidal, G. Padrón, C. Britto, P. Cuervo, and J. B. De Jesus. 2018. In-depth quantitative proteomic analysis of trophozoites and pseudocysts of *Trichomonas vaginalis*. J. Proteome Res. 17: 3704–3718.

Dill, K. A., and S. Bromberg. 2011. Molecular Driving Forces, 2nd edn. Garland Science, New York, NY.

Dill, K. A., K. Ghosh, and J. D. Schmit. 2011. Physical limits of cells and proteomes. Proc. Natl. Acad. Sci. USA 108: 17876–17882.

Dix, J. A., and A. S. Verkman. 2008. Crowding effects on diffusion in solutions and cells. Annu. Rev. Biophys. 37: 247–263.

Ellis, R. J. 2001. Macromolecular crowding: Obvious but underappreciated. Trends Biochem. Sci. 26: 597–604.

Elowitz, M. B., M. G. Surette, P. E. Wolf, J. B. Stock, and S. Leibler. 1999. Protein mobility in the cytoplasm of *Escherichia coli*. J. Bacteriol. 181: 197–203.

Fagerbakke, K. M., M. Heldal, and S. Norland. 1996. Content of carbon, nitrogen, oxygen, sulfur and phosphorus in native aquatic and cultured bacteria. Aquat. Microb. Ecol. 10: 15–27.

Fagerbakke, K. M., S. Norland, and M. Heldal. 1999. The inorganic ion content of native aquatic bacteria. Can. J. Microbiol. 45: 304–311.

Finlay, B. J., and G. Uhlig. 1981. Calorific and carbon values of marine and freshwater Protozoa. Helgol. Mar. Res. 34: 401–412.

Gifford, S. M., J. W. Becker, O. A. Sosa, D. J. Repeta, and E. F. DeLong. 2016. Quantitative transcriptomics reveals the growth- and nutrient-dependent response of a streamlined marine methylotroph to methanol and naturally occurring dissolved organic matter. mBio 7: e01279–16.

Gillooly, J. F., J. H. Brown, G. B. West, V. M. Savage, and E. L. Charnov. 2001. Effects of size and temperature on metabolic rate. Science 293: 2248–2251.

Guillon, L., S. Altenburger, P. L. Graumann, and I. J. Schalk. 2013. Deciphering protein dynamics of the siderophore pyoverdine pathway in *Pseudomonas aeruginosa*. PLoS One 8: e79111.

Guo, M., A. J. Ehrlicher, M. H. Jensen, M. Renz, J. R. Moore, R. D. Goldman, J. Lippincott-Schwartz, F. C. Mackintosh, and D. A. Weitz. 2014. Probing the stochastic, motor-driven properties of the cytoplasm using force spectrum microscopy. Cell 158: 822–832.

Hansen, P. J., P. K. Bjørnsen, and B. W. Hansen. 1997. Zooplankton grazing and growth: Scaling within the 2-2,000-μm body size range. Limnol. Oceanogr. 42: 687–704.

He, L., and B. Niemeyer. 2003. A novel correlation for protein diffusion coefficients based on molecular weight and radius of gyration. Biotechnol. Prog. 19: 544–548.

Heldal, M., D. J. Scanlan, S. Norland, F. Thingstad, and N. H. Mann. 2003. Elemental composition of single cells of various strains of marine *Prochlorococcus* and *Synechococcus* using Xray microanalysis. Limnol. Oceanogr. 48: 1732–1743.

Herendeen, S. L., R. A. VanBogelen, and F. C. Neidhardt. 1979. Levels of major proteins of *Escherichia coli* during growth at different temperatures. J. Bacteriol. 139: 185–194.

Ho, K. P., and W. J. Payne. 1979. Assimilation efficiency and energy contents of prototrophic bacteria. Bioctech. Bioeng. 21: 787–802.

Ho, T. Y., A. Quigg, Z. V. Finkel, A. J. Milligan, K. Wyman, P. G. Falkowski, and F. M. M. Morel. 2003. The elemental composition of some marine phytoplankton. J. Phycol. 39: 1145–1159.

Hong, L., and J. Lei. 2009. Scaling law for the radius of gyration of proteins and its dependence on hydrophobicity. J. Polym. Sci. B. Polym. Phys. 47: 207–214.

Huggett, M. J., D. H. Hayakawa, and M. S. Rappé. 2012. Genome sequence of strain HIMB624, a cultured representative from the OM43 clade of marine Betaproteobacteria. Stand. Genom. Sci. 6: 11–20.

Hunter, K., and A. H. Rose. 1972. Influence of growth temperature on the composition and physiology of microorganisms. J. Appl. Chem. Biotechnol. 22: 527–540.

Joyner, R. P., J. H. Tang, J. Helenius, E. Dultz, C. Brune, L. J. Holt, S. Huet, D. J. Müller, and K. Weis. 2016. A glucose-starvation response regulates the diffusion of macromolecules. eLife 5: e09376.

Kharasch, M. S., and B. Sher. 1925. The electronic conception of valence and heats of combustion of organic compounds. J. Phys. Chem. 29: 625–658.

Klumpp, S., M. Scott, S. Pedersen, and T. Hwa. 2013. Molecular crowding limits translation and cell growth. Proc. Natl. Acad. Sci. USA 110: 16754–16759.

Konopka, M. C., I. A. Shkel, S. Cayley, M. T. Record, and J. C. Weisshaar. 2006. Crowding and confinement effects on protein diffusion *in vivo*. J. Bacteriol. 188: 6115–6123.

Konopka, M. C., K. A. Sochacki, B. P. Bratton, I. A. Shkel, M. T. Record, and J. C. Weisshaar. 2009. Cytoplasmic protein mobility in osmotically stressed *Escherichia coli*. J. Bacteriol. 191: 231–237.

Kumar, M., M. S. Mommer, and V. Sourjik. 2010. Mobility of cytoplasmic, membrane, and DNA-binding proteins in *Escherichia coli*. Biophys. J. 98: 552–559.

Lange, H. C., and J. J. Heijnen. 2001. Statistical reconciliation of the elemental and molecular biomass composition of *Saccharomyces cerevisiae*. Biotechnol. Bioeng. 75: 334–344.

Linton, J. D., and R. J. Stephenson. 1978. A preliminary study on growth yields in relation to the carbon and energy content of various organic growth substrates. FEMS Microbiol. Letters 3: 95–98.

Luby-Phelps, K. 2000. Cytoarchitecture and physical properties of cytoplasm: Volume, viscosity, diffusion, intracellular surface area. Int. Rev. Cytol. 192: 189–221.

Lynch, M. 2007. The Origins of Genome Architecture. Sinauer Assocs., Inc. Sunderland, MA.

Lynch, M., and G. K. Marinov. 2015. The bioenergetic costs of a gene. Proc. Natl. Acad. Sci. USA 112: 15690–15695.

Lynch, M., and B. Trickovic. 2020. A theoretical framework for evolutionary cell biology. J. Mol. Biol. 432: 1861–1879.

Mairet, F., J. L. Gouzé, and H. de Jong. 2021. Optimal proteome allocation and the temperature dependence of microbial growth laws. NPJ Syst. Biol. Appl. 7: 14.

Menden-Deuer, S., and E. J. Lessard. 2000. Carbon to volume relationships for dinoflagellates, diatoms, and other protist plankton. Limnol. Oceanogr. 45: 569–579.

Mika, J. T., P. E. Schavemaker, V. Krasnikov, and B. Poolman. 2014. Impact of osmotic stress on protein diffusion in *Lactococcus lactis*. Mol. Microbiol. 94: 857–870.

Milo, R., and R. Phillips. 2016. Cell Biology by the Numbers. Garland Science, New York, NY.

Mohr, P. W., and S. Krawiec. 1980. Temperature characteristics and Arrhenius plots for nominal psychrophiles, mesophiles and thermophiles. J. Gen. Microbiol. 121: 311–317.

Montagnes, D. J. S., and D. J. Franklin. 2001. Effect of temperature on diatom volume, growth rate, and carbon and nitrogen content: Reconsidering some paradigms. Limnol. Oceanogr. 46: 2008–2018.

Montagnes, D. J. S., S. A. Kimmance, and D. Atkinson. 2003. Using Q_{10}: Can growth rates increase linearly with temperature? Aquatic Microb. Ecol. 32: 307–313.

Montero Llopis, P., O. Sliusarenko, J. Heinritz, and C. Jacobs-Wagner. 2012. *In vivo* biochemistry in bacterial cells using FRAP: Insight into the translation cycle. Biophys. J. 103: 1848–1859.

Mu, X., S. Choi, L. Lang, D. Mowray, N. V. Dokholyan, J. Danielsson, and M. Oliveberg. 2017. Physicochemical code for quinary protein interactions in *Escherichia coli*. Proc. Natl. Acad. Sci. USA 114: E4556–E4563.

Nenninger, A., G. Mastroianni, and C. W. Mullineaux. 2010. Size dependence of protein diffusion in the cytoplasm of *Escherichia coli*. J. Bacteriol. 192: 4535–4540.

Noll, P., L. Lilge, R. Hausmann, and M. Henkel. 2020. Modeling and exploiting microbial temperature response. Processes 8: 121.

Nozaki, Y. 1997. A fresh look at element distribution in the North Pacific. Eos 78: 221–223.

Payne, W. J. 1970. Energy yields and growth of heterotrophs. Annu. Rev. Microbiol. 24: 17–52.

Phillips, R., J. Kondev, J. Theriot, and H. Garcia. 2012. Physical Biology of the Cell, 2nd edn. Garland Science, New York, NY.

Pinho, N, J. R. Wiśniewski, G. Dias-Lopes, L. Saboia-Vahia, A. C. S. Bombaa, C. Mesquita-Rodrigues, R. Menna-Barreto, E. Cupolillo, J. B. de Jesus, G. Padrón, et al. 2020. In-depth quantitative proteomics uncovers species-specific metabolic programs in *Leishmania* (Viannia) species. PLoS Negl. Trop. Dis. 14: e0008509.

Potma, E. O., W. P. de Boeij, L. Bosgraaf, J. Roelofs, P. J. van Haastert, and D. A. Wiersma. 2001. Reduced protein diffusion rate by cytoskeleton in vegetative and polarized *Dictyostelium* cells. Biophys. J. 81: 2010–2019.

Ratkowsky, D. A., R. K. Lowry, T. A. McMeekin, A. N. Stokes, and R. E. Chandler. 1983. Model for bacterial culture growth rate throughout the entire biokinetic temperature range. J. Bacteriol. 154: 1222–1226.

Ratkowsky, D. A., J. Olley, and T. Ross. 2005. Unifying temperature effects on the growth rate of bacteria and the stability of globular proteins. J. Theor. Biol. 233: 351–362.

Raven, J. A., and R. J. Geider. 1988. Temperature and algal growth. New Phytol. 110: 441–461.

Redfield, A. C. 1934. The proportions of organic derivatives in sea water and their relation to the composition of plankton, pp. 176–192. James Johnstone Memorial Volume. University Press of Liverpool, Liverpool, UK.

Roels, J. A. 1980. Application of macroscopic principles to microbial metabolism. Biotech. Bioeng. 22: 2457–2514.

Savage, V. M., and G. B. West. 2006. Biological scaling and physiological time: biomedical applications, pp. 141–163. In T. S. Deisboeck and J. Y. Kresh (eds.) Complex Systems Science in Biomedicine. Springer Inc., New York, NY.

Schavemaker, P. E., A. J. Boersma, and B. Poolman. 2018. How important is protein diffusion in prokaryotes? Front. Mol. Biosci. 5: 93.

Schroeder, D. V. 2000. An Introduction to Thermal Physics. Addison Wesley, San Francisco, CA.

Smoluchowski, M. 1915. Über Brownsche Molekularbewegung unter Einwirkung äußerer Kräfte und den Zusammenhang mit der verallgemeinerten Diffusionsgleichung. Ann. Phys. 353: 1103–1112.

Soh, S., M. Banaszak, K. Kandere-Grzybowska, and B. A. Grzybowski. 2013. Why cells are microscopic: A transport-time perspective. J. Phys. Chem. Lett. 4: 861–865.

Swaminathan, R., C. P. Hoang, and A. S. Verkman. 1997. Photobleaching recovery and anisotropy decay of green fluorescent protein GFP-S65T in solution and cells: Cytoplasmic viscosity probed by green fluorescent protein translational and rotational diffusion. Biophys. J. 72: 1900–1907.

Tyn, M. T., and T. W. Gusek. 1990. Prediction of diffusion coefficients of proteins. Biotechnol. Bioeng. 35: 327–338.

von Stockar, U., and I. W. Marison. 2005. The use of calorimetry in biotechnology bioprocesses and engineering. Adv. Biochem. Eng. Biotech. Book Series 40: 93–136.

Wang, M., C. G. Kurland, and G. Caetano-Anollés. 2011. Reductive evolution of proteomes and protein structures. Proc. Natl. Acad. Sci. USA 108: 11954–11958.

Williams, K., F. Percival, J. Merino, and H. A. Mooney. 1987. Estimation of tissue construction cost from heat of combustion and organic nitrogen content. Plant Cell Environ. 10: 725–734.

Zimmerman, S. B., and S. O. Trach. 1991. Estimation of macromolecule concentrations and excluded volume effects for the cytoplasm of *Escherichia coli*. J. Mol. Biol. 222: 599–620.

CHAPTER 8

Evolutionary Scaling Relationships in Cell Biology

Cells vary widely in terms of shape, physiological properties, metabolic features, and internal architecture. Of particular importance is cell size, which influences myriad cellular features ranging from nutrient uptake to internal transport. Cell-size variation among species is likely driven by a variety of selective forces, including size-selective predators, buoyancy, resistance to flow dynamics, and osmotic pressure. Among the most well-studied unicellular species, cell volumes vary by approximately eleven orders of magnitude, 10^{-3} to 10^8 μm^3, across the Tree of Life, with up to 10^7-fold differences within major phylogenetic groups (Figure 8.1). By comparison, the range in size between the smallest and largest mammals, a bumblebee bat versus a blue whale, is eight orders of magnitude. On average, prokaryotic cells are smaller than those of eukaryotic species, but some eukaryotes have cell volumes smaller than the average bacterium.

There are a few striking exceptions at the large end of the scale not shown in Figure 8.1. For example, the unicellular green alga *Acetabularia* is up to 10 cm in length and contains just a single nucleus, but has a complex architecture similar to that of a vascular plant. A multinucleate green alga, *Caulerpa*, produces complex holdfasts, stalks, and fronds up to metres in length, despite being unicellular. *Gromia sphaerica*, a testate amoeba that lives on marine sediments at depths of > 1 km, produces cells up to 4 cm in diameter. There are also giant bacteria. The marine-sediment bacterium *Thiomargarita magnifica* approaches 10^{12} μm^3 in size, and *Epulopiscium*, a gut symbiont associated with surgeonfish, has a volume well over 10^6 μm^3.

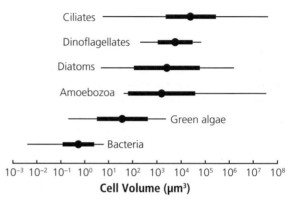

Figure 8.1 Distributions of species-specific cell volumes for major phylogenetic groups for which multiple measures are available. Solid points denote group means; bold and narrow horizontal bars denote standard deviations and ranges. Data from Lynch et al. (2022b).

Cell size is a major organizing factor in biology, with a wide array of cellular features scaling in predictable size-dependent manners. Not all such relationships are linear, but they often unfold in ways that transcend the boundaries between major phylogenetic groups, even between prokaryotes and eukaryotes. Commonly called 'laws of nature' or 'rules of life', such patterns identify strict limits on what evolution has been able to achieve in the natural world. What accounts for the empty regions in phenotypic space? Are missing combinations of trait values a consequence of biophysical and/or biochemical constraints? Are certain combinations simply too disharmonious to be promoted by selection? Or do both factors play a role?

Evolutionary Cell Biology. Michael Lynch, Oxford University Press. © Michael Lynch (2024). DOI: 10.1093/oso/9780192847287.003.0008

The most notable of cell biology's scaling laws define the ways in which bioenergetic features relate to cell volume. These constitute the primary subject matter of the following pages, although numerous relationships for other types of traits are explored in subsequent chapters. The focus here is on the *evolutionary* scaling of traits with size across species. There are equally compelling questions regarding scaling relationships on non-evolutionary timescales (Marshall 2020; Cadart and Heald 2022), e.g., cellular responses to nutritional status, temperature, and other physical/chemical factors. Ultimately, we wish to know whether long-term evolutionary trajectories reflect within-species developmental responses to the environment (Chapter 9).

Before proceeding, a simple overview of the ways in which scaling laws are expressed and interpreted mathematically is in order. Using this framework, we then summarize a number of general scaling relationships regarding energy acquisition and growth. This is followed by an overview of the possible evolutionary mechanisms driving such patterns and their implications for understanding the consequences of the prokaryote–eukaryote transition.

Describing Allometric Relationships

The description of a scaling relationship between two traits demands a statistical approach, as the twin goals are generally to quantify the average pattern and degree of noise in the response of one trait to a change in the other. The relationship may be positive or negative, but provided a proportional change in one trait is associated with a change in the other, the model can be succinctly written in the form of a simple, two-parameter power function,

$$z = \alpha S^{\beta} + e, \qquad (8.1a)$$

where in this case z is the measured phenotype of interest, S is a measure of organism size (usually mass or volume), α is the fitted scaling constant (giving the expected value of z when $S = 1$), and β is the fitted scaling coefficient. Equation 8.1a indicates that, on average, a twofold change in S is associated with a 2^{β}-fold change in z. The e term in

Equation 8.1a is usually left out of such expressions (and will be from here on), but is nonetheless relevant, as it is the deviation between observed and predicted values. With appropriate statistical analysis (below), the average value of e is zero, and the magnitude of the variance of e is a measure of the goodness-of-fit of the data to the model.

There is an elegant simplicity to power functions, as they exhibit linear form when z and S are jointly transformed logarithmically. On a log scale, Equation 8.1a becomes

$$\log(z) = \log(\alpha) + \beta \log(S), \qquad (8.1b)$$

providing a simple basis for estimating the parameters α and β with linear-regression analysis. The model is linear regardless of the logarithmic scale employed, e.g., base 10 as generally used here (denoted as log), or on the scale of natural logarithms (base $e \simeq 2.318$, denoted as ln). First popularized by Thompson (1917) and Huxley (1932), power-function scalings in biology are generally referred to as allometric functions, with $\beta = 1$ denoting an isometric relationship. If β is positive but < 1, then z becomes proportionally smaller with increasing S (sublinear or hypometric scaling), as $z/S = \alpha S^{\beta}/S = \alpha S^{\beta-1}$, with the exponent $\beta - 1$ being negative. In contrast, $\beta > 1$ implies supralinear or hypermetric scaling.

As we discuss later on, cell biology is well-endowed with features that are reasonably described by Equation 8.1b as a first-order approximation. In principle, although rarely relied upon, more complex functions are possible. For example, β could be a function of S. Note also that the scales on which biological traits are measured are generally arbitrary. For example, pH is measured on a log scale. Even when a particular measure does not strictly adhere to the form of Equation 8.1a, a variety of mathematical transformations to a new scale can often lead to behavior consistent with the simplest power-law form (Lynch and Walsh 1998; Frank 2016).

Regressions of trait values on organism size with slopes approximating multiples of $1/3$ are particularly intriguing, as they raise the possibility of simple geometric explanations. For example, when S is on the scale of mass or volume, $\beta = 1$ suggests a mechanism directly proportional to the mass of cellular material, $\beta = 2/3$ suggests a mechanism related to surface area (because area is a function of the square and mass is a function of the cube of a length measurement), and $\beta = 1/3$ suggests a mechanism related to a linear dimension of the organism. As early investigators found numerous regression coefficients to be in the neighborhood of $x/3$ (where x is an integer value, usually 2 or 3), there was a tendency to assume they were exactly $x/3$ and then embark on proposals of mechanistic explanations for the observed patterns.

Even at an early stage in these kinds of studies, discomfort was expressed with the generality of various geometry-based hypotheses (e.g., von Bertalanffy 1957), although the tradition of searching for them continues today, with a tendency to view significant deviations as little more than annoying secondary effects. As we discuss below, much attention has been given to ideas associated with fractal delivery systems that suggest allometric coefficients of the form $x/4$ rather than $x/3$. In light of the usual uncertainties in statistical analyses, however, it is often hard to justify one of these scalings versus the other when confronted with real data. Is, for example, an estimated coefficient of $\beta = 0.29$ more consistent with $1/3 \simeq 0.33$ or $1/4 = 0.25$ scaling, or should the strict adherence to either belief be avoided?

Scaling Laws in Cellular Bioenergetics

The vast majority of work on biological scaling relationships comes from ecologists striving to understand the basic energetic features of ecosystems, often with a focus on multicellular taxa (e.g., Burger et al. 2019; Hatton et al. 2019). Here, our attention is confined to the attributes of species that normally live as single cells, although (as discussed in Chapter 24) there are intriguing extensions of the results for unicellular organisms to multicellular lineages. Several power-law scalings of biological features with cell size were encountered in Chapter 7 – the slightly less than linear proportionality between cell dry weight and cell volume; the decline in the fractional contribution of DNA to total cellular biomass with increasing cell size; and

the sublinear scaling of the number of mRNA and protein molecules per cell with cell volume.

Strong correlational patterns imply strong constraints, and a key challenge for evolutionary cell biology is to determine their mechanistic basis. At least three classes of explanations always merit consideration: 1) inevitable outcomes of biophysical/biochemical limitations; 2) consequences of evolutionary channelling towards particular combinations of trait values that maximize fitness; and/or 3) reflections of drift barriers beyond which the efficiency of selection is compromised (Chapter 4).

Metabolic rate

In any discussion of size scaling of biological traits, it is appropriate to start with metabolic-rate data, as no trait has been more widely measured phylogenetically. In a statement that quickly became canonized as 'Kleiber's law', Kleiber (1932, 1947) argued that the total metabolic rate of an organism (typically measured as the rate of oxygen consumption) scales as the 3/4 power of body mass. His original analyses were largely derived from observations on vertebrates.

Although considerable subsequent research has led to a substantially altered view, West et al. (1997, 1999, 2002) have promoted the idea that 3/4 power-law scalings constitute universal laws relevant not just to metabolic rate but to a wide array of organismal features across the entire Tree of Life. The novelty of their work derives from a mechanistic view of fractal delivery systems (e.g., hierarchical branching networks of capillaries or leaf veins) for nutrients and respiratory gases. However, these details are not pursued here for several reasons. First, it is unclear how the features of a branching delivery network would apply to single cells. Second, a number of questionable mathematical assumptions underlying the fractal models have been highlighted (Dodds et al. 2001; Banavar et al. 2002; Kozłowski and Konarzewski 2004, 2005; Chaui-Berlinck 2006; Apol et al. 2008), and remain despite the originators' valiant efforts to dispute them (Brown et al. 2005; Savage et al. 2007). Third, although a study with protists yielded a power-law relationship with an exponent quite close to the theoretical prediction of 3/4 (Fenchel 2014), such behavior is a simple consequence of cells evolving flatter forms as species increase their average cell volumes, and is thus explainable with a surface-area constraint model. Although discrepancies with the 3/4 rule have been made repeatedly (Dodds et al. 2001; Kozłowski and Konarzewski 2005; Glazier 2015a,b), the universality of 3/4 power-law scaling continues to be promoted (West 2017).

Not only is the allometric coefficient for metabolic rate often inconsistent with $x/4$ power-law scaling, but taken across the Tree of Life, the regression appears to be non-linear (Zeuthen 1953). For example, with cell dry weight being the measure of size, DeLong et al. (2010) found the allometric slope for metabolic rate to be ∼2.0 for heterotrophic bacteria and ∼1.1 for unicellular eukaryotes. The same scaling was found whether cells were active and well-nourished or inactive and starved. Using updated cell-size measurements, the allometric slopes for the two groups are more on the order of 1.3 and 1.0 (Figure 8.2). However, the two estimates are not significantly different, and although there is little overlap in cell sizes between the two groups, a hypothesis of complete continuity of scaling across both groups cannot be ruled out. Assuming an isometric relationship for eukaryotic cells, the data in Figure 8.2 imply a simple relationship, with a mean metabolic rate ≃ 22 nL O_2 uptake/μg cell dry weight/day (at 20°C). Using a much smaller data set based on a mixture of just eight heterotrophs and phototrophs grown at a much colder temperature (5°C), Johnson et al. (2009) again observed an isometric relationship, and suggested a universal constant equivalent to 82 nL O_2 uptake/μg cell dry weight/day.

Finally, although metabolic rate is a classical physiological measurement, readily estimated as the rate of oxygen consumption or heat dissipation, its cell biological interpretation is generally far from clear. Total metabolic-rate measurements quantify the burning of carbon sources, but taken alone provide little information on the extent to which energy is converted to biomass production (growth and reproduction), the key targets of natural selection. Given the average constancy of the rate of energy utilization per unit mass noted above (independent of cell dry weight), were metabolic rate to be

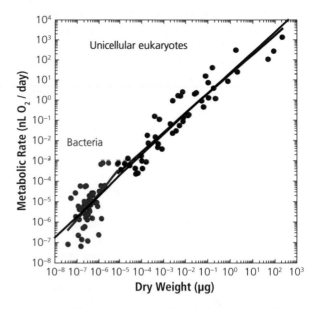

Figure 8.2 Allometric scalings of metabolic rate per cell (M) with cell volume (V) in heterotrophic unicellular species, with units given on the axes: $M = 23.3V^{1.01}$ (solid black line; standard error of exponent = 0.03); separate regressions for bacteria and unicellular eukaryotes (colored lines) have slopes of 1.26 and 0.97, respectively, although these are not significantly different. Data are from DeLong et al. (2010) with updated cell volumes derived from the survey in Lynch et al. (2022b). All metabolic rates are scaled to expected values at 20°C using a Q_{10} value of 2.5.

somehow proportional to the rate of biomass production, cell-division rates would also be expected to be nonresponsive to cell size. As discussed below, however, this expectation is far from fulfilled.

Lifetime energy requirements of a cell

Natural selection advances adaptations that enhance an organism's energetic capacity, either directly via growth and reproduction or indirectly via survivorship. However, adaptations themselves incur baseline construction and maintenance costs. To understand the capacity of selection to incorporate adaptive modifications, we need to scale the net energetic costs and benefits relative to the total cellular energy budget (the summed costs of cell construction and maintenance per cell lifetime), as this defines the visibility of trait modifications to selection. If the relative benefits do not sufficiently outweigh the costs, the trait will be liable to loss as selection is overwhelmed by the power of random genetic drift and deleterious-mutation accumulation (Chapters 6 and 17). If the costs sufficiently outweigh the benefits, loss will be actively promoted by selection.

The total cellular energy requirements per cell cycle partition into components associated with: 1) baseline maintenance and survival; and 2) construction of the essential parts of daughter cells (for growth and reproduction). Maintenance needs include energy invested in mRNA and protein processing, osmoregulation, intracellular transport, signal transduction, motility, and DNA repair. As the length of the cell cycle is prolonged, e.g., owing to resource limitation, the total maintenance requirements per cell cycle will grow approximately linearly with the cell-division time, whereas the contribution involving the construction of new parts (a roughly one-time investment) will remain approximately constant. As a consequence, the total lifetime energetic requirements of a cell (from birth to fission) will typically increase as growth conditions decline, eventually reaching a critical point where resources are just sufficient for maintenance (with nothing left for allocation to reproduction).

A powerful empirical approach allows the partitioning of the energy associated with maintenance vs. growth requirements of single-celled organisms. The method relies on estimates of the consumption rate of an energy-limiting resource at different cell-division rates (Foundations 8.1). For cells that can be grown on a defined medium in a continuous-flow chemostat (Figure 8.3), the rate of resource consumption per cell can be estimated from the difference in resource concentration between the inflow and outflow, the known cell density (which reaches an equilibrium in the growth chamber), and the flow rate. Different food resources vary in their energetic content, requiring a normalization of results across studies. However, conversion of resource consumption to units of ATP yield (the universal energy currency of cells) is readily accomplished if the metabolic pathways through which the substrate passes are known (Tempest and Neijssel 1984; Russell and Cook 1995).

The elegance of a continuous-flow culture is that an equilibrium cell-division rate is rapidly achieved, which is simply equal to the dilution rate of the chemostat. If the rate of resource consumption per cell is determined at several cell-division rates, a plot of the former versus the latter is expected to yield a straight line, with the slope providing an estimate of the amount of resource consumed to produce a new cell, and the intercept (denoting the point at which resource consumption is insufficient to support growth) providing a measure of baseline metabolic requirements (Figure 8.3). Pioneered by Bauchop and Elsden (1960), this regression approach is often called a Pirt (1982) plot.

The general procedure has been applied to enough organisms to reveal some broad generalizations (Figure 8.4). First, the basal metabolic rate (normalized to a constant temperature of 20°C for all species) scales almost linearly with cell volume across both bacteria and eukaryotes, with an allometric relationship of

$$C_M = 0.39V^{0.88}, \qquad (8.2a)$$

where C_M is in units of 10^9 molecules of ATP/cell/hour, and cell volume V is in units of μm^3. Care should be taken in the literal interpretation of C_M, as metabolic requirements of cells depend on their growth rates, and some maintenance activities can be strongly correlated with growth and hence, reside in the estimated component for construction costs (van Bodegom 2007; Biselli et al. 2020).

Second, the scaling of the growth requirement per cell is even closer to linearity with respect to cell volume (i.e., with an exponent near 1.0),

$$C_G = 26.9V^{0.96}, \qquad (8.2b)$$

where C_G is in units of 10^9 molecules of ATP/cell. If one further considers that a portion of eukaryotic cell volume is associated with vesicles and therefore relatively inert biologically, the regressions on active (or 'effective') cell volumes might yield modified allometric scaling coefficients. Unfortunately, little information is available on the scaling of vacuolar volume with total cell size, although an analysis for photosynthetic cells suggests ~89% active volume for a 1-μm^3 cell, declining to ~58%

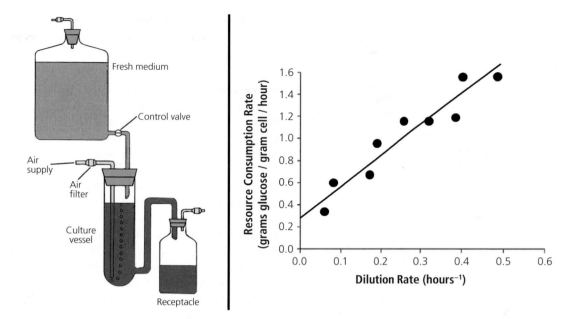

Figure 8.3 Estimating the costs of building and maintaining cells with chemostat cultures. **Left:** A chemostat (the central vessel) with a continuous input of resources (set by the valve on the upper reservoir containing sterile medium) and outflow into the bottom discard container. Aeration ensures an even distribution of cells within the culture and helps prevent wall growth. **Right:** A typical Pirt plot of the rate of glucose consumption by cultures of a bacterium grown at different dilution rates. Note that the measure of resource consumption is in units of glucose/total cell dry weight. This can be converted to ATP consumption per cell using information on the ATP equivalents derived per unit glucose consumed and the dry weight of individual cells. The y-intercept, with units of resource consumed/cell dry weight/time, is a measure of the basal maintenance requirement; the slope, with units of resource consumed/cell division (beyond basal metabolic requirements), is a measure of the total resource requirements for cell growth.

for a 10^4-μm^3 cell (Okie 2013). Such extremes are unlikely in heterotrophic species.

The total cost of building a cell is

$$C_T \simeq C_G + t_D C_M, \qquad (8.2c)$$

where t_D is the cell-division time in hours. The relationships in Equations 8.2a–c then imply that provided $t_D < 67V^{0.1}$ hours (assuming 20°C), which is generally the case in well-nourished cells, the contribution from cell growth dominates.

The preceding relationships will prove useful in subsequent chapters where attempts are made to determine the costs of various cellular features relative to a cell's entire energy budget. In Foundations 8.2, an entirely different approach to the problem, applicable to organisms that cannot be readily assayed in chemostats but for which metabolic-rate data are available, is shown to yield results that are quite compatible with Equations 8.2a,b.

Notably, both the maintenance and growth relationships scale in an apparently continuous fashion across bacteria and eukaryotes, despite the substantial difference in cell contents between the groups. On the one hand, eukaryotic cells contain internal lipid membranes, which are energetically expensive, but on the other hand, such cells are less densely packed with biomaterials (Chapter 7).

Finally, although almost all studies on scaling relationships in biology focus on the slopes of the regressions, an equally important matter concerns the magnitude of the normalization constant. Why, for example, is the cost of growth of a 1-μm^3 cell $\simeq 3 \times 10^{10}$ ATP hydrolyses (Equation 8.2b), and not substantially more or less? A crude calculation suggests that this is largely a consequence of biology's reliance on ATP as an energy source and the energetic content of the biomaterials from which cells are built. From Milo and Phillips (2016),

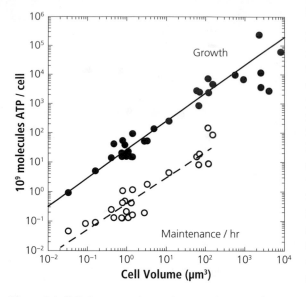

Figure 8.4 Cellular energetic requirements in units of numbers of ATP → ADP hydrolysis events, all scaled to 20°C to normalize results from studies involving different temperatures. The maintenance and growth costs for bacterial (black) and unicellular eukaryotic (blue) species are fitted with allometric regression lines. Growth data for cell cultures from a few multicellular eukaryotes (red) are not included in the regression analysis. From Lynch and Marinov (2015).

under average physiological conditions ~14,300 calories (cal) are released per mol of ATP hydrolyses, which translates into 2.4×10^{-20} cal/individual hydrolysis event. From a wide variety of sources, the energetic content of unicellular organisms is 4.7×10^{-6} cal/ng dry weight (Supplemental Table 8.1). Using these results, and converting to cell volume with the relationship in Figure 7.1, leads to an estimated cost of cell construction of $\sim 114 \times 10^{9}$ ATP hydrolyses/μm^3 cell volume, which is within a factor of four of the actual estimate based on elaborate work on resource consumption (Equation 8.2b).

Thus, the energetic cost of building cells is a historical legacy of the kinds of biomaterials (e.g., nucleic acids, proteins, and lipid membranes) that life has been dependent upon since the establishment of LUCA. The deeper question as to why life came to depend on ATP as a universal material for energy exchange remains unresolved.

The speed limit on cell-division rates

Natural selection feeds off of genotypic differences in rates of genome transmission on an absolute timescale. Thus, fitness ultimately depends not just on the rate of resource acquisition, but also on the rate at which assimilated resources are transformed into new cells, as opposed to being burned in non-productive activities. Thousands of studies have been performed on the growth rates of various species under a multitude of conditions, but given the diversity of approaches, the only fair comparison is to evaluate maximum known cell-division rates. Even then, the data must be normalized to a constant temperature (as the latter influences all aspects of biology; Chapter 7), and there is no guarantee that studies on any particular species have actually uncovered the optimal growth conditions.

With these caveats in mind, although the observations just discussed indicate near constant energy cost of cells per unit volume across the Tree of Life, the *rate* at which biomass is constructed is far from constant. Most notably, the evolutionary relationship between maximum cell-division rate and cell size (in units of dry weight per cell) is qualitatively different between heterotrophic bacteria and eukaryotes (Figure 8.5a). Here, the growth rate is measured as the maximum exponential rate of expansion $b_{max} = \ln(2)/t_D$, where t_D is the average cell-division time (in days). For bacteria, the scaling of this trait with cell size is positive, with an allometric coefficient of 0.28 (SE = 0.07). Although there is a more than a tenfold range of variation in cell-division times for any specific cell size, this is in part due to sampling error. Thus, bacteria certainly do not adhere to the idea that large cells suffer from reduced rates of cell division as a result of a surface area:volume constraint (below).

In contrast, unicellular eukaryotes universally exhibit weak, negative scaling of maximum growth rate with cell size. For amoeboid forms, ciliates, dinoflagellates, and other heterotrophic flagellates, the allometric scaling coefficients fall in the narrow range of −0.19 to −0.22. Despite this uniform scaling over six orders of magnitude of cell-size differences, the elevations of the power-law functions vary among groups, with ciliates having the highest growth rates and flagellates the lowest.

Unfortunately, there is very little overlap in the sizes of bacterial and eukaryotic cells in these analyses of heterotrophs. Thus, it is unclear whether the observed shift in scaling behavior is a consequence of fundamental biological differences between groups or a reflection of a more general scaling relationship, with a global optimum size for cell-division rates on the order of 10^{-6} μg dry weight (fortuitously near the approximate size boundary between prokaryotes and eukaryotes). Although it may seem puzzling why all bacteria don't evolve to very large sizes and all unicellular eukaryotes to very small sizes, it should be remembered that total fitness is determined by the difference between birth and death rates, and that the optimum size for survivorship may differ greatly among environments.

Do these observations generalize to phototrophs? In general reviews of marine phytoplankton, Raven (1986, 1994) argued (with fairly limited data) that there is a reversal in the scaling of cell-division times at a cell volume of $\sim$30 μm^3, which equates to an approximate cell dry weight of 10^{-5} μg, close to the area of overlap in size of bacterial and eukaryotic heterotrophs in Figure 8.5a. However, although a similar argument was made by Marañòn (2015), Marañón et al. (2013), and Ward et al. (2017), the broader comparative analysis in Figure 8.5b does not support this sort of non-monotonic scaling for phototrophs. There is no relationship between b_{max} and cell size in cyanobacteria, and for the two eukaryotic groups with data on several dozens of species, green algae and diatoms, the allometric scaling coefficient is −0.09, which is about half that found for heterotrophs.

It bears emphasizing that the cell-division rates summarized in Figure 8.5 are all derived from pure cultures grown under optimized laboratory conditions. In nature, organisms may rarely (if ever) experience such conditions, commonly dividing at least one to two orders of magnitude more slowly than maximum rates. Indeed, many microbes inhabiting aquatic sediments may have generation times on the order of several years, and in some cases even hundreds to thousands of years (Hoehler and Jørgensen 2013; Morono et al. 2020). In principle, such cells may often enter semi-dormant states with maintenance requirements orders of magnitude lower than

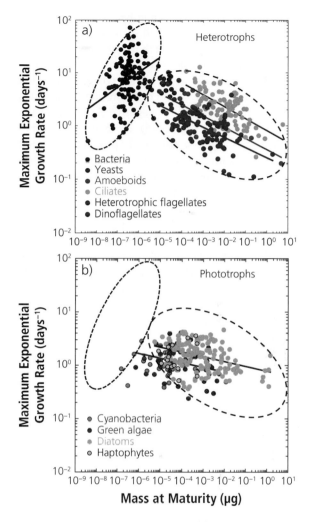

Figure 8.5 Maximum growth rates, ln(2)/doubling time, for **a)** heterotrophic and **b)** phototrophic unicellular species, scaled to expected values at 20°C. Fitted lines are given only for the cases in which the least-squares regression is significant; the slopes are 0.28 for heterotrophic bacteria, −0.21 for amoeboid eukaryotes, −0.22 for ciliates, −0.21 for heterotrophic flagellates (excluding dinoflagellates), −0.19 for dinoflagellates, and −0.09 for both green algae and diatoms. The dashed ellipses enveloping the data for bacterial and eukaryotic heterotrophs in the upper panel are transferred to the lower panel for comparative purposes. From Lynch et al. (2022b).

those implied in Figure 8.4 (Munder et al. 2016; Lennon et al. 2021).

To sum up, the preceding results permit three fairly general statements about the biology of cells.

First, in both eukaryotes and bacteria, *absolute* rates of biomass acquisition/cell increase with cell size. This follows from the fact that the energetic requirement for growth scales nearly linearly with cell volume, i.e., as $\sim V^{0.96}$ (Figure 8.4), while cell-division times (the inverse of the division rate) scale much more weakly and even negatively with V in the case of bacteria (Figure 8.5). For both bacteria and eukaryotes, the absolute rate of production per cell increases with cell size, but it does so supralinearly with size in bacteria (because the cell-division time declines with size) but sublinearly in eukaryotic cells (because the cell-division time increases weakly with size). On the other hand, the rate of productivity per unit biomass, b_{max}, which is more directly related to fitness, changes directionality between groups, being positive in bacteria but negative in eukaryotes, yielding the pattern in Figure 8.5a.

Second, returning to the results in the previous section, insight can be gained on the efficiency of conversion of assimilated energy into growth, C_G/C_T. The lifetime cellular energy budget (C_T) is a function of the cell-division time, t_D, as summarized in Equation 8.2c. Using the quantitative expressions for C_G (the cost of growth) and C_M (the cost of maintenance/day) as functions of cell volume (V) in Equations 8.2a,b, and defining t_D in units of days,

$$C_G/C_T \simeq \frac{1}{1 + 0.35 t_D V^{-0.08}}. \qquad (8.3)$$

Recalling from above that $t_D = \ln(2)/b_{max}$, noting from the regression in Figure 8.5 that bacterial $b_{max} \simeq 527 B^{0.28}$, where B is the dry weight per cell in μg, and using Equation 7.1 to express B in terms of V, leads to $t_D \simeq 0.055 V^{-0.26}$. Substitution of the latter expression into Equation 8.3 then leads to growth-efficiency estimates ranging from 0.92 to 0.99 for the range of bacterial cell volumes of 0.01 to 10 μm^3.

Thus, for bacterial cells growing at maximum rates, the vast majority of assimilated energy is allocated to growth, increasingly so in cells of larger size. With poorer growth conditions (larger t_D), these efficiencies will necessarily decline. For example, with a tenfold increase in t_D, the corresponding efficiencies become 0.79 to 0.92, and with a hundredfold increase in t_D, they reduce to 0.10 to 0.54. Thus, a lower limit to bacterial cell size

arguably results from the progressive increase in the fraction of energy intake that must be devoted to maintenance in the face of a relatively long cell-division time (Kempes et al. 2012).

Growth efficiencies are somewhat lower for eukaryotic cells. For heterotrophic eukaryotic flagellates, including dinoflagellates, growing at maximum rate, $t_D \simeq 0.18 V^{0.17}$ days, efficiencies decline from 0.93 to 0.83 for cell volumes of 10 to 10^6 μm^3. For amoeboid forms, the minimum cell division time is $t_D \simeq 0.094 V^{0.19}$ days, leading to a range for C_G/C_T of 0.95 to 0.84 for cell volumes of 100 to 10^8 μm^3. For ciliates, $t_D \simeq 0.036 V^{0.20}$ days, and C_G/C_T ranges from 0.97 to 0.92 for cell volumes of 10^3 to 10^7 μm^3. Thus, as in heterotrophic bacteria, when experiencing maximum growth capacity, most eukaryotic cells incorporate > 90% of assimilated energy into growth versus maintenance, although the scaling of efficiency with cell volume is negative.

A final noteworthy point is that the upper halves of the dashed ellipses in Figure 8.5a demarcate apparent absolute upper bounds to cell-division rates (normalized to 20°C) that natural selection has been able to achieve. For heterotrophic eukaryotes at this temperature, no cell divides in < 1.7 hours, and no cell > 1 μg in dry weight divides in < 8 hours. No phototroph of any size, bacterial or eukaryotic, divides in < 4 hours at 20°C. On the other hand, at the same temperature, some large bacterial heterotrophs can divide in just 15 minutes.

Can these upper limits be broken? When strong and prolonged periods of selection are imposed on growth rate in simple laboratory experiments with microbes, there can be an up to 30% increase in b_{max}, e.g., in the bacteria *E. coli* (Wiser et al. 2013; Marshall et al. 2022) and *Mycoplasma mycoides* (Moger-Reischer et al. 2023), the yeast *Kluyveromyces* (Groeneveld et al. 2009), and the ciliate *Tetrahymena* (Tarkington and Zufall 2021). Thus, natural selection appears to be somewhat stalled with respect to maximum growth rates, possibly because the strength of selection on maximum growth rates in nature is weaker than in laboratory settings where a premium can be put on this single trait. Notably, in both studies with bacteria, cell volume increased in parallel with cell growth rate, consistent with the scaling relationship in Figure 8.5, and suggesting a prior history of selection for small size (for reasons unassociated with growth rate). Although cell size also increased in the study with *Kluyveromyces*, this was accompanied by a dramatic change in cell shape, leading to an increase in surface area:volume ratio. Finally, although it might be argued that there is a trade-off between growth rate and competitive ability, the limited data in microbes suggests otherwise (Gounand et al. 2016).

The Limits to Natural Selection Imposed by the Drift Barrier

The classical dogma in physiological ecology is that scalings of bioenergetic features are unavoidable constraints of biochemistry and/or biophysics. As outlined in detail by West (2017), numerous arguments based on organism size lead to power-law behavior with exponents being multiples of 1/3 or 1/4 depending on one's mechanistic perspective. However, although such hypotheses are based on what may appear to be reasonable arguments, the inferred supportive evidence almost always derives from statistical analysis of patterns, rather than on direct experimental evidence of mechanistic constraints. Moreover, as noted earlier, a more fundamental problem is that neither 1/3 nor 1/4 power-law scalings provide general explanations for the phylogenetic patterning of bioenergetic traits in unicellular organisms.

Most striking are the opposite directions of scaling of maximum growth rate and efficiency with cell size in bacteria versus eukaryotes (Figure 8.5). For bacterial growth rates, the scaling is positive, although the data are noisy enough that the true exponent could be anywhere in the range of 0.14 to 0.42. In contrast, the scaling exponents for individual heterotrophic unicellular eukaryotic groups are not only negative, but more compatible with −1/5 than either −1/4 or −1/3 power-law scaling. For phototrophs, the growth-rate scaling exponent is even weaker than −1/5, being closer to −1/10.

This lack of robust support for biophysical-constraint hypotheses suggests a need to evaluate the problem of growth-rate scaling from an entirely different perspective. Shifting the view from biophysical constraints to limits on the evolutionary process, one possibility is that with increasing cell size, the efficiency of natural selection

declines, owing to the associated reduction in effective population size (N_e). The intention here is not to promote the idea of precise 1/5 (or 1/10) power-law scaling relationships. Further empirical study of the distribution of mutations with small effects in various phylogenetic lineages will be necessary for that level of resolution. The key point is that any reduction in N_e will diminish the ability of natural selection to purge deleterious mutations, thereby leading to a reduction in trait performance.

If this hypothesis is correct, arguments that attempt to explain scaling patterns across the Tree of Life purely on the basis of cell-geometric arguments will be incomplete, if not entirely misplaced. If, on the other hand, it could be shown that the population-genetic environment has no influence on scaling relationships, this would imply that the structure of cell biology is such that there is always a supply of mutations with sufficiently large, variable, and beneficial effects to bring things to their biophysical limits regardless of phylogenetic affinity.

Recall from Chapter 4 that the key determinant of whether natural selection can eradicate a deleterious mutation with effect s is whether the ratio of the power of selection to the power of drift $s/(1/N_e) = N_e s$ exceeds 1.0. The fact that N_e scales with the −0.20 power of organism size (Figure 4.3) implies that species with larger cell sizes have reduced capacities to promote growth-rate promoting mutations and to eradicate growth-rate reducing mutations of small effects. As noted in Chapter 5 and further elaborated on in Chapter 17, the evidence is compelling that a large fraction of mutations have fitness effects below $|s| = 10^{-5}$, extending down to 10^{-10} or even lower in species with larger cell sizes. Because N_e in unicellular species is typically in the range of 10^6 to 10^9 individuals, this means that a substantial number of mutations with individually very mildly deleterious (i.e., growth-reducing) effects are free to accumulate in species with relatively small N_e while still being subject to efficient purging in larger-N_e species.

As discussed in Chapter 4, several genetic features determine how the efficiency of selection against mildly deleterious mutations scales with the demographic features of a population. These are evaluated in a stepwise fashion in the next few paragraphs to show how the progressive incorporation of natural genomic features can plausibly lead to the kinds of negative scaling seen in Figure 8.5.

The simplest starting point assumes that selection operates on individual genetic loci independently of events occurring at other genomic locations. This requires either very high recombination rates or such small population sizes that co-segregating variants are never simultaneously present at multiple loci. Consider the situation in which each locus harbors two possible alleles, + (beneficial) and − (deleterious), with the mutation rates from + to −, and vice versa, being u_{10} and u_{01}, and the + allele having advantage s. At small enough population sizes that $N_e s \ll 1$, the long-term average frequency of the favorable + allele is simply a function of the ratio of mutation rates, $u_{01}/(u_{01} + u_{10})$, but as the condition $N_e s \gg 1$ is approached, the increased efficiency of selection drives the mean frequency to 1.0 (Figure 8.6a). Under this model, the transition between these two extremes occurs in a narrow (order of magnitude) range of N_e near $1/s$. Thus, this simple model is inadequate to explain a consistent scaling of mean performance across several orders of magnitude of population size.

Suppose, however, that there are multiple, completely linked loci with the same mutational and selection properties, with a haplotype's growth rate (performance) being determined in an additive fashion (equal to the fraction of loci occupied by + alleles), and fitness being defined by a multiplicative (independent effects) model, $(1 - s)^n \simeq e^{-sn}$, where n is the number of − alleles. In this case, it can take as many as five orders of magnitude of N_e to span the full range of equilibrium mean growth rate (Figure 8.6b). This shift in behavior is a consequence of selective interference among simultaneously segregating mutations, as described in Chapter 4. For populations of moderate size, there will be genetic variation among individuals in terms of the total number of + alleles harbored across loci, the result being that many new beneficial mutations will arise on suboptimal genetic backgrounds destined to eventual loss. At the largest population sizes, however, selection still keeps deleterious alleles at very low frequencies at all genomic sites, reducing the effects of background interference.

A third issue to consider is that fitness effects will vary among genomic sites, with sites with large effects likely being much rarer than sites with small effects (Chapter 5). This further flattens the scaling of mean trait performance with organism/population size. Even with free recombination, such behavior results because the sites with progressively smaller effects require increasingly high N_e for selection to promote their favorable alleles. As a consequence, the equilibrium mean performance will be a mixture of the responses found for the constant-effects model. This can be observed as the stepwise increment in mean performance in Figure 8.6c where there are sites with just three different effects. A smoother transition would arise with a more continuous distribution of site effects.

Finally, Figure 8.6d considers the situation in which different types of sites are completely linked. In this case, as a site with major effects becomes surrounded by increasing numbers of minor-effect sites, there is again a progressive decline in the rate of scaling of mean performance with N_e. Increasing numbers of minor-effect loci cause increased background interference for selection operating on major-effect sites, while also contributing more to the total maximum performance of the trait (diluting the overall influence of the major-effect sites).

Without a detailed understanding of the fine-scaled distribution of genomic sites with different magnitudes of mutational effects, a precise statement cannot be made on the expected scaling relationship between mean performance and N_e, other than that it should be negative. This might be viewed as a shortcoming of the theory. However, it should be recalled that virtually all biophysical explanations simply adhere to the assumption of a fixed power-law relationship, even when the latter can be shown to be not generalizable. The main limitation of the drift-barrier hypothesis is not an absence of generalizable theory, but rather a lack of precise information on mutations with small effects.

For a trait like growth rate, it can be expected that essentially every genomic site influences performance in at least a small way, and that there will be considerable variation among nucleotide sites in terms of average effects and recombination rates. Thus, because the plotted theoretical results encompass a wide range of plausible genetic

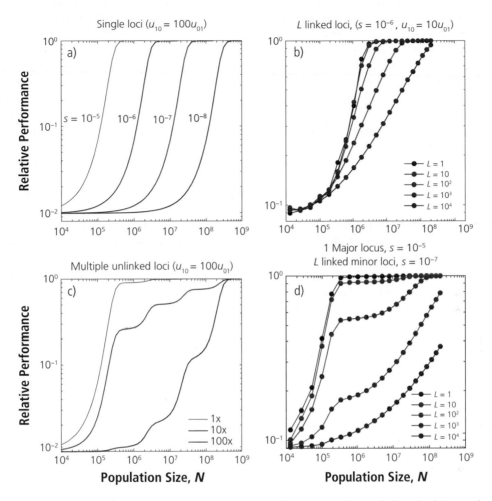

Figure 8.6 Evolutionary long-term average performance (under drift-mutation-selection balance) relative to the maximum value of a trait experiencing an exponential fitness function. Each locus has two possible alleles, + and −, with additive effects on the trait (denoted as performance), but with individual fitness declining as e^{-sn}, where s is the selective disadvantage, and n is the number of − alleles per individual. Reversible mutation operates between the two alternative alleles at rates u_{01} from − to +, and u_{10} from + to −. Populations consist of N haploid individuals. **a:** Results are given for four values of s for the situation in which there is free recombination between sites. Mean performance makes a rapid transition at the point at which $1/N = s$, from the neutral expectation $u_{01}/(u_{01} + u_{10})$ at small N to the domain in which the beneficial allele frequency $\simeq 1$ at large N. Analytical results are obtained from formulae in Kimura et al. (1963). **b:** Results for the situation in which $s = 10^{-6}$, under the assumption of L completely linked genomic sites, all with the same effects. With larger numbers of linked sites, the gradient in mean performance with N becomes more gradual. Here and in panel **d**, the results are obtained by computer simulation using the methods in Lynch (2020). **c:** Modification of the situation in panel **a** to allow for a mixture of unlinked genomic sites with different effects, with various weights (1×, 10×, and 100×) given to the four curves in panel **a**. As the population size increases, selection becomes progressively more efficient at promoting alleles with smaller effects. For example, $L = 10\times$ means that for each site with effect $s = 10^{-5}$, there are ten with $s = 10^{-6}$, 100 with $s = 10^{-7}$, and 1000 with $s = 10^{-8}$. **d:** The situation in which there are blocks of linked loci, one major-effect site with $s = 10^{-5}$, and increasing numbers (L) of linked loci with minor effects $s = 10^{-7}$. As L increases, the minor loci contribute proportionally more to total performance, and create background selection interference with the major locus and among themselves.

properties, it is clear that there is little justification for ignoring the possibility that the variation in the population-genetic environment plays a significant role in defining relationships between maximum performance, organism size, and N_e. For this not to be the case, all deleterious mutations would have to have effects smaller than the smallest $1/N_e$ (from Figure 4.3, on the order of 10^{-9}), and hence, to be impervious to selective removal in all organisms, and/or larger than the largest $1/N_e$ (on the order of 10^{-4}), and hence, impervious to fixation across the Tree of Life. It is highly implausible that no mutations would have fitness effects in the range of 10^{-9} to 10^{-4}.

What we can say is that, for heterotrophic eukaryotes, both b_{max} and N_e scale with the −0.2 power of body mass, which means that species-specific b_{max} is directly proportional to N_e. Thus, for every fractional decline in N_e, there is an equal fractional decline in b_{max}, and this is consistent over the entire span of N_e. For this to occur, within the window of mutations with selection coefficients in the range of 10^{-9} to 10^{-4}, there must be an inverse relationship between the number and effects of mutations, such that the product remains constant over this full range of s.

To see this, note that for the interval of N_e between 10^4 and 10^5, there will be an interval of deleterious mutations with s in the range of 10^{-5} to 10^{-4} that are vulnerable to fixation by drift, but would be efficiently purged by natural selection for $N_e > 10^5$. Likewise, for the interval of N_e between 10^5 and 10^6, there will be an interval of deleterious mutations with s in the range of 10^{-6} to 10^{-5} that are vulnerable to fixation by drift, but would be efficiently purged by natural selection at $N_e > 10^6$, and so on. Thus, for the interval-specific decline in b_{max} to remain constant across N_e, the number of relevant mutational effects in each interval must be inversely proportional to the effects, such that the product remains constant. Such a negative exponential distribution of fitness effects is at least qualitatively compatible with what we know about the subject within various species (Chapters 5 and 17). Note, however, that the concern here is with the cross-phylogeny balance of load-creating mutations, with the total load of mutations with effect $s \simeq 1/N_e$

remaining approximately constant as N_e changes across lineages.

Owing to insufficient information on N_e in phototrophs, this type of argument cannot be quantitatively extended to this group, except to say that the exact argument would apply if N_e happened to scale with the −0.1 power of body mass in phototrophs. There is a similar lack of information on the scaling of N_e with cell size in prokaryotes. However, given that b_{max} increases with cell size in bacteria, and assuming that N_e scales negatively with cell size (as in eukaryotes), the drift-barrier hypothesis cannot explain the size-specific growth scaling in bacteria. The possibility is then raised that adaptive evolution in bacteria is typically nearly invulnerable to compromises resulting from drift. Assuming an upper limit to $N_e \simeq 10^9$, this would require that the vast majority of deleterious mutations in such species have $s > 10^{-9}$, as suggested in Chapter 5.

Membrane Bioenergetics and the Prokaryote–Eukaryote Transition

As noted in Chapter 2, a peculiar feature of cell biology is the localization of the key machinery associated with energy production to lipid bilayers. The series of complexes known as the electron-transport chain (ETC) couple the extraction of electrons from the oxidation of organic compounds to the export of hydrogen ions, maintaining a concentration gradient of the latter across the membrane. The biochemical details of this process are covered in all biochemistry texts, and are not elaborated upon here. The salient issue is that the cross-membrane gradient in hydrogen ion concentration driven by the ETC causes chemiosmotic pressure for the return movement of hydrogen ions through membrane-embedded ATP synthase complexes, coupling this mechanical energy to the production of ATP (Chapter 2). One of the key differences between prokaryotes and eukaryotes is that in the former, all of these events take place on the inner cell membrane, whereas membrane bioenergetics has been relocated/restricted to the inner membranes of mitochondria in eukaryotes (where they would have been present from the outset in the primordial mitochondrion; Chapter 23).

Table 8.1 Geometric features for cells of common shapes. Abbreviations: $r < \ell$, radii or half-widths for spheroids; h, full length of a cylinder or rod; $\alpha = h/r$ or ℓ/r. Note that for a rod, h is the length from one rounded tip to the other. The formula for the surface area of a spheroid is known as Knud Thomsen's approximation.

Shape	Surface Area (S)	Volume (V)	S/V
Sphere	$4\pi r^2$	$(4/3)\pi r^3$	$3/r$
Cylinder	$2\pi r(r + h)$	$\pi r^2 h$	$2(1 + \alpha)/h$
Rod	$2\pi r h$	$\pi r^2[h - (2r/3)]$	$(2\alpha/r)/[\alpha - (2/3)]$
Prolate spheroid	$2.02\,\pi r^2(1 + 2\alpha^{1.61})^{0.63}$	$(4/3)\pi r^2 \ell$	$(1.5/\ell)(1 + 2\alpha^{1.61})^{0.63}$

Under the assumption that energy production is limited by the number of ATP synthase complexes, which in turn are assumed to be limited by the availability of plasma membrane surface area, Lane (2002, 2015, 2020; Lane and Martin 2010) argued that the endosymbiotic establishment of mitochondria freed eukaryotes of this constraint by providing effectively unlimited inner mitochondrial membranes. This assertion led to the further claim that the energetic boost made possible by mitochondria constituted a watershed moment in evolution by generating excess power essential to all things associated with eukaryogenesis. Under this view, the mitochondrion is not simply one of the many unique features of eukaryotes. Rather, it is the key feature that enabled the evolution of internal cell structure, large cell size, expanded genomes, multicellularity, sex, behavior, etc., all of which are viewed as hallmarks of organismal superiority.

Before evaluating the likelihood of this scenario, a brief consideration of the surface area problem is in order. The general formulae for several common cell shapes are provided in Table 8.1. Regardless of the shape, volume always increases with the cube of a linear dimension, whereas the surface area increases with the square of the linear measure. The surface area to volume ratio depends on shape, but it is always inversely related to a linear dimension of the cell.

Because the production of ATP in prokaryotes is highly dependent on complexes embedded in the plasma membrane, the geometric constraint argument implies that if the cell surface is a limiting resource, there should be a reduction in energy production per unit volume with increasing cell size. However, the analyses in the previous section already shed doubt on this assertion, showing that increased cell size is associated with higher, not lower, maximum rates of growth in bacterial species. In contrast, mitochondria-bearing eukaryotes have lower energetic capacities than prokaryotes on a volumetric basis, and growth rates decline with increasing cell size. This matters from an evolutionary perspective because it is the growth rate per unit volume, often called the specific growth rate, that determines the rate of gene flow into the next generation. Thus, observations on the growth-rate potential across the Tree of Life are uniformly inconsistent with the basic premise underlying the Lane hypothesis.

Energy production and the mitochondrion

A consideration of eukaryotic cell anatomy provides a more mechanistic view of why the total membrane energetic capacity of eukaryotic cells is nothing out of the ordinary. A key question is whether mitochondria endow eukaryotic cells with enhanced membrane surface area for the occupancy of ATP synthase. Although the situation at the time of first colonization of the mitochondrion is unknown, the iconic view of mitochondria being tiny, bean-shaped cellular inclusions is not generalizable. For example, many unicellular eukaryotes harbor just a single mitochondrion or one that developmentally moves among alternative reticulate states (e.g., Burton and Moore 1974; Rosen et al. 1974; Osafune et al. 1975; Biswas et al. 2003; Yamaguchi et al. 2011; Uwizeye et al. 2021). Such geometries necessarily have lower total surface areas than a collection of spheroids with similar total volumes. For the range of species that have been examined, many of which do have small individualized mitochondria, the total outer surface

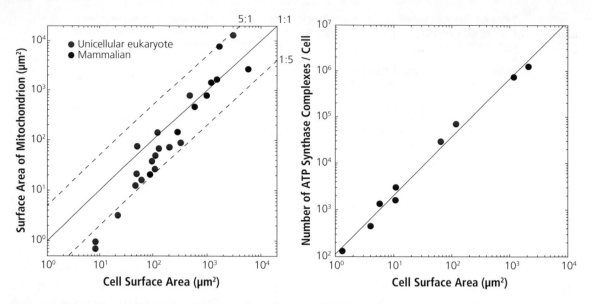

Figure 8.7 The scaling of mitochondrial features with cell size. **Left:** Relationship between the total outer surface area of mitochondria and that of the plasma membrane. Diagonal lines denote three idealized ratios of the two. Data are from Lynch and Marinov (2017) and Uwizeye et al. (2021). **Right:** The number of ATP synthase complexes per cell scales with cell surface area (S, in μm^2) as $113S^{1.26}$ ($r^2 = 0.99$); black data points are for bacteria.

area of mitochondria per cell is generally on the order of the total area of the plasma membrane, with no observed ratio exceeding 5:1, and ratios for the smallest species being less than 1:5 (Figure 8.7a). Given the putative archaeal nature of the cell that hosted the primordial mitochondrion (Chapters 3 and 23), it is likely that the starting condition resembled the situation in today's smallest eukaryotes.

Although the outer surface area of the mitochondrion has been the most common source of measurements, it is less relevant than the inner membrane, where the ATP synthase complex sits. However, for the few species with available data, the ratios of inner to outer membrane areas for mitochondria are modest, ~ 5.0, 2.4, 2.5, and 5.2, respectively, in mammals, the green alga *Ochromonas*, the plant *Rhus toxicodendron*, and the ciliate *Tetrahymena* (summarized in Lynch and Marinov 2017). Moreover, the total surface areas of mitochondria substantially overestimate the real estate allocated to ATP synthase complexes, which are often restricted to two rows on the narrow edges of the inner invaginations (called cristae), commonly comprising $\ll 10\%$ of the total internal membrane area (Kühlbrandt 2015).

Three additional observations raise further questions as to whether membrane surface area is a limiting factor in ATP synthesis. First, multiple observations on the developmental responses of organelles to cell growth indicate that the total mitochondrial volume remains proportional to cell volume (Atkinson et al. 1974; Grimes et al. 1974; Posakony et al. 1977; Pellegrini 1980; Rafelski et al. 2012), and the same has been observed in the comparative analysis of protist species (Fenchel 2014). From arguments in the preceding section, this implies that the surface area of mitochondria scales with the two-thirds power of cell volume. If mitochondrial surface area limits cellular energy production, to maintain mitochondrial generating power capacity per unit volume, the concentration of mitochondria would need to scale as $V^{1/3}$ rather than following the observed volume-independent pattern. Second, only a fraction of bacterial membranes appears to be allocated to bioenergetic functions (Magalon and Alberge 2015), again shedding doubt on whether membrane area is a limiting factor for prokaryotic energy production. Third, as is discussed more fully in Chapter 9, in every bacterial species for which data are available, growth in cell volume is exponential or close to it. Thus, as in the among-species pattern (Figure 8.5), growth rates of individual bacterial cells increase with cell volume despite the reduction in the surface area:volume ratio.

Still further insight into this issue derives from the average packing density of ATP synthase for the few species with sufficient proteomic data (Lynch and Marinov 2017). For example, the estimated number of complexes in *E. coli* is ~3000, and the surface area of the cell is ~16 μm^2. A single ATP synthase in this species occupies ~64 nm^2 (Lücken et al. 1990) of surface area, so the total set of complexes occupies ~2% of the cell membrane. Drawing from additional observations on four other diverse bacterial species, the overall average membrane occupancy of ATP synthase is just 1% in bacteria (Lynch and Marinov 2017).

This kind of analysis can be extended to the few eukaryotes for which data are available, noting that eukaryotic ATP synthases are slightly larger, with maximum surface area of ~110 nm^2 (Abrahams et al. 1994; Stock et al. 1999). Although ATP synthase resides in mitochondria in eukaryotes, it is relevant to evaluate the fractional area that would be occupied were they to be located in the cell membrane. Such hypothetical packing densities are 5–7% for yeast and mammalian cells (Lynch and Marinov 2017). (Adding in the membrane occupancy of the entire ETC would increase these proportions by ~4-fold, in both prokaryotes and eukaryotes). These observations suggest a ~5-fold increase in ATP synthase abundance relative to cell-surface area expectations in eukaryotes, although the data conform to a continuous allometric function with no dichotomous break between bacteria and eukaryotes (Figure 8.7b).

To sum up, these multifaceted observations are consistently contrary to the idea that the relocation of membrane bioenergetics endowed eukaryotes with enhanced growth efficiency beyond what would be expected of bacterial cells of similar size. Indeed, if there are any effects at all, they appear to be negative. One could perhaps argue that eukaryotes would be even poorer growers were it not for the presence of mitochondria, but extrapolations into the realm of unobserved data is inadvisable.

Cellular investment in ribosomes

The ribosome content of a cell provides another strong indicator of its bioenergetic potential, translation capacity in particular. Owing to the large number of proteins required to build the complex, ribosomes are energetically costly, and the number per cell within a species appears to be universally correlated with cellular growth rate. At low nutritional states, cells reduce their investments in ribosomes relative to components of the cell involved in nutrient uptake (Chapter 9). One might then expect variation in the translational capacity of cells of different species to reflect their intrinsic bioenergetic potential.

As noted in Chapter 7, the genome-wide total and mean number of transcripts per gene scale with cell volume as $V^{0.42}$ and $V^{0.26}$, respectively, and the analogous scalings for proteins are $V^{0.95}$ and $V^{0.65}$, with no dichotomous break between prokaryotes and eukaryotes (Lynch and Marinov 2015). As with the transcripts they process and the proteins they produce, the numbers of ribosomes per cell also appear to scale sublinearly with cell volume, $\propto V^{0.80}$, in a continuous fashion across bacteria and unicellular eukaryotes, including cultured cells from multicellular species (Figure 8.8). Note that, under the assumption that ribosomes produce proteins at approximately constant rates in different organisms, the scaling of protein production per volume would be $V^{0.80}/V = V^{-0.20}$. Thus, the cellular concentration of ribosomes matches the scaling of maximum growth rates with eukaryotic cell size outlined in Figure 8.5.

The mitochondrion as a driver of eukaryotic evolution

Lane and Martin (2010) and Lane (2015) proposed a scenario for how the mitochondrion became established by a series of adaptive steps, arguing that the eukaryotic leap to increased gene number and cellular complexity, and a subsequent adaptive cascade of morphological diversification 'was strictly dependent on mitochondrial power'. A similar argument was made by DeLong et al. (2010), and many others have repeated the narrative that eukaryogenesis and all of the associated

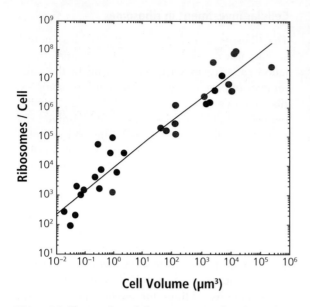

Figure 8.8 The number of ribosomes per cell scales with cell volume (V, in μm^3) as $8810 V^{0.80}$ ($r^2 = 0.92$; standard error of the exponent = 0.04). Colour coding as in the previous figure, with orange denoting green algae and land-plant cells. From Lynch and Marinov (2017, 2018) and a few additional data points from more recent literature (Supplemental Table 8.2). Although the estimated slope based on bacterial data alone is hypermetric, and that for eukaryotes alone is hypometric, neither is significantly different from 1.0, nor from each other.

downstream effects would be impossible without mitochondria.

However, as should now be clear, there is no evidence in support of this hypothesis. Diverse sets of comparative observations all lead to the opposite conclusion. Large bacterial cells do not suffer from reduced rates of biomass production, but eukaryotic cells do. There is not a quantum leap in the surface area of bioenergetic membranes in eukaryotes. The idea that ATP synthesis is limited by total membrane surface area is not supported. Moreover, the numbers of ribosomes and ATP synthase complexes per cell, joint indicators of a cell's capacity to convert energy into biomass, scale with cell size in a continuous fashion both within and between bacterial and eukaryotic groups. In addition, as noted in subsequent chapters, the absolute costs of producing not only ribosomes but also the remaining proteins

in cells (the members of the ETC in particular) are significantly higher in eukaryotes than in bacteria, owing to the substantial increase in complex size, gene lengths, investment in nucleosomes, etc. Finally, there is the additional matter of the expense of building mitochondria, associated with the high biosynthetic costs of lipid bilayers (Chapter 15).

Although further discussion on the origin of the mitochondrion will be presented in Chapter 23, the idea that the mitochondrion engendered a bioenergetics revolution can be put to rest for now. The relocation of membrane bioenergetics to inner mitochondrial membranes may have endowed eukaryotes with novel possibilities for further remodelling of cellular features. But an enhanced capacity for transforming energy into biomass was not one of them.

Summary

- Cell volumes vary by approximately eleven orders of magnitude across the Tree of Life, with most being in the range of 10^{-3} to 10^8 μm^3. Although most prokaryotic cells are < 10 μm^3 in size, a few giants exceed 10^9 μm^3, and a few unicellular eukaryotes with complex morphology are orders of magnitude larger.

- Species-specific mean phenotypes for various traits often scale in a continuous manner with cell size, implying substantial constraints on evolutionary diversification. A central goal of evolutionary cell biology is to determine the degree to which such patterns are consequences of biophysical constraints, selective disadvantages of discordant combinations, and/or outcomes of a reduction in the efficiency of selection that increases with cell size.

- One of the most studied physiological traits is metabolic rate, which scales positively with cell volume in a nearly isometric fashion in unicellular species. Such behavior is inconsistent with the 2/3 or 3/4 power-law scaling often invoked in the literature for multicellular species. Despite their ease of acquisition, metabolic-rate measures provide little insight into the basic currency of natural selection. However, when combined with growth-rate data, they can yield information on the efficiency of the rate of conversion of food resources into growth and reproduction.

- The energetic costs of constructing and maintaining cells scale nearly isometrically with cell volume across the Tree of Life, despite the significant differences in cellular architectures between prokaryotes and eukaryotes. The ultimate biophysical/evolutionary constraints on the total costs per unit volume remain to be determined, but it is relevant that the caloric content of biomass exhibits little phylogenetic variation.

- Maximum cell-division rates scale positively with cell size among heterotrophic bacterial species, but negatively among eukaryotic heterotrophs. Similarly, there is a directional shift in the efficiency of conversion of energy to growth in heterotrophic bacteria versus unicellular eukaryotes, with growth efficiency being lowest in large eukaryotic cells, and highest in large prokaryotic cells.

- The precise mechanisms that define the upper limits to growth rate remain unresolved, but the case can be made that the negative scaling with cell size in eukaryotes is (at least in part) a consequence of the reduced efficiency of natural selection operating on growth-rate related mutations in organisms with progressively larger cell size. As cell size increases, the effective population size decreases, and a larger number of mild growth-rate-reducing mutations are free to drift to fixation.

- It is commonly asserted that the establishment of the mitochondrion released the host eukaryotic cell from a surface area:volume constraint, precipitating a bioenergetic revolution. However, a diversity of observations, ranging from the scaling of energetic traits with cell size to the anatomy and cellular content of mitochondria, ATP synthase, and ribosomes, are inconsistent with this hypothesis.

Foundations 8.1 The cost of building a cell

Cell-division rates ultimately depend on the rate of acquisition of energy necessary to build a new cell. Arguably, the best currency to use in such analyses is units of ATP usage, as the hydrolysis of phosphate bonds in the conversion of ATP to ADP (and in some cases, ADP to AMP, or GTP to GDP) delivers the vast majority of energy for cellular functions. In principle, with a solid enough understanding of biosynthetic pathways and the various inputs of cellular resources, one could calculate the total energy required to build a cell by summing over the demands for the replacement of proteins, nucleic acids, lipids, etc. However, energy transformation is not 100% efficient, cellular components turn over on timescales less than the life of a cell, and energy must be invested into additional maintenance functions. Thus, the total energy utilized by a cell before giving rise to two daughters must exceed the cost of producing the standing levels of cellular components. This total level of investment (maintenance plus construction) represents the net cost of building a cell.

Determining the quantities of interest here is generally difficult for cells growing in natural environments, as most heterotrophic organisms consume a variety of resources varying in energy content. Thus, most knowledge in this area is derived from the growth of microorganisms on a

defined medium, with a single limiting carbon/energy source that enters a metabolic pathway with well-understood ATP-generating properties. If the organism can be grown in a chemostat (Figure 8.3a), it is straightforward to calculate both the rate of cell division and the rate of substrate consumption, and therefore to obtain the ratio, i.e., the yield of cells per unit consumption.

Data derived from such analyses were the source of the information presented in Chapter 7 on yields of biomass per unit carbon consumption (Figure 7.8). However, as previously noted, the level of yield depends on the nature of the carbon source, so a more meaningful measure focuses on a secondary conversion to the yield per unit ATP hydrolysis. Such a measure is more generalizable, as it accounts for differences in energetic contents among alternative carbon sources.

A chemostat (Figure 8.3a) consists of a closed environment in which sterilized medium is pumped in at a defined rate, with resource-depleted effluent (including the cells suspended within it) being eliminated at the same rate. If such a system is seeded with a pure population of a microbe, after several rounds of cell division, the population size will reach a steady state defined by the flow rate and the nutrient concentration. At this point, the population will have grown to a density at which the cell-division rate b equals the dilution rate d (i.e., the flow rate divided by the culture volume). The rate of resource consumption per cell is then equal to the rate of loss of substrate (the flow rate times the concentration difference between the inflow and outflow) divided by the number of cells in the steady-state culture.

Joint insight into maintenance and growth requirements is acquired by culturing cells under different flow rates. Such treatment imposes different nutritional states, as low and high dilution rates lead to high and low population densities (and hence, low and high nutrient availabilities per cell). Assuming a constant rate of resource consumption per cell necessary for maintenance (C_M), the consumption rate (per unit time) at cell division rate b (equivalent to the dilution rate d) can be written as

$$C = C_M + (b \cdot C_G) \qquad (8.1.1)$$

where C_G measures the total growth-related consumption per cell division (Tempest and Neijssel 1984; Russell and Cook 1995). From a fitted least-squares regression of observed consumption rates C against growth rates b, the intercept and slope respectively estimate the cellular requirements for maintenance per unit time and growth per cell division, C_M and C_G (Figure 8.3b).

The total cost of producing a cell at any growth rate b can be obtained by multiplying the consumption rate C by the mean cell division time, which is equivalent to the reciprocal of the cell division rate, $1/b$,

$$C_T = (C_M/b) + C_G. \qquad (8.1.2)$$

Provided the assumption of a constant rate of basal metabolism independent of the growth rate is correct, this means that lifetime resource requirements are higher in slower-growing cells owing to the increased cumulative maintenance requirements under a longer lifespan. (The first term in Equation 8.1.2 scales with the cell-division time).

In the preceding formula, C_T and C_G are expressed in units of the amount of resources consumed per cell. Here, however, we are interested in defining C_T to be the number of ATP hydrolyses required to yield a new cell, so appropriate conversions need to be made. C_G is then defined as the total number of ATP hydrolysis equivalents consumed in the production of building blocks leading to an offspring cell (independent of time), and C_M as the number of ATPs utilized per cell per unit time for maintenance. The quantity $1/C_G$ is often denoted as Y_{max}, as it represents the yield of cells per unit resource consumption that would occur in the absence of basal cellular requirements.

Although the general approach just taken assumes that metabolic requirements are constant, independent of the rate of growth, alternative formulations have been developed for the situation in which there is an additional metabolic cost to growth (Tempest and Neijssel 1984; Wieser 1994; Russell and Cook 1995). Note that if maintenance costs are linearly related to the growth rate, this additional contribution is simply contained within the term C_G in Equation 8.1.1, but in principle, a term that is a non-linear function (e.g., a quadratic) of b can be included. Although such alternative expressions can yield somewhat different interpretations on how energy is partitioned as a function of the growth environment, the total energy requirement observed at any particular growth rate remains unambiguous.

Foundations 8.2 Connecting metabolic rates with growth potential

The power of the preceding (Pirt plot) approach is that it can be applied to any sort of energy source, provided proper conversion to the level of ATP equivalence can be made. However, aside from the laborious nature of maintaining chemostats, the challenges become almost insurmountable with organisms requiring a complex diet, as is the case for the heterotrophic eukaryotes that make a living by consuming other cells. The problem here is that food resources consisting of undefined molecular mixtures cannot be converted to ATP equivalents based on basic biochemical principles.

For this reason, we turn to a second approach, motivated by the potential utility of joint information on metabolic rate and growth rate in unicellular species. If one has measures on both the rate of oxygen consumption (M, numbers of O_2 molecules consumed/cell/time) and the rate of cell division (R, the inverse of the cell doubling time), M/R provides an estimate of the number of O_2 molecules consumed per cell division. (Because metabolic rates are generally recorded as volumetric consumption rates, to obtain M, the appropriate conversion of units needs to be made. For example, at standard biological temperatures and pressures, 1 L (litre) $O_2 \simeq 0.045$ mol O_2, and multiplying by Avogadro's number yields 2.7×10^{22} molecules of O_2/litre). Basic observations from biochemistry (Foundations 17.2) indicate that the number of ATPs produced per oxygen atom consumed (the so-called P:O ratio; Mookerjee et al. 2017) $\simeq 2.5$, so under steady-state

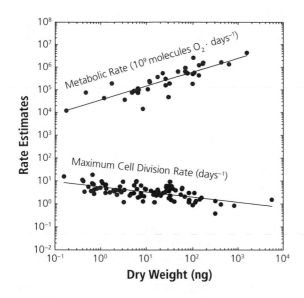

Figure 8.9 Allometric regressions of maximum growth rate and metabolic rate of ciliates, with all values normalized to a temperature of 20°C. From Lynch et al. (2022a).

assumptions, $5M/R$ provides an estimate of the number of ATP hydrolyses per cell division (as the number of ATPs per di-oxygen molecule, O_2, is 5).

A final key issue here is that $5M/R$ only represents the fraction of carbon resources allocated to energy production (θ), the remainder being used as carbon skeletons in biomass production. Analysis of the biosynthetic pathways of amino acids, nucleotides, and lipids leads to estimates of $\theta = 0.21$, 0.17, and 0.27, respectively (Chapter 17). As there are numerous, small additional ATP-consuming costs associated with biomass assembly, e.g., polymerization, protein folding, cargo delivery, mRNA replacement,

etc. (Lynch and Marinov 2015, 2017), and nucleotides comprise a minor fraction of total biomass, a reasonable first-order approximation for cell-wide θ is then 0.5, and dividing $5M/R$ by this leads to

$$C'_G \simeq \frac{10M}{R}, \qquad (8.2.1)$$

as an alternative measure of the total energetic requirements for daughter-cell construction (again, in units of ATP-hydrolysis equivalents). The utility of this approach is the lack of requirement for detailed knowledge of the nature of the food supply, which instead is reflected in the more easily observed metabolic rate.

As an explicit example of the utility of this approach, we consider a broad set of data for ciliate species with nearly a nearly 10,000-fold range of variation in cell size (Figure 8.9). The rate of cell division (per day) scales negatively with cell dry weight (B, in ng),

$$R \simeq 5.6B^{-0.22}, \qquad (8.2.2a)$$

whereas the metabolic rate (molecules O_2/cell/day) scales positively

$$M \simeq (3.4 \times 10^{13})B^{0.62}, \qquad (8.2.2b)$$

yielding

$$C'_G \simeq (6.1 \times 10^{13})B^{0.84}. \qquad (8.2.2c)$$

This can be compared to extrapolation from the regression based on Pirt-plot analyses (Equation 8.2b), which, after conversion of cell volume to cell dry weight, becomes

$$C_G \simeq (6.7 \times 10^{13})B^{1.05}. \qquad (8.2.3)$$

Thus, the two approaches give fairly similar results, although the Pirt-plot formula (to which only a single ciliate contributes) is closer to an isometric relationship.

Literature Cited

Abrahams, J. P., A. G. Leslie, R. Lutter, and J. E. Walker. 1994. Structure at 2.8 Å resolution of F1-ATPase from bovine heart mitochondria. Nature 370: 621–628.

Apol, M. E. F., R. S. Etienne, and H. Olff. 2008. Revisiting the evolutionary origin of allometric metabolic scaling in biology. Func. Ecol. 22: 1070–1080.

Atkinson, A. W. Jr., P. C. John, and B. E. Gunning. 1974. The growth and division of the single mitochondrion and other organelles during the cell cycle of *Chlorella*, studied by quantitative stereology and three-dimensional reconstruction. Protoplasma 81: 77–109.

Banavar, J. R., J. Damuth, A. Maritan, and A. Rinaldo. 2002. Supply-demand balance and metabolic scaling. Proc. Natl. Acad. Sci. USA 99: 10506–10509.

Bauchop, T., and S. R. Elsden. 1960. The growth of micro-organisms in relation to their energy supply. J. Gen. Microbiol. 23: 457–469.

Biselli, E., S. J. Schink, and U. Gerland. 2020. Slower growth of *Escherichia coli* leads to longer survival in carbon starvation due to a decrease in the maintenance rate. Mol. Syst. Biol. 16: e9478.

Biswas, S. K., M. Yamaguchi, N. Naoe, T. Takashima, and K. Takeo. 2003. Quantitative three-dimensional structural analysis of *Exophiala dermatitidis* yeast cells by freeze-substitution and serial ultrathin sectioning. J. Electron Microsc. (Tokyo) 52: 133–143.

Brown, J. H., G. B. West, and B. J. Enquist. 2005. Yes, West, Brown and Enquist's model of allometric scaling is both mathematically correct and biologically relevant. Func. Ecol. 19: 735–738.

Burger, J. R., C. Hou, and J. H. Brown. 2019. Toward a metabolic theory of life history. Proc. Natl. Acad. Sci. USA 116: 26653–26661.

Burton, M. D., and J. Moore. 1974. The mitochondrion of the flagellate, *Polytomella agilis*. J. Ultrastruct. Res. 48: 414–419.

Cadart, C., and R. Heald. 2022. Scaling of biosynthesis and metabolism with cell size. Mol. Biol. Cell 33: pe5.

Chaui-Berlinck, J. G. 2006. A critical understanding of the fractal model of metabolic scaling. J. Exp. Biol. 209: 3045–3054.

DeLong, J. P., J. G. Okie, M. E. Moses, R. M. Sibly, and J. H. Brown. 2010. Shifts in metabolic scaling, production, and efficiency across major evolutionary transitions of life. Proc. Natl. Acad. Sci. USA 107: 12941–12945.

Dodds, P. S., D. H. Rothman, and J. S. Weitz. 2001. Re-examination of the '3/4-law' of metabolism. J. Theor. Biol. 209: 9–27.

Fenchel, T. 2014. Respiration in heterotrophic unicellular eukaryotic organisms. Protist 165: 485–492.

Frank, S. A. 2016. The invariances of power law size distributions. F1000 Res. 5: 2074.

Glazier, D. S. 2015a. Body-mass scaling of metabolic rate: What are the relative roles of cellular versus systemic effects? Biology (Basel) 4: 187–199.

Glazier, D. S. 2015b. Is metabolic rate a universal 'pacemaker' for biological processes? Biol. Rev. Camb. Philos. Soc. 90: 377–407.

Gounand, I., T. Daufresne, D. Gravel, C. Bouvier, T. Bouvier, M. Combe, C. Gougat-Barbera, F. Poly, C. Torres-Barceló, and N. Mouquet. 2016. Size evolution in microorganisms masks trade-offs predicted by the growth rate hypothesis. Proc. Biol. Sci. 283: 20162272.

Grimes, G. W., H. R. Mahler, and R. S. Perlman. 1974. Nuclear gene dosage effects on mitochondrial mass and DNA. J. Cell Biol. 61: 565–574.

Groeneveld, P., A. H. Stouthamer, and H. V. Westerhoff. 2009. Super life – how and why 'cell selection' leads to the fastest-growing eukaryote. FEBS J. 276: 254–270.

Hatton, I. A., A. P. Dobson, D. Storch, E. D. Galbraith, and M. Loreau. 2019. Linking scaling laws across eukaryotes. Proc. Natl. Acad. Sci. USA 116: 21616–21622.

Hoehler, T. M., and B. B. Jørgensen. 2013. Microbial life under extreme energy limitation. Nat. Rev. Microbiol. 11: 83–94.

Huxley, J. S. 1932. Problems of Relative Growth. Methuen & Co. Ltd., London, UK.

Johnson, M. D., J. Völker, H. V. Moeller, E. Laws, K. J. Breslauer, and P. G. Falkowski. 2009. Universal constant for heat production in protists. Proc. Natl. Acad. Sci. USA 106: 6696–6699.

Kempes, C. P., S. Dutkiewicz, and M. J. Follows. 2012. Growth, metabolic partitioning, and the size of microorganisms. Proc. Natl. Acad. Sci. USA 109: 495–500.

Kimura, M., T. Maruyama, and J. F. Crow. 1963. The mutation load in small populations. Genetics 48: 1303–1312.

Kleiber, M. 1932. Body size and metabolism. Hilgardia 6: 315–353.

Kleiber, M. 1947. Body size and metabolic rate. Physiol. Rev. 27: 511–541.

Kozłowski, J., and M. Konarzewski. 2004. Is West, Brown and Enquist's model of allometric scaling mathematically correct and biologically relevant? Func. Ecol. 18: 283–289.

Kozłowski, J., and M. Konarzewski. 2005. West, Brown and Enquist's model of allometric scaling again: The same questions remain. Func. Ecol. 19: 739–743.

Kühlbrandt, W. 2015. Structure and function of mitochondrial membrane protein complexes. BMC Biol. 13: 89.

Lane, N. 2002. Power, Sex, Suicide: Mitochondria and the Meaning of Life. Oxford University Press, Oxford, UK.

Lane, N. 2015. The Vital Question. W. W. Norton & Co., Inc., New York, NY.

Lane N. 2020. How energy flow shapes cell evolution. Curr. Biol. 30: R471–R476.

Lane, N., and W. Martin. 2010. The energetics of genome complexity. Nature 467: 929–934.

Lennon, J. T., F. den Hollander, M. Wilke-Berenguer, and J. Blath. 2021. Principles of seed banks and the emergence of complexity from dormancy. Nat. Commun. 12: 4807.

Lücken, U., E. P. Gogol, and R. A. Capaldi. 1990. Structure of the ATP synthase complex (ECF1F0) of *Escherichia coli* from cryoelectron microscopy. Biochem. 29: 5339–5343.

Lynch, M. 2020. The evolutionary scaling of cellular traits imposed by the drift barrier. Proc. Natl. Acad. Sci. USA 117: 10435–10444.

Lynch, M., and G. K. Marinov. 2015. The bioenergetic costs of a gene. Proc. Natl. Acad. Sci. USA 112: 15690–15695.

Lynch, M., and G. K. Marinov. 2017. Membranes, energetics, and evolution across the prokaryote–eukaryote divide. eLife 6: e20437.

Lynch, M., and G. K. Marinov. 2018. Reply to Martin and colleagues: Mitochondria do not boost the bioenergetic capacity of eukaryotic cells. Biology Direct 13: 26.

Lynch, M., P. E. Schavemaker, T. J. Licknack, Y. Hao, and A. Pezzano. 2022a. Evolutionary bioenergetics of ciliates. J. Eukaryot. Microbiol. 65: e12934.

Lynch, M., B. Trickovic, and C. P. Kempes. 2022b. Evolutionary scaling of maximum growth rates with cell size. Sci. Rep. 12: 22586.

Lynch, M., and J. B. Walsh. 1998. Genetics and Analysis of Quantitative Traits. Sinauer Associates, Inc., Sunderland, MA.

Magalon, A., and F. Alberge. 2015. Distribution and dynamics of OXPHOS complexes in the bacterial cytoplasmic membrane. Biochim. Biophys. Acta 1857: 198–213.

Marañòn, E. 2015. Cell size as a key determinant of phytoplankton metabolism and community structure. Annu. Rev. Mar. Sci. 7: 241–264.

Marañón, E, P. Cermeño, D. C. López-Sandoval, T. Rodriguez-Ramos, C. Sobrino, M. Huete-Ortega, J. M. Blanco, and J. Rodriguez. 2013. Unimodal size scaling of phytoplankton growth and the size dependence of nutrient uptake and use. Ecol. Lett. 16: 371–379.

Marshall, D. J., M. Malerba, T. Lines, A. L. Sezmis, C. M. Hasan, R. E. Lenski, and M. J. McDonald. 2022. Long-term experimental evolution decouples size and production costs in *Escherichia coli*. Proc. Natl. Acad. Sci. USA 119: e2200713119.

Marshall, W. F. 2020. Scaling of subcellular structures. Annu. Rev. Cell Dev. Biol. 36: 219–236.

Milo, R., and R. Phillips. 2016. Cell Biology by the Numbers. Garland Science, New York, NY.

Moger-Reischer, R. Z., J. I. Glass, K. S. Wise, L. Sun, D. M. C. Bittencourt, B. K. Lehmkuhl, M. Lynch, and J. T. Lennon. 2023. Evolution of a minimal cell. Nature 620: 122–127.

Mookerjee, S. A., A. A. Gerencser, D. G. Nicholls, and M. D. Brand. 2017. Quantifying intracellular rates of glycolytic and oxidative ATP production and consumption using extracellular flux measurements. J. Biol. Chem. 292: 7189–7207.

Morono, Y., M. Ito, T. Hoshino, T. Terada, T. Hori, M. Ikehara, S. D'Hondt, and F. Inagaki F. 2020. Aerobic microbial life persists in oxic marine sediment as old as 101.5 million years. Nat. Commun. 11: 3626.

Munder, M. C., D. Midtvedt, T. Franzmann, E. Nüske, O. Otto, M. Herbig, E. Ulbricht, P. Müller, A. Taubenberger, S. Maharana, et al. 2016. A pH-driven transition of the cytoplasm from a fluid- to a solid-like state promotes entry into dormancy. eLife 5: e09347.

Okie, J. G. 2013. General models for the spectra of surface area scaling strategies of cells and organisms: Fractality, geometric dissimilitude, and internalization. Am. Nat. 181: 421–439.

Osafune, T., S. Mihara, E. Hase, and I. Ohkuro. 1975. Electron microscope studies of the vegetative cellular life cycle of *Chlamydomonas reinhardi* Dangeard in synchronous culture. III. Three-dimensional structures of mitochondria in the cells at intermediate stages of the growth phase of the cell cycle. J. Electron Microsc. (Tokyo) 24: 247–252.

Pellegrini, M. 1980. Three-dimensional reconstruction of organelles in *Euglena gracilis* Z. I. Qualitative and quantitative changes of chloroplasts and mitochondrial reticulum in synchronous photoautotrophic culture. J. Cell Sci. 43: 137–166.

Pirt, S. J. 1982. Maintenance energy: A general model for energy-limited and energy-sufficient growth. Arch. Microbiol. 133: 300–302.

Posakony, J. W., J. M. England, and G. Attardi. 1977. Mitochondrial growth and division during the cell cycle in HeLa cells. J. Cell Biol. 74: 468–491.

Rafelski, S. M., M. P. Viana, Y. Zhang, Y. H. Chan, K. S. Thorn, P. Yam, J. C. Fung, H. Li, L. da F. Costa, and W. F. Marshall. 2012. Mitochondrial network size scaling in budding yeast. Science 338: 822–824.

Raven, J. A. 1986. Physiological consequences of extremely small size for autotrophic organisms in the sea. In T. Platt and E. K. W. Li (eds.) Photosynthetic picoplankton. Can. Bull. Fish. Aquat. Sci. 214: 1–70. Ottawa.

Raven, J. A. 1994. Why are there no picoplanktonic O2 evolvers with volumes less than 10^{-19} m^3? J. Plank. Res. 16: 565–580.

Rosen, D., M. Edelman, E. Galun, and D. Danon. 1974. Biogenesis of mitochondria in *Trichoderma viride*: Structural changes in mitochondria and other spore constituents during conidium maturation and germination. Microbiol. 83: 31–49.

Russell, J. B., and G. M. Cook. 1995. Energetics of bacterial growth: Balance of anabolic and catabolic reactions. Microbiol. Rev. 59: 48–62.

Savage, V. M., B. J. Enquist, and G. B. West. 2007. Comment on 'A critical understanding of the fractal model of metabolic scaling'. J. Exp. Biol. 210: 3873–3874.

Stock, D., A. G. Leslie, and J. E. Walker. 1999. Molecular architecture of the rotary motor in ATP synthase. Science 286: 1700–1705.

Tarkington, J., and R. A. Zufall. 2021. Temperature affects the repeatability of evolution in the microbial eukaryote *Tetrahymena thermophila*. Ecol. Evol. 11: 13139–13152.

Tempest, D. W., and O. M. Neijssel. 1984. The status of YATP and maintenance energy as biologically interpretable phenomena. Annu. Rev. Microbiol. 38: 459–486.

Thompson, D. A. 1917. On Growth and Form. Cambridge University Press, Cambridge, UK.

Tyn, M. T., and T. W. Gusek. 1990. Prediction of diffusion coefficients of proteins. Biotechnol. Bioeng. 35: 327–338.

Uwizeye, C., J. Decelle, P.-H. Jouneau, S. Flori, B. Gallet, J.-B. Keck, D. Dal Bo, C. Moriscot, C. Seydoux,

F. Chevalier, et al. 2021. Morphological bases of phytoplankton energy management and physiological responses unveiled by 3D subcellular imaging. Nat. Comm. 12: 1049.

van Bodegom, P. 2007. Microbial maintenance: A critical review on its quantification. Microb. Ecol. 53: 513–523.

von Bertalanffy, L. 1957. Quantitative laws in metabolism and growth. Quart. Rev. Biol. 32: 217–231.

Ward, B. A., E. Marañòn, B. Sauterey, J. Rault, and D. Claessen. 2017. The size dependence of phytoplankton growth rates: A trade-off between nutrient uptake and metabolism. Am. Nat. 189: 170–177.

West, G. 2017. Scale: The Universal Laws of Growth, Innovation, Sustainability, and the Pace of Life in Organisms, Cities, Economies, and Companies. Penguin Press, London, UK.

West, G. B., J. H. Brown, and B. J. Enquist. 1997. A general model for the origin of allometric scaling laws in biology. Science 276: 122–126.

West, G. B., J. H. Brown, and B. J. Enquist. 1999. The fourth dimension of life: Fractal geometry and allometric scaling of organisms. Science 284: 1677–1679.

West, G. B., W. H. Woodruff, and J. H. Brown. 2002. Allometric scaling of metabolic rate from molecules and mitochondria to cells and mammals. Proc. Natl. Acad. Sci. USA 99 Suppl. 1: 2473–2478.

Wieser, W. 1994. Cost of growth in cells and organisms: General rules and comparative aspects. Biol. Rev. 69: 1–33.

Wiser, M. J., N. Ribeck, and R. E. Lenski. 2013. Long-term dynamics of adaptation in asexual populations. Science 342: 1364–1367.

Yamaguchi, M., Y. Namiki, H. Okada, Y. Mori, H. Furukawa, J. Wang, M. Ohkusu, and S. Kawamoto. 2011. Structome of *Saccharomyces cerevisiae* determined by freeze-substitution and serial ultrathin-sectioning electron microscopy. J. Electron Microsc. (Tokyo) 60: 321–335.

Zeuthen, E. 1953. Oxygen uptake as related to body size in organisms. Quart. Rev. Biol. 28: 1–12.

CHAPTER 9

Cell Growth and Division

The central mission of all cells – to survive and reproduce – is a product of the relentless operation of natural selection. For unicellular organisms, the matter of cellular reproduction naturally brings us into contact with the issue of cellular growth. Typically, cells double in size and then reproduce by binary fission, although there are cases in which offspring and adult cell sizes differ by more than twofold, e.g., budding in some yeast, and multiple internal fissions in some algae. The essential issue is that continuous proliferation of a population requires the growth and division of individual cells, which requires the intake and conversion of nutrients to biomolecules. Here, we focus on several general challenges that exist for any growth mechanism, deferring the molecular details on resource uptake and cell fission to Chapters 10 and 18, respectively.

First, cell growth requires coordination between the intake of resources and their conversion into cellular material. Even the simplest of cells consist of thousands of types of molecules, so the overall process may seem hopelessly beyond mechanistic interpretation. However, some aspects of cell growth can be understood in general terms using models incorporating a minimum level of molecular complexity.

Second, in relating growth in cellular biomass to the matter of reproduction, the issue arises as to how a cell decides when to undergo fission. In principle, cells might simply divide after a critical time period has passed, although this would require slowing the clock down in nutrient-poor environments. Alternatively, division might be delayed until a critical cell size (possibly environmentally determined) is reached. Still another possibility is that the license to divide is based on the attainment of a specific growth increment, in which case the size at division would be defined by the prior size at birth. Regardless of the target criterion, cells must generally possess compensatory mechanisms to prevent runaway growth or diminution in size in extreme individuals.

Third, cell division is not a perfect process. Some size variation among sister cells always results from binary fission, and this is inevitably accompanied by variance in the partitioning of the parental cell contents. Entirely a consequence of the limits to the perfection of cell-division mechanisms, such stochasticity generates phenotypic variation even in otherwise genetically uniform populations, and at a level that is potentially much higher than in multicellular species. Some have argued that the production of phenotypic variation has been promoted by natural selection as a bet-hedging strategy to cope with heterogeneous environments. However, as discussed below, non-genetic sources of phenotypic variation reduce the efficiency of natural selection and impose long-term fitness loads, leaving many open questions on this matter.

The following pages attempt to summarize what we know about these issues. Remarkably, however, the depth of understanding of the mechanistic determinants of the emergent properties of size and growth rate lags that for many of the more intricate and lower-level molecular features of cells.

Ribosomes and Cell Growth

Given that protein constitutes a large fraction of cell mass, before considering the more quantitative aspects of growth, an overview of the molecular machine dedicated to protein production is in order. Cells make enormous investments in ribosomes, with up to 50% of all transcription being devoted

Evolutionary Cell Biology. Michael Lynch, Oxford University Press. © Michael Lynch (2024). DOI: 10.1093/oso/9780192847287.003.0009

to the production of ribosomal RNA (rRNA, the catalytic heart of the ribosome) and up to 50% of messenger RNA (mRNA) production allocated to ribosomal proteins (Warner et al. 2001). Each ribosome can process only one mRNA at a time, and ribosomes are energetically expensive to produce, so strong regulatory associations between cellular growth rates and the number of ribosomes per cell can be expected. Overly low numbers of ribosomes relative to the cellular supply of nutrients compromises the rate of production of cellular biomass, but excess investment in ribosomes diverts energy from other cellular processes essential to resource acquisition.

Consistent with this expectation, for all species in which the issue has been addressed, there is a strong and essentially linear relationship between cell growth rate and the ratio of total RNA to total protein mass in the cell (Figure 9.1a). In other words, there is a predictable shift in the molecular contents of cells as they are exposed to more nutrient-rich environments. Generally, the RNA/protein mass ratio is in the range of 0.1 to 0.2 at low growth rates, and then increases to ~0.5 or even more in the fastest growing cells. These types of responses are retained when the growth-rate differences are created by varying the types of substrates (as opposed to altering the concentration of a single limiting nutrient) (Schaechter et al. 1958; Fraenkel and Neidhardt 1961). Thus, the level of RNA production is regulated by an indirect connection with the growth rate itself, rather than by direct resource-specific signals.

Although the patterns illustrated in Figure 9.1a refer to the total RNA in a cell, additional data suggest a coordinated response for mRNAs, tRNAs, and rRNAs, such that the number of ribosomes per cell also scales directly with the cellular growth rate. For most species that have been examined, the ratio of rRNA to total RNA in cells falls in the range of 0.55 to 0.88, typically not deviating by more than 0.15 between different growth rates (Figure 9.1b). Thus, with increasing nutrient availability, the number of ribosomes per cell increases in a coordinated way with the growth rate.

Such proportionality arises from various feedback mechanisms. Ribosome biogenesis is often controlled indirectly by the level of free rRNA in the cell, the production of which is in turn regulated via the amount of uncharged tRNAs (Liu et al. 2015). In *E. coli*, for example, an alarmone (guanosine tetraphosphate or ppGpp) is produced when uncharged tRNAs (transfer RNAs unattached to amino acids) accumulate in the face of an inadequate supply of amino acids, and the reduced translational rate suppresses rRNA production (Potrykus et al. 2011; Wu et al. 2022). When ribosomal proteins are in excess in the cell relative to the rRNAs with which they assemble, the former bind to their own mRNAs, thereby repressing their own production. In contrast, in the soil bacterium *Bacillus subtilis*, inhibition of rRNA production results from a drop in cellular GTP levels (a result of enhanced incorporation of GTP into ppGpp) (Krásný and Gourse 2004). A variety of other mechanisms exist in eukaryotes (Warner et al. 2001; Parenteau et al. 2019). In the following pages, we encounter many other examples of evolutionary wandering of mechanisms regulating otherwise highly conserved cellular functions.

A relatively simple theoretical argument potentially explains the linear relationship between investment in ribosomes and growth rate (Foundations 9.1). Assuming that all but a small fixed fraction of ribosomes is actively engaged in translation and that active ribosomes are generally saturated with mRNAs, the overall growth rate can only be enhanced by increasing the translation rate per ribosome and/or the number of ribosomes per cell. For species with available data, translation rates per ribosome generally change by no more than a factor of two over a scale in which the cell growth rate varies by up to two orders of magnitude (Figure 9.1c). Thus, elevated investment in ribosomes appears to be the dominant mechanism by which cells increase rates of protein production.

Under this view, a plot of the mass ratio of ribosomal protein to total protein against the cell growth rate (Equation 9.1.3) has a specific biological meaning – the y intercept is a measure of the investment in inactive ribosomes relative to the total pool of proteins, and the inverse of the slope is a measure of the ratio of protein mass produced per mass of ribosomal protein. Figure 9.1a provides such a plot for *E. coli*, except that the y-axis values need to be multiplied by 0.53 to convert to the ribosomal protein

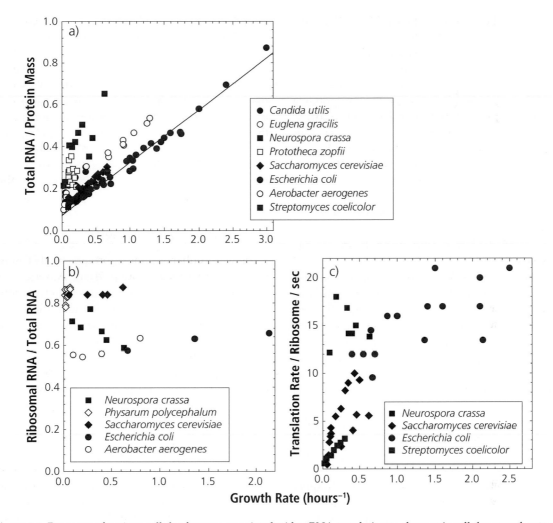

Figure 9.1 Response of various cellular features associated with mRNA translation to changes in cellular growth rate. References: *Candida utilis* (Brown and Rose 1969); *Euglena gracilis* (Freyssinet and Schiff 1974); *Neurospora crassa* (Alberghina et al. 1975); *Physarum polycephalum* (Plaut and Turnock 1975); *Prototheca zopfii* (Poyton 1973); *Saccharomyces cerevisiae* (Boehlke and Friesen 1975; Waldron and Lacroute 1975; Bonven and Gullov 1979; Metzl-Raz et al. 2017); *Escherichia coli* (Rosset et al. 1964; Forchhammer and Lindahl 1971; Dennis and Bremer 1974; Young and Bremer 1976; Scott et al. 2010; Zhu et al. 2016); *Aerobacter aerogenes* (Fraenkel and Neidhardt 1961; Tempest et al. 1965).

ratio for this species (Scott et al. 2010). Although *E. coli* will be used as an exemplar in the following analyses, this species has a distinctly lower intercept and slope for the response plot than in other species, meaning that *E. coli* achieves a maximum growth rate with a relatively low investment in RNA (and presumably ribosomes).

The total investment in ribosomal proteins in *E. coli* can be obtained by multiplying the total RNA/total protein mass ratio by the average

rRNA/total RNA ratio of 0.62 from Dennis and Bremer (1974), and then by the ratio of ribosomal protein to rRNA mass of 0.53. From Figure 9.1a, this leads to the conclusion that $\sim 3\%$ of the protein in a non-growing *E. coli* cell is associated with ribosomes. This is in reasonable agreement with a more direct estimate of $\sim 8\%$ associated with non-translating ribosomes in budding yeast (Metzl-Raz et al. 2017). Although little work has been done on the matter, in *E. coli*, and likely other bacteria, ribosomes dimerize and become translationally quiescent under nutritionally starved states (Vila-Sanjurjo 2008; Yoshida and Wada 2014). A non-zero reserve at near-zero growth rate is not too surprising, as complete ribosome loss sentences a cell to death by eliminating any possibility of responding to improved nutrient conditions (Mori et al. 2017).

The fractional investment in ribosomal proteins increases to 28% for *E. coli* cells growing at maximum rate, and when the total amount of accessory proteins associated with translation is added in, this number needs to be multiplied by ~ 1.7 to determine the total investment in the process of translation (Scott et al. 2010). Thus, a rapidly growing *E. coli* cell devotes nearly 50% of its protein to translation.

Although the data are less extensive, eukaryotes have higher total RNA/total protein and rRNA/total RNA mass ratios than bacteria (Figure 9.1a,b), as well as higher numbers of ribosomal proteins per ribosome (Chapter 6). This means that, to achieve equivalent growth rates, eukaryotic cells must make an even larger fractional investment in ribosome production than prokaryotes.

The results in Figure 9.1 can be used to estimate the absolute upper bound on the growth rate by considering the expected division time of a hypothetical cell consisting of nothing but translational machinery. The inverse of the slope in Figure 9.1a implies a rate of protein mass produced per unit ribosome mass per hour of 7.5. In other words, a healthy *E. coli* ribosome can replace its own protein constituents in about $60/7.5 = 8$ minutes.

The upper limit to the growth rate can also be calculated more directly by considering the number of amino acids per ribosome and the upper bound to the rate of translation (again, assuming that the cell consists of nothing but actively engaged translation machinery). The full set of bacterial ribosomal proteins comprising an individual ribosome contains $\sim 7,500$ amino acids, and the upper bound to the translation rate is ~ 20 amino acids/ribosome/second (Figure 9.1C). If one then liberally assumes that an extended ribosome (the complete translational machinery) contains twice as many amino acids as the ribosome itself, then the time required for complete replacement of an extended ribosome is $(2 \cdot 7500)/20 = 750$ seconds, or 12.5 minutes, in good agreement with the prior calculation (which ignored accessory proteins).

Thus, without an increase in the rate of translation or a decrease in the size of an extended ribosome, the cell-division time in an *E. coli*-like bacterium cannot be reduced below ~ 12.5 minutes, indicating that the massive cost of the ribosome itself imposes a significant limit on the rate of cell division. Under optimal growth conditions, some bacteria (e.g., *Vibrio natriegens*) have doubling times very close to this ultimate limit (Chapter 8).

Models for Cellular Growth

Natural selection promotes phenotypes that maximize the rate of entry of progeny into the subsequent generation, which requires both reproduction and survival. Here, we consider the issues in a very general sense, with an initial focus on simple expressions for the response of cell division rates to the concentration of a limiting nutrient, e.g., glucose for a laboratory-grown bacterium, or phosphorus for a planktonic alga. We then explore how cell size and division time are set and interrelated.

Below, we discuss how, even in a constant environment, substantial variation typically exists in the division times of individual cells, owing to internal stochastic processes. Nonetheless, an ensemble of cells can be described by the average population-level rate of increase r. Letting N_0 and N_t denote population sizes at two points in time, then assuming constant conditions,

$$N_t = N_0 e^{rt} \tag{9.1}$$

describes the trajectory of numbers of individuals over this period (Foundations 9.1). Defined in this way, r is a measure of the per-capita exponential growth rate (with units of time^{-1}). Taking logarithms and rearranging,

$$r = \frac{\ln N_t - \ln N_0}{t}. \tag{9.2}$$

The doubling time for population size, obtained by setting $N_t/N_0 = 2$, is

$$t_d = \ln(2)/r. \tag{9.3}$$

Like interest in a bank account, the doubling time of $\simeq 0.693/r$ is less than expected under linear growth.

The preceding expressions apply to the special situation in which a population is expanding in a nutritionally constant environment, but of course, no population can grow exponentially for an indefinite period of time. In more general applications in population biology, r is usually used to describe the actual rate of population growth, which reflects the net difference between birth and death rates, $r = b - d$. In this chapter, however, the focus is primarily on laboratory cultures, where there is typically very little direct cell death. In that case, r can be viewed as the rate of cell birth ($r \simeq b$ because $d \simeq 0$). With a constant steady-state distribution of cell sizes at division, b is equivalent to the exponential rate of increase in cellular biomass (Jun et al. 2018). In a laboratory chemostat where cells are being regularly drawn off (Figure 8.3), the birth rate can be kept indefinitely at a constant level equal to the dilution rate, d (because $r = b - d = 0$ within the growth chamber).

From observations on bacteria grown under constant conditions, Monod (1949) concluded that the growth-rate response to a limiting nutrient concentration (S) can be described by a simple hyperbolic function,

$$r = r_{max}\left(\frac{S}{K_r + S}\right), \qquad (9.4a)$$

where r_{max} is the maximum rate of growth (asymptotically approached as $S \rightarrow \infty$), and K_r is the half-saturation constant for growth (equivalent to the resource concentration at which $r = r_{max}/2$). As discussed in Chapter 18, this formula is identical in form to the commonly employed Michaelis–Menten equation for nutrient uptake and other enzymatic reactions,

$$u = u_{max}\left(\frac{S}{K_u + S}\right), \qquad (9.4b)$$

where K_u is the half-saturation constant for uptake, which is not necessarily equal to K_r.

Numerous other models have been proposed to link growth rate to nutrient availability. For example, with a focus on algal cells in continuous culture, Droop (1973, 1974) considered a construct in which the growth rate depends on the *internal* cellular concentration of the limiting nutrient (Q, commonly referred to as the cell quota),

$$r = r_{max}\left(1 - \frac{\phi}{Q}\right). \qquad (9.5)$$

Under this model, cell division ceases when Q drops below the critical internal concentration ϕ, and r asymptotically approaches the maximum possible value r_{max} as the internal nutritional state Q increases. An attractive feature of this expression is that cell growth is more naturally connected with internal than external nutrient levels. Although internal nutrient pools are not necessarily easy to estimate, measures of r and Q in cultures of single species of phytoplankton have repeatedly supported the general form of Equation 9.5 (Figure 9.2).

Despite its different functional underpinnings, the structure of Equation 9.5 is entirely compatible with the Monod growth equation. This can be seen by noting that for a steady-state system, the rate of nutrient uptake must equal the product of the cell quota and the rate of cell growth, i.e., $u = r \cdot Q$, which implies a cell quota equal to the ratio of rates of uptake and growth, $Q = u/r$. Substituting this expression into Equation 9.5 and rearranging yields

$$r = r_{max}\left(\frac{u}{(r_{max}\phi) + u}\right), \qquad (9.6)$$

which again has the form of a hyperbolic relationship, in this case between r and the rate of nutrient uptake. If Equation 9.4b is substituted for u here, a more complex expression is obtained in terms of S and the uptake parameters, but this is still hyperbolic with respect to the external nutrient concentration S, recovering the form of Equation 9.4a.

Equations 9.4a and 9.5 have been used to describe thousands of microbial growth responses and are often referred to as growth laws. However, the models are phenomenological in the sense that they do not explicitly describe any of the underlying cellular mechanisms connecting substrate uptake, utilization, and growth. They simply describe general growth responses to nutrient limitation with a minimum amount of detail. More complex models have been proposed. For example, Maitra and Dill (2015) and Weiße et al. (2015) presented formulations that include ribosomes, other RNAs, protein, and ATP as the underlying variables, in both cases

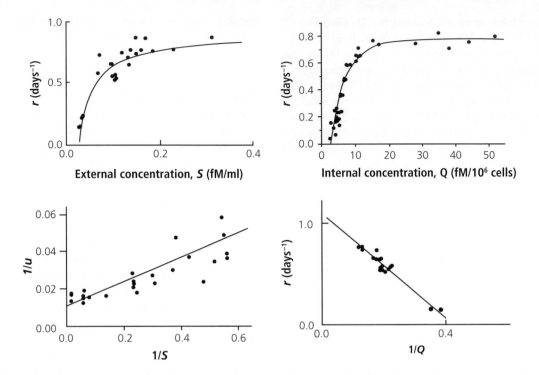

Figure 9.2 Various growth (r) and nutrient uptake (u, units of pg/10^6 cells/day) responses of the marine planktonic chrysophyte alga *Monochrysis lutheri* to various concentrations of vitamin B_{12}, recorded for populations of cells grown in a chemostat. The upper left and right graphs provide the relationships between the cell division rate and external and internal nutrient concentrations. The lower right graph illustrates the inverse linear relationship between r and $1/Q$, as predicted by Equation 9.5. For the lower left panel, the straight line relationship for the rate of nutrient uptake is derived from Equation 9.4b, which rearranges to $(1/u) = (1/u_{max}) + (K_u/u_{max})(1/S)$. The data are from Droop (1973, 1974, 1983).

generating predictions that are consistent with the Monod growth model and ribosome-growth coupling noted above. Models with an intermediate level of complexity, describing r_{max} and K_r in mechanistic terms associated with the translational capacity of ribosomes and the nutritional capacity of the environment are outlined in Foundations 9.2. These provide a satisfying explanation for the response of ribosome investment to increased nutrient availability noted in the preceding section.

Control of Cell Size

Chapter 8 discussed how cell volumes vary by approximately eleven orders of magnitude among unicellular species. Within-species variation also exists as a consequence of prevailing environmental conditions, stochastic variation in cell volume arising during division, and position in the cell cycle (age variation). Nonetheless, under any particular

environmental setting, the range of cell sizes within a species is generally fairly narrow, with standard deviations well below the mean. This implies the existence of homeostatic mechanisms for cell-size regulation.

Under constant conditions, the average rate of increase in cell size between divisions must equal the average rate of increase in cell number, i.e., the rate of cell division. If this were not the case, cell size would become progressively smaller or larger. In other words, at steady state, cells must double in size at the same rate that the population doubles in cell number. This, however, leaves open the possibility of a diversity of patterns of biomass growth within the life span of a cell. Resolving this issue is critical to understanding how cell size and division time are jointly determined.

As outlined above, the numbers of cells within expanding populations kept at constant environmental conditions increase exponentially in time. If the biomass of individual cells grew in a parallel manner, under steady-state conditions, cell volume would grow in accordance with Equation 9.1,

$$V_t = V_0 e^{rt}, \qquad (9.7)$$

where V_0 is the size of a new-born cell, and V_t is its size t time units later. Under this model, the proportional rate of change in cell volume is independent of cell size, although larger cells grow more rapidly in an absolute sense.

Exponential growth specifically implies that the metabolic features of growing cells remain constant, independent of size, with the ensemble of constituent molecules operating via a fixed set of reaction rates per unit cytoplasmic volume. However, exponential growth in cell size is not essential for balanced growth. The only requirement is that cumulative cellular biomass doubles from birth to death, i.e., matches the rate of increase in cell number. In principle, growth might be linear, with the rate of acquisition of biomass being independent of cell size, or sigmoidal, with the rate of cell growth initially accelerating and then decelerating as a critical size is approached.

Nonetheless, observations on the growth of individual cells for multiple bacterial species support the exponential cell-growth model (or something very close to it) (Voorn and Koppes 1998; Santi et al. 2013; Campos et al. 2014; Iyer-Biswas et al. 2014; Osella et al. 2014; Susman et al. 2018), with no known striking exceptions. This model also extends to eukaryotes. Godin et al. (2010) and Bryan et al. (2010) observed exponential growth not only in the bacteria *E. coli* and *B. subtilis*, but also in the yeast *S. cerevisiae* and mouse lymphoblast cells, and similar observations have been made on human osteosarcoma cells (Mir et al. 2011) and in the ciliate *Paramecium tetraurelia* (Kimball et al. 1959).

In terms of cell-size homeostasis, however, there remains a problem. Owing to stochasticities arising during division, not all cells have exactly the same size at birth. How then are the sizes of consecutive cells produced within a lineage regulated so as to prevent overly small/large cells from spawning ever more extreme descendants? If cells simply grew exponentially for a specified time before division, following a timer model, those that were larger at birth would grow more over the specified

duration, leading to a potentially runaway size distribution (Figure 9.3). Under an alternative sizer model, cells might be programmed to divide once a critical volume is reached.

For the best-studied organism, *E. coli*, both of these models come up short. Instead, for a given environment, cells appear to add an approximately constant volume (Δ) prior to division (Taheri-Araghi et al. 2015) (Figure 9.4). This adder model leads to a simple mechanism of cell-size homeostasis, with the steady-state expected offspring size being equal to Δ. Contrary to the sizer model, the adder model predicts that larger new-born cells will divide at larger sizes (with expected volume $V_0 + \Delta$), in effect being oblivious to their initial size. Nonetheless, with Δ remaining independent of size, the adder model also predicts that cells with extreme size will produce descendants progressively returning to the expected size Δ. The simple basis for such homeostasis can be seen by the following argument.

If a new-born cell is larger than Δ, say by an amount v, then cell division will occur at expected size $(v + \Delta) + \Delta = v + 2\Delta$, and the average offspring size will be half that, $(v/2) + \Delta$, and hence shifted back towards the long-term expected value Δ by an amount $v/2$. The opposite (a shift towards larger offspring size) occurs if an offspring cell happens to be slightly smaller than Δ. In both cases, the deviations from the expected newborn size decline over time, insuring rapid convergence back to Δ. These arguments ignore new deviations that arise at each subsequent division, and the actual damping process is less smooth than this simple description implies (Tanouchi et al. 2015; Susman et al. 2018). Nonetheless, the overall condition of homeostasis is retained.

Figure 9.4 shows how the behavior of *E. coli* cells at the extreme ends of the size spectrum deviates slightly from the expectations of this simple adder model, which assumes independence between the parental cell size and the subsequent growth increment. Instead, the latter appears to decline with increasing cell size. Such second-order effects can be accommodated by a simple modification of the growth-increment expression (Delarue et al. 2017; Jun et al. 2018; Susman et al. 2018). Consider, for example, the situation in which there is some memory of parental cell size at birth (V_p) such that the predicted offspring size is $V_0 = \alpha V_p + \Delta$. Setting

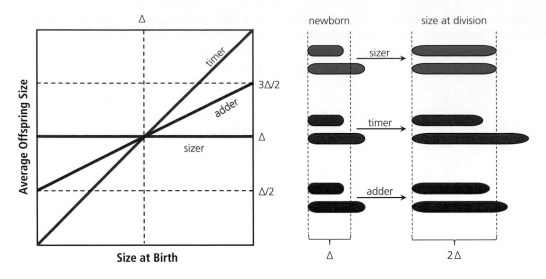

Figure 9.3 Expected relationship between cell volume at birth and at subsequent cell division under three alternative growth models. Δ serves as a reference point. Under the sizer model, regardless of the size at birth, the size at division always returns to 2Δ. Under the simplest form of the timer model, because of exponential growth, all cells grow by the same factor over a given duration, so individuals of extreme size produce offspring that are just as extreme; if there are stochastic deviations in offspring size or growth rates, the size distribution will diverge. Under the adder model, cells above or below the average still produce deviant progeny, but the average size of progeny is less extreme than that of parental cells, resulting in convergence of the cell-size distribution to an equilibrium.

$V_0 = V_p$ yields an equilibrium cell size of $\Delta/(1-\alpha)$, which implies homeostasis provided $-1 < \alpha < 1$. If $\alpha < -1$, cell size declines to zero, whereas $\alpha > 1$ leads to runaway cell growth.

This more general model accommodates a wide range of species. The one member of the archaea in which the growth properties have been investigated, *Halobacterium salinarum,* appears to adhere closely to the pure adder model (Eun et al. 2017). However, two bacterial species with asymmetric cell division, *Caulobacter crescentus* (Campos et al. 2014; Iyer-Biswas et al. 2014) and *Mycobacterium smegmatis* (Santi et al. 2013; Logsdon et al. 2017; Priestman et al. 2017) as well as the symmetrically dividing bacterium *Pseudomonas aeruginosa* (Deforet et al. 2015) have slightly positive values of $\alpha < 1$. In contrast, budding yeast *S. cerevisiae* (Di Talia et al. 2007; Soifer and Barkai 2014; Soifer et al. 2016; Chandler-Brown et al. 2017) and fission yeast *S. pombe* (Fantes 1977; Sveiczer et al. 1996) have slightly negative values of α.

Finally, the form of a growth model has implications for the response of cell-division time to size at birth. For example, under an adder model of cell-division, cells that are larger at birth divide at an earlier age because the additive increment Δ is achieved more rapidly. From the form of the exponential-growth model (Equation 9.7), the division time for a cell of initial size V_0 under the pure adder model is

$$t_d = \frac{\ln[1 + (\Delta/V_0)]}{r}, \qquad (9.8)$$

yielding a predicted decline in t_d with increasing V_0, consistent with observations in *E. coli* (Figure 9.4). We next discuss how, in a dynamically growing population, this may lead to an equilibrium mean offspring size > Δ, as offspring of large-size deviants will be promoted into the population at a higher rate than those of small-size deviants.

Molecular mechanisms of division-size determination

The simple models just outlined provide a statistical view of the features of cell division, but leave unanswered questions as to the molecular mechanisms enabling cells to determine when they have reached the critical threshold for division. Resolving these issues is essential to understanding how changes in

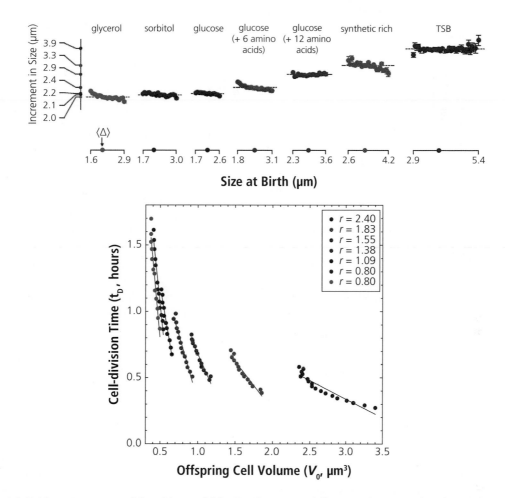

Figure 9.4 Evidence in support of the adder model for *E. coli* grown in different media. Proceeding from left to right, the growth media are increasingly rich sources of energy, carbon, and other nutrients. **Upper panel:** By looking at individual cells within each medium (at a constant concentration), it can be seen that the growth increments to maturity (Δ) are nearly independent of the size at birth. **Lower panel:** The negative relationship between the cell-division time and cell volume at birth becomes progressively stronger in media that support lower growth rates (r in units of hours^{-1}). In the upper and lower panels, size is presented as cell length and cell volume, respectively, although the two are directly proportional, given that *E. coli* cells are nearly cylindrical in shape. Note that the adder model only provides a first-order approximation, as cells at the extremes of the size range deviate from expectations of constant Δ (upper panel). From Taheri-Araghi et al. (2015).

cell sizes and division times might be accomplished by evolution.

One model for this decision-making process invokes a burst of cell-division inhibitor produced at the time of cell division, which then gradually becomes diluted as cell volume increases. An alternative model invokes the gradual build-up of an activator molecule to the point at which a threshold concentration is exceeded. Simple mathematical constructs exist to explain the features of these alternative models (Sompayrac and Maaløe 1973; Amir 2014; Deforet et al. 2015; Soifer et al. 2016). However, for species in which the molecular underpinnings of cell-division time have been sought, inhibitor mechanisms have generally come to the fore (Figure 9.5).

For example, the soil bacterium *Bacillus subtilis* determines the time of cell division by use of a

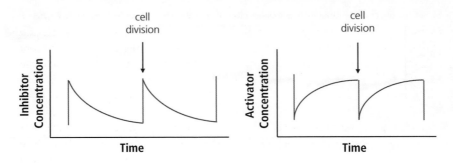

Figure 9.5 Two alternative models for the determination of cell size at division, presented in a generic fashion. On the left, a pulse of an inhibitor produced at the time of cell division experiences a progressive reduction in concentration as cell volume increases during growth; division occurs once the inhibitor concentration declines to a threshold level. On the right, an activator is continuously produced during cell growth, increasing in concentration until a critical level signals the time to division.

two-component system (Weart et al. 2007; Chien et al. 2012). The tubulin-like cell-division protein (FtsZ) acts as a central hub and has a nearly constant concentration under all nutritional conditions. At high nutrient conditions, an inhibitor molecule (UgtP) oligomerizes with FtsZ, preventing formation of the cytokinetic ring until a relatively large cell size dilutes UgtP and frees up FtsZ molecules. Under low nutrient conditions, UgtP is sequestered away from FtsZ, allowing division at a smaller cell size.

Escherichia coli utilizes a different inhibitor mechanism to determine the time of division. In this case, an inhibitor molecule oscillates back and forth between the cell poles, such that a minimum concentration exists at the cell midpoint. Once the inhibitor concentration drops below a critical point by growth dilution, cell division ensues (Lutkenhaus 2008).

Inhibitor mechanisms for division-time determination extend to yeasts. The fission yeast *S. pombe* utilizes a spatial gradient to sense its size – an activator of mitosis is centrally located, whereas an inhibitor of the activator has a gradient initiating at the cell poles; as the cell grows, the inhibitor concentration declines to the point at which mitosis is activated (Moseley et al. 2009). In contrast, in the budding yeast *S. cerevisiae*, a short burst of synthesis of a mitosis inhibitor is elicited shortly after cell division in a size-independent manner (Turner et al. 2012; Schmoller et al. 2015; Litsios et al. 2019).

Smaller cells, with a higher inhibitor concentration at birth, must then add more volume to reduce the inhibitor to its critical concentration to allow mitosis to proceed.

Remarkably, although all four of these systems rely on mechanisms of inhibition to determine cell-division time, the molecular details are essentially non-overlapping. As in the case of the regulation of ribosome biogenesis (Chapter 6), this implies that over evolutionary time the basic machinery dictating the key life-history features of cells – size and age of reproduction – has been rewired on multiple occasions. How such modifications are made without imperilling the fitness of individuals with intermediate states is unclear, and constitutes a major challenge for evolutionary cell biology. Regardless of the underlying evolutionary mechanisms, one should be wary of Jacob's (1998) proclamation that 'All that is true for *E. coli*, is true for the elephant.'

The described models invoke the monitoring of some sort of molecule to indirectly forecast the appropriate time for division, but do so in a way that is notably agnostic with respect to genome replication processes. A related mechanistic proposal, not necessarily unconnected with the processes noted above, is that bacterial cell division is somehow guided by the state of genome replication (Wallden et al. 2016; Amir et al. 2017; Jun et al. 2018; Si et al. 2019), as is the case with the eukaryotic cell cycle (Chapter 10). The idea here is that bacteria growing at rapid rates experience pileups of partially replicated nested genomes as the rate of genome replication lags behind the rate of production of the remaining cellular constituents. In one version of this model, the critical determining factor is the ratio between the number of genome origins of replication (the starting points of duplicating genomes) and cell volume (Jun et al. 2018; Si et al. 2019). This model fairly accurately fits the response of *E. coli* cell volume to cell growth rate, predicting an exponential response of cell size to growth rate, in accordance with some observations (discussed later). The model also predicts that the negative scaling of cell-division time and size at birth will become increasingly strong in nutrient-poor environments (Figure 9.4).

However, there are two caveats with respect to this model. First, any number of other underlying division-time determinants beyond the numbers of origins of replication (but highly correlated with them) might play a key role. Indeed, rather different models, with a focus on partitioning of resources between ribosomes and unspecified division proteins, fit the data just as well (Bertaux et al. 2020; Serbanescu et al. 2020), and additional work suggests that DNA replication and cytoplasmic growth jointly influence the time to division (Colin et al. 2021). Second, an origin-counting model cannot apply to eukaryotes, which always undergo a single genome replication per cell division.

Despite these rather unsettling uncertainties, these simple models (or variants of them) provide a clear path to understanding the molecular basis for evolutionary changes in cell size/division time via alterations in the concentrations and/or activities of the products of as few as two genes, e.g., an inhibitor molecule and its interacting partner. In principle, for example, larger cell size can be achieved by increasing the burst size of the mitotic inhibitor upon cell division or by reducing the sensitivity of its partner to inhibition.

Environmental determinants of cell size

Whatever the molecular mechanism of cell-size regulation, the key parameters associated with the system must vary with environmental conditions. For example, cell size typically increases with nutrient availability, which under the adder model implies an increase in Δ. Such patterns have been clearly documented in *E. coli* (Taheri-Araghi et al. 2015; Zheng et al. 2020), *Salmonella typhimurium* (Schaechter et al. 1958), and in many other bacteria (Jun et al. 2018). In *S. typhimurium* and *E. coli*, there is an ∼5-fold increase in cell volume over the full range of growth rates (Volkmer and Heinemann 2011; Si et al. 2017), whereas the full response in the photosynthetic cyanobacterium *Synechocystis* is a 1.5-fold increase in cell volume (Zavřel et al. 2019).

A positive relationship between cell volume and nutrient status has also been documented in unicellular eukaryotes. For example, the ciliate *Tetrahymena* exhibits a twofold increase in cell volume with nutrient availability (Zalkinder 1979), and the budding yeast *S. cerevisiae* (Tyson et al. 1979; Ferrezuelo et al. 2012) and the green alga *Chlorella pyrenoidosa* (Prokop and Ricica 1968) both have fivefold ranges. Given this near-universality of the positive physiological response of cell volume to nutrient availability, it too has often been christened as a 'growth law'.

What remains unclear is the extent to which this kind of cell size–growth rate relationship is driven by adaptive processes, i.e., whether increasing cell volume under high nutrient conditions somehow enhances the cell-division rate beyond that expected under constant cell size. Although it is certainly unlikely that such a universal response is maladaptive, an argument is made below as to how, under the adder model, cells phenotypically shifted to larger sizes might passively accumulate in cultures growing with higher rates of cell division.

Equally unclear is the degree to which the purely physiologically driven cell size–growth relationship associated with resource availability is related to the phylogenetic association between cell size and maximum growth rate outlined in Chapter 8 (Figure 8.5). On the one hand, the positive association between maximum cell-division rate and cell size among heterotrophic bacterial species is consistent with the developmental-plasticity pattern. Also consistent with this pattern is a long-term selection experiment for higher growth rate in *E. coli,* which yielded a parallel response in cell volume (Figure 9.6). On the

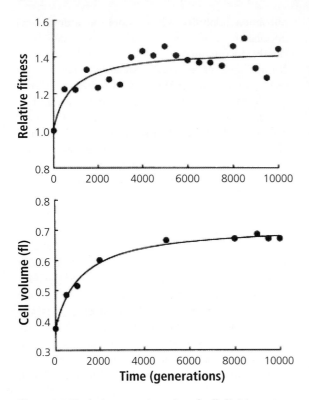

Figure 9.6 Evolutionary trajectories of cell-division rates and cell volume in a 10,000 generation experiment in which *E. coli* was subjected to persistent selection for higher growth rate. From Lenski and Travisano (1994).

other hand, among eukaryotic species, maximum cell-division rates decline with increasing cell size (Figure 8.5), contrary to within-species responses to a shift in nutrient availability (Figure 9.7a).

Why is there a conflict between responses at the evolutionary and physiological levels in eukaryotes, but not in prokaryotes? One possibility is that, despite retaining the physiological downshift in size under low-nutrient conditions, evolution of the maximum growth capacity of eukaryotic cells with increasing size is compromised owing to the reduction in efficiency of natural selection imposed by the increased power of random genetic drift (Chapter 8). By extension, if this hypothesis is correct, small- to moderately sized bacterial species should retain the flexibility to jointly evolve large cell size and high growth rates (Figure 9.6), whereas eukaryotic cells should be much more constrained.

Finally, we consider the effects of temperature, one of the most widely varying environmental parameters influencing cell physiology. Membrane fluidity, diffusion coefficients, and essentially all biochemical reaction rates (within the bounds of protein stability) increase with increasing temperature (Chapter 7). Given the positive association between cell size and growth rate in constant-temperature environments, one might expect that higher temperatures would promote both faster growth (within physiological limits) and larger cell volumes. Unfortunately, there is remarkably little information on this matter in prokaryotes, although in their seminal work, Schaechter et al. (1958) found that *Salmonella* cells grown at low temperature are substantially larger than those growing at identical rates (with lower nutrients) at higher temperatures. Their results suggest that temperature induces a different cell-size response to growth rate than does nutrient availability – to maintain a specific growth rate at lower temperatures, individual cells seemingly have to be larger (Figure 9.7b).

How generalizable is this sort of observation? Although there is a long history of thought on the relationship between organism size and temperature, the focus has mostly been on multicellular species. Here, the general idea is that organisms living in cooler environments have larger body sizes (within and among species), ostensibly because reduced surface area:volume ratios mitigate heat loss. This pattern has come to be known as Bergmann's rule (Bergmann 1847). Although its generality for multicellular organisms has been questioned (Riemer et al. 2018), the expected pattern does appear to hold for microbial eukaryotes, albeit likely for different reasons than proposed for homeothermic vertebrates.

In every study where the matter has been closely investigated, average cell volume declines with increasing temperature, while the growth rate increases. Such observations have been drawn from ciliates, flagellates, amoeboid heterotrophs, and diverse phototrophs, with the average response being an ∼ 25% decrease in cell volume accompanying a 10°C increase in temperature (Atkinson et al. 2003; Fu and Gong 2017; Zohary et al. 2020). If nothing else, such observations demonstrate that the

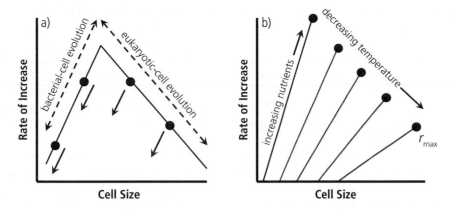

Figure 9.7 Idealized views of the response of cell growth rates to environmental effects. **a)** The solid black lines denote the phylogenetic relationship for the maximum rate of cellular growth, r_{max}, taken from Figure 8.5, where it is shown that the scaling is positive for bacteria but negative for eukaryotes. The four black dots denote hypothetical species, and the red arrows denote the universal reduction in cell size and growth rate in response to reduced nutrient supplies. **b)** The joint response of cell size and growth rate to changes in nutrient availability and temperature for an arbitrary genotype. The slopes and end points of the lines are arbitrarily placed, although it is known that r_{max} is reduced at lower temperatures (downward progression of dots), and the overall expectation is that when nutrient concentrations are manipulated so that cells are dividing at the same rates at different temperatures, cell size will be larger at the lower temperatures (rightward progression of coloured lines).

positive association between cell size and growth rate found in different nutritional environments is not generalizable to other environmental effects.

Again, whereas such a universal temperature response among diverse unicellular species might suggest the need for a general adaptive explanation for such behavior, the basis for such patterns remains unclear. Indeed, given the existence of size-selective predation and potential size-associated competitive interactions and physical–environmental effects, mortality rates are likely to be size dependent, so it is by no means clear that induction of large cell size is uniformly favorable in colder natural environments.

Is the physiological response to temperature change, running in the opposite direction to that induced by nutritional differences, an unavoidable by-product of the molecular mechanisms that set times to division? Future work in this area should look to the numerous experiments showing that when the translational capacity of ribosomes is compromised by chemical manipulation in *E. coli*, the phenotypic scaling between cell size and growth rate runs in opposite directions to the nutrient-based pattern (Scott et al. 2010; Jun et al. 2018;

Serbanescu et al. 2020). Perhaps the same underlying mechanism applies to temperature shifts.

Scaling of Intracellular Features

It is well known that various organs, tissues, and other body parts scale with body size as multicellular organisms grow (Thompson 1917), a phenomenon known as developmental allometry. Less clear is the extent to which internal cellular features (including transcript and protein numbers, organelle numbers and size, etc.) scale as cells grow from birth to maturity. A general positive relationship between absolute levels of cellular components and cell volume can be expected, as the organelles and molecular constituents of cells have functional roles whose total demands typically increase with the volume of cell. However, the expected quantitative patterns of scaling are less clear.

On the one hand, intracellular features might scale isometrically throughout growth (thereby keeping the intracellular environment relatively constant). Such scaling would be consistent with the exponential growth in cell volume noted earlier, which implies the maintenance of constant growth capacity per unit cell volume regardless of cell size. On the other hand, as cells grow and experience reductions in the surface area:volume ratio, the effective availability of nutrients per unit biomass might be reduced. If so, altered investments in the machinery associated with nutrient uptake and intracellular transport may be required, much like the responses of ribosome investment seen when cells are grown under different nutrient conditions.

For the few eukaryotic cellular traits with a modicum of data, isometric scaling with cell volume appears to be the norm. For example, in yeasts, mitochondrial volume constitutes $\sim 1\%$ of cell volume throughout life in *S. cerevisiae* (Rafelski et al. 2012), $\sim 10\%$ in *Candida albicans* (Tanaka et al. 1985), and $\sim 9\%$ in *Cryptococcus neoformans* (Mochizuki et al. 1998). Isometric scaling is also true in HeLa cells, with $\sim 10\%$ of the fractional volume of consisting of mitochondria (Posakony et al. 1977). Throughout growth in *Euglena gracilis*, the plastid constitutes $\sim 16\%$ and the mitochondrion $\sim 6\%$ of the total cell volume (Pellegrini 1980). Likewise, in the green alga *Chlorella fusca,* the volumetric contributions of plastids, mitochondria, and vacuoles remain nearly constant, at 40%, 3%, and 10% respectively (Atkinson et al. 1974). Total vacuole volume also scales nearly isometrically in *S. cerevisiae*, constituting $\sim 6\%$ of cell volume throughout the cell cycle (Chan and Marshall 2014; Chan et al. 2016). On the other hand, the essentially linear actin cables in yeast appear to scale with the length of the cell (McInally et al. 2021).

Compelling evidence for active cell volumetric control of organelle size derives from observations on the eukaryotic nucleus. In both *S. cerevisiae* and *S. pombe*, nuclear volume comprises a nearly constant ~ 6–8% of cell volume throughout cell growth (Jorgensen et al. 2007; Goehring and Hyman 2012). Transplants of nuclei from small to large cells reveal that the nucleus expands to the size expected given the host-cell volume. Such responses are not affected by the amount DNA in the nucleus, as they are even observed when DNA content is increased as much as 16-fold (Neumann and Nurse 2007). Similar responses have been seen in vertebrate cell cultures (Levy and Heald 2012), and it appears that the nuclear-to-cytoplasmic volume ratio is largely governed by a simple balance of osmotic pressures

associated with the numbers of protein molecules in the nucleus versus the cytoplasm (Deviri and Safran 2022; Lemière et al. 2022). Notably, across a wide range of prokaryotic species, the size of the nucleoid (the amorphous region of the genome, without nuclear envelopes or histone packaging of chromosomes) also grows nearly isometrically with cell volume within the growth cycle (Gray et al. 2019).

Although the molecular mechanisms underlying cytoplasmic composition homeostasis remain mostly unknown (Chan and Marshall 2010, 2012; Goehring and Hyman 2012; Brangwynne 2013), the emerging picture is that cells typically operate as bioreactors, with relatively constant internal compositions, until rapid remodelling takes place at the time of division. Still, this leaves open the question as to whether individual cellular features grow independently through time at roughly the same rate, or are somehow mutually guided via feedback associated with cell volume. These two alternative models make somewhat different predictions with respect to scaling relationships (Foundations 9.3). Passive homeostasis might simply arise from global changes in transcription rates in response to growth rate, thereby leading indirectly to coordinated assembly of subcellular compartments without the need for elaborate system-specific regulatory mechanisms.

It is also unclear how the ontogenetic patterns noted here relate to among-species scaling patterns observed at the phylogenetic level (Chapter 8). Returning to the questions relating to cell size and growth rate in the previous section, are the prevailing statistical relationships seen between pairs of characters during development recapitulated over evolutionary time with the divergence of phylogenetic lineages, or can evolution promote shifts in cellular composition in arbitrary directions? An organism's repertoire of developmental and phenotypic plasticities sets the range of phenotypic combinations that can be achieved and tested by natural selection prior to genetic change, so (in principle) genetic alterations that simply hardwire a plastic response into a constitutively expressed phenotype may provide a readily accessible route to multivariate evolution. This very old idea (Baldwin 1896; Waddington 1942) essentially suggests that natural selection typically exploits the paths of least resistance by genetically assimilating pre-existing possibilities. Such a view is not very different than the conventional vision of evolution as a process of descent with modification.

If this is the case for intracellular architecture, then the observations noted here suggest that iso-metric scaling of eukaryotic cell parts should also prevail at the phylogenetic level. Although the topic is largely unexplored, the kinds of phylogenetic scalings outlined in Chapters 7 and 8 provide com-pelling material for future investigation. Indeed, with its strong molecular basis, evolutionary cell biology provides a powerful platform for under-standing the mechanistic links (or lack thereof) between allometric scaling relationships at the onto-genetic (developmental), environmental (physio-logical), and phylogenetic levels.

Phenotypic Variation in Cell Size and Division Time

Although the preceding discussion has focused largely on the average behavior of cell-growth fea-tures, the stochasticity of events inherent in growth-related processes generates non-trivial variation in cell traits (Geiler-Samerotte et al. 2013). Sources of variation for cell size and growth rate include: 1) variation in birth size owing to imperfect par-titioning at cell division; 2) variation in numbers of ribosomes and of other critical molecules per cell, partly associated with variation in initial par-titioning, but also from subsequent events such as transcription and translation; 3) inaccuracies in the growth-increment target; and 4) extrinsic variation in the microenvironment.

Many attempts have been made to incorporate one or more of these factors into models of steady-state distributions of cell size and division time (e.g., Powell 1956; Scherbaum and Rasch 1957; Koch and Schaechter 1962; Tyson and Hannsgen 1985a,b; Taheri-Araghi et al. 2015; Jun et al. 2018). The math-ematical forms of cell-feature distributions are not uniformly agreed upon, and the details are not pur-sued here. However, it is worth noting that pre-dicted patterns are often closely related to formal distributions derived in the early days of statistics

for entirely different reasons. For example, the Yule (1925) distribution can be used to describe situations involving parallel (autonomous) growth of cellu-lar constituents, such that cell parts are duplicated during the cell-growth process independently with fixed probability per unit time, with cell division occurring at the time of duplication of the final part. In contrast, a Pearson type III distribution describes a situation in which cell division takes place only after the completion of a series of consecutive (inter-dependent) steps, with each initiated step com-pleted with a certain probability per unit time fol-lowing the exit from the preceding step (Kendall 1948). Although these models do not strictly incor-porate variability in size at birth, they do have features that are conceptually connected to the assumptions under the adder model, where a cer-tain amount of cellular biomass must accrue before the cell divides. They also generate skewed distribu-tions with long tails to the right, superficially similar to what is typically seen with real data (Figure 9.8).

Results from single-cell monitoring demonstrate that the magnitude of standing variation among genetically uniform cells is generally quite large. For example, observations from well-mixed labora-tory cultures of unicellular species commonly yield coefficients of variation (CVs, equal to the standard deviation divided by the mean) in the range of 0.1 to 0.5 for size at birth and maturity, incremental growth, and age at division (Table 9.1). As all of the studies in Table 9.1 involve single genotypes, the observed variance is due entirely to vagaries in the internal and external cellular environment.

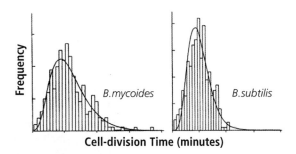

Figure 9.8 Pearson type III distributions fitted to data on the frequency distributions of cell-division times for two species of the bacterium *Bacillus*. From Powell (1958).

Table 9.1 Coefficients of variation (CV, standard deviation divided by the mean) for growth-related features of cells.

Species	Trait	CV	Reference
Bacteria:			
Aerobacter cloacae	Generation time	0.18	Powell 1958
Azotobacter agilis	Elongation rate	0.10	Harvey et al. 1967
	Generation time	0.22	Harvey et al. 1967
Bacillus mycoides	Generation time	0.48	Powell 1956
Bacillus subtilis	Generation time	0.54	Powell 1956
Bacterium aerogenes	Generation time	0.30	Powell 1956
Escherichia coli	Elongation rate	0.08	Taheri-Araghi et al. 2015
	Division length	0.14	Taheri-Araghi et al. 2015
		0.12	Harvey et al. 1967
	Birth length	0.16	Taheri-Araghi et al. 2015
	Generation time	0.21	Taheri-Araghi et al. 2015
		0.28	Harvey et al. 1967
		0.30	Kiviet et al. 2014
	Added length	0.24	Taheri-Araghi et al. 2015
Proteus vulgaris	Generation time	0.32	Powell 1956
Pseudomonas aeruginosa	Generation time	0.14	Powell 1958
Serratia marcescens	Generation time	0.17	Powell 1958
	Generation time	0.14	Tyson 1989
Streptococcus faecalis	Generation time	0.27	Powell 1956
Eukaryotes:			
Saccharomyces cerevisiae	Length of G1 phase	0.46	Di Talia et al. 2007
Schizosaccharomyces pombe	Division length	0.07	Tyson 1989
Tetrahymena pyriformis	Generation time	0.12	Scherbaum and Rasch 1957
	Division size	0.12	Scherbaum and Rasch 1957

Yet, CVs of this magnitude are substantially greater than those observed for morphometric traits in genetically variable samples of multicellular organisms, which are usually on the order of 0.05–0.10 (Lynch and Walsh 1998).

Owing to bursty transcription and translation (Rhee et al. 2014; Cao and Grima 2020; Chapter 21), high levels of cell-to-cell variation extend to the molecular level, and this likely feeds back in ways that contribute to variation in cell life-history traits. For a diversity of prokaryotes and eukaryotes, the CV for the number of molecules of particular proteins scales as $\bar{z}^{-0.2}$, where $\bar{z}$ is the mean number of proteins/cell (Vallania et al. 2014). As the average number of protein molecules per genetic locus per cell ranges from 10 to 10^5 from the smallest to the largest cell types (Figure 7.4), this implies CVs $\simeq 0.6$ to 0.1, with some evidence suggesting that 0.1 may be close to the asymptotic lower limit for highly expressed proteins (Keren et al. 2015). The CV for protein numbers also increases with decreasing cell division rates by a factor of ~ 3 over the whole range of growth rates (Keren et al. 2015).

Stochastic partitioning of cell contents at division

In most studies of variation in multicellular organisms, the relative contributions of different causal sources of variation are unknown (Lynch and Walsh 1998). However, for cellular traits, a number of insights can be gained from first principles. We start with the ways in which the basic features of molecular segregation during cell division generates variation among progeny. Such stochastic inheritance can have an equally (if not greater) overall effect than intrinsic transcriptional noise (Chapter 21) for the simple reason that upstream variation in molecular abundance can further generate downstream gene-expression noise, and vice versa. Such an outcome is a simple consequence of the structure of biology – numerous cellular products are responsible in one or more ways for their own production (Kiviet et al. 2014).

For the simplest case of a cell containing n molecules at the time of division, with each being independently and randomly distributed to the two daughter cells with probability $1/2$, the average number of molecules inherited per offspring cell is $\bar{n}_o = n/2$. From the binomial sampling formula, the variance (i.e., the square of the standard deviation) is $\sigma_{n_o}^2 = n(1/2)(1/2) = n/4$. The coefficient of variation is then $CV(n_o) = \sigma_{n_o}/\bar{n}_o = 1/\sqrt{n}$, showing that relative to the mean, the standard deviation is inversely related to the square root of the number of molecules being partitioned. This simple principle predicts elevated CVs in small cells containing smaller numbers of molecules. It may also, in part, explain the reduction in CVs in traits in multicellular species, as the noise from their constituent cells is averaged out in whole tissues.

Additional sources of stochastic inheritance during cell division can inflate the level of variation. The argument outlined in the previous paragraph assumes an ideal situation in which each daughter cell draws from an identical cytoplasmic pool. If, however, the cell volumes of daughter cells are unequal (owing to the imperfect positioning of the division septum), the coefficient of variation for offspring cells increases to

$$CV(n_o) = \left(\frac{1 - [CV(V)]^2}{\bar{n}} + \right.$$
$$\left. \left[[CV(V)]^2 \cdot \{ [CV(n)]^2 + 1 \} \right] \right)^{0.5}, \quad (9.9)$$

where $\bar{n}$ is the average number of molecules per adult cell, and $CV(V)$ and $CV(n)$ are, respectively, the coefficients of variation for offspring (sister-cell) volume and for the number of molecules per parental cell (Huh and Paulsson 2011). Empirical estimates of $CV(V) \simeq 0.05$ to 0.12 in *E. coli* (Trueba 1982), *Bacillus subtilis* (Nanninga et al. 1979), *Caulobacter crescentus* (Trueba 1982), and *Schizosaccharomyces pombe* (Johnson et al. 1979; Tyson 1989). $CV(n)$ is typically of similar magnitude to that for $CV(V)$ and relatively similar among species:

$\simeq 0.10$ in *E. coli* (Schaechter et al. 1962; Harvey et al. 1967), *Azotobacter agilis* (Harvey et al. 1967), and *Salmonella typhimurium* (Schaechter et al. 1962), 0.16 in the dinoflagellate *Gonyaulax polyedra* (Homma and Hastings 1989), and 0.07 in the yeast *S. pombe* (Tyson 1989).

Unless $\bar{n} < 100$, with $CV(V)$ and $CV(n)$ both < 0.1, it can be seen from Equation 9.9 that random partitioning of cell volume does not greatly elevate the level of variation in the inherited numbers of molecules beyond the binomial expectation, $1/\sqrt{n}$. On the other hand, if $CV(V) > 0.1$, the inflation can be greater than tenfold with small $\bar{n}$ (Figure 9.9).

Eukaryotic cells have an additional layer of stochasticity in that some molecules are segregated into vesicles or organelles prior to cell division, which are then randomly partitioned among offspring cells. Huh and Paulsson (2011) provide a general expression for the variation rendered under this model. If it assumed that the numbers

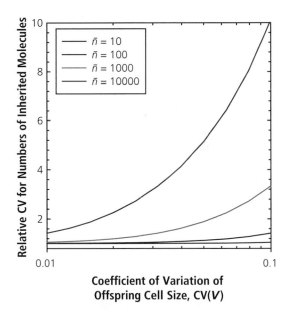

Figure 9.9 Coefficient of variation for the number of molecules inherited by daughter cells as a function of the coefficient of variation for volume of sister cells, $CV(V)$, resulting from slight asymmetries in division. Results are given for different mean numbers of molecules expected in parent cells, $\bar{n}$, relative to the case with simple binomial sampling, as calculated with Equation 9.9. The coefficient of variation for the number of molecules per parent cell, $CV(n_0)$, is assumed to equal 0.1 (based on empirical estimates described in the text).

of vesicles per daughter cell are independently distributed, and that the molecules are randomly distributed among vesicles,

$$\mathrm{CV}'(n_o) \simeq \left(\frac{1}{\bar{n}} + \frac{\{1 + [\mathrm{CV}(n_o)]^2\}\{1 + [\mathrm{CV}(v)]^2\}}{\bar{v}} \right)^{0.5},$$

$$(9.10)$$

where $\bar{v}$ and $\mathrm{CV}(v)$ are the mean and coefficient of variation of the number of vesicles per cell. From Equation 9.9, we know that $\mathrm{CV}(n_o) > 1/\sqrt{\bar{n}}$ and possibly as large as 10. Studies of mitochondrial inheritance in the fission yeast *S. pombe* (Jajoo et al. 2016) and of endosome inheritance in mammalian cell cultures (Bergeland et al. 2001) suggest that the partitioning of such organelles is only slightly less variable than the binomial expectation, which would imply $\mathrm{CV}(v) \simeq 1/\sqrt{\bar{v}}$. In addition, we expect the mean number of vesicles ($\bar{v}$) to be much lower than the mean number of molecules ($\bar{n}$) per cell.

Thus, it is clear that the stochastic partitioning of vesicles (described in the second fraction in Equation 9.10) can be a dominant source of intracellular variation unless there is some regulatory mechanism for controlling cargo partitioning among vesicles and vesicle partitioning among offspring cells. Moreover, variable organelle partitioning is likely to generate more phenotypic variation among cells than might be expected based just on organelle number. For example, because mitochondria are the sites of ATP production in eukaryotic cells, and ATP drives transcription and other cellular processes, mitochondrial partitioning during inheritance can have non-additive effects on offspring-cell performance (das Neves et al. 2010; Johnston et al. 2012).

Finally, it is worth noting that some cellular features can lead to a less variable pattern of inheritance of intracellular contents than expected by chance. For example, in *E. coli* (and many other bacteria) the genome is compacted into a centrally located nucleoid. The resultant mesh-like features serve as a barrier to the movement of ribosomes, which then become more concentrated towards cellular poles for purely physical reasons (Castellana et al. 2016). This may lead to a more even distribution of ribosome numbers in progeny cells than expected if each ribosome were drawn independently.

Phenotypic Variation and Adaptation

As explained in prior chapters, not all of evolutionary change is a product of natural selection. Adaptive as they might seem superficially, certain kinds of changes can only be efficiently promoted by selection under a narrow range of population-genetic conditions. Nonetheless, either unaware or unconvinced of such issues, numerous investigators have asserted that variation-inducing features are not simple consequences of biophysical constraints, but have been advanced by natural selection as strategies for survival in variable environments. In reality, there is a remarkable void of evidence for phenotypic variance (aside from regulated phenotypic plasticity) serving an adaptive purpose, and good reasons to think otherwise.

The following provides an overview of the general consequences of phenotypic variation for the process of natural selection. First, we consider how non-heritable environmental noise, such as that induced by cellular stochasticity, reduces the response to selection on a trait by obscuring the genetic differences among members of a population. Second, we demonstrate how, even in the absence of genetic variation, selection can yield a transient (and in some cases persistent) change in the phenotypic properties of a cell lineage, provided the environmental deviations among individuals are at least partially heritable, as will often be the case for growth-related traits. Finally, we return to the issue of whether phenotypic variation (within genotypes) is maintained by natural selection as a mechanism for coping with a variable environment.

Environmental variation and the efficiency of selection

One of the bedrock results of evolutionary theory concerns the nature of the underlying determinants of the resemblance between relatives. An appreciation of this matter is critical to understanding processes of adaptation for a very simple reason. Although the process of natural selection will always proceed provided there is fitness-associated phenotypic variation present, only the fraction of variation with a heritable genetic basis is relevant to permanent evolutionary change. As discussed in the following section, heritable environmental effects can lead to some response to directional selection, but any such response is transient, quickly decaying away once the prevailing selection pressure is terminated.

The central question here is the degree to which offspring phenotypes resemble those of their parents. For asexually reproducing cells, this is simply defined by the fraction of the phenotypic variation that is genetic in basis, a quantity known as the broad-sense heritability (or H^2) (Foundations 9.4). This key measure is readily estimated by taking a random sample of a population and performing a linear regression of offspring on parental phenotypes (Lynch and Walsh 1998). The best-fit slope, which is generally restricted to the range of 0.0 to 1.0, is equivalent to H^2 (Figure 9.10). Because total phenotypic variance is the sum of contributions from genetic and environmental effects, the higher the background noise from environmental causes, the lower the heritability of the trait.

The heritability of a trait is, in turn, directly related to the response to selection. Imagine a parental population with phenotypic mean $\overline{P}_p$ prior to selection, with directional selection then moving this to $\overline{P}_p'$, yielding a change of $S = \overline{P}_p' - \overline{P}_p$. This difference S in mean phenotypes prior to reproduction is called the selection differential. As

an example, Figure 9.10 shows a situation in which an initial phenotype distribution (black bell-shaped curve) is shifted to the right by viability selection (red curve). The diagonal line denotes the parent–offspring regression, i.e., the relationship between offspring and parent phenotypes. If there were perfect transmission of phenotypes across generations, i.e., if $H^2 = 1$, the mean offspring phenotype would be identical to that of the selected parent generation, and the response to selection (R) would equal the selection differential. However, if there is environmental variance for the trait, the slope of the parent–offspring regression will be < 1, and transmission will be less than perfect because the parental phenotypes deviate from their underlying genotypic values. This is often referred to regression towards mediocrity. If there is no genetic variation, there will be no permanent selection response at all.

Summing up, for a population of asexually reproducing cells, the response to selection is simply

$$R = H^2 S, \qquad (9.11)$$

showing that H^2 is equivalent to the fraction of the selection differential that is transmitted across generations. In a simple fashion, this result illustrates that the ability of natural selection to promote

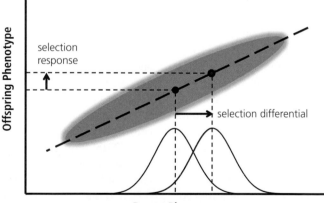

Figure 9.10 The response to directional selection for increased phenotypic values of a continuously distributed trait, represented by the transition of the black to the red bell-shaped curve. The difference between mean phenotypes after and before selection, but before reproduction, is known as the selection differential. The response to selection is determined by the degree to which offspring phenotypes resemble those of their parents, as illustrated by the diagonal dashed line through the blue cloud of parent/offspring values. The slope of this line is known as the heritability of the trait. In the absence of genetic variation, this regression line will have a slope equal to zero. Regardless of the heritability, in the absence of selection, the offspring mean phenotype is expected to equal that of the parental generation.

genetic change declines with increasing environmental variation.

Inheritance of environmental effects

Although a permanent response to directional selection requires the promotion of underlying genetic change, a transient response can sometimes be achieved in the absence of genetic variation. Because selection operates regardless of the source of phenotypic variation, if variation at the phenotypic level owing to intrinsic and/or extrinsic environmental effects is partly heritable across generations, the mean phenotype will still move in the direction of selection, even in the absence of genetic variation (Foundations 9.5). However, unlike the situation with genetic change, such a shift will not be permanent. Rather, under persistent directional selection, the population mean phenotype is expected to reach an alternative steady-state reflecting a balance between the directional force of selection operating on phenotypes and the erosion of progress each generation resulting from the dilution of inherited environmental effects. If selection is relaxed, all progress due to the inheritance of acquired environmental deviations will be quickly eroded away.

This sort of transient response to selection applies to any cellular feature that is partly inherited across generations. For example, any trait that is a function of the number of molecules within a cell (such as a metabolic rate) will naturally be subject to inheritance across cellular generations owing to the fact that the contents of progeny cells are derived immediately from parent-cell constituents, with the molecular composition subsequently undergoing turnover associated with continued production and degradation. Given that offspring in unicellular species inherit half of their parent-cell constituents, these kinds of effects are expected to be much more significant than in multicellular species. Indeed, the complete dissipation of maternal environmental effects can require up to ten generations in bacterial populations (Vashistha et al. 2021).

Such effects are of likely relevance to laboratory experiments that either intentionally or indirectly select for extreme phenotypes. For example, as noted for the adder growth model, large adult cells yield large progeny cells (although not as large, on average, as themselves), which more rapidly reach the point of cell division. Smaller cells take a longer time to reach the requisite cell-volume increase of Δ, and hence, lag in terms of their contribution to the growing population. Although the descendants of large, rapidly dividing cells will gradually move back towards the expected offspring size of Δ, with imperfect cell division, extreme cell sizes will continuously be produced anew, recreating the biasing process. This verbal model needs to be worked out in a more formal manner, but it provides a potentially simple and general explanation for the consistent observation of cells becoming larger in environments with higher nutritional status, even in the absence of genetic variation.

A selection experiment by Yoshida et al. (2014) may be relevant here. Using a cell sorter, they selected for smaller cell size in cultures of *E. coli* for 22 consecutive days by allowing only the smallest 1% of reproducing cells to propagate to the next generation. Overall, an $\sim 20\%$ decline of mean cell size was observed, with the variance in size decreasing only slightly (implying that sufficient opportunity for selection, but not necessarily genetic variance, remained throughout the experiment). Sequencing the entire genome of one selected population revealed only a single nucleotide change, the relevance of which remained unclear.

Although the logic just outlined provides a simple argument for why one expects resemblance between parents and offspring associated with transiently heritable environmental effects, there has been some suggestion of an even higher correlation between collateral relatives (i.e., relatives that are not direct descendants of each other) within genetically uniform cultures of cells. For example, Sandler et al. (2015) found that the correlation between cell-division times in parent and offspring lymphoblast cells is just 0.04, whereas that between sister cells is 0.71, and that between first cousins is 0.58. They call this elevated correlation among cousins relative to that between mother and offspring cells the 'cousin–mother inequality'. Cultures of mammalian cancer and embryonic stem cells exhibit similar behavior (Froese 1964; Kuchen et al. 2020).

These kinds of observations have also been made with cell-lineage studies of several bacterial species

(Powell 1958). For example, in *Aerobacter cloacae*, the correlation in cell-division time is −0.15 for mother–daughter cells, but 0.44 for sibs, and 0.19 for first cousins. Likewise, for *Serratia marcescens*, these correlations are, respectively, −0.20, 0.58, and 0.38. The reduced correlation between first cousins relative to that between sibs is consistent with a progressive dilution of shared effects, and as expected, an even further decline is observed for second cousins (Powell 1958).

Superficially, these results suggest a mechanism of inheritance that is lost for one generation, and then regained in the next, with subsequent erosion of the correlation occurring among the parallel descendants of maternal lineages. This led to the claim that such reappearance of heritability cannot be explained by stochastic inheritance, and requires an underlying deterministic mechanism (Pearl Mizrahi et al. 2015; Sandler et al. 2015). However, although a model can be set up in which an internal oscillator (putatively a circadian clock) operates with a periodicity such that first cousins are born at approximately the same time (Sandler et al. 2015), the following simple argument indicates that a deterministic mechanism need not be invoked.

Imagine that parent cells have their division times determined by physiological effects experienced early in life, but that en route to division, additional resources are gained (or lost) that influence the division times of their offspring, e.g., a burst of transcriptional/translational activity late in the maternal cell cycle. Upon fission, these resources are then approximately equally allocated to the two progeny cells, causing a sib correlation in the population, but having little effect on the maternal-offspring correlation. Although sibs share maternal effects, only a fraction of these are transmitted to the next generation (leading to a smaller first-cousin correlation, and a still smaller one for second cousins) (Staudte et al. 1996).

The adaptive value of phenotypic variation

Finally, we turn to the common argument that within-genotype phenotypic variation is molded by natural selection to provide a strategy for dealing with environmental variation (Thattai and van Oudenaarden 2004; Kussell and Leibler 2005;

Fraser and Kaern 2009; Eldar and Elowitz 2010; Zhuravel et al. 2010; Kiviet et al. 2014; Ackermann 2015; Jahn et al. 2015). As already noted, cellular features exhibit substantial, unavoidable variation owing simply to the intrinsic stochasticities of cellular processes. Here, we note several compelling theoretical reasons for why the further promotion of variation by selection should be the exception rather than the norm. The focus here is not on major discrete phenotypic changes induced by environmental triggers (such as spore formation, or transition to motility), which in many cases almost certainly represent adaptive survival mechanisms (Wolf et al. 2005; Losick and Desplan 2008). Rather, the question is whether the continuous range of variation typically associated with quantitative traits such as growth rate, cell size, and metabolic rates constitutes an adaptive bet-hedging strategy.

One of the most substantive reasons for questioning assertions about adaptive maintenance of phenotypic variation relates to the fact that selection is agnostic with respect to the underlying genetic/environmental determinants of variance. Following the logic outlined earlier, if selection favors an extreme phenotype, when individuals at the extreme are largely there as a consequence of non-genetic effects, the ability of selection to promote individuals with a genetic predisposition to extreme trait values will be compromised. This is because individuals with particularly extreme genetic values will compete for promotion by natural selection with those with more average genetic values but higher variance in expression (Bull 1987). Thus, selection for variance-producing genotypes is difficult when levels of stochastic phenotypic variance are already high. The likelihood of success is even lower if individuals with extreme genetic values have narrower conditional phenotype distributions around the expectation, as these would then be more visible to natural selection.

Of related importance is the fact that selection on phenotypic variation is a second-order effect, as individual genotypes are not promoted on the basis of their own expected genotypic values but via the distribution of their descendants' phenotypes. Unless there is continuing fluctuating selection for individuals at the opposite phenotypic extremes at a sufficiently high rate, the link between genotypic fitness and the ability to differentially generate variation will be weak. This will especially be the case for sexually reproducing species where recombination will progressively remove the disequilibrium between parental genotypic values and descendant phenotype distributions.

Although these arguments do not entirely rule out the possibility of direct selection for the production of broad phenotype distributions, they do lay out the substantial logical challenges confronting those who wish to invoke the existence of phenotypic variation as a direct product of adaptation. It is one thing to hypothesize on the optimality of a complex feature, but quite another to demonstrate that natural selection is actually capable of advancing such change.

As a more explicit example of interpretative difficulties here, consider a study in which single-cell monitoring methods were used to demonstrate that the rate of exponential growth of a culture of *E. coli* with the same average cell-division time is elevated if there is variance around the mean (Hashimoto et al. 2016). The authors argued that these results demonstrate a 'fundamental benefit of noise for population growth'. However, as we know that the rate of population expansion (r) is inversely related to cell doubling time (t_D) (Equation 9.2), this result was readily predictable in advance – for any absolute change in t_D, the increment in r with decreased t_D is greater than the decrease incurred with increased t_D. The behavior results simply because in a growing population, r is bounded above 0.0 and increases at an accelerating rate as t_D becomes small. No elaborate experiment was necessary to show this.

On the other hand, the outcome would have been completely different if the population was declining rather than increasing. In this case, a sublineage of cells with an absolute deviation in division time below the average will experience a greater change in the rate of decline than will a sublineage with a positive deviation of the same absolute amount. Here, variation in the underlying trait enhances the rate of decline of the sublineage. This is not a trivial example for the simple reason that, on average, populations ultimately must go through equal periods of growth and decline, else the population will either go extinct or fill the universe.

More generally, the relationship between the level of variation and the expansion of growth of a cell lineage can be seen to be a simple consequence of the form of the fitness function (Figure 9.11). If the relationship between phenotype and fitness is concave upwards, the average fitness of a variable population will be greater than that of a population having the same mean phenotype but no variance. In contrast, if the fitness–phenotype relationship is concave downwards, the opposite occurs – in this case, the boost in fitness from the upwardly deviating phenotypes is smaller than the loss of fitness in downwardly deviating phenotypes. An extreme case can be seen for the situation in which the trait is under stabilizing selection with the mean phenotype coinciding with the optimum – any deviation from the optimum will result in a decline in fitness. Only for the special situation in which the fitness function is perfectly linear is the influence of variation on fitness effectively neutral, owing to the fact that equal upward and downward phenotypic deviations have equivalent effects on fitness.

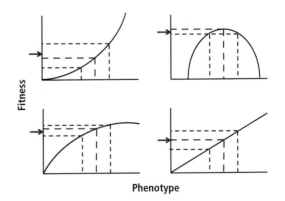

Figure 9.11 The influence of the form of the fitness function on the average fitness of a population of cells. The red lines are fitness functions denoting relationships between individual fitness and phenotype, and the dashed lines demarcate the expected fitnesses for three phenotypes, with the flanking two being equidistant from the one in the middle. In the absence of phenotypic variation, the mean fitness will be equal to the expectation for the middle phenotype (thick dashed lines), whereas in the presence of variation (here, assumed to be symmetrical around the mean), the average fitness for the population (blue arrows) will deviate in a direction depending on the curvature of the fitness function.

Finally, even these arguments are not ironclad, as they consider only the situation in which the phenotype distribution is symmetrically distributed about the mean. With asymmetric phenotype distributions, many alternative outcomes are possible, as the bulk of the phenotype distribution may reside in regions where the fitness function is either increasing or declining. The salient issue is that there is no general advantage to phenotypic variation. Although transient situations may arise in which variation is useful, the same may be said for periods in which it is detrimental.

The general conclusion then is that intrinsic variation in cellular processes alone results in high levels of phenotypic variation among individual cells, much higher than observed among individuals in multicellular species. As seductive as it is to attach an adaptive meaning to all things biological, the idea that phenotypic variance is generally promoted by selective processes appears to be a substantial overstatement, if not positively misleading. The very structure of biology makes the avoidance of phenotypic variation impossible.

Summary

- Observations from a diversity of organisms reveal a number of cellular responses to growth environments that are general enough to be labelled 'growth laws' by microbial physiologists. One universal relationship is the increase in relative investment in ribosomes with increasing cell-division rate, reflecting the conflict between the high energetic cost of ribosomes and their necessity for building cellular material.

- The response of cell-division rate to the concentration of a limiting nutrient can generally be described by a hyperbolic function similar to the Michaelis–Menten form for enzyme kinetics.

- The growth of cell volume within a cellular life cycle is typically exponential in form, as expected if reaction rates per cytoplasmic volume are nearly size independent. Consistent with this view, the ontogenetic response of cell composition to cell volume during individual growth appears generally to be isometric, such that the relative proportions of cell contents remain constant.

- Many prokaryotic and eukaryotic cells determine their division times by monitoring the total change in size, dividing only after a threshold amount of material has been added, rather than by targeting a specific size or time. Such behavior naturally leads to cell-size homeostasis.

- The models for growth patterns serve as first-order approximations and are phenomenological in nature, as the underlying mechanisms driving them remain uncertain and are variable among species. Nonetheless, evidence suggests that the determination of growth-size thresholds often involves the products of just two or three genes, implying relatively simple evolutionary paths for altering cell size and division time.

- In all species that have been studied closely, cell size increases with the nutrient status of the environment but decreases with increasing temperature. It remains unclear whether such shifts are adaptive in any way. They may simply be inevitable by-products of the underlying molecular mechanisms by which cells commit to division.

- Numerous sources of stochastic variation, ranging from sporadic transcription/translation to random partitioning of cellular contents at division, result in considerable phenotypic variation among genetically identical cells, even in well-mixed environments. The magnitude of such variation, which obscures the visibility of genetic differences to natural selection, is substantially greater in unicellular than in multicellular organisms.

- Owing to the fact that binary fission results in substantial sharing of the contents of parent, offspring, and sib cells, unicellular lineages are subject to significant inheritance of non-genetic effects, which can lead to transient shifts in phenotypic values in the absence of genetic change.

- Although there has been considerable speculation that the tendency to produce high levels of phenotypic variation has been advanced by natural selection as means for coping with variable environments, there is little empirical or theoretical support for this contention.

Foundations 9.1 The scaling of ribosome number and cell growth rate

Although cells in nature commonly experience fluctuations in resource availability on time scales shorter than the cell-division time, it is instructive to consider the steady-state situation in a constant environment, as when cells are grown in a continuous-flow chemostat (Chapter 8). Under such conditions, the production rate of every biomolecule (per existing molecule) in the cell must be identical to the rate of overall cell growth, ensuring a steady-state cellular composition.

The rate of translation per cell, and hence, the cellular growth rate, ultimately depends on the number of ribosomes and the number of mRNA transcripts that they engage with. Although translation also involves the use of accessory proteins (e.g., aminoacyl tRNA synthetases, elongation factor, and many others; Barenholz et al. 2016) and transfer RNAs, the abundance of such factors under steady-state growth will be in constant proportion to that of the ribosomes, leaving the latter as a quantifiable indicator of the rate of translation, and hence, cell growth. This argument assumes that cells are conservative with respect to the production of energetically expensive ribosomes, i.e., produce no more than needed to service the current mRNA pool. Here, we follow a derivation presented by Scott et al. (2010) to quantify this connection.

Letting M denote the total protein mass associated with a cell, and M_R denote the total protein mass associated with ribosomes and their affiliated proteins, i.e., 'extended ribosomes', then $f_R = M_R/M$ is the fractional allocation of proteins to translation. Letting m_R denote the protein mass of a single extended ribosome, which hereafter is simply abbreviated to ribosome, the number of ribosomes per cell is $N_R = M_R/m_R = f_R M/m_R$.

Assuming that all ribosomes are engaged in translation, letting k_T denote the rate of translation (i.e., the rate at which amino acids are added to elongating protein chains, here assumed to be constant), and letting m_{AA} be the average mass of an amino acid, the rate of increase in cellular protein mass is

$$\frac{dM}{dt} = m_{AA} \cdot k_T \cdot N_R = \left(\frac{m_{AA} \cdot k_T \cdot f_R}{m_R} \right) \cdot M. \quad (9.1.1a)$$

Because the mass of all components of the cell must increase at the same rate under steady-state conditions, and cell division must proceed at the same rate as growth in size, Equation 9.1.1a can also be written as

$$\frac{dM}{dt} = rM, \quad (9.1.1b)$$

with r denoting the per-capita rate of cell division. The solution of this expression is

$$M(t) = M(0) \cdot e^{rt}, \quad (9.1.1c)$$

where

$$r = m_{AA} \cdot k_T \cdot f_R/m_R, \quad (9.1.2a)$$

which can be condensed to a simpler form

$$r = K_R \cdot f_R, \quad (9.1.2b)$$

with $K_R = m_{AA}k_T/m_R$ being a measure of the translational capacity of the system (the rate of protein mass produced per unit mass of extended ribosomes).

Although the preceding derivation assumes that all ribosomes are actively engaged in translation, if a subfraction $f_{R,0}$ is inactive (independent of growth conditions), then

$$r = K_R \cdot (f_R - f_{R,0}), \quad (9.1.2c)$$

which rearranges to

$$f_R = f_{R,0} + \left(\frac{r}{K_R} \right). \quad (9.1.3)$$

The central assumptions in the preceding derivations are that the translation rate of engaged ribosomes (k_T) and the fraction of unoccupied ribosomes ($f_{R,0}$) are invariant with respect to growth rate. Under such conditions and subject to the constraint that $f_R \leq 1$, Equation 9.1.3 predicts a linear relationship between the fraction of protein invested in extended ribosomes and the rate of cell division, with the intercept being equivalent to the fraction of total cellular protein associated with unengaged ribosomes, and the slope ($1/K_R$) measuring the inverse of the translational capacity. If $f_{R,0}$ and K_R are functions of r, the scaling relationship in Equation 9.1.3 would be altered.

Foundations 9.2 Nutrient limitation and cell growth

In Foundations 9.1, an expression was derived for the rate of cellular growth in terms of ribosome processing. An alternative expression for the growth rate can be couched in terms of the rate of conversion of a limiting nutrient into biomass, again represented by the total mass of protein M. Under steady-state conditions, both approaches must yield equivalent answers for the rate of cell growth, as the rate of amino-acid uptake/biosynthesis must equal the rate at which amino acids are incorporated into proteins at steady state.

We first introduce this second approach, and then unify the two into a joint expression. Again following Scott et al. (2010), we let

$$\frac{dM}{dt} = c \cdot k_E \cdot M_E, \tag{9.2.1a}$$

where c is a constant representing the conversion of the nutrient into M, M_E is the summed mass of the proteins involved in nutrient acquisition and conversion into amino acids, and

$$k_E = k_{E,\max} \left(\frac{S}{K_S + S} \right), \tag{9.2.1b}$$

is the rate of nutrient acquisition per mass of enzyme protein, following the Michaelis–Menten form, which depends on the nutrient concentration (S), and the half-saturation constant (K_S).

We now assume that the total protein in a cell (M) can be partitioned into three sectors (Figure 9.12): a fraction taken to be quantitatively (although not necessarily qualitatively) invariant with respect to cell physiology; a fraction consisting entirely of ribosomal proteins and other proteins associated with translation (extended ribosomes, as in Foundations 9.1); and a fraction associated with metabolic features that respond to nutritional changes. Letting these three fractions be f_Q, f_R, and f_P, respectively, the system is constrained to obey

$$1 = f_Q + f_R + f_P. \tag{9.2.2}$$

Because f_Q is taken to be a constant, this means that increased investment in nutrient acquisition (f_P) necessitates a parallel reduction in investment in protein production (f_R), as the two must sum to $1 - f_Q$.

Further letting f_E denote the fraction of protein mass in sector P devoted to uptake of the limiting nutrient, i.e., $f_E = M_E/M_P$, and recalling from Foundations 9.1 that $f_P = M_P/M$, Equation 9.2.1a expands to

$$\frac{dM}{dt} = (c \cdot k_E \cdot f_E \cdot f_P) \cdot M. \tag{9.2.3}$$

As in Foundations 9.1, the product within the parentheses is equivalent to the rate of exponential growth, which can be further abbreviated to

$$r = K_N \cdot f_P, \tag{9.2.4}$$

where $K_N = c \cdot k_E \cdot f_E$ can be viewed as the nutritional capacity of the system.

We next wish to generate a more general growth-rate expression taking into joint consideration the underlying details about both translation (Foundations 9.1) and nutrient uptake. The key points are that under balanced growth: 1) the rate of nutrient conversion into biomass must be equivalent to the rate of protein production by ribosomes; and 2) the flexible fraction of the proteome must be apportioned into the fractions associated with translation (f_R) and nutrient provisioning (f_P).

As noted above, given that $f_R + f_P = 1 - f_Q$, there is an intrinsic trade-off between the two processes. The maximum possible fractional allocation to ribosomes (or to the remaining pool) is $f_{R,\max} = (1 - f_Q)$, or in other words,

$$f_R = f_{R,\max} - f_P. \tag{9.2.5}$$

Recalling Equation 9.1.2c and substituting for f_R from the preceding expression,

$$r = K_R \cdot (f_{R,\max} - f_{R,0} - f_P), \tag{9.2.6}$$

where, as in Foundations 9.1, K_R is a measure of translational capacity, and $f_{R,0}$ is the fraction of investment in inactive ribosomes. Further substitution for f_P from Equation 9.2.4 and some rearrangement leads to the overall solution

$$r = K_R \cdot (f_{R,\max} - f_{R,0}) \cdot \left(\frac{K_N}{K_R + K_N} \right), \tag{9.2.7}$$

This expression provides a mechanistic link between nutrient uptake and conversion to protein biomass by ribosomes. In effect, it describes the situation in which the allocation to R and P proteins, f_R and f_P, is mutually adjusted such that the rate of intake of critical nutrients is matched by the rate of conversion into protein, subject to the constraint that these must sum to $1 - f_Q$. The fraction in large parentheses on the right is a function of the translational and nutritional capacities of the system, with the cell growth rate $r \to 0$ as $K_N \to 0$, and r asymptotically approaching a maximum value of $K_R \cdot (f_{R,\max} - f_{R,0})$ as $K_N \to \infty$. Because the fraction on the right equals 0.5 when $K_R = K_N$, the ribosomal capacity can be viewed as the half-saturation constant for nutritional capacity. Thus, despite the added complexities, the overall expression for r retains the form of a Monod growth equation.

This kind of partitioning model can be taken in a number of other interesting directions. For example, it has long been known that cells under chronically high nutrient levels often switch to seemingly inefficient modes of energy production, e.g., engagement in fermentation processes, which leave incompletely oxidized products such as acetate or lactate, as opposed to the citric-acid cycle, which oxidizes glucose all the way down to CO_2. Such metabolic overflow, or energy spillage, at high resource levels can be explained by the fact that the machinery underlying fermentation processes involves many fewer enzymes than that required for the citric-acid cycle (Molenaar et al. 2009; Basan et al. 2015). The hypothesis here is that, when the external carbon supply is high, cells can increase the investment in the protein machinery necessary for biosynthesis by reducing the investment in the enzymes necessary for input into such pathways. In contrast, when the nutrient supply is low, investing more heavily in carbon metabolism allows cells to maximally direct flux towards biosynthesis.

Bertaux et al. (2020) and Serbanescu et al. (2020) have extended the preceding model to incorporate additional sector partitioning, e.g., cell division. These extensions allow for analysis of the cell size-growth rate relationship discussed in the text. There is room for caution in over-interpreting the good fits of models like these, as a large number of parameters are employed, not all of which are based on extrinsic estimates. The value of the approach resides in its ability to highlight the potential importance of broad classes of underlying mechanisms that can be followed up by further empirical study. Extensive details on the underlying proteomic responses of E. coli cells to changing nutritional conditions are provided in synthetic reviews by Belliveau et al. (2021) and Mori et al. (2021).

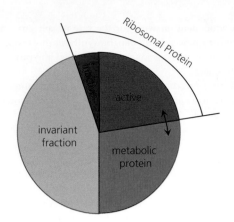

Figure 9.12 A conceptualized partitioning of total cellular proteins into three general sectors, one of which (blue) contributes an invariant proportion to the total pool regardless of the cellular nutritional state (after Scott et al. 2010). The total pool of ribosomal proteins is also considered to contain a small invariant fraction associated with inactive ribosomes (red). This leaves the pools of active ribosomal proteins (yellow) and metabolic proteins (peach) free to vary with respect to each other (i.e., the orange-yellow boundary associated with the double-headed arrow can move), but constrained to sum to the total area under the yellow and orange sectors.

Foundations 9.3 Scaling models for the development of cellular features

Given what little we know about the mechanisms driving the quantitative relationships between cells and their parts during cell growth, consideration of alternative models may be informative, particularly if they predict different patterns of scaling.

Here, following Equation 9.7, we evaluate two fairly general models, in both cases assuming exponential growth of the cell at rate r in terms of total volume, i.e., $V_t = V_0 e^{rt}$. First, consider the situation in which a cellular feature grows exponentially and autonomously (i.e., independent of cell volume, V) at rate β, such that the expected phenotypic value at time t is

$$z_t = z_0 e^{\beta t}, \tag{9.3.1}$$

where z_0 is the phenotypic value at cell birth. Log transforming Equations 9.7 and 9.3.1, solving the first expression for t, and substituting into the latter, we obtain

$$\log(z_t) = \left(\frac{\beta}{r}\right) \log(V_t) + c, \tag{9.3.2a}$$

where

$$c = \log(z_0) - \left(\frac{\beta}{r}\right) \log(V_0), \tag{9.3.2b}$$

can be viewed as the intercept of a log–log plot of z_t versus V_t throughout developmental progression. Noting that c is a constant determined by the size of the trait and cell volume at birth (as well as the growth parameters β and r), this model predicts an allometric (power-law; Chapter 8) relationship, with the slope providing an estimate of the ratio of growth rates (β/r). If the slope is equal to 1.0, then β must equal r, implying isometric growth.

Now consider the situation in which growth of the trait is directly linked to the growth in cell volume via some sort of regulatory mechanism (Harris and Theriot 2016), such that

$$\frac{dz}{dt} = \beta \cdot V_t = \beta \cdot V_0 e^{rt}, \tag{9.3.3}$$

the solution of which is

$$z_t = \left(\frac{\beta}{r}\right) V_t + c, \tag{9.3.4a}$$

with

$$c = z_0 - \left(\frac{\beta}{r}\right) V_0. \qquad (9.3.4b)$$

Note that the key scaling parameter is again the ratio of growth rates, β/r. However, in contrast to the volume-independent model, where there is linear scaling between the log-transformed values of z_t and V_t, when trait growth is coupled directly to cell volume, the scaling is linear on the original scale of measurement. If $\beta \simeq r$, which the data in the text suggest for volumetric traits, these two models will be difficult to distinguish based on growth-trajectory data alone, as in both cases, the relationship will be essentially linear on the original scale of measurement.

Foundations 9.4 Parent–offspring resemblance and the response to selection

The measure of a particular trait in a specific individual, P, can be viewed as the sum of its expected value given its genotypic composition, G, and a deviation from that expectation, E, owing to both internal effects associated with stochastic molecular behavior and external effects associated with physical, chemical, and biological aspects of the environment,

$$P = G + E. \qquad (9.4.1)$$

The genotypic value G can be thought of as the average phenotypic measure expected if a large number of individuals of the same genotype were monitored in an identical environmental setting. The environmental effect E summarizes the net positive or negative deviations around G, and has a mean (over all individuals) equal to zero and variance σ_E^2 among individuals (Lynch and Walsh 1998). Provided there is no genotype–environment covariance (i.e., environmental deviations are independent of the genetic background), the total phenotypic variance in the population is then the sum of the genetic and environmental variance components,

$$\sigma_P^2 = \sigma_G^2 + \sigma_E^2. \qquad (9.4.2)$$

These variance components relate directly to the resemblance between relatives. If the offspring of measured parents are allowed to develop to the same stage as the parents and then measured, one can produce a parent–offspring regression, which is equivalent to the straight line that best describes the overall relationship (Figure 9.10). The slope of a best fit line is known to be equal to the ratio of the covariance between x and y variables (denoted as $\sigma(x, y)$, with the two variables here being offspring and parent phenotypes) and the variance of the x variable (denoted as $\sigma^2(x)$, and here applying to parental phenotypes). (For those unfamiliar with statistics, a variance is the average squared deviation of measures from the mean, whereas a covariance is the average cross-product of x and y deviations from their respective means). Letting o and p denote offspring and parents, and assuming asexual reproduction, the covariance between offspring and parent pairs expands to

$$\sigma(P_o, P_p) = \sigma[(G_o + E_o), (G_p + E_p)]. \qquad (9.4.3a)$$

Although there are four potential cross-product terms in the covariance, assuming that the environmental deviations in different generations are uncorrelated (i.e., not inherited, an assumption relaxed in Foundations 9.5), there can only be covariance between the genetic values, so

$$\sigma(P_o, P_p) = \sigma(G_o, G_p), \qquad (9.4.3b)$$

and because parents and offspring have identical genetic values in an asexual population, the genetic covariance is the same as the genetic variance,

$$\sigma(P_o, P_p) = \sigma_G^2. \qquad (9.4.3c)$$

This follows because the covariance of a measure with itself is equal to the variance of the measure. The expected slope of the parent–offspring regression is then the ratio of Equations 9.4.3c and 9.4.2,

$$H^2 = \frac{\sigma_G^2}{\sigma_G^2 + \sigma_E^2}. \qquad (9.4.4)$$

This quantity, which is usually referred to as the broad-sense heritability, is simply the fraction of the total phenotypic variance attributable to genetic causes. Further aspects concerning the phenotypic covariances among clonal relatives can be found in Jun et al. (2018). Slight modifications are required under sexual reproduction, as parents only transmit half their genetic value to their progeny (Lynch and Walsh 1998).

Foundations 9.5 Transient response to selection without genetic change

Under the adder model for growth, Δ is equivalent to the expected increase in cell volume between cell divisions, and the expected size at birth is also Δ. This, however, is only strictly true in the absence of selection on cell size. Imagine a clonal (genetically uniform) population of cells with some variation in the realized value of Δ experienced by individual cells, owing to the vagaries in intracellular and external environments, and to the fact that cells do not divide with absolute symmetry.

With variation around the mean Δ, the size of an adult cell at the time of division can be expressed as

$$V_a = V_0 + \Delta + e_\Delta, \tag{9.5.1a}$$

where V_0 is the size at birth, Δ is the expected growth in size, and e_Δ is the deviation of the actual growth increment from Δ owing to background variation, assumed to have a mean value of zero and some variance σ_Δ^2. In the absence of selection, the expected value of V_0 is Δ, and the previous expression can be written as

$$V_a = 2\Delta + e_\Delta, \tag{9.5.1b}$$

with the average offspring cell size being $\overline{V}_0 = \overline{V}_a/2 = \Delta$ because the expected value of e_Δ (denoted by the overline) is equal to zero.

If, however, there is directional selection on cell size, the mean value $\overline{e}_\Delta$ is no longer equal to zero, as the cells with more extreme deviations are differentially promoted. Instead, in the first generation of selection, the average offspring cell size becomes

$$\overline{V}_0(1) = \Delta + (\overline{e}_\Delta/2)$$

assuming that on average half of the mean environmental deviation in the previous generation is transmitted to each offspring cell. If this same level of selection is continued for another generation, the mean becomes

$$\overline{V}_0(2) = \Delta + (\overline{e}_\Delta/2) + (\overline{e}_\Delta/4)$$

as a new deviation is added while half of the prior deviation is partially removed by 50% dilution. Using the series expansion

$$\sum_{i=1}^{t} x^i = \frac{x(1 - x^t)}{1 - x}, \tag{9.5.2}$$

with $x = 0.5$, after t generations of constant selection, the mean offspring size is

$$\overline{V}_0(t) = \Delta + \overline{e}_\Delta[1 - (1/2)^t] \tag{9.5.3}$$

which asymptotically approaches $\Delta + \overline{e}_\Delta$ as t increases.

This shows that the average size of cells in a population can quickly shift to a new value without any genetic change, with a deviation from the non-selection value Δ equal to the selection differential $\overline{e}_\Delta$. The central point is that, owing to the partial transmission of offspring deviations to subsequent generations, the mean phenotypes in a population are expected to change if directional selection persistently operates on a cellular trait, even if there is no genetic basis for the deviations.

Notably, however, this selection response is transient in that if selection is relaxed, the initial deviation $\overline{e}_\Delta$ declines by 50% each generation, rapidly returning the average offspring cell volume to Δ. In contrast, any genetic contribution to the selection response would remain following selection.

Finally, supposing extreme cells can sequester their excess endowment to a degree that allows greater than 50% retention, then the use of Equation 9.5.2 shows that an even greater transient boost can be obtained by selection on environmental deviations, e.g., with $x = 0.75$, $\overline{V}_0(t)$ has an asymptotic value of $\Delta + 3\overline{e}_\Delta$.

Literature Cited

Ackermann, M. 2015. A functional perspective on phenotypic heterogeneity in microorganisms. Nat. Rev. Microbiol. 13: 497–508.

Alberghina, F. A., E. Sturani, and J. R. Gohlke. 1975. Levels and rates of synthesis of ribosomal ribonucleic acid, transfer ribonucleic acid, and protein in *Neurospora crassa* in different steady states of growth. J. Biol. Chem. 250: 4381–4388.

Amir, A. 2014. Cell size regulation in bacteria. Phys. Rev. Lett. 112: 208102.

Amir A. 2017. Is cell size a spandrel? eLife 6: e22186.

Atkinson, A. W., Jr., P. C. John, and B. E. Gunning. 1974. The growth and division of the single mitochondrion and other organelles during the cell cycle of *Chlorella*, studied by quantitative stereology and three dimensional reconstruction. Protoplasma 81: 77–109.

Atkinson, D., B. J. Ciotti, and D. J. Montagnes. 2003. Protists decrease in size linearly with temperature: *ca.* 2.5% °C⁻¹. Proc. Biol. Sci. 270: 2605–2611.

Baldwin, J. M. 1896. A new factor in evolution. Amer. Natur. 30: 441–451.

Barenholz, U., L. Keren, E. Segal, and R. A. Milo. 2016. A minimalistic resource allocation model to explain ubiquitous increase in protein expression with growth rate. PLoS One 11: e0153344.

Basan, M., S. Hui, H. Okano, Z. Zhang, Y. Shen, J. R. Williamson, and T. Hwa. 2015. Overflow metabolism in *Escherichia coli* results from efficient proteome allocation. Nature 528: 99–104.

Belliveau, N. M., G. Chure, C. L. Hueschen, H. G. Garcia, J. Kondev, D. S. Fisher, J. A. Theriot, and R. Phillips. 2021. Fundamental limits on the rate of bacterial growth and their influence on proteomic composition. Cell Syst. 12: 924–944.e2.

Bergeland, T., J. Widerberg, O. Bakke, amd T. W. Nordeng. 2001. Mitotic partitioning of endosomes and lysosomes. Curr. Biol. 11: 644–651.

Bergmann, K. G. L. C. 1847. Über die Verhältnisse der wärmeokönomie der Thiere zu ihrer Grösse. Göttinger Studien 3: 595–708.

Bertaux, F., J. von Kügelgen, S. Marguerat, and V. Shahrezaei. 2020. A bacterial size law revealed by a coarse-grained model of cell physiology. PLoS Comput. Biol. 16: e1008245.

Boehlke, K. W., and J. D. Friesen. 1975. Cellular content of ribonucleic acid and protein in *Saccharomyces cerevisiae* as a function of exponential growth rate: Calculation of the apparent peptide chain elongation rate. J. Bacteriol. 121: 429–433.

Bonven, B., and K. Gullov. 1979. Peptide chain elongation rate and ribosomal activity in *Saccharomyces cerevisiae* as a function of the growth rate. Mol. Gen. Genet. 170: 225–230.

Brangwynne, C. P. 2013. Phase transitions and size scaling of membrane-less organelles. J. Cell Biol. 203: 875–881.

Brown, C. M., and A. H. Rose. 1969. Effects of temperature on composition and cell volume of *Candida utilis*. J. Bacteriol. 97: 261–270.

Bryan, A. K., A. Goranov, A. Amon, and S. R. Manalis. 2010. Measurement of mass, density, and volume during the cell cycle of yeast. Proc. Natl. Acad. Sci. USA 107: 999–1004.

Bull, J. J. 1987. Evolution of phenotypic variance. Evolution 41: 303–315.

Campos, M., I. V. Surovtsev, S. Kato, A. Paintdakhi, B. Beltran, S. E. Ebmeier, and C. Jacobs-Wagner. 2014. A constant size extension drives bacterial cell size homeostasis. Cell 159: 1433–1446.

Cao, Z., and R. Grima. 2020. Analytical distributions for detailed models of stochastic gene expression in eukaryotic cells. Proc. Natl. Acad. Sci. USA 117: 4682–4692.

Castellana, M., L. S. Hsin-Jung, and N. S. Wingreen. 2016. Spatial organization of bacterial transcription and translation. Proc. Natl. Acad. Sci. USA 113: 9286–9291.

Chan, Y.-H., and W. F. Marshall. 2010. Scaling properties of cell and organelle size. Organogenesis 6: 88–96.

Chan, Y.-H., and W. F. Marshall. 2012. How cells know the size of their organelles. Science 337: 1186–1189.

Chan, Y.-H., and W. F. Marshall. 2014. Organelle size scaling of the budding yeast vacuole is tuned by membrane trafficking rates. Biophys. J. 106: 1986–1996.

Chan, Y.-H., L. Reyes, S. M. Sohail, N. K. Tran, and W. F. Marshall. 2016. Organelle size scaling of the budding yeast vacuole by relative growth and inheritance. Curr. Biol. 26: 1221–1228.

Chandler-Brown, D., K. M. Schmoller, Y. Winetraub, and J. M. Skotheim. 2017. The adder phenomenon emerges from independent control of pre- and post-start phases of the budding yeast cell cycle. Curr. Biol. 27: 2774–2783.

Chien, A. C., S. K. Zareh, Y. M. Wang, and P. A. Levin. 2012. Changes in the oligomerization potential of the division inhibitor UgtP co-ordinate *Bacillus subtilis* cell size with nutrient availability. Mol. Microbiol. 86: 594–610.

Colin, A., G. Micali, L. Faure, M. Cosentino Lagomarsino, and S. van Teeffelen. 2021. Two different cell-cycle processes determine the timing of cell division in *Escherichia coli*. eLife 10: e67495.

das Neves, R. P., N. S. Jones, L. Andreu, R. Gupta, T. Enver, and F. J. Iborra. 2010. Connecting variability in global transcription rate to mitochondrial variability. PLoS Biol. 8: e1000560.

Deforet, M., D. van Ditmarsch, and J. B. Xavier. 2015. Cell-size homeostasis and the incremental rule in a bacterial pathogen. Biophys. J. 109: 521–528.

Delarue, M., D. Weissman, and O. Hallatschek. 2017. A simple molecular mechanism explains multiple patterns of cell-size regulation. PLoS One 12: e0182633.

Dennis, P. P., and H. Bremer. 1974. Macromolecular composition during steady-state growth of *Escherichia coli* B-r. J. Bacteriol. 119: 270–281.

Deviri, D., and S. A. Safran. 2022. Balance of osmotic pressures determines the nuclear-to-cytoplasmic volume ratio of the cell. Proc. Natl. Acad. Sci. USA 119: e2118301119.

Di Talia, S., J. M. Skotheim, J. M. Bean, E. D. Siggia, and F. R. Cross. 2007. The effects of molecular noise and size control on variability in the budding yeast cell cycle. Nature 448: 947–951.

Droop, M. R. 1973. Some thoughts on nutrient limitation in algae. J. Phycol. 9: 264–272.

Droop, M. R. 1974. The nutrient status of algal cells in continuous culture. J. Marine Biol. Assoc. UK 54: 825–855.

Droop, M. R. 1983. 25 years of algal growth kinetics: A personal view. Bot. Mar. 26: 99–112.

Eldar, A., and M. B. Elowitz. 2010. Functional roles for noise in genetic circuits. Nature 467: 167–173.

Eun, Y. J., P. Y. Ho, M. Kim, S. LaRussa, L. Robert, L. D. Renner, A. Schmid, E. Garner, and A. Amir. 2018.

Archaeal cells share common size control with bacteria despite noisier growth and division. Nat. Microbiol. 3: 148–154.

Fantes, P. A. 1977. Control of cell size and cycle time in *Schizosaccharomyces pombe*. J. Cell Sci. 24: 51–67.

Ferrezuelo, F., N. Colomina, A. Palmisano, E. Garí, C. Gallego, A. Csikász-Nagy, and M. Aldea. 2012. The critical size is set at a single-cell level by growth rate to attain homeostasis and adaptation. Nat. Commun. 3: 1012.

Forchhammer, J., and L. Lindahl. 1971. Growth rate of polypeptide chains as a function of the cell growth rate in a mutant of *Escherichia coli*. J. Mol. Biol. 55: 563–568.

Fraenkel, D. G., and F. C. Neidhardt. 1961. Use of chloramphenicol to study control of RNA synthesis in bacteria. Biochim. Biophys. Acta 53: 96–110.

Fraser, D., and M. Kaern. 2009. A chance at survival: Gene expression noise and phenotypic diversification strategies. Mol. Microbiol. 71: 1333–1340.

Freyssinet, G., and J. A. Schiff. 1974. The chloroplast and cytoplasmic ribosomes of *Euglena*: II. Characterization of ribosomal proteins. Plant Physiol. 53: 543–554.

Froese, G. 1964. The distribution and interdependence of generation times of HeLa cells. Exp. Cell Res. 35: 415–419.

Fu, R., and J. Gong. 2017. Single cell analysis linking ribosomal (r)DNA and rRNA copy numbers to cell size and growth rate provides insights into molecular protistan ecology. J. Eukaryot. Microbiol. 64: 885–896.

Geiler-Samerotte, K. A., C. R. Bauer, S. Li, N. Ziv, D. Gresham, and M. L. Siegal. 2013. The details in the distributions: Why and how to study phenotypic variability. Curr. Opin. Biotechnol. 24: 752–759.

Godin, M., F. F. Delgado, S. Son, W. H. Grover, A. K. Bryan, A. Tzur, P. Jorgensen, K. Payer, A. D. Grossman, M. W. Kirschner, et al. 2010. Using buoyant mass to measure the growth of single cells. Nat. Methods 7: 387–390.

Goehring, N. W., and A. A. Hyman. 2012. Organelle growth control through limiting pools of cytoplasmic components. Curr. Biol. 22: R330–R339.

Gray, W. T., S. K. Govers, Y. Xiang, B. R. Parry, M. Campos, S. Kim, and C. Jacobs-Wagner. 2019. Nucleoid size scaling and intracellular organization of translation across bacteria. Cell 177: 1632–1648.

Harris, L. K., and J. A. Theriot. 2016. Relative rates of surface and volume synthesis set bacterial cell size. Cell 165: 1479–1492.

Harvey, R. J., A. G. Marr, and P. R. Painter. 1967. Kinetics of growth of individual cells of *Escherichia coli* and *Azotobacter agilis*. J. Bacteriol. 93: 605–617.

Hashimoto, M., T. Nozoe, H. Nakaoka, R. Okura, S. Akiyoshi, K. Kaneko, E. Kussell, and Y. Wakamoto.

2016. Noise-driven growth rate gain in clonal cellular populations. Proc. Natl. Acad. Sci. USA 113: 3251–3256.

Homma, K., and J. W. Hastings. 1989. Cell growth kinetics, division asymmetry and volume control at division in the marine dinoflagellate *Gonyaulax polyedra*: A model of circadian clock control of the cell cycle. J. Cell Sci. 92: 303–318.

Huh, D., and J. Paulsson. 2011. Random partitioning of molecules at cell division. Proc. Natl. Acad. Sci. USA 108: 15004–15009.

Iyer-Biswas, S., C. S. Wright, J. T. Henry, K. Lo, S. Burov, Y. Lin, G. E. Crooks, S. Crosson, A. R. Dinner, and N. F. Scherer. 2014. Scaling laws governing stochastic growth and division of single bacterial cells. Proc. Natl. Acad. Sci. USA 111: 15912–15917.

Jacob, F. 1998. The Statue Within: An Autobiography. Cold Spring Harbor Laboratory Press, Cold Spring Harbor, NY.

Jahn, M., S. Günther, and S. Müller. 2015. Non-random distribution of macromolecules as driving forces for phenotypic variation. Curr. Opin. Microbiol. 25: 49–55.

Jajoo, R., Y. Jung, D. Huh, M. P. Viana, S. M. Rafelski, M. Springer, and J. Paulsson. 2016. Accurate concentration control of mitochondria and nucleoids. Science 351: 169–172.

Johnson, B. F., G. B. Calleja, I. Boisclair, and B. Y. Yoo. 1979. Cell division in yeasts. III. The biased, asymmetric location of the septum in the fission yeast cell, *Schizosaccharomyces pombe*. Exp. Cell Res. 123: 253–259.

Johnston, I. G., B. Gaal, R. P. Neves, T. Enver, F. J. Iborra, and N. S. Jones. 2012. Mitochondrial variability as a source of extrinsic cellular noise. PLoS Comput. Biol. 8: e1002416.

Jorgensen, P., N. P. Edgington, B. L. Schneider, I. Rupes, M. Tyers, and B. Futcher. 2007. The size of the nucleus increases as yeast cells grow. Mol. Biol. Cell 18: 3523–3532.

Jun, S., F. Si, R. Pugatch, and M. Scott. 2018. Fundamental principles in bacterial physiology-history, recent progress, and the future with focus on cell size control: A review. Rep. Prog. Phys. 81: 056601.

Kendall, D. G. 1948. On the role of variable generation time in the development of a stochastic birth process. Biometrika 35: 316–330.

Keren, L., D. van Dijk, S. Weingarten-Gabbay, D. Davidi, G. Jona, A. Weinberger, R. Milo, and E. Segal. 2015. Noise in gene expression is coupled to growth rate. Genome Res. 25: 1893–1902.

Kimball, R. F., T. O. Casperson, G. Svensson, and L. Carlson. 1959. Quantitative cytochemical studies on *Paramecium aurelia*. I. Growth in total dry weight measured by the scanning interference microscope and X-ray absorption methods. Exp. Cell. Res. 17: 160–172.

Kiviet, D. J., P. Nghe, N. Walker, S. Boulineau, V. Sunderlikova, and S. J. Tans. 2014. Stochasticity of metabolism and growth at the single-cell level. Nature 514: 376–379.

Koch, A. L., and M. Schaechter. 1962. A model for statistics of the cell division process. J. Gen. Microbiol. 29: 435–454.

Krásný, L., and R. L. Gourse. 2004. An alternative strategy for bacterial ribosome synthesis: *Bacillus subtilis* rRNA transcription regulation. EMBO J. 23: 4473–4483.

Kuchen, E. E., N. B. Becker, N. Claudino, and T. Höfer. 2020. Hidden long-range memories of growth and cycle speed correlate cell cycles in lineage trees. eLife 9: e51002.

Kussell, E., and S. Leibler. 2005. Phenotypic diversity, population growth, and information in fluctuating environments. Science 309: 2075–2078.

Lemière, J., P. Real-Calderon, L. J. Holt, T. G. Fai, and F. Chang. 2022. Control of nuclear size by osmotic forces in *Schizosaccharomyces pombe*. eLife 11: e76075.

Lenski, R. E., and M. Travisano. 1994. Dynamics of adaptation and diversification: A 10,000-generation experiment with bacterial populations. Proc. Natl. Acad. Sci. USA 91: 6808–6814.

Levy, D. L., and R. Heald. 2012. Mechanisms of intracellular scaling. Annu. Rev. Cell Dev. Biol. 28: 113–135.

Levy, S. F., N. Ziv, and M. L. Siegal. 2012. Bet hedging in yeast by heterogeneous, age-correlated expression of a stress protectant. PLoS Biol. 10: e1001325.

Litsios, A., D. H. E. W. Huberts, H. M. Terpstra, P. Guerra, A. Schmidt, K. Buczak, A. Papagiannakis, M. Rovetta, J. Hekelaar, G. Hubmann, et al. 2019. Differential scaling between G1 protein production and cell size dynamics promotes commitment to the cell division cycle in budding yeast. Nat. Cell Biol. 21: 1382–1392.

Liu, K., A. N. Bittner, and J. D. Wang. 2015. Diversity in (p)ppGpp metabolism and effectors. Curr. Opin. Microbiol. 24: 72–79.

Logsdon, M. M., P. Y. Ho, K. Papavinasasundaram, K. Richardson, M. Cokol, C. M. Sassetti, A. Amir, and B. B. Aldridge. 2017. A parallel adder coordinates mycobacterial cell-cycle progression and cell-size homeostasis in the context of asymmetric growth and organization. Curr. Biol. 27: 3367–3374.

Losick, R., and C. Desplan. 2008. Stochasticity and cell fate. Science 320: 65–68.

Lutkenhaus, J. 2008. Min oscillation in bacteria. Adv. Exp. Med. Biol. 641: 49–61.

Lynch, M., and J. B. Walsh. 1998. Genetics and Analysis of Quantitative Traits. Sinauer Associates, Inc., Sunderland, MA.

Maitra, A., and K. A. Dill. 2015. Bacterial growth laws reflect the evolutionary importance of energy efficiency. Proc. Natl. Acad. Sci. USA 112: 406–411.

McInally, S. G., J. Kondev, and B. L. Goode. 2021. Scaling of subcellular actin structures with cell length through decelerated growth. eLife 10: e68424.

Metzl-Raz, E., M. Kafri, G. Yaakov, I. Soifer, Y. Gurvich, and N. Barkai. 2017. Principles of cellular resource allocation revealed by condition-dependent proteome profiling. eLife 6: e28034.

Mir, M., Z. Wang, Z. Shen, M. Bednarz, R. Bashir, I. Golding, S. G. Prasanth, and G. Popescu. 2011. Optical measurement of cycle-dependent cell growth. Proc. Natl. Acad. Sci. USA 108: 13124–13129.

Mochizuki, T. 1998. Three-dimensional reconstruction of mitotic cells of *Cryptococcus neoformans* based on serial section electron microscopy. Nihon Ishinkin Gakkai Zasshi 39: 123–127.

Molenaar, D., R. van Berlo, D. de Ridder, and B. Teusink. 2009. Shifts in growth strategies reflect tradeoffs in cellular economics. Mol. Syst. Biol. 5: 323.

Monod, J. 1949. The growth of bacterial cultures. Ann. Rev. Microbiol. 3: 371–394.

Mori, M., S. Schink, D. W. Erickson, U. Gerland, and T. Hwa. 2017. Quantifying the benefit of a proteome reserve in fluctuating environments. Nat. Commun. 8: 1225.

Mori, M., Z. Zhang, A. Banaei-Esfahani, J. B. Lalanne, H. Okano, B. C. Collins, A. Schmidt, O. T. Schubert, D. S. Lee, G. W. Li, et al. 2021. From coarse to fine: The absolute *Escherichia coli* proteome under diverse growth conditions. Mol. Syst. Biol. 17: e9536.

Moseley, J. B., A. Mayeux, A. Paoletti, and P. Nurse. 2009. A spatial gradient coordinates cell size and mitotic entry in fission yeast. Nature 459: 857–860.

Nanninga, N., L. J. Koppes, and F. C. de Vries-Tijssen. 1979. The cell cycle of *Bacillus subtilis* as studied by electron microscopy. Arch. Microbiol. 123: 173–181.

Neumann, F. R., and P. Nurse. 2007. Nuclear size control in fission yeast. J. Cell Biol. 179: 593–600.

Osella, M., E. Nugent, and M. Cosentino Lagomarsino. 2014. Concerted control of *Escherichia coli* cell division. Proc. Natl. Acad. Sci. USA 111: 3431–3435.

Parenteau, J., L. Maignon, M. Berthoumieux, M. Catala, V. Gagnon, and S. Abou Elela. 2019. Introns are mediators of cell response to starvation. Nature 565: 612–617.

Pearl Mizrahi, S., O. Sandler, L. Lande-Diner, N. Q. Balaban, and I. Simon. 2016. Distinguishing between stochasticity and determinism: Examples from cell cycle duration variability. Bioessays 38: 8–13.

Pellegrini, M. 1980. Three-dimensional reconstruction of organelles in *Euglena gracilis* Z. II. Qualitative and quantitative changes of chloroplasts and mitochondrial reticulum in synchronous cultures during bleaching. J. Cell Sci. 46: 313–340.

Plaut, B. S., and G. Turnock. 1975. Coordination of macromolecular synthesis in the slime mould *Physarum polycephalum.* Mol. Gen. Genet. 137: 211–225.

Posakony, J. W., J. M. England, and G. Attardi. 1977. Mitochondrial growth and division during the cell cycle in HeLa cells. J. Cell Biol. 74: 468–491.

Potrykus, K., H. Murphy, N. Philippe, and M. Cashel. 2011. ppGpp is the major source of growth rate control in *E. coli.* Environ. Microbiol. 13: 563–575.

Powell, E. O. 1956. Growth rate and generation time of bacteria, with special reference to continuous culture. J. Gen. Microbiol. 15: 492–511.

Powell, E. O. 1958. An outline of the pattern of bacterial generation times. J. Gen. Microbiol. 18: 382–417.

Poyton, R. O. 1973. Effect of growth rate on the macromolecular composition of *Prototheca zopfii*, a colorless alga which divides by multiple fission. J. Bacteriol. 113: 203–211.

Priestman, M., P. Thomas, B. D. Robertson, and V. Shahrezaei. 2017. Mycobacteria modify their cell size control under sub-optimal carbon sources. Front. Cell Dev. Biol. 5: 64.

Prokop, A., and J. Ricica. 1968. *Chlorella pyrenoidosa* 7-11-05 in batch and in homogeneous continuous culture under autotrophic conditions. I. Growth characteristics of the culture. Folia Microbiol. (Praha) 13: 353–361.

Rafelski, S. M., M. P. Viana, Y. Zhang, Y.-H. Chan, K. S. Thorn, P. Yam, J. C. Fung, H. Li, L. da F. Costa, and W. F. Marshall. 2012. Mitochondrial network size scaling in budding yeast. Science 338: 822–824.

Rhee, A., R. Cheong, and A. Levchenko. 2014. Noise decomposition of intracellular biochemical signaling networks using nonequivalent reporters. Proc. Natl. Acad. Sci. USA 111: 17330–17335.

Riemer, K., R. P. Guralnick, and E. P. White. 2018. No general relationship between mass and temperature in endothermic species. eLife 7: e27166.

Rosset, R., R. Monier, and J. Julien. 1964. RNA composition of *Escherichia coli* as a function of growth rate. Biochem. Biophys. Res. Comm. 15: 329–333.

Sandler, O., S. P. Mizrahi, N. Weiss, O. Agam, I. Simon, and N. Q. Balaban. 2015. Lineage correlations of single cell division time as a probe of cell-cycle dynamics. Nature 519: 468–471.

Santi, I., N. Dhar, D. Bousbaine, Y. Wakamoto, and J. D. McKinney. 2013. Single-cell dynamics of the chromosome replication and cell division cycles in mycobacteria. Nat. Commun. 4: 2470.

Schaechter, M., O. Maaløe, and N. O. Kjeldgaard. 1958. Dependency on medium and temperature of cell size and chemical composition during balanced grown of *Salmonella typhimurium.* J. Gen. Microbiol. 19: 592–606.

Schaechter, M., J. P. Williamson, J. R. Hood, Jr., and A. L. Koch. 1962. Growth, cell and nuclear divisions in some bacteria. J. Gen. Microbiol. 29: 421–434.

Scherbaum, O., and G. Rasch. 1957. Cell size distribution and single cell growth in *Tetrahymena pyriformis* Gl1. Acta Pathol. Microbiol. 41: 161–182.

Schmoller, K. M., J. J. Turner, M. Kõivomägi, and J. M. Skotheim. 2015. Dilution of the cell cycle inhibitor Whi5 controls budding-yeast cell size. Nature 526: 268–272.

Scott, M., C. W. Gunderson, E. M. Mateescu, Z. Zhang, and T. Hwa. 2010. Interdependence of cell growth and gene expression: Origins and consequences. Science 330: 1099–1102.

Serbanescu, D., N. Ojkic, and S. Banerjee. 2020. Nutrient-dependent trade-offs between ribosomes and division protein synthesis control bacterial cell size and growth. Cell Rep. 32: 108183.

Si, F., G. Le Treut, J. T. Sauls, S. Vadia, P. A. Levin, and S. Jun. 2019. Mechanistic origin of cell-size control and homeostasis in bacteria. Curr. Biol. 29: 1760–1770.

Si, F., D. Li, S. E. Cox, J. T. Sauls, O. Azizi, C. Sou, A. B. Schwartz, M. J. Erickstad, Y. Jun, X. Li, et al. 2017. Invariance of initiation mass and predictability of cell size in *Escherichia coli.* Curr. Biol. 27: 1278–1287.

Soifer, I., and N. Barkai. 2014. Systematic identification of cell size regulators in budding yeast. Mol. Syst. Biol. 10: 761.

Soifer, I., L. Robert, and A. Amir. 2016. Single-cell analysis of growth in budding yeast and bacteria reveals a common size regulation strategy. Curr. Biol. 26: 356–361.

Sompayrac, L., and O. Maaløe. 1973. Autorepressor model for control of DNA replication. Nat. New Biol. 241: 133–135.

Staudte, R. G., J. Zhang, R. M. Huggins, and R. Cowan. 1996. A reexamination of the cell-lineage data of E. O. Powell. Biometrics 52: 1214–1222.

Susman, L., M. Kohram, H. Vashistha, J. T. Nechleba, H. Salman, and N. Brenner. 2018. Individuality and slow dynamics in bacterial growth homeostasis. Proc. Natl. Acad. Sci. USA 115: E5679–E5687.

Sveiczer, A., B. Novak, and J. M. Mitchison. 1996. The size control of fission yeast revisited. J. Cell Sci. 109: 2947–2957.

Taheri-Araghi, S., S. Bradde, J. T. Sauls, N. S. Hill, P. A. Levin, J. Paulsson, M. Vergassola, and S. Jun. 2015. Cell-size control and homeostasis in bacteria. Curr. Biol. 25: 385–391.

Tanaka, K., T. Kanbe, and T. Kuroiwa. 1985. Three-dimensional behaviour of mitochondria during cell division and germ tube formation in the dimorphic yeast *Candida albicans*. J. Cell Sci. 73: 207–220.

Tanouchi, Y., A. Pai, H. Park, S. Huang, R. Stamatov, N. E. Buchler, and L. You. 2015. A noisy linear map underlies oscillations in cell size and gene expression in bacteria. Nature 523: 357–360.

Tempest, D. W., J. R. Hunter, and J. Sykes. 1965. Magnesium-limited growth of *Aerobacter aerogenes* in a chemostat. J. Gen. Microbiol. 39: 355–366.

Thattai, M., and A. van Oudenaarden. 2004. Stochastic gene expression in fluctuating environments. Genetics 167: 523–530.

Thompson, D. A. 1917. On Growth and Form. Cambridge University Press, Cambridge, UK.

Trueba, F. J. 1982. On the precision and accuracy achieved by *Escherichia coli* cells at fission about their middle. Arch. Microbiol. 131: 55–59.

Turner, J. J., J. C. Ewald, and J. M. Skotheim. 2012. Cell size control in yeast. Curr. Biol. 22: R350–R359.

Tyson, J. J. 1989. Effects of asymmetric division on a stochastic model of the cell division cycle. Math. Biosci. 96: 165–184.

Tyson, J. J., and K. B. Hannsgen. 1985a. The distributions of cell size and generation time in a model of the cell cycle incorporating size control and random transitions. J. Theor. Biol. 113: 29–62.

Tyson, J. J., and K. B. Hannsgen. 1985b. Global asymptotic stability of the size distribution in probabilistic models of the cell cycle. J. Math. Biol. 22:61–68.

Tyson, C. B., P. G. Lord, and A. E. Wheals. 1979. Dependency of size of *Saccharomyces cerevisiae* cells on growth rate. J. Bacteriol. 138: 92–98.

Vallania, F. L., M. Sherman, Z. Goodwin, I. Mogno, B. A. Cohen, and R. D. Mitra. 2014. Origin and consequences of the relationship between protein mean and variance. PLoS One 9: e102202.

Vashistha, H., M. Kohram, and H. Salman. 2021. Non-genetic inheritance restraint of cell-to-cell variation. eLife 10: e64779.

Vila-Sanjurjo, A. 2008. Modification of the ribosome and the translational machinery during reduced growth due to environmental stress. EcoSal Plus 3(1). https://doi.org/10.1128/ecosalplus.2.5.6

Volkmer, B., and M. Heinemann. 2011. Condition-dependent cell volume and concentration of *Escherichia coli* to facilitate data conversion for systems biology modeling. PLoS One 6: e23126.

Voorn, W. J., and L. J. Koppes. 1998. Skew or third moment of bacterial generation times. Arch. Microbiol. 169: 43–51.

Waddington, C. 1942. Canalization of development and the inheritance of acquired characters. Nature 150: 563–565.

Waldron, C., and F. Lacroute. 1975. Effect of growth rate on the amounts of ribosomal and transfer ribonucleic acids in yeast. J. Bacteriol. 122: 855–865.

Wallden, M., D. Fange, E. G. Lundius, Ö. Baltekin, and J. Elf. 2016. The synchronization of replication and division cycles in individual *E. coli* cells. Cell 166: 729–739.

Warner, J. R., J. Vilardell, and J. H. Sohn. 2001. Economics of ribosome biosynthesis. Cold Spring Harb. Symp. Quant. Biol. 66: 567–574.

Weart, R. B., A. H. Lee, A. C. Chien, D. P. Haeusser, N. S. Hill, and P. A. Levin. 2007. A metabolic sensor governing cell size in bacteria. Cell 130: 335–347.

Weiße, A. Y., D. A. Oyarzún, V. Danos, and P. S. Swain. 2015. Mechanistic links between cellular trade-offs, gene expression, and growth. Proc. Natl. Acad. Sci. USA 112: E1038–E1047.

Wolf, D. M., V. V. Vazirani, and A. P. Arkin. 2005. Diversity in times of adversity: Probabilistic strategies in microbial survival games. J. Theor. Biol. 234: 227–253.

Wu, C., R. Balakrishnan, N. Braniff, M. Mori, G. Manzanarez, Z. Zhang, and T. Hwa. 2022. Cellular perception of growth rate and the mechanistic origin of bacterial growth law. Proc. Natl. Acad. Sci. USA 119: e2201585119.

Yoshida, M., S. Tsuru, N. Hirata, S. Seno, H. Matsuda, B. W. Ying, and T. Yomo. 2014. Directed evolution of cell size in *Escherichia coli*. BMC Evol. Biol. 14: 257.

Yoshida, H., and A. Wada. 2014. The 100S ribosome: Ribosomal hibernation induced by stress. Wiley Interdiscip. Rev. RNA 5: 723–732.

Young, R., and H. Bremer. 1976. Polypeptide-chain-elongation rate in *Escherichia coli* B/r as a function of growth rate. Biochem. J. 160: 185–194.

Yule, G. U. 1925. Mathematical theory of evolution, based on the conclusions of Dr. J. C. Willis, F.R.S. Phil. Trans. Royal Soc. London Series B 213: 21–87.

Zalkinder, V. 1979. Correlation between cell nutrition, cell size and division control. Part I. Biosystems 11: 295–307.

Zavřel, T., M. Faizi, C. Loureiro, G. Poschmann, K. Stühler, M. Sinetova, A. Zorina, R. Steuer, and J. Červený. 2019. Quantitative insights into the cyanobacterial cell economy. eLife 8: e42508.

Zheng, H., Y. Bai, M. Jiang, T. A. Tokuyasu, X. Huang, F. Zhong, Y. Wu, X. Fu, N. Kleckner, T. Hwa, and C. Liu. 2020. General quantitative relations linking cell growth and the cell cycle in *Escherichia coli*. Nat. Microbiol. 5: 995–1001.

Zhu, M., X. Dai, and Y. P. Wang. 2016. Real time determination of bacterial *in vivo* ribosome translation elongation speed based on LacZα complementation system. Nucleic Acids Res. 44: e155.

Zhuravel, D., D. Fraser, S. St.-Pierre, L. Tepliakova, W. L. Pang, J. Hasty, and M. Kaern. 2010. Phenotypic impact of regulatory noise in cellular stress-response pathways. Syst. Synth. Biol. 4: 105–116.

Zohary, T., G. Flaim, and U. Sommer. 2021. Temperature and the size of freshwater phytoplankton. Hydrobiol. 848: 143–155.

CHAPTER 10

The Cell Life Cycle

Having covered the general features of cell growth, we now focus on a number of key mechanistic and temporal aspects of cell life histories. Just as the soma of multicellular organisms undergo developmental changes, single cells progress through several stages from birth to division, some of which are very tightly defined and regulated. It is, for example, vitally important for cells to have properly duplicated their genomes and oriented each complement to their appropriate destinations at the time of division. This is a particular challenge for eukaryotic cells with multiple chromosomes. During mitosis, each chromosome must be replicated once, and only once, and parallel sets of chromosomes must be transmitted to each daughter cell.

Although most unicellular eukaryotes and all prokaryotic species reproduce in an effectively clonal manner, indefinite rounds of such propagation in the former are often punctuated by phases of sexual reproduction during which pairs of individuals exchange chromosomal segments by recombination. During such sexual phases, eukaryotic cells switch from mitotic to meiotic genome division, wherein a diploid phase is reduced to the haploid life-cycle stage. To return to the diploid state, haploid individuals must locate partners of the appropriate mating type and then undergo fusion with them. For most multicellular species the predominant growth stage is diploid. For a wide range of unicellular species the primary (vegetative) phase is haploid, while the diploid stage is simply a transient moment between the initiation of cell fusion and meiosis.

Sexual reproduction raises a number of functional and evolutionary issues, not all of which are fully understood. How and why did the complex process of meiosis, including organized modes of chromosomal segregation and recombination, evolve out of the already detailed orchestrations of mitosis? How do cells make 'decisions' to fuse only with appropriate partners? How are mating types determined, and why is the typical number of mating types within a species just two? Many of the proteins involved in various stages of sexual reproduction appear to diverge at unusually high rates, begging the question as to whether such evolution is the product of drive-like processes associated with the relentless operation of selection for successful gene transmission.

A secondary goal here is to introduce a breadth of comparative observations on the molecular basis of cellular features at a deeper level than encountered in previous chapters. Although this initial exploration is restricted to the diversification of life-history mechanisms, many of the underlying themes will reappear in subsequent chapters on other cellular features. Two key issues concern the evolution of complexity at the molecular and network levels. Proteins often consist of organized multimers of subunits. Sometimes the subunits are all encoded by the same locus (homomers), and other times they arise from different genetic loci (heteromers, often derived by gene duplication). Commonly, but not always, eukaryotic proteins take the second route, although there is little evidence (if any) that this increase in molecular complexity is driven by adaptive processes, a point already raised on prior pages with respect to ATP synthase and ribosomes. However, as such transitions elicit sustained, coordinated molecular coevolution at the binding interfaces of interacting partners, this molecular remodelling can passively lead

Evolutionary Cell Biology. Michael Lynch, Oxford University Press. © Michael Lynch (2024). DOI: 10.1093/oso/9780192847287.003.0010

to species reproductive-isolating barriers, as incompatibilities arise between the component parts residing in different lineages.

The regulatory networks of the cell cycle and the stages of mitosis and meiosis involve communication pathways between gene products. However, such systems are often endowed with seemingly excessive and arcane structures, the origins of which raise central evolutionary questions in themselves. Equally significant is the repeated observation that, even when a network structure remains constant, changes can occur in the underlying participating proteins, a process known as non-orthologous gene replacement. Recall that cellular systems with highly conserved functions but high levels of divergence of underlying control mechanisms have already been encountered in Chapter 9, e.g., ribosome biogenesis and division-time determination. An analogy is the legendary Ship of Theseus, whereby over time the Athenians gradually replaced every wooden plank, until none of the original components remained, raising the question as to whether the new construct is still equivalent to the original ship. In cell biology, the replacement planks are sometimes not even made from the original materials.

The Eukaryotic Cell Cycle

The life cycles of eukaryotic cells can be roughly subdivided into three phases, defined with respect to the genomic state: 1) a growth phase in which all cell contents other than the genome expand in number; 2) a period of genome replication in preparation for division (during which growth might continue); and 3) cell fission (cytokinesis) accompanied by transmission of separate genomes to each daughter cell. In practice, however, most cell biologists partition the eukaryotic cell cycle more finely into four or five genome-focused phases (Figure 10.1). The textbook model is: a brief (and sometimes undetectable) G_0 resting phase immediately following division (not shown in the figure); followed by a prolonged interphase, which is further divided into three stages – the G_1 (gap 1) phase during which cell size increases, the S (synthesis) phase during which the genome is replicated, and the G_2 (gap 2) phase during which the cell continues

to grow while containing a duplicated genome; and finally culminating in the M (mitotic) phase during which chromosomes are separated and cell division proceeds. We next discuss how the M phase is traditionally subdivided further into four or five subsections defined by chromosomal states.

Regulated checkpoints ensure that cells do not progress from one stage to the next unless all is in order. For example, a G_1/S checkpoint ensures that DNA synthesis does not initiate in the absence of sufficient cellular resources; a G_2/M checkpoint ensures that mitosis does not proceed until all chromosomes have been fully replicated; and an additional checkpoint within the M phase ensures that any problems in chromosome replication have been resolved prior to segregation.

This traditional scheme for classifying cell-cycle steps can be confusing, as the absolute and relative lengths of the cell-cycle phases vary greatly among organisms and cell types. The G_0 phase is often negligible in unicellular organisms (and consequently ignored in the overall scheme), but can be effectively indefinite in terminally differentiated cells of multicellular species. The G_1 and G_2 phases can be essentially absent in rapidly dividing cells in early metazoan development (as cells simply get progressively smaller); and this can also transiently occur in some unicellular species (such as the green

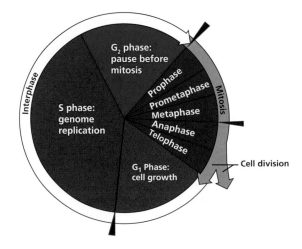

Figure 10.1 An idealized depiction of the eukaryotic cell cycle, divided into different phases (the absolute and relative lengths of which can vary dramatically among cell types). Checkpoints are denoted by the black arrows.

alga *Chlamydomonas*) that undergo multiple S/M cycles, without growth, prior to release of multiple progeny (e.g., 4, 8, or 16, following 2, 3, or 4 internal divisions).

Many overviews have been written on the complex web of interactions that constitute the eukaryotic cell cycle (e.g., Morgan 2007). However, the notation for the large number of participating proteins is often opaque and inconsistent across lineages, and this is not the place to recite the details. Suffice it to say that advancement through cell-cycle stages is generally governed by a multiplicity of cyclin proteins and their cyclin-dependent kinase (CDK) partners. Cyclins vary in concentration throughout the cell cycle, as they are actively degraded once deployed and then progressively resynthesized in the next cycle. Their functional role is to activate specific client CDKs, which then phosphorylate downstream target proteins, thereby directing entry into the next stage of cellular development. The following two sections highlight some surprising features of the underlying mechanisms.

Phylogenetic diversity

Despite the centrality of the cell cycle to all eukaryotes, the evolved diversity in the underlying regulatory machinery is striking. In effect, at the molecular level there is no standard eukaryotic cell cycle machinery. Among phylogenetic lineages, unrelated genes may carry out the same functions, and network topologies can change.

First, non-orthologous gene replacements of the proteins participating in the cell cycle are common (Jensen et al. 2006). For example, whereas most cell-cycle proteins in land plants and animals appear to be of common descent (Doonan and Kitsios 2009; Harashima et al. 2013), numerous fungi harbor key cell-cycle genes with no obvious relationship to those in the same network positions in plants and animals (Rhind and Russell 2000; Cross et al. 2011; Medina et al. 2016).

As one example, a fungal protein called E2F, which operates in the same pathway position as SBF in plants and animals (Figure 10.2), appears to be related to a viral protein, possibly acquired by horizontal transfer. The proteins with which E2F and SBF interact are also unrelated. The fact that the genomes of some basal fungal lineages harbor both E2F and SBF suggests the presence of a redundant regulatory system in basal fungi, with certain sublineages having lost SBF subsequently. Among the well-studied yeasts *Saccharomyces*, *Schizosaccharomyces*, and *Candida*, there are substantial non-orthologies in additional players in the cell cycle and their downstream regulated genes (Côte et al. 2009). Even among species within the genus *Saccharomyces*, differences exist in the interacting proteins at the G_1/S checkpoint (Drury and Diffley 2009).

Second, there are numerous examples in which the cell-cycle network topology itself has changed. For example, *Saccharomyces* deploys just a single CDK, whereas metazoan cells deploy at least four, while plants use two (Criqui and Genschik 2002). Substantial differences in the numbers of cyclins deployed in the cell cycle also exist among phylogenetic groups, e.g., three to four in yeasts, up to ten in plants and metazoans (Criqui and Genschik 2002; Cross and Umen 2015), and as many as two dozen in diatoms (Huysman et al. 2010) and ciliates (Stover and Rice 2011). In many cases, the additional genes originated via duplication. On the other hand, a broad phylogenetic survey suggests that many of the cell-cycle proteins observed across the eukaryotic phylogeny (and hence, by extrapolation were present in LECA) exhibit lineage-specific losses (Medina et al. 2016). For example, *Giardia intestinalis*, a single-celled parasite with two nuclei, has no anaphase-promoting complex (often used in targeting cyclins for degradation) and no checkpoint mechanism for mitotic entry (Gourguechon et al. 2013; Markova et al. 2016). Given that the cell-cycle network has been ascertained in only a few model organisms, and even then generally just partially so, many more variants are likely to be found.

A third key observation about the molecular basis of the cell cycle concerns the frequent redundancy in the underlying mechanisms. For example, during the DNA synthesis phase in yeast, at least three simultaneously acting mechanisms prevent secondary replication events (which would lead to chromosomal imbalance in progeny cells). The first of these involves proteolysis of the proteins that initiate replication; the second involves nuclear exclusion of key proteins; and the third involves

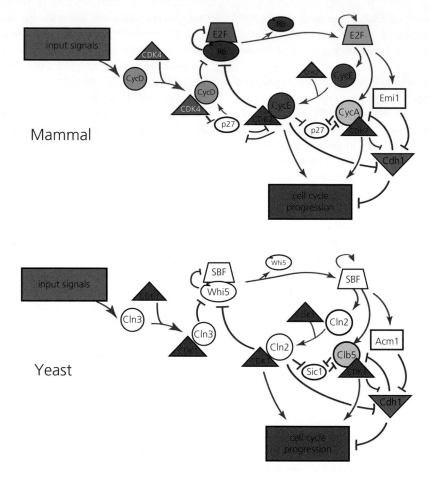

Figure 10.2 Nearly constant cell-cycle topology in the face of dramatic repatterning of the underlying components. The upper and lower panels denote the networks for mammal and budding yeast *S. cerevisiae*, respectively. Triangles denote CDKs, and circles cyclins. Green arrows denote activation steps; red lines with blunt ends denote inhibitory interactions. Proteins with coloured enclosures have orthologous sequences in other eukaryotes, and those with the same colour in mammal and yeast are orthologous to each other, whereas those in white are unrelated. The genes discussed in the text, E2F/SBF, appear as trapezoids at the top. From Cross et al. (2011).

direct binding at origins of replication. If deleted singly, none of these lead to unviability, implying that the three systems effectively back each other up (Drury and Diffley 2009).

A potential connection between such redundancy and the phylogenetic turnover of cell-cycle participants noted here can be seen as follows (with a more formal presentation appearing in Chapter 20). Imagine three layers of surveillance with error rates e_1, e_2, and e_3, operating in parallel so that the system fails only if all three layers fail to error-correct. The overall failure rate associated with the first

mechanism alone is e_1, with the first and second is $e_1 e_2$, and for all three is $e_1 e_2 e_3$. Because e_1, e_2, and e_3 are all < 1, multiple surveillance layers can greatly reduce the overall error rate. However, because natural selection operates on the cumulative error rate, $E = e_1 e_2 e_3$, and likely can only reduce it to some level defined by the power of random genetic drift (Chapter 8), there will typically be multiple degrees of freedom by which the overall minimum error rate can be achieved (Lynch 2012), i.e., a low value of e_1 can compensate for a high value of e_2 or e_3. This further implies that provided one or two components can together accomplish the target sum E, one (or even two) components are potentially free to be lost in individual lineages. In the long run, this may lead to a phylogenetic repatterning of the molecular mechanisms underlying a pathway through evolutionary cycles of emergence of redundancy followed by random loss of individual components (Figure 10.3).

Of further relevance to the repatterning of the underlying participants in the cell cycle is the observation that many such proteins can have additional cellular functions, including roles in transcription regulation and development in multicellular species. Multifunctional genes may be difficult to completely non-functionalize over evolutionary time, while still being subject to loss of individual subfunctions (such as participation in the cell cycle) when other redundant mechanisms remain (Chapter 6). Under the latter scenario, loss of connectivity of a particular gene with the cell cycle in a phylogenetic lineage might be followed by a regain in connectivity at a later point in time.

The key point here is that the cell cycle, one of the most central features of eukaryotic cells, provides a dramatic example of regulatory rewiring underlying an otherwise stable cellular attribute. Many more cases of this nature involving other aspects of cell biology appear in subsequent chapters. Although horizontal transfer (as implicated in yeast) can play a role in such evolutionary repatterning, the combination of gene duplication, transient redundancy, and multifunctionality of underlying participants further facilitates the opportunities for rewiring in an effectively neutral manner. Such neutral evolution apparently extends to key amino-acid residues in the final clients of the CDKs themselves, as even the phosphorylation sites appear to change locations among closely related taxa, while the regional clustering of sites within proteins is generally preserved (Moses et al. 2007; Holt et al. 2009).

Network complexity

It remains unclear why the cell cycle of most eukaryotes deploys such a large number of proteins with so many promoter and/or inhibitor activities (generally on the order of 20 or more). There is no evidence that such massive genomic investment is essential to an ordered cell-cycle progression, and all other things being equal, larger networks of proteins potentially impose a greater energetic burden on the cell, while also providing a larger target for mutational disruption.

Things may have been much simpler in the ancestral (pre-LECA) eukaryote. Prokaryotes do not have the elaborate mitotic cycles that are the hallmarks of eukaryotic genome replication, nor do they harbor any obvious orthologs of cyclins and CDKs. However, bacteria have loosely defined cell cycles governed by simple kinase-receptor systems that

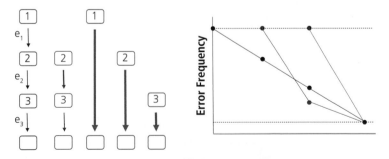

Figure 10.3 Three alternative surveillance-pathway architectures for perfecting a final outcome. There are three progressive steps in the black pathway. Under the red pathway, which has lost the first step, the same final outcome is accomplished by employing highly efficient surveillance in step 2, whereas the blue pathways (shown in three potential forms) deploy only a single, highly efficient step. The graph on the right illustrates the progressive removal of errors by successive steps. As shown on the left, the black pathway with three progressive steps can be simplified to three different one-step pathways (in blue), rendering non-overlapping mechanisms (with no shared molecular steps) in the resultant lineages.

enable a central response protein to cyclically dictate progression through growth, replication, and division stages (Biondi et al. 2006; Garcia-Garcia et al. 2016; Osella et al. 2017; Mann and Shapiro 2018). Moreover, the cell cycle of fission yeast (*S. pombe*) can be engineered to run with an extremely simplified control mechanism with just one CDK fused to a single cyclin (Coudreuse and Nurse 2010). If nothing else, these observations demonstrate the feasibility of an ancestral eukaryotic cell cycle driven by something as simple as a single self-oscillating module, and requiring no differential expression, interaction, and degradation of multiple participants.

The evolutionary mechanisms leading to the growth of network complexity, and how this can emerge by effectively neutral processes, were touched upon in Chapter 6. For now, we simply consider an observation from *S. cerevisiae*, a member of a yeast lineage that experienced an ancestral genome-duplication event, possibly resulting from interspecific hybridization (Wolfe and Shields 1997; Marcet-Houben and Gabaldón 2015). Although only a small fraction of duplicated gene pairs still survive in this species, a pair of genes involved in the *S. cerevisiae* mitotic cell cycle (*Bub1* and *Mad3*) is informative. In a number of eukaryotes (including some other yeasts and metazoans), a single gene

encodes for a key cell-cycle protein having two substantially different functions: binding to the kinetochore to ensure the proper segregation of sister chromosomes during mitosis, and regulating the spindle checkpoint. In *S. cerevisiae*, however, the duplicate versions of this gene partition up these tasks. Without a comparative perspective, one might conclude that either *Bub1* or *Mad3* evolved a new function, but instead this represents a clear example of subfunctionalization of the joint properties of an ancestral gene.

Remarkably, parallel patterns of subfunction partitioning by complementary degenerative mutation have occurred in multiple eukaryotic lineages following independent duplications of the same ancestral gene (Murray 2012; Suijkerbuijk et al. 2012) (Figure 10.4). Support for the idea that this increase in complexity of the underlying mitotic machinery does not involve the establishment of novel and/or beneficial functions is provided by an experimental replacement of the two *S. cerevisiae* genes by the single copy from a distantly related yeast species *Lachancea kluyveri*, which yielded negligible fitness consequences (Nguyen Ba et al. 2017). Taken together, these observations constitute a clear example of how the growth of network complexity can occur in the absence of any intrinsic selective advantages.

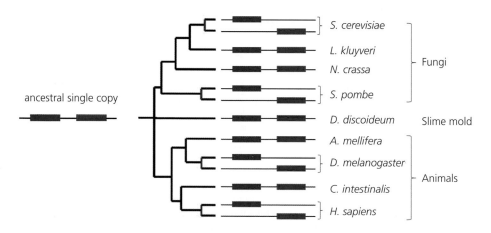

Figure 10.4 The phylogenetic distribution of genes related to two spindle-checkpoint genes in the yeast *S. cerevisiae*, *Bub1* (blue) and *Mad3* (red). In several species, e.g., *L. kluyveri* and *N. crassa,* a single gene is endowed with both the *Bub1* and *Mad3* subfunctions, but in four species (denoted by the red vertical bars on the phylogenetic tree on the left), including the reference *S. cerevisiae*, the bifunctional gene was duplicated, with each descendent copy then losing a complementary subfunction. The figure is simplified from Nguyen Ba et al. (2017).

Mitosis

Critical to the success of any cell lineage is the reliable production of progeny containing a full complement of the parental genome, i.e., the avoidance of chromosome loss during cell division. We focus first on issues related to mitosis, the process by which parental chromosomes are replicated and evenly transmitted to asexually produced daughter cells. In most bacterial species, mitosis is stereotypical – duplication of the genome (generally a single circular chromosome) starts from a single origin of replication, with DNA polymerases proceeding simultaneously down both sides of the circle until meeting at the single terminus. During this process, the newly emerging chromosomes begin to move towards opposite ends of the cell, and cell division

is completed as a constricting furrow pinches off the two daughter cells near the parental midpoint.

Eukaryotic mitosis is much more complicated. It always involves linear chromosomes, up to several dozen in number, each often longer than entire bacterial genomes (Lynch 2007). The choreographed process of eukaryotic mitosis is viewed conventionally as a series of five temporal stages familiar to all biology students (Figures 10.1 and 10.5): 1) interphase, wherein chromosomes duplicate into sisters; 2) prophase, wherein chromosomes condense and microtubule arrays begin to assemble; 3) metaphase, wherein the sister chromosomes, connected at centromeres and attached to kinetochore microtubules, line up in the middle of the spindle; 4) anaphase, wherein sister chromosomes detach from each other and move to opposite poles; and 5)

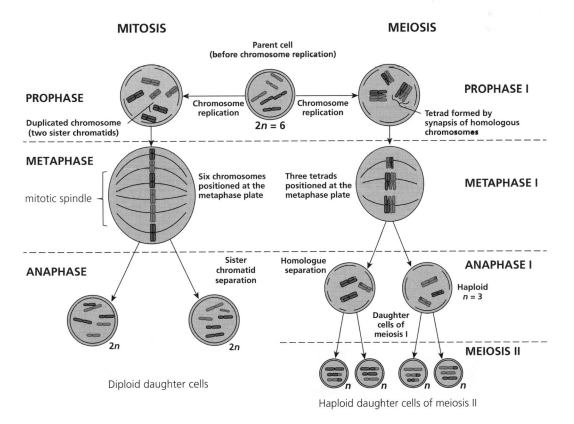

Figure 10.5 The contrast between the basic steps in eukaryotic mitosis and meiosis. Depicted are the assortment of three pairs of homologous chromosomes (red and blue denote chromosomes derived from paternal and maternal sources), which after duplication appear at metaphase as six pairs of sister chromatids in mitosis and as three pairs of tetrads in meiosis.

telophase, wherein chromosomes decondense into their new nuclear homes and the cell divides.

The evolutionary establishment of these sequential stages required numerous innovations not generally found in prokaryotes: 1) enclosure of the genome within a nuclear membrane perforated with nuclear-pore complexes to allow both export of mRNA to the cytoplasm for translation and import of key proteins into the nuclear environment; 2) the expanded use of nucleosomes (octomers involving four unique histone proteins, in contrast to tetrameric homomers in archaea) for spooling DNA; 3) substantial increases in numbers of origins of replication per chromosome and their parallel firing in the process of chromosome duplication; 4) capping of linear chromosomes with repeat-based telomeres and devotion of an enzyme (telomerase) for preventing end loss; 5) deployment of molecules for sister-chromatid cohesion prior to anaphase; 6) a switch from a membrane-based to a microtubule-based mechanism for segregating sister chromosomes; 7) establishment of centromeres for spindle attachment; and 8) the insertion of mitotic checkpoint mechanisms to ensure simultaneous and equitable migration of chromosomes to daughter cells.

Although these myriad features are shared by all of today's eukaryotes, the evolutionary order in which they appeared remains unknown. Moreover, as in the case of cell-cycle regulation, the molecular and cellular details of many aspects of eukaryotic mitosis have diverged so much among phylogenetic lineages that it is difficult to even specify the ancestral state of the underlying machinery. Phylogenetic analysis implies that at least 43 proteins involved

in genome replication were present in LECA, only 23 of which are found in all modern lineages, with others having been lost in lineage-specific manners (Aves et al. 2012). The following paragraphs attempt to highlight the diversity of mitotic mechanisms in a non-technical manner.

Numerous proteins used in eukaryotic mitosis have orthologs in archaea, with many of these experiencing duplication and functional divergence in eukaryotes (Aves et al. 2012; Lindås and Bernander 2013). To start the discussion, three key complexes involved in the initiation and progression of eukaryotic chromosome replication merit attention (Figure 10.6): 1) PCNA (proliferating cell nuclear antigen) is a trimeric ring that serves as a clamp to recruit DNA polymerase to single-stranded DNA; 2) RFC (replication factor, also known as the clamp loader) is a pentamer consisting of a chain of four similar subunits anchored to a larger component, which together endow the DNA polymerase with processivity; and 3) MCM (minichromosome maintenance complex) is a hexameric ring that unwinds DNA at the replication fork.

All three of these complexes exhibit substantial phylogenetic variation in terms of their underlying components (Chia et al. 2010). For example, in archaea, the trimeric PCNA can be a homomer (all three subunits encoded by the same locus) or a heteromer constructed from two or three distinct proteins. In contrast, eukaryotic PCNA is homomeric, with no evidence of functional superiority over the archaeal form (Fang et al. 2014). The RFC chain consists of one or two protein types in archaea, whereas each of the four subunits is encoded by a different

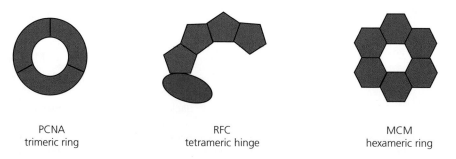

PCNA	RFC	MCM
trimeric ring	tetrameric hinge	hexameric ring

Figure 10.6 Simplified depictions of three of the multimeric proteins involved in DNA replication. The green subunits may all be encoded by the same genetic locus (homomers) or by two or more proteins (heteromers).

gene in eukaryotes. The MCM has one to five sub-unit types in archaea, whereas all six subunits are encoded by different genes in eukaryotes. In each of the three cases, the divergent eukaryotic compo-nents arose by gene duplication of ancestral compo-nents on the path from FECA to LECA, independent of the duplication events in archaea (Liu et al. 2009). Finally, another complex (GINS) that interacts with the MCM at origins of replication is generally a homotetramer in archaea but a heterotetramer in eukaryotes (Onesti and MacNeill 2013).

This collection of observations provides a first illustration of a recurrent theme of multimeric eukaryotic proteins often being more complex (in terms of number of gene products involved) than orthologous prokaryotic complexes (Chapter 13). The stochastic coevolution of interface residues among the partner proteins in heteromeric com-plexes can initiate and sharpen species boundaries, as the gene products from sister taxa diverge to the point of failing to interact in hybrid backgrounds (Zamir et al. 2012).

A second set of examples involving variation in the complexity of the components of the mitotic machinery involves the SMC (structural mainte-nance of chromosome) proteins, which are ubiqui-tous across the Tree of Life. All SMC complexes are dimers of coiled-coil proteins, with one end of the two members joining to make a flexible hinge, and the other end providing an opening that can be closed in certain contexts. In bacteria, the homodimeric molecules are involved in chro-mosome maintenance and compaction. In eukary-otes, the complexes are heterodimeric, and there are multiple copies with more diverse roles: SMC1/3 dimers form cohesins, which hold sister chromatids together during S phase; SMC2/4 dimers are part of the complex that condenses chromosomes to their metaphase state; and SMC5/6 dimers are recruited in some forms of DNA repair. Notably, although five gene duplications account for the six SMC proteins in eukaryotes, the components of indi-vidual complexes are not consistently each others' closest relatives. For example, *SMC1* and *SMC4* are sister genes, as are *SMC2* and *SMC3* (Cobbe and Heck 2004). All three heterodimers were estab-lished prior to LECA, again reflecting the deep roots of the components of the eukaryotic mitotic machinery.

Finally, although most readers will be familiar with the basic textbook version of mitosis (as illus-trated in Figure 10.5), this is a considerable over-simplification. As with the broader cell cycle, the mechanisms of eukaryotic chromosome segregation have diversified to an enormous extent from the standard model in almost every conceivable way. For example, in most taxa, DNA-attachment factors (known as kinetochores) assemble onto the cen-tromeres of sister chromosomes, connecting them to long polymeric proteins (the spindle microtubules) that guide chromosomes into daughter cells. Kine-tochore complexes consist of ~ 50 different proteins, many of which appear to have arisen by duplica-tion (Tromer et al. 2019). However, although the structure is thought to be relatively conserved, there are significant differences in component composi-tions among mammals, insects, and yeasts (Drin-nenberg et al. 2016). In the kinetoplastids (which include the parasitic trypanosomes), kinetochores are constructed out of 19 lineage-specific proteins (Akiyoshi and Gull 2014). Generally, the ends of spindle microtubules are anchored to cytoplasmic centrosomes during cell division, but the centro-some is absent in some groups such as planarians (Azimzadeh et al. 2012) and replaced by a non-homologous spindle pole body in budding yeast (Winey and Bloom 2012).

The most remarkable and visually obvious forms of variation in mitosis involve the behavior of the nuclear envelope and the locations of the micro-tubule organizing centres from which the spin-dles emerge (Sazer et al. 2014). In some lineages (e.g., metazoans and land plants) mitosis is open, with the nuclear envelope disappearing prior to metaphase, whereas in others (e.g., most fungi) the nuclear envelope remains intact throughout mito-sis. Across the phylogeny of eukaryotes, however, there is a complete continuum of intermediate forms of partially open mitosis. In addition, under closed mitosis, spindles can initiate inside or outside of the nuclear envelope (in the latter case penetrat-ing the membrane). The numbers of microtubules per kinetochore vary among lineages, as do the ways in which these are bundled. Most species have point centromeres on each chromosome, but numerous cases exist in which chromosomes are holocentric (with microtubules attaching along their full lengths).

Summaries of these and numerous other mitotic features across a wide range of phylogenetic lineages are provided in Kubai (1975), Heath (1980), and Raikov (1982). In one of the few studies to ever analyse genetic variation in internal cellular features, Farhadifar et al. (2015, 2020) revealed substantial levels of within- and between-species variation in spindle lengths, elongation rates, and centrosome sizes in *Caenorhabditis* nematodes, illustrating how mitosis-related traits can evolve both qualitatively and quantitatively.

In the Darwinian tradition, here we would be expected to speculate on, if not celebrate, the adaptive basis of the substantial lineage-specific diversification of mitosis across eukaryotes. However, there is no evidence that the emergence of mitosis or the downstream divergence of chromosome assortment mechanisms endowed their bearers with adaptive superiority. Moreover, the widespread existence (and success) of prokaryotes with consistently simple means of chromosome segregation, despite having had billions of years and thousands of lineages to have evolved alternative procedures further challenges the view that eukaryotic cell-division mechanisms are intrinsically advantageous. This being said, however, once mitosis had become established in LECA, it opened up novel pathways for further diversification by descent with modification, the primary innovation being the subject of the following section.

Meiosis

The most unique aspect of eukaryotic genome inheritance is meiotic cell division. Combined with the use of separate sexes, meiosis endows diploid eukaryotic cells with an organized mode of sexual reproduction and a capacity for generating substantial genetic variation among progeny. Via two nuclear divisions, only the first of which involves replication, meiosis reduces each pair of chromosomes in a diploid parental cell to single chromosomes in each of four haploid progeny cells (or

gametes) (Figure 10.5). Diploidy is subsequently restored by gamete fusion.

Although meiosis shares some physical processes with mitosis, two differences have major genetic consequences: 1) copies of homologous chromosomes in the diploid parent cell segregate independently into haploid progeny; and 2) prior to doing so, most chromosomes experience at least one crossover (with homologous chromosomes from the two parents swapping segments by recombination). Thus, unlike the situation in mitotic cell division, the products of meiosis are almost never identical to each other. The subsequent union of haploid gametes (generally from different parents) into diploids creates still more genetic diversity.

Almost certainly an evolutionary derivative of mitosis, the establishment of meiosis required four innovations (Figure 10.5): 1) physical pairing of homologous parental chromosomes during first-division prophase; 2) recombination between homologs (non-sisters) initiated by enzymatically induced double-strand breaks; 3) suppression of sister-chromatid separation (connected by centromeres) in the first division; and 4) the absence of chromosome replication during the second division. In effect, these modifications convert one-step mitosis into a two-step process by inserting the first meiotic division (and its associated peculiarities) into the mitotic cycle (Gerton and Hawley 2005; Wilkins and Holliday 2009).

During the first division, the genome is duplicated and rearranged, with the first two daughter cells being effectively genetically haploid (homozygous) for all DNA residing between the centromere and the proximal crossover, but potentially heterozygous for sites distal to the last crossover. For this reason, the first division is referred to as reductional. Cells enter the second meiotic division in the same way as in mitosis, with replicated chromosomes, which leads to complete haploidy for all chromosomal regions and is referred to as the equational division.

Origin and evolutionary modifications of meiosis

Meiosis is by no means a requirement for sexual reproduction, which simply requires cell fusion followed by cell division and separation. Meiotic sexual reproduction is unique in that it involves a cycling of haploid and diploid stages (even if one is fleetingly transient). Much speculation has been offered as to the order of events leading from mitosis to the more complex meiotic program (e.g., Maguire 1992; Kleckner 1996; Solari 2002; Egel and Penny 2007; Niklas et al. 2014). From a purely molecular perspective, what can be said is that as with mitosis, many of the molecular components of the meiotic machinery arose by gene duplication on the phylogenetic path from FECA to LECA. The members of one such pair, *Rad51/Dmc1*, are respectively used in mitotically and meiotically dividing cells and are further discussed below. Two pairs of proteins involved in mismatch repair and the processing of recombinant molecules resulting from single-strand invasion (Pms1/Mlh2 and Mlh1/Mlh3) also arose by duplication (Ramesh et al. 2005). In addition, Spo11 (which, as described below, creates double-strand breaks during meiosis) gave rise to two new genes by duplications prior to LECA (Malik et al. 2007).

Given that nearly all extant eukaryotes harbor a meiotic stage, we can be secure that the process is not maladaptive under most conditions, and is likely to be advantageous. However, care needs to be taken in assuming that any advantages that might be involved in the maintenance of modern meiosis reflect the factors involved in the origin of the process. For example, although meiosis is consistently associated with the production of variation in today's eukaryotes, it need not follow that the earliest evolutionary steps towards meiosis had anything to do with generating variation. Assuming a haploid ancestral state, a diploid (or higher-order polyploid) phase may have started as a simple form of endoreplication without cell division, as occurs in many rapidly dividing prokaryote cells. If that were the case, the subsequent addition of meiotic mechanisms for restoring haploidy (and avoiding chromosome imbalance) might have simply involved closed homozygous lineages and hence, evolved prior to sexual reproduction (Cleveland 1947). Even if this particular scenario seems unlikely, the central point remains – the fact that meiosis generates variation in most of today's

eukaryotes need not mean that its establishment was driven by selection for variation.

The fusion of two compatible haploid cells is a necessary condition for sexual reproduction. Even without recombination between homologs, the production of haploid progeny can generate some variation by independent segregation of chromosomes, provided multiple chromosomes are present. However, FECA almost certainly had a capacity for recombination, as most prokaryotes harbor systems for repairing accidentally broken chromosomes off of homologous sequence in another chromosomal copy or segment (Haldenby et al. 2009). In bacterial recombination, RecA protein forms a helical filament that coats single-stranded DNA at the ends of breaks and then plays a central role in searching for and transferring the strand to homologous double-stranded sequence. Eukaryotic Rad51 is related to bacterial RecA, and like the latter, coats single-stranded DNA and guides the initial stages of repair by homology search in mitotically dividing cells. The duplicate version of Rad51, called Dmc1, is specifically involved in inter-homolog pairing during meiosis (Ramesh et al. 2005).

Thus, although meiotic recombination provides an organized mechanism for the repair of double-strand breaks in today's eukaryotes (Bengtsson 1985; Bernstein et al. 1988; Hurst and Nurse 1991), given that the physical mechanism of recombination is highly conserved across the major domains of life, it is clear that meiosis was never a requirement for recombination. However, a unique feature of meiosis is that double-strand break repair does not just involve the resolution of prior physical accidents incurred by chromosomes. Rather, in most organisms breaks are actively promoted during meiosis by an enzyme called Spo11. It is tempting to conclude from this that Spo11 is maintained as a means for generating variation by recombining paired parental chromosomes, but as discussed below, this may simply be a by-product of an essential mechanism for holding homologs together during the first meiotic division.

Hickey and Rose (1988) proposed that rather than being a product of adaptive processes, cell fusion and self-inflicted meiotic recombination might have been forced upon an ancestral eukaryote by a selfish DNA element (e.g., a self-proliferating transposon or retrotransposon) as a means for the latter's transmission among host cells. Without such transmission, a mobile element is essentially confined to a single host-cell lineage, possibly driving its host to extinction by generating deleterious insertions if overly aggressive or itself succumbing to mutation load. This general idea is made plausible by the fact that some bacteria engage in a sort of sexual reproduction guided by the activities of plasmids. In *Enterococcus faecalis,* for example, strains containing a particular plasmid are attracted to pheromones produced by non-carrier strains, resulting in conjugation and transfer of the plasmid to the naive strain (Wirth 1994).

Finally, despite the canonical view of meiosis outlined in Figure 10.5, as with mitosis, the process has diversified in numerous ways across the eukaryotic domain (Loidl 2016; Zickler and Kleckner 2016). Textbook descriptions of meiosis are virtually always based on mammalian cells and/or the budding yeast *S. cerevisiae*, but almost every aspect of meiosis common to these organisms has been found to vary among other phylogenetic lineages. For example, several species of the yeast *Candida* lack multiple genes previously thought to be essential to meiosis, and yet still engage in the process, complete with Spo11-dependent recombination (Butler et al. 2009; Reedy et al. 2009). The missing components include members of the synaptonemal complex (a chromosome-length assemblage of polymeric proteins used to bind homologs together during meiosis I; Figure 10.7), the crossover-resolution pathway, and Dmc1 (the meiosis-specific cohesion protein). In the ciliate *Tetrahymena pyriformis,* chromosomes pair within a tube-like nucleus nearly twice the length

of the cell, with all of the telomeres grouped at one end and all of the centromeres at the other, again without a synaptonemal complex. Whereas most species deploy point centromeres, many cases exist in which the chromosomes are holocentric (tightly paired over the entire length), as in the nematode *C. elegans*. Cases also exist in which no recombination occurs during meiosis despite chromosome pairing, as in male *Drosophila*. In addition, certain proteins involved in determining the fates of meiotic double-strand breaks (crossovers vs. no crossovers) differ among phylogenetic groups (Zetka 2017).

Rapid evolution of meiosis-associated proteins

Despite their conserved functions and structural features, some meiosis-associated proteins have remarkably rapid rates of evolution at the amino-acid sequence level (Bogdanov et al. 2007; Bomblies et al. 2015; Bonner and Hawley 2019). Particularly notable is the synaptonemal complex (SC), which consists of a three-legged ladder-like structure, with lateral elements bound to each homolog, a central parallel element, and a series of transverse filaments connecting the lateral and central elements (Figure 10.7). The SC exhibits dramatically different protein sequences across metazoan lineages, to the point of there being questions as to homology (Fraune et al. 2012). Just within the genus *Drosophila*, the amino-acid sequences of several of the component proteins of the SC evolve at rates on the order of at least 40% that expected under neutrality, with a number of sites putatively being under positive selection for change (Anderson et al. 2009; Hemmer and Blumenstiel 2016). In the yeast *S. pombe*, the SC has been replaced by thread-like structures called linear elements, which are structurally different from conventional lateral elements and do not appear to engage with transverse filaments at all (Lorenz et al. 2004).

Although the iconic view of the SC invokes a rigid ladder-like structure, the underlying elements appear to be movable, with the overall structure behaving as a liquid crystal (Rog et al. 2017). This is of interest because liquid crystals are known to be highly sensitive to temperature variation (and for this reason, are often used as temperature sensors

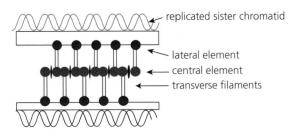

replicated sister chromatid

lateral element
central element
transverse filaments

Figure 10.7 Cartoon version of the synaptonemal complex holding two sister chromatids together.

in industrial applications). Meiotic processes tend to be highly sensitive to temperature (Bomblies et al. 2015; Lloyd et al. 2018), with the thermal optimum varying substantially among taxa, and the within-species temperature range for the faithful operation of meiosis often being only $\sim 5°C$. However, adaptation to environmental temperature variation cannot fully explain this extreme situation with the SC, as temperature influences the entire proteome, and most proteins are not this sensitive. It has been suggested that such rapid evolution is a consequence of a coevolutionary dance between interacting partners – with a slight modification of one member of the pair being met with a compensatory change in another (Bomblies et al. 2015), but as touched upon in Chapter 6, numerous factors determine whether coevolution between interacting molecules accelerate versus decelerate rates of sequence evolution.

One potential cause for high rates of evolution of the meiotic machinery is the relentless selection that must operate on parental chromosomes competing for successful transmission to gametes (Lindholm et al. 2016). Normally, one expects meiosis to give rise to four gametic products, all of which are free to contribute to the next generation. However, meiotic conflict can lead to situations in which one allelic type exhibits superiority with respect to another, as when one parental haplotype prevents the successful inheritance of another into gametes. Examples of such a meiotic-drive process are the spore-killer genes (often called sister killers) in a number of fungi, which increase their relative rates of transmission by killing haploid products that do not contain them (Turner and Perkins 1991; Vogan et al. 2019; López Hernández et al. 2021; Svedberg et al. 2021). Under such scenarios, parental cells will produce fewer than four haploid gametes, but the driving allele can still be brought to high frequency provided its success during meiosis exceeds the reduced production of successful progeny.

One cytological feature in particular naturally invites the emergence of exploitative chromosomes while incurring no fertility costs. For reasons that remain unclear, numerous phylogenetic groups exhibit a form of meiosis in females in which only one of the four meiotic products matures to a successful gamete, the other three being discarded. This so-called female meiosis has apparently evolved independently multiple times, being present in metazoans, land plants, ciliates, and a number of diatoms (Chepurnov et al. 2004), and naturally sets up a situation in which the four meiotic products compete for transmission to the one successful haploid egg. A drive-like process might then arise if centromere variants differ in their ability to successfully navigate to a particular location in the final meiotic tetrad (Figure 10.8). For example, an expansion of centromeric repeats leads to larger centromeres, which in principle can attract more kinetochores and spindle microtubules, and centromere location can also have effects (Chmátal et al. 2014; Iwata-Otsubo et al. 2017; Bracewell et al. 2019). As discussed in Chapter 4, even a 10^{-5} or so fitness advantage (on a scale of 1.0) would be adequate to drive such a centromere to high frequency. These kinds of observations motivate the centromere-drive hypothesis (Henikoff et al. 2001; Malik and Henikoff 2001).

Although there may be few side effects of a driving chromosome in female meiosis other than determining which homolog is promoted to the single egg cell, collateral problems may ensue in male meiosis (or even in mitosis), where there is an expected balanced outcome of cell division. This might then impose counter-selection for modifier mutations in centromeric proteins that restore normal meiotic segregation, thereby driving the rapid evolution of other genes involved in meiosis (Figure 10.8). In principle, once a such a modifier is driven to fixation, a new opportunity for another driving chromosome of a different nature might emerge, encouraging establishment of still another suppressor mutation.

Whereas the centromere-drive hypothesis provides a potentially simple explanation for the rapid evolution of the meiotic machinery, there are several uncertainties about the generality of the verbal model. First, the key requirement for a coevolutionary drive process is the maintenance of functionally significant polymorphisms in centromeric regions for a sufficiently long time to enable the centromeric proteins to respond by counter-selection. If a highly aggressive centromere rapidly goes to fixation, this will thwart the selective promotion

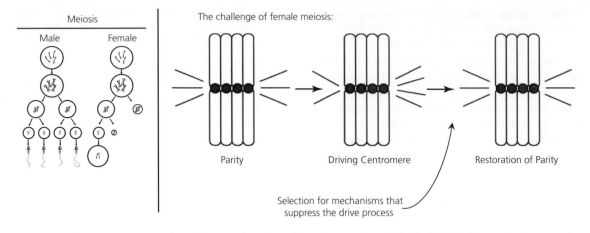

Figure 10.8 The centomere-drive hypothesis. **Left**: Male and female meiosis in diploid cells initially containing two pairs of homologous chromosomes, with the former leading to four haploid products and the latter to just one surviving gamete. **Right**: A driving centromere (red) is depicted as one with an excess number of attachments to spindle fibres during meiosis.

of modifier mutations, as homozygotes for driving centromeres might not experience problems with meiotic imbalance. Likewise, if the deleterious effects of a driving centromere on male fitness sufficiently exceed the power of the drive process, the driving centromere will simply be eliminated from the population too rapidly for the arrival of modifier mutations. Small population size might facilitate stochastic increases in the frequencies of mildly deleterious centromeres, but this will also reduce the rate of mutational origin of modifiers and the ability of natural selection to promote them. Thus, there must be a narrow range of population-genetic parameters conducive to centromeric drive.

A second concern with the centromere-drive hypothesis is that centromeric proteins must recognize the full set of centromere sequences across all chromosomes. There are no known chromosome-specific centromeric proteins. This means that in order to be successful, any modifier that restores parity at the problematical chromosome would have to do so without generating new difficulties with non-homologous chromosomes.

Finally, whereas the numerous examples noted here suggest a high rate of molecular evolution of components of the meiotic machinery, it is not clear whether such rapid evolution is fully general, let alone a consequence of driving centromeres. Of

particular interest is the centromeric variant of the histone H3 protein found in nucleosomes (CENP-A), which interacts with kinetochores and has been argued to evolve at an exceptionally high rate in *Drosophila* (Zwick et al. 1999; Malik and Henikoff 2001). In contrast, very distantly related plant species, which also have female meiosis, are able to accept transformations of CENP-A from each other, implying a relatively low level of functional divergence (Rosin and Mellone 2017). Moreover, the yeast *Saccharomyces*, which does not have female meiosis, nonetheless exhibits relatively high rates of centromere-sequence evolution (Bensasson et al. 2008; Bensasson 2011). These observations raise the caveat that, in addition to the population-genetic environment, aspects of the cellular environment dictate whether centromere drive (or any other meiotic-drive like process) can lead to accelerated rates of evolution of the participating genes.

One observation of potential relevance to this issue has been made in the ciliate *Tetrahymena*, where male meiosis is entirely absent. Both members of a conjugating pair undergo female-like meiosis, duplicate their single remaining haploid nucleus, and then pass one copy on to the other member. If rapid evolution of centromeric proteins in species with female meiosis is normally driven by deleterious side effects on male meiosis, accelerated rates of modifier evolution are unexpected in species without male meiosis. The observation of relative evolutionary stability of CENP-A in *Tetrahymena* is consistent with this idea (Elde et al. 2011), although this still leaves unexplained the counterexamples in the prior paragraph.

Recombination mechanisms

The key genetic features of meiosis are the independent segregation of non-homologous chromosomes and the production of chimeric daughter chromosomes by recombination between parental homologs. To avoid losses of chromosomal segments by deletion and/or duplication gains, recombination must be strictly confined to homologous chromosome regions. This requires proper chromosome alignment. In organisms that have been well-characterized, the initial search space for homology during early meiosis is usually greatly reduced by the clustering of telomeres near the nuclear periphery into a chromosomal bouquet (Scherthan 2001). Once juxtaposed, homologs are generally then held together by the SC, in combination with physical intercalations of single strands of DNA from one chromosome into the homologous regions of another (known as chiasmata), and perhaps pairing of non-coding RNAs (Barzel and Kupiec 2008; Ding et al. 2012). However, some species use one mechanism to the exclusion of another, or use entirely different mechanisms for chromosome pairing (Gerton and Hawley 2005).

As noted above, meiotic recombination events are not simple consequences of accidental chromosome breakage. Rather, they are specifically induced by the creation of double-strand breaks by Spo11, an enzyme related to a topoisomerase used in archaea to relieve supercoiling or tangled chromosomal regions (Robert et al. 2016; Vrielynck et al. 2016). Following a double-strand break, the ends of each fragment are partially digested, leaving single-stranded DNA overhangs, which then seek out, invade, and hybridize with the homologous region on the matching homolog. One or both members of the broken strand can invade the homolog, sometimes only transiently, and the ways in which the conjoined strands separate have consequences for the nature of the recombination event (Chapter 4). In almost all cases, there is a small patch of non-reciprocal exchange (called a gene conversion), but in a small fraction of cases a crossover occurs, yielding complete reciprocal exchange between homologs distal to the break (Figure 4.6).

Typically, no more than one or two double-strand breaks per chromosome arm are resolved as crossovers during an individual meiotic event (Chapter 4). However, the numbers of non-crossover events are tens to hundreds of times higher (De Muyt et al. 2009; de Massy 2013), and these transiently conjoined chromosomal regions keep the parental chromosomes in parallel during metaphase I. Removal of meiosis-specific cohesins distal to a crossover allows homologs to separate at meiosis I, whereas maintenance of cohesion proximal to the centromere keeps sisters joined until meiosis II (Watanabe 2012). Notably, the ciliate *Tetrahymena*, appears to utilize the same hinge during mitosis and meiosis (Howard-Till et al. 2013), and other species may use an entirely unrelated protein (Watanabe 2005).

As discussed in the following section, the typical adaptive view of homolog pairing is that such juxtaposition helps ensure a steady supply of recombinant chromosomes, providing useful variation upon which natural selection can act. A more structural view is that the primary role of homolog pairing is the inhibition of non-homologous recombination, which would lead to deleterious ectopic insertions, deletions, and chromosomal rearrangements (Wilkins and Holliday 2009). This being said, however, meiosis is far from a perfect process. For example, in humans $\sim 25\%$ of female meiotic products are aneuploid (Wang et al. 2017), and separation of sister rather than homologous chromosomes at meiosis I is not uncommon (Ottolini et al. 2015). With 23 chromosomes per human genome, this implies an error rate of $\sim 1\%$ per chromosome. The rate of non-disjunction for the X chromosome in *Drosophila* is estimated to be $\simeq 0.5\%$ (Zeng et al. 2010).

The evolutionary consequences of sexual reproduction

As outlined in subsequent chapters, the myriad novel cellular features in LECA opened up numerous avenues for eukaryotic cellular diversification. However, the onset of meiosis was unique in that it dictated a new mechanism for the inheritance of the genetic machinery itself, potentially defining new paths by which general evolutionary processes could proceed. For example, meiosis combined with conjugation (syngamy) ensures that, except in the case of self-fertilization, sexually produced progeny genomes are mixtures from two individual parents. Although the generation of variation may be viewed as beneficial from the standpoint of natural selection, it also means that, once obtained, an optimal parental genotype will generally not be perfectly transmitted to offspring.

Sexual reproduction is also costly in other ways. Cells of different mating types must locate each other and then fuse, but cell fusion also provides a vehicle for pathogen transmission. In species with separate sexes, females often contribute the bulk of the energetic investment in offspring, and the average number of progeny produced by individuals is typically reduced by a factor of two at the population level. The fact that asexual organisms pay none of these costs inspires the search for adaptive explanations for the evolutionary maintenance of sexual reproduction.

Obligate asexuality does not appear to be difficult to evolve from sexual reproduction, especially in organisms with mixed life cycles where phases of clonal reproduction alternate with sexual episodes (the usual situation for unicellular eukaryotes). Nearly every major phylogenetic group of multicellular eukaryotes harbors at least one obligately asexual lineage propagating via unfertilized eggs (Bell 1982). Although obligate asexuality is thought to be rare, this view is largely derived from observations of multicellular organisms, where the mating system is fixed and easily observed. The situation might be quite different in microbes, where induction of the sexual phase is often non-obvious. Nonetheless, with a typical focus on multicellular species, most evolutionary biologists assume that the rarity of transitions to propagation via unfertilized eggs (called parthenogenesis in animals, and apomixis in land plants) implies that there must be a large enough intrinsic advantage of sexual reproduction to offset the significant disadvantages just noted. The usual conclusion is that this must be

associated with the production of genetically variable offspring.

There are numerous ways in which the production of genetic variation by meiotic segregation and recombination might be advantageous (Williams 1975; Maynard Smith 1978; Kondrashov 1993; Barton and Charlesworth 1998). For example, outcrossing provides a means for promoting beneficial combinations of alleles from different genetic loci. Instead of waiting for two complementary mutations to sequentially arise in a single asexual lineage, single mutations contained within two different lineages can be combined, potentially reducing the waiting time for the emergence of a complex adaptation (Chapter 6). In addition, sexual reproduction can facilitate the purging of deleterious alleles (which constitute the bulk of spontaneously arising mutations). By expanding the range of variation in the numbers of deleterious mutations in offspring (some inheriting more and others less than the average parental number), sexual reproduction provides a more efficient route to reducing harmful mutation load by natural selection.

One concern with most arguments for the evolutionary maintenance of sex is their dependence on group-selection arguments – the inferred advantages are viewed through a long-term lens of the population. The fact that selection at the individual level is much more immediate than population turnover rates raises the question as to why, once established, sexual reproduction is resistant to invasion and eventual displacement by derived asexuals. One potential mechanism is purely genetic. With no known exceptions, diploid asexual cells are still capable of mitotic recombination and use this capacity to repair double-strand breaks off a homolog. However, because recombination generates local patches of homozygosity via gene conversion (Chapter 4), purely asexual lineages can be expected to experience progressive loss of heterozygosity, and hence, a relentlessly increasing exposure of deleterious recessive alleles carried in the original founder of the asexual lineage (as well as those subsequently arising), eventually leading to extinction. Thus, the capacity for complementation after each round of outcrossing may be a primary factor favoring at least periodic sexual reproduction in diploid organisms (Archetti 2004,

2005). However, this argument seems of minor significance for species with predominantly haploid life stages.

Meiotically reproducing species necessarily go through both haploid and diploid phases, so either might be the dominant state in the life cycle. In most multicellular species, diploidy is the rule, and the imagined genetic advantages of sexual outcrossing include the masking of deleterious recessive alleles, the ability to exploit any instances of heterozygote advantage, and the provisioning of secondary templates for double-strand break repair in mitotically reproducing cells. However, many unicellular eukaryotes are predominantly haploid, which reduces the investment in DNA, while also enhancing the exposure of recessive alleles to natural selection. If meiosis arose in a predominantly haploid organism, there is less justification for invoking diploid-specific genetic arguments (e.g., heterozygote superiority, or deleterious-mutation masking) for the origin of sexual reproduction. However, the latter factors could play a role in the maintenance of sexual reproduction in downstream lineages of diploids

Finally, to put things in a broader context, note that ameiotically reproducing prokaryotes are not strictly asexual, owing to the availability of multiple forms of gene transmission (i.e., the incorporation of exogenous DNA by direct uptake, plasmid transformation, or viral transfection). The archaea, in particular, are capable of cell fusion and bi-directional exchange of genomic material (Naor and Gophna 2013; van Wolferen et al. 2016; Wagner et al. 2017). Thus, strictly speaking, genetic mixing (and any evolutionary advantages that come with it) is not unique to eukaryotes, raising further doubts as to whether meiosis arose as a means for generating variation.

In summary, although an enormous amount of evolutionary theory has been devoted to understanding the adaptive significance of sexual reproduction, numerous observations raise questions about the traditional assumption that meiotic recombination originated and continues to be maintained by selection as a variance-generating mechanism. An alternative view is that recombination in eukaryotes is an inevitable consequence of the structural mechanisms of meiosis, with any variation generated being an indirect by-product, just as variance produced by mutation is a consequence of the inability of natural selection to reduce the replication error rate to zero (Chapter 4). The near-constancy of one crossover per chromosome arm among all eukaryotes (regardless of chromosome size; Chapter 4), and the absence of a correlation between chromosome number and ecological factors, do not inspire confidence in the idea that natural selection favors crossing over. Rather, it raises the possibility that selection reduces the latter to a near absolute minimum.

Mating Types

Aside from matters of genome processing, meiotic sexual reproduction introduces other novel evolutionary challenges that are absent from asexual lineages. Most notably, the necessity of cell fusion raises the issue of how cells avoid nonproductive interactions with inappropriate mating types and/or foreign species, and equally importantly, how cells efficiently attract conspecifics. The origin of mating types themselves and the factors that govern the numbers of such types within populations are also of key interest here.

In unicellular species, most sexual communication systems rely on pheromones, with contact between appropriate mating types initiating a cascade of effects starting with cell fusion and progressing to downstream meiotic control. The genetic bases for such systems are generally quite simple, usually involving just one or two linkage groups of a small number of genes, each segregating effectively as a single supergene. The norm is just two mating types per species, although exceptions exist.

Despite their centrality to organismal fitness, the components of mating-type systems sometimes evolve quite rapidly. Indeed, the phylogenetic breadth of mate-recognition systems implies that existing mechanisms are frequently taken over by entirely new processes. As already noted for the cell cycle and meiosis, this again poses the question as to how such shifts occur without causing massive internal incompatibilities in the lineages involved.

Mating type determination

The mere existence of mating types raises an evolutionary challenge, as each individual can only mate with a fraction of the members of the population (those with an alternative mating type). The situation is most extreme in the case of two mating types, where only half of the population is available to each individual (assuming a 1:1 sex ratio). In principle, multiple mating types will increase the opportunities for outcrossing. Yet, most sexual eukaryotic species have just two self-incompatible mating types.

Chemical recognition does not impose an absolute need for mating types, as all members of the population could in principle encode for the same signal and receptor proteins. Such a mutual recognition system is denoted as bipolar (Figure 10.9). However, an obvious limitation of such a system is the potential for an individual's receptors to be overwhelmed by its own pheromone molecules, obscuring the chemical gradients necessary to localize other members of the population. This might then endow a selective advantage to a genotype that loses the ability to either signal or receive. In a sea of bipolar cells, a mutant cell defective for pheromone production might also gain a selective advantage owing to the absence of expenditure on biosynthesis of the attractant.

In this sense, a bipolar recognition system is expected to be vulnerable to the emergence of a unipolar system (two unique mating types) by subfunctionalization, with one cell lineage retaining the signal-producing gene but losing the receptor, and vice versa for the second cell lineage. Once established, such a system might then be further refined by secondary novel gene acquisitions such that both mating types produce unique pheromones and receptors (Figure 10.9). However, maintenance of such a refined unipolar mating system requires that the receptor and signal genes be tightly linked chromosomally, as recombination would assort inappropriate mixes into the same gamete, thereby leading to non-functional mating capacities (Nei 1969; Hoekstra 1980).

Unfortunately, biology's descriptive language for mating systems is non-standardized, with different terms often used for functionally equivalent systems in fungi (and other unicellular species), land plants, and animals. Homothallism, equivalent to self-compatibility, refers to situations in which specific genotypes are capable of mating with other members of the same genotype. Heterothallism refers to self-incompatible systems requiring separate mating types (in land plants, systems with separate sexes are denoted as dioecious). These terms get blurred in organisms such as some yeast with internal mechanisms for switching mating types through genetic modifications; such species are homothallic, but could also be termed sequential hermaphrodites. In multicellular organisms, simultaneous hermaphroditism is possible, as in monoecious plants in which individuals produce male and female floral parts, but this is not known for unicellular species. Finally, the terms isogamous and anisogamous are used to refer to situations in which gamete types are morphologically indistinguishable versus distinct (as in eggs and sperm in land plants and animals); but even these terms can be a bit misleading, as isogamous species generally have different mating types and hence, underlying molecular differences.

A broader array of mating systems has been described in the fungi than in any other major eukaryotic lineage, although this could be a simple consequence of the magnitude of research focused on this group. In *S. cerevisiae, S. pombe,* and several other yeasts, there are two distinct mating types, each with unique pheromones and receptors, but these are achieved by mating-type switching (Hanson and Wolfe 2017), whereby cassettes of genes are swapped into a particular site by recombination. In this sense, mating-type determination involves a single tightly linked region (multigenic, but effectively segregating as a single locus). Individual genotypes are genetically hermaphrodites, but at the phenotypic level, individual cells mate in a unipolar manner. Once two complementary types are attracted to each other, the production of mating-type specific agglutinins (coagulants) is induced, and heterodimeric transcription factors constructed from components from each pair member elicit downstream meiotic activities.

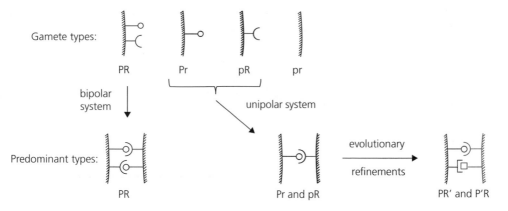

Figure 10.9 The evolution of mating-type determination by pheromone receptor systems, with P and p denoting the presence and absence of a gene associated with pheromone production, and R and r denoting the presence and absence of a gene for the receptor. **Upper row:** In the simplest system, there are four potential haplotypes, one having both functions, two having just a single function, and one being non-functional. In a system containing only haplotype PR, all individuals are capable of interacting with each other, whereas a system containing Pr and pR haplotypes (but no PR) will consist of two mating types. **Lower row:** Over time, the latter unipolar system may experience gains of novel pheromones and receptors, such that all participants contain non-interacting pairs of pheromones and receptors, further increasing the efficiency of recognition by compatible genotypes.

In the smuts, a group of plant pathogens within the mushroom family, the mating system sometimes involves two unlinked loci, although again each locus actually consists of linked blocks of genes (Bakkeren et al. 2008). In this case, one locus typically encodes for linked pheromones and receptors, while the second encodes for a transcription factor that governs downstream cellular events associated with syngamy and meiosis. As in the yeasts, different mating types recognize different pheromones, but in smuts four possible outcomes are possible, from fully compatible to fully incompatible, depending upon the allelic status at the two loci.

To further emphasize the diversity of evolved mating systems, just a few other examples of unicellular systems are noted here. Diatoms are known for their diversity of mating systems and sometimes rapid rates of evolution of underlying components (Armbrust and Galindo 2001; Chepurnov et al. 2004). Some diatoms have homothallic mating systems (capable of selfing), whereas others are heterothallic, and among these are cases of both isogamy and anisogamy. The diatom *Seminavis* has two mating types whose activities are coordinated by a two-step signalling system, the first involving a chemo-attractant that acts on a global basis, and

the second operating only after the perception of a mating partner and stimulating entry into cell-cycle arrest and gametogenesis (Moeys et al. 2016).

Not all species release mate-attraction pheromones. Although some ciliates, such as *Euplotes* do emit pheromones, others such as *Paramecium* simply deploy mating-type specific agglutinins upon contact. In *Paramecium tetraurelia*, which has a transcriptionally silent germline nucleus (the micronucleus) and a 'somatic' macronucleus (Figure 10.10), epigenetic events are involved in the maintenance of the two mating types (E and O, for even and odd), such that the latter are determined entirely by the maternal cytoplasm. In E-type cells, the mating-type gene (*mtA*, residing in the micronucleus) is passed on intact to the macronucleus, whereas in O-type cells, the promoter region is spliced out, rendering the macronuclear variant non-functional (Singh et al. 2014). In a related species, *P. septaurelia*, *mtA* is not differentially processed, but instead another gene (*mtB*, a transcription factor that regulates *mtA*) experiences a non-functionalizing deletion in the macronucleus of O-type cells. Thus, even members of the same genus can have substantially different mechanisms of mating-type determination.

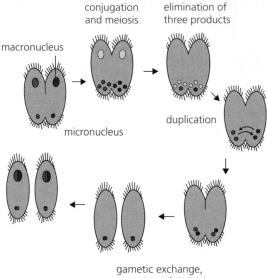

conjugation and meiosis

elimination of three products

macronucleus

duplication

micronucleus

gametic exchange, restoration of diploidy

Figure 10.10 The bi-nuclear endowment of a ciliate cell. Ciliates generally harbor a single, transcriptionally silent, germline micronucleus and a larger transcriptionally active macronucleus, which is comprised of hundreds of processed copies of micronuclear chromosomes, each with specific segments of DNA spliced out. The macronucleus is degraded after mating and replaced by a processed version of the new micronucleus. Both are transmitted intact during clonal phases of growth, the micronucleus by mitosis and the macronucleus by a less precise fission process. When two mating types conjugate, their micronuclei undergo meiosis, but three of the meiotic products are discarded by each cell ('female meiosis'); the remaining haploid nucleus is duplicated, with each conjugating cell exchanging a single gamete to restore diploidy; finally, the conjugating partners separate, and the new micronuclei serve as templates for building new macronuclei.

In *Chlamydomonas*, two mating types produce unique agglutinins on their flagella, which cross-react as recognition and adhesion mechanisms, leading to a cascade of events, again including the formation of a heterodimeric transcription factor composed of subunits derived from each mating type (Goodenough et al. 2007). The mating-type locus consists of a moderately sized (~ 300 kb) non-recombining linkage region. The agglutinins are very large (> 3300 amino acids in length), and the two types within a species are almost completely divergent in sequence, despite having very similar overall structures. The level of sequence divergence between *Chlamydomonas* species for orthologous agglutinins is also high, $\sim 2\times$ that for proteins used in cell-wall construction, which themselves are quite divergent (Lee et al. 2007).

Mating-type number

Given the relative simplicity of mating-type determination in most species, the establishment of more than two mating types is feasible, and as discussed below, rare mating types can sometimes have a strong selective advantage. However, unless a newly emergent mating type has a very high affinity towards the existing two, a two-type system can be very difficult to invade (Hadjivasiliou and Pomiankowski 2016), and indeed in multicellular species, there are virtually always just two distinct sexes.

The overall incidence of multiple mating-type systems in unicellular species is unclear, and for most single-celled organisms there is no information at all. Nonetheless, there are well-documented cases of unicellular species with more than two mating types. As one extreme example, two Basidiomycetes, *Schizophyllum commune* and *Coprinus cinereus*, have thousands of mating types (Kothe 1999; Riquelme et al. 2005). Some members of the green-algal genus *Closterium* have up to 15 mating types (Sekimoto et al. 2012). Three mating types are known in the slime mold *Dictyostelium discoideum*, and the number can be higher in other congeners, although there are also cases of homothallism (Bloomfield 2011). Two of the mating types in *D. discoideum* are specified by completely unrelated genes (and unknown to be related to those in any other species).

Although *Paramecium* species typically have two mating types, a number of ciliates (including *Euplotes*, *Tetrahymena*, *Glaucoma*, and *Stylonychia*) have up to twelve types (Phadke and Zufall 2009). In *Tetrahymena thermophila*, seven mating-type genes are tandemly arrayed in the germline micronucleus, but stochastic deletion events result in the macronucleus of progeny cells having only one complete gene (Cervantes et al. 2013), as in yeast mating-type switching.

Cell fusion

Once appropriate mating types have encountered each other, the climax of the sexual life cycle requires gamete fusion, which necessitates merging of the lipid bilayers of two cells that otherwise would be repellent under vegetative growth. For such purposes, a wide variety of unicellular eukaryotes utilize an integral membrane protein known as HAP2 (Wong and Johnson 2010; Speijer et al. 2015; Okamoto et al. 2016), although different, lineage-specific proteins appear to be involved in initial adhesion (Liu et al. 2015). Fungi and some animal species do not encode for HAP2 (although land plants do), and utilize alternative mechanisms for cell fusion.

Although quite divergent sequence-wise, HAP2 is highly similar in structure to the fusogen proteins used by lipid-bound viruses as entry mechanisms into host cells (Fédry et al. 2017; Pinello et al. 2017), implying either an extraordinary example of convergent evolution or an outcome of horizontal transfer. If eukaryotic HAP2 is derived from a virus (rather than the other way around), this would be compatible with sexual reproduction having arisen via the guidance of an ancient mobile element, as proposed by Hickey and Rose (1988).

HAP2 is generally expressed in both mating types of isogamous species such as the slime mold *Physarum* and the green algae *Chlamydomonas* and *Gonium* (reviewed in Cole et al. 2014), whereas in anisogamous species (e.g., animals and land plants), HAP2 is typically expressed by just one gamete type. In ciliates such as *Paramecium* and *Tetrahymena*, both members of a conjugating pair express HAP2, exchange meiotic products, and then disconnect (Cole et al. 2014; Orias 2014); this process is functionally equivalent to isogamy, but requires a mechanism for separation as well as merger.

Coevolution of pheromones and their receptors

Given the expense of biosynthesizing and exporting pheromones into the surrounding medium, and the potentially catastrophic outcome of mating with the wrong species, pheromone receptors can be expected to have a high specificity for their cognate pheromones. It has been argued that, below a critical cell size, active searching (swimming motility) for mates is energetically less costly than the recurrent production of released pheromones (Cox and Sethian 1985; Dusenbery and Snell 1995), but empirical work makes clear that chemical pheromones are widely used throughout unicellular eukaryotes. Mating pheromones may be simply attached to the cell surface, serving as final checkpoints in the decision to mate, or they may be released to the environment, with the resultant plume increasing the effective target size of a cell.

Mate-recognition systems in unicellular eukaryotes are best understood in yeast species (Michaelis and Barrowman 2012; Hanson and Wolfe 2017). In *S. cerevisiae*, each of the two mating types, α and *a*, produces a unique pheromone that attracts and elicits a developmental cascade in the other. That is, *a*-type cells secrete *a* pheromone while expressing the α-pheromone receptor, and vice versa. A similar system exists in *S. pombe*. Pheromones in yeast and other fungi are small peptides, on the order of a dozen amino acids in length (Urban et al. 1996), although the precursor molecules from which they are proteolytically cleaved prior to excretion are substantially larger (typically 40 to 160 amino acids). There are on the order of 8,000 receptors on the surface of *S. cerevisiae* cells (~ 80 per μm^2 of surface area), and these have a dissociation constant of 6×10^{-9} M (Jenness et al. 1986). The latter quantity is approximately equal to the substrate concentration at which the receptor is operating at 50% of its maximum rate (Chapters 18 and 19). For intracellular enzymes, dissociation constants for substrates are typically in the neighborhood of 10^{-4} M (Chapter 19), thus highlighting the exceptionally strong affinity of pheromone receptors for their signal molecules.

In the ciliate *Euplotes*, the pheromone and receptor are both encoded by the same gene, and alternative splicing results in one variant (the pheromone) being excreted into the environment and the other variant being a trans-membrane receptor (Luporini et al. 1996, 2005). The initial (pre-cleavage) protein is 70–140 amino acids in length, with the pheromone itself being reduced to 40–90 residues and excreted at a level of ~ 2–20 pg per cell per day (equivalent to $\sim 10^8$ to 10^9 proteins). There are 20–50 million receptors per cell surface, equivalent to ~ 1700 per

μm^2 of surface area, and dissociation constants are in the range of 0.6 to 10^{-8} M.

As in yeasts and *Euplotes*, the green alga *Closterium* deploys small peptides as pheromones (in this case, glycoproteins with ~ 150 amino acids) (Sekimoto et al. 2012), and these exhibit activity down to $\sim 10^{-10}$ M. The pheromone of *Volvox carteri*, also a glycoprotein, is one of the most potent effector molecules known, thought to operate with full effectiveness at 6×10^{-17} M, a sensitivity that may be made possible by secondary amplification involving the extracellular matrix (Hallmann et al. 1998; Hallmann 2008).

Not all pheromones are proteins. Those in the diatom *Seminavis* are small metabolites (Moeys et al. 2016), and those in the brown alga *Ectocarpus* consist of a blend of simple organic compounds derived from fatty acids and often containing pentane or hexane rings (Boland 1995; Pohnert and Boland 2002), which can be highly effective down to concentrations of 5×10^{-10} M.

These kinds of observations make clear that mating pheromones have independently evolved multiple times. However, once established, simple one-to-one signal-receptor systems are also subject to passive divergence by random genetic drift. This can happen if there are latent degrees of freedom in the signal and the receptor, as the sequence of the signalling molecule drifts slightly from its current state, opening up an opportunity for the receptor to change to a better match (or vice versa) (Lynch and Hagner 2015). Over a timescale sufficient for multiple mutations to accumulate, the basic mode of communication may then remain the same, while the communication language (i.e., the sequence motifs underlying the pheromone and its receptor) diverges (Figure 10.11).

By this means, coevolutionary drift in pheromones and their receptors can be expected to become so extreme in some cases as to lead to complete absence of interspecific recognition. A clear indication of this potential is provided by the engineering of a reproductively isolated strain of *S. pombe* with just a few amino-acid changes, effectively creating a new species (Seike et al. 2015). However, although comparative analysis suggests a combination of relaxed and positive selection in generating pheromone diversity among yeast species (Martin et al. 2011), there may also be constraints on such systems, as cross-talk is still possible between members of quite distant phylogenetic lineages. For example, *S. cerevisiae* still recognizes pheromones from species that have been separated for hundreds of millions of years (Rogers et al. 2015), and similar observations have been made for distantly related species of the yeast genus *Candida* (Lin et al. 2011). In addition, cases exist in smuts where pheromone receptor systems are cross-compatible between species that have been separated for hundreds of millions of years (Kellner et al. 2011; Xu et al. 2016), although again specific single amino-acid changes imposed on the pheromones and/or receptors can elicit large changes in specificity in other mushrooms (Fowler et al. 2001).

The limited studies that have been pursued in phylogenetic groups outside of the fungi provide indirect support for the widespread occurrence of considerable divergence in communication systems. In the ciliate *Euplotes*, the amino-acid sequences of excreted pheromones are far more variable among species than the peptides within the cleaved portions of the precursor molecules, to the point of being nearly completely divergent among congeners (Luporini et al. 1996, 2005). Within the genus *Closterium*, some species do not respond to the pheromones of others (Tsuchikane et al. 2008).

Taken together, these observations suggest that mating pheromone-receptor systems are capable of

Figure 10.11 Gradual change in a communication system involving a receptor and its signal. Upper and lower cases of the same letter imply a match, and the only requirement of the system is that at least two of three letters match. Six incremental changes lead to completely non-overlapping (and incompatible) words, although at each step the system is fully functional.

rapid interspecific divergence within at least some phylogenetic lineages. However, this is clearly an area in need of more mechanistically informative work. Whereas new receptors (with only single amino-acid changes) can be manufactured to discriminate against foreign pheromones, such specificity modifications may often come at the expense of efficiency of mating with conspecifics, reducing their likelihood of accumulating in nature.

Sexual Systems in Unicellular versus Multicellular Organisms

The vast majority of our knowledge of sex-determination systems and their genetic bases derives from studies on multicellular organisms. This gives a quite biased view of the general condition across the bulk of eukaryotic phylogeny. Three substantial differences, discussed next, emphasize the fact that similar selective challenges lead to radically different evolutionary responses in multicellular versus unicellular species, possibly as a consequence of the dramatic shifts in the population-genetic environment.

Isogamy versus anisogamy

First, as noted previously, sexual reproduction in most unicellular species involves isogamy, with the morphologically identical (but chemically different) gamete types being denoted as mating types. In contrast, virtually all complex multicellular organisms (animals and land plants) exhibit anisogamy, the operating definition being that females produce larger gametes than males. Notably, among freshwater phytoplankton, there is a moderate tendency for larger-celled species to be anisogamous (Madsen and Waller 1983). Once established, anisogamy is thought to secondarily facilitate the evolution of numerous other sexually dimorphic traits, as females with high investments per egg are selected to be choosy, and males with more numerous and individually cheap sperm are selected to be more indiscriminate and to acquire traits that enhance access to females (Maynard Smith 1978).

One explanation for the evolution of anisogamy invokes the concept of disruptive selection owing to an inherent evolutionary trade-off (Kalmus

and Smith 1960; Parker et al. 1972; Bell 1978; Charlesworth 1978; Parker 1978; Bulmer and Parker 2002; Iyer and Roughgarden 2008). With a fixed amount of resources at the time of gametogenesis (R), there is an inverse relationship between gamete size and gamete number – one large gamete of size R could be produced, or two of size $R/2$, four of size $R/4$, and so on. All other things being equal, larger numbers of gametes are advantageous, as they cumulatively have the potential to encounter more recipient partner gametes. However, if zygote survival increases with zygote size (the sum of the sizes of the two fusing gametes), there can also be a premium on producing a few large gametes. For the overall productivity of such a system to be elevated above that of isogamy, the increase in zygote survival with size need not be much greater than linear (Schuster and Sigmund 1982; Cox and Sethian 1984). Under this view, once two gamete sizes are established, strong preferential fusion between large and small gametes is expected to evolve secondarily, as the combination of two small gametes would have disproportionately low fitness, and the combination of two large gametes may also diminish fitness (owing to an inappropriately large size). Close linkage between a mating-type locus and a gamete-size locus can facilitate the maintenance of this kind of disassortative-mating system (Charlesworth 1978; Matsuda and Abrams 1999).

Given the universality of anisogamy in multicellular species, how can the opposite – the ubiquity of isogamy within unicellular species – be explained? Perhaps the most fundamental issue is that in multicellular organisms, there are no strict limits on the numbers of gametes that can be produced by the two sexes, as complex gonads can produce up to millions of meiotic products. In contrast, under unicellularity, all mating types produce the same number (four) of meiotic products, perhaps removing much of the potential pressure for alternative sexually selected traits. A second relevant issue, ignored by the models noted here, is stabilizing selection on the size of haploid cells themselves, a plausible scenario given that the predominant life stage for many unicellular organisms is haploid. As all mating types will typically be exposed to the same ecological conditions, and potentially so for long periods of clonal expansion, they will also typically be under the same size-selective forces.

Sex ratio

Simple population-genetic models suggest that separate sexes (or mating types) will typically lead to a stable 1:1 sex ratio as a consequence of frequency-dependent selection (Charlesworth and Charlesworth 2010). If + mating types are rare, then many − mating types will remain available, imposing selection in favor of genotypes producing more of the former to take advantage of this open resource (Figure 10.12). The opposite is expected if − mating types are disproportionately rare. For large populations with n unique mating types, assuming equal access to each other, the expectation is then that all types will have equilibrium frequencies of $\sim 1/n$. The well-known 1:1 sex ratio seen in most animals with separate sexes is in general accordance with this hypothesis, as are the limited data for mating-type frequencies in unicellular species.

In accordance with this hypothesis, in the ciliate *Tetrahymena thermophila*, which as noted above has a mating-type switching mechanism, all seven mating types typically coexist at roughly equal frequencies in the same ponds (Doerder et al. 1995). A study of another ciliate, *Paramecium bursaria*, which has a genetic sex-determination mechanism, revealed approximately equal frequencies of all four possible mating types (Kosaka 1991). The three mating types in the slime mold *D. discoideum* are also found in approximately equal frequencies in nature (Douglas et al. 2016). Likewise, natural populations of the ascomycete *Stagonospora nodorum* harbor approximately equal frequencies of the two possible mating types (Sommerhalder et al. 2006), and the same is true for the wheat blotch fungus *Mycosphaerella graminicola* (Gurung et al. 2011).

One nuance with respect to the predicted equilibration of mating-type frequencies is the assumption of equal costs of producing alternative mating types, which is likely usually met under isogamy and unicellularity. More generally, theory predicts an equal total expenditure on different mating types (sexes), e.g., if the production of individual daughters is energetically more expensive than that of sons, the equilibrium sex ratio is expected to be male biased (Charnov 1981; Charlesworth and Charlesworth 2010). It is, therefore, of interest that in the diatom *Cyclotella meneghiniana*, which is anisogamous (producing eggs and sperm) and has female meiosis, the sex ratio tends to be male biased (Shirokawa and Shimada 2013).

Sex chromosomes

The third major distinction between the genetics of mating systems in unicellular versus multicellular species concerns the nature of the genetic regions involved. Numerous animal and land-plant species have sex-determination systems based on fully differentiated sex chromosomes. In contrast, no unicellular species is known to harbor sex chromosomes. Instead, unicellular species rely on cassettes of a small number of tightly linked genes embedded within an otherwise freely recombining region. It is often argued that the somewhat

blue favored   stable equilibrium   red favored

Figure 10.12 Frequency-dependent selection for alternative mating types. On the left, the blue type is outnumbered by reds, leading to a selective advantage for blue, as not all reds will be capable of finding an appropriate mate. The opposite situation occurs to the right. In the centre, blues and reds have equal access to each other, leading to equivalent fitnesses for both types.

expanded sex-determination regions in some fungi and algae represent 'incipient sex chromosomes' in early stages of development towards full-fledged sex chromosomes. However, there is no obvious reason why sex-chromosome evolution should have been delayed in unicellular species, nor is there any reason why sex chromosomes should be viewed as an evolutionary advance. In short, the idea that organisms with separate sexes are destined to ultimately acquire such specialized sex chromosomes is less than compelling.

An alternative view is that the evolution of sex chromosomes is a pathological consequence of the low rates of recombination and high rates of random genetic drift experienced by multicellular species (Lynch 2007). Complete differentiation of sex chromosomes requires an outward expansion of recombination suppression around the primordial sex-determination (mating-type) locus, so as to allow the differential establishment and silencing of appropriate genes involved in sexual identity. Owing to their much higher rates of recombination per physical distance on chromosomes and to the diminished sensitivity to random genetic drift (Chapter 4), unicellular lineages may simply not provide the appropriate population-genetic environment for full sex-chromosome differentiation by degenerative mutation, regardless of the available time span. Although the isogamous mating systems of unicellular species also mitigate the opportunities for the evolution of differentiated sex chromosomes, a case can also be made for the opposite causal connection, i.e., that the origin of sex chromosomes facilitates the evolution of phenotypic sexual differentiation (Rice 1984).

Summary

- Cell biologists commonly subdivide the life of a cell into stages based on the state of growth and genome replication, the so-called cell cycle. However, the relative durations of stages vary greatly among phylogenetic lineages. In eukaryotes, check-points dictating the progression between stages are usually directed by elaborate networks of interacting proteins that cycle in expression levels.

- Despite the centrality of a well-coordinated cell cycle to all eukaryotes, there is remarkable evolutionary fluidity in the structure of the regulatory network, the nature of the participating proteins, and the positions of active sites in such proteins. In other words, there is no 'textbook' molecular description of the eukaryotic cell cycle.

- Variation in cell-cycle complexity and evolutionary rewiring of the overall network appear to be facilitated by duplication of ancestral component genes followed by subfunctionalization, providing striking examples of passive increases in network complexity in the absence of any intrinsic selective advantage of such structure.

- Establishment of the eukaryotic mitotic mechanism for replicating and evenly apportioning chromosomes to daughter cells involved the introduction of at least eight modifications not found in prokaryotes. The participating proteins typically assemble into multimers, which are commonly homomeric in archaea, but heteromeric in eukaryotes. Again, there is enormous phylogenetic variation in the structural features of mitosis among eukaryotes, but as yet no compelling evidence that the emergence of such variation, or of mitosis itself, was driven by adaptive processes.

- Unique to eukaryotes is meiosis, a two-stage modification of mitosis that reduces diploid genomes to haploids, which subsequently fuse as gametes to restore the diploid stage. Meiosis creates genetic variation via independent segregation of non-homologous chromosomes and the exchange of sequence between homologous parental chromosomes. As in the case of mitosis, much of the meiotic machinery appears to have arisen by gene duplication on the path from FECA to LECA, and a good deal of phylogenetic diversification of the underlying mechanisms has subsequently developed.

- Many of the proteins associated with meiosis appear to undergo relatively rapid sequence evolution. One popular argument for this, the centromere-drive hypothesis, postulates that meiosis sets up opportunities for centromeres to evolve so as to enhance their probability of appearing in haploid products, which induces secondary selective pressures on the meiotic machinery to eliminate negative cytogenetic side effects of the drive process. Empirical support for this idea remains mixed.

- It is commonly believed that meiosis increases the efficiency of natural selection in promoting beneficial mutations and purging deleterious load. However, it remains unclear whether this adaptive explanation for the maintenance of sexual reproduction is relevant to the question of why meiosis arose in the first place.

- Because it relies on the fusion of two haploid cells to produce the diploid substrate necessary for meiosis, sexual reproduction promoted the evolution of pheromone/receptor-based mating types to enhance the likelihood of mate acquisition. The mating-type determination systems of different phylogenetic lineages are highly diverse, often independently evolved, and include cases of more than two mating types.

- The simple one-to-one signal–receptor interactions in most mating-type recognition systems appear to be susceptible to coevolutionary drift, which over long timescales can lead to the passive emergence of reproductively isolated lineages.

- The sexual reproductive systems of unicellular species are substantially different from those in animals and land plants. Unlike the latter, the former generally have morphologically indistinguishable gamete types (as opposed to size-differentiated eggs and sperm) and short chromosomal segments involved in sex determination (as opposed to fully differentiated sex chromosomes). These features appear to be natural outcomes of the altered population-genetic environments of unicellular species.

Literature Cited

Akiyoshi, B., and K. Gull. 2014. Discovery of unconventional kinetochores in kinetoplastids. Cell 156: 1247–1258.

Anderson, J. A., W. D. Gilliland, and C. H. Langley. 2009. Molecular population genetics and evolution of *Drosophila* meiosis genes. Genetics 181: 177–185.

Archetti, M. 2004. Loss of complementation and the logic of two-step meiosis. J. Evol. Biol. 17: 1098–1105.

Archetti, M. 2005. Recombination and loss of complementation: A more than two-fold cost for parthenogenesis. J. Evol. Biol. 17: 1084–1097.

Armbrust, E. V., and H. M. Galindo. 2001. Rapid evolution of a sexual reproduction gene in centric diatoms of the genus *Thalassiosira*. Appl. Environ. Microbiol. 67: 3501–3513.

Aves, S. J., Y. Liu, and T. A. Richards. 2012. Evolutionary diversification of eukaryotic DNA replication machinery. Subcell. Biochem. 62: 19–35.

Azimzadeh, J., M. L. Wong, D. M. Downhour, A. Sánchez Alvarado, and W. F. Marshall. 2012. Centrosome loss in the evolution of planarians. Science 335: 461–463.

Bakkeren, G., Kämper, and J. Schirawski. 2008. Sex in smut fungi: Structure, function and evolution of mating-type complexes. Fungal Genet. Biol. 45 Suppl 1: S15–S21.

Barton, N. H., and B. Charlesworth. 1998. Why sex and recombination? Science 281: 1986–1990.

Barzel, A., and M. Kupiec. 2008. Finding a match: How do homologous sequences get together for recombination? Nat. Rev. Genet. 9: 27–37.

Bell, G. 1978. The evolution of anisogamy. J. Theor. Biol. 73: 247–270.

Bell, G. 1982. The Masterpiece of Nature: The Evolution and Genetics of Sexuality. University of California Press, Berkeley, CA.

Bengtsson, B. O. 1985. Biased conversion as the primary function of recombination. Genet. Res. 47: 77–80.

Bensasson, D. 2011. Evidence for a high mutation rate at rapidly evolving yeast centromeres. BMC Evol. Biol. 11: 211.

Bensasson, D., M. Zarowiecki, A. Burt, and V. Koufopanou. 2008. Rapid evolution of yeast centromeres in the absence of drive. Genetics 178: 2161–2167.

Bernstein, H., F. A. Hopf, and R. E. Michod. 1988. Is meiotic recombination an adaptation for repairing DNA, producing genetic variation, or both?, pp. 139–160. In H. Bernstein, F. Hopf, R. Michod, and B. R. Levin (eds.) The Evolution of Sex. Sinauer Assocs., Sunderland, MA.

Biondi, E. G., S. J. Reisinger, J. M. Skerker, M. Arif, B. S. Perchuk, K. R. Ryan, and M. T. Laub. 2006. Regulation of the bacterial cell cycle by an integrated genetic circuit. Nature 444: 899–904.

Bloomfield, G. 2011. Genetics of sex determination in the social amoebae. Dev. Growth Differ. 53: 608–616.

Bogdanov, Y. F., T. M. Grishaeva, and S. Y. Dadashev. 2007. Similarity of the domain structure of proteins as a basis for the conservation of meiosis. Int. Rev. Cytol. 257: 83–142.

Boland, W. 1995. The chemistry of gamete attraction: Chemical structures, biosynthesis, and (a)biotic degradation of algal pheromones. Proc. Natl. Acad. Sci. USA 92: 37–43.

Bomblies, K., J. D. Higgins, and L. Yant. 2015. Meiosis evolves: Adaptation to external and internal environments. New Phytol. 208: 306–323.

Bonner, A. M., and R. S. Hawley. 2019. Functional consequences of the evolution of Matrimony, a meiosis-specific inhibitor of polo kinase. Mol. Biol. Evol. 36: 69–83.

Bracewell, R., K. Chatla, M. J. Nalley, and D. Bachtrog. 2019. Dynamic turnover of centromeres drives karyotype evolution in *Drosophila*. eLife 8: e49002.

Bulmer, M. G., and G. A. Parker. 2002. The evolution of anisogamy: A game-theoretic approach. Proc. Biol. Sci. 269: 2381–2388.

Butler, G., M. D. Rasmussen, M. F. Lin, M. A. Santos, S. Sakthikumar, C. A. Munro, E. Rheinbay, M. Grabherr, A. Forche, J. L. Reedy, et al. 2009. Evolution of pathogenicity and sexual reproduction in eight *Candida genomes*. Nature 459: 657–662.

Cervantes, M. D., E. P. Hamilton, J. Xiong, M. J. Lawson, D. Yuan, M. Hadjithomas, W. Miao, and E. Orias. 2013. Selecting one of several mating types through gene segment joining and deletion in *Tetrahymena thermophila*. PLoS Biol. 11: e1001518.

Charlesworth, B. 1978. The population genetics of anisogamy. J. Theor. Biol. 73: 347–357.

Charlesworth, B., and D. Charlesworth. 2010. Elements of Evolutionary Genetics. Roberts and Company, Greenwood Village, CO.

Charnov. E. L. 1981. The Theory of Sex Allocation. Princeton University Press, Princeton, NJ.

Chepurnov, V. A., D. G. Mann, K. Sabbe, and W. Vyverman. 2004. Experimental studies on sexual reproduction in diatoms. Int. Rev. Cytol. 237: 91–154.

Chia, N., I. Cann, and G. J. Olsen. 2010. Evolution of DNA replication protein complexes in eukaryotes and Archaea. PLoS One 5: e10866.

Chmátal, L., S. I. Gabriel, G. P. Mitsainas, J. Martinez-Vargas, J. Ventura, J. B. Searle, R. M. Schultz, and M. A. Lampson. 2014. Centromere strength provides the cell biological basis for meiotic drive and karyotype evolution in mice. Curr. Biol. 24: 2295–2300.

Cleveland, L. R. 1947. The origin and evolution of meiosis. Science 105: 287–289.

Cobbe, N., and M. M. Heck. 2004. The evolution of SMC proteins: Phylogenetic analysis and structural implications. Mol. Biol. Evol. 21: 332–347.

Cole, E. S., D. Cassidy-Hanley, J. Fricke Pinello, H. Zeng, M. Hsueh, D. Kolbin, C. Ozzello, T. Giddings, Jr., M. Winey, and T. G. Clark. 2014. Function of the male-gamete-specific fusion protein HAP2 in a seven-sexed ciliate. Curr. Biol. 24: 2168–2173.

Côte, P., H. Hogues, and M. Whiteway. 2009. Transcriptional analysis of the *Candida albicans* cell cycle. Mol. Biol. Cell 20: 3363–3373.

Coudreuse, D., and P. Nurse. 2010. Driving the cell cycle with a minimal CDK control network. Nature 468: 1074–1079.

Cox, P. A., and J. A. Sethian. 1984. Search, encounter rates, and the evolution of anisogamy. Proc. Natl. Acad. Sci. USA 81: 6078–6079.

Criqui, M. C., and P. Genschik. 2002. Mitosis in plants: How far we have come at the molecular level? Curr. Opin. Plant Biol. 5: 487–493.

Cross, F. R., N. E. Buchler, and J. M. Skotheim. 2011. Evolution of networks and sequences in eukaryotic cell cycle control. Philos. Trans. R. Soc. Lond. B Biol. Sci. 366: 3532–3544.

Cross, F. R., and J. G. Umen. 2015. The *Chlamydomonas* cell cycle. Plant J. 82: 370–392.

de Massy, B. 2013. Initiation of meiotic recombination: how and where? Conservation and specificities among eukaryotes. Annu. Rev. Genet. 47: 563–599.

De Muyt, A., R. Mercier, C. Mézard, and M. Grelon. 2009. Meiotic recombination and crossovers in plants. Genome Dyn. 5: 14–25.

Ding, D. Q., K. Okamasa, M. Yamane, C. Tsutsumi, T. Haraguchi, M. Yamamoto, and Y. Hiraoka. 2012. Meiosis-specific noncoding RNA mediates robust pairing of homologous chromosomes in meiosis. Science 336: 732–736.

Doerder, F. P., M. A. Gates, F. P. Eberhardt, and M. Arslanyolu. 1995. High frequency of sex and equal frequencies of mating types in natural populations of the ciliate *Tetrahymena thermophila*. Proc. Natl. Acad. Sci. USA 92: 8715–8718.

Doonan, J. H., and G. Kitsios. 2009. Functional evolution of cyclin-dependent kinases. Mol. Biotechnol. 42: 14–29.

Douglas, T. E., J. E. Strassmann, and D. C. Queller. 2016. Sex ratio and gamete size across eastern North America in *Dictyostelium discoideum*, a social amoeba with three sexes. J. Evol. Biol. 29: 1298–1306.

Drinnenberg, I. A., S. Henikoff, and H. S. Malik. 2016. Evolutionary turnover of kinetochore proteins: A ship of Theseus? Trends Cell Biol. 26: 498–510.

Drury, L. S., and J. F. Diffley. 2009. Factors affecting the diversity of DNA replication licensing control in eukaryotes. Curr. Biol. 19: 530–535.

Dusenbery, D. B., and T. W. Snell. 1995. A critical body size for use of pheromones in mate location. J. Chem. Ecol. 21: 427–438.

Egel, R., and D. Penny. 2007. On the origin of meiosis in eukaryotic evolution: coevolution of meiosis and mitosis from feeble beginnings, pp. 249–288. In R. Egel and D-H. Lankenau (eds.) Genome Dynamics Stability. Springer, Heidelberg, Germany.

Elde, N. C., K. C. Roach, M. C. Yao, and H. S. Malik. 2011. Absence of positive selection on centromeric histones in *Tetrahymena* suggests unsuppressed centromere: Drive in lineages lacking male meiosis. J. Mol. Evol. 72: 510–520.

Fang, J., P. Nevin, V. Kairys, Č. Venclovas, J. R. Engen, and P. J. Beuning. 2014. Conformational analysis of processivity clamps in solution demonstrates that tertiary structure does not correlate with protein dynamics. Structure 22: 572–581.

Farhadifar, R., C. F. Baer, A. C. Valfort, E. C. Andersen, T. Müller-Reichert, M. Delattre, and D. J. Needleman. 2015. Scaling, selection, and evolutionary dynamics of the mitotic spindle. Curr. Biol. 25: 732–740.

Farhadifar, R., C. H. Yu, G. Fabig, H. Y. Wu, D. B. Stein, M. Rockman, T. Müller-Reichert, M. J. Shelley, and D. J. Needleman. 2020. Stoichiometric interactions explain spindle dynamics and scaling across 100 million years of nematode evolution. eLife 9: e55877.

Fédry, J., Y. Liu, G. Péhau-Arnaudet, J. Pei, W. Li, M. A. Tortorici, F. Traincard, A. Meola, G. Bricogne, N. V. Grishin, et al. 2017. The ancient gamete fusogen HAP2 is a eukaryotic Class II fusion protein. Cell 168: 904–915.

Fowler, T. J., M. F. Mitton, L. J. Vaillancourt, and C. A. Raper. 2001. Changes in mate recognition through alterations of pheromones and receptors in the multisexual mushroom fungus *Schizophyllum commune*. Genetics 158: 1491–1503.

Fraune, J., M. Alsheimer, J. N. Volff, K. Busch, S. Fraune, T. C. Bosch, and R. Benavente. 2012. *Hydra* meiosis reveals unexpected conservation of structural synaptonemal complex proteins across metazoans. Proc. Natl. Acad. Sci. USA 109: 16588–16593.

Garcia-Garcia, T., S. Poncet, A. Derouiche, L. Shi, I. Mijakovic, and M. F. Noirot-Gros. 2016. Role of protein phosphorylation in the regulation of cell cycle and DNA-related processes in bacteria. Front. Microbiol. 7: 184.

Gerton, J. L., and R. S. Hawley. 2005. Homologous chromosome interactions in meiosis: Diversity amidst conservation. Nat. Rev. Genet. 6: 477–487.

Goodenough, U., H. Lin, and J. H. Lee. 2007. Sex determination in *Chlamydomonas*. Semin. Cell Dev. Biol. 18: 350–361.

Gourguechon, S., L. J. Holt, and W. Z. Cande. 2013. The *Giardia* cell cycle progresses independently of the anaphase-promoting complex. J. Cell Sci. 126: 2246–2255.

Gurung, S., S. B. Goodwin, M. Kabbage, W. W. Bockus, and T. B. Adhikari. 2011. Genetic differentiation at microsatellite loci among populations of *Mycosphaerella*

graminicola from California, Indiana, Kansas, and North Dakota. Phytopathology 101: 1251–1259.

Hadjivasiliou, Z., and A. Pomiankowski. 2016. Gamete signalling underlies the evolution of mating types and their number. Philos. Trans. R. Soc. Lond. B Biol. Sci. 371(1706).

Haldenby, S., M. F. White, and T. Allers. 2009. RecA family proteins in archaea: RadA and its cousins. Biochem. Soc. Trans. 37: 102–107.

Hallmann, A. 2008. VCRPs, small cysteine-rich proteins, might be involved in extracellular signaling in the green alga *Volvox*. Plant Signal Behav. 3: 124–127.

Hallmann, A., K. Godl, S. Wenzl, and M. Sumper. 1998. The highly efficient sex-inducing pheromone system of *Volvox*. Trends Microbiol. 6: 185–189.

Hanson, S. J., and K. H. Wolfe. 2017. An evolutionary perspective on yeast mating-type switching. Genetics 206: 9–32.

Harashima, H., N. Dissmeyer, and A. Schnittger. 2013. Cell cycle control across the eukaryotic kingdom. Trends Cell Biol. 23: 345–356.

Heath, I. B. 1980. Variant mitoses in lower eukaryotes: Indicators of the evolution of mitosis. Int. Rev. Cytol. 64: 1–80.

Hemmer, L. W., and J. P. Blumenstiel. 2016. Holding it together: Rapid evolution and positive selection in the synaptonemal complex of *Drosophila*. BMC Evol. Biol. 16: 91.

Henikoff, S., K. Ahmad, and H. S. Malik. 2001. The centromere paradox: Stable inheritance with rapidly evolving DNA. Science 293: 1098–1102.

Hickey, D. A., and M. R. Rose. 1988. The role of gene transfer in the evolution of eukaryotic sex, pp. 161–175. In H. Bernstein, F. Hopf, R. Michod, and B. R. Levin (eds.) The Evolution of Sex. Sinauer Associates, Sunderland, MA.

Hoekstra, R. F. 1980. Why do organisms produce gametes of only two different sizes? Some theoretical aspects of the evolution of anisogamy. J. Theor. Biol. 87: 785–793.

Holt, L. J., B. B. Tuch, J. Villén, A. D. Johnson, S. P. Gygi, and D. O. Morgan. 2009. Global analysis of Cdk1 substrate phosphorylation sites provides insights into evolution. Science 325: 1682–1686.

Howard-Till, R. A., A. Lukaszewicz, M. Novatchkova, and J. Loidl. 2013. A single cohesin complex performs mitotic and meiotic functions in the protist *Tetrahymena*. PLoS Genet. 9: e1003418.

Hurst, L. D., and P. Nurse. 1991. A note on the evolution of meiosis. J. Theor. Biol. 150: 561–563.

Huysman, M. J., C. Martens, K. Vandepoele, J. Gillard, E. Rayko, M. Heijde, C. Bowler, D. Inzé, Y. Van de Peer, L. De Veylder, et al. 2010. Genome-wide analysis of the diatom cell cycle unveils a novel type of cyclins involved in environmental signaling. Genome Biol. 11: R17.

Iwata-Otsubo, A., J. M. Dawicki-McKenna, T. Akera, S. J. Falk, L. Chmátal, K. Yang, B. A. Sullivan, R. M. Schultz, M. A. Lampson, and B. E. Black. 2017. Expanded satellite repeats amplify a discrete CENP-A nucleosome assembly site on chromosomes that drive in female meiosis. Curr. Biol. 27: 2365–2373.

Iyer, P., and J. Roughgarden. 2008. Gametic conflict versus contact in the evolution of anisogamy. Theor. Popul. Biol. 73: 461–472.

Jenness, D. D., A. C. Burkholder, and L. H. Hartwell. 1986. Binding of alpha-factor pheromone to *Saccharomyces cerevisiae* a cells: Dissociation constant and number of binding sites. Mol. Cell Biol. 6: 318–320.

Jensen, L. J., T. S. Jensen, U. de Lichtenberg, S. Brunak, and P. Bork. 2006. Co-evolution of transcriptional and post-translational cell-cycle regulation. Nature 443: 594–597.

Kalmus, H., and C. A. Smith. 1960. Evolutionary origin of sexual differentiation and the sex-ratio. Nature 186: 1004–1006.

Kellner, R., E. Vollmeister, M. Feldbrügge, and D. Begerow. 2011. Interspecific sex in grass smuts and the genetic diversity of their pheromone-receptor system. PLoS Genet. 7: e1002436.

Kleckner, N. 1996. Meiosis: How could it work? Proc. Natl. Acad. Sci. USA 93: 8167–8174.

Kondrashov, A. S. 1993. Classification of hypotheses on the advantage of amphimixis. J. Hered. 84: 372–387.

Kosaka, T. 1991. Life cycle of *Paramecium bursaria* syngen 1 in nature. J. Protozool. 38: 140–148.

Kothe, E. 1999. Mating types and pheromone recognition in the Homobasidiomycete *Schizophyllum commune*. Fungal Genet. Biol. 27: 146–152.

Kubai, D. F. 1975. The evolution of the mitotic spindle. Int. Rev. Cytol. 43: 167–227.

Lee, J. H., S. Waffenschmidt, L. Small, and U. Goodenough. 2007. Between-species analysis of short-repeat modules in cell wall and sex-related hydroxyproline-rich glycoproteins of *Chlamydomonas*. Plant Physiol. 144: 1813–1826.

Lin, C. H., A. Choi, and R. J. Bennett. 2011. Defining pheromone-receptor signaling in *Candida albicans* and related asexual *Candida species*. Mol. Biol. Cell 22: 4918–4930.

Lindås, A. C., and R. Bernander. 2013. The cell cycle of archaea. Nat. Rev. Microbiol. 11: 627–638.

Lindholm, A. K., K. A. Dyer, R. C. Firman, L. Fishman, W. Forstmeier, L. Holman, H. Johannesson, U. Knief, H. Kokko, A. M. Larracuente, et al. 2016. The ecology and evolutionary dynamics of meiotic drive. Trends Ecol. Evol. 31: 315–326.

Liu, Y., J. Pei, N. Grishin, and W. J. Snell. 2015. The cytoplasmic domain of the gamete membrane fusion protein HAP2 targets the protein to the fusion site in *Chlamydomonas* and regulates the fusion reaction. Development 142: 962–971.

Liu, Y., T. A. Richards, and S. J. Aves. 2009. Ancient diversification of eukaryotic MCM DNA replication proteins. BMC Evol. Biol. 9: 60.

Lloyd, A., C. Morgan, F. C. Franklin, and K. Bomblies. 2018. Plasticity of meiotic recombination rates in response to temperature in *Arabidopsis*. Genetics 208: 1409–1420.

Loidl, J. 2016. Conservation and variability of meiosis across the eukaryotes. Annu. Rev. Genet. 50: 293–316.

López Hernández, J. F., R. M. Helston, J. J. Lange, R. B. Billmyre, S. H. Schaffner, M. T. Eickbush, S. McCroskey, and S. E. Zanders. 2021. Diverse mating phenotypes impact the spread of *wtf* meiotic drivers in *Schizosaccharomyces pombe*. eLife 10: e70812.

Lorenz, A., J. L. Wells, D. W. Pryce, M. Novatchkova, F. Eisenhaber, R. J. McFarlane, and J. Loidl. 2004. *S. pombe* meiotic linear elements contain proteins related to synaptonemal complex components. J. Cell Sci. 117: 3343–3351.

Luporini, P., C. Alilmenti, C. Ortenzi, and A. Vallesi. 2005. Ciliate mating types and their specific protein pheromones. Acta Protozool. 44: 89–101.

Luporini, P., C. Miceli, C. Ortenzi, and A. Vallesi. 1996. Ciliate pheromones. Prog. Mol. Subcell. Biol. 17: 80–104.

Lynch, M. 2007. The Origins of Genome Architecture. Sinauer Associates, Inc., Sunderland, MA.

Lynch, M. 2012. Evolutionary layering and the limits to cellular perfection. Proc. Natl. Acad. Sci. USA 109: 18851–18856.

Lynch, M., and K. Hagner. 2015. Evolutionary meandering of intermolecular interactions along the drift barrier. Proc. Natl. Acad. Sci. USA 112: E30–E38.

Madsen, J. D., and D. M. Waller. 1983. A note on the evolution of gamete dimorphism in algae. Amer. Natur. 121: 443–447.

Maguire, M. P. 1992. The evolution of meiosis. J. Theor. Biol. 154: 43–55.

Malik, H. S., and S. Henikoff. 2001. Adaptive evolution of Cid, a centromere-specific histone in *Drosophila*. Genetics 157: 1293–1298.

Malik, S. B., M. A. Ramesh, A. M. Hulstrand, and J. M. Logsdon, Jr. 2007. Protist homologs of the meiotic *Spo11* gene and topoisomerase VI reveal an evolutionary history of gene duplication and lineage-specific loss. Mol. Biol. Evol. 24: 2827–2841.

Mann, T. H., and L. Shapiro. 2018. Integration of cell cycle signals by multi-PAS domain kinases. Proc. Natl. Acad. Sci. USA 115: E7166–E7173.

Marcet-Houben, M., and T. Gabaldón. 2015. Beyond the whole-genome duplication: Phylogenetic evidence for an ancient interspecies hybridization in the baker's yeast lineage. PLoS Biol. 13: e1002220.

Markova, K., M. Uzlikova, P. Tumova, K. Jirakova, G. Hagen, J. Kulda, and E. Nohynkova. 2016. Absence of a conventional spindle mitotic checkpoint in the binucleated single-celled parasite *Giardia intestinalis*. Eur. J. Cell. Biol. 95: 355–367.

Martin, S. H., B. D. Wingfield, M. J. Wingfield, and E. T. Steenkamp. 2011. Causes and consequences of variability in peptide mating pheromones of ascomycete fungi. Mol. Biol. Evol. 28: 1987–2003.

Matsuda, H., and P. A. Abrams. 1999. Why are equally sized gametes so rare? The instability of isogamy and the cost of anisogamy. Evol. Ecol. Res. 1: 769–784.

Maynard Smith, J. 1978. The Evolution of Sex. Cambridge University Press, Cambridge, UK.

Medina, E. M., J. J. Turner, R. Gordân, J. M. Skotheim, and N. E. Buchler. 2016. Punctuated evolution and transitional hybrid network in an ancestral cell cycle of fungi. eLife 5: e09492.

Michaelis, S., and J. Barrowman. 2012. Biogenesis of the *Saccharomyces cerevisiae* pheromone a-factor, from yeast mating to human disease. Microbiol. Mol. Biol. Rev. 76: 626–651.

Moeys, S., J. Frenkel, C. Lembke, J. T. F. Gillard, V. Devos, K. Van den Berge, B. Bouillon, M. J. J. Huysman, S. De Decker, J. Scharf, et al. 2016. A sex-inducing pheromone triggers cell cycle arrest and mate attraction in the diatom *Seminavis robusta*. Sci. Rep. 6: 19252.

Morgan, D. O. 2007. The Cell Cycle: Principles of Control. Oxford University Press, Oxford, UK.

Moses, A. M., M. E. Liku, J. J. Li, and R. Durbin. 2007. Regulatory evolution in proteins by turnover and lineage-specific changes of cyclin-dependent kinase consensus sites. Proc. Natl. Acad. Sci. USA 104: 17713–17718.

Murray, A. W. 2012. Don't make me mad, Bub! Dev. Cell 22: 1123–1125.

Naor, A., and U. Gophna. 2013. Cell fusion and hybrids in Archaea: Prospects for genome shuffling and accelerated strain development for biotechnology. Bioengineered 4: 126–129.

Nei, M. 1969. Linkage modifications and sex difference in recombination. Genetics 63: 681–699.

Nguyen Ba, A. N., B. Strome, S. Osman, E. A. Legere, T. Zarin, and A. M. Moses. 2017. Parallel reorganization of protein function in the spindle checkpoint pathway through evolutionary paths in the fitness landscape that appear neutral in laboratory experiments. PLoS Genet. 13: e1006735.

Niklas, K. J., E. D. Cobb, and U. Kutschera. 2014. Did meiosis evolve before sex and the evolution of eukaryotic life cycles? Bioessays 36: 1091–1101.

Okamoto, M., L. Yamada, Y. Fujisaki, G. Bloomfield, K. Yoshida, H. Kuwayama, H. Sawada, T. Mori, and H. Urushihara. 2016. Two HAP2-GCS1 homologs responsible for gamete interactions in the cellular slime mold with multiple mating types: Implication for common mechanisms of sexual reproduction shared by plants and protozoa and for male-female differentiation. Dev. Biol. 415: 6–13.

Onesti, S., and S. A. MacNeill. 2013. Structure and evolutionary origins of the CMG complex. Chromosoma 122: 47–53.

Orias, E. 2014. Membrane fusion: HAP2 protein on a short leash. Curr. Biol. 24: R831–R833.

Osella, M., S. J. Tans, and M. C. Lagomarsino. 2017. Step by step, cell by cell: Quantification of the bacterial cell cycle. Trends Microbiol. 25: 250–256.

Ottolini, C. S., L. Newnham, A. Capalbo, S. A. Natesan, H. A. Joshi, D. Cimadomo, D. K. Griffin, K. Sage, M. C. Summers, A. R. Thornhill, et al. 2015. Genome-wide maps of recombination and chromosome segregation in human oocytes and embryos show selection for maternal recombination rates. Nat. Genet. 47: 727–735.

Parker, G. A. 1978. Selection on non-random fusion of gametes during the evolution of anisogamy. J. Theor. Biol. 73: 1–28.

Parker, G. A., R. R. Baker, and V. G. Smith. 1972. The origin and evolution of gamete dimorphism and the male–female phenomenon. J. Theor. Biol. 36: 529–553.

Phadke, S. S., and R. A. Zufall. 2009. Rapid diversification of mating systems in ciliates. Biol. J. Linnean Soc. 98: 187–197.

Pinello, J. F., A. L. Lai, J. K. Millet, D. Cassidy-Hanley, J. H. Freed, and T. G. Clark. 2017. Structure-function studies link Class II viral fusogens with the ancestral gamete fusion protein HAP2. Curr. Biol. 27: 651–660.

Pohnert, G., and W. Boland. 2002. The oxylipin chemistry of attraction and defense in brown algae and diatoms. Nat. Prod. Rep. 19: 108–122.

Raikov, I. B. 1982. Morphology of Eukaryotic Protozoan Nuclei. Cell Biology Monographs, Vol. 9. Springer-Verlag, Vienna, Austria.

Ramesh, M. A., S. B. Malik, J. M. Logsdon, Jr. 2005. A phylogenomic inventory of meiotic genes; Evidence for sex in *Giardia* and an early eukaryotic origin of meiosis. Curr. Biol. 15: 185–191.

Reedy, J. L., A. M. Floyd, and J. Heitman. 2009. Mechanistic plasticity of sexual reproduction and meiosis in the *Candida* pathogenic species complex. Curr. Biol. 19: 891–899.

Rhind, N., and P. Russell. 2000. Chk1 and Cds1: Linch-pins of the DNA damage and replication checkpoint pathways. J. Cell Sci. 113: 3889–3896.

Rice, W. R. 1984. Sex chromosomes and the evolution of sexual dimorphism. Evolution 38: 735–742.

Riquelme, M., M. P. Challen, L. A. Casselton, and A. J. Brown. 2005. The origin of multiple B mating specifici-ties in *Coprinus cinereus*. Genetics 170: 1105–1119.

Robert, T., A. Nore, C. Brun, C. Maffre, B. Crimi, H. M. Bourbon, and B. de Massy. 2016. The TopoVIB-like pro-tein family is required for meiotic DNA double-strand break formation. Science 351: 943–949.

Rog, O., S. Köhler, and A. F. Dernburg. 2017. The synap-tonemal complex has liquid crystalline properties and spatially regulates meiotic recombination factors. eLife 6: e21455.

Rogers, D. W., J. A. Denton, E. McConnell, and D. Greig. 2015. Experimental evolution of species recognition. Curr. Biol. 25: 1753–1758.

Rosin, L. F., and B. G. Mellone. 2017. Centromeres drive a hard bargain. Trends Genet. 33: 101–117.

Sazer, S., M. Lynch, and D. Needleman. 2014. Decipher-ing the evolutionary history of open and closed mitosis. Curr. Biol. 24: R1099–R1103.

Scherthan, H. 2001. A bouquet makes ends meet. Nat. Rev. Mol. Cell Biol. 2: 621–627.

Schuster, P., and K. Sigmund. 1982. A note on the evolution of sexual dimorphism. J. Theor. Biol. 94: 107–110.

Seike, T., T. Nakamura, and C. Shimoda. 2015. Molec-ular coevolution of a sex pheromone and its receptor triggers reproductive isolation in *Schizosaccharomyces pombe*. Proc. Natl. Acad. Sci. USA 112: 4405–4410.

Sekimoto, H., J. Abe, and Y. Tsuchikane. 2012. New insights into the regulation of sexual reproduc-tion in *Closterium*. Int. Rev. Cell. Mol. Biol. 297: 309–338.

Shirokawa, Y., and M. Shimada. 2013. Sex allocation pat-tern of the diatom *Cyclotella meneghiniana*. Proc. Biol. Sci. 280: 20130503.

Singh, D. P., B. Saudemont, G. Guglielmi, O. Arnaiz, J.-F. Gout, M. Prajer, A. Potekhin, E. Przybòs, A. Aubusson-Fleury, S. Bhullar, et al. 2014. Genome-defence small RNAs exapted for epigenetic mating-type inheritance. Nature 509: 447–452.

Solari, A. J. 2002. Primitive forms of meiosis: The possible evolution of meiosis. Biocell 26: 1–13.

Sommerhalder, R. J., B. A. McDonald, and J. Zhan. 2006. The frequencies and spatial distribution of mating types in *Stagonospora nodorum* are consistent with recurring sexual reproduction. Phytopath. 96: 234–239.

Speijer, D., J. Lukeš, and M. Eliás. 2015. Sex is a ubiquitous, ancient, and inherent attribute of eukaryotic life. Proc. Natl. Acad. Sci. USA 112: 8827–8834.

Stover, N. A., and J. D. Rice. 2011. Distinct cyclin genes define each stage of ciliate conjugation. Cell Cycle 10: 1699–1701.

Suijkerbuijk, S. J., T. J. P. van Dam, G. E. Karagöz, E. von Castelmur, N. C. Hubner, A. M. S. Duarte, M. Vleugel, A. Perrakis, S. G. D. Rüdiger, B. Snel, et al. 2012. The ver-tebrate mitotic checkpoint protein BUBR1 is an unusual pseudokinase. Dev. Cell 22: 1321–1329.

Svedberg, J., A. A. Vogan, N. A. Rhoades, D. Sarmara-jeewa, D. J. Jacobson, M. Lascoux, T. M. Hammond, and H. Johannesson. 2021. An introgressed gene causes mei-otic drive in *Neurospora sitophila*. Proc. Natl. Acad. Sci. USA 118: e2026605118.

Tromer, E. C., J. J. E. van Hooff, G. J. P. L. Kops, and B. Snel. 2019. Mosaic origin of the eukaryotic kinetochore. Proc. Natl. Acad. Sci. USA 116: 12873–12882.

Tsuchikane, Y., M. Ito, and H. Sekimoto. 2008. Reproduc-tive isolation by sex pheromones in the *Closterium lob-sterium peracerosum-strigosum-littrorale* complex (Zygne-matales, Charophycea). J. Phycol. 44: 1197–1203.

Turner, B. C., and D. D. Perkins. 1979. Meiotic drive in *Neurospora* and other fungi. Amer. Natur. 137: 416–429.

Urban, M., R. Kahmann, and M. Bölker. 1996. The bial-lelic mating type locus of *Ustilago maydis*: Remnants of an additional pheromone gene indicate evolution from a multiallelic ancestor. Mol. Gen. Genet. 250: 414–420.

van Wolferen, M., A. Wagner, C. van der Does, and S. V. Albers. 2016. The archaeal Ced system imports DNA. Proc. Natl. Acad. Sci. USA 113: 2496–2501.

Vogan, A. A., S. L. Ament-Velásquez, A. Granger-Farbos, J. Svedberg, E. Bastiaans, A. J. Debets, V. Coustou, H. Yvanne, C. Clavé, S. J. Saupe, et al. 2019. Combina-tions of *Spok* genes create multiple meiotic drivers in *Podospora*. eLife 8: e46454.

Vrielynck, N., A. Chambon, D. Vezon, L. Pereira, L. Chely-sheva, A. De Muyt, C. Mézard, C. Mayer, and M. Grelon. 2016. A DNA topoisomerase VI-like complex initiates meiotic recombination. Science 351: 939–943.

Wagner, A., R. J. Whitaker, D. J. Krause, J. H. Heilers, M. van Wolferen, C. van der Does, and S. V. Albers. 2017. Mechanisms of gene flow in archaea. Nat. Rev. Microbiol. 15: 492–501.

Wang, S., N. Kleckner, and L. Zhang. 2017. Crossover mat-uration inefficiency and aneuploidy in human female meiosis. Cell Cycle 4: 1–3.

Watanabe, Y. 2005. Shugoshin: Guardian spirit at the cen-tromere. Curr. Opin. Cell. Biol. 17: 590–595.

Watanabe, Y. 2012. Geometry and force behind kinetochore orientation: Lessons from meiosis. Nat. Rev. Mol. Cell Biol. 13: 370–382.

Wilkins, A. S., and R. Holliday. 2009. The evolution of meiosis from mitosis. Genetics 181: 3–12.

Williams, G. C. 1975. Sex and Evolution. Princeton University Press, Princeton, NJ.

Winey, M., and K. Bloom. 2012. Mitotic spindle form and function. Genetics 190: 1197–1224.

Wirth, R. 1994. The sex pheromone system of *Enterococcus faecalis*. More than just a plasmid-collection mechanism? Eur. J. Biochem. 222: 235–246.

Wolfe, K. H., and D. C. Shields. 1997. Molecular evidence for an ancient duplication of the entire yeast genome. Nature 387: 708–713.

Wong, J. L., and M. A. Johnson. 2010. Is HAP2-GCS1 an ancestral gamete fusogen? Trends Cell Biol. 20: 134–141.

Xu, L., E. Petit, and M. E. Hood. 2016. Variation in mate-recognition pheromones of the fungal genus *Microbotryum*. Heredity 116: 44–51.

Zamir, L., M. Zaretsky, Y. Fridman, H. Ner-Gaon, E. Rubin, and A. Aharoni. 2012. Tight coevolution of proliferating cell nuclear antigen (PCNA)-partner interaction networks in fungi leads to interspecies network incompatibility. Proc. Natl. Acad. Sci. USA 109: E406–E414.

Zeng, Y., H. Li, N. M. Schweppe, R. S. Hawley, and W. D. Gilliland. 2010. Statistical analysis of nondisjunction assays in *Drosophila*. Genetics 186: 505–513.

Zetka, M. 2017. When degradation spurs segregation. Science 355: 349–350.

Zickler, D., and N. Kleckner. 2016. A few of our favorite things: Pairing, the bouquet, crossover interference and evolution of meiosis. Semin. Cell Dev. Biol. 54: 135–148.

Zwick, M. E., J. L. Salstrom, and C. H. Langley. 1999. Genetic variation in rates of nondisjunction: Association of two naturally occurring polymorphisms in the chromokinesin nod with increased rates of nondisjunction in *Drosophila melanogaster*. Genetics 152: 1605–1614.

CHAPTER 11

Cellular Senescence

In all multicellular species, progressive deterioration of cellular features eventually gives rise to organismal breakdown, essentially guaranteeing mortality beyond a particular maximum lifespan. Although some forms of environmental modification, such as restricted feeding, can prolong longevity, in no case is a multicellular soma known to be immortal. Dreams of breaking this barrier pharmacologically notwithstanding, there is simply no way to prevent the relentless accumulation of deleterious somatic mutations, as most single-cell variants cannot be selectively eliminated from a terminally differentiated tissue. Here, we focus on mortality features intrinsic to unicellular species (i.e., independent of external factors such as predators and pathogens). Although individual cells do not commonly undergo senescent deterioration within the span of single cell divisions, a different view emerges on the timescale of cell lineages.

It can be difficult to contemplate the issue of senescence in unicellular species, as the concept normally applies to specific adult individuals, the identity of which is often blurred when reproduction involves binary fission. In the case of an asymmetrically reproducing cell, as in the budding yeast *S. cerevisiae*, it is natural to call the smaller product the daughter cell, although it need not always follow that the smallest member of a pair is the more rejuvenated of the two (in the case of *S. cerevisiae*, it is). Things become more ambiguous when fission is morphologically symmetrical, as even in this case there can still be asymmetrical transmission of damage. One might then view the least damaged of the two cells following a division as the offspring, although quantifying such differences between two daughter cells is difficult. An alternative strategy is to base the age of a cell on some key

inherited component, such as a specific pole of a parental cell.

When viewed in this way, quite possibly all cell lineages eventually succumb to senescent decline, even if nothing more than as a consequence of a series of unfortunate stochastic events. This means that senescence is an inherent feature of life, present since the beginning of biology. There is nothing intrinsically beneficial about death, but as is discussed later, the selective advantage of prolonging the life of individual cell lineages beyond extended periods of time may be negligible, in effect stalled by a drift barrier. What remains quite unclear is the degree to which senescence proceeds more rapidly in some unicellular species than in others.

Physiological Load

The gradual deterioration of the self-sustaining features of cells inevitably leads to enhanced mortality rates with increasing age. The relevant determinants of the rate of decline include: the preservation of membrane, messenger RNA, and protein integrity; clearance of cytotoxic features such as harmful metabolites and protein aggregates; and maintenance of large cellular complexes such as ribosomes and nuclear-pore complexes. All biomolecules are subject to harmful chemical and physical modifications. For example, lipid molecules can experience oxidative damage resulting from free radicals released during metabolism. The same is true of RNAs, which can lead to further downstream problems when erroneous messenger RNAs are translated. Proteins can accumulate damage in a number of ways, such as deamidation of Asn to Asp and Gln to Glu, both of which lead to a negatively

Evolutionary Cell Biology. Michael Lynch, Oxford University Press. © Michael Lynch (2024). DOI: 10.1093/oso/9780192847287.003.0011

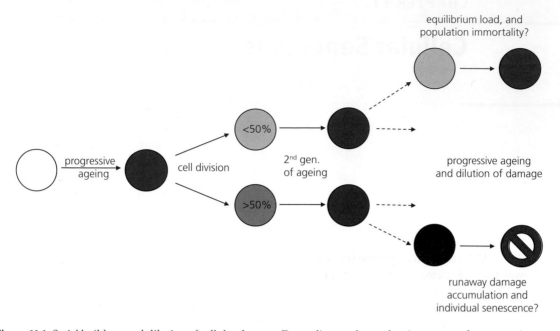

Figure 11.1 Serial build-up and dilution of cellular damage. Depending on the stochastic patterns of asymmetric damage inheritance among sublineages, the entire population of cells may remain effectively immortal, while some portions of the extended pedigree experience runaway damage and go extinct. The intensity of blue denotes the physiological load carried by a cell. The top sublineage experiences regular dilution of parental-cell damage, leading to periodic rejuvenation and persistence. The bottom sublineage represents the opposite extreme – a series of consecutive generations of inheritance of excess parental-cell damage, leading to cumulative physiological deterioration. Cell-lineage viability is retained as long as the rate of production of rejuvenated descendants exceeds the rate of production of permanently senescing sublineages.

charged amino acid with possible effects on three-dimensional protein structure.

The fundamental question is how the accumulation of damage over the life of a cell lineage can proceed at a low enough rate to avoid overwhelming the capacity for repair and eventual cell-lineage demise. Two potential routes for avoiding extinction by senescence at the population level are: 1) a rate of damage accumulation in an average cellular lifetime less than or equal to the rate at which such damage is diluted by cell division; and/or 2) asymmetric inheritance of parental-cell damage, such that one of the two 'daughter' products of cell division acquires more damage than its parent had at birth, while the other is rejuvenated (Figure 11.1).

Error catastrophe

Under the assumption that deteriorated proteins and macromolecular complexes accumulate during

cell lifetimes, Medvedev (1962) and Orgel (1963) proposed an error-catastrophe hypothesis for ageing, arguing in particular that progressive decline in the accuracy of the transcription and translation machinery would lead to still-further accumulation of erroneous proteins. The idea here is that because transcription and translation processes involve multiple proteins (Chapter 20), as these incur errors, the error rate itself will increase, leading to a downward spiral of cell fitness. Some contribution to error catastrophe might result from genomic mutations (discussed later), but the mechanisms envisioned here primarily concern phenotypic errors.

The vast majority of research on this matter has focused on the cells of multicellular species. Abnormal proteins accumulate in the cells of old organisms (Holliday and Tarrant 1972; Rothstein 1975), although it is unclear whether this is a consequence of declining accuracy of transcription and/or translation. Although ribosomes in rodents do not appear to decline in terms of translational accuracy with age, ribosomal activity does decline (Mori et al. 1979; Butzow et al. 1981; Filion and Laughrea 1985). There is some suggestion that transcript errors accumulate in old *S. cerevisiae* cells, although these results were obtained in cell lines with error-prone RNA polymerase, and it also remains unclear whether the error rate of transcription itself increases (as expected under the error-catastrophe hypothesis) or whether older transcripts simply accumulate more damage while circulating within cells (Vermulst et al. 2015).

Other results yield a mixed perspective on the error-catastrophe hypothesis. For example, Krisko and Radman (2013) found that progressive oxidative damage in *E. coli* magnifies the genomic mutation rate, and an *E. coli* strain engineered to be defective in translation experienced an increase in the mutation rate, but this was apparently a consequence of the activation of an error-prone DNA repair pathway (Bacher and Schimmel 2007). By applying the drug streptomycin, Edelmann and Gallant (1977) elevated the translation-error rate by $50\times$ in the same species, and found that although the misincorporation of amino acids into proteins increased substantially, cellular error levels reached a steady state, with growth potential reaching a plateau, rather than exhibiting a downward spiral. Thus, these studies demonstrate that physiological damage may not always lead to a relentless increase in genomic damage by positive feedback.

Orgel (1970) eventually realized that error catastrophe is not inevitable. As noted earlier, the key issue is whether the error rate grows sufficiently rapidly with the accumulation of damage to outpace the rate of removal. Letting $D(t)$ be the amount of cellular damage at time t, if λ is the baseline rate of production of new cellular damage (assumed to be constant), and δ is the fractional rate of removal of prior damage, the expected amount of damage in the next time unit is $\lambda + (1 - \delta)D(t)$ (Foundations 11.1). Assuming that λ and δ remain constant, if $0 < \delta \leq 1$, a stable equilibrium level of damage equal to λ/δ is predicted, such that the input of new damage is exactly balanced by the loss of old damage. Error catastrophe requires that prior damage accelerates the rate of accumulation of future damage.

Of course, a sufficiently high equilibrium level of damage might still ensure cell-lineage demise. An additional issue is that this simple computation is performed in a deterministic framework and assumes no variation among cells. As discussed below, variation in damage inheritance is the expectation, and this can yield a tail of extreme individuals with low fitness. Under some conditions, this might lead to a point of no return, i.e., a sort of ratchet effect towards lineage extinction. However, as noted in Foundations 11.1, among-individual variation can also lead to population immortality, as cells with exceptionally low damage loads are promoted over damage-laden cells.

Cellular versus population immortality

Because all progeny inherit their initial cytoplasm from their mothers, asymmetric transmission of damage provides a potentially powerful mechanism for avoiding population senescence – although maternal cells might progressively accumulate damage that reduces cell-division potential, at least one of their daughters might be rejuvenated (Ackermann et al. 2007; Erjavec et al. 2008; Chao 2010). Under this view, the gradual senescence that accrues in some cell lineages can be offset by rejuvenation in others (Figure 11.2; Foundations 11.1), ensuring population survival. This is no different than a population of a multicellular species remaining viable despite the senescence of old individuals.

The stalked bacterium *Caulobacter crescentus*, whose swimming progeny eventually settle down themselves, provides a clear example of such an effect. Cell division is asymmetrical in this taxon, with the sedentary maternal cells experiencing a progressive loss of cell-division potential, and near-complete loss of reproductive capacity after producing ~200 progeny (Ackermann et al. 2003, 2007). However, the species lives on, as the physiological decline of parental cells is offset by improved states in progeny.

In contrast, something close to an equilibrium growth rate is demonstrable with *E. coli* cell lineages grown in an apparatus within which mother cells are trapped and consecutive progeny are released. *E. coli* cells divide in a morphologically symmetrical manner, but in this 'mother machine', maternal cells retain one pole dating back to the onset of the

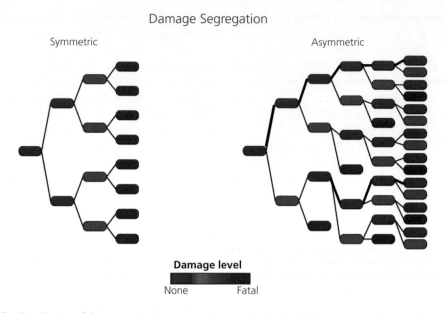

Figure 11.2 The distribution of damage inheritance in pedigrees of cells. **Left:** Transmission of parental damage is perfectly symmetrical, leading to invariance in the level of damage among descendent cells. In principle, the descendants can reach an equilibrium level of non-lethal damage if the rate of damage removal is not overwhelmed by damage accumulation each generation. **Right:** Asymmetric inheritance leads to phenotypic variance in the damage levels in populations of cells, with some nearly damage-free cells being maintained at the expense of damage-laden cells. Two extended single-cell lineages with low levels of damage (blue) are highlighted with bold connecting lines.

experiment, with the opposite pole being created anew at each cell division. When treated in this way, the growth rates of individual maternal cells remain constant for up to 150 divisions. Although there does appear to be an increase in the probability of mortality per division after ~100 generations (Wang et al. 2010), the evidence for progressive senescence is less than compelling. As discussed further below, other observations on the growth of single *E. coli* cells reveal that the cell-division times of descendant lineages eventually converge on equilibrium values (Chao 2010; Rang et al. 2011).

Molecular and Cellular Determinants

The previous section highlighted the conditions under which some or all cells within a population will experience senescence associated with physiological decline, and also how populations can remain effectively immortal when individual cell lineages are not. Here, we more explicitly consider the physical aspects of damage inheritance observed in various unicellular organisms. As noted

earlier, asymmetric damage inheritance provides a simple mechanism by which some sublineages of cells become rejuvenated (albeit at the expense of ageing in others).

In *E. coli*, protein aggregates that accumulate with age and are presumably harmful to cell health, are progressively moved to cell poles in an entirely passive manner (Lindner et al. 2008; Winkler et al. 2010). To see how this process can lead to lineage-specific senescence (or at least cell-to-cell heterogeneity in damage accumulation), consider an aggregate incapable of movement. If such a cellular inclusion develops at the pole of a cell, it will remain at that location in all descendants inheriting the pole (Figure 11.3). Likewise, if it appears near the midpoint of a symmetrically dividing cell, then it will also succumb to a permanent pole location in all future descendants. Finally, if the aggregate is near the 1/4 or 3/4 point in a rod-shaped cell, upon division at the midpoint, it will be present at the midpoint of one of the daughter cells, and hence, will sequester at the pole of one of the granddaughters. A similar mechanism appears

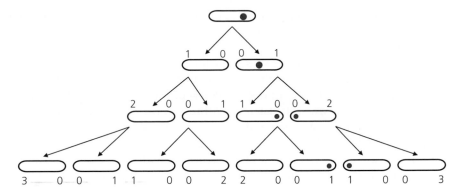

Figure 11.3 Passive relocation of a cellular protein aggregate (red sphere) to the poles of symmetrically dividing cells. Because the aggregate is immobile, regardless of the starting point, it will eventually be located near the poles of descendant cells, where it will then be retained indefinitely. The numbers at the ends of cells denote the ages of poles relative to the base individual. This schematic only shows the potential distribution of aggregates descendent from the single ancestral cell, but in reality, new damage aggregates will arise in each cell and be apportioned into its descendants in the same manner. Not shown is the possibility that aggregates may also dissipate over time. Modified from Lindner et al. (2008).

to result in the differential sequestration of outer membrane proteins to the older poles of cells (Bergmiller et al. 2017).

In contrast to the situation in bacteria, which generally lack significant internal structure, aggregates in eukaryotic cells can be actively transported along the cytoskeleton to destinations associated with protein-management activities, including aggregation disassembly and protein refolding by chaperones (Kawaguchi et al. 2003; Kaganovich et al. 2008; Malmgren Hill and Nyström 2016). Despite such activities, however, the few eukaryotic species to be evaluated still exhibit patterns of senescence of individual lineages.

One of the more dramatic examples involves the budding yeast, *S. cerevisiae*, which has asymmetric cell division, with the daughter (the bud) being substantially reduced in size relative to the mother. The evidence supports the long-term sequestration of damage within maternal *S. cerevisiae* cells (Aguilaniu et al. 2003), with ageing cells progressively experiencing higher mortality rates and higher degrees of sterility (inability to mate) (Kaeberlein 2010; Schlissel et al. 2017) (Figure 11.4). Although one of the most well-established methods for lifespan extension in animals is caloric restriction (imposition of a nutrient-sufficient diet low in calories) (Fontana et al. 2010), such treatment has essentially no effect on lifespan in *S. cerevisiae*, which produce

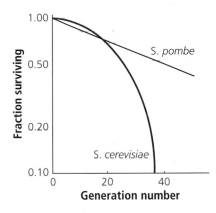

Figure 11.4 Individual cells of the budding yeast *S. cerevisiae* experience progressive decline in cell-division potential, and therefore an accelerated rate of mortality with age, whereas those of the fission yeast *S. pombe* experience a constant mortality rate, as revealed by the diagonal survivorship curve (on a logarithmic scale). From Spivey et al. (2017).

an average of 25 to 30 offspring per maternal cell regardless of conditions (Huberts et al. 2014).

Numerous mechanisms for ageing have been suggested for yeast, including the accumulation of extrachromosomal circular DNAs, malfunctioning mitochondria, and damaged proteins (Macara and Mili 2008; Kaeberlein 2010; Denoth-Lippuner et al. 2014; Kaya et al. 2021). The nuclei of older cells also progressively experience a genome-wide loss of nucleosomes, elevated rates of chromosomal breaks, and increases in translocations, insertions of mitochondrial DNA, and transposition of mobile elements (Hu et al. 2014). Several dozen proteins are either mother-enriched or daughter-enriched, the former being biased towards those localized to plasma membranes and vacuoles, and the latter towards those involved in bud construction and genome maintenance (Yang et al. 2015). Such asymmetric inheritance must be due to active transport and/or exclusion mechanisms, which include a diffusion barrier at the bud neck preventing the passage of large complexes (Shcheprova et al. 2008). Individual molecules of ~135 proteins are retained within maternal *S. cerevisiae* cells for up to 28 generations, consistent with the potential for long-term accumulation of protein damage (Thayer et al. 2014).

In contrast to *S. cerevisiae*, the fission yeast *S. pombe* has morphologically symmetrical division. Yet, there is still some asymmetric inheritance of damage, with 'maternal' cells (those retaining bud scars) inheriting more damaged proteins (Erjavec et al. 2008). Despite this partitioning, senescence appears not to occur under well-nourished conditions (Spivey et al. 2017) (Figure 11.4), instead arising primarily under stressful conditions, where cells accumulating large aggregates suffer higher mortality rates (Coelho et al. 2013). Whether such behavior reflects the predicted behavior of the model outlined in Foundations 11.1, wherein both sublineages of daughter cells evolve a finite equilibrium cell-division time and hence, retain 'immortality', remains unclear.

Finally, an issue of special interest with respect to senescence in eukaryotes is the biogenesis of complex macromolecular structures and organelles during cell division. Are pre-existing assemblies retained intact across cell divisions, with entirely new complexes being assembled from newly synthesized subcomponents as daughter cells grow? Or are such features frequently disassembled and reconstructed from mixtures of old and newly synthesized components?

For many organelles (e.g., endoplasmic reticulum, mitochondria, vacuoles, peroxisomes) in yeast, pre-existing structures and their protein constituents are symmetrically inherited, with new membrane complexes (comprised of components of similar ages) being incorporated with organelle growth. For example, old nuclear-pore complexes (consisting of dozens of proteins) are inherited as a unit in yeast and remain separate from those constructed from newly synthesized components (Shcheprova et al. 2008; Menendez-Benito et al. 2013). In post-mitotic rat brain cells, proteins in large complexes appear to remain in place for extended periods of time, with ~25% of proteins associated with nuclear pores remaining intact after an entire year (Toyama et al. 2013). In *C. elegans*, age-related deterioration of nuclear-pore complexes leads to increasingly leaky nuclei enabling the entry of increased numbers of inappropriate cytoplasmic proteins (D'Angelo et al. 2009).

These diverse observations make clear that, like multicellular organisms, unicells accrue damage over their individual lifespans. However, unlike the somas of a multicellular species, which eventually succumb to nonremovable damage accumulation, extended lineages of unicellular species can avoid permanent senescent decline as selection promotes cells with minimal damage. As is discussed in more detail later, these principles also apply to genomic damage, which inevitably accumulates as permanent mutations in lineages by cell-to-cell descent. Whereas specific mutant cells typically cannot be selectively removed from mixed tissues of multicellular species, leading to maladies such as cancer, they can be purged from unicellular populations by natural selection, as the cell is the unit of selection.

Evolution of Senescence

The preceding discussion makes clear that unicellular species can experience senescence, with sublineages stochastically incurring enough consecutive generations of excess damage accumulation eventually succumbing to a physiological load. Less clear, however, is the extent to which the rate of senescent decline is molded by the forces of natural selection, as opposed to being a simple passive response to selection on other cellular features.

Evolutionary theories of senescence have been developed primarily from the standpoint of multicellular organisms, focusing on the separation of the largely quiescent germline and the soma, which represents the manifestation of nearly all gene expression. As the latter is disposed after each sexual generation, its accumulated damage is expendable, whereas the expectation is that the germline will be effectively immortal.

Starting from this premise, the evolution of senescence is thought to be molded by two mechanisms (Finch 1991; Charlesworth 2000; Arking 2006). First, the pleiotropy hypothesis embraces the idea of trade-offs between alternative fitness traits, e.g., selection for energetic investment in early reproduction subtracting from what can be invested in survival (Williams 1957). A second hypothesis focuses on the diminishing fitness pay-offs of progeny produced late in life (Medawar 1952; Hamilton 1966). Like compound interest applied to a bank account, the pay-offs of which grow exponentially, progeny successfully produced at a young age enter the gene pool earlier and contribute to elevated rates of expansion of the genes contributing to such a life-history strategy. As a consequence, the efficiency of natural selection operating on age-specific reproductive and survival rates must eventually decline with increasing age. In the extreme situation in which individuals late in life produce no offspring at all, there can be no selection on survival-enhancing traits expressed only beyond that point (except in the case of cross-generational care, as in social animals). As a consequence, should they exist, deleterious late-acting mutations are expected to accumulate by a combination of mutation pressure and random genetic drift.

These two hypotheses are not mutually exclusive, as there is no reason that both mechanisms cannot be operating simultaneously. Moreover, as the number of genes influencing survival and reproductive rates must constitute nearly the entire genome, the genetic mechanisms of senescence are likely to be diverse among phylogenetic lineages due to the stochastic nature of the mutation process and the targets hit.

Unfortunately, because comparative studies of senescence in unicellular species are nearly nonexistent, the extent to which either of these hypotheses is relevant to microbes remains unclear. They are testable, however, and evidence of genetic variation for lifespan in yeast (Kaya et al. 2021) suggests a way forward. For example, the antagonistic-pleiotropy hypothesis predicts that species with higher cell-division rates would exhibit more rapid rates of demise in sublineages of cells that progressively inherit the most damage from their parental cells. The mutation-accumulation hypothesis predicts that alleles with deleterious effects on fitness will be enriched in classes of genes whose expression is confined to old mother cells. Whether genes with the postulated age-specific expression properties actually ever exist in unicellular species is an open question, but data from animal species support this argument (Rodriguez et al. 2017; Cui et al. 2019; Cheng and Kirkpatrick 2021).

Mutational Meltdown

One of the more dramatic observations from experimental gerontology is that isolated lineages of vertebrate cells generally have finite lifespans, regardless of the population size of cells. After a series of cell divisions of apparently unimpaired growth, culture loss often becomes certain. Fibroblast cell lines (derived from connective tissue) from long-lived vertebrates are capable of more divisions than those from short-lived species, and cells taken from older individuals have diminished numbers of remaining cell divisions (Martin et al. 1970; Hayflick 1977). As these results arise in cultures containing large numbers of cells, they suggest some kind of near-deterministic mechanism of sudden breakdown in cellular integrity, e.g., programmed cell death. However, observations of single-cell lineages suggest that this is not the case – the loss of cell-division potential is not abrupt but continuous throughout the lifespan of the extended lineage (Smith and Pereira-Smith 1977; Raes and Renmacle 1983; Karatza and Shall 1984; Angello and Prothero 1985) (Figure 11.5).

Although the relevance of cells having an 'out-of-body' experience to matters involving unicellular species is unclear, there are cases of single-celled organisms having finite numbers of cell divisions. Most notable is the situation in ciliates, many of which have characteristic culture lifespans. The substantial body of early research on these kinds of observations has been extensively reviewed (Nanney 1974; Smith-Sonneborn 1981; Bell 1988). For a number of species, extinction occurs in a predictable number of cell divisions (usually in 150–1500 cell divisions, depending on the isolate) with low intrastrain variance. Moreover, this can be true both for lineages maintained by single-cell descent

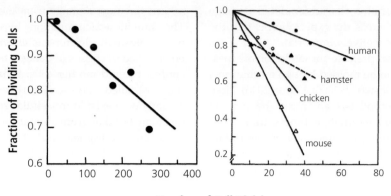

Number of Cell Divisions

Figure 11.5 Left: Continuous decline of cell-division potential in single-cell lines of the ciliate *Paramecium caudatum* (Takagi and Yoshida 1980). **Right:** Similar plots for fibroblasts cultured from various vertebrates: human (Smith and Pereira-Smith 1977); hamster (Raes and Remacle 1973); chicken (Angello and Prothero 1985); and mouse (Karatza and Shall 1984). The lines are least-squares regressions scaled to give a time-zero fitness equal to 1.0, except in the case of hamster. From Lynch and Gabriel (1990).

and in moderately large cultures. The results do not appear to be a consequence of deterioration of the lab environment, and nuclear transplantation experiments indicate that clonal decline is caused by nuclear, rather than cytoplasmic, factors (Aufderheide 1984, 1987). As in fibroblast cultures, close observations of the fitness features of individual *Paramecium* cells are inconsistent with a threshold reduction in cell-division potential (Figure 11.5). Rather, there is a steady decline in cell reproductive capacity with time (~0.1% per generation) (Smith-Sonneborn 1979; Takagi and Yoshida 1980; Lynch and Gabriel 1990).

Whereas the exact mechanisms underlying such behavior are not clear, the observed patterns are consistent with some sort of cumulative genomic deterioration. As first pointed out by Muller (1964), obligately asexual lineages are subject to a ratchet-like mechanism of mutation accumulation. With a high rate of genome-wide input of deleterious mutation (Chapter 4), populations of clonally reproducing cells are expected to contain a distribution of fitness classes defined by recurrent mutation pressure and selection against highly loaded individuals. This distribution is expected to be unimodal (and under many genetic models, close to Poisson in form; Chapters 3 and 23), with only a small fraction of the population contained within the best fitness class. In each generation, the possibility exists that,

by chance, all individuals in this class either leave no progeny at all or produce only progeny acquiring a new deleterious mutation. Either way, the best class will have been irreversibly lost, as in the absence of sexual reproduction, there is no way to recover the prior fitness class except in the extremely rare case of a precise back-mutation. Once the best class is lost, the previously second-best class advances to be the best in the population, but it too will eventually suffer the same fate, and so on.

The natural culmination of Muller's ratchet is extinction by a process known as the mutational meltdown (Lynch and Gabriel 1990; Lynch et al. 1993; Gabriel et al. 1994). Initially, a population will typically have excess reproductive capacity, but as mean individual fitness declines, a point will eventually be reached at which each individual is just able to replace itself on a per-generation basis. At that point, any further increase in the mutation load will lead to a reduction in population size, but with fewer individuals, natural selection is less efficient (Chapter 4). This then increases the rate at which the fitness ratchet clicks, continuously driving the population to a still smaller size, culminating in a downward spiral that leads to extinction via deleterious-mutation accumulation.

As this process can operate in populations containing many thousands of asexual individuals, yielding fairly deterministic extinction times despite the stochastic accumulation of mutations, it may provide an explanation for the above-noted results. The situation for ciliates likely relates to their unique binuclear form of genetic organization, involving a germline micronucleus and a somatic macronucleus. As discussed in Chapter 10, during macronuclear maturation, all chromosomes are duplicated to very high ploidy levels (typically, several hundreds), and it is from this genome that all gene expression occurs. Although the micronucleus replicates by mitosis during clonal propagation, macronuclear division is poorly understood, but is more along the lines of random fragmentation than symmetric mitosis.

Given this genetic system, loss of viability in a ciliate population during clonal propagation must be a consequence of deleterious changes in the macronuclear genome. However, such deterioration is unlikely to involve genomic mutations at the nucleotide level. In ciliates, point mutations involving single base-pair substitutions arise at a rate of $\sim 10^{-11}$/nucleotide site/cell division in the micronucleus (Sung et al. 2012; Long et al. 2016), the lowest known rate of any organism. Assuming the error rate is the same in the macronucleus, then because the macronuclear genome size is $\sim 10^8$ bp, only one mutation is expected to arise per haploid genome every 1000 cell divisions. Thus, the more likely path to cellular senescence in ciliates involves the amitotic (fission-like) nature of macronuclear propagation, which can lead to random drift in chromosome copy numbers, stoichiometric imbalance, and eventual loss of entire chromosomes. Assuming simple random sampling of chromosomes following macronuclear genome duplication, Kimura (1957) produced a stochastic model of chromosome loss that yields extinction dynamics reasonably consistent with the existing data. Notably, extinction can be averted entirely if pre-senescent clones are allowed to undergo sexual reproduction, which results in the development of a new macronucleus from the micronuclear template.

Mutation at the nucleotide level is also unlikely to account for the behavior of vertebrate fibroblast cultures. Although the mutation rate of such cells may be as high as 10^{-9} per bp per cell division (Lynch 2010) and the haploid genome size is of order 10^9 bp, at most 10% of the vertebrate genome is under significant selection, so it is likely that < 1 functionally relevant mutation (and maybe considerably less) arises per cell division. A more likely explanation for eventual culture demise is progressive loss of telomeres from chromosome ends. Indeed, immortalized mammalian cell lines are readily obtained from cancer cells with substantial genomic rearrangements, showing that mutation accumulation is an unlikely extinction mechanism.

Summary

- The molecular constituents of cells naturally deteriorate over time, raising the possibility of senescence in unicellular organisms analogous to the ageing that occurs in the somas of multicellular species. However, the asymmetric inheritance of damage by daughter cells, whether programmed or simply stochastic, can lead to the rejuvenation of one member of the pair at the expense of the other. This then ensures indefinite population survival, as in multicellular species that discard the senescing somas of ageing individuals.

- Two hypotheses have been suggested for the evolution of senescence: 1) a tradeoff associated with pleiotropic effects, whereby investment in early reproduction imposes costs on future survival; and 2) diminishing pay-offs resulting from the compound-interest effect of early contributions to the gene pool enhancing fitness more than progeny produced later in life. Both hypotheses were constructed with multicellular species in mind, and the extent to which they apply to unicellular species remains unexplored.

- Numerous examples exist in which single-cell lineages have well-defined finite lifespans, most notably in ciliates and vertebrate-cell cultures. The mechanisms driving extinction in these cases likely involve large-scale problems with chromosomal integrity and imbalance, such as telomere and/or chromosome loss, rather than the accumulation of point mutations. In such cases, rejuvenation can be accomplished by sexual outcrossing and/or inactivation of genes involved in telomere erosion.

Foundations 11.1 The physiological damage load in a cell lineage

Over time, whether stationary or in a period of active growth, a cell can be expected to acquire some reduction in physiological capacities owing to the accumulation of damaged molecules (here, assumed to accumulate at rate λ per unit time). Prior damage might also be fractionally eliminated at rate δ, such that with a starting level of damage at birth D_0, the amount of damage accumulated over time can be expressed with the recursion equation

$$D(t+1) = \lambda + (1-\delta)D(t). \qquad (11.1.1a)$$

Note that, provided $0 < \delta \leq 1$, for large t, an equilibrium level of damage is asymptotically approached. This is obtained by setting $D(t+1) = D(t)$,

$$\widehat{D} = \frac{\lambda}{\delta}. \qquad (11.1.1b)$$

A special case was considered by Chao (2010), who assumed $\delta t \ll 1$, so that damage simply accumulates linearly with time before cell division, and the above-noted equilibrium is never reached. Letting D_0 denote the damage of a particular cell at birth, its damage at the time of cell division T is then

$$D(T) \simeq D_0 + \lambda T. \qquad (11.1.2)$$

Suppose that some critical cell feature has to achieve a critical value P_c prior to cell division (e.g., time to achieve a threshold amount of some key cellular component), with cellular damage detracting from the rate of accumulation of P, such that

$$\frac{dP}{dt} = 1 - D(t). \qquad (11.1.3)$$

Equation 11.1.3 equates the rate of product accumulation to 1.0 in the absence of damage. The solution to Equation 11.1.3, obtained after substituting Equation 11.1.2, leads to a quadratic equation defining the time to cell division,

$$P_c = P(T) = (1-D_0)T - (\lambda/2)T^2. \qquad (11.1.4)$$

Given D_0 at birth and P_c at division (a fixed parameter), Equation 11.1.4 yields the cell-division time T as a function of λ. With no damage accumulation, $\lambda = D_0 = 0$, and Equation 11.1.4 leads to the definition $P_c = T$, i.e., P_c denotes the time required to linearly build up to the checkpoint in the absence of any damage accumulation.

The prior expressions assume cells growing deterministically, with each daughter cell sharing exactly half the damage in the maternal cell at the time of fission. However, with asymmetrical cell division, the population will be heterogeneous. Letting $a \leq 1/2$ and $1 - a \geq 1/2$ be the fractions of damage transmitted to the two offspring cells (indexed as 1 and 2, with the maternal cell denoted as 0), from Equation 11.1.2 the damage transmitted to the two newborns becomes

$$D_1 = a(D_0 + \lambda T_0), \qquad (11.1.5a)$$

$$D_2 = (1-a)(D_0 + \lambda T_0). \qquad (11.1.5b)$$

Substituting these expressions for D_1 and D_2 for D_0 in Equation 11.1.4, and solving yields the times to cell division for the two daughter cells as a function of the features of the maternal cell (D_0, T_0),

$$T_i = \frac{1 - D_i - \sqrt{(1-D_i)^2 - 2P_c\lambda}}{\lambda}, \qquad (11.1.6)$$

where $i = 1$ or 2.

This overall system of equations allows one to start with a given value of T_0 for a parental cell, and then obtain the cell division times for all subsequent cells (Figure 11.6). The critical issue with respect to senescence is whether T_1 and/or T_2 increase without limits. If this were true for both daughter-cell types, runaway damage and ageing of the entire population would ensue. In contrast, population survival in spite of some cell senescence would occur if T_2 stabilized, i.e., the least-loaded daughters do not experience ever-increasing damage.

Equilibrium requires that the reduction in damage by dilution from the mother cell be balanced by the build-up of new damage in the daughter. The equilibrium conditions are obtained by substituting $D_i = D_0$ into Equations 11.1.5a,b. Letting $\alpha = a/(1-a)$ and writing $T_0 = \widehat{T}_1, \widehat{T}_2$, yields the equilibrium physiological loads,

$$\widehat{D}_1 = \widehat{T}_1\lambda\alpha, \qquad (11.1.7a)$$

$$\widehat{D}_2 = \widehat{T}_2\lambda/\alpha. \qquad (11.1.7b)$$

Finally, substituting into Equation 11.1.4 yields the equilibrium solutions,

$$\widehat{T}_1 = \frac{1 - \sqrt{1 - 2P_c\lambda(1+2\alpha)}}{\lambda(1+2\alpha)}, \qquad (11.1.8a)$$

$$\widehat{T}_2 = \frac{1 - \sqrt{1 - 2P_c\lambda[1+(2/\alpha)]}}{\lambda[1+(2/\alpha)]}. \qquad (11.1.8b)$$

The conditions for equilibrium are that the terms within the square roots in the previous equations be non-negative, which requires

$$\alpha \leq \frac{1 - 2P_c\lambda}{4P_c\lambda}, \qquad (11.1.9a)$$

$$1/\alpha \leq \frac{1 - 2P_c\lambda}{4P_c\lambda}. \qquad (11.1.9b)$$

Thus, a key for the immortality of a cell line is that the composite quantity $P_c\lambda$ not be too large. Given the

definition of P_c noted above, $P_c\lambda$ can be thought of as the total amount of damage that would be built up in a cell with minimum possible division time. With symmetric division ($\alpha = 1$), the criterion for equilibrium reduces to $P_c\lambda \leq 1/6$. With asymmetric inheritance and $P_c\lambda$ sufficiently small, both $\widehat{T}_1$ and $\widehat{T}_2$ reach real equilibria, but with increasing levels of damage, at most one of the cell classes

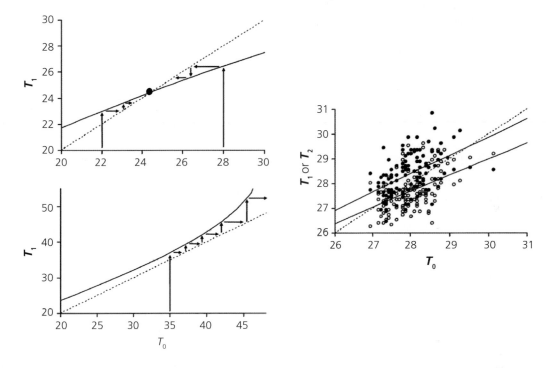

Figure 11.6 Left: Recursion plots for the relationship between cell-division times (T) in mother and daughter cells (the x and y axes, respectively) for the case of symmetrical cell division. Diagonal dotted lines denote potential points of equilibrium, with daughter and parental-cell division times being equal to each other. Starting with a particular value of T for a mother cell, Equation 11.1.4 defines the starting damage in that cell, which, when entered into Equation 11.1.5, generates the damage in the first-generation daughter cell at the time of its division; substitution of the latter into Equation 11.1.6 then leads to the cell-division time of this daughter cell; and this sequence of calculations can be done anew to yield the cell-division time in the next generation. The solid line denotes the expected relationship between mother and offspring cell-division times, which must intersect the dashed line for an equilibrium to exist, and moreover must intersect from above to the left for this equilibrium to be stable (upper left panel). The solid lines with arrows denote the successive projections of cell-division times for mothers and daughters from one generation to the next – starting with a particular maternal T_0, the offspring T_1 is determined by reading the appropriate value for the solid line off the y axis; this time value then becomes T_0 for the next generation, and so on. In the upper panel, $P_c = 2$ and $\lambda = 0.005$, so the key parameter $P_c\lambda = 0.10$, which (according to Equation 11.1.9) is sufficient to allow a stable equilibrium, as designated by the intersection of the solid and dashed lines; in this case, if T starts below the equilibrium, it progressively increases until reaching this point, and vice versa, if T starts above the equilibrium. In the lower panel, the rate of damage accumulation is too high for an equilibrium to be achieved; regardless of the starting point, T eventually increases without bound as descendants progressively accumulate more damage. **Right:** Actual observations on the cell-division times for parent and offspring pairs of *E. coli* cells, with the solid and open points designating the times for the daughter cells with the longest and shortest division times. Although the data are quite noisy, the regression lines for both sets of points intersect the solid diagonals, consistent with a non-senescing population. From Chao (2010).

can reach an equilibrium. Eventually, with high enough $P_c\lambda$, neither daughter is capable of sustained reproduction, and senescence of the entire population would ensue without some intervening rejuvenation mechanism.

Some insight into the level of fitness reduction that results from damage accumulation can be acquired by noting that, for the case of symmetrical division, a damage-free cell would reproduce at age P_c, doubling the population size in that interval. In contrast, in the presence of damage accumulation, cell doubling will require time T_1 (as defined by Equation 11.1.8a with $\alpha = 1$), so that at time P_c the population would have increased by a factor of $2^{P_c/\widehat{T}_1}$. Thus, relative to a damage-free cell, the reproductive rate become $(2^{P_c/\widehat{T}_1})/2$. The fractional reduction in

fitness, which can be viewed as the equilibrium physiological load of damage accumulation, is then

$$L = 1 - 2^{(P_c/\widehat{T}_1)-1}, \qquad (11.1.10a)$$

which for moderate amounts of damage, $P_c/\widehat{T}_1$, simplifies to

$$L \simeq \frac{1 - (P_c/\widehat{T}_1)}{\sqrt{2}}. \qquad (11.1.10b)$$

The damage load can be viewed as the amount of improvement in cell fitness that could be achieved by establishing mechanisms that completely eliminate damage, i.e., $\lambda = 0$, although this view ignores any bioenergetic cost of producing and maintaining damage-control mechanisms.

Literature Cited

Ackermann, M., A. Schauerte, S. C. Stearns, and U. Jenal. 2007. Experimental evolution of aging in a bacterium. BMC Evol. Biol. 7: 126.

Ackermann, M., S. C. Stearns, and U. Jenal. 2003. Senescence in a bacterium with asymmetric division. Science 300: 1920.

Aguilaniu, H., L. Gustafsson, M. Rigoulet, and T. Nyström. 2003. Asymmetric inheritance of oxidatively damaged proteins during cytokinesis. Science 299: 1751–1753.

Angello, J. C., and J. W. Prothero. 1985. Clonal attenuation in chick embryo fibroblasts. Cell Tissue Kinet. 18: 27–43.

Arking, R. T. 2006. The Biology of Aging, 3rd edn. Oxford University Press, New York, NY.

Aufderheide, J. 1984. Clonal aging in *Paramecium tetraurelia*. I. Absence of evidence for a cytoplasmic factor. Mech. Ageing Dev. 28: 57–66.

Aufderheide, J. 1987. Clonal aging in *Paramecium tetraurelia*. II. Evidence of functional changes in the macronucleus with age. Mech. Ageing Dev. 37: 265–279.

Bacher, J. M., and P. Schimmel. 2007. An editing-defective aminoacyl-tRNA synthetase is mutagenic in aging bacteria via the SOS response. Proc. Natl. Acad. Sci. USA 104: 1907–1912.

Bell, G. 1988. Sex and Death in Protozoa: The History of an Obsession. Cambridge University Press, Cambridge, UK.

Bergmiller, T., A. M. C. Andersson, K. Tomasek, E. Balleza, D. J. Kiviet, R. Hauschild, G. Tkačik, and C. C. Guet. 2017. Biased partitioning of the multidrug efflux pump AcrAB-TolC underlies long-lived phenotypic heterogeneity. Science 356: 311–315.

Butzow, J. J., M. G. McCool, and G. L. Eichhorn. 1981. Does the capacity of ribosomes to control translation fidelity change with age? Mech. Ageing Dev. 15: 203–216.

Chao, L. 2010. A model for damage load and its implications for the evolution of bacterial aging. PLoS Genet. 6: e1001076.

Charlesworth, B. 2000. Fisher, Medawar, Hamilton and the evolution of aging. Genetics 156: 927–931.

Cheng, C., and M. Kirkpatrick. 2021. Molecular evolution and the decline of purifying selection with age. Nat. Commun. 12: 2657.

Cui, R., T. Medeiros, D. Willemsen, L. N. M. Iasi, G. E. Collier, M. Graef, M. Reichard, and D. R. Valenzano. 2019. Relaxed selection limits lifespan by increasing mutation load. Cell 178: 385–399.

Coelho, M., A. Dereli, A. Haese, S. Kühn, L. Malinovska, M. E. DeSantis, J. Shorter, S. Alberti, T. Gross, and I. M. Tolić-Nørrelykke. 2013. Fission yeast does not age under favorable conditions, but does so after stress. Curr. Biol. 23: 1844–1852.

D'Angelo, M. A., M. Raices, S. H. Panowski, and M. W. Hetzer. 2009. Age-dependent deterioration of nuclear pore complexes causes a loss of nuclear integrity in postmitotic cells. Cell 136: 284–295.

Denoth Lippuner, A., T. Julou, and Y. Barral. 2014. Budding yeast as a model organism to study the effects of age. FEMS Microbiol. Rev. 38: 300–325.

Edelmann, P., and J. Gallant. 1977. On the translational error theory of aging. Proc. Natl. Acad. Sci. USA 74: 3396–3398.

Erjavec, N., M. Cvijovic, E. Klipp, and T. Nyström. 2008. Selective benefits of damage partitioning in unicellular systems and its effects on aging. Proc. Natl. Acad. Sci. USA 105: 18764–18769.

Filion, A. M., and M. Laughrea. 1885. Translation fidelity in the aging mammal: studies with an accurate *in vitro* system on aged rats. Mech. Ageing Dev. 29: 125–142.

Finch, C. E. 1991. Longevity, Senescence, and the Genome. University of Chicago Press, Chicago, IL.

Fontana, L., L. Partridge, and V. D. Longo. 2010. Extending healthy life span – from yeast to humans. Science 328: 321–326.

Gabriel, W., M. Lynch, and R. Bürger. 1994. Muller's ratchet and mutational meltdowns. Evolution 47: 1744–1757.

Hamilton, W. D. 1966. The moulding of senescence by natural selection. J. Theor. Biol. 12: 12–45.

Hayflick, L. 1977. The cellular basis for biological aging, pp. 159–188. In C. E. Finch and L. Hayflick (eds.) Handbook of the Biology of Aging. Van Nostrand Reinhold, New York, NY.

Holliday, R., and G. M. Tarrant. 1972. Altered enzymes in ageing human fibroblasts. Nature 238: 26–30.

Hu, Z., K. Chen, W. Li, and J. K. Tyler. 2014. Transcriptional and genomic mayhem due to aging-induced nucleosome loss in budding yeast. Microb. Cell. 1: 133–136.

Huberts, D. H., J. González, S. S. Lee, A. Litsios, G. Hubmann, E. C. Wit, and M. Heinemann. 2014. Calorie restriction does not elicit a robust extension of replicative lifespan in *Saccharomyces cerevisiae*. Proc. Natl. Acad. Sci. USA 111: 11727–11731.

Kaeberlein, M. 2010. Lessons on longevity from budding yeast. Nature 464: 513–519.

Kaganovich, D., R. Kopito, and J. Frydman. 2008. Misfolded proteins partition between two distinct quality control compartments. Nature 454: 1088–1095.

Karatza, C., and S. Shall. 1984. The reproductive potential of normal mouse embryo fibroblasts during culture *in vitro*. J. Cell Sci. 66: 401–409.

Kawaguchi, Y., J. J. Kovacs, A. McLaurin, J. M. Vance, A. Ito, and T. P. Yao. 2003. The deacetylase HDAC6 regulates aggresome formation and cell viability in response to misfolded protein stress. Cell 115: 727–738.

Kaya, A., C. Z. J. Phua, M. Lee, L. Wang, A. Tyshkovskiy, S. Ma, B. Barre, W. Liu, B. R. Harrison, X. Zhao, et al. 2021. Evolution of natural lifespan variation and molecular strategies of extended lifespan in yeast. eLife 10: e64860.

Kimura, M. 1957. Some problems of stochastic processes in genetics. Ann. Math. Stat. 28: 882–901.

Krisko, A., and M. Radman. 2013. Phenotypic and genetic consequences of protein damage. PLoS Genet. 9: e1003810.

Lindner, A. B., R. Madden, A. Demarez, E. J. Stewart, and F. Taddei. 2008. Asymmetric segregation of protein aggregates is associated with cellular aging and rejuvenation. Proc. Natl. Acad. Sci. USA 105: 3076–3081.

Long, H., D. J. Winter, A. Y.-C. Chang, W. Sung, S. H. Wu, M. Balboa, R. B. R. Azevedo, R. A. Cartwright, M. Lynch, and R. A. Zufall. 2016. Low base-substitution mutation rate in the germline genome of the ciliate *Tetrahymena thermophila*. Genome Biol. Evol. 8: 3629–3639.

Lynch, M. 2010. Evolution of the mutation rate. Trends Genetics 26: 345–352.

Lynch, R. Bürger, D. Butcher, and W. Gabriel. 1993. Mutational meltdowns in asexual populations. J. Heredity 84: 339–344.

Lynch, M., and W. Gabriel. 1990. Mutation load and the survival of small populations. Evolution 44: 1725–1737.

Macara, I. G., and S. Mili. 2008. Polarity and differential inheritance – universal attributes of life? Cell 135: 801–812.

Malmgren Hill, S., and T. Nyström. 2016. Selective protein degradation ensures cellular longevity. eLife 5: e17185.

Martin, G. M., C. A. Sprague, and C. J. Epstein. 1970. Replicative life-span of cultivated human cells: Effects of donor's age, tissue, and genotype. Lab Invest. 23: 86–92.

Medawar, P. B. 1952. An Unsolved Problem of Biology. H. K. Lewis, London, UK.

Medvedev, Z. A. 1962. A hypothesis concerning the way of coding interaction between transfer RNA and messenger RNA at the later stages of protein synthesis. Nature 195: 38–39.

Menendez-Benito, V., S. J. van Deventer, V. Jimenez-Garcia, M. Roy-Luzarraga, F. van Leeuwen, and J. Neefjes. 2013. Spatiotemporal analysis of organelle and macromolecular complex inheritance. Proc. Natl. Acad. Sci. USA 110: 175–180.

Mori, N., D. Mizuno, and S. Goto. 1979. Conservation of ribosomal fidelity during ageing. Mech. Ageing Dev. 10: 379–398.

Muller, H. J. 1964. The relation of recombination to mutational advance. Mutat. Res. 1: 2–9.

Nanney, D. L. 1974. Aging and long-term temporal regulation in ciliated protozoa: A critical review. Mech. Ageing Dev. 3: 81–105.

Orgel, L. E. 1963. The maintenance of the accuracy of protein synthesis and its relevance to ageing. Proc. Natl. Acad. Sci. USA 49: 517–521.

Orgel, L. E. 1970. The maintenance of the accuracy of protein synthesis and its relevance to ageing: A correction. Proc. Natl. Acad. Sci. USA 67: 1476.

Raes, M., and J. Renmacle. 1983. Ageing of hamster embryo fibroblasts as the result of both differentiation and stochastic mechanisms. Exp. Gerontol. 18: 223–240.

Rang, C. U., A. Y. Peng, and L. Chao. 2011. Temporal dynamics of bacterial aging and rejuvenation. Curr. Biol. 21: 1813–1816.

Rodríguez, J. A., U. M. Marigorta, D. A. Hughes, N. Spataro, E. Bosch, and A. Navarro. 2017. Antagonistic pleiotropy and mutation accumulation influence human senescence and disease. Nat. Ecol. Evol. 1: 55.

Rothstein, M. 1975. Aging and the alteration of enzymes: A review. Mech. Ageing Dev. 4: 325–338.

Schlissel, G., M. K. Krzyzanowski, F. Caudron, Y. Barral, and J. Rine. 2017. Aggregation of the Whi3 protein, not loss of heterochromatin, causes sterility in old yeast cells. Science 355: 1184–1187.

Shcheprova, Z., S. Baldi, S. B. Frei, G. Gonnet, and Y. Barral. 2008. A mechanism for asymmetric segregation of age during yeast budding. Nature 454: 728–734.

Smith, J. R., and O. Pereira-Smith. 1977. Colony size distribution as a measure of age in cultured human cells. A brief note. Mech. Ageing Dev. 6: 283–286.

Smith-Sonneborn, J. 1979. DNA repair and longevity assurance in *Paramecium tetraurelia*. Science 203: 1115–1117.

Smith-Sonneborn, J. 1981. Genetics and aging in protozoa. Int. Rev. Cytol. 73: 319–354.

Spivey, E. C., S. K. Jones, J. R. Rybarski, F. A. Saifuddin, and I. J. Finkelstein. 2017. An aging-independent replicative lifespan in a symmetrically dividing eukaryote. eLife 6: e20340.

Sung, W., A. Tucker, T. G. Doak, J. Choi, W. K. Thomas, and M. Lynch. 2012. Extraordinary genome stability in the ciliate *Paramecium tetraurelia*. Proc. Natl. Acad. Sci. USA 109: 19339–19344.

Takagi, Y., and M. Yoshida. 1980. Clonal death associated with the number of fissions in *Paramecium caudatum*. J. Cell Sci. 41: 177–191.

Thayer, N. H., C. K. Leverich, M. P. Fitzgibbon, Z. W. Nelson, K. A. Henderson, P. R. Gafken, J. J. Hsu, and D. E. Gottschling. 2014. Identification of long-lived proteins retained in cells undergoing repeated asymmetric divisions. Proc. Natl. Acad. Sci. USA 111: 14019–14026.

Toyama, B. H., J. N. Savas, S. K. Park, M. S. Harris, N. T. Ingolia, J. R. Yates, and M. W. Hetzer. 2013. Identification of long-lived proteins reveals exceptional stability of essential cellular structures. Cell 154: 971–982.

Vermulst, M., A. S. Denney, M. J. Lang, C.-W. Hung, S. Moore, M. A. Moseley, J. W. Thompson, V. Madden, J. Gauer, K. J. Wolfe, et al. 2015. Transcription errors induce proteotoxic stress and shorten cellular lifespan. Nat. Commun. 6: 8065.

Wang, P., L. Robert, J. Pelletier, W. L. Dang, F. Taddei, A. Wright, and S. Jun. 2010. Robust growth of *Escherichia coli*. Curr. Biol. 20: 1099–1103.

Williams, G. C. 1957. Pleiotropy, natural selection and the evolution of senescence. Evolution 11: 398–411.

Winkler, J., A. Seybert, L. König, S. Pruggnaller, U. Haselmann, V. Sourjik, M. Weiss, A. S. Frangakis, A. Mogk, and B. Bukau. 2010. Quantitative and spatio-temporal features of protein aggregation in *Escherichia coli* and consequences on protein quality control and cellular ageing. EMBO J. 29: 910–923.

Yang, J., M. A. McCormick, J. Zheng, Z. Xie, M. Tsuchiya, S. Tsuchiyama, H. El-Samad, Q. Ouyang, M. Kaeberlein, B. K. Kennedy, et al. 2015. Systematic analysis of asymmetric partitioning of yeast proteome between mother and daughter cells reveals 'aging factors' and mechanism of lifespan asymmetry. Proc. Natl. Acad. Sci. USA 112: 11977–11982.

PART 4

Structural Evolution

CHAPTER 12

The Protein World

The vast majority of cellular functions involve the use of one or more proteins. Although these biomolecules have myriad specialized functions, many of which are covered in subsequent chapters, we focus here on general issues. Proteins are composed of linear strings of amino-acid residues, parts of which generally assemble into simple secondary structures, e.g., helices and sheets held together by hydrogen bonds. These, in turn, arrange into tertiary (three-dimensional and frequently globular) architectures. Quaternary structures, the subject of Chapter 13, arise when separate proteins associate into higher-order assemblages via binding interfaces. Whereas most genomes encode for several thousand proteins, only a few hundred protein-coding genes are shared across all species (Harris et al. 2003; Koonin 2003), implying that lineage-specific gains and losses of genes are common.

We explore three general topics in this chapter. First, we consider the fundamental biochemical and biophysical properties of the 20 major amino acids that serve as the building blocks of virtually all proteins. It is highly unlikely that all 20 amino acids entered the biological world at the same moment of time, so it is of interest to consider the potential order of evolutionary entry, as well as the consequences of a presumably simpler early amino-acid alphabet.

Second, one of the central problems of protein science concerns the stable folding of proteins into their so-called native states. Levinthal (1968) famously pointed out that proteins longer than a few dozen residues cannot possibly examine all feasible configurations en route to final assembly, concluding that folding pathways must be guided by information in the primary amino-acid sequence. The underlying guidelines must operate on timescales short enough to enable rapid responses to gene-expression demands and accurate enough to ensure the proper assembly of catalytic sites and avoidance of the energetic wastage associated with the disposal of improper assemblies. Poorly folded proteins impose the additional risk of initiating inappropriate aggregations with self and non-self proteins.

Third, in light of these features, we review the evolutionary constraints on the amino-acid sequences found in different proteins, in different regions of proteins, and in different phylogenetic lineages. The central questions here concern the degree to which various pairs of amino acids are substitutable for each other, the extent to which evolution at one particular site is independent of that of others, and the overall capacity of natural selection to counter the relentless input of amino acid-altering mutations.

The Essential Features of Proteins

Proteins are composed of variable amino-acid chain lengths, parts of which typically fold into more compact localized domains. Average domain sizes are roughly constant across prokaryotes and eukaryotes, but the linkers between domains tend to be several-fold longer in eukaryotes, leading to ~50% longer total chain lengths in the latter (an average of ~530 residues in eukaryotes, and ~350 in prokaryotes; Wang et al. 2011; Rebeaud et al. 2021). Given that each amino acid is chemically unique with respect to molecular weight, charge, hydrophobicity, polarity, etc. (Table 12.1), when primary sequences are further combined into three-dimensional forms, the resultant combinatorial diversity comprises an essentially boundless array of protein structures and functions.

Evolutionary Cell Biology. Michael Lynch, Oxford University Press. © Michael Lynch (2024). DOI: 10.1093/oso/9780192847287.003.0012

Each amino acid consists of a central carbon atom attached to one hydrogen atom, one NH_2 (amino) group, one CO_2 (carbonyl) group, and a unique cognate side chain (Figure 12.1). Peptide chains are assembled by mRNA-translating ribosomes, with the amide group of each consecutive amino acid reacting with the carboxyl group of the adjacent member of the growing chain (Figure 12.2). Glycine has the simplest side chain, just a single hydrogen atom, and is therefore symmetrical and quite flexible. Two residues contain sulfur (cysteine and methionine), whereas several have side chains containing nitrogen. Serine and threonine side chains uniquely carry an OH group. Proline is exceptional in that the side chain is covalently bonded to the nitrogen atom of the peptide backbone, and as a consequence is the only amino acid lacking an amide hydrogen atom for use in hydrogen bonding. Alanine, valine, leucine, and isoleucine have simple side chains ending in CH_3, and like glycine and proline are highly hydrophobic.

Amino-acid composition

To a large extent, the different properties of amino acids dictate where they are deployed within proteins, and this has direct implications for the biochemical and structural consequences of mutations. To acquire functionality, proteins need to achieve proper folds, which are strongly dependent upon backbone hydrogen bonds between residues (often located distantly on the polypeptide chain). In addition, hydrophobic residues, which avoid water molecules, tend to be buried within the cores of proteins. Exposure of hydrogen bonds and hydrophobic residues of protein cores leads to folding instability, which increases stickiness and the potential to engage in inappropriate protein–protein interactions. Thus, the surfaces of proteins are typically well wrapped with hydrophilic residues so as to minimize the intrusion of water molecules into the core.

It is unlikely that the amino-acid alphabet had reached the current 20-residue state when the first protein-based cells emerged some 4 billion years ago. This then raises the question as to how much functional protein diversity might have been achieved in a setting involving a smaller number of amino acids. The potential seems large, given that many proteins with diverse functions in today's world do not contain the full set of 20 amino acids. An extreme case is an antifreeze protein in a flounder fish that contains only seven different residues (Sicheri and Yang 1995), and a number of proteins in prokaryotes are entirely devoid of basic residues (McDonald and Storrie-Lombardi 2010). Moreover, gene-sequence manipulations of modern-day proteins show that, provided the catalytic site is not compromised, substantial reductions in the number of distinct amino acids used in primary sequences can be achieved without loss of function. For example, Akanuma et al. (2002) were able to modify a 213-residue protein involved in pyrimidine biosynthesis to function in the absence of seven amino acids, with 188 positions being occupied by just nine amino acids. A bovine pancreatic trypsin inhibitor modified to contain >33% alanine residues retained its native fold and functions (Islam et al. 2008). In addition, a simplified version of an archaeal chorismate mutase has been engineered to contain just nine amino acids (MacBeath et al. 1998; Walter et al. 2005). Several other such studies are reviewed in Longo and Blaber (2012) and Longo et al. (2013).

Although a diverse protein repertoire can be derived from a restricted set of amino acids, laboratory evolution experiments also suggest that enhanced enzyme efficiency would have been promoted by expansion of the amino-acid alphabet (Müller et al. 2013). Indeed, in experiments where the 20 canonical amino acids are supplemented with non-canonical forms, enzymes can be engineered to have still higher catalytic rates than found in natural populations (Windle et al. 2017; Zhao et al. 2020; de la Torre and Chin 2021). Thus, the canonical set of 20 amino acids upon which all life depends constitutes a constraint on natural selection's ability to promote proteins with optimal features.

Origin of amino acids

Given that the substantial differences among amino-acid features (e.g., positive vs. negative charge, hydrophilic vs. hydrophobic) define their potential contributions to various cellular transactions, an understanding of the temporal order of

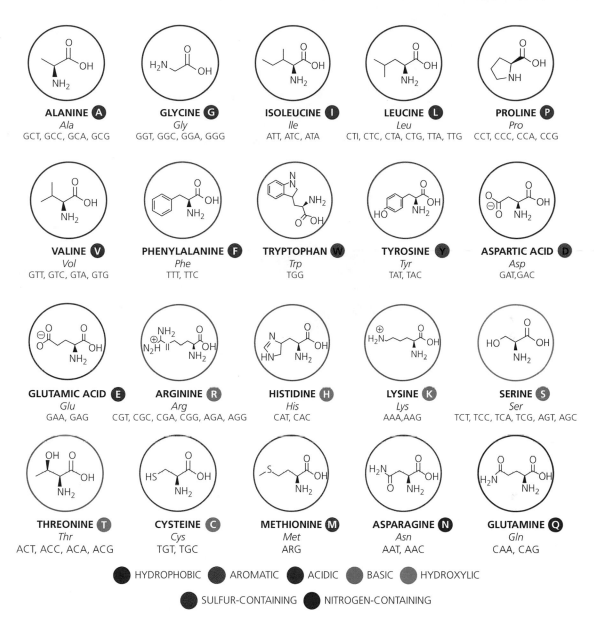

Figure 12.1 A pictorial guide to the structures of the twenty amino acids generally used in proteins. The side-chains unique to each amino acid are shown to the left of the amino (NH$_2$) group; all unlabelled nodes are carbon atoms, each of which has four bonds (all of which are to hydrogens, unless labelled otherwise). The genetic (triplet DNA) codes for each amino acid are those for the so-called universal genetic code; slight variants of this code exist in the nuclear genomes of some unicellular eukaryotes, as well as in the mitochondrial genomes of many eukaryotes. As denoted by the index at the bottom, the colour-coded circles denote some of the unique biochemical properties of individual residues. Aromatic residues contain closed carbon rings. Acidic residues are negatively charged, whereas basic residues are positively charged.

evolutionary incorporation of the amino acids into the early proteome might help clarify the origin of cellular features. All of the numerous attempts devoted to such inference rely on assumptions with tenuous validity, and the initial functions of some amino acids may have been totally unrelated to

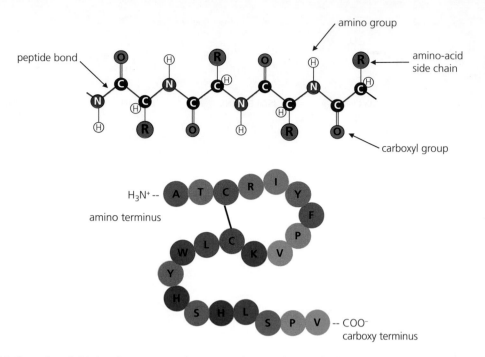

Figure 12.2 Second- and third-order structure of proteins. **Above:** Polypeptide chains are built by covalent bonding between the amide and carboxyl groups of adjacent amino acids (blue lines). The specific features of each amino acid are represented as side chains (R) off the overall backbone. **Below:** The emerging chain typically goes through a folding process, which in some cases allows for further covalent bonding between the sulfide residues of two cysteines. Unbound amide and carboxyl groups reside at the first and final amino acids.

their use in today's proteins (e.g., charged amino acids might have been deployed to cell surfaces to improve adhesion to counter-charged surfaces). With these caveats in mind, the following is a brief survey of the conclusions reached by various approaches.

Davis (1999) postulated that the earliest arriving amino acids would have the simplest production mechanisms, i.e., with the fewest steps in today's biosynthetic pathways. Most amino-acid biosynthesis initiates at hubs of central metabolism – the citric-acid cycle, the pentose-phosphate cycle, or the central trunk that connects the two, allowing the derivation of proximity measures for all 20 amino acids (Figure 12.3). For example, alanine, aspartic acid, asparagine, glutamic acid, and glutamine are just one to two steps removed from their metabolic by-product precursors, whereas biosynthesis of histidine, lysine, phenylalanine, tryptophan, and tyrosine requires 10–13 additional steps. Under Davis's hypothesis, the earliest amino acids were

aspartic acid, glutamic acid, asparagine, and glutamine (for a variety of reasons, he viewed alanine as a later addition). These four building blocks are often referred to as the 'nitrogen-fixing' amino acids as the first two are, respectively, produced by the addition of an amino group to oxaloacetate and α-ketoglutarate (both components of the citric-acid cycle), with secondary amino additions then leading to asparagine and glutamine. One concern with this type of reasoning is that variation exists in the pathways used in amino-acid biosynthesis by different species (Chapter 19), leaving uncertain the number of steps required in the production of various amino acids in ancestral pre-LUCA species.

An alternative approach to inferring the temporal ordering of amino-acid appearance relies on phylogenetic analysis. For example, if one is willing to assume that the amino-acid content of the most strongly conserved protein sequences across the Tree of Life reflects the availability of amino acids

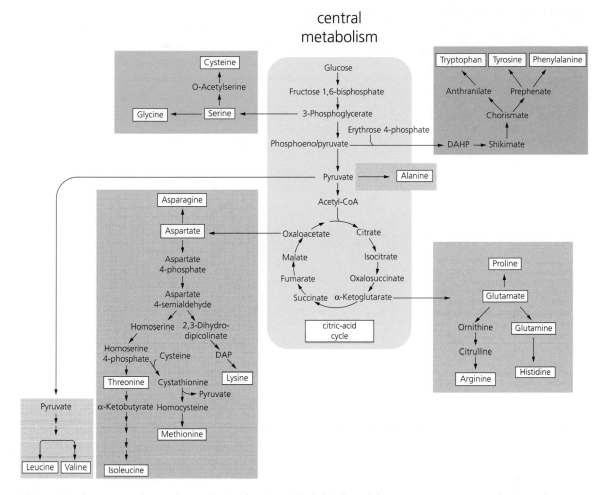

Figure 12.3 Common pathways for synthesis of amino acids (white boxes) from precursors generated in central metabolism.

at the times of protein origin, one is led to conclude that alanine, glycine, aspartic acid, and valine were early arrivals, with cysteine, tryptophan, tyrosine, phenylalanine, glutamine, and glutamic acid being among the late arrivals (Brooks and Fresco 2002; Brooks et al. 2002). Using a rather different approach, involving simple pairwise comparisons of sequences in sister taxa with an outgroup to infer the directionality of amino-acid substitutions, Jordan et al. (2005) suggested a universal trend across the Tree of Life toward an increase in cysteine, methionine, histidine, serine, and phenylalanine, and a decrease in proline, alanine, glycine, and glutamic acid. The underlying assumption here is

that amino acids that are declining in frequency represent the pool of early arrivals. A clear concern with these approaches is the assumption that there has been insufficient time in the history of life for the complete erasure of information on amino-acid compositions at the pre-LUCA stage. It is difficult to reconcile this view with the vast stretch of post-LUCA time and known rates of mutation (Chapter 4).

Despite the uncertainties in our ability to project backwards to the primordial amino-acid pool, by integrating the above sorts of analyses with empirical observations on the ease of synthesizing amino acids in potential settings for the origin of

Table 12.1 Properties of the 20 major amino acids (and their conventional abbreviations). MW denotes the molecular weight in grams/mol. Hydropathy is the log of a coefficient that measures the propensity of a molecule to dissociate from water into a nonpolar solvent (Wolfenden et al. 2015). Interface denotes the log of the ratio of the incidence of an amino acid on interfaces to that on exposed surfaces of proteins; these numbers are taken from *E. coli*, although similar results are obtained in other species (Levy et al. 2012). GC is the average fractional G/C content within codons in the primary genetic code (Figure 12.1). Cost is the biosynthetic cost of a single amino acid in units of ATP hydrolyses, which assumes a starting point of glucose, and includes both the loss of ATP generation due to the diversion of precursors (with ATP-generating power) and the direct use of ATP in biosynthesis (derived in Chapter 17).

Amino acid	Polarity	Charge	MW	Hydropathy	Interface	GC	Cost
Alanine (Ala, A)	non-polar	0	89	2.11	0.01	0.83	16
Arginine (Arg, R)	polar	+	174	−4.32	−0.09	0.83	31
Asparagine (Asn, N)	polar	0	132	−4.88	−0.27	0.17	16
Aspartic acid (Asp, D)	polar	−	133	−3.29	−0.75	0.50	14
Cysteine (Cys, C)	non-polar	0	121	1.53	1.04	0.50	17
Glutamic acid (Glu, E)	polar	−	147	−2.26	−0.79	0.50	20
Glutamine (Gln, Q)	polar	0	146	−4.07	−0.41	0.50	21
Glycine (Gly, G)	non-polar	0	75	0.20	−0.18	0.83	12
Histidine (His, H)	polar	+	155	−3.49	0.12	0.50	32
Isoleucine (Ile, I)	non-polar	0	131	4.24	1.11	0.11	39
Leucine (Leu, L)	non-polar	0	131	4.24	0.91	0.38	44
Lysine (Lys, K)	polar	+	146	−0.27	−1.18	0.17	36
Methionine (Met, M)	non-polar	0	149	1.91	1.01	0.33	25
Phenylalanine (Phe, F)	non-polar	0	165	2.64	1.27	0.17	62
Proline (Pro, P)	non-polar	0	115	3.75	−0.18	0.83	26
Serine (Ser, S)	polar	0	105	−2.82	0.14	0.50	14
Threonine (Thr, T)	polar	0	119	−1.83	0.10	0.50	20
Tryptophan (Trp, W)	non-polar	0	204	1.83	0.79	0.68	71
Tyrosine (Tyr, Y)	polar	0	181	−0.31	0.88	0.17	57
Valine (Val, V)	non-polar	0	117	4.09	0.76	0.50	31

life, a loose argument has been made for a limited set of 10 prebiotic amino acids: alanine, aspartic acid, glutamic acid, glycine, isoleucine, leucine, proline, serine, threonine, and valine (Higgs and Pudritz 2009; Longo and Blaber 2014). Notably absent from this list are two of the earliest arrivers under Davis's hypothesis, asparagine and glutamine. If roughly correct, this list of early amino acids has implications for the temporal ordering of the emergence of cellular biochemistry and the features of early proteins. For example, an absence of the basic, positively charged amino acids (arginine and lysine) likely would have limited the potential for intimate relationships between proteins and acidic nucleic acids (with negatively charged backbones).

Protein Folding and Stability

To acquire their enzymatic or structural features, individual polypeptide chains generally must fold into specific three-dimensional configurations. The overall architecture of an entire amino-acid chain is referred to as its tertiary structure, and the most appropriate functional configuration is referred to as its native state.

In the process of complete folding, numerous substructures are initially formed, the most common of which are α helices and β sheets. In α helices, the amide group of every amino acid donates a hydrogen bond to the backbone carboxyl group of the amino acid four residues earlier in the polypeptide chain. Total helix-chain lengths are typically on the order of 10–15 residues (Figure 12.4). Methionine, alanine, leucine, glutamic acid, and lysine have high helix-forming propensities, whereas glycine is poor in this regard. A proline residue will break or kink a helix because it cannot donate an amide hydrogen bond. In contrast, β sheets consist of sets of chains (each chain typically being 3–10 residues long) held together by backbone hydrogen bonds. Such sheets can consist of parallel or anti-parallel chains, usually four to five but as many as 10, with the physical distance between hydrogen-bonding residues in the primary sequence depending on the length of the strands within the sheet.

Higher-order structures are commonly assembled from α helices and β sheets. For example, coiled coils result from the interlacing of two or three adjacent α helices. Helix–loop–helix repeats can yield a variety of higher-order geometric forms, depending on the angular features of the loop. The $(\beta\alpha)_8$ barrel, one of the most common enzyme folds throughout the Tree of Life, consists of eight alternating units of β strands and α helices folded into an internal curved β sheet surrounded by α helices.

The reliance of almost all proteins on a moderate number of fold types is unlikely to simply be an evolutionary fossil of common ancestry. Rather, commonly observed folds appear to be natural outcomes of the fundamental features of peptide

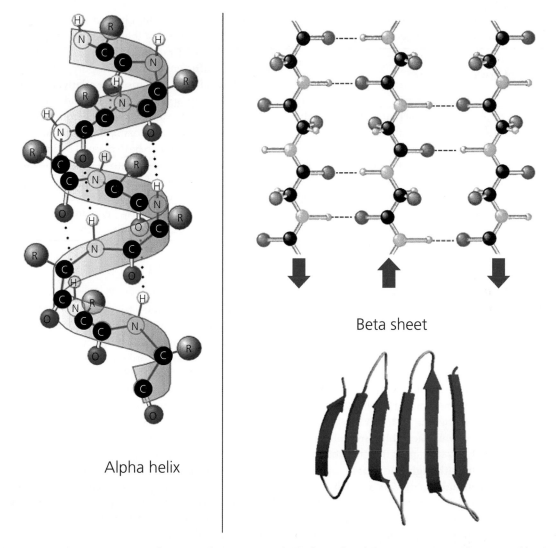

Beta sheet

Alpha helix

Figure 12.4 Two common types of protein substructures. **Left:** Hydrogen bonds between amino acids separated by three steps, shown as dotted lines, hold alpha helices together. **Right:** Here the bonds are between more distant amino acids on adjacent chains, creating beta sheets. The ribbon diagram to the lower right depicts an anti-parallel beta sheet.

chains, including the intrinsic ability to hydrogen bond and form hydrophobic associations with each other. Indeed, random sequences of amino acids (even those involving reduced sets of amino acids, including homopolymers) commonly generate stably folded proteins (Doi et al. 2005; Zhang et al. 2006; López de la Osa et al. 2007; Go et al. 2008; Labean et al. 2011). This suggests that the compact globular nature of proteins is an expectation based on physical properties, and need not be entirely a product of the guiding hand of natural selection (Alva et al. 2015). Thus, the majority of common folds in today's proteins may have been present even before the establishment of the full genetic code.

The rate of protein folding

Given the large number of fold types in the protein world, the specific folding pathways utilized by different proteins must be extraordinarily diverse. Nonetheless, considerable effort has gone towards identifying general solutions to the 'protein folding problem' that transcend the details of secondary structure. This is a highly technical field, but enough quantitative information now exists to yield general insight into the typical timescales and energetic forces involved in productive protein folding. We start by focusing on folding unassisted by outside factors, deferring until Chapter 14 consideration of the cellular mechanisms (chaperones) that have evolved to assist with the process.

One conceptual solution to the Levinthal paradox invokes the metaphor of a folding funnel, with an energetically favorable bias in the landscape of possible folds acting to progressively channel a protein towards the relatively stable native state (Dill and McCallum 2012; Englander and Mayne 2014, 2017; Wolynes 2015; Neupane et al. 2016). The hallmark of a stable protein is a well-packed hydrophobic core, resulting from the self-aggregation of non-polar surfaces in water, although the actual underlying molecular mechanisms driving such clustering remain unclear (Ball 2008; Snyder et al. 2011). Additional factors involved in protein folding and stability include hydrogen bonds in α helices and β sheets, electrostatic interactions between residues with different charges, and disulfide bonds between cysteine residues. Consistent with there being a multifactorial basis, the folding times of most proteins are quite resilient to sequence changes, with random mutagenesis (sometimes involving multiple residues) generally causing no more than a tenfold increase in the mean folding time, and as many as half of residue changes causing reduced folding times (Kim et al. 1998; Plaxco et al. 2000).

Under this general model of folding, the approach to the native state can be viewed as a series of stochastic samplings of alternative states, with the initial folding of local domain structures causing a progressive reduction in the multiplicity of routes to the final native state. Lin and Zewail (2012) go so far as to suggest that the force resulting from the mere presence of random hydrophobic residues is generally sufficient to induce a polypeptide chain of < 200 residues to collapse to a relatively compact form within an appropriate biological time frame. Indeed, despite the apparent complexity of the process, as a first-order approximation, the known folding rates of proteins can be explained by knowledge of just the total chain length (i.e., the number of amino-acid residues, L). At least in the range of $L = 20$ to 300, there is a dramatic reduction in the spontaneous folding rate (k_f, given in units of sec^{-1}) with increasing L, with the function

$$k_f \simeq (1.1 \times 10^8)e^{-1.3\sqrt{L}} \qquad (12.1)$$

explaining ~78% of the variance in the folding rate among proteins (Dill et al. 2011). Over an order of magnitude increase in chain length, k_f declines approximately seven orders of magnitude from 3×10^4 to 2×10^{-3} sec^{-1} (Figure 12.5). Notably, the protein-folding rates used to derive this formula have been almost universally estimated with *in vitro* methods. This raises concerns because the macromolecular crowding within cells (Dhar et al. 2010) and the attachment of nascent chains to the ribosome during translation (Kaiser et al. 2011) can modulate folding by reducing the formation of inappropriate folds. Unfortunately, the technical difficulty of quantifying protein assembly *in vivo* remains formidable.

Numerous attempts have been made to improve the prediction of folding rates by incorporating additional information. For example, Ivankov and

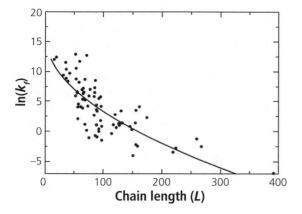

Figure 12.5 Reduction in the folding rate (k_f) with increasing chain length (L), as described by Equation 12.1. The fit is based on 80 proteins known to engage in two- and multi-state folding. From Dill et al. (2011).

Finkelstein (2004) suggested a refinement that subtracts the subsets of residues incorporated into α helices from L. However, the resultant regression yields only a marginal improvement over the preceding expression with the added expense of requiring a detailed understanding of the protein's secondary structure. Incorporation of further information on secondary structure, amino-acid composition, and/or the number of chain contacts has little added effect on the accuracy of prediction (Grantcharova et al. 2001; Ivankov and Finkelstein 2004; Prabhu and Bhuyan 2006; Galzitskaya 2008; Huang et al. 2012). Given that k_f itself is subject to measurement error, there may be little room for improvement beyond the pattern summarized in Equation 12.1.

Finally, it bears emphasizing that, although chain length alone is a fairly good predictor of the folding rate, this need not exclude the importance of other factors, but simply means that any additional factors must be either tightly correlated with chain length or of minor significance. In other words, chain length may provide an overall summary measure with good predictive ability but possibly with little mechanistic relevance. In addition, not too much should be read into the significance of the $L^{0.5}$ scaling in Equation 12.1, as exponents in the range of 0.1–0.7 yield fits that are nearly equally as good.

How many contortions might a protein go through en route to achieving a proper fold? Because the mean time for a chain to switch from one configuration to another is estimated to be $\sim 10^{-9}$ sec (Zana 1975), taking the reciprocal of Equation 12.1 as the approximate mean time to complete a search, the average number of configurations sampled prior to finding the proper fold solution $\simeq 10 e^{\sqrt{L}}$. For $L = 200$, this implies an average of 13×10^6 configurations searched in a time span of ~ 0.013 seconds. For $L = 300$, this jumps to $\sim 33 \times 10^7$ configurations searched over 0.33 seconds, and with $L = 500$ to $\sim 51 \times 10^9$ searches over 51 seconds. This implies that beyond chain lengths of 200–300 residues, unassisted folding times rapidly approach biologically unrealistic levels, a point to which we return in Chapter 14. Thus, it may not be a coincidence that protein domains exceeding 300 residues in length are rare (Wheelan et al. 2000; Wang et al. 2011; Lin and Zewail 2012), and that average lengths of entire proteins are commonly on the order of 300 residues in most species.

To what extent do observed folding rates approach the maximum rates possible from the standpoint of biophysical limitations? Following the suggestion that an upper bound to the rate of folding of a single-domain protein $\simeq (10^8/L)$ sec^{-1} (Kubelka et al. 2004), by comparison with Equation 12.1, it can be seen that empirically observed folding rates are typically far below the maximum. For example, for a 100-residue protein, the maximum folding rate is predicted to be $\simeq 10^6$/sec, whereas Equation 12.1 implies an average observed rate of only 249/sec. Even the most rapidly folding proteins do so one to two orders of magnitude more slowly than this proposed protein-folding speed limit (Kubelka et al. 2004). Thus, it appears that natural selection has generally been unable to achieve perfection in folding rates.

Does the level of refinement of folding rates vary among species? Although an ideal comparison of orthologous proteins in different phylogenetic lineages has not been performed, it has been argued that proteins of equivalent length fold at least 10 times more rapidly in bacteria than in eukaryotes (Galzitskaya et al. 2011). Proteins also tend to be longer in eukaryotes than in prokaryotes, which will further exacerbate the protein-folding challenges in the former group. In one of the only comparative studies of protein-folding pathways, Lim et al. (2018; Lim and Marqusee 2018) found for the protein ribonuclease H that, although different bacterial lineages use the same type of folding intermediate, the pathways to get there differ; and another study of a protease revealed dramatic differences in the folding mechanism (Nixon et al. 2021). Thus, even within bacteria, there are apparently evolutionary paths open to divergence of folding mechanisms without compromising folding rate.

Stability of folding

As with protein-folding rates, folding stability (the tendency to remain folded after achieving the native state) is largely a function of the total chain length, with additional information on sequence and secondary structure again not greatly improving predictability (Robertson and Murphy 1997; Ghosh and Dill 2009; Khan and Vihinen 2010; Dill et al. 2011; Jarzab et al. 2020). The diverse mechanisms by which stability is achieved include packing effects of hydrophobic residues, backbone hydrogen bonds, and favorable electrostatic interactions (Miller et al. 2010). Thus, not surprisingly, protein-folding rates and stability are not independent attributes (Plaxco et al. 2000; Sato et al. 2001). Proteins that fold rapidly are often also quite stable, but conflicts can also exist. For example, whereas proteins are positively selected to fold into their proper native states, negative selection may operate to avoid folding too rapidly and/or too stably into misfolded states. Random mutagenesis with a model protein demonstrated that, although a substantial fraction of mutations result in faster folding times, nearly all of these have the side effect of reducing stability, suggesting that natural selection places a premium on the latter (Kim et al. 1998).

Indirect evidence supports the idea that selection on folding stability plays a central role in amino-acid sequence evolution. For example, there is a strong correlation between the thermostability of individual proteins and the optimal growth temperature of bacterial species, and random amino-acid substitutions in protein cores are more deleterious than those for surface residues (Dehouck et al. 2008; Tripathi et al. 2016; Jarzab et al. 2020).

In particular, the total usage of seven amino acids – four hydrophobic (Ile, Val, Trp, and Leu), one polar (Tyr), and two charged (Arg and Glu) – is highly correlated with optimal growth temperature (Zeldovich et al. 2007). The usage of this particular mix of residues may represent a compromise between the conflicting challenges of folding rapidly and avoiding stable misfolded configurations (Berezovsky et al. 2007). Under this hypothesis, the reduced incidence of these seven residues at lower temperatures is viewed as a by-product of the relaxed intensity of selection on folding mechanisms in less-extreme thermal backgrounds.

Protein stability is deemed to be positively associated with fitness in the sense that destabilized proteins are prone to loss of function, aggregation, and/or direct toxicity. Nonetheless, most proteins sit on the 'margin of stability' in the sense that only one or two mutations are often sufficient to induce complete loss of stability. Although it is commonly argued that marginal stability is required for proper protein function, with excess stability somehow reducing protein performance, this has not held up to close scrutiny. It is relatively easy to create more stable proteins by mutagenesis (Matsuura et al. 1999; Sullivan et al. 2012; Bershtein et al. 2013), and the individual residues contributing to stability typically interact in an additive fashion (Wells 1990; Serrano et al. 1993; Zhang et al. 1995). Numerous proteins have been engineered to have increased stability with few, if any, consequences for enzyme efficiency (e.g., Giver et al. 1998; van den Berg et al. 1998; Taverna and Goldstein 2002; Borgo and Havranek 2012; Moon et al. 2014). Indeed, when wide phylogenetic comparisons are used to generate consensus sequences of proteins, the resultant synthesized peptides are commonly more stable than the extant proteins, each of which has a subset of the overall stabilizing consensus residues (Sternke et al. 2019).

An alternative explanation for all of these observations is that marginal stability evolves as a simple consequence of the diminishing benefits of increased stability. This would be the case, for example, if fitness is a hyperbolic function of the energy associated with the forces holding a protein together (Govindarajan and Goldstein

1997; Taverna and Goldstein 2002; Bloom et al. 2005; Wylie and Shakhnovich 2011; Serohijos and Shakhnovich 2014). Under this model, proteins are expected to be pushed by natural selection to more stable configurations until reaching the point where any further fitness improvement is small enough to be offset by the vagaries of random genetic drift and/or mutation pressure towards less-stable states (the drift barrier; Chapter 5). In essence, under any particular population-genetic environment, a quasi-steady-state distribution of stability is expected to evolve to the point at which the rates of fixation of beneficial and deleterious mutations are equal (Figure 12.6). The overall prediction of this hypothesis is that the mean folding stability of proteins will evolve to higher values in populations with larger effective population sizes. This same hypothesis may explain the higher folding rates in prokaryotes than in eukaryotes.

A more mechanistic view of these issues can be acquired by considering the typical features of evolved proteins. The folding stability of proteins is often on the order of $\Delta G = -3$ to -20 kcal/mol (Plaxco et al. 2000; Dill et al. 2011). To put this in perspective, the average energy associated with single hydrogen bonds in peptides is thought to be on the order of -2 kcal/mol (Sheu et al. 2003; Wendler et al. 2010). With the expected fraction of folded proteins being $\simeq e^{-\Delta G/RT}/(1 + e^{-\Delta G/RT})$ at thermodynamic equilibrium, where $RT \simeq 0.6$ kcal/mol, a protein with $\Delta G = -3$ is expected to be folded $> 99\%$ of the time. This diminishes to 96.5 and 84% for $\Delta G = -2$ and -1, respectively.

A survey of experimental assays of mutational effects suggests an average $\Delta\Delta G \simeq 0.6$ kcal/mol (SD = 1.1) associated with individual surface residues, and higher destabilizing effects (1.4 kcal/mol; SD = 1.7) for core residues. The distributions of both kinds of effects are roughly normal (bell-shaped; Figure 12.7), so the overall distribution of site-specific effects for an entire protein is essentially a mixture of normals. Because

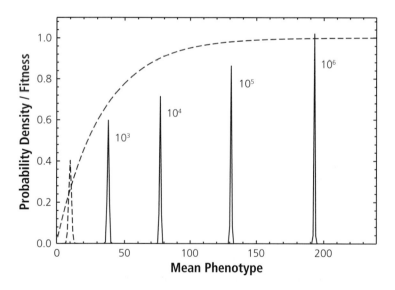

Figure 12.6 Evolution of population mean phenotypes on a hyperbolic fitness function (given by the dashed red line). Shown are steady-state distributions of mean phenotypes under a balance between the forces of mutation, selection, and random genetic drift. It is assumed that a large number of factors contribute to the trait, in this case folding stability, with each factor having two alternative states (positive or negative with respect to stability). The dashed black line gives the distribution under effective neutrality (i.e., when the effective population size is small enough that the power of selection is overcome by random genetic drift), the position and width of the distribution being a function of the relative rates of mutations to stabilizing versus destabilizing mutations. The distribution shifts to the right with larger effective population sizes (given as the four labelled peaks, $N_e = 10^3$ to 10^6), as the efficiency of natural selection becomes greater. Details on the methods used to generate these results are given in Chapter 5 and in Lynch (2018).

smaller proteins have a higher fraction of surface residues, the average $\Delta\Delta G$ is expected to be smaller.

If the drift-barrier hypothesis is indeed the explanation for the evolution of marginal stability, the distribution of $\Delta\Delta G$ values seen in such surveys must reflect the natural outcome of the joint forces of mutation, selection, and drift in positioning a population on the fitness-stability landscape (Wylie and Shakhnovich 2011). Unfortunately, as in most areas of cell biology, there are few comparative studies bearing on this issue. However, an *in vitro* evaluation of the folding stability of the dihydrofolate reductase enzyme from 36 species of mesophilic bacteria revealed a substantial range of variation among species, with the standard deviation being roughly 10% of the mean (Figure 12.8).

Determinants of Protein-Sequence Evolution

There can be substantial variation in the evolutionary rates of substitution among amino-acid sites within a given protein. Unsurprisingly, positions involved in catalytic sites are generally under strong purifying selection, and as a consequence, proteins often retain the ability to function appropriately in foreign cellular backgrounds after very long periods of evolutionary divergence. For example, a survey of the performance of over 400 different human proteins in yeast (lineages that separated over a billion years ago) revealed that nearly half were able to complement the absence of the native yeast gene (Kachroo et al. 2015). Similar results were obtained when bacterial genes were substituted for their orthologs in yeast (Kachroo et al. 2017). Remarkably, however, although between 60% and 90% of genes involved in various aspects of metabolism were able to complement (along with ~50% of genes involved in transcription, ~65% involved in DNA replication and repair), nearly all involved in cell growth were unable to complement. Thus, proteins whose functions are closely related to fitness need not remain highly conserved at the protein-sequence level. Here, we explore a wide range of issues bearing on the mechanisms responsible for the substantial evolutionary-rate variation that exists among proteins and among sites within them.

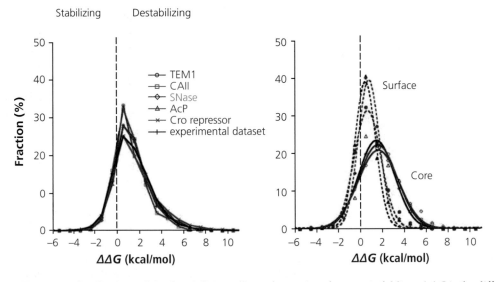

Figure 12.7 Frequency distributions of the destabilizing effects of mutations for protein folding. $\Delta\Delta G$ is the difference in ΔG (a measure of folding stability) between the mutant and native protein. Positive values of $\Delta\Delta G$ denote destabilizing mutations; negative values denote stabilizing mutations. Plots on the left map the distribution of $\Delta\Delta G$ for random amino acid exchanges at sites across the entire protein; results are given for five specific proteins and a larger set of experimental data. In the right panel, mutations are subdivided into those affecting surfaces and internal cores of proteins. From Tokuriki et al. (2007) and Tokuriki and Tawfik (2009).

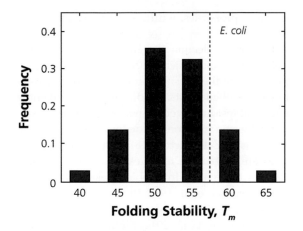

Figure 12.8 The distribution of folding stabilities of the enzyme dihydrofolate reductase for 36 species of mesophilic bacteria, measured as the temperature (°C) required for 50% unfolding *in vitro*. The position for *E. coli* is given as a reference point. From Bershtein et al. (2015).

Lessons from phylogenetic comparisons

Comparisons of the sequences of orthologous protein-coding genes over a diversity of species have left little doubt that amino-acid sequences undergo slow but relentless change over evolutionary time. Not all amino-acid substitutions are acceptable in all contexts, and there is substantial variation in evolutionary rates among different proteins and different phylogenetic lineages, but only a tiny fraction of amino-acid sites is invariant across the entire Tree of Life.

The most common approach to estimating protein evolutionary rates starts at the level of DNA-sequence analysis, and compares the rates of nucleotide substitution at amino-acid replacement and silent sites where, respectively, nucleotide substitutions do or do not elicit a change at the amino-acid level. Owing to the nature of the genetic code, ~25% of nucleotide sites in a protein-coding gene are typically silent. For example, third-positions in codons for the eight amino acids for which A, C, G, or T lead to the same residue (Figure 12.1) are referred to as fourfold redundant sites. The usual assumption is that such sites evolve in a neutral fashion, owing to their invisibility at the amino-acid level. If this is the case, then the rate of nucleotide substitution (i.e., the rate at which

one nucleotide type is displaced by another at the population level) at silent sites is expected to equal the mutation rate per site per generation (u) in accordance with the neutral theory (Chapter 4). The total expected silent-site divergence between two lineages separated by t time units (in this case, generations) would then be $2\,tu$ mutational changes per site, the 2 appearing because mutations accumulate independently down each lineage.

Having such a benchmark of neutral divergence is informative, as it provides a means for interpreting rates of amino-acid substitution by factoring out the contribution of mutation pressure from that associated with selection. If substitutions at replacement sites are selected against, which is generally the case (Kimura 1983; Nei and Kumar 2000; Yang 2014), their rate of divergence should be lower than the expected neutral benchmark. Letting ϕ_f be the probability of fixation of a newly arisen replacement mutation, the rate of divergence at such sites has expectation $2t \cdot (2Nu) \cdot \phi_f$, where N is the absolute population size, and $2Nu$ is the rate at which mutations arise within each population (assumed to be diploid) per nucleotide site. If mutations at replacement sites are neutral, the fixation probability is simply equal to the initial frequency of a mutation, $1/(2N)$, and the overall amount of divergence is again equal to the neutral expectation $2tu$.

Generally, we do not have an accurate measurement of the divergence time t between two species, nor of the mutation rate u. However, if one simply takes the ratio of the observed divergences at replacement and silent sites (denoted d_N and d_S, respectively, with the N referring to non-synonymous or amino-acid replacement sites), the resultant ratio has expectation $[2t \cdot (2Nu) \cdot \phi_f]/(2tu) = 2N \cdot \phi_f$, assuming silent sites do indeed evolve in a neutral fashion. When rewritten as $\phi_f/[1/(2N)]$, this ratio is seen to be equivalent to the fixation probability at replacement sites relative to the neutral expectation. Thus, under appropriate conditions, d_N/d_S provides a biologically interpretable measure of the degree of selective constraint on a protein-coding gene. Assuming that the majority of mutations are either neutral or deleterious, d_N/d_S is equivalent to the fraction of amino acid-altering mutations that evade the eyes of natural selection, and for that reason it is sometimes referred to as the width of the selective sieve.

There are many caveats with respect to this sort of analysis. First, it is assumed that silent sites are neutral, whereas we know that such sites can experience some selection at various levels, e.g., from preferential tRNA recognition of certain nucleotides in third positions, mRNA structural constraints, and influence of translation speed on folding efficiency (Sharp et al. 2005; Zhou et al. 2010; Lawrie et al. 2013; Long et al. 2018; Walsh et al. 2020). Second, d_N is generally measured as an average over multiple sites within a gene, obscuring the fact that, although many substitutions can be strongly selected against, a minority may nonetheless be advanced by positive selection. Third, there is the difficult matter of accurately estimating d_N and d_S from highly divergent sequences, as multiple substitutions at individual sites will lead to an undercounting of the actual numbers of changes that have accrued, especially at more rapidly evolving silent sites.

These and many other matters have been taken up in detail in the technical field of DNA-sequence analysis, but justified or not, the d_N/d_S ratio remains a central parameter determined in almost all molecular-evolution studies. With few exceptions, proteome-wide studies involving d_N/d_S between moderately related species yield average ratios on the order of 0.05–0.25 (Kuo et al. 2009; Stanley and Kulathinal 2016; Lynch et al. 2017). This implies that on *average* only 5–25% of amino-acid alterations of proteins are typically acceptable in nature in the long term. There is, however, a wide range of variation among proteins and among species. Given the possibility of selection on silent sites, especially in large-N_e species, such differences also need to be cautiously interpreted as implying lineage-specific differences in the efficiency of natural selection. Moreover, d_N/d_S analyses leave unresolved the degree to which neutral versus beneficial nonsynonymous mutations contribute to the pool of fixed amino-acid replacement substitutions.

This being said, comparative analyses have led to a number of other general observations that leave little doubt that the majority of amino acid-altering mutations are removed by purifying selection in nature. First, most amino-acid substitutions that do occur involve exchanges of amino acids with similar chemical properties, with radical exchanges being more common in low-N_e species with reduced efficiency of selection (Bergman and Eyre-Walker 2019; Weber and Whelan 2019). Second, there is a premium on the use of amino acids with relatively low biosynthetic costs, conditional on maintaining a level of residue diversity necessary for maintaining stable and functional proteins (Krick et al. 2014; Venev and Zeldovich 2018). Third, substitution rates are higher for residues on protein surfaces than for those in hydrophobic cores (Suckow et al. 1996; Goldman et al. 1998; Bustamente et al. 2000; Ramsey et al. 2011; Roscoe et al. 2013; Firnberg et al. 2014; Sarkisyan et al. 2016; Moutinho et al. 2019). Fourth, membrane proteins exposed to the external environment evolve relatively rapidly, perhaps in response to adaptive challenges, compared to cytosolic proteins confined to the more homeostatic internal cellular environment (Sojo et al. 2016).

Observations from experimental mutagenesis

As an alternative to deriving indirect inferences from distantly diverged proteins, the degree of constraint on protein sequences can be directly evaluated by comparing the performance of randomly mutagenized sequences. However, although such an approach has the advantage of avoiding problems of sequence saturation, silent-site selection, etc., it has the strong limitation of only being able to identify residue changes with major effects. The central issue here is that selection in nature is capable of eradicating deleterious mutations with selective disadvantages down to order $1/N_e$, where N_e (the effective population size) is typically in the range of 10^4 to 10^9 (Chapter 4), whereas lab experiments are generally unable to detect deleterious mutations with fitness effects smaller than 10^{-3}. Thus, random mutagenesis experiments invariably underestimate the fraction of amino-acid substitutions that are eliminated by purifying selection in nature.

This being said, such experiments have been illuminating in a number of ways. For example, Yampolsky and Stoltzfus (2005) summarized the relative exchangeabilities of amino-acid pairs observed in such studies. Hydrophobic residues tend to be most substitutable with other hydrophobic residues, and

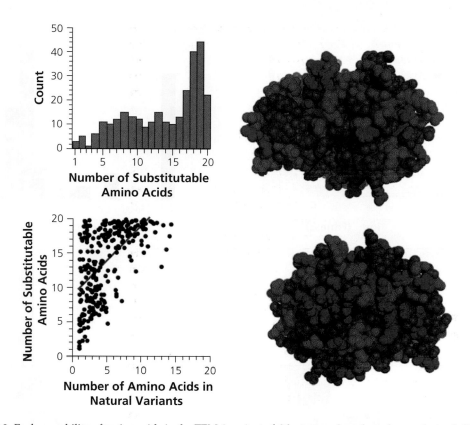

Figure 12.9 Exchangeability of amino acids in the TEM-1 variant of β-lactamase, based on observations of all 19 possible amino-acid alterations at each position in the protein. **Upper left:** The distribution of the number of possible amino acids residing at each site that retain enzyme function. **Lower left:** Weak relationship between the number of amino acids observed at particular positions in sequences of natural variants of β-lactamase and the number of the 20 possible amino acids that retain enzyme function in laboratory manipulations. **Right:** Spatial pattern of amino-acid exchangeability on the surface residues, with red denoting four or fewer, yellow five to nine, and green 10 or more amino acids conferring a functional enzyme. The upper panel illustrates the residues on the side of the molecule containing the active site (marked by the black arrow), whereas the lower panel illustrates the opposite side of the molecule. Note that these are examinations of single amino-acid changes, and some exchanges become possible on different genetic backgrounds. From Deng et al. (2012) and Firnberg et al. (2014).

hydrophilic residues with each other, whereas exchanges between these two extreme groups tend to be unacceptable.

The protein most extensively studied in this way is β-lactamase, a bacterial enzyme that hydrolyses antibiotics such as penicillin. The functional consequences of every possible amino-acid substitution at every position in β-lactamase have been characterized (Deng et al. 2012; Jacquier et al. 2013; Firnberg et al. 2014). In agreement with evolutionary divergence data, this work reveals that the number of acceptable amino acids (out of 20) at individual sites

is frequently below 15 (and often much lower), with surface residues generally being more receptive to change (Figure 12.9). However, a number of surface positions far from the active site are highly sensitive to mutations, ruling out the generality that all surface residues are under relatively relaxed selection. Several residues can be altered in ways that increase molecular stability, and the overall distribution of effects is bimodal, with most acceptable variants having functionality just slightly below the norm and a small fraction being nonfunctional (Figure 12.10).

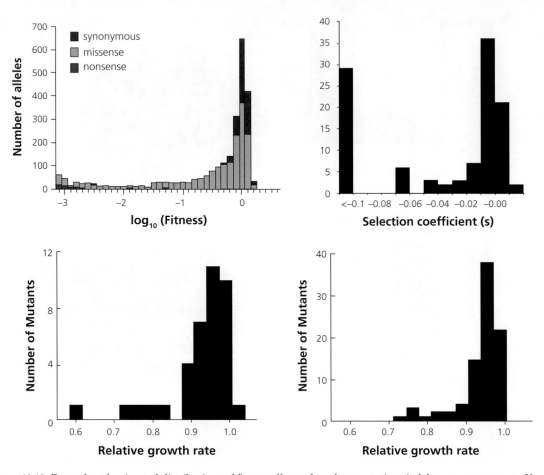

Figure 12.10 Examples of estimated distributions of fitness effects of random mutations in laboratory constructs. **Upper left:** β-lactamase, with fitness given on a log scale for three types of single amino-acid substitutions (from Firnberg et al. 2014). **Upper right:** Nonsynonymous substitutions in genes involved in arabinose metabolism in *Salmonella,* with fitness scored as the relative deviation in performance from the native protein sequence (from Lind et al. 2016). **Lower left and right:** Respectively, synonymous and nonsynonymous substitutions in ribosomal protein genes in *Salmonella* (from Lind et al. 2010).

Surprisingly, a number of sites that are known to vary among β-lactamase sequences from natural isolates are intolerant to amino-acid substitutions, an observation that has been seen in other proteins (Mishra et al. 2016). This raises questions about the common assumption that sites with high natural levels of variability experience low functional constraints. It also suggests the importance of context dependence (Chapter 5), with certain sites being more or less accepting of alterations depending on the state of other sites within the protein (Bershtein et al. 2006; Salverda et al. 2011), a point

to which we return later. A strong role for contingency is implicated from the observation that when mutations at two sites that are acceptably exchangeable on their own are combined in this protein, they commonly lead to nonfunctional molecules (Axe 2000). Moreover, chimeric molecules obtained by splicing together halves of different natural variants are often completely nonfunctional (Axe 2000).

Numerous assays from random mutagenesis experiments with other proteins are in general agreement with the preceding results. For example,

Guo et al. (2004) examined the performance of $\sim 10^5$ single amino-acid substitutions in 3-methyladenine DNA glycosylase, a DNA-repair enzyme in humans, and found that 34% of exchanges led to enzyme inactivation. Substitutions in α-helices were about twice as exchangeable as those in β-strands (as seen in other studies; Silverman et al. 2001; Firnberg et al. 2014), and those in turns and loops were still more acceptable. Similar analyses with different proteins have yielded estimates of 30–80% for fractions of nonfunctionalizing mutations (Guo et al. 2004; Axe et al. 1996; Materon and Palzkill 2001). Again, although the definition of nonfunctional varies among studies (with most incapacitated enzymes actually retaining at least a small amount of functionality), owing to measurement limitations, all such studies must underestimate the total fraction of mutations that would be removed by purifying selection in nature.

From an evolutionary standpoint, it is more desirable to know the net consequences of mutations not simply for molecular function but for overall organismal fitness, and to have such measurements on a continuous rather than a yes/no scale with an arbitrary cut-off. Although the data are more limited here, when available they generally point in the same direction. For example, in the case of β-lactamase, the distribution of fitness effects for single amino-acid substitutions has a mode near zero, with a small fraction being favorable, and a long tail to the left (denoting deleterious effects), with $\sim 40\%$ of these having selection coefficients $0 < s < 0.1$, and only $\sim 6\%$ completely obliterating enzyme function (Figure 12.10).

Direct fitness assays of random mutations in other genes are generally consistent with the observations for β-lactamase (Figure 12.10; see also Roscoe et al. 2013; Lind et al. 2016; Sarkisyan et al. 2016; Lundin et al. 2018). Typically, the main peak in the distribution of fitness effects is near zero, with only a small fraction of mutations improving fitness, the majority reducing fitness by no more than 10% (on average just 1%), and with a secondary peak associated with mutations lacking entirely in activity. Notably, in a number of cases, a few silent-site substitutions have discernible negative effects, implying effects on transcription, translation, and/or folding efficiency. Ribosomal-protein genes are particularly pronounced in this regard, having similar distributions of fitness effects for both silent and replacement substitutions (Figure 12.10).

Expression level and the propensity for sequence change

It has long been thought that the evolutionary rate of a protein is inversely related to its functional significance – the higher the relevance to fitness, the lower the acceptability of amino-acid changes. However, there is no formal way to rank functional significance, and simply invoking low d_N/d_S introduces a circularity. To avoid falling into this seductive trap, an exploration of alternative explanations is warranted.

As noted above, residues buried within a protein generally evolve at substantially lower rates than those exposed on protein surfaces. Buried polar residues involved in hydrogen bonding are especially conserved, although the constraint on molecular evolution declines with increasingly large internal cores of proteins, presumably because stability is distributed across more residues (Franzosa and Xia 2009; Worth and Blundell 2009, 2010). Proteins with especially low rates of substitution for surface residues have even more exceptionally low rates for the core residues, leading to the suggestion that alterations in surface residues facilitate the acceptance of mutations in the core (Tóth-Petróczy and Tawfik 2011). However, it remains unclear whether this is a causal relationship or simply a consequence of some proteins being under greater overall constraint at all positions.

A plausible argument for reduced rates of evolution in core positions is the intimate involvement of backbone hydrogen bonds and hydrophobic effects in the maintenance of folding stability. However, the actual mechanisms may be more complicated, as other features are correlated with interior versus exterior residues. For example, the tendency to engage in unproductive aggregations with other proteins is a function of surface residues, and amino-acid substitutions in such regions might sometimes even be driven by positive selection to avoid aggregation (Wright et al. 2005). Moreover, the relative packing density of residues is strongly correlated with solvent accessibility, and when these two are jointly controlled for in a multiple regression, the former accounts for more of the variance in evolutionary rate than the latter (Toft and Fares 2010; Yeh et al. 2014).

These observations on the consequences of mutations for protein stability and adhesivity help explain a general observation on relative rates of protein evolution. As noted above, it was long thought that variation in evolutionary rates would be dictated by the functional significance of a protein, but as gene-knockout studies raised questions about this interpretation, it became clear that the best evolutionary rate predictor is the expression level of a protein (Pál et al. 2001; Zhang and Yang 2015).

One interpretation of this pattern invokes the idea that natural selection operates to minimize the likelihood of improper folding and of instability once properly folded (Serohijos et al. 2012). This idea is further motivated by the suspicion that misfolded proteins commonly arise as a consequence of erroneous protein sequences resulting from transcriptional or translational errors (Drummond and Wilke 2008; Yang et al. 2010). That the latter is a significant challenge is made plausible by the fact that proteins with low amino-acid substitution rates also have low substitution rates at silent sites, which might reflect selection to avoid amino-acid misloading by non-cognate transfer RNAs at non-optimal codons (Chapter 20). The fact that most mutations influencing protein performance do so by eliciting changes in protein folding and stability, rather than by directly compromising the catalytic core, provides further support for the misfolding hypothesis (Bloom et al. 2007; Shi et al. 2012). Under this view, the consequences of misfolding are more significant in an abundant protein simply because the absolute number of problematical molecules is greater. However, it could be argued that individual molecules are more expendable when proteins are highly abundant.

There are alternative (and not necessarily mutually exclusive) explanations for low rates of evolution in highly expressed protein-coding genes. For example, the misinteraction hypothesis postulates purifying selection for surface residues that avoid promiscuous interactions/aggregations with inappropriate proteins (Levy et al. 2012; Yang et al. 2012). With a focus on surface residues, this explanation differs from the emphasis of the misfolding hypothesis on the importance of protein-core residues to folding and stability. That inappropriate protein–protein interactions are a non-trivial selective challenge is highlighted by the fact that ∼20% of protein molecules are typically bound with non-specific partners in yeast and metazoan cells (Zhang et al. 2008).

Under the misinteraction hypothesis, the efficiency of selection against amino-acid substitutions in surface residues is again expected to be especially elevated in more highly expressed proteins. Thus, it is of interest that in *E. coli*, the more abundant proteins have a lower tendency to aggregate (de Groot and Ventura 2010), apparently because of their reduced surface hydrophobicity (Ishihama et al. 2008). The same is true for human proteins (Tartaglia et al. 2007).

To investigate this idea further, Levy et al. (2012) ordered the full set of amino acids with respect to their tendency to adhere to other molecules, using information on their degree of participation in natural interfaces. They found a negative correlation between a protein's cellular abundance and the predicted adhesiveness of its surface. This effect diminishes from *E. coli* to yeast to human, consistent with an expected reduction in the efficiency of selection against mildly deleterious mutations in species with reduced effective population sizes. Notably, in bacteria, the disparity in evolutionary rates between highly and lowly expressed genes is greatest in species with rapid cell-division rates (Vieira-Silva et al. 2011), which might reflect the latter species having larger effective population sizes and hence, a higher level of efficiency of natural selection.

Mutation pressure and biased amino acid usage

The particular amino acids deployed within a protein need not simply be outcomes of selection. As discussed in Chapter 5, the likelihood of occupancy of any position within a protein by a particular residue is a joint function of the mutation biases involving individual allelic variants and the ratio of the power of selection to drift. To understand the relative roles of these two determinants, it is necessary to consider why genome-wide G+C nucleotide compositions range from ~0.25 to ~0.80 among different species (Lynch 2007), and whether such biases have cascading effects on encoded amino-acid composition.

If genomic G+C composition reflects the prevailing pressure of mutation, it ought to be correlated with the expectations based on known mutational spectra, as recorded in mutation-accumulation experiments (Long et al. 2018). Letting u be the mutation pressure of A+T nucleotides to C+G, and v be the reciprocal rate, the expected equilibrium frequency of G+C under mutation pressure alone is simply $u/(u + v)$. Across the Tree of Life, average genome-wide G+C compositions are strongly correlated with this neutral expectation (Figure 12.11). Nonetheless, despite the positive correlation, almost all genomes also have an excess G+C content relative to the neutral expectation. Notably, the deviations of G+C composition from the neutral expectations are particularly large at silent sites, supporting the idea that, contrary to popular belief, such sites do not generally evolve in a neutral fashion. The general interpretation of these results is that, whereas there is biased mutation pressure towards A+T content in most species (most neutral G+C expectations being < 0.5), there is near universal selection for G+C.

For the organisms in Figure 12.11, the ratio of mutation rates from G+C → A+T to the reverse ranges from 16 (very low G/C-content genomes) to 0.4 (moderately high G/C-content genomes). The mechanisms driving such vastly different mutation spectra remain unclear. One suggestion is that nucleotide-composition bias is a product of lineage-specific selection pressures, e.g., to generate base compositions conducive to producing the amino-acid compositions of proteins most compatible with the challenges of specific environments (Mendez et al. 2010). However, as noted in Chapter 4, optimization arguments for phylogenetic variation in the mutation rate itself have not been successful, and explanations for a fine-tuning of the molecular spectrum are even more challenging. There is no direct evidence that mutation spectra are driven by selection, and the possibility that the substantial level of divergence may have been governed largely by effectively neutral processes cannot be ruled out (Haywood-Farmer and Otto 2003). Selection

operates on the genome-wide mutation rate, driving this down to some level beyond which further advantages are offset by the power of random genetic drift, but conditional on any particular overall rate, the mutational spectrum may be free to wander over evolutionary time.

Why is there near-universal selection for G+C composition, regardless of the magnitude of mutation pressure towards A+T? The bioenergetic costs of all four nucleotides are very similar (Chapter 17), so selective discrimination on this basis is unlikely. High-temperature environments might impose selection for higher G+C composition because G:C pairs involve three hydrogen bonds (as opposed to two for A:T), rendering a higher degree of DNA (and RNA) stability (Musto et al. 2004; Basak and Ghosh 2005). However, although there are correlations between G+C content and optimal growth temperatures within narrow phylogenetic groups of bacteria, this is not true on a broader phylogenetic scale. Moreover, the G+C composition of silent sites does not exhibit such correlations, contrary to expectations if there is genome-wide selection for duplex stability (Hurst and Merchant 2001). Adenine and guanine (purine) nucleotides contain three more nitrogen atoms than do pyrimidines, so long-term residence in nitrogen-limiting environments might select for genomes enriched with Cs and Ts (Rocha and Danchin 2002; Luo et al. 2015), but as DNA consists of A:T and G:C bonds, such selection would have to occur at the strand-specific RNA level and also be efficient enough to discriminate a difference of two nitrogen atoms against a total-cell backdrop of billions of these. Finally, gene conversion (a local product of recombination between two non-identical sequences; Chapter 4) is thought to be weakly biased towards Cs and Gs (when mismatches with As and Ts arise) across the Tree of Life (Lassalle et al. 2015), providing still another pressure on nucleotide composition, depending on the level of recombination.

Regardless of the mechanisms driving genome-wide nucleotide composition, the central question here is whether such biases have repercussions at the level of amino-acid utilization across phylogenetic lineages. Owing to the structure of the genetic code, the codons for some amino acids are much richer in GC content than others (Figure 12.1);

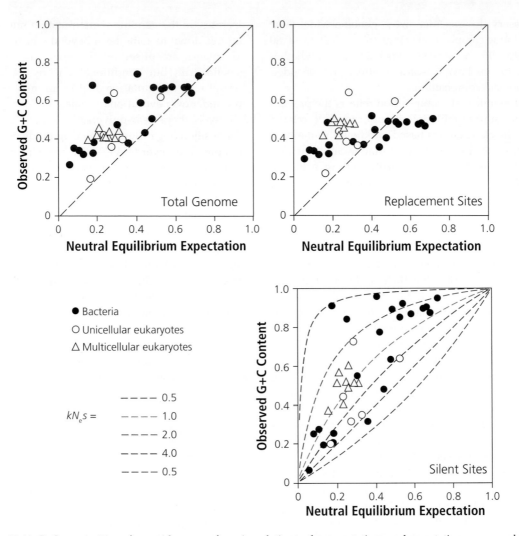

Figure 12.11 G+C compositions for a wide range of species relative to the expectations under mutation pressure alone (the neutral equilibrium expectation, derived from directly estimated spectra of spontaneous mutations obtained from mutation-accumulation experiments). The dashed diagonal lines give the null expectation for the situation in which G+C content is entirely driven by mutation. Replacement sites are positions within codons for which all mutations lead to an amino-acid substitution, whereas silent sites are those for which mutations have no effect on the encoded amino acid. As revealed by the distribution of points above the diagonal, nearly all species have excess G+C content, and the pattern is even more extreme for silent sites than for replacement sites in protein-coding genes. The quantity kN_es, where $k = 2$ for haploids and 4 for diploids, is a measure of the strength of selection relative to drift (Chapter 5), with $|kN_es| \ll 1$ implying effective neutrality. From Long et al. (2018).

e.g., 83% for alanine, arginine, glycine, and proline, but ≤17% for asparagine, isoleucine, lysine, phenylalanine, and tyrosine. Thus, we wish to know whether certain population-genetic environments promote the use of particular amino acids independent of their immediate functional significance.

From Figure 12.11, this can be seen to be the case – species with very strong mutation pressure towards A+T also gravitate to codons with high A+T composition. Among species, genome-wide G+C content at first and second positions of codons (which mostly consist of amino-acid replacement sites) is correlated with that at third positions (which are largely silent sites) (Gu et al. 1998; D'Onofrio et al. 1999; Bastolla et al. 2004; Chen et al. 2004). Moreover, the proteome-wide usage of amino acids with GC-rich codons in different species more than doubles across the range of genome-wide GC composition at silent sites, whereas that of the AT-rich group declines by more than 50% (Knight et al. 2001; Li et al. 2015). Thus, although there is strong selection for amino-acid composition at key sites within most proteins, it appears that mutation pressure is frequently sufficient to overcome the weak selection for amino-acid usage in a substantial fraction of less functionally significant sites.

These observations are of relevance to the question as to whether isolated lineages evolve completely independently at the molecular level. When two separate lineages independently acquire the same novel phenotype from the same starting state, the change is said to be parallel, whereas independent acquisition of the same state from different initial conditions represents convergent evolution (Zhang and Kumar 1997; Storz 2016).

Evolutionary substitutions of certain types of amino acids at particular sites within proteins occur more frequently than expected by chance (e.g., Bazykin et al. 2007; Rokas and Carroll 2008; Elias and Tawfik 2012; Ayuso-Fernández et al. 2018; Cano et al. 2022), and there is little question that lineages do occasionally respond to the same selective challenge in parallel manners. A dramatic example was revealed in replicated *E. coli* populations exposed to an increasing gradient of the antibiotic trimethoprim, which exhibited a similar temporal ordering of mutations conferring resistance in the dihydrofolate reductase gene (Toprak et al. 2011). However, demonstrating that convergent/parallel evolution is an outcome of shared selective pressures is difficult without rigorous statistical and/or empirical analysis, and numerous examples exist in which parallel evolution has inspired arguments about the channelling of molecular adaptations, only to be overturned by subsequent evaluation (Storz 2016).

This being said, it has been consistently observed that, relative to the neutral expectation, the incidence of amino-acid convergence events becomes progressively less common with more distantly related lineages (Goldstein et al. 2015; Shah et al. 2015; Zou and Zhang 2015). A compelling explanation for such behavior is that, as a protein accepts amino-acid changes at a variety of sites in different lineages, the local selective environments at other sites are altered, thereby diminishing the likelihood of subsequent mutations being channelled to the same set of residues. Such a model implies a predominance of both effectively neutral substitutions (allowing change to occur at individual sites) and of epistasis (interaction effects between individual sites).

Epistasis and compensatory mutation

The preceding sections provided numerous examples in which the effects of mutations in protein-coding sequences are epistatic with respect to fitness. That is, the fitness effects of individual mutations often depend on local context. Direct evidence for such interactions derives from experimental mutagenesis experiments, such as that of Bank et al. (2015), who in an analysis of >1000 double mutants in the binding domain of a yeast heat-shock protein found a preponderance of pairs with negative combined effects on fitness (beyond the additive expectations based on single-mutational effects). In that study, very few mutational pairs exhibited positive epistatic effects.

In a somewhat different study, Lunzer et al. (2010) substituted (one at a time) 168 amino acids in the isopropymalate dehydrogenase protein in *E. coli* to match the differences in the orthologous protein in *Pseudomonas aeruginosa*. On the *E. coli* background, 63 of these single substitutions were functionally compromised, whereas only one had improved performance. In another comparative study, Starr et al. (2018) reconstructed estimated ancestral states in a yeast heat-shock protein (Hsp70) and then laboriously substituted amino acids from the modern-day sequence into the ancestral form, and vice versa. Although Hsp70 has retained a highly conserved function over a billion years of evolution, > 75% of these single-residue exchanges were deleterious, even though they must have been acceptable over the course of evolution. All of these observations are consistent with stochastic lineage-specific additions of mutations conditional upon earlier changes progressively altering the permissive environment for substitution.

Further indirect evidence for the long-term evolutionary significance of epistasis derives from a number of different comparative analyses. For example, in a study of 16 eukaryotic proteins, each with >1000 sequences available from a wide variety of phylogenetic lineages, Breen et al. (2012) found that the average amino-acid site is occupied by just eight different amino acids, even though ample evolutionary time has elapsed for all mutation types to have appeared at each site, i.e., on average each site can shift to only seven alternative amino acids. The authors reasoned that d_N/d_S ought to be $7/19 = 0.36$ if amino-acid altering mutations accumulate in a noninteractive way, i.e., 64% of amino-acid replacements would be expected to be unacceptable. However, the average observed d_N/d_S ratio (measured from sequence divergence between species) for these proteins is $\sim 7\times$ lower than this expectation, leading to the conclusion that negative epistatic fitness effects must be pervasive among mutations. The implication is that if a particular amino acid fixes at one particular site, it creates a local environmental shift within the protein that prevents the fixation of the majority of amino acid-altering mutations at other sites, leaving an average of only ~ 1 permissible change per site at any particular point in evolutionary time.

A second compelling line of evidence for the role of epistasis in protein evolution derives from the observation that many amino-acid changes that cause human pathologies (and are therefore rare in the human population) are nonetheless well-established (with no pathogenic effects) in other mammalian species (Kondrashov et al. 2002; Gao and Zhang 2003). Very similar observations have been made with mutations known to be pathogenic in *Drosophila*, but established in other insect species (Kulathinal et al. 2004). As the frequency of such compensated deviations does not increase with the evolutionary distance of a lineage, this suggests that they accrue relatively rapidly, rather than awaiting long-term protein remodelling. The effects of such mutations must be context dependent.

Finally, amino-acid changes in proteins tend to be clustered within a sequence, generally on a chain-length scale of <10 residues, and also tend to preserve the local charge of the protein (Callahan et al. 2011). Notably, the average physical distance between central carbon atoms of amino acids in folded proteins plateaus at chain distances >10 residues, implying that on average residues separated by <10 positions have a high likelihood of physical interaction. The fact that silent-site substitutions are not clustered argues against the pattern being a result of regional mutational hot spots. Additional work shows that long-range epistatic interactions are not uncommon (Sharir-Ivry and Xia 2018).

A General Model for Protein Evolution

A key point from the previous discussion is that, more often than not, many cumulative amino-acid changes have little impact on the immediate functionality of a gene. Rather, much of protein evolution appears to reflect little more than a restricted random walk down nearly neutral pathways (Figure 12.12). Some of these pathways may involve the fixation of effectively neutral but slightly deleterious mutations (as outlined in Chapter 5), which then allow the fixation of a compensatory mutation that was insignificantly favorable (or even deleterious) on the prior ancestral background but now more favorable in its new context. Such compensatory changes are not necessarily epistatic with respect to the long-term enhancement of total fitness, although they are epistatic with respect to the physical structure of the protein.

Thus, an emerging view of protein-sequence evolution is that at any point in time the number of degrees of freedom for change at individual amino-acid sites is small, with the identities of exchangeable amino acids shifting with fortuitous prior fixations elsewhere in the molecule (Goldstein and Pollock 2017; Xie et al. 2021; Park et al. 2022). In part, restricted sequence walks are governed by the nature of the genetic code, as single mutations at each replacement nucleotide site can generate at most three alternative amino acids. More generally, however, the structural environment of the protein itself will dictate the subset of permissible (effectively neutral) amino-acid exchanges. Over time, slight shifts in the amino-acid constitution of the protein, each nearly neutral incrementally,

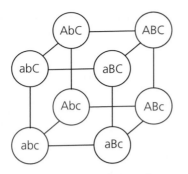

Figure 12.12 A cartoon view of the restricted paths to evolution for a three-residue protein with two alternative states at each site. Of the eight possible sequences, four (in black) are effectively inviable owing to their deleterious fitness effects being large enough for selection to overcome random genetic drift. The four states in red are either effectively neutral, or involve the substitution of a selectively beneficial change on the adjacent background. For example, A is permissible only on a bc, Bc, or BC background, and B is only permissible on a background containing A. Under this view, a lineage is free to wander back and forth along the red route.

alter the local protein structural environment, further restricting the degrees of freedom for future changes, but in a progressively divergent way, allowing the long-term degrees of freedom for change at a large fraction of sites to wander.

Such cycles of modification of the background environment, and subsequent channelling of permissible mutations allows for an expansive set of paths open to evolutionary change across the Tree of Life, while rendering individual lineages victims of historical contingency. Under this model, because slightly deleterious mutations can sometimes fix, such events also pave the way for the subsequent fixation of compensatory beneficial mutations without significant consequences for long-term adaptation in terms of protein function. Moreover, by this cumulative process, amino-acid changes that were originally effectively neutral may sometimes become entrenched to the point of being essential to protein functionality and hence, nearly irreversible evolutionarily. If this view is correct, then experimental studies of the fitness consequences of amino-acid substitutions at individual sites may poorly reflect the effects of individual residues when first established in the historical development of a protein.

Summary

- Proteins consist of chains of amino acids, generally folded into subunits, such as helices and sheets, which then further arrange into tertiary structures essential for function.

- At the dawn of the protein world, only a fraction of the twenty amino acids used in today's organisms would have been in play, and other non-canonical amino acids might have been used. Nonetheless, enormous functional diversity of proteins can still be generated by a reduced amino-acid alphabet, although an expanded vocabulary would allow for further refinement in catalytic activity and efficiency.

- One of the major challenges of proteins is their initial need to fold into three-dimensional structures to achieve functionality. Although there are a number of important substructural influences, folding rates are largely determined by the amino-acid chain length, and those in excess of ~250 residues are generally incapable of folding on their own on reasonable time scales.

- Despite the high level of refinement, the functionality of proteins has not reached the limits set by biophysics. Catalytic rates can be improved by the use of noncanonical amino acids. Folding rates and stability are also less than their maximum possible values, and potentially more so in eukaryotes than prokaryotes. These observations suggest that the efficiency of natural selection is stalled by a drift barrier.

- Based on phylogenetic comparisons of sequence data, only 5–25% of amino acid-altering mutations are acceptable in nature, although experimental substitutions of random amino acids consistently indicate that most such mutations do not entirely eliminate protein function. The distribution of fitness effects associated with amino-acid exchanges generally has a mode not significantly different from zero, a long tail towards deleterious effects, and only a small tail containing favorable changes. Although the overall conclusion is that the majority of mutations at the protein level are mildly deleterious, the details of the distribution in the range of very small effects, which is most critical to evolutionary theory, remains uncertain.

- One of the primary determinants of the rate of evolution of a protein is its level of expression. This is thought to be a consequence of strong purifying selection for the maintenance of folding stability to avoid the production of wasted or harmful by-products and/or selection for surface residues that minimize misinteractions with other key proteins.

- The magnitude of mutation bias varies widely among phylogenetic lineages, but is usually in the direction of A and T nucleotides. As a consequence of the structure of the genetic code, this can sometimes drive the biased deployment of particular amino acids in the proteome, leading to parallel evolution in different lineages with little involvement of selection.

- Amino acid-altering mutations frequently have context-dependent fitness effects, whereby the incorporation of earlier mutations can dictate whether specific subsequent substitutions are deleterious, beneficial, or effectively neutral. As a consequence, the fixation of effectively neutral (but mildly deleterious) mutations can pave the way for the future fixation of compensatory mutations that otherwise would not be beneficial. Over time, a series of such subtle remodelling events can lead to the entrenchment of previously neutral amino-acid substitutions to the point of becoming near essential to protein functionality. In retrospect, although such progressive changes may appear to involve adaptive fixations and strong epistatic effects, the entire process may unfold with only minor consequences for overall fitness. This view of protein evolution is entirely compatible with long-term wandering of amino acid sequences along the drift barrier.

Literature Cited

Akanuma, S., T. Kigawa, and S. Yokoyama. 2002. Combinatorial mutagenesis to restrict amino acid usage in an enzyme to a reduced set. Proc. Natl. Acad. Sci. USA 99: 13549–13553.

Alva, V., J. Söding, and A. N. Lupas. 2015. A vocabulary of ancient peptides at the origin of folded proteins. eLife 4: e09410.

Axe, D. D. 2000. Extreme functional sensitivity to conservative amino acid changes on enzyme exteriors. J. Mol. Biol. 301: 585–595.

Axe, D. D., N. W. Foster, and A. R. Fersht. 1996. Active barnase variants with completely random hydrophobic cores. Proc. Natl. Acad. Sci. USA 93: 5590–5594.

Ayuso-Fernández, I., F. J. Ruiz-Dueñas, and A. T. Martínez. 2018. Evolutionary convergence in lignin-degrading enzymes. Proc. Natl. Acad. Sci. USA 115: 6428–6433.

Ball, P. 2008. Water as an active constituent in cell biology. Chem. Rev. 108: 74–108.

Bank, C., R. T. Hietpas, J. D. Jensen, and D. N. Bolon. 2015. A systematic survey of an intragenic epistatic landscape. Mol. Biol. Evol. 32: 229–238.

Basak, S., and T. C. Ghosh. 2005. On the origin of genomic adaptation at high temperature for prokaryotic organisms. Biochem. Biophys. Res. Commun. 330: 629–632.

Bastolla, U., A. Moya, E. Viguera, and R. C. van Ham. 2004. Genomic determinants of protein folding thermodynamics in prokaryotic organisms. J. Mol. Biol. 343: 1451–1466.

Bazykin, G. A., F. A. Kondrashov, M. Brudno, A. Poliakov, I. Dubchak, and A. S. Kondrashov. 2007. Extensive parallelism in protein evolution. Biol. Direct 2: 20.

Berezovsky, I. N., K. B. Zeldovich, and E. I. Shakhnovich. 2007. Positive and negative design in stability and thermal adaptation of natural proteins. PLoS Comput. Biol. 3: e52.

Bergman, J., and A. Eyre-Walker. 2019. Does adaptive protein evolution proceed by large or small steps at the amino acid level? Mol. Biol. Evol. 36: 990–998.

Bershtein, S., W. Mu, A. W. Serohijos, J. Zhou, and E. I. Shakhnovich. 2013. Protein quality control acts on folding intermediates to shape the effects of mutations on organismal fitness. Mol. Cell 49: 133–144.

Bershtein, S., M. Segal, R. Bekerman, N. Tokuriki, and D. S. Tawfik. 2006. Robustness–epistasis link shapes the fitness landscape of a randomly drifting protein. Nature 444: 929–932.

Bershtein, S., A. W. Serohijos, S. Bhattacharyya, M. Manhart, J. M. Choi, W. Mu, J. Zhou, and E. I. Shakhnovich. 2015. Protein homeostasis imposes a barrier on functional integration of horizontally transferred genes in bacteria. PLoS Genet. 11: e1005612.

Bloom, J. D., A. Raval, and C. O. Wilke. 2007. Thermodynamics of neutral protein evolution. Genetics 175: 255–266.

Bloom, J. D., M. M. Meyer, P. Meinhold, C. R. Otey, D. MacMillan, and F. H. Arnold. 2005. Evolving strategies for enzyme engineering. Curr. Opin. Struct. Biol. 15: 447–452.

Borgo, B., and J. J. Havranek. 2012. Automated selection of stabilizing mutations in designed and natural proteins. Proc. Natl. Acad. Sci. USA 109: 1494–1499.

Breen, M. S., C. Kemena, P. K. Vlasov, C. Notredame, and F. A. Kondrashov. 2012. Epistasis as the primary factor in molecular evolution. Nature 490: 535–538.

Brooks, D. J., and J. R. Fresco. 2002. Increased frequency of cysteine, tyrosine, and phenylalanine residues since the last universal ancestor. Mol. Cell. Proteomics 1: 125–131.

Brooks, D. J., J. R. Fresco, A. M. Lesk, and M. Singh. 2002. Evolution of amino acid frequencies in proteins over deep time: Inferred order of introduction of amino acids into the genetic code. Mol. Biol. Evol. 19: 1645–1655.

Bustamante, C. D., J. P. Townsend, and D. L. Hartl. 2000. Solvent accessibility and purifying selection within proteins of *Escherichia coli* and *Salmonella enterica*. Mol. Biol. Evol. 17: 301–308.

Callahan, B., R. A. Neher, D. Bachtrog, P. Andolfatto, and B. I. Shraiman. 2011. Correlated evolution of nearby residues in drosophilid proteins. PLoS Genet. 7: e1001315.

Cano, A. V., H. Rozhoňová, A. Stoltzfus, D. M. McCandlish, and J. L. Payne. 2022. Mutation bias shapes the spectrum of adaptive substitutions. Proc. Natl. Acad. Sci. USA 119: e2119720119.

Chen, S. L., W. Lee, A. K. Hottes, L. Shapiro, and H. H. McAdams. 2004. Codon usage between genomes is constrained by genome-wide mutational processes. Proc. Natl. Acad. Sci. USA 101: 3480–3485.

Davis, B. K. 1999. Evolution of the genetic code. Prog. Biophys. Mol. Biol. 72: 157–243.

de la Torre, D., and J. W. Chin. 2021. Reprogramming the genetic code. Nat. Rev. Genet. 22: 169–184.

de Groot, N. S., and S. Ventura. 2010. Protein aggregation profile of the bacterial cytosol. PLoS One 5: e9383.

Dehouck, Y., B. Folch, and M. Rooman. 2008. Revisiting the correlation between proteins' thermoresistance and organisms' thermophilicity. Protein Eng. Des. Sel. 21: 275–278.

Deng, Z., W. Huang, E. Bakkalbasi, N. G. Brown, C. J. Adamski, K. Rice, D. Muzny, R. A. Gibbs, and T. Palzkill. 2012. Deep sequencing of systematic combinatorial libraries reveals β-lactamase sequence constraints at high resolution. J. Mol. Biol. 424: 150–167.

Dhar, A., A. Samiotakis, S. Ebbinghaus, L. Nienhaus, D. Homouz, M. Gruebele, and M. S. Cheung. 2010. Structure, function, and folding of phosphoglycerate kinase are strongly perturbed by macromolecular crowding. Proc. Natl. Acad. Sci. USA 107: 17586–17591.

Dill, K. A., K. Ghosh, and J. D. Schmit. 2011. Physical limits of cells and proteomes. Proc. Natl. Acad. Sci. USA 108: 17876–17882.

Dill, K. A., and J. L. MacCallum. 2012. The protein-folding problem, 50 years on. Science 338: 1042–1046.

Doi, N., K. Kakukawa, Y. Oishi, and H. Yanagawa. 2005. High solubility of random-sequence proteins consisting of five kinds of primitive amino acids. Protein Eng. Des. Sel. 18: 279–284.

D'Onofrio, G., K. Jabbari, H. Musto, and G. Bernardi. 1999. The correlation of protein hydropathy with the base composition of coding sequences. Gene 238: 3–14.

Drummond, D. A., and C. O. Wilke. 2008. Mistranslation-induced protein misfolding as a dominant constraint on coding-sequence evolution. Cell 134: 341–352.

Elias, M., and D. S. Tawfik. 2012. Divergence and convergence in enzyme evolution: Parallel evolution of paraoxonases from quorum-quenching lactonases. J. Biol. Chem. 287: 11–20.

Englander, S. W., and L. Mayne. 2014. The nature of protein folding pathways. Proc. Natl. Acad. Sci. USA 111: 15873–15880.

Englander, S. W., and L. Mayne. 2017. The case for defined protein folding pathways. Proc. Natl. Acad. Sci. USA 114: 8253–8258.

Firnberg, E., J. W. Labonte, J. J. Gray, and M. Ostermeier. 2014. A comprehensive, high-resolution map of a gene's fitness landscape. Mol. Biol. Evol. 31: 1581–1592.

Franzosa, E. A., and Y. Xia. 2009. Structural determinants of protein evolution are context-sensitive at the residue level. Mol. Biol. Evol. 26: 2387–2395.

Galzitskaya, O. V., N. S. Bogatyreva, and A. V. Glyakina. 2011. Bacterial proteins fold faster than eukaryotic proteins with simple folding kinetics. Biochemistry (Moscow) 76: 225–235.

Galzitskaya, O. V., D. C. Reifsnyder, N. S. Bogatyreva, D. N. Ivankov, and S. O. Garbuzynskiy. 2008. More compact protein globules exhibit slower folding rates. Proteins 70: 329–332.

Gao, L., and J. Zhang. 2003. Why are some human disease-associated mutations fixed in mice? Trends Genet. 19: 678–681.

Ghosh, K., and K. A. Dill. 2009. Computing protein stabilities from their chain lengths. Proc. Natl. Acad. Sci. USA 106: 10649–10654.

Giver, L., A. Gershenson, P. O. Freskgard, and F. H. Arnold. 1998. Directed evolution of a thermostable esterase. Proc. Natl. Acad. Sci. USA 95: 12809–12813.

Go, A., S. Kim, J. Baum, and M. H. Hecht. 2008. Structure and dynamics of *de novo* proteins from a designed superfamily of 4-helix bundles. Protein Sci. 17: 821–832.

Goldman, N., J. L. Thorne, and D. T. Jones. 1998. Assessing the impact of secondary structure and solvent accessibility on protein evolution. Genetics 149: 445–458.

Goldstein, R. A., S. T. Pollard, S. D. Shah, and D. D. Pollock. 2015. Nonadaptive amino acid convergence rates decrease over time. Mol. Biol. Evol. 32: 1373–1381.

Goldstein, R. A., and D. D. Pollock. 2017. Sequence entropy of folding and the absolute rate of amino acid substitutions. Nat. Ecol. Evol. 1: 1923–1930.

Govindarajan, S., and R. A. Goldstein. 1997. Evolution of model proteins on a foldability landscape. Proteins 29: 461–466.

Grantcharova, V., E. J. Alm, D. Baker, and A. L. Horwich. 2001. Mechanisms of protein folding. Curr. Opin. Struct. Biol. 11: 70–82.

Gu, X., D. Hewett-Emmett, and W.-H. Li. 1998. Directional mutational pressure affects the amino acid composition and hydrophobicity of proteins in bacteria. Genetica 102/103: 383–931.

Guo, H. H., J. Choe, and L. A. Loeb. 2004. Protein tolerance to random amino acid change. Proc. Natl. Acad. Sci. USA 101: 9205–9210.

Harris, J. K., S. T. Kelley, G. B. Spiegelman, and N. R. Pace. 2003. The genetic core of the universal ancestor. Genome Res. 13: 407–412.

Haywood-Farmer, E., and S. P. Otto. 2003. The evolution of genomic base composition in bacteria. Evolution 57: 1783–1792.

Higgs, P. G., and R. E. Pudritz. 2009. A thermodynamic basis for prebiotic amino acid synthesis and the nature of the first genetic code. Astrobiol. 9: 483–490.

Huang, J. T., D. J. Xing, and W. Huang. 2012. Relationship between protein folding kinetics and amino acid properties. Amino Acids 43: 567–572.

Hurst, L. D., and A. R. Merchant. 2001. High guanine-cytosine content is not an adaptation to high temperature: A comparative analysis amongst prokaryotes. Proc. Biol. Sci. 268: 493–497.

Ishihama, Y., T. Schmidt, J. Rappsilber, M. Mann, F. U. Hartl, M. J. Kerner, and D. Frishman. 2008. Protein abundance profiling of the *Escherichia coli* cytosol. BMC Genomics 9: 102.

Islam, M. M., S. Sohya, K. Noguchi, M. Yohda, and Y. Kuroda. 2008. Crystal structure of an extensively simplified variant of bovine pancreatic trypsin inhibitor in which over one-third of the residues are alanines. Proc. Natl. Acad. Sci. USA 105: 15334–15339.

Ivankov, D. N., and A. V. Finkelstein. 2004. Prediction of protein folding rates from the amino acid sequence-predicted secondary structure. Proc. Natl. Acad. Sci. USA 101: 8942–8944.

Jacquier, H., A. Birgy, H. Le Nagard, Y. Mechulam, E. Schmitt, J. Glodt, B. Bercot, E. Petit, J. Poulain, G. Barnaud, et al. 2013. Capturing the mutational landscape of the β-lactamase TEM-1. Proc. Natl. Acad. Sci. USA 110: 13067–13072.

Jarzab, A., N. Kurzawa, T. Hopf, M. Moerch, J. Zecha, N. Leijten, Y. Bian, E. Musiol, M. Maschberger, G. Stoehr, et al. 2020. Meltome atlas – thermal proteome stability across the Tree of Life. Nat. Methods 17: 495–503.

Jordan, I. K., F. A. Kondrashov, I. A. Adzhubei, Y. I. Wolf, E. V. Koonin, A. S. Kondrashov, and S. Sunyaev. 2005. A universal trend of amino acid gain and loss in protein evolution. Nature 433: 633–638.

Kachroo, A. H., J. M. Laurent, A. Akhmetov, M. Szilagyi-Jones, C. D. McWhite, A. Zhao, and E. M. Marcotte. 2017. Systematic bacterialization of yeast genes identifies a near-universally swappable pathway. eLife 6: e25093.

Kachroo, A. H., J. M. Laurent, C. M. Yellman, A. G. Meyer, C. O. Wilke, and E. M. Marcotte. 2015. Systematic humanization of yeast genes reveals conserved functions and genetic modularity. Science 348: 921–925.

Kaiser, C. M., D. H. Goldman, J. D. Chodera, I. Tinoco, Jr., and C. Bustamante. 2011. The ribosome modulates nascent protein folding. Science 334: 1723–1727.

Khan, S., and M. Vihinen. 2010. Performance of protein stability predictors. Hum. Mutat. 31: 675–684.

Kim, D. E., H. Gu, and D. Baker. 1998. The sequences of small proteins are not extensively optimized for rapid folding by natural selection. Proc. Natl. Acad. Sci. USA 95: 4982–4986.

Kimura, M. 1983. The Neutral Theory of Molecular Evolution. Cambridge University Press, Cambridge, UK.

Knight, R. D., S. J. Freeland, and L. F. Landweber. 2001. A simple model based on mutation and selection explains trends in codon and amino acid usage and GC composition within and across genomes. Genome Biol. 2: RESEARCH0010.

Kondrashov, A. S., S. Sunyaev, and F. A. Kondrashov. 2002. Dobzhansky–Muller incompatibilities in protein evolution. Proc. Natl. Acad. Sci. USA 99: 14878–14883.

Koonin, E. V. 2003. Comparative genomics, minimal genesets and the last universal common ancestor. Nature Rev. Microbiol. 1: 127–136.

Krick, T., N. Verstraete, L. G. Alonso, D. A. Shub, D. U. Ferreiro, M. Shub, and I. E. Sánchez. 2014. Amino acid metabolism conflicts with protein diversity. Mol. Biol. Evol. 31: 2905–2912.

Kubelka, J., J. Hofrichter, and W. A. Eaton. 2004. The protein folding 'speed limit'. Curr. Opin. Struct. Biol. 14: 76–88.

Kulathinal, R. J., B. R. Bettencourt, and D. L. Hartl. 2004. Compensated deleterious mutations in insect genomes. Science 306: 1553–1554.

Kuo, C. H., N. A. Moran, and H. Ochman. 2009. The consequences of genetic drift for bacterial genome complexity. Genome Res. 19: 1450–1454.

Labean, T. H., T. R. Butt, S. A. Kauffman, and E. A. Schultes. 2011. Protein folding absent selection. Genes (Basel) 2: 608–626.

Lassalle, F., S. Périan, T. Bataillon, X. Nesme, L. Duret, and V. Daubin. 2015. GC-Content evolution in bacterial genomes: The biased gene conversion hypothesis expands. PLoS Genet. 11: e1004941.

Lawrie, D. S., P. W. Messer, R. Hershberg, and D. A. Petrov. 2013. Strong purifying selection at synonymous sites in *D. melanogaster*. PLoS Genet. 9: e1003527.

Levinthal, C. 1968. Are there pathways for protein folding? J. Chim. Physique 65: 44–45.

Levy, E. D., S. De, and S. A. Teichmann. 2012. Cellular crowding imposes global constraints on the chemistry and evolution of proteomes. Proc. Natl. Acad. Sci. USA 109: 20461–20466.

Lim, S. A., E. R. Bolin, and S. Marqusee. 2018. Tracing a protein's folding pathway over evolutionary time using ancestral sequence reconstruction and hydrogen exchange. eLife 7: e38369.

Lim, S. A., and S. Marqusee. 2018. The burst-phase folding intermediate of ribonuclease H changes conformation over evolutionary history. Biopolymers 109: e23086.

Li, J., J. Zhou, Y. Wu, S. Yang, and D. Tian. 2015. GC-content of synonymous codons profoundly influences amino acid usage. G3 (Bethesda) 5: 2027–2036.

Lin, M. M., and A. H. Zewail. 2012. Hydrophobic forces and the length limit of foldable protein domains. Proc. Natl. Acad. Sci. USA 109: 9851–9856.

Lind, P. A., L. Arvidsson, O. G. Berg, and D. I. Andersson. 2016. Variation in mutational robustness between different proteins and the predictability of fitness effects. Mol. Biol. Evol. 34: 408–418.

Lind, P. A., O. G. Berg, and D. I. Andersson. 2010. Mutational robustness of ribosomal protein genes. Science 330: 825–827.

Long, H., W. Sung, S. Kucukyildirim, E. Williams, S. F. Miller, W. Guo, C. Patterson, C. Gregory, C. Strauss, C. Stone, et al. 2018. Evolutionary determinants of genome-wide nucleotide composition. Nat. Ecol. Evol. 2: 237–240.

Longo, L. M., and M. Blaber. 2012. Protein design at the interface of the pre-biotic and biotic worlds. Arch. Biochem. Biophys. 526: 16–21.

Longo, L. M., and M. Blaber. 2014. Prebiotic protein design supports a halophile origin of foldable proteins. Front. Microbiol. 4: 418.

Longo, L. M., J. Lee, and M. Blaber. 2013. Simplified protein design biased for prebiotic amino acids yields a foldable, halophilic protein. Proc. Natl. Acad. Sci. USA 110: 2135–2139.

López de la Osa, J., D. A. Bateman, S. Ho, C. González, A. Chakrabartty, and D. V. Laurents. 2007. Getting specificity from simplicity in putative proteins from the prebiotic earth. Proc. Natl. Acad. Sci. USA 104: 14941–14946.

Lundin, E., P. C. Tang, L. Guy, J. Näsvall, and D. I. Andersson. 2018. Experimental determination and prediction of the fitness effects of random point mutations in the biosynthetic enzyme HisA. Mol. Biol. Evol. 35: 704–718.

Lunzer, M., G. B. Golding, and A. M. Dean. 2010. Pervasive cryptic epistasis in molecular evolution. PLoS Genet. 6: e1001162.

Luo, H., L. R. Thompson, U. Stingl, and A. L. Hughes. 2015. Selection maintains low genomic GC content in marine SAR11 lineages. Mol. Biol. Evol. 32: 2738–2748.

Lynch, M. 2007. The Origins of Genome Architecture. Sinauer Associates, Inc., Sunderland, MA.

Lynch, M. 2018. Phylogenetic diversification of cell biological features. eLife 7: e34820.

Lynch, M., M. Ackerman, K. Spitze, Z. Ye, and T. Maruki. 2017. Population genomics of *Daphnia pulex*. Genetics 206: 315–332.

MacBeath, G., P. Kast, and D. Hilvert. 1998. A small, thermostable, and monofunctional chorismate mutase from the archaeon *Methanococcus jannaschii*. Biochemistry 37: 10062–10073.

Materon, I. C., and T. Palzkill. 2001. Identification of residues critical for metallo-β-lactamase function by codon randomization and selection. Protein Sci. 10: 2556–2565.

Matsuura, T., K. Miyai, S. Trakulnaleamsai, T. Yomo, Y. Shima, S. Miki, K. Yamamoto, and I. Urabe. 1999. Evolutionary molecular engineering by random elongation mutagenesis. Nat. Biotechnol. 17: 58–61.

McDonald, G. D., and M. C. Storrie-Lombardi. 2010. Biochemical constraints in a protobiotic earth devoid of basic amino acids: The 'BAA(-) world'. Astrobiol. 10: 989–1000.

Mendez, R., M. Fritsche, M. Porto, and U. Bastolla. 2010. Mutation bias favors protein folding stability in the evolution of small populations. PLoS Comput. Biol. 6: e1000767.

Miller, C., M. Davlieva, C. Wilson, K. I. White, R. Couñago, G. Wu, J. C. Myers, P. Wittung-Stafshede, and Y. Shamoo. 2010. Experimental evolution of adenylate kinase reveals contrasting strategies toward protein thermostability. Biophys. J. 99: 887–896.

Mishra, P., J. M. Flynn, T. N. Starr, and D. N. Bolon. 2016. Systematic mutant analyses elucidate general and client-specific aspects of Hsp90 function. Cell Rep. 15: 588–598.

Moon, S., D. K. Jung, G. N. Phillips, Jr., and E. Bae. 2014. An integrated approach for thermal stabilization of a mesophilic adenylate kinase. Proteins 82: 1947–1959.

Moutinho, A. F., F. F. Trancoso, and J. Y. Dutheil. 2019. The impact of protein architecture on adaptive evolution. Mol. Biol. Evol. 36: 2013–2028.

Müller, M. M., J. R. Allison, N. Hongdilokkul, L. Gaillon, P. Kast, W. F. van Gunsteren, P. Marliére, and D. Hilvert. 2013. Directed evolution of a model primordial enzyme provides insights into the development of the genetic code. PLoS Genet. 9: e1003187.

Musto, H., H. Naya, A. Zavala, H. Romero, F. Alvarez-Valín, and G. Bernardi. 2004. Correlations between genomic GC levels and optimal growth temperatures in prokaryotes. FEBS Lett. 573: 73–77.

Nei, M. and S. Kumar. 2000. Molecular Evolution and Phylogenetics. Oxford University Press, Oxford, UK.

Neupane, K., D. A. Foster, D. R. Dee, H. Yu, F. Wang, and M. T. Woodside. 2016. Direct observation of transition paths during the folding of proteins and nucleic acids. Science 352: 239–242.

Nixon, C. F., S. A. Lim, Z. R. Sailer, I. N. Zheludev, C. L. Gee, B. A. Kelch, M. J. Harms, and S. Marqusee. 2021. Exploring the evolutionary history of kinetic stability in the α-lytic protease family. Biochemistry 60: 170–181.

Pál, C., B. Papp, and L. D. Hurst. 2001. Highly expressed genes in yeast evolve slowly. Genetics 158: 927–931.

Park, Y., B. P. H. Metzger, and J. W. Thornton. 2022. Epistatic drift causes gradual decay of predictability in protein evolution. Science 376: 823–830.

Plaxco, K. W., K. T. Simons, I. Ruczinski, and D. Baker. 2000. Topology, stability, sequence, and length: Defining the determinants of two-state protein folding kinetics. Biochemistry 39: 11177–11183.

Prabhu, N. P., and A. K. Bhuyan. 2006. Prediction of folding rates of small proteins: Empirical relations based on length, secondary structure content, residue type, and stability. Biochemistry 45: 3805–3812.

Ramsey, D. C., M. P. Scherrer, T. Zhou, and C. O. Wilke. 2011. The relationship between relative solvent accessibility and evolutionary rate in protein evolution. Genetics 188: 479–488.

Rebeaud, M. E., S. Mallik, P. Goloubinoff, and D. S. Tawfik. 2021. On the evolution of chaperones and cochaperones and the expansion of proteomes across the Tree of Life. Proc. Natl. Acad. Sci. USA 118: e2020885118.

Robertson, A. D., and K. P. Murphy. 1997. Protein structure and the energetics of protein stability. Chem. Rev. 97: 1251–1268.

Rocha, E. P., and A. Danchin. 2002. Base composition bias might result from competition for metabolic resources. Trends Genet. 18: 291–294.

Rokas, A., and S. B. Carroll. 2008. Frequent and widespread parallel evolution of protein sequences. Mol. Biol. Evol. 25: 1943–1953.

Roscoe, B. P., K. M. Thayer, K. B. Zeldovich, D. Fushman, and D. N. Bolon. 2013. Analyses of the effects of all ubiquitin point mutants on yeast growth rate. J. Mol. Biol. 425: 1363–1377.

Salverda, M. L., E. Dellus, F. A. Gorter, A. J. Debets, J. van der Oost, R. F. Hoekstra, D. S. Tawfik, and J. A. de Visser. 2011. Initial mutations direct alternative pathways of protein evolution. PLoS Genet. 7: e1001321.

Sarkisyan, K. S., D. A. Bolotin, M. V. Meer, D. R. Usmanova, A. S. Mishin, G. V. Sharonov, D. N. Ivankov, N. G. Bozhanova, M. S. Baranov, O. Soylemez, et al. 2016. Local fitness landscape of the green fluorescent protein. Nature 533: 397–401.

Sato, S., S. Xiang, and D. P. Raleigh. 2001. On the relationship between protein stability and folding kinetics: A comparative study of the N-terminal domains of RNase HI, *E. coli* and *Bacillus stearothermophilus* L9. J. Mol. Biol. 312: 569–577.

Serohijos, A. W., Z. Rimas, and E. I. Shakhnovich. 2012. Protein biophysics explains why highly abundant proteins evolve slowly. Cell Rep. 2: 249–256.

Serohijos, A. W., and E. I. Shakhnovich. 2014. Contribution of selection for protein folding stability in shaping the patterns of polymorphisms in coding regions. Mol. Biol. Evol. 31: 165–176.

Serrano, L., A. G. Day, and A. R. Fersht. 1993. Stepwise mutation of barnase to binase. A procedure for engineering increased stability of proteins and an experimental analysis of the evolution of protein stability. J. Mol. Biol. 233: 305–312.

Shah, P., D. M. McCandlish, and J. B. Plotkin. 2015. Contingency and entrenchment in protein evolution under purifying selection. Proc. Natl. Acad. Sci. USA 112: E3226–E3235.

Sharir-Ivry, A., and Y. Xia. 2018. Nature of long-range evolutionary constraint in enzymes: Insights from comparison to pseudoenzymes with similar structures. Mol. Biol. Evol. 35: 2597–2606.

Sharp, P. M., E. Bailes, R. J. Grocock, J. F. Peden, and R. E. Sockett. 2005. Variation in the strength of selected codon usage bias among bacteria. Nucleic Acids Res. 33: 1141–1153.

Sheu, S. Y., D. Y. Yang, H. L. Selzle, and E. W. Schlag. 2003. Energetics of hydrogen bonds in peptides. Proc. Natl. Acad. Sci. USA 100: 12683–12687.

Shi, Z., J. Sellers, and J. Moult. 2012. Protein stability and *in vivo* concentration of missense mutations in phenylalanine hydroxylase. Proteins 80: 61–70.

Sicheri, F., and D. S. Yang. 1995. Ice-binding structure and mechanism of an antifreeze protein from winter flounder. Nature 375: 427–431.

Silverman, J. A., R. Balakrishnan, and P. B. Harbury. 2001. Reverse engineering the β/α_8 barrel fold. Proc. Natl. Acad. Sci. USA 98: 3092–3097.

Snyder, P. W., J. Mecinovic, D. T. Moustakas, S. W. Thomas, 3rd, M. Harder, E. T. Mack, M. R. Lockett, A. Héroux, W. Sherman, and G. M. Whitesides. 2011. Mechanism of the hydrophobic effect in the biomolecular recognition of arylsulfonamides by carbonic anhydrase. Proc. Natl. Acad. Sci. USA 108: 17889–17894.

Sojo, V, C. Dessimoz, A. Pomiankowski, and N. Lane. 2016. Membrane proteins are dramatically less conserved than water-soluble proteins across the Tree of Life. Mol. Biol. Evol. 33: 2874–2884.

Stanley, C. E., Jr., and R. J. Kulathinal. 2016. flyDIVaS: A comparative genomics resource for *Drosophila* divergence and selection. G3 (Bethesda) 6: 2355–2363.

Starr, T. N., J. M. Flynn, P. Mishra, D. N. A. Bolon, and J. W. Thornton. 2018. Pervasive contingency and entrenchment in a billion years of Hsp90 evolution. Proc. Natl. Acad. Sci. USA 115: 4453–4458.

Sternke, M., K. W. Tripp, and D. Barrick. 2019. Consensus sequence design as a general strategy to create hyperstable, biologically active proteins. Proc. Natl. Acad. Sci. USA 116: 11275–11284.

Storz, J. F. 2016. Causes of molecular convergence and parallelism in protein evolution. Nat. Rev. Genet. 17: 239–250.

Suckow, J., P. Markiewicz, L. G. Kleina, J. Miller, B. Kisters-Woike, and B. Müller-Hill. 1996. Genetic studies of the Lac repressor. XV: 4000 single amino acid substitutions and analysis of the resulting phenotypes on the basis of the protein structure. J. Mol. Biol. 261: 509–523.

Sullivan, B. J., T. Nguyen, V. Durani, D. Mathur, S. Rojas, M. Thomas, T. Syu, and T. J. Magliery. 2012. Stabilizing proteins from sequence statistics: The interplay of conservation and correlation in triosephosphate isomerase stability. J. Mol. Biol. 420: 384–399.

Tartaglia, G. G., S. Pechmann, C. M. Dobson, and M. Vendruscolo. 2007. Life on the edge: A link between gene expression levels and aggregation rates of human proteins. Trends Biochem. Sci. 32: 204–206.

Taverna, D. M., and R. A. Goldstein. 2002. Why are proteins marginally stable? Proteins 46: 105–109.

Toft, C., and M. A. Fares. 2010. Structural calibration of the rates of amino acid evolution in a search for Darwin in drifting biological systems. Mol. Biol. Evol. 27: 2375–2385.

Tokuriki, N., F. Stricher, J. Schymkowitz, L. Serrano, and D. S. Tawfik. 2007. The stability effects of protein mutations appear to be universally distributed. J. Mol. Biol. 369: 1318–1332.

Tokuriki, N., and D. S. Tawfik. 2009. Chaperonin overexpression promotes genetic variation and enzyme evolution. Nature 459: 668–673.

Toprak, E., A. Veres, J. B. Michel, R. Chait, D. L. Hartl, and R. Kishony. 2011. Evolutionary paths to antibiotic resistance under dynamically sustained drug selection. Nat. Genet. 44: 101–105.

Tóth-Petróczy, A., and D. S. Tawfik. 2011. Slow protein evolutionary rates are dictated by surface–core association. Proc. Natl. Acad. Sci. USA 108: 11151–11156.

Tripathi, A., K. Gupta, S. Khare, P. C. Jain, S. Patel, P. Kumar, A. J. Pulianmackal, N. Aghera, and R. Varadarajan. 2016. Molecular determinants of mutant phenotypes, inferred from saturation mutagenesis data. Mol. Biol. Evol. 33: 2960–2975.

van den Berg, P. A., A. van Hoek, C. D. Walentas, R. N. Perham, and A. J. Visser. 1998. Flavin fluorescence dynamics and photoinduced electron transfer in *Escherichia coli* glutathione reductase. Biophys. J. 74: 2046–2058.

Venev, S. V., and K. B. Zeldovich. 2018. Thermophilic adaptation in prokaryotes is constrained by metabolic costs of proteostasis. Mol. Biol. Evol. 35: 211–224.

Vieira-Silva, S., M. Touchon, S. S. Abby, and E. P. Rocha. 2011. Investment in rapid growth shapes the evolutionary rates of essential proteins. Proc. Natl. Acad. Sci. USA 108: 20030–20035.

Walsh, I. M., M. A. Bowman, I. F. Soto Santarriaga, A. Rodriguez, and P. L. Clark. 2020. Synonymous codon substitutions perturb cotranslational protein folding *in vivo* and impair cell fitness. Proc. Natl. Acad. Sci. USA 2020 117: 3528–3534.

Walter, K. U., K. Vamvaca, and D. Hilvert. 2005. An active enzyme constructed from a 9-amino acid alphabet. J. Biol. Chem. 280: 37742–37746.

Wang, M., C. G. Kurland, and G. Caetano-Anollés. 2011. Reductive evolution of proteomes and protein structures. Proc. Natl. Acad. Sci. USA 108: 11954–11958.

Weber, C. C., and S. Whelan. 2019. Physicochemical amino acid properties better describe substitution rates in large populations. Mol. Biol. Evol. 36: 679–690.

Wells, J. A. 1990. Additivity of mutational effects in proteins. Biochemistry 29: 8509–8517.

Wendler, K., J. Thar, S. Zahn, and B. Kirchner. 2010. Estimating the hydrogen bond energy. J. Phys. Chem. A. 114: 9529–9536.

Wheelan, S. J., A. Marchler-Bauer, and S. H. Bryant. 2000. Domain size distributions can predict domain boundaries. Bioinformatics 16: 613–618.

Windle, C. L., K. J. Simmons, J. R. Ault, C. H. Trinh, A. Nelson, A. R. Pearson, and A. Berry. 2017. Extending enzyme molecular recognition with an expanded amino acid alphabet. Proc. Natl. Acad. Sci. USA 114: 2610–2615.

Wolfenden, R., C. A. Lewis, Jr., Y. Yuan, and C. W. Carter Jr. 2015. Temperature dependence of amino acid hydrophobicities. Proc. Natl. Acad. Sci. USA 112: 7484–7488.

Wolynes, P. G. 2015. Evolution, energy landscapes and the paradoxes of protein folding. Biochimie 119: 218–230.

Worth, C. L., and T. L. Blundell. 2009. Satisfaction of hydrogen-bonding potential influences the conservation of polar side chains. Proteins 75: 413–429.

Worth, C. L., and T. L. Blundell. 2010. On the evolutionary conservation of hydrogen bonds made by buried polar amino acids: The hidden joists, braces and trusses of protein architecture. BMC Evol. Biol. 10: 161.

Wright, C. F., S. A. Teichmann, J. Clarke, and C. M. Dobson. 2005. The importance of sequence diversity in the aggregation and evolution of proteins. Nature 438: 878–881.

Wylie, C. S., and E. I. Shakhnovich. 2011. A biophysical protein folding model accounts for most mutational fitness effects in viruses. Proc. Natl. Acad. Sci. USA 108: 9916–9921.

Xie, V. C, J. Pu, B. P. Metzger, J. W. Thornton, and B. C. Dickinson. 2021. Contingency and chance erase necessity in the experimental evolution of ancestral proteins. eLife 10: e67336.

Yampolsky, L. Y., and A. Stoltzfus. 2005. The exchangeability of amino acids in proteins. Genetics 170: 1459–1472.

Yang, J. R., B. Y. Liao, S. M. Zhuang, and J. Zhang. 2012. Protein misinteraction avoidance causes highly expressed proteins to evolve slowly. Proc. Natl. Acad. Sci. USA 109: E831–E840.

Yang, J. R., S. M. Zhuang, and J. Zhang. 2010. Impact of translational error-induced and error-free misfolding on the rate of protein evolution. Mol. Syst. Biol. 6: 421.

Yang, Z. 2014. Molecular Evolution: A Statistical Approach. Oxford University Press, Oxford, UK.

Yeh, S. W., J. W. Liu, S. H. Yu, C. H. Shih, J. K. Hwang, and J. Echave. 2014. Site-specific structural constraints on protein sequence evolutionary divergence: Local packing density versus solvent exposure. Mol. Biol. Evol. 31: 135–139.

Zana, R. 1975. On the rate-determining step for helix propagation in the helix–coil transition of polypeptides in solution. Biopolymers 14: 2425–2428.

Zeldovich, K. B., P. Chen, and E. I. Shakhnovich. 2007. Protein stability imposes limits on organism complexity and speed of molecular evolution. Proc. Natl. Acad. Sci. USA 104: 16152–16157.

Zhang, J., and S. Kumar. 1997. Detection of convergent and parallel evolution at the amino acid sequence level. Mol. Biol. Evol. 14: 527–536.

Zhang, J., S. Maslov, and E. I. Shakhnovich. 2008. Constraints imposed by non-functional protein–protein interactions on gene expression and proteome size. Mol. Syst. Biol. 4: 210.

Zhang, J., and J. R. Yang. 2015. Determinants of the rate of protein sequence evolution. Nat. Rev. Genet. 16: 409–420.

Zhang, X. J., W. A. Baase, B. K. Shoichet, K. P. Wilson, and B. W. Matthews. 1995. Enhancement of protein stability by the combination of point mutations in T4 lysozyme is additive. Protein Eng. 8: 1017–1022.

Zhang, Y., I. A. Hubner, A. K. Arakaki, E. Shakhnovich, and J. Skolnick. 2006. On the origin and highly likely completeness of single-domain protein structures. Proc. Natl. Acad. Sci. USA 103: 2605–2610.

Zhao, J., A. J. Burke, and A. P. Green. 2020. Enzymes with noncanonical amino acids. Curr. Opin. Chem. Biol. 55: 136–144.

Zhou, T., W. Gu, and C. O. Wilke. 2010. Detecting positive and purifying selection at synonymous sites in yeast and worm. Mol. Biol. Evol. 27: 1912–1922.

Zou, Z., and J. Zhang. 2015. Are convergent and parallel amino acid substitutions in protein evolution more prevalent than neutral expectations? Mol. Biol. Evol. 32: 2085–2096.

CHAPTER 13

Multimerization

In all organisms, proteins commonly operate by binding to other proteins, either transiently or in more permanent structures. Here, we consider multimeric assemblages of polypeptide subunits held together at binding interfaces in a non-covalent manner. Virtually all biological pathways and cell structural features involve one or more multimers, called dimers when there are two subunits, trimers when there are three, tetramers when there are four, etc. (Figure 13.1). Higher-order complexes built out of lower-order multimers include flagella, nuclear-pore complexes, and centrosomes, all of which are discussed in separate chapters.

Before exploring the evolution of the basal building blocks of protein complexes, a brief overview of nomenclature will be useful. When the subunits of a multimer are encoded by the same genetic locus, the complex is referred to as a homomer. Complexes consisting of subunits from two or more loci are heteromers, the extreme situation being the case in which each subunit is encoded by a different genetic locus. When considering homomers, it is further useful to distinguish between those using isologous versus heterologous forms of interfaces. In the former case, both participants deploy the same surface patch in binding (aligning 'face-to-face', although generally rotated), whereas in the latter case, each binding partner utilizes different interface residues (aligning 'face-to-back'). By definition, all heteromeric interfaces are heterologous.

Odd-mers (e.g., trimers, pentamers) rely almost exclusively on heterologous interfaces, as these are required to create closed structures (Figure 13.2). Closed structures with just one type of heterologous interface are referred to as cyclical multimers. A special case is the domain-swapping homodimer, which has a geometric configuration allowing two symmetric heterologous interfaces to form a closed loop. In contrast, dihedral multimers have multiple axes of symmetry and require more than one interface type, with each usually being isologous (Figure 13.2).

How and why multimeric structures originate constitute two of the more significant issues in evolutionary cell biology. It is often assumed that the kinds of organized diversity embodied in multimers could only be a product of natural selection, with higher-order multimers sometimes being referred to as the 'end points' or 'pinnacles' of stepwise evolution. Should this be true, then an obvious question is why only a subset of lineages have been able to achieve such lofty heights. In fact, the distribution of multimeric states in prokaryotes is not much different from that in eukaryotes, and there are numerous examples of evolutionary reversions to lower oligomeric states. Moreover, although eukaryotes may have a higher incidence of heteromeric than homomeric structures relative to their orthologs in bacteria (Figure 13.3; Reid et al. 2010; Nishi et al. 2013; Marsh and Teichmann 2014), there is no compelling evidence that the former are functionally superior (e.g., Chapters 10, 20, and 24).

Sometimes a multimeric complex achieves a new function by virtue of special features associated with the interface itself. For example, there are numerous cases in which higher-order complexes assemble into containers (proteasomes, some chaperones, and cages for vesicles) or fibrils (actins and tubulins). However, with enzymes, it is frequently the case that all subunits retain their original monomeric functions, in which case, for example, a dimeric enzyme would simply have two

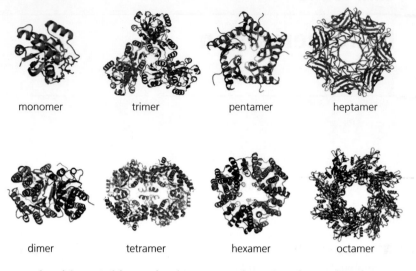

monomer   trimer   pentamer   heptamer

dimer   tetramer   hexamer   octamer

Figure 13.1 Some examples of the varied forms of multimeric complexes. In each case, the subunits are presented in different colours, although each subunit might be encoded by the same genetic locus (homomers).

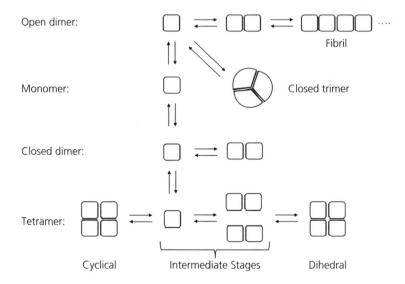

Figure 13.2 Structures of monomers, dimers, and tetramers for the case of homomeric proteins. Blue-blue and red-red denote isologous interfaces, and blue-red denote heterologous interfaces. For a closed dimer, with one isologous interface, there is only one possible topology. A dimer with a heterologous interface (top) is subject to the development of open fibrils, unless it is geometrically configured to form a closed loop, in which case an odd-mer (e.g., a closed trimer) would have to be produced (as an even-mer would necessitate one interface being isologous). With two interfaces, a tetramer can take on two possible topologies: cyclical when heterologous interfaces attract, and dihedral when isologous interfaces attract. The latter case is sometimes referred to as a dimer of dimers.

catalytic sites but no new function; this does not preclude subtle changes in the kinetic features of the monomeric subunits.

Because the extent to which multimeric structures are promoted or maintained by selection remains unclear, it is useful to think of them as 'biology's snowflakes', to remind us that beauty and diversity can often arise for purely physical reasons. We first consider the incidence of various multimeric forms of molecules across the Tree of Life, highlighting the apparent lack of association between molecular and organismal complexity. We then review some of the

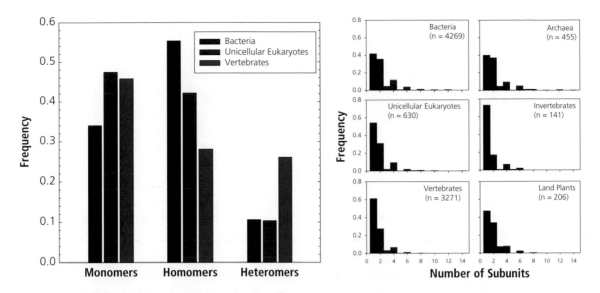

Figure 13.3 Left: Relative frequencies of the three main classes of protein structures in three phylogenetically broad groups. **Right:** Frequency distributions of levels of multimeric complexity (including monomers) for homomers for six major taxonomic groups; although the data are more limited, the distributions are similar for heteromers. The data are drawn from a diversity of proteins that have been structurally characterized in a broad range of organisms. From Lynch (2012).

biophysical considerations relevant to transitions between alternative oligomeric forms, showing that the paths to multimerization can be remarkably simple. Following this entrée into the biophysics of aggregation and the general features of known interfaces, we review the basic theory for understanding how multimeric architectures evolve.

The Incidence and Architectural Features of Multimers

There is astounding diversity in the higher-order structure of proteins. Surveys of known quaternary structures indicate that multimers comprise ~60% of all characterized proteins, with homomers being about twice as frequent as heteromers (Marianayagam et al. 2004; Levy et al. 2006; Lynch 2012). Because complex proteins are more difficult to characterize structurally, it is likely that the incidence of multimers is even greater than 60%. The vast majority of multimers are dimers, with a roughly exponential pattern of decline in frequencies with elevated numbers of subunits, and odd-mers tending to be under-represented relative to even-mers (Figure 13.3; Goodsell and Olson 2000; Mei et al. 2004). Strikingly, these distributions are essentially

independent of phylogenetic context, with those for prokaryotes being quite similar to those for vertebrates and land plants (Figure 13.3).

Although these distributions are potentially biased by the nature of proteins that have been characterized at the structural level, a comparison of orthologous metabolic proteins across taxa supports the conclusion that there is no gradient of molecular complexity (number of subunits) with organismal complexity. Moreover, within and among most major lineages, there is substantial variation in the multimeric states of the same proteins with the same functions (Reid et al. 2010; Lynch 2013). For example, it is not uncommon for the same enzyme to operate as a monomer in some bacteria, a dimer in others, and a tetramer in still others. Haemoglobins provide a textbook view of the extraordinary diversity of complexes that can exist (ranging from monomers to structures containing more than 100 subunits) (Shionyu et al. 2001). Levy et al. (2006) suggest that at the point of 30–40% amino-acid sequence divergence between orthologs, the probability of two species sharing quaternary structure drops to ~70%.

Numerous studies also reveal that orthologous proteins with the same numbers of subunits often

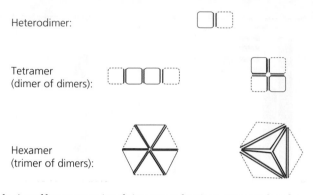

Heterodimer:

Tetramer
(dimer of dimers):

Hexamer
(trimer of dimers):

Figure 13.4 Structural topologies of heteromers involving two subunits per repeat (one bounded by black solid lines and the other by dashed lines). All interfaces in the structures shown here are heterologous, except the green-green pair for the tetramer on the left.

utilize different binding interfaces in different phylogenetic lineages. For example, in the case of glycosyltransferases (enzymes involved in the transfer of sugars to acceptor molecules), a dramatic diversity of monomeric, homodimeric, and homotetrameric states exists across the Tree of Life, with many of the independently evolved dimers utilizing different binding interfaces, sometimes even on opposite sides of the monomeric subunit (Hashimoto et al. 2011). Examples are also known of multimers in which some species use isologous and others use heterologous interfaces (Vassylyev et al. 2006). For proteins with known multimeric structures in more than one species, ~36% of cases exhibit variants with different interfaces, with ~4% of cases having more than five different binding modes across species (Dayhoff et al. 2010).

Ahnert et al. (2015) produced a 'periodic table' of multimer types based on the total number of repeating units, the number of subunit types (gene products) per repeat, and the number of ways in which these can assemble into multimers. Figure 13.4 shows just a small set of the possible topologies for heteromers, with two to three repeats and two subunit types per repeat. When there are four or more repeats, the final complexes can be cyclical or dihedral in form. For example, a cyclic homomeric tetramer has just one type of face-to-back interface repeated four times in a closed loop, whereas a dihedral homomeric tetramer utilizes two different isologous interfaces twice (Figure 13.2). If there are two subunits per repeat and three repeats, the

resultant structure is a hexamer, which can take on two configurations (Figure 13.4). A wide variety of higher-order topologies beyond those outlined in Figures 13.2 and 13.4 are known to exist in biology (Ahnert et al. 2015), but whenever there exists the possibility of either a dihedral or cyclical closed structure (as with tetramers; Figure 13.2), dihedral forms are ~10× more frequent than cyclical forms (Levy et al. 2008). This suggests a greater ease of evolution towards isologous binding interfaces, a point discussed further below.

For heteromeric even-mers beyond dimers (e.g., tetramers and hexamers), there is the possibility of uneven stoichiometry from the contributing proteins (e.g., a tetrameric complex might be AAAB as opposed to AABB). Uneven stoichiometry is thought to be nearly twice as common in bacteria as in eukaryotes, and the incidence of unevenness increases with the number of contributing proteins, for example ~10% when just two or three proteins participate, but ~30% when there are five or more contributing proteins (Marsh et al. 2015).

Propensity to Aggregate

The reliance on multimers imposes costs and risks upon cells. To achieve multimeric architectures, cellular concentrations of monomeric subunits must be kept at sufficiently high levels to allow reasonable encounter and aggregation rates between partner molecules. Because proteins come at a biosynthetic cost, this may then entail an excess investment that would otherwise not be required for monomers. If, on the other hand, multimers are less subject to degradation, the overall cost of monomer production could be mitigated. However, an additional issue is that adhesive interfaces can promote promiscuous interactions with non-cognate molecules and/or lead to the runaway production of harmful self-aggregates, including fibrils.

The proximity of proteins to the edge of misaggregation propensity is demonstrated by experimental work in *E. coli* showing that single amino-acid substitutions to hydrophobic surface residues often shift proteins into multi-molecular states (Garcia-Seisdedos et al. 2017). In a well-known case in human biology, a single glutamate-to-valine substitution in the haemoglobin molecule induces the fibril formation that causes sickle-cell anaemia. Experimental observations in yeast suggest that promiscuous binding leads to ~25% of proteins being at least transiently bound to inappropriate partners (Zhang et al. 2008).

The clear implication here is that any involvement of natural selection in the evolution of higher-order protein structures must include not just the enforcement of productive self-binding but the elimination of surface features that promote harmful structures (Zabel et al. 2019). Mechanisms for reducing problems with inappropriate aggregation may include the spatial configuration of adhesive residues on polypeptide chains in ways that influence the relative rates of monomer folding and subunit aggregation. For example, locating interface residues on the C-terminal end of a protein (the last to emerge from the ribosome) reduces the likelihood of premature aggregation of partially folded proteins (Natan et al. 2018; Gartner et al. 2020). The spatially restricted nature of translation also reduces the likelihood of off-target binding. For example, co-translation of mRNAs by groups of ribosomes (so-called polysomes jointly processing the same mRNA) creates a situation in which monomers are initially co-localized at substantially higher concentrations than expected under a random intracellular distribution, increasing the association rate, and sometimes even leading to co-translational assembly (Wells et al. 2015). In the case of bacterial operons, spatial clustering even allows for enhanced assembly of heteromers (Wells et al. 2016).

Theory of association

Acknowledging that there are many underlying molecular factors, we now consider in the simplest and most general terms the principles underlying the rate and stability of molecular aggregation, building on the concept of molecular diffusion introduced in Chapter 7. After emerging from a ribosome, amino-acid chains will initially be monomeric in form, and will retain that state until encountering a binding partner. Even then, the maintenance of an appropriate multimeric form requires sufficient binding energy across the interface. In principle, given steady-state conditions (e.g., a constant concentration of protein per unit cell volume), an equilibrium will be reached within the cytoplasm reflecting equal rates of association into multimers and dissociation of the latter back to monomers. In this sense, few proteins can be viewed as having a single form, and when one is referred to, this should simply be viewed as the dominant phase.

The equilibrium distribution of multimeric forms within a cell depends on at least three factors: 1) the cellular concentration of proteins, which dictates the encounter rate of monomeric subunits; 2) the rate of particle movement and binding, which dictates the assembly rate; and 3) the stability of the binding interface, which dictates the longevity of a newly formed multimer. The basic principles can be understood by focusing on a system allowing for just monomers and dimers.

Owing to the destabilizing forces of molecular motion, two subunits will have a thermodynamically determined probability of being in complex depending on the rates of association of monomers and dissociation of dimers. Letting [A] and [AA] denote the cellular concentrations of solo A molecules and AA complexes, the fraction of complexes versus singletons is

$$p_{AA} = \frac{[AA]}{[A] + [AA]}. \tag{13.1a}$$

By multiplying the numerator and denominator by [A]/[AA], the expected fraction of dimers at equilibrium can then be expressed as

$$p_{AA} = \frac{[A]}{K_D + [A]}, \tag{13.1b}$$

where

$$K_D = \frac{[A][A]}{[AA]} \qquad (13.2a)$$

is the equilibrium dissociation constant, which depends on the strength of binding (Foundations 13.1).

Written in this way, K_D is seen to be equivalent to the cellular concentration of monomers at which aggregation leads to equivalent concentrations of dimeric and monomeric complexes, i.e., $p_{AA} = 0.5$ when $[A] = K_D$, which requires $[AA] = [A]$. Note that $[A]$, $[AA]$, and K_D all have the same units of concentration, usually written below as μM (where $1\ \mu M = 10^{-6}$ M) because cytoplasmic protein concentrations are typically in the μM range (Chapter 7). In a more mechanistic sense, K_D is equivalent to the ratio of dissociation and association rates, where the latter can be expressed in terms of the random encounter rate (Chapter 7) and the probability of proper binding (Foundations 13.1). Because the rate of encounter is a function of the molecular concentration, proteins with lower binding affinities are expected to require higher cellular concentrations to achieve an effective level of complexation.

To put Equation 13.2a into a more biophysical perspective, the dissociation constant can be written in energetic terms as

$$K_D = e^{-\Delta E/RT}, \qquad (13.2b)$$

where ΔE denotes the excess energy required for dissociation (in kcal/mol), R is the Boltzmann constant K_B (encountered in previous chapters) scaled up to the molar equivalent, and RT denotes the standard background energy associated with Brownian molecular motion (where for most biological temperatures, $RT \simeq 0.6$ kcal/mol) (Foundations 13.1). Thus, as the binding energy across the interface between two subunits increases, K_D asymptotically approaches zero and the probability of complexation approaches 1.0.

We emphasize that the quantitative value of the dissociation constant is context dependent. As defined in Equation 13.2b, K_D has a rather precise meaning from a pure chemistry perspective, and is typically recorded in an aqueous solution containing only A molecules. Written in the form of Equation 13.1b, however, K_D can be thought of in much more general terms, as a simple indicator of the degree of affinity of two A molecules for each other in a system of arbitrary complexity.

Within the cellular environment, which can contain thousands of proteins encoded by other loci, a monomeric subunit is confronted not just with the challenge of adhering to its binding partner, but also with the additional problem of avoiding promiscuous engagement with non-cognate molecules. Assuming that the molecule of interest is sufficiently adhesive that it is essentially always in complex with either a self or foreign molecule, the methods in Foundations 13.1 can be used to derive a modified expression for the fraction of molecules bound up in appropriate homodimeric complexes. Maintaining the structure of the preceding formula,

$$K_D' = K_D \cdot \phi e^{\Delta E'/(RT)}. \qquad (13.3)$$

Here, ϕ is the effective concentration of foreign proteins with a capacity for promiscuous binding, and $\Delta E'$ is the strength of binding associated with promiscuous liaisons. In effect, the terms to the right of K_D amount to a weighting factor that increases the level of dissociation as a function of background interference. Because ϕ can easily be on the order of 10 or greater, $\Delta E'$ must be much smaller than RT if a high incidence of engagement in non-productive (and possibly harmful) complexes is to be avoided.

We shortly discuss that a typical value of ΔE for a binding interface is 10 kcal/mol, which implies $K_D = e^{-10/0.6} \simeq 0.058\ \mu M$. Under these conditions and assuming no promiscuous binding, Equation 13.1b indicates that, when equilibrium monomeric concentrations are 0.1, 1.0, and 10.0 μM, the expected fractions of dimers are $\simeq 0.63$, 0.94, and 0.99, respectively. Supposing, however, that $\phi = 10$ and that $\Delta E'/(RT)$ is 1 kcal/mol, then K_D is modified to $K_D' \simeq 1.6\ \mu M$, and for equilibrium monomeric concentrations of 0.1, 1.0, and 10.0 μM, the fraction of dimers is now reduced to 0.06, 0.38, and 0.86 respectively, owing to the fact that many monomers will be sequestered in inappropriate complexes.

The physical features of interfaces

Geometric constraints generally ensure that the interface between two globular molecules constitutes only a moderate fraction of the total surface area. Direct observations from a large number of proteins suggest a range of ~5 to 30%, with a weak positive linear scaling between interface size and total monomeric surface area (Chothia and Janin 1975; Jones and Thornton 1996; Bordner and Abagyan 2005; Lynch 2013). Whether such scaling is a simple consequence of geometry or a result of larger proteins requiring larger interfaces for stabilization remains unclear.

Although little work has been done on the adhesive features of homomeric interfaces, surveys of a diverse set of transient protein–protein interactions (e.g., antigen-antibodies and enzyme-inhibitors from a variety of taxa; Horton and Lewis 1992; Jones and Thornton 1996; Bogan and Thorn 1998; Kastritis et al. 2011) lead to two conclusions (Figure 13.5). First, the average interfacial binding strength is ~18RT with a standard deviation of 5RT (i.e., about 18× greater than background thermal energy). We can expect the binding strength of more permanent multimers to be somewhat greater, and

Brooijmans et al. (2002) suggest an average of 22RT for homodimer interfaces. Second, there is only a weak relationship between binding strength and total interface size, with the minimum observed interface area being ~1000 Å^2 (equivalent to 10 nm^2) and the minimum binding energy being ~6 kcal/mol, i.e., 10RT. Drawing from a diverse set of observations, Bahadur et al. (2003) find an average interface area of ~4000 Å^2 for homodimers and ~2000 Å^2 for hetero-complexes.

An approximate mechanistic argument for the lower limit to a functional interface area can be derived as follows (Day et al. 2012). From Figure 13.5, it can be seen that a rough upper limit to the binding strength per Å^2 is ~0.012 kcal/mol. For two proteins to have any affinity at all, the total strength of the interaction must be at least equal to the energetic cost of simply holding the two interfaces in a particular orientation, which is on the order of 8.0 kcal/mol (Janin 1995; Zhang et al. 2008; Day et al. 2012). The minimum interface size necessary to cover this cost, ~8.0/0.012 = 667 Å^2, is also in accord with Figure 13.5. Taking an average binding strength of a homodimeric interface to be 22RT, Equation 13.2b then implies $K_D \simeq 0.3$ nM, which is

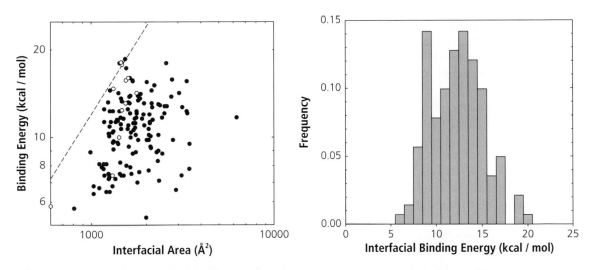

Figure 13.5 Binding strengths associated with interfaces of heterodimeric complexes. **Left:** The weak relationship between binding energy of a complex and the area of the interface. Solid circles (Kastritis et al. 2011); open circles (Horton and Lewis 1992). The dashed line denotes an approximate upper limit to the binding energy of ~0.012 kcal·mol^{-1}·Å^{-2}, where Å denotes Angstroms (equivalent to 0.1 nm). **Right:** An approximately normal distribution of interface binding strengths, with mean = 18.3RT and SD = 4.8RT, where RT is the energy associated with background thermal motion (Foundations 13.1). Data are extracted from a wide variety of sources (summarized in Kastritis et al. 2011).

of the order of magnitude of protein concentrations often seen in cells (Chapter 7).

The binding properties noted here are generally functions of no more than a few dozen amino-acid residues (Chothia and Janin 1975). For example, surveys over diverse organisms suggest means of ~80–100 for homodimer interfaces and 50–60 for heterodimers, with strong asymmetries in the distributions yielding lower modes than means (Bahadur et al. 2003; Zhanhua et al. 2005), such that most interfaces are in the range of 10 to 40 (Bordner and Abagyan 2005). However, these total sizes exaggerate the number of actual residues involved in adhesion, as the total binding energy is usually concentrated in a small number of hot spots, with < 10 residues typically involved (Bogan and Thorn 1998; Hochberg et al. 2020; Pillai et al. 2020).

Binding can come about in multiple ways, although utilization of hydrophobic residues and hydrogen bonds are the predominant factors. Interface hydrophobicity is generally intermediate to that of interior cores and exterior surfaces of proteins, with a typical enrichment of the more hydrophobic and polar uncharged residues, most notably F, C, L, M, I, Y, W, and V (Bordner and Abagyan 2005; Levy 2010). A summary from diverse proteins and organisms indicates that the average numbers of hydrogen bonds are ~11 and 18 for heterodimer and homodimer interfaces, respectively, with ~0.2 per interface residue (Jones and Thornton 1995; Bahadur et al. 2003; Zhanhua et al. 2005).

Why are homomeric structures so common? Many homomers utilize isologous interfaces rotated 180° with respect to each other, a feature that may reflect a simple geometric 'two for the price of one' argument made by Monod et al. (1965). Supposing that residue A binds with a complementary residue B on a parallel flat interface, then rotating around this affinity pair by 180° will yield a complementary B–A match, thereby doubling the strength of binding with just a single mutational change. Extensions of this argument lead to the conclusion that extant proteins have an innate propensity for self-assembly relative to random surfaces (Ispolatov et al. 2005; Lukatsky et al. 2007; André et al. 2008). The flip-side of this view is that an isologous interface is also expected to have a 50% reduction in the mutation rate to binding residues (because the mutational target size is reduced by 50%), and that any mutation eliminating a binding residue will have double the effect (because two binding pairs of sites are lost simultaneously). Thus, it remains unclear whether Monod's argument is a sufficient explanation for the excess abundance of isologous interfaces (André et al. 2008; Plaxco and Gross 2009).

Notably, the structural evolution of multimeric proteins involves more than the conjoining interfaces. For example, Marsh and Teichmann (2014) found that the subunits of multimeric assemblies are significantly more flexible than proteins operating exclusively as monomers. The average flexibility of subunits increases with the number of nonhomologous subunits in a heteromeric complex, and as a protein complex acquires new subunits over time, the consecutive additions tend to be increasingly flexible (Marsh and Teichmann 2014). One interpretation of these patterns is that flexibility facilitates the conformational changes necessary for successful binding. There are, however, open questions of cause versus effect here, as selection for monomer rigidity may be relaxed in a protein that is secondarily stabilized by a binding partner.

As in the case of folding stability (Chapter 12), once a highly refined level of binding efficiency has evolved, the fitness advantages of further improvement are expected to become increasingly negligible, rendering a situation in which excess capacity for binding remains unutilized. For studies in which alanine residues are individually substituted for the amino-acid constituents of interfaces, the effects on binding strength typically have distributions in which the highest density is near zero, with ~10% having slightly positive effects on binding (Bogan and Thorn 1998). A number of examples also exist in which multimers have been engineered to have stronger interfaces and increased thermostabilities (Griffin and Gerrard 2012), consistent with the hypothesis of excess capacity.

Given that there are typically unutilized residues at most binding interfaces, these observations suggest that a large number of alternative and nearly energetically equivalent amino-acid compositions exist at individual interfaces. This further implies the potential for considerable interface sequence wandering over evolutionary time (Figure 13.6).

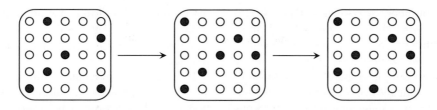

Figure 13.6 Effectively neutral evolution of new binding configurations at an interface patch. There are 25 potential binding residues in the patch, but at any point in time here, there are only six adhesive residues (in red). Over time, the specific residues involved in binding are free to wander over the surface, so long as six remain actively engaged. In principle, the binding number could wander above six by mutation and drift, although here it is assumed that six are sufficient to confer maximum fitness.

Consistent with this view, amino-acid sequences at interfaces do not evolve at unusually slow rates – on average not more than 50% more slowly, and sometimes slightly more rapidly, than residues on exposed surfaces and in internal cores (Grishin and Phillips 1994; Caffrey et al. 2004; Bordner and Abagyan 2005; Mintseris and Weng 2005). Over time, this combination of multiple degrees of evolutionary freedom and the diminishing fitness advantages of increased binding can lead to effectively neutral interface evolution. The eventual result of such divergence is a situation in which the interfaces on molecules derived from different phylogenetic lineages evolve to become non-functional in a cross-species constructs, owing to the lack of cross-recognition.

The few attempts to evaluate heterologous compatibilities have led to mixed results, but interpretations are made difficult by the fact that most examples focus on constructs where the catalytic function is built from the interface (which will reduce the opportunities for divergence). Cross-species molecular hybrids of thymidylate synthase from *E. coli* and *Lactobacillus casei*, two very distantly related bacteria, appear to be fully functional (Greene et al. 1993). Functionality has also been demonstrated for hybrid ornithine decarboxylase dimers between the monomeric subunits of *Trypanosoma brucei* and mouse, whereas cross-species dimers between *T. brucei* and *Leishmania donovani* (both trypanosomes) are non-functional (Osterman et al. 1994). For triose phosphate isomerase, cross-species hybrid dimers involving mammals, chicken, and yeast have lower levels of catalytic efficiency and enzyme stability than within-species dimers (in this case, the catalytic

site is not created by the interface) (Sun et al. 1992). Careful comparative analyses of this sort, involving a gradient of phylogenetic relationships, and extended to situations in which catalysis is independent of interface binding, will be essential to furthering our understanding of how interfaces evolve.

Evolutionary Considerations

Given the penchant for most biologists to assume that virtually every aspect of biology owes its origin to natural selection, it will come as no surprise that most attempts to explain the existence of multimers start with this implicit assumption (e.g., Monod et al. 1965; Goodsell and Olson 2000; Marianayagam et al. 2004; Mei et al. 2004; Hashimoto et al. 2011; Griffen and Gerrard 2012). The proposed advantages are diverse. First, as noted in Chapter 12, it is generally easier to fold multiple small proteins than a single long one, and cases are known in which multimerization can actually enhance the folding rates and stability of the monomeric subunits (Zheng et al. 2012).

Second, the encounter rate of an enzyme and a small substrate is proportional to the effective radius of the enzyme (Chapter 7), and the elimination of extraneous protein surface may further enhance the frequency of productive encounters between catalytic sites and their substrates. However, these considerations need to be tempered by the fact that multimerization reduces the number of enzymatic complexes subject to diffusion. For example, a dimer with double the volume of a monomer would be expected to have a radius (and hence, encounter rate) $\simeq 2^{1/3} \times$ that of the monomer but half the number of particles, which would reduce the encounter rate by a factor of $2^{-2/3}$, unless there were additional favorable factors involved.

Third, multimerization may have secondary advantages. For example, a smaller surface area-to-volume ratio might reduce a protein's vulnerability to denaturation or engagement in promiscuous interactions (Bershstein et al. 2012). Higher-order structures might also reduce the sensitivity of catalytic sites to internal motions, thereby increasing substrate specificity. In addition, complexation offers increased opportunities for allosteric regulation of protein activity (with structural changes in one subunit induced by substrate binding being transmitted to and altering the catalytic capacity of another).

Finally, multimerization can sometimes lead to the creation of an entirely new function by inducing secondary structural changes that alter the catalytic site. For example, members of the archaea have a CS_2 anhydrase that converts CS_2 into H_2S by a process similar to carbonic anhydrase's conversion of CO_2 to HCO_3, but whereas the latter is a monomer, the former has a hexadecameric architecture that prevents access of CO_2 to the catalytic site (Smeulders et al. 2013).

Weighted against these potential advantages, one must also consider the negative side effects that can result from proteins with a tendency to multimerize. First, unless a newly emerging dimer has a single isologous interface, concatenations into indefinite fibrils can arise. Human disorders involving the production of amyloid fibrils, such as Alzheimer's and Parkinson's diseases, amyotrophic lateral sclerosis, and sickle-cell anaemia, are prominent examples of the negative consequences of overly adhesive proteins (Chiti and Dobson 2009; Jucker and Walker 2013).

Second, as noted above, to achieve a critical concentration of an active multimer, the abundance of monomeric subunits must be raised to a high enough level to ensure an adequate number of encounters for successful complex assembly. Any increase in subunit production will entail an energetic cost, and highly expressed proteins are also vulnerable to promiscuous interactions (Semple et al. 2008; Vavouri et al. 2009). If, however, multimeric proteins are less vulnerable to

thermo-instabilities and degradation, multimerization might actually result in a lower energetic demand, as the cost of protein replacement is reduced.

Third, even if a particular multimeric structure is advantageous, it need not follow that a new mutation to such a form can be easily promoted by selection, as the fixation process critically depends on the background conditions in which the mutation first appears. As we will discuss below for diploid species, some dimerizing mutations may have deleterious effects when combined with ancestral monomeric proteins in heterozygous individuals (where potentially detrimental heterotypic complexation can occur). Because new mutations in diploid species are always present on a heterozygous background, such a process has the potential to greatly reduce the probability of establishment of a dimerizing allele.

In summary, although the large pool of multimeric structures in today's organisms cannot possibly be strongly maladaptive, this need not imply that they have arisen or are currently maintained by adaptive processes. Indeed, despite the plausibility of many of the mentioned hypotheses, empirical evidence for the adaptive value of alternative multimeric structures is near non-existent. Moreover, a number of examples can be pointed to in which a more complex structure seemingly operates no more efficiently in its lineage than a simpler structure in others.

For example, the mismatch-repair system, which plays important roles in replication fidelity, DNA repair, and recombination, is comprised of monomeric proteins in bacteria but dimers in eukaryotes (Kunkel and Erie 2005; Iyer et al. 2006), yet the repair efficiency of eukaryotic systems appears to be lower than that in prokaryotes (Lynch 2011). The overall fidelity of amino-acid loading of class I tRNA synthetase enzymes (which load specific amino acids onto cognate tRNAs) tends to be much greater than that of the class II enzymes, despite the fact that the former operate as monomers and the latter as dimers (Freist et al. 1998). The sliding clamps used in DNA replication are homodimeric in bacteria but homotrimeric in eukaryotes, with both structures having very similar overall architecture (Kelman and O'Donnell

1995), yet replication-fork progression rates are nearly an order of magnitude faster in prokaryotes (Chapter 20). A more practical example involves the insulin hormone, which normally operates as a homohexamer in humans, but has been engineered to become a monomer by incorporating just one or two amino-acid changes in the interface (Brange et al. 1988). In the treatment of diabetes, the monomers are absorbed much more rapidly than the multimers and have equivalent efficacy, although they can sometimes lead to the production of amyloid fibres at the site of injection. Finally, although the ribosome has a much more complex protein repertoire in eukaryotes than in prokaryotes (Chapter 6), there is no evidence that translation fidelity is elevated in the former (Chapter 20).

Motivated by these observations, we now consider the more formal evolutionary theory essential to understanding how the evolution of alternative multimeric structures proceeds. The focus is primarily on situations in which the gene function is independent of the number of subunits in the complex, i.e., protein function is not a structural outcome of the interface. Quite different outcomes are expected when the interfaces between subunits are essential to function, as this strongly constrains the acceptability amino-acid substitutions (Abrusán and Marsh 2018). Although the basic theoretical issues appear clear, we are still a long way from formally testing the theory, which will require a combination of comparative analysis and experimental work on orthologous proteins across diverse phylogenetic taxa.

Transitions from monomeric to higher-order states

In one of the few experimental studies of the causes and consequences of dimerization, Hochberg et al. (2020) compared the features of oestrogen and steroid receptors. These hormone-activated transcription factors diverged following duplication of an ancestral gene at the base of the chordate lineage, but the former operate as dimers, and the latter as monomers. By comparing the amino-acid sequences from diverse species, the authors were able to predict, reconstruct, and evaluate the functional properties of ancestral sequences. The key

finding was that the alternative functional features of monomeric and dimeric family members are likely not a consequence of multimerization per se, but that following its initial establishment, the dimeric form gradually became more entrenched into this structural form as mutations to hydrophobic residues accumulated in the interface.

This proposed scenario is very similar to the kind of evolutionary trajectory postulated under the constructive neutral-evolution model (Chapter 6), which envisions higher-order complexes arising following the fortuitous appearance of effectively neutral interfaces, which in turn confer a permissive environment for the downstream accumulation of mutations that are only deleterious if exposed on the surface of a molecule. What remains to be achieved, however, is the ground-truthing essential to any evolutionary hypothesis – an understanding of the underlying population-genetic processes essential to this kind of transition from monomers to dimers (and vice versa).

Guided by knowledge on the physical features of interfaces mentioned above, we start with a relatively simple null model in which the architectural features of a homomeric protein can be described as a linear series of alternative structural states, with increasing binding strength of a potentially dimeric interface. Each class has a finite probability (per unit of evolutionary time) of transitioning to an adjacent class by gaining or losing a binding residue (Figure 13.7). The alternative molecular phenotypes range from the extreme case of a pure monomer (state $i = 1$) through a series of states ($i > 1$) in which the dimeric interface becomes increasingly stable owing to the establishment of adhesive amino acids.

Under this model, over evolutionary time, the molecular state will wander across the landscape of alternative phenotypes to a degree that depends on the evolutionary transition rates between states. However, provided the transition possibilities are bidirectional, given enough time and constant rates, a quasi-steady-state distribution is expected to emerge. At this point, the long-term average evolutionary rates of movement into and out of each class are equal, which further implies that transitions towards a more dimeric state occur at exactly the same rate as transitions towards a more monomeric state. Such a long-term steady-state

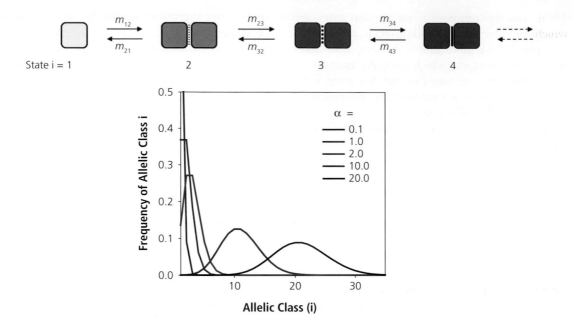

Figure 13.7 An idealized model for evolution along a linear array of alternative quaternary states of a protein. Here, the states describe a series of molecular variants for a homodimer with increasing numbers of binding sites at the interface, so that the state on the left is essentially a pure monomer, and moving to the right, the states have progressively stronger binding interfaces. The coefficients m_{ij} are evolutionary rates of transition from state i to state j, the values of which will depend on the power of mutation, random genetic drift, and selection, which together govern the composite parameter α in Equation 13.4. In effect, α is a measure of the net pressure from mutation and selection in the forward direction, as described in the text. The lower panel gives the steady-state probability distributions of states (i) for different values of α for the model described in the text.

condition results from the fact that more abundant states are so because they have lower rates of export to adjacent states, whereas less abundant states have higher rates of export. Given a constant set of conditions, the resultant steady-state distribution describes both the long-term expected history of a protein in a particular phylogenetic lineage and the diversification across independent lineages.

A useful feature of this model is that the equilibrium probability of being in any particular state is proportional to the product of all of the rate coefficients pointing towards the state from both the upward and downward directions (Foundations 5.3). In addition, these transition coefficients have a relatively simple interpretation, with each being equivalent to the product of the number of mutations arising in the population per generation and the probability of fixation, which in turn depend on the directional powers of mutation and selection

relative to the stochastic force of random genetic drift.

Here, to illustrate the key points in as simple a manner as possible, we assume that mutations destined to fixation arise infrequently enough that populations are almost always in a pure state of one form or another. We also assume a constant upward mutation rate between classes, i.e., $\mu_{i,i+1} = u$ for all i, as would be closely approximated if the number of potential surface residues that can mutate to adhesive states is large. Letting v be the mutational rate of loss of an adhesive residue, the downward mutation rate must increase linearly with the allelic state, i.e., $\mu_{i,i-1} = (i-1)v$, owing to the increase in the number of adhesive residues at the interface subject to loss with increasing i. As u is an aggregate mutation rate over a large number of sites, and v is a per-site mutation rate, we expect u/v to be > 1, and perhaps greatly so.

The logical starting point is a neutral model in which each alternative state (i) has equivalent fitness (i.e., all dimers operate with equal efficiencies as each other and as monomers). Three features are immediately apparent under such conditions. First, because each adjacent pair of states is separated by a pair of upwards and downwards coefficients and all fixation probabilities are equal under neutrality (Chapter 4), the steady-state distribution depends only on the ratios of mutation rates (i.e., not on their absolute values). Second, because such ratios decrease with increasing numbers of binding sites, the equilibrium frequencies of the higher states must eventually diminish toward zero, ensuring the existence of a quasi-steady-state probability distribution of the array of possible states. Third, because the number of mutations arising each generation is a function of the population size, and the probability of fixation equals the inverse of the population size (again assuming neutrality), the transition rates are entirely independent of population size. From Foundations 5.3, the equilibrium probability distribution simplifies to

$$\widetilde{P}_i = \frac{e^{-\alpha}\alpha^{i-1}}{(i-1)!}. \tag{13.4}$$

This is a Poisson probability distribution, with a single parameter equal to the ratio of mutation rates, $\alpha = u/v$.

Under neutrality, the probability of the extreme monomeric state ($i = 1$) is simply $\widetilde{P}_1 = e^{-u/v}$, which with $u/v = 0.01, 0.1, 1.0$, and 10.0 becomes $\widetilde{P}_1 = 0.99, 0.90, 0.37$, and 0.000045, respectively. Thus, even in the absence of adaptive differences among allelic states, the probability of a dimeric structure $(1 - \widetilde{P}_1)$ can be substantial. In addition, because the variance of a Poisson distribution is equal to the mean, for large u/v substantial variation in interface binding strength is expected among lineages (Figure 13.7), consistent with the observations mentioned above regarding variation for interfacial areas and binding strengths. Finally, given the population-size independence of these results, assuming the mutation ratio u/v is reasonably constant across phylogenetic lineages, this neutral model predicts that the probability distributions of monomers and dimers should be independent of phylogenetic context, i.e., approximately the same

in bacteria as in multicellular eukaryotes, which is consistent with the observations in Figure 13.3.

What are the consequences of selection towards one extreme or the other? Because the probability of fixation depends on the effective population size (N_e) when selection operates, we can anticipate a shift in the form of the distribution from the neutral expectation as the efficiency of selection increases with increasing N_e. The key modification that must be made is the weighting of the mutation pressure in each transition coefficient by the probability of fixation of the mutant allele, which is no longer the initial frequency.

There are numerous ways in which fitness might change over the phenotypic array displayed in Figure 13.7. Nonetheless, regardless of the fitness function, for each pair of upward and downward coefficients, the ratio of fixation probabilities in the two directions is equal to $e^{2N_e s}$ assuming haploidy (and $e^{4N_e s}$ assuming diploidy), where s is the relative fitness difference between the two adjacent states (Foundations 5.3). If one assumes weak positive selection in the direction of increasing i, such that the difference in selective advantage between adjacent states remains constant, Equation 13.4 still holds but with $\alpha = (u/v)e^{2N_e s}$, i.e., the Poisson distribution is maintained, except that the mutation-pressure ratio is multiplied by the selection-pressure ratio. If the prevailing selection pressure is in the direction of the monomeric state, then $\alpha = (u/v)e^{-2N_e s}$. Thus, the ratio of the power of selection to the power of drift, $s/(1/N_e) = N_e s$ simply enters as an exponential weighting factor, pushing the system further in the favored direction than would be expected on the basis of mutation pressure alone. If the power of drift is substantially greater than that of selection, $1/N_e \gg s$, then $N_e s \ll 1$ and $e^{2N_e s} \simeq 1$, and the distribution of state frequencies is still closely approximated by the neutral expectation.

This model can accommodate any alternative pattern of fitness variation among sites. For example, weakly dimerizing states might create structural defects not experienced by monomers, which could induce a sign change in s with increasing i. Moreover, as noted above, one could argue that any incremental selective advantages with increasing numbers of binding residues will progressively decline, owing to the non-linear relationship between binding energy and the dissociation constant. Any such modifications can be implemented readily through appropriate changes to the selection-pressure weighting terms, but would invariably lead to situations in which the long-term evolutionary distribution is no longer exactly Poisson.

The key point is that, regardless of the selection scheme, the model outlined in Figure 13.7 is sufficiently general to provide mechanistic insight into the driving forces leading to the so-called entrenchment of multimeric states caused by the accumulation of multiple binding residues. Indeed, the theory suggests that entrenchment is a natural expectation in the sense that the cumulative fixed mutations that promote a stronger interface will result in substantial structural defects if all are simultaneously exposed. This would happen, for example, in an experimental manipulation that suddenly exposed an entire hydrophobic patch on the interface (in contrast to more natural single amino-acid substitutions that are more likely to occur in evolution, as envisioned in Figure 13.7).

The basic structure of this model can also be extended to the secondary evolution of homotetramers from homodimers (Lynch 2013). Notably, dimers appear to often constitute the initial step in the evolution of higher-order multimers. For example, Levy et al. (2008) found that whenever a tetramer in some lineage is related to a dimer in another, the dimer interface is usually conserved in the tetramer. This observation motivates the suggestion that tetramers typically evolve via an intermediate dimeric state, with the dominant dimeric interface evolving first (Levy et al. 2008; Dayhoff et al. 2010), followed by the joining of a potentially weaker interface to form a complex of four. Consistent with this view, the order of assembly events for higher-order multimers within cells appears to reflect the postulated order of evolutionary emergence of different interfaces (Marsh et al. 2013; Marsh and Teichmann 2015), a putative case of ontogeny recapitulating phylogeny. Nonetheless, given the steady-state distributional properties noted, there is need for care in interpretation here, as the sharing of interfaces between orthologous dimers and tetramers could in many cases reflect reversion of the latter to the former. This caveat is motivated by the theoretical expectation that,

under a steady-state situation, the net evolutionary flux rate from dimer to tetramer is equal to that in the opposite direction.

Regardless of the exact details of the model, two central points emerge from the preceding analyses, relevant not just to the issue of quaternary protein structure but also to the distributions of all complex traits. First, as already noted for the case of neutrality, substantial phenotypic variation can arise among lineages experiencing identical intensities of selection, demonstrating how risky it can be to assume adaptive explanations for phenotypic divergence among lineages. For example, if the composite parameter $\alpha = 1$, the probability of being in the pure monomeric state is 0.37, and that of being in the remaining (at least partially) dimeric categories is 0.63. Thus, as substantial phenotypic diversity can exist among phylogenetic lineages confronted with identical evolutionary pressures, any attempt to explain these differences in terms of imagined lineage-specific selection pressures might be quite misplaced.

Second, the most common state is not necessarily the optimum. Even with negative selection against multimers, they will still be common provided the mutational bias towards binding affinity is sufficiently large. The mode of the distribution is entirely determined by the composite parameter α, and if $2N_e s \ll 1$, the prevailing molecular phenotype will be essentially defined by the mutation spectrum. Notably, mutation pressure in many phylogenetic lineages is biased toward the production of A/T nucleotides (Hershberg and Petrov 2010; Hildebrand et al. 2010; Lynch 2010; Long et al. 2018; Chapter 12). This encourages a bias toward more hydrophobic, and hence, more adhesive, surface residues (which, owing to the nature of the genetic code, have more A/T-rich codons) (Knight et al. 2001; Bastolla et al. 2004; Hochberg et al. 2020; Table 12.1).

In summary, the preceding analyses suggest that the substantial phylogenetic variation that exists in the multimeric states of proteins is not necessarily a consequence of idiosyncratic modes of selection in different lineages. Rather, it is an expected outcome of the stochastic evolutionary dynamics that arise in finite populations when the combined pressures of mutation and selection are not overwhelmingly large in one direction. If this hypothesis is correct, and one had the ability to sample a single evolutionary lineage over a very long period of time, orthologous proteins in different phylogenetic lineages would be observed to occupy various multimeric states in frequencies reflecting the underlying transition probabilities.

Although we do not have the luxury of making such observations directly, provided enough time has elapsed for the Tree of Life to have reached the steady-state distribution, a corollary can be tested – the number of transitions from a monomeric to a dimeric state on the branches of a phylogeny should equal that in the opposite direction, and symmetry should also hold for dimer–tetramer transitions, etc. Unfortunately, owing to the huge imbalance in the taxa and proteins with existing structural data, such a test cannot yet be made, and if such efforts are to be pursued, the phylogenetic sampling depth will need to be substantially greater than the expected transition times between alternative states. For example, if the likely transition rate between states is on the order of 10^{-8} per year, a focus on a lineage that diverged more recently than 10^8 years ago would be uninformative unless it contained large numbers of independent lineages.

The domain-swapping model

The preceding section focused on a gradualistic model for evolution along a gradient of adjacent allelic classes. However, transitions between monomeric and dimeric states can sometimes be precipitated by a major structural mutation that is fundamentally different than the single amino-acid substitutions envisioned previously (e.g., an insertion or deletion; Hashimoto and Panchenko 2010; Plach et al. 2017).

One specific and frequently invoked mechanism for the origin of homodimers is encapsulated in the domain-swapping model (Bennett et al. 1994; Kuriyan and Eisenberg 2007), whereby a monomeric protein forming a closed loop with two interfacing domains (within the same polypeptide chain) is physically altered in such a way that intramolecular binding is no longer possible. This can be caused, for example, by a major deletion in the linker between the two binding domains that prevents them from meeting. In principle, such a modification can be accomodated by reciprocal domain swapping between monomeric subunits,

the result being a dimer with two, rather than one, heterologous interfaces (Figure 13.8).

An attractive feature of this model is that the well-endowed binding domains already present in the ancestral protein do not have to go through a phase of incremental improvement. In addition, depending on the configuration of mutant monomers, the process envisioned here is not restricted to the origin of dimers, but extends to the establishment of higher-order multimers as well (Ogihara et al. 2001). Finally, as in the model introduced in the previous section, it is plausible that the process is bidirectional, with insertion mutations in the linker sequence sometimes causing reversion of a domain-swapping dimer to the monomeric condition.

There are plenty of seemingly plausible examples of domain-swapping proteins (Liu and Eisenberg 2002). However, the conditions required for such evolution are particularly sensitive to the population-genetic environment. Here, we consider the simplest case in which an allele for the domain-swapping protein arises by a single deletion mutation that denies self-accessibility within the ancestral monomer. If the dimer is beneficial, such a mutant allele can readily proceed to fixation by positive selection in a haploid species. However, in a diploid species, the mutant allele will initially be present exclusively in heterozygotes, raising potential challenges for establishment.

A key issue is whether heterozygote fitness is compromised by the production of malfunctioning composites of the two alternative monomeric subunits, e.g., chimeras between proteins with and without a deletion. Presumably, the magnitude of any heterozygote disadvantage will depend on the rate of folding of the ancestral monomer and the overall cellular concentration of both allelic products, as slow folding and/or high concentration should magnify the likelihood of chimeric assemblies, including the potential concatenation into harmful fibrils. The details matter here because reduced fitness in heterozygotes presents a barrier to the spread of a mutant allele unless it can somehow rise to a high enough frequency that beneficial homozygotes become frequent enough to offset any heterozygote disadvantages.

Here, we simply provide a heuristic guide to the most salient issue – the low likelihood of an evolutionary sojourn through a bottleneck in mean

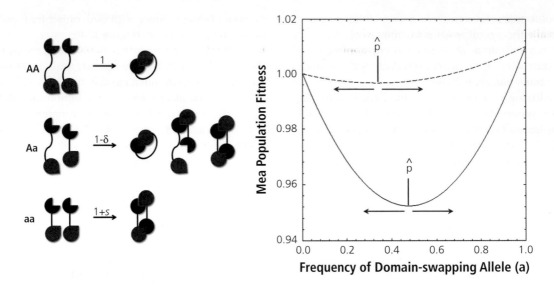

Figure 13.8 The domain-swapping model in a diploid population. **Left:** Relative to the ancestral monomeric type **A**, the domain-swapping allele **a** has a fitness deficit of δ in the heterozygous state, but advantage s in the homozygous state ($s = 0$ meaning that the alternative homozygous states are equivalent in fitness). The configurations following the arrows denote the assembly states of the proteins within diploid cells. **Right:** Such a scenario results in a fitness surface (as a function of the frequency of the domain-swapping allele) with a valley at an intermediate frequency $\widehat{p}$ for the domain-swapping allele, owing to the fact that low-frequency alleles are present almost exclusively in deleterious **Aa** heterozygotes. If the frequency of allele **a** is to the left of the bottom of the valley, the prevailing pressure of selection is to remove the allele from the population, whereas to the right of the valley, selection promotes allele **a**, potentially driving it to fixation. Two examples are shown, both with $s = 0.01$, and with $\delta = 0.01$ for the dashed curve and 0.10 for the solid curve.

population fitness unless the heterozygote disadvantage is overwhelmed by the power of random genetic drift (Lynch 2012). Of special interest is the critical effective population size (N_e^*) beyond which the efficiency of selection is so strong that there is effectively no possibility of making a transition to a domain-swapping allele.

Letting heterozygotes have a fractional fitness reduction δ and domain-swapping homozygotes an advantage s, under the assumption of random mating, mean population fitness reaches a minimum when the population frequency of the domain-swapping allele is $\widehat{p} = \delta/(s + 2\delta)$ (Figure 13.8). This is an unstable equilibrium point. When the domain-swapping allele has frequency $< \widehat{p}$, it will be found essentially exclusively in heterozygotes, and will therefore act like a deleterious mutation being removed from the population at rate δ, i.e., there will be net selection against the domain-swapping allele. However, if the domain-swapping allele has frequency $> \widehat{p}$, homozygotes will be sufficiently frequent that there will be net selection in favor of

this allele. Thus, because the initial frequency of a novel domain-swapping allele is very low (on the order of the reciprocal of the absolute population size), the key issue is whether the mutant allele can drift against a gradient of negative selection up to frequency $\widehat{p}$, whereupon it becomes subject to net positive selection.

The population-size barrier to the establishment of the domain-swapping protein is

$$N_e^* \simeq \frac{s + 2\delta}{\delta^2},\qquad(13.5)$$

which reduces to $N_e^* \simeq 2/\delta$ if there is no selective advantage of the domain-swapping homozygote. In this case, for example, if the deleterious effect in heterozygotes is just 0.002, there is essentially no chance of establishment of an otherwise neutral domain-swapping allele unless the effective population size is smaller than 1000. Even if the domain-swapping homozygote had a 1% advantage ($s = 0.01$), the barrier is still a very small $N_e^* = 3500$. Thus, a small heterozygote disadvantage is a very strong impediment to the establishment of an allele even if it is advantageous when fixed.

To sum up, under the domain-swapping model, a transition from a monomeric to a dimeric state is most plausible under two sets of conditions: 1) a haploid population, in which case heterozygote disadvantage is never experienced; or 2) a diploid population, in which selection against heterozygotes is inefficient, either because the effective population size is small (which allows selection to be overcome by drift) or because the reduction in heterozygote fitness is negligible. Unfortunately, although a knowledge of the fitness consequences of mixtures of the products of ancestral and derived alleles is essential to resolving how readily domain-swapping can evolve in diploid populations, there appear to be no data on this key issue or even on whether domain-swapping dimers confer greater or lesser fitness than monomers.

Notably, the theory presented here is entirely general in that a simple change in definition of terms is all that is required for considering the reverse transition of domain-swapping homodimer to monomer, a scenario that certainly cannot be ruled out on the basis of existing data. Indeed, for the simplest case in which there is no heterozygote disadvantage, if u is the rate of mutation to dimers and v is the reverse mutation rate, and s_d is the selective advantage of dimers (negative if dimers are disadvantageous), the steady-state probability of being in the dimeric state is

$$\widetilde{P} = \frac{\alpha}{1 + \alpha},\qquad(13.6)$$

where $\alpha = (u/v)e^{4N_e s_d}$, and $1 - \widetilde{P}$ is the probability of monomers. Note that this formula follows directly from the theory discussed in the preceding section, being the special case in which there are just two possible states (Foundations 5.3).

Heteromers from homomers

Although heteromers can, in principle, arise from promiscuous interactions among non-orthologous proteins, most seem to originate from interactions between paralogs arising from gene duplication. For example, Mcm1 is a transcriptional regulator that operates as a homodimer in many fungal species. Following duplication in *S. cerevisiae*, the paralogous copies acquired complementary mutations that cause heterodimer assembly; the loss of either duplicate is lethal, but can be rescued by constructs of ancestral homodimers (Baker et al. 2013). Following whole-genome duplication in yeast, many other homomeric complexes made a transition to heteromeric states (Pereira-Leal et al. 2007). Likewise, experimental work involving historical reconstructions suggests that haemoglobin, which is deployed as a heterotetramer in a number of metazoan species, evolved from a homodimer, with the transition following gene duplication, and just two subsequent amino-acid substitutions being sufficient to confer a new binding interface (Pillai et al. 2020). Many other examples of homomer-to-heteromer transitions in eukaryotes are summarized in Chapter 24.

Transitions to heteromeric structures can emerge in a variety of ways. For example, duplication might occur first in an ancestral gene with no intrinsic tendency to form dimers, with secondary complementary mutations resulting in complexation of the paralogous products. However, an alternative, and perhaps more likely, scenario involves the situation in which an ancestral gene already engages in homodimerization and therefore has a well-established interface at the outset. The initial steps in developing a heteromeric interface would then require the accumulation of unique interface mutations in both paralogs so as to encourage heterodimerization while discouraging homodimer formation. Such a process might be facilitated if there was no intrinsic benefit to dimerizing, as this would eliminate any negative consequences of relinquishing homodimerization. On the other hand, without some form of reinforcement by selection, the long-term maintenance of the heterodimer would also be evolutionarily unstable owing to the fact that each locus would be subject to loss by degenerative mutations (Chapter 6).

There are several mechanisms by which reinforcement might occur. Suppose, for example, that each monomeric subunit from the ancestral gene had multiple, independently mutable subfunctions. Then, gene duplication followed by complementary degenerative mutations (the process of subfunctionalization) would lead to the joint preservation of both paralogs, with the evolved heterodimer still carrying out the combined subfunctions of the ancestral gene, but with the subfunctions partitioned to each subunit. Alternatively, if for other structural reasons an evolved heterodimer outperformed the homodimeric products of each individual locus, this could lead to positive selection for heterodimeric complexation provided the structural changes necessary for avoiding self-recognition were compatible with those for promoting heterodimerization (Marchant et al. 2019).

Notably, the latter scenario need not always await the arrival of new mutations affecting function (Lynch 2012). Consider, for example, the situation in which an ancestral locus encoding a homodimer harbors two alleles, **A** and **a**, such that the cross-product dimer created within heterozygotes elevates fitness beyond that for either of the two pure types produced in homozygotes. Assuming random assembly of allelic products, heterozygous cells would be expected to produce three types of dimeric constructs in a binomial 1:2:1 ratio, i.e., half of the dimers would consist of products from the two alleles. Letting the fitnesses of the three diploid genotypes (**AA**, **Aa**, and **aa**) at the ancestral locus be $1 - s_1$, 1, and $1 - s_2$, respectively, owing to heterozygote superiority, the two alleles will have been maintained in the ancestral (pre-duplication) population by balancing selection, with frequencies $p_1 = s_1/(s_1 + s_2)$ and $p_2 = s_2/(s_1 + s_2)$. There is, however, an intrinsic constraint with such a balanced polymorphism. Because the individuals with highest fitness (heterozygotes) always segregate equal numbers of both alleles into the next generation, there is no possibility for all members of the population to have a pure hetero-allelic state – the frequency of heterozygous individuals will be kept near the Hardy–Weinberg expectation of $2p_1p_2$.

Gene duplication removes this barrier by providing the opportunity for each locus to become fixed for a different ancestral allele (Lynch et al. 2001). Once this point has been reached, then every member of the population would have the expression pattern found in the ancestral heterozygote (Figure 13.9) – fitter than the average member of the ancestral population, but in the early stages, with

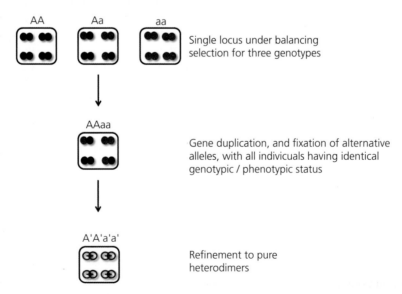

AA   Aa   aa

Single locus under balancing
selection for three genotypes

AAaa

Gene duplication, and fixation of alternative
alleles, with all indivicuals having identical
genotypic / phenotypic status

A'A'a'a'

Refinement to pure
heterodimers

Figure 13.9 A potential path to the evolution of heterodimers from a homodimeric state following gene duplication at a heterozygous locus. Solid circles represent individual proteins derived from alternative alleles, which together make dimers consisting of **A** (blue) and/or **a** (red) subunits (four such complexes are shown per cell); ancestral heterozygotes (prior to duplication) produce three types of dimers in a 1:2:1 ratio, assuming random assembly. Following gene duplication, each locus becomes fixed for an alternative allele, and the encoded products at each locus subsequently diverge to the point that self-assembly is avoided, leading to 'fixed heterozygosity'.

every individual still producing the three dimeric types in a 1:2:1 ratio. Following the establishment of this duplication state, subsequent mutational modifications involving the interfaces of one or both loci could then facilitate heterodimerization, eventually to the point at which homodimer assembly no longer occurs. This particular model is, of course, irrelevant for haploid species, which cannot harbor ancestral heterozygosity.

All of this being said, duplication of a homomerizing gene product need not always lead a heteromer. For example, Billerback et al. (2013) created subunit variants of the normally homomeric barrel-shaped, bacterial chaperone, GroEL (Chapter 14), and found that, instead of assembling as heteromeric complexes, the resultant assemblies were individualized homomers. Moreover, Hochberg et al. (2018) argued that following gene duplication most homomers actually evolve to avoid the construction of heteromeric complexes, potentially becoming preserved by either subfunctionalization or neofunctionalization of different homomeric complexes.

Summary

- Across the Tree of Life, at least 60% of proteins assemble into higher-order multimeric structures, with homomers being about twice as frequent as heteromers. Dimers are more common than tetramers, which are more common than hexamers, and so on, with odd-mers being underrepresented relative to even-mers.

- These distributions are very similar across all phylogenetic groups of prokaryotes and eukaryotes, indicating a minimal gradient of molecular complexity (number of subunits) with organismal complexity. Moreover, orthologous proteins often have different numbers of subunits in different phylogenetic groups, and even when the level of multimerization is conserved, different taxa commonly utilize different binding interfaces between monomeric subunits.

- A substantial contributor to these patterns is the tendency for proteins to be naturally self-adhesive. This leads to a situation in which monomeric proteins are often just one or two amino-acid substitutions away from switching to a dimeric state or to the production of harmful open-ended fibrils.

- The binding interfaces of multimers are relatively simple – typically involving fewer than 10 key residues, and having binding strengths generally 15–25 × the level of background thermal energy.

- Binding interfaces usually have an excess subpopulation of non-adhesive residues. This enables the specific binding sites of any particular lineage to wander in an effectively neutral fashion over evolutionary time, which in turn can lead to incompatibilities between monomeric subunits from divergent taxa.

- Although numerous adaptive explanations have been proposed for the widespread use of multimers, other than for the cases in which a new catalytic function is conferred by the interface or in which a functional cage or fibre is produced, there is very little direct evidence that multimeric proteins are intrinsically advantageous. Moreover, any proposed advantages must be weighed against several costs of relying on multimers, such as the engagement in promiscuous binding and the necessity of producing elevated numbers of monomers essential for complex formation. Resolution of all of these issues will require comparative, experimental work on orthologous proteins with different multimeric structures but conserved functions across the Tree of Life.

- Because of mutation bias towards adhesive amino-acid residues, there is an innate tendency for monomeric proteins to move in the direction of becoming homomeric multimers. This can gradually lead to a situation in which such complexes appear to be entrenched by reinforcing binding sites, even though the process need not have been driven by selection. As a consequence of such directionality and stochastic gains and losses, there can be a broad distribution of molecular phenotypes in different lineages exposed to identical processes of selection, mutation, and drift.

- A celebrated mode of origin of homodimers is domain-swapping, wherein a monomer containing two internal binding domains incurs a deletion in the linker that prevents self-binding and encourages assembly into closed dimers with two heterologous binding interfaces. The reciprocal route is also possible, i.e., an insertion in a linker separating two dimer-interface domains. A major challenge of such transitions in diploid species is the possibility of harmful chimeric complexes between the two allelic products in heterozygous individuals, which imposes a barrier in mean population fitness that can only be overcome by a sufficiently small population size to enable drift across the fitness valley.

- Transitions from homomeric to heteromeric structures are commonly observed, although more so in eukaryotes, and usually following gene duplication with the sister genes then becoming specialized binding partners. Such transitions may initiate when there is a balanced ancestral polymorphism at the locus, with the heterozygote having superior fitness, and effective fixation of the hetero-complex only becoming possible after duplication enables each locus to adopt a particular ancestral allelic type.

Foundations 13.1 Association/dissociation equilibria

To understand a variety of issues involving reaction dynamics and equilibria, knowledge of a few basic features of molecular thermodynamics is required. Consider two molecules, A and B, with the potential to join together to form a non-covalent complex AB, e.g., a dimer. In a steady-state environment, nearly all such systems will reach an equilibrium containing fixed relative concentrations of A, B, and AB. In this particular chapter, the focus is often on the special case in which A = B, i.e., the two molecules are of the same type, forming a homodimer, but the more general solution is given here. There are two ways to obtain the equilibrium solution.

The first approach takes a macroscopic view of the problem, using only information on the concentrations of the system components and their rates of interchange. Letting k_{on} be the association rate of A and B to form AB, and k_{off} be the reciprocal dissociation rate of AB to A and B, at steady state the rate of formation of AB must equal its rate of dissociation,

$$k_{on}[A][B] = k_{off}[AB], \qquad (13.1.1)$$

where the quantities in brackets denote equilibrium concentrations. This general relationship is known as the law of mass action.

The dissociation constant K_D (not to be confused with the dissociation *rate* k_{off}) is the ratio of the reverse and forward rates, which in turn relates to the ratio of reactant molecules under equilibrium conditions. Rearranging Equation 13.1.1,

$$K_D = \frac{k_{off}}{k_{on}} = \frac{[A][B]}{[AB]}, \qquad (13.1.2)$$

defined at 1 M total concentrations of A and B. Although the underlying on/off rates, k_{on} and k_{off}, are not considered further here, note that these are dictated by the structural features of the molecular participants, which determine the rates of encounter and efficiency of binding (critical to k_{on}) and degree of complex stability (critical to k_{off}) (Kastritis and Bonvin 2012).

Letting p_C denote the fraction of A molecules that are in complex, and rearranging and substituting from Equation 13.1.1, at equilibrium,

$$p_C = \frac{[AB]}{[A] + [AB]} = \frac{[B]}{([A][B]/[AB]) + [B]} = \frac{[B]}{K_D + [B]}. \qquad (13.1.3a)$$

This expression shows that K_D is equivalent to the equilibrium concentration of B at which half of the A molecules are in complex with B. If A = B, then

$$p_C = \frac{2[AA]}{[A] + 2[AA]} = \frac{2[A]}{([A][A]/[AA]) + 2[A]}$$
$$= \frac{2[A]}{K_D + 2[A]}, \qquad (13.1.3b)$$

because there are two A molecules within each complex.

The second approach takes a more detailed, thermodynamical view of the alternative states of the system. The key here is that, from the perspective of a single A molecule, there are a number of potential microstates (the set of all possible configurations of the entire population of A and B molecules) involving the states of all B molecules in the system (some of which include a B molecule in complex with A). A classical result from the field of statistical mechanics is that the probability of a particular microstate i of a molecular system is proportional to the function $e^{E_i/(K_B T)}$, where E_i is the energy associated with the state (a more positive number implying a more energetically favorable state), K_B is the

Boltzmann constant, and T is the temperature in degrees Kelvin. At the molecular level, the thermodynamic stabilities of microstates are expressed relative to the background energy related to thermal motion of the solvent molecules $(k_B T)$.

Consider the situation in which there are N molecules of B for each molecule of A, with a B molecule having energy E_{on} when bound to A and E_{off} when free. Focusing on a specific molecule of A, the total energy of the system will then be NE_{off} if A is unbound, and $E_{on} + (N-1)E_{off}$ if a particular A molecule is bound to a single B. The one significant remaining issue is the number of ways in which this one particular A molecule can be bound with various alternative molecules of B, relative to the number of microstates in which none of the local B molecules is bound. This turns out be simply equal to the concentration of B; see Phillips et al. (2013, p. 237–244) for an explicit derivation. To account for this effect, the probability of an individual AB association must be multiplied by [B], whereas the weighting factor for the situation in which A is unbound is just 1. Thus, an alternative way of expressing the probability that a molecule of A is complexed with B is

$$p_C = \frac{[B]e^{[E_{on}+(N-1)E_{off}]/(RT)}}{e^{NE_{off}/(RT)} + [B]e^{[E_{on}+(N-1)E_{off}]/(RT)}} \quad (13.1.4a)$$

$$= \frac{[B]e^{\Delta E/(RT)}}{1 + [B]e^{\Delta E/(RT)}} \quad (13.1.4b)$$

where $\Delta E = E_{on} - E_{off}$.

Note that there has been a change in notation here. The usual convention is to express molecular concentrations and energies associated with them in terms of molar quantities (mol/litre), and so K_B has been scaled up to its molar equivalent R, which is simply K_B times Avogadro's constant (6.02×10^{23} molecules/mol). With R being equal to 1.987 cal $\cdot$ mol^{-1} $\cdot$ K^{-1}, at standard temperature 25°C (equivalent to 298 K), $RT \simeq 0.6$ kcal/mol. Throughout, we adhere to this as an approximate constant, as even a 25°C change in temperature alters RT by <10%. For

Equation 13.1.4b to work properly, ΔE must also have units of kcal/mol, and [B] must be the molar concentration of B.

Comparing Equations 13.1.3 and 13.1.4b shows that setting

$$K_D = e^{-\Delta E/(RT)} \quad (13.1.5)$$

provides an alternative definition of the dissociation constant in thermodynamic terms. The binding-energy differential ΔE is positive for a pair of molecules with an energetically favorable interaction, so that with increasing affinity, $K_D \to 0$ and $p_C \to 1$.

The general approach leading to Equation 13.1.4b can be used to estimate the frequency of alternative states in any localized molecular system at equilibrium. In the example here, there are only two alternative states for any particular molecule, so the solution is relatively simple, but with multiple reactants, the bookkeeping for alternative, combinatorial states becomes increasingly complex. The sum of terms in the denominator of Equation 13.1.4a is known as the partition function, as it insures that the probabilities of all possible microstates sum to 1.0. The overall set of probabilities for alternative states is generally referred to as the Boltzmann distribution. In this particular example, there are just two alternatives, A being bound to B with probability p_C, and A being unbound with probability $(1 - p_C)$.

Note that Equation 13.1.4b, for the probability of one of two particular molecular states in a population of molecules, is identical in form to Equation 13.6, which expresses the probability of one particular allelic state in a population of individuals (fully derived in Foundations 5.3). There is, thus, a remarkable convergence in the form of these statistical-mechanic and evolutionary-genetic equations, with the prefix terms ([B] and u/v, respectively) being measures of the intrinsic pressure towards the state (owing to molecular concentration and mutation bias, respectively), and the exponential terms denoting the added pressure associated with energetic favorabilities and selective advantages.

Literature Cited

Abrusán, G., and J. A. Marsh. 2018. Ligand binding site structure influences the evolution of protein complex function and topology. Cell Rep. 22: 3265–3276.

Ahnert, S. E., J. A. Marsh, H. Hernández, C. V. Robinson, and S. A. Teichmann. 2015. Principles of assembly reveal a periodic table of protein complexes. Science 350: aaa2245.

André, I., C. E. Strauss, D. B. Kaplan, P. Bradley, and D. Baker. 2008. Emergence of symmetry in homooligomeric biological assemblies. Proc. Natl. Acad. Sci. USA 105: 16148–16152.

Bahadur, R. P., P. Chakrabarti, F. Rodier, and J. Janin. 2003. Dissecting subunit interfaces in homodimeric proteins. Proteins 53: 708–719.

Baker, C. R., V. Hanson-Smith, and A. D. Johnson. 2013. Following gene duplication, paralog interference constrains transcriptional circuit evolution. Science 342: 104–108.

Bastolla, U., A. Moya, E. Viguera, and R. C. van Ham. 2004. Genomic determinants of protein folding thermodynamics in prokaryotic organisms. J. Mol. Biol. 343: 1451–1466.

Bennett, M. J., S. Choe, and D. Eisenberg. 1994. Domain swapping: Entangling alliances between proteins. Proc. Natl. Acad. Sci. USA 91: 3127–3131.

Bershtein, S., W. Mu, and E. I. Shakhnovich. 2012. Soluble oligomerization provides a beneficial fitness effect on destabilizing mutations. Proc. Natl. Acad. Sci. USA 109: 4857–4862.

Billerbeck, S., B. Calles, C. L. Müller, V. de Lorenzo, and S. Panke. 2013. Towards functional orthogonalisation of protein complexes: Individualisation of GroEL monomers leads to distinct quasihomogeneous single rings. Chembiochem. 14: 2310–2321.

Bogan, A. A., and K. S. Thorn. 1998. Anatomy of hot spots in protein interfaces. J. Mol. Biol. 280: 1–9.

Bordner, A. J., and R. Abagyan. 2005. Statistical analysis and prediction of protein–protein interfaces. Proteins 60: 353–366.

Brange, J., U. Ribel, J. F. Hansen, G. Dodson, M. T. Hansen, S. Havelund, S. G. Melberg, F. Norris, K. Norris, L. Snel, et al. 1988. Monomeric insulins obtained by protein engineering and their medical implications. Nature 333: 679–682.

Brooijmans, N., K. A. Sharp, and I. D. Kuntz. 2002. Stability of macromolecular complexes. Proteins 48: 645–653.

Caffrey, D. R., S. Somaroo, J. D. Hughes, J. Mintseris, and E. S. Huang. 2004. Are protein–protein interfaces more conserved in sequence than the rest of the protein surface? Protein Sci. 13: 190–202.

Chiti, F., and C. M. Dobson. 2009. Amyloid formation by globular proteins under native conditions. Nat. Chem. Biol. 5: 15–22.

Chothia, C., and J. Janin. 1975. Principles of protein–protein recognition. Nature 256: 705–708.

Day, E. S., S. M. Cote, and A. Whitty. 2012. Binding efficiency of protein–protein complexes. Biochemistry 51: 9124–9136.

Dayhoff, J. E., B. A. Shoemaker, S. H. Bryant, and A. R. Panchenko. 2010. Evolution of protein binding modes in homooligomers. J. Mol. Biol. 395: 860–870.

Freist, W., H. Sternbach, I. Pardowitz, and F. Cramer. 1998. Accuracy of protein biosynthesis: Quasi-species nature of proteins and possibility of error catastrophes. J. Theor. Biol. 193: 19–38.

Garcia-Seisdedos, H., C. Empereur-Mot, N. Elad, and E. D. Levy. 2017. Proteins evolve on the edge of supramolecular self-assembly. Nature 548: 244–247.

Gartner, F. M., I. R. Graf, P. Wilke, P. M. Geiger, and E. Frey. 2020. Stochastic yield catastrophes and robustness in self-assembly. eLife 9: e51020.

Goodsell, D. S., and A. J. Olson. 2000. Structural symmetry and protein function. Annu. Rev. Biophys. Biomol. Struct. 29: 105–153.

Greene, P. J., F. Maley, J. Pedersen-Lane, and D. V. Santi. 1993. Catalytically active cross-species heterodimers of thymidylate synthase. Biochemistry 32: 10283–10288.

Griffin, M. D., and J. A. Gerrard. 2012. The relationship between oligomeric state and protein function. Adv. Exp. Med. Biol. 747: 74–90.

Grishin, N. V., and M. A. Phillips. 1994. The subunit interfaces of oligomeric enzymes are conserved to a similar extent to the overall protein sequences. Protein Sci. 3: 2455–2458.

Hashimoto, K., T. Madej, S. H. Bryant, and A. R. Panchenko. 2010. Functional states of homooligomers: Insights from the evolution of glycosyltransferases. J. Mol. Biol. 399: 196–206.

Hashimoto, K., H. Nishi, S. Bryant, and A. R. Panchenko. 2011. Caught in self-interaction: Evolutionary and functional mechanisms of protein homooligomerization. Phys. Biol. 8: 035007.

Hashimoto, K., and A. R. Panchenko. 2010. Mechanisms of protein oligomerization, the critical role of insertions and deletions in maintaining different oligomeric states. Proc. Natl. Acad. Sci. USA 107: 20352–20357.

Hershberg, R., and D. A. Petrov. 2010. Evidence that mutation is universally biased towards AT in bacteria. PLoS Genet. 6: e1001115.

Hildebrand, F., A. Meyer, and A. Eyre-Walker. 2010. Evidence of selection upon genomic GC-content in bacteria. PLoS Genet. 6: e1001107.

Hochberg, G. K. A., Y. Liu, E. G. Marklund, B. P. H. Metzger, A. Laganowsky, and J. W. Thornton. 2020. A hydrophobic ratchet entrenches molecular complexes. Nature 588: 503–508.

Hochberg, G. K. A., D. A. Shepherd, E. G. Marklund, I. Santhanagoplan, M. T. Degiacomi, A. Laganowsky, T. M. Allison, E. Basha, M. T. Marty, M. R. Galpin, et al. 2018. Structural principles that enable oligomeric small heat-shock protein paralogs to evolve distinct functions. Science 359: 930–935.

Horton, N., and M. Lewis. 1992. Calculation of the free energy of association for protein complexes. Protein Sci. 1: 169–181.

Ispolatov, I., A. Yuryev, I. Mazo, and S. Maslov. 2005. Binding properties and evolution of homodimers in protein–protein interaction networks. Nucleic Acids Res. 33: 3629–3635.

Iyer, R. R., A. Pluciennik, V. Burdett, and P. L. Modrich. 2006. DNA mismatch repair: Functions and mechanisms. Chem. Rev. 106: 302–323.

Janin, J. 1995. Protein–protein recognition. Prog. Biophys. Mol. Biol. 64: 145–166.

Jones, S., and J. M. Thornton. 1995. Protein–protein interactions: A review of protein dimer structures. Prog. Biophys. Mol. Biol. 63: 31–65.

Jones, S., and J. M. Thornton. 1996. Principles of protein–protein interactions. Proc. Natl. Acad. Sci. USA 93: 13–20.

Jucker, M., and L. C. Walker. 2013. Self-propagation of pathogenic protein aggregates in neurodegenerative diseases. Nature 501: 45–51.

Kastritis, P. L., and A. M. Bonvin. 2012. On the binding affinity of macromolecular interactions: Daring to ask why proteins interact. J. R. Soc. Interface 10: 20120835.

Kastritis, P. L., I. H. Moal, H. Hwang, Z. Weng, P. A. Bates, A. M. Bonvin, and J. Janin. 2011. A structure-based benchmark for protein–protein binding affinity. Protein Sci. 20: 482–491.

Kelman, Z., and M. O'Donnell. 1995. Structural and functional similarities of prokaryotic and eukaryotic DNA polymerase sliding clamps. Nucleic Acids Res. 23: 3613–3620.

Knight, R. D., S. J. Freeland, and L. F. Landweber. 2001. A simple model based on mutation and selection explains trends in codon and amino acid usage and GC composition within and across genomes. Genome Biol. 2: RESEARCH0010.

Kunkel, T. A., and D. A. Erie. 2005. DNA mismatch repair. Annu. Rev. Biochem. 74: 681–710.

Kuriyan, J., and D. Eisenberg. 2007. The origin of protein interactions and allostery in colocalization. Nature 450: 983–990.

Levy, E. D. 2010. A simple definition of structural regions in proteins and its use in analyzing interface evolution. J. Mol. Biol. 403: 660–670.

Levy, E. D., E. Boeri Erba, C. V. Robinson, and S. A. Teichmann. 2008. Assembly reflects evolution of protein complexes. Nature 453: 1262–1265.

Levy, E. D., J. B. Pereira-Leal, C. Chothia, and S. A. Teichmann. 2006. 3D complex: A structural classification of protein complexes. PLoS Comput. Biol. 2: e155.

Liu, Y., and D. Eisenberg. 2002. 3D domain swapping: As domains continue to swap. Protein Sci. 11: 1285–1299.

Long, H., W. Sung, S. Kucukyildirim, E. Williams, S. F. Miller, W. Guo, C. Patterson, C. Gregory, C. Strauss, C. Stone, et al. 2018. Evolutionary determinants of genome-wide nucleotide composition. Nat. Ecol. Evol. 2: 237–240.

Lukatsky, D. B., B. E. Shakhnovich, J. Mintseris, and E. I. Shakhnovich. 2007. Structural similarity enhances interaction propensity of proteins. J. Mol. Biol. 365: 1596–1606.

Lynch, M. 2010. Rate, molecular spectrum, and consequences of spontaneous mutations in man. Proc. Natl. Acad. Sci. USA 107: 961–968.

Lynch, M. 2011. The lower bound to the evolution of mutation rates. Gen. Biol. Evol. 3: 1107–1118.

Lynch, M. 2012. The evolution of multimeric protein assemblages. Mol. Biol. Evol. 29: 1353–1366.

Lynch, M. 2013. Evolutionary diversification of the multimeric states of proteins. Proc. Natl. Acad. Sci. USA 110: E2821–E2828

Lynch, M., M. O'Hely, B. Walsh, and A. Force. 2001. The probability of fixation of a newly arisen gene duplicate. Genetics 159: 1789–1804.

Marchant, A., A. F. Cisneros, A. K Dubé, I. Gagnon-Arsenault, D. Ascencio, H. Jain, S. Aubé, C. Eberlein, D. Evans-Yamamoto, N. Yachie, and C. R. Landry. 2019. The role of structural pleiotropy and regulatory evolution in the retention of heteromers of paralogs. eLife 8: e46754.

Marianayagam, N. J., M. Sunde, and J. M. Matthews. 2004. The power of two: Protein dimerization in biology. Trends Biochem. Sci. 29: 618–625.

Marsh, J. A., H. Hernández, Z. Hall, S. E. Ahnert, T. Perica, C. V. Robinson, and S. A. Teichmann. 2013. Protein complexes are under evolutionary selection to assemble via ordered pathways. Cell 153: 461–470.

Marsh, J. A., H. A. Rees, S. E. Ahnert, and S. A. Teichmann. 2015. Structural and evolutionary versatility in protein complexes with uneven stoichiometry. Nat. Commun. 6: 6394.

Marsh, J. A., and S. A. Teichmann. 2014. Protein flexibility facilitates quaternary structure assembly and evolution. PLoS Biol. 12: e1001870.

Marsh, J. A., and S. A. Teichmann. 2015. Structure, dynamics, assembly, and evolution of protein complexes. Annu. Rev. Biochem. 84: 551–575.

Mei, G., A. Di Venere, N. Rosato, and A. Finazzi-Agrò. 2005. The importance of being dimeric. FEBS J. 272: 16–27.

Mintseris, J., and Z. Weng. 2005. Structure, function, and evolution of transient and obligate protein–protein interactions. Proc. Natl. Acad. Sci. USA 102: 10930–10935.

Monod, J., J. Wyman, and J. P. Changeux. 1965. On the nature of allosteric transitions: A plausible model. J. Mol. Biol. 12: 88–118.

Natan, E., T. Endoh, L. Haim-Vilmovsky, T. Flock, G. Chalancon, J. T. S. Hopper, B. Kintses, P. Horvath, L. Daruka, G. Fekete, et al. 2018. Cotranslational protein assembly imposes evolutionary constraints on homomeric proteins. Nat. Struct. Mol. Biol. 25: 279–288.

Nishi, H., K. Hashimoto, T. Madej, and A. R. Panchenko. 2013. Evolutionary, physicochemical, and functional mechanisms of protein homooligomerization. Prog. Mol. Biol. Transl. Sci. 117: 3–24.

Ogihara, N. L., G. Ghirlanda, J. W. Bryson, M. Gingery, W. F. DeGrado, and D. Eisenberg. 2001. Design of three-dimensional domain-swapped dimers and fibrous oligomers. Proc. Natl. Acad. Sci. USA 98: 1404–1409.

Osterman, A., N. V. Grishin, L. N. Kinch, and M. A. Phillips. 1994. Formation of functional cross-species heterodimers of ornithine decarboxylase. Biochemistry 33: 13662–13667.

Pereira-Leal, J. B., E. D. Levy, C. Kamp, and S. A. Teichmann. 2007. Evolution of protein complexes by duplication of homomeric interactions. Genome Biol. 8: R51.

Phillips, R., J. Kondev, J. Theriot, and H. Garcia. 2013. Physical Biology of the Cell. Garland Science, New York, NY.

Pillai, A. S., S. A. Chandler, Y. Liu, A. V. Signore, C. R. Cortez-Romero, J. L. P. Benesch, A. Laganowsky, J. F. Storz, G. K. A. Hochberg, and J. W. Thornton. 2020. Origin of complexity in haemoglobin evolution. Nature 581: 480–485.

Plach, M. G., F. Semmelmann, F. Busch, M. Busch, L. Heizinger, V. H. Wysocki, R. Merkl, and R. Sterner. 2017. Evolutionary diversification of protein–protein interactions by interface add-ons. Proc. Natl. Acad. Sci. USA 114: E8333–E8342.

Plaxco, K. W., and M. Gross. 2009. Protein complexes: The evolution of symmetry. Curr. Biol. 19: R25–R26.

Reid, A. J., J. A. Ranea, and C. A. Orengo. 2010. Comparative evolutionary analysis of protein complexes in *E. coli* and yeast. BMC Genomics 11: 79.

Semple, J. I., T. Vavouri, and B. Lehner. 2008. A simple principle concerning the robustness of protein complex activity to changes in gene expression. BMC Syst. Biol. 2: 1.

Shionyu, M., K. Takahashi, and M. Go. 2001. Variable subunit contact and cooperativity of hemoglobins. J. Mol. Evol. 53: 416–429.

Smeulders, M. J., A. Pol, H. Venselaar, T. R. Barends, J. Hermans, M. S. Jetten, and H. J. Op den Camp. 2013. Bacterial CS_2 hydrolases from *Acidithiobacillus thiooxidans* strains are homologous to the archaeal catenane CS_2 hydrolase. J. Bacteriol. 195: 4046–4056.

Sun, A. Q., K. U. Yüksel, and R. W. Gracy. 1992. Interactions between the catalytic centers and subunit interface of triosephosphate isomerase probed by refolding, active site modification, and subunit exchange. J. Biol. Chem. 267: 20168–20174.

Vassylyev, D. G., H. Mori, M. N. Vassylyeva, T. Tsukazaki, Y. Kimura, T. H. Tahirov, and K. Ito. 2006. Crystal structure of the translocation ATPase SecA from *Thermus thermophilus* reveals a parallel, head-to-head dimer. J. Mol. Biol. 364: 248–258.

Vavouri, T., J. I. Semple, R. Garcia-Verdugo, and B. Lehner. 2009. Intrinsic protein disorder and interaction promiscuity are widely associated with dosage sensitivity. Cell 138: 198–208.

Wells, J. N., L. T. Bergendahl, and J. A. Marsh. 2015. Cotranslational assembly of protein complexes. Biochem. Soc. Trans. 43: 1221–1226.

Wells, J. N., L. T. Bergendahl, and J. A. Marsh. 2016. Operon gene order is optimized for ordered protein complex assembly. Cell Rep. 14: 679–685.

Zabel, W. J., K. P. Hagner, B. J. Livesey, J. A. Marsh, S. Setayeshgar, M. Lynch, and P. G. Higgs. 2019. Evolution of protein interfaces in multimers and fibrils. J. Chem. Physics 150: 225102.

Zhanhua, C., J. G. Gan, L. Lei, M. K. Sakharkar, and P. Kangueane. 2005. Protein subunit interfaces: Heterodimers versus homodimers. Bioinformation 1: 28–39.

Zhang, J., S. Maslov, and E. I. Shakhnovich. 2008. Constraints imposed by non-functional protein–protein interactions on gene expression and proteome size. Mol. Syst. Biol. 4: 210.

Zheng, W., N. P. Schafer, A. Davtyan, G. A. Papoian, and P. G. Wolynes. 2012. Predictive energy landscapes for protein–protein association. Proc. Natl. Acad. Sci. USA 109: 19244–19249.

CHAPTER 14

Protein Management

Some of the fundamental kinetic and structural features of proteins were reviewed in the two preceding chapters. We now move on to additional issues relevant to the life-histories of proteins, most notably matters associated with folding assistance, post-translational modifications, and protein disposal through degradation. Whereas much attention has been given to transcriptional control of gene expression (Chapter 21), these three protein-management processes are also central to gene performance.

As noted in Chapter 12, many small proteins are capable of folding on their own without any external physical assistance. Such proteins must be endowed with amino-acid sequences carrying all of the 'information' essential to acquiring proper three-dimensional structures. However, in all organisms, numerous proteins (particularly large ones) require some form of folding assistance from helper proteins called chaperones. Even in the presence of chaperones, some proteins fail to ever achieve their native states, and these must be disposed of to avoid misinteractions with other proteins and the generation of cellular malfunctions. In addition, some proteins, such as those involved in the cell cycle, need to be conditionally active and efficiently eliminated after completing their missions. Such selective protein removal often relies upon particular markings directing delivery to the cellular degradation machinery. Still other post-translational markings on proteins confer particular subcellular functions, e.g., in signal-transduction pathways (Chapter 22).

In addition to outlining general aspects of protein management, this chapter also provides numerous examples that reinforce the principles regarding the evolutionary aspects of cellular features outlined in preceding and subsequent chapters. For example, much of the machinery associated with protein-folding assistance and selective degradation consists of higher-order multimers that have frequently changed with respect to subunit number and type (Chapter 13). The coevolution of chaperones and their client-gene products raise issues of how a cellular feature with multiple substrates might become constrained by a 'jack of all trades, master of none' syndrome (Chapter 21). Finally, the sites of post-translational markings appear to be nearly free to wander evolutionarily over protein surfaces, providing a means for the effectively neutral rewiring of regulatory mechanisms (Chapters 6, 10, and 21).

Chaperone Assistance

A common mechanism by which proper folding of proteins is achieved involves chaperone guidance into a protective environment for confining the ways in which captured protein molecules can move. This lowers the energetic barrier necessary to achieve a stable folding configuration by a newly formed protein, while also minimizing the potential for harmful interactions of misfolded proteins with others in the cell.

The widespread use of molecular chaperones (themselves proteins) across the Tree of Life motivates numerous evolutionary questions. First, what are the mechanisms by which chaperones recognize their cognate client proteins? Second, do certain classes of chaperones coevolve with individual client proteins in ways that make them less effective with other potential clients? Third, once a protein becomes reliant on chaperone assistance for proper folding, does this act as an evolutionary trap

Evolutionary Cell Biology. Michael Lynch, Oxford University Press. © Michael Lynch (2024). DOI: 10.1093/oso/9780192847287.003.0014

by further relaxing selection on features essential to unassisted self-folding, thereby facilitating the effectively neutral accumulation of otherwise deleterious mutations? Fourth, does chaperone dependence facilitate the evolution of adaptations that would otherwise not be possible because of their negative effects on self-folding? Fifth, given that chaperones consume ATP in the folding process, what is the energetic cost to the cell of producing and relying upon such assistance?

Phylogenetic diversity of chaperones

Because orthologs of some chaperones are found in all three domains of life, they may have been present in LUCA (Rebeaud et al. 2021), paving the way to the establishment of proteins too large to self-fold. However, the substantial diversity of chaperone types within organismal lineages also leads to the conclusion that these helper molecules originated more than once, often converging on similar molecular structures and mechanisms (Schilke et al. 2006, Stirling et al. 2006). Further functional diversification of chaperones followed gene duplication and sub/neofunctionalization on multiple occasions in eukaryotes (Abascal et al. 2013; Carretero-Paulet et al. 2013) and prokaryotes (Bittner et al. 2007; Wang et al. 2013; Weissenbach et al. 2017), possibly driven by adaptive conflicts imposed by alternative client-protein pools. Moreover, as outlined next, striking examples exist of evolutionary transitions between homomeric and heteromeric chaperone structures.

The following paragraphs provide a brief overview of chaperone diversity in the bacterial, archaeal, and eukaryotic domains, although only the most well-studied chaperone families will be introduced. Unfortunately, chaperone-family nomenclature is difficult to navigate, as the labelling of orthologous genes is often inconsistent among organismal lineages. To avoid this notational morass, an attempt is made below to simplify discussion via a slight abuse of taxon-specific notation. Many chaperones are referred to as heat-shock (or heat-stress) proteins, owing to their induced overexpression at extreme temperatures, and such labels are often post-scripted by a number

referring to the approximate size in kiloDaltons (a measure of mass, with one kDa $\simeq 7.5$ amino acids), a notation that will be adhered to in a number of cases discussed here. Adding to the complexity of classification, not all heat-shock proteins are exclusively involved in protein folding, with some being more associated with protein degradation and/or disaggregation.

Within the bacteria, there are three major classes of chaperones: 1) Trigger Factor; 2) a consortium of Hsp40, Hsp70, and a Nucleotide Exchange Factor (NEF); and 3) GroEL/GroES. All three classes have divergent molecular architectures and are deployed in substantially different ways. Trigger Factor is a monomeric protein that binds to nascent peptides as they emerge from the ribosome, effectively producing a preliminary folding space without requiring ATP for function. Hsp40 (a tweezers-like dimeric protein) acts as a co-chaperone, binding exposed hydrophobic patches on unfolded proteins and targeting them to Hsp70 (a monomeric protein), which stabilizes the client protein in an ATP-dependent manner. Hsp70 can also operate as an 'unfoldase', consuming ~5 ATPs per protein in the process, and has many other housekeeping roles (Sharma et al. 2010; Rosenzweig et al. 2019). NEF plays a regulatory role in these processes. Although this tripartite Hsp40/Hsp70/NEF system is widespread among bacteria, at least one lineage appears to have lost it (Warnecke 2012).

The best-studied bacterial chaperone is GroEL (more generally known as chaperonin 60, with the name GroEL being used for the *E. coli* protein). GroEL has a large barrel-like structure, consisting of two heptameric rings stacked back to back (with all 14 subunits encoded by the same gene) (Figure 14.1). Each ring comprises a separate chamber within which the folding of individual client proteins proceeds after closure of the GroEL cavity by a co-chaperone lid (heptameric GroES). Substrate proteins are captured via interactions with their hydrophobic residues and then stretched and remodelled within the folding cage. The mechanics of GroEL/ES involve a form of allostery, with cycles of enclosure and release – the binding of ATPs to one ring results in the release of the GroES cap from the other. Each round of turnover of a protein

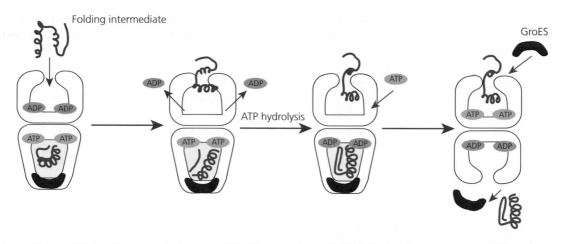

Figure 14.1 An idealized cross-sectional view of GroEL-assisted protein folding. The GroEL chaperone has a double barrel (back-to-back) shape. Loading of the top barrel with a client protein (blue) combined with seven ATPs leads to binding of the lid (GroES) and release of the processed protein from the bottom barrel.

requires ~11 seconds and consumes seven ATPs, one ATP for each of the subunits of the ring (Keskin et al. 2002; Ueno et al. 2004; Horwich et al. 2009). However, in *E. coli*, the half-time for completion of assisted folding is ~45 sec (Kerner et al. 2005), suggesting that an average client protein engages in $45/11 \simeq 4$ GroEL-assisted folding attempts before success is achieved, and consuming ~28 ATPs in doing so. Although some *E. coli* proteins require an average of ~40 cycles (~280 ATPs) to achieve proper folding (Santra et al. 2017), this is still a relatively small price to pay, as the biosynthetic cost of a single amino acid is ~30 ATPs (Chapter 17), and proteins typically contain many dozens of amino acids.

GroEL is present throughout the entire bacterial phylogeny, with some species harboring multiple variants that are likely subfunctionalized with respect to client proteins (Lund 2009; Henderson et al. 2013). However, a few bacterial species (e.g., *Mycoplasma* and *Ureaplasma*; Wong and Houry 2004) seem to have lost GroEL. Random mutagenesis studies indicate that *Ureaplasma* proteins are just one or two mutations removed from GroEL dependence (Ishimoto et al. 2014), further evidence for the point made in Chapter 12 that proteins commonly evolve to be just beyond the margin of stability. There is also some evidence that not all bacterial

GroELs follow the *E. coli* model of oligomeric structure, with dimeric or tetrameric structures likely present in some taxa. It is difficult to see how such reduced structures could serve as chaperones, and they may have entirely different functions, as GroEL is known to engage in different activities in a number of species, e.g., adhesion to host cells, secretion, DNA binding, cell–cell communication, and even toxicity (Henderson et al. 2013).

Like bacteria, many archaea deploy chaperones in the Hsp40/Hsp70/NEF group, suggesting that this particular family dates back to LUCA. However, Hsp40 and Hsp70 are apparently absent from the most thermophilic archaea, which is surprising given the negative effects of high temperature on folding stability. Those archaeal species containing Hsp40 and Hsp70 appear to have acquired them by lateral transfer from bacteria (Macario et al. 2006). Archaea do not harbor Trigger Factor (Laksanalamai et al. 2004), although there is an apparently unrelated mechanism for stabilizing nascent proteins emerging from ribosomes (Spreter et al. 2005).

Although GroEL/GroES is absent from archaea except in rare cases of horizontal transfer (Hirtreiter et al. 2009), there is a chaperonin (CCT, also known as the thermosome) with substantial structural similarity (Foundations 14.1). As with bacterial

GroEL, the archaeal CCT forms a double-ringed barrel structure, but instead of there being a separate GroES-like cap, each monomeric subunit contains a built-in apical loop. These apical lids close like a camera iris, leaving a small opening. This may enable proteins too large to enter the chamber in their entirety to experience progressive folding by threading (Rüßmann et al. 2012). Despite the similarity of the double-barrel architecture of CCT to the form of GroEL, it remains unclear whether the two are derived from a common ancestor, as there is only ~20% sequence similarity. In addition, the GroEL ring contains seven subunits, whereas CCT contains eight or nine (Archibald et al. 1999). Finally, archaeal prefoldin is a heterohexameric co-chaperone, comprised of two monomeric subunit types, that serves to transfer proteins to CCT.

Eukaryotes deploy several chaperones in protein-folding assistance, the major ones being a ribosome-associated complex, consisting of Hsp40 and Hsp70 partners; Hsp90 (a dimer involved in both folding and aggregation suppression); prefoldin; and CCT. The eukaryotic prefoldin hexamer consists of six subunit types, as opposed to two in archaea, and all of the monomeric subunits of eukaryotic CCT are also encoded by different genes (Foundations 14.1). Moreover, unlike the situation with bacterial GroEL, where proteins are processed in a chamber with hydrophilic walls, CCT-folding assistance involves binding of the substrate to the apical domains of the internal chamber. Eukaryotic organelles (mitochondria and chloroplasts) utilize bacterial-derived orthologs of GroEL and GroES, called Hsp60 and Hsp10, respectively, but unlike bacterial GroEL, mitochondrial Hsp60 operates as a single, rather than a double, ring.

As in bacteria, the eukaryotic Hsp70 proteins are monomers, containing one domain for protein binding and another for ATPase activity. Hsp70 has commonly diversified into a dozen or more copies in various eukaryotes, and Hsp40 even more so (Craig and Marszalek 2011; Bogumil et al. 2014). Moreover, the ancestry of these gene families is mixed, with some members showing greater phylogenetic affinities to bacteria and others to archaea. As in bacteria, the eukaryotic system initiates with a Hsp40 protein recruiting a client protein and then stimulating Hsp70 ATPase activity to assist in folding of the client protein. Eukaryotic Hsp70s commonly operate with several different Hsp40 proteins, but specialization also occurs.

As with bacterial GroEL, eukaryotic CCTs often have accessory functions (Henderson et al. 2013). For example, the specific system operating as a chaperone in the mitochondrion is also involved in mitochondrial genome maintenance and protein import. Hsp90 proteins are found throughout eukaryotes, with separate families operating in the cytoplasm, the endoplasmic reticulum, the mitochondrion, and the plastid (in plants), and these interact with a diversity of co-chaperones, with numerous secondary functions, patchily distributed among various branches of the eukaryotic tree (Johnson and Brown 2009; Taipale et al. 2010; Johnson 2012).

If any generality emerges from this morass of complexity, it is that nearly all of cellular life depends on protein complexes specifically assigned to protein folding. Despite the increased proteome complexity in eukaryotes, there has been no major expansion in the core types of chaperones, although the numbers of gene copies and assisting co-chaperones did expand (Powers and Balch 2013; Rebeaud et al. 2021), as did the heteromeric complexity of the individual types. Such alterations must have been accomplished in such a way that the basic folding capacities of cells remained uncompromised during the transitions. There is no evidence that the systems established in any particular lineage are fundamentally superior in any ways to those in others, but as discussed next, each system must be specifically tuned to its resident client proteins.

Client–chaperone coevolution

Unlike most enzymes, chaperones typically have a wide variety of client substrates. For example, ~20% of the ~4000 encoded proteins in *E. coli* are chaperone dependent, and of these, at least 250 appear to rely on GroEL for proper folding, while another 400 or so rely on Hsp40/70, and about 170 are serviceable by both (Kerner et al. 2005; Fujiwara et al. 2010; Niwa et al. 2012). In yeast, ~20% of proteins are clients of Hsp90 alone (Taipale et al. 2010).

Such a vast repertoire of substrates raises questions about the degree to which the features of individual chaperone systems are compromised by the numbers of client genes, one issue being that any evolutionary movement toward a better fit of one client may diminish the effectiveness with others (Lynch and Hagner 2015; Figure 14.2). Wang et al. (2002) gained insight into this matter by

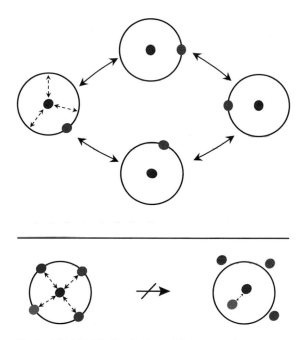

Figure 14.2 An idealized view of the recognition 'sequence space' for client proteins of a chaperone. The central red dot represents the position of the chaperone relative to the recognition profiles of its various client proteins. The closer the chaperone is to a hypothetical client protein within this space, the better the recognition, with the radius of the black circle denoting the minimum distance necessary for recognition. In the upper panel, the chaperone has just one client protein, so the pair is free to wander through sequence space, so long as the matching specificity is kept within the minimum limit (denoted by the dashed arrows). In the lower panel, the chaperone has four client proteins, and this prevents the chaperone sequence from wandering, as any movement with respect to one client protein is likely to reduce the affinity for others, e.g., movement of the red dot (chaperone) towards a particular client protein (pink).

engineering *E. coli* to carry a foreign protein (green fluorescent protein, GFP, encoded on a plasmid) and then imposing a selective challenge on cultures to improve the folding of GFP into functional molecules (readily revealed as fluorescing cells). This resulted in the evolution of a novel GroEL variant with substantially improved GFP folding but a reduced ability to fold normal client proteins, consistent with the chaperone being intrinsically constrained by the need to simultaneously satisfy the needs of multiple interactors.

Another potential example of such a compromise is the reliance of eukaryotic actins and tubulins on chaperones for the production of properly folded monomeric subunits. Together, these two molecules form the cytoskeleton, serve as highways for the transport of various cargoes, and have roles in numerous other eukaryotic cell functions (Chapter 16). However, despite their relatively simple and highly conserved structures, and a history extending back to at least LECA, in no case have the monomers of either protein been found to be capable of self-folding. Instead, both are major clients of CCT, which appears to have evolved specialized features for such processing (Llorca et al. 2001). Although bacterial GroEL will bind eukaryotic actin and tubulin, it is incapable of guiding them to their native conformations (Tian et al. 1995). Given the high intracellular concentrations of actins and tubulins, it is plausible that fine-tuning for processing these key client proteins imposes constraints on the capacity of CCT to service alternative substrates.

A third potential example of chaperones coevolving with client features involves bacterial species experiencing deleterious-mutation accumulation as a consequence of serial bottlenecks. Notably, GroEL comprises up to 70% of the protein in some insect endosymbiotic bacteria, which are thought to experience increased random genetic drift owing to their vertical transmission from maternal to daughter insect. This has led to the suggestion that the elevated investment in chaperones arose as a mechanism to accommodate the accumulation of mildly deleterious mutations in the endosymbiont's protein-coding genes (Moran 1996; Fares et al. 2004). However, such drift-prone bacterial lineages experience even more accelerated rates of sequence evolution in the chaperones themselves than in other proteins (Herbeck et al. 2003; Warnecke and Rocha 2011). This coincident elevation of amino-acid substitutions in both chaperones and client proteins then raises questions as to whether the force driving these changes is adaptive remodelling of key chaperone motifs in response to mutations in specific client genes, and/or whether chaperone over-expression is an evolutionary compensation for its own reduced catalytic capacity. Whatever the mechanism, it is notable that substantial overexpression of chaperone genes also occurs in nucleomorphs (remnants of the nuclear genomes contained within photosynthetic endosymbionts of some algae) that also exhibit elevated rates of protein evolution (Hirakawa et al. 2014).

Given their large size, high expression levels, and reliance on ATP, chaperones such as GroEL comprise a significant fraction of the energy budget of a cell. Thus, elevated chaperone expression may come at a considerable cost that is only warranted under extreme genetic conditions. Experimental data do suggest that bacterial cells respond physiologically to the genome-wide accumulation of deleterious mutations by up-regulating GroEL expression. For example, Maisnier-Patin et al. (2005) observed such a response in mutation-accumulation lines of *Salmonella*, with additional artificial enhancement of GroEL expression resulting in still further increase in fitness. Similarly, Fares et al. (2002) found that *E. coli* lines allowed to accumulate enough mutations to reduce fitness by ~50% were restored to ~90% fitness following overexpression of GroEL; this type of observation extends to Hsp70 as well (Aguilar-Rodríguez et al. 2016). In contrast, *E. coli* cultures maintained at large population sizes often evolve reduced GroEL expression, possibly as a consequence of selection for mutations that reduce unnecessary energetic expenditure (Sabater-Muñoz et al. 2015).

This being said, selection for improved client protein folding is not the only evolutionary determinant of the architectural features of chaperones. Most notably, owing to their roles as safe havens for protein assembly, chaperones are vulnerable to exploitation by pathogens. For example, the genome of bacteriophage T4 (a virus of *E. coli*) encodes for a protein that is a molecular mimic of GroES and uses this feature to assemble its head proteins with GroEL (Keppel et al. 2002). Many other bacteriophage are dependent on host-encoded chaperones for proper development (Nakonechny and Teschke 1998; Karttunen et al. 2015). In fact, it was a serendipitous study of bacteriophage that led to the discovery of GroEL/GroES – the finding of *E. coli* mutants that promoted defective bacteriophage capsid assembly (Georgopoulos 2006). Many eukaryotic viruses also rely on host-cell chaperones to complete their life cycles (Geller et al. 2012). Thus, the degree to which selection to avoid cellular parasites directly conflicts with selection for efficient handling of a cell's endogenous proteins by chaperones merits further study.

Many other open questions remain with respect to the coevolution of chaperones and their client proteins, including the extent to which clients become evolutionarily addicted to assisted folding once reliance on a chaperone has become initiated. Following the sort of constructive neutral-evolution scenario outlined in Chapter 6, with a reliable mechanism of assisted folding in place, mutations that would otherwise prevent self-folding of a protein might be expected to accumulate. A phylogenetic analysis of the clients of human Hsp90 suggest that this is not the case, with both gains and losses of chaperone dependence being common (Taipale et al. 2012), so this is another area ripe for further investigation.

Much of the uncertainty here is a consequence of the low level of understanding of the precise mechanisms by which chaperones identify their client proteins, the details of which are central to all aspects of coevolutionary engagement and escape. Rousseau et al. (2006) suggest that 10–20% of the residues within proteomes across the Tree of Life are contained within segments with a capacity for aggregation if left unfolded, but that such regions tend to be flanked with positively charged amino acids (arginine, lysine, and proline) that are targets of chaperones. Less clear, however, is whether the latter sequences arose in response to the accumulation of aggregation-prone sequences, or appeared first and simply paved the way for the safe accumulation of otherwise adhesive amino-acid residues.

If the former is involved, this then raises the challenging question as to why selection should not minimize the accumulation of aggregation features to start with, as opposed to accepting such properties and then making compensatory modifications to minimize their effects.

Chaperone-mediated phenotypic evolution

Given that chaperones modulate protein quantity and quality, the question arises as to whether such activity can influence individual phenotypes in ways that might modify the course of evolution. In the search for adaptive purposes of traits, one particularly extreme view has been promoted: that chaperones facilitate adaptive evolution by buffering the normally deleterious effects of mutant alleles, thereby encouraging the effectively neutral build-up of a load of hidden but latent phenotypic effects. The idea that such variation might be exposed if a chaperone system becomes overwhelmed in a stressful environment lead to the suggestion that chaperones can act as 'capacitors' for evolutionary change by promoting the expression of conditionally beneficial effects (Rutherford and Lindquist 1998). Further imagining that stressful environments are the ones where aberrant phenotypes are most likely to have utility led to speculation that chaperones (in actuality, their liability to becoming overwhelmed) enhance the ability of populations to adapt to extreme selective challenges. If sustained, this might somehow eventually lead to the constitutive expression of the previously suppressed variant, moving the population into an entirely new phenotypic domain.

Multiple arguments shed doubt on the correctness of these ideas (Levy and Siegal 2008; Tomala and Korona 2008; Siegal and Masel 2012; Charlesworth et al. 2017). First, because chaperones service hundreds of client proteins, for adaptive capacitance to work, the exposure of any single transiently beneficial variant must outweigh the consequences of a likely vast array of other exposed deleterious variants. Second, there is the issue of how a variant that is not expressed for considerable periods of time can avoid the neutral accumulation of still more deleterious, condition-dependent mutations, thereby eventually being rendered non-functional when exposed. Third, if some mechanism does exist by which transient exposure could lead to the expression of a novel protein function, then what becomes of the original function? Fourth, the suppression of chaperone activity can lead to the release of mobile-element activity (Specchia et al. 2010) and/or elevated rates of production of aneuploid progeny (Chen et al. 2012), imposing additional negative consequences. Fifth, for the entire scenario to work, chaperone stress must keep the extreme phenotype exposed to selection for a long enough time to enable new mutations to produce a mechanism for constitutive expression, but short enough to avoid population extinction.

Finally, implicit in the argument that compromised chaperone capacity leads to a release of latent variation is the assumption that chaperones do indeed buffer the effects of new mutations. In fact, the empirical evidence suggests otherwise. In yeast, the effects of standing variation are muted by chaperone activity, but the phenotypic effects of *de novo* mutations are actually magnified on average (Geiler-Samerotte et al. 2016). The overall implication is that natural selection differentially promotes alleles whose effects are buffered by chaperones, not the other way around.

This being said, although it is unlikely that chaperones have been advanced to enhance long-term evolvability, they may nonetheless play indirect roles in short-term evolutionary processes. An example of how chaperones might mediate the evolution of a novel protein function is provided by an experiment in which phosphotriesterase, an expendable protein in *Pseudomonas aeruginosa*, was selected for a novel arylesterase function (Wyganowski et al. 2013). In the experimental system, by controlling the expression of GroEL, it was possible to select for protein function under conditions of either high or low chaperone activity. High chaperone levels allowed the advancement of protein variants with elevated catalytic activity but low folding stability, whereas subsequent return to a low level of GroEL imposed strong selection for compensatory mutations against destabilizing mutations. Several rounds of such selection eventually led to a 10^4-fold increase in arylesterase activity and a near absence of GroEL dependency.

Additional experiments of this nature have led to the improvement of the catalytic performance of other enzymes at the expense of self-folding capacity (Tokuriki and Tawfik 2009).

This kind of experimental result, reliant on a contrived experimental setup – alternating periods of high and low GroEL expression, and selection on a non-essential protein, needs to be tempered vis-á-vis the patterns actually seen with natural GroEL clients. Contrary to expectations under the hypothesis that chaperones lead to a relaxation of selection on protein evolution and/or facilitate movement into new adaptive domains, the client proteins of GroEL tend to be slowly evolving (Williams and Fares 2010). Although the subset of clients that are obligately dependent on GroEL and Hsp70 do evolve somewhat more rapidly at the protein-sequence level (Bogumil and Dagan 2010; Williams and Fares 2010; Aguilar-Rodríguez et al. 2016; Kadibalban et al. 2016; Alvarez-Ponce et al. 2019), such a pattern could also exist for reasons unassociated with folding.

In summary, all of the preceding observations strongly support the view that the function of chaperones is to suppress the negative phenotypic consequences of problematic protein folding, rather than to store away hopeful monsters. Pushing most organisms beyond their physiological capacities invariably leads to aberrant, pathological phenotypes, so there is nothing particularly unique about the phenotypic consequences of overtaxed chaperones. More generally, the broader idea that various biological features have emerged specifically to enhance the long-term evolvability of species is without support and largely incompatible with evolutionary theory. Conveniently ignored here is the fact that selection operates on individuals in the present and has no capacity to see into the future. There is no known evolutionary mechanism to advance a cellular feature for the specific purpose of allowing the long-term accumulation of suppressed variation with conditionally beneficial effects in some future environment. Indeed, population-genetic theory demonstrates that a release of hidden genetic variation on a mutant background (or in a stressful environment) is a generic property of complex genetic systems, regardless of the prior state of buffering, and not an indicator of the prior evolution of a mechanism for ensuring robustness (Hermisson and Wagner 2004).

It is true that any mechanism that can sufficiently increase the robustness of an organism to perturbations can be selectively favored (de Visser et al. 2003), provided that the strength of selection exceeds that of random genetic drift and that the energetic cost is not too great. Chaperones do indeed expand the capacity of organisms to survive through stressful conditions. However, it does not follow that the assimilation of such a mechanism into a species owes its existence to selection for the long-term evolutionary flexibility of the lineage, nor even that there are any long-term benefits. More likely, there are disadvantages. Although selection for a robustness-enhancing feature may hide background defects in the short term, in the long run, a new load of defects is expected to bring the population back to the previous fitness state, but with the added expense of maintaining a new layer of surveillance machinery (Frank 2007; Gros and Tenaillon 2009; Lynch 2012). In this sense, the idea that natural selection produces fundamentally superior organisms by adding layers and layers of buffering mechanisms to stabilize high fitness is an illusion (Chapter 20).

Protein Disposal

All organisms are confronted with the challenge of eliminating proteins that are structurally aberrant (owing to improper folding), functionally unnecessary or inappropriate (owing to the completion of prior tasks), or damaged by a wide variety of intracellular effects (such as thermal denaturation and oxidation). To accomplish such tasks, most prokaryotes and possibly all eukaryotes harbor a special molecular machine, the proteasome, which carries out processive protein degradation in an ATP-consuming process. The proteasome consists of a barrel-like structure, reminiscent of that found for the CCT, which provides a safe compartment for restricting protease activity to target proteins and protecting other desirable proteins from proteolysis.

The proteasome exhibits a phylogenetic gradient in complexity similar to that seen for CCT chaperones. In archaea, eukaryotes, and a few bacteria, the barrel consists of four layers of heptameric rings, with the outer rings forming pores through which cargoes are delivered. In most archaea, the two inner (β) rings are homomeric, comprised of catalytic subunits encoded by a single locus, whereas the outer (α) scaffold rings are homomers of another gene product. In contrast, in eukaryotes each of the fourteen subunits (seven for the α and β rings, respectively) are encoded separately (Pühler et al. 1993). Deviating substantially from the situation in archaea and eukaryotes, the bacterial proteasome is generally comprised of two homomeric rings with six subunits, although archaeal-like structures with seven subunits are found sporadically throughout the bacterial domain (Valas and Bourne 2008; Fuchs et al. 2017, 2018).

Thus, although the proteasome dates back to LUCA, we are again confronted with both an increase in complexity and an expansion in the number of subunits of the eukaryotic version, which must have involved an evolutionary alteration of binding interfaces (Foundations 14.1). Based on their phylogenetic distribution, the origins of all fourteen distinct eukaryotic subunits predate LECA (Bouzat et al. 2000). Along with this shift in proteasome complexity, the regulator proteins that control the entry of cargo proteins into the proteasome consist of at least six different subunit types throughout eukaryotes but only one in archaea (Fort et al. 2015).

In parallel with the proteasome, numerous other proteases operate in both prokaryotic and eukaryotic cells (e.g., Clausen et al. 2011). Many of these complete the degradational process, as proteasome degradation only reduces substrates to short oligopeptides, not single amino acids. Additional machinery, the exosome and its regulatory proteins, is deployed in the selective degradation of specific RNAs (Makino et al. 2013). As in the case of the proteasome, the nine subunit barrel of the exosome has experienced an increase in complexity from archaea to eukaryotes (three vs. nine distinct contributing proteins).

The selective targeting of proteins for disposal is generally orchestrated by pathways dedicated to marking molecules with specific degradation signals. In eukaryotes, the most prominent mechanism by far is the ubiquitylation pathway (Mogk et al. 2007; Sriram et al. 2011; Varshavsky 2011, 2019). In a series of three enzymatically guided steps, ubiquitin is delivered and ligated to specific lysine residues on target molecules in an ATP-dependent manner (Figure 14.3). From this starting point, chains of polyubiquitin are then grown, providing a signal for proteasome delivery. Deubiquitylation occurs prior to entry into the proteasome, sparing the ubiquitin molecules from degradation. The presence of all components of this pathway in some lineages of archaea implies a pre-LECA origin, apparently with independent expansions and specialization of component parts occurring in animals and land plants (Grau-Bové et al. 2015). Pathways with essentially the same features but quite different molecular participants exist in bacteria (Mogk et al. 2007; Mukherjee and Orth 2008), so an even earlier origin cannot be ruled out. In a related eukaryotic pathway, acetylation of specific residues provides another signal for degradation (Hwang et al. 2010; Shemorry et al. 2013).

In addition to its central role in protein degradation, the eukaryotic ubiquitylation/deubiquitylation pathway provides a means for dynamically switching proteins between alternative activity states in a wide variety of cellular functions. These include the cell cycle, DNA repair, vesicle trafficking, and signal transduction (e.g., Hirsch et al. 2009; Raiborg and Stenmark 2009; Ulrich and Walden 2010). Remarkably, a number of pathogenic bacteria have independently evolved molecular mimics of ubiquitin ligases, enabling them to commandeer various aspects of the machinery of host cells (Hicks et al. 2010).

The ubiquitin–proteasome degradation system provides still another example of the importance of intracellular molecular languages in guiding key cellular events. In this case: 1) specific amino-acid residues at the termini of proteins (usually the N-terminal ends) define their susceptibility to ubiquitylation; 2) specific internal sites of the target molecules (usually lysine residues) are post-translationally modified by the covalent conjugation of ubiquitin (usually as polyubiquitin chains); and 3) the resultant linked ubiquitin moieties serve as indicators for delivery of the modified protein to the proteasome. Ubiquitylation is

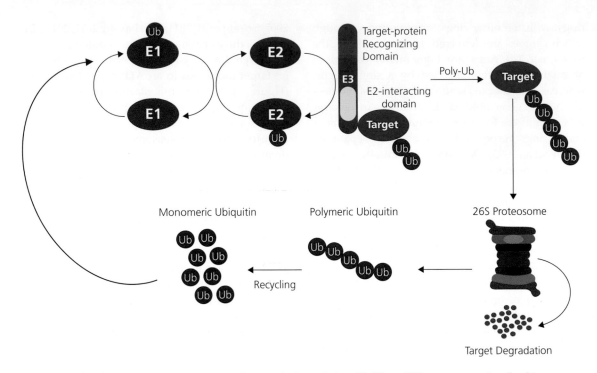

Figure 14.3 The ubiquitin–proteasome system for protein degradation. E1, E2, and E3 are enzymes involved in sequestering and covalently conjugating ubiquitin moieties to specific sites on a target protein, with the build-up of polyubiquitin chains serving as a signal for the recognition and processive degradation of the protein by the proteasome.

mediated by ubiquitin ligases, which rely on specific amino-acid sequence motifs for precise ubiquitin conjugation. Hundreds of such ligases with unique recognition sequences are often encoded within individual genomes, providing both specificity and an immense functional reach of the overall system.

The recognition determinants for protein degradation generally consist of specific amino-acid residues at the N- or C-termini of proteins, referred to as N-degrons or C-degrons (Figure 14.4). The exact nature of degrons (i.e., the recognition language) can differ among major groups of organisms (e.g., bacteria, land plants, and animals; Mogk et al. 2007). Further complicating things is the presence of enzymes for removing the initial methionine residues on polypeptide chains, other endopeptidases for severing small N-terminal peptide chains (thereby exposing new degradation determinants), and still others for converting some

N-terminal amino-acid residues to others (e.g., Asn to Asp and Gln to Glu in eukaryotes). The latter residues can be viewed as secondary/tertiary destabilizing N-terminal residues, as they are only effective after modification, and even then, often only after covalent attachment of yet another amino acid that serves as the primary determinant (Arg in the case of eukaryotes). In the case of *E. coli*, N-terminal Arg and Lys serve as secondary destabilizing factors residues, which become active after terminal attachment of Leu. It has been suggested that the Arg-transferase utilized in eukaryotes is related to one of the Leu-transferases in bacteria (Graciet et al. 2006).

Together then, two signals, one for denoting stabilization/destabilization status and the other for ligase-mediated attachment of ubiquitin to specific sites, largely determine the half-lives of individual proteins. Notably, the internal lysine sites involved in ubiquitylation are only slightly more conserved over evolutionary time than other adjacent lysine

Primary destabilizing and stabilizing residues:

Modification and addition of destabilizing residue:

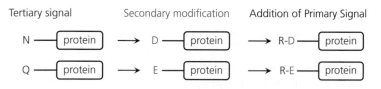

Figure 14.4 Some of the N-end rules for the acquisition of polyubiquitin signals that mediate protein degradation by the proteasome (known from yeast, land plant, and mammalian cells). In the lower panel, the tertiary signal asparagine (N) is converted to the secondary signal aspartic acid (D) by deamidation, and likewise for glutamine (Q) to glutamic acid (E), and then a transferase adds arginine (R, the primary destabilizing residue), thereby enabling the recognition and marking of the resultant protein by ubiquitin ligases. From Graciet and Wellmer (2010) and Varshavsky (2011, 2019).

residues, suggesting a high degree of redundancy with respect to location combined with stabilizing selection on the total numbers of such sites per protein (Hagai et al. 2012; Lu et al. 2017), a point highlighted below for phosphorylation sites.

Finally, although bacteria have a pathway for protein disposal that is similar to that of eukaryotes, including the use of N-end rules, the bacterial pathway is substantially simpler than that in eukaryotes. Notably, however, a number of the destabilizing N-terminal amino acids in eukaryotes are the same as those in bacteria, suggesting a common ancestry of this N-degron system that emerged prior to LUCA. Varshavsky (2011) has suggested that, despite the increased complexity of the system that mediates processive proteolysis in eukaryotes, the eukaryotic system is no more efficient than that in bacteria, with 'overdesign' in the former having arisen by effectively neutral processes operating during phases of reduced effective population sizes.

Post-Translational Modification

An additional stage in the life histories of many proteins involves post-translational covalent linkage of small molecules to certain amino-acid residues. The diversity of molecular moieties that can be conjugated to proteins is substantial, ranging from small phosphoryl, adenyl, acetyl, and amide groups to larger molecules, such as sugars and fatty acid chains, and even to entire proteins, such as ubiquitin (just discussed). Although nearly all amino-acid residues can participate in such modifications, the exact residue marked in any situation depends on the organism and cellular context. The precise functions of such markings are known in just a fraction of cases, but there is little question that post-translational modifications can lead to changes in structure, stability, and/or localization of the affected proteins, thereby modulating their functions. Two cases stand out in particular – the eukaryotic cell cycle (Chapter 10) and signal-transduction systems used in environmental sensing (Chapter 22).

Thus, whereas the classical view of gene regulation focuses on gene-expression modification at the level of transcription (Chapter 21), post-translational modifications yield additional dimensions to the overall complexity of regulation in both prokaryotes and eukaryotes. Although the mechanics differ dramatically, there are several similarities between the evolutionary features of transcriptional and post-translational regulation: both involve *trans*-regulating proteins with interactions targeting simple binding sites (DNA in the first case, and proteins in the second); both are subject to divergence with non-functional consequences; and both are typically under some form of purifying selection.

Although post-translational modification is largely uncharted territory for the field of evolutionary biology, one major target of study involves phosphorylation (Bradley 2022). Phosphoryl-group (PO_4^{3-}) additions are generally restricted to serine, threonine, and tyrosine residues in animals, and to arginine, aspartate, cysteine, and histidine in bacteria (Chapter 22). Covalent attachment of phosphoryl groups is generally carried out by specialized enzymes called kinases, most of which have simple recognition sites comprised of a substrate amino-acid residue plus just two to four flanking residues (Ubersax and Ferrell 2007; Miller and Turk 2018; Ochoa et al. 2018). Such simplicity raises the potential for substantial promiscuity, often rendering inferences on the functional significance of specific sites quite uncertain. Gratuitous phosphorylation may be difficult to select against, as the cost of just a few extra ATP hydrolyses is relatively small compared to the total cost of building a protein (the average cost of an amino acid being ~30 ATPs; Chapter 17). However, certain forms of inappropriate phosphorylation may have negative functional consequences (Brunk et al. 2018; Cantor et al. 2018; Viéitez et al. 2022).

The immediate effect of phosphorylation is the addition of a negative charge to the acceptor residue. Such a change can often have downstream effects such as protein activation or inhibition. In addition, protein phosphorylation can be rapidly reversed by use of specific phosphatases (Chapter 22). Substantial numbers of eukaryotic genes are dedicated to post-translational modifications of this sort. For example, ~2% of the yeast genome encodes for protein kinases, with ~40,000 phosphosites distributed throughout the proteome (Zhu et al. 2001; Lanz et al. 2021). More than 500 kinases and 200 phosphatases are encoded in the human genome (Manning et al. 2002; Alonso et al. 2004).

Proteome-wide data provide insights into the long-term evolutionary stability of phosphorylation sites. For example, comparative studies in yeasts and mammals indicate that many phosphorylated serines and threonines are under purifying selection to retain their phosphosite status (Gray and Kumar 2011; Levy et al. 2012), and sites known to have functionally relevant phosphorylation are more conserved than those with no known function. There is also evidence that sets of phosphorylation sites undergo subfunctionalization following gene duplication (Amoutzias et al. 2010; Freschi et al. 2011; Kaganovich and Snyder 2012), with each member of a paralogous pair partitioning up the ancestral sites, although the functional significance of this remains unclear. While a large fraction of such sites appears free to vary among species in terms of status and location, not all phosphosites are simply evolving neutrally (Moses et al. 2007; Holt et al. 2009; Landry et al. 2009; Nguyen Ba and Moses 2010; Freschi et al. 2014; Studer et al. 2016). For example, ~5% of all phosphosites appear to have been conserved across the entire yeast lineage (dating back ~700 million years), but even when the same phosphorylatable residue is present in two moderately related species, their phosphorylation status may differ.

Phosphosites tend to be clustered on the surface of a protein or in disordered regions, and the critical feature may simply be the production of functionally appropriate local charge. Notably, the negatively charged aspartate and glutamate residues often serve as replacements (and/or sources) for their phosphorylatable counterparts (although the amino-acid interconversions require two nucleotide substitutions per codon), i.e., phosphosites often evolve from phosphomimetic Asp and Glu sites, and vice versa (Kurmangaliyev et al. 2011; Pearlman et al. 2011; Diss et al. 2012).

Taken together, these observations suggest a scenario whereby the degree of a protein's phosphorylation is under stabilizing selection for an appropriate total negative charge, with the specific locations of many of the affected residues relatively free to wander in a quasi-neutral fashion (Lienhard 2008; Landry et al. 2014). That is, the level of phosphorylation of individual proteins appears to operate as a sort of quantitative trait, with the total number of phosphorylated residues being conserved, but also with enough degrees of freedom that there can be considerable turnover of specific phosphosites on evolutionary timescales (Foundations 14.2).

Summary

- All organisms harbor subsets of proteins whose proper folding requires assistance from chaperones. These molecular guardians appear to have been present in LUCA, and likely paved the way for the establishment of long proteins incapable of self-folding. Despite their critical functions, the families of chaperones have diversified substantially with respect to multimeric structures, with expansions in complexity being most extreme in eukaryotes.

- The process of chaperone-assisted folding is relatively cheap, on the order of the biosynthetic cost of one to a few amino acids per protein molecule, although the chaperones themselves are often complex and can constitute a substantial fraction of the total protein within a cell.

- As the number of chaperone systems per cell is dwarfed by the number of their client proteins, coevolutionary conflicts arise with respect to the recognition of specific clients, with the fine-tuning to any one particular client reducing the affinity to others.

- Chaperones are commonly exploited by viruses as assembly chambers for viral capsids, imposing still additional constraints on the evolution of chaperone recognition capacities. The very high evolutionary rates of some eukaryotic chaperones may be a consequence of host–pathogen coevolutionary arms races.

- The extent to which client proteins become addicted to the safe havens of chaperones and embark on a path of no return to self-folding is unclear, but the relatively low cost of such dependence may mean that many proteins are not far from drifting down a path of chaperone dependence by effectively neutral processes.

- It has been argued that chaperones serve as capacitors of adaptive evolution by masking the deleterious effects of mutations in benign environments but releasing novel phenotypes when stressful environments overwhelm surveillance systems. There is, however, no direct support for the idea that chaperones are maintained to enhance the evolvability of species, and multiple lines of evidence are inconsistent with it.

- Essentially all organisms have systems for selectively destroying damaged or superfluous proteins, largely via a barrel-like machine called the proteasome. Selective protein degradation relies on a detailed set of communication rules involving sequence motifs on target molecules and a system of enzymes for marking specific sites as indicators for disposal. The baseline system for regulated protein degradation dates back to LUCA, although the complexity of the processes has expanded in eukaryotes.

- Across the Tree of Life, the structures and functions of numerous proteins are influenced by post-translational modifications involving the covalent conjugation of various side groups to specific amino-acid residues. Best studied is the case of phosphorylation, wherein the specific locations of many target phosphosites appear relatively free to wander in an effectively neutral fashion over evolutionary timescales provided their local density does not change significantly.

Foundations 14.1 The CCT complex

Transition steps from a homomeric to a heteromeric even-mer (8 subunits):

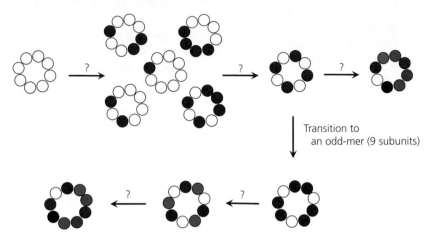

Figure 14.5 A simplified view of some of the challenges to the evolution of a heteromeric ring molecule. When the first (red) variant of the monomeric subunit appears (either as an allelic variant, or as a duplicate gene), prior to sufficient sequence divergence, the two types are likely to form a diversity of hetero-oligomeric structures within a cell. If a pair of sufficiently distinct interfaces can be established, an organized architecture involving alternating subunits might be acquired, e.g., alternating white and red subunits in the case of an even-mer. A ring with an odd number of subunits imposes additional challenges; for example, in the first step (with two subunit types), the positions cannot be evenly divided between two monomeric subtypes. A final structure involving eight or nine distinct members of the ring requires several additional gene duplications followed by the evolution of two distinct binding interfaces by each monomeric subunit, with each step introducing stoichiometric challenges.

The CCT (chaperonin containing tailless) complex presents a striking example of a transition of a multimeric protein from a homomeric to a complex heteromeric state. Restricted to archaea and eukaryotes, CCT chaperonins are generally double-barrelled hexadecamers (occasionally octodecamers), i.e., with eight or nine monomeric subunits per barrel (Archibald et al. 1999, 2001).

In archaea, the overall structure is homomeric or heteromeric with two or three alternating subunits (in eight- or nine-component barrels, respectively). The evidence suggests that conditionally deleterious mutations have accumulated in the contact regions between paralogous subunits in heteromeric archaeal CCTs, with compensatory mutations then serving to create a sort of evolutionary entrapment (Ruano-Rubio and Fares 2007). Under this hypothesis, the ancestral CCT was a homo-oligomer that then diversified in architecture following gene duplication, via an effectively neutral evolutionary pathway and with no significant change at the functional level (Archibald et al. 1999). Nonetheless,

the evolution of complexity is not unidirectional in CCT, as there are examples of the reversion of heteromeric complexes to homomers.

In contrast, all of the CCT subunits in eukaryotes are encoded by separate genes. With eight different subunits per ring in the eukaryotic version of CCT, there are thousands of possible arrangements under random assembly, and yet it is thought that just one assembly is consistently achieved in the cell (Kalisman et al. 2012), i.e., there are precisely calibrated binding affinities between the eight subunits. The underlying duplication and divergence of subunits occurred early in eukaryotic history, apparently pre-LECA, as the different subunits within a species are more divergent from each other than are orthologous subunits across major phylogenetic lineages (Fares and Wolfe 2003). Moreover, the eight eukaryotic duplicates appear to diverge at the amino-acid sequence level at rates exceeding the neutral expectation, thus suggesting positive selection for diversification in function, potentially with each subunit being relatively specialized to a different set of client proteins (Fares and Wolfe 2003; Joachimiak et al. 2014).

Understanding the evolution of an initially homomeric ring into such a complex heteromeric state imposes several challenges. At each evolutionary step, a mechanism is required to permanently preserve both the new and the old members of the complex, either via the gain of a beneficial function or complementary losses of subfunctions (Chapter 13). Moreover, the addition of each new member of the ring likely requires the fixation of at least two mutations, as ring architectures necessitate that each

subunit be involved in two distinct interfaces. Each step of the process also raises the noted problem of hetero-oligomerization – the assembly of heterogeneous mixtures of subunits in individual complexes that is likely to persist until a high level of interface specificity has evolved (Figure 14.5). The need for understanding of these kinds of issues is not confined to chaperone evolution, as numerous other cellular features have ring-like structures, e.g., the nuclear pore (Chapter 15), the proteasome (this chapter), and a number of DNA-binding proteins in eukaryotes (Chapter 10).

Foundations 14.2 The evolution of a digital trait

Phosphorylation and other post-translational modifications are examples of digital traits, in the sense that they have a simple molecular basis with the resultant phenotypes taking on integer values (e.g., equal to the number of modified amino-acid residues). Many other cellular features have this property, e.g., the number of residues involved in binding of a protein to its substrate, and the number of saturated and unsaturated bonds in a lipid molecule.

Such restriction of simple molecular traits to discontinuous values may impose unique evolutionary consequences. For example, the optimum binding energy for a particular trait may be unattainable unless it coincides with an integer multiple of the underlying granularity. If this is not the case, two allelic states straddling the optimum may have nearly the same fitness, resulting in an essentially neutral process of molecular evolution combined with a permanent state of suboptimal fitness. In addition, if certain suboptimal allelic states are more accessible by mutation, this can compete with the ability of natural selection to promote higher-fitness states. As we discuss below, such conflicts can even be present in the absence of mutation bias. Finally, if a system has excess capacity, such that the typical state (e.g., number of modified residues) is well below the maximum possible value, substantial drift is possible among alternative phenotypes with equivalent effects.

Drawing upon an approach introduced in Chapter 5, here we consider a simple model for exploring these issues, with ℓ equivalent sites (factors), each with two alternative functional states, $+$ and $-$, contributing positively and negatively to the trait. Under this model, a multiplicity of functionally equivalent classes exists with respect to the number of positive alleles (m). As an example, for

the case of $\ell = 4$, there are five genotypic classes ($m = 0$, 1, 2, 3, and 4), with multiplicities 1, 4, 6, 4, and 1, respectively (Figure 14.6). These multiplicities are equivalent to the coefficients of a binomial expansion, e.g., $(x+y)^m$. With equivalent fitness for all members within a particular class, this variation in multiplicity of states plays an important role in determining the long-term evolutionary distribution of alternative classes – all other things being equal, classes with higher multiplicities are more accessible over evolutionary time.

As discussed in Foundations 5.2, a system like this yields an equilibrium distribution of a population occupying alternative states over a long evolutionary time period, given constancy of the population-genetic environment. That is, over time the mean phenotype is expected to wander within limits dictated by the strength of selection for alternative classes, the degree of mutation bias, and the power of random genetic drift. Justification of this quasi-steady-state view derives from the fact that many cellular traits have functions (and cytoplasmic environments) that may have remained relatively stable for tens to thousands of millions of years (even in the face of a changing external environment).

The probabilities of alternative states depend on the relative magnitudes of the transition coefficients between adjacent classes (Figure 14.6). Each of these coefficients is equal to the product of a multiplicity, a per-site mutation rate, and a probability of fixation of a new mutation. The per-generation mutation rates from the $-$ to $+$ state, and vice versa, are defined to be u_{01} and u_{10}, respectively. The probability of fixation is given by the standard expression outlined in Chapter 4. A haploid, non-recombining population is assumed here, so that each set of functionally equivalent states comprises a genotypic class.

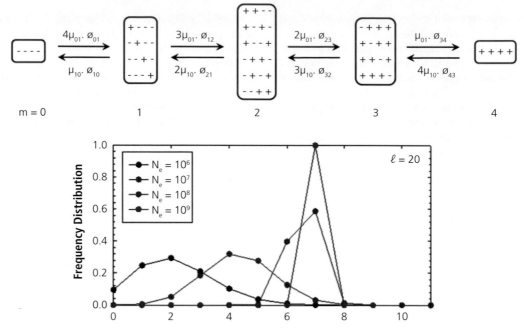

Figure 14.6 Top: Schematic for the transition rates (terms on arrows) between adjacent classes under the sequential-fixation model for the case of $\ell = 4$ sites. Under this model, transitions are rare enough (owing to small enough population sizes and/or mutation rates) that populations essentially always reside in pure states relative to the much less common polymorphic transition periods. Mutation rates towards + and– alleles are denoted by u_{01} and u_{10}, respectively, and ϕ_{xy} denotes the probability of fixation of a new mutation of state y arising in a population of state x. **Bottom:** Equilibrium haplotype (genotype) distributions for four effective population sizes (N_e), given for the situation in which the capacity of the system is $\ell = 20$ sites, and selection is of a stabilizing nature with optimum genotypic value (for the number of + alleles) $\theta = 7.0$ and width of the fitness function $\omega = 5000$. The mutation rate in the direction of – alleles is assumed to be $10 \times$ that in the opposite direction. Results are derived by use of Equations 14.2.2 and 14.2.4. As $N_e \to \infty$, the mean genotypic value converges on the optimum (in this case, $\theta = 7$), although even at very large N_e, substantial variation among population states can remain. Note that only integer values of the average value of m (dots) are possible under the sequential-fixation model, and these are connected by lines merely for visualization.

In the limiting case of neutrality, the equilibrium probability of any site being occupied by a + allele is simply $\eta = u_{01}/(u_{01} + u_{10})$, the fraction of the summed mutation rates in the + direction, and the states of all sites will be independent. (Here, the probability of fixation, $1/N$, factors out because it is identical for all mutations). The neutral probability of a population residing in state m is then simply defined by the binomial distribution,

$$\widetilde{P}_{n,m} = \binom{\ell}{m} \eta^m (1 - \eta)^{\ell - m}. \tag{14.2.1}$$

Thus, in this limiting case, the probability distribution for the class types only depends on: 1) the ratio of mutation

rates, not on their absolute values; and 2) the binomial coefficient associated with each class, which defines the multiplicity of equivalent states in the class. The long-term mean and variance of the trait under neutrality, defined by the properties of the binomial distribution, are $\mu_n = \ell\eta$ and $\sigma_n^2 = \ell\eta(1 - \eta)$, respectively.

Selection alters this baseline distribution by weighting each class by the factor $e^{2N_e s_m}$ (with a 4 being substituted for the 2 under diploidy), where N_e is the effective population size, and s_m is a measure of the class-specific deviation of fitness from some reference point (e.g., from the fitness of the optimal phenotype). The quantity $N_e s_m = s_m/(1/N_e)$ is equivalent to the ratio of the strength of selection relative to that of drift. The basis for this weighting term has already been discussed in Foundations 5.2 – it is the ratio of fixation probabilities from class $m-1$ to m, and vice versa.

The overall distribution can then be written as

$$\widetilde{P}_m = \widetilde{P}_{n,m} \cdot e^{2N_e s_m} = C \cdot \binom{\ell}{m} \beta^m e^{2N_e s_m}, \qquad (14.2.2)$$

where $\beta = u_{01}/u_{10}$ is the mutation bias (the ratio of mutation rates in both directions), and the normalization constant C is equal to the reciprocal of the sum of the terms to the right of C for $m=0$ to ℓ, which ensures that the frequencies sum to 1.0. The term $(1-\eta)^\ell$ from Equation 14.2.1 has been absorbed into C, as it is a constant independent of m, and the specific reference from which the class-specific fitness deviations are measured does not matter either, as it cancels out through the normalization constant. The mean phenotype is

$$\mu_m = \sum_{m=0}^{\ell} m \cdot \widetilde{P}_m, \qquad (14.2.3)$$

which reduces to $\ell\eta$ in the case of neutrality.

As a specific example of the application of Equation 14.2.2, consider the case of a trait under stabilizing selection, such that the fitness of an individual in genotypic class m is denoted by the Gaussian function,

$$W_m = e^{-(m-\theta)^2/(2\omega^2)}, \qquad (14.2.4)$$

where θ is the optimum phenotypic value, and ω is a measure of the width of the fitness function (analogous to the standard deviation of a normal distribution). Selection is purely directional if $m=0$ or ℓ, and neutrality is approached as $\omega \to \infty$. Although m is confined to integer values, θ need not be. The selection coefficient can be arbitrarily defined as $s_m = W_m - W_0$.

An application of the Gaussian fitness function to Equation 14.2.2, shown in Figure 14.6, illustrates several general points. First, a gradient in the average class value (e.g., the number of phosphorylation sites) is expected with respect to the effective population size, the exact location on the phenotypic scale depending on the strength of selection. When the fitness function is sufficiently flat that $N_e \ll \omega^2$, selection is overwhelmed by the power of drift, and the distribution converges on the neutral expectation, Equation 14.2.1. Only when $N_e \gg \omega^2$ does the force of selection overwhelm the power of drift to the extent that the population will almost always reside near the optimum. The actual optimum will only be achievable if θ is an integer. If this is not the case, the two attainable phenotypes straddling the optimum will be present as alternative states with frequencies depending on their relative fitnesses and the pattern of mutation.

Second, there will frequently be two or more classes with probabilities substantially greater than zero, and sometimes with nearly equivalent values. The fact that populations will frequently have different phenotypic states even in a constant population-genetic environment raises significant reservations about the common practice of assuming that phenotypic differences are a consequence of different forms of selection.

Finally, because of the multiplicity of alternative, functionally equivalent states within each class, populations residing within the same class will commonly have different configurations of $-$ and $+$ states. For example, for the case of $\ell = 10$ and two populations in state $m = 3$, the probability of no overlapping use of $+$ sites is $[1 - (3/10)][1 - (3/9)][1 - (3/8)] \simeq 0.29$. At equilibrium in state $m > 0$, the probability of any specific $+$ site in one population being $-$ in another is $(\ell - m)/\ell$. Each of these points is relevant to a diversity of situations in cellular evolution where there are multiple solutions to the same problem, e.g., the specific amino-acid residues serving as phospho-sites on a post-translationally modified protein, or serving as binding residues on the interfaces in protein complexes.

Literature Cited

Abascal, F., A. Corpet, Z. A. Gurard-Levin, D. Juan, F. Ochsenbein, D. Rico, A. Valencia, and G. Almouzni. 2013. Subfunctionalization via adaptive evolution influenced by genomic context: The case of histone chaperones ASF1a and ASF1b. Mol. Biol. Evol. 30: 1853–1866.

Aguilar-Rodríguez, J., B. Sabater-Muñoz, R. Montagud-Martínez, V. Berlanga, D. Alvarez-Ponce, A. Wagner, and M. A. Fares. 2016. The molecular chaperone DnaK is a source of mutational robustness. Genome Biol. Evol. 8: 2979–2991.

Alonso, A., J. Sasin, N. Bottini, I. Friedberg, A. Osterman, A. Godzik, T. Hunter, J. Dixon, and T. Mustelin. 2004. Protein tyrosine phosphatases in the human genome. Cell 117: 699–711.

Alvarez-Ponce, D, J. Aguilar-Rodríguez, and M. A. Fares. 2019. Molecular chaperones accelerate the evolution of their protein clients in yeast. Genome Biol. Evol. 11: 2360–2375.

Amoutzias, G. D., Y. He, J. Gordon, D. Mossialos, S. G. Oliver, and Y. Van de Peer. 2010. Posttranslational regulation impacts the fate of duplicated genes. Proc. Natl. Acad. Sci. USA 107: 2967–2971.

Archibald, J. M., T. Cavalier-Smith, U. Maier, and S. Douglas. 2001. Molecular chaperones encoded by a reduced nucleus: The cryptomonad nucleomorph. J. Mol. Evol. 52: 490–501.

Archibald, J. M., J. M. Logsdon, and W. F. Doolittle. 1999. Recurrent paralogy in the evolution of archaeal chaperonins. Curr. Biol. 9: 1053–1056.

Bittner, A. N., A. Foltz, and V. Oke. 2007. Only one of five GroEL genes is required for viability and successful symbiosis in *Sinorhizobium meliloti*. J. Bacteriol. 189: 1884–1889.

Bogumil, D., D. Alvarez-Ponce, G. Landan, J. O. McInerney, and T. Dagan. 2014. Integration of two ancestral chaperone systems into one: The evolution of eukaryotic molecular chaperones in light of eukaryogenesis. Mol. Biol. Evol. 31: 410–418.

Bogumil, D., and T. Dagan. 2010. Chaperonin-dependent accelerated substitution rates in prokaryotes. Genome Biol. Evol. 2: 602–608.

Bouzat, J. L., L. K. McNeil, H. M. Robertson, L. F. Solter, J. E. Nixon, J. E. Beever, H. R. Gaskins, G. Olsen, S. Subramaniam, M. L. Sogin, et al. 2000. Phylogenomic analysis of the alpha proteasome gene family from early-diverging eukaryotes. J. Mol. Evol. 51: 532–543.

Bradley, D. 2022. The evolution of post-translational modifications. Curr. Opin. Genet. Dev. 76: 101956.

Brunk, E., R. L. Chang, J. Xia, H. Hefzi, J. T. Yurkovich, D. Kim, E. Buckmiller, H. H. Wang, B. K. Cho, C. Yang, et al. 2018. Characterizing posttranslational modifications in prokaryotic metabolism using a multiscale workflow. Proc. Natl. Acad. Sci. USA 115: 11096–11101.

Cantor, A. J., N. H. Shah, and J. Kuriyan. 2018. Deep mutational analysis reveals functional trade-offs in the sequences of EGFR autophosphorylation sites. Proc. Natl. Acad. Sci. USA 115: E7303–E7312.

Carretero-Paulet, L., V. A. Albert, and M. A. Fares. 2013. Molecular evolutionary mechanisms driving functional diversification of the HSP90A family of heat-shock proteins in eukaryotes. Mol. Biol. Evol. 30: 2035–2043.

Charlesworth, D., N. H. Barton, and B. Charlesworth. 2017. The sources of adaptive variation. Proc. R. Soc. B 284: 20162864.

Chen, G., W. D. Bradford, C. W. Seidel, and R. Li. 2012. Hsp90 stress potentiates rapid cellular adaptation through induction of aneuploidy. Nature 482: 246–250.

Clausen, T., M. Kaiser, R. Huber, and M. Ehrmann. 2011. HTRA proteases: Regulated proteolysis in protein quality control. Nat. Rev. Mol. Cell Biol. 12: 152–162.

Craig, E. A., and J. Marszalek. 2011. Hsp70 chaperones. In Encyclopedia of Life Sciences [online]. https://onlinelibrary.wiley.com/doi/book/10.1002/047001590X John Wiley & Sons, Ltd., Chichester, UK.

de Visser, J. A., J. Hermisson, G. P. Wagner, L. A. Meyers, H. Bagheri-Chaichian, J. L. Blanchard, L. Chao, J. M. Cheverud, S. F. Elena, W. Fontana, et al. 2003. Evolution and detection of genetic robustness. Evolution 57: 1959–1972.

Diss, G., L. Freschi, and C. R. Landry. 2012. Where do phosphosites come from and where do they go after gene duplication? Int. J. Evol. Biol. 2012: 843167.

Fares, M. A., E. Barrio, B. Sabater-Muñoz, and A. Moya. 2002. The evolution of the heat-shock protein GroEL from *Buchnera*, the primary endosymbiont of aphids, is governed by positive selection. Mol. Biol. Evol. 19: 1162–1170.

Fares, M. A., A. Moya, and E. Barrio. 2004. GroEL and the maintenance of bacterial endosymbiosis. Trends Genet. 20: 413–416.

Fares, M. A., and K. H. Wolfe. 2003. Positive selection and subfunctionalization of duplicated CCT chaperonin subunits. Mol. Biol. Evol. 20: 1588–1597.

Fort, P., A. V. Kajava, F. Delsuc, and O. Coux. 2015. Evolution of proteasome regulators in eukaryotes. Genome Biol. Evol. 7: 1363–1379.

Frank, S. A. 2007. Maladaptation and the paradox of robustness in evolution. PLoS One 2: e1021.

Freschi, L., M. Osseni, and C. R. Landry. 2014. Functional divergence and evolutionary turnover in mammalian phosphoproteomes. PLoS Genet. 10: e1004062.

Freschi, L., M. Courcelles, P. Thibault, S. W. Michnick, and C. R. Landry. 2011. Network rewiring by gene duplication. Mol. Syst. Biol. 7: 504.

Fuchs, A. C. D., V. Alva, L. Maldoner, R. Albrecht, M. D. Hartmann, and J. Martin. 2017. The architecture of the Anbu complex reflects an evolutionary intermediate at the origin of the proteasome system. Structure 25: 834–845.

Fuchs, A. C. D., L. Maldoner, K. Hipp, M. D. Hartmann, and J. Martin. 2018. Structural characterization of the bacterial proteasome homolog BPH reveals a tetradecameric double-ring complex with unique inner cavity properties. J. Biol. Chem. 293: 920–930.

Fujiwara, K., Y. Ishihama, K. Nakahigashi, T. Soga, and H. Taguchi. 2010. A systematic survey of *in vivo* obligate chaperonin-dependent substrates. EMBO J. 29: 1552–1564.

Geiler-Samerotte, K. A., Y. O. Zhu, B. E. Goulet, D. W. Hall, and M. L. Siegal. 2016. Selection transforms the landscape of genetic variation interacting with Hsp90. PLoS Biol. 14: e2000465.

Geller, R., S. Taguwa, and J. Frydman. 2012. Broad action of Hsp90 as a host chaperone required for viral replication. Biochim. Biophys. Acta 1823: 698–706.

Georgopoulos, C. 2006. Toothpicks, serendipity and the emergence of the *Escherichia coli* DnaK (Hsp70) and GroEL (Hsp60) chaperone machines. Genetics 174: 1699–1707.

Graciet, E., R. G. Hu, K. Piatkov, J. H. Rhee, E. M. Schwarz, and A. Varshavsky. 2006. Aminoacyl-transferases and the N-end rule pathway of prokaryotic/eukaryotic specificity in a human pathogen. Proc. Natl. Acad. Sci. USA 103: 3078–3083.

Graciet, E., and F. Wellmer. 2010. The plant N-end rule pathway: Structure and functions. Trends Plant Sci. 15: 447–453.

Grau-Bové, X., A. Sebé-Pedrós, and I. Ruiz-Trillo. 2015. The eukaryotic ancestor had a complex ubiquitin signaling system of archaeal origin. Mol. Biol. Evol. 32: 726–739.

Gray, V. E., and S. Kumar. 2011. Rampant purifying selection conserves positions with posttranslational modifications in human proteins. Mol. Biol. Evol. 28: 1565–1568.

Gros, P. A., and O. Tenaillon. 2009. Selection for chaperone-like mediated genetic robustness at low mutation rate: impact of drift, epistasis and complexity. Genetics 182: 555–564.

Hagai T, Á. Tóth-Petróczy, A. Azia, and Y. Levy. 2012. The origins and evolution of ubiquitination sites. Mol. Biosyst. 8: 1865–1877.

Henderson, B., M. A. Fares, and P. A. Lund. 2013. Chaperonin 60: A paradoxical, evolutionarily conserved protein family with multiple moonlighting functions. Biol. Rev. Camb. Phil. Soc. 88: 955–987.

Herbeck, J. T., D. J. Funk, P. H. Degnan, and J. J. Wernegreen. 2003. A conservative test of genetic drift in the endosymbiotic bacterium *Buchnera*: Slightly deleterious mutations in the chaperonin groEL. Genetics 165: 1651–1660.

Hermisson, J., and G. P. Wagner. 2004. The population genetic theory of hidden variation and genetic robustness. Genetics 168: 2271–2284.

Hicks, S. W., and J. E. Galán. 2010. Hijacking the host ubiquitin pathway: Structural strategies of bacterial E3 ubiquitin ligases. Curr. Opin. Microbiol. 13: 41–46.

Hirakawa, Y., S. Suzuki, J. M. Archibald, P. J. Keeling, and K. Ishida. 2014. Overexpression of molecular chaperone genes in nucleomorph genomes. Mol. Biol. Evol. 31: 1437–1443.

Hirsch, C., R. Gauss, S. C. Horn, O. Neuber, and T. Sommer. 2009. The ubiquitylation machinery of the endoplasmic reticulum. Nature 458: 453–460.

Hirtreiter, A. M., G. Calloni, F. Forner, B. Scheibe, M. Puype, J. Vandekerckhove, M. Mann, F. U. Hartl, and M. Hayer-Hartl. 2009. Differential substrate specificity of group I and group II chaperonins in the archaeon *Methanosarcina mazei*. Mol. Microbiol. 74: 1152–1168.

Holt, L. J., B. B. Tuch, J. Villén, A. D. Johnson, S. P. Gygi, and D. O. Morgan. 2009. Global analysis of Cdk1 substrate phosphorylation sites provides insights into evolution. Science 325: 1682–1686.

Horwich, A. L., A. C. Apetri, and W. A. Fenton. 2009. The GroEL/GroES cis cavity as a passive anti-aggregation device. FEBS Lett. 583: 2654–2662.

Hwang, C. S., A. Shemorry, and A. Varshavsky. 2010. N-terminal acetylation of cellular proteins creates specific degradation signals. Science 327: 973–977.

Ishimoto, T., K. Fujiwara, T. Niwa, and H. Taguchi. 2014. Conversion of a chaperonin GroEL-independent protein into an obligate substrate. J. Biol. Chem. 289: 32073–32080.

Joachimiak, L. A., T. Walzthoeni, C. W. Liu, R. Aebersold, and J. Frydman. 2014. The structural basis of substrate recognition by the eukaryotic chaperonin TRiC/CCT. Cell 159: 1042–1055.

Johnson, J. L. 2012. Evolution and function of diverse Hsp90 homologs and cochaperone proteins. Biochim. Biophys. Acta 1823: 607–613.

Johnson, J. L., and C. Brown. 2009. Plasticity of the Hsp90 chaperone machine in divergent eukaryotic organisms. Cell Stress Chaperones 14: 83–94.

Kadibalban, A. S., D. Bogumil, G. Landan, and T. Dagan. 2016. DnaK-dependent accelerated evolutionary rate in prokaryotes. Genome Biol. Evol. 8: 1590–1599.

Kaganovich, M., and M. Snyder. 2012. Phosphorylation of yeast transcription factors correlates with the evolution of novel sequence and function. J. Proteome Res. 11: 261–268.

Kalisman, N., C. M. Adams, and M. Levitt. 2012. Subunit order of eukaryotic TRiC/CCT chaperonin by cross-linking, mass spectrometry, and combinatorial homology modeling. Proc. Natl. Acad. Sci. USA 109: 2884–2889.

Karttunen, J., S. Mäntynen, T. O. Ihalainen, J. K. Bamford, and H. M. Oksanen. 2015. Non-structural proteins P17 and P33 are involved in the assembly of the internal membrane-containing virus PRD1. Virology 482: 225–233.

Keppel, F., M. Rychner, and C. Georgopoulos. 2002. Bacteriophage-encoded cochaperonins can substitute for *Escherichia coli*'s essential GroES protein. EMBO Rep. 3: 893–898.

Kerner, M. J., D. J. Naylor, Y. Ishihama, T. Maier, H.-C. Chang, A. P. Stines, C. Georgopoulos, D. Frishman, M. Hayer-Hartl, M. Mann, et al. 2005. Proteome-wide analysis of chaperonin-dependent protein folding in *Escherichia coli*. Cell 122: 209–220.

Keskin, O., I. Bahar, D. Flatow, D. G. Covell, and R. L. Jernigan. 2002. Molecular mechanisms of chaperonin GroEL–GroES function. Biochemistry 41: 491–501.

Kurmangaliyev, Y. Z., A. Goland, and M. S. Gelfand. 2011. Evolutionary patterns of phosphorylated serines. Biol. Direct 6: 8.

Laksanalamai, P., T. A. Whitehead, and F. T. Robb. 2004. Minimal protein-folding systems in hyperthermophilic archaea. Nat. Rev. Microbiol. 2: 315–324.

Landry, C. R., L. Freschi, T. Zarin, and A. M. Moses. 2014. Turnover of protein phosphorylation evolving under stabilizing selection. Front. Genet. 5: 245.

Landry, C. R., E. D. Levy, and S. W. Michnick. 2009. Weak functional constraints on phosphoproteomes. Trends Genet. 25: 193–197.

Lanz, M. C., K. Yugandhar, S. Gupta, E. J. Sanford, V. M. Faça, S. Vega, A. M. N. Joiner, J. C. Fromme, H. Yu, and M. B. Smolka. 2021. In-depth and 3-dimensional exploration of the budding yeast phosphoproteome. EMBO Rep. 22: e51121.

Levy, E. D., S. W. Michnick, and C. R. Landry. 2012. Protein abundance is key to distinguish promiscuous from functional phosphorylation based on evolutionary information. Philos. Trans. R. Soc. Lond. B Biol. Sci. 367: 2594–2606.

Levy, S. F., and M. L. Siegal. 2008. Network hubs buffer environmental variation in Saccharomyces cerevisiae. PLoS Biol. 6: e264.

Lienhard, G. E. 2008. Non-functional phosphorylations? Trends Biochem. Sci. 33: 351–352.

Llorca, O., J. Martín-Benito, J. Grantham, M. Ritco-Vonsovici, K. R. Willison, J. L. Carrascosa, and J. M. Valpuesta. 2001. The 'sequential allosteric ring' mechanism in the eukaryotic chaperonin-assisted folding of actin and tubulin. EMBO J. 20: 4065–4075.

Lu, L., Y. Li, Z. Liu, F. Liang, F. Guo, S. Yang, D. Wang, Y. He, J. Xiong, D. Li, et al. 2017. Functional constraints on adaptive evolution of protein ubiquitination sites. Sci. Rep. 7: 39949.

Lund, P. A. 2009. Multiple chaperonins in bacteria – why so many? FEMS Microbiol. Rev. 33: 785–800.

Lynch, M. 2012. Evolutionary layering and the limits to cellular perfection. Proc. Natl. Acad. Sci. USA 109: 18851–18856.

Lynch, M., and K. Hagner. 2015. Evolutionary meandering of intermolecular interactions along the drift barrier. Proc. Natl. Acad. Sci. USA 112: E30–E38.

Macario, A. J., L. Brocchieri, A. R. Shenoy, and E. Conway de Macario. 2006. Evolution of a protein-folding machine: Genomic and evolutionary analyses reveal three lineages of the archaeal hsp70(dnaK) gene. J. Mol. Evol. 63: 74–86.

Maisnier-Patin, S., J. R. Roth, A. Fredriksson, T. Nyström, O. G. Berg, and D. I. Andersson. 2005. Genomic buffering mitigates the effects of deleterious mutations in bacteria. Nat. Genet. 37: 1376–1379.

Makino, D. L., F. Halbach, and E. Conti. 2013. The RNA exosome and proteasome: Common principles of degradation control. Nat. Rev. Mol. Cell. Biol. 14: 654–660.

Manning, G., D. B. Whyte, R. Martinez, T. Hunter, and S. Sudarsanam. 2002. The protein kinase complement of the human genome. Science 298: 1912–1934.

Miller, C. J., and B. E. Turk. 2018. Homing in: Mechanisms of substrate targeting by protein kinases. Trends Biochem. Sci. 43: 380–394.

Mogk, A., R. Schmidt, and B. Bukau. 2007. The N-end rule pathway for regulated proteolysis: Prokaryotic and eukaryotic strategies. Trends Cell Biol. 17: 165–172.

Moran, N. A. 1996. Accelerated evolution and Muller's rachet in endosymbiotic bacteria. Proc. Natl. Acad. Sci. USA 93: 2873–2878.

Moses, A. M., M. E. Liku, J. J. Li, and R. Durbin. 2007. Regulatory evolution in proteins by turnover and lineage-specific changes of cyclin-dependent kinase consensus sites. Proc. Natl. Acad. Sci. USA 104: 17713–17718.

Mukherjee, S., and K. Orth. 2008. A protein pupylation paradigm. Science 322: 1062–1063.

Nakonechny, W. S., and C. M. Teschke. 1998. GroEL and GroES control of substrate flux in the in vivo folding pathway of phage P22 coat protein. J. Biol. Chem. 273: 27236–27244.

Nguyen Ba, A. N., and A. M. Moses. 2010. Evolution of characterized phosphorylation sites in budding yeast. Mol. Biol. Evol. 27: 2027–2037.

Niwa, T., T. Kanamori, T. Ueda, and H. Taguchi. 2012. Global analysis of chaperone effects using a reconstituted cell-free translation system. Proc. Natl. Acad. Sci. USA 109: 8937–8942.

Ochoa, D., D. Bradley, and P. Beltrao. 2018. Evolution, dynamics and dysregulation of kinase signalling. Curr. Opin. Struct. Biol. 48: 133–140.

Pearlman, S. M., Z. Serber, and J. E. Ferrell Jr. 2011. A mechanism for the evolution of phosphorylation sites. Cell 147: 934–946.

Powers, E. T., and W. E. Balch. 2013. Diversity in the origins of proteostasis networks – a driver for protein function in evolution. Nat. Rev. Mol. Cell Biol. 14: 237–248.

Pühler, G., F. Pitzer, P. Zwickl, W. Baumeister. 1993. Proteasomes: Multisubunit proteinases common to Thermoplasma and eukaryotes. Syst. Appl. Microbiol. 16: 734–741.

Raiborg, C., and H. Stenmark. 2009. The ESCRT machinery in endosomal sorting of ubiquitylated membrane proteins. Nature 458: 445–452.

Rebeaud, M. E., S. Mallik, P. Goloubinoff, and D. S. Tawfik. 2021. On the evolution of chaperones and cochaperones and the expansion of proteomes across the Tree of Life. Proc. Natl. Acad. Sci. USA 118: e2020885118.

Rosenzweig, R., N. B. Nillegoda, M. P. Mayer, and B. Bukau. 2019. The Hsp70 chaperone network. Nat. Rev. Mol. Cell Biol. 20: 665–680.

Rousseau, F., L. Serrano, and J. W. Schymkowitz. 2006. How evolutionary pressure against protein aggregation shaped chaperone specificity. J. Mol. Biol. 355: 1037–1047.

Ruano-Rubio, V., and M. A. Fares. 2007. Testing the neutral fixation of hetero-oligomerism in the archaeal chaperonin CCT. Mol. Biol. Evol. 24: 1384–1396.

Rüßmann, F., M. J. Stemp, L. Mönkemeyer, S. A. Etchells, A. Bracher, and F. U. Hartl. 2012. Folding of large multidomain proteins by partial encapsulation in the chaperonin TRiC/CCT. Proc. Natl. Acad. Sci. USA 109: 21208–21215.

Rutherford, S. L., and S. Lindquist. 1998. Hsp90 as a capacitor for morphological evolution. Nature 396: 336–342.

Sabater-Muñoz, B., M. Prats-Escriche, R. Montagud-Martínez, A. López-Cerdán, C. Toft, J. Aguilar-Rodríguez, A. Wagner, and M. A. Fares. 2015. Fitness trade-offs determine the role of the molecular chaperonin GroEL in buffering mutations. Mol. Biol. Evol. 32: 2681–2693.

Santra, M., D. W. Farrell, and K. A. Dill. 2017. Bacterial proteostasis balances energy and chaperone utilization efficiently. Proc. Natl. Acad. Sci. USA 114: E2654–E2661.

Schilke, B., B. Williams, H. Knieszner, S. Pukszta, P. D'Silva, E. A. Craig, and J. Marszalek. 2006. Evolution of mitochondrial chaperones utilized in Fe-S cluster biogenesis. Curr. Biol. 16: 1660–1665.

Sharma, S. K., P. De los Rios, P. Christen, A. Lustig, and P. Goloubinoff. 2010. The kinetic parameters and energy cost of the Hsp70 chaperone as a polypeptide unfoldase. Nat. Chem. Biol. 6: 914–920.

Shemorry, A., C. S. Hwang, and A. Varshavsky. 2013. Control of protein quality and stoichiometries by N-terminal acetylation and the N-end rule pathway. Mol. Cell 50: 540–551.

Siegal, M. L., and J. Masel. 2012. Hsp90 depletion goes wild. BMC Biol. 10: 14.

Specchia, V., L. Piacentini, P. Tritto, L. Fanti, R. D'Alessandro, G. Palumbo, S. Pimpinelli, and M. P. Bozzetti. 2010. Hsp90 prevents phenotypic variation by suppressing the mutagenic activity of transposons. Nature 463: 662–665.

Spreter, T., M. Pech, and B. Beatrix. 2005. The crystal structure of archaeal nascent polypeptide-associated complex (NAC) reveals a unique fold and the presence of a ubiquitin-associated domain. J. Biol. Chem. 280: 15849–15854.

Sriram, S. M., B. Y. Kim, and Y. T. Kwon. 2011. The N-end rule pathway: Emerging functions and molecular principles of substrate recognition. Nat. Rev. Mol. Cell. Biol. 12: 735–747.

Stirling, P. C., S. F. Bakhoum, A. B. Feigl, and M. R. Leroux. 2006. Convergent evolution of clamp-like binding sites in diverse chaperones. Nat. Struct. Mol. Biol. 13: 865–870.

Studer, R. A., R. A. Rodriguez-Mias, K. M. Haas, J. I. Hsu, C. Viéitez, C. Solé, D. L. Swaney, L. B. Stanford, I. Liachko, R. Böttcher, et al. 2016. Evolution of protein phosphorylation across 18 fungal species. Science 354: 229–232.

Taipale, M., D. F. Jarosz, and S. Lindquist. 2010. HSP90 at the hub of protein homeostasis: Emerging mechanistic insights. Nat. Rev. Mol. Cell Biol. 11: 515–528.

Taipale, M., I. Krykbaeva, M. Koeva, C. Kayatekin, K. D. Westover, G. I. Karras, and S. Lindquist. 2012. Quantitative analysis of HSP90-client interactions reveals principles of substrate recognition. Cell 150: 987–1001.

Tian, G., I. E. Vainberg, W. D. Tap, S. A. Lewis, and N. J. Cowan. 1995. Specificity in chaperonin-mediated protein folding. Nature 375: 250–253.

Tokuriki, N., and D. S. Tawfik. 2009. Chaperonin overexpression promotes genetic variation and enzyme evolution. Nature 459: 668–673.

Tomala, K., and R. Korona. 2008. Molecular chaperones and selection against mutations. Biol. Direct 3: 5.

Ubersax, J. A., and J. E. Ferrell Jr. 2007. Mechanisms of specificity in protein phosphorylation. Nat. Rev. Mol. Cell. Biol. 8: 530–541.

Ueno, T., H. Taguchi, H. Tadakuma, M. Yoshida, and T. Funatsu. 2004. GroEL mediates protein folding with a two successive timer mechanism. Mol. Cell 14: 423–434.

Ulrich, H. D., and H. Walden. 2010. Ubiquitin signalling in DNA replication and repair. Nat. Rev. Mol. Cell. Biol. 11: 479–489.

Valas, R. E., and P. E. Bourne. 2008. Rethinking proteasome evolution: Two novel bacterial proteasomes. J. Mol. Evol. 66: 494–504.

Varshavsky, A. 2011. The N-end rule pathway and regulation by proteolysis. Protein Sci. 20: 1298–1345.

Varshavsky, A. 2019. N-degron and C-degron pathways of protein degradation. Proc. Natl. Acad. Sci. USA 116: 358–366.

Viéitez, C., B. P. Busby, D. Ochoa, A. Mateus, D. Memon, M. Galardini, U. Yildiz, M. Trovato, A. Jawed, A. G. Geiger, et al. 2022. High-throughput functional characterization of protein phosphorylation sites in yeast. Nat. Biotechnol. 40: 382–390.

Wang, J. D., C. Herman, K. A. Tipton, C. A. Gross, and J. S. Weissman. 2002. Directed evolution of substrate-optimized GroEL/S chaperonins. Cell 111: 1027–1039.

Wang, Y., W. Y. Zhang, Z. Zhang, J. Li, Z. F. Li, Z. G. Tan, T. T. Zhang, Z. H. Wu, H. Liu, and Y. Z. Li. 2013. Mechanisms involved in the functional divergence of duplicated GroEL chaperonins in *Myxococcus xanthus* DK1622. PLoS Genet. 9: e1003306.

Warnecke, T. 2012. Loss of the DnaK-DnaJ-GrpE chaperone system among the Aquificales. Mol. Biol. Evol. 29: 3485–3495.

Warnecke, T., and E. P. Rocha. 2011. Function-specific accelerations in rates of sequence evolution suggest predictable epistatic responses to reduced effective population size. Mol. Biol. Evol. 28: 2339–2349.

Weissenbach, J., J. Ilhan, D. Bogumil, N. Hülter, K. Stucken, and T. Dagan. 2017. Evolution of chaperonin gene duplication in stigonematalean cyanobacteria (subsection V). Genome Biol. Evol. 9: 241–252.

Williams, T. A., and M. A. Fares. 2010. The effect of chaperonin buffering on protein evolution. Genome Biol. Evol. 2: 609–619.

Wong, P., and W. A. Houry. 2004. Chaperone networks in bacteria: Analysis of protein homeostasis in minimal cells. J. Struct. Biol. 146: 79–89.

Wyganowski, K. T., M. Kaltenbach, and N. Tokuriki. 2013. GroEL/ES buffering and compensatory mutations promote protein evolution by stabilizing folding intermediates. J. Mol. Biol. 425: 3403–3414.

Zhu, H., M. Bilgin, R. Bangham, D. Hall, A. Casamayor, P. Bertone, N. Lan, R. Jansen, S. Bidlingmaier, T. Houfek, et al. 2001. Global analysis of protein activities using proteome chips. Science 293: 2101–2105.

CHAPTER 15

Lipids and Membranes

Although much of cell biology focuses on proteins and the machines constructed from them, thousands of molecular forms of lipids are utilized across the Tree of Life, with dozens to hundreds of types frequently being used within individual species (Fahy et al. 2005; Oger and Cario 2013; Brügger 2014; Buehler 2016; Sohlenkamp and Geiger 2016). Lipids are used for multiple cell functions, including energy storage and occasionally as cofactors for protein function, but we focus here specifically on their deployment in membranes. Cell envelopes provide a barrier to the external environment, and in doing so ensure the co-localization of genomes with the products they produce, thereby conferring individuality, a critical requirement for heritable evolutionary processes. In eukaryotes, membranes also circumscribe a wide variety of intracellular organelles, including the endoplasmic reticulum (ER), the Golgi, the nuclear envelope, mitochondria and plastids, and transport vesicles.

Consisting of millions to billions of non-covalently linked molecules, lipid membranes are typically highly fluid, constituting an effectively two-dimensional liquid, with an intrinsic biophysical capacity for both flexibility and resistance to breakage and leakage. Given their multifaceted structural and functional roles, the universal use of lipids is unlikely to be simply a frozen accident in biology. Indeed, it is difficult to see how the establishment and diversification of cellular life would have been possible without them.

Membranes also provide platforms for the residence of key proteins with diverse functions. Most notable are the trans-membrane channels, importers, and exporters used for ion and nutrient acquisition and balance (Chapter 18), electron transfer chains and ATP synthases used for energy production (Chapter 23), and components of signal-transduction pathways used for environmental sensing and communication (Chapter 22). Taken together, the proteins involved in these diverse functions typically comprise 10–30% of the total set of proteins encoded in the genomes of species.

In the latter part of this chapter, we review how establishment of the intricate system of vesicle trafficking in eukaryotes entailed a significant investment in a diverse repertoire of proteins required in vesicle formation, transport, and localization. Given these additional investments, the energetic costs of lipids are particularly germane to understanding the evolution of eukaryotic cells, and these are covered in detail in Chapter 17.

Molecular Structure

Marsh (2013) provides an encyclopaedic coverage of the various classes of membrane lipids. Here, we provide a simple overview of the key issues from an evolutionary perspective. Most membrane lipids in bacteria and eukaryotes reside in two families: the glycerophospholipids and the sphingolipids. In both cases, a polar (hydrophilic) head group is attached to a negatively charged phosphate, which in turn connects to a linker, glycerol in the case of glycerophospholipids and sphingosine in the case of sphingolipids (Figure 15.1). Glycerophospholipids have two fatty acid chains attached to the glycerol linker, whereas sphingosine provides one built-in chain, which joins with another fatty acid in sphingolipids.

Such modular structure allows for enormous diversity of lipid types through the exchange of variable parts, including the head groups. The most common glycerophospholipid head groups

Evolutionary Cell Biology. Michael Lynch, Oxford University Press. © Michael Lynch (2024). DOI: 10.1093/oso/9780192847287.003.0015

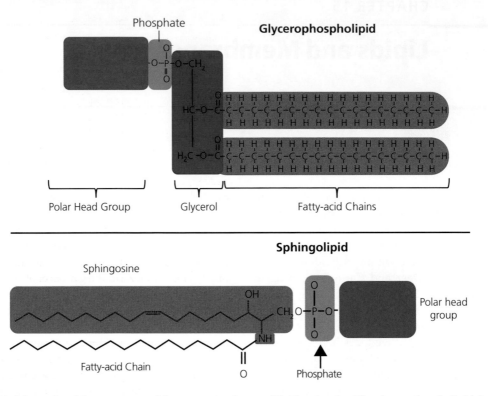

Figure 15.1 Schematics of the structures of the two major classes of lipid molecules. The glycerophospholipid depicted here is saturated, as the fatty-acid tails contain only single carbon-carbon (C–C) bonds. The sphinoglipid has a single double bond, denoted by the double line in the tail, where each kink denotes a C (and the side hydrogens are not shown). A third common group of membrane lipids (not shown) consists of a diverse array of sterols, which lack head groups and intercalate between the fatty-acid tails of membranes.

are choline, ethanolamine, serine, glycerol, inositol, and phosphatidyl glycerol. The cognate lipids are known, respectively, as phosphatidylcholines, phosphatidylethanolamines, phosphatidylserines, phosphatidylglycerols, phosphatidylinositols, and cardiolipins. Additional structural diversity is associated with the number of carbon atoms and the numbers and locations of double C=C bonds in the fatty-acid chains. (Double bonds are referred to as unsaturated, as the carbon atoms are bound to only single hydrogen atoms). The lengths of fatty-acid chains are typically in the range of 14 to 22 carbons, whereas the number of C=C bonds is usually between 0 and 5, and these features have a strong influence on membrane width and flexibility. In various phylogenetic groups, there are still other layers of combinatorial complexity, with the head groups of some lipids being modified

by additions of various small molecules, and some fatty acids containing methyl side branches and/or ring structures at the ends (Geiger et al. 2010; Buehler 2016). Although the precise functions of most such variants are unknown, they may play roles in thermal stability, permeability, and/or protection from various damaging agents.

In contrast to the water soluble head-groups of membrane lipids, the fatty-acid tails are highly hydrophobic. As a consequence of this amphipathic (or synonymously, amphiphilic) structure, the roughly cylinder-shaped lipid molecules naturally self-associate into organized aggregates, with their hydrophobic tails lying parallel to each other in single sheets (Figure 15.2). Moreover, the most thermodynamically stable state is one in which two sheets (leaflets) align with their tails contra-posed, minimizing the contact of hydrophobic tails with water, and leaving flexible walls of head groups on the sides exposed to water. The internal hydrophobic environment of lipid bilayers makes

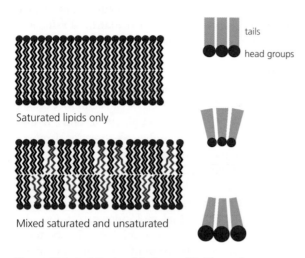

Figure 15.2 Architectural features of lipid membranes. **Left:** Lipid bilayers, with the hydrophilic head groups in red and the hydrophobic fatty-acid tails in black. In the lower left, some individual molecules (blue) have C=C (unsaturated) bonds, yielding slightly kinked tails and a more open membrane. **Right:** The width of the tail region relative to the head group determines the tendency of a membrane to curve inwardly versus outwardly.

them extremely impermeable to charged ions, which must then be transported through gated protein channels (Chapter 18).

Most classes of phospholipids are shared among bacteria and eukaryotes, although their relative usages can vary dramatically (Table 15.1), even across leaflets. The maintenance of such lipid diversity over billions of years of evolution may relate to the variation in structural flexibility endowed by alternative head groups and fatty-acid chains. Indeed, microbial species are generally phenotypically plastic with respect to the lipid profiles of their membranes, shifting their composition in response to environmental change, e.g., using phosphorus-free lipids instead of phospholipids when phosphorus is limiting (Benning et al. 1995; Van Mooy et al. 2009; Zavaleta-Pastor et al. 2010; Carini et al. 2015). With increasing temperature, many cells physiologically remodel their membranes to contain lipids with longer and more saturated fatty acids or to incorporate different head groups (Holm et al. 2022). By this means, membrane fluidity and permeability are kept relatively constant, a process known as homeoviscous adaptation (Sinensky 1974; van de Vossenberg et al. 1995). Without a shift in lipid composition, increased temperature would

Table 15.1 Fractional contributions of lipid molecules to cell membranes in select species. The surveys exclude contributions from sterols and proteins, and are generally given for optimal growth conditions. The central point is that distantly related species often utilize the same types of lipids, although in different proportions.

Organism	PC	PE	PG	PI	PS	C	LPG	O
Bacteria:								
Bacillus subtilis	0.00	0.24	0.35	0.00	0.00	0.18	0.23	0.00
Caulobacter crescentus	0.00	0.00	0.88	0.00	0.00	0.12	0.00	0.00
Escherichia coli	0.00	0.75	0.19	0.00	0.00	0.06	0.00	0.00
Staphylococcus aureus	0.00	0.00	0.53	0.00	0.00	0.07	0.40	0.00
Zymomonas mobilis	0.13	0.62	0.20	0.00	0.00	0.01	0.00	0.03
Eukaryotes:								
Mus musculus, thymocytes	0.57	0.21	0.00	0.07	0.10	0.00	0.00	0.06
Vigna radiata, seedlings	0.47	0.35	0.05	0.05	0.08	0.00	0.00	0.00
Dictyostelium discoideum	0.29	0.55	0.01	0.08	0.02	0.02	0.00	0.03
Dunaliella salina	0.15	0.41	0.15	0.06	0.00	0.00	0.00	0.22
Saccharomyces cerevisiae	0.17	0.18	0.00	0.23	0.21	0.03	0.00	0.19
Schizosaccharomyces pombe	0.42	0.23	0.00	0.25	0.03	0.06	0.00	0.02

Abbreviations: PC, phosphatidylcholine; PE, phosphatidylethanolamine; PG, phosphatidylglycerol; PI, phosphatidylinsoitol; PS, phosphatidylserine; C, cardiolipin; LPG, lysophosphatidylglycerol; O, other.
References: *Bs*: den Kamp (1969); López et al. (1998); *Cc*: Contreras et al. (1978); *Ec*: Raetz et al. (1979); Rietveld et al. (1993); *Sa*: Haest et al. (1972); Mishra and Bayer (2013); *Zm*: Carey and Ingram (1983). *Mm*: Van Blitterswijk et al. (1982); *Vr*: Yoshida and Uemura (1986); *Dd*: Weeks and Herring (1980); *Ds*: Peeler et al. (1989); *Sc*: Zinser et al. (1991); Tuller et al. (1999); Blagović et al. (2005); *Sp*: Koukou et al. (1990).

magnify membrane permeability and fluidity, eventually leading to the loss of cell homeostasis. Homeoviscous adaptation has been observed in all domains of life (Haest et al. 1969; Arthur and Watson 1976; Hazel 1995; Toyoda et al. 2009; Nozawa 2011; Oger and Cario 2013; Ernst et al. 2016), and can be especially refined in organisms such as mammalian pathogens that regularly experience large shifts in temperature (external environment vs. host) (Li et al. 2012). Monitoring mechanisms, essential for an adaptive physiological response, involve proteins that regularly probe membranes for levels of fluidity (Harayama and Riezman 2018).

Finally, as noted in Chapter 3, the structures of lipid molecules in archaea differ significantly from those of eukaryotes and bacteria (Koga and Morii 2007; Chong 2010; Oger and Cario 2013; Buehler 2016). Most notably, archaea generally utilize isoprenoid hydrocarbon chains (with methyl side groups branching off the tails, rather than simple hydrogen atoms). Despite these differences, however, most of the head groups utilized in phospholipids in eukaryotes and bacteria are also deployed in archaea. A particularly unique aspect of archaeal membranes is the partial use of bipolar lipids, which span the entire width of the membrane.

Membrane Structure

Despite their flexibility and fluidity, membranes have a high capacity for maintaining stable sheet-like structures. Owing to the difficulties of moving a polar headgroup through the hydrophobic interior of a bilayer, swaps of molecules between leaflets are negligible unless promoted by specialized transport proteins. However, as the individual molecules are held together by non-covalent forces, lateral diffusive movement of molecules within a leaflet is essentially unavoidable.

This being said, lipid variants are not homogeneously distributed within membranes. Rather, molecules tend to aggregate with their own types as they encounter each other by diffusion, leading to a sort of self-organized phase separation that generates patchy variation in membrane properties. Larger patches, referred to as lipid rafts, are themselves capable of diffusive lateral movement across membranes. Such variation is relevant to the distribution of membrane proteins, as a stable platform for any particular membrane-spanning protein requires a good match between the membrane thickness and the protein's hydrophobic transmembrane domains. Hence, specific types of proteins are associated with particular lipid rafts, further adding to membrane heterogeneity (Mitra et al. 2004).

The lateral diffusion coefficients of individual glycerophospholipid molecules in a bilayer are $D \simeq 2$ to $4 \, \mu m^2 / sec$ at $25°C$ (Devaux and McConnell 1972; Wu et al. 1977; Jin et al. 1999). Thus, letting $D = 3$, and assuming unbiased directional movement such that the mean squared distance travelled over a two-dimensional surface is $4Dt$, where t is measured in seconds (Chapter 7), the mean absolute distance travelled is $\sim 2\sqrt{Dt}$, or $\sim 3.5 \, \mu m / sec$. This implies that individual molecules can diffuse the equivalent of full lengths of small cells in a matter of seconds. Rates of diffusion of lipid rafts are one to two orders of magnitude lower, declining with the size of the raft (Schütte et al. 2017; Zeno et al. 2018).

To put this into perspective, recall from Chapter 7 that diffusion rates of proteins are on the order of $10-40 \, \mu m^2 / sec$ within a cytoplasmic environment. Thus, diffusion inhibition from molecular crowding within membranes is substantially greater than in the cytoplasm. The lateral diffusion coefficients of membrane proteins are even lower than those for lipids, e.g., $\sim 0.04-0.3 \, \mu m^2 / sec$ for mitochondrial membrane proteins (Gupte et al. 1984). In *E. coli*, such coefficients decline from ~ 0.2 to $0.02 \, \mu m^2 / sec$ as the number of membrane-spanning helices in proteins increases from 3 to 14 (Kumar et al. 2010; Schavemaker et al. 2018). Thus, although membrane proteins are mobile in an absolute sense, they are effectively stationary from the perspective of cytoplasmic proteins.

Lipid molecules are not strictly cylindrical in shape. Rather, depending on the size of the head group relative to the tail width, the overall form can be closer to a cone, with unsaturated fatty-acid tails tending to fan out. As a consequence, curvature is induced when molecules of particular geometric shapes associate with each other (Figure 15.2). This simple structural mechanism reduces the energy necessary to mold membranes into particular shapes, and in part explains the differential distribution of lipid types on the inner versus outer leaflets of membranes.

Generation of stronger curvature typically requires additional sources of bending energy derived from ATP- or GTP-hydrolysing processes (Helfrich 1973). For example, motor proteins moving along microtubules or actin filaments (Chapter 15) can pull membranes into tubular forms. In addition, a wide variety of membrane proteins have functions specifically associated with the bending and sculpting of membranes into specific forms (Shibata et al. 2009; Jarsch et al. 2016). Insertions of hydrophobic protein wedges naturally cause a membrane to bend towards the narrower end of the inserted protein, as occurs when ATP synthase molecules inhabit the tips of cristae on the internal membranes of mitochondria. Scaffolding proteins with natural curvature and an affinity for specific lipid head groups can force lipid bilayers to conform to the same curvature, and are widely used in the formation of vesicles (as described in more detail below). Transmembrane proteins, which traverse the space between two membranes, help maintain specific distances between leaflet layers.

Eukaryotes and the Endogenous Organellar Explosion

The proliferation of internal membrane-bound organelles is one of the hallmark features distinguishing eukaryotes from prokaryotes. Prominent in almost all eukaryotic cells are the ER (the site of production of many proteins and lipids), the Golgi (the site of post-translational processing and transport), and lysosomes and peroxisomes (devoted to degradation). Based on their phylogenetic distributions across eukaryotes, all of these embellishments likely date to LECA. With two exceptions, all eukaryotic organelles are believed to be endogenous in origin, having developed by descent with modification and containing no internal genomes. In Chapter 24, special attention is given to mitochondria and chloroplasts, which arose exogenously via endosymbiotic events dating back to known bacterial lineages. Many organelles have membrane-contact sites (e.g., covering 2–5% of the surface areas between the ER and contacted mitochondria;

Phillips and Voeltz 2016), with functions in inter-organellar communication, further contributing to the complex interior of eukaryotic cells.

At both the cell-biological and evolutionary levels, there are numerous unsolved problems as to how individual organelle types achieve their distinctive shapes, identities, and interactions. Organelle assembly and identity may in part be an intrinsic consequence of the self-assembly features of the component molecules described above. However, other mechanisms must be involved. For example, major portions of the core ER can have a layered, spiralling architecture, resembling a parking structure (Terasaki et al. 2013). The ER is also continuous with the nuclear envelope (Foundations 15.1), and the peripheral ER often exhibits a matrix-like structure involving narrow tubules (Nixon-Abell et al. 2016). These and other morphological features appear to be generated by the relative concentrations of just two types of membrane-shaping proteins, one encouraging flat sheets and the other curvature (Shemesh et al. 2014). Phylogenetic diversification extends to the Golgi, the central hub for vesicle trafficking and post-translational modification. Many eukaryotes lack classical stacked Golgi, and yet contain the genes associated with Golgi trafficking, suggesting independent loss of this morphology at least eight times (Dacks et al. 2003; Mironov et al. 2007; Mowbrey and Dacks 2009).

A common view is that the emergence of organelles led to fundamentally superior organisms (Lane and Martin 2010; Gould 2018). This embrace of the assumption that increased cellular complexity is always a positive development ignores the fact that the currency of natural selection is the rate of progeny production, not the preponderance of offspring embellishments. While one can marvel at the many intricacies associated with eukaryotic organelles, as noted in several prior chapters, numerous lines of evidence are inconsistent with the idea that eukaryotic cell structure is intrinsically advantageous relative to that of prokaryotes, and from several perspectives, it is notably worse (e.g., growth rates). Organelles do enable eukaryotes to accomplish cellular tasks in novel ways relative to prokaryotes, but as discussed in Chapter 17, the investment in internal membranes comprises a substantial energetic burden on cells. In addition, although the body plan of the eukaryotic cell allows for novel functions, such as vesicle transport, such elaborations can also impose liabilities. For example, the eukaryotic endocytic pathway provides a direct route for cellular entry and exit by numerous pathogens (e.g., Heuer et al. 2009; Szumowski et al. 2014; Renard et al. 2015; Shen et al. 2015; Shi et al. 2016).

Notably, the ability to evolve internal membranes is not an exclusive feature of eukaryotes. Although most prokaryotes are channelled down pathways of morphological simplicity, many of them have internal cellular structures (Kerfeld et al. 2018; Greening and Lithgow 2020). For example, the planctomycetes, a group of aquatic bacteria, are endowed with substantial tubular networks of internal membranes (Fuerst and Sagulenko 2011; Acehan et al. 2014; Boedecker et al. 2017), reminiscent of the endomembrane system of eukaryotes, but likely independently evolved. Although the functions of such membranes are not fully resolved, one structure (the anammoxosome) sequesters a reaction that converts nitrite and ammonium ions to nitrogen gas (van Niftrik and Jetten 2012). Some members of the planctomycetes are capable of phagocytosis and reproduce by budding (Shiratori et al. 2019), possibly using eukaryote-like mechanisms, and related groups of bacteria (e.g., verrucomicrobia and chlamydiae) also have endomembranes. Simple mutations in the membrane-binding protein MreB (Chapter 16) can induce striking invaginations in *E. coli* cells (Salje et al. 2011). Finally, the archaeon *Igniococcus hospitalis* deploys two membranes in a way that is quite distinct from the double membranes of Gram-negative bacteria, with carbon metabolism sequestered to the voluminous intermembrane space and the genome and RNA processing kept separate inside the internal membrane (Flechsler et al. 2021).

There are many other examples of compartmentalized organelles in bacteria. For example, the photosynthetic machinery in cyanobacteria is sequestered within a carboxysome (Savage et al. 2010), and numerous cyanobacteria regulate their buoyancy by use of gas vacuoles (Walsby 1972). A microcompartment for ethanolamine metabolism consisting of hexameric protein subunits is present in *E. coli* (Tanaka et al. 2010), and *Salmonella* harbors another such structure for propanediol utilization (Chowdhury et al. 2015). In these particular cases, the intracellular compartment consists of an assembly of protein multimers, much like the capsids of viruses. However, magnetotactic bacteria contain crystals of magnetite or iron phosphate enclosed by phospholipid membranes (Byrne et al. 2010; Jogler et al. 2011). The giant cells of *Epulopiscium,* a symbiotic bacterium (up to 1 mm in length) inhabiting triggerfish guts, contain stacked 'vesicles' of unknown function near the cell membrane (Robinow and Angert 1998), and the even larger *Thiomargarita magnifica*, which attains lengths greater than 1 cm, sequesters its genome and ribosomes within a membrane-bound organelle (Volland et al. 2022).

These examples suffice to demonstrate that prokaryotes are free to evolve internal cell structures when the selective pressures to do so are present, which by extension implies an absence of such selection, or even counter-selection, in most microbial species. This leaves the evolutionary conditions leading to the widespread proliferation of internal membranes in the ancestral eukaryote as one of the greatest mysteries of evolutionary cell biology.

Not ruling out an early role for adaptation, little more can be said than that unknown historical contingencies led to the adoption and apparent permanent retention of the eukaryotic cell plan at some point on the path from FECA to LECA. It has been suggested that the internal membrane system arose in an ancestral lineage with naturally invaginated cell membranes, with the protrusions merging gradually over evolutionary time to create internal vesicles and the nuclear envelope (Baum and Baum 2014; Imachi et al. 2020), but the evolutionary incentive to do so remains unclear. Nonetheless, once established, internal cell membranes may have promoted downstream changes, such as the colonization of introns in nuclear genomes, which essentially eliminated the possibility of ever relinquishing the nuclear envelope (as described more fully below).

Vesicle Trafficking

Although most molecular interactions in prokaryotes are governed by diffusion-like processes, eukaryotic cells rely extensively on active transport of macromolecules. Such transport pathways include the endocytic internalization of extracellularly derived cargoes, vesicle transport of molecules from one organelle to another, and the translocation of proteins and RNA molecules across the nuclear envelope. Each of them involves one or more modes of intermolecular communication. Correct substrates must be identified to the exclusion of erroneous and sometimes harmful cargoes, and individual vesicles must be delivered to their appropriate destinations. Thus, intracellular transport raises many of the same issues encountered with metabolic (Chapter 19), transcription (Chapter 21), and signal-transduction networks (Chapter 22), most notably the evolutionary origin and maintenance of the specificity of the languages underlying intermolecular interactions.

Eukaryotes deploy lipid-bound vesicles in a wide range of trafficking activities, including endocytosis and exocytosis, digestion, and transport between the ER and the Golgi. The life cycle of a vesicle begins with assembly at sites of initiation, usually by pinching off a parental membrane, proceeds through a period of delivery through the intracellular domain, and ends with docking and fusion to another lipid-bound compartment at the site of delivery (Figure 15.3). As these processes all occur simultaneously and bi-directionally, the quantitative partitioning of lipid membranes throughout the cell can remain in a roughly steady-state condition, such that membrane areas lost by donors are balanced by those gained by recipients, despite substantial traffic between compartments. The rate of membrane flux can be quite high. For example, amoeboid cells can internalize the equivalent of the entire surface membrane in the form of endocytic vesicles in just an hour (Ryter and de Chastellier 1977; Bowers et al. 1981; Steinman et al. 1983), and the blood parasite *Trypanosoma* can do so four times per hour (Engstler et al. 2004).

The origin of the vesicle-transport system remains obscure, as there are few obvious orthologs of any components known in prokaryotes. However, the planctomycetes (noted in the prior section) lend credence to the idea that some aspects of an endomembrane system may have been present in the primordial eukaryote, i.e., in FECA (Lonhienne et al.

2010). Such a hypothesis is consistent with phylogenetic analyses suggesting an origin of various aspects of endocytosis as well as the ER and the secretory system prior to LECA (Jékely 2003; Dacks et al. 2008; Podar et al. 2008; Makarova et al. 2010; Wideman et al. 2014; Klinger et al. 2016; Zaremba-Niedzwiedzka et al. 2017; Kontou et al. 2022).

Pointing out that many bacteria release outer-membrane vesicles into the extracellular environment, Gould et al. (2016) suggested that the eukaryotic endomembrane system originated with such processes contained within the mitochondrial endosymbiont. One obvious concern with this hypothesis is the absence of any mechanistic evolutionary argument for how or why the simple production of vesicles by the primordial mitochondrion would have become transformed into a highly organized and nuclear-encoded vesicle transport system by the host cell. The same unknowns exist for the invagination hypothesis of Baum and Baum (2014). Given that any such modifications must have involved incremental evolutionary processes, future understanding in this area would profit from an evolutionary genetic perspective.

The following two subsections provide a brief and simplified overview of what is known about the various steps from cargo uptake to delivery, the focus being on general principles. The enormously detailed molecular mechanisms can be explored further in many specialized publications.

Vesicle production

Lipid membranes are constantly recycled via fission and fusion processes. Rather than forming *de novo*, vesicles are typically derived via the invagination (endocytosis) or budding (inter-organelle transport) of a pre-existing membrane. Vesicle birth generally involves the recruitment of specific proteins dedicated to inducing membrane curvature. In most cases, cage-like lattices are used to support developing vesicles before they are eventually pinched off from parental membranes (Field et al. 2011): 1) clathrin-coated vesicles import cargoes across the cell membrane in the form of endosomes, and are also deployed in the trans-Golgi network; 2) COPI (coat protein I)-coated vesicles carry cargoes between different Golgi compartments and from

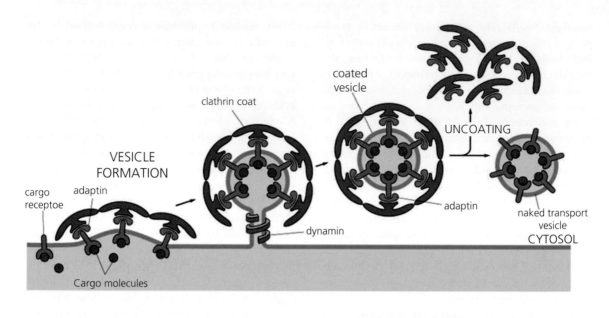

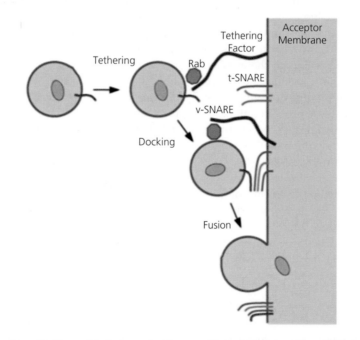

Figure 15.3 Generic schematic of a vesicle-transport pathway with several stages. **Above:** Cargo capture: external cargo molecules (red) are initially bound by specific cargo-transport proteins (blue). Vesicle budding and coating: specific adaptor proteins (light green) bind to the cargo receptors, and in turn recruit vesicle-coat proteins (dark green), which induce membrane curvature (dark grey). Vesicle scission: coat proteins continue to be recruited, and the stem is eventually squeezed off with a concatamer of dynamin molecules (yellow coil). Vesicle uncoating: the coat proteins are removed, leaving the lipid-bound vesicle free to bind to a recipient membrane. **Below:** Tethering: a specific RAB protein (blue) provides recognition between the vesicle and a tethering effector molecule (green), and vesicle and target SNARE proteins (orange and lavender) join to seal the final connection. Fusion: after docking with the recipient membrane, the cargo is unloaded.

the Golgi to the ER; and 3) COPII-coated vesicles export cargoes from the ER. In all cases, large protein lattices assemble from lower-order trimers (clathrin and COPI) or dimers (COPII), which then co-assemble into structures with distinct geometric shapes and sizes (Figure 15.4).

Given the widespread presence of clathrin throughout the eukaryotic domain, the logical conclusion is that LECA also deployed clathrin-coated vesicles (Field et al. 2007). Nevertheless,

substantial diversification of clathrin-coated vesicles has occurred, and their diameters range from ~30 to 200 nm among observed species (McMahon and Boucrot 2011; Kaksonen and Roux 2018). Such size differences are in part a function of the architecture of the clathrin molecule itself, e.g., the numbers of α helices constituting the long connecting arms, but turgor-pressure differences among cell types (which would influence membrane bendability) might be involved as well. Size variation is also

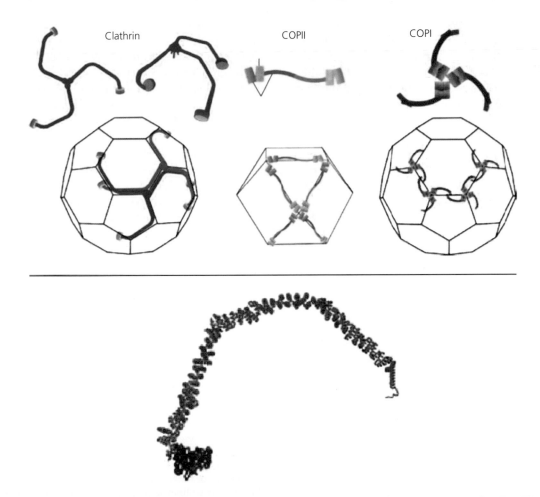

Figure 15.4 Schematics of the higher-order structure of the proteinaceous coats of eukaryotic lipid vesicles. **Top:** The basic structures of clathrin and COPI coats are homotrimeric subunits, whereas COPII is a heterodimer. These subunits organize into lattices with distinct geometric shapes, the specific dimensions of which are defined by the lengths of the domains of the monomeric subunits. **Bottom:** A monomeric subunit of clathrin. Each linear domain of the arm consists of a long series of α helices, the numbers of which define the overall dimensions of the lattice. It remains to be seen whether such structures vary in any meaningful way phylogenetically. From Edeling et al. (2006) and Harrison and Kirchhausen (2010).

known for COPI- and COPII-coated vesicles (Faini et al. 2012).

Although many details still need clarification, clathrin-vesicle formation initiates when specialized proteins bound to the source membrane recruit the coat proteins (Boettner et al. 2011; McMahon and Boucrot 2011; Kirchausen et al. 2014). A primary group of such proteins called adaptors (or adaptins) are thought to recognize specific cargo-recruitment molecules, which in turn have affinities to specific cargo types. In this sense, adaptins serve as an informational link between cargoes and coat recruitment, although this is a simplified view in that other ancillary proteins can be involved in clathrin recruitment, some of which appear to be lineage specific (Adung'a et al. 2013). In addition, the finer details on how clathrin-coated pits come to contain their cargoes or even whether cargoes are essential to trigger vesicle formation remain unclear (Kaksonen and Roux 2018). Pits may stochastically develop and abort, and cargoes with higher affinities for such settings will naturally accumulate to a greater extent (Weigel et al. 2013).

With one exception, adaptor proteins are heterotetramers comprised of two large, one medium, and one small subunit. Each subunit has orthologs across all adaptors and is also related to a particular protein involved in the COPI coat (Schledzewski et al. 1999). Moreover, the two large subunits appear to have arisen by a gene duplication that preceded the origin of the different adaptor complexes, and the same is true of the medium and small subunits. These observations suggest that the ancestral adaptor (in LECA) may have been a heterodimer of just single small and large subunits (Schledzewski et al. 1999). Under this hypothesis, subsequent duplication of both subunits followed by divergence led to the heterotetrameric state, with further duplications and divergence of all subunits leading to the various classes of adaptors.

Because all five known adaptor proteins as well as COPI-coated vesicles are found throughout the eukaryotic domain, their diversification must have preceded LECA, as in the case of clathrin. Using the form of the genealogical relationships among the various complexes then provides a potential means for ordering events in the diversification of

vesicle-trafficking pathways on the path between FECA and LECA (Figure 15.5). Such a perspective leads to the suggestion that an early ancestral adaptor diverged from the COPI coat, with the former then undergoing a series of duplications leading to five different adaptors, possibly prior to the emergence of clathrin (Hirst et al. 2011). The form of relationship between the gene-family members suggests that the deployment of adaptors in endosomes emerged prior to the expansion of their use in the trans-Golgi network. Likely, other adaptor-like complexes remain to be discovered, given that a distantly related ortholog has recently been found in scattered lineages (Hirst et al. 2014). Unlike the complexes described previously, this new complex is a heterohexamer.

The protein dynamin is central to the completion of vesicle formation, at least in metazoans (Praefcke and McMahon 2004). After the development of clathrin-coated invagination begins, dynamin assembles into collar-like helical structures and uses mechanical energy derived from GTP hydrolysis to pinch off the neck. Once a critical length of the collar has been achieved by oligomerization of dynamin, a chain reaction generates the mechanical force necessary for vesicle release. Dynamin

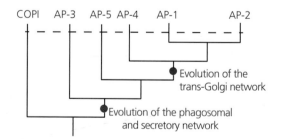

Figure 15.5 A view of the timing of the evolutionary diversification of the five known adaptor proteins (AP) and the COPI coat subunit. All of the nodes on the tree emerged prior to the last eukaryotic common ancestor (LECA, dashed line), as all components are distributed throughout the entire eukaryotic phylogeny. Thus, the structure of the tree yields a hypothesis about the order of events in the functional diversification of adaptor proteins, e.g., implying a likely early presence of COPI, but a relatively late recruitment of adaptor proteins to the trans-Golgi network. From Hirst et al. (2011).

appears to be absent from many eukaryotic lineages, which nonetheless often harbor a separate clade of dynamin-like proteins that likely serve a similar function (Liu et al. 2012b; Briguglio and Turkewitz 2014).

An extreme form of membrane-mediated ingestion is the process of phagocytosis. Aided by their extensive cytoskeletons, most eukaryotes without cell walls are able to ingest large particles, including other cells. They do this by invagination of the surrounding cell membrane, followed by internalization and fusion with digestive vacuoles. Phylogenetic analysis suggests that LECA harbored many of the genes underlying the core machinery employed in the phagosome of today's species (Yutin et al. 2009; Boulais et al. 2010), with considerable independent additions and diversifications occurring in different descendant lineages, and complete losses in a few cases (e.g., chlorophytes and fungi). The logical conclusion is that LECA had no cell wall, and if not capable of phagocytosis, was primed for its subsequent emergence. Whether this capacity enabled the primordial eukaryote to ingest the ancestral mitochondrion, or came later, remains unclear (Chapters 3 and 23).

Feeding by phagocytosis demands considerable membrane recycling. For example, the digestive system of ciliates consists of a steady-stream of food vacuoles. *Paramecium* and *Tetrahymena* cells produce several hundred to a few thousand of these each day, resulting in the recycling of the plasma membrane 5–50× during complete cell cycles (Lee 1942; Rasmussen 1976; Smith-Sonneborn and Rodermel 1976; Fok et al. 1988; Ramoino and Franceschi 1992; Gangar et al. 2015; Chan et al. 2016). When the predatory ciliate *Euplotes* feeds on the smaller ciliate *Tetrahymena*, food-vacuole membrane equivalent to the entire surface area of the predator's cell can be ingested every five minutes (Kloetzel 1974). This is also approximately the case for amoebae feeding on ciliates or other prey (Marshall and Nachmias 1965; Wetzel and Korn 1969).

Vesicle delivery

Once formed, vesicles must find their way to an appropriate recipient, and in doing so, avoid fusing with inappropriate membranes. The entire process entails multiple layers of information exchange, but central to such navigation are members of the RAB GTPase family of proteins, which help specify the locations to which vesicles are delivered. RABs act as switches by undergoing conformational changes when bound by GDP (inactive state) versus GTP (active state), via processes that involve two other diverse sets of proteins. Specific GEFs (guanine exchange factors, which promote GDP release) catalyse conversion from the GDP- to GTP-bound forms, leading to activation, whereas GAPs (GTPase-activating proteins) do the reverse, leading to GTP hydrolysis and inactivation. Still other proteins are involved in RAB activation/deactivation cycles; e.g., RAB-escort proteins deliver their cognate RABs to specific cellular locations, whereas after inactivation, RABs are recycled back to their membranes of origin via other specific proteins (called GDP-dissociation inhibitors). The N-terminal residues of RABs contain vesicle-specificity information, whereas the C-terminals are involved in targeting and adhesion to destination lipid membranes. Still other enzymes endow these regions with post-translational modifications that confer specificity (Pylypenko and Goud 2012).

The main point here is that the transport of specific kinds of vesicles to precise target locations involves an elaborate choreography of several layers of specialized protein–protein interactions, involving multiple gene-family expansions that are essentially unique to eukaryotes. The genomes of eukaryotic species typically harbor 10–100 distinct RABs, and phylogenetic analysis suggests that LECA may have contained up to 23 RAB genes (Elias et al. 2012; Klöpper et al. 2012), with some lineages then experiencing losses of distinct family members. Fungi commonly encode no more than a dozen (Brighouse et al. 2010).

Also involved in vesicle delivery to specific sites are a large set of SNARE (soluble N-ethylmaleimide-sensitive factor-attachment protein receptor) proteins, which act in a zipper-like fashion, with coordination between specific sets on vesicles and recipient membranes. The appropriate recognition groups attach to each other in bundles known as SNARE pins. Like the coat proteins, adaptins, and RABs, the main SNARE types diversified into subfamilies prior to LECA (Klöpper et al. 2007). Many of the subfamilies have expanded in lineage-specific ways, but with no obvious relationship to organismal complexity (Sanderfoot 2007; Kienle et al. 2009).

Evolutionary issues

Although many details still require clarification, the information summarized above indicates that essentially all of the components of the vesicle-trafficking system of eukaryotes diversified through multiple gene-duplication events prior to LECA. As discussed further in Chapter 24, one or more whole-genome duplication events on the path from FECA to LECA may have contributed to such diversification.

To this end, to help explain the diversification of transport pathways, the organelle-paralogy hypothesis (Figure 15.6) invokes repeated rounds of gene duplication and joint coevolution of clusters of components towards more specialized functions (Dacks and Field 2007; Mast et al. 2014). However, although such descent with modification provides a logical argument for diversification (Ramadas and Thattai 2013), many issues remain unresolved. Gene duplication alone does not ensure diversification in function, especially in a multi-layered system that requires coordinated behavior of hundreds of component parts. At the very least, such evolution requires a series of sub- and/or neofunctionalizing events to ensure the coordinated preservation of mutually interacting components (Foundations 15.1). The population-genetic conditions permissive for such specialization have not been worked out and seem rather formidable, as the subcomponents of each descendent pathway must not only evolve pathway-specific features but also relinquish pre-duplication features to avoid pathway crosstalk.

Equally challenging is understanding how the multiple layers of communication necessary for specialized trafficking pathways evolve. Adaptor proteins provide the interface between various cargoes and the specific coat proteins of vesicles; different RAB proteins specify unique types of vesicles and convey information on subcellular localization; and specific pairs of vesicle and target SNARE proteins recognize each other to ensure vesicle delivery to proper destinations. Although such a layered system might be viewed as exquisitely intricate, it

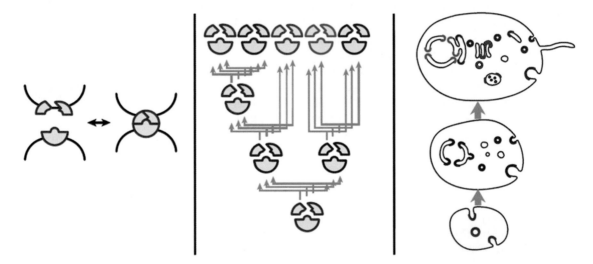

Figure 15.6 The organelle-paralogy hypothesis (Dacks and Field 2007; Mast et al. 2014). **Left:** An ancestral communication mechanism between two molecular structures, e.g., an adaptor protein and its cargo receptor; a RAB protein and a tethering molecule; or a v-SNARE and a t-SNARE. (See Figure 15.3). **Centre:** Nested sets of duplications of the genes for both participants, followed by molecular coevolution, eventually lead to interacting pairs with specific functions (represented by different colours) isolated from other such pairs. **Right:** These specializations are proposed to lead in turn to partitioning with respect to subcellular functions and localizations.

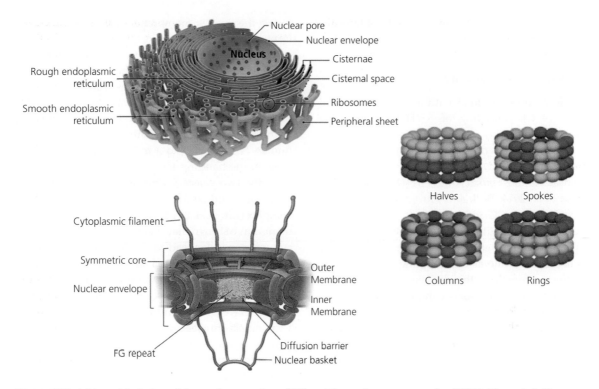

Figure 15.7 A hierarchical view of the nuclear envelope (NE) and the nuclear-pore complex (NPC). **Upper left:** The outer layer of the NE is continuous with the endoplasmic reticulum. Perforating both sides of the NE are nuclear pores. From Blackstone and Prinz (2016). **Lower left:** The NPC, through which cargoes must be transported, is a complex structure consisting of hundreds of proteins encoded by multiple loci. From Jena (2020). **Right:** The central core of the NPC consists of eight spokes, each consisting of two columns, for a total of 16 columns per pore, all of which are organized into a cylinder containing four layers.

comes at a substantial cost in terms of bioenergetic demand and mutational vulnerability of the large number of components.

The Nuclear Envelope

If there is an iconic feature of the eukaryotic cell, it is the housing of the genome inside the nucleus. Rather than floating freely in the cytoplasm, the nucleus is surrounded by a double membrane (involving two lipid bilayers), with the outer layer being continuous with the ER and periodically bending around at nuclear pores to form the inner nuclear membrane (Figure 15.7). There is also a proteinaceous support layer interior to the nuclear envelope, consisting of lamins in metazoa and amoebozoa (Simon and Wilson 2011; Burke and

Stewart 2013) and apparently unrelated proteins in plants and other organisms (Cavalier-Smith 2005).

Among other things, genomic sequestration behind the nuclear envelope separates transcription (intranuclear) from translation (extranuclear), paving the way for the emergence of introns that must be spliced out of pre-messenger RNAs before meeting the cytosolic ribosomes (Lynch 2007). It is through the nuclear pores that mRNAs and partially assembled ribosomes are actively exported to the cytoplasm and nuclear proteins (e.g., transcription factors, histones, and DNA-repair enzymes) are imported. There is evidence in flies that clusters of proteins are sometimes exported as particles to the cytoplasm by budding of the inner nuclear membrane (Speese et al. 2012), but it remains unclear whether this is a common phenomenon, and it is virtually certain that the bulk of transport proceeds through pores.

Nuclear-pore architecture

Nuclear pores are lined with a nuclear-pore complex (NPC), consisting of ∼500–1000 individual Nup (nucleoporin) proteins encoded by ∼30 separate genes. Exceeding the mass of a ribosome by more than tenfold, the NPC is the largest protein complex in most eukaryotic cells (Devos et al. 2014; Field et al. 2014; Beck and Hurt 2017). A brief discourse on the NPC will reinforce the contention that large complexes within eukaryotic cells are typically grown out of a series of gene-duplication events (Chapter 6), while also illustrating that despite its conserved functions, the NPC has experienced considerable diversification at the architectural level. There are interesting lessons in coevolution to be learned as well, as pathogens that require entry into the nucleus (e.g., for replication and/or transcription) must successfully navigate the NPC.

The core of the NPC is both vertically and radially symmetrical, consisting of four stacked rings (two on the nuclear side and two on the cytoplasmic side), each comprised of eight spokes, which in turn consist of two parallel columns of several proteins (Figure 15.7). The proteins in adjacent columns are related as pairs, each derived by gene duplication (Alber et al. 2007a,b). This one-to-one correspondence of multiple pairs of duplicates is again consistent with a massive amount of duplication activity in the ancestor leading to LECA.

Comparison of the parts lists from diverse species suggests that LECA had an NPC structure very much like that in today's species. Moreover, its evolutionary roots appear to be associated with the proteins involved in vesicle production. Most notably, the core proteins of the inner rings appear to be related to the membrane-bending proteins involved in vesicle formation (COPI, COPII, and clathrin), motivating the hypothesis that all of these molecules are derived from a common ancestral protein, deemed the protocoatomer (Devos et al. 2004; Mans et al. 2004; Alber et al. 2007a,b; Brohawn et al. 2008). The protein used in the sculpting of internal vesicles of cells (ESCRT) is also used in

fusing the nuclear membranes at the pore junctions (Vietri et al. 2015).

Despite these common roots, there are differences in NPC components among lineages (Mans et al. 2004; Bapteste et al. 2005; DeGrasse et al. 2009; Neumann et al. 2010; Devos et al. 2014; Akey et al. 2022; Zhu et al. 2022). As one example, the overall mass of the *S. cerevisiae* NPC is only ∼50% of that of the human NPC, owing to a reduction in the number of subunits banding together in the ring in yeast. Even the two yeasts *S. cerevisiae* and *S. pombe* have different numbers of subunits in the multimeric complex (Liu et al. 2012a; Stuwe et al. 2015). Experiments have shown that changes in the expression of subunit genes can lead to an alteration in the overall structure, suggesting a simple path to variation in pore composition/size within and among species (Rajoo et al. 2018). Larger compositional changes are known as well. For example, the deployment of proteins on the nuclear and cytoplasmic sides of the pore is asymmetrical in the case of yeast, animals, and land plants, but relatively symmetrical in the case of trypanosomes (Obado et al. 2016a,b).

Finally, it is worth noting that the NPC has evolved a number of secondary functions, including involvement in chromosome organization and positioning and in mediating of transcription of tRNAs and mRNAs (Fahrenkrog et al. 2004; Xu and Meier 2008; Strambio-De-Castillia et al. 2010; Vaquerizas et al. 2010; Ikegami and Lieb 2013). In yeast, and likely many other species, a number of genes have short motifs that target their location to the nuclear periphery via interactions with the NPC (Ahmed et al. 2010). Thus, the NPC evolved to become the hub of many activities beyond cargo transport.

Nuclear transport

The nuclear pore is lined with a large number of phenylalanine-glycine (FG) Nups, each containing up to 50 FG repeats (Figure 15.7). These highly unstructured molecules can be viewed as a spaghetti-like sieve through which cargoes bound by appropriate nuclear-transporter proteins (importins and exportins) are actively delivered while inappropriate molecules are excluded (Sorokin et al. 2007; Grünwald et al. 2011; Hülsmann

2012; Vovk et al. 2016). Such selective filtering relies on molecular communication between the FG Nups and the transporters, as well as between the latter and their specific cargoes. Cargo recognition typically involves a nuclear-localization signal on cargo proteins, which attracts a nuclear-transporter protein. Such signals are generally quite simple, typically involving three or four consecutive basic amino acids (arginine or lysine), although the consensus sequence appears to vary among species (Kosugi et al. 2009). A separate set of transporter proteins is assigned to mRNA export.

The exact mechanisms facilitating cargo transport are not fully resolved, but the process is fast, allowing the delivery of up to 1000 molecules per second per pore (Yang et al. 2004). Transport is governed by gradients of Ran-GTP and Ran-GDP, associated with the transporter-cargo complexes. Specific enzymes devoted to this Ran-GDP/GTP cycle define still another means of molecular communication in the nuclear-transport pathway.

Although the basic mechanism of communication between the transport machinery and FG Nups is conserved across taxa, there is drift in the language used across lineages. For example, human transport substrates are not imported into the nucleus of *Amoeba proteus* unless they are co-injected with human importins (Feldherr et al. 2002). Among yeast species, the FG Nups have diverged at the sequence level at much higher rates than other genes, with the greatest elevation arising in sequences interspersed between the Nup repeats (Denning and Rexach 2007). The ciliate *Tetrahymena thermophila* harbors two nuclei (the transcriptionally silent micronucleus and the transcriptionally active macronucleus), one of which has pores lined with FG Nups, while the other has Nups with novel NIFN repeats, implying distinct permeability of the two nuclear membranes (Iwamoto et al. 2017).

Evolutionary considerations

The universal presence of a nuclear envelope in eukaryotes tells us that it was present in LECA, and this motivates two major evolutionary questions. First, what were the driving forces underlying the emergence of the nucleus? Second, once

established, what secondary evolutionary challenges/opportunities did the nuclear envelope impose on other aspects of cellular evolution?

It is not clear that genome sequestration would have any intrinsic advantage in a prokaryote-like ancestor, and the failure of nearly all prokaryotes to make such a transition over billions of years suggests that there is none. Nonetheless, two hypotheses attempt to explain the origin of a nuclear envelope based on a selection scenario surmised to be unique to eukaryotes.

First, Martin and Koonin (2006) proposed that the evolution of the nuclear envelope was forced by the origin of introns (intervening sequences of messenger RNAs that must be spliced out to yield a productive mRNA) as a mechanism for preventing the early translation of inappropriate (not yet spliced) messages in the cytosol. Second, Jékely (2008) suggested that the origin of the mitochondrion forced the sequestration of nuclear-encoded genes. Here, the idea is that once the ribosomal protein-coding genes of the primordial mitochondrion were transferred to the nucleus of the host cell (Chapter 23), there would have been a risk of constructing chimeric (and potentially malfunctional) ribosomes consisting of mixtures of proteins with host and endosymbiont functions. In principle, this problem could be avoided by assembling the cytosolic ribosomes prior to nuclear export, and addressing the mRNAs for the mitochondrial ribosomal protein-coding genes to the mitochondrion.

A key difficulty with both of these arguments is the assumption of a self-inflicted, pre-established harmful condition from which the host cell was unable to escape. If a problem was deleterious enough to encourage a massive repatterning of cellular architecture, why weren't the original mutational variants that created such a bleak situation simply removed from the population by purifying selection prior to fixation? One potential scenario that might enable the imposition of such a dire setting is an extremely deep and prolonged population bottleneck, on the road from FECA to LECA (Chapter 24).

However, a more plausible alternative is that the nuclear envelope evolved prior to the establishment of introns and mitochondria by some form of positive selection, thereby paving the way for the latter changes secondarily. Alternative possibilities include the protection of the genome from shearing forces in cells with cytoplasmic streaming and/or from invasive self-proliferating genomic parasites. Most bacterial genomes are largely devoid of mobile-genetic elements, in principle because the typically large effective population sizes of such species enables them to resist the fixation of harmful insertions (Lynch 2007). In contrast, few eukaryotes are able to cleanse themselves entirely of such elements, with a large fraction of many eukaryotic genomes being a result of the activities of parasitic DNAs (Chapter 24).

Although the nuclear envelope provides a physical barrier to invasive genomic parasites, it is by no means perfect, and most such elements depend on access to the host genome for survival. Among the most prominent of these are the mobile-genetic elements that literally reside within the nuclear genome – transposons and retrotransposons. To produce their encoded mobilization factors necessary for proliferation, the genetic material of such elements must be transcribed in the nucleus and translated in the cytoplasm, and the resultant products must be able to return to susceptible genomic territories to produce daughter copies.

The FG Nup-gated nuclear pores serve as a primary guardian against uncontrolled element spread, as illustrated by dozens of examples of the coevolution of Nups and genomic parasites. For example, two inner-channel Nups in *Drosophila* play a central role in a pathway for transposon silencing (Munafó et al. 2021). Yeast retrotransposons have a requirement for the host-cell FxFG repeats in Nup124 (Dang and Levin 2000; Kim et al. 2005), although nuclear pore-associated factors also have inhibitory effects on retrotransposition (Irwin et al. 2005). Notably, the same Nup protein is exploited by the human retrovirus HIV-1 (Varadarajan et al. 2005; Woodward et al. 2009; Lee et al. 2010). On the other hand, Nup124 prevents entry of hepatitis B virus, specifically via the FxFG repeats (Schmitz et al. 2010). Many other exogenous viruses have been found to engage in genetic conflicts with Nups of their host species (e.g., Gallay et al. 1995, 1997; Strunze et al. 2005; Satterly et al. 2007; Bardina et al. 2009; Porter and Palmenberg 2009).

Given the potentially high evolutionary rates of the nuclear-pore components driven by infectious agents, and the NPC's involvement in chromosome organization and interactions with the spindle during meiosis, it would not be surprising if the sequence divergence of the NPC sequence played a central role in the emergence of species-isolating barriers. This could happen if coevolutionary changes of interacting NPC components within species lead to cross-species assembly incompatibilities. Thus, it is of interest that, although few reproductive isolating barriers have been elucidated at the molecular level in any species, in one of the major engines of speciation research, the genus *Drosophila*, negative interactions between heterospecific Nups have a direct role in hybrid incompatibility. Moreover, the causal genes have evolved at highly elevated rates, apparently driven by positive selection (Presgraves 2007; Presgraves and Stephan 2007; Tang and Presgraves 2009).

Finally, there is the matter of genome size and its relationship to nuclear architecture. Under the assumption that the rate of export of transcripts from the nucleus is limited by the surface area of the nuclear envelope, Cavalier-Smith (1978, 2005) suggested the need for a strong coordination between nuclear and cell volumes. Drawing from observations of an association between genome size and nuclear volume (mostly in land plants; Price et al. 1973), his nucleoskeletal hypothesis postulates that organisms with large cells evolve large genome sizes as a means to support a large nuclear membrane. Under this view, DNA has a structural role, independent of its coding content, with a larger nuclear envelope leading to an associated increase in the number of pores, which in turn support an enhanced flow of mRNAs to maintain the needs imposed by large cell size. The limited amount of comparative data suggests a roughly constant scaling of total nuclear pore number with nuclear surface area, with a pore density generally between 5 and $15/\mu m^2$ (Figure 15.8).

However, a number of observations shed doubt on the nucleoskeletal hypothesis. First, it is unclear that flow rates through pores is the limiting factor in material transport, e.g., as opposed to association rates between cargoes and transporters. Empirical studies suggest the latter,

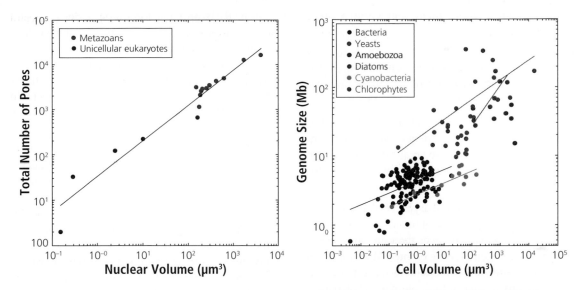

Figure 15.8 Phylogenetic relationships between genome size and nuclear and cell-volume features. **Left:** Phylogenetic scaling of the total number of nuclear pores with nuclear volume, $33.1V^{0.78}$ ($r^2 = 0.94$). Because area scales with $V^{2/3}$, this implies that the mean number of pores per area scales very weakly with an increase in nuclear volume, i.e., with the ~0.1 power of the latter. Data are from Keddie and Barajas (1969), Atkinson et al. (1974), Maul and Deaven (1977), Maul (1977), Winey et al. (1997), and Henderson et al. (2007). **Right:** The scaling of genome size with cell volume. For both heterotrophic bacteria and cyanobacteria, there is a significant positive scaling with genome size (in megabases), $4.3V^{0.17}$ ($r^2 = 0.19$) and $2.7V^{0.17}$ ($r^2 = 0.71$), respectively. In eukaryotes, only the regressions for chlorophytic algae and diatoms are significant, with respective slopes of 0.29 and 0.59. Data are contained in Supplemental Table 15.1.

with transporter efficiency being greatly compromised by off-binding to non-specific substrates (Riddick and Macara 2005; Timney et al. 2006). Second, as pointed in out in Chapter 9, nuclear volume does not appear to be regulated by the amount of DNA in a cell. Third, evolutionary increases in genome size in organisms with larger cell sizes may simply be an indirect consequence of the latter experiencing the passive expansion of excess DNA owing to higher levels of random genetic drift (Chapter 6). The most notable source of genome expansion is mobile-element activity (Lynch 2007), a highly mutationally hazardous enterprise and hence, less than ideal substrate for building a nuclear support structure.

Perhaps the key problem with the nucleoskeletal hypothesis is that the data do not strongly support a general relationship between genome and cell sizes (Figure 15.8). Although some groups of eukaryotes do exhibit an increase in genome size with cell volume, the slopes of the scaling relationship are far below the value of 1.5 expected if the

ratio of nuclear surface area to cell volume is kept constant by changes in bulk DNA. Moreover, there is a weak but significantly positive scaling between genome size and cell volume for both heterotrophic and photosynthetic bacteria, neither of which have nuclear envelopes. The latter pattern is largely due to the fact that bacteria with larger cells generally have genomes with larger numbers of genes.

Summary

- Lipids are an essential ingredient of life. All cells, and all organelles within eukaryotic cells, are surrounded by bilayers of tightly packed lipid molecules. The hydrophobic tails of lipids face the interior of membranes, whereas the hydrophilic head groups face the outside.

- There is enormous combinatorial diversity of lipid molecules within and among species, involving variation in head-group types, lengths of tails and numbers and locations of double carbon bonds within them, and various other embellishments.

- Most species can developmentally alter the composition of lipid profiles as a mechanism of physiological acclimation in response to environmental change.

- Individual lipid molecules are free to diffuse laterally in a two-dimensional manner within membranes, although they do so at rates that are an order of magnitude lower than diffusion rates for proteins within the cytoplasm.

- The hallmark of the eukaryotic cell plan is the presence of a network of internal membrane-bound organelles. However, there are no absolute barriers to the emergence of internal membranes in prokaryotes, and some actually have them. Nor is there any evidence that internal cell structure endows eukaryotes with fundamental superiority in fitness.

- Rather than forming *de novo*, vesicles are typically pinched off from source membranes, with the assistance of cage-like assemblies of coat proteins. They are then delivered to source membranes, where they fuse, with the overall gain/loss dynamics leading to an approximately steady-state distribution of cell constituents.

- Specificity in the vesicle-trafficking system is a function of several layers of intermolecular crosstalk, including adaptor proteins for selecting cargo, RAB proteins for guiding delivery, and SNARE proteins for promoting appropriate membrane fusion. Although gene duplication is known to underlie diversification in specificity, the precise population-genetic requirements for the stable emergence of such partitioning remain to be worked out.

- The nuclear envelope of eukaryotic cells is a continuous elaboration of the ER, and harbors the pores through which nuclear–cytoplasmic transport occurs. Embedded within these openings, nuclear-pore complexes consist of several hundreds of proteins that guide the bidirectional passage of appropriate RNAs and proteins. Nuclear transport is a selective process involving proteins contained within the pore and transporter proteins that selectively bind to particular cargoes.

- Why the nuclear envelope evolved is not entirely clear, but once established it altered the intracellular environment in ways that created a permissive setting for the colonization of genomic elements previously forbidden by natural selection, e.g., introns and mobile-genetic elements. Ongoing molecular arms races dictate the success of intracellular pathogens that require access to host molecules residing within the nucleus, and as a by-product, nuclear-pore proteins appear to diverge in ways that contribute to the establishment of species-isolating barriers.

Foundations 15.1 Probability of preservation and subdivision of labor by duplicated interactions

Chapter 6 introduced the concept of subfunctionalization of the two members of a duplicated gene pair. To recap the central point, genes often have multiple, independently mutable subfunctions, and after gene duplication, these can become reciprocally silenced, leading to more specialized daughter genes. The question of interest here is how frequently pairs of interacting genes (e.g., members of a transport pathway) can partition up their functions after both members of the pair are simultaneously duplicated (as would occur following a whole-genome duplication event). In the extreme, assuming both genes are capable of subfunctionalization, this can lead to two independently operating pathways. The approach taken here is not fully general, in that it assumes a situation in which

key mutations fix sequentially in the population, although it does highlight the basic principles that will need to be accounted for in a fuller development.

Before addressing the two-gene model, it will be useful to understand the quantitative expectations for the single-gene situation. Consider the case illustrated in Figure 15.9, where initially a single protein-coding gene has a coding region and two regulatory elements for different subfunctions. It will be assumed here that all mutations with significant effects on gene activity are degenerative in nature, with loss of single subfunctions occurring at rate μ_s for each regulatory element, and mutations that eliminate whole-gene function arising at rate μ_n. Under this model, there are two possible fates following gene duplication:

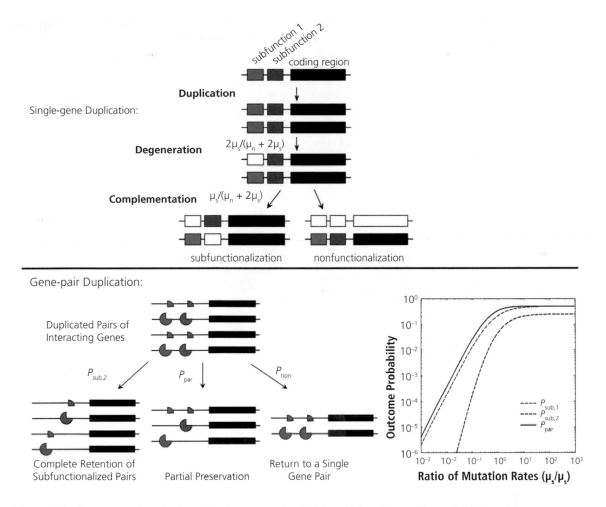

Figure 15.9 The preservation of pairs of duplicate genes by division of labor. **Above:** The probabilities of the individual steps involved in the preservation of two duplicate genes, each with two independently mutable subfunctions. μ_s and μ_n denote mutation rates for losses of subfunctions and complete gene silencing. **Below:** The alternative fates of duplicated pairs of interacting genes, and their probabilities, as described in Foundations 15.1. In both cases, the fate probabilities are functions of the ratio of rates of subfunctionalizing to non-functionalizing mutations (μ_s/μ_n).

one of the genes will become completely silenced (nonfunctionalization), returning the system to the initial state of a single active gene, or the two genes will become mutually preserved by subfunctionalization, as in this case joint retention is necessary to retain the full complement of gene activity. The loss of single gene features is assumed to be a neutral process owing to the redundancy of the two-gene system, so that each step of permissible mutations proceeds at a rate equivalent to the determining mutation rate. Note that

the two subfunctions need not be regulatory elements and could instead be protein domains.

If subfunctionalization is to occur, the first mutation to fix must be of the subfunctionalizing type, the probability of which is $2\mu_s/(\mu_n + 2\mu_s)$. This expression follows from the fact that there are three ways to mutate each fully endowed gene, two of which eliminate single subfunctions. Conditional upon arriving at this initial state, the remaining fully intact gene cannot be non-functionalized, as this would fully eliminate one subfunction entirely, although it can lose the remaining redundant subfunction, implying a permissible mutation rate of μ_s. However, the partially incapacitated copy can be completely silenced by either a non-functionalizing mutation or by a mutation to the remaining subfunction, giving a total rate of $\mu_n + \mu_s$. The total permissible mutation rate during the second step is then $\mu_n + 2\mu_s$, with the probability that the second mutation leads to joint subfunctionalization being $\mu_s/(\mu_n + 2\mu_s)$. The total probability of subfunctionalization is equal to the product of the two stepwise probabilities,

$$P_{\text{sub},1} = \frac{2\mu_s^2}{(\mu_n + 2\mu_s)^2},$$ (15.1.1a)

with the probability of non-functionalization being

$$P_{\text{non},1} = 1 - P_{\text{sub},1}.$$ (15.1.1b)

Now consider the situation in which a pair of interacting genes (e.g., a donor and its recipient) is duplicated simultaneously, with each pair having two independently mutable interactions (as indicated by the different colours and complementary shapes in Figure 15.9). Following the same mutation scheme, there are four possible final fates of this system: 1) complete subfunctionalization and the preservation of two specialized single-subfunction interactions; 2) one fully endowed donor gene, and two specialized recipients; 3) two specialized donors, and one fully endowed recipient (not shown); and 4) nonfunctionalization of one donor and one recipient and return to the single-pair (ancestral) situation.

Multiple paths involving multiple steps lead to each of these final outcomes, rendering the bookkeeping tedious, so only a few of the results are sketched out. It is relatively straightforward to obtain the probability of complete subfunctionalization, as this requires that a series of four subfunctionalizing mutations occur before any gene is completely nonfunctionalized. Moreover, specific subfunctions must be retained in each gene – the two donor genes must preserve alternative subfunctions, as must the two recipient genes. The probability of each specific subfunctionalizing mutation is $\mu_s/(\mu_n + 2\mu_s)$, and because there are two ways by which the donor copies can be resolved (blue in one, and green in the other, in either order), and likewise for the recipient genes, the probability of preservation of the four-gene set by subfunctionalization is

$$P_{\text{sub},2} = \frac{4\mu_s^4}{(\mu_n + 2\mu_s)^4},$$ (15.1.2)

which is equivalent to the square of the single-gene-result, $P_{\text{sub},1}^2$.

We next consider the probability of return to a single-pair system, which requires the complete loss of function

of one donor and one recipient gene. There are three ways by which this endpoint can come about. First, if the initial mutation is nonfunctionalizing, which occurs with probability $\mu_n/(\mu_n + 2\mu_s)$, the system effectively returns to a one-gene system, as only the remaining pair of duplicates is not capable of further evolution. The net probability of return to the ancestral state by this path is then simply

$$P_{\text{non},a} = \frac{\mu_n P_{\text{non},1}}{\mu_n + 2\mu_s}.$$ (15.1.3a)

Second, there are two additional paths to a one-pair system if the first mutation is of the subfunctionalizing type, probability $2\mu_s/(\mu_n + 2\mu_s)$, and the second is non-functionalizing. When there are three fully functional and one subfunctionalized genes, the total rate of permissible mutations in the next step is $d = 3(\mu_n + 2\mu_s)$. The probability that the subfunctionalized copy is silenced in the next step is then $(\mu_n + \mu_s)/d$, and this returns the system to the identical situation noted in the previous paragraph – one fully endowed gene of one type and two of the other, with a probability of nonfunctionalization of $P_{\text{non},1}$ in the final step. Alternatively, a member of the pair of fully endowed genes will be non functionalized with probability $2\mu_n/d$, in which case the remaining single-subfunction gene will be lost with probability $(\mu_n + \mu_s)/(\mu_n + 2\mu_s)$, leaving one fully endowed donor and recipient gene. Collecting terms, the probability of complete nonfunctionalization by these two path types is

$$P_{\text{non},b} = \frac{2\mu_s(\mu_n + \mu_s)}{3(\mu_n + 2\mu_s)^2}\left(P_{\text{non},1} + \frac{2\mu_n}{\mu_n + 2\mu_s}\right).$$ (15.1.3b)

The third potential path to complete nonfunctionalization follows when the first two mutations are of the subfunctionalizing type. This can only occur if one of each such mutations is allocated to a donor and the other to a recipient gene (as otherwise, both members of donor and/or recipient would be permanently preserved by subfunctionalization). The probability of this starting point is $[2\mu_s/(\mu_n + 2\mu_s)] \cdot (4\mu_s/d)$. Completion of the path to complete nonfunctionalization then requires that one of the single-subfunction genes is silenced by the next mutation, the probability of which is $(\mu_n + \mu_s)/(\mu_n + 2\mu_s)$, and that the final single-subfunction gene is also silenced in the remaining step, which also occurs with probability $(\mu_n + \mu_s)/(\mu_n + 2\mu_s)$. Collecting terms,

$$P_{\text{non},c} = \frac{8\mu_s^2(\mu_n + \mu_s)^2}{3(\mu_n + 2\mu_s)^4}.$$ (15.1.3c)

Summing up terms, the total probability of return to a single-pair system by random silencing of one donor and one recipient gene is

$$P_{\text{non},2} = P_{\text{non},a} + P_{\text{non},b} + P_{\text{non},c}.$$ (15.1.4)

The probability of partial preservation is

$$P_{par} = 1 - P_{sub,2} - P_{non,2},\qquad(15.1.5)$$

with half of these cases involving two specialized donors and one two-subfunction recipient, and the other half the reciprocal situation.

The solutions of the above formulae, given in Figure 15.9, are simple functions of the ratio μ_s/μ_n. As noted, the probability of complete subfunctionalization of a two-component pathway is substantially smaller than that of subfunctionalization of a single two-function gene, being equivalent to the square of the latter. On the other hand, the probability of partial preservation of a pathway (involving just one member of the pair) is $\simeq 2P_{sub,1}$, provided $\mu_s < \mu_n$ (which is likely to be the case). Thus, given that there are two genes involved, only one of which will be preserved by subfunctionalization, the probability of preservation per gene is very nearly the same with the duplication of a two-gene system as in the case of single-gene duplication.

Literature Cited

Acehan, D., R. Santarella-Mellwig, and D. P. Devos. 2014. A bacterial tubulovesicular network. J. Cell Sci. 127: 277–280.

Adung'a, V. O., C. Gadelha, and M. C. Field. 2013. Proteomic analysis of clathrin interactions in trypanosomes reveals dynamic evolution of endocytosis. Traffic 14: 440–457.

Ahmed, S., D. G. Brickner, W. H. Light, I. Cajigas, M. McDonough, A. B. Froyshteter, T. Volpe, and J. H. Brickner. 2010. DNA zip codes control an ancient mechanism for gene targeting to the nuclear periphery. Nat. Cell Biol. 12: 111–118.

Akey, C. W., D. Singh, C. Ouch, I. Echeverria, I. Nudelman, J. M. Varberg, Z. Yu, F. Fang, Y. Shi, J. Wang, et al. 2022. Comprehensive structure and functional adaptations of the yeast nuclear pore complex. Cell 185: 361–378.

Alber, F., S. Dokudovskaya, L. M. Veenhoff, W. Zhang, J. Kipper, D. Devos, A. Suprapto, O. Karni-Schmidt, R. Williams, B. T. Chait, et al. 2007a. Determining the architectures of macromolecular assemblies. Nature 450: 683–694.

Alber, F., S. Dokudovskaya, L. M. Veenhoff, W. Zhang, J. Kipper, D. Devos, A. Suprapto, O. Karni-Schmidt, R. Williams, B. T. Chait, et al. 2007b. The molecular architecture of the nuclear pore complex. Nature 450: 695–701.

Arthur, H., and K. Watson. 1976. Thermal adaptation in yeast: Growth temperatures, membrane lipid, and cytochrome composition of psychrophilic, mesophilic, and thermophilic yeasts. J. Bacteriol. 128: 56–68.

Atkinson, A. W., Jr., P. C. John, and B. E. Gunning. 1974. The growth and division of the single mitochondrion and other organelles during the cell cycle of *Chlorella*, studied by quantitative stereology and three-dimensional reconstruction. Protoplasma. 81: 77–109.

Bapteste, E., R. L. Charlebois, D. MacLeod, and C. Brochier. 2005. The two tempos of nuclear pore complex evolution: Highly adapting proteins in an ancient frozen structure. Genome Biol. 6: R85.

Bardina, M. V., P. V. Lidsky, E. V. Sheval, K. V. Fominykh, F. J. van Kuppeveld, V. Y. Polyakov, amd V. I. Agol. 2009. Mengovirus-induced rearrangement of the nuclear pore complex: Hijacking cellular phosphorylation machinery. J. Virol. 83: 3150–3161.

Baum, D. A., and B. Baum. 2014. An inside-out origin for the eukaryotic cell. BMC Biol. 12: 76.

Beck, M., and E. Hurt. 2017. The nuclear pore complex: Understanding its function through structural insight. Nat. Rev. Mol. Cell. Biol. 18: 73–89.

Benning, C., Z. H. Huang, and D. A. Gage. 1995. Accumulation of a novel glycolipid and a betaine lipid in cells of *Rhodobacter sphaeroides* grown under phosphate limitation. Arch. Biochem. Biophys. 317: 103–111.

Blackstone, C., and W. A. Prinz. 2016. Keeping in shape. eLife. 5: e20468.

Blagović, B., J. Rupčić, M. Mesarić, and V. Marić. 2005. Lipid analysis of the plasma membrane and mitochondria of brewer's yeast. Folia Microbiol. (Praha) 50: 24–30.

Boedeker, C., M. Schüler, G. Reintjes, O. Jeske, M. C. van Teeseling, M. Jogler, P. Rast, D. Borchert, D. P. Devos, M. Kucklick, et al. 2017. Determining the bacterial cell biology of Planctomycetes. Nat. Commun. 8: 14853.

Boettner, D. R., R. J. Chi, and S. K. Lemmon. 2011. Lessons from yeast for clathrin-mediated endocytosis. Nat. Cell Biol. 14: 2–10.

Boulais, J., M. Trost, C. R. Landry, R. Dieckmann, E. D. Levy, T. Soldati, S. W. Michnick, P. Thibault, and M. Desjardins. 2010. Molecular characterization of the evolution of phagosomes. Mol. Syst. Biol. 6: 423.

Bowers, B., T. E. Olszewski, and J. Hyde. 1981. Morphometric analysis of volumes and surface areas in membrane compartments during endocytosis in *Acanthamoeba*. J. Cell. Biol. 88: 509–515.

Brighouse, A., J. B. Dacks, and M. C. Field. 2010. Rab protein evolution and the history of the eukaryotic endomembrane system. Cell. Mol. Life Sci. 67: 3449–3465.

Briguglio, J. S., and A. P. Turkewitz. 2014. *Tetrahymena thermophila*: A divergent perspective on membrane traffic. J. Exp. Zool. B Mol. Dev. Evol. 322: 500–516.

Brohawn, S. G., N. C. Leksa, E. D. Spear, K. R. Rajashankar, and T. U. Schwartz. 2008. Structural evidence for common ancestry of the nuclear pore complex and vesicle coats. Science 322: 1369–1373.

Brügger, B. 2014. Lipidomics: Analysis of the lipid composition of cells and subcellular organelles by electrospray ionization mass spectrometry. Annu. Rev. Biochem. 83: 79–98.

Buehler, L. K. 2016. Cell Membranes. Garland Science, New York, NY.

Burke, B., and C. L. Stewart. 2013. The nuclear lamins: Flexibility in function. Nat. Rev. Mol. Cell. Biol. 14: 13–24.

Byrne, M. E., D. A. Ball, J. L. Guerquin-Kern, I. Rouiller, T. D. Wu, K. H. Downing, H. Vali, and A. Komeili. 2010. *Desulfovibrio magneticus* RS-1 contains an iron- and phosphorus-rich organelle distinct from its bullet-shaped magnetosomes. Proc. Natl. Acad. Sci. USA 107: 12263–12268.

Carey, V. C., and L. O. Ingram. 1983. Lipid composition of *Zymomonas mobilis*: Effects of ethanol and glucose. J. Bacteriol. 154: 1291–1300.

Carini, P., B. A. Van Mooy, J. C. Thrash, A. White, Y. Zhao, E. O. Campbell, H. F. Fredricks, and S. J. Giovannoni. 2015. SAR11 lipid renovation in response to phosphate starvation. Proc. Natl. Acad. Sci. USA 112: 7767–7772.

Cavalier-Smith, T. 1978. Nuclear volume control by nucleoskeletal DNA, selection for cell volume and cell growth rate, and the solution of the DNA C-value paradox. J. Cell. Sci. 34: 247–278.

Cavalier-Smith, T. 2005. Economy, speed and size matter: Evolutionary forces driving nuclear genome miniaturization and expansion. Ann. Bot. 95: 147–175.

Chan, M., A. Pinter, and I. Sohi. 2016. The effect of temperature on the rate of food vacuole formation in *Tetrahymena thermophila*. The Expedition 6.

Chong, P. L. 2010. Archaebacterial bipolar tetraether lipids: Physico-chemical and membrane properties. Chem. Phys. Lipids 163: 253–265.

Chowdhury, C., S. Chun, A. Pang, M. R. Sawaya, S. Sinha, T. O. Yeates, and T. A. Bobik. 2015. Selective molecular transport through the protein shell of a bacterial microcompartment organelle. Proc. Natl. Acad. Sci. USA 112: 2990–2995.

Contreras, I., L. Shapiro, and S. Henry. 1978. Membrane phospholipid composition of *Caulobacter crescentus*. J. Bacteriol. 135: 1130–1136.

Dacks, J. B., L. A. Davis, A. M. Sjögren, J. O. Andersson, A. J. Roger, and W. F. Doolittle. 2003. Evidence for Golgi bodies in proposed 'Golgi-lacking' lineages. Proc. Biol. Sci. 270 Suppl 2: S168–S171.

Dacks, J. B., and M. C. Field. 2007. Evolution of the eukaryotic membrane-trafficking system: Origin, tempo and mode. J. Cell Sci. 120: 2977–2985.

Dacks, J. B., P. P. Poon, and M. C. Field. 2008. Phylogeny of endocytic components yields insight into the process of nonendosymbiotic organelle evolution. Proc. Natl. Acad. Sci. USA 105: 588–593.

Dang, V. D., and H. L. Levin. 2000. Nuclear import of the retrotransposon Tf1 is governed by a nuclear localization signal that possesses a unique requirement for the FXFG nuclear pore factor Nup124p. Mol. Cell. Biol. 20: 7798–7812.

DeGrasse, J. A., K. N. DuBois, D. Devos, T. N. Siegel, A. Sali, M. C. Field, M. P. Rout, and B. T. Chait. 2009. Evidence for a shared nuclear pore complex architecture i.e., conserved from the last common eukaryotic ancestor. Mol. Cell. Proteomics 8: 2119–2130.

den Kamp, J. A., P. P. Bonsen, and L. L. van Deenen. 1969. Structural investigations on glucosaminyl phosphatidylglycerol from *Bacillus megaterium*. Biochim. Biophys. Acta 176: 298–305.

Denning, D. P., and M. F. Rexach. 2007. Rapid evolution exposes the boundaries of domain structure and function in natively unfolded FG nucleoporins. Mol. Cell. Proteomics 6: 272–282.

Devaux, P., and H. M. McConnell. 1972. Lateral diffusion in spin-labeled phosphatidylcholine multilayers. J. Am. Chem. Soc. 94: 4475–4481.

Devos, D., S. Dokudovskaya, F. Alber, R. Williams, B. T. Chait, A. Sali, and M. P. Rout. 2004. Components of coated vesicles and nuclear pore complexes share a common molecular architecture. PLoS Biol. 2: e380.

Devos, D. P., R. Gräf, and M. C. Field. 2014. Evolution of the nucleus. Curr. Opin. Cell Biol. 28: 8–15.

Edeling, M. A., C. Smith, and D. Owen. 2006. Life of a clathrin coat: Insights from clathrin and AP structures. Nat. Rev. Mol. Cell Biol. 7: 32–44.

Elias, M., A. Brighouse, C. Gabernet-Castello, M. C. Field, and J. B. Dacks. 2012. Sculpting the endomembrane system in deep time: High resolution phylogenetics of Rab GTPases. J. Cell Sci. 125: 2500–2508.

Engstler, M., L. Thilo, F. Weise, C. G. Grünfelder, H. Schwarz, M. Boshart, and P. Overath. 2004. Kinetics of endocytosis and recycling of the GPI-anchored variant surface glycoprotein in *Trypanosoma brucei*. J. Cell Sci. 117: 1105–1115.

Ernst, R., C. S. Ejsing, and B. Antonny B. 2016. Homeoviscous adaptation and the regulation of membrane lipids. J. Mol. Biol. 428: 4776–4791.

Fahrenkrog, B., J. Köser, and U. Aebi. 2004. The nuclear pore complex: A jack of all trades? Trends Biochem. Sci. 29 175–182.

Fahy, E., S. Subramaniam, H. A. Brown, C. K. Glass, A. H. Merrill Jr., R. C. Murphy, C. R. Raetz, D. W. Russell, Y. Seyama, W. Shaw, et al. 2005. A comprehensive classification system for lipids. J. Lipid Res. 46: 839–861.

Faini, M., S. Prinz, R. Beck, M. Schorb, J. D. Riches, K. Bacia K, Brügger, F. T. Wieland, and J. A. Briggs. 2012. The structures of COPI-coated vesicles reveal alternate coatomer conformations and interactions. Science 336: 1451–1454.

Feldherr, C., D. Akin, T. Littlewood, and M. Stewart. 2002. The molecular mechanism of translocation through the nuclear pore complex is highly conserved. J. Cell Sci. 115: 2997–3005.

Field, M. C., C. Gabernet-Castello, and J. B. Dacks. 2007. Reconstructing the evolution of the endocytic system: Insights from genomics and molecular cell biology. Adv. Exp. Med. Biol. 607: 84–96.

Field, M. C., L. Koreny, and M. P. Rout. 2014. Enriching the pore: Splendid complexity from humble origins. Traffic 15: 141–156.

Field, M. C., A. Sali, and M. P. Rout. 2011. On a bender – BARs, ESCRTs, COPs, and finally getting your coat. J. Cell Biol. 193: 963–972.

Flechsler, J., T. Heimerl, H. Huber, R. Rachel, and I. A. Berg. 2021. Functional compartmentalization and metabolic separation in a prokaryotic cell. Proc. Natl. Acad. Sci. USA 118: e2022114118.

Fok, A. K., B. C. Sison, M. S. Ueno, and R. D. Allen. 1988. Phagosome formation in *Paramecium*: Effects of solid particles. J. Cell Sci. 90: 517–524.

Fuerst, J. A., and E. Sagulenko. 2011. Beyond the bacterium: Planctomycetes challenge our concepts of microbial structure and function. Nat. Rev. Microbiol. 9: 403–413.

Gallay, P., T. Hope, D. Chin, and D. Trono. 1997. HIV-1 infection of nondividing cells through the recognition of integrase by the importin/karyopherin pathway. Proc. Natl. Acad. Sci. USA 94: 9825–9830.

Gallay, P., S. Swingler, J. Song, F. Bushman, and D. Trono. 1995. HIV nuclear import is governed by the phosphotyrosine-mediated binding of matrix to the core domain of integrase. Cell 83: 569–576.

Gangar, S., S. Kanageswaran, S. Lai, and A. Persson. 2015. Effect of temperature and time on the ciliary function of

Tetrahymena thermophila based on food vacuole formation. The Expedition 5:1–15.

Geiger, O., N. González-Silva, I. M. López-Lara, and C. Sohlenkamp. 2010. Amino acid-containing membrane lipids in bacteria. Prog. Lipid Res. 49: 46–60.

Gould, S. B. 2018. Membranes and evolution. Curr. Biol. 28: R381–R385.

Gould, S. B., S. G. Garg, and W. F. Martin. 2016. Bacterial vesicle secretion and the evolutionary origin of the eukaryotic endomembrane system. Trends Microbiol. 24: 525–534.

Greening, C., and T. Lithgow. 2020. Formation and function of bacterial organelles. Nat. Rev. Microbiol. 18: 677–689.

Grünwald, D., R. H. Singer, and M. Rout. 2011. Nuclear export dynamics of RNA-protein complexes. Nature 475: 333–341.

Gupte, S., E. S. Wu, L. Hoechli, M. Hoechli, K. Jacobson, A. E. Sowers, and C. R. Hackenbrock. 1984. Relationship between lateral diffusion, collision frequency, and electron transfer of mitochondrial inner membrane oxidation-reduction components. Proc. Natl. Acad. Sci. USA 81: 2606–2610.

Haest, C. W., J. de Gier, J. A. den Kamp, P. Bartels, and L. L. van Deenen. 1972. Changes in permeability of *Staphylococcus aureus* and derived liposomes with varying lipid composition. Biochim. Biophys. Acta 255: 720–733.

Haest, C. W., J. de Gier, and L. L. van Deenen. 1969. Changes in the chemical and the barrier properties of the membrane lipids of *E. coli* by variation of the temperature of growth. Chem. Phys. Lipids 3: 413–417.

Harayama, T., and H. Riezman. 2018. Understanding the diversity of membrane lipid composition. Nat. Rev. Mol. Cell Biol. 19: 281–296.

Harrison, S. C., and T. Kirchhausen. 2010. Conservation in vesicle coats. Nature 466: 1048–1049.

Hazel, J. R. 1995. Thermal adaptation in biological membranes: Is homeoviscous adaptation the explanation? Annu. Rev. Physiol. 57: 19–42.

Helfrich, W. 1973. Elastic properties of lipid bilayers: Theory and possible experiments. Z. Naturforsch. C. 28: 693–703.

Henderson, G. P., L. Gan, and G. J. Jensen. 2007. 3-D ultrastructure of *O. tauri*: Electron cryotomography of an entire eukaryotic cell. PLoS One 2: e749.

Heuer, D., A. Rejman Lipinski, N. Machuy, A. Karlas, A. Wehrens, F. Siedler, V. Brinkmann, and T. F. Meyer. 2009. *Chlamydia* causes fragmentation of the Golgi compartment to ensure reproduction. Nature 457: 731–735.

Hirst, J., L. D. Barlow, G. C. Francisco, D. A. Sahlender, M. N. Seaman, J. B. Dacks, and M. S. Robinson. 2011. The fifth adaptor protein complex. PLoS Biol. 9: e1001170.

Hirst, J., A. Schlacht, J. P. Norcott, D. Traynor, G. Bloomfield, R. Antrobus, R. R. Kay, J. B. Dacks, and M. S. Robinson. 2014. Characterization of TSET, an ancient and widespread membrane trafficking complex. eLife 3: e02866.

Holm, H. C., H. F. Fredricks, S. M. Bent, D. P. Lowenstein, J. E. Ossolinski, K. W. Becker, W. M. Johnson, K. Schrage, and B. A. S. Van Mooy. 2022. Global ocean lipidomes show a universal relationship between temperature and lipid unsaturation. Science 376: 1487–1491.

Hülsmann, B. B., A. A. Labokha, and D. Görlich. 2012. The permeability of reconstituted nuclear pores provides direct evidence for the selective phase model. Cell 150: 738–751.

Ikegami, K., and J. D. Lieb. 2013. Integral nuclear pore proteins bind to Pol III-transcribed genes and are required for Pol III transcript processing in *C. elegans*. Mol. Cell 51: 840–849.

Imachi, H., M. K. Nobu, N. Nakahara, Y. Morono, M. Ogawara, Y. Takaki, Y. Takano, K. Uematsu, T. Ikuta, M. Ito, et al. 2020. Isolation of an archaeon at the prokaryote–eukaryote interface. Nature 577: 519–525.

Irwin, B., M. Aye, P. Baldi, N. Beliakova-Bethell, H. Cheng, Y. Dou, W. Liou, and S. Sandmeyer. 2005. Retroviruses and yeast retrotransposons use overlapping sets of host genes. Genome Res. 15: 641–654.

Iwamoto, M., H. Osakada, C. Mori, Y. Fukuda, K. Nagao, C. Obuse, Y. Hiraoka, and T. Haraguchi. 2017. Compositionally distinct nuclear pore complexes of functionally distinct dimorphic nuclei in the ciliate *Tetrahymena*. J. Cell Sci. 130: 1822–1834.

Jarsch, I. K., F. Daste, and J. L. Gallop. 2016. Membrane curvature in cell biology: An integration of molecular mechanisms. J. Cell Biol. 214: 375–387.

Jékely, G. 2003. Small GTPases and the evolution of the eukaryotic cell. Bioessays 25: 1129–1138.

Jékely, G. 2008. Origin of the nucleus and Ran-dependent transport to safeguard ribosome biogenesis in a chimeric cell. Biol. Direct 3: 31.

Jena, B. P. 2020. Cellular Nanomachines. Springer, Cham.

Jin, A. J., M. Edidin, R. Nossal, and N. L. Gershfeld. 1999. A singular state of membrane lipids at cell growth temperatures. Biochemistry 38: 13275–13278.

Jogler, C., G. Wanner, S. Kolinko, M. Niebler, R. Amann, N. Petersen, M. Kube, R. Reinhardt, and D. Schüler. 2011. Conservation of proteobacterial magnetosome genes and structures in an uncultivated member of the deep-branching Nitrospira phylum. Proc. Natl. Acad. Sci. USA 108: 1134–1139.

Kaksonen, M., and A. Roux. 2018. Mechanisms of clathrin-mediated endocytosis. Nat. Rev. Mol. Cell Biol. 19: 313–326.

Keddie, F. M., and L. Barajas. 1969. Three-dimensional reconstruction of *Pityrosporum* yeast cells based on serial section electron microscopy. J. Ultrastruct. Res. 29: 260–275.

Kerfeld, C. A., C. Aussignargues, J. Zarzycki, F. Cai, and M. Sutter. 2018. Bacterial microcompartments. Nat. Rev. Microbiol. 16: 277–290.

Kienle, N., T. H. Klöpper, and D. Fasshauer. 2009. Differences in the SNARE evolution of fungi and metazoa. Biochem. Soc. Trans. 37: 787–791.

Kim, M. K., K. C. Claiborn, and H. L. Levin. 2005. The long terminal repeat-containing retrotransposon Tf1 possesses amino acids in Gag that regulate nuclear localization and particle formation. J. Virol. 79: 9540–9555.

Kirchhausen, T., D. Owen, and S. C. Harrison. 2014. Molecular structure, function, and dynamics of clathrin-mediated membrane traffic. Cold Spring Harb. Perspect. Biol. 6: a016725.

Klinger, C. M., A. Spang, J. B. Dacks, and T. J. Ettema. 2016. Tracing the archaeal origins of eukaryotic membrane-trafficking system building blocks. Mol. Biol. Evol. 33: 1528–1541.

Kloetzel, J. A. 1974. Feeding in ciliated protozoa. I. Pharyngeal disks in *Euplotes*: A source of membrane for food vacuole formation? J. Cell Sci. 15: 379–401.

Klöpper, T. H., C. N. Kienle, and D. Fasshauer. 2007. An elaborate classification of SNARE proteins sheds light on the conservation of the eukaryotic endomembrane system. Mol. Biol. Cell 18: 3463–3471.

Klöpper, T. H., N. Kienle, D. Fasshauer, and S. Munro. 2012. Untangling the evolution of Rab G proteins: Implications of a comprehensive genomic analysis. BMC Biol. 10: 71.

Koga, Y., and H. Morii. 2007. Biosynthesis of ether-type polar lipids in archaea and evolutionary considerations. Microbiol. Mol. Biol. Rev. 71: 97–120.

Kontou, A., E. K. Herman, M. C. Field, J. B. Dacks, and V. L. Koumandou. 2022. Evolution of factors shaping the endoplasmic reticulum. Traffic 23: 462–473.

Kosugi, S., M. Hasebe, M. Tomita, and H. Yanagawa. 2009. Systematic identification of cell cycle-dependent yeast nucleocytoplasmic shuttling proteins by prediction of composite motifs. Proc. Natl. Acad. Sci. USA 106: 10171–10176.

Koukou, A. I., D. Tsoukatos, and C. Drainas. 1990. Effect of ethanol on the phospholipid and fatty acid content of *Schizosaccharomyces pombe* membranes. J. Gen. Microbiol. 136: 1271–1277.

Kumar, M., M. S. Mommer, and V. Sourjik. 2010. Mobility of cytoplasmic, membrane, and DNA-binding proteins in *Escherichia coli*. Biophys. J. 98: 552–559.

Lane, N., and W. Martin. 2010. The energetics of genome complexity. Nature 467: 929–934.

Lee, J. W. 1942. The effect of temperature on food-vacuole formation in *Paramecium*. Physiol. Zool. 15: 453–458.

Lee, K., Z. Ambrose, T. D. Martin, I. Oztop, A. Mulky, J. G. Julias, N. Vandegraaff, J. G. Baumann, R. Wang, W. Yuen, et al. 2010. Flexible use of nuclear import pathways by HIV-1. Cell Host Microbe 7: 221–233.

Li, Y., D. A. Powell, S. A. Shaffer, D. A. Rasko, M. R. Pelletier, J. D. Leszyk, A. J. Scott, A. Masoudi, D. R. Goodlett, X. Wang, et al. 2012. LPS remodeling is an evolved survival strategy for bacteria. Proc. Natl. Acad. Sci. USA 109: 8716–8721.

Liu, X., J. M. Mitchell, R. W. Wozniak, G. Blobel, and J. Fan. 2012a. Structural evolution of the membrane-coating module of the nuclear pore complex. Proc. Natl. Acad. Sci. USA 109: 16498–16503.

Liu, Y. W., A. I. Su, and S. L. Schmid. 2012b. The evolution of dynamin to regulate clathrin-mediated endocytosis: Speculations on the evolutionarily late appearance of dynamin relative to clathrin-mediated endocytosis. Bioessays 34: 643–647.

Lonhienne, T. G., E. Sagulenko, R. I. Webb, K. C. Lee, J. Franke, D. P. Devos, A. Nouwens, B. J. Carroll, and J. A. Fuerst. 2010. Endocytosis-like protein uptake in the bacterium *Gemmata obscuriglobus*. Proc. Natl. Acad. Sci. USA 107: 12883–1288.

López, C. S., H. Heras, S. M. Ruzal, C. Sánchez-Rivas, and E. A. Rivas. 1998. Variations of the envelope composition of *Bacillus subtilis* during growth in hyperosmotic medium. Curr. Microbiol. 36: 55–61.

Lynch, M. 2007. The Origins of Genome Architecture. Sinauer Associates, Inc., Sunderland, MA.

Makarova, K. S., N. Yutin, S. D. Bell, and E. V. Koonin. 2010. Evolution of diverse cell division and vesicle formation systems in Archaea. Nat. Rev. Microbiol. 8: 731–741.

Mans, B. J., V. Anantharaman, L. Aravind, and E. V. Koonin. 2004. Comparative genomics, evolution and origins of the nuclear envelope and nuclear pore complex. Cell Cycle 3: 1612–1637.

Marsh, D. 2013. Handbook of Lipid Bilayers, 2nd edn. CRC Press, Boca Raton, FL.

Marshall, J. M., and V. T. Nachmias. 1965. Cell surface and pinocytosis. J. Histochem. Cytochem. 13: 92–104.

Martin, W., and E. V. Koonin. 2006. Introns and the origin of nucleus-cytosol compartmentalization. Nature 440: 41–45.

Mast, F. D., L. D. Barlow, R. A. Rachubinski, and J. B. Dacks. 2014. Evolutionary mechanisms for establishing eukaryotic cellular complexity. Trends Cell Biol. 24: 435–442.

Maul, G. G. 1977. The nuclear and the cytoplasmic pore complex: Structure, dynamics, distribution, and evolution. Int. Rev. Cytol. Suppl. 6:75–186.

Maul, G. G., and L. Deaven. 1977. Quantitative determination of nuclear pore complexes in cycling cells with differing DNA content. J. Cell Biol. 73:748–760.

McMahon, H. T., and E. Boucrot. 2011. Molecular mechanism and physiological functions of clathrin-mediated endocytosis. Nat. Rev. Mol. Cell Biol. 12: 517–533.

Mironov, A. A., V. V. Banin, I. S. Sesorova, V. V. Dolgikh, A. Luini, and G. V. Beznoussenko. 2007 Evolution of the endoplasmic reticulum and the Golgi complex, pp. 61–72. In G. Jékely (ed.) Eukaryotic Membranes and Cytoskeleton: Origins and Evolution. Landes Bioscience, Austin, TX.

Mishra, N. N., and A. S. Bayer. 2013. Correlation of cell membrane lipid profiles with daptomycin resistance in methicillin-resistant *Staphylococcus aureus*. Antimicrob. Agents Chemother. 57: 1082–1085.

Mitra, K., I. Ubarretxena-Belandia, T. Taguchi, G. Warren, and D. M. Engelman. 2004. Modulation of the bilayer thickness of exocytic pathway membranes by membrane proteins rather than cholesterol. Proc. Natl. Acad. Sci. USA 101: 4083–4088.

Mowbrey, K., and J. B. Dacks. 2009. Evolution and diversity of the Golgi body. FEBS Lett. 583: 3738–3745.

Munafó, M., V. R. Lawless, A. Passera, S. MacMillan, S. Bornelöv, I. U. Haussmann, M. Soller, G. J. Hannon, and B. Czech. 2021. Channel nuclear pore complex subunits are required for transposon silencing in *Drosophila*. eLife 10: e66321.

Neumann, N., D. Lundin, and A. M. Poole. 2010. Comparative genomic evidence for a complete nuclear pore complex in the last eukaryotic common ancestor. PLoS One 5: e13241.

Nixon-Abell, J., C. J. Obara, A. V. Weigel, D. Li, W. R. Legant, C. S. Xu, H. A. Pasolli, K. Harvey, H. F. Hess, E. Betzig, et al. 2016. Increased spatiotemporal resolution reveals highly dynamic dense tubular matrices in the peripheral ER. Science 354: aaf3928.

Nozawa, Y. 2011. Adaptive regulation of membrane lipids and fluidity during thermal acclimation in *Tetrahymena*. Proc. Jpn. Acad. Ser. B Phys. Biol. Sci. 87: 450–462.

Obado, S. O., M. Brillantes, K. Uryu, W. Zhang, N. E. Ketaren, B. T. Chait, M. C. Field, and M. P. Rout. 2016a. Interactome mapping reveals the evolutionary history of the nuclear pore complex. PLoS Biol. 14: e1002365.

Obado, S. O., L. Glover, and K. W. Deitsch. 2016b. The nuclear envelope and gene organization in parasitic protozoa: specializations associated with disease. Mol. Biochem. Parasitol. 209: 104–113.

Oger, P. M., and A. Cario. 2013. Adaptation of the membrane in Archaea. Biophys. Chem. 183: 42–56.

Peeler, T. C., M. B. Stephenson, K. J. Einspahr, and G. A. Thompson. 1989. Lipid characterization of an enriched plasma membrane fraction of *Dunaliella salina* grown in media of varying salinity. Plant Physiol. 89: 970–976.

Phillips, M. J., and G. K. Voeltz. 2016. Structure and function of ER membrane contact sites with other organelles. Nat. Rev. Mol. Cell. Biol. 17: 69–82.

Podar, M., M. A. Wall, K. S. Makarova, and E. V. Koonin. 2008. The prokaryotic V4R domain is the likely ancestor of a key component of the eukaryotic vesicle transport system. Biol. Direct 3: 2.

Porter, F. W., and A. C. Palmenberg. 2009. Leader-induced phosphorylation of nucleoporins correlates with nuclear trafficking inhibition by cardioviruses. J. Virol. 83: 1941–1951.

Praefcke, G. J., amd H. T. McMahon. 2004. The dynamin superfamily: Universal membrane tubulation and fission molecules? Nat. Rev. Mol. Cell Biol. 5: 133–147.

Presgraves, D. C. 2007. Does genetic conflict drive rapid molecular evolution of nuclear transport genes in *Drosophila*? Bioessays 29: 386–391.

Presgraves, D. C., and W. Stephan. 2007. Pervasive adaptive evolution among interactors of the *Drosophila* hybrid inviability gene, Nup96. Mol. Biol. Evol. 24: 306–314.

Price, H. J., A. H. Sparrow, and A. F. Nauman. 1973. Correlations between nuclear volume, cell volume and DNA content in meristematic cells of herbaceous angiosperms. Experientia 29: 1028–1029.

Pylypenko, O., and B. Goud. 2012. Posttranslational modifications of Rab GTPases help their insertion into membranes. Proc. Natl. Acad. Sci. USA 109: 5555–5556.

Raetz, C. R., G. D. Kantor, M. Nishijima, and K. F. Newman. 1979. Cardiolipin accumulation in the inner and outer membranes of *Escherichia coli* mutants defective in phosphatidylserine synthetase. J. Bacteriol. 139: 544–551.

Rajoo, S., P. Vallotton, E. Onischenko, and K. Weis. 2018. Stoichiometry and compositional plasticity of the yeast nuclear pore complex revealed by quantitative fluorescence microscopy. Proc. Natl. Acad. Sci. USA 115: E3969–E3977.

Ramadas, R., and M. Thattai. 2013. New organelles by gene duplication in a biophysical model of eukaryote endomembrane evolution. Biophys. J. 104: 2553–2563.

Ramoino, P., and T. C. Franceschi. 1992. Food vacuole formation in the culture life of *Paramecium primaurelia* mating type I and mating type II lines. Ital. J. Zool. 59: 401–405.

Rasmussen, L. 1976. Nutrient uptake in *Tetrahymena pyriformis*. Carlsberg Res. Comm. 41: 143–167.

Renard, H. F., M. Simunovic, J. Lemière, E. Boucrot, M. D. Garcia-Castillo, S. Arumugam, V. Chambon, C. Lamaze, C. Wunder, A. K. Kenworthy, et al. 2015. Endophilin-A2 functions in membrane scission in clathrin-independent endocytosis. Nature 517: 493–496.

Riddick, G., and I. G. Macara. 2005. A systems analysis of importin-alpha-beta mediated nuclear protein import. J. Cell Biol. 168: 1027–1038.

Rietveld, A. G., J. A. Killian, W. Dowhan, and B. de Kruijf. 1993. Polymorphic regulation of membrane phospholipid composition in *Escherichia coli*. J. Biol. Chem. 268: 12427–12433.

Robinow, C., and E. R. Angert. 1998. Nucleoids and coated vesicles of '*Epulopiscium*' spp. Arch. Microbiol. 170: 227–235.

Ryter, A., and C. de Chastellier. 1977. Morphometric and cytochemical studies of *Dictyostelium discoideum* in vegetative phase. Digestive system and membrane turnover. J. Cell Biol. 75: 200–217.

Salje, J., F. van den Ent, P. de Boer, and J. Löwe. 2011. Direct membrane binding by bacterial actin MreB. Mol. Cell 43: 478–487.

Sanderfoot, A. 2007. Increases in the number of SNARE genes parallels the rise of multicellularity among the green plants. Plant Physiol. 144: 6–17.

Satterly, N., P. L. Tsai, J. van Deursen, D. R. Nussenzveig, Y. Wang, P. A. Faria, A. Levay, D. E. Levy, and B. M. Fontoura. 2007. Influenza virus targets the mRNA export machinery and the nuclear pore complex. Proc. Natl. Acad. Sci. USA 104: 1853–1858.

Savage, D. F., B. Afonso, A. H. Chen, and P. A. Silver. 2010. Spatially ordered dynamics of the bacterial carbon fixation machinery. Science 327: 1258–1261.

Schavemaker, P. E., A. J. Boersma, and B. Poolman. 2018. How important is protein diffusion in prokaryotes? Front. Mol. Biosci. 5: 93.

Schledzewski, K., H. Brinkmann, and R. R. Mendel. 1999. Phylogenetic analysis of components of the eukaryotic vesicle transport system reveals a common origin of adaptor protein complexes 1, 2, and 3 and the F subcomplex of the coatomer COPI. J. Mol. Evol. 48: 770–778.

Schmitz, A., A. Schwarz, M. Foss, L. Zhou, B. Rabe, J. Hoellenriegel, M. Stoeber, N. Panté, and M. Kann. 2010. Nucleoporin 153 arrests the nuclear import of hepatitis B virus capsids in the nuclear basket. PLoS Pathog. 6: e1000741.

Schütte, O. M., I. Mey, J. Enderlein, F. Savić, B. Geil, A. Janshoff, and C. Steinem. 2017. Size and mobility of lipid domains tuned by geometrical constraints. Proc. Natl. Acad. Sci. USA 114: E6064–E6071.

Shemesh, T., R. W. Klemm, F. B. Romano, S. Wang, J. Vaughan, X. Zhuang, H. Tukachinsky, M. M. Kozlov, and T. A. Rapoport. 2014. A model for the generation and interconversion of ER morphologies. Proc. Natl. Acad. Sci. USA 111: E5243–E5251.

Shen, Q. T., X. Ren, R. Zhang, I. H. Lee, and J. H. Hurley. 2015. HIV-1 Nef hijacks clathrin coats by stabilizing AP-1: Arf1 polygons. Science 350: aac5137.

Shi, X., P. Halder, H. Yavuz, R. Jahn, and H. A. Shuman. 2016. Direct targeting of membrane fusion by SNARE mimicry: Convergent evolution of *Legionella* effectors. Proc. Natl. Acad. Sci. USA 113: 8807–8812.

Shibata, Y., J. Hu, M. M. Kozlov, and T. A. Rapoport. 2009. Mechanisms shaping the membranes of cellular organelles. Annu. Rev. Cell Dev. Biol. 25: 329–354.

Shiratori, T., S. Suzuki, Y. Kakizawa, and K. I. Ishida. 2019. Phagocytosis-like cell engulfment by a planctomycete bacterium. Nat. Commun. 10: 5529.

Simon, D. N., and K. L. Wilson. 2011. The nucleoskeleton as a genome-associated dynamic 'network of networks'. Nat. Rev. Mol. Cell Biol. 12: 695–708.

Sinensky, M. 1974. Homeoviscous adaptation – a homeostatic process that regulates the viscosity of membrane lipids in *Escherichia coli*. Proc. Natl. Acad. Sci. USA 71: 522–525.

Smith-Sonneborn, J., and S. R. Rodermel. 1976. Loss of endocytic capacity in aging *Paramecium*. The importance of cytoplasmic organelles. J. Cell Biol. 71: 575–588.

Sohlenkamp, C., and O. Geiger. 2016. Bacterial membrane lipids: Diversity in structures and pathways. FEMS Microbiol. Rev. 40: 133–159.

Sorokin, A. V., E. R. Kim, and L. P. Ovchinnikov. 2007. Nucleocytoplasmic transport of proteins. Biochemistry (Mosc). 72: 1439–1457.

Speese, S. D., J. Ashley, V. Jokhi, J. Nunnari, R. Barria, Y. Li, B. Ataman, A. Koon, Y.-T. Chang, Q. Li, et al. 2012. Nuclear envelope budding enables large ribonucleoprotein particle export during synaptic Wnt signaling. Cell 149: 832–846.

Steinman, R. M., I. S. Mellman, W. A. Muller, and Z. A. Cohn. 1983. Endocytosis and the recycling of plasma membrane. J. Cell Biol. 96: 1–27.

Strambio-De-Castillia, C., M. Niepel, and M. P. Rout. 2010. The nuclear pore complex: Bridging nuclear transport and gene regulation. Nat. Rev. Mol. Cell. Biol. 11: 490–501.

Strunze, S., L. C. Trotman, K. Boucke, and U. F. Greber. 2005. Nuclear targeting of adenovirus type 2 requires CRM1-mediated nuclear export. Mol. Biol. Cell 16: 2999–3009.

Stuwe, T., C. J. Bley, K. Thierbach, S. Petrovic, S. Schilbach, D. J. Mayo, T. Perriches, E. J. Rundlet, Y. E. Jeon, L. N. Collins, et al. 2015. Architecture of the fungal nuclear pore inner ring complex. Science 350: 56–64.

Szumowski, S. C., M. R. Botts, J. J. Popovich, M. G. Smelkinson, and E. R. Troemel. 2014. The small GTPase RAB-11 directs polarized exocytosis of the intracellular pathogen *N. parisii* for fecal–oral transmission from *C. elegans*. Proc. Natl. Acad. Sci. USA 111: 8215–8220.

Tanaka, S., M. R. Sawaya, and T. O. Yeates. 2010. Structure and mechanisms of a protein-based organelle in *Escherichia coli*. Science 327: 81–84.

Tang, S., and D. C. Presgraves. 2009. Evolution of the *Drosophila* nuclear pore complex results in multiple hybrid incompatibilities. Science 323: 779–782.

Terasaki, M., T. Shemesh, N. Kasthuri, R. W. Klemm, R. Schalek, K. J. Hayworth, A. R. Hand, M. Yankova, G. Huber, J. W. Lichtman, et al. 2013. Stacked endoplasmic reticulum sheets are connected by helicoidal membrane motifs. Cell 154: 285–296.

Timney, B. L., J. Tetenbaum-Novatt, D. S. Agate, R. Williams, W. Zhang, B. T. Chait, and M. P. Rout. 2006. Simple kinetic relationships and nonspecific competition govern nuclear import rates *in vivo*. J. Cell Biol. 175: 579–593.

Toyoda, T., Y. Hiramatsu, T. Sasaki, and Y. Nakaoka. 2009. Thermo-sensitive response based on the membrane fluidity adaptation in *Paramecium multimicronucleatum*. J. Exp. Biol. 212: 2767–2772.

Tuller, G., T. Nemec, C. Hrastnik, and G. Daum. 1999. Lipid composition of subcellular membranes of an FY1679-derived haploid yeast wild-type strain grown on different carbon sources. Yeast 15: 1555–1564.

Van Blitterswijk, W. J., G. De Veer, J. H. Krol, and P. Emmelot. 1982. Comparative lipid analysis of purified plasma membranes and shed extracellular membrane vesicles from normal murine thymocytes and leukemic GRSL cells. Biochim. Biophys. Acta 688: 495–504.

van de Vossenberg, J. L., T. Ubbink-Kok, M. G. Elferink, A. J. Driessen, and W. N. Konings. 1995. Ion permeability of the cytoplasmic membrane limits the maximum growth temperature of bacteria and archaea. Mol. Microbiol. 18: 925–932.

Van Mooy, B. A., H. F. Fredricks, B. E. Pedler, S. T. Dyhrman, D. M. Karl, M. Koblízek, M. W. Lomas, T. J. Mincer, L. R. Moore, T. Moutin, et al. 2009. Phytoplankton in the ocean use non-phosphorus lipids in response to phosphorus scarcity. Nature 458: 69–72.

van Niftrik, L., and M. S. Jetten. 2012. Anaerobic ammonium-oxidizing bacteria: Unique microorganisms with exceptional properties. Microbiol. Mol. Biol. Rev. 76: 585–596.

Vaquerizas, J. M., R. Suyama, J. Kind, K. Miura, N. M. Luscombe, and A. Akhtar. 2010. Nuclear pore proteins nup153 and megator define transcriptionally active regions in the *Drosophila* genome. PLoS Genet. 6: e1000846.

Varadarajan, P., S. Mahalingam, P. Liu, S. B. Ng, S. Gandotra, D. S. Dorairajoo, and D. Balasundaram. 2005. The functionally conserved nucleoporins Nup124p from fission yeast and the human Nup153 mediate nuclear import and activity of the Tf1 retrotransposon and HIV-1 Vpr. Mol. Biol. Cell 16: 1823–1838.

Vietri, M., K. O. Schink, C. Campsteijn, C. S. Wegner, S. W. Schultz, L. Christ, S. B. Thoresen, A. Brech, C. Raiborg, and H. Stenmark. 2015. Spastin and ESCRT-III coordinate mitotic spindle disassembly and nuclear envelope sealing. Nature 522: 231–235.

Volland, J.-M., S. Gonzalez-Rizzo, O. Gros, T. Tyml, N. Ivanova, F. Schulz, D. Goudeau, N. H. Elisabeth, N. Nath, D. Udwary, et al. 2022. A centimeter-long bacterium with DNA compartmentalized in membrane-bound organelles. Science 376: 1453–1458.

Vovk, A., C. Gu, M. G. Opferman, L. E. Kapinos, R. Y. Lim, R. D. Coalson, D. Jasnow, and A. Zilman. 2016. Simple biophysics underpins collective conformations of the intrinsically disordered proteins of the nuclear pore complex. eLife 5: e10785.

Walsby, A. E. 1972. Structure and function of gas vacuoles. Bacteriol. Rev. 36: 1–32.

Weeks F. G., and G. Herring. 1980. The lipid composition and membrane fluidity of *Dictyostelium discoideum* plasma membranes at various stages during differentiation. J. Lipid Res. 21: 681–686.

Weigel, A. V., M. M. Tamkun, and D. Krapf. 2013. Quantifying the dynamic interactions between a clathrin-coated pit and cargo molecules. Proc. Natl. Acad. Sci. USA 110: E4591–E4600.

Wetzel, M. G., and E. D. Korn. 1969. Phagocytosis of latex beads by *Acanthamoeba castellanii* (Neff). 3. Isolation of the phagocytic vesicles and their membranes. J. Cell Biol. 43: 90–104.

Wideman, J. G., K. F. Leung, M. C. Field, and J. B. Dacks. 2014. The cell biology of the endocytic system from an evolutionary perspective. Cold Spring Harb. Perspect. Biol. 6: a016998.

Winey, M., D. Yarar, T. H. Giddings, Jr., and D. N. Mastronarde. 1997. Nuclear pore complex number and distribution throughout the *Saccharomyces cerevisiae* cell cycle by three-dimensional reconstruction from electron micrographs of nuclear envelopes. Mol. Biol. Cell. 8: 2119–2132.

Woodward, C. L., S. Prakobwanakit, S. Mosessian, and S. A. Chow. 2009. Integrase interacts with nucleoporin NUP153 to mediate the nuclear import of human immunodeficiency virus type 1. J. Virol. 83: 6522–6533.

Wu, E. S., K. Jacobson, and D. Papahadjopoulos. 1977. Lateral diffusion in phospholipid multibilayers measured by fluorescence recovery after photobleaching. Biochemistry 16: 3936–3941.

Xu, X. M., and I. Meier. 2008. The nuclear pore comes to the fore. Trends Plant Sci. 13: 20–27.

Yang, W., J. Gelles, and S. M. Musser. 2004. Imaging of single-molecule translocation through nuclear pore complexes. Proc. Natl. Acad. Sci. USA 101: 12887–12892.

Yoshida, S., and M. Uemura. 1986. Lipid composition of plasma membranes and tonoplasts isolated from etiolated seedlings of mung bean (*Vigna radiata* L.). Plant Physiol. 82: 807–812.

Yutin, N., M. Y. Wolf, Y. I. Wolf, and E. V. Koonin. 2009. Origins of phagocytosis and eukaryogenesis. Biol. Direct 4: 9.

Zaremba-Niedzwiedzka, K., E. F. Caceres, J. H. Saw, D. Bäckström, L. Juzokaite, E. Vancaester, K. W. Seitz, K. Anantharaman, P. Starnawski, K. U. Kjeldsen, et al. 2017. Asgard archaea illuminate the origin of eukaryotic cellular complexity. Nature 541: 353–358.

Zavaleta-Pastor, M., C. Sohlenkamp, J. L. Gao, Z. Guan, R. Zaheer, T. M. Finan, C. R. Raetz, I. M. López-Lara, and O. Geiger. 2010. *Sinorhizobium meliloti* phospholipase C required for lipid remodeling during phosphorus limitation. Proc. Natl. Acad. Sci. USA 107: 302–307.

Zeno, W. F., M. O. Ogunyankin, and M. L. Longo. 2018. Scaling relationships for translational diffusion constants applied to membrane domain dissolution and growth. Biochim. Biophys. Acta 1860: 1994–2003.

Zhu, X., G. Huang, C. Zeng, X. Zhan, K. Liang, Q. Xu, Y. Zhao, P. Wang, Q. Wang, Q. Zhou, et al. 2022. Structure of the cytoplasmic ring of the *Xenopus laevis* nuclear pore complex. Science 376: eabl8280.

Zinser, E., C. D. Sperka-Gottlieb, E. V. Fasch, S. D. Kohlwein, F. Paltauf, and G. Daum. 1991. Phospholipid synthesis and lipid composition of subcellular membranes in the unicellular eukaryote *Saccharomyces cerevisiae*. J. Bacteriol. 173: 2026–2034.

CHAPTER 16

Cytoskeleton, Cell Shape, and Motility

As in multicellular organisms, single cells are confronted with challenges associated with structural support and delivery of biomolecules, albeit at a different scale. Most cells rely on endoskeletons (filaments and tubules) and/or exoskeletons (cell walls) to maintain cell shape integrity. Cell division also requires membrane-deforming proteins. In eukaryotes, various modes of internal cellular movement require cytoskeletal highways for molecular motors, which transport large cargoes using ATP hydrolysis as fuel. Central to all of these cellular features are protein fibrils and sheets comprised of long concatenations of monomeric subunits held together by non-covalent forces. How fundamental features such as cell division were carried out prior to the origin of filament-forming proteins is unknown, but the emergence of self-assembling fibrils would have been a watershed moment for evolution, providing new opportunities for cellular features requiring structural support systems.

This chapter continues an exploration of the internal anatomy and natural history of cellular components and looks at the evolutionary diversification of cytoskeletal proteins and their varied functions. It also covers the diverse sets of eukaryote-specific molecular motors and their roles in intracellular transport. Although prokaryotes seem to be devoid of such machines, fibrillar proteins do exist in prokaryotes, playing contrasting roles in structural support and cell division relative to what is seen in eukaryotes.

In eukaryotes, fibrillar proteins are also central to swimming and crawling, so this chapter explores a few generalities with respect to cellular locomotion. Both prokaryotes and eukaryotes use flagella to swim, and such structures are sometimes suggested to be so complex as to defy an origin by normal evolutionary processes. However, not only are there plausible routes for the emergence of flagella via modifications of pre-existing cellular features, but flagella have evolved more than once. Notably, prokaryotic and eukaryotic flagella evolved independently and operate in completely different manners. Nonetheless, despite these differences, the limits to motility of single-celled organisms follow some general scaling laws across the Tree of Life.

The Basic Cytoskeletal Infrastructure

The three major groups of fibrillar proteins are discussed in the following sections. The two most celebrated members, actins and tubulins, are often viewed as eukaryotic innovations. However, relatives of both families have diverse functions in various prokaryotic lineages, including cell-shape sculpting, cell division, and plasmid partitioning during replication (Ozyamak et al. 2013). This phylogenetic distribution suggests that the antecedents of both actins and tubulins were present as early as LUCA. Moreover, it bears noting that unique types of filament-forming proteins, not discussed here, are sequestered in various prokaryotic lineages (Wagstaff and Löwe 2018). This supports the idea that fibrils are not particularly difficult to evolve, the main requirements being the emergence of a heterologous interface for fibril formation from monomeric subunits and a presumed selective advantage. Indeed, for many proteins, the avoidance of mutations leading to the production of open chains is a major evolutionary challenge (Chapter 13).

Evolutionary Cell Biology. Michael Lynch, Oxford University Press. © Michael Lynch (2024). DOI: 10.1093/oso/9780192847287.003.0016

Despite their sequence divergence, fibrillar proteins from very distant species faithfully assemble and function in novel host species (Horio and Oakley 1994; Osawa and Erickson 2006), indicating both a self-organizational capacity and a conserved architecture of the monomeric subunits. General reviews can be found in Wickstead and Gull (2011), Michie and Löwe (2006), Erickson (2007), and Löwe and Amos (2009). An overview of the energetic costs of producing cytoskeletal proteins in eukaryotes is provided in Foundations 16.1.

Actins

Comprised of one of the most abundant proteins within eukaryotic cells, actin polymers have diverse roles, including cell-cortex formation, vesicle trafficking, cell division, endocytosis, and amoeboid movement. Actin filaments are homopolymeric, double-stranded, and helical, ~7 nm in diameter (Figure 16.1). Free actin monomers are generally bound to ATP, and upon joining a filament undergo ATP hydrolysis. Actin-filament assembly is polar, with the monomeric subunits being added at the plus (barbed) end. Combined with removal at the

minus end, this can result in the 'apparent' movement of a filament in a treadmilling-like process (Carlier and Shekhar 2017). There are different critical concentrations of activated monomers for the plus and minus ends, and depending on the end-specific rates of addition and removal, total filament length can either grow or shrink.

Multiple actin families exist within most eukaryotes, with the variants often having different functions (e.g., Sehring et al. 2007; Joseph et al. 2008; Velle and Fritz-Laylin 2019), although yeasts and *Giardia* encode only a single copy. Numerous additional proteins are associated with the assembly of actin cables and networks in eukaryotes. For example, at least eight families of actin-related proteins (ARPs), related to actins by gene duplication, are involved in the nucleation of new filaments and patterning of branches off parental actins (Goodson and Hawse 2002). Profilins and formins are involved in the regulated delivery of monomers to growing filaments. Proteins in the WASP (Wiscott–Aldrich syndrome protein) family are involved in context-specific catalysis of ARP-induced branch growth, and appear to be essential in crawling motility (Fritz-Laylin et al. 2017). At least three major families of WASP proteins are inferred to have been present in LECA, and these have in turn diversified enormously among some phylogenetic lineages, while also being lost from others in coordination with ARP loss (Veltman and Insall 2010; Kollmar et al. 2012).

Bacterial actin-like proteins share the basic structural features of eukaryotic actin (including the contact sites at monomer interfaces and the ATP binding site involved in polymerization/depolymerization). Although this suggests phylogenetic relatedness (Ghoshdastider et al. 2015), the sequences of eukaryotic and prokaryotic actins are substantially different (sometimes to the point of not being recognizable by this means; Bork et al. 1992), leaving the issue unresolved. Actin-like proteins with diverse functions are sporadically distributed over the bacterial phylogeny, usually with only one or two types per species. For example, MreB is involved in the maintenance of shape in rod-shaped cells, forming helical shapes around the cell periphery (Margolin 2009). ParR acts like a eukaryotic centromeric-binding protein, elongating

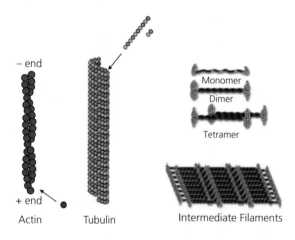

Figure 16.1 Examples of the structural forms derived from the three key forms of fibril-forming proteins in eukaryotes. Actin filaments are solid, double-stranded helices. Individual tubulin filaments are tubes comprised of heterodimers of two elemental forms of tubulin (purple and blue). Intermediate filaments can assemble into higher-order structures such as cables and sheets.

and pushing plasmid copies to the opposite ends of the dividing parental cell using an actin-like concatomer of ParM (Salje et al. 2010). In some bacteria, magnetite crystals organize along a MamK filament, creating a magnet used in orientation (Bazylinski and Frankel 2004). Expression of DivB in punctate foci during times of stress induces *Streptomyces* to grow into branching, filamentous mats.

Although usually consisting of double-stranded helices, these broad families of bacterial filaments vary widely with respect to the periodicity of twists and even the direction of helix winding (handedness). A protein harbored by the archaeon *Pyrobaculum calidifontis*, crenactin, has close sequence similarity to eukaryotic actins and again shares similar helical structure (Ettema et al. 2011; Braun et al. 2015; Izoré et al. 2016). All of these observations strongly suggest that an actin-like protein was present in LUCA.

Tubulins

The second major group of eukaryotic fibrils is comprised of tubules, which assemble through the stepwise addition of heterodimers of α- and β-tubulin subunits. Typically, 13 protofilaments are assembled into hollow cylindrical microtubules, approximately 25 nm in diameter (Figure 16.1), $\sim 2.5\times$ that of actin filaments. Unlike actins, which depend on ATP hydrolysis for assembly, tubulins use GTP hydrolysis. Like actin filaments, microtubules are polar in nature. The minus ends are generally anchored at a microtubule-organizing centre, such as a centrosome, a spindle-pole body, or a basal body. Growth and contraction occur at the opposite end in a process known as dynamic instability, which plays a central role in various aspects of cell motility (Mitchison and Kirschner 1984; Gardner et al. 2013; Akhmanova and Steinmetz 2015).

As with actins, there is clear evidence that the antecedents of tubulins were present in LUCA, the most telling of which is FtsZ, a bacterial protein that assembles into homomeric filaments that can form bundles (but not organized microtubules). Despite having very low sequence identity with tubulin, FtsZ subunits are nearly identical in structure to tubulin monomers, and the few sites conserved

between FtsZ and tubulin are almost all involved in GTP binding (Löwe and Amos 2009). FtsZ produces rings that guide bacterial cell division, although at least one bacterial group (the Planctomycetes) deploys an unrelated protein for such purposes (Van Niftrik et al. 2009). Notably, FtsZ-related proteins are also present in numerous eukaryotes, where they form the division ring responsible for fission of mitochondria and chloroplasts. This is presumably a vestige of these organelles having arisen by bacterial endosymbiosis (Chapter 23), although the loss of such proteins from a number of eukaryotic lineages (including metazoans) implies alternative modes of organelle division (Kiefel et al. 2004; Bernander and Ettema 2010).

α- and β-tubulin are products of an ancient, pre-LUCA gene duplication. In addition to these two types, at least six other distinct tubulin families are known in eukaryotes (although only γ seems to be present in all lineages), and these are generally deployed in different cellular constructs, e.g., basal bodies out of which cilia and flagella grow, rails for vesicle transport, and mitotic/meiotic spindles used for chromosome separation. Such diversification of function is sometimes called the multi-tubulin hypothesis (McKean et al. 2001; Dutcher 2003). Divergent variants even exist within the α and β subtypes. Consider, for example, two closely related amoeboid protists with extreme forms of pseudopodia. Foraminiferans produce branching networks (reticulopodia) that assist in prey capture, whereas radiolarians produce stiff, spine-like structures (axopodia) that assist in floating and capturing prey. Each group deploys a uniquely modified β tubulin to construct the helical filaments involved in such structures (Hou et al. 2013).

Tubulins can also assemble into higher-order structures. For example, most eukaryotes harbor a pair of barrel-shaped organelles called centrioles, comprised largely of microtubules. With few exceptions, centrioles consist of rings of nine microtubule triplets, with the overall structure resembling a cartwheel. Throughout eukaryotes, centrioles serve as the basal body from which flagella and cilia grow. Phylogenetic analysis implies their presence in the ancestral eukaryote, with losses in a few lineages, including yeasts and land plants (Carvalho-Santos et al. 2010, 2011; Hodges et al. 2010).

In animals and some fungi, centrioles have a second role, assembling into centrosomes that serve as the organizing centre from which the mitotic spindle expands during cell division (Chapter 10). Some insects have evolved centrioles with much more elaborate structures than the conventional nine-triplet cartwheel, whereas a few protists have simpler structures (Gönczy 2012). The mechanisms driving such change remain unknown. During sexual reproduction in animals, the centriole is excluded from the egg and introduced via the sperm, possibly an evolutionary outcome of sexual conflict and/or sperm competition driving the high rate of evolution of centriolar protein sequences (Carvalho-Santos et al. 2011; Ross and Normark 2015).

Although the 13-protofilament microtubule appears to be unique to eukaryotes, variants of this form exist in some prokaryotes. For example, some members of the Verrucomicrobiales genus *Prosthecobacter* contain two tubulin-like genes, BtubA and BtubB, that form heterodimers, which in turn polymerize into four-filament microtubules capable of dynamic instability and are also joined by a third protein seemingly related to a eukaryotic motor protein (Pilhofer et al. 2011; Deng et al. 2017). The emergent picture is that FtsZ and tubulins are all members of a common family of proteins, with Btub possibly having arisen by horizontal transfer from a eukaryote and evolving a heterodimeric form after gene duplication.

One of the more interesting aspects of actin and tubulin-related proteins is their exchanged roles in cell division in eukaryotes vs. prokaryotes. As noted above, cytokinesis involves a tubulin-like (FtsZ) ring in bacteria but an actin ring in eukaryotes. In contrast, whereas microtubules are universally used to move chromosomes apart in eukaryotes, actin-like molecules are involved in the partitioning of plasmid genomes in most bacteria (Larsen et al. 2007; Salje et al. 2010; Szewczak-Harris and Löwe 2018), although tubulin-like molecules are used in still others (Larsen et al. 2007). This differential utilization of fibrillar proteins provides a clear example of convergent evolution of the same function from substantially different starting points. Notably, phylogenetic subgroups of the archaea also appear to deploy diverse sets of filaments in cytokinesis, some FtsZ-like and others actin-like (Lindås et al. 2008; Samson and Bell 2009; Makarova et al. 2010; Pelve et al. 2011).

Finally, as discussed below, eukaryotic actins and tubulins act in consortia with motor proteins to accomplish a wide array of intracellular functions. However, fibrils are also capable of work on their own, creating pushing forces that can be used to move various organelles and to deform the cell membrane. How can this occur if the fibrils grow at the tips that are in contact with their targets? Although not all of the details are worked out, it appears that continued fibril elongation occurs by a sort of Brownian-motion ratchet (Mogilner and Oster 1996). Slight fluctuations at points of contact allow occasional room for a new monomer to join the pushing end of the fibril, thereby ratcheting the point of contact forward.

Intermediate filaments

A third group of fibril forming proteins known as intermediate filaments are structurally quite unlike tubulins and actins, forming unpolarized cables and sheets that are not dynamic. Although such structures are ultimately assembled from dimeric subunits, unlike the situation in actins and tubulins, these organize into coiled coils before further development into higher-order structures (Figure 16.1). Intermediate filaments are mostly confined to metazoans and slime molds, where they have a variety of functions involved in structural support, e.g., lamins, keratins, and desmins (Preisner et al. 2018). This restricted distribution, sequestered to one clade, suggests a post-LECA origin. However, a potential homolog, crescentin, determines cell shape in the bacterium *Caulobacter crescentus*.

Cell Shape

Cells come in a wide variety of sizes and shapes. Indeed, unicellular eukaryotes often have such characteristic morphologies that they are often used as diagnostics for identification to the genus or species levels, e.g., as in diatoms and dinoflagellates. Bacteria also have a wide range of possible cell shapes (Young 2006), although only a small number of these are commonly utilized. Most bacteria are either spherical or rod-shaped with an average axial (length:width) ratio of ~ 3, with non-motile cells being spheres more frequently than are flagellated cells (Dusenbery 1998; Young 2006). When confined to constricted settings or forced to bend, bacteria can mold into novel shapes, but upon release to more natural settings, they return to their characteristic shapes (Männik et al. 2009; Amir et al. 2014), demonstrating both a capacity for plasticity and a significant degree of genetic determinism. Cytoskeletal fibrils and cell walls provide support structures essential for maintaining shape.

The main challenge here from an evolutionary perspective is understanding the selective pressures that promote the maintenance of particular shapes in specific ecological contexts. Although the evolutionary advantages of alternative cell shapes are unknown in most cases, a variety of factors have been envisioned (Pirie 1973; Marshall 2011; Pančić et al. 2019). For example, given the known behavior of different shapes within fluids, e.g., spheres, oblate spheroids (flat ellipsoids), and prolate spheroids (elongate ellipsoids), the advantages of alternative forms can be postulated from a purely biophysical perspective (Dusenbery 1998; Schuech et al. 2019). If there is a premium on increased surface area (as would be the case if cell-surface area limited nutrient uptake), disc-shaped cells would be expected to be common. However, such forms are almost never found among bacteria (one exception being *Haloquadratum*, which forms flat squares), although some diatoms have them. Spheres (which have a minimum surface:volume ratio) have the highest rate of diffusion through a liquid, and would be selectively promoted if random dispersal (with no direct motility) were to be advantageous. On the other hand, the sedimentation rate of spheres exceeds that of all other ellipsoids, which may be disadvantageous under many conditions, e.g., phytoplankton sinking out of the photic zone. Up to a $2.4\times$ reduction in sinking rate occurs with a rod-like form, but this maximum effect requires an axial ratio of ~ 30, far beyond what is typically observed (some needle-shaped diatoms being an exception). Swimming efficiency is greatest for a rod with an axial ratio of 2 (Dusenbery 1998), which minimizes drag, but this relative length is smaller than what is typically seen.

Why then are rod-shaped bacteria so common? Rods may be advantageous in environments where sheer forces are high, as they are able to attach to solid surfaces better than spheres. Environmental sensing (Chapter 22) may also be enhanced in elongate cells, which can have a 100- to 600-fold enhanced capacity for inferring the direction of chemical gradients (Dusenbery 1998). This can be accomplished by simultaneously sensing the environment at both ends of the cell or by performing a temporal survey by swimming across environmental gradients. Spheres are much more subject to rotational diffusion (and hence, loss of directionality) than rods.

Despite the uncertainties regarding ecological selective forces, rod-shaped bacterial cells appear to follow specific scaling patterns. Rods generally grow only in length (not width), and this keeps the surface area:volume (SA:V) roughly constant over the cell cycle, as the end caps make only a small contribution. The SA:V ratio is maintained near $2\pi V^{-1/3}$ at both the level of individual species when grown under different nutritional levels as well as at the phylogenetic (between-species) level (Ojkic et al. 2019). This specific SA:V scaling is accomplished by maintaining the aspect ratio (length:width) at a constant value $\simeq 4$. Under this scaling pattern, as a cell doubles in volume, the SA:V ratio decreases by a factor of 1.26.

Although the specific underlying selective forces are not always clear, cellular dimensions can be responsive to relatively simple modifications at the molecular level. The molecular toolbox for cell-shape determination in bacteria primarily involves two cytoskeletal molecules noted above: the tubulin-like FtsZ and the actin-like MreB. For rod-shaped cells, MreB and its relatives direct the synthesis of the sidewalls in ways that ensure uniform width (Margolin 2009; Typas et al. 2011; Ursell et al. 2014; Dion et al. 2019). On timescales of just a few years, laboratory cultures of *E. coli* have been found to evolve increases in cell volume via a single mutation in the MreB gene. Additional studies show that substitutions of the 20 alternative amino acids at just one amino-acid site in the MreB gene generate a range of cell sizes, with an increase in growth rate accompanying an increase in cell volume up to a point (Monds et al. 2014). Numerous other examples exist in which simple manipulations of the protein-coding regions of MreB and FtsZ elicit radical shifts in cell form, e.g., spheres to rods to long filaments (Young 2010).

When MreB is deleted entirely, rod-shaped cells generally shift to more spherical forms (Margolin 2009), and indeed bacteria that are naturally coccoid have invariably lost MreB. However, some rod-shaped bacteria have also lost MreB, implying the presence of alternative methods for the maintenance of this shape. Moreover, even when MreB is utilized in the formation of the same cell shape in different species, the end point can be achieved by rather different mechanisms. For example, the two model bacteria, *E. coli* and *B. subtilis*, use MreB to maintain their rod shapes, but the helical movement of this molecule proceeds in left-hand fashion in the former and in right-hand fashion in the latter. In addition, the deployment of MreB in wall growth differs. *E. coli* regulates the amount of peptidoglycan loaded per MreB patch, whereas *B. subtilis* regulates the speed of growth of individual MreB patches (Wang et al. 2012). The two species also differ in the ways in which they integrate subunits into the cell wall, the thickness of which in *E. coli* is only ∼10% that in *B. subtilis* (Billaudeau et al. 2017).

Lactococcus lactis lacks MreB and is typically an ovoid cell, but under certain environmental conditions the cells can become rod-shaped, even extending to long filaments. FtsZ is involved in these transformations (Pérez-Núñez et al. 2011). As noted above, this protein generally directs the production of the cross-wall septum during cell division, but in *L. lactis* multiple rings of the division proteins develop along the filaments, inhibiting complete septation. This kind of transition from spherical (coccoid) to rod-shaped cells has been seen in other bacteria (Lleo et al. 1990), but cases are also known in which shifts from rod to coccoid shapes are elicited by deletions of genes other than FtsZ and MreB (Veyrier et al. 2015). Thus, although it has been suggested that the ancestral bacterium was rod shaped (Siefert and Fox 1998; Tamames et al. 2001), with coccoid forms being derived evolutionary dead-end states, the ease of both transitions leaves little room for such a simple interpretation.

Cell Walls

Most prokaryotes maintain their cell shapes via external cell-wall structures. The same is true for fungi, green plants, and many other lineages of unicellular algae. In addition to their roles in cell-shape determination, such outer coverings can have other functions, such as protection against pathogens and consumers. Structural support also allows freshwater cells to resist the turgor pressure that inevitably results from high intracellular molarity, which encourages the influx of water molecules from the environment (Chapter 18). Cells lacking such pressure resistance would blow up without active mechanisms for the continuous efflux of water. Raven (1982) makes the point that cell walls represent a one-time investment against the osmolarity problem, whereas an active mechanism of water removal, e.g., a contractile vacuole, incurs costs that must be paid throughout the life of the cell. Assuming negligible costs of cell-wall maintenance, the latter expense must eventually surpass the former in slower-growing cells.

Bacteria

One of the most striking variants of cell-wall architecture distinguishes Gram-negative from Gram-positive bacteria (originally identified by a difference in staining intensity of the cells). Bacterial cell walls are constructed primarily out of peptidoglycan (often called murein), which consists of a mesh of polysaccharide (glycan) chains linked together by short peptides (Figure 16.2). Gram-positive bacteria (also known as monoderms) have a single cell membrane surrounded by a thick cell wall, whereas Gram-negative bacteria (diderms) sandwich a thin peptidoglycan layer between two lipid membranes (often with an additional layer of lipopolysaccharides on the cell surface). Although the basic structure of peptidoglycan is universal, there is substantial variation among bacterial lineages with respect to peptide sequence, location of crosslinks between the peptide chains, and secondary modifications of the components (Vollmer and Seligman 2010).

How a dual-membrane form could have evolved from an ancestor with a single membrane (assuming this was, in fact, the ancestral bacterial state)

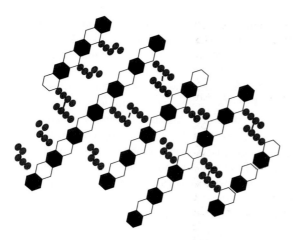

Figure 16.2 A cartoon version of a peptidoglycan layer. The paired hexagons represent dimers of NAM and NAG, whereas the small chains represent cross-linked peptides.

is an unsolved problem. One idea derives from the observation that monoderms often produce double-membrane endospores. These are produced by engulfment within the maternal cell, with the plasma membrane of the mother becoming an external membrane of the spore (Vollmer 2011; Tocheva et al. 2016). Some diderm bacteria can also produce endospores wherein the outer membrane of the maternal cell is displaced by the inner membrane (Tocheva et al. 2011). These observations show how a cell can undergo special kinds of divisions to establish an altered cell envelope. However, such changes are terminal and lost upon spore germination, so such repatterning is a far cry from demonstrating how a stable developmental pattern of two-layer biogenesis is acquired in an evolutionary sense. One interpretation of extant bacterial phylogeny is that the ancestral bacterium was a sporulating diderm (with monoderms losing the outer membrane) (Tocheva et al. 2016). If correct, this would be another example of the regression of complexity to simpler forms (while still leaving the origin of the diderm phenotype unexplained).

Like Gram-positive bacteria, archaea nearly always have single lipid membranes, often surrounded by a variant of peptidoglycan (pseudomurein) and then by an outer crystalline shell called the S layer (Albers and Meyer 2011; Visweswaran et al. 2011; Oger and Cario 2013). S layers consist

of highly organized, essentially crystalline, protein lattices, with substantial variation in design among species. As there is significant variation in the biosynthetic pathways leading to murein (bacteria) and pseudomurein (archaea), and little homology between the genes involved, it has been suggested that the cell walls of these two groups evolved independently (Hartmann and König 1990; Steenbakkers et al. 2006).

Although cell walls provide obvious advantages in many ecological settings, they have been lost in a few bacterial lineages (e.g., *Mycoplasma* and *Ureaplasma*). Numerous eukaryotic lineages do not have cell walls, and it remains unclear as to whether LECA did. Multiple lines of evidence suggest that cell walls comprise a substantial fraction of a cell's energy budget, imposing a 'use-it-or-lose-it' situation, i.e., a strong advantage to loss during prolonged periods in which the investment in biosynthesis outweighed the fitness advantages.

The energetic costs of wall construction can be made for the well-understood bacterium *E. coli* in the following way (Lynch and Trickovic 2020), using methods developed in Chapter 17. The carbohydrate subunit of peptidoglycan consists of two joined molecules, NAG (N-acetylglucosamine) and NAM (N-acetylmuramic acid), with short peptide chains fused to the latter for cross-bridging. The total cost of synthesis of each peptidoglycan unit can be shown to be $\sim$209 ATP hydrolyses, and there are $\sim$3.5 × 10^6 disaccharides in the entire cell wall (Wientjes et al. 1991; Gumbart et al. 2014), implying a total cost of $\sim$7.3 × 10^8 ATP hydrolyses (which constitutes $\sim$2.7% of the total cell-growth budget). The *E. coli* envelope also contains $\sim$10^6 molecules of Braun's lipoprotein (Li et al. 2014; Asmar and Collet 2018), which attach the outer membrane to the cell wall. Each of these molecules consists of a chain of 58 amino acids (worth $\sim$34 ATPs each, including chain elongation) and three chains of palmitate ($\sim$122 ATPs each) for a total cost of $(58 \times 34) + (3 \times 122) = 2338$ ATPs per unit, implying $\sim$2.3 × 10^9 ATPs in total, or $\sim$8.5% of the total cell budget. There are in addition $\sim$1.2 × 10^6 lipopolysaccharide molecules on the surface of the cell (Neidhardt et al. 1990), with a cost of 1048 ATPs per molecule, accounting for another 3.6% of the cell's construction budget. Finally, the total cost of lipids in the cell membranes of *E. coli* comprises ~34% of the cell budget (Chapter 17). Thus, summing all of these components, the entire envelope constitutes ~49% of the energy budget of an *E. coli* cell. This is likely a downwardly biased estimate, as the costs of surface antigens attached to the ends of lipopolysaccharides (and present in variable amounts in different strains) have been ignored.

Extension of this sort of analysis to the Gram-positive bacterium *B. subtilis* leads to the conclusion that, despite substantial envelope organization differences relative to *E. coli*, about a third of the total cell energy budget is allocated to the envelope (Lynch and Trickovic 2020). Thus, for species with cell walls, lipid membranes are generally still the most expensive components of the envelope. Most significantly, the overall cost of the bacterial cell envelope constitutes a larger fraction of a cell's energy budget than any other single component, although this declines with increasing cell volume as a consequence of the reduction in SA:V ratio (Chapter 17).

Eukaryotes

Numerous wall types exist in eukaryotes. One macro-scale classification uses categories related to whether walls are internal or external to membranes, neither, or both (Becker 2000), although such a scheme belies the underlying diversity in chemical makeup (Niklas 2004). For example, structural support in diatoms is provided by silicious shells. The thecal plates of dinoflagellates, and the scales of haptophytes, chrysophytic algae, and other phytoplankton species are constructed from diverse substrates, including calcite, chitin, and cellulose (Okuda 2002).

One of the most common constituents of eukaryotic cell walls, possibly the most abundant biomolecule in the world, is a linear polymer of glucose residues called cellulose. Found throughout plants, some slime molds, some bacteria, and even a group of animals (tunicates), this simple molecule has variants among phylogenetic lineages, with the glucose residues sometimes carrying modified side chains. The phylogenetic distribution of cellulose biosynthetic capacity, along with the conserved structure and sequence

of the underlying machinery, strongly suggests its movement into and throughout eukaryotes by horizontal transfer events. The ultimate agent may have been the endosymbiotic colonist that formed the chloroplast. However, it remains a mystery as to why eukaryotes have completely abandoned the use of peptidoglycans, if they ever had them. Notably, biosynthesis of peptidoglycan in bacteria starts with UDP-acetyl glucosamine (where UDP denotes uridine diphosphate), whereas cellulose biosynthesis starts with UDP-glucose, implying either an ancient common ancestry for the production of such building blocks, or another case of convergent evolution.

Contrary to the situation with bacterial peptidoglycan, which is held together by peptide cross-linking, cellulose consists of long aggregated filaments of the underlying glucose derivatives held together by intra- and inter-chain hydrogen bonds. Cellulose production proceeds in a spatially organized fashion via large, membrane-bound cellulose synthase complexes, organized in rows, arrays, or rosettes that define the variable dimensions of the resultant microfibrils and their mesh-like forms in different species (Niklas 2004; Saxena and Brown 2005). However, in seemingly no cases do eukaryotic cell walls consist solely of cellulose, with a wide variety of glycoproteins, lignins, pectins, uronic acids, xyloglucans, and many other subunits being utilized in various forms of cross-linking in individual lineages (Niklas 2004; Domozych et al. 2014). In the brown algae, cellulose is a relatively minor component of cell walls, the majority consisting of other forms of linked polysaccharides called alginates and sulfated fucans, both thought to have been acquired by horizontal transfer from bacteria (Michel et al. 2010).

Even rough estimates of the costs of these components would be useful. Using methods outlined in Chapter 17, based on the known biosynthetic pathway, the cost of each glucan subunit of cellulose is equivalent to ~34 ATP hydrolyses, although the total investment that cells make in cellulose (which requires estimates of the total number of subunits per wall) remains an open question. Deposition of silicon into the frustules of diatoms may be relatively cheap, involving the expenditure of perhaps just a single ATP per incorporated molecule, with the total cost to a cell being only $\sim 2\%$ of the total energy budget (Raven 1983), although this ignores the underlying organic wall components.

Explicit insight into the substantial costs of walls can be achieved for the well-studied budding yeast by considering the following numbers. Yin et al. (2007) estimate that $\sim 21\%$ of the biomass of a *S. cerevisiae* cell consists of the cell wall ($\sim 4\%$ protein, 30% mannose, 60% glucan, and 1% chitin). The glucan subunits are derived from $\sim 7.7 \times 10^9$ glucose molecules, which would otherwise be available for cellular energy or as carbon skeletons for other biosynthetic pathways, and the $\sim 1.7 \times 10^6$ protein molecules per adult cell surface constitute $\sim 2\%$ of the total protein content of the cell (Klis et al. 2014). Similar numbers were obtained for another yeast species, *Candida albicans* (Klis et al. 2014).

Given these observations, a rough estimate of the cost of the cell wall in *S. cerevisiae* can be calculated as follows. Mannose is a simple derivative of glucose and comprises half the biomass of the glucans, so together these two constituents account for the equivalent of $\sim 1.5 \times 7.7 \times 10^9 = 1.2 \times 10^{10}$ glucose molecules. Noting that *S. cerevisiae* has an average cell volume $\simeq 70\ \mu m^3$, Equation 8.2b implies that the cost of building a yeast cell $\simeq 1.6 \times 10^{12}$ ATP hydrolyses. Assuming that a glucose molecule is equivalent to 32 ATPs under aerobic respiration (Chapter 17), this translates to a cost of 5×10^{10} in units of glucose molecules. Thus, mannose and glucans together account for $\sim 26\%$ of the energetic growth requirements in this species. Assuming an average cost of synthesizing and concatenating amino acids of ~ 34 ATP hydrolyses each (Chapter 17), and an average protein length of 400 amino acids, the cost of wall proteins $\simeq 1.7 \times 10^6 \times 400 \times 34 = 23 \times 10^9$ ATPs, which is $\sim 1.4\%$ of the total energy budget. With the cell membrane in this species constituting an additional 4.6% of the cell budget (Chapter 17), the entire cell envelope in *S. cerevisiae* comprises $\sim 32\%$ of the total cell energy budget.

Molecular Motors

In prokaryotes, most intracellular molecular transport is passively driven by random diffusion resulting from background thermally induced molecular motion. In contrast, a good deal of molecular movement in eukaryotes involves active transport, which comes at a cost. Actins and tubulins can exert mechanical pressure on their own, as they push against larger structures and thermal fluctuations allow the stochastic insertion of monomeric subunits and progressive extension. However, molecular-motor proteins supplement the cytomotive capacity of the cytoskeletal filaments, using them as highways for dragging along attached cargoes (Figure 16.3). Such motors transduce chemical energy into mechanical force via ATP hydrolysis, engaging in diverse functions, ranging from cargo transport to the movement of flagella (Howard 2001).

Given their apparent absence from prokaryotes, all three major families of cytoskeleton-associated motors – dynein, kinesin, and myosin – likely emerged on the path between FECA and LECA. Myosins travel exclusively on actin filaments,

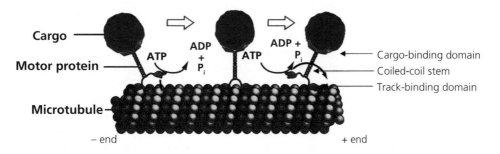

Figure 16.3 An example of a motor protein interfacing with a microtubule, along which it walks fuelled by energy derived from ATP hydrolysis, carrying its cargo (in this case a transport vesicle).

whereas kinesins and dyneins operate on microtubules (tubulin filaments). Although their physical structures vary, each type of molecule works by the same mechanism. Each motor molecule has an ATP binding site, a track binding site, and a tail domain involved in cargo attachment. Some details remain unclear (Hwang and Karplus 2019), but the molecules effectively walk along their cytoskeletal roadways, with each ATP hydrolysis resulting in one step forward (in some cases two hydrolyses occur per step; Zhang et al. 2015).

Ubiquitous to all eukaryotes, kinesins underwent massive diversification into an estimated eleven families prior to LECA (Lawrence et al. 2004; Wickstead et al. 2010). Prominent roles include vesicle transport, spindle assembly, chromosome segregation, and intraflagellar transport of flagellar components. Monomeric, dimeric, trimeric, and tetrameric forms exist, and the multimers may be either homomeric or heteromeric. The motor domains generally operate like walking feet paired together at one end of the molecular structure, although tetramers have motor domains on both ends. Kinesin movement is generally unidirectional, towards the + ends of microtubules (usually moving from the centre to the edges of cells). Typical rates of movement are on the order of 0.5 to 1 μm/sec and stride lengths $\simeq$8 nm (Block et al. 1990; Howard 2001; Muthukrishnan et al. 2009; Scholey 2013). Motors involved in meiosis and mitosis are much slower, with speeds on the order of 0.05 μm/sec. Different kinesins vary in neck lengths, and these and other molecular features determine the degree of processivity of the motors along tubulin filaments (Hariharan and Hancock 2009; Shastry and Hancock 2010; Cochran 2015).

Myosins are nearly ubiquitous among eukaryotes, but apparently have been lost from the red algal, diplomonad, and trichomonad lineages (Richards and Cavalier-Smith 2005; Foth et al. 2006). Although myosins and kinesins differ in size and exhibit little sequence similarity, given the similar three-dimensional structures of their functional domains, they may have evolved from a common ancestor (Kull et al. 1998). The most likely ancestral state prior to eukaryotic diversification is something on the order of three myosin classes, all derived from a single common ancestral protein en route from FECA to LECA. However, like kinesins, myosins can be further diversified into a large number of subclasses (as many as 24; Foth et al. 2006; Goodson and Dawson 2006), again with numerous functions, including cellular motility, and transport of organelles, vesicles, and mRNAs. An argument has been made that diversification of myosin functions followed paths of subfunctionalization, from generalized to more specialized forms (Mast et al. 2012).

Although most myosins are dimeric with parallel coiled-coil domains, monomeric forms exist, as do dimers with anti-parallel constructs with actin-binding domains at each end (Quintero and Yengo 2012). Almost all myosins move in the + direction of actin filaments. Each type of molecule has a specific affinity for certain types of filaments, a particular step length (usually 5 to 40 nm per step), and speeds often in the range of 0.3 to 0.5 μm/sec (Elting et al. 2011). However, a myosin involved in cytoplasmic streaming in plant cells can move at rates up to 60 μm/sec (Titus 2016). Exceptionally high rates occur when myosins run rather than walk, jumping from point to point with step lengths greater than those that can be accomplished when one motor domain is always in contact with the filament.

The third major class of motor proteins, dyneins, has a substantially different structure than kinesins and myosins, including an intramolecular hexameric ring consisting of AAA (ATPases Associated with cellular Activities) domains, all physically linked in a single gene (Kardon and Vale 2009). Containing >4500 amino acid residues, dyneins are among the largest known proteins. In total, there are at least nine deeply diverging dynein lineages, most of which trace back to LECA, although they were independently lost in the lineages leading to land plants and red algae. Dyneins maintain the attribute of having two walking feet, with each step fuelled by ATP hydrolysis. However, contrary to the situation with kinesins and myosins, dyneins always walk towards the − ends of microtubules, typically towards the cell centre, and with variable speeds in the range of those noted earlier (Walter and Diez 2012; Sweeney and Holzbaur 2016). Dyneins are also unique in that all but one cytoplasmic family member is associated with the cilium/flagellum (Wickstead and Gull 2007; Wilkes et al. 2008). The latter operate in a team-like fashion to cause microtubules to slide past each other, thereby eliciting ciliary movement.

Although molecular motors have been subject to numerous biochemical and biophysical studies, few questions have been asked from an evolutionary perspective. The three central phenotypic features of motor proteins are their step length, step rate, and processivity (distance progressed before falling off a substrate). All three parameters are readily estimated in laboratory constructs, and the rate of movement is equal to the product of the length and rate of steps. However, almost nothing is known about how these traits vary among species or among motor family members within species. In addition, the total drain on a cell's energy budget owing to motor activity remains unclear.

Motility

All cells are capable of movement, even if simply as a consequence of the physical forces associated with cell division. Here, our concern is with active movement within the lifespan of a cell, as in swimming or crawling. The range of cellular motility mechanisms is diverse, and Miyata et al. (2020) catalogue 18 distinct modes of single-cell movement. Moreover, even seemingly similar, complex structural features such as flagella have evolved independently in different lineages. Regardless of the mechanism, all forms of motility require an investment of energy, either a gradient of hydrogen ions driving a turbine-like apparatus or active hydrolysis of ATP used in the conversion of chemical to mechanical energy for flagellar flexing. In addition, the construction of the motility machinery itself can require a substantial energetic investment. Thus, unless active motility provides a boost in fitness, the mechanism will be liable to rapid loss, as this can free up substantial energy for other cellular functions.

The details of how simple cells can utilize sensory mechanisms to direct motion will be addressed in Chapter 18. The primary focus here is on the evolutionary origins of motility machineries and the costs of building and running them. Flagella and cilia have received considerably more attention than mechanisms of crawling, and accordingly are given the most attention in the following sections. A broad overview of bacterial motility systems is given in Wadhwa and Berg (2022).

Crawling

Numerous bacteria can move on solid surfaces (Jarrell and McBride 2008; Nan and Zusman 2016; Wadhwa and Berg 2022). For example, some bacteria, such as *Neisseria*, crawl over surfaces by twitching of Type IV pili, whereas members of the Bacteroidetes glide by use of surface adhesins, and still others move with an inchworm-like mechanism involving membrane-embedded proteins. In addition, crawling is achieved by similar enough molecular mechanisms over diverse phylogenetic groups of eukaryotes that it has been argued that LECA was capable of such movement (Fritz-Laylin et al. 2017). Thus, given that a similar conclusion has been reached on the origins of the eukaryotic flagellum, and the fact that numerous eukaryotic lineages have cells capable of both flagellar swimming and amoeboid movement (Fulton 1977; Brunet et al. 2021; Prostak et al. 2021), LECA may have been capable of both forms of motility.

Despite the approximate uniformity of crawling mechanisms in eukaryotes, most notably the shared use of actins and actin-associated proteins to generate expansion and retraction of cellular protrusions, a variety of structures are used. These can generally be classified into two major groups: 1) pseudopods, which involve broad protrusions underlain by branched-actin networks, allow cells to drag themselves across surfaces; and 2) filopods, which involve narrower, linear extensions of unbranched actin bundles, enabling a more walking-like motion. Unlike more permanent flagella, all such structures are dynamic, allowing rapid changes in shape via local assembly and disassembly of the underlying cytoskeleton. One of the primary side activities associated with crawling is the ingestion of food particles by phagocytosis.

In general, crawling on surfaces proceeds at much lower speeds than swimming through liquids. For example, the amoeba *Naegleria gruberi* crawls at a rate of 1 to 2 μm/sec (Preston and King 2003), which is one to two orders of magnitude lower than typical swimming speeds for prokaryotes and eukaryotes of similar size (below). Gliding motility in *Myxococcus* is even slower, on the order of 0.03 μm/sec (Tchoufag et al. 2019).

Little is known about the energetic cost of crawling in unicellular species. However, based on the cost of actin polymerization, Flamholz et al. (2014) estimated that the crawling motion in a goldfish keratocyte (a cell type that normally colonizes the cornea in vertebrates) consumes $\sim$4 $\times$ 10^5 ATPs/second. Recalling from Chapter 8 that the basal metabolic requirement of a vertebrate cell $\simeq$5 $\times$ 10^7 ATPs/second, and that, assuming a one-day cell division time, the cost of building such a cell is $\sim$6 $\times$ 10^8 ATPs/second, the cost of crawling in this particular case is on the order of 1% of such a cell's operational budget (and $\sim$0.1% of the total energy budget).

Prokaryotic flagella

Some planktonic cyanobacteria, such as *Synechococcus*, can swim without flagella (Waterbury et al. 1985), and others are capable of buoyancy regulation via gas vacuoles (Walsby 1994), but the vast majority of bacterial movement in open fluids involves a flagellum, one of the most complex molecular machines within bacterial cells (Figure 16.4). The dozens of proteins contributing to overall flagellar structure can be divided into five major components: the basal body (which includes the membrane-embedded stator within which the motor rotates), the switch, the hook, the filament, and the export apparatus. Bacterial flagellar assembly starts at the basal body, anchoring the overall structure to the cell membrane, and then proceeds to the construction of the hook, and finally to the filament. The latter is a hollow, tubular structure, consisting of tens of thousands of flagellin molecules, making this one of the most abundant proteins in swimming bacterial cells. Its assembly proceeds by the export of flagellin proteins through the central pore. Specific chaperones are assigned to each exported protein, which have to remain unfolded while in transit to the tip. The monomeric

flagellin subunits are then configured into helical forms, which vary in structural arrangements both within and between species (Turner et al. 2000; Kreutzberger et al. 2022).

Unicellular species live in a world with a low Reynold's number (Foundations 16.2), where viscous forces dominate to such a degree that there is essentially no inertia upon cessation of swimming. Thus, bacterial flagella essentially operate like cork screws, pulling the cell forward through an effectively syrup-like environment. Generally driven by a proton-motive force (but in some species, by a sodium-motive force or even by divalent cations; Ito and Takahashi 2017), the helical filament rotates within the membrane-embedded hub (Manson et al. 1977, 1980; Meister et al. 1987). The overall structure is reminiscent of a rotor guided by a surrounding stationary stator in an electrical motor, except that the primary flagellar motor is run by series of still smaller motors (Figure 16.5). As the protons pass through channels, the associated rotational force is applied to the internal shaft, analogous to the coordinated pairing of gears associated with a bicycle chain. Speed and orientation can be regulated through various signal-transduction systems that modify flagellar activity (Wadhams and Armitage 2004; Boehm et al. 2010; Chapter 22).

Despite the intricacies of the bacterial flagellum, there is substantial phylogenetic variation in its basic features, including the angularity and periodicity of the helices, clockwise versus counterclockwise rotation, the number of distinct flagellin proteins incorporated into the filament, and the number of protofilaments per flagellum (Pallen and Matzke 2006; Galkin et al. 2008; Chaban et al. 2018; Kaplan et al. 2019). Variation also exists in the structure and number of components in the basal body and in the cellular positions and numbers of flagella (Chen et al. 2011; Moon et al. 2016; Chaban et al. 2018; Rossmann and Beeby 2018). Although the flagella of most bacteria are not surrounded by membranes, such sheaths are present in numerous genera including *Helicobacter*, *Vibrio*, and *Bdellovibrio* (Seidler and Starr 1968; Geis et al. 1993; Zhu et al. 2017). In spirochaetes, the flagella are not even external, but reside within the periplasmic space (between the two membranes), where their twisting

Prokaryote   Eukaryote

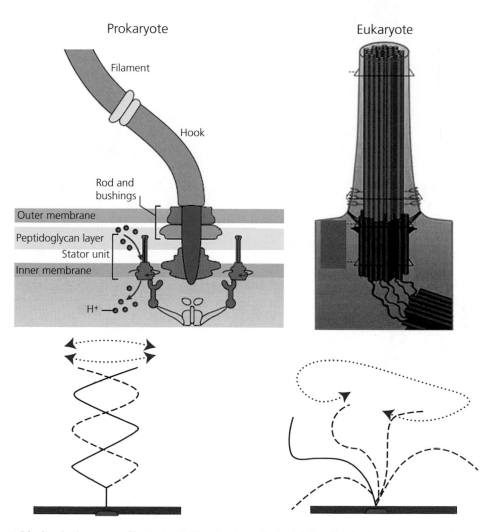

Figure 16.4 Idealized schematics of bacterial (left) and eukaryotic (right) flagella. The former emanates from a complex apparatus embedded in the double membrane, and operates in a corkscrew-like manner. The latter grows out of a basal body (centriole), has an interior consisting of tubulins along which motor proteins move (not shown), and operates in a whip-like manner.

contorts the entire cell. The degree to which the scattered presence of flagella throughout the bacterial phylogeny is due to lineage-specific losses or gains by horizontal gene transfer remains unclear (Snyder et al. 2009).

How might a molecular machine as complex as a bacterial flagellum have arisen? Paths of descent with modification of pre-existing structures are supported by several lines of evidence. For example, the subcomponent that drives flagellar-component export appears to be related to the catalytic subunits of ATP synthase. In addition, the remainder of the system appears to be related to cell-surface projections used to transfer infection-determining proteins into host cells, so-called Type-III secretion systems (Pallen et al. 2005). At least 10 components of such secretion systems have strong homologies to components of the bacterial flagellum, and the overall architecture is similar. The main difference is that the secretion system harbors a stiff injection apparatus, rather than a flagellum (Blocker et al. 2003; Egelman 2010).

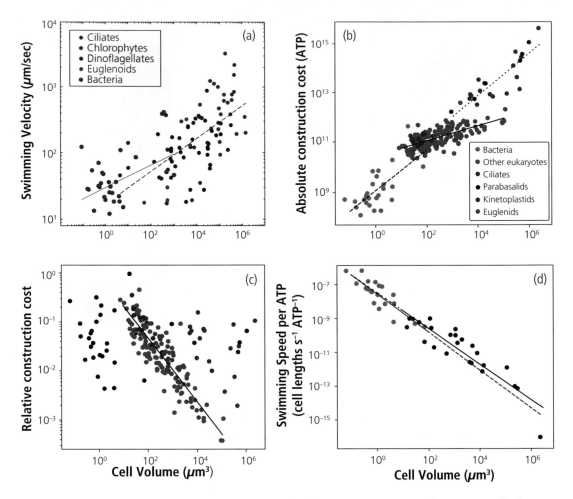

Figure 16.5 Scaling relationships for swimming speeds and flagellar construction costs with respect to cell volume. **a:** Swimming speeds (v) vs. cell volume (V) in unicellular species. For eukaryotes, $v = 16.6V^{0.25}$, $r^2 = 0.35$); and for bacteria, $v = 29.5V^{0.17}$, $r^2 = 0.17$) (as the data are very noisy, the data for all phylogenetic groups were pooled for the eukaryote analysis). The measured values are corrected for temperature assuming a Q_{10} of 2.5. From Lynch and Trickovic (2020). **b:** Absolute costs of total flagellar construction. In units of 10^9 ATP hydrolyses, the fitted power-law relationships are: $1.1V^{0.81}$ for bacteria, and $2.6V^{0.31}$ for eukaryotic flagellates excluding ciliates and parabasalids, which are described by $2.8V^{0.86}$. **c:** Flagellar construction costs as a fraction of total-cell construction costs. Color-coding of points as described in panel **b.** **d:** Swimming speed (in units of cell length) per ATP invested in flagellar construction. In units of 10^{-8}, the regressions for bacteria (grey) and eukaryotes (black) are, respectively, $2.7V^{-1.13}$ and $3.1V^{-1.06}$. The data for the final three panels are taken from Schavemaker and Lynch (2022).

Liu and Ochman (2007) provide evidence that least 24 of the >50 genes whose products comprise the bacterial flagellar system were likely present in the common ancestor to all bacteria, and that many of the components were derived by gene duplication. They further suggest that the order of inferred duplications imply an 'inside-out' sequence of evolutionary steps that in turn reflects the assembly process, i.e., from basal body to hook to junction to filament, a potential example of ontogeny recapitulating phylogeny. These observations suggest that the flagellum arose from something like a Type-III secretion system, rather than the other way around, although this is not an iron-clad conclusion. Just as the flagellum may have evolved from bacterial structures with alternative functions, partial loss of the flagellum sometimes leads to modified structures with novel functions. For example, *Buchnera aphidicola*, an endosymbiont that inhabits the cells of aphids, has lost the flagellar filament, but the cells still harbor hundreds of hook/basal-body structures, with likely roles in secretion into host cells, much like Type-III secretion systems (Maezawa et al. 2006; Toft and Fares 2008).

Finally, although fairly similar in overall structure, the flagella of archaea appear to be completely unrelated to those of bacteria, providing a remarkable example of convergent evolution (Kreutzberger et al. 2022). As in bacteria, the archaeal flagellum (often called the archaellum) rotates via an embedded membrane structure. However, rather than being driven by a proton-motive force, ATP hydrolysis directly drives rotation of the archaellum (Thomas et al. 2001; Desmond et al. 2007; Lassak et al. 2012; Daum et al. 2017; Albers and Jarrell 2018). The components of the archaeal flagellum appear to be unrelated to bacterial flagellins, and unlike the situation in bacteria, archaeal flagella are not hollow, have no hook structure, and appear most closely related to bacterial Type-IV pilus systems (Albers and Pohlschröder 2009), which are used as adhesive structures and in twitching motility, although this too might be a matter of convergence.

Eukaryotic flagella

This theme of diversity extends to the eukaryotic flagellum, which differs from anything utilized in the prokaryotic world. Unlike the rotational flagella of prokaryotes, eukaryotic flagella operate by a whip-like mechanism driven by ATP-consuming motor proteins, which cause doublets of internal microtubules to slide past each other within the lumen of the whip. Only a few eukaryotic lineages lack the ability to produce flagella, most notably slime molds, yeasts, and land plants, motivating the conclusion that the origin of the eukaryotic flagellum preceded LECA (Cavalier-Smith 2002; Nevers et al. 2017). In some protists, such as ciliates and parabasalids, enormous proliferations of short flagella coat large portions of the cell surface. The word cilia (singular cilium) is generally reserved for these

cases, although aside from length differences relative to cell size, the basic structure remains the same.

Eukaryotic flagella are much larger than those in bacteria, generally on the order of 200 nm in diameter as compared to 10–30 nm in prokaryotes. They almost always consist of rings of nine peripheral microtubule doublets surrounding a central pair (Ginger et al. 2008; Ishikawa and Marshall 2011, 2017). This overall core structure, known as an axoneme, is surrounded by an extension of the cell membrane. The doublets grow out of cylinders known as basal bodies, which consist of triplets of δ- and ϵ-tubulins that arose by gene duplication prior to LECA (Dutcher 2003). In most eukaryotes, the basal bodies are recycled centrioles, which are used as mitotic-spindle organizing centres during cell division.

Although bacterial flagella are comprised of a few dozen distinct proteins, on the order of 250–500 proteins contribute to the eukaryotic flagellum, up to 3% of the proteins encoded in a eukaryotic genome (Avidor-Reiss et al. 2004; Pazour et al. 2005; Smith et al. 2005). For the few organisms for which the flagellar proteome has been evaluated (the green alga *Chlamydomonas*, the ciliate *Tetrahymena*, and mammals), ~200 proteins are flagellar-specific and not found in non-ciliated species (Avidor-Reiss et al. 2004; Smith et al. 2005).

Numerous activities occur within the lumens of eukaryotic flagella. As there are no ribosomes within the flagellum, all materials must be selectively imported by intraflagellar transport (IFT). Eukaryotic flagella also contain metabolic machinery for generating ATPs to fuel the molecular motors involved in transport and motility (Ginger et al. 2008). Kinesins move structural components forward on one member of each microtubule doublet, while dyneins move cargo in the opposite direction on the other member, preventing bi-directional traffic collisions (Stepanek and Pigino 2016). Transport rates are on the order of 1–3 μm/sec (Ishikawa and Marshall 2017).

The most convincing hypothesis for the evolution of the eukaryotic flagellum invokes an autogenous origin, starting as protruding microtubule bundle emanating from a microtubule organizing centre (as used in mitotic spindles) (Cavalier-Smith 1978; Jékely and Arendt 2006). The initial structure may have had little to do with motility at all, operating instead as an environmental sensing organ. Indeed, flagella also function secondarily as sense organs, harboring numerous signal-transduction systems (environmental sensors that initiate information cascades within cells; Chapter 22). The central pair of microtubules is typically absent in derived structures called primary cilia, which are used as sensing organs in cells of metazoa.

Eventually an IFT system would have to evolve, along with the ninefold symmetric structure of the axoneme, and the recruitment of molecular motors. Such a scenario may have involved little more than a series of gene duplication and modification events associated with the tubulins and motor proteins deployed in the cellular interior (Hartman and Smith 2009). An alternative hypothesis postulates an origin via a virus that somehow convinced a host cell to produce a primitive basal body for extruding viral particles (Satir et al. 2007; Alliegro and Satir 2009). An obvious problem with this idea is its failure to address the cost of maintaining such a harmful pathogen for the eons required for the emergence of the complex ciliary structure.

Further insight into the origins of the components of the eukaryotic flagellum derives from similarities to various features of the nuclear-pore complex and vesicle-scaffolding proteins. Fibres at the base of the flagellum may operate as gateways to admission of appropriate protein cargoes for intraflagellar transport, making use of RAN-GTP cycles and importin molecules as in the case of transport through nuclear pores (Rosenbaum and Witman 2002; Dishinger et al. 2010; Kee et al. 2012). Notably, the two IFT protein complexes that bind to cargoes (and are in turn dragged by motors) appear to have arisen by duplication from a common ancestor related to vesicle-coat proteins (van Dam et al. 2013). As noted in Chapter 15, the latter proteins are in turn related to the scaffold of the nuclear-pore complex (NPC). Left unresolved here is whether the NPC preceded the evolution of the flagellum, or vice versa, but these observations again illustrate how the complex features of cells often evolve by duplication and modification of pre-existing structures, rather than by *de novo* establishment.

As in bacteria, there are many variants on the structure and deployment of eukaryotic flagella. For example, the green alga *Chlamydomonas* can swim either by a breast stroke or by undulatory waves (Tam and Hosoi 2011). The cryptophyte *Ochromonas* has perpendicular hairs (mastigonemes) radiating from the flagellum, creating a feather-like appearance. Combined with the use of flagella, euglenoids are capable of additional cell contortions (called metaboly), shifting the positions of cellular contents progressively along its axis like food moving through an animal's gut (Noselli et al. 2019). In trypanosomes, a single flagellum is embedded along the side of the cell.

The eukaryotic flagellum has served as a model for understanding how size homeostasis is maintained for cellular organelles, one of the key questions in developmental biology focused at the cellular level (Chapter 9). Inappropriate flagellum lengths (including unequal lengths in species with paired flagella) lead to aberrant swimming patterns (Tam et al. 2003; Khona et al. 2013). Because eukaryotic flagella undergo constant assembly and disassembly of tubulin subunits (Marshall and Rosenbaum 2001), maintenance of a constant length and symmetry implies that the two rates must be balanced. This further requires that at least one of the two reaction rates depends on length, such that the rate of disassembly exceeds that of assembly above the equilibrium length, and vice versa below the equilibrium. Such regulation is supported by experiments in the bi-flagellated *Chlamydomonas*, showing that when one flagellum is severed, it grows back while the other shrinks until the two equilibrate in length (Ludington et al. 2012, 2013). The data appear to support an active-disassembly model, with depolymerization increasing with flagellum length (Fai et al. 2019), and this is also true for *Giardia*, which has four pairs of flagella of different lengths (McInally et al. 2019). Random fluctuations in the assembly/disassembly process leads to natural variation between flagellar lengths within and among cells (Bauer and Marshall 2021).

The costs of swimming

Although many cells can swim, and it is easy to imagine adaptive reasons for doing so, e.g., directive movement towards patchy resources and avoidance of predators, not all cells living in aquatic environments can self-propel. The latter include many planktonic bacteria, diatoms, and green algae. This suggests that the energetic cost of swimming in certain settings may exceed the benefits. Drawing from prior work (Lynch and Trickovic 2020; Schavemaker and Lynch 2022), rough estimates are now provided for the costs for both the act of swimming and of building the mobility apparatus.

The minimum power requirement for swimming can be estimated as the work needed to move an object at a particular velocity (Foundations 16.2). In biology, however, the mechanical force essential for movement is obtained from chemical energy, i.e., ATP hydrolysis. As the conversion of energy is a less than perfect process, the ultimate cost of swimming must exceed the expectations based on mechanical energy alone. From comparisons of cost predictions based on physical theory with the actual costs of moving flagella, the universal conclusion from a diversity of organisms is that the conversion of chemical energy into motion is quite low—the power requirement for swimming is $\sim 100\times$ that expected based solely on the physics of the process (Foundations 16.2). Brownian-motion jostling of cells, rotational diffusion, helical swimming patterns, and flexibility of the flagellum are among the reasons for these two efficiencies.

It has been suggested that the total cost of swimming is still trivially small (Purcell 1977), but this sort of statement needs context, especially from an evolutionary perspective. Only a few attempts have been made to estimate the cost of swimming, but Raven and Richardson (1984) concluded that the cost of running a flagellum for an idealized dinoflagellate is about equal to the basal metabolic rate, whereas Katsu-Kimura et al. (2009) suggested 70% for the ciliate *Paramecium*. Crawford (1992) estimated that 1–10% of ciliate and dinoflagellate *total* energy budgets are allocated to swimming, whereas Fenchel and Finlay (1983) estimated fractional total cellular costs of $\sim 0.4\%$ in the cryptomonad *Ochromonas* and the ciliate *Didinium*. These quantitative differences in results are not necessarily incompatible, as metabolic rates are generally just a small fraction of total cellular energy budgets.

Although fractional energy investments as low as 0.1 to 1% may seem trivial in an absolute sense,

as discussed in Chapters 4 and 5, the power of selection is almost always sufficient to perceive a cost as low as 0.01% (and orders of magnitude lower in unicellular species). Thus, there is little question that the cost of swimming is sufficiently large that such a trait exists in an evolutionary 'use it or lose it' context. That is, the promotion of non-motile mutants is highly likely if the benefits of motility do not exceed the costs.

A more general perspective on the cost of swimming can be obtained from the phylogenetic scaling relationship between swimming velocity (v in units of μm/sec) and cell volume (V in units of μm^3) (Figure 16.5a). For eukaryotic species, average $v \simeq 17V^{0.25}$, and noting that the radius of a sphere is $r = (3V/4\pi)^{1/3}$, this suggests a swimming speed in units of effective cell diameters ($2r$) of $\sim 13V^{-0.08}$/sec. The average estimates are then near 4 and 13 lengths/sec, respectively, for the largest ($V = 10^6$) and smallest ($V = 1$) swimming cells of eukaryotes. For bacteria, average swimming speeds in units of cell length are even more extreme: scaling as $\sim 24V^{-0.16}$/sec, or 17 and 35 lengths/sec for the largest ($V = 10$) and smallest ($V = 0.1$) bacterial cells (Some ciliates are capable of 'jumping' behavior, which yields rates of 130–150 lengths/sec (Gilbert 1994)).

Scaled by organismal length, swimming speeds for eukaryotic cells are comparable to those observed for the larvae of marine invertebrates, which generally fall in the range of 5 to 10 lengths/sec (Chia et al. 1984), and much higher than those for swimming vertebrates (fish, birds, seals, and whales), nearly all of which fall in the range of 0.5 to 2.5 m/sec (Sato et al. 2007). The peak speeds of the fastest fish (marlins and sailfish) are ~ 15 body lengths/sec, and the maximum speed for an Olympic swimmer is ~ 2 body lengths/sec. Despite their small size and simplicity, bacteria are clear champions of the swimming world.

How much energy does such activity demand? Equation 16.2.2 shows that the power required for swimming scales with the product of the viscosity of the medium (η), the radius of the cell (r), and the squared swimming velocity (v^2), and to account for the inefficient conversion from chemical energy to motion, this must further be multiplied by 100 (from above). Assuming spherical cells, the scaling relationship for eukaryotes in Figure 16.4a implies $rv^2 \simeq 171 V^{0.83}$ $\mu m^3/sec^2$. From Chapter 3, the viscosity of water is 10^{-2} g $\cdot$ cm^{-1} $\cdot$ sec^{-1} (assuming 20°C), and after applying this to Equation 16.2.2 and appropriate changes of units, the average power requirement of swimming is estimated to be (1.6 $\times$ $10^{-16})V^{0.83}$ kg $\cdot$ m^2 $\cdot$ sec^{-3}, which has equivalent units of joules/sec (here, cell volume V is in units of μm^3).

To put this in more familiar terms, note that a rough estimate of the energy associated with the hydrolysis of one mole of ATP at physiological conditions is 50 kilojoules. After allowing for Avogadro's number of molecules per mole, the power requirement for swimming in units of ATP molecules hydrolyzed/hour is then $\sim(17 \times 10^6)V^{0.83}$ for eukaryotic cells, and similar extrapolation leads to $(54 \times 10^6)V^{0.67}$ for bacteria. Recalling that average basal metabolic rates scale isometrically with cell volume, $\simeq 0.4 \times 10^9 V$ ATP hydrolyses/hour at 20° C (Chapter 8), this suggests a fractional energetic investment in swimming (relative to total maintenance costs) in the range of 1–4% for the largest and smallest eukaryotic cells (assuming the cell is constantly swimming), but substantially higher in bacteria, ranging from 6% in the largest to 29% in the smallest swimming species.

Notably, these estimates do not include the cost of building the swimming apparatus. As outlined in Foundations 16.3, the total biosynthetic cost of the proteins comprising each *E. coli* flagellum and its basal protein parts is equivalent to $\sim 10^8$ ATP hydrolyses. Recalling from Chapter 8 that the total cost of building an *E. coli* cell is $\sim 2.7 \times 10^{10}$ ATP hydrolyses, the bioenergetic cost of building each of the multiple flagella is then on the order of 0.4% of the total cellular energy budget. The flagella of some bacteria are surrounded by a lipid membrane, and although *E. coli* is not one of them, were it to do so, the additional cost would be $\sim 3.6 \times 10^8$ ATP hydrolyses per flagellum, raising the total construction cost by $\sim 5\times$ relative to that for a naked flagellum. Recalling from above that the cost of swimming is $\sim(54 \times 10^6)V^{0.67}$ ATP hydrolyses per hour for bacteria, because the average volume of an *E. coli* cell is $V \simeq 1$ μm^3, the cell division time is typically no more than a few hours, and an average

cell harbors several flagella, more than 50% of the cost of swimming will typically be associated with building (as opposed to operating) the apparatus. There will, of course, be quantitative differences among species after variation in flagellar lengths and numbers and cell size are accounted for. However, a larger analysis of ~ 40 bacterial species reveals that the absolute costs of flagellar construction scale isometrically with cell volume, with a median cost of $\sim 4\%$ relative to total cell construction costs (Figure 16.5b,c).

Calculations for eukaryotes (Foundations 16.3) suggest roughly similar relative costs of swimming. Noting that the volume of a *C. reinhardtii* cell is ~ 150 μm^3, and allowing for a 24-hour cell-division time implies an energetic cost of swimming of $\sim 2.6 \times 10^{10}$ ATP per cell life time, whereas the costs of flagellar construction are $\sim 3 \times 10^{10}$ ATP hydrolyses for the protein content and $\sim 10^{10}$ for the lipid membrane (for each of the two flagella). With the total cost of building a *C. reinhardtii* cell being $\simeq 3.2 \times 10^{12}$ ATP hydrolyses (Chapter 8), the relative costs of swimming and constructing the flagellar pair are 0.8% and 2.5%, respectively.

A broader phylogenetic analysis for eukaryotes with small numbers of flagella indicates that, unlike the situation with bacteria, total flagellar construction costs scale hypometrically with cell volume, with a slope $\simeq 0.3$, so that the relative costs decline with increasing cell volume (Figure 16.5b). In contrast, for ciliates and parabasalids (with cells typically coated with small cilia), the scaling relationship is again isometric. Overall, the median total cost of the eukaryotic swimming apparatus relative to the total cell energy budget is $\sim 3\%$, although these costs can be substantially higher for ciliates and parabasalids than for other eukaryotic flagellates (Figure 16.5c).

Despite the substantial differences in flagellar architecture among prokaryotes and eukaryotes, there is a consistent inverse relationship between the swimming speed (in units of cell length) per construction cost and cell volume across all cellular organisms (Figure 16.5d), with the smallest bacteria having an $\sim 10^4$-fold elevation in this measure of performance relative to the largest ciliates. Thus, as in the case of replication fidelity (Chapter 4) and growth-rate potential (Chapter 8), swimming efficiency is substantially greater in bacteria than in eukaryotes.

Taken together, these results suggest that from an evolutionary perspective, the cost of swimming constitutes a significant fraction of a cell's total energy budget. In accordance with this view, many bacteria have evolved mechanisms for selective use of flagella, restricting their use only to environments providing significant benefits. When confronted with nutrient depletion, some species shed all external features of their flagella, leaving behind just a plugged remnant of the flagellar motor, whereas others cease assembly, leading to a reduction in flagellum number as the cells divide (Ferreira et al. 2019).

Prolonged periods of silencing of flagellar genes would be expected to eventually lead to the accumulation of deactivating mutations. As such a cascade progresses, there must eventually be a point at which mutation accumulation is so high that reversion to motility is impossible without horizontal gene transfer, and this presumably explains the absence of flagella/cilia from numerous branches in the Tree of Life. However, a remarkable case of the resurrection of a flagellum was observed in an experimental construct of the bacterium *Pseudomonas fluorescens* engineered to be non-motile by deletion of a key regulatory gene for flagellar gene expression (Taylor et al. 2015). After just four days of strong selection for motility, this bacterium regained flagella as mutations redirected a regulatory gene for nitrogen assimilation towards the promotion of expression of still intact flagellar genes (albeit at the expense of nitrogen uptake).

Finally, there is the issue of why prokaryotes and eukaryotes evolved such dramatically different flagellum types. As noted by Schavemaker and Lynch (2022), the cost of eukaryotic flagella per flagellum unit length substantially exceeds that for prokaryotes. Were bacteria to expend their normal flagellar construction costs on a flagellum of the eukaryotic form, the latter would be extremely short and unlikely to elicit much motility. This raises the possibility that the eukaryotic flagellum arose in a pre-LECA ancestor devoid of a bacterial-like flagellum, absolving it from competing with such a form, while also requiring an expansion in cell size to a level that could support the magnified construction costs.

Summary

- Eukaryotic cells rely heavily on polymers of monomeric (actin) or heterodimeric (tubulin) protein subunits concatenated into linear fibrils. Such constructs are used in a variety of cellular functions, ranging from structural support to intracellular vesicle trafficking to motility.

- Although less apparent visually, evolutionary relatives of such filaments are also found in prokaryotes. This implies their presence at the time of the last universal common ancestor, with subsequent functional modifications arising in the eukaryotic domain. Shifts in the deployment of tubulins versus actins for different functions between the major domains of life provide dramatic examples of convergent/divergent evolution of cellular functions at the molecular level.

- Scaled to the total costs of constructing cells, the investment in fibrillar proteins in eukaryotes ($\simeq$0.5 to 4%) may be only twofold greater than it is in prokaryotes.

- Cell sizes and shapes are sustained by endoskeletons (fibrillar proteins) and/or exoskeletons (cell walls). Although the detailed genetic mechanisms underlying cell shape are largely unknown in eukaryotes, substantial shifts in the geometric features of bacterial cells can be accomplished by just single amino-acid changes in one or two proteins.

- Many species deploy hard cell coverings, either internal or external to the plasma membrane. These provide resistance to turgor pressure and likely have other functions such as avoiding predation. Although peptidoglycans (prokaryotes) and cellulose and chitin (eukaryotes) are common components of cell walls, numerous variants are deployed across the eukaryotic phylogeny, including the use of inorganic components (e.g., silicon and calcium) in some lineages.

- The investment in cell walls can be quite substantial, accounting for as much as half the costs of the complete cell envelope, with the latter (including lipid membranes) comprising on the order of 20–40% of a cell's lifetime energy budget. These numbers are strikingly similar to the realtor's rule-of-thumb that one should spend no more than one-third of one's income on a home.

- Whereas most intracellular transport in bacteria is accomplished by passive diffusion, eukaryotic cells deploy molecular motors to carry large cargoes, such as vesicles, using the cytoskeleton as a structural highway. All three major classes of molecular motors within eukaryotes (kinesins, myosins, and dyneins) diversified into multiple subfamilies with different subfunctions prior to the last eukaryotic common ancestor.

- The most complex morphological structure of most prokaryotic cells is the flagellum, whose origin seems rooted to structures used in ATP synthase and in secretion systems, with the overall elaborations having arisen in part by gene duplication. Numerous eukaryotes also swim with flagella, but the eukaryotic flagellum evolved independently of that in prokaryotes and operates in a fundamentally different way, using cytoskeletal and motor proteins.

- In units of body lengths, swimming rates are several times higher in prokaryotes than eukaryotes, but their operating costs per unit volume are also higher. The efficiency of conversion of chemical to mechanical energy used in motion is quite low (of order 1%). As a fraction of a cell's total energy budget, the costs of flagella are on the order of 0.1–1.0%, with the costs of construction and operation being of similar magnitudes. Thus, there is a strong selective premium for eliminating flagella unless the advantages outweigh the costs.

Foundations 16.1 The eukaryotic cellular investment in the cytoskeleton

Given the central roles that cytoskeletal proteins (actin and tubulin) play in eukaryotic cell biology, it is of interest to evaluate the fraction of the cell's energy budget devoted to their production. Estimates are available for the average number of monomers of each protein within the cells of a few species, and this information combined with the cost of protein biosynthesis (to be covered in more detail in Chapter 17) and the cost of building an entire cell (Chapter 8) can be used to obtain a rough estimate of the relative cost of the cytoskeleton.

For the two model yeast species, the average numbers of actin and tubulin monomers per cell are, respectively: 88,600 and 22,400 for *S. cerevisiae* (Norbeck and Blomberg 1997; Kulak et al. 2014); and 731,500 and 125,100 for *S. pombe* (Wu and Pollard 2005; Marguerat et al. 2012; Kulak et al. 2014). Together, these two molecules account for ~0.24% and 0.77% of the total number of proteins per cell in *S. cerevisiae* and *S. pombe*, respectively.

These numbers can be converted into bioenergetic cost estimates by noting that: 1) actin and tubulin monomers contain ~375 and 450 amino acids, respectively; and 2) the average cost of amino-acid biosynthesis is 30 ATP hydrolyses per amino acid, with another 4 hydrolyses necessary for polypeptide-chain elongation (Chapter 17). Taking the total cost (in ATP hydrolyses per cell cycle) to be the product of the number of monomers, the number of amino acids per monomer, and 34 ATP hydrolyses per amino-acid residue leads to estimates of 1.1×10^9 and 3.3×10^8 for actin and tubulin in *S. cerevisiae*, and 9.1×10^9 and 1.9×10^9 in *S. pombe*. There are additional costs associated with transcription of the genes for these proteins, but as will be more fully discussed in Chapter 17, about 90% of the total cost of running genes is associated with protein production, so these numbers are only slight underestimates.

To put things into broader perspective, recall that Equation 8.2b provides an estimate of the cost of building an entire cell. Given cell volumes of $\sim 70 \, \mu m^3$ for *S. cerevisiae* and $130 \, \mu m^3$ for *S. pombe*, the number of ATP hydrolyses required for the construction of cells in these two species $\simeq 1.5 \times 10^{12}$ and 2.9×10^{12}, respectively. Thus, just 0.1% and 0.4% of the total energy budgets for cell construction in these species is devoted to the cytoskeleton, with ~80% of such costs being associated with actin. Similar calculations using data from mouse fibroblast cells (Schwanhäusser et al. 2011) lead to estimates of 3.8% for actins and 1.9% for tubulins, and for human HeLa cells (Kulak et al. 2014) of 0.1% for actins and 0.5% for tubulins.

What can be said of such investments in prokaryotes? Although few quantitative data are available, the two major cytoskeletal proteins in *E. coli*, FtsZ (related to actin) and MreB (related to tubulin), have average abundances of 3,450 and 1,060 protein monomers per cell (Pla et al. 1991; Rueda et al. 2003; Lu et al. 2007; Taniguchi et al. 2010; Wiśniewski and Rakus 2014; Vischer et al. 2015; Ouzounov et al. 2016; Bratton et al. 2018), with respective contents of 383 and 347 amino acids/monomer. With the cost of building an *E. coli* cell being $\simeq 27 \times 10^9$ ATP hydrolyses (Chapter 8), the fractional contributions of these two molecules to the total cell budget are 0.16% and 0.04%, respectively.

Thus, although the common view is that eukaryotes invest substantially more in cytoskeletal infrastructure than do prokaryotes, this simple comparison suggests comparable costs in yeasts and *E. coli*, with an inflation in metazoan cells.

Foundations 16.2 The physical challenges to cellular locomotion

A key to understanding the relative advantages/disadvantages of motility in single-celled organisms is in quantifying the way in which the resistance of a fluid to the motion of an object scales with the size of the object (Purcell 1977). The central concept is related to the definition of a dimensionless index known as the Reynolds number, which equals the ratio of inertial to viscous forces,

$$\mathrm{Re} = \frac{\rho v L}{\eta}, \qquad (16.2.1)$$

where ρ is the fluid density (g/cm^3), v is the velocity of the object (cm/sec), L is the characteristic linear dimension of the object (which depends on the shape; cm), and η is the fluid viscosity (g/cm·sec). For water, $\eta/\rho \simeq 10^{-2}$ cm^2/sec. Although the source of Equation 16.2.1 may not be intuitive, the derivation follows from the fact that the inertial force of an object (the numerator in Re) is equal to mass times acceleration (the rate of change of velocity), i.e., $(\rho L^3) \cdot (v/t)$, where t denotes time. The denominator follows from the definition of the coefficient of viscosity (η) as the viscous force per unit area per the velocity gradient (change in velocity per distance), which gives a total viscosity force of $\eta \cdot L^2 \cdot (v/L)$. Noting that $(v/t)/(v/L) = L/t$ is equivalent to v reduces $[(\rho L^3) \cdot (v/t)]/[\eta \cdot L^2 \cdot (v/L)]$ to Equation 16.2.1.

The Reynolds number is a convenient index of the challenges confronted by an object moving in a fluid. When Re < 1, viscous forces dominate, and in the limiting case of Re $\ll$ 1, the motion at any particular moment is essentially independent of all prior action. In the latter case, the resistance of the fluid is so great that movement of the object ceases nearly instantaneously if the force of motion is stopped. Almost all aspects of cellular movement are in this low Reynolds number range. For example, as noted for bacterial motility, v is almost always <100 μm/sec, and most bacterial cells have lengths of order 1 μm, so given that 1 μm = 10^{-4} cm, Re is on the order of 10^{-4}.

A key definition from physics is that power (the rate of doing work, or equivalently the rate of energy utilization) is equal to the product of force and velocity, $P = F \cdot v$. From

Stokes's law, which specifically applies to low Reynolds-number situations, the inertial drag force for a sphere with radius r is $F = 6\pi r \eta v$, so

$$P = 6\pi r \eta v^2. \qquad (16.2.2)$$

The formula differs for different shapes (see pages 56–57 in Berg (1993) for some approximations, and Perrin (1934, 1936) for more general results), although the scaling of P with ηv^2 remains. One has to be careful with units here. If F has units of kg-m/sec^2, and v has units of m/sec, then P has units of watts or joules/sec.

To determine the total metabolic power required to maintain velocity v, the preceding expression must be divided by the efficiency of conversion of electrochemical energy to directional motion (i.e., the ratio of propulsive power output to rotary power input), which in *E. coli* is estimated to be 0.017 (Chattopadhyay et al. 2006). Estimates from several other species have similarly low values: ~0.1% for the ciliate *Paramecium* (Katsu-Kimura et al. 2009); 1.3% for the euglenoid *Eutreptiella* (Arroyo et al. 2012); 0.8%, 1.5%, and 1.4%, respectively, for the green algae *Chlamydomonas* (Tam and Hosoi 2011), *Polytoma uvella* (Gittleson and Noble 1973), and *Tetraflagellochloris mauritanica* (Barsanti et al. 2016); 8% for the archaeon *Halobacterium* (Kinosita et al. 2016); and 1% and 5%, respectively, for the bacterium *Streptococcus* (Meister et al. 1987).

To determine the potential benefits of swimming in terms of resource acquisition, consider the overall process as being equivalent to a measure of effective diffusion of the cell through the medium. Suppose a cell swims for time τ with constant velocity v in a three-dimensional environment, pauses for an infinitesimally short time, and then randomly starts off in a new direction, with the swimming durations being exponentially distributed. Although the average physical length of each individual bout is $v\tau$, the runs are distributed over three dimensions (Chapter 7). Because the times are exponentially distributed, with variance $(\overline{\tau^2} - \overline{\tau}^2)$ equal to the square of the mean, the average squared value of run times is $\overline{\tau^2} = 2\overline{\tau}^2$. Thus, the mean-squared length of movement per bout is

$2v^2\bar{\tau}^2/3$. By definition, the mean-squared deviation of movement in τ time units is $2D\tau$ (Chapter 7), where D is the diffusion coefficient. Equating these two expressions and factoring out 2τ leads to

$$D = \frac{v^2\tau}{3} \qquad (16.2.3a)$$

under the assumption that the directions of movement between adjacent steps are random.

If there is some memory of the swimming process such that successive bouts tend to go roughly in the same direction, the average distance travelled is expected to increase. If, however, there is a tendency to switch to opposing directions, the cumulative distance travelled will be decreased. Such correlations of movement can be accommodated by dividing the previous expression by $(1-\bar{c})$, where $\bar{c}$ is the mean cosine of the angle of switching (Lovely and Dahlquist 1975),

$$D = \frac{v^2\bar{\tau}}{3(1-\bar{c})}. \qquad (16.2.3b)$$

If the angle of switching is small, $(1-\bar{c}) \simeq \overline{\theta^2}/2$, where θ is in radians. The mean-squared angular (rotational) deviation $\overline{\theta^2}/2$ is analogous to the mean-squared linear (translational) movement encountered in Chapter 7, leading to the concept of rotational diffusion. By definition, $\overline{\theta^2}/2 = 2D_r\bar{\tau}$, where D_r is the rotational diffusion coefficient.

Even in the absence of behaviorally induced switching of angles, the jostling of fluid molecules causes rotational diffusion of the cell, just as in the case of translational diffusion. It is known that under these conditions $D_r = k_BT/(8\pi\eta r^3)$ (Berg 1993; Equation 6.6). This inverse relationship between the rate of rotational diffusion and cell size means that directional swimming becomes increasingly challenging in small cells that are continually being reoriented by physical forces (Mitchell 1991). Using the preceding approximation for $(1-\bar{c})$ leads to an effective diffusion rate of

$$D = \frac{v^2\bar{\tau}}{3(1-\bar{c})} = \frac{2v^2\bar{\tau}}{3\overline{\theta^2}} = \frac{v^2}{6D_r} = \frac{4\pi\eta r^3 v^2}{3k_BT}. \qquad (16.2.4a)$$

for a cell at the mercy of random thermal jostling. From Chapter 2, we know that $k_BT \simeq 4.1 \times 10^{-14} \mathrm{cm}^2 \cdot \mathrm{g} \cdot \mathrm{sec}^{-2}$, and for water $\eta \simeq 10^{-3}\mathrm{g} \cdot \mathrm{cm}^{-1} \cdot \mathrm{sec}^{-1}$, so for a cell in a freshwater environment, subject to translational and rotational diffusion

$$D \simeq 10^{12} \cdot r^3 v^2, \qquad (16.2.4b)$$

where r and v have units of cm and cm/sec, respectively, yielding D with units of $\mathrm{cm}^2/\mathrm{sec}$.

The preceding expressions can be used to estimate the influence of swimming on the encounter rate of molecules.

For example, if one considers a bacterium with typical $r \simeq 10^{-4}$ cm, and $v \simeq 3 \times 10^{-3}$ cm/sec, then assuming randomly directed swimming, the effective diffusion coefficient for the cell is $D \simeq 10^{-5}$ cm^2/sec. From Figure 7.6, the average diffusion coefficient for a typical anion/cation is also $\simeq 10^{-5}$ cm^2/sec, with that for a protein containing 100 amino acids being $\simeq 10^{-6}$ cm^2/sec. The encounter rate between two different types of diffusing particles is proportional to the sum of their diffusion coefficients. Thus, without direct behavioral modifications of swimming direction, a swimming bacterial cell can encounter randomly distributed resources at a rate at least double that expected for a non-motile cell. Larger, more rapidly swimming cells will achieve more. There are, however, some subtle complications not dealt with here, most notably the fact that a boundary layer of nutrient-depleted fluid adheres to swimming cells, thereby reducing the effective concentration of resources to a degree that declines with swimming speed (Chapter 18).

Behavioral mechanisms can substantially magnify average levels of resource availability by biasing movement towards resource-rich patches in a heterogeneous environment. For example, an ability to respond to local resource abundance (e.g., by chemotaxis; Chapter 22) can enhance encounter rates by fostering a positive directional bias of movement up a resource gradient. Likewise, a negative bias can lead to local retention in a nutrient-rich patch. Consider, for example, a strong positive correlation of directional movement, $\bar{c} = 0.9$. Applying this to Equation 16.2.3b, with a bacterial swimming speed of $v \simeq 3 \times 10^{-3}$ cm/sec and $\bar{\tau} = 1$ sec, leads to $D \simeq 3 \times 10^{-5}$ cm^2/sec, whereas $\bar{c}=0$ (random movement) leads to $D \simeq 3 \times 10^{-6}$, and $\bar{c} = -0.8$ leads to $D \simeq 1.7 \times 10^{-6}$ cm^2/sec.

Arguments like these provide a starting point for estimating costs and benefits of locomotion and chemoreception. Based on a metabolic-scaling argument, Dusenbery (1997) estimated that there is a minimal size limit of $\sim$0.8 μm below which there is no advantage to motility, although his arguments ignore the cost of building the motility apparatus (Foundations 16.3). If one computes the Sherwood number (a measure of the encounter rate of nutrients in a swimming cell relative to that expected by passive diffusion in a non-motile cell; Foundations 18.1) and compares that to the ratio of construction costs of swimming and non-motile cells, it is found that the flagella for bacterial-sized cells cannot be paying for themselves in terms of increased nutrient accessibility (Schavemaker and Lynch 2022). Whereas swimming does have such an advantage in moderate- to large-sized eukaryotic cells, motility in bacteria must be associated with other advantages such as location of nutrient-rich patches.

Foundations 16.3 The construction costs of flagella

The cost of building a molecular machine such as a flagellum can be estimated using the general methodology outlined in Foundations 16.1 (with further elaborations on the underpinnings in Chapter 17). For a bacterial flagellum, the biosynthetic costs can be roughly computed as follows. From Berg (2003), an average *E. coli* flagellar filament contains ~ 5340 flagellin protein molecules, each containing ~ 500 amino acids, implying a total investment of $\sim 2.7 \times 10^6$ amino acids / flagellar filament. The remaining proteins associated with the basal body, rod, and hook comprise only about 5% of the protein in the flagellum (Sosinsky et al. 1992; Berg 2003), so in total there are $\sim 3 \times 10^6$ amino acids involved. As noted in Foundations 17.2, the total translation-associated cost per amino acid (including synthesis and polymerization) is ~ 34 ATP hydrolyses, and additional costs of genes at the DNA and mRNA levels only inflate these estimates by $\sim 10\%$. This implies a protein biosynthetic cost per *E. coli* flagellum of $\sim 10^8$ ATP hydrolyses. Supplementing this estimate with the costs of the flagellar export-apparatus proteins, Schavemaker and Lynch (2022) obtained an estimate of 2×10^8.

For some bacteria, there is an additional cost of wrapping the filament with a lipid membrane. Although not one of these, *E. coli* will nonetheless be used here to examine the consequences of such an elaboration. Again, the details underlying the computations will be given in Chapter 17. The essential points are that with an average *E. coli* flagellum radius and length of 0.01 μm and 6 μm, respectively, the cylindrical surface area of the flagellum is $\sim 0.4 \, \mu m^2$. As the average membrane area occupied by a bacterial lipid molecule is $0.65 \times 10^{-6} \, \mu m^2$, after accounting for the two leaflets of the lipid bilayer, there would

be an estimated 1.2×10^6 lipid molecules surrounding this flagellum. The biosynthetic costs of lipids are much greater than those for proteins, averaging 300 ATP hydrolyses per lipid molecule, leading to a total biosynthetic cost of the hypothetical flagellar membrane of $\sim 3.6 \times 10^8$ ATP hydrolyses, $\sim 2\times$ that of the protein components.

To extend these calculations to eukaryotes, consider the green alga *Chlamydomonas reinhardtii*, which generally has two flagella with approximate radii and lengths of 0.075 μm and 20 μm. After accounting for the slightly elevated cost of lipid biosynthesis in eukaryotes ($\simeq 350$ ATPs per molecule), this implies a total membrane cost surrounding each flagellum of $\sim 10^{10}$ ATP. Raven and Richardson (1984) estimate there to be $\sim 48,000$ tubulin monomers (each ~ 450 amino acids in length) per μm of flagellum, implying a cost for this major molecule of $\sim 1.5 \times 10^{10}$ ATP. There are also ~ 600 copies of the large motor protein dynein (each $\sim 15,000$ amino acids in length) per μm, implying an additional construction cost of $\sim 6.1 \times 10^9$ ATP. There are numerous other proteins within eukaryotic flagella, including the smaller motor protein kinesin, but their summed number is unlikely to rival that for tubulin and dynein, so the total cost of protein synthesis associated with each *C. reinhardtii* flagellum is $\sim 3 \times 10^{10}$ ATP. This is $\sim 3\times$ the lipid cost, opposite the situation for a bacterial flagellum, but expectedly so, given the lower surface area:volume ratio of the thicker eukaryotic flagellum. A much more detailed accounting in Schavemaker and Lynch (2022) yields similar estimates.

For further information on these matters, including the costs of archaeal flagella, see Schavemaker and Lynch (2022).

Literature Cited

Akhmanova, A., and M. O. Steinmetz. 2015. Control of microtubule organization and dynamics: two ends in the limelight. Nat. Rev. Mol. Cell. Biol. 16: 711–726.

Albers, S. V., and K. F. Jarrell. 2018. The archaellum: An update on the unique archaeal motility structure. Trends Microbiol. 26: 351–362.

Albers, S. V., and B. H. Meyer. 2011. The archaeal cell envelope. Nat. Rev. Microbiol. 9: 414–426.

Albers, S. V., and M. Pohlschröder. 2009. Diversity of archaeal type IV pilin-like structures. Extremophiles 13: 403–410.

Alliegro, M. C., and P. Satir. 2009. Origin of the cilium: Novel approaches to examine a centriolar evolution hypothesis. Methods Cell Biol. 94: 53–64.

Amir, A., F. Babaeipour, D. B. McIntosh, D. R. Nelson, and S. Jun. 2014. Bending forces plastically deform growing bacterial cell walls. Proc. Natl. Acad. Sci. USA 111: 5778–5783.

Arroyo, M., L. Heltai, D. Millán, and A. DeSimone. 2012. Reverse engineering the euglenoid movement. Proc. Natl. Acad. Sci. USA 109: 17874–17879.

Asmar, A. T., and J. F. Collet. 2018. Lpp, the Braun lipoprotein, turns 50 – Major achievements and remaining issues. FEMS Microbiol. Lett. 365 (18).

Avidor-Reiss, T., A. M. Maer, E. Koundakjian, A. Polyanovsky, T. Keil, S. Subramaniam, and C. S. Zuker. 2004. Decoding cilia function: Defining specialized genes required for compartmentalized cilia biogenesis. Cell 117: 527–539.

Barsanti, L., P. Coltelli, V. Evangelista, A. M. Frassanito, and P. Gualtieri. 2016. Swimming patterns of the quadriflagellate *Tetraflagellochloris mauritanica* (Chlamydomonadales, Chlorophyceae). J. Phycol. 52: 209–218.

Bauer, D., H. Ishikawa, K. A. Wemmer, N. L. Hendel, J. Kondev, and W. F. Marshall. 2021. Analysis of biological noise in the flagellar length control system. iScience 24: 102354.

Bazylinski, D. A., and R. B. Frankel. 2004. Magnetosome formation in prokaryotes. Nat. Rev. Microbiol. 2: 217–230.

Becker, B. 2000. The cell surface of flagellates, pp. 110–123. In B. S. C. Leadbeater and J. C. G. Taylor (eds.) The Flagellates: Unity, Diversity, and Evolution. Taylor and Francis, New York, NY.

Berg, H. C. 1993. Random Walks in Biology. Princeton University Press, Princeton, NJ.

Berg, H. C. 2003. The rotary motor of bacterial flagella. Annu. Rev. Biochem. 72: 19–54.

Bernander, R., and T. J. Ettema. 2010. FtsZ-less cell division in archaea and bacteria. Curr. Opin. Microbiol. 13: 747–752.

Billaudeau, C., A. Chastanet, Z. Yao, C. Cornilleau, N. Mirouze, V. Fromion, and R. Carballido-López. 2017. Contrasting mechanisms of growth in two model rod-shaped bacteria. Nat. Commun. 8: 15370.

Block, S. M., L. S. Goldstein, and B. J. Schnapp. 1990. Bead movement by single kinesin molecules studied with optical tweezers. Nature 348: 348–352.

Blocker, A., K. Komoriya, and S. Aizawa. 2003. Type III secretion systems and bacterial flagella: Insights into their function from structural similarities. Proc. Natl. Acad. Sci. USA 100: 3027–3030.

Boehm, A., M. Kaiser, H. Li, C. Spangler, C. A. Kasper, M. Ackermann, V. Kaever, V. Sourjik, V. Roth, and U. Jenal. 2010. Second messenger-mediated adjustment of bacterial swimming velocity. Cell 141: 107–116.

Bork, P., C. Sander, and A. Valencia. 1992. An ATPase domain common to prokaryotic cell cycle proteins, sugar kinases, actin, and hsp70 heat-shock proteins. Proc. Natl. Acad. Sci. USA 89: 7290–7294.

Bratton, B. P., J. W. Shaevitz, Z. Gitai, and R. M. Morgenstein. 2018. MreB polymers and curvature localization are enhanced by RodZ and predict *E. coli*'s cylindrical uniformity. Nat. Commun. 9: 2797.

Braun, T., A. Orlova, K. Valegård, A. C. Lindås, G. F. Schröder, and E. H. Egelman. 2015. Archaeal actin from a hyperthermophile forms a single-stranded filament. Proc. Natl. Acad. Sci. USA 112: 9340–9345.

Brunet, T., M. Albert, W. Roman, M. C. Coyle MC, D. C. Spitzer, and N. King. 2021. A flagellate-to-amoeboid switch in the closest living relatives of animals. eLife 10: e61037.

Carlier, M. F., and S. Shekhar. 2017. Global treadmilling coordinates actin turnover and controls the size of actin networks. Nat. Rev. Mol. Cell. Biol. 18: 389–401.

Carvalho-Santos, Z., J. Azimzadeh, J. B. Pereira-Leal, and M. Bettencourt-Dias. 2011. Evolution: Tracing the origins of centrioles, cilia, and flagella. J. Cell Biol. 194: 165–175.

Carvalho-Santos, Z., P. Machado, P. Branco, F. Tavares-Cadete, A. Rodrigues-Martins, J. B. Pereira-Leal, and M. Bettencourt-Dias. 2010. Stepwise evolution of the centriole-assembly pathway. J. Cell Sci. 123: 1414–1426.

Cavalier-Smith, T. 1978. The evolutionary origin and phylogeny of microtubules, mitotic spindles and eukaryote flagella. Biosystems 10: 93–114.

Cavalier-Smith, T. 2002. The phagotrophic origin of eukaryotes and phylogenetic classification of Protozoa. Internat. J. Syst. Evol. Microbiol. 52: 297–354.

Chaban, B., I. Coleman, and M. Beeby. 2018. Evolution of higher torque in *Campylobacter*-type bacterial flagellar motors. Sci. Rep. 8: 97.

Chattopadhyay, S., R. Moldovan, C. Yeung, and X. L. Wu. 2006. Swimming efficiency of bacterium *Escherichia coli*. Proc. Natl. Acad. Sci. USA 103: 13712–13717.

Chen, S., M. Beeby, G. E. Murphy, J. R. Leadbetter, D. R. Hendrixson, A. Briegel, Z. Li, J. Shi, E. I. Tocheva, A. Müller, et al. 2011. Structural diversity of bacterial flagellar motors. EMBO J. 30: 2972–2981.

Chia, F.-S., J. Buckland-Nicks, and C. M. Young. 1984. Locomotion of marine invertebrate larvae: A review. Can. J. Zool. 62: 1205–1222.

Cochran, J. C. 2015. Kinesin motor enzymology: Chemistry, structure, and physics of nanoscale molecular machines. Biophys. Rev. 7: 269–299.

Crawford, D. W. 1992. Metabolic cost of motility in planktonic protists: Theoretical considerations on size scaling and swimming speed. Microb. Ecol. 24: 1–10.

Daum, B., J. Vonck, A. Bellack, P. Chaudhury, R. Reichelt, S. V. Albers, R. Rachel, and W. Kühlbrandt. 2017. Structure and *in situ* organisation of the *Pyrococcus furiosus* archaellum machinery. eLife 6: e27470.

Deng, X., G. Fink, T. A. M. Bharat, S. He, D. Kureisaite-Ciziene, and J. Löwe. 2017. Four-stranded mini microtubules formed by *Prosthecobacter* BtubAB show dynamic instability. Proc. Natl. Acad. Sci. USA 114: E5950–E5958.

Desmond, E., C. Brochier-Armanet, and S. Gribaldo. 2007. Phylogenomics of the archaeal flagellum: Rare horizontal gene transfer in a unique motility structure. BMC Evol. Biol. 7: 106.

Dion, M. F., M. Kapoor, Y. Sun, S. Wilson, J. Ryan, A. Vigouroux, S. van Teeffelen, R. Oldenbourg, and E. C. Garner. 2019. *Bacillus subtilis* cell diameter is determined by the opposing actions of two distinct cell wall synthetic systems. Nat. Microbiol. 4: 1294–1305.

Dishinger, J. F., H. L. Kee, P. M. Jenkins, S. Fan, T. W. Hurd, J. W. Hammond, Y. N. Truong, B. Margolis, J. R. Martens, and K. J. Verhey. 2010. Ciliary entry of the kinesin-2 motor KIF17 is regulated by importin-beta2 and RanGTP. Nat. Cell Biol. 12: 703–710.

Domozych, D. S., I. Sørensen, Z. A. Popper, J. Ochs, A. Andreas, J. U. Fangel, A. Pielach, C. Sacks, H. Brechka, P. Ruisi-Besares, et al. 2014. Pectin metabolism and assembly in the cell wall of the charophyte green alga *Penium margaritaceum*. Plant Physiol. 165: 105–118.

Dusenbery, D. B. 1997. Minimum size limit for useful locomotion by free-swimming microbes. Proc. Natl. Acad. Sci. USA 94: 10949–10954.

Dusenbery, D. B. 1998. Fitness landscapes for effects of shape on chemotaxis and other behaviors of bacteria. J. Bacteriol. 180: 5978–5983.

Dutcher, S. K. 2003. Long-lost relatives reappear: Identification of new members of the tubulin superfamily. Curr. Opin. Microbiol. 6: 634–640.

Egelman, E. H. 2010. Reducing irreducible complexity: Divergence of quaternary structure and function in macromolecular assemblies. Curr. Opin. Cell Biol. 22: 68–74.

Elting, M. W., Z. Bryant, J. C. Liao, and J. A. Spudich. 2011. Detailed tuning of structure and intramolecular communication are dispensable for processive motion of myosin VI. Biophys. J. 100: 430–439.

Erickson, H. P. 2007. Evolution of the cytoskeleton. Bioessays 29: 668–677.

Ettema, T. J., A. C. Lindås, and R. Bernander. 2011. An actin-based cytoskeleton in archaea. Mol. Microbiol. 80: 1052–1061.

Fai, T. G., L. Mohapatra, P. Kar, J. Kondev, and A. Amir. 2019. Length regulation of multiple flagella that self-assemble from a shared pool of components. eLife 8: e42599.

Fenchel, T., and B. J. Finlay. 1983. Respiration rates in heterotrophic, free-living protozoa. Microb. Ecol. 9: 99–122.

Ferreira, J. L., F. Z. Gao, F. M. Rossmann, A. Nans, S. Brenzinger, R. Hosseini, A. Wilson, A. Briegel, K. M. Thormann, P. B. Rosenthal, et al. 2019. γ-proteobacteria eject their polar flagella under nutrient depletion, retaining flagellar motor relic structures. PLoS Biol. 17: e3000165.

Flamholz, A., R. Phillips, and R. Milo. 2014. The quantified cell. Mol. Biol. Cell 25: 3497–3500.

Foth, B. J., M. C. Goedecke, and D. Soldati. 2006. New insights into myosin evolution and classification. Proc. Natl. Acad. Sci. USA 103: 3681–3686.

Fritz-Laylin, L. K., S. J. Lord, and R. D. Mullins. 2017. WASP and SCAR are evolutionarily conserved in actin-filled pseudopod-based motility. J. Cell Biol. 216: 1673–1688.

Fulton, C. 1977. Cell differentiation in *Naegleria gruberi*. Annu. Rev. Microbiol. 31: 597–629.

Galkin, V. E., X. Yu, J. Bielnicki, J. Heuser, C. P. Ewing, P. Guerry, and E. H. Egelman. 2008. Divergence of quaternary structures among bacterial flagellar filaments. Science 320: 382–385.

Gardner, M. K., M. Zanic, and J. Howard. 2013. Microtubule catastrophe and rescue. Curr. Opin. Cell. Biol. 25: 14–22.

Geis, G, Suerbaum S, Forsthoff B, Leying H, and Opferkuch W. 1993. Ultrastructure and biochemical studies of the flagellar sheath of *Helicobacter pylori*. J. Med. Microbiol. 38: 371–377.

Ghoshdastider, U., S. Jiang, D. Popp, and R. C. Robinson. 2015. In search of the primordial actin filament. Proc. Natl. Acad. Sci. USA 112: 9150–9151.

Gilbert, J. J. 1994. Jumping behavior inthe oligotrich ciliates *Strobilidium velox* and *Halteria grandinella*, and its significance as a defense against rotifer predators. Microb. Ecol. 27: 189–200.

Ginger, M. L., N. Portman, and P. G. McKean. 2008. Swimming with protists: Perception, motility and flagellum assembly. Nat. Rev. Microbiol. 6: 838–850.

Gittleson, S. M., and R. M. Noble. 1973. Locomotion in *Polytomella agilis* and *Polytoma uvella*. Trans. Amer. Micro. Soc. 92: 122–128.

Gönczy, P. 2012. Towards a molecular architecture of centriole assembly. Nat. Rev. Mol. Cell Biol. 13: 425–435.

Goodson, H. V., and S. C. Dawson. 2006. Multiplying myosins. Proc. Natl. Acad. Sci. USA 103: 3498–3499.

Goodson, H. V., and W. F. Hawse. 2002. Molecular evolution of the actin family. J. Cell Sci. 115: 2619–2622.

Gumbart, J. C., M. Beeby, G. J. Jensen, and B. Roux. 2014. *Escherichia coli* peptidoglycan structure and mechanics as predicted by atomic-scale simulations. PLoS Comput. Biol. 10: e1003475.

Hariharan, V., and W. O. Hancock. 2009. Insights into the mechanical properties of the kinesin neck linker domain from sequence analysis and molecular dynamics simulations. Cell. Mol. Bioeng. 2: 177–189.

Hartman, H., and T. F. Smith. 2009. The evolution of the cilium and the eukaryotic cell. Cell Motil. Cytoskeleton 66: 215–219.

Hartmann, E., and H. König. 1990. Comparison of the biosynthesis of the methanobacterial pseudomurein and the eubacterial murein. Naturwissenschaften 77: 472–475.

Hodges, M. E., N. Scheumann, B. Wickstead, J. A. Langdale, and K. Gull. 2010. Reconstructing the evolutionary history of the centriole from protein components. J. Cell Sci. 123: 1407–1413.

Horio, T., and B. R. Oakley. 1994. Human gamma-tubulin functions in fission yeast. J. Cell. Biol. 126: 1465–1473.

Hou, Y., R. Sierra, D. Bassen, N. K. Banavali, A. Habura, J. Pawlowski, and S. S. Bowser. 2013. Molecular evidence for β-tubulin neofunctionalization in Retaria (Foraminifera and radiolarians). Mol. Biol. Evol. 30: 2487–2493.

Howard, J. 2001. Mechanics of Motor Proteins and the Cytoskeleton. Sinauer Associates, Inc. Sunderland, MA.

Hwang, W., and Karplus M. 2019. Structural basis for power stroke vs. Brownian ratchet mechanisms of motor proteins. Proc. Natl. Acad. Sci. USA 116: 19777–19785.

Ishikawa, H., and W. F. Marshall. 2011. Ciliogenesis: Building the cell's antenna. Nat. Rev. Mol. Cell Biol. 12: 222–234.

Ishikawa, H., and W. F. Marshall. 2017. Testing the time-of-flight model for flagellar length sensing. Mol. Biol. Cell. 28: 3447–3456.

Ito, M., and Y. Takahashi. 2017. Nonconventional cation-coupled flagellar motors derived from the alkaliphilic *Bacillus* and *Paenibacillus* species. Extremophiles 21: 3–14.

Izoré, T., D. Kureisaite-Ciziene, S. H. McLaughlin, and J. Löwe. 2016. Crenactin forms actin-like double helical filaments regulated by arcadin-2. eLife 5: e21600.

Jarrell, K. F., and M. J. McBride. 2008. The surprisingly diverse ways that prokaryotes move. Nat. Rev. Microbiol. 6: 466–476.

Jékely, G., and D. Arendt. 2006. Evolution of intraflagellar transport from coated vesicles and autogenous origin of the eukaryotic cilium. Bioessays 28: 191–198.

Joseph, J. M., P. Fey, N. Ramalingam, X. I. Liu, M. Rohlfs, A. A. Noegel, A. Müller-Taubenberger, G. Glöckner, and M. Schleicher. 2008. The actinome of *Dictyostelium discoideum* in comparison to actins and actin-related proteins from other organisms. PLoS One 3: e2654.

Kaplan, M., D. Ghosal, P. Subramanian, C. M. Oikonomou, A. Kjaer, S. Pirbadian, D. R. Ortega, A. Briegel, M. Y. El-Naggar, and G. J. Jensen. 2019. The presence and absence of periplasmic rings in bacterial flagellar motors correlates with stator type. eLife 8: e43487.

Kardon, J. R., and R. D. Vale. 2009. Regulators of the cytoplasmic dynein motor. Nat. Rev. Mol. Cell. Biol. 10: 854–865.

Katsu-Kimura, Y., F. Nakaya, S. A. Baba, and Y. Mogami. 2009. Substantial energy expenditure for locomotion in ciliates verified by means of simultaneous measurement of oxygen consumption rate and swimming speed. J. Exp. Biol. 212: 1819–1824.

Kee, H. L., J. F. Dishinger, T. L. Blasius, C. J. Liu, B. Margolis, and K. J. Verhey. 2012. A size-exclusion permeability barrier and nucleoporins characterize a ciliary pore complex that regulates transport into cilia. Nat. Cell Biol. 14: 431–437.

Khona, D. K., V. G. Rao, M. J. Motiwalla, P. C. Sreekrishna Varma, A. R. Kashyap, K. Das, S. M. Shirolikar, L. Borde, J. A. Dharmadhikari, A. K. Dharmadhikari, et al. 2013. Anomalies in the motion dynamics of long-flagella mutants of *Chlamydomonas reinhardtii*. J. Biol. Phys. 39: 1–14.

Kiefel, B. R., P. R. Gilson, and P. L. Beech. 2004. Diverse eukaryotes have retained mitochondrial homologues of the bacterial division protein FtsZ. Protist 155: 105–115.

Kinosita, Y., N. Uchida, D. Nakane, and T. Nishizaka. 2016. Direct observation of rotation and steps of the archaellum in the swimming halophilic archaeon *Halobacterium salinarum*. Nat. Microbiol. 1: 16148.

Klis, F. M., C. G. de Koster, and S. Brul. 2014. Cell wall-related bionumbers and bioestimates of *Saccharomyces cerevisiae* and *Candida albicans*. Eukaryot. Cell 13: 2–9.

Kollmar, M., D. Lbik, and S. Enge. 2012. Evolution of the eukaryotic ARP2/3 activators of the WASP family WASP, WAVE, WASH, and WHAMM, and the proposed new family members WAWH and WAML. BMC Res. Notes 5: 88.

Kreutzberger, M. A. B., R. R. Sonani, J. Liu, S. Chatterjee, F. Wang, A. L. Sebastian, P. Biswas, C. Ewing, W. Zheng, F. Poly, et al. 2022. Convergent evolution in the supercoiling of prokaryotic flagellar filaments. Cell 185: 3487–3500.e14.

Kulak, N. A., G. Pichler, I. Paron, N. Nagaraj, and M. Mann. 2014. Minimal, encapsulated proteomic-sample processing applied to copy-number estimation in eukaryotic cells. Nat. Methods 11: 319–324.

Kull, F. J., R. D. Vale, and R. J. Fletterick. 1998. The case for a common ancestor: Kinesin and myosin motor proteins and G proteins. J. Muscle Res. Cell Motil. 19: 877–886.

Larsen, R. A., C. Cusumano, A. Fujioka, G. Lim-Fong, P. Patterson, and J. Pogliano. 2007. Treadmilling of a prokaryotic tubulin-like protein, TubZ, required for plasmid stability in *Bacillus thuringiensis*. Genes Dev. 21: 1340–1352.

Lassak, K., T. Neiner, A. Ghosh, A. Klingl, R. Wirth, and S. V. Albers. 2012. Molecular analysis of the crenarchaeal flagellum. Mol. Microbiol. 83: 110–124.

Lawrence, C. J., R. K. Dawe, K. R. Christie, D. W. Cleveland, S. C. Dawson, S. A. Endow, L. S. B. Goldstein, H. V. Goodson, N. Hirokawa, J. Howard, et al. 2004. A standardized kinesin nomenclature. J. Cell Biol. 167: 19–22.

Li, G. W., D. Burkhardt, C. Gross, and J. S. Weissman. 2014. Quantifying absolute protein synthesis rates reveals principles underlying allocation of cellular resources. Cell 157: 624–635.

Lindås, A. C., E. A. Karlsson, M. T. Lindgren, T. J. Ettema, and R. Bernander. 2008. A unique cell division machinery in the Archaea. Proc. Natl. Acad. Sci. USA 105: 18942–18946.

Liu, R., and H. Ochman. 2007. Stepwise formation of the bacterial flagellar system. Proc. Natl. Acad. Sci. USA 104: 7116–7121.

Lleo, M. M., P. Canepari, and G. Satta. 1990. Bacterial cell shape regulation: Testing of additional predictions unique to the two-competing-sites model for peptidoglycan assembly and isolation of conditional rod-shaped mutants from some wild-type cocci. J. Bacteriol. 172: 3758–3771.

Lovely, P. S., and F. W. Dahlquist. 1975. Statistical measures of bacterial motility and chemotaxis. J. Theor. Biol. 50: 477–496.

Löwe, J., and L. A. Amos. 2009. Evolution of cytomotive filaments: The cytoskeleton from prokaryotes to eukaryotes. Int. J. Biochem. Cell Biol. 41: 323–329.

Lu, P., C. Vogel, R. Wang, X. Yao, and E. M. Marcotte. 2007. Absolute protein expression profiling estimates the relative contributions of transcriptional and translational regulation. Nat. Biotechnol. 25: 117–124.

Ludington, W. B., L. Z. Shi, Q. Zhu, M. W. Berns, and W. F. Marshall. 2012. Organelle size equalization by a constitutive process. Curr. Biol. 22: 2173–2179.

Ludington, W. B., K. A. Wemmer, K. F. Lechtreck, G. B. Witman, and W. F. Marshall. 2013. Avalanche-like behavior in ciliary import. Proc. Natl. Acad. Sci. USA 110: 3925–3930.

Lynch, M., and B. Trickovic. 2020. A theoretical framework for evolutionary cell biology. J. Mol. Biol. 432: 1861–1879.

Maezawa, K., S. Shigenobu, H. Taniguchi, T. Kubo, S. Aizawa, and M. Morioka. 2006. Hundreds of flagellar basal bodies cover the cell surface of the endosymbiotic bacterium *Buchnera aphidicola* sp. strain APS. J. Bacteriol. 188: 6539–6543.

Makarova, K. S., N. Yutin, S. D. Bell, and E. V. Koonin. 2010. Evolution of diverse cell division and vesicle formation systems in Archaea. Nat. Rev. Microbiol. 8: 731–741.

Männik, J., R. Driessen, P. Galajda, J. E. Keymer, and C. Dekker. 2009. Bacterial growth and motility in submicron constrictions. Proc. Natl. Acad. Sci. USA 106: 14861–14866.

Manson, M. D., P. M. Tedesco, and H. C. Berg. 1980. Energetics of flagellar rotation in bacteria. J. Mol. Biol. 138: 541–561.

Manson, M. D., P. Tedesco, H. C. Berg, F. M. Harold, and C. Van der Drift. 1977. A proton motive force drives bacterial flagella. Proc. Natl. Acad. Sci. USA 74: 3060–3064.

Margolin, W. 2009. Sculpting the bacterial cell. Curr. Biol. 19: R812–R822.

Marguerat, S., A. Schmidt, S. Codlin, W. Chen, R. Aebersold, and J. Bähler. 2012. Quantitative analysis of fission yeast transcriptomes and proteomes in proliferating and quiescent cells. Cell 151: 671–683.

Marshall, W. F. 2011. Origins of cellular geometry. BMC Biol. 9: 57.

Marshall, W. F., and J. L. Rosenbaum. 2001. Intraflagellar transport balances continuous turnover of outer doublet microtubules: Implications for flagellar length control. J. Cell Biol. 155: 405–414.

Mast, F. D., R. A. Rachubinski, and J. B. Dacks. 2012. Emergent complexity in myosin V-based organelle inheritance. Mol. Biol. Evol. 29: 975–984.

McInally, S. G., J. Kondev, and S. C. Dawson. 2019. Length-dependent disassembly maintains four different flagellar lengths in *Giardia*. eLife 8: e48694.

McKean, P. G., S. Vaughan, and K. Gull. 2001. The extended tubulin superfamily. J. Cell Sci. 114: 2723–2733.

Meister, M., G. Lowe, and H. C. Berg. 1987. The proton flux through the bacterial flagellar motor. Cell 49: 643–650.

Michel, G., T. Tonon, D. Scornet, J. M. Cock, and B. Kloareg. 2010. The cell wall polysaccharide metabolism of the brown alga *Ectocarpus siliculosus*. Insights into the evolution of extracellular matrix polysaccharides in Eukaryotes. New Phytol. 188: 82–97.

Michie, K. A., and J. Löwe. 2006. Dynamic filaments of the bacterial cytoskeleton. Annu. Rev. Biochem. 75: 467–492.

Mitchell, J. G. 1991. The influence of cell size on marine bacterial motility and energetics. Microbial Ecol. 22: 227–238.

Mitchison, T., and M. Kirschner. 1984. Dynamic instability of microtubule growth. Nature 312: 237–242.

Miyata, M., R. C. Robinson, T. Q. P. Uyeda, Y. Fukumori, S.-I. Fukushim, S. Haruta, M. Homma, K. Inaba, M. Ito, C. Kaito, et al. 2020. Tree of motility – a proposed history of motility systems in the tree of life. Genes Cells 25: 6–21.

Mogilner, A., and G. Oster. 1996. Cell motility driven by actin polymerization. Biophys. J. 71: 3030–3045.

Monds, R. D., T. K. Lee, A. Colavin, T. Ursell, S. Quan, T. F. Cooper, and K. C. Huang. 2014. Systematic perturbation of cytoskeletal function reveals a linear scaling relationship between cell geometry and fitness. Cell Rep. 9: 1528–1537.

Moon, K. H., X. Zhao, A. Manne, J. Wang, Z. Yu, J. Liu, and M. A. Motaleb. 2016. Spirochetes flagellar collar protein FlbB has astounding effects in orientation of periplasmic flagella, bacterial shape, motility, and assembly of motors in *Borrelia burgdorferi*. Mol. Microbiol. 102: 336–348.

Muthukrishnan, G., Y. Zhang, S. Shastry, and W. O. Hancock. 2009. The processivity of kinesin-2 motors suggests diminished front-head gating. Curr. Biol. 19: 442–447.

Nan, B., and D. R. Zusman. 2016. Novel mechanisms power bacterial gliding motility. Mol. Microbiol. 101: 186–193.

Neidhardt, F. C., J. L. Ingraham, and M. Schaechter. 1990. Physiology of the Bacterial Cell: A Molecular Approach. Sinauer Associates Inc., Sunderland, MA.

Nevers, Y., M. K. Prasad, L. Poidevin, K. Chennen, A. Allot, A. Kress, R. Ripp, J. D. Thompson, H. Dollfus, O. Poch, and O. Lecompte. 2017. Insights into ciliary genes and evolution from multi-level phylogenetic profiling. Mol. Biol. Evol. 34: 2016–2034.

Niklas, K. J. 2004. The cell walls that bind the Tree of Life. BioScience 54: 831–841.

Norbeck, J., and A. Blomberg. 1997. Two-dimensional electrophoretic separation of yeast proteins using a non-linear wide range (pH 3–10) immobilized pH gradient in the first dimension; Reproducibility and evidence for isoelectric focusing of alkaline (pI > 7) proteins. Yeast 13: 1519–1534.

Noselli, G., A. Beran, M. Arroyo, and A. DeSimone. 2019. Swimming *Euglena* respond to confinement with a behavioral change enabling effective crawling. Nature Phys. 15: 496–502.

Oger, P. M., and A. Cario. 2013. Adaptation of the membrane in Archaea. Biophys. Chem. 183: 42–56.

Ojkic, N., D. Serbanescu, and S. Banerjee. 2019. Surface-to-volume scaling and aspect ratio preservation in rod-shaped bacteria. eLife 8: e47033.

Okuda, K. 2002. Structure and phylogeny of cell coverings. J. Plant Res. 115: 283–288.

Osawa, M., and H. P. Erickson. 2006. FtsZ from divergent foreign bacteria can function for cell division in *Escherichia coli*. J. Bacteriol. 188: 7132–7140.

Ouzounov, N., J. P. Nguyen, B. P. Bratton, D. Jacobowitz, Z. Gitai, and J. W. Shaevitz. 2016. MreB orientation correlates with cell diameter in *Escherichia coli*. Biophys. J. 111: 1035–1043.

Ozyamak, E., J. M. Kollman, and A. Komeili. 2013. Bacterial actins and their diversity. Biochemistry 52: 6928–6939.

Pallen, M. J., and N. J. Matzke. 2006. From *The Origin of Species* to the origin of bacterial flagella. Nat. Rev. Microbiol. 4: 784–790.

Pallen, M. J., C. W. Penn, and R. R. Chaudhuri. 2005. Bacterial flagellar diversity in the post-genomic era. Trends Microbiol. 13: 143–149.

Pančić, M., R. R. Torres, R. Almeda, and T. Kiø rboe. 2019. Silicified cell walls as a defensive trait in diatoms. Proc. Biol. Sci. 286: 20190184.

Pazour, G. J., N. Agrin, J. Leszyk, and G. B. Witman. 2005. Proteomic analysis of a eukaryotic cilium. J. Cell. Biol. 170: 103–113.

Pelve, E. A., A. C. Lindås, W. Martens-Habbena, J. R. de la Torre, D. A. Stahl, and R. Bernander. 2011. Cdv-based cell division and cell cycle organization in the thaumarchaeon *Nitrosopumilus maritimus*. Mol. Microbiol. 82: 555–566.

Pérez-Núñez, D., R. Briandet, B. David, C. Gautier, P. Renault, B. Hallet, P. Hols, R. Carballido-López, and E. Guédon. 2011. A new morphogenesis pathway in bacteria: Unbalanced activity of cell wall synthesis machineries leads to coccus-to-rod transition and filamentation in ovococci. Mol. Microbiol. 79: 759–771.

Perrin, F. 1934. Mouvement Brownien d'un ellipsoide. I. Dispersion diélectrique pour des molécules ellipsoidales. J. Phys. Radium 5: 497–511.

Perrin, F. 1936. Mouvement Brownien d'un ellipsoide. II. Rotation libre et dépolarisation des fluorescences. Translation et diffusion de molécules ellipsoidales. J. Phys. Radium 7: 1–11.

Pilhofer, M., M. S. Ladinsky, A. W. McDowall, G. Petroni, and G. J. Jensen. 2011. Microtubules in bacteria: Ancient tubulins build a five-protofilament homolog of the eukaryotic cytoskeleton. PLoS Biol. 9: e1001213.

Pirie, N. W. 1973. On being the right size. Annu. Rev. Microbiol. 27: 119–132.

Pla, J., M. Sanchez, P. Palacios, M. Vicente, and M. Aldea. 1991. Preferential cytoplasmic location of FtsZ, a protein essential for *Escherichia coli* septation. Mol. Microbiol. 5: 1681–1686.

Preisner, H., J. Habicht, S. G. Garg, and S. B. Gould. 2018. Intermediate filament protein evolution and protists. Cytoskeleton 75: 231–243.

Preston, T. M., and C. A. King. 2003. Locomotion and phenotypic transformation of the amoeboflagellate *Naegleria gruberi* at the water–air interface. J. Eukaryot. Microbiol. 50: 245–251.

Prostak, S. M., K. A. Robinson, M. A. Titus, and L. K. Fritz-Laylin. 2021. The actin networks of chytrid fungi reveal evolutionary loss of cytoskeletal complexity in the fungal kingdom. Curr. Biol. 31: 1192–1205.

Purcell, E. M. 1977. Life at low Reynolds number. Amer. J. Physics 45: 3–11.

Quintero, O. A., and C. M. Yengo. 2012. Myosin X dimerization and its impact on cellular functions. Proc. Natl. Acad. Sci. USA 109: 17313–17314.

Raven, J. A. 1982. The energetics of freshwater algae; energy requirements for biosynthesis and volume regulation. New Phytol. 92: 1–20.

Raven, J. A. 1983. The transport and function of silicon in plants. Biol. Rev. 58: 179–207.

Raven, J. A., and K. Richardson. 1984. Dinophyte flagella: A cost–benefit analysis. New Phytol. 98: 259–276.

Richards, T. A., and T. Cavalier-Smith. 2005. Myosin domain evolution and the primary divergence of eukaryotes. Nature 436: 1113–1118.

Rosenbaum, J. L., and G. B. Witman. 2002. Intraflagellar transport. Nat. Rev. Mol. Cell Biol. 3: 813–825.

Ross, L., and B. B. Normark. 2015. Evolutionary problems in centrosome and centriole biology. J. Evol. Biol. 28: 995–1004.

Rossmann, F. M., and M. Beeby. 2018. Insights into the evolution of bacterial flagellar motors from high-throughput *in situ* electron cryotomography and subtomogram averaging. Acta Crystallogr. D Struct. Biol. 74: 585–594.

Rueda, S., M. Vicente, and J. Mingorance. 2003. Concentration and assembly of the division ring proteins FtsZ, FtsA, and ZipA during the *Escherichia coli* cell cycle. J. Bacteriol. 185: 3344–3351.

Salje, J., P. Gayathri, and J. Löwe. 2010. The ParMRC system: Molecular mechanisms of plasmid segregation by actin-like filaments. Nat. Rev. Microbiol. 8: 683–692.

Samson, R. Y., and S. D. Bell. 2009. Ancient ESCRTs and the evolution of binary fission. Trends Microbiol. 17: 507–513.

Satir, P., C. Guerra, and A. J. Bell. 2007. Evolution and persistence of the cilium. Cell Motil. Cytoskeleton 64: 906–913.

Sato, K., Y. Watanuki, A. Takahashi, P. J. O. Miller, H. Tanaka, R. Kawabe, P. J. Ponganis, Y. Handrich, T. Akamatsu, Y. Watanabe, et al. 2007. Stroke frequency, but not swimming speed, is related to body size in free-ranging seabirds, pinnipeds and cetaceans. Proc. Biol. Sci. 274: 471–477.

Saxena, I. M., and R. M. Brown Jr. 2005. Cellulose biosynthesis: Current views and evolving concepts. Ann. Bot. 96: 9–21.

Schavemaker, P. E., and M. Lynch. 2022. Flagellar energy costs across the Tree of Life. eLife 11: e77266.

Scholey, J. M. 2013. Kinesin-2: A family of heterotrimeric and homodimeric motors with diverse intracellular transport functions. Annu. Rev. Cell Dev. Biol. 29: 443–469.

Schuech, R., T. Hoehfurtner, D. J. Smith, and S. Humphries. 2019. Motile curved bacteria are Pareto-optimal. Proc. Natl. Acad. Sci. USA 116: 14440–14447.

Schwanhäusser, B., D. Busse, N. Li, G. Dittmar, J. Schuchhardt, J. Wolf, W. Chen, and M. Selbach. 2011. Global quantification of mammalian gene expression control. Nature 473: 337–342.

Sehring, I. M., C. Reiner, J. Mansfeld, H. Plattner, and R. Kissmehl. 2007. A broad spectrum of actin paralogs in *Paramecium tetraurelia* cells display differential localization and function. J. Cell Sci. 120: 177–190.

Seidler, R. J., and M. P. Starr. 1968. Structure of the flagellum of *Bdellovibrio bacteriovorus*. J. Bacteriol. 95: 1952–1955.

Shastry, S., and W. O. Hancock. 2010. Neck linker length determines the degree of processivity in kinesin-1 and kinesin-2 motors. Curr. Biol. 20: 939–943.

Siefert, J. L., and G. E. Fox. 1998. Phylogenetic mapping of bacterial morphology. Microbiology 144: 2803–2808.

Smith, J. C., J. G. Northey, J. Garg, R. E. Pearlman, and K. W. Siu. 2005. Robust method for proteome analysis by MS/MS using an entire translated genome: Demonstration on the ciliome of *Tetrahymena thermophila*. J. Proteome Res. 4: 909–919.

Sosinsky, G. E., N. R. Francis, D. J. DeRosier, J. S. Wall, M. N. Simon, and J. Hainfeld. 1992. Mass determination and estimation of subunit stoichiometry of the bacterial hook-basal body flagellar complex of *Salmonella typhimurium* by scanning transmission electron microscopy. Proc. Natl. Acad. Sci. USA 89: 4801–4805.

Steenbakkers, P. J., W. J. Geerts, N. A. Ayman-Oz, and J. T. Keltjens. 2006. Identification of pseudomurein cell wall binding domains. Mol. Microbiol. 62: 1618–1630.

Stepanek, L., and G. Pigino. 2016. Microtubule doublets are double-track railways for intraflagellar transport trains. Science 352: 721–724.

Snyder, L. A., N. J. Loman, K. Fütterer, and M. J. Pallen. 2009. Bacterial flagellar diversity and evolution: Seek simplicity and distrust it? Trends Microbiol. 17: 1–5.

Sweeney, H. L., and E. L. F. Holzbaur. 2016. Motor proteins, pp. 69–86. In T. D. Pollard and R. D. Goldman (eds.) The Cytoskeleton. Cold Spring Harbor Laboratory Press, Cold Spring Harbor, NY.

Szewczak-Harris, A., and J. Löwe. 2018. Cryo-EM reconstruction of AlfA from *Bacillus subtilis* reveals the structure of a simplified actin-like filament at 3.4-Å resolution. Proc. Natl. Acad. Sci. USA 115: 3458–3463.

Tam, L. W., W. L. Dentler, and P. A. Lefebvre. 2003. Defective flagellar assembly and length regulation in LF3 null mutants in *Chlamydomonas*. J. Cell Biol. 163: 597–607.

Tam, D., and A. E. Hosoi. 2011. Optimal feeding and swimming gaits of biflagellated organisms. Proc. Natl. Acad. Sci. USA 108: 1001–1006.

Tamames, J., M. Gonzalez-Moreno, J. Mingorance, A. Valencia, and M. Vicente. 2001. Bringing gene order into bacterial shape. Trends Genet. 17: 124–126.

Taniguchi, Y., P. J. Choi, G. W. Li, H. Chen, M. Babu, J. Hearn, A. Emili, and X. S. Xie. 2010. Quantifying *E. coli* proteome and transcriptome with single-molecule sensitivity in single cells. Science 329: 533–538.

Taylor, T. B., G. Mulley, A. H. Dills, A. S. Alsohim, L. J. McGuffin, D. J. Studholme, M. W. Silby, M. A. Brockhurst, L. J. Johnson, and R. W. Jackson. 2015. Evolutionary resurrection of flagellar motility via rewiring of the nitrogen regulation system. Science 347: 1014–1017.

Tchoufag, J., P. Ghosh, C. B. Pogue, B. Nan, and K. K. Mandadapu. 2019. Mechanisms for bacterial gliding motility on soft substrates. Proc. Natl. Acad. Sci. USA 116: 25087–25096.

Thomas, N. A., S. L. Bardy, and K. F. Jarrell. 2001. The archaeal flagellum: A different kind of prokaryotic motility structure. FEMS Microbiol. Rev. 25: 147–174.

Titus, M. A. 2016. Myosin-driven intracellular transport, pp. 265–279. In T. D. Pollard and R. D. Goldman (eds.) The Cytoskeleton. Cold Spring Harbor Laboratory Press, Cold Spring Harbor, NY.

Tocheva, E. I., E. G. Matson, D. M. Morris, F. Moussavi, J. R. Leadbetter, and G. J. Jensen. 2011. Peptidoglycan remodeling and conversion of an inner membrane into an outer membrane during sporulation. Cell 146: 799–812.

Tocheva, E. I., D. R. Ortega, and G. J. Jensen. 2016. Sporulation, bacterial cell envelopes and the origin of life. Nat. Rev. Microbiol. 14: 535–542.

Toft, C., and M. A. Fares. 2008. The evolution of the flagellar assembly pathway in endosymbiotic bacterial genomes. Mol. Biol. Evol. 25: 2069–2076.

Turner, L., W. S. Ryu, and H. C. Berg. 2000. Real-time imaging of fluorescent flagellar filaments. J. Bacteriol. 182: 2793–2801.

Typas, A., M. Banzhaf, C. A. Gross, and W. Vollmer. 2011. From the regulation of peptidoglycan synthesis to bacterial growth and morphology. Nat. Rev. Microbiol. 10: 123–136.

Ursell, T. S., J. Nguyen, R. D. Monds, A. Colavin, G. Billings, N. Ouzounov, Z. Gitai, J. W. Shaevitz, and K. C. Huang. 2014. Rod-like bacterial shape is maintained by feedback between cell curvature and cytoskeletal localization. Proc. Natl. Acad. Sci. USA 111: E1025–E1034.

van Dam, T. J., M. J. Townsend, M. Turk, A. Schlessinger, A. Sali, M. C. Field, and M. A. Huynen. 2013. Evolution of modular intraflagellar transport from a coatomer-like progenitor. Proc. Natl. Acad. Sci. USA 110: 6943–6948.

Van Niftrik, L., W. J. C. Geerts, E. G. van Donselaar, B. M. Humbel, R. I. Webb, H. R. Harhangi, H. J. M. Op den Camp, J. A. Fuerst, A. J. Verkleij, M. S. M. Jetten, and M. Strous. 2009. Cell division ring, a new cell division protein and vertical inheritance of a bacterial organelle in anammox planctomycetes. Mol. Microbiol. 73: 1009–1019.

Velle, K. B., and L. K. Fritz-Laylin. 2019. Diversity and evolution of actin-dependent phenotypes. Curr. Opin. Genet. Dev. 58/59: 40–48.

Veltman, D. M., and R. H. Insall. 2010. WASP family proteins: Their evolution and its physiological implications. Mol. Biol. Cell 21: 2880–2893.

Veyrier, F. J., N. Biais, P. Morales, N. Belkacem, C. Guilhen, S. Ranjeva, O. Sismeiro, G. Péhau-Arnaudet, E. P. Rocha, C. Werts, et al. 2015. Common cell shape evolution of two nasopharyngeal pathogens. PLoS Genet. 11: e1005338.

Vischer, N. O., J. Verheul, M. Postma, B. van den Berg van Saparoea, E. Galli, P. Natale, K. Gerdes, J. Luirink, W. Vollmer, M. Vicente, T. den Blaauwen, et al. 2015. Cell age dependent concentration of Escherichia coli divisome proteins analyzed with ImageJ and ObjectJ. Front. Microbiol. 6: 586.

Visweswaran, G. R., B. W. Dijkstra, and J. Kok. 2011. Murein and pseudomurein cell wall binding domains of bacteria and archaea – a comparative view. Appl. Microbiol. Biotechnol. 92: 921–928.

Vollmer, W. 2011. Bacterial outer membrane evolution via sporulation? Nat. Chem. Biol. 8: 14–18.

Vollmer, W., and S. J. Seligman. 2010. Architecture of peptidoglycan: More data and more models. Trends Microbiol. 18: 59–66.

Wadhams, G. H., and J. P. Armitage. 2004. Making sense of it all: Bacterial chemotaxis. Nat. Rev. Mol. Cell Biol. 5: 1024–1037.

Wadhwa, N., and H. C. Berg. 2022. Bacterial motility: Machinery and mechanisms. Nat. Rev. Microbiol. 20: 161–173.

Wagstaff, J., and J. Löwe. 2018. Prokaryotic cytoskeletons: Protein filaments organizing small cells. Nat. Rev. Microbiol. 16: 187–201.

Walter, W. J., and S. Diez. 2012. Molecular motors: A staggering giant. Nature 482: 44–45.

Walsby, A. E. 1994. Gas vesicles. Microbiol. Rev. 58: 94–144.

Wang, S., L. Furchtgott, K. C. Huang, and J. W. Shaevitz. 2012. Helical insertion of peptidoglycan produces chiral ordering of the bacterial cell wall. Proc. Natl. Acad. Sci. USA 109: E595–E604.

Waterbury, J. B., J. M. Willey, D. G. Franks, F. W. Valois, and S. W. Watson. 1985. A cyanobacterium capable of swimming motility. Science 230: 74–76.

Wickstead, B., and K. Gull. 2007. Dyneins across eukaryotes: A comparative genomic analysis. Traffic 8: 1708–1721.

Wickstead, B., and K. Gull. 2011. The evolution of the cytoskeleton. J. Cell Biol. 194: 513–525.

Wickstead, B., K. Gull, and T. A. Richards. 2010. Patterns of kinesin evolution reveal a complex ancestral eukaryote with a multifunctional cytoskeleton. BMC Evol. Biol. 10: 110.

Wientjes, F. B., C. L. Woldringh, and N. Nanninga. 1991. Amount of peptidoglycan in cell walls of gram-negative bacteria. J. Bacteriol. 173: 7684–7691.

Wilkes, D. E., H. E. Watson, D. R. Mitchell, and D. J. Asai. 2008. Twenty-five dyneins in *Tetrahymena*: A re-examination of the multidynein hypothesis. Cell Motil. Cytoskeleton 65: 342–351.

Wiśniewski, J. R., and D. Rakus. 2014. Quantitative analysis of the *Escherichia coli* proteome. Data Brief 1: 7–11.

Wu, J. Q., and T. D. Pollard. 2005. Counting cytokinesis proteins globally and locally in fission yeast. Science 310: 310–314.

Yin, Q. Y., P. W. de Groot, L. de Jong, F. M. Klis, and C. G. De Koster. 2007. Mass spectrometric quantitation of covalently bound cell wall proteins in *Saccharomyces cerevisiae*. FEMS Yeast Res. 7: 887–896.

Young, K. D. 2006. The selective value of bacterial shape. Microbiol. Mol. Biol. Rev. 70: 660–703.

Young, K. D. 2010. Bacterial shape: Two-dimensional questions and possibilities. Annu. Rev. Microbiol. 64: 223–240.

Zhang, P., W. Dai, J. Hahn, and S. P. Gilbert. 2015. *Drosophila* Ncd reveals an evolutionarily conserved powerstroke mechanism for homodimeric and heterodimeric kinesin-14s. Proc. Natl. Acad. Sci. USA 112: 6359–6364.

Zhu, S., T. Nishikino, B. Hu, S. Kojima, M. Homma, and J. Liu. 2017. Molecular architecture of the sheathed polar flagellum in *Vibrio alginolyticus*. Proc. Natl. Acad. Sci. USA 114: 10966–10971.

PART 5

Energetics and Metabolism

CHAPTER 17

The Costs of Cellular Features

The pervasiveness of natural selection has been demonstrated thousands of times by measuring the fitness of individuals with different phenotypic values of a trait (Manly 1985; Endler 1986; Hoekstra et al. 2001; Walsh and Lynch 2018). However, it is one thing to identify optimum phenotypic values, conditional on having a trait, but quite another to determine the overall consequences of developing and exhibiting the trait at all. Virtually all organismal features require energetic expenditures for construction and maintenance, and the total benefit of a trait needs to be considered in the light of this baseline investment. The problem is similar to the challenges any company faces when developing a new product – how much profit will the new product earn versus the costs of developing and producing it? The explorations of the costs of cell walls and of swimming motility in Chapter 16 are applications of this idea.

To more broadly appreciate the relevant investments that cells make in different types of functions, and to understand the resultant barriers to trait establishment and maintenance, a quantifiable measure of the costs of cellular attributes is essential. In principle, such costs might be measured as the change in cell growth rate were the trait to be manipulated so as to be expressed versus unexpressed (or overexpressed vs. normally expressed) while conveying no benefits. In reality, however, such a measure is nearly impossible for many well-established traits for the simple reason that once integrated into critical cellular pathways, a trait will accumulate secondary side effects over evolutionary time. For example, increases in gene-expression levels will often have side effects (e.g., promiscuous binding or aggregation) that are irrelevant to the basic construction/maintenance costs of gene products.

Thus, we require an indirect way of summarizing the baseline costs of simply expressing and maintaining a trait that does not require invasive manipulation of the cell; and beyond this, we require a way to quantify the fitness consequences of such costs. To provide a quantitative framework for addressing these issues, we first outline the conceptual link between cell-biological expenditures and evolution. This is followed by applications to several key features exhibiting a substantial gradient in complexity across the Tree of Life, most notably the expansion of gene- and genome-architectural complexity and of the investment in membrane-bound organelles in the eukaryotic domain. These analyses show that such elaborations need not have been driven by positive selection for cellular complexity, but rather may be inevitable passive consequences of directional mutation biases and of the diminished efficiency of selection resulting from a reduction in effective population size. Although such a scenario might suggest the cumulative development of a long-term fitness drag on species with reduced population sizes, this need not always be the case. This is because embellishments incorporated into the genomic real estate and/or cellular repertoires by nonadaptive mechanisms can also serve as novel substrate for future adaptive evolution by descent with modification.

The Bioenergetic Cost of a Cellular Feature

Regardless of the nature of a cellular feature, be it a non-coding RNA, a protein molecule or complex, or a membrane associated with an organelle, all cellular traits entail baseline expenditures on construction and maintenance. This motivates the

Evolutionary Cell Biology. Michael Lynch, Oxford University Press. © Michael Lynch (2024). DOI: 10.1093/oso/9780192847287.003.0017

need for a universal currency by which costs can be measured in a way that generalizes across phylogenetic lineages and trait types. For such purposes, there seems to be little alternative than a measure of energy utilization. In the traditional fields of ecology and evolutionary biology, maximization of energy flow through biomass production has long been viewed as a target of natural selection, given that all processes devoted to survival and offspring production require energy (e.g., Lotka 1922; Van Valen 1976, 1980). However, justification for a focus on energy is based on more than tradition.

Although cost analyses might be carried out with alternative currencies (e.g., the mass of carbon or some other key nutrient), the nature of limiting nutrients can vary among phylogenetic lineages and even within species growing in different environments, restricting the general utility of such metrics. For example, silicon availability can be critical to diatom growth but is irrelevant to most other organisms. Likewise, nitrogen will rarely be limiting to a species capable of nitrogen fixation, although molybdenum may be. In any event, regardless of the substrates and elemental building blocks being consumed, their entry into all aspects of cellular maintenance and biomass production must ultimately be coupled with energy consumption. In addition, most operational and maintenance costs of cellular features can only be measured in terms of energy utilization.

As to the specific energetic currency to be used, there seems to be little alternative than the consumption of ATP molecules. Across the Tree of Life, hydrolysis of ATP to ADP is universally deployed in the vast majority of energy-requiring processes, including conserved metabolic pathways, modes of amino-acid, nucleotide, and lipid biosynthesis, and mechanisms of assembly of cellular polymers (e.g., DNA, RNA, and protein). Some processes involve the hydrolysis of other nucleotides such as CTP and GTP, but the energy release in these cases can still be counted in terms of ATP equivalents, as can the use of coenzymes such as NADH and NADPH in electron-transport chains (ETCs) used to produce ATP.

Thus, in keeping with previous efforts in microbial bioenergetics (Bauchop and Elsden 1960; Atkinson 1970; Stouthamer 1973; Tempest and Neijssel 1984; Russell and Cook 1995), all costs described here are defined in units equivalent to numbers of ATP hydrolyses. Additional justification for this approach derives from the observation that, although the yields of microbes grown on alternative substrates (per unit carbon consumed) can vary substantially with the nature of the substrate, once the appropriate energetic conversions are made, the number of ATP hydrolyses necessary to build an offspring cell are found to be relatively constant (Chapter 8).

The direct cost of a trait can be subdivided into three components: 1) biosynthesis of the basic building blocks; 2) assembly of the building blocks into the full structure; and 3) maintenance (Figure 17.1). First, with respect to biosynthesis, nearly all cellular features are assembled from four broad classes of monomeric subunits: amino acids, nucleotides, lipids, and carbohydrates. If not provided by the outside environment, such molecules must be synthesized within the cell by processes requiring carbon skeletons and energy consumption (derived, for example, from transformations of glucose or acetate to precursor metabolites such as pyruvate and acetyl CoA and then to downstream products). When such building blocks are available externally (a situation enjoyed by predators), the reliance on *de novo* biosynthesis will be diminished, but costs of resource acquisition and of various transformations of the precursors (e.g., branch-point metabolites) will still exist. Second, the assembly cost of a cellular feature is the sum of requirements for construction from its monomeric building blocks. For example, protein assembly requires polymerization of the constituent amino acids, addition of post-translational modifications, and folding of the amino-acid chain into the appropriate globular form; and the construction of higher-order structures such the cytoskeleton requires additional expenditures underlying multimerization. Third, cellular traits almost always experience maintenance costs, e.g., accommodation of molecular turnover, and identification and elimination of cumulative errors.

Given the near universality of many biosynthetic pathways and enzyme-reaction mechanisms, all three sets of costs can often be calculated to a good approximation from information residing in the biochemistry, biophysics, and cell-biological literature. Nonetheless, these direct-cost components do not fully describe the cellular consequences of a trait's presence. Even if the trait under consideration pays for itself by endowing the cell with increased fitness (in excess of the direct baseline costs), trait construction and maintenance will still impose an additional drain on resources that could have been allocated to other essential cellular functions. For example, when metabolic precursors that are generally processed for ATP production are instead allocated to the production of a focal trait (i.e., as carbon skeletons), the loss of availability for other purposes represents an opportunity cost (Figure 17.1). This follows from the fact that the production of an additional building block (or entire trait) adds to a cell's lifetime energy-budget requirements, while subtracting from the prior pool of resources available for other functions, thereby extending the cell cycle. Atkinson (1970) defined opportunity costs as the 'prices of metabolites', and seemingly independent

of him, Craig and Weber (1998) and Akashi and Gojobori (2002) used this approach to partition the costs of amino acids into components associated with the utilization of metabolites for carbon skeletons and with subsequent energy investments used in modifying these to final products.

Summing up, the total cost of synthesizing and maintaining a trait is the sum of the direct and opportunity costs,

$$c_T = c_D + c_O, \tag{17.1}$$

where all costs represent the cumulative expenditures over the entire lifespan of the cell. The direct cost, c_D, reflects actual ATP hydrolysis (or related) reactions resulting in heat dissipation in the cell. However, as a diversion of metabolic precursors, c_O will not be manifested in heat production, given that no ATP is produced or consumed. For this reason, it may be argued that c_O should not be included in cost analyses in situations where energy is a non-limiting resource. If, for example, the energy

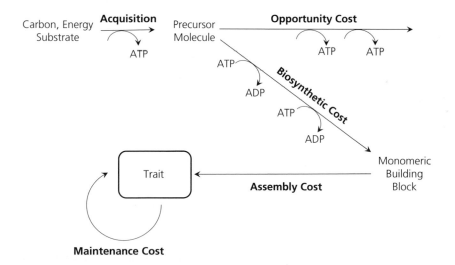

Figure 17.1 The distinction between direct and opportunity costs associated with synthesizing molecular building blocks (e.g., an amino acid). As the energy resource (e.g., glucose) is partially metabolized into precursor metabolites (carbon skeletons) necessary for synthesis of the building block, the additional energy that could have been captured from the complete metabolism of the resource is the opportunity cost. The conversion of precursor metabolites to some molecular building blocks can also consume electron-carrier molecules such as NADH, which, if not used in building-block synthesis, would have generated ATP, and thus represent an additional source of opportunity cost (Foundations 17.2). The consumption of ATP in the biosynthetic process defines the direct cost of building-block synthesis. The assembly of macromolecules such as proteins and nucleic acids from building blocks requires additional post-synthesis investment, such as the costs of assembly (e.g., polymerization) and maintenance (e.g., associated with molecular turnover, or DNA repair).

extracted from the food source is in excess supply relative to a key elemental resource (e.g., carbon, nitrogen, phosphorus, or iron), c_O might be largely irrelevant to the fitness of the cell, reducing the energetic cost from a fitness perspective to $c_T \simeq c_D$. However, it remains unclear how commonly such conditions arise, particularly when one considers that energy is used not only in the production of biomass, but also is the key contributor to all activities involved in cell maintenance.

An additional observation relevant to this issue derives from studies of microbes growing on defined media with all carbon and energy being provided by a single compound and all other nutrients being in excess supply (Chapter 7). Notably, in such settings, growth yields per carbon consumed increase linearly with the substrate heat of combustion up until a threshold value, and thereafter level off (Figure 7.8). This suggests that biology is constructed in such a way that below a critical substrate value of $\simeq 10$ kcal/g carbon, growth is typically limited by energy, whereas food supplies above this threshold contain excess energy relative to carbon content required for growth. Fortuitously, the most common substrate used in growth experiments with microbes, glucose, has a heat of combustion of 9.3 kcal/g carbon, close to the threshold at which growth is equally limited by carbon and energy. Very few commonly used substrates have heats of combustion much beyond the apparent threshold (values being 11.0, 13.6, and 14.8 for glycerol, ethanol, and methanol, respectively), providing further justification for a focus of energy as a basis for cost measures.

The Evolutionary Cost of a Cellular Feature

Given a measure of the absolute cost of a cellular feature, the overall implications depend on the cell's lifetime energy requirements. From a cell physiological perspective, the energetic cost of the feature must be scaled relative to the total cost of building and maintaining the cell (Chapter 8), and evolutionary considerations further require that this relative measure be translated to an appropriate metric of the impact on fitness.

Suppose the cell has a baseline total energy budget per cell cycle of C_T (which includes the costs of growth and maintenance, with a capital C denoting a whole-cell cost). The presence of the trait under consideration (c_T) influences the lifetime energy budget such that $C_T = C_T' + c_T$, where C_T' is the total energy budget in the absence of the trait. Note that the view here is that C_T is the sum of direct and opportunity costs, which includes the direct energetic expenditure of ATP on biosynthesis and the indirect diversion of energy-containing carbon skeletons (opportunity loss), as would be reflected in the consumption of glucose molecules consumed in a chemostat and their conversion to ATP equivalents.

Owing to the additional amount of resource acquisition necessary to complete the cell cycle, investment in the trait is expected to increase the cell-division time by a factor equal to the ratio c_T/C_T (ignoring for the time being any direct advantages conferred by the trait, and assuming $c_T \ll C_T$). The baseline fitness effects of building and maintaining the trait then impose a selection disadvantage $\simeq \ln(2)(c_T/C_T)$, with $\ln(2) \simeq 0.69$ simply scaling to the continuous-growth scale (Foundations 17.1). Thus, aside from the factor $\ln(2)$, the evolutionary cost of a trait in terms of energy investment is simply equal to its fractional contribution to the total-cell energy budget.

One caveat is that this definition assumes that the addition of the trait does not somehow alter the cell's basic metabolic makeup in other ways that would modify the total baseline energy budget C_T. In some instances, there may be non-additive side effects associated with a trait, as for example, when a novel protein promiscuously interacts with inappropriate substrates, aggregates with other cellular components, and/or excessively occupies cellular volume or membrane real estate. However, even if this does occur, the result given in Foundations 17.1 will be modified only slightly if the fractional alteration to C_T is small, as is likely for cellular modifications involving just one or two genes. The general point remains – as there will always be some baseline cost of expressing a trait, the net benefit needs to be derived by subtracting the construction/maintenance costs from the direct advantages accrued from increased survival and/or reproduction (Figure 17.2).

It further follows that if a trait is to be maintained by natural selection, it must pay for itself in terms of fitness enhancement – the baseline costs must be sufficiently small that the net benefit s_n is greater than the power of random genetic drift ($1/N_e$ for a haploid, and $1/2N_e$ for a diploid population, where N_e is the effective population size). If this condition is not met, the trait will generally be vulnerable to loss by degenerative mutations, unless it is so entrenched into other aspects of organismal function that removal is impossible. Likewise, it follows that in this domain of effective neutrality ($|s| < 1/N_e$), mutation pressure can lead to the slow accumulation of cellular modifications that impose a net drain on a cell's energetic capacity, as long as the incremental deleterious fitness effects are $< 1/N_e$. We next discuss this issue in the context

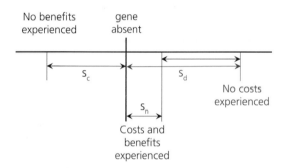

Figure 17.2 The evolutionary distinction between the construction/maintenance cost of a trait, here denoted as the product of a particular gene, and the direct benefits. s_c represents the reduction in fitness that would be expected from the presence of the trait in the absence of any direct benefits; it may be viewed as the selective advantage of a mutant relieved of the trait in an environment in which no advantages of the trait are experienced. s_d is a measure of the increase in fitness that would accrue in the absence of any assembly/maintenance costs. The difference $s_n = s_d - s_c$ denotes the net fitness advantage of the trait; if this value is negative, the trait is selectively disadvantageous despite any ecological benefits accrued. Clearly, a gene will be selected against if $s_d < s_c$, but $s_d > s_c$ is not a sufficient condition for gene preservation by natural selection, as the absolute value of s_n must exceed the power of random genetic drift, $1/N_e$ in a haploid species and $1/(2N_e)$ in a diploid, to be readily perceived by natural selection (Chapter 4).

of alterations to genome architecture and gene structure in various eukaryotic lineages.

Biosynthetic Costs of Nucleotides and Amino Acids

Much of the infrastructure of cells consists of DNA, RNA, and protein. Although there are many energetic expenses associated with assembly, processing, and maintenance of these molecules, the primary costs are associated with synthesis of the monomeric building blocks of the polymeric chains. Because the biosynthetic pathways for nucleotides and amino acids are generally conserved across the Tree of Life (Chapter 19), it is relatively straightforward, albeit tedious, to estimate the direct and opportunity costs of these basic units (Foundations 17.2). As discussed above, such costs are quantified in terms of ATP usage, specifically the number of phosphorus atoms released via ATP hydrolyses, the primary source of energy in most cellular reactions. Biosynthetic-pathway steps not involving ATP, but relying on different reactions including electron transfers via coenzyme conversions (e.g., NADH to NAD$^+$, NADPH to NADP$^+$, and FADH$_2$ to FAD) can be translated to ATP equivalents using conventions in biochemistry based on the known pathway for ATP production (Foundations 17.2).

As a first-order approximation, with glucose as the starting point, the average opportunity and direct costs of building the four ribonucleotides (used in RNA molecules) are $\simeq 43$ and 6 ATPs, respectively, with the total costs ranging from 43 for UTP to 55 for ATP (Foundations 17.2). Deoxyribonucleotides (used in DNA), which are made from the former by ribonucleotide reductase (and thymidylate synthase for U $\rightarrow$ T), have the same opportunity costs, but average direct costs elevated to 8 ATPs. Although the opportunity costs for both A:T and C:G bonds in double-stranded DNA are 85.5 ATP equivalents, the direct costs of the former are higher than the latter (19 vs. 17 ATPs), so that the total cost of an A:T bond is elevated by $\sim 2\%$.

The average opportunity and direct costs per amino acid are ~ 24 and 6 ATPs (Foundations 17.2), with the total average cost per residue being

~61% of the cost of an average nucleotide. However, contrary to the situation with nucleotides, all of which have similar biosynthetic costs, the total costs for different amino acids range sixfold, from 12 ATPs for glycine to 71 for tryptophan, in strong correlation with the molecular weights of the individual residues (Seligmann et al. 2003). It has been suggested that these cost differences among amino acids may be large enough to be discriminated by natural selection (Akashi and Gojobori 2002), and averaged across the Tree of Life, there is indeed a nearly thirtyfold reduction in the use of the most expensive relative to the cheapest amino acids (Krick et al. 2014). However, for selection to be capable of driving differences in amino-acid utilization, the cost differential between alternative amino acids at single codon sites must be sufficiently large (relative to the cell's entire energy budget) to overcome the background noise associated with random genetic drift. Close scrutiny of the matter is warranted, given that the maximum cost differential per amino-acid exchange is (71–12) = 59 ATP equivalents (tryptophan vs. glycine).

Why the focus on amino acids at single sites as units of selection rather than the full genome content? For selection to be effective at discriminating differences in the amino-acid contents of stretches of sequence longer than single codons, multiple codons for expensive versus cheap types of amino-acid variants would have to be associated within chromosomal segments, and recombination among sites would have to be sufficiently rare for such outlier segments to be a reliable target of selection. Given that most populations have levels of heterozygosity < 0.01, and that ~75% of nucleotide sites within a gene are replacement sites (Chapters 4 and 12), then for a gene of average length (~1000 bp), the number of amino-acid polymorphisms segregating simultaneously will generally be < 5, many of which by chance will involve amino acids with small cost differences. Moreover, because the rate of recombination per nucleotide site is roughly equal to that of mutation (Chapter 4), considerable decoupling of segregating amino-acid substitutions will occur over time.

Consider first the situation for an *E. coli*-sized cell, which with volume 1 μm^3 requires $\sim 3 \times 10^{10}$ ATP hydrolyses to build (Chapter 8). The mean

number of protein molecules associated with an average gene in a cell of this size is ~ 1700, with almost all genes falling in the range of 10 to 10^4 protein copies/cell (Chapter 7). Thus, relative to the total cell budget, the maximum energetic impact of the substitution of a single amino-acid residue (assuming a highly expressed gene with 10^4 protein copies/cell) $\simeq (59 \times 10^4)/(3 \times 10^{10}) \simeq 2 \times 10^{-5}$. As this quantity times ln(2) translates into a selection coefficient (Foundations 17.1), then recalling that selection is effective in a haploid population if $2N_e s > 1$ (Chapter 8), the effective population size N_e need only exceed $\sim 4 \times 10^4$ for selection to promote this extreme an amino-acid change in a highly expressed gene. For a gene with just 10 protein copies per cell, the critical N_e increases to 4×10^7, which is near the typical N_e for microbial populations (Chapter 4).

This rough analysis suggests that the selective promotion of amino acids with low biosynthetic costs (assuming they do not compromise protein function) can indeed be quite effective in prokaryotes, although with diminishing strength in lowly expressed genes. Consistent with these arguments, Akashi and Gojobori (2002) found a nearly 10% decline in the average cost per amino-acid residue in proteins with increasing gene-expression levels in *E. coli* and *B. subtilis*, and a similar conclusion was reached for other bacterial species (Heizer et al. 2006; Raiford et al. 2012). Thus, at least in bacteria, amino-acid substitutions that may be neutral with respect to protein function may nonetheless be advanced via their relative energetic demands.

Now consider a yeast-sized cell, $\sim 100~\mu m^3$ in volume, with a range of 100 to 10^6 protein copies per gene (Chapter 7). Given the near-linear scaling of lifetime cellular energy budgets with cell volume (Chapter 8), the construction cost for this larger cell is $\sim 100 \times$ that for a typical bacterium, but so is the upper limit for protein number. Thus, assuming similar protein length, the upper limit to the relative cellular expense (c_T / C_T) is the same as for *E. coli*, whereas the lower limit is $\sim 10 \times$ smaller. The critical N_e values then become 4×10^4 to 4×10^8 for highly versus lowly expressed genes, with the latter being near the upper bound seen in unicellular eukaryotes (Chapter 4). Similar calculations for a metazoan cell size of $\sim 1000~\mu m^3$ in volume yield a range of critical N_e values of 4×10^4 to 4×10^9 for the most highly to most lowly expressed genes. The latter critical point is orders of magnitude greater than any N_e estimate for a multicellular species.

Consistent with these disparities, indirect analyses suggest that selection may promote energetically cheap amino acids in highly expressed genes in unicellular eukaryotes, and perhaps even in some genes in multicellular eukaryotes (Swire 2007; Heizer et al. 2011). However, the overall pattern is substantially weaker than in bacteria – only a 1% reduction in the average amino-acid cost in highly versus lowly expressed genes in *S. cerevisiae* (Raiford et al. 2008), and 3% in the flour beetle *Tribolium castaneum* (Williford and Demuth 2012).

These observations suggest that in eukaryotes, selection for usage of energetically cheap amino acids approaches effective neutrality for a large fraction of genes with lower expression, and increasingly so in organisms that are larger in size. Even this, however, is a liberal conclusion, in that the preceding computations were carried out with the most extreme bioenergetic cost difference between amino acids, 59 ATPs. All but five amino acids have ATP costs in the range of 12 to 36 ATPs, with most pairs having differences < 10 ATPs (Foundations 17.2). Amino-acid cost differences on the order of 6 ATPs (as opposed to 59) require effective population sizes to be $\sim$ 10-fold higher than those noted previously for selection to be efficient, greatly reducing the likelihood of selective promotion of amino acids based on their energetic demands in species with large cell sizes. Thus, biased amino-acid usage in species with insufficient N_e to enable selection based on cost differences must have alternative explanations, such as mutation bias or simple functional constraints. Notably, in *E. coli* and yeast, amino-acid sites containing expensive residues tend to evolve more slowly than those harbored by the cheapest residues, suggesting that in these relatively high-N_e species, energetically costly residues are primarily relied upon for key structural or functional reasons (Seligmann et al. 2003; Barton et al. 2010).

A further issue of concern is the nature of amino-acid acquisition. Although it is often the case that most (and in some cases all) amino acids are synthesized within the cell, many species (e.g., metazoans) are unable to synthesize one or more

amino acids and must acquire them from external food sources. As noted above, the evolutionary loss of a biosynthetic pathway is expected when the payoffs of direct biosynthesis are sufficiently small relative to the cost of constructing and maintaining the pathway, assuming a readily available external source of the building block. Remarkably, however, patterns of reduced usage of expensive amino acids also occur in microbes that are auxotrophic (unable to synthesize) for such residues (Swire 2007; Raiford et al. 2012). Although there is no direct cost of biosynthesis of externally acquired amino acids, Swire (2007) argues that there is still an opportunity cost – an amino acid taken up by a heterotroph can either be directly incorporated into a protein or degraded to produce ATP that can be utilized in other cellular processes. Given that the amount of energy extracted from the breakdown of an amino acid is about the same as the energy for building one, that most of the total cost of amino-acid biosynthesis involves opportunity loss, and that energy must be expended for amino-acid uptake, the differences in cost for directly acquired versus internally synthesized amino acids may be relatively minor. A clearer resolution of this matter is obviously desirable.

An Empirical Shortcut to Cost Estimates

There are two major challenges to applying the preceding theory to non-model organisms. First, for those who have worked through the bookkeeping contortions in Foundations 17.2 and 17.5, the complexity of estimating the costs of various cellular components from known biosynthetic pathways will be apparent. Moreover, although many such pathways are highly conserved across the Tree of Life, variants do exist within and among species (Chapter 19), and for the vast majority of organisms, the precise nature of the underlying reactions may be uncertain even for the simplest of building blocks. In addition, for biological compounds for which biosynthetic mechanisms are completely unknown, pathway analysis is not an option.

Second, as outlined in the preceding section, to make a connection to evolutionary theory, all costs of cellular components need to be normalized by the entire cellular energy budget. Effective methods for estimating the latter have been applied to a wide variety of microbes, leading to a general expression for the cost of building and maintaining a cell as a function of cell volume (Chapter 8). However, the methods involved are not broadly utilizable, as they require the culturing of organisms on a defined medium in chemostats, and/or the joint estimation of rates of growth and metabolism. Implementation of such methods can be highly impractical for organisms that are long-lived, consume other organisms with unknown chemical compositions, and/or are difficult to culture in laboratory environments.

Thus, there is a need for at least approximate methods that can be applied in the absence of detailed knowledge of an organism's biochemical pathways or growth features. One possible approach invokes the Kharasch and Sher (1925) formula for the degree of reduction of an organic compound, defined as N_E in Equation 7.14. For a wide range of organic compounds, N_E is almost perfectly correlated with the heats of combustion, making it a reliable indicator of the energy content of a substance. N_E is a simple function of the number of carbon, hydrogen, and oxygen atoms in a compound, and from the standpoint of the costs of building blocks, a key missing element would seem to be phosphorus, which enters into the energetic values of a number of precursor compounds, nucleic acids, phospholipids, etc. However, modification of the expression for N_E to

$$N_{ATP} = 4N_C + N_H - 2N_O + 10N_P, \qquad (17.2)$$

where the terms to the right are numbers of the subscripted element in the molecular formula for the compound, provides an excellent first-order approximation to the total costs of a wide range of cellular compounds (Figure 17.3). This means that, from the standpoint of cell biology, the total cost of an organic substance, in terms of ATP equivalents, can be closely approximated using only the atomic numbers in the chemical formula. By extension, it also implies that, as a first-order approximation, the total cost of building a cell, C_T, might be measurable using simple information on the cellular elemental composition for C, H, O, and P alone.

To gain some appreciation for the potential utility of this approach, recall that chemostat results

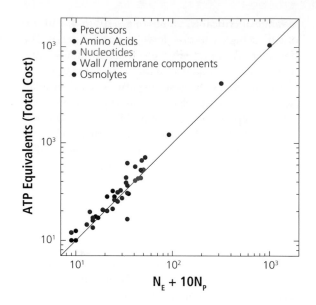

Figure 17.3 Total costs associated with various cellular building blocks (estimated from biochemical pathways) in comparison to the modified index for the degree of chemical reduction, Equation 17.2.

suggest that the cost of constructing a cell in a species with average volume 1 μm^3 is approximately 2.7×10^{10} ATPs (Chapter 8). For a cell with this average size, the expected newborn size is 0.67 μm^3, which then grows to size 1.33 μm^3 at the time of cell division. From Equation 7.1, a volume increase of 0.67 μm^3 is equivalent to a dry weight increase of $\simeq 0.00039$ ng, approximately 50% of which is carbon (Chapter 7). Noting that carbon has a molecular weight of 12 g/mol, and applying Avogadro's number for the number of molecules/mol leads to $N_C \simeq 10^{10}$ atoms of carbon required for the construction of a newborn cell of this size. Averaging over elemental analyses for two bacteria and two yeasts, Duboc et al. (1985) estimated C : H : O : P molar ratios of 1.00 : 1.73 : 0.57 : 0.02. Applying the elemental numbers to Equation 17.2 then leads to a total cost estimate of 4.7×10^{10}, within a factor of two of the direct estimate. An analysis with ciliates yielded a similar conclusion (Lynch et al. 2022).

Given the potential sources of inaccuracies in these kinds of estimates, this level of disagreement should not be viewed as too serious, particularly because for most downstream evolutionary interpretations, an estimate of C_T to an order of magnitude accuracy is generally sufficient. Equation 17.2 is actually expected to underestimate more direct estimates of construction costs, as this indirect approach only estimates the ATP equivalents tied up in the actual standing biomass in a cell. Not included are the costs of acquisition of resource molecules (nutrient import through channels), of assembling building blocks into higher-order structures, of molding such structures into their appropriate forms, etc. However, the sum of these additional costs is generally small relative to the costs of synthesizing the basic building blocks.

Despite its promising practical utility, a theoretical basis for Equation 17.2 remains to be developed. However, a crude understanding for why N_{ATP} provides a reasonable approximation to cell structural costs derives from the fact that $N_E = 24$ for glucose, whereas the latter has an energetic content equivalent to 32 ATPs (based on known biochemical pathways; Fundamentals 17.2). With the latter transformation, this suggests that at least at the whole-cell level, multiplication of the right side of Equation 17.2 by 1.33 should yield the predicted number of ATPs. If the cell were capable of extracting fewer than 32 ATPs per glucose molecule, the weighting factor would be reduced, becoming 1.0 for the case in which glucose equivalence in ATPs is just 24 (and 0.5 in the case of just 12).

In prior attempts to draw connections like that just mentioned, Williams et al. (1987) ignored phosphorus, whereas Duboc et al. (1985) advocated a weighting factor of 5. However, both approaches lead to poorer fits of the data for the cellular building blocks outlined in Figure 17.3, especially for precursor molecules. Nonetheless, because phosphorus is such a small component of overall cellular biomass, none of these differences in treatment has much impact in estimating the total cost of a cell, C_T. For example, in the preceding example, ignoring N_P entirely only reduces the estimate of N_{ATP} by 3%. Accounting for nitrogen and/or sulfur content does not improve the situation.

The Energetic Cost of a Gene

As a first application of the preceding ideas, consider the total cost of maintaining and operating a

gene, which involves up to three levels of investment. First, even for an unexpressed genomic segment, there are DNA-level costs in terms of replication and chromosome maintenance (which, in eukaryotes, include the costs of nucleosomes around which the DNA is wrapped). Second, transcription and transcript processing impose additional costs. Third, for protein-coding genes, there are major additional costs in terms of amino acid biosynthesis and polypeptide processing. The sum of investments at these three levels constitutes the total cost of any genomic sequence.

As mentioned above, individual genes face a 'use it or lose it' challenge. If the net fitness advantage of a gene is smaller than the power of random genetic drift, it will be vulnerable to passive inactivation by the accumulation of degenerative mutations, and if there is a strong enough net selective disadvantage of the remnant pseudogene, physical removal will be accelerated by directional selection for deletion mutations. On the other hand, biased mutation pressure towards insertions of excess and even nonfunctional DNA can facilitate genomic expansion, provided the cost of the excess genomic material is smaller than the power of random genetic drift.

One of the mysteries of genome evolution concerns the number of genes contained within genomes and the mechanisms responsible for the lineage-specific expansions of such numbers in eukaryotes, especially in multicellular species (Lynch 2007a; Lynch et al. 2011; Chapter 24). The genomes of most prokaryotes contain < 5000 protein-coding genes, whereas those of most eukaryotes harbor $> 10^4$ genes, with genomes of most multicellular eukaryotes (metazoans and land plants) containing 15,000 to 30,000. Most significantly, the expansion of total genome sizes from prokaryotes (generally 1 to 10 million base pairs) to unicellular eukaryotes (generally 10 to 100 million base pairs) to multicellular eukaryotes (10^8 to 10^{10} base pairs) is much less a consequence of an increase in gene number than of the proliferation of noncoding DNA. Most prokaryotic genomes are highly streamlined, typically containing $< 5\%$ intergenic DNA and devoid of introns and mobile-genetic elements. In contrast, the genomes of multicellular eukaryotes often contain $< 5\%$ coding DNA and harbor massive numbers of large introns, mobile elements, and other forms of DNA insertions. This raises the question as to whether expansions in genome size and gene-structural complexity are driven by adaptive processes as opposed to being inevitable consequences of the increased power of random genetic drift and biased mutational insertion pressure in organisms of larger size.

A common view is that that there is an intrinsic advantage to both cellular complexity and multicellularity (Chapter 24), but such a stance is nothing more than an assumption made by the one multicellular species that dominates the biological and intellectual worlds (Lynch 2007b; Booth and Doolittle 2015). In fact, there is no direct evidence that what we regard as complexity has been directly promoted by natural selection. If the advancement of complexity is a general goal of selection, given that all extant organisms are temporally equidistant from the last universal common ancestor, the more astounding observation is that complex multicellularity (involving large numbers of cell types) is largely confined to just two recent phylogenetic lineages, metazoans and land plants.

To help explain the gradient from extreme streamlining of genomes in prokaryotes to the extraordinary bloating of genomes in metazoans and land plants, we now draw from a wide variety of observations from cell biology and biochemistry to evaluate how expensive a gene (or segment of DNA) is from an energetic perspective, drawing heavily from Lynch and Marinov (2015).

Chromosome-associated costs

We first consider the baseline cost of harboring a segment of DNA, regardless of its expression level. Genome replication requires the synthesis of two new DNA strands from each parental double-helix DNA molecule. From Foundations 17.2, the average total biosynthetic cost per deoxyribonucleotide is 52 ATPs. Of the numerous additional costs of a gene at the DNA level, the major one is the unwinding of the parental double-helix, which requires ~ 1 ATP per nucleotide. All other replication-related costs – opening of origins of replication, clamp loading, proofreading, production of the RNA primers used for replicate-strand extension, ligation of Okazaki fragments, and DNA repair – are

an order of magnitude or so smaller. Thus, noting the double-stranded nature of DNA, as a first-order approximation, the total cost of replicating a gene L_n nucleotides in length is $\sim 105 L_n$.

There is, however, one additional major chromosome-level cost specific to eukaryotes – the highly ordered, dense coverage of nucleosomes, each of which contains two histone complexes followed by a linker histone. Throughout eukaryotes, each nucleosome wraps ~ 147 bp, and with an average linker length between nucleosomes of 33 bp, there is on average one nucleosome per 180-bp interval. Weighting by the cost of synthesizing the amino acids that comprise histone proteins and the cost of translating such proteins (Foundations 17.2), the total nucleosome-associated cost $\simeq 190 L_n$ ATPs (using slightly modified building costs from Lynch and Marinov 2015), which is more than the DNA itself.

Taking all of these issues into consideration, the total chromosome-level cost of a bacterial DNA segment (in units of ATP hydrolyses) is

$$c_{\text{DNA},b} \simeq 105 L_n, \qquad (17.2a)$$

whereas for a haploid eukaryote,

$$c_{\text{DNA},h} \simeq 295 L_n, \qquad (17.2b)$$

and doubling the preceding cost for a diploid eukaryote yields

$$c_{\text{DNA},d} \simeq 590 L_n. \qquad (17.2c)$$

These expressions provide a quantitative basis for understanding the evolutionary mechanisms underlying the dramatic differences in gene structure and genomic architecture that exist between prokaryotes and eukaryotes. Because replication is essentially a one-time investment in the life of a cell, the maximum fractional contribution of the DNA-level cost of a gene to a cell's total energy budget occurs at minimum cell-division times (which minimize maintenance costs; Chapter 8). The maximum chromosome-level cost can then be approximated as c_{DNA}/C_G, where C_G is total cost of constructing a cell. Thus, a prokaryotic cell with a representative volume of 1 μm^3 (and associated cellular construction cost of $\sim 3 \times 10^{10}$ ATPs) has a replication-associated cost of DNA $\simeq (3 \times 10^{-9}) L_n$, which implies a fractional drain on the total cellular energy budget of (3×10^{-9}) for a 1-bp insertion and (3×10^{-6}) for a gene-sized insertion of 1000 bp. Thus, because free-living prokaryotes typically have effective population sizes $\simeq 10^8$ (Chapter 4), when growing at maximum rates, such organisms experience efficient enough selection to remove insertions as small as 10 bp (and even 1 bp when N_e approaches the apparent upper bound of 10^9).

In contrast, for a unicellular haploid eukaryote with a moderately sized 100 μm^3 cell (e.g., a yeast), the fractional cost of DNA is $\simeq 10^{-10} L_n$, yielding maximum relative chromosome-level costs of 10^{-9} and 10^{-7} for 10 bp and 1000 bp segments of DNA, respectively. Because unicellular eukaryotes often have $N_e < 10^8$, sometimes ranging down to 10^6, these results imply that energetic costs of small- to moderately sized insertions in such species will frequently be indiscernible by natural selection.

For a larger cell size of 2500 μm^3, more typical of a multicellular eukaryote, and a diploid genome, the relative cost of DNA declines further to $\simeq 10^{-11} L_n$, so even a 10^5-bp segment of DNA has a relative cost of just 10^{-6}. The effective population sizes of invertebrates tend to be in the neighborhood of 10^6, with that of some vertebrates (including humans, historically) ranging down to 10^4. Thus, even though the chromosome-level cost of a DNA insertion in a diploid multicellular eukaryote is ~6× that in a prokaryote, the disparity in total cellular energy budgets dilutes the effect, such that the power of random genetic drift is sufficient to overwhelm the ability of selection to prevent the accumulation of quite large insertions on the basis of energetic costs at the chromosome level.

These results help explain the evolutionary maintenance of the highly streamlined genomes of prokaryotes relative to eukaryotes. As outlined in Foundations 17.3, however, there is an additional cost of excess DNA, unassociated with bioenergetics – all excess DNA, even when nonfunctional, is dangerous in that it increases the substrate for mutations to gene malfunction (Lynch 2007a). On a per-nucleotide basis, this too is typically a weak evolutionary cost, strong enough to be perceived in many microbes but often effectively neutral in eukaryotes (multicellular species in particular).

Transcription-associated costs

Although the costs of transcription are numerous, and not all of them can be quantified with certainty, the major contributors are well understood. Thus, it is possible to achieve order of magnitude estimates of the investments required to produce individual transcripts, again adhering to the strategy summarized in Lynch and Marinov (2015).

Because transcripts are typically degraded (and must be replaced) on time scales much shorter than cell-division intervals, the total cost of transcription per cell cycle depends on the lifespan of a cell (T). If we consider a cell containing an average number of transcripts N_r at birth and a degradation rate per transcript of δ_r, during its lifetime, a cell must produce N_r additional surviving transcripts to create a daughter cell at the same steady-state level as well as an additional $\delta_r N_r T / \ln(2)$ replacement molecules to offset molecular degradation (Foundations 17.4). Here, it will be assumed that the set of N_r transcripts necessary to provision the equivalent of a daughter cell require *de novo* synthesis of nucleotides, whereas the remaining $\delta_r N_r T / \ln(2)$ replacement molecules are produced from recycled ribonucleotides.

Several forms of transcription-associated costs are general across prokaryotes and eukaryotes, but only two of these are quantitatively significant enough to be of concern here. The primary investment is the synthesis of ribonucleotides, with an average cost of ~45 ATP per base (Foundations 17.2). Adding in the 2 ATP equivalents required for each chain-elongation step, the total cost of *de novo* ribonucleotide synthesis associated with a gene with transcript length L_r is then $\simeq 47 N_r L_r$. The second major cost involves the replacement of degraded transcripts within the life-span of the cell, and here the total expenditure is simply taken to be the two ATPs that must be expended per nucleotide for each chain-elongation step, which leads to a cost of $2 \delta_r N_r L_r T / \ln(2)$.

A third cost, associated with helix unwinding, is < 5% of that associated with ribonucleotide recycling. A fourth cost is associated with aborted transcripts (as not all transcription-initiation events lead to completed transcripts), but because such events generally occur within the first ten nucleotides, this cost is even smaller than that for helix unwinding. Still smaller is the cost of activating and initiating transcription. Thus, to a close approximation, the sum of the two predominant costs of transcription – expenditure on ribonucleotide synthesis and chain elongation closely approximates the total cost of transcribing a gene in the lifetime of a bacterial cell (in units of ATP),

$$c_{\mathrm{RNA},b} \simeq N_r L_r (50 + 2.9 \delta_r T). \tag{17.3a}$$

Eukaryotes incur several additional energy-consuming features of transcription, but only two of them are quantitatively relevant here. First, eukaryotic mRNAs are terminated by extended poly(A) tails, with initial lengths of ~250 nucleotides. Taking into consideration the full costs of As necessary for the standing pool of mRNAs and the two ATPs per bp necessary for chain elongation associated with excess degraded transcripts, the total cost of poly(A) tails is $\sim 250 N_r (45 + 2.9 \delta_r T)$. Second, as noted above, eukaryotic DNA is populated by regularly spaced nucleosomes, and in order for RNA polymerases to proceed, the DNA must be unwrapped from these; this and other related processing entails a total energetic cost $\simeq 0.25 N_r L_r \delta_r T$. For intron-containing genes, there is a small additional cost of splicing, and there are also costs associated with transcript termination, 5' mRNA capping, phosphorylation cycles associated with the RNA polymerase II, and nuclear export, but all of these are quite small relative to the two costs noted.

Summing the eukaryotic-specific components with Equation 17.3a, the total cost associated with transcription for a eukaryotic gene is

$$c_{\mathrm{RNA},e} \simeq N_r [(12,500 + 50 L_r) + (725 + 3.2 L_r) \delta_r T], \tag{17.3b}$$

where L_r is the length of the primary transcript (before intron removal and splicing). Note that Equations 17.3a,b are written so that the total cost associated with transcription is subdivided into two components, the first defining the baseline requirement for building a cell, and the second being a linear function of the cell division time.

Observations from single-cell methodologies provide quantitative insight into some of the key parameters in these formulations. As noted in Chapter 7, standing numbers of transcripts per gene (N_r) are generally quite small. For example, for *E. coli*, average $N_r \simeq 5$ (with a range of 0 to 100 among genes) (Lu et al. 2007; Li and Xie 2011). The mean N_r is 10 per gene in *S. cerevisiae* (Lu et al. 2007; Zenklusen et al. 2008), and the median is ~20 in mammalian cells (Islam et al. 2011; Schwanhäusser et al. 2011; Marinov et al. 2014). In all cases, there is a broad distribution around the mean, so that genes with numbers of transcripts deviating tenfold from the mean are not uncommon (Golding et al. 2005; Raj et al. 2006; Taniguchi et al. 2010; Csárdi et al. 2015).

Estimates of transcript-decay rates suggest that the half-lives of mRNAs are typically much shorter than cell division times. In *E. coli*, $\sim 80\%$ of mRNAs have decay rates (δ_r) in the range of 7 to 20/hour, with a median of 12/hour (Bernstein et al. 2002; Taniguchi et al. 2010). The mRNAs in *Bacillus subtilis* have a median decay rates of 8/hour (Hambraeus et al. 2003); and in *Lactococcus lactis*, mean and median mRNA decay rates are in the range of 3 to 7/hour, decreasing with decreasing cellular growth rates (Dressaire et al. 2013). For eukaryotes, median mRNA decay rates range from 3 to 6/hour in *S. cerevisiae* (Wang et al. 2002; Neymotin et al. 2014), and average decay rates are 5/hour in *S. pombe* (Eser et al. 2016) and 0.1/hour in mouse fibroblast cells (Schwanhäusser et al. 2011). To a first-order approximation, these results suggest that $\delta_r T$ (the number of replacement transcripts per gene per cell lifetime) is generally within the range of 10 to 100, which from Equations 17.3a,b further implies that mRNA decay typically inflates the total cost of transcription by a factor 2–6× relative to that expected on the basis of the steady-state number of transcripts per cell, $50 N_r L_r$ and $N_r [(12,500 + 50 L_r)]$, for bacteria and eukaryotes, respectively.

Recalling the preceding results for chromosome-level costs, it can be seen that the costs at the level of transcription will often be several times higher. Considering bacteria, for example, and noting that the length of a transcript is very close to the length of a gene ($L_n \simeq L_r$), the ratio of Equations 17.3a and 17.2a, $N_r (0.5 + 0.029 \delta_r T)$, always exceeds 1.0 provided the steady-state number of transcripts is > 2, and can be several times higher for genes with higher expression levels and/or large numbers of transcript half-lives per cell cycle.

For the yeast *S. cerevisiae*, >95% of genes are intron-free, and the remaining 5% contain only a single small intron, so as a first-order approximation, it can again be assumed that genes and transcripts have essentially the same lengths $L_n \simeq L_r = 2000$ bp. As noted above, the mean number of mRNAs per gene per yeast cell is $\overline{N}_r \simeq 10$, and with an average decay rate of 4.5/hour and a doubling time $\simeq 1.5$ hours under optimal growth conditions, $\delta_r T \simeq 7$. From Equation 17.3b, the cost of transcription for a typical yeast gene is then on the order of $10 \cdot \{(12, 500 + 100, 000) + [(725 + 6, 400) \cdot 7]\} \simeq 1.6 \times 10^6$ ATPs. By contrast, from Equation 17.2b, the chromosome-level cost of a gene in this species $\simeq 0.6 \times 10^6$ ATPs.

Finally, we consider the situation for a typical human gene, where the median number of mRNAs per gene $\simeq 20$, and the average mRNA decay rate is ~ 1.4/day. Assuming a cell division time of one day and an average primary transcript length of 47 kilobases (owing to the burden of numerous, large introns, which are transcribed before being spliced), and ignoring the small contribution from the cost of splicing, the cost of transcription per gene is then on the order of $20 \cdot \{(12, 500 + 2, 350, 000) + [(725 + 150, 400) \cdot 1.4]\} \simeq 5 \times 10^7$ ATPs. Total gene lengths are difficult to define in metazoans, but a 50% inflation relative to the pre-mRNA (~ 70 kilobases) is not unreasonable. Equation 17.2c then implies a typical cost at the chromosome level of 4×10^7 ATPs. Taken together, all of these results suggest that transcription-associated costs in mammalian cells are typically of the same order of magnitude as those at the chromosome level, although the former can greatly exceed the latter for highly expressed genes.

Translation-associated costs

The conceptual approach employed in the preceding section can be extended to the protein level by again assuming that the cost of production of the steady-state number of proteins must be covered by *de novo* synthesis of amino acids, with the excess molecules needed to compensate for protein decay being acquired from salvaged amino acids. Although several sources of costs underlie protein

production and subsequent management, the overwhelming contributions are associated with just three functions, the biochemical details of which are summarized in Lynch and Marinov (2015).

First, the cost associated with the production of the standing level of protein for a particular gene necessary for an offspring cell is $N_p L_p \bar{c}_{AA}$, where N_p is the number of protein molecules per newborn cell, L_p is the number of amino-acids per protein, and $\bar{c}_{AA}$ is the average total cost of synthesis per amino-acid residue (assumed to be equivalent to 30 ATP hydrolyses; Foundations 17.2). Second, the total cost associated with chain elongation of all proteins produced during the cell cycle is $4N_p L_p [1 + (\delta_p T / \ln(2))]$, where δ_p is the rate of protein decay, and the 4 results from the 2 ATPs required for activating the cognate tRNA, an additional 1 for transferring the tRNA to the ribosome, and 1 more for the movement of the ribosome to an adjacent mRNA triplet). Third, degradation of proteins imposes an approximate cost of $N_p L_p \delta_p T / \ln(2)$ ATP hydrolyses. Additional costs small enough be ignored are associated with translation initiation and termination, post-translational modification, protein folding, and degradation. Summing up the three primary expenses, the total protein-level cost of a gene in both bacteria and eukaryotes is

$$c_{PRO} \simeq N_p L_p (34 + 7\delta_p T), \qquad (17.4)$$

where again the first term represents a one-time cost incurred regardless of the length of the cell cycle, and the second term grows linearly with the cell-division time owing to the cumulative costs of protein turnover and replacement.

Insight into the relative magnitudes of the two terms in Equation 17.4 requires information on protein-degradation rates, which are typically much lower than those of their cognate mRNAs. In the bacterium *L. lactis*, the vast majority of protein decay rates are in the range of 0.04 to 6.0/hour, with the median being 0.1 to 0.9/hour depending on the growth rate (Lahtvee et al. 2014). Those for other bacteria are commonly in the range of 0.05 to 0.20/hour (Trötschel et al. 2013). In *S. cerevisiae*, the median and mean decay rate is ~ 1.4/hour, with most values for individual proteins falling in the range of 0.2 to 5.5/hour under optimal growth conditions (Belle et al. 2006), and the median declining to 0.1/hour in nutrient limiting conditions (Shahrezaei and Swain 2008; Helbig et al. 2011). In mouse fibroblast cells, the median decay rate of a protein is ~ 0.02/hour (with a range of 0.002 to 0.3/hour) (Schwanhäusser et al. 2011), and in a human cancer-cell line, the range is from 0.04 to 1.3/hour (Eden et al. 2011). Given the known division times for the cell types noted earlier, $\delta_p T$ will commonly be < 1, and seldom > 10. This implies that the second term in Equation 17.4, the cost of protein degradation, will generally be of the same order of magnitude as the *de novo* protein-synthesis cost or smaller.

Cellular abundances of proteins (N_p) are much higher than those for their cognate mRNAs, with the average ratio of the two per gene being 450 in *E. coli*, 5100 in *S. cerevisiae*, and 2800 in mammalian fibroblasts (Ghaemmaghami et al. 2003; Lu et al. 2007; Schwanhäusser et al. 2011). How do these numbers translate into the total protein-level cost of a gene? To keep the computations simple but still accurate to a first-order approximation, it will be assumed here that $7\delta_p T \simeq 6$, so that $c_{PRO} \simeq 40 N_p L_p$.

In *E. coli*, the average number of proteins per gene is ~ 2250, and the average protein contains ~ 300 residues, implying an average $c_{PRO} \simeq 3 \times 10^7$ ATP/protein-coding gene. By comparison, as noted above, the average chromosome-level cost is $\sim 10^5$, and the average transcription-associated cost is only a few times greater than 10^5. Thus, the vast majority of the energetic cost of a protein-coding gene in bacteria is associated with translation. For the yeast *S. cerevisiae*, there is an average of $\sim 50,000$ proteins per genetic locus per cell, and the average protein length is $\sim 50\%$ greater than in *E. coli*, yielding an average total cost of translation of $\sim 8 \times 10^8$ ATP per protein-coding gene, which is again approximately two orders of magnitude greater than the summed costs at the chromosome and transcription levels.

Evolutionary implications

The full slate of data necessary to estimate the total cost of a gene is only available for a few species (Lynch and Marinov 2015). However, the existing data are in full accord with the hypothesis that baseline selective consequences (s_c) of such costs tend to exceed the power of random genetic drift in microbes and then progressively become smaller than the power of genetic drift in larger eukaryotic species (Figure 17.4).

For almost all genes in the bacterium *E. coli*, s_c falls in the range of 10^{-6} to 10^{-3}, far above the likely minimum values that can be perceived by selection in this large-N_e species (Figure 17.4). Such fitness costs have been directly corroborated in microbial constructs engineered to have additional gene copies and/or enhanced expression levels (Dean et al. 1986; Stoebel et al. 2008; Scott et al. 2010; Adler et al. 2014; Bienick et al. 2014). If such genes were to find themselves in an environment where their functions were no longer useful, inactivating mutations would be strongly selected for.

Within eukaryotes, some lowly expressed genes have roughly the same absolute costs of *E. coli* genes. However, owing to the increased total cellular energy budgets, the fractional costs are reduced, rendering s_c for many eukaryotic genes < 10^{-6} and in some cases as low as 10^{-9}. This is significant because the reduction in the effective population sizes of such species alters the location of the drift barrier. For the majority of genes in *S. cerevisiae* (yeast), *C. elegans* (nematode), and *A. thaliana* (land plant), the costs at the chromosomal and transcriptional levels are below or near the drift barrier, implying that without translation most gene-sized insertions will be essentially invisible to the eyes of selection. That is, the cost of translation is the major factor pushing the baseline s_c of eukaryotic genes past the drift barrier.

A more general analysis over a large number of species indicates that the total cost of a gene (relative to the cell energy budget) declines with increasing cell size across the Tree of Life (Figure 17.5). Average estimates of all three cost measures in bacteria are generally substantially greater than those in eukaryotes, although there is continuity in the scaling between groups. In addition, as mentioned above, there is a consistent ranking of s_{DNA} and s_{RNA} < s_{PRO}, with a one to two order of magnitude increase from the former to the latter.

These results suggest that evolutionary increases in cell size (and the associated increased power of random genetic drift in larger organisms) promote a shift in the selective environment such that

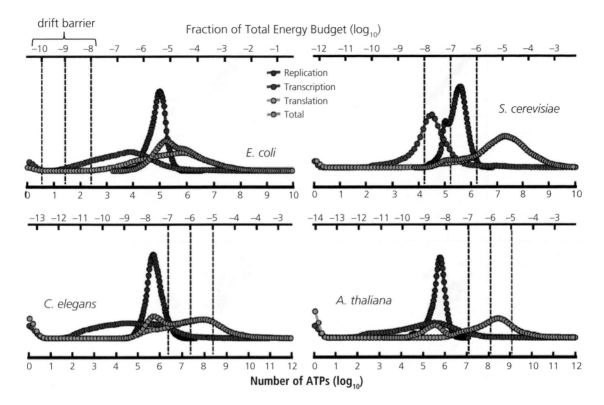

Figure 17.4 Frequency distributions of the total costs for all genes (and their three components, associated with replication, transcription, and translation) in four species for which the detailed transcriptomic and proteomic data necessary for full analyses are available. The data are presented as frequency histograms, so that the peaks demarcate the mode of the costs over the full set of genes. The lower axis (in $\log_{10}$ units) denotes the total (direct plus opportunity) costs, whereas the upper axis is obtained by dividing these numbers by the total cost of building a cell (from Chapter 8), yielding a measure of the selective cost of the maintenance and operation of a gene relative to the cell's total energy budget. The vertical dashed lines denote the drift barrier, with the middle line approximately demarcating $1/N_e$ for the species, and the two flanking lines simply providing order-of-magnitude margins for error. The fact that the relative costs of virtually all genes in *E. coli* are far to the right of the drift barrier implies that their baseline costs can easily be perceived by selection – if such genes do not pay their way by endowing the cell with benefits in excess of such costs, they would be rapidly purged by degenerative mutation and negative selection. Moving into eukaryotes, the effective population size declines with organism size, moving the drift barrier to the right, whereas the relative costs of genes move to the left owing to the predominant effect of an increase in the total cost of the larger cells (note the changes in scale). As a consequence, with increasing organism size, the baseline costs of genes tend to move to the left of the drift barrier, implying that gene-specific costs are too small, particularly at the DNA and RNA levels, to be perceived by selection. From Lynch and Marinov (2015).

gene-sized insertions in eukaryotes, particularly in multicellular species, are commonly effectively neutral from a bioenergetic perspective. Although costs at the RNA level are frequently greater than those at the DNA level, these are often still not large enough to overcome the power of random genetic drift in eukaryotic cells. This means that many non-functional DNAs that are inadvertently transcribed in eukaryotes (especially in multicellular species) still cannot be opposed by selection. On the other hand, with the cost at the protein level generally being much greater than that at the RNA level, segments of DNA that are translated can sometimes impose large enough energetic costs to be susceptible to selection, even in multicellular species.

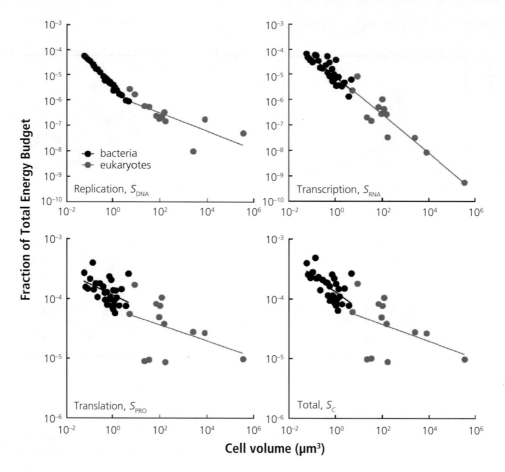

Figure 17.5 Estimated costs (relative to estimated total cellular energy budgets) for average genes in 44 species of prokaryotes and eukaryotes. For all components of cost, there is a clear negative scaling with increasing cell volume, both within and among groups, with no discontinuity in the scaling behavior between prokaryotes and eukaryotes. From Lynch and Marinov (2015); related analyses appear in Chiyomaru and Takemoto (2020).

Such observations are relevant to the idea that an enhanced ability to generate energy, made possible by the origin of the mitochondrion, was a prerequisite for the evolution of increased gene numbers, protein lengths, protein folds, protein–protein interactions, and regulatory elements in eukaryotic cells (Lane and Martin 2010). As already noted in Chapter 8, there is no dichotomous break in the size-scaling of the metabolic properties in prokaryotic versus eukaryotic cells, and here we see that increased cell size does not induce a condition in which gene addition becomes an increasing selective burden, but quite the contrary. In terms of the fractional contribution to a cell's energy budget,

which ultimately determines whether selection can oppose genome expansion, the cost of an average gene decreases with increasing cell size at the DNA, RNA, and protein levels.

Thus, population-genetic arguments based on both the mutational-hazard hypothesis (Foundations 17.2) and on known features of cellular energetics lead to the conclusion that passive increases in genome size naturally arise in organisms with increased cell sizes (which, by correlation, have reduced effective population sizes). This supports the view that variation in the power of random genetic drift has played a central role in the historical diversification of genome size and possibly cellular architecture across the Tree of Life (Chapter 24).

Finally, note that genes may have costs beyond those mentioned here. For example, there may be additional opportunity costs with respect to transcription and translation, as RNA polymerases, tRNAs, and ribosomes must be deployed in gene expression, reducing their availability to service other genes. High gratuitous expression of proteins can lead to significant misfolding problems that induce secondary biosynthesis costs associated with the up-regulation of chaperones (Geiler-Samerotte et al. 2011; Frumkin et al. 2017). However, experimental work suggests that the predominant cost of genes is indeed associated with the biosynthesis of the elemental building blocks, rather than with toxic problems associated with harmful misfolding and protein–protein misinteractions (Stoebel et al. 2008; Plata et al. 2010; Eguchi et al. 2018), with the quantitative effects being in reasonable accord with the numbers cited here (Tomala and Korona 2013; Adler et al. 2014).

The Cost of Lipids and Membranes

Eukaryotes distinguish themselves from prokaryotes by relying on membrane-bound organelles with diverse functions, including sequestration and gated access to the genomic material, vesicle transport of a multiplicity of cargoes, power production, and platforms for molecular assembly. As outlined in Chapter 15, numerous features of these complex membrane systems appear to have evolved by repeated rounds of gene duplication and divergence. Given that the rate of *de novo* gene duplication in prokaryotes is comparable to that in eukaryotes (Lynch 2007a), and that some prokaryotes have organelles and internal membranes, why then do almost all prokaryotes remain devoid of membranes?

One possibility is that the evolution of internal membranes is the null state, driven by mutational bias and basic biophysical forces (Ramadas and Thattai 2013; Mani and Thattai 2016), but that the construction and maintenance of such embellishments is energetically expensive enough that their incremental accrual is thwarted by the efficiency

of natural selection in prokaryotes. This could then lead to the emergence of internal cellular complexity as a simple consequence of nearly neutral, drift-like processes in larger eukaryotic cells with reduced N_e, rather than by direct promotion by positive selection (Lynch 2007a).

Because eukaryotic cells are typically larger than those of prokaryotes, often substantially so, there is an increase in the absolute investment of lipids based on the cell membrane alone. However, given a constant shape, because the surface area of a cell increases with the square of cell length, whereas the volume increases with the cube of length, the relative investment in the external membrane declines with cell volume (Table 8.1). Nonetheless, when the investment in internal membranes in eukaryotes is added in, a larger fraction of cellular biomass is allocated to lipids than in prokaryotes. The mean fractional contribution of lipids to total dry weight is ~ 0.06 for bacteria, ~ 0.08 for yeast species, and ~ 0.15 for unicellular photosynthetic algae (Chapter 7), and the fractional costs at the bioenergetic level are even greater than this.

To understand why this is so, we require information on the numbers of lipid molecules required for membrane production over the life of the cell. Owing to the absence of information on the rate of lipid-molecule turnover, it will be necessary here to rely on N_l, the number of molecules per newborn cell, as an estimate of the minimum lifetime requirement for lipids. However, as there is no evidence that membrane lipids are rapidly degraded, the bias of the resultant estimates is not expected to be large. N_l can be determined by dividing cellular membrane areas by the head-group areas of membrane lipid molecules, most of which are within 10% of an average $a_l = 0.65$ nm^2 (Nagle and Tristram-Nagle 2000; Petrache et al. 2000; Kučerka et al. 2011). The thickness of the bilayer (h) is approximately twice the radius of the head-group area, which $\simeq 0.5$ nm in all cases, plus the total length of the internal hydrophobic tail domains (Lewis and Engelman 1983; Mitra et al. 2004), generally $\simeq 3.0$ nm, and so sums to $h \simeq 4.0$ nm. There are slight increases in bilayer thickness with the length of the fatty-acid chain deployed (Rand and Parsegian 1989; Wieslander et al. 1995; Rawicz et al. 2000), as each single and double carbon–carbon bond adds ~ 0.15 and 0.13 nm, respectively.

To gain some appreciation for the number of lipid molecules per cell, consider a spherical cell with radius r, with a surface area for the outer side of the bilayer of $4\pi r^2$, and $4\pi (r - h)^2$ for the inner layer. The summed area is then $4\pi[r^2 + (r - h)^2]$, which $\simeq 8\pi r^2$ for $h \ll r$. Dividing by a_l gives the total number of lipid molecules in the bilayer. With $a_l = 0.65$ nm^2 and $h = 4.0$ nm, for a bacterial-sized cell with radius $r = 1000$ nm (1 μm), there are then $\sim 4 \times 10^7$ total lipid molecules in the cell membrane. This increases to $\sim 4 \times 10^9$ for a sphere with radius 10 μm and $\sim 4 \times 10^{11}$ for a 100 μm sphere. For Gram-negative bacteria (which include *Escherichia*, *Pseudomonas*, *Helicobacter*, and *Salmonella*), there are two cell membranes, which roughly doubles these numbers.

These values might need to be modified somewhat to account for the typical 20–50% occupancy of membranes by proteins (Lindén et al. 2012). If, for example, 50% of the membrane surface is occupied by proteins, the preceding values for lipids would need to be diminished by 50%, although this would be offset with costs associated with membrane proteins. Regardless, it remains clear that the number of lipids in the cell membrane alone is almost always greater than the number of nucleotides in a cell's genome, often by orders of magnitude.

Costs of individual molecules

We now consider the total (direct plus opportunity) costs of biosynthesizing individual lipid molecules, the details of which are presented in Foundations 17.5. Most cellular membranes are predominantly comprised of glycerophospholipids (Table 17.1), which despite containing a variety of head groups (e.g., glycerol, choline, serine, and inositol), all have biosynthetic costs per molecule (in units of ATP hydrolyses) of

$$c_L \simeq 283 + [30 \cdot (n_L - 16)] + (5 \cdot n_U), \qquad (17.5a)$$

$$c_L \simeq 291 + [32 \cdot (n_L - 16)] + (5 \cdot n_U), \qquad (17.5b)$$

in bacteria and eukaryotes respectively, where n_L is the mean fatty-acid chain length (number of backbone carbons), and n_U is the mean number of

unsaturated carbons per fatty-acid chain. Although variants on glycerophospholipids are utilized in a variety of species (Guschina and Harwood 2006; Geiger et al. 2010), these are structurally similar enough that the preceding expressions should still provide good first-order approximations. The direct costs, which ignore the opportunity loss of ATP-generating potential from the diversion of metabolic precursors, are

$$c_L' \simeq 88 + [10 \cdot (n_L - 16)] + (5 \cdot n_U), \qquad (17.6a)$$

$$c_L' \simeq 104 + [12 \cdot (n_L - 16)] + (5 \cdot n_U), \qquad (17.6b)$$

in bacteria and eukaryotes, respectively.

For most lipids in biological membranes, $14 \leq n_L \leq 22$ and $0 \leq n_U \leq 6$, so the total cost per lipid molecule is generally in the range of $c_L \simeq 220$ to 500 ATP, although the average over the pooled population of lipids deployed in species-specific membranes is much narrower. Cardiolipin, which rarely comprises more than 20% of membrane lipids, is exceptional because it derives from a fusion of two phosphatidylglycerols, and has total and direct costs of ~ 570 and 190 ATPs/molecule. The key point is that, on an individual molecule basis, lipids are an order of magnitude more expensive than the other two major monomeric building blocks of cells (nucleotides and amino acids).

Application of the preceding expressions to known membrane compositions indicates that the biosynthetic costs of eukaryotic lipids are somewhat higher than those in bacteria, e.g., on average 7% per lipid molecule in the plasma membrane and for whole eukaryotic cells, although the average cost of mitochondrial lipids is especially high (Table 17.1). These elevated expenses in eukaryotes are a consequence of two factors: 1) the added cost of mitochondrial export of oxaloacetate to generate acetyl-CoA necessary for lipid biosynthesis in the cytoplasm; and 2) the tendency for eukaryotic lipids to have longer chains containing more desaturated carbons.

Total cellular investment

We are now in position to estimate the total cell bioenergetic cost associated with membrane lipids. Recalling the surface occupancy of individual lipid molecules and the bilayer nature of membranes, the biosynthetic cost of a membrane with surface area A (in units of μm^2, and ignoring the unknown contribution from lipid turnover) is

$$C_L \simeq 2A \cdot \bar{c}_L/(0.65 \times 10^{-6}), \qquad (17.7)$$

where $\bar{c}_L$ is the average total cost of a lipid molecule (e.g., as given in Table 17.1). Dividing C_L by the total cost of building a progeny cell (Chapter 8) yields the proportion of a cell's total growth budget allocated to a membrane.

There are only a few cell types for which the internal anatomy has been scrutinized well enough to achieve an accounting for the full set of membrane types. However, the limited data uniformly suggest allocations of ~20–60% of a cell's growth budget to membrane lipids (Table 17.2). As expected based on the surface area:volume relationship, the plasma membrane constitutes a diminishing cost with increasing cell size, from >16% in bacterial-sized cells with volumes <2 μm^3 to <7% in moderately sized eukaryotic cells. The majority of the total cost of membrane production in eukaryotic cells is associated with internal membranes. For example, 90% of the membrane budget of the green alga *Dunaliella* is associated with organelles.

Enough information is available on the total investment in mitochondrial lipid membranes that a general statement can be made for this particular organelle. Over the eukaryotic domain, the total surface area of mitochondria (inner plus outer membranes, summed over all mitochondria, in μm^2) scales with cell volume (V, in units of μm^3) as $A \simeq 3V$ (Figure 17.6a). Applying this to Equation 17.7, with

the average total cost of mitochondrial lipids from Table 17.1, and letting the total growth requirements of a cell be $(27 \times 10^9)V$ from Chapter 8, implies a relative cost of mitochondrial membrane lipids of 0.13 (not discounting for membrane proteins), essentially independent of eukaryotic cell size.

These results indicate that the construction of mitochondrial membranes, critical to ATP synthase operation in eukaryotes, amounts to a ~13% drain on the cellular energy budget beyond what would be necessary had ATP synthase remained in operation on the plasma membrane (with mitochondria being absent). These first-order approximations apply to rapidly growing cells, for which the contributions of cell maintenance and lipid-molecule turnover to the total cellular energy budget will be minor. For slowly growing cells, the costs will be higher or lower depending on whether the cost of mitochondrial-membrane maintenance is above or below that for total cellular maintenance. However, the central point remains – the costs of mitochondrial membranes represent a substantial baseline price, not incurred by prokaryotes, resulting from the relocation of bioenergetics from cell envelopes to the interior of eukaryotic cells.

Although the data are less extensive, this approach can be extended to show that different cost scalings exist for other types of internal membranes. For example, across the Tree of Life, the outer surface area of the nuclear envelope $\simeq 2.7V^{0.5}$ with units as given above (Figure 17.6b). Assuming the average cost per lipid molecule in these organelles to be about the same as for the entire cell

Table 17.1 Bioenergetic costs for the synthesis of lipid molecules, in units of ATP hydrolyses. Data are provided for species with lipid composition measurements of the parameters needed to solve Equations 17.5a–17.6b, along with additional information on the contribution from cardiolipin. PL denotes glycerophospholipid, and C cardiolipin, with the cost for the remaining small fraction per molecule being taken to be the average of the preceding two. Total cost denotes the opportunity plus direct cost per molecule incorporated into the membrane. Mean costs/molecule are obtained by weighting the PL and cardiolipin costs by their fractional contributions. Standard deviations among species are given in parentheses. Modified from Lynch and Marinov (2017).

Source	PL Cost		Composition		Mean Cost	
	Total	Direct	PL	C	Total	Direct
Bacteria, whole cell	299 (22)	94 (8)	0.89 (0.09)	0.09 (0.06)	326 (14)	99 (6)
Eukaryotes whole cell	326 (21)	124 (9)	0.95 (0.03)	0.04 (0.03)	346 (19)	128 (8)
Eukaryotes plasma membrane	338 (16)	125 (7)	0.95 (0.05)	0.03 (0.03)	348 (19)	124 (7)
Eukaryotes mitochondrion	345 (42)	129 (18)	0.85 (0.08)	0.11 (0.05)	376 (37)	134 (17)

Table 17.2 Contributions of membranes to total cellular growth costs (in units of ATP equivalents). Cell volumes (Vol) and total membrane surface areas (SA) are in units of μm^3 and μm^2, respectively. Fractional contributions to the total cell growth requirements are given for the plasma membrane (Pm), mitochondrial membranes (inner + outer, Mt), nuclear envelope (Nu), endoplasmic reticulum and Golgi (ER/G), vesicles and vacuoles (V), and Total. The fraction of the total cell growth budget allocated to membranes is obtained by applying Equation 17.7, using the species-specific lipid biosynthesis costs (Table 17.1) where possible (and otherwise applying the averages for eukaryotic species), and normalizing by the allometric equation for ATP growth requirements given by Equation 8.2b. The SA given for *E. coli* is twice that of the cell, as this species has two membranes; the results for the two algae, *O. tauri* and *D. salina*, do not include the investment in plastid membranes. From Lynch and Marinov (2017).

			Fractional contributions to total cell growth					
Organism	Vol	SA	Pm	Mt	Nu	ER/G	V	Total
Bacteria:								
Staphylococcus aureus	0.29	2.1	0.240					0.240
Escherichia coli	0.98	8.6	0.337					0.337
Bacillus subtilis	1.41	6.0	0.161					0.161
Eukaryotes:								
Ostreococcus tauri	0.9	14	0.364	0.030	0.149	0.033	0.036	0.612
Saccharomyces cerevisiae	44	211	0.066	0.061	0.034	0.022	0.023	0.206
Dunaliella salina	591	2326	0.028	0.035	0.014	0.065	0.065	0.207

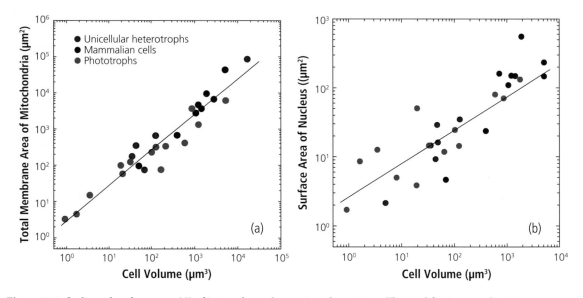

Figure 17.6 Scaling of surface areas (A) of internal membranes in eukaryotes. **a:** The total for inner and outer mitochondrial membranes scales nearly isometrically as $A = 3.0V^{0.98}$ ($r^2 = 0.92$), where A and V have units of μm^2 and μm^3, with little variation among functionally different types of organisms (phototrophs include land plants and various phylogenetic groups of phytoplankton). **b:** Total surface area of the nucleus scales across phylogenetic groups as $A = 2.7V^{0.48}$ ($r^2 = 0.68$). Data from Lynch and Marinov (2017), Uwizeye et al. (2021), and a few other references.

(~ 346 ATPs, from Table 17.1), and recalling that the nuclear envelope has a double membrane, the total cost of the nuclear envelope (scaled to the total cell budget) $\simeq 0.1V^{-0.5}$. Thus, the fractional cost of the nuclear envelope (relative to a cell's total energy budget) declines with increasing cell volume, from $\sim 3\%$ for a small eukaryote with $V = 10$ μm^3 to $\sim 0.3\%$ when $V = 1000$ μm^3.

Unfortunately, there are insufficient data to estimate scaling relationships for other organelle surface areas. However, multiple studies make clear that in all but the smallest eukaryotic cells, the summed contributions from membranes associated with the endoplasmic reticulum, Golgi, and assorted vesicles and vacuoles exceed those associated with the nuclear envelope, often by more than tenfold (Table 17.2). Thus, adding in the mitochondrial investment, it appears that 10–20% of a eukaryotic cell's energy budget is typically invested in internal membranes.

Although there are a number of other costs associated with membranes and their processing, as in the case of nucleic acids and proteins, the major costs appear to be associated with the biosynthesis of the basic building blocks noted earlier. Consider, for example, the cost of molding membranes into specific shapes. The problem is most simply evaluated for spherical vesicles, for which the bending energy is $\sim 400\ k_B T$ per vesicle. This energetic requirement is independent of vesicle size because, although there is more surface area to bend in larger vesicles, the curvature is reduced, resulting in cancelling of the two effects (Phillips et al. 2012). Knowing the rate of membrane flux for a cell then allows a rough estimate of the total bending energy required per unit time. For example, the entire cell membranes of mammalian fibroblast and macrophage cells are internalized by pinocytosis in about 0.5–2.0 hours (Steinman et al. 1976). The same is true for *Dictyostelium* (slime mold) cells (Thilo and Vogel 1980). Knowing the surface area of the cell and the average surface area of a vesicle, one can then estimate that about 1000 vesicles must be produced per minute. Assuming that the bending energy is acquired via ATP hydrolysing reactions in the production of protein-coating cages (Chapter 15), and that an energy of $\sim 20\ k_B T$ is associated with each ATP hydrolysis, the total energy demand for bending $\simeq 400 \times 1000/20 = 20{,}000$ ATP/minute. The cost of membrane fusion is smaller, being equivalent to $\sim 20\ k_B T$, or about one ATP hydrolysis, per fusion event (François-Martin et al. 2017). For a *Dictyostelium*-sized cell with $V = 600\ \mu m^3$, the total bending energy associated with endocytosis is quite small, $\simeq 0.001\%$ of the cell's total maintenance requirements.

Summary

- Understanding the baseline energetic costs of constructing and maintaining cellular features is essential to evaluating the relative investments of total cellular budgets in alternative functions, to determining the benefits that must be accrued if a trait is to pay for itself, and to learning whether certain aspects of a cell's biology are liable to accumulate by biased mutation pressure despite being mildly deleterious.

- The assessment of cell budgets in units of energy, rather than elemental constitution, is desirable because the latter does not contribute uniformly to structural costs and is often of minor relevance to maintenance and operational activities. In addition, most organic food substrates are more limiting with respect to energy than carbon content. As the universal currency of bioenergetics across the Tree of Life, the number of ATP hydrolyses provides a natural measure of energy expenditure.

- The total energetic cost of a cellular feature can be viewed as the sum of three components: the direct costs invested in construction and in operation/maintenance; and the opportunity costs resulting from the diversion of resources (e.g., carbon skeletons) to the trait of interest and away from other cellular functions. The fitness cost of a cellular feature, exclusive of downstream phenotypic advantages, is obtained by scaling the absolute cost of construction and maintenance by the total cellular energy budget.

- Quantification of the costs of constructing cellular constituents requires information on the biosynthetic costs of basic building blocks such as nucleotides (used in RNAs and DNA), amino acids (proteins), and lipids (membranes). Although individual amino acids are, on average, less energetically expensive than nucleotides, there is a sixfold range of variation among the 20 amino-acid types. For highly expressed protein-coding genes in species with large effective population sizes (e.g., bacteria), such cost differences can be significant enough for selection to discriminate among alternative amino-acid types.

- The energetic costs of a gene (or genomic segment) can be partitioned into three levels of investment: chromosomal, transcriptional, and translational. The half-lives of transcripts are typically much shorter than the lifespan of a cell, so these mRNAs must be continually replaced to keep the cell at steady state. As a consequence of such recycling, the overall cost of transcription of a gene is typically slightly more than the cost at the chromosomal level. Owing to the much larger number of proteins per cell, the average cost of translation per gene is $\sim 100\times$ that of transcripts.

- Although the absolute costs of genes increase with increasing organism size, the proportional cost relative to entire cell energy budgets declines. As a consequence, because effective population sizes decline with increasing organism size, larger organisms experience increased vulnerability to the passive accumulation of extraneous genomic material. Whereas a genomic insertion of just a few bases can be purged by natural selection in prokaryotic species with high N_e, in larger eukaryotic species with relatively small N_e, insertions as large as several kilobases can be invisible to the eyes of selection. This variation in the efficiency of selection helps explain the streamlined genomes of microbes versus the bloated genomes of multicellular eukaryotes.

- Membrane biosynthesis constitutes a substantial fraction of the total energy budgets of cells, generally > 20%. The costs of individual lipid molecules are 6 to 8× that of individual nucleotides, and depending on the size of the cell, there will generally be on the order of 10^7 to 10^{12} lipid molecules in the cell membrane alone, typically well in excess of the number of nucleotides per genome.

- The hallmark of eukaryotic cells, membrane-bound organelles, imposes a substantial increase in the energetic cost per cell, not incurred by prokaryotes. Independent of cell volume, $\sim 13\%$ of eukaryotic cell budgets are associated mitochondrial membrane lipids.

Foundations 17.1 The relationship between bioenergetic costs and the strength of selection

Understanding the baseline fitness consequences of the total energetic investment in a trait requires a quantitative definition of the effects on the cell's reproductive rate (exclusive of any downstream changes in fitness owing to phenotypic effects). The selective cost associated with investment is defined as

$$s = r - r', \qquad (17.1.1a)$$

where $r = \ln(2)/\tau$ denotes an exponential rate of growth, with τ being the mean cell-division time (or population doubling time; Chapter 9), and r' and r denoting the growth rates in the presence and absence of the attribute under consideration. Denoting the increase in cell-division time as $\Delta_\tau = (\tau' - \tau)$,

$$s = \ln(2)\left(\frac{1}{\tau} - \frac{1}{\tau'}\right) \simeq \frac{\ln(2) \cdot \Delta_\tau}{\tau}. \qquad (17.1.1b)$$

Further assuming that the cost of the trait is much less than the lifetime energetic expenditure of a cell, $c_T \ll C_T$, so that the increment in cell-division time scales linearly with the proportional increase in investment,

$$\tau' \simeq \tau\left(1 + \frac{c_T}{C_T}\right). \qquad (17.1.2)$$

Noting that $\Delta_\tau \simeq \tau c_T / C_T$ then leads to

$$s \simeq \frac{\ln(2) \cdot c_T}{C_T}. \qquad (17.1.3)$$

This shows that the intrinsic selective disadvantage associated with the bioenergetic cost of a trait scales directly with the proportional increase in the total energy demand per cell cycle (Lynch and Marinov 2015; Ilker and Hinczewski 2019).

Foundations 17.2 The biosynthetic costs of nucleotides and amino acids

Critical to understanding the energetic costs of nucleic acids and proteins are the biosynthetic costs of the basic building blocks from which these polymers are built. The numerous published accounts for such costs (Atkinson 1970; McDermitt and Loomis 1981; Williams 1987; Craig and Weber 1998; Akashi and Gojobori 2002; van Milgen 2002; Wagner 2005; Barton et al. 2010; Arnold et al. 2015) are mostly presented without detail or reference to prior work, and none are in entire agreement. Thus, given the morass of technical literature on biochemical pathways underlying such computations, it is desirable to have steps involved in the cost computations laid out in enough detail that the reader can readily make modifications should the biochemistry interpretations be deemed problematic.

As a unit of energetic currency, we will rely on ATP usage, specifically, the number of phosphorus atoms released via ATP hydrolyses, the primary source of energy in most cellular reactions. There are, however, two complications to deal with. First, instead of ATP, CTP and GTP are utilized in a few cellular reaction steps (e.g., lipid biosynthesis), and these will be treated as equivalent to ATP. Second, electron transfers resulting from conversions of coenzyme NADH to NAD^+, NADPH to $NADP^+$, and $FADH_2$ to FAD drive the delivery of hydrogen ions (H^+) that contribute to the proton-motive force used to produce ATP and are also involved in direct transactions. As there is a precise recipe for converting hydrogen ions flowing through ATP synthase into ATP production, coenzyme use can be converted to ATP equivalents, as now described.

ATP / coenzyme equivalents

Recall from Chapter 2 that ATP is produced by ATP synthase complexes via the loading of protons (H^+) into the subunits of the rotating c ring in the F_O complex, which sits in mitochondrial membranes of eukaryotes and in the plasma membranes of prokaryotes. Rotation of the ring requires that each subunit be loaded with a proton, and each rotation universally leads to the production of three ATPs. Thus, if the c ring contains $n = 12$ subunits, $12/3 = 4$ protons are required to produce each ATP. In eukaryotes, each NADH drives the pumping of ten protons via the ETC, while $FADH_2$ has a lower energy state and delivers only six. This is the basis for the common textbook assertion that the ATP value of a NADH is on the order of 2.5 to 3.0, as $10/4 = 2.5$. (Here, we assume negligible proton leakage, although if this is known, it could be readily factored in). However, the number of c subunits actually

varies from $n = 8$ to 15 (e.g., $n = 10$ for yeast and *E. coli*, and $n = 8$ for mammals, with most species being near the lower end; Chapter 2).

For eukaryotes, there is an additional small complication in that mitochondrially produced ATP must be exported to the cytoplasm and ADP imported back. This requires the use of one extra H^+ per ATP exported, so the H^+ to ATP ratio in eukaryotes is $(n + 3)/3$, and the ATP/NADH ratio becomes $10 \cdot 3/(n + 3) = 30/(n + 3)$, i.e., 2.7, 2.3, and 2.0, for $n = 8$, 10, and 12, respectively. For prokaryotes, ATP/ADP transport is not required, and ETCs are shorter, with eight rather than ten, protons released by an NADH-initiated chain reaction (Nicholls and Ferguson 2013), so the ATP/NADH ratio is on the order of $8 \cdot 3/n = 24/n$. In this case, $n = 8$, 10, and 12 lead to ratios of 3.0, 2.4, and 2.0, respectively, similar to those for eukaryotes.

Thus, under the assumption that $n = 8$ to 12 or so in most organisms, and acknowledging that the exact value of n (determination of which requires structural work) is unknown for most species, an ATP/NADH conversion factor of 2.5 appears to be a well-justified approximation for both eukaryotes and prokaryotes. To keep things simple, the following analyses will adhere to this value. Given that $FADH_2$ generates only 6 H^+, then $(6/10) \times 2.5 = 1.5$, which is commonly stated for the ATP/$FADH_2$ equivalence, seems also to be justified. Should the data for a particular taxon warrant modifications of these ratios, the following calculations are readily modifiable. An excellent summary of the issues underlying the confusing use of different conversion ratios in the literature is given by Silverstein (2005).

Costs of precursor molecules

All cellular building blocks are constructed out of carbon skeletons that are ultimately derived from an organic carbon source, here assumed to be glucose. More generally, however, organic molecules are sources of both carbon and energy. As a carbon resource progressively moves down energy-generating metabolic pathways, when the modified intermediate products are diverted into side pathways for biosynthesis, the precursor molecule is no longer available for downstream energy production, the loss of which is viewed as an opportunity cost. The following paragraphs provide explanations for the quantitative derivation of the opportunity costs associated with eleven precursor molecules necessary for amino-acid and nucleotide biosynthesis.

All of these precursors reside in key positions in the TCA cycle, glycolytic pathway, or pentose-phosphate shunt. The key steps and products generated are outlined in Figure 17.7, omitting numerous non-energy-generating steps that are covered in detail in all biochemistry textbooks. Using the shown conversion factors, these are then transformed into ATP equivalents in the following table.

We adopt the strategy of Atkinson (1970), which starts with the assumption that, under aerobic metabolism, after descending through glycolysis and the citric-acid (also known as TCA or Krebs) cycle (Figure 17.7), a single glucose molecule would yield a net 4 ATP, 10 NADH, and 2 FADH$_2$ molecules, for a total value of $4 + (10 \times 2.5) + (2 \times 1.5) = 32$ ATP equivalents. In the first steps of glycolysis, two phosphates are added, yielding the loss of two ATPs, so the bi-phosphorylated product (fructose-1, 6-bisphosphate, not shown) has actually gained energy and has a value of 34 ATPs (as the initial subtraction of two ATPs is no longer necessary). The first precursor molecules of interest are the 3-carbon metabolites dhap and g3p,

which are derived by splitting the prior 6-carbon molecule, and each has a value of 34/2. Thereafter, the descending metabolites have decreasing values, as ATPs and/or NADHs have been produced at higher steps in the chain. An exception arises upon arrival at the citric-acid cycle, as acetyl-CoA joins oaa, yielding a summed value of $10 + 10 = 20$ ATPs. The early production of 1 NADH in the cycle results in αkg having a reduced value of $20 - 2.5 = 17.5$ ATP equivalents, and further energy extraction reduces oaa to 10.

The remaining precursor molecules of interest are associated with the pentose-phosphate shunt, which diverts glucose derivatives from the glycolytic pathway to produce nucleotides (below). The first component in this pathway, ribose-5-phosphate (penP), is derived from g6p at the expense of 2 NADHs, and therefore has an energetic value of $33 - (2 \cdot 2.5) = 28$ ATP equivalents. Production of eryP is energy neutral, whereas pRpp arises after an ATP → AMP conversion, and therefore has a value elevated by 2 ATP equivalents.

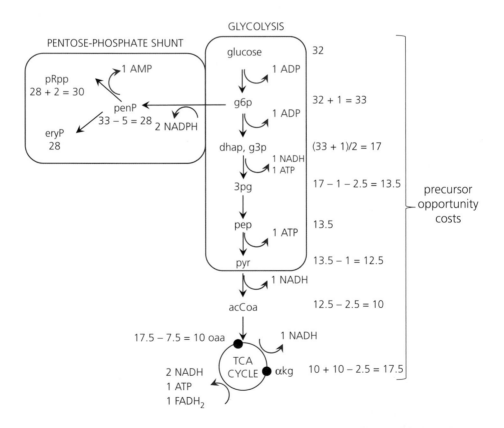

Figure 17.7 Summary of the three major cellular pathways for the production of the three major carriers of energy and reducing power (ATP, NADH, and FADH$_2$), and the routes to the primary precursor molecules leading to the biosynthetic pathways for amino acids and nucleotides. Computations are given in units of ATP equivalents. Note that below g6p in glycolysis, there are actually two copies of each listed derivative, as the 6-carbon glucose has been split into two 3-carbon compounds, and hence, the division by 2 at the dhap/g3p position. For simplicity, numerous intermediate steps extraneous to the computations are left out; details on these can be found in all basic biochemistry text books.

Precursor	Abbrev.	ATP	NADH	FADH$_2$	Total
Ribose 5-phosphate	penP	5	8	2	28.0
5-Phosphoribosyl pyrophosphate	pRpp	7	8	2	30.0
Erythrose 4-phosphate	eryP	5	8	2	28.0
Dihydroxyacetone phosphate	dhap	3	5	1	17.0
Glyceraldehyde-3-phosphate	g3p	3	5	1	17.0
3-Phosphoglycerate	3pg	2	4	1	13.5
Phosphoenolpyruvate	pep	2	4	1	13.5
Pyruvate	pyr	1	4	1	12.5
Acetyl-CoA	acCoA	1	3	1	10.0
Oxaloacetate	oaa	1	3	1	10.0
α-ketoglutarate	αkg	2	5	2	17.5

The formula Total ATP Cost = $(1 \times$ ATP$)$ + $(2.5 \times$ NADH$)$ + $(1.5 \times$ FADH$_2)$ yields the total costs for each precursor, summarized in the table above. Modifications can become necessary with sources of carbon other than glucose, e.g., see Akashi and Gojobori (2002) for acetate and malate, although some of their computations stray from the above scheme.

Amino-acid biosynthesis
As an entrée into the logic of estimating the cost of a basic building block, we now turn to the amino acids used in protein production. All of these are derived from the

above-noted precursor molecules, and although variants exist, the downstream biosynthetic paths are conserved in most species (Chapter 19), and the energetics associated with variant paths will be similar. Acknowledging that future researchers may wish to make modifications in some cases, we retain a focus on the situation in which entry into metabolism is based on glucose.

The basic calculations start with the costs of the precursor molecules used for carbon skeletons, which then descend down various modifying steps. Many of these steps require an investment loss in energy of at least the same order of magnitude as an ATP, and so must enter into the final bookkeeping. In a few cases, a step is energy producing. Detailed information on these matters can be found in most biochemistry and/or bioenergetics text books, and a simple listing of relevant reactions and their values in terms of ATP hydrolyses is given here without further explanation.

Reaction	ATP
1) NADH $\rightarrow$ NAD$^+$	2.5
1) NADPH $\rightarrow$ NADP$^+$	2.5
2) Trans-aminotransferase reaction	3.5
3) Glutamine $\rightarrow$ Glutamate	1.0
4) Aspartate $\rightarrow$ Fumarate	1.0
5) Acetyl-CoA $\rightarrow$ CoA-SH	2.5
6) ATP $\rightarrow$ AICAR	3.0
7) Succinyl-CoA $\rightarrow$ CoA	1.0
8) H$_4$ folate $\rightarrow$ N^5,N^{10}-methylene H$_4$ folate	−2.5
9) N^5-methyl H$_4$ folate $\rightarrow$ H$_4$ folate	5.0
10) N^{10}-formyl H$_4$ folate $\rightarrow$ H$_4$ folate	1.0

Amino acid	Precursor	ATP	Reactions	Opportunity	Direct	Total
Alanine	pyr		1	12.5	2.5	15.0
Arginine	αkg	1	1(2), 2, 4, 5	17.5	13.0	30.5
Asparagine	oaa	2	2, 3	10.0	6.5	16.5
Aspartate	oaa		2	10.0	3.5	13.5
Cysteine	3pg		1(−1), 2, 5	13.5	3.5	17.0
Glutamate	αkg		1	17.5	2.5	20.0
Glutamine	αkg	1	1	17.5	3.5	21.0
Glycine	3pg		1(−1), 2, 8	13.5	−1.5	12.0
Histidine	pRpp		1(−2), 2, 3, 6	30.0	2.5	32.5
Isoleucine	pyr, oaa	2	1(3), 2(2)	22.5	16.5	39.0
Leucine	2 pyr, acCoA	2	2(2)	35.0	9.0	44.0
Lysine	pyr, oaa	1	1(2), 2(2)	22.5	13.0	35.5
Methionine	oaa, cysteine, -pyr	1	1(2), 2	9.5	15.5	25.0
Phenylalanine	2 pep, eryP	1	2	55.0	4.5	59.5
Proline	αkg	1	1(3)	17.5	8.5	26.0
Serine	3pg		1(−1), 2	13.5	1.0	14.5
Threonine	oaa	2	1(2), 2	10.0	10.5	20.5
Tryptophan	2 pep, eryP, pRpp, -pyr, serine, -g3p	1	3	69.0	2.0	71.0
Tyrosine	2 pep, eryP	1	1(−1), 2	55.0	2.0	57.0
Valine	2 pyr		1, 2	25.0	6.0	31.0

The preceding tables provide a list of the precursors and the reaction steps involved in the synthesis of each amino acid. The total precursor costs, obtained by weighting each component by its total ATP content and summing, constitute the opportunity costs for each amino acid. The direct costs are estimated by weighting each step in the downstream biosynthetic pathway with its associated energy and summing. Numbers in parentheses under Reactions denote the net number of reactions of the indexed type; if not noted, this number is 1, and a negative number implies an energy gain. All costs are in units of numbers of ATP hydrolysis equivalents. For some amino acids, there can be shifts to alternative biosynthetic pathways under certain environmental conditions, although it remains unclear whether the cost estimates outlined are greatly altered (Du et al. 2018).

Nucleotide biosynthesis

The general strategy for estimating the costs of amino acids is readily extended to nucleotide biosynthesis, again facilitated by the fact that the basic biosynthetic pathways are nearly universal across the Tree of Life. All nucleotide synthesis starts with the precursor pRpp (phosphoribosyl pyrophosphate) as a carbon skeleton. This 5-carbon ring is an early derivative of glucose-6-phosphate, which, if not diverted, would eventually descend down the citric-acid cycle (Figure 17.7). Amino-acid molecules also contribute carbon skeletons, and here it is assumed that they must first be synthesized. As these could have been utilized for other purposes (or not produced at all, thereby saving energy), they (along with pRpp) will be viewed as opportunity costs. The direct costs include additional investments of ATP and other cofactors in the downstream steps of molding and modifying the precursor carbon skeletons en route to producing nucleotides.

Consider first the production of purines (adenine and guanine). From the amino-acid table shown above, it can be seen that the diversion of pRpp deprives the cell of 30 ATP hydrolysis equivalents. One glycine molecule is also consumed in the early stages of biosynthesis, the production of which results in an opportunity loss of 12 ATPs. This leads to a total opportunity loss for purines of 30 + 12 = 42 ATPs per utilized pRpp molecule (Figure 17.8).

As pRpp and glycine merge and descend down 10 steps in the production of the final purine precursor (IMP), direct costs consume the equivalent of ∼9 ATPs; these derive from: 4 ATP → ADP; 2 glutamine → glutamate; 2 N^{10}-formyl H_4 folate → H_4 folate; and 1 aspartate → fumarate reactions. Conversion of IMP to AMP consumes a GTP and an aspartate to fumarate exchange, for an additional direct cost of 2 ATPs in AMP production. Conversion of IMP to GMP involves 1 ATP → AMP reaction, production of 1 NADH, and 1 glutamine → glutamate

exchange for an additional direct cost of 0.5 ATPs in GMP production.

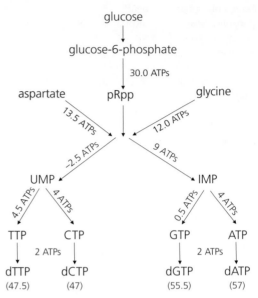

Figure 17.8 Key steps in the biosynthesis of ribonucleotides and deoxyribonucleotides, and summations of the energy demands (in units of ATP equivalents per molecule produced). Red and black denote opportunity- and direct-cost components.

Pyrimidine (C and T) production also starts with pRpp, but consumes an aspartate (rather than glycine) molecule with an opportunity cost of 13.5 ATPs, yielding a total opportunity cost of 43.5 ATPs. The route to the final pyrimidine precursor UMP results in the production of 1 NADH, for a direct cost of −2.5 ATP equivalents. Conversion of UMP to UTP consumes 2 ATPs, and then conversion of UTP to CTP consumes an additional ATP, and involves a glutamine → glutamate exchange, for an additional direct cost of 4 ATP equivalents. Conversion of UMP to TMP involves a N^5,N^{10}-methylene H_4 folate → H_4 folate exchange, which is equivalent to a direct cost of 2.5 ATP equivalents, and further conversion to TTP requires a final expenditure of 2 ATPs.

The costs of the four ribonucleotides are summed up in the following table.

Nucleotide	Opportunity	Direct	Total
Adenine (ATP)	42.0	13.0	55.0
Guanine (GTP)	42.0	11.5	53.5
Cytosine (CTP)	43.5	1.5	45.0
Uracil (UTP)	43.5	-0.5	43.0
Thymine (TTP)	43.5	2.0	45.5

Foundations 17.3 The mutational hazard of excess DNA

All genes have a deleterious-mutational target size equal to the number of nucleotide sites for which alteration of the nucleotide identity has the potential to influence fitness (Lynch 2007a). This number will include most amino-acid replacement sites in the coding region, and to a much lesser (but not always insignificant) extent silent sites (for which the nucleotide identity has no impact on the encoded amino acid, but may influence aspects of mRNA folding and/or the rate of translation). Here, we consider the impact of larger segments of gene-associated non-coding DNA that are relevant to successful gene expression. These include introns, which commonly populate eukaryotic genes, transcription-factor binding sites, and numerous other regulatory sequences in upstream (5') untranslated regions (UTRs) of transcripts.

What is the contribution of such elements for the vulnerability of a gene to mutational inactivation? Introns are transcribed into pre-mRNAs, and must be properly spliced out to yield a productive mature mRNA, with accurate recognition by the splicing machinery depending on diffuse motifs at both intron ends comprising a total ~ 20 to 30 nucleotides. Transcription-factor binding sites typically consist of motifs 8–16 bases in length, and other elements in UTRs are usually of this same size. Even entirely nonfunctional DNA can magnify a gene's vulnerability to deleterious mutations, as such material can acquire detrimental gain-of-function mutations. To simplify discussion, we will consider an embellishment to a gene that magnifies the mutational target size by ten nucleotides.

The mutation rate per nucleotide site per generation ranges from $\sim 10^{-10}$ in prokaryotes and many unicellular eukaryotes to 10^{-8} in vertebrates (Chapter 4), so the addition of a gene-structural embellishment that introduces 10-bp of sequence critical to gene function is equivalent to increasing the mutation rate to null alleles to 10^{-9} to 10^{-7}. This type of mutation-rate inflation for an embellished gene operates in a manner effectively identical to selection, as it is a measure of the excess rate of removal of such alleles from the population by conversion to null alleles.

How do mutational hazards compare with the energetic costs of nucleotides from a gene-level perspective? Returning to the results in the text, a 10-bp segment of DNA imposes an energetic penalty (relative to the total cost of building a cell) of $\sim 10^{-8}$ in a typical bacterium, $\sim 10^{-9}$ in a unicellular eukaryote, and $\sim 10^{-10}$ in a multicellular eukaryote. These rough calculations suggest that, for bacteria, the primary selective disadvantage of excess DNA is associated with energetic costs (this, at most, being of the same order of magnitude as the mutational hazard). In contrast, the mutational cost starts to exceed the energetic penalty in unicellular eukaryotes, and greatly exceeds it in multicellular species. In the latter case, however, even the mutational cost of a gene-structural embellishment is insufficient to overcome the power of random genetic drift. Hence, from the perspectives of both energetics and genetics, we expect a gradient in genome size and gene-structural complexity from prokaryotes to multicellular eukaryotes, provided there is a mutational bias towards insertions.

Foundations 17.4 Numbers of molecules required in a cellular lifespan

From the time of birth to the time of cell division, for any particular cellular feature, a cell must accumulate new constituent molecules to a level consistent with the contents of a new cell, and in doing so will often have to acquire replacement molecules to balance any decay processes. Here, we consider the situation in which a newborn cell contains N_0 molecules of the type being considered (e.g., the number of transcripts or protein molecules associated with a particular gene, or the number of lipid molecules in a cell membrane). This number must then double to $2N_0$ molecules to allow for binary fission. Letting β be the rate of production of the molecule and δ be the rate of decay, so that $r = \beta - \delta$ is the net growth rate in cell size, then assuming exponential growth

$$\frac{dN}{dt} = rN \qquad (17.4.1)$$

denotes the rate of increase in molecule number. This expression integrates to

$$N_t = N_0 e^{rt}, \qquad (17.4.2)$$

so with the cell dividing when $N_t = 2N_0$, the cell-division time is $T = \ln(2)/r$.

The average number of molecules in the parental cell over its entire lifespan is obtained by integrating over Equation 17.4.2 from 0 to T,

$$\overline{N} = \frac{N_0}{T} \int_0^T e^{rt} \cdot dt = \frac{N_0(e^{rT} - 1)}{rT} = \frac{N_0}{\ln(2)}, \qquad (17.4.3)$$

with the final simplification following from the fact that $rT = \ln(2)$. The total number of molecules produced per cellular lifespan (N_p) is then the product of the average number of molecules ($\overline{N}$), the production rate per molecule (β), and the cell division time (T),

$$N_p = \frac{\beta N_0 T}{\ln(2)} \qquad (17.4.4)$$

which by using $\beta = r + \delta$ and $r = \ln(2)/T$ becomes

$$N_p = N_0 \left(1 + \frac{\delta T}{\ln(2)}\right). \qquad (17.4.5)$$

A simple interpretation of this expression is that during its lifespan, a cell must produce N_0 new, surviving molecules (to yield an offspring cell) and $\delta N_0 T / \ln(2)$ replacement molecules to offset molecular degradation/loss. Note that this second (maintenance) term increases linearly with the cell-division time.

Foundations 17.5 The biosynthetic costs of lipid molecules

Estimation of the total cellular expenditure on the synthesis of a lipid molecule requires separate consideration of the investments in the three subcomponents of these molecules: the fatty-acid tails, head groups, and linkers (Lynch and Marinov 2017; Lynch and Trickovic 2020). As in the applications for amino acids and nucleotides, we will quantify such costs in units of ATP-hydrolysis equivalents. CTP, which is utilized in a few reaction steps in lipid biosynthesis, will be treated as equivalent to ATP, and electron transfers resulting from conversions of NADH to NAD$^+$ and NADPH to NADP$^+$ will again be assumed equivalent to 2.5 ATPs, and FADH$_2$ to FAD to 1.5 ATPs (the resultant computations are slightly different than those in Lynch and Trickovic (2020) owing to slightly different assumptions about these coenzyme conversions).

The starting point for the synthesis of most fatty acids is the production of one particular linear chain, palmitate, containing 16 carbon atoms. Production of this molecule takes place within a large complex, known as fatty-acid synthase. In bacteria, biosynthesis of each molecule requires the consumption of 8 acetyl-CoA molecules, 7 ATPs, and reductions involving 14 NADPH. As noted in Foundations 17.2, each molecule of acetyl-CoA is equivalent to a net opportunity loss of 10 ATPs. Thus, the total cost of production of one molecule of palmitate is $(8 \times 10) + (7 \times 1) + (14 \times 2.5) = 122$ ATP in bacteria. Fatty-acid production is slightly more expensive in non-photosynthetic eukaryotes, where acetyl-CoA is produced in the mitochondrion and reacts with oxaloacetate to produce citrate, which must then be exported. Cleavage of oxaloacetate in the cytosol yields acetyl-CoA at the expense of 1 ATP, and a series of reactions serve to return oxaloacetate to the citric-acid cycle in an effectively ATP neutral way. Thus, the cost of palmitate increases to 122 + $(8 \times 1) = 130$ ATP. Each additional pair of carbons added to this primary fatty-acid chain requires 1 additional acetyl-CoA and ATP, and 2 additional NADPHs, or an equivalent of 15 ATPs, and each subsequent desaturation of a bond consumes one NADPH, or 2.5 ATP equivalents.

To evaluate the total cost of a lipid molecule, we first consider the situation for glycerophospholipids, for

which glycerol-3-phosphate serves as the linker between the fatty-acid chains and the headgroup. Conversion of glycerol-3-phosphate from dhap (within the glycolytic pathway, and having a cost of 17 ATPs; Figure 17.7) consumes 1 NADH, so the use of this molecule as a linker in a lipid molecule has an opportunity cost of 17 + 2.5 = 19.5 ATPs. Linking each of the two fatty-acid tails requires 1 ATP, and linking the head group involves two CTP hydrolyses, yielding a total cost of $(19.5 + 2 + 2) = 23.5$.

All that remains is the cost of synthesis of the head group. In the case of phosphatidylglycerol, the head group is glycerol-3-phosphate, the cost of which is 19.5 ATP, as just noted, so the total cost of this molecule in a bacterium is $\simeq (2 \cdot 122) + 23.5 + 19.5 = 287$ ATP. From Foundations 17.2, the cost of a serine is 13.5 ATP, six fewer than for glycerol-3-phosphate, so the total cost of a phosphatidylserine is 281 ATP, and because ethanolamine and choline are simple derivatives of serine, this closely approximates the costs of both phosphatidylethanolamine and phosphatidylcholine. The headgroup of phosphatidylinositol is inosital, which is derived from glucose-6-phosphate (Figure 17.7), diverting the latter from glycolysis and depriving the cell of the equivalent of 33 ATPs, so the total cost of production of this molecule is 300.5 ATP. Finally, cardiolipin is synthesized by the fusion of two phosphatidylglycerols and the release of one glycerol, so taking the return from the latter to be 15 ATP, the total cost per molecule produced is $(2 \times 287) - 15 = 559$ ATP for bacteria (and 575 for eukaryotes), and the respective direct costs are 176 and 208 ATPs.

Estimation of the cost of biosynthesis of sphingolipids follows many of the steps just outlined. Construction of the linker molecule requires a palmitate molecule and the expenditure of 1 NADPH, for a total of 139 and 147 ATP in bacteria and eukaryotes, respectively. Then, a single fatty-acid chain is added, so assuming this is palmitate, this requires the expenditure of another 125.5 or 133.5 ATP. Finally, there are the costs of synthesizing and adding the head group, both of which are outlined in the preceding paragraphs.

Literature Cited

Adler, M., M. Anjum, O. G. Berg, D. I. Andersson, and L. Sandegren. 2014. High fitness costs and instability of gene duplications reduce rates of evolution of new genes by duplication-divergence mechanisms. Mol. Biol. Evol. 31: 1526–1535.

Akashi, H., and T. Gojobori. 2002. Metabolic efficiency and amino acid composition in the proteomes of *Escherichia coli* and *Bacillus subtilis*. Proc. Natl. Acad. Sci. USA 99: 3695–3700.

Arnold, A., M. Sajitz-Hermstein, and Z. Nikoloski. 2015. Effects of varying nitrogen sources on amino acid synthesis costs in *Arabidopsis thaliana* under different light and carbon-source conditions. PLoS One 10: e0116536.

Atkinson, D. E. 1970. Adenine nucleotides as universal stoichiometric metabolic coupling agents. Adv. Enzyme Regul. 9: 207–219.

Barton, M. D., D. Delneri, S. G. Oliver, M. Rattray, and C. M. Bergman. 2010. Evolutionary systems biology of amino acid biosynthetic cost in yeast. PLoS One 5: e11935.

Bauchop, T., and S. R. Elsden. 1960. The growth of microorganisms in relation to their energy supply. J. Gen. Microbiol. 23: 457–469.

Belle, A., A. Tanay, L. Bitincka, R. Shamir, and E. K. O'Shea. 2006. Quantification of protein half-lives in the budding yeast proteome. Proc. Natl. Acad. Sci. USA 103: 13004–13009.

Bernstein, J. A., A. B. Khodursky, P. H. Lin, S. Lin-Chao, and S. N. Cohen. 2002. Global analysis of mRNA decay and abundance in *Escherichia coli* at single-gene resolution using two-color fluorescent DNA microarrays. Proc. Natl. Acad. Sci. USA 99: 9697–9702.

Bienick, M. S., K. W. Young, J. R. Klesmith, E. E. Detwiler, K. J. Tomek, and T. A. Whitehead. 2014. The interrelationship between promoter strength, gene expression, and growth rate. PLoS One 9: e109105.

Booth, A., and W. F. Doolittle. 2015. Eukaryogenesis, how special really? Proc. Natl. Acad. Sci. USA 112: 10278–10285.

Chiyomaru, K., and K. Takemoto. 2020. Revisiting the hypothesis of an energetic barrier to genome complexity between eukaryotes and prokaryotes. R. Soc. Open Sci. 7: 191859.

Craig, C. L., and R. S. Weber. 1998. Selection costs of amino acid substitutions in ColE 1 and colIa gene clusters harbored by *Escherichia coli*. Mol. Biol. Evol. 15: 774–776.

Csárdi, G., A. Franks, D. S. Choi, E. M. Airoldi, and D. A. Drummond. 2015. Accounting for experimental noise reveals that mRNA levels, amplified by post-transcriptional processes, largely determine steady-state protein levels in yeast. PLoS Genet. 11: e1005206.

Dean, A. M., D. E. Dykhuizen, and D. L. Hartl. 1986. Fitness as a function of beta-galactosidase activity in *Escherichia coli*. Genet. Res. 48: 1–8.

Dressaire, C., F. Picard, E. Redon, P. Loubière, I. Queinnec, L. Girbal, and M. Cocaign-Bousquet. 2013. Role of mRNA stability during bacterial adaptation. PLoS One 8: e59059.

Du, B., D. C. Zielinski, J. M. Monk, and B. O. Palsson. 2018. Thermodynamic favorability and pathway yield as evolutionary tradeoffs in biosynthetic pathway choice. Proc. Natl. Acad. Sci. USA 115: 11339–11344.

Duboc, P., N. Schill, L. Menoud, W. van Gulik, and U. von Stockar. 1995. Measurements of sulfur, phosphorus and other ions in microbial biomass: Influence on correct determination of elemental composition and degree of reduction. J. Biotechnol. 43: 145–158.

Eden, E., N. Geva-Zatorsky, I. Issaeva, A. Cohen, E. Dekel, T. Danon, L. Cohen, A. Mayo, and U. Alon. 2011. Proteome half-life dynamics in living human cells. Science 331: 764–768.

Endler, J. A. 1986. Natural Selection in the Wild. Princeton University Press, Princeton, NJ.

Eguchi, Y., K. Makanae, T. Hasunuma, Y. Ishibashi, K. Kito, and H. Moriya. 2018. Estimating the protein burden limit of yeast cells by measuring the expression limits of glycolytic proteins. eLife 7: e34595.

Eser, P., L. Wachutka, K. C. Maier, C. Demel, M. Boroni, S. Iyer, P. Cramer, and J. Gagneur. 2016. Determinants of RNA metabolism in the *Schizosaccharomyces pombe* genome. Mol. Syst. Biol. 12: 857.

François-Martin, C., J. E. Rothman, and F. Pincet. 2017. Low energy cost for optimal speed and control of membrane fusion. Proc. Natl. Acad. Sci. USA 114: 1238–1241.

Frumkin, I., D. Schirman, A. Rotman, F. Li, L. Zahavi, E. Mordret, O. Asraf, S. Wu, S. F. Levy, and Y. Pilpel. 2017. Gene architectures that minimize cost of gene expression. Mol. Cell 65: 142–153.

Geiger, O., N. González-Silva, I. M. López-Lara, and C. Sohlenkamp. 2010. Amino acid-containing membrane lipids in bacteria. Prog. Lipid Res. 49: 46–60.

Geiler-Samerotte, K. A., M. F. Dion, B. A. Budnik, S. M. Wang, D. L. Hartl, and D. A. Drummond. 2011. Misfolded proteins impose a dosage-dependent fitness cost and trigger a cytosolic unfolded protein response in yeast. Proc. Natl. Acad. Sci. USA 108: 680–685.

Ghaemmaghami, S., W. K. Huh, K. Bower, R. W. Howson, A. Belle, N. Dephoure, E. K. O'Shea, and J. S. Weissman. 2003. Global analysis of protein expression in yeast. Nature 425: 737–741.

Golding, I., J. Paulsson, S. M. Zawilski, and E. C. Cox. 2005. Real-time kinetics of gene activity in individual bacteria. Cell 123: 1025–1036.

Guschina, I. A., and J. L. Harwood. 2006. Lipids and lipid metabolism in eukaryotic algae. Prog. Lipid Res. 45: 160–186.

Hambraeus, G., C. von Wachenfeldt, and L. Hederstedt. 2003. Genome-wide survey of mRNA half-lives in *Bacillus subtilis* identifies extremely stable mRNAs. Mol. Genet. Genomics 269: 706–714.

Helbig, A. O., P. Daran-Lapujade, A. J. van Maris, E. A. de Hulster, D. de Ridder, J. T. Pronk, A. J. Heck, and M. Slijper. 2011. The diversity of protein turnover and abundance under nitrogen-limited steady state conditions in *Saccharomyces cerevisiae*. Mol. Biosyst 7: 3316–3326.

Heizer, E. M., Jr., D. W. Raiford, M. L. Raymer, T. E. Doom, R. V. Miller, and D. E. Krane. 2006. Amino acid cost and codon-usage biases in 6 prokaryotic genomes: A whole-genome analysis. Mol. Biol. Evol. 23: 1670–1680.

Heizer, E. M., Jr., M. L. Raymer, and D. E. Krane. 2011. Amino acid biosynthetic cost and protein conservation. J. Mol. Evol. 72: 466–473.

Hoekstra, H. E., J. M. Hoekstra, D. Berrigan, S. N. Vignieri, A. Hoang, C. E. Hill, P. Beerli, and J. G. Kingsolver. 2001. Strength and tempo of directional selection in the wild. Proc. Natl. Acad. Sci. USA 98: 9157–9160.

Ilker, E., and M. Hinczewski. 2019. Modeling the growth of organisms validates a general relation between metabolic costs and natural selection. Phys. Rev. Lett. 122: 238101.

Islam, S., U. Kjällquist, A. Moliner, P. Zajac, J. B. Fan, P. Lönnerberg, and S. Linnarsson. 2011. Characterization of the single-cell transcriptional landscape by highly multiplex RNA-seq. Genome Res. 21: 1160–1167.

Kharasch, M. S., and B. Sher. 1925. The electronic conception of valence and heats of combustion of organic compounds. J. Phys. Chem. 29: 625–658.

Krick, T., N. Verstraete, L. G. Alonso, D. A. Shub, D. U. Ferreiro, M. Shub, and I. E. Sánchez. 2014. Amino acid metabolism conflicts with protein diversity. Mol. Biol. Evol. 31: 2905–2912.

Kučerka, N., M. P. Nieh, and J. Katsaras. 2011. Fluid phase lipid areas and bilayer thicknesses of commonly used phosphatidylcholines as a function of temperature. Biochim. Biophys. Acta 1808: 2761–2771.

Lahtvee, P. J., A. Seiman, L. Arike, K. Adamberg, and R. Vilu. 2014. Protein turnover forms one of the highest maintenance costs in *Lactococcus lactis*. Microbiology 160: 1501–1512.

Lane, N., and W. Martin. 2010. The energetics of genome complexity. Nature 467: 929–934.

Lewis, B. A., and D. M. Engelman. 1983. Lipid bilayer thickness varies linearly with acyl chain length in fluid phosphatidylcholine vesicles. J. Mol. Biol. 166: 211–217.

Li, G. W., and X. S. Xie. 2011. Central dogma at the single-molecule level in living cells. Nature 475: 308–315.

Lindén, M., P. Sens, and R. Phillips. 2012. Entropic tension in crowded membranes. PLoS Comput. Biol. 8: e1002431.

Lotka, A. J. 1922. Contribution to the energetics of evolution. Proc. Natl. Acad. Sci. USA 8: 147–151.

Lu, P., C. Vogel, R. Wang, X. Yao, and E. M. Marcotte. 2007. Absolute protein expression profiling estimates the relative contributions of transcriptional and translational regulation. Nat. Biotechnol. 25: 117–124.

Lynch, M. 2007a. The Origins of Genome Architecture. Sinauer Associates, Inc., Sunderland, MA.

Lynch, M. 2007b. The frailty of adaptive hypotheses for the origins of organismal complexity. Proc. Natl. Acad. Sci. USA 104 (Suppl.): 8597–8604.

Lynch, M., L.-M. Bobay, F. Catania, J.-F. Gout, and M. Rho. 2011. The repatterning of eukaryotic genomes by random genetic drift. Ann. Rev. Genomics Hum. Genet. 12: 347–366.

Lynch, M., and G. K. Marinov. 2015. The bioenergetic costs of a gene. Proc. Natl. Acad. Sci. USA 112: 15690–15695.

Lynch, M., and G. K. Marinov. 2017. Membranes, energetics, and evolution across the prokaryote–eukaryote divide. eLife 6: e20437.

Lynch, M., P. Schavemaker, T. Licknack, and Y. Hao, and A. Pezzano. 2022. Evolutionary bioenergetics of ciliates. J. Euk. Microbiol. 69: e12934.

Lynch, M., and B. Trickovic. 2020. A theoretical framework for evolutionary cell biology. J. Mol. Biol. 432: 1861–1879.

Mani, S., and M. Thattai. 2016. Stacking the odds for Golgi cisternal maturation. eLife 5: e16231.

Manly, B. F. J. 1985. The Statistics of Natural Selection. Chapman & Hall, London, UK.

Marinov, G. K., B. A. Williams, K. McCue, G. P. Schroth, J. Gertz, R. M. Myers, and B. J. Wold. 2014. From single-cell to cell-pool transcriptomes: Stochasticity in gene expression and RNA splicing. Genome Res. 24: 496–510.

McDermitt, D. K., and R. S. Loomis. 1981. Elemental composition of biomass and its relation to energy content, growth efficiency, and growth yield. Ann. Bot. 48: 275–290.

Mitra, K., I. Ubarretxena-Belandia, T. Taguchi, G. Warren, and D. M. Engelman. 2004. Modulation of the bilayer thickness of exocytic pathway membranes by membrane proteins rather than cholesterol. Proc. Natl. Acad. Sci. USA 101: 4083–4088.

Nagle, J. F., and S. Tristram-Nagle. 2000. Structure of lipid bilayers. Biochim. Biophys. Acta 1469: 159–195.

Neymotin, B., R. Athanasiadou, and D. Gresham. 2014. Determination of *in vivo* RNA kinetics using RATE-seq. RNA 20: 1645–1652.

Nicholls, D. G., and S. J. Ferguson. 2013. Bioenergetics, 4th edn. Academic Press, New York, NY.

Petrache, H. I., S. W. Dodd, and M. F. Brown. 2000. Area per lipid and acyl length distributions in fluid phosphatidylcholines determined by ^{2}H NMR spectroscopy. Biophys. J. 79: 3172–3192.

Phillips, R., J. Kondev, J. Theriot, and H. Garcia. 2012. Physical Biology of the Cell, 2nd edn. Garland Science, New York, NY.

Plata, G., M. E. Gottesman, and D. Vitkup. 2010. The rate of the molecular clock and the cost of gratuitous protein synthesis. Genome Biol. 11: R98.

Raiford, D. W., E. M. Heizer, Jr., R. V. Miller, H. Akashi, M. L. Raymer, and D. E. Krane. 2008. Do amino acid biosynthetic costs constrain protein evolution in *Saccharomyces cerevisiae*? J. Mol. Evol. 67: 621–630.

Raiford, D. W., E. M. Heizer, Jr., R. V. Miller, T. E. Doom, M. L. Raymer, and D. E. Krane. 2012. Metabolic and translational efficiency in microbial organisms. J. Mol. Evol. 74: 206–216.

Raj, A., C. S. Peskin, D. Tranchina, D. Y. Vargas, and S. Tyagi. 2006. Stochastic mRNA synthesis in mammalian cells. PLoS Biol. 4: e309.

Ramadas, R., and M. Thattai. 2013. New organelles by gene duplication in a biophysical model of eukaryote endomembrane evolution. Biophys. J. 104: 2553–2563.

Rand, R. P., and V. A. Parsegian. 1989. Hydration forces between phospholipid bilayers. Biochim. Biophys. Acta 988: 351–376.

Rawicz, W., K. C. Olbrich, T. McIntosh, D. Needham, and E. Evans. 2000. Effect of chain length and unsaturation on elasticity of lipid bilayers. Biophys. J. 79: 328–339.

Russell, J. B., and G. M. Cook. 1995. Energetics of bacterial growth: Balance of anabolic and catabolic reactions. Microbiol. Rev. 59: 48–62.

Schwanhäusser, B., D. Busse, N. Li, G. Dittmar, J. Schuchhardt, J. Wolf, W. Chen, and M. Selbach. 2011. Global quantification of mammalian gene expression control. Nature 473: 337–342.

Scott, M., C. W. Gunderson, E. M. Mateescu, Z. Zhang, and T. Hwa. 2010. Interdependence of cell growth and gene expression: Origins and consequences. Science 330: 1099–1102.

Seligmann, H. 2003. Cost-minimization of amino acid usage. J. Mol. Evol. 56: 151–161.

Shahrezaei, V., and P. S. Swain. 2008. Analytical distributions for stochastic gene expression. Proc. Natl. Acad. Sci. USA 105: 17256–17261.

Silverstein, T. 2005. The mitochondrial phosphate-to-oxygen ratio is not an integer. Biochem. Mol. Biol. Educ. 33: 416–417.

Steinman, R. M., S. E. Brodie, and Z. A. Cohn. 1976. Membrane flow during pinocytosis: A stereologic analysis. J. Cell Biol. 68: 665–687.

Stoebel, D. M., A. M. Dean, and D. E. Dykhuizen. 2008. The cost of expression of *Escherichia coli* lac operon proteins is in the process, not in the products. Genetics 178: 1653–1660.

Stouthamer, A. H. 1973. A theoretical study on the amount of ATP required for synthesis of microbial cell material. Antonie Van Leeuwenhoek 39: 545–65.

Swire, J. 2007. Selection on synthesis cost affects interprotein amino acid usage in all three domains of life. J. Mol. Evol. 64: 558–571.

Taniguchi, Y., P. J. Choi, G. W. Li, H. Chen, M. Babu, J. Hearn, A. Emili, and X. S. Xie. 2010. Quantifying *E. coli* proteome and transcriptome with single-molecule sensitivity in single cells. Science 329: 533–538.

Tempest, D. W., and O. M. Neijssel. 1984. The status of YATP and maintenance energy as biologically interpretable phenomena. Annu. Rev. Microbiol. 38: 459–486.

Thilo, L., and G. Vogel. 1980. Kinetics of membrane internalization and recycling during pinocytosis in *Dictyostelium discoideum*. Proc. Natl. Acad. Sci. USA 77: 1015–1019.

Tomala, K., and R. Korona. 2013. Evaluating the fitness cost of protein expression in *Saccharomyces cerevisiae*. Genome Biol. Evol. 5: 2051–2060.

Trötschel, C., S. P. Albaum, and A. Poetsch. 2013. Proteome turnover in bacteria: Current status for *Corynebacterium glutamicum* and related bacteria. Microb. Biotechnol. 6: 708–719.

Uwizeye, C., J. Decelle, P.-H. Jouneau, S. Flori, B. Gallet, J.-B. Keck, D. Dal Bo, C. Moriscot, C. Seydoux, F. Chevalier, et al. 2021. Morphological bases of phytoplankton energy management and physiological responses unveiled by 3D subcellular imaging. Nature Comm. 12: 1049.

van Milgen, J. 2002. Modeling biochemical aspects of energy metabolism in mammals. J. Nutr. 132: 3195–3202.

Van Valen, L. 1976. Energy and evolution. Evol. Theory 1: 179–229.

Van Valen, L. 1980. Evolution as a zero-sum game for energy. Evol. Theory 4: 289–300.

Wagner, A. 2005. Energy constraints on the evolution of gene expression. Mol. Biol. Evol. 22: 1365–1374.

Walsh, J. B., and M. Lynch. 2018. Selection and Evolution of Quantitative Traits. Oxford University Press, Oxford, UK.

Wang, Y., C. L. Liu, J. D. Storey, R. J. Tibshirani, D. Herschlag, and P. O. Brown. 2002. Precision and functional specificity in mRNA decay. Proc. Natl. Acad. Sci. USA 99: 5860–5865.

Wieslander, A., S. Nordström, A. Dahlqvist, L. Rilfors, and G. Lindblom. 1995. Membrane lipid composition and cell size of *Acholeplasma laidlawii* strain A are strongly influenced by lipid acyl chain length. Eur. J. Biochem. 227: 734–744.

Williams, K., F. Percival, J. Merino, and H. A. Mooney. 1987. Estimation of tissue construction cost from heat of combustion and organic nitrogen content. Plant Cell Environ. 10: 725–734.

Williford, A., and J. P. Demuth. 2012. Gene expression levels are correlated with synonymous codon usage, amino acid composition, and gene architecture in the red flour beetle, *Tribolium castaneum*. Mol. Biol. Evol. 29: 3755–3766.

Zenklusen, D., D. R. Larson, and R. H. Singer. 2008. Single-RNA counting reveals alternative modes of gene expression in yeast. Nat. Struct. Mol. Biol. 15: 1263–1271.

CHAPTER 18

Resource Acquisition and Homeostasis

The most fundamental challenge of any organism is the acquisition of resources necessary for cell growth and replication, without which the transmission of genes to the next generation is impossible. Cells must also avoid predation, infection, and other sources of mortality, but this requires resource investment as well. The goal here is to consider some of the evolved strategies that single-celled organisms exploit to harvest nutrients, often with extremely low environmental concentrations.

Nutrient acquisition takes many forms, from the direct uptake of small molecules from the environment to the fixation of carbon by photosynthesis to the engulfment of smaller cells by phagocytosis to the scavenging of host nutrients by intracellular parasites. Rather than provide an encyclopaedic coverage of the topic, only the first two of these strategies is explored in detail here.

Although the molecular details can be quite complicated, relatively simple models describe in mechanistic terms the basic responses of uptake rates to ambient nutrient concentrations. At the very least, these models highlight the key cellular features that need to be quantified to understand the basis for interspecies differences in nutrient-uptake parameters. Unfortunately, they also point to substantial gaps in our knowledge. Photosynthesis is of particular interest as it fuels the biological world, and yet has unexplained inefficiencies.

Finally, because all cellular machinery has optimal operating conditions, homeostatic mechanisms exist to keep internal physiological conditions (e.g., ion balance) relatively constant in the face of a variable external world. Such challenges impose a recurrent energy cost to the maintenance budget of all cells. Thus, in this chapter we also provide a brief overview of two key homeostatic mechanisms: osmoregulation and the maintenance of an internal physiological clock.

Adaptive Fine-Tuning of Elemental Composition

As outlined in Chapter 7, ~ 20 elements are essential to the growth of most cells. The bulk of biomass consists of carbon, hydrogen, oxygen, nitrogen, phosphorus, and sulfur atoms (diminishing in that order), and to a lesser extent the cations potassium, magnesium, calcium, and sodium, and ~ 10 trace metals. Different environments can provide quite different blends of these elements, and this can promote the growth of some species over others, depending on their elemental requirements and uptake efficiencies (Tilman 1982).

Motivated by the fact that the basic building blocks of life vary in elemental content (C, N, P, and S in particular), several authors have suggested that natural selection fine-tunes the elemental composition of cells to prevailing environmental conditions through genome-wide shifts in the use of alternative nucleotides and/or amino acids. Under this view, organisms living in environments depleted for one of these elements are expected to evolve towards reduced reliance on the same element in their building block repertoires. For the same reason, enzymes involved in the metabolism of a scarce element are expected to have nucleotides and/or amino acids depleted for the same element (Baudouin-Cornu et al. 2001; Alves and Savageau 2005; Acquisti et al. 2009; Grzymski and Dussaq 2012; Fasani and Savageau 2014). It has even been suggested that the genetic code is structured in such a way that the nitrogen content of the nucleotides within codons

Evolutionary Cell Biology. Michael Lynch, Oxford University Press. © Michael Lynch (2024). DOI: 10.1093/oso/9780192847287.003.0018

is correlated with that of the encoded amino acids, generating a reinforcing effect at the level of both building blocks (Bragg and Hyder 2004; Shenhav and Zeevi 2020).

Consistent with this view, Baudouin-Cornu et al. (2001) found that enzymes involved in the assimilation of sulfur and carbon in *E. coli* and *S. cerevisiae* contain amino acids that are, respectively, depleted with sulfur and carbon (e.g., methionine and cysteine in the case of sulfur). Similar claims have been made based on the absence of methionine and cysteine residues in cyanobacterial proteins with enhanced expression during times of sulfur depletion (Mazel and Marliére 1989), and on the differential utilization of nitrogen in proteins involved in anabolism versus catabolism (Acquisti et al. 2009). On the other hand, Baudouin-Cornu et al. (2001) found no such depletion for sulfur metabolizing enzymes in mammals, arguing that this is because mammals are less deprived of sulfur.

Two strong and often unstated assumptions underly these causal arguments: 1) that the genomic attributes of extant species have been molded evolutionarily by the current features of the environments that they inhabit; and 2) that natural selection is capable of promoting incremental modifications at the codon level. The first assumption ignores the possibility that matching between genomic requirements and the environment may be a simple consequence of competitive assortment of species, with differences in the former having evolved for entirely different reasons. The second assumption relies on the idea that natural selection is powerful enough to discriminate among alternative building blocks differing in just one or two atoms of a particular type. Whether the population-genetic conditions necessary for such selection are likely to be met can be evaluated by applying the same strategy outlined in Chapter 17 for the differential costs of amino acids, but in this case by measuring costs in units of elemental composition, rather than ATP equivalents.

Consider first the selective usage of alternative nucleotides on the basis of their carbon and nitrogen contents. Shenhav and Zeevi (2020) suggested that the vertical distribution of the C:N ratio in marine environments has driven the evolutionary divergence of genome-wide nucleotide usage in the bacteria inhabiting different depth strata. To determine whether the strength of selection required for such genome-wide remodelling is feasible, consider that over a range of growth conditions, the dry weight of *E. coli* and other bacterial cells consists of 45–47% carbon and 10–13% nitrogen (Folsom and Carlson 2015; Chapter 7). The dry weight of an average cell in this species $\simeq 0.00038$ ng, which given the molecular weights of C and N, implies $\sim 9 \times 10^9$ C and $\sim 2 \times 10^9$ N atoms per cell. An A:T nucleotide pair in DNA contains 10 C and 7 N atoms, whereas a G:C pair contains 9 and 8, respectively. Thus, exchanges between A:T and G:C pair add/subtract just a single atom of each of these elements, changing the fractional elemental usages of C and N per cell by $\sim 10^{-10}$ and 0.5×10^{-9}, respectively.

Given that no known species has an effective population size of $N_e > 10^{10}$, such a single genomic base-pair shift in a non-transcribed genomic region is expected to be barely detectable by natural selection, if at all. However, as most genes are represented by multiple copies of mRNA per cell, these numbers need to be multiplied accordingly for transcribed genes. Nonetheless, because the average number of mRNA copies per cell per gene is <10 in *E. coli* (Chapter 3), and unlike DNA, mRNAs are single-stranded, this still leaves the elemental cost very close to, if not below, the drift barrier. Thus, it appears doubtful that nucleotide usage can be selected on the basis of C:N composition.

Now consider the situation at the amino-acid level. The maximum difference in carbon atom content among the 20 amino acids is 9 (tryptophan vs. glycine) and for nitrogen is 3 (arginine vs. several other residues). The maximum fractional impact of a single amino-acid substitution is then $\sim 10^{-9}$ for both C and N for a gene with just a single protein copy per cell for *E. coli*. However, for *E. coli*, the mean number of proteins per cell per expressed gene $\simeq 2000$, with a very large standard deviation ($\simeq 5000$) (Lynch and Marinov 2015). As this will frequently place the maximum fractional impact in the range of 10^{-6} to 10^{-5}, this suggests that the elemental costs of some amino-acid substitutions are in the realm of being detectable by natural selection, at least in bacteria.

The conditions necessary for effective selection on elemental composition are less conducive in eukaryotes. Because the average number of proteins per gene per cell increases less than linearly with cell volume (with the $\sim 2/3$ power; Chapter 7), the cost/benefit of an amino-acid substitution is expected to decline with cell volume. Thus, because N_e declines by two to five orders of magnitude from bacteria to eukaryotes, and cell volume increases by one to three orders of magnitude, it is questionable as to whether cellular elemental composition can be driven by natural selection in eukaryotes, especially in multicellular species with large cells. This provides a simple alternative explanation for the mammalian results noted earlier.

In summary, despite the strong arguments that have been made for the global significance of selection on elemental composition, even extending to the evolution of the genetic code (see Rozhoňová and Payne (2021) and Xu and Zhang (2021) for compelling counterpoints), it is striking that almost all such studies are devoid of formal evolutionary analyses. Notably, in one of the only applications of this kind, Günther et al. (2013) found that high-frequency derived alleles in *Arabidopsis* populations have elevated nitrogen content, contrary to expectations for a plant expected to be under nitrogen limitation.

Given the tiny selective pressures involved and the extremely long timescales required for genome-wide repatterning of elemental composition, to the extent that matchings in organismal and environmental features occur, a more likely explanation is that other evolutionary forces drive cellular elemental composition, with species then assorting into environments that are most compatible with elemental demands. One such force is mutational bias, which plays a strong role in guiding nucleotide content of genomes across the Tree of Life, often operating in a direction contrary to that of natural selection (Long et al. 2017). Future work in this area would profit from a more holistic conceptual approach incorporating general evolutionary principles.

Nutrient Uptake Kinetics

Nutrient uptake rates are concentration-dependent. At low substrate concentrations, consumers are limited by resource encounter rates, whereas at high concentrations, uptake mechanisms become

saturated. One of the simplest models for nutrient-uptake kinetics captures both of these effects, and has the Michaelis–Menten form (identical in structure to the Monod growth equation introduced in Chapter 9, and derived for more general enzyme kinetics in Chapter 19),

$$V = \frac{V_{\max}S}{K_S + S},\qquad(18.1)$$

where V is the rate of uptake of the nutrient (substrate) with concentration S, and K_S is the concentration at which V is 50% of its maximum value, $V_{\max}$. Written in this way, the rate of nutrient is a hyperbolic function of the substrate concentration, which closely approximates empirical results from hundreds of studies. The ratio $V_{\max}/K_S$ is close to the uptake rate per unit nutrient for $S \ll K_S$ and is commonly referred to as the uptake affinity.

The model parameters $V_{\max}$ and K_S are simple summary descriptors of potentially complex cellular mechanisms, leaving unexplained the actual determinants of such patterns. A more mechanistic model (Foundations 18.1) directly accounts for random encounter rates of the substrate by molecular diffusion and active transport across the cell membrane, while retaining the hyperbolic form of Equation 18.1.

Substantial attention has been given to nutrient-uptake parameters in phytoplankton species, particularly for nitrogen and phosphorus, which commonly limit growth in freshwater and marine ecosystems. The key parameters appear to be cell-volume dependent. Summarizing over a wide phylogenetic range of species, Edwards et al. (2012) found power-law exponents of 0.82 and 0.94 for $V_{\max}$ (per cell) for N and P uptake, respectively, when scaled against cell volume, and exponents of 0.33 and 0.53 for K_S for these same elements. Uptake affinities ($V_{\max}/K_S$) for both elements also increase with cell size, with allometric scaling powers on the order of 0.75 to 0.85. Similar results were obtained by Lomas et al. (2014) and Marañón (2015). Note, however, that although uptake affinities have positive allometric coefficients, implying that larger cells (on average) have higher rates of nutrient uptake on a per-cell basis, because the coefficients are < 1.0, on a per-biomass basis, uptake rates actually decline with increasing cell size.

What are the underlying determinants of these scaling features? Foundations 18.1 shows that V_{max} is equivalent to the product of the number of transporters on the cell surface (per cell) and the rate of handling of substrate molecules by an engaged transporter. As the surface area of a cell is proportional to the square of the linear dimension, whereas cell volume is proportional to the cube of the latter, if the density of transporters per unit surface area remains constant, one would expect V_{max} to scale with the 2/3 power of cell volume (with variation around this expectation owing to shape differences). Some suggestion that such scaling occurs derives from observations on uptake rates for several amino acids in a polyploid series of yeast cells (from haploidy to tetraploidy), which increase in cell volume but are otherwise isogenic (Hennaut et al. 1970). However, the fact that the exponents of V_{max} and V_{max}/K_S (0.82 to 0.94, noted earlier) for interspecific comparisons are substantially higher than 0.67 suggests an increase in transporter density and/or a decrease in handling time with cell volume.

The half-saturation constant K_S is also expected to increase with transporter numbers and rate of handling (Foundations 18.1), albeit in a somewhat different way, so a positive scaling of this parameter with cell volume is again expected. Why the allometric scaling of K_S (noted above) is reduced twofold relative to that for V_{max} is unclear, but this could happen if the capture rates of transporters declined with increasing cell volume. In principle, the theory in Foundations 18.1 might be brought in closer alignment with empirical observations by directly evaluating the relationship between transporter number and the geometric features of cells, but the necessary empirical work remains to be done.

There has been considerable speculation on the adaptive designs of various sized cells for particular environments (e.g., Chakraborty et al. 2017), with Litchman et al. (2007) arguing that the uptake kinetics of different phytoplankton groups were essentially permanently molded by the physical/chemical environments in which the groups emerged phylogenetically (generally hundreds of millions of years ago). However, this frozen-phenotype view belies the rapidity with which metabolic processes can evolve in microbes (Helling et al. 1987; Maharjan et al. 2007; Blount et al. 2012; Samani and Bell 2016). As noted for nucleotide and amino-acid composition, a more likely explanation for any phenotype-environment matching is simple competitive assortment in temporally and spatially varying environments.

Channels and transporters

Lipid membranes are essentially impermeable to the kinds of ions and organic molecules that cells harvest and export for nutritional and homeostatic reasons. Thus, for nearly every type of essential molecule, specialized trans-membrane proteins are dedicated to selective transmission across the lipid bilayer. Relative to cytoplasmic proteins, such molecules can be viewed as being inside-out, in the sense that the exterior of the transmembrane domain is hydrophobic (to enable embedding into the lipid environment), whereas the channel interior is hydrophilic.

Membrane pores can be roughly partitioned into two categories. Channels are selective, sometimes gated, sieves that allow the transmission of specific molecules, while excluding others. As channels do not bind molecules directly, their ion transfer can approach the rate of diffusion. In contrast, transporters bind cargo directly, reducing rates of molecular transmission to levels in the range of 50 to 250 molecules per second, although most of the data are restricted to *E. coli* (Waygood and Steeves 1980; Naftalin et al. 2007) and yeast (Kruckeberg et al. 1999; Ye et al. 2001). Whereas channels operate in a passive manner and can only transport solutes down a concentration gradient, traffic through transporters requires energy to induce the structural changes necessary for opening and closing the pores on opposite sides of the membrane.

Cellular investments in membrane proteins are enormous. Approximately 20–35% of genes in prokaryotes and eukaryotes encode for them (Wallin and von Heijne 1998; Stevens and Arkin 2000), and they typically comprise on the order of 25% of the total protein mass in cells (Itzhak et al. 2016), and up to 50% of membrane surface areas (Chapter 17). Because many types of channels are found across the Tree of Life (Greganova et al. 2013), e.g., potassium channels

(Loukin et al. 2005), and aquaporins (Abascal et al. 2014), they must have originated very early (pre-LUCA). On the other hand, many channel types have been lost from specific lineages but dramatically expanded in others. For example, potassium channels are absent from fission yeast, and only present as one copy in budding yeast, but have expanded to ~ 300 copies in *Paramecium*. Losses of transporters are particularly common in parasite lineages, which can, for example, obtain nitrogen-containing amino acids and nucleotides from their host cells, eliminating the need to import ammonia for biosynthesis.

How difficult is it to evolve membrane-embedded proteins? Although hydrophobic transmembrane domains are a key requirement for reliable membrane insertion, the barrier to such a condition may not be great. For example, numerous studies have found that small random peptides often spontaneously develop into α-helices that adhere to the surfaces of lipid bilayers and under some conditions even insert as bundles of several helices (Lear et al. 1988; Pohorille et al. 2005; Mulkidjanian et al. 2009). Part of the reason for the enhanced stability of α-helices when aligned along membranes is the partial sheltering of their hydrogen bonds from open water. Thus, for biophysical reasons alone, protochannels may have been quite feasible in the earliest stages of cellular evolution.

There are many ways by which membrane pores use energy to drive the passage of substrates. Many transporters couple the power from downhill diffusion of one molecule to the uphill work required for the import of a second molecule. Others cotransport protons or sodium ions (by the proton/sodium motive force) to drive cargo import. ABC (ATP binding cassette) transporters, found throughout bacteria and eukaryotes, utilize the energy from ATP hydrolysis to pump substrates against a concentration gradient. The exact number of ATP hydrolyses per substrate molecule is not firmly established, but it is thought to be near two, the same as the number of ATP binding domains per transporter (Higgins 1992; Stefan et al. 2020). Some evidence suggests that transporters have ATPase activity even in the absence of substrate, leading to up to dozens of ATP hydrolyses per substrate-molecule uptake (Bock et al. 2019). Bacteria can also produce solute-binding proteins (SBPs) that sequester substrate molecules and then pass them on to cognate transporters (Driessen et al. 2000). In Gram-negative bacteria (with two membranes), the SBPs float freely in the periplasm, whereas in Gram-positives they are anchored to the cell surface.

Given the energy requirement for substrate transporters, as well as the expense of producing the membrane proteins themselves, it is desirable to know the total costs involved in import, as this must influence the success/failure of specific organisms in different environments. Although the requisite information for making such calculations appears to be lacking, what needs to be done is clear. Knowing the number of atoms of a particular element comprising a newborn cell and the cell-division time, the total rate of cellular import can be computed, and if one is willing to assume an energetic cost per imported molecule (e.g., two ATP hydrolyses per molecule), the total cost of the import process can be calculated. If one knows the rate of transport per channel, one can also estimate the number of channels necessary for such transport, and given a knowledge of the proteins involved, the cost of synthesizing the transporters themselves can be estimated.

Using this approach, Phillips and Milo (2009) estimated that at saturating glucose concentrations (with an assumed transport rate of 100 glucose molecules/transporter/sec), $\sim 4\%$ of the membrane areas of *E. coli* cells must be allocated to glucose transporters. Assuming a 40-minute lifespan, and $\sim 2 \times 10^9$ glucose molecules to build a new cell, this implies: 1) a total cellular uptake rate of $\sim 833{,}000$ glucose molecules/sec, which at a transport rate of 100 molecules/sec, suggests the presence of $\sim 8{,}000$ glucose transporters per cell; and 2) a cost of glucose uptake per cell division of $\sim 4 \times 10^9$ ATP hydrolyses (assuming a cost of two ATPs per imported glucose molecule), which is $\sim 15\%$ of the total cost of building an *E. coli* cell. These numbers further imply a cost of import per transporter of 500,000 ATP hydrolyses per cell division, which greatly exceeds the cost of building the transporters, as even a 1000 residue protein costs $\sim 30{,}000$ ATP hydrolyses (Chapter 17). All of these calculations are extremely rough, but they will hopefully motivate the need for future observations on numbers of transporters per cell, transport rates, and energetic costs of transport per cargo molecule.

Physiological acclimation

Care should be taken in the interpretation of estimates of nutrient-uptake parameters, as these are often derived from cells in various states of physiological acclimation, and it is known that uptake affinities can be developmentally variable. For example, when cells are acclimated to different nutrient conditions, and then assayed over a full range of nutrient concentrations, K_S is typically lower for cells previously exposed to nutrient-depleted medium (Collos et al. 2005), i.e., substrate affinity increases as the acclimation concentration decreases. In principle, such acclimation might be accomplished by regulation of transporter abundance per unit surface area. However, in many species, these physiological responses are associated with a developmental shift between the use of high- and low-affinity transporters, with the former being up-regulated in low-nutrient conditions (Eide 2012).

Empirical observations with *S. cerevisiae* led Levy et al. (2011) to suggest that dual-transporter systems enable cells to respond more effectively to shifts to low-nutrient conditions. The idea here is that low-affinity transporters activate a starvation response more readily than do high-affinity transporters, and in doing so enhance the expression of high-affinity transporter genes, thereby enabling the cell to endure a longer period of nutrient scarcity. What remains unclear, however, is why the same dynamical response could not be achieved with a single high-affinity transporter system by simply up-regulating the transporter concentration in response to decreasing nutrient abundance.

An alternative explanation that may solve this puzzle invokes a trade-off between the maximum rate of nutrient uptake and K_S in different iso-forms (Gudelj et al. 2007), such that with limited membrane space available for transporters, the one with the highest affinity at the current substrate concentration is utilized. Such a trade-off might arise if transporters operate in a bi-directional manner (Figure 18.1). At high nutrient concentrations, high-affinity transporters can be saturated on both sides of the cell membrane, causing nutrient efflux to compete with influx, whereas low-affinity transporters experience asymmetry, facilitating the net rate of influx (Bosdriesz et al. 2018). The resultant model, which allows the transporter binding-site conformation to switch from one side of the membrane to the other, leads to a modification of the normal Michaelis–Menten uptake kinetics. In contrast to uptake always increasing with the external substrate concentration, bi-directional flow results in a situation where, for any particular transporter K_S, there is a substrate concentration at which the uptake rate is maximized. With the optimum concentration increasing with decreasing affinity (Figure 18.1), the model then predicts that lower-affinity transporters have higher rates of influx at higher external nutrient concentrations simply because they experience less conflict with internal cell concentrations.

Because genomes typically only encode for a small number of transporters for particular substrates, often no more than two, some peculiar behavior can arise under this model. Due to the existence of an optimal K_S for peak influx at any external nutrient concentration, the model predicts that, for two transporters with K_S flanking the peak, there will always be a nutrient concentration at which both have equivalent influx rates. Notably, Wykoff et al. (2007) suggest that dual-transporter systems can exhibit bistable states, wherein at particular nutrient concentrations, a polymorphic (but clonally uniform) population of cells can be maintained. Although these authors view this sort of scenario to be the result of natural selection favoring a strategy for anticipating different types of nutrient-level changes, such an outcome may be nothing more than a physical consequence of the nature of the system.

Advantages of motility

Average swimming speeds in bacteria range from 10 to 100 μm/sec, whereas those in eukaryotic microbes are more on the order of 100–1000 μm/sec (Figure 16.4). There are numerous contexts in which swimming might confer an advantage, including predator avoidance and mate acquisition. In predatory species, motility will also increase the

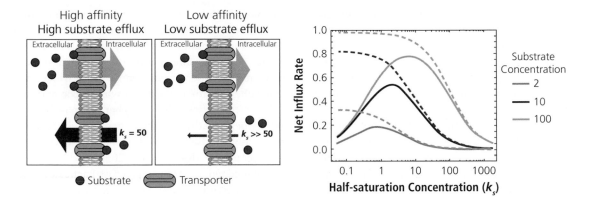

Figure 18.1 A structural explanation for the use of dual high/low-affinity transporters for nutrient uptake. **Left:** Transporters can be bi-directional, in the sense that they can bind to substrates on both sides of the membrane. In that case, when nutrient concentrations are high internally and externally, rates of influx will be nearly equal to those of efflux. However, low-affinity transporters require higher concentrations to become saturated, and therefore maintain larger differences between influx and eflux rates at high nutrient concentrations. **Right:** Net influx rates for transporters with different affinities (k_S, x axis) are given for three substrate concentrations (solid lines) and assuming constant V_{max}. Low k_S (half-saturation constant) implies high affinity for the substrate. Net influx is low when k_S is low because saturation occurs at very low substrate concentrations, and the high affinity leads to near-equal levels of binding on both sides of the membrane. On the other hand, with very high k_S, although there is little competitive binding, the overall influx rate is low simply because of the low absolute affinity on the external side of the membrane. Overall rates of influx increase with nutrient concentration regardless of k_S, but the optimum k_S shifts to the right (meaning that the optimum affinity is lowered) with higher concentrations. The dashed lines show the behavior that would occur with purely Michaelis–Menten uptake dynamics – in this case, the optimal k_S is always zero, as there is no competitive binding on the inside of the membrane. Derivations and other details can be found in Bosdriesz et al. (2018).

encounter rate of prey (Foundations 18.2), and when integrated with chemosensory systems, motility allows for habitat selection (Chapter 22).

Here, we simply explore the extent to which swimming can increase the rate of uptake of passively diffusing molecules. Recall that small ions have typical diffusion coefficients in the range of $D = 700$ to $2000\ \mu m^2/sec$, with an average value of $1300\ \mu m^2/sec$ (Figure 7.6). If swimming is to substantially increase the rate of nutrient uptake, the product of the velocity (v) and the average distance between particles ($\bar{\ell}$), which has units of $\mu m^2/sec$, needs to exceed the diffusion coefficient of the harvested particles. Purcell (1977) called the ratio $\bar{\ell}v/D$ the stirring number (see also Zehr et al. 2017).

Assuming a random distribution, the mean inter-particle distance is approximately $[3/(4\pi S)]^{1/3}$, which is equivalent to the radius of a conceptual sphere having a volume per particle equal to the inverse of the substrate concentration, $1/S$. For

example, a concentration of 1 mM, roughly the situation for inorganic carbon in seawater (Table 7.1), contains $\sim$600,000 particles/μm^3, which implies an average distance between particles of $\bar{\ell} = 0.007\ \mu m$. To match the average rate of ionic diffusion noted earlier, such that $\bar{\ell}v = D$, the swimming velocity ($v = D/\bar{\ell}$) would then need to be $v \simeq 186,000\ \mu m/sec$, far beyond the normal range in unicellular species. Similar calculations for a concentration of 1 μM (2 to 30× lower than average N and P concentrations in the ocean) lead to a critical speed of $\sim$18,600 $\mu m/sec$, and for 1 nM (similar to the concentrations of a number of trace metals) the critical speed is $\sim$1,860 $\mu m/sec$. Because cells swimming in low Reynolds number environments drag along most of their boundary layer as they move (Berg and Purcell 1977), these are lower-bound estimates of critical swimming speeds. In addition, no consideration has been given to the energetic cost of swimming.

The overall implication here is that swimming by microbes generally does not greatly magnify the rate of intake of randomly distributed inorganic nutrients. A more formal analysis in Foundations 18.2 suggests a < 5% enhancement in bacteria, but potentially an up to twofold inflation in large eukaryotic cells. Large organic compounds such as proteins have diffusion coefficients roughly 10 × smaller than those of inorganic ions, which would reduce the critical swimming velocities tenfold, so depending on the environmental concentrations there may be some significant increases in their encounter rate. Although the largest, fastest swimming eukaryotic cells may experience significant gains in the uptake rate of nutrients by enhanced stirring alone (Short et al. 2006; Solari et al. 2011), motility may have a greater impact in facilitating active searches for resource gradients when coupled with chemoreception (Chapter 22).

Photosynthesis

Whereas the acquisition of dissolved inorganic molecules is important to cells, only a tiny fraction of organisms can survive on such resources alone. Almost all of the global ecosystem ultimately depends on the conversion of solar energy into chemical energy used in the biosynthesis of organic materials. Organisms capable of such transformations are called phototrophs, in contrast to heterotrophs, which acquire their reduced carbon compounds from the former, either via uptake as single molecules or by bulk ingestion (phagocytosis or herbivory). Five other carbon fixation mechanisms, not requiring light, are exploited by various groups of anaerobic prokaryotes (Thauer 2007; Berg et al. 2010; Fuchs 2011). Collectively, all of these nonheterotrophs are called autotrophs. Approximately 99% of total global primary production is generated by RuBisCO, the key carbon fixation enzyme in phototrophs (Raven 2009).

Photosynthesis is known to occur in seven bacterial phyla, but only in the cyanobacteria and photosynthetic eukaryotes does it result in oxygen release. In these cases of oxygenic photosynthesis, water is used as an electron donor in ATP recharging. The remaining groups of phototrophic prokaryotes

are anoxygenic, relying on reduced inorganic compounds such as ferrous iron, hydrogen, and sulfur as electron suppliers. In today's highly oxidized world, such substances are only locally abundant in a small number of environments. However, the atmosphere of the early Earth was anoxic and had CO_2 concentrations 100 × higher than in today's atmosphere, leading to the suggestion that anoxygenic photosynthesis was the first established form of phototrophy (Raven et al. 2017). Indeed, given its phylogenetic distribution across bacteria, it may even have been present in the ancestral bacterium (Woese et al. 1985; Woese 1987), with many bacterial lineages subsequently experiencing loss and transition to heterotrophic life styles.

The origin of oxygenic photosynthesis was a key moment in evolutionary history, providing a means for exploiting a permanent and reliable supply of light energy to extract electrons from highly plentiful water molecules, using the resultant energy for the downstream synthesis of organic matter, and releasing O_2 in the process. The emergence and phylogenetic spread of oxygenic photosynthesis had a profound effect on the Earth's biogeochemistry, ultimately leading to a 10^5-fold rise of atmospheric O_2 concentration (Figure 18.2). The first big phase of oxygenation occurred $\sim$ 2.5 billion years ago (BYA),

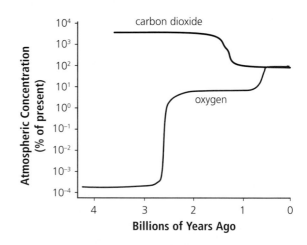

Figure 18.2 The historical record of the Earth's atmospheric CO_2 and O_2 concentrations, given as a crude idealization of patterns outlined in Shih (2015) and Raven (2017). Note that the vertical axis is on a logarithmic scale.

presumably the result of the origin of photosynthetic cyanobacteria. As a result, the atmospheric O_2 concentration rose to a level $\simeq 5\%$ of today's atmosphere. It is thought that this set the stage for the evolution of oxygenic metabolism by aerobic heterotrophs (Lyons et al. 2014).

This new atmospheric state then remained relatively constant until ~ 750 million years ago, at which point there was a further elevation of O_2 concentration to today's level. This second quantum shift in atmospheric O_2 content may have been a consequence of the origin of diverse groups of photosynthetic eukaryotes (including chlorophytes, red algae, diatoms, dinoflagellates, euglenoids, and haptophytes). All of these lineages acquired their photosynthetic capacities (embodied in chloroplasts) via horizontal transfers of endosymbionts ultimately of cyanobacterial origin (Shih 2015; Fischer et al. 2016; Chapter 23). During this second transition, there was a gradual draw-down of atmospheric CO_2 to today's level (which, despite the current uptick, is far lower than that 1 BYA).

The transformation of solar to chemical energy

As almost all of today's photosynthesis is carried out by aerobes, the remaining focus is on oxygenic photosynthesis. The details of this complex system are outlined in most biochemistry texts, and only a gross overview is given here. A key point is that the transition to oxygenic photosynthesis involved the emergence of two major innovations: the water-oxidizing photosystem II, which releases electrons, protons, and oxygen; and photosystem I, which utilizes the electron/proton output from photosystem II to drive the production of the energy-carrying compounds ATP and NADPH. (The numbering system here is a historical artifact of the timing of discoveries).

Both photosystems are thought to date to a duplication event deep in the bacterial phylogeny, although the sequence similarity of their molecular constituents is so low as to raise considerable uncertainty on this interpretation. An additional issue is that the two photosystems coexist only in cyanobacteria and in the derived plastids of photosynthetic eukaryotes, although each individual system resides alone in a few other non-cyanobacterial groups. Thus, a key question is whether photosynthetic cyanobacteria uniquely joined the two systems via a horizontal transfer, or whether the dual system is ancestral, with other lineages containing just one having experienced loss of the other. Because all lineages of oxygenic photosynthetic bacteria appear to be derived within their phylogenetic groups, with no evidence for ancestral phototrophism within any group (even cyanobacteria) (Fischer et al. 2016; Shih et al. 2017), the most parsimonious explanation is that multiple horizontal-transfer events occurred deep in bacterial phylogeny, one of which led to the dual I/II system in cyanobacteria.

Both photosystems are comprised of complexes of ~ 20 protein subunits and cofactors, which jointly carry out processes of light-harvesting, photoprotection, and transfer of electrons and protons. Two of the major photosystem-II proteins evolved by gene duplication prior to the diversification of the cyanobacteria and operate as heterodimers, whereas all bacterial photosystem-I complexes operate as homodimers, except in cyanobacteria, where they are heterodimeric (Cardona 2015; Cardona et al. 2015).

The overall process of photosynthesis is further subdivided into the light reactions (carried out by the photosystems just noted) and the downstream carbon-fixation dark reactions (somewhat of a misnomer, as they can occur in light and dark). The quantum efficiency of the light reaction of photosynthesis, defined to be the number of moles of photons absorbed per mole of oxygen produced, is sometimes viewed as a universal constant. However, there is continuing disagreement as to its actual value (Melis 2009; Hill and Govindjee 2014), and although most estimates are in the range of 8 to 10 photons per O_2, it remains possible that this number differs among phylogenetic lineages. Taking the average, ~ 9 photons are required to convert two H_2O and two $NADP^+$ into one O_2, two protons, and two NADPH. The protons generated by this reaction are used to drive the production of ATP from ADP using a membrane-embedded ATP synthase (in essentially the same way that the analogous machine operates on the plasma membrane of bacteria and the inner mitochondrial membranes of eukaryotes; Chapters 2 and 17).

The carbon-fixation (dark) reactions, also known as the Calvin–Benson cycle, consume three molecules of CO_2 to produce one triose phosphate molecule (glyceraldehyde-3-phosphate) that can then be used in the synthesis of higher-order organic compounds. The process is a cycle in that a molecule of ribulose-1,5-bisphosphate (RuBP) is initially consumed, broken down, and then regenerated, yielding a net release of one glyceraldehyde-3-phosphate. This is an energy-intensive process, with the production of each new triose phosphate requiring the investment of 6 NADPHs and 9 ATPs provided by the light reaction. This means that for the production of the 6 NADPHs needed to generate each triose phosphate, ~ 27 photons (4.5 for each NADPH, from the preceding paragraph) are required. If we assume that ~ 3 protons are required for the production of each ATP (Chapters 2 and 17), then 27 protons and ~ 122 photons (4.5 per proton, from above) are required per triose phosphate. Thus, in total, $\sim 27 + 122 = 149$ photons are required for the production of each triose phosphate, i.e., ~ 50 photons per fixed carbon atom.

Not all wavelengths of light are utilized by photosynthetic organisms. Using chlorophylls a and b, along with a set of accessory pigments, all arranged within a reaction centre in the large photosystem complexes, most photosynthesis relies on photons in the 400–700 nm wavelength range (essentially the visible light spectrum). However, not all wavelengths of light are utilized with equal efficiency, and substantial energy in the infrared range of sunlight goes unharvested by most plants. A few photosynthesizers such as the cyanobacterium *Acaryochloris* have an accessory pigment that expands the range of availability up to 750 nm (Chen and Blankenship 2011), but it remains unclear why the vast majority of photosynthesizers leave this resource untouched.

The world's most abundant enzyme

The centrepiece of the Calvin–Benson cycle in photosynthesis is the enzyme RuBisCO (ribulose-1,5-bisphosphate carboxylase/oxygenase), which catalyses the joining of RuBP, a five-carbon molecule, and CO_2 into the 6-carbon product that

is subsequently split into two 3-carbon molecules (glyceraldehyde-3-phosphates). The latter are then deployed in downstream biosynthetic pathways as well as in the recycling of RuBP for another round of carbon fixation. As the major generator of primary organic material in the biosphere, RuBisCO is thought to be the most abundant protein on Earth, summing to $\sim 0.7 \times 10^{12}$ kg, 95% of which is in terrestrial plants (Bar-On and Milo 2019).

Given the enzyme's conserved role across biology, RuBisCO exhibits a surprising level of structural diversity. Phylogenetically, the enzyme consists of two major variants. Form I (a hexadecamer, consisting of a ring of eight large subunits, capped off on both ends by rings of four small subunits) is found in most eukaryotic algae and cyanobacteria. Form II (no small subunits) is found in anaerobic photosynthetic proteobacteria and dinoflagellates, each of which has multiple structural forms (Morse et al. 1995; Erb and Zarzycki 2018). Moreover, there are additional variants in prokaryotes. In the photosynthetic bacterium *Rhodopseudomonas palustris*, RuBisCO is a homohexamer, whereas in *Rhodospirillum rubrum*, it is a homodimer. For the two Archaea in which a form of RuBisCO is known, in one case it is an homooctamer (*Pyrococcus horikoshii*), and in another case it is a homodecamer (*Thermococcus kodakaraensis*) (Maeda et al. 2002). The evolutionary roots of RuBisCO remain unclear, but the presence of RuBisCO-like proteins with other functions in various bacterial and archaeal lineages suggests that CO_2-fixing RuBisCO may be derived from a protein with quite different features, possibly nucleotide assimilation (Ashida et al. 2005; Erb and Zarzycki 2018).

Despite the energetic expense of photosynthesis, RuBisCO is a surprisingly inefficient enzyme. As there are many applied reasons for understanding the mechanisms of photosynthesis, especially in agriculture, the uptake kinetics of RuBisCO have been studied in a broader phylogenetic context than any other enzyme, usually by evaluating *in vitro* performances of isolated complexes. Estimates of k_{cat} (a measure of the maximum processing rate per enzyme molecule; Chapter 19) for CO_2 range from 1 to 5/sec in photosynthetic eukaryotes, with those in prokaryotes ranging from 5 to 14/sec (Uemura et al. 1997; Tcherkez et al. 2006; Savir et al. 2010;

Galmés et al. 2014a,b), orders of magnitude below the possible values based on biophysical limitations. Likewise, k_{cat}/k_S (a measure of efficiency of processing at non-saturating conditions) for CO_2 ranges from 10^5 to 5×10^5 $mol^{-1}sec^{-1}$ in eukaryotes, and from 10^4 to 2×10^5 $mol^{-1}sec^{-1}$ in bacteria, in both cases far below the diffusion limit expected for a perfect enzyme (Chapter 19). These low turnover rates per catalytic site necessitate that RuBisCO be maintained at high intracellular concentrations, further magnifying the cost of relying on this large enzyme. Moreover, the *in vitro* measures belie the actual situation in nature. Relative to maximum-capacity measurements of RuBisCO in the laboratory, the global time-average performance of the enzyme in nature is just 1% for terrestrial and 15% for marine environments (Bar-On and Milo 2019).

A large part of the problem here is that RuBisCO is a remarkably error-prone enzyme. As implied by its designation as a carboxylase/oxygenase, the enzyme competitively binds both CO_2 and O_2, with the latter event resulting in the production of a toxic by-product that is eliminated by a process called photorespiration. Assuming that substrate molecule concentrations are less than saturating, the error rate of RuBisCO resulting from encounters with oxygen molecules is a function of the specificity ratio, which is equivalent to the ratio of the uptake affinities, k_{cat}/k_S, for the two substrates,

$$R_{O_2/CO_2} = \frac{k_{cat,O_2}/k_{S,O_2}}{k_{cat,CO_2}/k_{S,CO_2}}. \qquad (18.2a)$$

Estimates of R_{O_2/CO_2} for RuBisCO from eukaryotes fall in the range of 0.005 to 0.015, whereas those for prokaryotes range from 0.02 to 0.10. Thus, if confronted with equal concentrations of the two gases, the error rate would be in the range of a few per cent. Whereas this may not seem like an overly large problem, the total error rate is a function of the product of the specificity ratio and the ratio of substrate concentrations, C_{O_2/CO_2},

$$E_{O_2/CO_2} = \frac{R_{O_2/CO_2} \cdot C_{O_2/CO_2}}{1 + (R_{O_2/CO_2} \cdot C_{O_2/CO_2})} \simeq R_{O_2/CO_2} \cdot C_{O_2/CO_2}, \qquad (18.2b)$$

where the final approximation applies if $R_{O_2/CO_2} \cdot C_{O_2/CO_2} \ll 1$. Today's atmosphere imposes much higher concentrations of O_2 than CO_2, ~250 and 7 μM in saturated water, respectively, so $C_{O_2/CO_2} \simeq$ 36. This implies astoundingly high error rates, in the range of 0.15 to 0.35 for eukaryotes. As discussed later, cellular mechanisms for concentrating CO_2 intracellularly have evolved to mitigate this problem, but such embellishments impose additional layers of cellular investment.

The underlying factors responsible for these suboptimal features of RuBisCO remain unclear. Phylogenetic reconstructions of ancestral states and evaluation of their biochemical features suggest that the addition of the small capping subunits to produce Form I enzymes led to both increased affinity and specificity for CO_2 prior to the great oxygenation event (Schulz et al. 2022). However, whereas such results suggest that Form I RuBisCOs may have been preadapted to changing atmospheric conditions, the enduring oxygen sensitivity of the enzyme for the past two billion years is left unexplained.

Tortell (2000) suggested an adaptive basis for an observed gradient between R_{O_2/CO_2} and the estimated geological age of a phylogenetic group, with a decrease from ~0.02 in ancient cyanobacterial lineages to 0.005 in the more recent diatom lineage (although red algae are a clear outlier). They argue that the enzyme evolved higher CO_2 specificity in more recent lineages as atmospheric CO_2/O_2 ratios declined (Figure 18.2). However, the assumption here is that RuBisCO initially evolved in bacteria inhabiting anoxic environments, and then remained frozen in the ancestral form in descendent cyanobacteria, whereas it became free to evolve in new directions in eukaryotic lineages following the great oxidation event (~2.5 BYA). Recalling the points made earlier for nutrient uptake, i.e., the well-known ability of microbes to evolve metabolically on timescales of thousands of generations, the likelihood of enzymatic features remaining frozen with suboptimal features for thousands of millennia is very small. Moreover, the frozen-accident hypothesis is inconsistent with broad phylogenetic analyses suggesting substantial adaptive amino-acid substitutions in RuBisCO, including on the dimeric interfaces near the active sites of the enzyme (Kapralov and Filatov 2007; Young et al. 2012; Poudel et al. 2020).

A common, alternative narrative is that the low efficiency and error-prone nature of RuBisCO is an inevitable consequence of structural constraints. Under this view, the enzyme actually has near-optimal performance conditional on unavoidable compromises. Most notably, a suggested negative correlation between the catalytic affinity for CO_2 and the ability of the enzyme to discriminate against O_2 motivates the view that simultaneous improvement in both features is structurally difficult (Tcherkez et al. 2006; Savir et al. 2010; Shih et al. 2016; Flamholz et al. 2019). Catalytic affinities and specificities for CO_2 vary by more than an order of magnitude across species but do so in a phylogenetically dependent manner. Cyanobacteria are at one extreme, having high maximum rates of carboxylation but low specificities for CO_2, whereas the red algae are at the opposite end of the spectrum, with chlorophytes having intermediate values of both. However, it has been argued that confidence in the constraint argument has been inflated by a failure to account for the phylogenetic structure in the data, with the statistical correlations between RuBisCO kinetic parameters being largely due to differences between (rather than within) eukaryotic and prokaryotic groups (Bouvier et al. 2021). In addition, the proposed mechanistic constraint has not yet been demonstrated empirically.

As alluded to above, there is an additional determinant of CO_2-utilization efficiency by RuBisCO – the ratio of O_2 to CO_2 concentrations of the two substrates at the active site of the enzyme, C_{O_2/CO_2}. Many species reduce the latter via a CO_2-concentrating mechanism (CCM) for increasing the internal availability of CO_2 (Beardall and Giordano 2002; Falkowski and Raven 2007). Such mechanisms are best known in land plants harboring C_4 and CAM (crassulacean acid) metabolisms. In the latter cases, CO_2 is incorporated into malate in spatially or temporally separate contexts from RuBisCO, which subsequently receives its CO_2 payload after malate dissociation. Malate formation appears to have been evolutionarily modified into a CCM many dozens of times independently in land plants (Brown et al. 2011; Heyduk et al. 2019), as well as in various unicellular lineages (Raven et al. 2017). For example, cyanobacteria concentrate CO_2 locally into microcompartments housing the photosynthetic apparatus (carboxysomes).

Because an efficient CCM will offset a higher intrinsic error rate, RuBisCO evolution provides another potential example of a bivariate drift barrier. Indeed, there is a strong inverse relationship between the specificity factor and CCM factor across species (Badger et al. 1998), i.e., species with a strong capacity to concentrate CO_2 have less discriminating RuBisCOs. In addition, species with high specificity factors produce less RuBisCO (Hobson and Guest 1983). Thus, the overall patterns are consistent with the evolutionary-layering hypothesis (Lynch and Hagner 2014), which postulates that an increase in complexity need not ultimately yield an increase in overall system performance, as the addition of a layer in a molecular pathway alleviates the intensity of selection operating on prior components.

If nothing else, these observations on one of the biosphere's most important proteins provide some cautionary notes for those who fully subscribe to the adaptive paradigm that all biological features have been molded to absolute perfection by the relentless pressure of natural selection. Are certain regions of phenotypic space empty (e.g., high CO_2 affinity and specificity in the case of RuBisCO) because such phenotypes are absolutely unattainable on biophysical grounds, or because once produced the selective advantages are so weak or the mutational biases in the opposite direction so strong that they cannot be maintained in natural populations?

Biotechnology industries are built on the assumption that nature can be improved upon. With respect to engineering more efficient RuBisCO molecules with greater catalytic efficiency and speed, some improvements have been made with as few as one or two amino-acid substitutions, although these have not yet translated into increased productivity of recipient (crop) species (Spreitzer and Salvucci 2002; Greene et al. 2007; Gomez-Fernandez et al. 2018; Wilson et al. 2018; Zhou and Whitney 2019). Whereas such progress may seem unimpressive, it should be kept in mind that the period of investigation extends over just 20 years, in contrast to > 2 billion years of evolution of RuBisCO in nature. Notably, there are success stories with respect to the laboratory improvement of other enzymes, e.g., a substantial increase in the efficiency of superoxide dismutase (Stroppolo et al. 2001), as will be discussed further in Chapter 19.

Osmoregulation

Although cells often experience dramatically different environments on within- and between-generation timescales, they strive to ensure that internal physiological operating conditions remain relatively constant. Because the structure and function of proteins depends on their surrounding chemistry, to operate in an efficient manner, cells must regulate the concentrations of their constituents, including hydrogen ions and all other dissolved substances. Among other things, this raises the necessity of osmoregulation. Hyperosmotic (saline) environments promote cell dehydration, excess molecular crowding, and reduced levels of molecular diffusion. In contrast, hypoosmotic environments induce cell swelling and potential membrane rupture.

Water can pass through cell membranes at potentially very rapid rates, owing to the presence of membrane-spanning pores called aquaporins, which passively admit water molecules while excluding other ions. Because aquaporins are present across the entire Tree of Life, with the apparent exception of some thermophilic bacteria (Abascal et al. 2014), they were likely present in LUCA. In eukaryotes, they have diversified into two major groups, one of which engages in glycerol transport for reasons discussed later. However, osmoregulation is not a simple matter of regulating the number aquaporin channels. Much of it involves the active transport and/or production of solutes.

For bacterial cells completely surrounded by a cell wall, the natural tendency for cell-volume expansion by water entry is countered by wall resistance. However, the resultant equilibrium turgor pressure by no means eliminates problems with osmotic stress, particularly when the environment is hyperosmotic. Because there are no active pumps for water molecules in bacteria, osmoregulation requires continual environmental sensing and, when necessary, rapid fluxes of ions and/or the production of osmolytes (Bremer and Krämer 2019). Commonly synthesized solutes are small organic compounds such as glycerol, proline, glycine betaine, and sucrose, which have no effects on the performance of active molecules and hence, are called compatible solutes (Figure 18.3). Such molecules can also be expelled rapidly if the external medium is too hypoosmotic.

The importance of cell walls for reducing the costs of osmotic balance is illustrated by a class of phenotypic variants in numerous bacterial species called L-form cells (Errington et al. 2016; Errington 2017; Claessen and Errington 2019). Such wall-less cells are induced by chronic exposure to antibiotics that interfere with peptidoglycan production. However, the maintenance of L-form cultures requires an osmoprotective medium (usually achieved by the addition of sucrose, which is not metabolized). Many eukaryotic lineages have abandoned cell walls, and long-term evolution experiments to gradually wean L-forms from dependence on a hyperosmotic environment would help reveal how secondary accommodations emerge to affect such a radical change in cellular structure. Given the high energetic costs of cell walls (Chapters 16 and 17), the development of stable cultures of such cells may also have utility in biotechnological applications.

Halophilic bacteria often pay the costs of osmoregulation by using compatible solutes by active instead of by active ionic balance. The compatible-solute strategy requires the intracellular synthesis of organic osmolytes, whereas the ion-balance strategy requires the flux of appropriate ions across the plasma membrane. Mixtures of the two strategies occur, and such regulation is especially important in marine and other hypersaline environments.

The biosynthetic costs per compatible-solute molecule (in units of ATP hydrolyses) range from 20 to 66, increasing with increasing molecular weight of the compound (Figure 18.3). The molarity of compatible solutes in cells of marine microbes led Oren (1999) to suggest that up to 85% of the construction cost of a halophilic cell may be associated with osmoprotection. Given these substantial investments for organisms inhabiting saline environments, it remains unclear why different organisms deploy different compatible solutes. Glycerol is the cheapest solute to produce, and also has the highest solubility. However, although relied upon

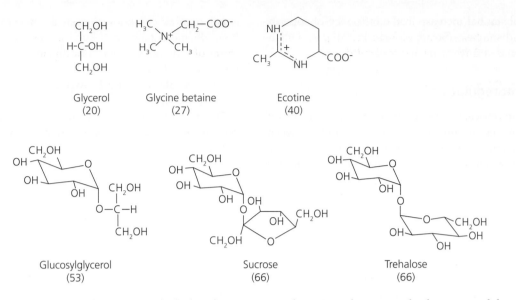

Figure 18.3 Compatible solutes commonly deployed in osmotic regulation in prokaryotes and eukaryotes, and the biosynthetic costs per molecule (in parentheses, and in units of ATP hydrolyses). Taken from Oren (1999; his figure 2), with the costs modified to be compatible with the recipes outlined in Foundations 17.2.

by yeast and the marine alga *Dunaliella*, which has specialized membranes, glycerol is a challenge because of its ability to permeate most membranes. Many non-halophilic bacteria utilize sucrose or trehalose.

Equally unclear is why species rely on compatible solutes at all (as opposed to maintaining ionic balance). Although the data are scant, as noted above, most specialized transporters appear to require one to two ATP hydrolyses for the movement of each transported molecule (Oren 1999; Patzlaff et al. 2003; Lycklama à Nijeholt 2018). Species that regulate osmotic pressure with ion balance often do so by modulating the cellular concentration of KCl. Potassium and chloride ions are imported through specific transporters, coupled with sodium export and proton input. The cost of import appears to be on the order of 0.5–0.67 ATP hydrolyses per KCl imported, 35–160× lower than the cost of synthesizing a compatible-solute molecule. These calculations do not include the costs of the machinery necessary to carry out these processes, which are likely greater than the costs for the biosynthetic machinery for compatible solutes.

Wall-less cells (most eukaryotes) face additional problems with osmotic pressure in hypoosmotic environments. To retain high levels of intracellular solutes, there is a need for a mechanism for removal of water molecules that work themselves through the membrane. This problem is magnified greatly in cells engaging in phagocytosis and the inevitable consumption of the external medium. Among flagellated eukaryotes, even walled cells have a water problem, as the flagella are surrounded by lipid bilayers (but not by walls), providing an entry for water molecules. Many eukaryotes deal with this relentless import of water by use of a contractile vacuole, which grows in volume by accumulating excess cell water, and then periodically ejects its contents into the surrounding medium. How this is accomplished in a way that selectively retains key cytoplasmic constituents remains unclear, although a number of hypotheses have been suggested (Raven and Doblin 2014).

The amount of water expelled by eukaryotic cells is impressive. For example, Lynn (1982) found a positive scaling of the volume of the contractile vacuole and cell volume in ciliates, with the total rate of volumetric output (μm^3/day) scaling as $460V^{0.88}$, where V is the cell volume (in μm^3). For the range of ciliate cell sizes studied, $\sim 10^4$ to 10^6 μm^3, between 152 and 87 cell volumes of water are ingested (and expelled) per day. Even higher rates have been found in other organisms. For the green alga *Chlamydomonas reinhardtii*, with a cell volume of $\sim 140\,\mu m^3$ and an average expulsion rate of 38,000 μm^3/day (Raven 1982; Buchmann and Becker 2009), the implied number of cell volumes ingested (and expelled) per day is 270. Another green alga *Mesostigma viride*, which harbors eight contractile vacuoles, expels 1150 equivalents of its volume per day (Buchmann and Becker 2009).

A deeper understanding of the magnitude of these demands can be obtained from expressions for the flow rate of water across a membrane of known porosity separating two liquids with different osmolarities (Foundations 18.3). Such analyses lead to the conclusion that a very substantial fraction of eukaryotic cell-maintenance requirements must be devoted to osmoregulation. The overall implication is that, owing to different osmotic demands and the evolved mechanisms for dealing with them, environments with different salinities will substantially alter the growth performance of organisms even in the face of equivalent levels of critical resources.

Circadian Rhythms

Most multicellular organisms use internal molecular clocks to generate circadian rhythms. In effect, such timekeepers enable organisms to predict the short-term future and make gene-expression and physiological changes appropriate to the demands of day/night cycles. The definition of a circadian system is somewhat subjective, but the central feature is some form of robust rhythmicity, with a near 24-hour periodicity that can be accurately entrained by an external factor (such as a light:dark cycle) but can also remain free-running for extended periods in a constant environment. Lack of sensitivity to absolute temperature (unlike most biochemical properties) is generally taken to be a hallmark of a true circadian rhythm.

Most multicellular organisms have generation lengths of weeks to years, so such diurnal cycles occur many times throughout the life of the individual. However, in unicellular species, minimum cell-division times can be much less than 24 hours. In such species, individual cells commonly complete their entire lives in very different periods of the diurnal cycle, raising questions as to the utility of a molecular clock. Nevertheless, they do exist.

Circadian rhythms are known in numerous unicellular phototrophs, including green and red algae, dinoflagellates, and euglenoids (Roenneberg and Merrow 2001; Brunner and Merrow 2008; Noordally and Millar 2015). Eelderink-Chen et al. (2010) found that imposition of a temperature cycle can entrain a circadian rhythm in the yeast *S. cerevisiae*, although the cycle is damped within just two days of constant temperature. A day:night circadian rhythm is also loosely coupled to the mitotic cell cycle in the filamentous fungus *Neurospora*, albeit with cell division being far from synchronous (Hong et al. 2014). Although few of the details have been worked out, the timekeeping mechanism in all of these cases is thought to involve rhythmicity of transcription in clock-component genes, as is known in land plants and metazoans. In multiple eukaryotes (humans, green algae, and filamentous fungi), rhythmicity appears to be governed by intracellular levels of magnesium, which operates as a cofactor with ATP and can therefore globally influence cellular features such as translation rates (Feeney et al. 2016).

Although the existence of entrainable clocks in heterotrophic bacteria is circumstantial at best (Sartor et al. 2019), the cyanobacterium *Synechococcus* has a well-characterized circadian clock, with a simple mechanism quite unlike that in eukaryotes (Figure 18.4). Indeed, the core oscillator can be made to operate *in vitro* in a solution containing just three proteins (KaiA, KaiB, and KaiC) and ATP (Nakajima et al. 2005; Phong et al. 2013), which engage in a cycle of post-translational modifications. The central hub, KaiC, is phosphorylated in the presence of KaiA, but once this occurs, KaiB forms a complex with KaiC that inhibits KaiA, thereby promoting dephosphorylation of KaiC, starting the cycle anew. This cycle further elicits a signal-transduction cascade (Chapter 22) of gene-expression events leading to growth-inhibition in the absence of light. Also of note, KaiC is a homo-hexamer. Although the phosphorylation of individual KaiC molecules is a stochastic process, the hexamers dissociate and reassociate, leading to homogenization of their

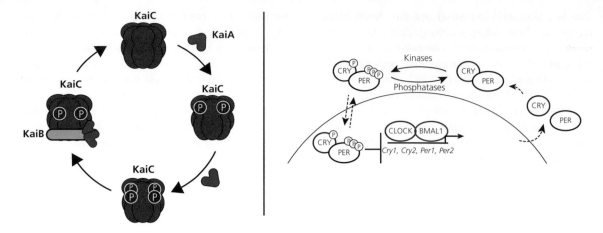

Figure 18.4 Left: The molecular network for the circadian clock in the cyanobacterium *Synechococcus*. Hexameric KaiC is autophosphorylated in the presence of KaiA (right), until it is joined by KaiB (left), which encourages dephosphorylation. Driven by ATP, this cyclical pattern results in stable oscillations. From Phong et al. (2013). **Right:** A simplified view of the vertebrate molecular clock. The curved line denotes the nuclear envelope. The heterodimeric transcription factor Clock/Bmal1 activates transcription of the Per and Cry genes, the mRNAs of which are translated in the cytoplasm. The proteins of the latter two genes then heterodimerize, return to the nucleus, and suppress their own expression. The kinetics lead to stable 24-hour cycles. From Partch et al. (2014). The only commonality between the two systems is the use of phosphorylation/dephosphorylation cycles.

phosphorylation states and hence a more coherent signal (Kageyama et al. 2006).

The protein-driven *Synechococcus* clock is different from that typically observed in eukaryotic systems (Figure 18.4), which rely on negative feedback involving activation of genes whose products ultimately repress their own expression. The cycle seems to be set by metabolic oscillations determined by diurnal resource availability, and hence is only indirectly correlated with the light:dark cycle itself (Pattanayak et al. 2015). As with the clocks of land plants and animals, however, the *Synechococcus* circadian rhythm can track changes in the day:night cycle even in an *in vitro* setting, providing a built-in response to seasonal changes (Leypunskiy et al. 2017). On the other hand, exposure to short pulses of darkness reset the clock in inappropriate ways, decreasing cell-division times (Lambert et al. 2016), and this raises a second issue.

A key challenge for circadian clocks is the sensitivity to internal and external noise. The former derives from small numbers of molecules per cell and bursty transcription, while the latter results from weather effects such as cloud coverage. Internal noise is expected to be more significant for species with smaller cells, which naturally harbor smaller numbers of proteins. Thus, it is of interest that *Prochlorococcus*, a marine cyanobacterium that is significantly smaller than *Synechococcus*, does not have a free-running circadian clock. Instead, *Prochlorococcus* deploys an hourglass-like clock, which responds directly to each daily change in light:dark and does not run freely in the absence of a daily signal. This shift is a consequence of the absence of the negative-feedback component KaiA in *Prochlorococcus* (Holtzendorff et al. 2008).

It has been suggested that such shifts in clock structure may be adaptive in small-celled species. When the internal noise is sufficiently large, free-running clocks are perturbed enough by noise amplification in the negative-feedback loop to become less reliable than hourglass clocks that are reset each day. On the other hand, free-running clocks appear to be less vulnerable to large-amplitude fluctuations in external signals (Chew et al. 2018; Pittayakanchit et al. 2018). The effects here are not large, however, and it remains to be shown whether the stochasticity of the clocks in these alternative systems have adaptive consequences.

How much does it cost to run and maintain a molecular clock? Although the details remain to be worked out, progress can be made in the cyanobacterial example just given, as it appears that the vast majority of the cost is associated with the biosynthesis of the proteins involved. The Kai system is driven by ATP hydrolysis, but the 24-hour cycle involves the hydrolysis of just 60 ATP molecules per hexamer (Terauchi et al. 2007). Because each monomer of the KaiC hexamer consists of 518 amino acids, with a biosynthetic cost of ~ 30 ATP hydrolyses per amino acid (Chapter 17), provided the cell-division time is a day or less, it is clear that the biosynthesis of the Kai components constitutes almost all of the cost.

Chew et al. (2018) estimate there to be $\sim 4{,}000$ KaiA molecules/cell (284 amino acids each), $11{,}000$ for KaiB (102 amino acids), and $8{,}000$ for KaiC. Assuming a total cost of ~ 30 ATPs per amino acid, implies a total biosynthetic cost $\simeq 2 \times 10^8$, as compared to the $\sim 5 \times 10^5$ for running the clock. Using Equation 8.2b, the cost of building a *Synechococcus* cell (with volume $\sim 0.5~\mu m^3$) is $\sim 14 \times 10^9$ ATP hydrolyses, so on the order of 1.5% of the construction energy budget is consumed by building and running the clock.

A similar computation can be made for the diminutive marine green alga *Ostreococcus*, which has a cell volume $\simeq 1.7~\mu m^3$, and therefore requires $\sim 45 \times 10^9$ ATP hydrolyses to build a cell. The clock in this species runs by a simple transcriptional loop mechanism, with one protein (CCA1) repressing the transcription of another (TOC1), and the latter activating transcription of the former (Bouget et al. 2014). The circadian cycle is driven by protein degradation, with the number of CCA1 protein copies cycling from 100 to 400 per day, and of TOC1 proteins cycling from 10 to 150 per cell (van Ooijen et al. 2011). Thus, the protein degradation rates are 0.75 and 0.93, respectively. Letting the proteins be ~ 300 and 500 in length, and assuming one cell division per day, using Equation 17.4, which assumes that amino acids are recycled, leads to estimates of total costs of protein production of 3.6×10^6 and 2.2×10^6 ATP hydrolyses for CCA1 and TOC1. Together, these constitute just 0.01% of the cell's total energy budget. A number of costs are ignored here, but their inclusion would be unlikely to increase this estimate by more than an order of magnitude. For example, although the cost of transcription has been ignored, this is known to be on the order of 10% of the protein-level cost (Chapter 17). The clock in this species seems to run via nontranscriptional rhythms of a light-sensitive protein (O'Neill et al. 2011), which could further increase the cost several fold. However, the central point seems to remain: the overall cost of running a clock in this species is kept very small because of the low number of proteins involved. Recall, however, that were there to be an advantage of losing a clock, a bioenergetic advantage as low as 0.01% would be quite easily promoted by natural selection.

Finally, it is worth contrasting the elegant simplicity of the few molecular clocks that have been dissected in microbial species with the extraordinarily baroque multi-loop constructs deployed by animals (not fully shown in Figure 18.4). Not only are the pathways more complex, elongated, and mechanistically diverse, but metazoans appear to have multiple clocks for different subcellular features (Mofatteh et al. 2021). Combined with the basic Cdk/kinase oscillator deployed in the cell cycle (Chapter 10), the latter can be intertwined, often with one master clock locking in the phasing of others. Some subsidiary metazoan clocks have cycle lengths that are harmonics of the usual 24-hour periodicity, e.g., 8- or 12-hour periods, and some can be made to run autonomously with respect to others.

There is no evidence that the embellishments of complex clocks in multicellular species are more efficient, and they are certainly more energetically costly. This lead one author to dub them an example of a 'Rube Goldberg' construct (Sancar 2008), after the cartoonist famous for his creation of extraordinarily complex machines to carry out simple tasks. As discussed in prior chapters and in those to come, this seemingly gratuitous increase in the complexity of molecular mechanisms is a hallmark of multicellular organisms. Rather than being driven by adaptive processes, many such embellishments have likely accrued in a nonadaptive manner as a passive consequence of the change in the population-genetic environment of such lineages. The emergent traits remain critical to organismal well-being, but gradually accrue aspects of overdesign.

Summary

- Of the ~ 20 chemical elements essential for growth, almost all are enriched intracellularly relative to environmental levels. It has been suggested that low environmental concentrations of elements can encourage selection for genome-wide enrichment of particular nucleotides depauperate for such elements. However, the selection differentials at the single-nucleotide level are so small that this is unlikely to occur in even the largest prokaryotic populations. On the other hand, selection for/against usage of particular amino acids with extreme elemental compositions is feasible in some unicellular species with very large effective populations sizes.

- Species-specific nutrient-uptake affinities increase with cell volume on a per-cell basis, but decline on a per-biomass basis. Although the data are limited, $> 10\%$ of cell energy budgets is allocated to the physical process of nutrient uptake, and many species acclimate to their surrounding nutrient levels by deploying membrane-bound transporters with different levels of substrate affinity.

- In large eukaryotic cells, swimming can increase rates of nutrient uptake by overcoming diffusion limitations. However, the effects are small in bacterial-sized cells, suggesting that the primary function of motility involves other factors such as the movement up resource gradients by chemotaxis.

- The process of oxygenic photosynthesis, responsible for the fixation of almost all of Earth's carbon, is an energetically expensive process. It is carried out by RuBisCO, probably the world's most abundant protein, but also one of the least efficient and most error-prone of all known enzymes. It remains unclear whether the latter features are consequences of biophysical constraints.

- To keep their internal conditions at near-optimal states for physiological function, all cells have a variety of homeostatic mechanisms. Osmoregulation is a perpetual problem for cells, as internal solute concentrations are generally unequal to those in the external environment. Cell walls counter the osmotic stress confronted by many species, but wall-less species (many eukaryotes) must continuously pump water out, often with a daily export equivalent to 100–1000 cell volumes. The relative costs of structural (wall) versus dynamical (pump) solutions to the osmoregulation problem remain to be worked out, but both consume a considerable fraction of cellular resources.

- Despite having cell-division times that are typically < 1 day, a number of unicellular species have independently evolved internal clocks with 24-hour periodicities, which provide a basis for predicting and preparing for diurnal environmental changes. Although only understood in a few species, the molecular mechanisms appear to consume no more than 2% of a cell's daily energy budget, and most of this is associated with construction rather than running of the clock. The molecular details of clocks in unicellular species are striking in their simplicity relative to the baroque mechanisms deployed in multicellular species.

Foundations 18.1 The response of uptake rate to nutrient concentration

Because the uptake of a molecule is typically carried out by specific membrane-bound nutrient transporters, the total rate of uptake by a cell is expected to scale positively with both the number of transporters residing in the cell membrane and the nutrient concentration at the surface of the cell. The latter, however, is not the same as the concentration in the bulk medium. As nutrients at the cell surface are constantly being taken up, there is a natural concentration gradient away from the cell surface governed by the rate of nutrient diffusion. Thus, one way of achieving a mechanistic understanding of the cellular rate of nutrient uptake is to first consider the transporter-based uptake kinetics conditional on the nutrient concentration at the cell surface, and then to relate the latter to the external diffusion process (Armstrong 2008).

Letting f denote the fraction of transporters on the cell surface engaged with a substrate molecule, t_h be the handling time of a captured molecule (the time to transport the molecule to the cytoplasm), k_c be the capture rate of a substrate molecule by an unengaged transporter, and S_0 be the nutrient concentration at the cell surface, the rate of change in transporter occupancy is

$$\frac{df}{dt} = k_c S_0 (1 - f) - f/t_h. \quad (18.1.1a)$$

At steady-state, $df/dt = 0$, and rearrangement of Equation 18.1.1a yields a mean fraction of occupied transporters

$$\widetilde{f} = \frac{S_0}{[1/(k_c t_h)] + S_0}, \quad (18.1.1b)$$

which asymptotically approaches 1.0 as $S_0 \to \infty$. With n_T transporters per cell surface, and each engaged transporter handling its cargo at rate $1/t_h$, the total rate of nutrient uptake is then

$$V = \frac{n_T \widetilde{f}}{t_h} = \frac{V_{max} S_0}{k_T + S_0}, \quad (18.1.2)$$

with $V_{max} = n_T/t_h$ being the maximum uptake rate (achieved when all transporters are engaged), and $k_T = [1/(k_c t_h)]$ being the nutrient concentration at the cell surface at which the uptake rate is half the maximum value.

Although Equation 18.1.2 has the convenient Michaelis–Menten form, as in Equation 18.1, the substrate concentration is inconveniently that at the cell surface, S_0. The next step is to determine the expected nutrient concentration at the cell surface, S_0, conditional on the measurable concentration in the bulk medium, S_∞. Pasciak and Gavis (1974) noted that at steady state, the transport-limited flux rate into the cell (noted previously) must equal the delivery rate of the nutrient to the cell surface.

The latter can be obtained by use of Fick's law (Foundations 7.2), which states that the flux rate of a diffusing substrate across a point is equal to the product of the concentration gradient at that point and the diffusion coefficient,

$$F = D \cdot (\partial S_d / \partial d)|_{d=r_e}, \quad (18.1.3a)$$

where D is the diffusion constant for the substrate, r_e is the effective radius of the cell (which can be obtained from the cell volume, assuming a sphere), and d is the distance from the center of the cell. To obtain the concentration gradient, we assume an asymptotic approach to the bulk fluid concentration (S_∞) with distance d, starting with the concentration at the cell surface, S_0,

$$S_d = S_0 + (S_\infty - S_0)\left(1 - \frac{r_e}{d}\right), \quad (18.1.3b)$$

which has the properties of $S_d = S_0$ when $d = r_e$, and $S_d \to S_\infty$ as $r_e/d \to 0$. The concentration gradient is then

$$\frac{\partial S_d}{\partial d} = \frac{r_e(S_\infty - S_0)}{d^2}. \quad (18.1.3c)$$

Substituting into Equation 18.1.3a, with $d = r_e$, the flux per unit surface area of the sphere is found to be

$$F = D(S_\infty - S_0)/r_e, \quad (18.1.3d)$$

and multiplication by the cell surface area $4\pi r_e^2$ yields

$$V = (4\pi r_e)D(S_\infty - S_0). \quad (18.1.3e)$$

This shows that the total flux rate to the cell is proportional to r_e, rather than to the surface area of the cell, because of the inverse relationship between the flux rate per surface area and r_e.

Often Equation 18.1.3e is multiplied by a dimensionless Sherwood number (Sh), which accounts for effects on the diffusion rate resulting from the deviation of the actual cell shape from a sphere and from other active processes that magnify the movement of substrate molecules relative to the case of pure diffusion (e.g., swimming). In other words, Sh is defined as the ratio of the realized rate of uptake to the expectation under diffusion alone. Defining the Péclet number as $Pe = 2r_e v/D$, where $2r_e$ is the length of the cell, v is the swimming velocity, and D is the diffusion coefficient, $Sh \simeq [1 + (1 + 2Pe)^{1/3}]/2$ (Guasto et al. 2012).

Equation 18.1.3e assumes $Sh = 1$, and for the remaining formulae, we suppress the use of Sh, noting that where desired it can simply be inserted as a prefactor for D. However, to gain some appreciation for the degree to which swimming can enhance nutrient uptake, recall from Figure 7.6 that $D \simeq 10^3 \mu m^2/sec$ for small ions. From Figure 16.4 a typical swimming speed for a bacterial cell with volume 1 μm^3 (which implies $2r_e \simeq 1.2$ μm) is $v = 50$ $\mu m/sec$. This implies a Péclet number of 0.06, and a Sherwood number equal to 1.02, i.e., a 2% increase in the rate of nutrient uptake. For a eukaryotic cell of volume 10^4 μm^3 ($2r_e \simeq 120$ μm) and velocity 300 $\mu m/sec$, the Péclet number becomes 36, yielding a Sherwood number of 2.6. Thus, by breaking down the diffusion barrier by swimming, large (phytoplankton-sized) cells are capable of a substantial increase in the rate of nutrient uptake. This, of course, needs to be tempered by the increased cost of swimming. Finally, rearranging Equation 18.1.3e to solve for S_0 and substituting the latter into Equation 18.1.2 yields the expected uptake rate

$$V = \frac{V_{max}(V - D' S_\infty)}{V + (k_T - S_\infty)D'}, \quad (18.1.4a)$$

where $D' = D(4\pi r_e)$. Although Equation 18.1.4a is a quadratic equation, making the solution for V somewhat complicated, Armstrong (2008) found that a close approximation is provided by

$$V = \frac{V_{max}S_\infty}{k_T + [V_{max}/(4\pi D)] + S_\infty}. \quad (18.1.4b)$$

Note that this expression has the same form as Equation 18.1.2, with the relevant nutrient concentration now being that in the bulk medium S_∞, and the half-saturation constant being expanded to $k_T + [V_{max}/(4\pi D)]$.

The overall utility of this approach is that it provides a mechanistic understanding of uptake kinetics in terms of both the cellular features (number, affinity, and handling time of transporters, n_T, k_c, and t_h, respectively), and the additional factors influencing the magnitude of the diffusive boundary around the cell (embodied in the parameter D). Summarizing from earlier,

$$V_{max} = \frac{n_T}{t_h}, \quad (18.1.5a)$$

$$K_S = \frac{1}{t_h}\left(\frac{1}{k_c} + \frac{n_T}{4\pi D}\right), \quad (18.1.5b)$$

which further implies an uptake affinity of

$$\frac{V_{max}}{K_S} = \frac{(4\pi D)n_T}{(4\pi D/k_c) + n_T}. \quad (18.1.5c)$$

Foundations 18.2 Encounter and capture rates

Ingestion rates are functions of both the rates at which cells physically encounter molecules (or larger prey items) and the subsequent efficiency of particle capture. The preceding theory simply treated the capture rate (per unit substrate concentration) as a fixed constant, but here we go further in mechanistic terms by considering the roles played by the rate of movement of the consumer and its substrate, the size of the cell, and the number of receptors on the surface (transporters) used in capturing the substrate. We first consider the encounter rate, and then the efficiency of capture, the product of which can be viewed as the capture rate k_c.

Foundations 7.2 showed that the rate of encounter between two passively diffusing particles is of the form

$$k_e = 4\pi RD, \quad (18.2.1a)$$

where R is the sum of the radii of the two particles, and D is the sum of their diffusion coefficients. After defining the diffusion coefficients in terms of particle radii, and assuming one particle (in this case, the consumer) is much larger than the other, this leads to

$$k_e \simeq \left(\frac{2k_BT}{3\eta}\right)\left(\frac{r_c}{r_n}\right), \quad (18.2.1b)$$

where k_B is the Boltzmann constant, T is the temperature (in Kelvins), η is the viscosity of the medium, and r_c and r_n are the radii of the consumer cell and the nutrient particle. In this case of passive diffusion, the encounter rate is directly proportional to the width of the consumer cell (here assumed to be a sphere).

Suppose now that the consumer and its prey are capable of swimming at rates well beyond the diffusion rates.

Then, each particle type (predator and prey) can be viewed as sweeping through a cylinder at a rate proportional to the square of its radius and the respective swimming velocity (v_c and v_n). Under the assumption of random swimming patterns,

$$k_e = \frac{\pi R^2[(v_c + v_n)^3 - (v_c - v_n)^3]}{6v_cv_n}, \quad (18.2.2a)$$

where R is now taken to be the encounter radius of consumer and substrate, which with a sensory system in the former might exceed the sum $(r_c + r_n)$ (Gerritsen and Strickler 1977). Assuming the prey is a passive nutrient molecule, such that $v_c \gg v_n$, this expression reduces to

$$k_e = \pi R^2 v_c, \quad (18.2.2b)$$

the area of a circle of radius R times the speed of the consumer. Thus, active swimming greatly increases the encounter rate, as this is now proportional to the square of the perception radius of the consumer. If turbulence contributes to movement of the consumer and/or substrate, this can be accommodated by addition of an appropriate constant to v_c (Rothschild and Osborn 1988; Evans 1989).

One limitation of these expressions is the assumption of no back-tracking on the part of the consumer, such that the same nutrient concentration is always being encountered, as might be the case if the consumer never changed direction. If, on the other hand, the direction of movement changes stochastically, the consumer will occasionally encounter an already explored patch. In this case, Equation 18.2.1a again applies, but with D redefined to be an effective diffusion coefficient. For example, in the case of run-and-tumble motion (Foundations 16.2), $D = v_c^2 \bar{\tau}/(3(1 - \bar{c}))$, where $\bar{\tau}$ is the mean time between switching directions, and $\bar{c}$ is the average cosine of the angular change in direction. Equation 18.2.1a then expands to

$$k_e = \frac{4\pi R v_c^2 \bar{\tau}}{3(1 - \bar{c})}, \tag{18.2.2c}$$

with the encounter rate again being linear in R (Visser and Kiørboe 2006).

If the consumer cell were 100% efficient at capturing particles upon encounter, the consumption rate would simply be proportional to the encounter rate and the substrate concentration (assuming non-saturating conditions). However, nutrient uptake generally can only be accomplished when substrate molecules encounter particular receptors on the cell surface. Thus, we must further consider the capture efficiency, an issue first tackled in the pioneering work of Berg and Purcell (1977); additional details can be found in Mitchell (1991, 2002).

The central idea is that conditional upon an encounter, if a particle is not captured by a receptor, it will still diffuse around in the vicinity of the cell for quite some time, providing additional opportunities for proper engagement. Letting r_r be the radius of a receptor, with N_r of these distributed over the surface of the cell, the total area of receptors is $\pi N_r r_r^2$, and the probability that a random encounter does not hit a receptor is one minus the fraction of cell surface occupied by receptors,

$$P_0 = 1 - \frac{N_r r_r^2}{4r_c^2}. \tag{18.2.3}$$

Assuming that each fresh (independent) encounter of the cell starts at approximate distance r_r from the cell surface, the probability that the particle will be captured anew and not lost forever is

$$P_s = \frac{r_c}{r_c + r_r}. \tag{18.2.4}$$

The probability that the particle eventually escapes after a series of encounters is then

$$P_{esc} = \sum_{i=0}^{\infty} P_0^i P_s^i (1 - P_s) = \frac{1 - P_s}{1 - P_0 P_s}, \tag{18.2.5}$$

which leads to the probability of capture

$$P_{cap} = 1 - P_{esc} = \frac{N_r r_r}{4r_c + N_r r_r}. \tag{18.2.6}$$

The rate of successful encounters (the capture rate) is then the product $k_e P_{cap}$.

This result provides insight into the degree to which a cell membrane needs to be populated by receptors to maintain a high probability of capture of encountered particles. Equation 18.2.6 shows that the probability of capture reaches 50% when the number of receptors per cell is $N_r = 4r_c/r_r$, which would occupy $(\pi r_r^2)(4r_c/r_r)$ of the total cell surface area $(4\pi r_c^2)$, implying a fractional coverage of r_r/r_c. Thus, as receptor diameters are $\ll r_c$, only a small fraction of the cell surface needs to be occupied by receptors to ensure a high capture efficiency.

Foundations 18.3 The cost of osmoregulation

The cytoplasm of cells is often hyperosmotic with respect to the surrounding fluid, causing a tendency for water molecules to enter through the partially permeable cell membrane. Here, we attempt to calculate the energy required to maintain the hyperosmotic states of cells. The pressure difference across a membrane can be computed with the van't Hoff equation,

$$\Delta P = \Delta C \cdot k_B T, \tag{18.3.1}$$

where ΔC is the difference in the concentration of solute molecules across the membrane, k_B is the Boltzmann constant, and T is the temperature (in Kelvins). For most biological temperatures $k_B T \simeq 4.1 \times 10^{-21}$ Joules (1 Joule $= 1 \, kg \cdot M^2 \cdot sec^{-2} = 0.00024$ kcal).

As an example, the approximate osmolarity of freshwater (summed over all solutes) is 2 mM, or equivalently 2 osmol/M^3. Considering a typical cell with a cytoplasmic osmolarity of 100 osmol/M^3, the concentration difference is 98 osmol/M^3. Multiplying by Avogadro's number of molecules/mol, $\Delta C = 590 \times 10^{23}$ molecules/M^3, and multiplying by $k_B T$ yields $\Delta P = 242{,}000$ Joules/M^3.

The flow rate across a membrane is equal to the product of the pressure difference, the hydraulic conductivity of the membrane (L_p), and the membrane surface area (A),

$$\Delta F = \Delta P \cdot L_p \cdot A, \tag{18.3.2}$$

where L_p has units of $M^4 \cdot sec^{-1} \cdot Joules^{-1}$ when ΔP is in units of Joules/M^3 and A is in units of M^2, yielding ΔF in units of M^3/sec.

At this point, these formulations may be non-informative to many biologists. However, to understand the challenge that osmotic pressure imposes upon cells, consider that an average freshwater eukaryote has a cell-membrane conductivity $L_p \simeq 7 \times 10^{-15}$ M^4 · sec^{-1} · Joules^{-1} (Raven 1982; Hellebust et al. 1989). Assuming a spherical cell with volume 1000 μm^3, the surface area would be $A = 482$ μm^2, and substitution into Equation 18.3.2 leads to a predicted water intake of $\sim$7 $\times 10^4$ μm^3/day, i.e., a daily volumetric intake equal to 70$\times$ the cell volume. To maintain cell homeostasis, this is also the amount of water that must be pumped out of the cell per day.

As the mechanism by which cells expel water is not fully understood, the energy required to maintain osmotic balance remains uncertain. However, Raven (1982) reasoned out an energetic cost of $\sim$10^{-16} mol ATP hydrolyses/ μm^3. Using this as an approximation, after multiplying by Avogadro's number, the 1000 μm^3 cell just noted (with a water intake rate of 7×10^4 μm^3/day) would require $\sim$4 $\times 10^{12}$ ATP hydrolyses/day for pumping. The daily total maintenance cost for a cell of this size is also $\sim$4 $\times 10^{12}$ ATP hydrolyses (Chapter 8). There are a number of uncertainties in these calculations, but assuming they are roughly correct, the implication is that a substantial fraction of cellular maintenance costs in eukaryotic cells derives from the regulation of osmotic balance. For ciliates, it has been estimated that 2–40% of total cell energy budgets are associated with contractile-vacuole pumping (increasing with cell volume), with values for *Tetrahymena* and *Paramecium* being in the range of 10 to 15% (Lynch et al. 2022).

Literature Cited

Abascal, F., I. Irisarri, and R. Zardoya. 2014. Diversity and evolution of membrane intrinsic proteins. Biochim. Biophys. Acta 1840: 1468–1481.

Acquisti, C., S. Kumar, and J. J. Elser. 2009. Signatures of nitrogen limitation in the elemental composition of the proteins involved in the metabolic apparatus. Proc. Biol. Sci. 276: 2605–2610.

Alves, R., and M. A. Savageau. 2005. Evidence of selection for low cognate amino acid bias in amino acid biosynthetic enzymes. Mol. Microbiol. 56: 1017–1034.

Armstrong, R. A. 2008. Nutrient uptake rate as a function of cell size and surface transporter density: A Michaelis-like approximation to the model of Pasciak and Gavis. Deep Sea Research Part I: Oceanogr. Res. Papers 55: 1311–1317.

Ashida, H., A. Danchin, and A. Yokota. 2005. Was photosynthetic RuBisCO recruited by acquisitive evolution from RuBisCO-like proteins involved in sulfur metabolism? Res. Microbiol. 156: 611–618.

Badger, M. R., T. J. Andrews, S. M. Whitney, M. Ludwig, D. C. Yellowlees, W. Leggat, and G. D. Price. 1998. The diversity and coevolution of Rubisco, plastids, pyrenoids, and chloroplast-based CO$_2$-concentrating mechanisms in algae. Can. J. Botany 76: 1052–1071.

Bar-On, Y. M., and R. Milo. 2019. The global mass and average rate of rubisco. Proc. Natl. Acad. Sci. USA 116: 4738–4743.

Baudouin-Cornu, P., Y. Surdin-Kerjan, P. Marlière, and D. Thomas. 2001. Molecular evolution of protein atomic composition. Science 293: 297–300.

Beardall, J., and M. Giordano. 2002. Ecological implications of microalgal and cyanobacterial CO$_2$ concentrating mechanisms, and their regulation. Func. Plant Biol. 29: 335–347.

Berg, H. C., and E. M. Purcell. 1977. Physics of chemoreception. Biophys. J. 20: 193–219.

Berg, I. A., D. Kockelkorn, W. H. Ramos-Vera, R. F. Say, J. Zarzycki, M. Hügler, B. E. Alber, and G. Fuchs. 2010. Autotrophic carbon fixation in archaea. Nat. Rev. Microbiol. 8: 447–460.

Blount, Z. D., J. E. Barrick, C. J. Davidson, and R. E. Lenski. 2012. Genomic analysis of a key innovation in an experimental *Escherichia coli* population. Nature 489: 513–518.

Bock, C., T. Zollmann, K. A. Lindt, R. Tampé, and R. Abele. 2019. Peptide translocation by the lysosomal ABC transporter TAPL is regulated by coupling efficiency and activation energy. Sci. Rep. 9: 11884.

Bosdriesz, E., M. T. Wortel, J. R. Haanstra, M. J. Wagner, P. de la Torre Cortés, and B. Teusink. 2018. Low affinity uniporter carrier proteins can increase net substrate uptake rate by reducing efflux. Sci. Rep. 8: 5576.

Bouget, F. Y., M. Lefranc, Q. Thommen, B. Pfeuty, J. C. Lozano, P. Schattab, H. Botebo, and V. Vergé. 2014. Transcriptional versus non-transcriptional clocks: A case study in *Ostreococcus*. Mar. Genomics 14: 17–22.

Bouvier, J. W., D. M. Emms, T. Rhodes, J. S. Bolton, A. Brasnett, A. Eddershaw, J. R. Nielsen, A. Unitt, S. M. Whitney, and S. Kelly. 2021. Rubisco adaptation is more limited by phylogenetic constraint than by catalytic trade-off. Mol. Biol. Evol. 38: 2880–2896.

Bragg, J. G., and C. L. Hyder. 2004. Nitrogen versus carbon use in prokaryotic genomes and proteomes. Proc. Biol. Sci. 271 Suppl 5: S374–S377.

Bremer, E., and R. Krämer. 2019. Responses of microorganisms to osmotic stress. Annu. Rev. Microbiol. 73: 313–334.

Brown, N. J., C. A. Newell, S. Stanley, J. E. Chen, A. J. Perrin, K. Kajala, and J. M. Hibberd. 2011. Independent and parallel recruitment of preexisting mechanisms underlying C_4 photosynthesis. Science 331: 1436–1439.

Brunner, M., and M. Merrow. 2008. The green yeast uses its plant-like clock to regulate its animal-like tail. Genes Dev. 22: 825–831.

Buchmann, K., and B. Becker. 2009. The system of contractile vacuoles in the green alga *Mesostigma viride* (Streptophyta). Protist 160: 427–443.

Cardona, T. 2015. A fresh look at the evolution and diversification of photochemical reaction centers. Photosynth. Res. 126: 111–134.

Cardona, T., J. W. Murray, and A. W. Rutherford. 2015. Origin and evolution of water oxidation before the last common ancestor of the cyanobacteria. Mol. Biol. Evol. 32: 1310–1328.

Chakraborty, S., L. T. Nielsen, and K. H. Andersen. 2017. Trophic strategies of unicellular plankton. Am. Nat. 189: E77–E90.

Chen, M., and R. E. Blankenship. 2011. Expanding the solar spectrum used by photosynthesis. Trends Plant Sci. 16: 427–431.

Chew, J., E. Leypunskiy, J. Lin, A. Murugan, and M. J. Rust. 2018. High protein copy number is required to suppress stochasticity in the cyanobacterial circadian clock. Nat. Commun. 9: 3004.

Claessen, D., and J. Errington. 2019. Cell wall deficiency as a coping strategy for stress. Trends Microbiol. 27: 1025–1033.

Collos, Y., A. Vaquer, and P. Souchou. 2005. Acclimation of nitrate uptake by phytoplankton to high substrate levels. J. Phycol. 41: 466–478.

Driessen, A. J., B. P. Rosen, and W. N. Konings. 2000. Diversity of transport mechanisms: Common structural principles. Trends Biochem. Sci. 25: 397–401.

Edwards, K. F., M. K. Thomas, C. A. Klausmeier, and E. Litchman. 2012. Allometric scaling and taxonomic variation in nutrient utilization traits and maximum growth rate of phytoplankton. Limnol. Oceanogr. 57: 554–566.

Eelderink-Chen, Z., G. Mazzotta, M. Sturre, J. Bosman, T. Roenneberg, and M. Merrow. 2010. A circadian clock in *Saccharomyces cerevisiae*. Proc. Natl. Acad. Sci. USA 107: 2043–2047.

Eide, D. J. 2012. An 'inordinate fondness for transporters' explained? Sci. Signal. 5: 5.

Erb, T. J., and J. Zarzycki. 2018. A short history of RubisCO: The rise and fall (?) of nature's predominant CO(2) fixing enzyme. Curr. Opin. Biotechnol. 49: 100–107.

Errington, J. 2017. Cell wall-deficient, L-form bacteria in the 21st century: A personal perspective. Biochem. Soc. Trans. 45: 287–295.

Errington, J., K. Mickiewicz, Y. Kawai, and L. J. Wu. 2016. L-form bacteria, chronic diseases and the origins of life. Philos. Trans. R. Soc. Lond. B Biol. Sci. 371: 20150494.

Evans, G. T. 1989. The encounter speed of moving predator and prey. J. Plank. Res. 11: 415–417.

Falkowski, P. G., and J. A. Raven. 2007. Aquatic Photosynthesis, 2nd edn Princeton University Press, Princeton, NJ.

Fasani, R. A, and M. A. Savageau. 2014. Evolution of a genome-encoded bias in amino acid biosynthetic pathways is a potential indicator of amino acid dynamics in the environment. Mol. Biol. Evol. 31: 2865–2878.

Feeney, K. A., L. L. Hansen, M. Putker, C. Olivares-Yañez, J. Day, L. J. Eades, L. F. Larrondo, N. P. Hoyle, J. S. O'Neill, and G. van Ooijen. 2016. Daily magnesium fluxes regulate cellular timekeeping and energy balance. Nature 532: 375–379.

Fischer, W. W., J. Hemp, and J. E. Johnson. 2016. Evolution of oxygenic photosynthesis. Annu. Rev. Earth Planetary Sci. 44: 647–683.

Flamholz, A. I., N. Prywes, U. Moran, D. Davidi, Y. M. Bar-On, L. M. Oltrogge, R. Alves, D. Savage, and R. Milo. 2019. Revisiting trade-offs between Rubisco kinetic parameters. Biochemistry 58: 3365–3376.

Folsom, J. P., and R. P. Carlson. 2015. Physiological, biomass elemental composition and proteomic analyses of *Escherichia coli* ammonium-limited chemostat growth, and comparison with iron- and glucose-limited chemostat growth. Microbiology 161: 1659–1670.

Fuchs, G. 2011. Alternative pathways of carbon dioxide fixation: Insights into the early evolution of life? Annu. Rev. Microbiol. 65: 631–658.

Galmés, J., P. J. Andralojc, M. V. Kapralov, J. Flexas, A. J. Keys, A. Molins, M. A. Parry, and À. M. Conesa. 2014a. Environmentally driven evolution of Rubisco and improved photosynthesis and growth within the C3 genus *Limonium* (Plumbaginaceae). New Phytol. 203: 989–999.

Galmés, J., M. V. Kapralov, P. J. Andralojc, M. À. Conesa, A. J. Keys, M. A. Parry, and J. Flexas. 2014b. Expanding knowledge of the Rubisco kinetics variability in plant species: Environmental and evolutionary trends. Plant Cell. Environ. 37: 1989–2001.

Gerritsen, J., and J. R. Strickler. 1977. Encounter probabilities and community structure in zooplankton: A mathematical model. J. Fish. Res. Bd. Canada 34: 73–82.

Gomez-Fernandez, B. J., E. Garcia-Ruiz, J. Martin-Diaz, P. Gomez de Santos, P. Santos-Moriano, F. J. Plou, A. Ballesteros, M. Garcia, M. Rodriguez, V. A. Risso, et al. 2018. Directed *in vitro* evolution of Precambrian and extant Rubiscos. Sci. Rep. 8: 5532.

Greene, D. N., S. M. Whitney, and I. Matsumura. 2007. Artificially evolved *Synechococcus* PCC6301 Rubisco variants exhibit improvements in folding and catalytic efficiency. Biochem. J. 404: 517–524.

Greganova, E., M. Steinmann, P. Mäser, and N. Fankhauser. 2013. *In silico* ionomics segregates parasitic from free-living eukaryotes. Genome Biol. Evol. 5: 1902–1909.

Grzymski, J. J, and A. M. Dussaq. 2012. The significance of nitrogen cost minimization in proteomes of marine microorganisms. ISME J. 6: 71–80.

Guasto, J. S., R. Rusconi, and R. Stocker. 2012. Fluid mechanics of planktonic microorganisms. Ann. Rev. Fluid Mechanics 44: 373–400.

Gudelj, I., R. E. Beardmore, S. S. Arkin, and R. C. MacLean. 2007. Constraints on microbial metabolism drive evolutionary diversification in homogeneous environments. J. Evol. Biol. 20: 1882–1889.

Günther, T., C. Lampei, and K. J. Schmid. 2013. Mutational bias and gene conversion affect the intraspecific nitrogen stoichiometry of the *Arabidopsis thaliana* transcriptome. Mol. Biol. Evol. 30: 561–568.

Hellebust, J. A., T. Mérida, and I. Ahmad. 1989. Operation of contractile vacuoles in the euryhaline green flagellate *Chlamydomonas pulsatilla* (Chlorophyceae) as a function of salinity. Mar. Biol. 100: 373–379.

Helling, R. B., C. N. Vargas, and J. Adams. 1987. Evolution of *Escherichia coli* during growth in a constant environment. Genetics 116: 349–358.

Hennaut, C., F. Hilger, and M. Grenson. 1970. Space limitation for permease insertion in the cytoplasmic membrane of *Saccharomyces cerevisiae*. Biochem. Biophys. Res. Commun. 39: 666–671.

Heyduk, K., J. J. Moreno-Villena, I. S. Gilman, P. A. Christin, and E. J. Edwards. 2019. The genetics of convergent evolution: Insights from plant photosynthesis. Nat. Rev. Genet. 20: 485–493.

Higgins, C. F. 1992. ABC transporters: From microorganisms to man. Annu. Rev. Cell Biol. 8: 67–113.

Hill, J. F., and Govindjee. 2014. The controversy over the minimum quantum requirement for oxygen evolution. Photosynth. Res. 122: 97–112.

Hobson, L.A., and K. P. Guest. 1983. Values of net compensation irradiation and their dependence on photosynthetic efficiency and respiration in marine unicellular algae. Mar. Biol. 74: 1–7.

Holtzendorff, J., F. Partensky, D. Mella, J. F. Lennon, W. R. Hess, and L. Garczarek. 2008. Genome streamlining results in loss of robustness of the circadian clock in the marine cyanobacterium *Prochlorococcus marinus* PCC 9511. J. Biol. Rhythms 23: 187–199.

Hong, C. I., J. Zámborszky, M. Baek, L. Labiscsak, K. Ju, H. Lee, L. F. Larrondo, A. Goity, H. Siong Chong, W. J. Belden, et al. 2014. Circadian rhythms synchronize mitosis in *Neurospora crassa*. Proc. Natl. Acad. Sci. USA 111: 1397–1402.

Itzhak, D. N., S. Tyanova, J, Cox, and G. H. Borner. 2016. Global, quantitative and dynamic mapping of protein subcellular localization. eLife 5: e16950.

Kageyama, H., T. Nishiwaki, M. Nakajima, H. Iwasaki, T. Oyama, and T. Kondo. 2006. Cyanobacterial circadian pacemaker: Kai protein complex dynamics in the KaiC phosphorylation cycle *in vitro*. Mol. Cell. 23: 161–171.

Kapralov, M. V., and D. A. Filatov. 2007. Widespread positive selection in the photosynthetic Rubisco enzyme. BMC Evol. Biol. 7: 73.

Kruckeberg, A. L., L. Ye, J. A. Berden, and K. van Dam. 1999. Functional expression, quantification and cellular localization of the Hxt2 hexose transporter of *Saccharomyces cerevisiae* tagged with the green fluorescent protein. Biochem. J. 339: 299–307.

Lambert, G., J. Chew, and M. J. Rust. 2016. Costs of clock-environment misalignment in individual cyanobacterial cells. Biophys. J. 111: 883–891.

Lear, J. D., Z. R. Wasserman, and W. F. DeGrado. 1988. Synthetic amphiphilic peptide models for protein ion channels. Science 240: 1177–1181.

Levy, S., M.Kafri, M. Carmi, and N. Barkai. 2011. The competitive advantage of a dual-transporter system. Science 334: 1408–1412.

Leypunskiy, E., J. Lin, H. Yoo, U. Lee, A. R. Dinner, and M. J. Rust. 2017. The cyanobacterial circadian clock follows midday *in vivo* and *in vitro*. eLife 6: e23539.

Litchman, E., C. A. Klausmeier, O. M. Schofield, and P. G. Falkowski. 2007. The role of functional traits and trade-offs in structuring phytoplankton communities: Scaling from cellular to ecosystem level. Ecol. Letters 10: 1170–1181.

Lomas, M. W., J. A. Bonachela, S. A. Levin, and A. C. Martiny. 2014. Impact of ocean phytoplankton diversity on phosphate uptake. Proc. Natl. Acad. Sci. USA 111: 17540–17545.

Long, H., W. Sung, S. Kucukyildirim, E. Williams, S. F. Miller, W. Guo, C. Patterson, C. Gregory, C. Strauss, C. Stone, et al. 2017. Evolutionary determinants of genome-wide nucleotide composition. Nature Ecol. Evol. 2: 237–240.

Loukin, S. H., M. M. Kuo, X. L. Zhou, W. J. Haynes, C. Kung, and Y. Saimi. 2005. Microbial K+ channels. J. Gen. Physiol. 125: 521–527.

Lycklama à Nijeholt, J. A., R. Vietrov, G. K. Schuurman-Wolters, and B. Poolman. 2018. Energy coupling efficiency in the Type I ABC transporter GlnPQ. J. Mol. Biol. 430: 853–866.

Lynch, M., and K. Hagner. 2014. Evolutionary meandering of intermolecular interactions along the drift barrier. Proc. Natl. Acad. Sci. USA 112: E30–E38.

Lynch, M., and G. K. Marinov. 2015. The bioenergetic costs of a gene. Proc. Natl. Acad. Sci. USA 112: 15690–15695.

Lynch, M., P. Schavemaker, T. Licknack, Y. Hao, and A. Pezzano. 2022. Evolutionary bioenergetics of ciliates. J. Euk. Microbiol. 69: e12934.

Lynn, D. H. 1982. Dimensionality and contractile vacuole function in ciliated protozoa. J. Exp. Zool. 223: 219–229.

Lyons, T. W., C. T. Reinhard, and N. J. Planavsky. 2014. The rise of oxygen in Earth's early ocean and atmosphere. Nature 506: 307–315.

Maeda, N., T. Kanai, H. Atomi, and T. Imanaka. 2002. The unique pentagonal structure of an archaeal Rubisco is essential for its high thermostability. J. Biol. Chem. 277: 31656–31662.

Maharjan, R. P., S. Seeto, and T. Ferenci. 2007. Divergence and redundancy of transport and metabolic rate-yield strategies in a single *Escherichia coli* population. J. Bacteriol. 189: 2350–2358.

Marañón, E. 2015. Cell size as a key determinant of phytoplankton metabolism and community structure. Ann. Rev. Mar. Sci. 7: 241–264.

Mazel, D., and P. Marlière. 1989. Adaptive eradication of methionine and cysteine from cyanobacterial light-harvesting proteins. Nature 341: 245–248.

Melis, A. 2009. Solar energy conversion efficiencies in photosynthesis: Minimizing the chlorophyll antennae to maximize efficiency. Plant Sci. 177: 272–280.

Mitchell, J. G. 1991. The influence of cell size on marine bacterial motility and energetics. Microbial Ecol. 22: 227–238.

Mitchell, J. G. 2002. The energetics and scaling of search strategies in bacteria. Amer. Nat. 160: 727–740.

Mofatteh, M., F. Echegaray-Iturra, A. Alamban, F. Dalla Ricca, A. Bakshi, and M. G. Aydogan. 2021. Autonomous clocks that regulate organelle biogenesis, cytoskeletal organization, and intracellular dynamics. eLife 10: e72104.

Morse, D., P. Salois, P. Markovic, and J. W. Hastings. 1995. A nuclear-encoded form II RuBisCO in dinoflagellates. Science 268: 1622–1624.

Mulkidjanian, A. Y., M. Y. Galperin, and E. V. Koonin. 2009. Co-evolution of primordial membranes and membrane proteins. Trends Biochem. Sci. 34: 206–215.

Naftalin, R. J., N. Green, and P. Cunningham. 2007. Lactose permease H+-lactose symporter: Mechanical switch or Brownian ratchet? Biophys. J. 92: 3474–3491.

Nakajima, M., K. Imai, H. Ito, T. Nishiwaki, Y. Murayama, H. Iwasaki, T. Oyama, and T. Kondo. 2005. Reconstitution of circadian oscillation of cyanobacterial KaiC phosphorylation *in vitro*. Science 308: 414–415.

Noordally, Z. B., and A. J. Millar. 2015. Clocks in algae. Biochemistry 54: 171–183.

O'Neill, J. S., G. van Ooijen, L. E. Dixon, C. Troein, F. Corellou, F. Y. Bouget, A. B. Reddy, and A. J. Millar. 2011. Circadian rhythms persist without transcription in a eukaryote. Nature 469: 554–558.

Oren, A. 1999. Bioenergetic aspects of halophilism. Microbiol. Mol. Biol. Rev. 63: 334–348.

Partch, C. L., C. B. Green, and J. S. Takahashi. 2014. Molecular architecture of the mammalian circadian clock. Trends Cell Biol. 24: 90–99.

Pasciak, W. J., and J. Gavis. 1974. Transport limitation of nutrient uptake in phytoplankton. Limnol. Oceanogr. 19: 881–888.

Pattanayak, G. K., G. Lambert, K. Bernat, and M. J. Rust. 2015. Controlling the cyanobacterial clock by synthetically rewiring metabolism. Cell Rep. 13: 2362–2367.

Patzlaff, J. S., T. van der Heide, and B. Poolman. 2003. The ATP/substrate stoichiometry of the ATP-binding cassette (ABC) transporter OpuA. J. Biol. Chem. 278: 29546–29551.

Phillips, R., and R. Milo. 2009. A feeling for the numbers in biology. Proc. Natl. Acad. Sci. USA 106: 21465–21471.

Phong, C., J. S. Markson, C. M. Wilhoite, and M. J. Rust. 2013. Robust and tunable circadian rhythms from differentially sensitive catalytic domains. Proc. Natl. Acad. Sci. USA 110: 1124–1129.

Pittayakanchit, W., Z. Lu, J. Chew, M. J. Rust, and A. Murugan. 2018. Biophysical clocks face a trade-off between internal and external noise resistance. eLife 7: e37624.

Pohorille, A., K. Schweighofer, and M. A. Wilson. 2005. The origin and early evolution of membrane channels. Astrobiology 5: 1–17.

Poudel, S., D. H. Pike, H. Raanan, J. A. Mancini, V. Nanda, R. E. M. Rickaby, and P. G. Falkowski. 2020. Biophysical analysis of the structural evolution of substrate specificity in RuBisCO. Proc. Natl. Acad. Sci. USA 117: 30451–30457.

Purcell, E. M. 1977. Life at low Reynolds number. Amer. J. Physics 45: 3.

Raven, J. A. 1982. The energetics of freshwater algae: Energy requirements for biosynthesis and volume regulation. New Phytol. 92: 1–20.

Raven, J. A. 2009. Contributions of anoxygenic and oxygenic phototrophy and chemolithotrophy to carbon and oxygen fluxes in aquatic environments. Aquatic Microb. Ecol. 56: 177–192

Raven, J. A. 2017. The possible roles of algae in restricting the increase in atmospheric CO_2 and global temperature. European J. Phycol. 52: 506–522.

Raven, J. A., J. Beardall, and P. Sánchez-Baracaldo. 2017. The possible evolution and future of CO_2-concentrating mechanisms. J. Exp. Bot. 68: 3701–3716.

Raven, J. A., and M. A. Doblin. 2014. Active water transport in unicellular algae: Where, why, and how. J. Exp. Bot. 65: 6279–6292.

Roenneberg, T., and M. Merrow. 2001. Circadian systems: Different levels of complexity. Philos. Trans. R. Soc. Lond. B Biol. Sci. 356: 1687–1696.

Rothschild, B. J., and T. R. Osborn. 1988. Small-scale turbulence and plankton contact rates. J. Plank. Res. 10: 465–474.

Rozhoňová, H., and J. L. Payne. 2021. Little evidence the standard genetic code is optimized for resource conservation. Mol. Biol. Evol. 38: 5127–5133.

Samani, P., and G. Bell. 2016. Experimental evolution of the grain of metabolic specialization in yeast. Ecol. Evol. 6: 3912–3922.

Sancar, A. 2008. The intelligent clock and the Rube Goldberg clock. Nat. Struct. Mol. Biol. 15: 23–24.

Sartor, F., Z. Eelderink-Chen, B. Aronson, J. Bosman, L. E. Hibbert, A. N. Dodd, Á. T. Kovács, and M. Merrow. 2019. Are there circadian clocks in non-photosynthetic bacteria? Biology (Basel) 8: E41.

Savir, Y., E. Noor, R. Milo, and T. Tlusty. 2010. Cross-species analysis traces adaptation of Rubisco toward optimality in a low-dimensional landscape. Proc. Natl. Acad. Sci. USA 107: 3475–3480.

Schulz, L., Z. Guo, J. Zarzycki, W. Steinchen, J. M. Schuller, T. Heimerl, S. Prinz, O. Mueller-Cajar, T. J. Erb, and G. K. A. Hochberg. 2022. Evolution of increased complexity and specificity at the dawn of form I Rubiscos. Science 378: 155–160.

Shenhav, L., and D. Zeevi. 2020. Resource conservation manifests in the genetic code. Science 370: 683–687.

Shih, P. M. 2015. Photosynthesis and early Earth. Curr. Biol. 25: R855–R859.

Shih, P. M., A. Occhialini, J. C. Cameron, P. J. Andralojc, M. A. Parry, and C. A. Kerfeld. 2016. Biochemical characterization of predicted Precambrian RuBisCO. Nat. Commun. 7: 10382.

Shih, P. M., L. M. Ward, and W. W. Fischer. 2017. Evolution of the 3-hydroxypropionate bicycle and recent transfer of anoxygenic photosynthesis into the Chloroflexi. Proc. Natl. Acad. Sci. USA 114: 10749–10754.

Short, M. B., C. A. Solari, S. Ganguly, T. R. Powers, J. O. Kessler, and R. E. Goldstein. 2006. Flows driven by flagella of multicellular organisms enhance long-range molecular transport. Proc. Natl. Acad. Sci. USA 103: 8315–8319.

Solari, C. A., K. Drescher, S. Ganguly, J. O. Kessler, R. E. Michod, and R. E. Goldstein. 2011. Flagellar phenotypic plasticity in volvocalean algae correlates with Péclet number. J. R. Soc. Interface 8: 1409–1417.

Spreitzer, R. J., and M. E. Salvucci. 2002. Rubisco: Structure, regulatory interactions, and possibilities for a better enzyme. Annu. Rev. Plant Biol. 53: 449–475.

Stefan, E., S. Hofmann, and R. Tampé. 2020. A single power stroke by ATP binding drives substrate translocation in a heterodimeric ABC transporter. eLife 9: e55943.

Stevens, T. J., and I. T. Arkin. 2000. Do more complex organisms have a greater proportion of membrane proteins in their genomes? Proteins 39: 417–420.

Stroppolo, M. E., M. Falconi, A. M. Caccuri, and A. Desideri. 2001. Superefficient enzymes. Cell. Mol. Life Sci. 58: 1451–1460.

Tcherkez, G. G., G. D. Farquhar, and T. J. Andrews. 2006. Despite slow catalysis and confused substrate specificity, all ribulose bisphosphate carboxylases may be nearly perfectly optimized. Proc. Natl. Acad. Sci. USA 103: 7246–7251.

Terauchi, K., Y. Kitayama, T. Nishiwaki, K. Miwa, Y. Murayama, T. Oyama, and T. Kondo. 2007. ATPase activity of KaiC determines the basic timing for circadian clock of cyanobacteria. Proc. Natl. Acad. Sci. USA 104: 16377–16381.

Thauer, R. K. 2007. A fifth pathway of carbon fixation. Science 318: 1732–1733.

Tilman, G. D. 1982. Resource Competition and Community Structure. Monographs in Population Biology, Vol. 17. Princeton University Press, Princeton, NJ.

Tortell, P. D. 2000. Evolutionary and ecological perspectives on carbon acquisition in phytoplankton. 45: 744–750.

Uemura, K., Anwaruzzaman, S. Miyachi, and A. Yokota. 1997. Ribulose-1,5-bisphosphate carboxylase/oxygenase from thermophilic red algae with a strong specificity for CO_2 fixation. Biochem. Biophys. Res. Commun. 233: 568–571.

van Ooijen, G., L. E. Dixon, C. Troein, and A. J. Millar. 2011. Proteasome function is required for biological timing throughout the twenty-four hour cycle. Curr. Biol. 21: 869–875.

Visser, A. W., and T. Kiørboe. 2006. Plankton motility patterns and encounter rates. Oecologia 148: 538–546.

Wallin, E., and G. von Heijne. 1998. Genome-wide analysis of integral membrane proteins from eubacterial, archaean, and eukaryotic organisms. Protein Sci. 7: 1029–1038.

Waygood, E. B., and T. Steeves. 1980. Enzyme I of the phosphoenolpyruvate: Sugar phosphotransferase system of *Escherichia coli*. Purification to homogeneity and some properties. Can. J. Biochem. 58: 40–48.

Wilson, R. H., E. Martin-Avila, C. Conlan, and S. M. Whitney. 2018. An improved *Escherichia coli* screen for Rubisco identifies a protein–protein interface that can enhance CO_2-fixation kinetics. J. Biol. Chem. 293: 18–27.

Woese, C. R. 1987. Bacterial evolution. Microbiol. Rev. 51: 221–271.

Woese, C. R, B. A. Debrunner-Vossbrinck, H. Oyaizu, E. Stackebrandt, and W. Ludwig. 1985. Gram-positive bacteria: Possible photosynthetic ancestry. Science 229: 762–765.

Wykoff, D. D., A. H. Rizvi, J. M. Raser, B. Margolin, and E. K. O'Shea. 2007. Positive feedback regulates switching of phosphate transporters in *S. cerevisiae*. Mol. Cell. 27: 1005–1013.

Xu, H., and J. Zhang. 2021. Is the genetic code optimized for resource conservation? Mol. Biol. Evol. 38: 5122–5126.

Ye, L., J. A. Berden, K. van Dam, and A. L. Kruckeberg. 2001. Expression and activity of the Hxt7 high-affinity hexose transporter of *Saccharomyces cerevisiae*. Yeast 18: 1257–1267.

Young, J. N., R. E. Rickaby, M. V. Kapralov, and D. A. Filatov. 2012. Adaptive signals in algal Rubisco reveal a history of ancient atmospheric carbon dioxide. Philos. Trans. R. Soc. Lond. B Biol. Sci. 367: 483–492.

Zehr, J. P., J. S. Weitz, and I. Joint. 2017. How microbes survive in the open ocean. Science 357: 646–647.

Zhou, Y., and S. Whitney. 2019. Directed evolution of an improved Rubisco; *in vitro* analyses to decipher fact from fiction. Int. J. Mol. Sci. 20: 5019.

CHAPTER 19

Enzymes and Metabolic Pathways

Virtually every aspect of cell biology relies on intermolecular interactions, some of which extend into long, multistep pathways. In many pairwise interactions, the partners represent a coevolutionary loop with each participant constituting part of the selective environment of the other. This is true, for example, of crosstalk involving DNA-protein interactions, such as transcription factors and their binding sites (Chapter 21). It also applies to protein-protein interactions underlying the cell cycle (Chapter 9), conferring specificity in vesicle trafficking in eukaryotes (Chapter 15), and transmitting environmental information (Chapter 22). Metabolic pathways are different in that there is typically no direct physical contact between the major players, with each simply imposing a chemical modification on a substrate, which is then passed on to another enzyme. Most cellular processes involving resource acquisition and biosynthesis involve multiple steps of this sort.

Catabolic pathways break down complex organic molecules into simpler ones for a variety of reasons: the release of energy for use in other cellular processes; conversion to simpler precursors for use in specific biosynthetic pathways; and detoxification of harmful substances. Sometimes an intermediate metabolite in a pathway has no function other than to serve as a link between two compounds that are useful. Frequently, a link serves multiple purposes, e.g., generating a useful precursor molecule for a biosynthetic pathway, and in the process regenerating ATP and/or NADH. Anabolic pathways, which require energy input, are responsible for the synthesis and elaboration of the constituents of proteins, nucleic acids, lipids, and other molecules from which cells are built.

The features of metabolic pathways motivate several fundamental evolutionary questions. First, what limits the level of performance of their enzymatic constituents? Given that biomass and energy production are central to all of biology, and given that the costs of protein machineries are substantial, one would expect metabolic enzymes to have high efficiencies and specificities so as to minimize the infrastructure costs. Second, what are the evolutionary sources of the enzymes that enable the emergence of novel metabolic features? One expects a descent-with-modification scenario, with the emergence of new functions requiring moderate numbers of amino-acid substitutions, but how does a specialized enzyme already integrated into metabolism become free to take on a new activity? Third, what determines the structure of metabolic pathways, and how recalcitrant are their component parts to replacement? Again, one might expect such features to be among the most conserved aspects of cell biology.

Comparative analyses of metabolic machineries provide surprising answers to all of these questions. Despite the impressive performance of enzymes, their catalytic efficiencies are generally far below the biophysical limits, whereas inherent inaccuracies provide latent potential for the emergence of new catalytic functions. Despite the centrality of metabolic pathways to all of life, both their structure and the identities of the underlying participants exhibit phylogenetic variation. As with many other cell biological features, numerous aspects of the metabolic machinery appear to be influenced by the intervention of nonadaptive processes.

Evolutionary Cell Biology. Michael Lynch, Oxford University Press. © Michael Lynch (2024). DOI: 10.1093/oso/9780192847287.003.0019

Enzymes

Essentially all metabolic pathways are driven by enzymatic proteins whose primary roles are to facilitate the transformation of substrates into products. Thus, any attempt to understand the mechanisms of metabolic-pathway evolution must begin with a basic appreciation of the fundamental features of enzymes. First, enzymes generally do not create entirely novel biochemical reactions, but simply magnify transformation rates beyond what happens spontaneously. The level of enhancement of catalytic rates can be enormous, typically in the range of 10^7 to $10^{26} \times$ the corresponding rates in the absence of enzyme (Wolfenden and Snider 2001; Edwards et al. 2012). Second, enzymes are recyclable – after they disengage with a substrate/product molecule, they are generally free to enter into a new reaction. Third, most enzymes carry out very simple reactions, usually involving no more than two substrates. Such simplicity likely results from the diminishingly small probabilities of joint encounters of multiple sets of molecules. Fourth, there are usually no more than one or two substrate modifications (in the form of chemical-bond changes) per enzymatic step. A natural consequence of such simplicity is the necessity of multistep pathways for the complete breakdown or synthesis of complex molecules.

Basic enzymology

Although most enzymes catalyse simple chemical reactions, these are almost always targeted to highly specialized substrates. Biochemists generally classify enzymes into six categories according to the reaction types in which they engage, although layers of subdivision also enter the overall classification scheme. Oxidoreductases (e.g., oxidases and dehydrogenases) transfer electrons from one substrate to another, whereas transferases (e.g., transaminases and kinases) transfer larger functional groups (e.g., amine or phosphate groups) from one substrate to another. Hydrolases (e.g., lipases, phosphatases, and peptidases) cleave chemical bonds by addition of water, leading to the breakdown of one substrate into two product molecules, whereas lyases (e.g., decarboxylases)

break bonds in a nonhydrolytic fashion. Isomerases make molecular changes within a single molecule, whereas ligases (e.g., synthetases) join two molecules.

Although the structural features of enzymes dictate their chemical potential and specificity, the actual reaction rates are highly dependent on the substrate concentration. The basic issues are outlined in Foundations 19.1, which derives the famous two-parameter formulation of Michaelis and Menten (1913) describing this relationship. Under this model, an enzymatic reaction rate is viewed as the net outcome of two processes: 1) the rate of productive encounters between enzyme and substrate, which is mainly a function of physical factors and molecular densities; and 2) the rate of the ensuing chemical transformation, which is a function of enzyme structure and the nature of the interaction. These features define two key quantitative parameters for every enzyme–substrate interaction: 1) the maximum reaction rate (per active site of the enzyme) at saturating substrate concentration, k_{cat}, which is a measure of catalytic efficiency once engaged with a substrate molecule (often also referred to as the enzyme turnover number); and 2) the substrate concentration at which the reaction rate equals half the maximum rate, K_S, which is a function of the physical rates of association and dissociation between enzyme and substrate (Figure 19.1). The hyperbolic Michaelis–Menten relationship between the rate of product formation and substrate concentration arises in many other contexts in biology, including nutrient uptake and growth responses to nutrient levels (Chapters 9 and 18).

To understand how the kinetic parameters of enzyme performance are guided by natural selection, it is useful to know the typical substrate concentrations within the intracellular environment. At sufficiently high concentrations, such that the time intervals between substrate encounter rates are negligible, an enzymatic reaction rate is defined by k_{cat} alone, whereas at sufficiently low concentrations the reaction rate is a function of the composite parameter, k_{cat}/K_S, which is the slope of the rate response to low substrate concentrations (generally called the kinetic efficiency; Foundations 19.1). If substrate concentrations are typically $\gg K_S$, selection

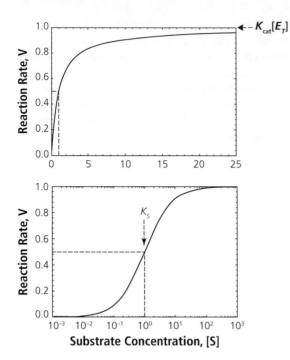

Figure 19.1 Michaelis–Menten enzyme kinetics as a function of substrate concentration, [S], as defined in Equation 19.1.4. The maximum rate $k_{cat}[E_T]$ is arbitrarily set to equal 1.0, with k_{cat} being the maximum catalytic rate, $[E_T]$ the enzyme concentration, and K_S the half-saturation constant (here also set equal to 1.0). The plot appears hyperbolic on an arithmetic scale and sigmoid on a logarithmic scale.

for higher reaction rates would largely be focused on the catalytic efficiency (k_{cat}), but at low substrate concentrations, a reduction in K_S may also be favored.

Selection can enhance the flux rate of a reaction in three ways: 1) by increasing the rate of association of enzyme and substrate (k_a); 2) by decreasing the backwards rate of disengagement (k_d); and 3) by increasing the forward rate of catalysis (k_{cat}). Crowley (1974) refers to such changes as influences on affinity, efficiency, and velocity. Selection is expected to act in a directional fashion on each of these three parameters. However, there are biophysical limits to what can be accomplished. Moreover, as can be seen from Equations 19.1.1c and 19.1.2, changes in k_{cat} influence both the maximum reaction rate and the

half-saturation constant, so these two features are not entirely independent (see also Foundations 18.1).

Only a few attempts have been made to estimate the intracellular concentrations of metabolites. The bacterium *E. coli* is one of the few species for which kinetic parameters are available for a large number of enzymes, and in this case for cells in exponential-growth phase, most metabolite concentrations are in excess of the K_S values of their associated enzymes (Bennett et al. 2009). Indeed, ~60% of metabolites (including ATP and NAD^+) have concentrations more than tenfold higher than their associated K_S, implying near saturation-level flux rates. Given the costs of protein biosynthesis, at least for this set of enzymes in this species, this suggests that selection puts a premium on active-site features that maximize the flux per unit enzyme (i.e., high k_{cat}). Nonetheless, as we discuss next, k_{cat} is often far below what seems possible biophysically.

Degree of molecular perfection

A perfect enzyme would have such a high catalytic efficiency that the reaction rate is limited only by the rate of encounters with substrate molecules. In this case, the reaction would be effectively instantaneous following a productive co-localization of enzyme and substrate molecules. Thus, understanding the determinants of enzyme–substrate encounter rates is an essential starting point to evaluating the limits of what is possible versus what natural selection has actually accomplished.

The rate at which randomly and homogeneously distributed particles engage with each other in an aqueous medium can be viewed as the product of three factors: 1) the rate of encounter by diffusion-like processes (k_e); 2) a potential multiplier to account for further attractive (or repulsive) charges between enzyme and substrate ($m_e > 1$ or < 1, respectively); and 3) the fraction of encounters that are in the correct orientation for proper engagement (p_e),

$$k_{enc} = k_e m_e p_e. \tag{19.1}$$

Assuming that encounters between enzyme (E) and substrate (S) occur via a three-dimensional diffusion process, and that both types of molecules can be approximated as spheres, with the latter being much smaller than the former, the diffusion limit is

$$k_e \simeq (1.7 \times 10^9) \left(\frac{r_E}{r_S} \right), \qquad (19.2)$$

in units of $M^{-1}sec^{-1}$, where r_E and r_S are the effective radii of enzyme and substrate particles, and M denotes the number of moles per litre (Foundations 7.2). The assumption that $r_E \gg r_S$ is entirely reasonable for metabolic enzymes. For example, the 20 amino acids have individual masses ranging from 75 to 204 g/mol, whereas the molecular weight (g/mol) of glucose is 180, of NADH is 663, of ATP is 507, and of nucleosides A, G, C, and T is 135, 151, 111, and 126 g/mol, respectively. Most enzymes contain on the order of a few hundred amino acids. Thus, noting that a radius is roughly proportional to the one-third power of mass, depending on the enzyme and substrate, for metabolic reactions r_E/r_S will typically be in the neighborhood of 10–100, implying a k_e on the order of 10^{10} to $10^{11} M^{-1}sec^{-1}$.

Although numerous factors can cause the overall encounter rate, k_{enc}, to deviate from the expectation based on diffusion alone, much less is known about the two other components in Equation 19.1. Early attempts to estimate the fraction of productive encounters, p_e, generally assumed that overlap of the substrate molecule with a small circular patch on the surface of the enzyme is critical for proper adhesion (Berg and von Hippel 1985; Janin 1997; Camacho et al. 2000). However, simply multiplying the overall encounter rate by the fraction of properly oriented molecules yields rates that are too low (Berg 1985; Schlosshauer and Baker 2004), probably because some initially unfavorable encounters undergo rotational diffusion, with the two molecules sliding around each other rather than completely disengaging. If at least one contact point is correct, rotational movement may quickly locate a second favorable contact, thereby facilitating the remaining fine tuning (Northrup and Erickson 1992). All of these things taken together, orientation limitations can still reduce k_{enc} by as much as 100 to 1000 fold (Schlosshauer and Baker 2004), i.e., p_e is commonly <0.01.

Less clear is the degree to which electrostatic-charge effects enhance an enzyme's likelihood of productive encounters (Zhou 1993). For a very well-studied enzyme, barnase, and its inhibitor, barstar, Janin (1997) estimated m_e to be $\simeq 10^5$, but because $p_e \simeq 10^{-5}$ for this particular interaction, electrostatic effects and hydrophobic steering essentially compensate for each other, rendering $k_{enc} \simeq k_e$. One final point to consider is that the internal milieu of a cell is hardly the open-water environment assumed in most diffusion theory, with molecular crowding reducing the rate of diffusion of proteins by at least tenfold (Chapter 7).

A summary of the preceding considerations allows a rough statement on the maximum achievable rate of an enzymatic reaction, conditioned on the internal state of today's cells. As noted above, the ideal situation would be one in which the overall reaction rate is dictated only by the rate of encounter, which is expected to be on the order of 10^{10} to $10^{11} M^{-1}sec^{-1}$ in a purely aqueous environment, assuming spherical particles with all contacts being productive regardless of orientation. However, after accounting for inappropriate orientations and the effects of molecular crowding in a cellular environment, the best that may be physically achievable inside cells is an association rate of order 10^8 to $10^9 M^{-1}sec^{-1}$.

How does this biophysical limit to catalytic rates compare with the evolved capacities of enzymes? Returning to the Michaelis–Menten formula, Equation 19.1.4, we see that at low substrate concentrations (the situation under diffusion limitation), the reaction rate is closely approximated by $k_{cat}[E_T][S]/K_S$, where the quantities in brackets denote the molar concentrations (M, in moles per litre) of enzyme and substrate. Because $K_S \simeq k_{cat}/k_e$ (Foundations 19.1), $k_{cat}[E_T][S]/K_S$ is equivalent to $k_e[E_T][S]$, showing that the preceding arguments on the upper limit to k_e also apply to the kinetic efficiency, k_{cat}/K_S, an observable composite parameter.

Using this logic and early observations on triosephosphate isomerase, Albery and Knowles (1976a,b) argued that many enzymes have been pushed to perfection by natural selection, and hence, have reached the end of their 'evolutionary development'. However, summarizing kinetic data for thousands of enzymes in dozens of species, Bar-Even et al. (2011) and Davidi et al. (2018) reached a substantially different conclusion. The average

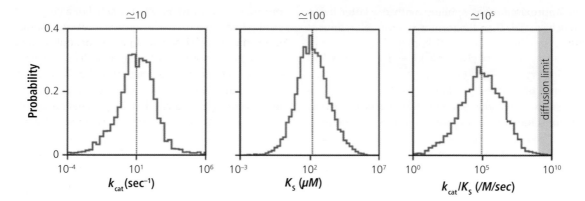

Figure 19.2 Distributions of kinetic features of enzymes surveyed across the Tree of Life. Sample sizes are 3500, 12576, and 4451 for the three parameters, almost all determined *in vitro*. Median values are denoted by the vertical dashed lines. From Davidi et al. (2018).

enzyme has a $k_{cat}/K_S \simeq 10^5 \, M^{-1}sec^{-1}$, several orders of magnitude below the apparent diffusion limit (Figure 19.2). Although most rate estimates are derived from *in vitro* aqueous environments, they are highly correlated with *in vivo* estimates (Davidi et al. 2016), with the average k_{cat}/K_S of the latter for specific enzymes being somewhat lower than the *in vitro* estimates but generally by no more than 10×.

The simplest interpretation of these observations is that, on average, only one in 10^3 to 10^4 collisions between an enzyme and a substrate molecule typically results in a productive interaction (Bar-Even et al. 2015). A similar conclusion is reached if one considers k_{cat} alone. Based on a number of biochemical considerations (e.g., electrostatic and hydrogen-bond interactions, bond cleavage, exclusion of water molecules), the approximate upper bound of k_{cat} for metabolic reactions $\simeq 10^6 \, sec^{-1}$ (Hammes 2002; Benkovic and Hammes-Schiffer 2003). However, in *E. coli*, average $k_{cat} \simeq 10 \, sec^{-1}$, with an approximate range of 10^{-2} to 10^3, and a survey based on a smaller number of enzymes in the plant *Arabidopsis* yields k_{cat} estimates of a similar order of magnitude (Davidi et al. 2016, 2018; Chen and Nielsen 2021).

Taken together, these observations suggest that enzyme efficiencies are generally far below what should be physically possible, one of the most dramatic examples being the RuBisCO protein discussed in Chapter 18. The most compelling explanation for such striking disparities is a form of the drift-barrier hypothesis (Hartl et al. 1985; Heckmann et al. 2018), which postulates a sort of diminishing-returns epistasis, such that as catalytic rates reach higher and higher levels, additional improvements enhance organismal fitness to a lesser and lesser extent. This requires that the mapping of fitness on catalytic rate eventually reaches a plateau, much like the Michaelis–Menten relationship itself. One reason why such a relationship is likely to exist is discussed below in the context of the sensitivity of the fluxes through enzymatic pathways – such networks dilute the selection intensity on the individual components.

Consistent with this theory, Bar-Even et al. (2011) found that enzymes involved in secondary metabolism are on average ∼30× less efficient than those involved in central metabolism, as expected if selection operates less efficiently on downstream enzymes. The assumption here is that variants associated with secondary metabolism have diminished fitness effects owing to their lower degree of entrenchment in cellular biochemistry. In addition, prokaryotic enzymes have slightly better kinetics than those from eukaryotes (Bar-Even et al. 2011; Hanson et al. 2021), as expected for species with higher effective population sizes.

An additional issue of relevance to the drift-barrier hypothesis for the low catalytic rates of enzymes is that the flux rate of a reaction is equal to the product of the catalytic rate per enzyme molecule and the enzyme concentration. Proteins are energetically costly (Chapter 17), and any increase in the catalytic rate per enzyme molecule would presumably reduce the number of molecules required for reactions, thereby freeing up resources for other cellular functions. This further emphasizes the surprisingly low catalytic efficiency of enzymes. An alternative argument for low catalytic rates is that enzymes face biophysical compromises, such that any improvement in efficiency comes at the cost of enzyme stability, solubility, or accuracy. However, although such trade-offs surely exist, they are not universal (Nguyen et al. 2017; Klesmith et al. 2017), as indicated by numerous observations on protein chemistry (Chapter 12). Thus, the relatively low efficiencies of enzymes remains one of the big unsolved mysteries of evolutionary biochemistry.

Finally, it is worth noting that enzyme molecules have finite life spans, owing to self-inflicted damage generated by reaction chemistry. The working hypothesis is that enzyme failure is not so much due to a progressive erosion in functionality (i.e., molecular senescence) as to random but permanent inactivation per enzymatic reaction (Hanson et al. 2021). The mean number of catalytic cycles in the lifespan of an enzyme is on the order of 10^4 to 10^5. This imposes a cost in terms of molecular replacement to maintain steady-state conditions (Chapter 17).

Enzyme promiscuity

In addition to being far from perfect in terms of catalytic rates and damage avoidance, most enzymes also bind to inappropriate substrates at fairly high frequencies, the case of RuBisCO discussed in Chapter 18 providing a dramatic example. Although quantitative analyses have been largely confined to bacteria, error rates at normal cellular concentrations of off-target substrates are commonly in the range of 0.1% to well over 1% (e.g., Wilks et al. 1988; Yano et al. 1998; Rothman and Kirsch 2003; Rakus et al. 2008; Ge et al. 2014; Fernandes et al. 2015). A review of ∼250 enzymes in *E. coli* suggests an average error rate of 0.5% (Notebaart et al. 2014), with 37% of metabolic enzymes engaging with multiple substrates at biologically significant rates (Nam et al. 2012)

Additional large-scale studies in *E. coli* illustrate the dramatic reservoir of latent activities of proteins. For example, ∼20% of single-gene knockouts in this species can be rescued by overexpression of at least one non-cognate gene (Patrick et al. 2007). Starting with a full library of *E. coli* constructs, each overexpressing a single endogenous protein residing on a plasmid, Soo et al. (2011) found measurable resistance to 237 novel (and likely never-before encountered) toxins and antibiotics in ∼0.4% of cases. In addition, using repeated cycles of mutagenic polymerase chain reaction, cloning, transformation, and selective challenge, Aharoni et al. (2005) generated mutants for phosphotriesterase capable of hydrolysing substrates that had been rarely, if ever, encountered before. In this latter study, depending on the substrate selected for, the ancestral gene function sometimes declined but in other cases it increased, the end result being proteins with substantially remodelled catalytic sites with their own novel profiles of promiscuous functions.

The nature of enzyme promiscuity ranges from substrate ambiguity (catalysis of the same reaction with alternative substrates) to catalytic ambiguity (catalysis of different reactions with the same substrate) (Copley 2003; Khersonsky et al. 2006; Tokuriki and Tawfik 2009; Babtie et al. 2010; Jia et al. 2013). A dramatic example of catalytic ambiguity derives from observations on errors in the production of the energy storage polymer glycogen. Glycogen polymers are built as glycogen synthase progressively incorporates glycosyl moieties from the precursor uridine diphosphate-glucose, but a fraction ∼10^{-4} of the time a phosphate (from uridine diphosphate) is erroneously inserted (Tagliabracci et al. 2011). In some cases promiscuity can be condition dependent, with the specific reactions and substrates differing depending on the local chemistry dictated by cell physiological states.

Although the cost of enzyme promiscuity for cell performance and fitness remains unclear, its ubiquity implies that, when confronted with new biochemical challenges, adaptation need not await the mutational origin of new functions from scratch (Copley 2020). Indeed, despite the relatively low kinetic rates of promiscuous reactions, they are still generally orders of magnitude greater than rates of non-catalysed reactions, thereby providing a well-endowed starting point for adaptive exploitation (even if far from perfect). Even completely randomly designed proteins have a high probability of having some (albeit weak) functions, as they often harbor surface pockets capable of a variety of catalytic activities (Skolnick and Gao 2013).

Not surprisingly, it has been repeatedly found that given sufficient selection pressure and a large enough population size, cells are capable of evolving novel metabolic features in remarkably short periods of time (often in < 1000 cell divisions in laboratory evolution experiments). These sorts of experiments typically start with complete deletions of a key enzyme, and then reveal the emergence of compensatory structural and/or regulatory changes at seemingly unrelated loci (McLoughlin and Copley 2008; Kim et al. 2010, 2019; Blank et al. 2014). Evolutionary advancement frequently initiates with the fortuitous appearance of just a few mutations enhancing a baseline level of promiscuity followed by a relatively rapid remodelling of the catalytic core (Matsumura and Ellington 2001; Copley 2014, 2020; Kim et al. 2019; Morgenthaler et al. 2019). Many long-term evolution studies with *E. coli* have established novel pathways for the uptake and utilization of non-native carbon resources, through broadened substrate utilization mechanisms involving specific enzymes (Boronat et al. 1983; Lee and Palsson 2010) and/or by the emergence of novel transporters for resource uptake (Blount et al. 2012; Quandt et al. 2014).

With a > 3-billion-year legacy of evolution distributed over countless proteins in an enormous number of phylogenetically independent lineages, there may be few enzymatic reactions whose evolution is absolutely dependent on *de novo* mutations. Nowhere is the evidence for this more striking than in the field of industrial enzymology, where substantial progress has been made in developing enzymes with novel specificities, activities, and stability, again typically starting with pre-existing proteins with promiscuous functions and progressively modifying these via strategies involving random mutagenesis (Johannes and Zhao 2006; Hult and Berglund 2007; Brustad and Arnold 2011; Bornscheuer et al. 2012; Abatemarco et al. 2013).

Pathway Flux Control

Metabolite processing often involves a series of steps carried out by different enzymes, the glycolytic pathway being one of many notable examples (Figure 19.3). This raises questions about the degree to which the overall flux rates of final products are dictated by the properties of the individual pathway components. Some aspects of these issues can be evaluated by treating the flux of metabolites through a pathway as a steady-state process with the metabolite entering the pathway having a constant concentration (as would be the case with a stable rate of nutrient intake) and no feedback inhibition from the final product. Under such a framework, the internal metabolites in the pathway will also be maintained at steady-state concentrations, so that the rate of entry into the pathway must equal the rates of flux through each subsequent step, including the final output from the system.

As outlined in Foundations 19.2, the total flux rate through a linear pathway is a function of the enzyme kinetics operating at each step. The transition rates between all pairs of adjacent metabolites are in turn defined by both the concentrations and the biochemical features of each enzyme in the pathway. The general expression for the equilibrium flux rate is a fairly complicated function of the features at individual steps, showing that there are innumerable ways to regulate overall pathway flux. However, one fairly simple question can be asked (and potentially answered). How dependent is the overall flux rate on the features of enzymes at successive positions in the pathway?

Pathway position and the strength of selection

Consider a linear metabolic pathway of the form $S_1 \rightarrow S_2 \rightarrow S_3 \rightarrow \cdots \rightarrow P$, where the terms S_x denote a linear array of modified metabolites, eventually leading to the final product P. The arrows on the pathway denote potential points of control associated with the enzyme at each position. Mathematical analysis leads to the prediction that the sensitivity of the overall flux rate to P declines dramatically (exponentially) with increasing downstream position of enzymes in

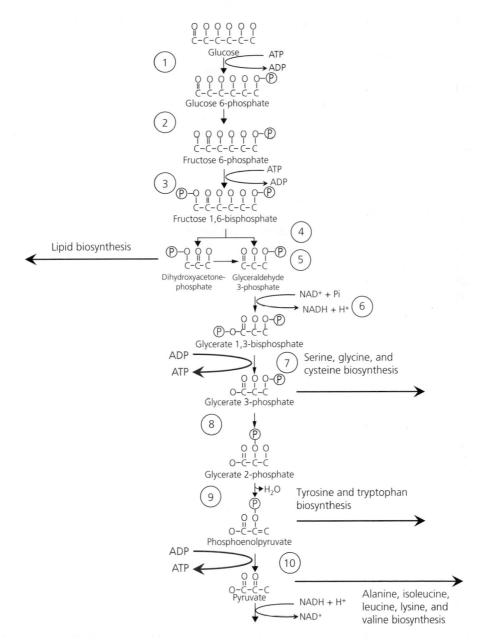

Figure 19.3 The canonical pathway of glycolysis, with 10 different enzymes operating in a stepwise fashion to produce two pyruvates from a glucose molecule (only one pyruvate is shown). As noted by the side arrows, a number of the intermediate metabolites are used as entry points to various biosynthetic pathways. (Note that the fate of only one of the 3-carbon products is given below step 5). Two ATPs are consumed early in the process, but four are produced in downstream steps (two for each pyruvate), giving a net yield of two ATPs per glucose molecule consumed.

the pathway (Foundations 19.2). This prediction appears to also hold for segments of branched pathways, with enzymes appearing earlier in branches being under stronger selective constraint (Rausher 2012). If this hypothesis is correct, enzymes higher up in a pathway should exhibit signs of stronger selection for the level of molecular refinement.

Although there is need for more comparative work, the existing data are at least superficially in accord with this expectation. For example, consistent with the flux-control model, for the linear pathways for amino-acid biosynthesis in *E. coli*, the response rates and the maximum-response levels to amino-acid depletion decline with the position of enzymes in the pathway (Zaslaver et al. 2004). In the anthocyanin-production pathway in plants, upstream enzymes evolve more slowly than those downstream (Rausher et al. 1999; Lu and Rausher 2003), and this pattern has been observed with multiple biosynthetic pathways in plants and other eukaryotes (Vitkup et al. 2006; Ramsay et al. 2009; Wright and Rausher 2010). Arguably, these differences reflect a reduction in the efficiency of purifying selection operating to remove mildly deleterious mutations from downstream genes. Genes whose products are associated with pathways with higher flux rates also exhibit low rates of amino-acid sequence evolution (Colombo et al. 2014).

The significant influence of pathway position is also illustrated by numerous studies in bacteria, yeast, and mammals showing that genes whose products have more interacting partners evolve more slowly than those with fewer interactors, i.e., are more constrained from accepting amino-acid changes (Fraser et al. 2002; Vitkup et al. 2006; Yang et al. 2009; Montanucci et al. 2011; Aguilar-Rodríguez and Wagner 2018). Such patterns are consistent with the general flux-control model, given that highly connected genes must on average enter earlier into pathways than more downstream genes with fewer interacting partners. Note, however, that enzymes entering into a pathway upstream of a large number of branchpoints will also tend to have more pleiotropic effects, which could magnify the overall strength of purifying selection for reasons other than pathway location and flux control (Ramsay et al. 2009).

Finally, the prediction of greater control of overall pathway flux by upstream enzymes is not absolute. All other things being equal, changes in upstream enzymes are likely to be more effective. But all other things may not always be equal. For example, some enzymes may have architectures that make them more or less vulnerable to change for reasons unassociated with pathway position, and enzymes that are utilized in multiple contexts may be constrained with respect to overall cellular concentrations.

Speed versus efficiency

Beyond the issue of the relatively low catalytic rates of many enzymes, most catabolic pathways are much less efficient at extracting energy from substrate molecules than seems possible. For example, glycolysis, a significant mechanism for ATP production in most cells, yields a net gain of just two ADP-to-ATP conversions per glucose molecule consumed, whereas the energy contained within glucose is sufficient for up to four such conversions. As a consequence, as glycolysis converts single glucose molecules to two lactic acid molecules, only about half of the energy released is stored in the form of ATPs produced, the remainder being lost as heat to the environment (as demonstrated by any well-nourished compost pile). Given that energy can be hard to come by, why is so much change left on the counter? One possibility, outlined in Foundations 19.3, relates to the intrinsic trade-off between the rate of a chemical reaction and the efficiency of product formation.

As noted above, assuming constant concentrations of substrate and product molecules, a system of metabolic reactions will reach a steady-state flux rate, with the rate of resource consumption being balanced by the rate of product output. Just as the rate of diffusion of molecules across a boundary layer is proportional to the concentration gradient, the rate of flux through a chemical reaction is proportional to the energy gradient across the pathway. With increasing efficiency of energy extraction, there is reduced flux through the overall system, the result being that the total rate of energy harvesting (the product of the flux rate of the carbon source and the efficiency of conversion to ATP per molecule consumed) is maximized at some intermediate level of efficiency. In the case of glycolysis, the rate of ATP production is maximized when $\sim$2 ATPs are produced per glucose consumed (Foundations 19.3), essentially what is observed within cells. Whereas, in principle, 4 ATPs could be produced from the energy of one glucose molecule, this would reduce the total rate of ATP production by 90%. Thus, energy dissipation in terms of heat loss seems to be an intrinsic requirement if natural selection is focused on maximizing the rate of flux rather than on the efficiency of resource utilization.

Note that this sort of pathway outcome illustrates the use of digital traits in cell biology. One molecule of glucose might yield 1, 2, 3, or 4 ATPs, but not 1.5 or 2.5, even if such an output might be mathematically optimal. Glycolytic conversion of glucose to lactic acid yields $\sim$205 kJoules/mol glucose consumed, whereas each conversion of ADP to ATP stores 50 kJoules/mol. Fortuitously perhaps, the optimal (but not possible) rate of ATP production of 2.05 per glucose molecule (Foundations 19.3) is very close to the realized value of 2, and $\sim$25% better (in terms of the ATP production rate) than would be the case if 1 or 3 ATPs were produced.

Although concepts of evolutionary trade-offs or constraints pervade the evolutionary ecology literature, most invocations are at best based on intuition (which need not be correct) or at worst entirely *ad hoc*. Justified or not, such arguments are predicated on the assumption that natural selection has unlimited power, save for biophysical constraints. Here, however, we are confronted with real limitations of biological systems. In principle, depending on the overall energetics of a pathway and the nature of the substrate molecule, the optimum strategy could be further from the integer possibilities than the outcome observed for the glycolytic pathway (e.g., an optimum of 2.50 rather than 2.05).

These kinds of speed versus efficiency trade-offs exist in a number of other contexts in cell biology. For example, microbes that exhibit high growth yields per carbon molecule consumed also have lower levels of heat dissipation (von Stockar and Liu 1999). Wagoner and Dill (2019) have considered this kind of trade-off for various molecular machines,

finding that, for cases such as ATP synthase, optimization has placed a premium on efficiency, perhaps not surprisingly given that the main mission of this machine is energy (ATP) production. (In contrast, glycolysis produces intermediate metabolites that serve as precursors for various biosynthetic pathways).

Pathway Expansion and Contraction

One of the most striking aspects of metabolism, at least to a non-biochemist, is the large number of incremental pathway steps often used to accomplish what superficially seem to be relatively simple molecular alterations. Many enzymes produce intermediate metabolites whose sole role is to be passed on to another type of enzyme for further processing.

Consider, for example, glycolysis, which requires 10 separate enzymes to convert one 6-carbon/6-oxygen glucose molecule to two 3-carbon/3-oxygen pyruvates (Figure 19.3). As just noted, a few of the intermediates do serve as precursors for other pathways for building-block production, but pyruvate is a key entry point for cellular energy production, and the ten steps to get there might seem excessive. However, Bar-Even et al. (2012) argue that intrinsic constraints of extant biochemistry (thermodynamic limitations, limited availability of enzymatic mechanisms, and physicochemical properties of pathway intermediates) make any alternative routes from glucose to pyruvate implausible. In other words, their view, which was extended to all of central metabolism by Noor et al. (2010), is that given the evolved structure of biochemistry and the intermediate metabolites that must be relied upon for downstream biosynthetic pathways, there is no shorter pathway between glucose and pyruvate.

Of course, this begs the question as to why biochemistry evolved to have such a structure. As we discuss next, cases do exist in which individual proteins are capable of carrying out multiple reactions. Moreover, as it is highly unlikely that the extended catabolic and anabolic pathways known in today's organisms were present at the moment of life's origin, deeper questions arise as to how complex metabolic pathways become assembled evolutionarily. As an entrée to this particular problem, we start with a consideration of the theoretical issues, and then transition to empirical data bearing on the theory.

Stochastic meandering of pathway architecture

Consider the situation outlined in Figure 19.4, where a focal resource, S, is potentially obtainable either directly from the environment or indirectly via some other cellular activity that yields a precursor metabolite P, which in the presence of an appropriate enzyme can be converted into S (Figure 19.4). If P is present in sufficient quantity, the shuttling of a fraction of an appropriately converted product (S) to a downstream pathway might be an advantageous strategy. The addition of an enzymatic mechanism for producing and feeding P into a pre-existing pathway would then be viewed as a forwards or backwards extension, depending on one's perspective. Either way, if the initial ability to obtain S directly from the environment is lost, the pathway will have increased in length.

As discussed in other contexts in previous chapters, provided there are non-zero probabilities of forward and reverse evolutionary transitions between all adjacent pathway states, for any set of conditions there is a steady-state probability

that a population will reside in one of the alternative states if it were to be sampled at enough intervals over evolutionary time (Foundations 19.4). Borenstein et al. (2008) provide substantial evidence for such evolution among microbes, showing that gains and losses of external precursor compounds (upstream metabolites that cannot be synthesized and are exogenously acquired from the environment) are common, with phylogenetic shifts likely driven by environmental availabilities of alternative substrates.

To determine the population-genetic and ecological conditions that are most conducive to the evolution of the three alternative pathways in Figure 19.4, two basic issues need to be considered: 1) the relative rates of origin and loss of pathway links via mutational mechanisms; and 2) the relative selective advantages of the alternative pathways. As outlined in Chapter 5, the transition rate between two adjacent states is equal to the product of the numbers of relevant mutations arising per generation in the population and their probability of fixation. The long-term probabilities of occupancy of the alternative states are then functions of the relative biases of mutation and selection. If the bias in mutation pressure in one particular direction is sufficiently strong relative to any opposing selection pressures,

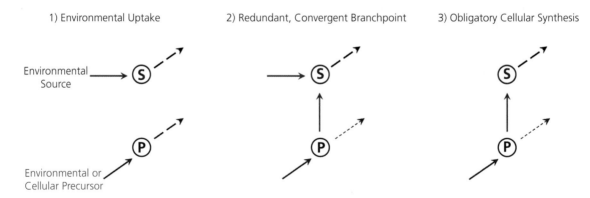

Figure 19.4 Alternative pathway topologies for acquisition of a key metabolite S. **1:** At one extreme, the sole source of S is environmental (blue). **2:** From this perspective, upstream pathway growth occurs if a mechanism (green arrow) arises that converts another substrate (precursor P) to the focal metabolite (S). Initially, this results in two (at least partially redundant) pathways to metabolite S, and the output of the pre-existing pathway descending from P (dashed red lines which may or may not be relevant to cellular fitness) will be reduced accordingly. **3:** If the environmental uptake mechanism for S is then lost, acquisition of S will have become obligatorily dependent on the precursor pathway, with the linear pathway now having an additional step. In principle, this evolutionary trajectory can be reversed if a population in the third state acquires a capacity for direct environmental uptake of S.

the state of the population will be driven primarily by mutation, with the most frequently occupied state not necessarily being that which yields the highest fitness.

Three factors can contribute to the differential fitness of the pathways illustrated in Figure 19.4. First, the direct effects of resource S on fitness must be considered. If R_S is the amount of resource available for direct uptake, and R_P is the additional amount that can be added by the novel enzymatic step, then (R_S+R_P) is the total amount of resource available to a variant possessing both input mechanisms. Assuming that fitness, $W(R)$, is an increasing function of resource availability, approaching an asymptotic maximum value at high R (Figure 19.5), then pathway 2 has the highest direct pay-off, whereas that of pathway 3 can be greater or smaller than that of pathway 1, depending on the relative availability and convertibility of precursor P. Because natural selection operates on relative fitnesses, it is sufficient to denote the direct selective advantages of pathways 1, 2, and 3 as deviations from the maximum (state 2) value: $-r_1$, 0, and $-r_3$. The magnitudes of the two negative coefficients will depend on the resource availability via the alternative pathways,

asymptotically approaching zero as both resources approach saturating levels, and $W(R)$ plateaus.

Second, each pathway will incur a baseline cost in terms of the required energetic investments for the production and maintenance of the enzymatic machinery and/or acquisition of any additional cofactors needed for uptake/production of S. These costs, which can be denoted $-c_1$, $-(c_1+c_3)$, and $-c_3$, will necessarily be greater for pathway 2.

Third, each pathway can be silenced by degenerative mutations. Let u_1 be the rate of mutational loss of the enzymatic connection with the environmental source of S, and u_3 be the rate of mutational loss of the connection to P. Assuming the final product to be essential, a fraction of u_1 and u_3 newborn individuals are eliminated by selection for pathways 1 and 3, respectively, as they incur a loss of access to S. On the other hand, single deactivating mutations incurred by the redundant state (2) do not lead to the inviability of individuals, but simply transform state 2 either state 1 or 3.

Taking all three factors into consideration, the selective disadvantages of the three pathways are $r_1 + c_1 + u_1$, $c_1 + c_3$, and $r_3 + c_3 + u_3$. The key point is that the relative fitnesses of alternative mechanisms of resource acquisition are functions of: 1) the amounts of resource directly and indirectly available; 2) the bioenergetic costs of building and maintaining the two alternative mechanisms; and 3) the mutation rates to defective phenotypes. The redundant pathway (2) is advantageous with respect to resource acquisition and mutational vulnerability, but more costly with respect to construction and maintenance. Not included in this expression is the possibility that the precursor molecule P might have a pre-existing function in the cell, in which case diversion of a fraction of P towards the production of S could impose an additional cost via loss of some of the primary function of P, further detracting from the fitnesses of states 2 and 3.

There are two additional points to consider. First, as noted in Chapters 4 and 5, a key issue is the ratio of the magnitude of the strength of selection to the power of drift, $N_e s_i$, where N_e is the effective population size, and s_i is the selective advantage of type i (Foundations 19.4). This term influences the probability distribution of alternative states exponentially, via the multiplier $e^{N_e s_i}$, which is the ratio

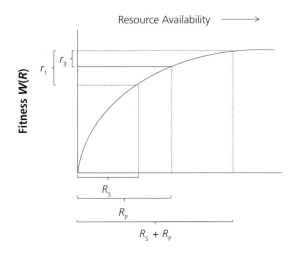

Figure 19.5 A hyperbolic fitness function, with R_S and R_P, respectively, denoting amounts of resources available directly from the environment and via a precursor molecule, and r_1 and r_3 denoting reductions in fitness relative to the case in which both sources of substrate molecule are available. Note that R_P might be smaller than R_S.

of fixation probabilities for mutations entering and leaving class *i*. Second, the steady-state probabilities of alternative pathways also depend on the ratios of rates of gain and loss of the end states by mutation. The essential result is that relative to the intermediate (redundant) state 2, the probability of each of the end states is simply defined by the net pressure of mutation and selection producing versus exiting these states (Foundations 19.4). In the extreme case of effective neutrality (all $|N_e s_i| \ll 1$), then $e^{N_e s_i} \simeq 1.0$, showing that selection is ineffective unless the effective population size is sufficiently large; in that case, the probabilities of alternative states depend only on the degree of mutation bias.

Based on these considerations, several conclusions can be drawn (Figure 19.6). First, because the rate of mutational production of a new mechanism of resource acquisition is likely to be substantially smaller than the mutational rate of inactivation, there will be a tendency for mutation to drive the redundant pathway to the two extreme states. This means that unless the selective advantage of the redundant state is much larger than the power of drift, and also large enough to offset mutational pressure, populations will most likely reside in one of the end states. Second, the non-redundant pathway that is most likely to emerge (uptake-dependent vs. precursor-dependent) depends on the relative magnitudes of the ratios of loss:gain rates of the two types.

Finally, it is important to note that all of the preceding arguments assume a non-recombining genome, such that offspring have genotypes identical to those of their parents in the absence of mutation. Suppose, however, that recombination can occur between the genes underlying the two alternative pathways, with the alternative types having a very simple genetic basis. Of the three types of recombinational encounters between alternative genotypes, 1–2, 1–3, and 2–3, only the 1–3 combination results in novel progeny genotypes. This is because a redundantly encoded genome recombined with a single-feature genome still results in progeny with a 1:1 ratio of parental

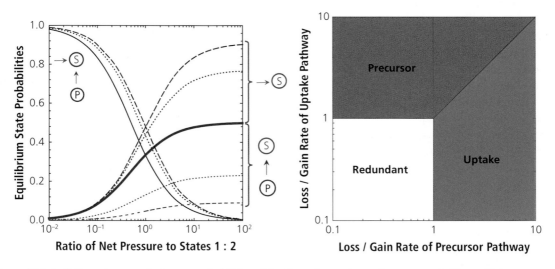

Figure 19.6 Left: Long-term evolutionary probabilities of the three alternative pathway states in Figure 19.4, as a function of the relative net forces of mutation and selection from and to the intermediate (redundant) state 2. The *x* axis is the ratio of the net pressure towards state 1 from state 2 relative to the reverse (2 to 1). Solid lines denote the situation in which the two extreme states (1 and 3) experience identical mutation/selection pressures (in this case, the incidences of states 1 and 3 are identical, and given as the thick cyan line). Dotted and dashed lines respectively denote situations in which the biases toward the extended-pathway state (3) are 0.3 and 0.1× that toward the external dependent state (1); with ratios of 3.33 and 10.0, the positions of the state-1 (blue) and state-3 (green) curves would simply be reversed. All curves follow directly from Equations 19.4.4a–c. **Right:** The phase diagram, denoting which of the three alternative pathway types will have the highest likelihood of occurrence, is entirely a function of two composite parameters, the ratios of the net rates of loss/gain of the two end states.

features, whereas a 1–3 recombination event results in one genome of type 2 and one completely lacking in function. This potential for recombinational breakdown between single-feature genotypes might promote the maintenance of genotypes with redundant acquisition pathways. However, such an effect can be partially or entirely offset if the redundant pathway has sufficiently low fitness, or if the population is small enough that the essential ingredient for recombinational production (the simultaneous presence of state 1 and 3 genotypes) is absent (Lynch 2007).

Although this model provides a heuristic understanding of how incremental changes might lead to among-species differentiation in metabolic pathways, many such pathways go far beyond the simple systems just explored. Consider, for example, the mysteries posed by the amino-acid biosynthetic pathways (Figure 12.3). Only three amino acids (glutamate, aspartate, and alanine) are a single step removed from a metabolic precursor, with several others being derived from these primary products via one or more additional steps. For example, asparagine is one step removed from aspartate, whereas lysine, methionine, and threonine are several steps removed, and isoleucine is still more steps removed from threonine (Chapters 12 and 17). Many of the intermediate products leading to the production of downstream amino acids simply serve as pathway stepping stones, with no other function. If this was always the case, how could the steps leading to such products be established evolutionarily prior to completion of the pathway, assuming only the end product is of utility?

As a potential solution to this dilemma, Horowitz (1945) proposed a retrograde model for the evolution of biosynthetic pathways, postulating that upstream steps to a pathway are added as the environmental availability of downstream metabolites becomes limited. Under this hypothesis, the final (downstream) step in a pathway is the first to have been acquired, with the preceding step being established second, and so on. Such a model assumes that molecular intermediates were at one time freely available in the environment (as in the progression in Figure 19.4), but is challenged by the fact that intermediate metabolites often exhibit high levels of chemical instability.

The origin of novel enzymes

Although the theory outlined above clarifies the population-genetic requirements for shortening or lengthening metabolic pathways in a quite generic way, a deeper appreciation of the actual biological underpinnings of such transitions is necessary to achieve a mechanistic understanding of how pathways expand. In the preceding example, a transition from state 1 to 2 requires the emergence of an enzymatic function for converting precursor molecule P to the final resource S, whereas a transition from state 3 to 2 requires the appearance of a mechanism for direct acquisition of S. As it is quite unlikely that new molecular mechanisms arise *de novo* in highly refined states, more gradual mechanisms must be sought.

As already noted, all enzymes make errors (i.e., bind to inappropriate substrates) at some low frequency. A fortuitous consequence of such promiscuity is the predisposition of many enzymes to respond evolutionarily to selection pressure to utilize alternative substrates. On the other hand, because the refinement of a pre-existing side effect may come at the expense of the primary function, additional genomic changes may often be essential to allow the emergence of an efficient novel function. Gene duplication provides a powerful route for remodelling an enzyme evolutionarily without relinquishing key ancestral functions, by allowing each copy to become more specialized to individual tasks (Copley 2012).

For the simplest case in which a gene is completely duplicated, it will go through an initial phase of functional redundancy, and unless maintained by selection for increased dosage, will generally suffer one of three fates: loss via a nonfunctionalizing mutation; preservation by subfunctionalization (with the two sister copies becoming specialized to independently mutable subfunctions); or preservation by neofunctionalization (evolution of an entirely novel function by one copy, and retention of the original key functions by the other copy) (Chapter 6).

Given the promiscuity of individual enzymes, in the case of metabolic-function evolution, there may be very few cases of pure neofunctionalization. This would mean that almost all cases of long-term preservation of duplicate metabolic genes are likely driven by either dosage demands or selection for refinements and losses of pre-existing functions (i.e, enzyme diversification). The latter mechanism was, in fact, suggested by Kacser and Beeby (1984; see also Pfeiffer et al. 2005) as a potential explanation for how the earliest enzymes (presumably with much lower substrate fidelity than today's enzymes) might have diversified in function. When a gene is amplified in copy number, this provides additional mutational targets (free of compromising constraints) for the refinement of at least one copy to alternative, more specialized subfunctions. Quite similar arguments for the origin of new functions were subsequently promoted by Piatigorsky and Wistow (1991), Hughes (1994), and Bergthorsson et al. (2007).

There are, however, some subtle distinctions between alternative verbal models in this area. For example, Hughes's (1994) adaptive-conflict hypothesis envisions situations in which an ancestral gene carries out two or more essential but mutually compromised functions that then become specialized after duplication. In contrast, Bergthorsson et al. (2007) focuses more on situations in which a specialized ancestral gene acquires a novel mutation inducing weak functional promiscuity, which then undergoes adaptive refinement following a duplication event that increases its exposure to natural selection by amplifying its expression level. Both hypotheses contain an element of neofunctionalization, and both involve a sort of subfunctionalization, the primary difference being the time-zero degree of essentiality of the alternative gene functions.

Voordeckers et al. (2012) obtained evidence for an innovation-amplification-divergence scenario in yeast wherein an ancestral glucosidase protein, active primarily with maltose, had weak promiscuous attraction to isomaltose. Upon duplication, via just three key amino-acid changes, one copy lost maltose activity while becoming optimized for isomaltose utilization. Likewise, Blount et al. (2012) found that the evolution of the ability to utilize citrate in an experimental population of *E. coli* was initiated by a regulatory change that increased the expression of a citrate transporter, followed by amplification to a multicopy array of the modified locus. In a comparative study that involved the resurrection of ancestral states, Rauwerdink et al. (2016) found that plant cyanohydrin lyases, which produce insecticidal cyanides, evolved from an initially weak promiscuous side effect of plant esterases, only becoming refined after duplication. Significantly, all of these studies illustrate dramatic shifts in metabolic functions being driven by just a few mutational changes.

Pathway Participant Remodelling

Although most of the pathways associated with central metabolism are typically presented in biochemistry textbooks as invariant, formal analyses have generally been carried out in no more than a handful of taxa. Nonetheless, existing comparisons amply demonstrate that the structures and constituents of key metabolic pathways sometimes vary dramatically among species. As an entrée into this area, we now consider how a cellular process even as fundamental as amino-acid biosynthesis can be subject to substantial modifications among phylogenetic lineages.

Two tyrosine-biosynthesis pathways are known, both starting with prephenate as a precursor, but having different intermediates (Jensen and Pierson 1975; Song et al. 2005). Both pathways involve transamination and dehydrogenation steps, but in reverse order (Figure 19.7, top). Among bacteria, many different lineages use one or the other pathway, whereas species that are capable of both may employ a single broad-specificity enzyme. The cofactors deployed in the key reactions, NAD^+ versus $NADP^+$, are also subject to phylogenetic changes.

Cysteine is produced by at least three different pathways in various organisms, two of which involve completely non-overlapping reactions. In animals, the sulfhydryl group of cysteine is derived from methionine, whereas in plants and prokaryotes that have been studied, inorganic sulfide provides the source (Figure 19.7, middle). In the budding yeast *S. cerevisiae*, there is a modification of the animal pathway such that the methionine → homocysteine step is replaced by two others. However, there is also a good deal of variation among fungi, with the fission yeast *S. pombe* having the

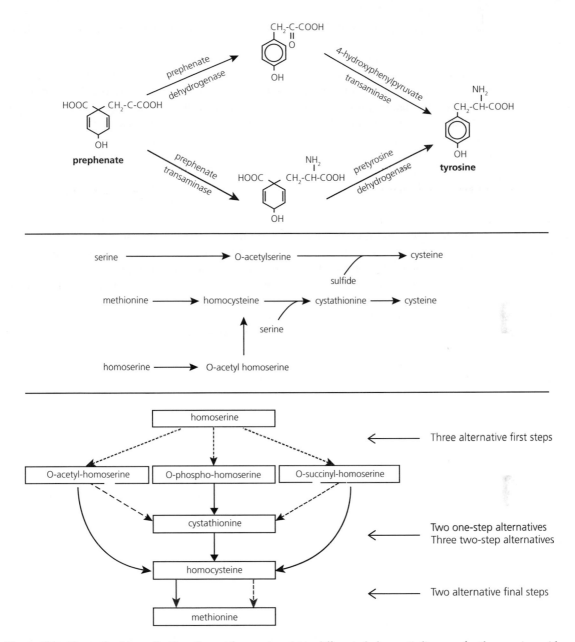

Figure 19.7 Alternative biosynthetic pathways known to exist in different phylogenetic lineages for three amino acids.

pathway initiating with serine, and many species having both types of pathways (Hébert et al. 2011).

Lysine biosynthesis provides still another example of pathway diversification. Two major modes of production exist for this amino acid: the diaminopimelate (DAP) pathway and the α-aminoadipate (AAA) pathway, the first starting from aspartate and the second from α-ketoglutarate, and with no overlap in the intermediate products. Both pathways are quite long, typically with eight to nine steps, although short-length variants that bypass some internal steps are known (Nishida et al. 1999; Hudson et al. 2006; Curien et al. 2008; Torruella et al. 2009). The DAP pathway (and variants of it) is present in bacteria, plants, and some algae, and the AAA pathway has been found in fungi, a few bacteria, and some protists. A few taxa may have both pathways, and lateral gene transfer has been suggested (Nishida and Nishiyama 2012).

Perhaps the most spectacular example of diversification of amino-acid biosynthetic pathways involves methionine production (Figure 19.7, bottom). In all cases, the starting point is homoserine, but three alternative first-step metabolites can be produced (acetylated, phosphorylated, or succinylated homoserine). From these points, the addition of cysteine can lead to cystathione (a sulfur source) in a single step, which is then converted to homocysteine in the penultimate step to methionine. However, single-step pathways from both acetyl-homoserine and succinyl-homoserine to homocysteine also exist, utilizing H_2S and bypassing cystathione production. Finally, there are two alternative final-step mechanisms for converting homocysteine to methionine. Taken together, these observations suggest the possibility of as many as ten partially overlapping pathways. At least eight of these have been observed (Gophna et al. 2005), and some microbial species are capable of multiple production pathways. As some archaea appear to be completely lacking proteins for homoserine activation, additional pathways likely await discovery. Moreover, evolutionary shifts between alternative methionine biosynthetic pathways do not appear to be difficult. For example, a single amino-acid change is sufficient to alter the first step of homoserine activation, and not surprisingly, the first-step enzymes are polyphyletic with respect to function (Zubieta et al. 2008).

These four examples are by no means an exclusive list of variations in amino-acid biosynthesis pathways. For example, although glutamate is one of the simpler amino acids, being just one step (the addition of an NH_3) removed from α-ketoglutarate, the source of NH_3 in the reaction can vary, and the enzyme involved (glutamate synthase) utilizes different cofactors in different phylogenetic contexts (Dincturk and Knaff 2000; Suzuki and Knaff 2005). Just within the bacterial genus *Pseudomonas*, two non-overlapping pathways are known for

phenylalanine production, with some species having one of the pathways and others having both (Byng et al. 1983). The first step in the biosynthesis of arginine is the acetylation of glutamine, but this is done by apparently unrelated enzymes in various phylogenetic lineages (Xu et al. 2007). The branchpoint intermediate in the path leading to methionine versus threonine production is O-phosphohomoserine in plants, whereas it is one step higher in *E. coli* and *S. cerevisiae*, at homoserine (Curien et al. 2008). There are at least two largely non-overlapping pathways for serine production (Melcher et al. 1995; Ho et al. 1998; Shimizu et al. 2008), three for isoleucine (Hochuli et al. 1999), and two for glycine (Sakuraba et al. 1996; Ramazzina et al. 2010). Finally, the tryptophan operon, which is often used as a textbook example of gene regulation, shows extensive variation in modes of regulation among bacterial species, with some having fusions of the various genes in the pathway (Xie et al. 2003; Merino et al. 2008).

This broad set of examples, mostly drawn from prokaryotes but by no means exclusive to amino-acid biosynthesis (e.g., Hoshino and Gaucher 2018), highlights the flexibility of cells to evolutionary modifications of the machinery underlying key metabolic processes, presumably while keeping the overall performance of the pathway nearly constant. Several additional points are clear. First, even the pathways that one might expect to be highly conserved are subject to complete remodelling to the point of sharing no reaction steps. Second, even when species share the same pathway structure, the participating enzymes can be non-orthologous. Third, redundant pathways can often coexist in the same organism. Finally, alternative pathways often differ in efficiency (Du et al. 2018), e.g., in utilization of ATP, and it remains unclear as to why energetically suboptimal variants exist. Textbook pathways used for biosynthetic-cost estimates (Chapter 17) may need adjustment if different paths impose different energetic expenses.

These general observations, combined with the common existence of enzyme promiscuity and gene duplication, lead to a plausible set of scenarios for metabolic-pathway evolution involving intermediate stages of functional redundancy, fully consistent with the model outlined above (Figures 19.4–19.6).

Many variants of such a model can be envisioned, all of which fit well within the general theme of the patchwork model of pathway evolution (Ycas 1974; Jensen 1976; Copley 2000), under which enzymes with substrate ambiguity can sometimes be stitched together to establish novel pathways and/or to confer redundancy with pre-existing molecular mechanisms. We now consider the underlying issues in somewhat greater detail, again motivating each aspect of the discussion with a series of examples, primarily to emphasize just how common pathway remodelling is.

Non-orthologous gene replacement

As alluded to in the previous section, in addition to experiencing the rewiring of pathway structure, metabolic pathways can remain stable in form, while the underlying participants turn over. As a case in point, consider glycolysis, the major path to sugar breakdown in the cells of eukaryotes and many prokaryotes (Figure 19.3). Despite the centrality of glycolysis, extraordinary variation exists for all of the enzymatic components, with comparative analyses suggesting a very complex evolutionary history both within and among the major domains of life (Canback et al. 2002; Oslancová and Janecek 2004; Siebers and Schönheit 2005).

For example, bacterial phosphofructose kinase (step 2) and phosphoglycerate kinase (step 7) appear to be non-orthologous to the homologous enzymes used in eukaryotes (Galperin et al. 1998), and most of the pathway enzymes in archaea appear to be non-homologous to those in either eukaryotes or bacteria (Verhees et al. 2001; Kawai et al. 2005; Siebers and Schönheit 2005). Two types of fructose bisphosphate aldolase (step 4) are known, each deploying a different molecular mechanism, one group in animals and land plants, and the other in fungi, some algae, and bacteria. Various protist lineages harbor a diversity of sequences from both classes, with the complexity of the phylogenetic distribution strongly suggesting substantial ancient horizontal transfer (Siebers et al. 2001; Sánchez et al. 2002; Rogers and Keeling 2004; Allen et al. 2012). Clear examples of inter-kingdom horizontal transfers also exist for glyceraldehyde-3-phosphate dehydrogenase (step

6) (Qian and Keeling 2001; Takishita and Inagaki 2009). Finally, phosphoglycerate mutase (PGM, step 8) is found in two forms, apparently unrelated to each other and also differing in cofactor dependency (Liapounova et al. 2006). Several members of the archaea and bacteria harbor both types of PGM, consistent with the idea of the ancient origin of the two types (van der Oost et al. 2002; Johnsen and Schönheit 2007).

This list could be expanded, but by now the central point should be clear – as with many of the amino-acid biosynthetic pathways, probably all of the enzymes involved in glycolysis have evolved more than once across the Tree of Life from apparently unrelated ancestral enzymes and then been substituted for their non-orthologous analogues. Multiple mechanisms exist for such non-orthologous gene replacement. As noted above, phylogenetic analyses clearly implicate horizontal gene transfer in a number of instances, although most of the well-established examples are restricted to quite ancient periods in the Tree of Life. Although inter-kingdom transfers must be confronted with a number of selective challenges, these may be eased for enzymes universally conserved to have the same function. Indeed, occasions might arise in which the enzyme from a potential donor species has acquired slightly beneficial kinetic, assembly, and/or stability properties relative to those in the recipient, e.g., because of a population-genetic environment more conducive to efficient selection.

One situation that might encourage non-orthologous gene replacement (including that via horizontal transfer) arises during periods in which a resource that must normally be synthesized becomes freely available in the environment (e.g., Figure 19.4). Such a condition might relax selection on the maintenance of the internal catabolic pathway, thereby leading to the effectively neutral loss of one or more components via degenerative mutations. The resultant condition, known as auxotrophy (Chapter 18), is an inability to synthesize an essential organic compound. Reintroduction to an environment lacking the resource then imposes strong selection for pathway reconstruction, e.g., replacement of a missing step with a duplication from a non-orthologous gene with a suitably promiscuous function.

Consistent with such a scenario, it is common for some isolates within bacterial species to be missing parts of particular pathways and hence, to be deficient with respect to certain kinds of metabolism (Monk et al. 2014). This phenomenon has been repeatedly seen in endosymbiotic bacteria inhabiting the well-provisioned cells of insects (McCutcheon and Moran 2007), a dramatic example being the emergence of chimeric biosynthetic pathways in which different enzyme components are encoded in different genomes of the mutualists (Bublitz et al. 2019). The enzymes most likely to be lost involve steps (usually downstream) with minimal interactions with other pathways (Hittinger et al. 2004; Pál et al. 2006).

As noted above, further opportunities for pathway diversification exist when a broad-specificity enzyme is duplicated, particularly if different subfunctions are independently mutable, as this provides a route by which a gene from one pathway might move to another. By this means, through progressive gains and losses of subfunctions, genes might be free to wander across pathways, provided this is accomplished in ways that do not erode the operation of individual pathways. For example, *Mycobacterium tuberculosis* deploys a bifunctional enzyme used in both histidine and tryptophan biosynthesis (Due et al. 2011), and experimental work demonstrates that simple amino-acid substitutions can establish tryptophan-pathway activity from histidine-pathway genes (Leopoldseder et al. 2004; Näsvall et al. 2012). In a number of species that have experienced genome reductions, one or the other amino-acid biosynthetic path has been lost, with the bifunctional enzyme reverting to monofunctionality (Juárez-Vázquez et al. 2017).

Notably, indirect evidence suggests that entire pathways have sometimes arisen by duplication of most or all constituent proteins. For example, Velasco et al. (2002) suggest that the DAP pathway for lysine production (noted earlier) is related to the arginine-biosynthesis pathway, in that many of the constituent enzymes appear to be related. A link between the AAA pathway for lysine and the leucine-biosynthesis pathway has also been suggested, leading to the idea that pathways for the production of all three amino acids – leucine, arginine, and lysine – share a common ancestry,

possibly derived from a state in which all three were synthesized by enzymes with lower specificity (Nishida et al. 1999; Miyazaki et al. 2001; Fondi et al. 2007). Remarkable similarities between key steps in the TCA cycle and the paths for lysine, isoleucine, and leucine biosynthesis have also been suggested (Jensen 1976). Once a redundant set of pathways arises, reciprocal loss by subfunctionalization can occur in descendant sister taxa, leaving little or no record of past redundancy. For example, Jensen (1985) highlights hydrogenases in bacteria whose ancestral state involved joint NAD^+ and $NADP^+$ cofactor specificity but became fixed for one or the other alternatives in descendant sister taxa.

Returning to the situation with the glycolytic pathway (Figure 19.3), the potential to shift pathway affiliations seems particularly large, given the widespread moonlighting of such enzymes in functions completely unrelated to glycolysis (Kim and Dang 2005). Just a few examples are noted here. Glyceraldehyde 3-phosphate dehydrogenase (G3PDH; step 6) is known to bind DNA and telomeres, plays a role in cell-cycle regulation, and has many other functions associated with nuclear export, membrane fusion, phosphorylation, and DNA repair (Seidler 2013). Many examples are known in which fructose 1,6-bisphosphate (FBP; step 4) and triosephosphate isomerase (TPI, step 5) are involved in membrane fusion in pathogenic bacteria (Gozalbo et al. 1998; Modun and Williams 1999; Alvarez et al. 2003; Tunio et al. 2010). One of the two hexokinase (HEX, step 1) genes in *S. cerevisiae* also acts as a repressor of glucose metabolism; and HEX has been implicated in anti-apoptotic activity in mammals. Enolase (also known as phosphopyruvate hydratase or PPH, step 9) has also been found to have numerous secondary functions – it is a major lens protein in some vertebrates (Wistow and Piatigorsky 1988), plays a role in thermal tolerance in yeast (Iida and Yahara 1985), and can be involved in cell-cell communication (Miles et al. 1991; Castaldo et al. 2009) and transcriptional regulation (Chang et al. 2003; Kim and Dang 2005).

These kinds of observations extend to each of the remaining enzymes involved in glycolysis, as well as to many other enzymes involved in central metabolism. For example, all organisms harbor mRNA degradation machineries, but there are well-documented examples of different bacteria deploying unrelated ribonucleolytic enzymes in such processes (Kaberdin et al. 2011). Bacteria, archaea, and eukaryotes all have a mevalonate pathway (used in the production of isoprenoids), but the full set of enzyme participants appear to be non-orthologous (Lombard and Moreira 2010). Similarly, the participants in the production of coenzyme M (used in methane metabolism) are completely different in bacteria versus archaea (Wu et al. 2022). These extreme cases of complete non-overlap in enzyme participants often lead to the conclusion that such pathways have independently evolved in different lineages, but the alternative 'ship of Theseus' hypothesis is that such endpoints are long-term outcomes of the complete serial replacement of parts. Although often ignored, the types of gene loss and substitution implied by non-orthologous gene replacement also have significant implications for the common practice of using gene-sequence homology in model organisms to infer commonality of gene functions in more distantly related species (Gabaldón and Koonin 2013).

Internal pathway expansion via multifunctional enzymes

A fundamental feature of biochemical pathways is that consecutive enzymes must interact with one of the same ligands – the product of an upstream enzymatic reaction becomes the substrate for the next. Some enzymes exploit this linkage by engaging in two adjacent interactions in a pathway. A good example of such a bifunctional enzyme is fructose-1,6-bisphosphate aldolase/phosphatase, which in most archaea and many bacteria is used in gluconeogenesis (the reverse of the glycolytic pathway) to first convert dihydroxyacetone phosphate (DHAP) to FBP and then to dephosphorylate the latter to fructose-6-phosphate (Du et al. 2011). Unlike many bifunctional enzymes, both reactions are carried out by the same catalytic domain, leading to an efficient system in which the intermediate metabolite (FBP) is never released.

Situations like this open up the opportunity for pathway growth when an ancestral protein capable of carrying out two pathway steps becomes duplicated into two daughter copies, each of which loses an alternative function, locking the two into preservation by subfunctionalization. A possible example of such a situation exists in *E. coli*, where two consecutive enzymes in the methionine-biosynthetic pathway appear to have been derived from a common gene by duplication that then underwent subsequent specialization (Belfaiza et al. 1986; Parsot et al. 1987). Here, what is done in two steps in *E. coli* is accomplished by one related enzyme in *S. cerevisiae*. It has also been argued that a single enzyme converting O-acetyl-serine to cysteine is related to the two taking O-acetyl-homoserine to homocysteine in the methionine pathway (Figure 19.8; Parsot et al. 1987). Consecutive genes in the threonine-biosynthetic pathway also appear to have arisen by gene duplication in bacteria (Parsot 1986; Parsot et al. 1987), and similar structures for several of the sequential proteins in tryptophan biosynthesis suggest a common ancestry (Wilmanns et al. 1991; List et al. 2011). Arguments have also been made that a number of the enzymes in the purine-biosynthesis pathway have arisen by duplication and divergence (Kanai and Toh 1999; Zhang et al. 2008).

A few enzymes are known to be capable of catalysing more than two reactions (Roy 1999), e.g., dehydroquinate synthase (Carpenter et al. 1998) and carbamoyl phosphate synthetase (Raushel et al. 1998), either in the same site or with active sites connected by molecular tunnels. In principle, such enzymes provide rare opportunities for an entire pathway to be carried out by a single protein, as well as for repeated processes of duplication and subfunctionalization to lead to pathway extension. Of course, there seems to be no reason to rule out the reverse – pathway simplification, as a monofunctional enzyme becomes bifunctional by taking on the task of an adjacent step.

Summary

- Virtually all metabolic reactions are carried out by enzymes. Although such molecules enhance reaction rates by many orders of magnitude above spontaneous background levels, their catalytic capacities are still generally orders of magnitude below the biophysical limits, suggesting that the degree of performance is stalled at a drift barrier.

- Usually highly specialized for particular functions, most enzymes are also quite promiscuous, engaging with inappropriate substrate molecules up to 1% of the time. This must impose a burden on cells. But in the long run it also provides a launching pad for the evolution of novel enzyme functions through the refinement of latent capacities, commonly via just a few key amino-acid substitutions.

- Because individual enzymes typically carry out simple chemical reactions, typically involving only single chemical bonds, this generally necessitates the use of complex, stepwise metabolic pathways in the transformation of organic material.

- Theory suggests that the sensitivities of metabolic-pathway flux rates decline exponentially from the entry level to final steps in the pathway, and this appears to be reflected in a relaxation of selection on the amino-acid sequences of downstream enzymes.

- Catabolic pathways are commonly highly inefficient energetically in the sense that a substantial amount of energy derived from the breakdown of substrate molecules is simply released as heat rather than converted to bioenergetic currency (ATP). Flux-rate theory suggests that such inefficiency is an intrinsic requirement for maximizing the net rate of product formation. The digital nature of molecular transactions is also an intrinsic constraint that necessitates energy loss.

- Despite the near universality of many resource-utilization and biosynthetic pathways, both the structure and underlying enzyme participants of such pathways can differ dramatically among phylogenetic lineages. This suggests the existence of multiple degrees of freedom for pathway remodelling without significant negative fitness consequences. Gene duplication, combined with partitioning and divergent refinement of ancestral gene functions, plays a central role in such reconfigurations.

Foundations 19.1 Michaelis–Menten enzyme kinetics

The simplest approach to enzyme kinetics starts with the assumption that molecules of enzyme (E) and substrate (S) interact in two steps to yield a product (P). The first step involves the production of a non-covalent intermediate complex (ES) and is assumed to be reversible, governed by association–dissociation kinetics (Foundations 13.1). If successful, the second step transforms the substrate into the final product, after which the released enzyme is free to engage in another reaction. The overall reaction flow can be written as

$$E + S \rightleftharpoons ES \rightarrow E + P.$$

The intent here is to describe the relationship between the net conversion rate of S to P and the concentration of S. The rate of product formation is a function of the three rates associated with the arrows in the preceding formulation. Let k_a be the rate of association of E and S to form ES, k_d be the rate of dissociation of ES back to E and S,

and k_{cat} be the catalytic rate of production of P from ES (also known as the turnover rate). Under the assumption of a steady-state process (e.g., with the concentrations of all components remaining constant, owing to the constant replenishment of substrate and recycling of enzyme), the rates of gain and loss of the intermediate state must be equal, so that

$$k_a[E][S] = (k_d + k_{cat})[ES], \qquad (19.1.1a)$$

where brackets denote concentrations. Note that the rate of production of ES is a function of the concentrations of the two constituents and the rate of association per unit concentration. Rearranging, we find that

$$[ES] = \frac{[E][S]}{K_S}, \qquad (19.1.1b)$$

where

$$K_S = (k_d + k_{cat})/k_a \qquad (19.1.1c)$$

is a composite function of three parameters. Denoting the rate of production of P as $V = k_{cat}[ES]$, substitution of Equation 19.1.1b for [ES] implies

$$V = \frac{k_{cat}[E][S]}{K_S}. \tag{19.1.2}$$

There is one final complication in that [E] is the concentration of free enzyme, whereas the more easily measured total amount of enzyme in the system, $[E_T]$, is the sum of the amounts both freely circulating and tied up in the enzyme–substrate complex. Again using Equation 19.1.1b,

$$[E_T] = [E] + [ES] = [E]\left(1 + \frac{[S]}{K_S}\right). \tag{19.1.3}$$

Solving this equation for [E], and substituting into Equation 19.1.2 yields an equation of the form presented by Michaelis and Menten (1913),

$$V = \frac{k_{cat}[E_T][S]}{K_S + [S]}. \tag{19.1.4}$$

Notably, the derivation 19.1.4 was first presented by Briggs and Haldane (1925), which although yielding the same mathematical structure, involves some subtle distinctions from the approach used by Michaelis and Menten (1913). For interesting historical commentary on this and other matters, see Gunawardena (2012), as well as Johnson and Goody (2011), who also generated an English translation of Michaelis and Menten (1913). Formulae with the same form appear in many other areas of cell biology (Wong et al. 2018).

There are several notable features of this expression. First, for a constant level of total enzyme, Equation 19.1.4 predicts a hyperbolic relationship between the substrate concentration, [S], and the rate of product formation (Figure 19.1). At high substrate concentrations, V asymptotically approaches a maximum level $V_{max} = k_{cat}[E_T]$,

which shows that k_{cat} is the maximum rate of product formation per unit of total enzyme concentration.

Second, K_S is equivalent to the substrate concentration at which the reaction rate is half the maximum, as can be seen by substituting K_S for [S] in Equation 19.1.4. Thus, K_S is usually referred to as the half-saturation constant, although as can be seen from Equation 19.1.1c, K_S is not mathematically independent of k_{cat}.

Third, if the catalytic rate k_{cat} is much smaller than the dissociation rate k_d, then the former can be ignored in Equation 19.1.1c, yielding $K_S \simeq k_d/k_a$. This ratio of dissociation to association rates is usually referred to as the dissociation constant and often denoted by K_D.

Finally, as $[S] \to 0$, $V \to k_{cat}[E_T][S]/K_S$. The ratio $\phi_E = k_{cat}/K_S$ is often referred to as the kinetic efficiency, as it defines the innate capacity of an enzyme when the substrate is at nonsaturating levels.

Two additional points are worth noting about the Michaelis–Menten equation. First, inspired by their own kinetic observations and those made earlier by others, Michaelis and Menten postulated a hypothetical intermediate enzyme–substrate complex in their formulation as a means for achieving the known hyperbolic relationship between enzyme reaction rates and substrate concentration. However, several decades passed before any such complexes were actually observed. Thus, this early exercise in biochemical modelling is a beautiful example of a theoretical construct predicting a previously unseen phenomenon. Second, the enzyme pathway deployed assumes that product formation is irreversible, unlike the ES complex. Although this is a reasonable approximation if P remains rare (as would be the case if P were consumed in some other cellular process), as P becomes increasingly common, negative feedback may cause a reduction in the net forward reaction rate, necessitating further modification of the Michaelis–Menten formula (Gunawardena 2014).

Foundations 19.2 Evolutionary sensitivity of pathway steps

Metabolic pathways often consist of a linear series of steps, each transforming a metabolite to a new product that then serves as a substrate for a subsequent enzyme. Such a chain can be represented as follows:

$$S_0 \rightleftharpoons S_1 \rightleftharpoons S_2 \cdots\cdots\cdots \rightleftharpoons S_{n-1} \rightleftharpoons S_n,$$

with S_0 denoting the initial substrate, and the S_i denoting the intermediate metabolites en route to the final product

S_n. The right/left arrows between metabolites denote the forward and reverse reactions, which occur at rates k_{+i} and k_{-i}. In the following, we assume that the initial substrate is kept at a constant intracellular concentration $[S_0]$, and that the final metabolite is utilized (and hence, removed from the cytosol) at rate γ (per unit concentration). Assuming that the rates on the transition arrows in the reaction scheme remain constant, such a system will yield a set of steady-state concentrations for each of the metabolites.

Consider a simple two-enzyme system, for which the time course of concentration changes can be written as

$$\frac{d[S_1]}{dt} = k_{+1}[S_0] - (k_{-1} + k_{+2})[S_1] + k_{-2}[S_2] \qquad (19.2.1a)$$

$$\frac{d[S_2]}{dt} = k_{+2}[S_1] - (k_{-2} + \gamma)[S_2]. \qquad (19.2.1b)$$

The three terms in the first equation respectively denote the conversion of the initial substrate to S_1, the loss of S_1 to the two alternative substrates (S_0 and S_2), and the gain from the reverse reaction involving S_2. The steady-state concentrations are found by setting the derivatives equal to zero and solving the pair of equations,

$$[\widetilde{S}_1] = \frac{K_1[S_0]\{1 + (\gamma/k_{-2})\}}{1 + (\gamma K_2/k_{-1}) + (\gamma/k_{-2})} \qquad (19.2.2a)$$

$$[\widetilde{S}_2] = \frac{K_1 K_2[S_0]}{1 + (\gamma K_2/k_{-1}) + (\gamma/k_{-2})}. \qquad (19.2.2b)$$

The terms involving upper-case Ks are the ratios of forward and reverse reaction rates, often referred to as the equilibrium constants,

$$K_i = k_{+i}/k_{-i}, \qquad (19.2.3)$$

as they would define the equilibrium concentrations of two adjacent metabolites in the absence of any other steps in the system,

$$k_{+i}[\widetilde{S}_{i-1}] = k_{-i}[\widetilde{S}_i]. \qquad (19.2.4)$$

The rate of flux through the system (i.e., the amount of final product drawn off for use in cellular functions) is equal to the product of the concentration of the final metabolite and its utilization rate, $\gamma[\widetilde{S}_2]$,

$$F = \frac{\gamma K_1 K_2[S_0]}{1 + (\gamma K_2/k_{-1}) + (\gamma/k_{-2})}. \qquad (19.2.5)$$

The two other fluxes in the system, from S_0 to S_1 and from S_1 to S_2, occur at net rates $(k_{+1}[S_0] - k_{-1}[\widetilde{S}_1])$ and $(k_{+2}[\widetilde{S}_1] - k_{-2}[\widetilde{S}_2])$, respectively, when the system is in steady state. After substituting the equilibrium concentrations from Equations 19.2.2a,b, these flux rates can be shown to be identical to the steady-state exit rate from the pathway, Equation 19.2.5, as must be the case for a steady-state system.

Although the algebra gets more tedious with longer pathways, Heinrich and Rapoport (1974) obtained the general solution for a pathway of length n,

$$[\widetilde{S}_i] = [S_0] \prod_{j=1}^{i} K_j \cdot \frac{1 + \gamma \sum_{l=i+1}^{n}(1/k_{-l}) \prod_{m=l+1}^{n} K_m}{1 + \gamma \sum_{l=1}^{n}(1/k_{-l}) \prod_{m=l+1}^{n} K_m}, \qquad (19.2.6)$$

with the product terms being set equal to 1.0 when $m > n$. With this expression, the total flux rate can be written as the product of γ and the steady-state concentration of S_n (the final component of the pathway),

$$F = \frac{\gamma[S_0] \prod_{j=1}^{n} K_j}{1 + \gamma \sum_{l=1}^{n}(1/k_{-l}) \prod_{m=l+1}^{n} K_m}, \qquad (19.2.7)$$

which reduces to Equation 19.2.5 when $n = 2$. Similar results were obtained independently by Kacser and Burns (1973).

We are now in a position to address some key questions regarding the degree to which the overall flux through a pathway is influenced by the individual component reactions. To simplify the analysis, we assume that all of the enzymatic reactions have equal reversion rates k_- and equilibrium constants K. Because the summation in the denominator of Equation 19.2.7 is expected to be much greater than 1 (owing to the forwards nature of the pathway), the 1 can be dropped, and substitution of the constant parameters leads to

$$F \simeq \frac{k_- K^n[S_0]}{\sum_{i=0}^{n-1} K^i}. \qquad (19.2.8a)$$

Still further simplification is possible because the equilibrium constant K is likely to be $\gg 1$, in which case the final term in the series in the denominator dominates (because the terms progressively increase by the factor K), leading to

$$F \simeq k_- K[S_0]. \qquad (19.2.8b)$$

This shows that provided K is larger than 10 or so, the steady-state flux rate is nearly independent of the number of steps in the pathway.

Finally, we consider the sensitivity of the flux rate to changes in the enzymes (e.g., amounts or catalytic efficiencies) responsible for the individual steps. Suppose the equilibrium constant for a single step is multiplied by the factor x. Equation 19.2.8a defines F_* as the pre-perturbation flux rate, and F_j (the post-perturbation rate) is obtained from Equation 19.2.7 with all equilibrium constants equal to K, except that for the jth step, which is set to xK. The sensitivity coefficient for step j is then defined to be the fractional change in the total flux rate relative to its initial value, $(F_j - F_*)/F_*$, scaled by the fractional change in the equilibrium constant relative to its initial value $(x - 1)$,

$$C_j = \frac{F_j - F_*}{F_*(x - 1)}, \qquad (19.2.9a)$$

Substitution of the expressions for F_* and F_j into Equation 19.2.9a, followed by some algebra, yields

$$C_j = \frac{\sum_{i=0}^{n-j} K^i}{\sum_{i=0}^{n-1} K^i}, \qquad (19.2.9b)$$

and again assuming $K \gg 1$, so that only the final terms in each summation predominate, this further simplifies to

$$C_j \simeq \frac{1}{K^{j-1}}. \qquad (19.2.9c)$$

This shows that the relative sensitivity of the flux rate to a change in features influencing K at a particular step (e.g., enzyme concentration, substrate affinity, rate of reactivity) declines by a factor of $1/K$ with each increasing step.

Foundations 19.3 Optimization of the glycolytic flux rate

Glycolysis, a key energy-generating process found in most cells, involves a linear pathway of 10 steps starting with a glucose molecule and culminating in the production of two pyruvates (which can be further reduced to lactic acid) and two ATPs (Figure 19.3). At typical physiological concentrations of the participating molecules, the glucose-to-lactic acid reaction releases ~205 kJoules/mol, whereas each of the ADP-to-ATP conversions stores 50 kJoules/mol, meaning that 105 kJoules/mol is simply lost as heat.

Why has such an apparently wasteful process been universally retained, rather than the production of four ATPs (equivalent to a total of $4 \times 50 = 200$ kJoules/mol), or even three, being squeezed out of the reaction? One explanation is that, instead of maximizing the efficiency of resource utilization, natural selection puts a premium on the total rate of ATP production, which is a function of both the efficiency and speed of substrate utilization.

Letting $\Delta G_{gly} = -205$ kJoules/mol be the energy released by a glycolytic reaction, $\Delta G_{ATP} = 50$ kJoules/mol be the amount retained per net ATP produced (the negative and positive signs denote energy release and gain, respectively), and n be the number of ATPs produced per glucose molecule, the efficiency of the energy recapture can be written as

$$\eta = \frac{n \cdot \Delta G_{ATP}}{|\Delta G_{gly}|}. \qquad (19.3.1)$$

Thus, in terms of efficiency of resource utilization, just 49% of the energy endowment of glucose is retained by the organism.

The speed of a reaction (the flux rate) depends on the concentrations of reactants and products, as well as on the net energy differential between the two. In this case, the differential is equal to $|\Delta G_{gly} + (n \cdot \Delta G_{ATP})|$ (recalling that

ΔG_{gly} is negative). The flux rate through the system is

$$F_{glu} = L \cdot |\Delta G_{gly} + (n \cdot \Delta G_{ATP})|, \qquad (19.3.2)$$

where L is a constant that scales the flux rate to the net energy differential. Note that Equation 19.3.2 has the same form as the diffusion equation discussed in Chapter 7, where the diffusion rate was equal to the product of a diffusion coefficient and a local concentration gradient. Here, L is analogous to the diffusion coefficient, and the term in brackets is analogous to the concentration gradient. The actual value of L can depend on a number of cellular features, but is irrelevant to the following analysis.

The rate of ATP production is the product of the efficiency of conversion, Equation 19.3.1, and the glucose flux rate, Equation 19.3.2,

$$F_{ATP} = L \cdot \frac{n \cdot \Delta G_{ATP}}{\Delta G_{gly}} \cdot [\Delta G_{gly} + (n \cdot \Delta G_{ATP})]. \quad (19.3.3)$$

Here, the absolute signs have been removed, as the negatives cancel out. Note, however, that the first (efficiency) term increases with n, whereas the second (flux-rate) term becomes closer to zero with increasing n (because ΔG_{gly} is negative). This implies that there must be some intermediate value of n that maximizes F_{ATP}, the rate of ATP production. This behavior reflects the intrinsic tradeoff between speed and efficiency.

By taking the first derivative of Equation 19.3.3 with respect to n, setting this equal to zero (the peak of a non-linear function), and rearranging, we obtain

$$n^* = \frac{|\Delta G_{gly}|}{2 \cdot \Delta G_{ATP}} \qquad (19.3.4)$$

as the ratio of ATP to glucose molecules that maximizes the total rate of ATP production. Substituting from above for the energy terms yields $n^* = 2.05$, remarkably close to the actual glycolytic ratio of 2.

Using Equation 19.3.3, we can further inquire as to the consequences of extracting different numbers of ATPs in this reaction. For n = 1, 2, 3, and 4, F_{ATP} = 37.8L, 51.2L, 40.3L, and 4.9L, respectively. The number of ATPs produced per glucose molecule, n, can only take on integer values, but given that the adopted value of 2 is very close to the optimum, it follows that the alternatives of n = 1, 3, and 4 are expected to yield 26, 21, and 90% reductions in flux rates.

Further details on these issues can be found in Heinrich et al. (1997, 1999), with particularly lucid explanations given in Waddell et al. (1997) and Aledo and del Valle (2002).

Foundations 19.4 Extension/contraction of a metabolic pathway

Here we consider a situation in which there are two alternative pathways by which an organism can acquire an essential molecule S. Under state 1, S is only obtained by direct environmental uptake, whereas under state 3, S can only be obtained indirectly by conversion of an upstream precursor P. Both mechanisms of uptake are possible under state 2 (Figure 19.4). Under this model, evolutionary transitions occur between the alternative states by gains/losses of particular molecular mechanisms, with the probability of long-term population occupancy of each site being dictated by its selective advantage and mutation accessibility.

Following the scheme

$$\text{Pathway 1} \rightleftharpoons \text{Pathway 2} \rightleftharpoons \text{Pathway 3,}$$

the equilibrium probabilities of the reliance on these alternative pathways in a phylogenetic lineage depend on the evolutionary transition rates between adjacent states. Given constant and non-zero transition rates over a long evolutionary time period, a lineage would be expected to slowly wander from state to state. With sufficient time, regardless of the starting state, the time-averaged state probabilities would then reach an equilibrium defined by their relative selective advantages and the mutational interconversion rates (Foundations 5.2 and 5.3). A very simple model is outlined here for heuristic purposes.

Each of the four transition rates (associated with the arrows in the preceding scheme) is equal to the product of the rate of origin of the appropriate mutation and the subsequent probability of fixation. For example, transitions from state 1 to 2 depend on: the rate of origin of mutations that allow conversion of P to S, Nv_1, where N is the number of individuals in the population (assumed to be haploid), v_1 is the per-individual mutational rate of origin of the upstream pathway; and the probability of fixation is denoted by $\phi(s_2 - s_1)$, where $(s_2 - s_1)$ is the selective advantage of pathway 2 over that of pathway 1. Extending this approach to all four coefficients leads to the four transition rates

$$1 \rightarrow 2: \qquad \theta_{12} = Nv_1\phi(s_2 - s_1) \qquad (19.4.1a)$$

$$2 \rightarrow 1: \qquad \theta_{21} = Nu_1\phi(s_1 - s_2) \qquad (19.4.1b)$$

$$2 \rightarrow 3: \qquad \theta_{23} = Nu_3\phi(s_3 - s_2) \qquad (19.4.1c)$$

$$3 \rightarrow 2: \qquad \theta_{32} = Nv_3\phi(s_2 - s_3) \qquad (19.4.1d)$$

where u_1 is the mutational rate of loss of the precursor pathway, and v_3 and u_3 denote the mutational rates of gain and loss of the environmental uptake mechanism. (Here we assume that state transitions are rare enough that the simultaneous presence of all three states within a population is negligible, i.e., the sequential model in the parlance of Chapter 5).

Two mathematical simplifications lead to a relatively straightforward solution. First, the steady-state frequencies of each of the alternative states in a linear array can be determined by multiplying the coefficients on all of the arrows pointing toward the state from above and below (Foundations 5.4). This means that

$$\widetilde{P}_1 \propto \theta_{21}\theta_{32} \qquad (19.4.2a)$$

$$\widetilde{P}_2 \propto \theta_{12}\theta_{32} \qquad (19.4.2b)$$

$$\widetilde{P}_3 \propto \theta_{12}\theta_{23} \qquad (19.4.2c)$$

where $\propto$ means 'proportional to'. Second, the fixation probabilities are defined by the standard formula for newly arising mutations (Equation 4.1b), and have the property that the ratio of these probabilities for advantageous and deleterious mutations with the same absolute effects is

$$\phi(s)/\phi(-s) = e^{2N_es} \qquad (19.4.3)$$

for a haploid population with effective size N_e (Foundations 5.4). Dividing Equations 19.4.2a–c by $\theta_{12}\theta_{32}$, and applying Equation 19.4.3 leads to

$$\widetilde{P}_1 = C \cdot \beta_1 e^{2N_e(s_1 - s_2)} \qquad (19.4.4a)$$

$$\widetilde{P}_2 = C \cdot 1 \qquad (19.4.4b)$$

$$\widetilde{P}_3 = C \cdot \beta_3 e^{2N_e(s_3 - s_2)} \qquad (19.4.4c)$$

where $\beta_i = u_i/v_i$, and C is a normalization constant equal to the reciprocal of the sum of the three terms to the right of

C in Equations 19.4.4a,b,c (usage of C insures that $\widetilde{P}_1 + \widetilde{P}_2 + \widetilde{P}_3 = 1$). As β_i defines the ratio of the mutational rates of loss to gain of the two alternative mechanisms of substrate acquisition, these expressions show that the probabilities of alternative states are simple functions of the joint directional flux into the end states by mutation and selection processes.

Literature Cited

Abatemarco, J., A. Hill, and H. S. Alper. 2013. Expanding the metabolic engineering toolbox with directed evolution. Biotechnol. J. 8: 1397–1410.

Aguilar-Rodríguez, J., and A. Wagner. 2018. Metabolic determinants of enzyme evolution in a genome-scale bacterial metabolic network. Genome Biol. Evol. 10: 3076–3088.

Aharoni, A., L. Gaidukov, O. Khersonsky, S. M. Gould, C. Roodveldt, and D. S. Tawfik. 2005. The 'evolvability' of promiscuous protein functions. Nat. Genet. 37: 73–76.

Albery, W. J., and J. R. Knowles. 1976a. Evolution of enzyme function and the development of catalytic efficiency. Biochemistry 15: 5631–5640.

Albery, W. J., and J. R. Knowles. 1976b. Free-energy profile of the reaction catalyzed by triosephosphate isomerase. Biochemistry 15: 5627–5631.

Aledo, J. C., and A. E. del Valle. 2002. Glycolysis in Wonderland: The importance of energy dissipation in metabolic pathways. J. Chem. Educ. 79: 1336–1339.

Allen, A. E., A. Moustafa, A. Montsant, A. Eckert, P. G. Kroth, and C. Bowler. 2012. Evolution and functional diversification of fructose bisphosphate aldolase genes in photosynthetic marine diatoms. Mol. Biol. Evol. 29: 367–379.

Alvarez, R. A., M. W. Blaylock, and J. B. Baseman. 2003. Surface localized glyceraldehyde-3-phosphate dehydrogenase of *Mycoplasma genitalium* binds mucin. Mol. Microbiol. 48: 1417–1425.

Babtie, A., N. Tokuriki, and F. Hollfelder. 2010. What makes an enzyme promiscuous? Curr. Opin. Chem. Biol. 14: 200–207.

Bar-Even, A., A. Flamholz, E. Noor, and R. Milo. 2012. Rethinking glycolysis: On the biochemical logic of metabolic pathways. Nat. Chem. Biol. 8: 509–517.

Bar-Even, A., R. Milo, E. Noor, and D. S. Tawfik. 2015. The moderately efficient enzyme: Futile encounters and enzyme floppiness. Biochemistry 54: 4969–4977.

Bar-Even, A., E. Noor, Y. Savir, W. Liebermeister, D. Davidi, D. S. Tawfik, and R. Milo. 2011. The moderately efficient enzyme: Evolutionary and physicochemical trends shaping enzyme parameters. Biochemistry 50: 4402–4410.

Belfaiza, J., C. Parsot, A. Martel, C. B. de la Tour, D. Margarita, G. N. Cohen, and I. Saint-Girons. 1986. Evolution in biosynthetic pathways: Two enzymes catalyzing consecutive steps in methionine biosynthesis originate from a common ancestor and possess a similar regulatory region. Proc. Natl. Acad. Sci. USA 83: 867–871.

Benkovic, S. J., and S. Hammes-Schiffer. 2003. A perspective on enzyme catalysis. Science 301: 1196–1202.

Bennett, B. D., E. H. Kimball, M. Gao, R. Osterhout, S. J. Van Dien, and J. D. Rabinowitz. 2009. Absolute metabolite concentrations and implied enzyme active site occupancy in *Escherichia coli*. Nat. Chem. Biol. 5: 593–599.

Berg, O. G. 1985. Orientation constraints in diffusion-limited macromolecular association: The role of surface diffusion as a rate-enhancing mechanism. Biophys J. 47: 1–14.

Berg, O. G., and P. H. von Hippel. 1985. Diffusion-controlled macromolecular interactions. Annu. Rev. Biophys. Biophys. Chem. 14: 131–160.

Bergthorsson, U., D. I. Andersson, and J. R. Roth. 2007. Ohno's dilemma: Evolution of new genes under continuous selection. Proc. Natl. Acad. Sci. USA 104: 17004–17009.

Blank, D., L. Wolf, M. Ackermann, and O. K. Silander. 2014. The predictability of molecular evolution during functional innovation. Proc. Natl. Acad. Sci. USA 111: 3044–3049.

Blount, Z. D., J. E. Barrick, C. J. Davidson, and R. E. Lenski. 2012. Genomic analysis of a key innovation in an experimental *Escherichia coli* population. Nature 489: 513–518.

Borenstein, E., M. Kupiec, M. W. Feldman, and E. Ruppin. 2008. Large-scale reconstruction and phylogenetic analysis of metabolic environments. Proc. Natl. Acad. Sci. USA 105: 14482–14487.

Bornscheuer, U. T., G. W. Huisman, R. J. Kazlauskas, S. Lutz, J. C. Moore, and K. Robins. 2012. Engineering the third wave of biocatalysis. Nature 485: 185–194.

Boronat, A., E. Caballero, and J. Aguilar. 1983. Experimental evolution of a metabolic pathway for ethylene glycol utilization by *Escherichia coli*. J. Bacteriol. 153: 134–139.

Briggs, G. E., and J. B. Haldane. 1925. A note on the kinetics of enzyme action. Biochem. J. 19: 338–339.

Bublitz, D. C., G. L. Chadwick, J. S. Magyar, K. M. Sandoz, D. M. Brooks, S. Mesnage, M. S. Ladinsky, A. I. Garber, P. J. Bjorkman, V. J. Orphan, et al. 2019. Peptidoglycan production by an insect-bacterial mosaic. Cell 179: 703–712.

Brustad, E. M., and F. H. Arnold. 2011. Optimizing nonnatural protein function with directed evolution. Curr. Opin. Chem. Biol. 15: 201–210.

Byng, G. S., R. J. Whitaker, and R. A. Jensen. 1983. Evolution of L-phenylalanine biosynthesis in rRNA homology group I of *Pseudomonas*. Arch. Microbiol. 136: 163–168.

Camacho, C. J., S. R. Kimura, C. DeLisi, and S. Vajda. 2000. Kinetics of desolvation-mediated protein–protein binding. Biophys. J. 78: 1094–1105.

Canback, B., S. G. Andersson, and C. G. Kurland. 2002. The global phylogeny of glycolytic enzymes. Proc. Natl. Acad. Sci. USA 99: 6097–6102.

Carpenter, E. P., A. R. Hawkins, J. W. Frost, and K. A. Brown. 1998. Structure of dehydroquinate synthase reveals an active site capable of multistep catalysis. Nature 394: 299–302.

Castaldo, C., V. Vastano, R. A. Siciliano, M. Candela, M. Vici, L. Muscariello, R. Marasco, and M. Sacco. 2009. Surface displaced α-enolase of *Lactobacillus plantarum* is a fibronectin binding protein. Microb. Cell Fact. 8: 14.

Chang, Y. S., W. Wu, G. Walsh, W. K. Hong, and L. Mao. 2003. Enolase-α is frequently down-regulated in non-small cell lung cancer and predicts aggressive biological behavior. Clin. Cancer Res. 9: 3641–3644.

Chen, Y., and J. Nielsen. 2021. *In vitro* turnover numbers do not reflect *in vivo* activities of yeast enzymes. Proc. Natl. Acad. Sci. USA 118: e2108391118.

Colombo, M., H. Laayouni, B. M. Invergo, J. Bertranpetit, and L. Montanucci. 2014. Metabolic flux is a determinant of the evolutionary rates of enzyme-encoding genes. Evolution 68: 605–613.

Copley, S. D. 2000. Evolution of a metabolic pathway for degradation of a toxic xenobiotic: The patchwork approach. Trends Biochem. Sci. 25: 261–265.

Copley, S. D. 2003. Enzymes with extra talents: Moonlighting functions and catalytic promiscuity. Curr. Opin. Chem. Biol. 7: 265–272.

Copley, S. D. 2012. Toward a systems biology perspective on enzyme evolution. J. Biol. Chem. 287: 3–10.

Copley, S. D. 2014. An evolutionary perspective on protein moonlighting. Biochem. Soc. Trans. 42: 1684–1691.

Copley, S. D. 2020. The physical basis and practical consequences of biological promiscuity. Phys. Biol. 17: 051001.

Crowley, P. H. 1975. Natural selection and the Michaelis constant. J. Theor. Biol. 50: 461–475.

Curien, G., V. Biou, C. Mas-Droux, M. Robert-Genthon, J. L. Ferrer, and R. Dumas. 2008. Amino acid biosynthesis: New architectures in allosteric enzymes. Plant Physiol. Biochem. 46: 325–339.

Davidi, D., L. M. Longo, J. Jabłońska, R. Milo, and D. S. Tawfik. 2018. A bird's-eye view of enzyme evolution: Chemical, physicochemical, and physiological considerations. Chem. Rev. 118: 8786–8797.

Davidi, D., E. Noor, W. Liebermeister, A. Bar-Even, A. Flamholz, K. Tummler, U. Barenholz, M. Goldenfeld, T. Shlomi, and R. Milo. 2016. Global characterization of *in vivo* enzyme catalytic rates and their correspondence to *in vitro* k_{cat} measurements. Proc. Natl. Acad. Sci. USA 113: 3401–3406.

Dincturk, H. B., and D. B. Knaff. 2000. The evolution of glutamate synthase. Mol. Biol. Rep. 27: 141–148.

Du, B., D. C. Zielinski, J. M. Monk, and B. O. Palsson. 2018. Thermodynamic favorability and pathway yield as evolutionary tradeoffs in biosynthetic pathway choice. Proc. Natl. Acad. Sci. USA 115: 11339–11344.

Du, J., R. F. Say, W. Lü, G. Fuchs, and O. Einsle. 2011. Active-site remodelling in the bifunctional fructose-1,6-bisphosphate aldolase/phosphatase. Nature 478: 534–537.

Due, A. V., J. Kuper, A. Geerlof, J. P. von Kries, and M. Wilmanns. 2011. Bisubstrate specificity in histidine/tryptophan biosynthesis isomerase from *Mycobacterium tuberculosis* by active site metamorphosis. Proc. Natl. Acad. Sci. USA 108: 3554–3559.

Edwards, D. R., D. C. Lohman, and R. Wolfenden. 2012. Catalytic proficiency: The extreme case of S-O cleaving sulfatases. J. Am. Chem. Soc. 134: 525–531.

Fernandes, P., H. Aldeborgh, L. Carlucci, L. Walsh, J. Wasserman, E. Zhou, S. T. Lefurgy, and E. C. Mundorff. 2015. Alteration of substrate specificity of alanine dehydrogenase. Protein Eng. Des. Sel. 28: 29–35.

Fondi, M., M. Brilli, G. Emiliani, D. Paffetti, and R. Fani. 2007. The primordial metabolism: An ancestral interconnection between leucine, arginine, and lysine biosynthesis. BMC Evol. Biol. 7 Suppl. 2: S3.

Fraser, H. B., A. E. Hirsh, L. M. Steinmetz, C. Scharfe, and M. W. Feldman. 2002. Evolutionary rate in the protein interaction network. Science 296: 750–752.

Gabaldón, T., and E. V. Koonin. 2013. Functional and evolutionary implications of gene orthology. Nat. Rev. Genet. 14: 360–366.

Galperin, M. Y., D. R. Walker, and E. V. Koonin. 1998. Analogous enzymes: Independent inventions in enzyme evolution. Genome Res. 8: 779–790.

Ge, Y. D., P. Song, Z. Y. Cao, P. Wang, and G. P. Zhu. 2014. Alteration of coenzyme specificity of malate dehydrogenase from *Streptomyces coelicolor* A3(2) by site-directed mutagenesis. Genet. Mol. Res. 13: 5758–5766.

Gophna, U., E. Bapteste, W. F. Doolittle, D. Biran, and E. Z. Ron. 2005. Evolutionary plasticity of methionine biosynthesis. Gene 355: 48–57.

Gozalbo, D., I. Gil-Navarro, I. Azorín, J. Renau-Piqueras, J. P. Martínez, and M. L. Gil. 1998. The cell wall-associated glyceraldehyde-3-phosphate dehydrogenase of *Candida albicans* is also a fibronectin and laminin binding protein. Infect. Immun. 66: 2052–2059.

Gunawardena, J. 2012. Some lessons about models from Michaelis and Menten. Mol. Biol. Cell 23: 517–519.

Gunawardena, J. 2014. Time-scale separation – Michaelis and Menten's old idea, still bearing fruit. FEBS J. 281: 473–488.

Hammes, G. G. 2002. Multiple conformational changes in enzyme catalysis. Biochemistry 41: 8221–8228.

Hanson, A. D., D. R. McCarty, C. S. Henry, X. Xian, J. Joshi, J. A. Patterson, J. D. García-García, S. D. Fleischmann, N. D. Tivendale, and A. H. Millar. 2021. The number of catalytic cycles in an enzyme's lifetime and why it matters to metabolic engineering. Proc. Natl. Acad. Sci. USA 118: e2023348118.

Hartl, D. L., D. E. Dykhuizen, and A. M. Dean. 1985. Limits of adaptation: The evolution of selective neutrality. Genetics 111: 655–674.

Hébert, A., S. Casaregola, and J. M. Beckerich. 2011. Biodiversity in sulfur metabolism in hemiascomycetous yeasts. FEMS Yeast Res. 11: 366–378.

Heckmann, D., D. C. Zielinski, and B. O. Palsson. 2018. Modeling genome-wide enzyme evolution predicts strong epistasis underlying catalytic turnover rates. Nat. Commun. 9: 5270.

Heinrich, R., E. Meléndez-Hevia, F. Montero, J. C. Nuño, A. Stephani, and T. G. Waddell. 1999. The structural design of glycolysis: An evolutionary approach. Biochem. Soc. Trans. 27: 294–298.

Heinrich, R., F. Montero, E. Klipp, T. G. Waddell, and E. Meléndez-Hevia. 1997. Theoretical approaches to the evolutionary optimization of glycolysis: Thermodynamic and kinetic constraints. Eur. J. Biochem. 243: 191–201.

Heinrich, R., and T. A. Rapoport. 1974. A linear steady-state treatment of enzymatic chains: General properties, control and effector strength. Eur. J. Biochem. 42: 89–95.

Hittinger, C. T., A. Rokas, and S. B. Carroll. 2004. Parallel inactivation of multiple GAL pathway genes and ecological diversification in yeasts. Proc. Natl. Acad. Sci. USA 101: 14144–14149.

Ho, C. L., M. Noji, M. Saito, M. Yamazaki, and K. Saito. 1998. Molecular characterization of plastidic phosphoserine aminotransferase in serine biosynthesis from *Arabidopsis*. Plant J. 16: 443–452.

Hochuli, M., H. Patzelt, D. Oesterhel, K. Wüthrich, and T. Szyperski. 1999. Amino acid biosynthesis in the halophilic archaeon *Haloarcula hispanica*. J. Bacteriol. 181: 3226–3237.

Horowitz, N. H. 1945. On the evolution of biochemical syntheses. Proc. Natl. Acad. Sci. USA 31: 153–157.

Hoshino, Y., and E. A. Gaucher. 2018. On the origin of isoprenoid biosynthesis. Mol. Biol. Evol. 35: 2185–2197.

Hudson, A. O., B. K. Singh, T. Leustek, and C. Gilvarg. 2006. An LL-diaminopimelate aminotransferase defines a novel variant of the lysine biosynthesis pathway plants. Plant Physiol. 140: 292–301.

Hughes, A. L. 1994. The evolution of functionally novel proteins after gene duplication. Proc. Roy. Soc. Lond. B 256: 119–124.

Hult, K., and P. Berglund. 2007. Enzyme promiscuity: Mechanism and applications. Trends Biotechnol. 25: 231–238.

Iida, H., and I. Yahara. 1985. Yeast heat-shock protein of Mr 48,000 is an isoprotein of enolase. Nature 315: 688–690.

Janin, J. 1997. The kinetics of protein–protein recognition. Proteins 28: 153–161.

Jensen, R. A. 1976. Enzyme recruitment in evolution of new function. Annu. Rev. Microbiol. 30: 409–425.

Jensen, R. A. 1985. Biochemical pathways in prokaryotes can be traced backward through evolutionary time. Mol. Biol. Evol. 2: 92–108.

Jensen, R. A., and D. L. Pierson. 1975. Evolutionary implications of different types of microbial enzymology for L-tyrosine biosynthesis. Nature 254: 667–671.

Jia, B., G. W. Cheong, and S. Zhang. 2013. Multifunctional enzymes in archaea: Promiscuity and moonlight. Extremophiles 17: 193–203.

Johannes, T. W., and H. Zhao. 2006. Directed evolution of enzymes and biosynthetic pathways. Curr. Opin. Microbiol. 9: 261–267.

Johnsen, U., and P. Schönheit. 2007. Characterization of cofactor-dependent and cofactor-independent phosphoglycerate mutases from Archaea. Extremophiles 11: 647–657.

Johnson, K. A., and R. S. Goody. 2011. The original Michaelis constant: Translation of the 1913 Michaelis–Menten paper. Biochemistry 50: 8264–8269.

Juárez-Vázquez, A. L., J. N. Edirisinghe, E. A. Verduzco-Castro, K. Michalska, C. Wu, L. Noda-García, G. Babnigg, M. Endres, S. Medina-Ruíz, J. Santoyo-Flores, et al.

2017. Evolution of substrate specificity in a retained enzyme driven by gene loss. eLife 6: e22679.

Kaberdin, V. R., D. Singh, and S. Lin-Chao. 2011. Composition and conservation of the mRNA-degrading machinery in bacteria. J. Biomed. Sci. 18: 23.

Kacser, H., and R. Beeby. 1984. Evolution of catalytic proteins or on the origin of enzyme species by means of natural selection. J. Mol. Evol. 20: 38–51.

Kacser, H., and J. A. Burns. 1973. The control of flux. Symp. Soc. Exp. Biol. 27: 65–104.

Kanai, S., and H. Toh. 1999. Identification of new members of the GS ADP-forming family from the *de novo* purine biosynthesis pathway. J. Mol. Evol. 48: 482–492.

Kawai, S., T. Mukai, S. Mori, B. Mikami, and K. Murata. 2005. Hypothesis: Structures, evolution, and ancestor of glucose kinases in the hexokinase family. J. Biosci. Bioeng. 99: 320–330.

Khersonsky, O., C. Roodveldt, and D. S. Tawfik. 2006. Enzyme promiscuity: Evolutionary and mechanistic aspects. Curr. Opin. Chem. Biol. 10: 498–508.

Kim, J., J. J. Flood, M. R. Kristofich, A. B. Morgenthaler, T. Fuhrer, U. Sauer, D. Snyder, V. S. Cooper, C. C. Ebmeier, et al. 2019. Hidden resources in the *Escherichia coli* genome restore PLP synthesis and robust growth after deletion of the essential gene pdxB. Proc. Natl. Acad. Sci. USA 116: 24164–24173.

Kim, J., J. P. Kershner, Y. Novikov, R. K. Shoemaker, and S. D. Copley. 2010. Three serendipitous pathways in *E. coli* can bypass a block in pyridoxal-5′-phosphate synthesis. Mol. Syst. Biol. 6: 436.

Kim, J. W., and C. V. Dang. 2005. Multifaceted roles of glycolytic enzymes. Trends Biochem. Sci. 30: 142–150.

Klesmith, J. R., J. P. Bacik, E. E. Wrenbeck, R. Michalczyk, and T. A. Whitehead. 2017. Trade-offs between enzyme fitness and solubility illuminated by deep mutational scanning. Proc. Natl. Acad. Sci. USA 114: 2265–2270.

Lee, D. H., and B. Ø. Palsson. 2010. Adaptive evolution of *Escherichia coli* K-12 MG1655 during growth on a non-native carbon source, L-1,2-propanediol. Appl. Environ. Microbiol. 76: 4158–4168.

Leopoldseder, S., J. Claren, C. Jürgens, and R. Sterner. 2004. Interconverting the catalytic activities of $(\beta\alpha)(8)$-barrel enzymes from different metabolic pathways: Sequence requirements and molecular analysis. J. Mol. Biol. 337: 871–879.

Liapounova, N. A., V. Hampl, P. M. Gordon, C. W. Sensen, L. Gedamu, and J. B. Dacks. 2006. Reconstructing the mosaic glycolytic pathway of the anaerobic eukaryote *Monocercomonoides*. Eukaryot. Cell 5: 2138–2146.

List, F., R. Sterner, and M. Wilmanns. 2011. Related $(\beta\alpha)8$-barrel proteins in histidine and tryptophan biosynthesis: A paradigm to study enzyme evolution. Chembiochem. 12: 1487–1494.

Lombard, J., and D. Moreira. 2011. Origins and early evolution of the mevalonate pathway of isoprenoid biosynthesis in the three domains of life. Mol. Biol. Evol. 28: 87–99.

Lu, Y., and M. D. Rausher. 2003. Evolutionary rate variation in anthocyanin pathway genes. Mol. Biol. Evol. 20: 1844–1853.

Lynch, M. 2007. The evolution of genetic networks by nonadaptive processes. Nat. Rev. Genet. 8: 803–813.

Matsumura, I., and A. D. Ellington. 2001. *In vitro* evolution of beta-glucuronidase into a beta-galactosidase proceeds through non-specific intermediates. J. Mol. Biol. 305: 331–339.

McCutcheon, J. P., and N. A. Moran. 2007. Parallel genomic evolution and metabolic interdependence in an ancient symbiosis. Proc. Natl. Acad. Sci. USA 104: 19392–19397.

McLoughlin, S. Y., and S. D. Copley. 2008. A compromise required by gene sharing enables survival: Implications for evolution of new enzyme activities. Proc. Natl. Acad. Sci. USA 105: 13497–13502.

Melcher, K., M. Rose, M. Künzler, G. H. Braus, and K. D. Entian. 1995. Molecular analysis of the yeast SER1 gene encoding 3-phosphoserine aminotransferase: Regulation by general control and serine repression. Curr. Genet. 27: 501–508.

Merino, E., R. A. Jensen, and C. Yanofsky. 2008. Evolution of bacterial trp operons and their regulation. Curr. Opin. Microbiol. 11: 78–86.

Michaelis, L., and M. L. Menten. 1913. Die Kinetik der Invertinwirkung. Biochem. Z. 49: 333–369.

Miles, L. A., C. M. Dahlberg, J. Plescia, J. Felez, K. Kato, and E. F. Plow. 1991. Role of cell-surface lysines in plasminogen binding to cells: Identification of alpha-enolase as a candidate plasminogen receptor. Biochemistry 30: 1682–1691.

Miyazaki, J., N. Kobashi, M. Nishiyama, and H. Yamane. 2001. Functional and evolutionary relationship between arginine biosynthesis and prokaryotic lysine biosynthesis through alpha-aminoadipate. J. Bacteriol. 183: 5067–5073.

Modun, B., and P. Williams. 1999. The staphylococcal transferrin-binding protein is a cell wall glyceraldehyde-3-phosphate dehydrogenase. Infect. Immun. 67: 1086–1092.

Monk, J., J. Nogales, and B. O. Palsson. 2014. Optimizing genome-scale network reconstructions. Nat. Biotechnol. 32: 447–452.

Montanucci, L., H. Laayouni, G. M. Dall'Olio, and J. Bertranpetit. 2011. Molecular evolution and network-level analysis of the N-glycosylation metabolic pathway across primates. Mol. Biol. Evol. 28: 813–823.

Morgenthaler, A. B., W. R. Kinney, C. C. Ebmeier, C. M. Walsh, D. J. Snyder, V. S. Cooper, W. M. Old, and

S. D. Copley. 2019. Mutations that improve efficiency of a weak-link enzyme are rare compared to adaptive mutations elsewhere in the genome. eLife 8: e53535.

Nam, H., N. E. Lewis, J. A. Lerman, D. H. Lee, R. L. Chang, D. Kim, and B. O. Palsson. 2012. Network context and selection in the evolution to enzyme specificity. Science 337: 1101–1104.

Näsvall, J., L. Sun, J. R. Roth, and D. I. Andersson. 2012. Real-time evolution of new genes by innovation, amplification, and divergence. Science 338: 384–387.

Nguyen, V., C. Wilson, M. Hoemberger, J. B. Stiller, R. V. Agafonov, S. Kutter, J. English, D. L. Theobald, and D. Kern. 2017. Evolutionary drivers of thermoadaptation in enzyme catalysis. Science 355: 289–294.

Nishida, H., and M. Nishiyama. 2012. Evolution of lysine biosynthesis in the phylum deinococcus-thermus. Int. J. Evol. Biol. 2012: 745931.

Nishida, H., M. Nishiyama, N. Kobashi, T. Kosuge, T. Hoshino, and H. Yamane. 1999. A prokaryotic gene cluster involved in synthesis of lysine through the amino adipate pathway: A key to the evolution of amino acid biosynthesis. Genome Res. 9: 1175–1183.

Noor, E., E. Eden, R. Milo, and U. Alon. 2010. Central carbon metabolism as a minimal biochemical walk between precursors for biomass and energy. Mol. Cell 39: 809–820.

Northrup, S. H., and H. P. Erickson. 1992. Kinetics of protein–protein association explained by Brownian dynamics computer simulation. Proc. Natl. Acad. Sci. USA 89: 3338–3342.

Notebaart, R. A., B. Szappanos, B. Kintses, F. Pál, Á. Györkei, B. Bogos, V. Lázár, R. Spohn, B. Csörgo, A. Wagner, et al. 2014. Network-level architecture and the evolutionary potential of underground metabolism. Proc. Natl. Acad. Sci. USA 111: 11762–11767.

Oslancová, A., and S. Janecek. 2004. Evolutionary relatedness between glycolytic enzymes most frequently occurring in genomes. Folia Microbiol. (Praha) 49: 247–258.

Pál, C., B. Papp, M. J. Lercher, P. Csermely, S. G. Oliver, and L. D. Hurst. 2006. Chance and necessity in the evolution of minimal metabolic networks. Nature 440: 667–670.

Parsot, C. 1986 Evolution of biosynthetic pathways: A common ancestor for threonine synthase, threonine dehydratase and D-serine dehydratase. EMBO J. 5: 3013–3019.

Parsot, C., I. Saint-Girons, and G. N. Cohen. 1987. Enzyme specialization during the evolution of amino acid biosynthetic pathways. Microbiol. Sci. 4: 260–262.

Patrick, W. M., E. M. Quandt, D. B. Swartzlander, and I. Matsumura. 2007. Multicopy suppression underpins metabolic evolvability. Mol. Biol. Evol. 24: 2716–2722.

Pfeiffer, T., O. S. Soyer, and S. Bonhoeffer. 2005. The evolution of connectivity in metabolic networks. PLoS Biol. 3: e228.

Piatigorsky, J., and G. Wistow. 1991. The recruitment of crystallins: New functions precede gene duplication. Science 252: 1078–1079.

Qian, Q., and P. J. Keeling. 2001. Diplonemid glyceraldehyde-3-phosphate dehydrogenase (GAPDH) and prokaryote-to-eukaryote lateral gene transfer. Protist 152: 193–201.

Quandt, E. M., D. E. Deatherage, A. D. Ellington, G. Georgiou, and J. E. Barrick. 2014. Recursive genomewide recombination and sequencing reveals a key refinement step in the evolution of a metabolic innovation in *Escherichia coli*. Proc. Natl. Acad. Sci. USA 111: 2217–2222.

Rakus, J. F., A. A. Fedorov, E. V. Fedorov, M. E. Glasner, B. K. Hubbard, J. D. Delli, P. C. Babbitt, S. C. Almo, and J. A. Gerlt. 2008. Evolution of enzymatic activities in the enolase superfamily: L-rhamnonate dehydratase. Biochemistry 47: 9944–9954.

Ramazzina, I., R. Costa, L. Cendron, R. Berni, A. Peracchi, G. Zanotti, and R. Percudani. 2010. An aminotransferase branch point connects purine catabolism to amino acid recycling. Nat. Chem. Biol. 6: 801–806.

Ramsay, H., L. H. Rieseberg, and K. Ritland. 2009. The correlation of evolutionary rate with pathway position in plant terpenoid biosynthesis. Mol. Biol. Evol. 26: 1045–1053.

Raushel, F. M., J. B. Thoden, G. D. Reinhart, and H. M. Holden. 1998. Carbamoyl phosphate synthetase: A crooked path from substrates to products. Curr. Opin. Chem. Biol. 2: 624–632.

Rausher, M. D. 2012. The evolution of genes in branched metabolic pathways. Evolution 67: 34–48.

Rausher, M. D., R. E. Miller, and P. Tiffin. 1999. Patterns of evolutionary rate variation among genes of the anthocyanin biosynthetic pathway. Mol. Biol. Evol. 16: 266–274.

Rauwerdink, A., M. Lunzer, T. Devamani, B. Jones, J. Mooney, Z. J. Zhang, J. H. Xu, R. J. Kazlauskas, and A. M. Dean. 2016. Evolution of a catalytic mechanism. Mol. Biol. Evol. 33: 971–979.

Rogers, M., and P. J. Keeling. 2004. Lateral transfer and recompartmentalization of cycle enzymes of plants and algae. J. Mol. Evol. 58: 367–375.

Rothman, S. C., and J. F. Kirsch. 2003. How does an enzyme evolved *in vitro* compare to naturally occurring homologs possessing the targeted function? Tyrosine aminotransferase from aspartate aminotransferase. J. Mol. Biol. 327: 593–608.

Roy, S. 1999. Multifunctional enzymes and evolution of biosynthetic pathways: Retro-evolution by jumps. Proteins 37: 303–309.

Sakuraba, H., S. Fujiwara, and T. Noguchi. 1996. Metabolism of glyoxylate, the end product of purine degradation, in liver peroxisomes of fresh water fish. Biochem. Biophys. Res. Commun. 229: 603–606.

Sánchez, L., D. Horner, D. Moore, K. Henze, T. Embley, and M. Müller. 2002. Fructose-1,6-bisphosphate aldolases in amitochondriate protists constitute a single protein subfamily with eubacterial relationships. Gene 295: 51–59.

Schlosshauer, M., and D. Baker. 2004. Realistic protein–protein association rates from a simple diffusional model neglecting long-range interactions, free energy barriers, and landscape ruggedness. Protein Sci. 13: 1660–1669.

Seidler, N. W. 2013. Functional diversity. Adv. Exp. Med. Biol. 985: 103–147.

Shimizu, Y., H. Sakuraba, K. Doi, and T. Ohshima. 2008. Molecular and functional characterization of D-3-phosphoglycerate dehydrogenase in the serine biosynthetic pathway of the hyperthermophilic archaeon Sulfolobus tokodaii. Arch. Biochem. Biophys. 470: 120–128.

Siebers, B., H. Brinkmann, C. Dörr, B. Tjaden, H. Lilie, J. van der Oost, and C. H. Verhees. 2001. Archaeal fructose-1,6-bisphosphate aldolases constitute a new family of archaeal type class I aldolase. J. Biol. Chem. 276: 28710–28718.

Siebers, B., and P. Schönheit. 2005. Unusual pathways and enzymes of central carbohydrate metabolism in Archaea. Curr. Opin. Microbiol. 8: 695–705.

Skolnick, J., and M. Gao. 2013. Interplay of physics and evolution in the likely origin of protein biochemical function. Proc. Natl. Acad. Sci. USA 110: 9344–9349.

Song, J., C. A. Bonner, M. Wolinsky, and R. A. Jensen. 2005. The TyrA family of aromatic-pathway dehydrogenases in phylogenetic context. BMC Biol. 3: 13.

Soo, V. W., P. Hanson-Manful, and W. M. Patrick. 2011. Artificial gene amplification reveals an abundance of promiscuous resistance determinants in Escherichia coli. Proc. Natl. Acad. Sci. USA 108: 1484–1489.

Suzuki, A., and D. B. Knaff. 2005. Glutamate synthase: Structural, mechanistic and regulatory properties, and role in the amino acid metabolism. Photosynth. Res. 83: 191–217.

Tagliabracci, V. S., C. Heiss, C. Karthik, C. J. Contreras, J. Glushka, M. Ishihara, P. Azadi, T. D. Hurley, A. A. DePaoli-Roach, and P. J. Roach. 2011. Phosphate incorporation during glycogen synthesis and Lafora disease. Cell Metab. 13: 274–282.

Takishita, K., and Y. Inagaki. 2009. Eukaryotic origin of glyceraldehyde-3-phosphate dehydrogenase genes in Clostridium thermocellum and Clostridium cellulolyticum genomes and putative fates of the exogenous gene in the subsequent genome evolution. Gene 441: 22–27.

Tokuriki, N., and D. S. Tawfik. 2009. Protein dynamism and evolvability. Science 324: 203–207.

Torruella, G., H. Suga, M. Riutort, J. Peretó, and I. Ruiz-Trillo. 2009. The evolutionary history of lysine biosynthesis pathways within eukaryotes. J. Mol. Evol. 69: 240–248.

Tunio, S. A., N. J. Oldfield, A. Berry, D. A. Ala'Aldeen, K. G. Wooldridge, and D. P. Turner. 2010. The moonlighting protein fructose-1, 6-bisphosphate aldolase of Neisseria meningitidis: Surface localization and role in host cell adhesion. Mol. Microbiol. 76: 605–615.

van der Oost, J., M. A. Huynen, and C. H. Verhees. 2002. Molecular characterization of phosphoglycerate mutase in archaea. FEMS Microbiol. Lett. 212: 111–120.

Velasco, A. M., J. I. Leguina, and A. Lazcano. 2002. Molecular evolution of the lysine biosynthetic pathways. J. Mol. Evol. 55: 445–459.

Verhees, C. H., J. E. Tuininga, S. W. Kengen, A. J. Stams, J. van der Oost, and V. M. de Vos. 2001. ADP-dependent phosphofructokinases in mesophilic and thermophilic methanogenic archaea. J. Bacteriol. 183: 7145–7153.

Vitkup, D., P. Kharchenko, and A. Wagner. 2006. Influence of metabolic network structure and function on enzyme evolution. Genome Biol. 7: R39.

von Stockar, U., and J. Liu. 1999. Does microbial life always feed on negative entropy? Thermodynamic analysis of microbial growth. Biochim. Biophys. Acta 1412: 191–211.

Voordeckers, K., C. A. Brown, K. Vanneste, E. van der Zande, A. Voet, S. Maere, and K. J. Verstrepen. 2012. Reconstruction of ancestral metabolic enzymes reveals molecular mechanisms underlying evolutionary innovation through gene duplication. PLoS Biol. 10: e1001446.

Waddell, T. G., P. Repovic, E. Meléndez-Hevia, R. Heinrich, and F. Montero. 1997. Optimization of glycolysis: A new look at the efficiency of energy coupling. Biochem. Education 25: 204–205.

Wagoner, J. A., and K. A. Dill. 2019. Opposing pressures of speed and efficiency guide the evolution of molecular machines. Mol. Biol. Evol. 36: 2813–2822.

Wilks, H. M., K. W. Hart, R. Feeney, C. R. Dunn, H. Muirhead, W. N. Chia, D. A. Barstow, T. Atkinson, A. R. Clarke, and J. J. Holbrook. 1988. A specific, highly active malate dehydrogenase by redesign of a lactate dehydrogenase framework. Science 242: 1541–1544.

Wilmanns, M., C. C. Hyde, D. R. Davies, K. Kirschner, and J. N. Jansonius. 1991. Structural conservation in parallel beta/alpha-barrel enzymes that catalyze three sequential reactions in the pathway of tryptophan biosynthesis. Biochemistry 30: 9161–9169.

Wistow, G. J., and J. Piatigorsky. 1988. Lens crystallins: The evolution and expression of proteins for a highly specialized tissue. Annu. Rev. Biochem. 57: 479–504.

Wolfenden, R., and M. J. Snider. 2001. The depth of chemical time and the power of enzymes as catalysts. Acc. Chem. Res. 34: 938–945.

Wong, F., A. Dutta, D. Chowdhury, and J. Gunawardena. 2018. Structural conditions on complex networks for the Michaelis–Menten input-output response. Proc. Natl. Acad. Sci. USA 115: 9738–9743.

Wright, K. M., and M. D. Rausher. 2010. The evolution of control and distribution of adaptive mutations in a metabolic pathway. Genetics 184: 483–502.

Wu, H. H., M. D. Pun, C. E. Wise, B. R. Streit, F. Mus, A. Berim, W. M. Kincannon, A. Islam, S. E. Partovi, D. R. Gang, et al. 2022. The pathway for coenzyme M biosynthesis in bacteria. Proc. Natl. Acad. Sci. USA 119: e2207190119.

Xie, G., N. O. Keyhani, C. A. Bonner, and R. A. Jensen. 2003. Ancient origin of the tryptophan operon and the dynamics of evolutionary change. Microbiol. Mol. Biol. Rev. 67: 303–342.

Xu, Y., B. Labedan, and N. Glansdorff. 2007. Surprising arginine biosynthesis: A reappraisal of the enzymology and evolution of the pathway in microorganisms. Microbiol. Mol. Biol. Rev. 71: 36–47.

Yang, Y. H., F. M. Zhang, and S. Ge. 2009. Evolutionary rate patterns of the gibberellin pathway genes. BMC Evol. Biol. 9: 206.

Yano, T., S. Oue, and H. Kagamiyama. 1998. Directed evolution of an aspartate aminotransferase with new substrate specificities. Proc. Natl. Acad. Sci. USA 95: 5511–5515.

Ycas, M. 1974. On earlier states of the biochemical system. J. Theor. Biol. 44: 145–160.

Zaslaver, A., A. E. Mayo, R. Rosenberg, P. Bashkin, H. Sberro, M. Tsalyuk, M. G. Surette, and U. Alon. 2004. Just-in-time transcription program in metabolic pathways. Nat. Genet. 36: 486–491.

Zhang, Y., M. Morar, and S. E. Ealick. 2008. Structural biology of the purine biosynthetic pathway. Cell. Mol. Life Sci. 65: 3699–3724.

Zhou, H. X. 1993. Brownian dynamics study of the influences of electrostatic interaction and diffusion on protein–protein association kinetics. Biophys J. 64: 1711–1726.

Zubieta, C., K. A. Arkus, R. E. Cahoon, and J. M. Jez. 2008. A single amino acid change is responsible for evolution of acyltransferase specificity in bacterial methionine biosynthesis. J. Biol. Chem. 283: 7561–7567.

Information Processing

CHAPTER 20

Intracellular Errors

As noted repeatedly in prior chapters, few (if any) cellular processes have been pushed to the limits of perfection dictated by the laws of physics. The barrier to natural selection imposed by random genetic drift, combined with the recurrent introduction of deleterious mutations, insures that this will never happen, although such bounds may be closely approached in sufficiently large populations. Some would argue that biophysical trade-offs with other traits prevent selection from attaining univariate optima, but this simply extends the drift barrier to two dimensions.

One consequence of these limitations is that cells make errors, which, if not removed, can lead to progressive damage, resulting in elongated cell-division times and/or shortened lifespans. The challenges are often quite multifaceted. For example, the production of properly constructed proteins requires the avoidance of potential problems arising via dozens of cellular processes (Figure 20.1). A wide array of other cellular processes, including protein folding (Chapter 12), interactions of enzymes with inappropriate substrates (Chapter 19), and faulty assembly of proteins into higher-order structures (Chapter 13), are subject to error.

Cells have evolved multiple mechanisms that seemingly ameliorate the physiological consequences of error proliferation. For example, the incidence of errors arising during the replication of new DNA strands (heritable mutations) is reduced by a proofreading domain in the primary DNA polymerases (Chapter 4). Problems arising in non-replicating DNA are dealt with by a variety of repair mechanisms. Furthermore, mechanisms also exist for the detection and elimination of some types of erroneous transcripts, and some of the stages leading to translation involve proofreading mechanisms.

This chapter focuses entirely on the rates at which errors arise at the levels of transcription and translation, the mechanisms by which these are mitigated, the energetic burden of error surveillance, and the magnitude of selection operating to increase the fidelity of the underlying processes. Transcription and translation errors arise at rates that are orders of magnitude higher than those incurred during replication (Chapter 4), and this is likely in part an evolutionary consequence of the transient nature of such errors. Unlike replication errors, which create cumulative damage in genomic regions linked to mutator alleles, errors arising during transcription and translation will generally become decoupled from their source in no more than a single generation, reducing the strength of purifying selection against mechanisms of error production.

This scenario of diminished selection intensity raises the question as to how cells have evolved an array of mechanisms for error surveillance at the transcript and translational levels, and given this, why error rates are still so high. A key point made here is that, although multiple layers of surveillance lead to the impression of a highly refined system and encourage the common assertion that cells are robust to perturbations, the overall level of performance is likely no greater than that possible with a much simpler system. Such a conclusion is entirely consistent with the idea that natural selection operates on the total performance of a system distributed over multiple traits, each of which is subject to a drift barrier.

Evolutionary Cell Biology. Michael Lynch, Oxford University Press. © Michael Lynch (2024). DOI: 10.1093/oso/9780192847287.003.0020

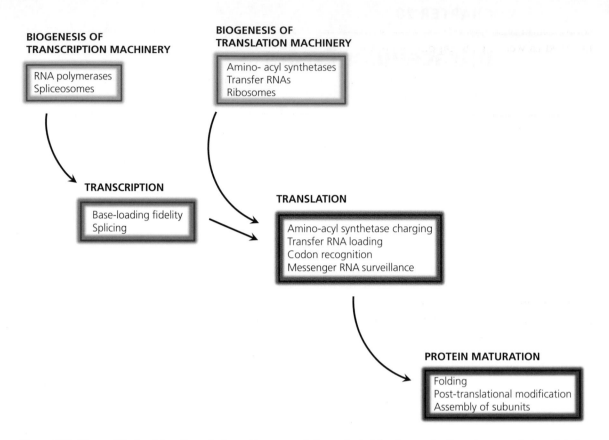

Figure 20.1 The multitude of functions that must be successfully navigated for the production of a properly translated and folded protein.

Transcript Fidelity

The first step in the successful development of a gene product is the generation of an appropriate RNA transcript from the underlying genomic sequence. The RNA polymerases responsible for transcription are typically comprised of several protein subunits. In eukaryotes, one of these complexes (Pol II) is reserved for the production of messenger RNAs and micro RNAs, another (Pol I) for the synthesis of most cytoplasmic ribosomal RNA subunits, a third (Pol III) for transfer RNA production, and still another for the mitochondrial genome (Werner and Grohmann 2011). Land plants deploy two additional RNA polymerases to generate small RNAs used in transcriptional silencing (Wierzbicki et al. 2009; Haag and Pikaard 2011); these seem to be derived from Pol II, but are highly divergent in sites that are otherwise conserved in Pols I–III, suggesting the possibility of reduced fidelity (Luo and Hall 2007; Landick 2009). In contrast, prokaryotes use just a single RNA polymerase (designated here as RNAP) to service all genes.

Despite their shared functions, the complexity of these enzymes is quite variable, with bacterial and archaeal RNAPs consisting of 5 and 12 subunits, respectively, eukaryotic Pols I and III containing 14 and 17 subunits, respectively (Carter and Drouin 2010), and Pols II, IV, and V all comprised of 12 subunits (Haag and Pikaard 2011). Yet, despite the fact that eukaryotic RNA polymerases contain roughly twice the number of components as those from bacteria, there is no evidence that the former carry out their tasks more efficiently.

Like replication, transcription involves phases of initiation, elongation, and termination, but several additional transcript-processing steps are involved as well (e.g., 5′-capping, intron removal, and addition of poly-A to 3′ ends of mRNAs in eukaryotes). Problems may arise at each of these stages: 1) transcription initiation at an inappropriate location, which can be particularly disastrous if it occurs downstream from the translation-initiation site; 2) inaccurate removal of introns (a problem largely confined to eukaryotes, and especially in multicellular lineages, which typically average five or more introns per protein-coding gene); 3) premature transcriptional termination (prior to the translation-termination codon being particularly detrimental); and 4) base-substitution and insertion/deletion errors.

Errors of the first three types are common (Suzuki et al. 2000; Frith et al. 2006), with Struhl (2007) suggesting that 90% of transcription initiation by Pol II in yeast is nothing more than transcriptional noise. It is sometimes difficult to know whether all 'aberrant' transcripts are actually errors as opposed to being products of various kinds of regulatory pathways, e.g., alternative transcription initiation sites and/or RNA editing. However, most variants involve single-site changes sporadically distributed across transcripts, often with downstream consequences with no obvious net benefits. The remaining discussion focuses on these erroneous base substitutions.

Although the mechanisms underlying transcription fidelity are not fully understood, they differ from those involved in replication (Chapter 8) in that RNAPs do not have separate domains set aside for proofreading. The potential for error correction remains, as incorrect base incorporation results in polymerase pausing, providing time for loss of the dangling base (Zenkin et al. 2006; Sydow and Cramer 2009; Sydow et al. 2009; Kaplan 2010; Yusenkova et al. 2010). However, in contrast to the situation with replication, there is no known mechanism for correcting errors after chain elongation is complete (e.g., by recognizing mismatches in the DNA–RNA hybrid molecule). Erroneous transcripts may typically arise by simple copying errors, and these may sometimes be exacerbated by base damage within the DNA template itself, as transcription often proceeds across damaged sites

by substituting an inappropriate base, often an A (Brégeon et al. 2003; Clauson et al. 2010; Fritsch et al. 2020). However, as post-transcriptional errors (e.g., damaged bases) cannot be ruled out, the term transcript error, rather than transcriptional error, will be adhered to below.

As the lifespans of individual transcripts are generally substantially shorter than the lifespans of cells (on the order of just minutes; Chapter 17), and most genes are represented in multiple mRNA copies per cell, transcription must progress at fairly high rates to meet cellular demands. Thus, unlike DNA, which only experiences one copying event per cell cycle, there can be dozens to hundreds of opportunities for error proliferation per transcribed site.

Owing to their singular occurrences, transcript errors can be challenging to identify. One might think that such errors could easily be enumerated by simply comparing the sequences of RNAs to the expectations based on their genomic sources, but because the sequencing error rate is generally substantially greater than the transcript-error rate, this approach will not work. Thus, until recently almost all information on transcript-error rates was obtained by indirect *in vitro* methods, generally by measuring the relative incorporation rates of two competing nucleotides across a specified template.

The average of three *E. coli* estimates obtained with this kind of approach (which themselves exhibit a 40-fold range of variation) is 1.4×10^{-4} per nucleotide site (Springgate and Loeb 1975; Blank et al. 1986; Goldsmith and Tawfik 2009), whereas the thermophilic bacterium *Thermus aquaticus* has an estimated error rate of 6.5×10^{-4} per nucleotide site (Yuzenkova et al. 2010). The sole reporter-construct estimate for budding yeast *S. cerevisiae* is 2.0×10^{-6} (Kireeva et al. 2008; Walmacq et al. 2009), whereas a single estimate for wheat is 2.4×10^{-4} per nucleotide site (de Mercoyrol et al. 1992). These early estimates, potentially subject to considerable experimental biases, suggested that transcript-error rates fall in the broad range of 10^{-6} to 10^{-3} per nucleotide site.

More recently, it has become possible to estimate genome-wide *in vivo* error rates of transcripts by directly sequencing individual mRNA molecules multiple times (Gout et al. 2013; Traverse and Ochman 2016; Fritsch et al. 2020; Li and Lynch 2020;

Li et al. 2022). These rates mostly fall in the range of 10^{-6} and 10^{-4} and reveal no obvious phylogenetic pattern (McCandlish and Plotkin 2013; Li et al. 2022). The highest direct estimates are all from one study (Traverse and Ochman 2016), possibly a consequence of methodological issues (Gout et al. 2017; Li and Lynch 2020; Meer et al. 2020), and if these are ignored, the range in known transcription error rates shrinks to $\sim 10^{-6}$ to 10^{-5} (Figure 20.2a). Thus, the earlier *in vitro* estimates are anomalously high.

To put these rates into perspective, the estimated transcript-error rate in *E. coli* exceeds the known genomic mutation rate per nucleotide site in this species (Lee et al. 2012) by a factor of 30,000. For all other species for which both rates are known, the transcript-error rate is inflated by factors of 10^3 to 10^4 (Figure 20.2). Thus, there is little question that rates of transcript-error production are substantially elevated relative to those arising during DNA replication, as previously suggested by Ninio (1991a,b). The probability of a base-substitution error in a small mRNA of ~ 1000 bp in length is on the order 0.1–1.0%.

Notably, these high error rates exist despite the proofreading capacity of RNAPs (Alic et al. 2007; Sydow and Cramer 2009), and they are not a consequence of an exceptionally high speed of copying. The few estimates of average speeds of transcript progression exhibit only a small range of phylogenetic variation: 46 bp/sec in *E. coli* (Golding and Cox 2004; Proshkin et al. 2010); 20–60 in yeasts (Mason and Struhl 2005; Larson et al. 2012; Eser et al. 2016; Lisica et al. 2016; Ucuncuoglu et al. 2016); 21 in *Drosophila*, 23 in rats, and 56 in humans (Ardehali and Lis 2009). In contrast, replication rates are typically 100–1000 bp/sec in prokaryotes (Hiriyanna and Ramakrishnan 1986; Stillman 1996; Myllykallio et al. 2000), but just 10–50 bp/sec in yeast, flies, and mammals (summarized in Lynch 2007). Thus, transcription is much slower than replication in prokaryotes, whereas both processes proceed at comparable rates in eukaryotes.

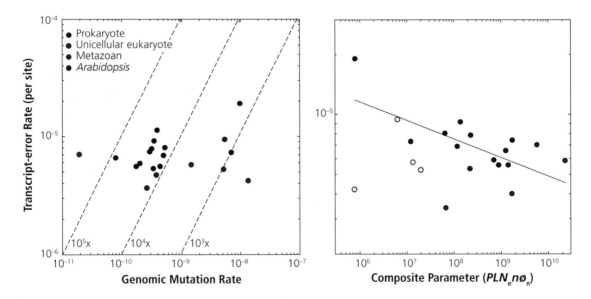

Figure 20.2 The distribution of transcript-error rates across the Tree of Life. **Left:** In comparison to mutation rates at the DNA level (per nucleotide site per generation), transcript-error rates are orders of magnitude higher and relatively constant. Diagonal dashed lines denote where data would lie with three different levels of inflation. **Right:** Negative association between transcript-error rates and a composite measure expected to be proportional to the efficiency of downward-selection (based on the effective population size, the effective number of codons expected to be under selection, and the degree of expression of individual errors, as described in the text). Only the data for unicellular species are used in the regression, and the line shown denotes the best fit, which is obtained when $x = 1.0$, i.e., additivity of effects is assumed. Data are from Li et al. (2022).

Translational Fidelity

Even when an mRNA emerges error-free prior to translation, several additional challenges must be met if a faithful protein product is to be synthesized (Parker 1989; Zaher and Green 2009; Han et al. 2020). First, specific amino-acyl synthetase proteins (AARSs), each assigned to a single amino acid, must initially harvest their cognate amino acids. Second, charged AARSs must then pass their cargo on to the appropriate tRNA. There is considerable room for error in both of these steps because the structural differences between some amino acids are quite minimal, e.g., valine and isoleucine differ by the presence of just a single methyl group. Most AARSs are endowed with proofreading mechanisms to minimize misloading errors (Hussain et al. 2010; Reynolds et al. 2010), although some species of *Mycoplasma* have lost the capacity for proofreading in multiple synthetases (Li et al. 2011; Yadavalli and Ibba 2013). Third, at the ribosome, each codon in an mRNA must be recognized by its cognate tRNA via proper codon:anticodon recognition. Proofreading appears to occur twice after initial tRNA loading, involving processes that require GTP hydrolysis (Ieong et al. 2016).

Infidelities at any one of these steps can lead to a diversity of errors in translated products. For example, misreads of sense codons can lead to alterations in protein structure/function. Misreads of sense codons as termination codons lead to prematurely truncated amino-acid chains, whereas misreading of termination codons as sense codons leads to termination read-through. Despite these immediate functional consequences, translation-error rates are even higher than transcript-error rates.

Most attempts to estimate the rate at which AARSs are mischarged have involved *in vitro* competition experiments between cognate and non-cognate amino acids. These measures are only rough estimates of likely *in vivo* error rates for two reasons. First, such evaluations almost always involve simple binary-choice experiments, leaving questions as to the total error rate expected in a more natural setting in which all 20 amino acids are present simultaneously. Second, most binary tests have focused on the loading of erroneous amino acids with physical features most similar to the cognate substrate of the focal AARS, raising the additional caveat that such pairwise estimates of misloading rates may be upwardly biased. Using this approach, the average pairwise misloading rate for a variety of species ranges from 0.0004 to 0.0055 (Table 20.1). As these estimates have been obtained with different methods, different AARSs, and different pairs of cognate and non-cognate amino acids, no conclusions can be drawn with respect to phylogenetic differences in AARS loading fidelity.

Data on the rate of mischarging of tRNAs (by inappropriate AARSs) are scant, but the few available estimates are of order 10^{-3} per tRNA (Yadavalli and Ibba 2013; Shepherd and Ibba 2014). Thus, given the sum of known AARS and tRNA misloading rates, it is clear that the potential for translation error is far higher than the transcript-error rate. Additional errors in translation will arise at the level of codon reading, although the only detailed estimates at this stage fall in the range of 10^{-7} to 10^{-4}, all involving *E. coli* constructs (Zhang et al. 2016). Thus, 10^{-3} would appear to be the lower limit to the total error rate per codon for the species that have been examined.

Several attempts have been made to estimate the total *in vivo* rate of amino-acid misincorporation into protein (which summarizes the net consequence of errors in all preceding steps, including transcription). As it is not easy to sequence single amino-acid chains, such studies have often been performed with target genes devoid of codons for a particular amino acid, and then searching for the incidence of that amino acid in synthesized proteins. In other cases, genes have been engineered to produce defective products unless a particular codon is misloaded by a specific amino acid, with the degree of rescue providing insight into the specific error rate at that one codon. As both of these methods are limited with respect to the amino-acid misloadings that can be detected, to obtain the total translation-error rate, correction needs to be made to account for the likely errors involving all amino acids not monitored. There are potential biases associated with such correction, but with these caveats in mind, average *in vivo* translation-error estimates (per codon) fall in the range of 0.005 to 0.017 (Table 20.1), substantially greater than the AARS misloading rates. An attempt to estimate

Table 20.1 Estimated error rates associated with translation (rate of amino-acid misincorporation per codon). Standard errors are in parentheses.

Species	Synthetase Loading	Total Translation	Translation Read-Through
Prokaryotes:			
Escherichia coli	0.0011 (0.0004)	0.0052 (0.0027)	0.0050 (0.0017)
Mycobacterium smegmatis		0.0168 (0.0147)	
Salmonella typhimurium			0.0034 (0.0014)
Five prokaryotic species	0.0017 (0.0009)		
Eukaryotes:			
Saccharomyces cerevisiae	0.0055 (0.0018)	0.0162 (0.0124)	0.0111 (0.0047)
Lupinus luteus	0.0009 (0.0005)		
Mus musculus		0.0152 (0.0065)	0.0042
Homo sapiens	0.0004 (0.0001)	0.0147 (0.0146)	

References: *E. coli*: Strigini and Brickman 1973; Edelmann and Gallant 1977; Fersht and Dingwall 1979; Fersht et al. 1980; Tsui and Fersht 1981; Bouadloun et al. 1983; Miller and Albertini 1983; Khazaie et al. 1984; Laughrea et al. 1987; Toth et al. 1988; Mikkola and Kurland 1992; Freist et al. 1998; Weickert and Apostol 1998; Chen et al. 2000; Beuning and Musier-Forsyth 2001; Tang and Tirrell 2002; Korencic et al. 2004; Kramer and Farabaugh 2007; Lue and Kelley 2007; SternJohn et al. 2007; Guo et al. 2009; Reynolds et al. 2010; Zhang et al. 2013; Manickam et al. 2014. *M. smegmatis*: Javid et al. 2014; Leng et al. 2015. *S. typhimurium*: Roth 1970. Five prokaryotes: Fersht et al. 1980; Beuning and Musier-Forsyth 2001; Korencic et al. 2004; Li et al. 2011. *S. cerevisiae*: Igloi et al. 1978; Firoozan et al. 1991; Bonetti et al. 1995; Freist et al. 1998; Stansfield et al. 1998; Williams et al. 2004; Salas-Marco and Bedwell 2005; Plant et al. 2007; SternJohn et al. 2007; Kiktev et al. 2009; Kramer et al. 2010; Reynolds et al. 2010; Torabi and Kruglyak 2011; Loenarz et al. 2014. *L. luteus*: Jakubowski 1980. *M. musculus*: Mori et al. 1979; Cassan and Rousset 2001; Azpurua et al. 2013. *H. sapiens*: Beuning and Musier-Forsyth 2001; Lue and Kelley 2007; Chen et al. 2011; Zhang et al. 2013.

For the estimation of the total translation-error rate by misreading of a single codon to yield one specific amino acid, the observed error rate must be divided by the detectability of all possible misreads. Here, it is assumed that misreads only involve codon misevaluations at a single nucleotide site, so for any specific codon there are nine possible misreads, of which some number $x < 9$ yield an amino-acid substitution (the remaining misreads leading to a stop, or owing to redundancy in the code, to no amino-acid substitution). A number $y \leq x$ of these reads leads to the misread being detected, so the detectability is y/x, and the total translation-error rate is taken to be the observed rate to the monitored amino acid divided by the detectability.

translation-error rates by mass spectrometry of individual protein fragments led to inexplicably low values in the approximate range of 10^{-4} to 10^{-3} for *E. coli* (Mordret et al. 2019).

A rough check on these numbers can be acquired from observations on another type of translational error – misreading of a termination codon as a sense codon (Parker 1989). Typically, such studies monitor the expression of reporter constructs containing premature termination codons that completely abrogate gene function unless experiencing read-through. The read-through error rate is then estimated as the fraction of protein expression relative to that for an intact gene copy. Although there can be substantial variation in the read-through rate depending on the local context of the non-sense codon, most studies average over several such sites, with reported rates of nonsense-codon misreading ranging from 0.003 to 0.011 (Table 20.1). Thus, despite the estimation uncertainties involved, the translation-error rate per codon appears to be $\sim 10,000\times$ greater than the transcript error rate per

nucleotide site. This also means that only a small fraction of errors at the protein level is associated with transcription. Moreover, these high error rates arise even though the rate of translation is not much greater, and often lower, than that of transcription, e.g., 10–50 codons/sec in *E. coli* (with the rate increasing with increasing growth rate; Lovmar and Ehrenberg 2006), and 4–7 codons/sec in mouse (Gerashchenko et al. 2021).

Assuming an average translation-error rate of 0.01 per codon, a newly synthesized protein of moderate size, 300 amino acids, would contain an average of three erroneous amino acids, and assuming a Poisson distribution of errors, only 5% of proteins of this size would be error free. For large complexes, such as the ribosome involving $\sim 10,000$ amino acids summed over all subunits, essentially every composite structure would be expected to contain errors.

The implications here are that, even within a population of genetically uniform cells, each cell will harbor a statistically and transiently unique distribution of variants for most proteins. Recalling Equation 7.2a, which predicts the numbers of protein molecules within cells, and again assuming an average of 300 amino acids per protein, a bacterial-sized cell of $\sim 1 \ \mu m^3$ is expected to contain $\sim 5 \times 10^6$ protein-sequence errors. Average protein lengths are more on the order of 500 amino acids in eukaryotes, so a yeast-sized cell of $\sim 100 \ \mu m^3$ can be expected to harbor $\sim 6 \times 10^8$ errors, and a larger eukaryotic cell of $\sim 10^5 \ \mu m^3$ harbors $\sim 4 \times 10^{10}$ errors.

Biophysics of Substrate Discrimination and the Cost of Proof-Reading

As discussed in prior chapters, there are two potential limits to the evolutionary achievement of molecular perfection. The biophysical barrier represents the ultimate goal that might be attained by a supreme biochemist, capable of developing constructs constrained only by diffusion limitations and energetic features of substrate-binding mechanisms. The evolutionary barrier is the boundary set by the degree to which random genetic drift and deleterious mutation reduces the efficiency of natural selection. The biophysical issues are the subject of this section, with matters related to the drift barrier following.

Errors arise during transcription and translation as a consequence of random diffusion and attachment of alternative substrates to catalytic sites. The frequency of inappropriate substrate usage depends on the relative concentrations and binding affinities involved. As the adhesion strengths of two alternative substrates become arbitrarily close, both substrates become equally likely to bind to an enzyme, provided the substrate concentrations are the same. As the difference in binding energies increases, the relative rate of an enzyme engaging with an inappropriate substrate declines exponentially. With unequal concentrations, the balance of competitive binding is tipped towards the more abundant substrate.

These issues can be addressed more formally as follows. Under a simple competitive-binding situation, the error rate can be evaluated by considering two competing Michaelis–Menten reactions and the resultant ratio of rates of engagement

with wrong (W) and right (R) substrate molecules (Foundations 20.1),

$$E = \frac{[W]}{[R]} \cdot \frac{k_{d,R}}{k_{d,W}}, \qquad (20.1a)$$

where the first term represents the ratio of substrate concentrations (denoted by brackets), and the second term is the ratio of dissociation constants. The latter are functions of the binding energies between substrates and enzyme, and their ratio can be represented in statistical-mechanic terms relative to the background energy of the system,

$$E = \frac{[W]}{[R]} \cdot e^{-\Delta E/(k_B T)}, \qquad (20.1b)$$

where ΔE is the difference of binding strengths involving correct and incorrect substrate molecules, k_B is the Boltzmann constant, and T is the temperature in degrees Kelvin (see Foundations 7.3; and pp. 1011-1016 in Phillips et al. 2013). The actual error rate (the fraction of incorrect reactions, ε) is equivalent to $E/(1+E)$, but this is essentially the same as E provided $E \ll 1$. Assuming that this condition is met and that concentrations of alternative substrates are equal,

$$\epsilon \simeq e^{-\Delta E/(k_B T)}. \qquad (20.2)$$

To gain an appreciation for the biological limits to accuracy under this simple model, note that most enzymes bind their specific substrates with energies of 12–24 $k_B T$ (Kuntz et al. 1999). The strength of a single hydrogen bond, 5–15 $k_B T$ depending on the context (Fersht 1999), puts this in perspective. For example, a G:C pairing in DNA involves three hydrogen bonds, whereas an A:T pairing involves two. It then follows that binding-strength differentials between preferred and non-preferred substrates (ΔE), especially those involving DNA and RNA, are generally on the order of 10 $k_B T$ or smaller, rendering most such biological processes error-prone. For example, taking 2 or 5 $k_B T$ to be binding-energy differences between two substrates yields error rates of $\epsilon \simeq 0.13$ and 0.007, respectively, and extreme differences of 10 and 15 $k_B T$ still yield $\epsilon \simeq 5 \times 10^{-5}$ and 3×10^{-7}, respectively. Even the latter is substantially higher than known DNA replication error rates (Chapter 8), showing that replication fidelity must involve processes beyond simple competitive binding of alternative nucleotides to single-stranded DNA.

Hopfield (1974) and Ninio (1975) realized how an indirect form of proofreading, emerging from a simple deviation from Michaelis–Menten enzyme kinetics, can lead to a dramatic amplification in substrate fidelity (Foundations 20.1). The key point is that if an enzyme can pause long enough for substrate molecules to dissociate before completing a reaction, this opens the opportunity for the repeated interrogation of alternative substrates. Molecules that are less likely to dissociate before conversion to the final product will then be utilized more frequently. However, this path to increased accuracy comes at an energetic cost in the form of ATP or GTP hydrolyses, as well as in time to complete the forwards reaction. For example, an ATP hydrolysis is required each time an amino acid is attached to an AARS, and this must be repeated whenever a substrate molecule is rejected prior to tRNA attachment.

The energetic cost to proofreading was first shown directly without any detailed knowledge of the underlying mechanism. For example, Hopfield et al. (1976) considered an *in vitro* system involving the transfer of either isoleucine or valine to the isoleucine tRNA via isoleucyl–tRNA synthetase. When isoleucine was the sole substrate, 1.6 ATP hydrolyses occurred per charged tRNA (implying that a correct substrate molecule is examined an average 1.6 times prior to permanent attachment). In contrast, 270 ATPs were consumed per charged tRNA when only valine was presented. Assuming equal substrate concentrations, these results suggest an error rate of $\sim 1.6/270 = 0.006$ resulting from the differential rejection of the two residue types.

Additional work revealed that 25–40 ATPs are consumed when properly charged AARSs are forced to deliver an amino acid to a non-cognate tRNA, implicating energy-consuming proofreading at the tRNA stage (Yamane and Hopfield 1977). Likewise, Muzyczka et al. (1972) found that bacteriophage DNA polymerases with mutations in their proofreading domains hydrolysed more or less nucleotides relative to wild-type when they were antimutators versus mutators, respectively. In addition, when properly charged tRNAs encounter inappropriate codons during translation, GTP is hydrolysed, implicating additional proofreading at the codon–anticodon recognition step on the ribosome (Thompson and Stone 1977; Yates 1979). Finally, hyper-accurate ribosomes in *E. coli* require about twice the number of GTPs to produce a peptide bond as in wild-type cells, presumably as a consequence of the increased number of rejection cycles per accepted amino acid (Andersson et al. 1986).

One can view the energetic cost of proofreading at two levels: the baseline cost of multiple interrogations of correct substrate molecules; and the additional cost incurred by engaging with inappropriate substrates. As an example of the baseline cost, consider the prior example in which 1.6 ATP hydrolyses are experienced per correct substrate molecule. This implies an intrinsic cost of proofreading of 0.6 ATPs per residue incorporated, as just 1 ATP would be consumed if the preferred substrate was never rejected.

The total energetic cost of erroneous substrate removal cannot be inferred without a detailed knowledge of the relative concentrations of alternative substrates and their kinetic coefficients. However, insight into this matter can be gained by extending the example from the preceding paragraph. Assuming equal concentrations of isoleucine and valine molecules within the cytosol implies an additional consumption of $(1.6/270) \cdot (270 - 1.6) \simeq 1.6$ ATP molecules per fixed valine. Thus, assuming valine is the only erroneous substrate recognized by this particular AARS, the total cost of proofreading is approximately 3.2 ATP hydrolyses per amino-acid incorporation, i.e., a doubling of the ATP consumption were isoleucine to be the only substrate. The cost of engaging with any other erroneous substrate will still be $\simeq 1.6$ per substrate type, provided the number of cycles is $\gg 1.6$. These observations imply that ATP-dependent template copying mechanisms, which arose very early in cellular evolution, likely imposed an energetic cost equivalent to at least a doubling or tripling in the consumption of ATP molecules for each proofreading step involving DNA/RNA transactions.

To put this in the context of a cell's total energy budget, recall from Chapter 8 that the cost of building a cell scales nearly linearly with cell volume, averaging $\sim 27V$ billion hydrolysed ATPs (where V is cell volume in units of μm^3), whereas from Chapter 7, the number of protein molecules (in millions) per cell is $\simeq 1.6V$. Assuming 400 amino acids per protein (an approximate average over prokaryotes and eukaryotes), two ATPs consumed by proofreading per amino-acid incorporation, and ignoring protein turnover, the fraction of a cell's total growth budget allocated to proofreading is $\sim 5\%$. Using a rather different approach, and less certain numbers, Savageau and Freter (1979) obtained an estimate of 2%. These rough calculations indicate that surveillance at the level of translation consumes a substantial fraction of a cell's energy budget. In principle, there are no limits to the level of accuracy that can be achieved by kinetic-proofreading mechanisms (Foundations 20.1), but each increment in accuracy involves additional energetic costs on the cell, hence reducing cellular rates of reproduction.

The Limits to Selection on Error Rates

Although an increase in phenotypic errors can have negative fitness consequences, natural selection has not driven error rates down to their biophysical limits. For example, the substantial room for improvement in translational fidelity is amply revealed by the fact that hyperaccurate ribosomes are readily obtained in microbial systems (Gorini 1971; Piepersberg et al. 1979; Bouadloun et al. 1983; Andersson et al. 1986; Zaher and Green 2010). Mikkola and Kurland (1992) found that natural isolates of E. coli have a tenfold range of translation-error rates, bracketing the values found for long-established lab populations. Although these authors found no correlation between the growth rates and translation accuracies of different strains, this is perhaps not surprising given the difficulties of measuring both parameters to a high degree of resolution. At high-nutrient levels, wild-type E. coli grow more rapidly than those with hyperaccurate ribosomes, but growth rates are approximately the same under low-nutrient conditions (Andersson et al. 1986).

Why has natural selection not been able to reduce transcript and translation-error rates to the levels observed for replication? One obvious difference is that genomic errors remain associated with mutator alleles until separated by recombination, which reinforces the fitness effects indirectly induced by a mutator, whereas transcription and translation errors are transient. The effects of a DNA-level mutator allele gradually build up over time with the accumulation of linked deleterious mutations. In contrast, the full effects of an allele reducing transcription accuracy are immediate, but not cumulative. The half-lives of individual transcripts are typically 5–10 minutes in bacteria, 10–20 minutes in yeast, and a few hours in mammalian cells, shorter than the time necessary for cell division in all cases (Chapter 17).

It need not follow, however, that the damage from transcript errors is quickly erased. Rather, a roughly steady-state density of erroneous transcripts can be expected within the cell, reflecting a balance between the decay of old mRNAs and the transcriptional production of new ones, and the same will be true of erroneous proteins. Thus, regardless of the transience of individual errors, one can expect a relatively constant number of total errors per cell, although the specific errors will vary temporally. For each expressed protein-coding gene, the total expected number of erroneous amino acids per cell at steady-state will equal the product of the error rate per codon, the number of codons per mRNA, and the total number of mRNAs (for transcript errors) or protein molecules (for translation errors). As noted in Chapter 10, the conditions for equilibrium might sometimes be violated if the machinery responsible for error minimization becomes compromised with cell age.

An additional issue here is that, although the total number of errors in a cell must increase with the number of molecules, the fitness effect of any single error in a particular protein may be diluted out with increasing copy numbers of the protein free of the specific error. The cellular setting for phenotypic mutations is fundamentally different than that for genetic mutations, which are either fully expressed (in haploids or homozygous diploids), or 50% expressed in diploid heterozygotes (assuming additivity). For a gene with a steady-state number n of transcripts or proteins per cell, the phenotypic manifestation of error expression will be a function of the product of the number of transcripts (proteins) and a dilution factor, $\phi(n)$ (Foundations 20.2). In the case of additive effects, the dilution factor is simply $1/n$, and the net effect of errors is independent of the number of molecules per cell, as $n \cdot (1/n) = 1$.

More generally, to obtain the total burden of errors on fitness, we require the total number of erroneous amino acids within proteins, each discounted by the appropriate dilution effect. Let P be the number of expressed protein-coding genes, L be the mean number of codons per protein, and u be the fraction of codons with errors (which is the translation-error rate in the case of proteins, and roughly three-quarters of the transcript-error rate for mRNAs, owing to the redundancy of the genetic code). The expected number of erroneous amino acids incorporated into proteins by a particular route (transcription or translation) is then $uP\overline{Ln}$, where $\overline{Ln}$ is the mean number of nucleotide sites in expressed transcripts per gene. Further, letting δ be the average reduction in fitness if a mutation were to be fully expressed (as in an encoded genomic error in a haploid organism), the total reduction in fitness associated with a particular type of error is $uP\overline{Ln\phi(n)\delta}$, where the overline denotes an average value over all protein-coding genes. This implies that the selection coefficient associated with a modifier of the transcript- or translation-error rate scales with $\overline{PLn\phi(n)\delta}$ or just $\overline{PL\delta}$ if the dilution factor is $1/n$.

As in the case of replication fidelity (Chapter 4), theory suggests that the efficiency of selection on transcript and translational fidelity should scale approximately inversely with the effective size of a population (N_e), as $1/N_e$ dictates the power of genetic drift (Foundations 20.2). If this general theoretical framework is correct, we then expect u to scale inversely with $N_e \overline{PLn\phi(n)}$, provided δ is fairly constant across species, which further reduces to $N_e \overline{PL}$ if $\phi(n) \simeq 1/n$. However, because all of the underlying cellular factors can covary with each other, as well as with N_e (e.g., organisms with small N_e tend to have relatively large P and L; Chapter 24), it is possible that no simple scaling with single parameters will emerge.

Nonetheless, the data are consistent with a weak negative scaling of u with the composite parameter. From Chapter 7, we know how the number of molecules n scales with cell volume, which enters $\overline{n\phi(n)}$ (Foundations 20.2). Estimates of P and $\overline{L}$ are generally available from genome sequencing data,

and N_e from population-level polymorphism data (Chapter 4). For the 15 unicellular species for which data are available for these parameters, estimates of the transcript-error rate scale negatively with the composite parameter $N_e \overline{PLn\phi(n)}$ (Figure 20.2, right). The best fit to the data is obtained when $\phi(n) = 1/n$, i.e., additivity. Although the slope of the overall regression is much shallower than the expectation of -1, this might be a consequence of variation in fitness effects associated with transcript/translational fidelity, e.g., if $\overline{\delta}$ (which has been ignored here) were to decline with increasing $N_e \overline{PLn}$ (Li et al. 2022).

Another suggested factor in the evolution of high transcription- and translation-error rates is the energetic cost of kinetic proofreading (Ehrenberg and Kurland 1984). The idea here is that, whereas increasingly accurate transcription and translation should improve cell health, these advantages might become offset by the costs of proofreading, which as noted above consumes ATP/GTP and magnifies the time to progression to successful polymerization. Under this view, too high or too low of a level of accuracy leads to reduced fitness, motivating the idea that the fidelity of transcription/translation may be kept at some intermediate optimal state by natural selection (Ehrenberg and Kurland 1984; Kurland and Ehrenberg 1984, 1987).

This kind of speed/accuracy trade-off argument has been made in numerous contexts in the biophysical literature (Foundations 19.3, 20.1, and 22.2). However, such optimization would only be possible if the supposed optimum error rate was accessible relative to the drift barrier. Otherwise, the latter would take precedence. Another implicit assumption in this optimization argument is that increases in accuracy can only be achieved via proofreading improvement rather than through modifications of the basic efficiency of the pre-proofreading steps in transcription and translation, i.e., in the kinetic coefficients. Finally, the optimization hypothesis does not explain the phylogenetic range (or lack thereof) in error rates, given the substantial differences in cell-division times among study species, which might alter the premium put on the speed of transcription.

One alternative explanation for elevated transcript-error rates (relative to DNA-level mutation) is that the translation error rate is so high that selection at the level of transcription accuracy is relaxed. The kind of scenario is the subject of the following section.

The Evolutionary Consequences of Surveillance-Mechanism Layering

The accuracy-demanding processes of replication, transcription, and translation all involve layers of mechanisms that improve fidelity. For example, genome replication involves highly selective DNA polymerases, with the small fraction of initial base misincorporations being subject to correction by subsequent proofreading, and the still smaller fraction of errors that escape proofreading being subject to mismatch repair.

Layering of defense mechanisms also occurs with transcript surveillance. In a number of organisms, certain subsets of erroneous mRNAs can be removed by mRNA-surveillance mechanisms that occur after the initiation of translation. These include: the nonsense-mediated decay (NMD) pathway, which eliminates a fraction of inappropriate mRNAs carrying premature termination codons (PTCs); no-go decay, which degrades mRNAs associated with stalled ribosomes; and non-stop decay, which removes mRNAs lacking a stop codon (Graille and Séraphin 2012; Kervestin and Jacobson 2012).

The central point is that some fraction of erroneous mRNAs is removed from the cell in the very first round of translation. This may help explain why *in vivo* transcript-error rate estimates are lower than those obtained by *in vitro* methods, which exclude translation-associated processes. Notably, many of these surveillance mechanisms are either restricted to or substantially elaborated on in the eukaryotic lineage.

The eukaryotic NMD process is mechanistically associated with spliceosomal introns, which are unique to eukaryotes (often exceeding an average of five per protein-coding gene) and must be spliced out of precursor mRNAs to produce productive transcripts prior to translation. Failure to remove an intron leads to a downstream frameshift two-thirds of the time, typically resulting in the appearance

of PTCs. If not removed from the cytoplasm, such aberrant transcripts will yield truncated proteins, which will generally be harmful to cell health. The NMD process removes many such transcripts during their first round of translation, distinguishing erroneous from true termination codons by use of information on the distance from the 3′ end of the transcript, including information laid down in the form of proteins marking intron-exon junctions (Hentze and Kulozik 1999; Gonzalez et al. 2001; Lykke-Andersen 2001; Mango 2001; Maquat and Carmichael 2001; Wilusz et al. 2001; Maquat 2004, 2006; Zhang et al. 2010).

Although not all PTC-containing mRNAs are detectable by the NMD process, the importance of NMD is illustrated by experiments in which the pathway has been silenced, which show substantial increases in PTC-containing mRNAs (Mendell et al. 2004; Mitrovich and Anderson 2000; Gout et al. 2017). Knockouts of the NMD pathway have small phenotypic effects in the yeasts *S. cerevisiae* and *S. pombe* (Leeds et al. 1992; Dahlseid et al. 1998; Mendell et al. 2000), moderate fitness effects in the nematode *C. elegans* (Hodgkin et al. 1989), and lethal effects in mice (Medghalchi et al. 2001). The enhanced sensitivity of multicellular species to NMD inactivation may simply be a consequence of greater rates of production of erroneous transcripts in complex genomes with more opportunities for splicing errors, although this conclusion is clouded by the fact that some of the proteins in the NMD pathway have additional cellular functions (Maquat 2006; Isken and Maquat 2008).

Mechanisms for surveillance and removal of faulty products also exist for noncoding RNAs, such as rRNAs and tRNAs. For example, eukaryotes (but apparently not bacteria) have a decay pathway capable of detecting and removing a subset of rRNAs with erroneous sequences (LaRiviere et al. 2006; Andersen et al. 2008; Sarkar et al. 2017; Lirussi et al. 2021). Multiple mechanisms for the removal of aberrant tRNAs also exist in prokaryotes and eukaryotes (Wilusz et al. 2011; Kramer and Hopper 2013). In addition, the accuracy of translation depends on a series of surveillance mechanisms for proper loading of tRNA synthetases by their cognate amino acids, proper recognition of tRNAs by their cognate synthetases, and proper codon recognition by tRNAs, with all three steps incorporating proofreading mechanisms.

It is common to view such layered security systems as reflections of adaptive evolutionary processes, the idea being that, once selection has brought a particular mechanism to perfection, a second layer can emerge, yielding a quantum leap in accuracy, with still other layers being subsequently added. Under this view, copying fidelity can be continuously pushed to higher levels with the addition of more and more layers of error correction. However, this view of ever-improving fidelity is likely incorrect. It fails to explain, for example, why eukaryotes have more elaborate surveillance mechanisms for errors in DNA, RNAs, and proteins than do bacteria, and yet still exhibit equal or higher net rates of error production.

Suppose that prior to the addition of a secondary line of defense, the primary mechanism is not constrained by biophysical limits, but rather by the

drift barrier. In that case, the fortuitous addition of a second layer of defense (with a large enough immediate effect) might lead to a larger boost in accuracy than possible under incremental changes made to the primary mechanism, thereby vaulting over the prior limits to natural selection. However, the initial boost in accuracy need not be permanent, as incremental reductions in the efficiency of both layers result in decay back to the drift barrier, rendering the overall system no more accurate than the previous single-layered system (Figure 20.3). The end result is a more complex and bioenergetically expensive system, which superficially looks robust, but is in fact no more accurate than the simpler ancestral state (Gros and Tenaillon 2009; Lynch 2012). Frank (2007) called this phenomenon the 'paradox of robustness'.

In effect, the combination of multiple lines of defense results in the relaxation of selection on the efficiency of individual layers, and hence,

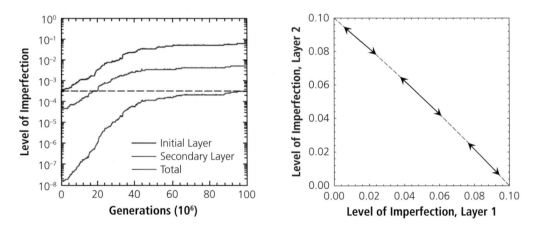

Figure 20.3 Left: Evolutionary addition of a secondary line of defense. Initially, the genome encodes for a single mechanism for error prevention (solid black line) leading to a level of perfection (at the starting point) denoted by the dashed line (this is the drift barrier, the exact position of which is defined by the power of random genetic drift, and the magnitude of bias in the spectrum of mutations improving or degrading trait efficiency). A secondary adaptation (blue line) arises that then removes all but a fraction of the errors remaining after level one. The total error rate (red line) is given by the product of the error rates at each individual level, which act sequentially in the cell. Although there is an initial quantum leap in accuracy, slight imperfections (too small to be removed by selection) gradually accumulate at each level, returning the overall system to the same level as experienced prior to establishment of the second layer. From Lynch (2012). **Right:** A bivariate drift barrier. Natural selection is capable of maintaining the two-layered system at a fitness level of $W = (1 - w_1)(1 - w_2) \simeq 1 - w_1 - w_2$, where the approximation assumes $w_1, w_2 \ll 1$, and w_1 and w_2 are the levels of inaccuracy at each layer. Here, w_1 and w_2 are able to jointly move along the diagonal line of constant fitness W, with an improvement in one trait being balanced by a reduction in the other. Ultimately, the system may collapse to a single-trait state, as the less efficient trait approaches a state of nonfunctionality so closely that it is highly vulnerable to loss by further degenerative mutations and random genetic drift.

the eventual degeneration of earlier established mechanisms. This is because natural selection operates on the output of an entire system, leaving multiple degrees of freedom for change in individual components so long as the summed level of efficiency remains at the drift barrier. With two layers, the bivariate drift barrier will have a line of equivalence, with pairs of phenotypes on the line having identical overall accuracies (Figure 20.3), and with three layers, there will be a trivariate drift barrier (i.e., a plane of equivalence). Such systems will be vulnerable to eventual loss of one component, provided such loss can be compensated by improved performance in the other(s).

Observations on the mutational properties of microbes support this view. For example, the elimination of mismatch repair from *E. coli, S. cerevisiae*, and other organisms generally results in an ~100× increase in the mutation rate. However, *Mycobacterium smegmatis*, a bacterium lacking the mismatch-repair pathway, and *Deinococcus radiodurans*, a bacterium in which mismatch repair only improves accuracy by a few-fold, both have mutation rates similar to those in other microbes (Long et al. 2015; Kucukyildirim et al. 2016). In addition, mycobacteria and other species lack the DNA-polymerase proofreading domain deployed by *E. coli* and many other bacteria, but harbor an entirely different mechanism for such purposes (Rock et al. 2015). The fact that bacterial populations founded with a mutator genotype frequently evolve lower mutation rates on relatively short timescales through compensatory molecular changes at genomic sites not involved in the initial mutator construct provides further evidence that individual fidelity mechanisms are not limited by biophysical constraints (McDonald et al. 2012; Turrientes et al. 2013; Wielgoss et al. 2013; Williams et al. 2013; Wei et al. 2022). Finally, different yeast species with very similar mutation rates have substantially different mutation spectra, implying variation in the underlying mechanisms of mutation and repair that nonetheless have the same net effects on the mutation rate (Long et al. 2016; Nguyen et al. 2020).

From the standpoint of evolutionary layering and translational fidelity, one can also point to exemplary cases involving AARSs, the enzymes responsible for sequestering cognate amino acids prior to passing them on to their appropriate tRNAs. Many AARSs are capable of post-transfer editing of mischarged tRNAs. However, the phenylalanyl-tRNA synthetase (PheRS) in *Mycoplasma mobile* has lost the capacity to edit, and instead simply relies on pre-transfer kinetic proofreading for discriminating against non-cognate amino acids (Yadavalli and Ibba 2013). Although this AARS is not sufficient to support *E. coli* growth, presumably owing to problems associated with mistranslation, the introduction of just two amino-acid changes into the evolutionarily deactivated editing domain removes this deficiency, increasing the level of accuracy by several fold. In budding yeast (*S. cerevisiae*), cytosolic PheRS is capable of editing a tRNA mischarged with an erroneous amino acid, whereas the mitochondrial PheRS is incapable of editing but nonetheless has a comparable error rate (Reynolds et al. 2010). Thus, the accuracy of the latter is solely dependent on a high level of initial specificity, which has apparently offset the loss of post-transfer editing in the cytosolic version. Many other examples are known in which tRNA-charging accuracy depends on a mixture of pre- and post-transfer fidelity, with a switch to strong reliance on just a single mechanism being conferred by no more than two amino-acid substitutions in some lineages (Martinis and Boniecki 2010).

Adaptive Significance of Errors

Given that the vast majority of amino-acid altering mutations have negative fitness effects (Chapter 12), it is reasonable to expect the same to be true of translation errors. Indeed, high translation-error rates are known to lead to malfunctioning cells (Lee et al. 2006; Nangle et al. 2006; Bacher and Schimmel 2007; Schimmel and Guo 2009), and as noted above, the removal of the surveillance capacity for aberrant mRNAs also causes fitness loss. By evaluating the growth rates of *Salmonella* cells containing various translation-fidelity mutations, Hughes (1991) and Hughes and Tubulekas (1993) found that a tenfold increase in the error rate generates a twofold reduction in growth rate, accompanied by substantial reductions in enzyme activity and protein stability, with no associated change in protein abundance. Similarly, in comparisons of wild-type

E. coli and a mutant line with enhanced translational fidelity, Musa et al. (2016) found an ∼ 30% decrease in enzyme activity in the former.

Despite these observations, following the grand tradition of assuming that everything biological must be optimized to maximize organismal fitness, several authors have argued that translational errors have been wrongly labelled as deleterious (Peltz et al. 1999; Santos et al. 1999; Pezo et al. 2004; Bacher et al. 2007; Pan 2013; Ribas de Pouplana 2014; Fan et al. 2015; Wang and Pan 2016). This adaptationist argument asserts that mistranslation is a regulated phenomenon, with organisms 'deliberately' making errors in order to expand the chemical diversity of the cell. The view here is that an intermediate level of mistranslation, fine-tuned by natural selection, yields populations of variant molecules, some of which will have large enough fitness-enhancing functions to offset the deleterious effects of others. This line of thinking derives from a rather liberal interpretation of multiple lines of observation.

First, if one engineers an *E. coli* cell line to be auxotrophic (unable to synthesize a particular nutrient) by introducing a missense mutation in a gene required for the synthesis of the nutrient, an editing defective tRNA synthetase can rescue the line, presumably by promoting the production of a small fraction of proteins with reversions to phenotypic function (Min et al. 2003). However, such an extreme starting point provides little (if any) evidence for the adaptive significance of error production, as auxotrophic mutants are expected to be rapidly purged from populations by natural selection except in cases where the nutrient is freely available in the environment, in which case there would be no selection for phenotypic reversion.

Second, cells under stress often have increased translation-error rates. For example, in bacteria exposed to antibiotics and/or oxidative damage, error rates can increase by 10 to 100× (Bacher and Schimmel 2007; Kramer and Farabaugh 2007; Javid et al. 2014; Fan et al. 2015; Leng et al. 2015; Vargas-Rodriguez et al. 2021). The mischarging of non-methionine tRNAs with methionine is particularly common in stressful situations, with up to 10% of incorporated methionine being erroneous in mammalian, yeast, and bacterial cells under

some conditions (Jones et al. 2011; Wiltrout et al. 2012; Schwartz and Pan 2017). In the hyperthermophilic archeon *Aeropyrum pernix*, growth at low temperatures is accompanied by global mischarging of leucine tRNAs by methionine AARS, and in *E. coli*, excess carbon leads to stop-codon readthrough (Zhang et al. 2020). It has been argued that conservation of a particular type of 'misfunction' must imply maintenance by natural selection as a mechanism for the adaptive exploitation of errors (Netzer et al. 2009; Pan 2013; Schwartz et al. 2016). However, cases are known in which stress leads to increased translational accuracy (Steiner et al. 2019), suggesting the possibility of reporting bias.

In principle, widespread methionine misincorporation might yield advantages beyond the direct effects on protein functions. There is, for example, biochemical evidence that methionine residues can serve as scavengers for reactive oxygen species via conversion to methionine sulfoxide (Levine et al. 1999; Stadtman and Levine 2003; Wang and Pan 2016), which might in turn protect the proteins containing them. Moreover, a common enzyme (methionine sulfoxide reductase) confers the ability to revert the modified amino acid back to methionine, implying that such residues can be recycled as antioxidants.

Nonetheless, these observations leave many questions unanswered. Although methionine misincorporation can alter the properties of individual proteins in potentially beneficial ways (Wang and Pan 2015; Schwartz and Pan 2016), this is not a demonstration of an overall induced selective advantage, as the majority of such variants at other loci are likely deleterious. Given that there are no known disadvantages of methionine in non-stressful conditions, if methionine serves such a useful function, why then is the cytoplasm not populated with higher free amounts of this amino acid, and why are more methionines not directly encoded into the proteome?

Third, when cells are extremely starved for one particular amino acid, mischarging of the cognate tRNA synthetase can increase misincorporation rates by up to an order of magnitude (Feeney et al. 2013). This is expected to arise naturally from the reduction in competitive binding between substrate types (Foundations 20.1). In some cases, such misloadings can provide rescue from an otherwise lethal situation. In *Acinetobacter baylyi*, for example, a mutation that allows the isoleucine AARS to mischarge with valine can increase the growth rate when isoleucine is strongly limiting (Bacher et al. 2007). Should this be surprising given the alternative outcome of no translation and death? In nature, a more common situation would be generic shortage of all amino acids, which as discussed in Foundations 20.1 is expected to reduce the rate of translation but not magnify the error rate.

Fourth, some organisms, such as members of the genus *Mycoplasma*, have one or more AARSs with editing defective domains (Li et al. 2011). The microsporidian *Vavraia culicis* harbors a defective leucine AARS, which misuses a variety of alternative amino acids up to 6% of the time (Melnikov et al. 2018). The fungal pathogen *Candida albicans* has experienced a reassignment of one particular leucine codon to serine (an alteration of the genetic code), but still misincorporates 1–6% leucines at such codons (Rocha et al. 2011; Bezerra et al. 2013). This codon ambiguity reduces fitness in normal environments, but by inducing the expression of stress-response proteins can create a competitive advantage in stressful environments (Santos et al. 1999). Again, however, although rare accidents can occasionally be useful, that does not mean that the proclivity to incur errors is promoted by natural selection.

In summary, although there is clear evidence that translation inaccuracy can increase during times of stress, on occasion even stochastically creating protein variants capable of improving a precarious situation, there is no direct evidence that error-prone translation has been promoted and/or maintained by selection as a strategy for adaptively generating variant protein pools. When cells are stressed, cellular functions go wrong, and there is no obvious reason why translation should be immune to pathological behavior. Notably, the examples promoted as poster children of adaptive translation inaccuracy are highly idiosyncratic in that they involve different amino acids – leucine and serine in the case of *Candida*, phenylalanine and leucine in the case of *Mycoplasma*, asparagine and aspartate in the case of *Mycobacteria*, and methionine in the case of *E. coli*, yeast, and mammals. There is

no obvious reason why such lineage-specific variation would be driven by phylogenetically specific adaptive requirements. Notably, many organisms with intrinsically defective translation efficiency are pathogens, which may be highly vulnerable to random genetic drift and loss of non-essential functions.

Summary

- The production of proteins and functional RNAs with correct sequences relies on the avoidance of error proliferation at myriad steps, ranging from transcription and translational fidelity to mRNA splicing to protein folding and assembly.

- Transcript-error rates are universally orders of magnitude greater than rates of mutation at the DNA level, and translation-error rates are still orders of magnitude higher. Thus, for any given genetic locus, cells typically contain populations of variant molecules.

- Errors at the level of transcription and translation are mitigated by kinetic proofreading, which relies on the competitive binding and release of correct and incorrect substrate molecules. Such processes consume considerable energy and/or extend the time to complete polymerization processes, thereby leading to intrinsic trade-offs between speed, energetic expense, and accuracy.

- Theory and observation suggest that the ability of selection to reduce transcript-error rates is a composite function of the effective population size, proteome size, and number of transcripts per cell, but the negative scaling is weaker than that observed for mutation rates at the DNA level for reasons that remain unclear.

- Although evolution has resulted in multiple layers of error-correcting mechanisms, particularly in eukaryotes, this does not appear to result in a long-term increase in accuracy. Rather, stepwise increases in fidelity are eventually dissipated evolutionarily, eventually resulting in a more complex system with no overall level of improvement. This leads to the false impression that multiple lines of defense imply an evolutionary enhancement of organismal robustness to error production.

- Arguments have been made that high rates of transcript and translation errors are promoted for their positive effects, but such conclusions are extreme examples of unconstrained pan-adaptationist thinking. Benefits can sometimes be found in particular settings, but significant deleterious effects are more pervasive. As the particular kinds of errors produced are idiosyncratic across species, and most species in which they have been identified are pathogens with small effective population sizes, an alternative view is that high rates of transient errors are simply a consequence of the compromised power of natural selection.

Foundations 20.1 Kinetic proof-reading

Numerous cell biological processes, including steps involved in DNA replication, transcription, and mRNA translation, involve various forms of proofreading. As first suggested by Hopfield (1974) and Ninio (1975), such processes exploit weak binding energies and the even smaller differences between correct and incorrect substrates to repeatedly interrogate bound substrates until passing them on to the next biochemical stage. In principle, there are no limits to the level of accuracy that can be achieved by such mechanisms, but any increase in accuracy comes at the cost of increased reaction times and energy consumption.

To gain an appreciation for the mechanisms by which proofreading can lead to an increase in accuracy, we first consider the null situation in which two alternative substrates are engaged in the same Michaelis–Menten reaction (Chapter 19), albeit at different rates (Figure 20.4). The right and wrong substrates will be designated R and W or as X when non-specified, with both being processed by the same enzyme E. Recall that under the Michaelis–Menten model, before the final product is arrived at, an enzyme–substrate complex EX is formed (Foundations 19.1), which then either returns to the prior state (E and X) by dissociation or proceeds to product formation, at rates $k_{d,X}$ and $k_{cat,X}$, respectively.

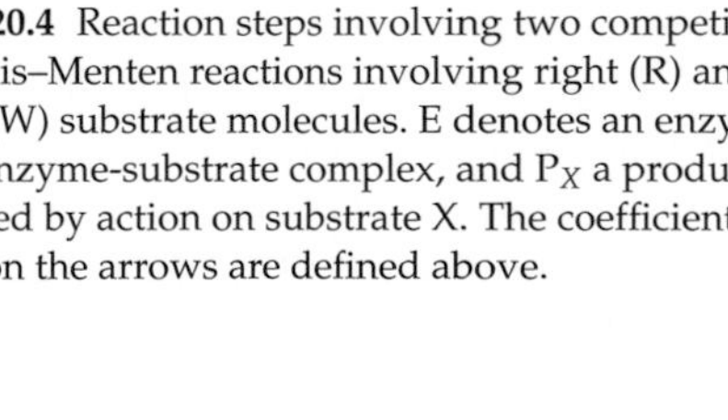

$$E + R \underset{k_{d,R}}{\overset{k_{a,R}}{\rightleftharpoons}} ER \xrightarrow{k_{cat,R}} E + P_R$$

$$E + W \underset{k_{d,W}}{\overset{k_{a,W}}{\rightleftharpoons}} EW \xrightarrow{k_{cat,W}} E + P_W$$

Figure 20.4 Reaction steps involving two competing Michaelis–Menten reactions involving right (R) and wrong (W) substrate molecules. E denotes an enzyme. EX an enzyme-substrate complex, and P_X a product generated by action on substrate X. The coefficients (rates) on the arrows are defined above.

The rate of production of the correct intermediate per unit enzyme is $k_{a,R}[R]$, where the brackets denote concentration, and once formed the intermediate is converted to final product with probability $\lambda_R = k_{cat,R}/(k_{d,R} + k_{cat,R})$, with similar expressions following for the incorrect substrate. Provided the error rate is $\ll 1$, it will be very closely approximated by the ratio of the forward rates for the incorrect and correct products (see main text),

$$\epsilon = \frac{k_{a,W} \cdot \lambda_W \cdot [W]}{k_{a,R} \cdot \lambda_R \cdot [R]}, \qquad (20.1.1a)$$

which can be seen to depend on ratios involving both substrate concentrations and kinetic coefficients. Assuming equal concentrations of the two substrates, and further supposing the association rates of the two substrates to be the same, which is reasonable if encounters are based on diffusion of two similar sized particles,

$$\epsilon = \frac{\lambda_W}{\lambda_R} = \frac{k_{cat,W} \cdot (k_{d,R} + k_{cat,R})}{k_{cat,R} \cdot (k_{d,W} + k_{cat,W})}. \qquad (20.1.1b)$$

If the catalytic rates greatly exceed the dissociation rates, the error rate may be high enough to violate the assumption of the preceding derivation, but will approach

$$\epsilon \simeq \frac{k_{cat,W}}{k_{cat,R} + k_{cat,W}}, \qquad (20.1.1c)$$

which is equal to one half when the two catalytic rates are equivalent. On the other hand, if both dissociation rates are large relative to the catalytic rates, and the latter are equivalent for both substrate molecules, Equation 20.1.1b reduces to the ratio of dissociation rates,

$$\epsilon \simeq \frac{k_{d,R}}{k_{d,W}}. \qquad (20.1.1d)$$

This is the limit to what can be achieved by an enzyme that discriminates solely on the basis of the sticking times (the inverse of the dissociation rates) of the substrates.

Now consider the situation in which a proofreading step is inserted into the previous scheme, designating this as the creation of a secondary complex EX* from EX at rate m_X. Under such conditions, the secondary complex is rejected and returned to state EX with decay rate $k_{d*,X}$ (Figure 20.5). This has the effect of creating a recurrent loop in the progression of a substrate molecule to the final-product step, with the number of excursions back to EX depending on the magnitude of the dissociation constant. The combination of repetitive interrogation and enhanced rejection of loosely bound complexes leads to an elevated level of fidelity to the appropriate substrate. A full exposition of the model can be found in Hopfield (1974) and Ninio (1975), but to focus on the central point, the simplest case is examined here: the situation in which all rate coefficients, except for the dissociation constants, are equal for both substrates.

The rate of production of product P_X can be determined by first partitioning the series of intermediate events into net rates/probabilities associated with four subcategories,

$$\lambda_{1,X} = k_{on}[X] \qquad \text{for } E + X \rightarrow EX \qquad (20.1.2a)$$

$$\lambda_{2,X} = \frac{m}{m + k_{d,X}} \qquad \text{for } EX \rightarrow EX^* \qquad (20.1.2b)$$

$$\lambda_{3,X} = \frac{k_{d*,X}}{k_{cat} + k_{d*,X}} \qquad \text{for } EX^* \rightarrow EX \qquad (20.1.2c)$$

$$\lambda_{4,X} = \frac{k_{cat}}{k_{cat} + k_{d*,X}} \qquad \text{for } EX^* \rightarrow E + P_X. \quad (20.1.2d)$$

The overall forward rate of production of P_X can be summarized as the series

$$\Lambda_{1,X} = \lambda_{1,X}\lambda_{2,X}\lambda_{4,X}\left[1 + (\lambda_{2,X}\lambda_{3,X}) + (\lambda_{2,X}\lambda_{3,X})^2 + \cdots\right]$$

$$(20.1.3a)$$

$$= \frac{\lambda_{1,X}\lambda_{2,X}\lambda_{4,X}}{1 - \lambda_{2,X}\lambda_{3,X}}. \qquad (20.1.3b)$$

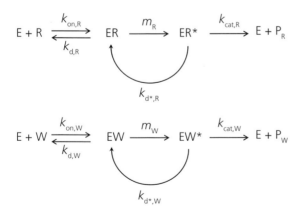

Figure 20.5 Modified reaction dynamics (compared to Figure 20.4) for the situation in which there is proof-reading of an intermediate complex (denoted by asterisks), again for two competing substrates, R and W.

Note that from the standpoint of the intrinsic error rate, $\lambda_{1,X}$ is irrelevant as k_{on} is the same for both substrates, so the ratio $\lambda_{1,W}/\lambda_{1,R} = [W]/[R]$. Again, factoring out the concentration effect by assuming $[W] = [R]$, the error rate is

$$\epsilon = \frac{\lambda_{2,W}\lambda_{4,W}(1 - \lambda_{2,R}\lambda_{3,R})}{\lambda_{2,R}\lambda_{4,R}(1 - \lambda_{2,W}\lambda_{3,W})}, \qquad (20.1.4)$$

and substituting from above yields

$$\epsilon = \frac{mk_{cat} + k_{d,R}k_{cat} + k_{d,R}k_{d*,R}}{mk_{cat} + k_{d,W}k_{cat} + k_{d,W}k_{d*,W}}. \qquad (20.1.5a)$$

The assumption that m and k_{cat} are the same for both substrate molecules can be relaxed to allow for inequalities in Equations 20.1.3b and 20.1.4, but this is not pursued here.

Comparison to Equation 20.1.1c shows that the key difference between the error-rate function with an intermediate error-checking step is the presence of products of terms, most notably of the dissociation constants associated with each substrate. If the dissociation coefficients are large relative to m and k_{cat}, and the ratios of dissociation coefficients at both steps are the same, i.e., $k_{d,X} \propto k_{d*,X}$,

$$\epsilon \simeq \left(\frac{k_{d,R}}{k_{d,W}}\right)^2. \qquad (20.1.5b)$$

Thus, in this limiting case, a proofreading step can reduce the error rate down to the square of that expected in the absence of proofreading, Equation 20.1.1d. These specific results, which make various assumptions about the equality of some pairs of rates for R and W substrates and relative magnitudes of different classes of coefficients, lead to a maximum level of error reduction. More general formulae can be found in Hopfield (1974), Ninio (1975), and Murugan et al. (2012).

By increasing the number of steps from substrate to final product, all other things being equal, proof-reading also increases the reaction time (along with the energetic cost), implying an intrinsic trade-off between accuracy and speed. Expanding from the logic underlying Equation 20.1.3a, the average number of intermediate steps that are cycled through before reaching the final product is

$$n_c = \sum_{i=1}^{\infty} i \cdot (\lambda_{2,X}\lambda_{3,X})^{i-1}$$

$$= \frac{1}{[1 - \lambda_{2,X}\lambda_{3,X}]^2} \qquad (20.1.6)$$

Note that $\lambda_{2,X}\lambda_{3,X}$ is a measure of the probability of transition from $EX \rightarrow EX^* \rightarrow EX$, i.e., of recycling. The time to complete a reaction is proportional to n_c.

This being said, it can also be shown that an increase in proofreading rates (which in part influence the overall reaction time) can lead to an increase in both reaction speed and accuracy (by promoting correct substrates more rapidly to the final product before dissociation) (Banerjee et al. 2017), so it is not inevitable that there is a trade-off between speed and accuracy, contrary to common opinion (Johansson et al. 2008; Wohlgemuth et al. 2011). Rather, the directionality of this relationship depends on the full set of parameters in Figure 20.5, some of which are jointly favorable for both traits and others of which are not. A lucid examination of these models, including an extension to an arbitrary number of correction steps is provided by Qian (2006).

Foundations 20.2 The evolutionary bounds on the transcription-error rate

Given that natural selection relentlessly promotes the sequences of protein-coding genes towards their optimal functional state, most errors in transcripts are expected to be deleterious (Chapter 12). Here, we consider the expected selective advantage of a genomic variant that improves transcription fidelity (or conversely the disadvantage of a variant that exacerbates the transcript-error rate). With slight modifications, the same approach can be used to ascertain the magnitude of selection operating on a variant that influences the translation-error rate. To achieve such an understanding, several factors must be considered: 1) the expected number of errors per molecule produced (transcript or protein) at the amino-acid sequence level (where fitness effects are manifest); 2) the total number of molecules per cell associated with each gene; and 3) the fitness effects of such errors.

Letting u be the rate of missense/nonsense transcript error per codon, and L_i be the number of amino acids in a protein of type i, the number (j) of errors in individual protein molecules of this type will be Poisson distributed with expectation uL_i,

$$P(j|uL_i) = \frac{e^{-uL_i}(uL_i)^j}{j!}. \tag{20.2.1}$$

From the standpoint of translation errors, u is defined to be the error rate per codon, whereas if μ were to be the transcript-error rate per nucleotide site, as it is here, because there are three nucleotide sites per codon, and $\sim 3/4$s of nucleotide substitutions cause an amino-acid substitution, $u \simeq 3\mu \cdot (3/4) = 9\mu/4$, i.e., the error rate per codon $\simeq 2.25 \times$ the error rate per nucleotide site.

Because transcription and translation errors are singular events, individual variant proteins within a cell will generally be just a fraction of the total pool of molecules for specific genetic loci. This raises the question of the degree to which the fitness effects of single molecules are manifest at the cellular level. As with variant alleles at a genetic locus, transcription and translation errors might behave in an additive, recessive, or dominant fashion, with the magnitude of the latter two conditions depending on the number of molecules per cell. Letting n_i be the number of molecules per cell for protein i, a flexible function that allows for alternative modes of dilution effects is

$$f(n_i) = \frac{1}{n_i^x}, \tag{20.2.2}$$

which equals 1.0 when $n_i = 1$ (effects are fully penetrant), and converges to 0.0 (effects are completely masked) as $n_i \to \infty$ at a rate that depends on the exponent x. When $x = 1$, $f(n_i) = 1/n_i$, and the number of copies of a protein has no effect, as the number of error-containing proteins, which is proportional to n_i, is compensated by the dilution effect, i.e., $n_i \cdot f(n_i) = 1$. Values of $0 < x < 1$ result in a relatively slow decline in the dilution effect with increasing n_i (with $x = 0$ implying complete dominance of errors). Positive synergistic effects are implied by $x < 0$, whereas $x > 1$ results in a relatively rapid decline in $f(n_i)$ (i.e., relatively recessive effects of errors).

From Chapter 12, the average reduction in fitness associated with single, fully expressed deleterious mutations in proteins is generally < 0.1, and based on known transcript- and translation-error rates and typical gene lengths, the number of errors per protein will generally be $\ll 10$. Thus, letting δ_i be the average fitness loss per single error in a single molecule if fully revealed, unless there are very strong epistatic effects, the total reduction in fitness associated with a protein containing j errors $\simeq j\delta_i$. Provided $j\delta_i \ll 1$, it is useful to use the further approximation, $j\delta_i \simeq 1 - e^{-j\delta_i}$.

For each locus i, the expected fractional reduction in fitness associated with the error burden (s_i) will then depend on the number of proteins per cell over which the errors are distributed, n_i, the degree of expression of individual errors, $f(n_i)$, and the distribution of the numbers of errors per protein $P(j|uL_i)$,

$$s_i = n_i \cdot f(n_i) \cdot \sum_{j=1}^{L_i} P(j|uL_i) \cdot (1 - e^{-j\delta_i}), \tag{20.2.3a}$$

$$\simeq n_i \cdot f(n_i) \cdot \left[1 - \exp\left(-\frac{uL_i\delta_i}{1+\delta_i}\right)\right]. \tag{20.2.3b}$$

Further simplification is obtained by noting that because δ_i and $uL_i\delta_i \ll 1$,

$$s_i \simeq uL_i \cdot n_i \cdot [f(n_i) \cdot \delta_i]. \qquad (20.2.3c)$$

Equation 20.2.3c shows that the fitness consequences of mistranslation are a function of three quantities: 1) the total mistranslation rate per expressed protein (uL_i); 2) the steady-state number of proteins per gene (n_i); and 3) the average effect of an amino-acid substitution (the final term in brackets, which is a function of the deleterious effect of a fully penetrant mutation and the dilution effect).

Finally, we require an expression for the entire burden of errors throughout the proteome for cell fitness. Here, letting P denote the number of protein-coding loci, we will assume that the effects of errors arising at each genetic locus act independently, such that total mean fitness can be written as

$$W(u) = \prod_{i=1}^{P}(1 - s_i) \simeq \exp\left(-\sum_{i=1}^{P} s_i\right)$$
$$\simeq 1 - \left\{(u P\overline{L}) \cdot [\overline{n \cdot f(n)}] \cdot \overline{\delta}\right\}, \qquad (20.2.4)$$

where an overline denotes an average over loci. The first approximation follows from $s_i \ll 1$, and the second from the assumption of independent n_i, L_i, and δ_i. The quantity $[1 - W(u)]$ defines the fractional selective disadvantage of error rate u relative to the situation in which $u = 0$. The difference in this quantity for two alleles resulting in error rates u_1 and u_2, $[W(u_1) - W(u_2)]$, provides a measure of the selective advantage of the first allele over the second.

As noted in Chapter 4, the magnitude of a selection coefficient dictates the capacity for natural selection to improve a trait. For a haploid population with genetic effective size N_e, the absolute value of $[W(u_1) - W(u_2)]$ must exceed $1/N_e$ for natural selection to discriminate between alternative alleles. For $|W(u_1) - W(u_2)| < 1/N_e$, drift overwhelms the power of selection, and hence this point is referred to as the drift barrier. Thus, letting Δ denote the fractional decline in the error rate between consecutive states in a hierarchy of mutationally connected alleles with effects on the error rate, the drift barrier is the error rate u^* that satisfies

$$N_e = \frac{1}{W[(1 - \Delta)u^*] - W[u^*]}. \qquad (20.2.5)$$

For $u > u^*$, natural selection is capable of driving the error rate to a lower value, whereas a situation in which $u < u^*$ is expected to be transient as selection is unable to maintain such a state.

Rearranging, and substituting from above, the lower bound to the error rate achievable by selection is

$$u^* \simeq \frac{1}{\Delta \cdot N_e \cdot P\overline{L} \cdot \overline{n}\phi(\overline{n}) \cdot \overline{\delta}}. \qquad (20.2.6)$$

where $\phi(\overline{n}) = \{1 - [x(1 - x)C_n^2/2]\}/\overline{n}^x$ is the average dilution factor, with $\overline{n}$ and C_n denoting the mean and coefficient of variation in expression level (obtained by Taylor-series expansion of Equation 20.2.2). This shows that, all other things being equal, there is an expected inverse relationship between the drift barrier and N_e. However, there is also an inverse scaling with the total number of codons in the proteome $P\overline{L}$, with the copy-number effect $\overline{n}\phi(\overline{n})$, and with the average fitness effect of mutations $\overline{\delta}$. The granularity of mutational changes in alleles influencing the error rate (Δ) operates as a simple scaling factor, but does not change the form of the relationship – the higher the value of Δ, the greater the difference of allelic effects, and hence the greater the efficiency of selection for a lower error rate.

Literature Cited

Alic, N., N. Ayoub, E. Landrieux, E. Favry, P. Baudouin-Cornu, M. Riva, and C. Carles. 2007. Selectivity and proofreading both contribute significantly to the fidelity of RNA polymerase III transcription. Proc. Natl. Acad. Sci. USA 104: 10400–10405.

Andersen, K. R., T. H. Jensen and D. E. Brodersen. 2008. Take the 'A' tail – quality control of ribosomal and transfer RNA. Biochim. Biophys. Acta 1779: 532–537.

Andersson, D. I., H. W. van Verseveld, A. H. Stouthamer, and C. G. Kurland. 1986. Suboptimal growth with hyper-accurate ribosomes. Arch. Microbiol. 144: 96–101.

Ardehali, M. B., and J. T. Lis. 2009. Tracking rates of transcription and splicing *in vivo*. Nat. Struct. Mol. Biol. 16: 1123–1124.

Azpurua, J., Z. Ke, I. X. Chen, Q. Zhang, D. N. Ermolenko, Z. D. Zhang, V. Gorbunova, and A. Seluanov. 2013. Naked mole-rat has increased translational fidelity compared with the mouse, as well as a unique 28S ribosomal RNA cleavage. Proc. Natl. Acad. Sci. USA 110: 17350–17355.

Bacher, J. M., and P. Schimmel. 2007. An editing-defective aminoacyl-tRNA synthetase is mutagenic in aging bacteria via the SOS response. Proc. Natl. Acad. Sci. USA 104: 1907–1912.

Bacher, J. M., W. F. Waas, D. Metzgar, V. de Crécy-Lagard, and P. Schimmel. 2007. Genetic code ambiguity confers a selective advantage on *Acinetobacter baylyi*. J. Bacteriol. 189: 6494–6496.

Banerjee, K., A. B. Kolomeisky, and O. A. Igoshin. 2017. Elucidating interplay of speed and accuracy in biological error correction. Proc. Natl. Acad. Sci. USA 114: 5183–5188.

Beuning, P. J., and K. Musier-Forsyth. 2001. Species-specific differences in amino acid editing by class II prolyl-tRNA synthetase. J. Biol. Chem. 276: 30779–30785.

Bezerra, A. R., J. Simões, W. Lee, J. Rung, T. Weil, I. G. Gut, M. Gut, M. Bayés, L. Rizzetto, D. Cavalieri, et al. 2013. Reversion of a fungal genetic code alteration links proteome instability with genomic and phenotypic diversification. Proc. Natl. Acad. Sci. USA 110: 11079–11084.

Blank, A., J. A. Gallant, R. R. Burgess, and L. A. Loeb. 1986. An RNA polymerase mutant with reduced accuracy of chain elongation. Biochemistry 25: 5920–5928.

Bonetti, B., L. Fu, J. Moon, and D. M. Bedwell. 1995. The efficiency of translation termination is determined by a synergistic interplay between upstream and downstream sequences in *Saccharomyces cerevisiae*. J. Mol. Biol. 251: 334–345.

Bouadloun, F., D. Donner, and C. G. Kurland. 1983. Codon-specific missense errors *in vivo*. EMBO J. 2: 1351–1356.

Brégeon, D., Z. A. Doddridge, H. J. You, B. Weiss, and P. W. Doetsch. 2003. Transcriptional mutagenesis induced by uracil and 8-oxoguanine in *Escherichia coli*. Mol. Cell 12: 959–970.

Carter, R., and G. Drouin. 2010. The increase in the number of subunits in eukaryotic RNA polymerase III relative to RNA polymerase II is due to the permanent recruitment of general transcription factors. Mol. Biol. Evol. 27: 1035–1043.

Cassan, M., and J. P. Rousset. 2001. UAG readthrough in mammalian cells: Effect of upstream and downstream stop codon contexts reveal different signals. BMC Mol. Biol. 2: 3.

Chen, J. F., N. N. Guo, T. Li, E. D. Wang, and Y. L. Wang. 2000. CP1 domain in *Escherichia coli* leucyl-tRNA synthetase is crucial for its editing function. Biochemistry 39: 6726–6731.

Chen, X., J. J. Ma, M. Tan, P. Yao, Q. H. Hu, G. Eriani, and E. D. Wang. 2011. Modular pathways for editing non-cognate amino acids by human cytoplasmic leucyl-tRNA synthetase. Nucleic Acids Res. 39: 235–247.

Clauson, C. L., K. J. Oestreich, J. W. Austin, and P. W. Doetsch. 2010. Abasic sites and strand breaks in DNA cause transcriptional mutagenesis in *Escherichia coli*. Proc. Natl. Acad. Sci. USA 107: 3657–3662.

Dahlseid, J. N., J. Puziss, R. L. Shirley, A. L. Atkin, P. Hieter, and M. R. Culbertson. 1998. Accumulation of mRNA coding for the ctf13p kinetochore subunit of *Saccharomyces cerevisiae* depends on the same factors that promote rapid decay of nonsense mRNAs. Genetics 150: 1019–1035.

de Mercoyrol, L., Y. Corda, C. Job, and D. Job. 1992, Accuracy of wheat-germ RNA polymerase II. General enzymatic properties and effect of template conformational transition from right-handed B-DNA to left-handed Z-DNA. Eur. J. Biochem. 206: 49–58.

Edelmann, P., and J. Gallant. 1977. Mistranslation in *E. coli*. Cell 10: 131–137.

Ehrenberg, M., and C. G. Kurland. 1984. Costs of accuracy determined by a maximal growth rate constraint. Quart. Rev. Biophys. 17: 45–82.

Eser, P., L. Wachutka, K. C. Maier, C. Demel, M. Boroni, S. Iyer, P. Cramer, and J. Gagneur. 2016. Determinants of RNA metabolism in the *Schizosaccharomyces pombe* genome. Mol. Syst. Biol. 12: 857.

Fan, Y., J. Wu, M. H. Ung, N. De Lay, C. Cheng, and J. Ling. 2015. Protein mistranslation protects bacteria against oxidative stress. Nucleic Acids Res. 43: 1740–1748.

Feeney, L., V. Carvalhal, X. C. Yu, B. Chan, D. A. Michels, Y. J. Wang, A. Shen, J. Ressl, B. Dusel, and M. W. Laird. 2013. Eliminating tyrosine sequence variants in CHO cell lines producing recombinant monoclonal antibodies. Biotechnol. Bioeng. 110: 1087–1097.

Fersht, A. R. 1999. Structure and Mechanism in Protein Science. W. H. Freeman & Co., New York, NY.

Fersht, A. R., and C. Dingwall. 1979. An editing mechanism for the methionyl-tRNA synthetase in the selection of amino acids in protein synthesis. Biochemistry 18: 1250–1256.

Fersht, A. R., J. S. Shindler, and W. C. Tsui. 1980. Probing the limits of protein-amino acid side chain recognition with the aminoacyl-tRNA synthetases. Discrimination against phenylalanine by tyrosyl-tRNA synthetases. Biochemistry 19: 5520–5524.

Firoozan, M., C. M. Grant, J. A. Duarte, and M. F. Tuite. 1991. Quantitation of readthrough of termination codons in yeast using a novel gene fusion assay. Yeast 7: 173–183.

Frank, S. A. 2007. Maladaptation and the paradox of robustness in evolution. PLoS ONE 2: e1021.

Freist, W., H. Sternbach, I. Pardowitz, and F. Cramer. 1998. Accuracy of protein biosynthesis: Quasi-species nature of proteins and possibility of error catastrophes. J. Theor. Biol. 193: 19–38.

Fritsch, C., J.-F. Gout, S. Haroon, A. Towheed, X. Zhang, Y. Song, S. Simpson, D. Wallace, K. Thomas, M. Lynch, et al. 2020. Genome-wide surveillance of transcription errors in response to genotoxic stress. Proc. Natl. Acad. Sci. USA 118: e2004077118.

Frith, M. C., J. Ponjavic, D. Fredman, C. Kai, J. Kawai, P. Carninci, Y. Hayashizaki, and A. Sandelin. 2006. Evolutionary turnover of mammalian transcription start sites. Genome Res. 16: 713–722.

Gerashchenko, M. V., Z. Peterfi, S. H. Yim, and V. N. Gladyshev. 2021. Translation elongation rate varies among organs and decreases with age. Nucleic Acids Res. 49: e9.

Golding, I., and E. C. Cox. 2004. RNA dynamics in live *Escherichia coli* cells. Proc. Natl. Acad. Sci. USA 101: 11310–11305.

Goldsmith, M., and D. S. Tawfik. 2009. Potential role of phenotypic mutations in the evolution of protein expression and stability. Proc. Natl. Acad. Sci. USA 106: 6197–6202.

Gonzalez, C. I., A. Bhattacharya, W. Wang, and S. W. Peltz. 2001. Nonsense-mediated mRNA decay in *Saccharomyces cerevisiae*. Gene 274: 15–25.

Gorini, L. 1971. Ribosomal discrimination of tRNAs. Nat. New Biol. 234: 261–264.

Gout, J.-F., W. Li, C. Fritsch, A. Li, S. Haroon, L. Singh, D. Hua, H. Fazelinia, S. Seeholzer, M. Lynch, et al. 2017. The landscape of transcription errors in eukaryotic cells. Science Advances 3: e1701484.

Gout, J. F., W. K. Thomas, Z. Smith, K. Okamoto, and M. Lynch. 2013. Large-scale detection of *in vivo* transcription errors. Proc. Natl. Acad. Sci. USA 110: 18584–18589.

Graille, M., and B. Séraphin. 2012. Surveillance pathways rescuing eukaryotic ribosomes lost in translation. Nat. Rev. Mol. Cell Biol. 13: 727–735.

Gros, P. A., and O. Tenaillon. 2009. Selection for chaperone-like mediated genetic robustness at low mutation rate: Impact of drift, epistasis and complexity. Genetics 182: 555–564.

Guo, M., Y. E. Chong, R. Shapiro, K. Beebe, X. L. Yang, and P. Schimmel. 2009. Paradox of mistranslation of serine for alanine caused by AlaRS recognition dilemma. Nature 462: 808–812.

Haag, J. R., and C. S. Pikaard. 2011. Multisubunit RNA polymerases IV and V: Purveyors of non-coding RNA for plant gene silencing. Nat. Rev. Mol. Cell. Biol. 12: 483–492.

Han, N. C., P. Kelly, and M. Ibba. 2020. Translational quality control and reprogramming during stress adaptation. Exp. Cell Res. 394: 112161.

Hentze, M. W., and A. E. Kulozik. 1999. A perfect message: RNA surveillance and nonsense-mediated decay. Cell 96: 307–310.

Hiriyanna, K. T., and T. Ramakrishnan. 1986. Deoxyribonucleic acid replication time in *Mycobacterium tuberculosis* H37 Rv. Arch. Microbiol. 144: 105–109.

Hodgkin, J., A. Papp, R. Pulak, V. Ambros, and P. Anderson. 1989. A new kind of informational suppression in the nematode *Caenorhabditis elegans*. Genetics 123: 301–313.

Hopfield, J. J. 1974. Kinetic proofreading: a new mechanism for reducing errors in biosynthetic processes requiring high specificity. Proc. Natl. Acad. Sci. USA 71: 4135–4139.

Hopfield, J. J., T. Yamane, V. Yue, and S. M. Coutts. 1976. Direct experimental evidence for kinetic proofreading in amino acylation of tRNA-Ile. Proc. Natl. Acad. Sci. USA 73: 1164–1168.

Hughes, D. 1991. Error-prone EF-Tu reduces *in vivo* enzyme activity and cellular growth rate. Mol. Microbiol. 5: 623–630.

Hughes, D., and I. Tubulekas. 1993. Ternary complex-ribosome interaction: Its influence on protein synthesis and on growth rate. Biochem. Soc. Trans. 21: 851–857.

Hussain, T., V. Kamarthapu, S. P. Kruparani, M. V. Deshmukh, and R. Sankaranarayanan. 2010. Mechanistic insights into cognate substrate discrimination during proofreading in translation. Proc. Natl. Acad. Sci. USA 107: 22117–22121.

Ieong, K. W., U. Uzun, M. Selmer, and M. Ehrenberg. 2016. Two proofreading steps amplify the accuracy of genetic code translation. Proc. Natl. Acad. Sci. USA 113: 13744–13749.

Igloi, G. L., F. von der Haar, and F. Cramer. 1978. Aminoacyl-tRNA synthetases from yeast: Generality of chemical proofreading in the prevention of misaminoacylation of tRNA. Biochemistry 17: 3459–3468.

Isken, O., and L. E. Maquat. 2008. The multiple lives of NMD factors: Balancing roles in gene and genome regulation. Nat. Rev. Genet. 9: 699–712.

Jakubowski, H. 1980. Valyl-tRNA synthetase form yellow lupin seeds: Hydrolysis of the enzyme-bound noncognate aminoacyl adenylate as a possible mechanism of increasing specificity of the aminoacyl-tRNA synthetase. Biochemistry 19: 5071–5508.

Javid, B., F. Sorrentino, M. Toosky, W. Zheng, J. T. Pinkham, N. Jain, M. Pan, P. Deighan, and E. J. Rubin. 2014. Mycobacterial mistranslation is necessary and sufficient for rifampicin phenotypic resistance. Proc. Natl. Acad. Sci. USA 111: 1132–1137.

Johansson, M., M. Lovmar, and M. Ehrenberg. 2008. Rate and accuracy of bacterial protein synthesis revisited. Curr. Opin. Microbiol. 11: 141–147.

Jones, T. E., R. W. Alexander, and T. Pan. 2011. Misacylation of specific nonmethionyl tRNAs by a bacterial methionyl-tRNA synthetase. Proc. Natl. Acad. Sci. USA 108: 6933–6938.

Kaplan, C. D. 2010. The architecture of RNA polymerase fidelity. BMC Biol. 8: 85.

Kervestin, S., and A. Jacobson. 2012. NMD: A multifaceted response to premature translational termination. Nat. Rev. Mol. Cell Biol. 13: 700–712.

Khazaie, K., J. H. Buchanan, and R. F. Rosenberger. 1984. The accuracy of Q beta RNA translation. 1. Errors during the synthesis of Q beta proteins by intact *Escherichia coli* cells. Eur. J. Biochem. 144: 485–489.

Kiktev, D., S. Moskalenko, O. Murina, A. Baudin-Baillieu, J. P. Rousset, and G. Zhouravleva. 2009. The paradox of viable sup45 STOP mutations: A necessary equilibrium between translational readthrough, activity and stability of the protein. Mol. Genet. Genomics 282: 83–96.

Kireeva, M. L., Y. A. Nedialkov, G. H. Cremona, Y. A. Purtov, L. Lubkowska, F. Malagon, Z. F. Burton, J. N. Strathern, and M. Kashlev. 2008. Transient reversal of RNA polymerase II active site closing controls fidelity of transcription elongation. Mol. Cell 30: 557–566.

Korencic, D., I. Ahel, J. Schelert, M. Sacher, B. Ruan, C. Stathopoulos, P. Blum, M. Ibba, and D. Söll. 2004. A free-standing proofreading domain is required for protein synthesis quality control in Archaea. Proc. Natl. Acad. Sci. USA 101: 10260–10265.

Kramer, E. B., and P. J. Farabaugh. 2007. The frequency of translational misreading errors in *E. coli* is largely determined by tRNA competition. RNA 13: 87–96.

Kramer, E. B., and A. K. Hopper. 2013. Retrograde transfer RNA nuclear import provides a new level of tRNA quality control in *Saccharomyces cerevisiae*. Proc. Natl. Acad. Sci. USA 110: 21042–21047.

Kramer, E. B., H. Vallabhaneni, L. M. Mayer, and P. J. Farabaugh. 2010. A comprehensive analysis of translational missense errors in the yeast *Saccharomyces cerevisiae*. RNA 16: 1797–1808.

Kucukyildirim, S., H. Long, W. Sung, S. F. Miller, T. G. Doak, and M. Lynch. 2016. The rate and spectrum of spontaneous mutations in *Mycobacterium smegmatis*, a bacterium naturally devoid of the post-replicative mismatch repair pathway. G3 (Bethesda) 6: 2157–2163.

Kuntz, I. D., K. Chen, K. A. Sharp, and P. A. Kollman. 1999. The maximal affinity of ligands. Proc. Natl. Acad. Sci. USA 96: 9997–10002.

Kurland, C. G., and M. Ehrenberg. 1984. Optimization of translation accuracy. Prog. Nucleic Acid Res. Mol. Biol. 31: 191–219.

Kurland, C. G., and M. Ehrenberg. 1987. Growth-optimizing accuracy of gene expression. Annu. Rev. Biophys. Biophys. Chem. 16: 291–317.

Landick, R. 2009. Functional divergence in the growing family of RNA polymerases. Structure 17: 323–325.

LaRiviere, F. J., S. E. Cole, D. J. Ferullo, and M. J. Moore. 2006. A late-acting quality control process for mature eukaryotic rRNAs. Mol. Cell 24: 619–626.

Larson, M. H., J. Zhou, C. D. Kaplan, M. Palangat, R. D. Kornberg, R. Landick, and S. M. Block. 2012. Trigger loop dynamics mediate the balance between the transcriptional fidelity and speed of RNA polymerase II. Proc. Natl. Acad. Sci. USA 109: 6555–6560.

Laughrea, M., J. Latulippe, A. M. Filion, and L. Boulet. 1987. Mistranslation in twelve *Escherichia coli* ribosomal proteins. Cysteine misincorporation at neutral amino acid residues other than tryptophan. Eur. J. Biochem. 169: 59–64.

Lee, H., E. Popodi, H. Tang, and P. L. Foster. 2012. Rate and molecular spectrum of spontaneous mutations in the bacterium *Escherichia coli* as determined by whole-genome sequencing. Proc. Natl. Acad. Sci. USA 109: E2774–E2783.

Lee, J. W., K. Beebe, L. A. Nangle, J. Jang, C. M. Longo-Guess, S. A. Cook, M. T. Davisson, J. P. Sundberg, P. Schimmel, and S. L. Ackerman. 2006. Editing-defective tRNA synthetase causes protein misfolding and neurodegeneration. Nature 443: 50–55.

Leeds, P., J. M. Wood, B. S. Lee, and M. R. Culbertson. 1992. Gene products that promote mRNA turnover in *Saccharomyces cerevisiae*. Mol. Cell. Biol. 12: 2165–2177.

Leng, T., M. Pan, X. Xu, and B. Javid. 2015. Translational misreading in *Mycobacterium smegmatis* increases in stationary phase. Tuberculosis 95: 678–681.

Levine, R. L., B. S. Berlett, J. Moskovitz, L. Mosoni, and E. R. Stadtman. 1999. Methionine residues may protect proteins from critical oxidative damage. Mech. Ageing Dev. 107: 323–332.

Li, L., M. T. Boniecki, J. D. Jaffe, B. S. Imai, P. M. Yau, Z. A. Luthey-Schulten, and S. A. Martinis. 2011. Naturally occurring aminoacyl-tRNA synthetases editing-domain mutations that cause mistranslation in *Mycoplasma* parasites. Proc. Natl. Acad. Sci. USA 108: 9378–9383.

Li, W., and M. Lynch. 2020. Universally high transcript error rates in bacteria. eLife 9: e54898.

Li, W., S. Baehr, M. Marasco, L. Reyes, D. Brister, C. S. Pikaard, J.-F. Gout, M. Vermulst, and M. Lynch. 2022. Evolution of transcript-error rates. (in prep.)

Lirussi, L., Ö. Demir, P. You, A. Sarno, R. E. Amaro, and H. Nilsen. 2021. RNA metabolism guided by RNA modifications: The role of SMUG1 in rRNA quality control. Biomolecules 11: 76.

Lisica, A., C. Engel, M. Jahnel, É. Roldán, E. A. Galburt, P. Cramer, and S. W. Grill. 2016. Mechanisms of backtrack recovery by RNA polymerases I and II. Proc. Natl. Acad. Sci. USA 113: 2946–2951.

Loenarz, C., R. Sekirnik, A. Thalhammer, W. Ge, E. Spivakovsky, M. M. Mackeen, M. A. McDonough, M. E. Cockman, B. M. Kessler, P. J. Ratcliffe, et al. 2014. Hydroxylation of the eukaryotic ribosomal decoding center affects translational accuracy. Proc. Natl. Acad. Sci. USA 111: 4019–4024.

Long, H., M. G. Behringer, E. Williams, R. Te, and M. Lynch. 2016. Similar mutation rates but highly diverse mutation spectra in ascomycete and basidiomycete yeasts. Genome Biol. Evol. 8: 3815–3821.

Long, H., S. Kucukyildirim, W. Sung, E. Williams, M. Ackerman, T. G. Doak, and M. Lynch. 2015. Background mutational features of the radiation-resistant bacterium *Deinococcus radiodurans*. Mol. Biol. Evol. 32: 2383–2392.

Lovmar, M., and M. Ehrenberg. 2006. Rate, accuracy and cost of ribosomes in bacterial cells. Biochimie 88: 951–961.

Lue, S. W., and S. O. Kelley. 2007. A single residue in leucyl-tRNA synthetase affecting amino acid specificity and tRNA aminoacylation. Biochemistry 46: 4466–4472.

Luo, J., and B. D. Hall. 2007. A multistep process gave rise to RNA polymerase IV of land plants. J. Mol. Evol. 64: 101–112.

Lykke-Andersen, J. 2001. mRNA quality control: Marking the message for life or death. Curr. Biol. 11: R88–R91.

Lynch, M. 2007. The Origins of Genome Architecture. Sinauer Associates, Inc., Sunderland, MA.

Lynch, M. 2012. Evolutionary layering and the limits to cellular perfection. Proc. Natl. Acad. Sci. USA 109: 18851–18856.

Mango, S. E. 2001. Stop making nonSense: The *C. elegans* smg genes. Trends Genet. 17: 646–653.

Manickam, N., N. Nag, A. Abbasi, K. Patel, and P. J. Farabaugh. 2014. Studies of translational misreading *in vivo* show that the ribosome very efficiently discriminates against most potential errors. RNA 20: 9–15.

Maquat, L. E. 2004. Nonsense-mediated mRNA decay: A comparative analysis of different species. Curr. Genomics 5: 175–190.

Maquat, L. E. (ed.) 2006. Nonsense-mediated mRNA Decay. Landes Bioscience, Georgetown, TX.

Maquat, L. E., and G. G. Carmichael. 2001. Quality control of mRNA function. Cell 104: 173–176.

Martinis, S. A., and M. T. Boniecki. 2010. The balance between pre- and post-transfer editing in tRNA synthetases. FEBS Lett. 584: 455–459.

Mason, P. B., and K. Struhl. 2005. Distinction and relationship between elongation rate and processivity of RNA polymerase II *in vivo*. Mol. Cell 17: 831–840.

McCandlish, D. M., and J. B. Plotkin. 2016. Transcriptional errors and the drift barrier Proc. Natl. Acad. Sci. USA 113: 3136–3138.

McDonald, M. J., Y. Y. Hsieh, Y. H. Yu, S. L. Chang, and J. Y. Leu. 2012. The evolution of low mutation rates in experimental mutator populations of *Saccharomyces cerevisiae*. Curr. Biol. 22: 1235–1240.

Medghalchi, S. M., P. A. Frischmeyer, J. T. Mendell, A. G. Kelly, A. M. Lawler, and H. C. Dietz. 2001. Rent1, a trans-effector of nonsense-mediated mRNA decay, is essential for mammalian embryonic viability. Hum. Mol. Genet. 10: 99–105.

Meer, K. M., P. G. Nelson, K. Xiong, and J. Masel. 2020. High transcriptional error rates vary as a function of gene expression level. Genome Biol. Evol. 12: 3754–3761.

Melnikov, S. V., K. D. Rivera, D. Ostapenko, A. Makarenko, N. D. Sanscrainte, J. J. Becnel, M. J. Solomon, C. Texier, D. J. Pappin, and D. Söll. 2018. Error-prone protein synthesis in parasites with the smallest eukaryotic genome. Proc. Natl. Acad. Sci. USA 115: E6245–E6253.

Mendell, J. T., S. M. Medghalchi, R. G. Lake, E. N. Noensie, and H. C. Dietz. 2000. Novel Upf2p orthologues suggest a functional link between translation initiation and nonsense surveillance complexes. Mol. Cell. Biol. 20: 8944–8957.

Mendell, J. T., N. A. Sharifi, J. L. Meyers, F. Martinez-Murillo, and H. C. Dietz. 2004. Nonsense surveillance regulates expression of diverse classes of mammalian transcripts and mutes genomic noise. Nat. Genet. 36: 1073–1078.

Mikkola, R., and C. G. Kurland. 1992. Selection of laboratory wild-type phenotype from natural isolates of *Escherichia coli* in chemostats. Mol. Biol. Evol. 9: 394–402.

Miller, J. H., and A. M. Albertini. 1983. Effects of surrounding sequence on the suppression of nonsense codons. J. Mol. Biol. 164: 59–71.

Min, B., M. Kitabatake, C. Polycarpo, J. Pelaschier, G. Raczniak, B. Ruan, H. Kobayashi, S. Namgoong, and D. Söll. 2003. Protein synthesis in *Escherichia coli* with mischarged tRNA. J. Bacteriol. 185: 3524–3526.

Mitrovich, Q. M., and P. Anderson. 2000. Unproductively spliced ribosomal protein mRNAs are natural targets of mRNA surveillance in *C. elegans*. Genes Dev. 14: 2173–2184.

Mordret, E., O. Dahan, O. Asraf, R. Rak, A. Yehonadav, G. D. Barnabas, J. Cox, T. Geiger, A. B. Lindner, and Y. Pilpel. 2019. Systematic detection of amino acid substitutions in proteomes reveals mechanistic basis of ribosome errors and selection for translation fidelity. Mol. Cell 75: 427–441.e5.

Mori, N., D. Mizuno, and S. Goto. 1979. Conservation of ribosomal fidelity during ageing. Mech. Ageing Dev. 10: 379–398.

Murugan, A., D. A. Huse, and S. Leibler. 2012. Speed, dissipation, and error in kinetic proofreading. Proc. Natl. Acad. Sci. USA 109: 12034–12039.

Musa, M., M. Radman, and A. Krisko. 2016. Decreasing translation error rate in *Escherichia coli* increases protein function. BMC Biotechnol. 16: 28.

Muzyczka, N., R. L. Poland, and M. J. Bessman. 1972. Studies on the biochemical basis of spontaneous mutation. I. A comparison of the deoxyribonucleic acid polymerases of mutator, antimutator, and wild type strains of bacteriophage T4. J. Biol. Chem. 247: 7116–7122.

Myllykallio, H., P. Lopez, P. López-García, R. Heilig, W. Saurin, Y. Zivanovic, H. Philippe, and P. Forterre. 2000. Bacterial mode of replication with eukaryotic-like machinery in a hyperthermophilic archaeon. Science 288: 2212–2215.

Nangle, L. A., C. M. Motta, and P. Schimmel. 2006. Global effects of mistranslation from an editing defect in mammalian cells. Chem. Biol. 13: 1091–1100.

Netzer, N., J. M. Goodenbour, A. David, K. A. Dittmar, R. B. Jones, J. R. Schneider, D. Boone, E. M. Eves, M. R. Rosner, J. S. Gibbs, et al. 2009. Innate immune and chemically triggered oxidative stress modifies translational fidelity. Nature 462: 522–526.

Nguyen, D. T., B. Wu, H. Long, N. Zhang, C. Patterson, S. Simpson, K. Morris, W. K. Thomas, M. Lynch, and W. Hao. 2020. Variable spontaneous mutation and loss of heterozygosity among heterozygous genomes in yeast. Mol. Biol. Evol. 37: 3118–3130.

Ninio, J. 1975. Kinetic amplification of enzyme discrimination. Biochimie 57: 587–595.

Ninio, J. 1991a. Connections between translation, transcription and replication error-rates. Biochimie 73: 1517–1523.

Ninio, J. 1991b. Transient mutators: A semiquantitative analysis of the influence of translation and transcription errors on mutation rates. Genetics 129: 957–962.

Pan, T. 2013. Adaptive translation as a mechanism of stress response and adaptation. Annu. Rev. Genet. 47: 121–137.

Parker, J. 1989. Errors and alternatives in reading the universal genetic code. Microbiol. Rev. 53: 273–298.

Peltz, S. W., A. B. Hammell, Y. Cui, J. Yasenchak, L. Puljanowski, and J. D. Dinman. 1999. Ribosomal protein L3 mutants alter translational fidelity and promote rapid loss of the yeast killer virus. Mol. Cell Biol. 19: 384–391.

Pezo, V., D. Metzgar, T. L. Hendrickson, W. F. Waas, S. Hazebrouck, V. Döring, P. Marlière, P. Schimmel, and V. De Crècy-Lagard. 2004. Artificially ambiguous genetic code confers growth yield advantage. Proc. Natl. Acad. Sci. USA 101: 8593–8597.

Phillips, R., J.Kondev, J. Theriot, and H. Garcia. 2013. Physical Biology of the Cell, 2nd edn. Garland Science, New York, NY.

Piepersberg, W., V. Noseda, and A. Böck. 1979. Bacterial ribosomes with two ambiguity mutations: Effects of translational fidelity, on the response to aminoglycosides and on the rate of protein synthesis. Mol. Gen. Genet. 171: 23–34.

Plant, E. P., P. Nguyen, J. R. Russ, Y. R. Pittman, T. Nguyen, J. T. Quesinberry, T. G. Kinzy, and J. D. Dinman. 2007. Differentiating between near- and non-cognate codons in Saccharomyces cerevisiae. PLoS One 2: e517.

Proshkin, S., A. R. Rahmouni, A. Mironov, and E. Nudler. 2010. Cooperation between translating ribosomes and RNA polymerase in transcription elongation. Science 328: 504–508.

Qian, H. 2006. Reducing intrinsic biochemical noise in cells and its thermodynamic limit. J. Mol. Biol. 362: 387–392.

Reynolds, N. M., J. Ling, H. Roy, R. Banerjee, S. E. Repasky, P. Hamel, and M. Ibba. 2010. Cell-specific differences in the requirements for translation quality control. Proc. Natl. Acad. Sci. USA 107: 4063–4068.

Ribas de Pouplana, L., M. A. Santos, J. H. Zhu, P. J. Farabaugh, and B. Javid. 2014. Protein mistranslation: Friend or foe? Trends Biochem. Sci. 39: 355–362.

Rocha, R., P. J. Pereira, M. A. Santos, and S. Macedo-Ribeiro. 2011. Unveiling the structural basis for translational ambiguity tolerance in a human fungal pathogen. Proc. Natl. Acad. Sci. USA 108: 14091–14096.

Rock, J. M., U. F. Lang, M. R. Chase, C. B. Ford, E. R. Gerrick, R. Gawande, M. Coscolla, S. Gagneux, S. M. Fortune, and M. H. Lamers. 2015. DNA replication fidelity in Mycobacterium tuberculosis is mediated by an ancestral prokaryotic proofreader. Nat. Genet. 47: 677–681.

Roth, J. R. 1970. UGA nonsense mutations in Salmonella typhimurium. J. Bacteriol. 102: 467–475.

Salas-Marco, J., and D. M. Bedwell. 2005. Discrimination between defects in elongation fidelity and termination efficiency provides mechanistic insights into translational readthrough. J. Mol. Biol. 348: 801–815.

Santos, M. A., C. Cheesman, V. Costa, P. Moradas-Ferreira, and M. F. Tuite. 1999. Selective advantages created by codon ambiguity allowed for the evolution of an alternative genetic code in Candida spp. Mol. Microbiol. 31: 937–947.

Sarkar, A., M. Thoms, C. Barrio-Garcia, E. Thomson, D. Flemming, R. Beckmann, and E. Hurt. 2017. Preribosomes escaping from the nucleus are caught during translation by cytoplasmic quality control. Nat. Struct. Mol. Biol. 24: 1107–1115.

Savageau, M. A., and R. R. Freter. 1979. On the evolution of accuracy and cost of proofreading tRNA aminoacylation. Proc. Natl. Acad. Sci. USA 76: 4507–4510.

Schimmel, P., and M. Guo. 2009. A tipping point for mistranslation and disease. Nat. Struct. Mol. Biol. 16: 348–349.

Schwartz, M. H., and T. Pan. 2016. Temperature dependent mistranslation in a hyperthermophile adapts proteins to lower temperatures. Nucleic Acids Res. 44: 294–303.

Schwartz, M. H., and T. Pan. 2017. tRNA misacylation with methionine in the mouse gut microbiome *in situ*. Microb. Ecol. 74: 10–14.

Schwartz, M. H., J. R. Waldbauer, L. Zhang, and T. Pan. 2016. Global tRNA misacylation induced by anaerobiosis and antibiotic exposure broadly increases stress resistance in *Escherichia coli*. Nucleic Acids Res. 44: 10292–10303.

Shepherd, J., and M. Ibba. 2014. Relaxed substrate specificity leads to extensive tRNA mischarging by *Streptococcus pneumoniae* class I and class II aminoacyl-tRNA synthetases. MBio 5: e01656–14.

Springgate, C. F., and L. A. Loeb. 1975. On the fidelity of transcription by *Escherichia coli* ribonucleic acid polymerase. J. Mol. Biol. 97: 577–591.

Stadtman, E. R., and R. L. Levine. 2003. Free radical-mediated oxidation of free amino acids and amino acid residues in proteins. Amino Acids 25: 207–218.

Stansfield, I., K. M. Jones, P. Herbert, A. Lewendon, W. V. Shaw, and M. F. Tuite. 1998. Missense translation errors in *Saccharomyces cerevisiae*. J. Mol. Biol. 282: 13–24.

Steiner, R. E., A. M. Kyle, and M. Ibba. 2019. Oxidation of phenylalanyl-tRNA synthetase positively regulates translational quality control. Proc. Natl. Acad. Sci. USA 116: 10058–10063.

SternJohn, J., S. Hati, P. G. Siliciano, and K. Musier-Forsyth. 2007. Restoring species-specific posttransfer editing activity to a synthetase with a defunct editing domain. Proc. Natl. Acad. Sci. USA 104: 2127–2132.

Stillman, B. 1996. Cell cycle control of DNA replication. Science 274: 1659–1664.

Strigini, P., and E. Brickman. 1973. Analysis of specific misreading in *Escherichia coli*. J. Mol. Biol. 75: 659–672.

Struhl, K. 2007. Transcriptional noise and the fidelity of initiation by RNA polymerase II. Nat. Struct. Mol. Biol. 14: 103–105.

Suzuki, Y., D. Ishihara, M. Sasaki, H. Nakagawa, H. Hata, T. Tsunoda, M. Watanabe, T. Komatsu, T. Ota, T. Isogai, et al. 2000. Statistical analysis of the 5′ untranslated region of human mRNA using 'oligo-capped' cDNA libraries. Genomics 64: 286–297.

Sydow, J. F., F. Brueckner, A. C. Cheung, G. E. Damsma, S. Dengl, E. Lehmann, D. Vassylyev, and P. Cramer. 2009. Structural basis of transcription: Mismatch-specific fidelity mechanisms and paused RNA polymerase II with frayed RNA. Mol. Cell 34: 710–721.

Sydow, J. F., and P. Cramer. 2009. RNA polymerase fidelity and transcriptional proofreading. Curr. Opin. Struct. Biol. 19: 732–739.

Tang, Y., and D. A. Tirrell. 2002. Attenuation of the editing activity of the *Escherichia coli* leucyl-tRNA synthetase allows incorporation of novel amino acids into proteins *in vivo*. Biochemistry 41: 10635–10645.

Thompson, R. C., and P. J. Stone. 1977. Proofreading of the codon–anticodon interaction on ribosomes. Proc. Natl. Acad. Sci. USA 74: 198–202.

Torabi, N., and L. Kruglyak. 2011. Variants in SUP45 and TRM10 underlie natural variation in translation termination efficiency in *Saccharomyces cerevisiae*. PLoS Genet. 7: e1002211.

Toth, M. J., E. J. Murgola, and P. Schimmel. 1988. Evidence for a unique first position codon-anticodon mismatch in vivo. J. Mol. Biol. 201: 451–454.

Traverse, C. C., and H. Ochman. 2016. Conserved rates and patterns of transcription errors across bacterial growth states and lifestyles. Proc. Natl. Acad. Sci. USA 113: 3311–3316.

Tsui, W. C., and A. R. Fersht. 1981. Probing the principles of amino acid selection using the alanyl-tRNA synthetase from *Escherichia coli*. Nucleic Acids Res. 9: 4627–4637.

Turrientes, M. C., F. Baquero, B. R. Levin, J.-L. Martínez, A. Ripoll, J.-M. González-Alba, R. Tobes, M. Manrique, M.-R. Baquero, M.-J. Rodríguez-Domínguez, et al. 2013. Normal mutation rate variants arise in a mutator (mut S) *Escherichia coli* population. PLoS ONE 8: e72963.

Ucuncuoglu, S., K. L. Engel, P. K. Purohit, D. D. Dunlap, D. A. Schneider, and L. Finzi. 2016. Direct characterization of transcription elongation by RNA polymerase I. PLoS One 11: e0159527.

Vargas-Rodriguez, O., A. H. Badran, K. S. Hoffman, M. Chen, A. Crnković, Y. Ding, J. R. Krieger, E. Westhof, D. Söll, and S. Melnikov. 2021. Bacterial translation machinery for deliberate mistranslation of the genetic code. Proc. Natl. Acad. Sci. USA 118: e2110797118.

Walmacq, C., M. L. Kireeva, J. Irvin, Y. Nedialkov, L. Lubkowska, F. Malagon, J. N. Strathern, and M. Kashlev. 2009. Rpb9 subunit controls transcription fidelity by delaying NTP sequestration in RNA polymerase II. J. Biol. Chem. 284: 19601–19612.

Wang, X., and T. Pan. 2015. Methionine mistranslation bypasses the restraint of the genetic code to generate mutant proteins with distinct activities. PLoS Genet. 11: e1005745.

Wang, X., and T. Pan. 2016. Stress response and adaptation mediated by amino acid misincorporation during protein synthesis. Adv. Nutr. 7: 773S–779S.

Wei, W., W.-C. Ho, M. Behringer, and M. Lynch. 2022. Rapid evolution of mutation rate and spectrum in response to environmental and population-genetic challenges. Nat. Comm. 13: 4752.

Weickert, M. J., and I. Apostol. 1998. High-fidelity translation of recombinant human hemoglobin in *Escherichia coli*. Appl. Environ. Microbiol. 64: 1589–1593.

Werner, F., and D. Grohmann. 2011. Evolution of multisubunit RNA polymerases in the three domains of life. Nat. Rev. Microbiol. 9: 85–98.

Wielgoss, S., J. E. Barrick, O. Tenaillon, M. J. Wiser, W. J. Dittmar, S. Cruveiller, B. Chane-Woon-Ming, C. Médigue, R. E. Lenski, and D. Schneider. 2013. Mutation rate dynamics in a bacterial population reflect tension between adaptation and genetic load. Proc. Natl. Acad. Sci. USA 110: 222–227.

Wierzbicki, A. T., T. S. Ream, J. R. Haag, and C. S. Pikaard. 2009. RNA polymerase V transcription guides ARGONAUTE4 to chromatin. Nat. Genet. 41: 630–634.

Williams, I., J. Richardson, A. Starkey, and I. Stansfield. 2004. Genome-wide prediction of stop codon readthrough during translation in the yeast *Saccharomyces cerevisiae*. Nucleic Acids Res. 32: 6605–6616.

Williams, L. N., A. J. Herr, and B. D. Preston. 2013. Emergence of DNA polymerase antimutators that escape error-induced extinction in yeast. Genetics 193: 751–770.

Wiltrout, E., J. M. Goodenbour, M. Fréchin, and T. Pan. 2012. Misacylation of tRNA with methionine in *Saccharomyces cerevisiae*. Nucleic Acids Res. 40: 10494–10506.

Wilusz, C. J., W. Wang, and S. W. Peltz. 2001. Curbing the nonsense: The activation and regulation of mRNA surveillance. Genes Dev. 15: 2781–2785.

Wilusz, J. E., J. M. Whipple, E. M. Phizicky, and P. A. Sharp. 2011. tRNAs marked with CCACCA are targeted for degradation. Science 334: 817–821.

Wohlgemuth, I., C. Pohl, J. Mittelstaet, A. L. Konevega, and M. V. Rodnina. 2011. Evolutionary optimization of speed and accuracy of decoding on the ribosome. Philos. Trans. R. Soc. Lond. B Biol. Sci. 366: 2979–2986.

Yadavalli, S. S., and M. Ibba. 2013. Selection of tRNA charging quality control mechanisms that increase mistranslation of the genetic code. Nucleic Acids Res. 41: 1104–1112.

Yamane, T., and J. J. Hopfield. 1977. Experimental evidence for kinetic proofreading in the aminoacylation of tRNA by synthetase. Proc. Natl. Acad. Sci. USA 74: 2246–2250.

Yates, J. L. 1979. Role of ribosomal protein S12 in discrimination of aminoacyl-tRNA. J. Biol. Chem. 254: 11550–11554.

Yuzenkova, Y., A. Bochkareva, V. R. Tadigotla, M. Roghanian, S. Zorov, K. Severinov, and N. Zenkin. 2010. Stepwise mechanism for transcription fidelity. BMC Biol. 8: 54.

Zaher, H. S., and R. Green. 2009. Fidelity at the molecular level: Lessons from protein synthesis. Cell 136: 746–762.

Zaher, H. S., and R. Green. 2010. Hyperaccurate and error-prone ribosomes exploit distinct mechanisms during tRNA selection. Mol. Cell. 39: 110–120.

Zenkin, N., Y. Yuzenkova, and K. Severinov. 2006. Transcript-assisted transcriptional proofreading. Science 313: 518–520.

Zhang, H., Z. Lyu, Y. Fan, C. R. Evans, K. W. Barber, K. Banerjee, O. A. Igoshin, J. Rinehart, and J. Ling. 2020. Metabolic stress promotes stop-codon readthrough and phenotypic heterogeneity. Proc. Natl. Acad. Sci. USA 117: 22167–22172.

Zhang, J., K. W. Ieong, H. Mellenius, and M. Ehrenberg. 2016. Proofreading neutralizes potential error hotspots in genetic code translation by transfer RNAs. RNA 22: 896–904.

Zhang, Z., B. Shah, and P. V. Bondarenko. 2013. G/U and certain wobble position mismatches as possible main causes of amino acid misincorporations. Biochemistry 52: 8165–8176.

Zhang, Z., L. Zhou, L. Hu, Y. Zhu, H. Xu, Y. Liu, X. Chen, X. Yi, X. Kong, and L. D. Hurst. 2010. Nonsense-mediated decay targets have multiple sequence-related features that can inhibit translation. Mol. Syst. Biol. 6: 442.

CHAPTER 21

Transcription

All life relies on heritable information encoded in DNA, with the majority of cellular functions being derived from products resulting from transcription of DNA into RNA. Some RNAs, such as those in the ribosome, have direct functions, but for protein-coding loci, the first-step messenger RNAs (mRNAs) must be translated subsequently into strings of amino acids. Gene expression is almost always regulated by specific DNA binding proteins that activate and/or repress their target genes by binding to adjacent regulatory regions. Most genomes encode for hundreds to thousands of such transcription factors (hereafter, TFs), each with unique DNA binding-motif requirements called TF-binding sites (hereafter, TFBSs). One can certainly imagine simpler mechanisms for using DNA-level information to make proteins (e.g., the use of no RNA intermediates at all), but these are the cards that were dealt to LUCA, and there is now no way to erase this legacy of the earliest stages of evolution.

Usually, TFs are referred to as *trans*-acting, in the sense that the genetic loci encoding for them are generally unlinked to (or at least physically distant from) their regulatory targets. In contrast, TFBSs are generally referred to as *cis*-acting, as they are physically adjacent to the affected coding regions. This distinction can be blurred in prokaryotes, where a TF is sometimes encoded in the same multilocus transcriptional unit (called an operon) as its target gene, and all genes are linked in non-recombining genomes.

Because TFs must bind to the regulatory sites of their target proteins with high affinity relative to off-target sites, gene regulation provides an excellent example of coevolution at the molecular level. A number of questions immediately arise. How long and accurate does a TFBS have to be to ensure high specificity with respect to its cognate TF? Are both the TF and the TFBS free to wander in sequence space, provided an adequate level of joint matching is maintained? What happens when a TF services increasingly large numbers of genes? How do new TF–TFBS interactions arise?

Gene transcription is generally not an autonomous, invariant process, but rather is driven by extra- or intracellular information. The signals range from small inorganic molecules to simple metabolites, sometimes produced by other gene products. These, in turn, activate (or suppress) specific TFs, typically by modifying intermediary regulatory proteins (themselves often TFs) in functionally significant ways. Central to all of biology, this transfer of environmental information via TFs to downstream gene expression, called signal transduction, is the topic of Chapter 22. In addition, although TF proteins and their binding sites provide the dominant mechanism for regulating gene expression, they are by no means the only intervening factor. For example, post-transcriptional regulation can occur in the form of small complementary RNAs that can bind to transcripts, and post-translational modifications (Chapter 14) can further modify the operational features of gene products. As an overall entrée into the overall field of gene regulation, however, this chapter focuses on processes driven by TF proteins.

A central issue with respect to understanding transcription and its consequences is stochasticity. Genes are generally present just once (haploids) or twice (diploids) within cells, and as outlined in Chapter 7, mRNAs of active genes are often present in just a dozen or fewer copies, with proteins typically being an order of magnitude or more abundant. Owing to the small numbers of molecules

Evolutionary Cell Biology. Michael Lynch, Oxford University Press. © Michael Lynch (2024). DOI: 10.1093/oso/9780192847287.003.0021

of individual types relative to the vast space within cells, intermolecular encounters are by no means certain, and as a consequence there can be considerable cell-to-cell variation in gene expression, even in a genetically homogeneous population. Thus, before discussing the biology and evolution of transcription, we need to understand some simple quantitative principles regarding molecules in single cells.

Molecular Stochasticity in Single Cells

The fitness of a cell ultimately depends on the quality, quantity, and stoichiometric relationships of its underlying functional constituents. However, with stochastic dynamics of transcription and translation at play, the numbers of individual proteins vary among cells, even in a completely homogeneous external environment and with genetically identical and uniformly aged cells. The many factors governing the probability distributions of numbers of molecules per cell can be subsumed into six coefficients: the rates at which an inactive gene enters the transcriptionally active state, and vice versa, k_{on} and k_{off} respectively; the rate at which an active gene transcribes mRNAs, k_m; the rate at which an mRNA is translated into proteins, k_p; and the rates of degradation of mRNAs and proteins, δ_m and δ_p respectively (Figure 21.1). We now consider the ways in which these various factors influence the distributions of numbers of transcripts and proteins within different cells. For heuristic purposes, the following assumes a simple system in which gene expression is a function of the binding of a single activating TF, as is the case for many bacterial genes.

Cellular mRNA abundances

We first consider the numbers of mRNAs found in cells, n_m, as this has cascading effects on protein numbers. As noted in Figure 21.2, active cells gain mRNAs by transcription and lose them by degradation, whereas inactive cells (with no TF engaged with the target TFBS) can only lose mRNAs. Cells can also move back and forth between active and inactive states. The total transition rates to smaller numbers of mRNAs increase linearly with increasing n_m simply because there are more targets for degradation, and as a consequence, such systems converge on a steady-state distribution regardless of the starting point. As outlined in Foundations 21.1, a particularly simple outcome is obtained when a gene is constitutively turned on ($k_{off} = 0$). In this case, n_m is Poisson distributed, with both the mean and the variance of the number of mRNAs per cell equalling the ratio of gain and loss rates, k_m/δ_m. The Poisson distribution is dominated by the zero (mRNA-free) class when the mean is smaller than 1.0, has maximum and equal probability in classes 0 and 1 when the mean is equal to 1.0, and converges on a normal (bell-shaped or Gaussian) distribution with larger mean (Figure 21.2).

Regulated genes are not continuously expressed, but instead are in the active state only a fraction of

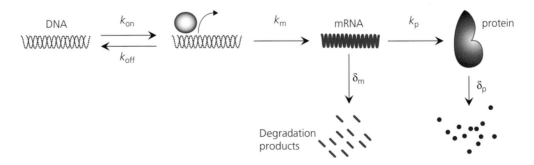

Figure 21.1 Flow chart for the key determinants of the number of proteins associated with a particular gene within a cell. As TFs (blue ball) join and depart their target binding sites, the associated gene makes transitions to the active and inactive states at respective rates k_{on} and k_{off}. An actively transcribing gene produces fully functional mRNAs (orange) at rate k_m, which are in turn translated into proteins (purple) at rate k_p. Messenger RNAs and proteins are degraded at rates δ_m and δ_p, respectively (small dashes and spheres).

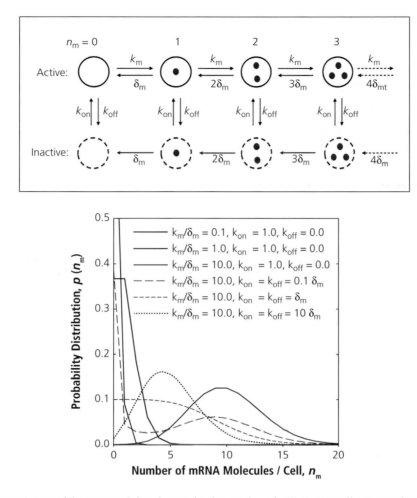

Figure 21.2 Determinants of the statistical distribution for the number of mRNAs per cell associated with a particular gene (n_m). **Above:** Flow chart for the transitions of numbers of mRNAs within individual cells (n_m). The rates are defined as in Figure 21.1. Circles with solid and dashed lines denote transcribing and non-transcribing cells, respectively, with the red dots denoting the numbers of transcripts in cells of various states. Note that cells with inactive genes can only lose (not gain) mRNAs. **Below:** Probability distributions for the steady-state number of mRNAs for a particular gene present in cells, as a function of the ratio of transcription to degradation rates, k_m/δ_m, and the rates of transition of cells from the off to on states, and vice versa, k_{on} and k_{off}, respectively. Solid lines are Poisson distributions, with mean k_m/δ_m for genes that are constitutively on, whereas the black dashed and dotted lines represent situations in which the transition rates to on and off states are equal, with the mode of the distribution shifting to the right with increasing rates of switching. The functions are obtained with Equation 21.1.7.

the time. Averaging over a sufficiently long period, this fractional time is a simple function of the ratio of association and dissociation rates of the TF,

$$P_{on} = k_{on}/(k_{on} + k_{off}). \qquad (21.1)$$

This result arises because under steady-state conditions, $P_{on}k_{off}$ must equal $(1 - P_{on})k_{on}$. The number of mRNAs per cell is then necessarily more complex than Poisson, as it involves a mixture of the distributions of n_m in active and inactive cells (see Foundations 21.1 for the full expression).

Contrary to common belief, gene regulation magnifies the variance of mRNA numbers among cells (Figure 21.2). This occurs because transient switches from active to inactive states result in a heavier weighting towards the categories with small numbers of mRNAs. Indeed, if the rate of switching among active and inactive states is sufficiently slow relative to the rate of degradation of mRNAs, a bimodal distribution can result, with one fraction (the inactive cells) carrying few mRNAs and the remaining active cells having an mRNA-number distribution close to that expected under the constitutive-expression model. Further insight into the likely specific forms of the distribution of mRNA molecules per cell requires quantitative estimates for the four key parameters: k_m, k_{on}, k_{off}, and δ_m.

Based on observed rates of mRNA chain elongation (Chapter 20), and assuming an average transcript length of ~ 1 kb, an activated *E. coli* gene is capable of producing as many as 50–150 transcripts/hour (Golding and Cox 2004; Proshkin et al. 2010). Approximate estimates of k_m for the yeasts *S. cerevisiae* and *S. pombe*, again largely based on chain elongation rates, and assuming an average transcript length of 2 kb, fall in the range of ~ 10 to 35/hour (Lynch 2007a; Zenklusen et al. 2008; Sun et al. 2012; Miguel et al. 2013). For mammalian cells grown in the lab, transcription rates across the genome have a roughly log-normal distribution, with a median of 2–3 mRNAs/hour and an approximate range of 0.1 to 30/hour (Darzacq et al. 2007; Schwanhäusser et al. 2011; Danko et al. 2013). Vertebrate genes typically contain multiple large introns, which are transcribed prior to removal, and this contributes substantially to these reduced rates. However, as the latter rates do not account for the time genes spend in the off state, and a substantial fraction of transcription events abort prior to complete elongation (>90% in mammals; Darzacq et al. 2007), they must underestimate k_m.

Transcript degradation rates are often estimated by inhibiting transcription and following the subsequent decline in mRNA numbers. The half-life of a molecule, $T_{0.5}$, denotes the time required for an initial concentration to decline by 50% and is related to the degradation rate δ by assuming random, exponential decay:

$$0.5 = e^{-\delta T_{0.5}}. \tag{21.2}$$

In *E. coli*, $\sim 80\%$ of mRNAs have half-lives between 3 and 8 mins, with a range of 1 to 15 mins and median ~ 5 mins (Bernstein et al. 2002; Taniguchi et al. 2010). Estimates of median half-lives of mRNAs in *S. cerevisiae* (Wang et al. 2002) and

mouse fibroblast cells (Schwanhäusser et al. 2011) are larger, 22 mins and 9 hours, respectively. Using this expression, degradation rates of 0.14, 0.034, 0.012, and 0.0013 per min are implied for half-lives of 5 mins, 20 mins, 60 mins, and 9 hours.

The rates at which genes turn on and off transcriptionally, k_{on} and k_{off}, dictate the dynamics of gene activation/inactivation. For example, the average time between bursts of transcription at a particular locus, which is equivalent to the mean time that a silent gene remains off, is equal to $1/k_{on}$. Once turned on, a gene remains transcriptionally active for an average interval of $1/k_{off}$, so the average number of transcripts produced during a bout of activity is k_m/k_{off}.

Unfortunately, little is known about the on and off rates, although So et al. (2011) estimate k_{on} to average about 0.003/sec in *E. coli*, with k_{off} often being about two to tenfold lower. Rates of a similar order of magnitude have been observed in mammalian cells (Darzacq et al. 2007). Given that mRNA burst sizes (per engaged gene) are generally in the range of 1 to 20 (Sanchez and Golding 2013), it follows that k_m must typically be on the order of 1–20 × larger than k_{off}, which implies k_m of the same order of magnitude noted above for this species. A primary determinant of the on–off process in highly expressed genes appears to involve alternating periods of engagement and dissociation of gyrase, a molecule used to relieve positive supercoiling of the DNA that results from the progression of the transcription machinery (Chong et al. 2014). In more lowly expressed genes, the stochastic engagement with TFs is likely to be involved.

The preceding survey provides a mechanistic understanding of why the per-cell numbers of mRNA molecules associated with individual genes are generally quite small (Chapter 7). Suppose, for example, that the rate of full mRNA production by an *E. coli* cell in the on state is $k_m \simeq 50$ mRNAs per active gene per hour, with a median degradation rate of $\delta_m \simeq 8$/hour, and that the gene is turned on only a fraction of the time ($P_{on} < 1$). At equilibrium, the average number of mRNAs per cell equals the ratio of the production and elimination rates, $\mu_m = P_{on}k_m/\delta_m$ (Foundations 21.1), so the theory predicts that genes in this species should commonly be represented by fewer than 10 mRNAs per cell. This qualitative prediction is consistent with an observed average number of 5 mRNAs/gene/cell (and a range of 0 to 100) in *E. coli* (Lu et al. 2007; Li and Xie 2011).

Modifications of the theory are necessary for eukaryotes, where there can be a negative feedback between mRNA synthesis and degradation (Sun et al. 2012; Haimovich et al. 2013), and where genes can have much more complex modes of regulation involving multiple TFs with higher-order interactions, long-distance enhancer elements, etc., especially in multicellular species (Hafner and Boettiger 2023). Nonetheless, it remains clear that even in large eukaryotic cells, the numbers of mRNAs per cell can be quite small, with a mean of just 10/gene in *S. cerevisiae* (Lu et al. 2007; Zenklusen et al. 2008), and medians in the vicinity of 20 in mammalian cells (Schwanhäusser et al. 2011; Marinov et al. 2014). In all cases, there is a broad distribution around the mean, with the variance typically exceeding the mean by several fold (Golding et al. 2005; Raj et al. 2006; Taniguchi et al. 2010), as expected for genes that are not constitutively active.

Cellular protein abundances

The ultimate manifestation of transcription is the number of protein molecules per cell. Because protein production depends on the presence of mRNAs, the kinds of transcriptional noise described previously naturally transmits to the level of translation. However, if a large number of proteins are translated per mRNA, the degree of noise propagation can be reduced, owing to the fact that the life span of a protein is typically greater than that of its associated mRNA. For example, in *S. cerevisiae*, most proteins outlive their maternal mRNAs, with the average ratio of half-lives being ~ 3 (Shahrezaei and Swain 2008; Martin-Perez and Villén 2017). Likewise, in mouse fibroblast cells, the median half-life of a protein, ~ 2 days (with a range of 3 to 500 hours), is $\sim 5\times$ greater than that of mRNAs (Schwanhäusser et al. 2011). The latter study also shows that translation rates per mRNA are $100\times$ greater than transcription rates, with a mode of ~ 200 and a range of 1 to 10^4 proteins/mRNA/hour. Thus, the temporal variation in protein numbers per cell is expected to be dampened in comparison to that for mRNAs (Figure 21.3).

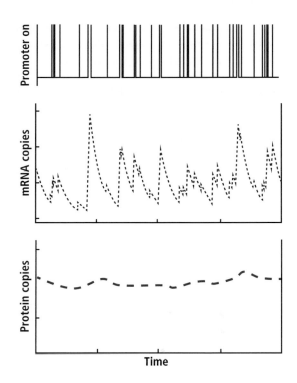

Figure 21.3 Idealized view of the temporal variation in gene expression within a cell. The gene is stochastically turned on (blue vertical bars) or off at points depending on the binding of the cognate TFs. Messenger RNAs are produced during the on periods, but decline at an exponential rate during off periods. Protein numbers also vary within the cell, rising during periods of mRNA abundance, but then declining via degradation during periods of mRNA rarity. However, the fluctuations in protein numbers are damped, owing to their greater longevities than mRNA molecules.

As a consequence of their greater half-lives and higher rates of production, proteins also tend to be much more abundant in cells than their cognate mRNAs (Chapter 7). For example, the average ratios are 450 in *E. coli*, 5100 in *S. cerevisiae*, and 2800 in mammalian fibroblasts (Ghaemmaghami et al. 2003; Lu et al. 2007; Schwanhäusser et al. 2011). Bacterial cells have protein-copy numbers typically ranging from 10 to 20,000 per gene (Ishihama et al. 2008; Malmström et al. 2009; Taniguchi et al. 2010). Yeast proteins fall in the range of 100 to 10^6 copies per cell with a median of ~ 4000 (Ghaemmaghami et al. 2003; Newman et al. 2006; Lu et al. 2007), and mammalian proteins range from 10 to 10^7 copies per cell with a median of 50,000 per expressed gene

(Schwanhäusser et al. 2011). Notably, TFs tend to be the rarest proteins within cells (Ghaemmaghami et al. 2003; Li et al. 2014; Marinov et al. 2014).

Two key determinants of the level of protein production are the number of mRNAs produced by active genes over the typical life span of a protein,

$$a = k_m/\delta_p, \tag{21.3a}$$

and the average number of proteins translated per life span of an mRNA,

$$b = k_p/\delta_m, \tag{21.3b}$$

where in both cases the average life span of a molecule is equal to the inverse of the decay rate. Together with the transcriptional activation and deactivation rates, k_{on} and k_{off}, these parameters define the distribution of protein numbers among cells (Shahrezaei and Swain 2008). The mean number of proteins per cell is a simple extension of the expected value for mRNAs ($\mu_m = P_{on}k_m/\delta_m$, from Foundations 21.1),

$$\mu_p = \frac{\mu_m k_p}{\delta_p} = P_{on} \cdot a \cdot b. \tag{21.4a}$$

The variance in number of protein molecules among cells is described by

$$\sigma_p^2 = \mu_p \left(1 + b + \frac{ab(1 - P_{on})\delta_p}{\delta_p + k_{on} + k_{off}} \right), \tag{21.4b}$$

which reduces to $\sigma_p^2 = \mu_p(1 + b)$ for a constitutively expressed gene (Thattai and van Oudenaarden 2001). Thus, for individual genes, the dispersion of protein numbers among cells is broader than that expected under a Poisson distribution.

These composite expressions define the wide variety of ways in which cells might control their steady-state numbers of active proteins, e.g., by adjusting rates of engagement, elongation, and decay. Several observations are suggestive as to how such alterations in protein expression are actually brought about. For example, Schwanhäusser et al. (2011) noted a strong correlation between the number of protein molecules per cell and the translation rate, and Wang et al. (2002) found that decay rates of mRNAs in yeast are coordinated among protein-coding loci whose products interact stoichiometrically. In *E. coli* and *S. cerevisiae*,

for proteins that assemble in complexes, the relative rates of protein production associated with each locus are directly proportional to the relative numbers of molecules required in each assembly, suggesting coordinated expression so as to maintain stoichiometric balance (Li et al. 2014). Orthologous genes in closely related yeast species often achieve similar levels of overall expression via compensatory changes in rates of transcription and mRNA degradation (Dori-bachash et al. 2011). Such a pattern is compatible with a bivariate drift barrier in which a particular phenotype can be achieved by interchangeable mechanisms, and draws empirical support from studies in *E. coli* jointly modifying regulatory and coding regions (Cisneras et al. 2023).

Expression noise and adaptation

The preceding overview makes clear that phenotypic noise is an inevitable consequence of the structure of biology, resulting from the stochastic features of TF binding and high rates of mRNA decay. Although muted somewhat, noise created at the level of bursty transcription still has cascading effects at the level of translation. There is essentially no way to completely eliminate such between-cell variation, and based on thermodynamic principles, any process for regulating gene expression stochasticity must require an energetic investment. Nonetheless, several authors have suggested that expression noise may be promoted by natural selection as a means for coping with a variable environment (Fraser et al. 2004; Tănase-Nicola and ten Wolde 2008; Wang and Zhang 2011; Levy et al. 2012; Liu et al. 2015; Wolf et al. 2015), operating as a sort of bet-hedging strategy (Chapter 22).

Part of the motivation for this argument is shown in Figure 21.4 – if the expected phenotype (in this case, mean expression level) of a particular genotype is far from the optimum dictated by environmental factors, only genotypes producing sufficiently variable progeny will have some hope of gene transmission to the next generation, as the offspring of more noise-suppressed genotypes will all have near-zero fitness. Note, however, that once the mean phenotype evolves to be in accordance with the environmental optimum, the opposite occurs: all members of noise-suppressed genotypes have

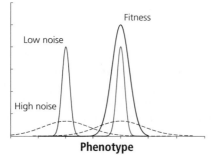

Figure 21.4 Phenotypic distributions for two alternative genotypes with low and high levels of expression noise, given by the solid and dashed lines respectively, relative to the fitness function. The red lines illustrate a situation in which two alternative genotypes with the same average expression level deviate far from the optimum phenotype (denoted by the peak of the black curve). The blue lines denote the situation when the genotypes have mean expression levels coinciding with the optimum.

high fitness, whereas individuals in the tails of the distribution for highly variable genotypes will have near-zero fitness. The point has been demonstrated in an experiment in yeast in which both the mean and variance of expression were altered in various alleles for a key gene (Duveau et al. 2018).

There are, however, multiple reasons for skepticism as to whether selection can modulate expression noise on a gene-by-gene basis. First, it is far from clear whether environmental optima exhibit sufficiently large fluctuations to encourage the evolution of an intermediate level of noise, and it is equally unclear whether environments are typically constant enough (or constantly variable enough) to promote the evolution of noise modulation. Second, any modifier for expression noise would need to be tightly linked to the modulated gene, most likely in the gene body itself. Otherwise any benefit associated with the modifier would be quickly disconnected after a few generations of recombination. Third, as outlined in detail at the end of Chapter 9, selection on the level of phenotypic variance production is a second-order effect and, at best, a very weak evolutionary force, as individual genotypes are not filtered on the basis of their own genetic merits but via the partly non-heritable features of their offspring. Most notably, phenotypic noise actually

reduces the response to selection by diminishing the relationship between individual phenotype and the underlying genotype. Related issues on this particular topic are covered by Matsumoto et al. (2015) and Mineta et al. (2015).

Finally, and perhaps most significantly, gene-expression noise is an intrinsic function of average expression, so any selection on the former, and vice versa, will naturally have cascading effects on both the target locus and other cellular participants. Taking the ratio of the standard deviation in protein number relative to the mean (i.e., the coefficient of variation, or CV), it can be seen from the leading term in Equation 21.4b (which defines the square of the standard deviation), that the CV is roughly inversely proportional to the square root of the mean number of proteins. Lestas et al. (2010) showed more generally that the CV is inversely proportional to the 0.25 to 0.5 power of the number of proteins produced per mRNA. This means that reducing the variation in protein expression by 50% requires roughly a 4–16× increase in the number of proteins relative to mRNA molecules. Thus, noise suppression comes at the expense of added protein production, whereas noise enhancement generally requires a reduction in protein number, which may compromise basic aspects of cell biology.

A related issue is that all of the results introduced here were derived under the assumption of haploidy (Foundations 21.1). Diploidy (common in eukaryotes) might further reduce the level of noise depending upon the degree to which expression is coordinated between the two copies of a gene. If allelic expression is completely independent, as some evidence suggests (Sepúlveda et al. 2016; Skinner et al. 2016), then having two genes expressed simultaneously may give a ∼15% reduction in the level of noise. On the other hand, with their more complex modes of regulation (often involving multiple TFs and/or accessory proteins), eukaryotic genes may commonly exhibit a gradient of transcriptional states (rather than simple on/off) (Corrigan et al. 2016), which further influences the noise process. These matters seem not to have entered the conversation over the many potential factors that might favor diploidy (Chapter 10).

The Basic Biology of Transcription

Gene transcription is carried out by multi-subunit DNA-dependent RNA polymerases (Chapter 20), which we simply call RNAPs. However, such complexes are generally non-autonomous in the sense that one to several accessory proteins, including TFs, must be present simultaneously for transcription activation (Jolma et al. 2015; Haberle and Stark 2018; Cramer 2019). The core promoters upon which the transcription machinery assembles typically reside within 100 or so bp of transcription initiation sites, whereas enhancer elements containing the TFBSs are generally located further upstream (in multicellular species sometimes up to 100,000 bp away; Hafner and Boettiger 2023). Individual TFs often service multiple genes, which facilitates coregulation of gene expression, but specialized one-to-one relationships between TFs and their client genes are not uncommon. For example, seven TFs control the expression of $\sim 50\%$ of regulated genes in *E. coli*, whereas ~ 60 TFs (about one fifth of the TFs in this species) service single genes (Martínez-Antonio and Collado-Vides 2003).

Liaisons between TFs and their target TFBSs on the DNA are usually governed by hydrogen bonds and van der Waals attractions between the two molecules. However, as a consequence of the negatively charged phosphate backbones of the DNA and positively charged residues on the protein, all TFs also inevitably engage in promiscuous interactions with off-target sites. Furthermore, these unavoidable non-specific interactions impose a substantial challenge for any TF, which must avoid too great of a burden of sequestration in nonfunctional locations while retaining a high enough affinity to its own specific binding sites.

Eukaryotic transcription raises additional issues in that the chromosomes are regularly wrapped around nucleosomes, formed from histones, and often further packed into higher-order structures. On the one hand, such structures can reduce the accessibility of a TF to a hidden TFBS, but the occlusion of TFs from non-regulatory DNA can also reduce the time spent on non-productive searching (Charoensawan et al. 2012; Thurman et al. 2012). In addition, some proteins such as cohesins, which encircle sister chromosomes during cell division

(Chapter 10), help recruit TFs to localized regions (Yan et al. 2013).

Many dozens of TF families exist across the Tree of Life, each structurally reliant on different DNA-binding domains. However, although each TF has maximum affinity for a specific DNA motif, there is no general regulatory code in TFs, i.e., no specific language involving one-to-one recognition matching between the amino-acid sequence of a TF and the nucleotide sequence of its binding site. Typically, 10–50 amino-acid residues in the TF are involved in contacts with the DNA, whereas TFBS motifs generally consist of 6–30 nucleotides, usually near the lower end, especially in eukaryotes (Luscombe and Thornton 2002).

A physical model for TF binding

The universal mode of transcription, involving the interaction of a specific protein (the TF) with a specific DNA binding site (the TFBS), provides a compelling platform for developing an evolutionary theory of gene expression couched in terms of the biophysics of intermolecular associations (Bintu et al. 2005). However, because many TFs are commonly present in fewer than a few dozen copies per cell, an understanding of transcription requires insight into the consequences of stochastic aspects of single-cell biology (as opposed to measuring just the average features of entire populations). Thus, to move forwards, we require a probabilistic framework for understanding the likelihood of TFs being bound to their specific TFBS targets in individual cells.

Consider a TF that recognizes an optimal binding motif containing ℓ key nucleotide sites. Empirical data from a variety of sources suggest that the average energetic cost of a single base mismatch is $\simeq 2$ in Boltzmann units of $k_B T$ (which $\simeq 0.6$ kcal/mol at most biological temperatures) (Table 21.1). Thus, under the assumption that binding strength scales linearly with the degree of correspondence between a TFBS and the optimal binding motif of its TF, the relevant phenotype from the perspective of binding efficiency of a target site can be viewed as the number of matches with the optimal recognition sequence ($m \leq \ell$). Numerous empirical studies support the additivity assumption as a first-order

Table 21.1 Features of the motifs of well-studied transcription factors. Motif sizes are based on consensus sequences. The estimated costs of mismatches are obtained from binding strength experiments in which single-base changes were made in motifs. Costs of single-base mismatches are in units of kcal/mol associated with background thermal motion; these average to 1.4 across the full set of studies, or in terms of Boltzmann units ($k_B T \simeq 0.6$ kcal/mol) to 2.3.

TF	Species	Motif (bp)	Cost of Mismatch Mean (Range)	References
CI	Lambda phage	17	1.4 (0.5–3.5)	Sarai and Takeda (1989)
Cro	Lambda phage	9	1.4 (0.5–2.5)	Takeda et al. 1989
Mnt	*Salmonella* phage P22	21	1.0 (0.3–1.6)	Fields et al. (1997); Berggrun and Sauer (2001)
CRP	*Escherichia coli*	22	1.7 (0.9–2.5)	Gunasekera et al. (1992); Kinney et al. (2010)
CRP	*Synechocystis* sp.	22	1.8 (0.7–3.0)	Omagari et al. (2004)
ArcA	*Shewanella oneidensis*	15	1.3 (0.1–3.4)	Schildbach et al. (1999); Wang et al. (2008)
Gcn4	*Saccharomyces cerevisiae*	11	1.0 (0.5–1.7)	Nutiu et al. (2011)
c-Myb	*Homo sapiens*	6	1.6 (0.6–2.8)	Oda et al. (1998)

approximation (von Hippel and Berg 1986; Sarai and Takeda 1989; Takeda et al. 1989; Fields et al. 1997; Shultzaberger et al. 2010), although higher-order effects involving the shape of TFBSs can also contribute to the overall binding energy (Yang et al. 2014; Le et al. 2018).

Given a potential TFBS sequence with a particular binding energy for a specific TF, we wish to know the probability of occupancy by the cognate TF, P_{on}, as this is a minimal requirement for expression of the associated gene. Recall from Equation 21.1 that P_{on} can be expressed in terms of association/dissociation rates. Here we take a related but more mechanistic approach, treating P_{on} as a function of both the binding site and the features of the intracellular environment that restrict the site's access to cognate TFs. Clearly, P_{on} will increase with the number of TF molecules in the cell (N_{tf}), but equally important are the ways in which individual TF molecules can become sidetracked by binding to alternative genomic sites. Other genes serviced by the TF (numbering N_{ot}, where 'ot' denotes off-target) will compete for the pool of TFs, but non-specific binding of TFs across the genome can be numerically more important.

Letting G denote the total genome size (in bp), because $\ell \ll G$, there are essentially G such non-specific sites in a haploid cell (with varying degrees of affinity). Letting the excess scaled binding energy of a target TFBS be $2m$, from Foundations 21.2 the probability that a specific target TFBS is occupied by its TF is

$$P_{on} \simeq \frac{1}{1 + B e^{-2m}}, \qquad (21.5)$$

where $B = G/(N_{tf} - N_{ot})$ is a measure of the concentration of background (non-specific) binding sites relative to the number of TF molecules available for the specific target site. As $B \to \infty$, $P_{on} \to 0$, whereas as $m \to \infty$, $P_{on} \to 1$.

A rough idea of the magnitude of B can be inferred by noting that G is generally in the range of 10^6 to 10^{10} bp, with prokaryotes falling at the lower end and multicellular eukaryotes at the higher end of the range (Lynch 2007a). For the model bacterium *E. coli*, the numbers of molecules per cell for particular TFs, N_{tf}, are often in the range of 100 to 1000, with just a few cases ranging as high as 50,000 (Robison et al. 1998). Somewhat lower numbers are estimated for another bacterium *Leptospira interrogans* (Malmström et al. 2008). In such species, it is unusual for the number of genes serviced by a particular TF, to exceed 100. Thus, taken together, these observations suggest a range for B on the order of 10^3 to 10^6 for prokaryotes. For the yeast *S. cerevisiae*, proteomic data suggest that the average number of molecules for individual TFs is on the order of 8000 per cell (Ghaemmaghami et al. 2003), so with a genome size of 12 Mb, B should be in the vicinity of 10^3 to 10^4. Proteins within mammalian cells appear to be about

10× as numerous as those in yeast (Schwanhäusser et al. 2011), but with a genome size of ~ 3000 Mb, B can be expected to be $\gg 10^3$.

All of these estimates of background interference assume that the primary mechanism reducing TF accessibility is non-specific binding on DNA. If other sources of interference exist (such as promiscuous binding to other proteins), B would be accordingly higher. On the other hand, DNA-binding proteins, such as histones in eukaryotes, could reduce B by restricting access of a TF to only a fraction of the genome. Thought of in a more general way, the composite parameter B is a composite measure of the totality of cellular features working against the binding of a TF to a specific cognate TFBS.

Equation 21.5 provides insight into the conditions necessary for a high probability of binding. For example, $m = 0.5 \ln(B)$ represents a key pivot point below which background interference results in $P_{on} < 0.5$. If the binding probability is to exceed 0.99, the number of matches (m) must exceed 6 for $B = 10^3$ and 11 for $B = 10^7$ (Figure 21.5). Thus, unless the level of background interference greatly exceeds $B = 10^7$, there is little to be gained in terms of binding probability for a motif in excess of a dozen bases. This means that a considerable amount of mismatching can be tolerated for a TFBS motif more than a dozen nucleotides in length.

The results in Figure 21.5 highlight two physical constraints on the basic process of transcription regulation by the binding of TFs to DNA. First, because mismatches come in approximately discrete packets (with relative binding energy $\simeq 2$/site), the opportunities for fine-tuning gene expression by altering the numbers of mismatches in a TFBS may be limited, although variation around this expectation (e.g., from not all mismatches having exactly the same consequences) will provide some flexibility. Modulation of gene expression might also be accomplished by altering the numbers of TFs residing inside cells (which will influence B). However, a secondary consequence of altering the concentration of a TF is that different client genes will also be affected.

Second, life's transcription mechanism comes at a significant energetic cost, in that to ensure that a particular gene is turned on, a substantial excess number of TF molecules must be produced

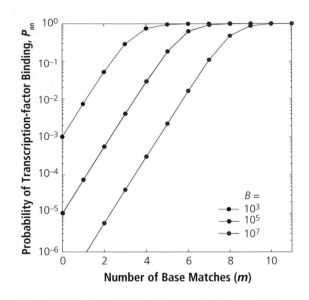

Figure 21.5 The probability that a particular TF binding site (TFBS) is bound by a cognate TF, given the level of background interference (B) and the number of nucleotides at the site (m) matching the optimal recognition sequence of the TF. The curves, obtained from Equation 21.5, cover the range of biologically plausible values of B (as described in the text).

to compensate for the unproductive engagements occurring at non-specific sites. For example, rearrangement of Equation 21.5 shows that for a TFBS motif with $m = 8$ matching bases to achieve a 0.9 probability of being bound to its cognate TF, 10 and 1000 TF molecules are required in cells with genome sizes of 10^7 and 10^9 bp, respectively, and $P_{on} = 0.99$ elevates these numbers to ~ 110 and 11,000. Thus, an unavoidable consequence of biology's mode of gene expression is that far more TF proteins must be produced than the numbers of genes to be serviced. This cost of living with a system that relies on mRNA production for gene expression necessarily increases in eukaryotic cells with larger genomes.

Encounter rates between TFs and their binding sites

The preceding analyses implicitly assume that the distribution of TFs within a cell is typically in a dynamic steady state of bound and unbound molecules. At first glance, the chances of a TF locating a specific cognate TFBS in a reasonable amount of time would seem to be daunting. However, as outlined in Foundations 21.3, the biophysical properties of cells are such that localization can generally be achieved in a matter of a few seconds or less in bacterial cells.

Despite their passive transport, TFs locate their target sites at rates exceeding the three-dimensional diffusion limit (Riggs et al. 1970), an observation that motivated the facilitated-diffusion model (von Hippel and Berg 1989). Given the minute sizes of individual TFBSs, a newly arisen TF molecule will essentially always first encounter a non-specific site on a chromosome before locating a proper, more energetically favorable target. The search process involves repeated association–dissociation events involving one-dimensional sliding along DNA molecules interspersed with three-dimensional jumping to new locations. During such episodes of intersegmental transfer, TF molecules are kept in the vicinity of the DNA, thereby avoiding the much larger and unproductive search space of the entire cytoplasm/nucleoplasm. Such three-dimensional wandering also minimizes the redundant interrogation of localized chromosomal space that would occur if a non-directed one-dimensional diffusion process followed first contact. Finally, it appears that the search for appropriate DNA binding sites is facilitated by protein–protein interactions within clusters of TFs, not just by direct DNA binding processes (Brodsky et al. 2020). Reviews covering many of the technical issues can be found in Gowers et al. (2005), Halford and Marko (2004), Halford (2009), Kolomeisky (2011), Zhou et al. (2011), Normanno et al. (2012), and Staller (2022).

The extent to which various species alter the locations of their chromosomal DNA to further assist in the search process remains unclear. However, the spatial issues incurred by the large cells of eukaryotes are of particular interest. The volumes of the nuclei of eukaryotic cells are typically larger than entire prokaryotic cells, and this can result in mean search times of individual TF molecules for a target TFBS of 1–200 minutes within the nuclear environment alone (Foundations 21.3). Although the overall search process can be sped up by producing more TF molecules, there is the additional issue of the cytoplasmic cell volume (within which TFs arise by translation), which is commonly 10–100× that of the nucleus. All other things being equal, this would result in an increase in the search time by 10–100× were the genome not concentrated within a nuclear envelope. Thus, although a number of hypotheses have been proposed for the evolution of the nuclear envelope and its relevance to the expanded sizes of eukaryotic cells (Chapter 15), the challenges of gene expression should be included in this list. The rate of gene expression in large cells might be extremely compromised if the genome were not confined to the restricted space of the nucleus.

Coevolution of Transcription Factors and Their Binding Sites

To be expressed, essentially every gene in every genome requires interaction with at least one TF. This implies that the TF mode of gene regulation must have been present in LUCA. Nonetheless, because many TFs service multiple genes, a fairly small fraction of most genomes is allocated to TF production, typically 1–5% of the protein-coding genes within a genome. Among prokaryotic species, the number of TF genes ranges from ∼5 to 500, scaling quadratically with the total number of protein-coding genes. Eukaryotic genomes generally encode for at least 100 TFs, with well over 1000 being harbored in multicellular species, and the scaling with total gene number being closer to linear (Riechmann et al. 2000; van Nimwegen 2003; Aravind et al. 2005; Iyer et al. 2008; Charoensawan et al. 2010).

Across the Tree of Life, many dozens of TF families have been identified based on the unique physical structures of their DNA-binding domains. However, changes in the regulatory vocabulary and the reading machinery have evolved on various time scales. Only 2% of specific DNA-binding domain families are shared across bacteria, archaea, and eukaryotes, and no clear TF orthologs are known across these three superkingdoms (Charoensawan et al. 2010). Dramatic differences appear among the major eukaryotic lineages as well (Riechmann et al. 2000). These observations alone suggest a substantial turnover in the specific TFs used in various lineages, a pattern that repeats itself at lower levels of phylogenetic organization (discussed later).

There has been much speculation, especially among those doing comparative developmental biology in animals, that eukaryotic morphological diversity has been driven by the exploitation of novel TF families and their recruitment to specific sets of genes. However, although there is no question that developmental evolution must involve modifications in gene regulation, it does not follow that the origin of multicellular complexity is an inevitable outcome of transcriptional complexity. As noted above, eukaryotes do not generally invest proportionately more in their TF repertoires at the genomic level than do prokaryotes. Moreover, many of the key TFs deployed in complex development are present in the unicellular relatives of animals and land plants (de Mendoza et al. 2013; Richter et al. 2018).

Another common argument is that most changes in gene regulation are a consequence of alterations in the *cis*-regulatory logic residing upstream of genes rather than a result of modifications in the agents of transcription, with some going so far as to claim that *cis*-regulatory modifications are the units of evolutionary change (Carroll et al. 2001; Davidson 2001; Wray 2007). The usual logic underlying this assertion is that, because individual TFs often service multiple genes, alterations of binding-site specificities of TFs are likely to have large-scale, negative pleiotropic consequences for fitness. Under this view, a change in expression pattern with minimal pleiotropic effects on multiple traits can only be achieved by recruiting, modifying, or eliminating TFBSs on a gene-by-gene basis. However, not all mutations arising in a gene with pleiotropic effects need themselves be pleiotropic, and as discussed further below, considerable evidence suggests that functional changes in TFs often have minimal side effects (Hsia and McGinnis 2003; Lynch and Wagner 2008; Wagner and Lynch 2008). Moreover, the target sizes for *trans*-acting mutations can be hundreds of times larger than those for *cis*-acting mutations (Gruber et al. 2012; Metzger et al. 2016), meaning that by sheer numerical dominance, such mutations can be quantitatively important.

General observations

Although high-throughput methodologies for genome-wide identification of TFs and their corresponding TFBSs promise to substantially expand our understanding of how such systems diversify (e.g., Berger and Bulyk 2009; Carey et al. 2012; Furey 2012; Ding et al. 2013; Smith et al. 2013; Levo and Segal 2014; Hill et al. 2021), most current insight into the mechanisms of gene-regulatory evolution still derives from observations from the usual key model systems – the bacterium *E. coli*, the yeast *S. cerevisiae*, the fly *D. melanogaster*, mouse, and human. From this limited set of taxa, several generalizations have started to emerge.

First, prokaryotes typically harbor substantially longer consensus TFBSs than do eukaryotes (Stewart et al. 2012). Moreover, unlike many eukaryotic TFBSs, prokaryotic binding sites are often palindromic in nature, with each half sequence being 7–11 bases in length and recognized by one of the two members of a homodimeric TF.

Second, the evolutionary features of TFs appear to depend on the number of host genes serviced. For example, from comparisons of multiple gammaproteobacteria, Rajewsky et al. (2002) found that TFs with larger numbers of target genes are more evolutionarily conserved at the amino-acid sequence level and with respect to the TFBS-recognition sequence. Nevertheless, Sengupta et al. (2002) observed a decline in binding-site specificity with increasing numbers of genes serviced by a TF in both *E. coli* and yeast. In principle, the latter condition may evolve so as to minimize the mutational burden on an organism, as a large number of TFBSs increases the overall mutational target size. However, an alternative explanation is that TFs with low specificity are recruited more frequently into various regulatory pathways over evolutionary time.

Third, in eukaryotes, multiple motifs for a particular TF are frequently present in the upstream regions of client genes (e.g., Gotea et al. 2010). Although it is commonly argued that such redundancy is maintained by natural selection, TFBS clustering can also arise naturally by small-scale duplication processes (Lusk and Eisen 2010; Nourmohammad and Lässig 2011). Thus, while the presence of multiple binding sites might help ensure that an adjacent gene will be activated, there is as yet no formal evidence that such configurations are anything more than a simple consequence of physical processes.

Evolutionary distributions of binding site motifs

TFs and their binding sites provide an explicit framework for evolutionary analysis, in that specific DNA-level features can be directly related to fitness (Gerland and Hwa 2002; Berg et al. 2004; Lässig 2007; Stewart et al. 2012; Lynch and Hagner 2015; Tuğrul et al. 2015). A common approach to understanding the evolution of binding motifs is to consider individual fitness to be a linear function of the fraction of time that a TFBS with m matching sites is expected to be bound by its cognate TF, e.g.,

$$W(m) = 1 + \alpha P_{on}(m), \qquad (21.6)$$

where α is a scaling factor relating binding probability to fitness, and $P_{on}(m)$ is defined in Equation 21.5. As $\alpha \rightarrow 0$, $W(m) \rightarrow 1$, implying neutrality. Equation 21.6 is often referred to as a mesa fitness function because fitness increases asymptotically from 1 to $(1+\alpha)$ as the probability of gene activation increases from 0 to 1.

Alternative models relating m to fitness are certainly possible, but once $W(m)$ is defined, and additional information is available on rates of mutational movement between alternative TFBS states, a number of basic issues regarding TFBS evolution can be examined using the methods outlined in Foundations 21.4. For example, it is well known that the genomic set of binding sites associated with a particular TF often exhibit variable motif sequences. Although such variation might partially result from selection for alternative levels of locus-specific gene expression, because of the diminishing-returns nature of the fitness function (Figure 21.5), variation in motif matching is also expected to arise naturally as selection pushes a population towards the drift barrier, where alternative high-m states are selectively equivalent (Berg and von Hippel 1987).

Over evolutionary time, the frequency distribution of the number of matches in the various TFBSs serviced by a particular TF is expected to reach an equilibrium between the mutational forces causing mismatches and the selective forces favoring mutant alleles with higher specificity. As always, the efficiency of selection is modulated by the power of genetic drift, which is inversely proportional to the effective population size (N_e). From Foundations 21.4, provided all nucleotides mutate to all others at equal rates, the equilibrium distribution takes on a simple form,

$$\widetilde{P}(m) = C\left[\binom{\ell}{m} 3^{\ell-m}\right] e^{2N_e W(m)}, \qquad (21.7)$$

where C is simply a normalization constant that ensures that the full set of probabilities, $\widetilde{P}(m)$, sum to one.

The equilibrium distribution $\widetilde{P}(m)$ can be viewed as either the long-term average probability of states at a particular TFBS as it wanders through evolutionary time, or as the expected distribution of m for a full set of equivalent TFBSs (for different genes within a particular host genome) at any one point in time. The exponential term in Equation 21.7 is a constant when $W(m)$ is invariant, and so the term within brackets is equivalent to the neutral distribution expected in the absence of selection, $\widetilde{P}_n(m)$. Thus, the evolutionary distribution of binding-site matching is equal to the neutral expectation weighted by an exponential gradient of the fitness surface relative to the power of random genetic drift, i.e., $W(m)$ divided by $1/(2N_e)$.

Solution of Equation 21.7 illustrates several general principles (Figure 21.6). First, the equilibrium distribution is completely independent of the mutation rate. The factor of three enters because it is assumed that there are three ways for a matching nucleotide to mutate to a mismatch but only one way for a reversion to arise. Because the former is simply a multiplicative function of the latter, the actual mutation rate cancels out.

Second, regardless of the set of parameter values, substantial variation in m is almost always expected among sites. Unless the motif size is small (e.g., $\ell = 8$) and levels of background interference and selection pressures are very high, most motifs are expected to contain mismatches. This behavior arises because the alternative states in the upper range of m are selectively equivalent with respect

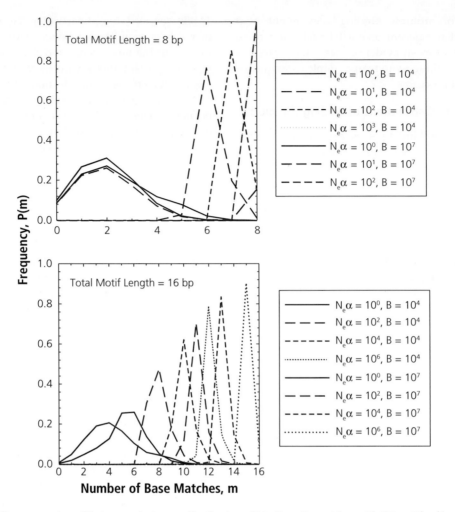

Figure 21.6 The expected equilibrium evolutionary distribution of binding site matches with TF motifs of lengths $\ell = 8$ and 16. Results are given for various levels of the strength of selection relative to the power of genetic drift ($N_e\alpha$), and two levels of background interference (B).

to each other owing to the plateau of $P_{on}(m)$ at high m. Indeed, with a motif size of 16 bp, essentially no TFBS is expected to be perfect, unless the power of selection is unrealistically high ($N_e\alpha > 10^6$). The exact form of $\widetilde{P}(m)$ will vary with different forms of the fitness function, $W(m)$, but provided the upper end of $W(m)$ becomes progressively flatter, the drift barrier combined with the multiplicity of sequences with identical matching levels will encourage substantial motif variation. Thus, the theory provides an explicit nonadaptive explanation for the high level of interspecific divergence in TFBS motifs

routinely seen in comparative studies as well as for the substantial variation in motif sequences at the intraspecific level (Zheng et al. 2011; Heinz et al. 2013).

Third, with relatively weak selection pressure ($N_e\alpha \ll 1$), $\widetilde{P}(m)$ is very heavily skewed towards small (but non-zero) numbers of matches (essentially the neutral expectation). This intrinsic weighting towards low numbers of matches is due to both biased mutation pressure and the high multiplicity of configurations leading to the same m with increasing numbers of mismatches.

Fourth, because the neutral distribution is heavily weighted toward low m, there can be a sharp 'phase transition' as $N_e\alpha$ crosses the threshold value of ~ 1.0. Notably, cases can even exist in which $\widetilde{P}(m)$ is bimodal, with a peak to the left resulting from the high multiplicity of motif configurations driven by mutation and a peak to the right driven by selection pressure. As the motifs within the different peaks of such distributions will deviate in both length and sequence, this result may help explain the widespread use of secondary TFBS motifs by TFs (noted earlier).

Fifth, although the preceding results have been derived for an interaction in which the TF is evolutionarily invariant (e.g., due to pleiotropic constraints associated with its use with other genetic substrates), when the TF coevolves with its TFBS, the overall results noted earlier largely remain except that the average equilibrium degree of matching declines (Lynch and Hagner 2015). This results because mutations in both the TF and the TFBS present a constant stream of changes in each other's selective landscape, in effect preventing strong specialization. A side consequence of this behavior is that, when both components of a TF/TFBS system are free to evolve, the underlying recognition motif is free to explore all of sequence space, conditional on the constraint of maintaining an adequate degree of matching at all points of time. This results in the origin of incompatibilities of TF/TFBS pairs in different phylogenetic lineages as the two systems drift apart to the extent that they no longer recognize each other in heterospecific combinations.

Sixth, in the case of a one-to-many scenario in which the TF interacts with multiple TFBSs, there can be a substantial degree of asymmetry in the rates of evolution of the two components (Lynch and Hagner 2015). Owing to its need to satisfy multiple partners, the TF experiences the strongest selective constraints, with the overall rate of evolutionary divergence declining with increasing numbers of partners. In effect, the master controlling element is no longer free to coevolve with single interacting partners, becoming increasingly constrained to accept only the small subset of mutations that is either effectively neutral for all partners, or the even-smaller subset with a net overall positive impact. In contrast, the TFBSs themselves continue to evolve in an essentially independent fashion, with distribution and rate features identical to what would be expected in a highly specific system, e.g., Equation 21.7. These results appear to be consistent with the observations, noted above, that TFs with larger numbers of target genes are more evolutionarily conserved at the amino-acid sequence level and evolve lower levels of binding-site specificities.

Finally, the theory helps clarify why TF motifs are typically so small. Owing to the saturating binding potential embodied in Equation 21.5, even the strongest levels of selection are unlikely to lead to mean binding-motif lengths in excess of 12 bp. Although an overly short TF-recognition motif may lead to excessive spurious binding to off-target sites, the challenges here are not too severe. Assuming equal nucleotide usage, the expected number of appearances of any particular sequence of length ℓ in a genome containing G bp is $G(1/4)^{\ell}$. Setting this equal to one, and rearranging, we find that less than one random motif is expected to be present by chance on each strand if the motif size exceeds $\ell^* = \ln(G)/\ln(4)$. For genome sizes of 10–1000 Mb, $\ell^* = 12$ to 15 bp. Combined, these two points provide a simple explanation for why TFBSs are generally shorter than 15 bp in length.

Although Stewart et al. (2012) have argued that TFBS evolution reflects an inherent trade-off between specificity to enhance the stability of gene expression (which increases with matching motif length) and robustness to mutational breakdown (which decreases with increasing length), it is clear that less than maximum matching lengths arise naturally as a consequence of mutation–selection balance. Direct selection for mutational robustness, a second-order effect, need not be invoked. In addition, all of these results indicate that, without direct empirical validation, the presence of motif variation in genomes, at both the intraspecific and interspecific levels, should not be taken as evidence of adaptive fine-tuning of individual loci.

Application of the models

The general model embodied in Equation 21.7 can be used for more than simply predicting the features of TF/TFBS systems. Assuming that the set of TFBSs in a particular species has evolved to its equilibrium distribution of motifs, one can compare the observed distribution of usage to the neutral expectation to estimate the strength of selection on functional binding sites, $2N_e s(m)$, necessary to account for the deviations between the two. This follows by rearranging Equation 21.7 to

$$W(m) = \left(\frac{1}{2N_e}\right) \ln \left(\frac{\widetilde{P}(m)}{\widetilde{P}_n(m)}\right). \qquad (21.8)$$

where $\widetilde{P}_n(m)$ is expected distribution under neutrality, which can be obtained from random nucleotide motifs exclusive of known binding sites, i.e., as a fraction of random genomic stretches of length ℓ containing m matches to the optimal motif. Note that this sort of application makes no assumptions about the form of the fitness function, and instead relies on the data to infer $W(m)$.

Mustonen and Lässig (2005) performed such an analysis for the cAMP receptor protein (CRP) in *E. coli*, showing that $2N_e s(m)$ for known TFBS sites for this factor is often in the range of 5 to 10, with the strength of negative selection declining monotonically with increasing binding affinity (Figure 21.7). An analysis of Abf1 binding sites in yeast yielded similar results (Mustonen et al. 2008), and an analysis of 12 additional TFs revealed a positive relationship between fitness and binding energy in each case (Haldane et al. 2014). Thus, all existing analyses appear to support the use of a fitness model that assumes a positive association with binding affinity, as in Equation 21.7. This type of analysis also harbors substantial potential for TFBS discovery in a genome using thermodynamic principles rather than consensus sequence motifs (Djordjevic et al. 2003; Mustonen and Lässig 2005; Lässig 2007; Mustonen et al. 2008).

As noted, $\widetilde{P}(m)$ is best described as a quasi-equilibrium, in that each individual motif is expected to wander across the entire distribution over evolutionary time, as described in Equation 21.4.1, with the entire ensemble of motifs retaining the steady-state pattern. This general principle leads to the prediction that, if the model is correct, and individual motifs are not being kept in their specific states by locus-specific selective

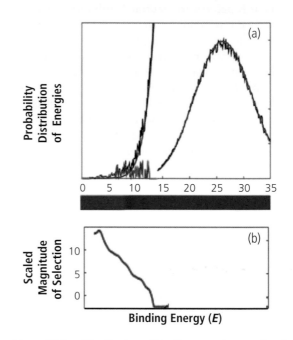

Figure 21.7 a: Distribution of binding energies associated with the transcription factor CRP in *E. coli*. Note that contrary to the approach in the text, the binding sites are characterized with respect to energy rather than mismatches, although the two scales are entirely interchangeable. Energies are computed using sliding windows of 22-bp (the length of the consensus TFBS for CRP) sequences across the entire *E. coli* genome. The energy scale is set such that $E = 0$ denotes the strongest possible binding site, with all other (more weakly binding) motif sequences simply being measured as the deviation from this value (and appearing further towards the right). The rapidly rising left curve is the tail of the remainder of the energy distribution (blue bell-shaped curve to the right) multiplied by 30 to enhance visualization. The solid lines illustrate the expected distribution based on the full set of possible 22-bp sequences under a random model using the known distribution of nucleotide types in the *E. coli* genome; these fit very well in the right portion of the distribution, which represents non-specific binding sites. The red line is the excess of motifs in the left tail from this neutral expectation. Motifs in the red region are viewed as true binding sites, whereas all others denote the background resulting from non-specific binding. **b:** As discussed in the text, for TFBS motifs deemed to be functional, the logarithm of the ratio of observed abundance relative to that expected under neutrality (the red line), $\widetilde{P}/\widetilde{P}_n$, provides an estimate of $2N_e s$, which is equivalent to the selective advantage of each site relative to the power of drift. From Mustonen and Lässig (2005).

pressures, comparison of the differences in binding energies among orthologous sites in different species should yield variances in motif binding consistent with the diffusion model. Observations on the Abf1 TF in four species of *Saccharomyces* are consistent with these expectations (Mustonen et al. 2008). Thus, consistent with theory, the specific sequences of functional TFBSs appear to be conserved only to the extent that they yield levels of matching consistent with the relevant domain of drift–mutation–selection equilibrium. Due to the multiplicity of binding site configurations deviating from the optimum, there is room for substantial sequence change via compensatory mutations.

Evolution of Pathway Architecture

Despite the centrality of TFBSs to gene expression and the common conservation of motifs between distant lineages (Nitta et al. 2015), consistent with the theory just noted, a diverse set of observations indicates that TFBS locations and motif sequences can vary dramatically among closely related lineages, often with no apparent phenotypic consequences (Borneman et al. 2007; Doniger and Fay 2007; Dowell 2010). However, these sorts of changes are not always simply due to random wandering of binding-site sequences, but to functional changes in the TFs themselves. For example, Nakagawa et al. (2013) found that the sequence specificities of members of the forkhead family of TFs have changed over time in the eukaryotic tree, with some evolving bi-specificity (i.e., using two different motifs), and others subsequently losing the ancestral specificity.

One of the most thoroughly analysed metazoan promoters is that for the Endo16 gene in the sea urchin *Strongylocentrotus purpuratus*, which is bound by seven different TFs and forms the heart of a complex developmental cascade (Yuh et al. 1998). The regulatory pathways associated with this gene were revealed through several years of study using multiple individuals from a diverse natural population, but this lack of genetic background control likely influenced the generality of the results. For example, it was subsequently determined that the TFBSs for Endo16 and for other regulatory genes in *S. purpuratus* harbor as much (and in some cases

more) within-species sequence variation as surrounding, presumably non-functional, nucleotides (Balhoff and Wray 2005; Garfield et al. 2012). Moreover, although the expression patterns of Endo16 appear to be conserved between different sea-urchin genera, there is virtually no similarity between the regulatory regions (Romano and Wray 2003). Similar kinds of observations have been made on the regulatory regions of developmental genes in different ascidian species (Oda-Ishii et al. 2005).

Additional examples of apparent stability of gene expression across species with little apparent regulatory-region sequence continuity have been noted in the congeneric nematodes *C. elegans* and *C. briggsae* (Barrière et al. 2011, 2012; Reece-Hoyes et al. 2013). Likewise, multiple studies on developmental genes in *Drosophila* indicate up to 5% turnover of TFBSs among closely related species (Moses et al. 2006; Crocker et al. 2008; Hare et al. 2008; He et al. 2011), again with conservation of gene expression patterns being maintained despite the underlying changes in the regulatory regions (Ludwig et al. 2011; Paris et al. 2013).

What remains unclear is whether the observed regulatory sequence changes in these studies are accompanied by modifications in the DNA-binding domains of the associated TFs. There are, however, clear examples of TF-associated changes in vertebrates. For example, Yokoyama and Pollock (2012) found that a single amino-acid change in transcription factor SP1, which occurred independently in birds and mammals, is associated with orchestrated TFBS motif changes in hundreds of genes in each lineage. Moreover, across the different orders of mammals, which diverged ~ 100 million years ago, at least a third of TFBSs appear not to be shared (Dermitzakis and Clark 2002; Schmidt et al. 2010; Yokoyama et al. 2011). These changes involve alterations in both the TFs utilized in gene expression and the motifs to which they bind. Substantial changes in the TFs bound to the regulatory regions of orthologous genes have even been observed in closely related mouse species (Stefflova et al. 2013). Small changes in TF amino-acid sequence are also known to be associated with changes in binding-site specificities in plants (Sayou et al. 2014).

As in the previous examples with invertebrates, the changes in vertebrate regulatory mechanisms again appear to often occur without noticeable effects on patterns of gene expression. For example, Fisher et al. (2006) found that the control region for the human RET-receptor kinase gene drives expression within zebrafish even though there is no obvious sequence similarity. Wilson et al. (2008) found that when human chromosome 11 is put into mouse cells, the pattern of transcription is very similar to that in human cells. However, in contrast to the situation for messenger RNAs for protien-coding genes, neither species background is sufficient to drive expression of the ribosomal RNA genes from the other, a phenomenon known as nucleolar dominance (Arnheim 1986).

Taken together, the preceding observations suggest the common existence of evolutionary pathways whereby the underlying mechanisms of gene regulation can be altered with no apparent modification in the outward phenotype. Such regulatory repatterning provides further evidence for evolution at the cellular level by effectively neutral mechanisms, a topic we revisit in the final section of the chapter.

We now move on to higher-order issues, in particular, the wide diversity of topological structures of gene-regulation pathways, much of which is unexplained from an evolutionary perspective. There is much to be considered here, including the numbers and types of steps in regulatory pathways, branching patterns, and the degree to which both pathway topologies and the individual participants remain constant over evolutionary time. As expected, considerable attention has been given to the idea that regulatory pathways are optimally structured to yield particular performance levels, response times, and stability. However, these conclusions are often reached from the starting assumption of an all-powerful hand of natural selection. We argue here that, as with the coevolution of TFs and their binding sites, numerous features of pathway-structure evolution are seemingly guided by nonadaptive mechanisms.

Activators versus suppressors

Before considering the higher-order architecture of pathways, it is essential to note that, although all of the preceding discussion has focused on gene activation by TFs, in their simplest form TFs can operate as activators or repressors (Figure 21.8). In the former case, via signal transduction (Chapter 22) the gene for the TF is activated, and the TF then activates the gene of interest. This mechanism of double-positive (++) control ensures that the gene is only active in the face of appropriate demand. However, the same end result can be obtained with double-negative (−−) control, whereby the TF operates as a repressor of transcription until it is released upon receiving an appropriate signal for gene-usage demand.

In a broad study of the regulation of *E. coli* genes, Savageau (1974, 1977, 1998) found that genes whose products are needed continuously tend to be subject to ++ regulation, whereas those that are only sporadically needed are generally under −− regulation. To explain this pattern, he proposed a 'use it or lose it' hypothesis. The simple basis of this idea is that proteins not carrying out a function are subject to the neutral accumulation of degenerative mutations during such periods of activity. Under this view, an activator TF that is rarely used will be subject to a high rate of pseudogenization, whereas a repressor TF used in this context will only rarely be subject to deleterious-mutation accumulation. In contrast, for a gene whose products are in high and frequent demand, a repressor TF would be unutilized most of the time, and hence, subject to degradation. Thus,

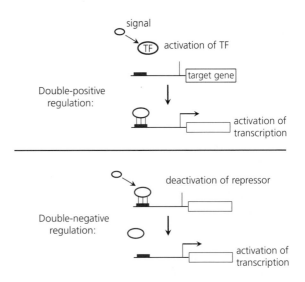

Figure 21.8 Gene expression regulation by an activator versus repressor TF.

under this hypothesis, there is a selective premium on the mode of regulation that involves a regulatory protein that is kept at the highest level of utilization. Supplementing this mutation-load argument is the idea that bound TFs reduce the likelihood of inadvertent transcription owing to non-specific binding by other TFs, which amounts to an error-minimization scenario (Shinar et al. 2006).

As pointed out by Gerland and Hwa (2009), the validity of these genetic load arguments depends on the population-genetic environment and the timescale of environmental shifts. They argue that the 'use it or lose it' principle is most likely to hold if populations are sufficiently small that conditionally deleterious mutations can drift to high frequency during periods of inactivity. At sufficiently large population sizes, deleterious mutations may rarely have time to rise to high frequencies between bouts of use/non-use (from Chapter 4, the time to fix a neutral mutation is approximately equal to twice the effective population size). Arguing that all deleterious mutations accumulating for an inactive TF gene will be immediately purged from the population upon demand for the gene product, they invoke a 'wear and tear' principle, whereby the least-used regulator can actually incur a slightly lower long-term average mutation load.

The issues are a bit subtle here, but the essential point is that the fitness difference between alternative modes of regulation under this second model is less than the mutation rate. This actually makes it highly unlikely that a domain in which the least-used mechanism will be most advantageous will ever be entered, as the mutation rate is weaker than the power of random genetic drift (Chapter 4). Thus, Savageau's hypothesis appears to be quite robust, and is well worth exploring in future studies with other organisms, especially given that it already draws support from observations on a high-N_e species, *E. coli*.

Regulatory rewiring

As outlined above, there are numerous examples in which the regulatory motifs associated with specific traits vary among species. However, this only touches the surface of the known ways in which regulatory mechanisms change over evolutionary time. Given the large numbers of TFs in most cells and their reliance on simple binding sites subject to stochastic mutational turnover, there are many plausible mechanisms for the emergence of novel intracellular transactions by effectively neutral processes (Johnson and Porter 2000; Force et al. 2005; Haag and Molla 2005; Lynch 2007b).

Several well-dissected examples demonstrate the complete rewiring of TF/TFBS associations, mostly in the yeast *S. cerevisiae*, where the study of gene regulation has been especially intense. Such studies strongly support the counter-intuitive idea that, even when under strong selection to maintain a stable phenotype, complex regulatory systems are subject to substantial modifications in their underlying structure.

Drawing on earlier work by Tanay et al. (2005) and Hogues et al. (2008), Lavoie et al. (2010) found massive differences in the regulatory machinery associated with the ribosomal-protein genes in *S. cerevisiae* and another yeast *Candida albicans*. Indeed, nearly every TF used in *S. cerevisiae* is utilized in a different way in the latter species and shifts in the consensus motifs for orthologous TFs occur as well. Some of this rewiring appears to be associated with a whole-genome duplication known to have occurred in ancestral *S. cerevisiae* (Wolfe and Shields 1997). For example, an activator and repressor that control ribosomal-protein expression in normal and stress conditions in *S. cerevisiae* are actually subfunctionalized duplicates of an ancestral gene inferred to have had both functions. Moreover, the various TFs involved have associations with novel functions in one or both species, showing expanded/contracted assignments.

Extending this theme, Martchenko et al. (2007) found that although *S. cerevisiae* and *C. albicans* have similar patterns of expression for genes associated with galactose metabolism, the underlying regulatory circuitry is completely different. Based on phylogenetic analysis, the ancestral species appears to have had shared (and perhaps redundant) regulatory motifs, with each of the two descendent lineages then going on to divergently utilize just one. Interestingly, the regulatory TF in *S. cerevisiae* (Gal4) is still retained and has similar binding properties in *C. albicans*, but is used in other processes (Askew et al. 2009). Even the regulatory mechanisms for the expression of histone proteins, some of the most evolutionary conserved proteins across eukaryotes, are dramatically altered across yeast species, both in terms of the TFs deployed and their binding motifs (Mariño-Ramírez et al. 2006)

The regulatory wiring for the mating-type locus is also dramatically changed in yeast (Baker et al. 2011, 2012; Sorrells et al. 2015; Britton et al. 2020). Two mating-type cells exist in these species, *a* and *α* (Chapter 10). In *C. albicans* and other basal yeast lineages, *a*-specific genes are activated by a regulatory protein only present in *a* cells, which keeps the *a*-specific genes off in *α* cells without any investment in TFs. This mode of regulation in *α* cells diverged in *S. cerevisiae*, where the *a*-specific genes remain constitutively active in *a* cells, as in *C. albicans*, but are kept silent in *α* cells by a specific repressor protein (i.e., requiring an added investment in gene regulation in such cells). The alterations responsible for these differences again appear to have arisen from an intermediate ancestral state in which two sets of regulation were used simultaneously, and then divergently resolved in descendent lineages.

Many additional examples of regulatory rewiring have been uncovered in comparative analyses of the gene modules of *S. cerevisiae* and *C. albicans* (Tuch et al. 2008; Sarda and Hannenhalli 2015; Nocedal et al. 2017), but these kinds of observations are by no means restricted to yeasts. For example, a number of studies have suggested substantial regulatory rewiring among bacterial species (Babu et al. 2004; Lozada-Chávez et al. 2006; Price et al. 2007), with the general conclusion being that TFs are much less conserved than their target genes, although detailed examples of closely related species are lacking (see Perez and Groisman 2009a,b).

A remarkable feature of all of these examples of the evolution of different control mechanisms is that they involve coordinated TFBS changes at multiple target loci. How might multiple genes acquire the same sets of regulatory changes without an intermediate state of massive fitness loss? The simplest routes appear to require an intermediate phase of redundancy with respect to the TF (Force et al. 2005; Tanay et al. 2005; Tuch et al. 2008) (Figure 21.9). If, for example, an ancestral TF exhibited bi-specificity,

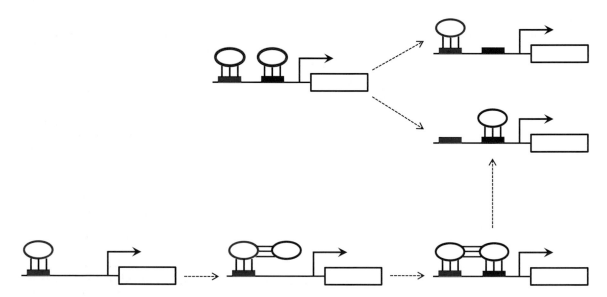

Figure 21.9 Two ways in which regulatory mechanisms might diverge between species in an effectively neutral fashion. **Above:** An ancestral transcription factor (orange) is capable of binding to two different, but functionally equivalent and hence, redundant, sites (blue and red rectangles). Following isolation by speciation, the TF in each descendent taxon then loses an alternative TFBS, hence, becoming more specialized. **Below:** A pre-existing TF (blue) fortuitously acquires binding affinity to another protein (red) that may or may not have DNA-binding activity. This opens up the opportunity for the emergence of a TFBS site with affinity to the recruited protein, resulting in redundant gene regulation, and allowing for the eventual loss of the ancestral regulatory mechanism (not shown).

i.e., was able to recognize two alternative TFBS motifs, random genetic drift (possibly accompanied by divergent mutation pressures) might result in the gradual loss of a different TFBS motif in each lineage. Such a condition would lead to relaxed selection on bi-specificity, with the TF then being free to lose a complementary motif in each lineage. The net effect of such a scenario would be the continued use of the same TF, but a change in the underlying regulatory language.

An apparent example of such evolutionary divergence is provided by LEAFY, a major regulator of flower development and cell division in land plants. Despite its presence in just a single copy per genome, the recognition motif of this TF differs substantially between mosses and the clade containing almost all other land plants. However, hornworts, which are basal with respect to the rest of land plants, utilize a third consensus motif while also harboring a capacity to promiscuously recognize the two motifs relied upon by other land plants (Sayou et al. 2014). This reciprocal focusing of a bi-specific ancestral TF may be a common mechanism of regulatory rewiring, at least in multicellular species, as roughly half of the TFs in mice and land plants recognize secondary motifs (Badis et al. 2009; Jolma et al. 2013; Franco-Zorrilla et al. 2014; Morgunova et al. 2018).

Divergence of TFBS motifs can also be achieved by an effectively neutral process of subfunctionalization within a single genome, when an ancestral TF gene with two regulatory motifs becomes duplicated, with the two copies then retaining just single, complementary recognition motifs. In this case, the overall biology of the organism will again remain the same, although the regulatory network will have become more complex, owing to the specialization of the individual TFs. Analyses in the nematode *C. elegans* (Reece-Hoyes et al. 2013) and the budding yeast *S. cerevisiae* (Pougach et al. 2014) provide considerable support for this model of regulatory rewiring.

Finally, the TF used in one particular lineage might fortuitously recruit an unrelated TF through a spurious protein–protein interaction (Figure 21.9). Although initially neutral, this interaction might then encourage the gradual evolution of local binding sites complementary to the second TF, at which point the first TF might become superfluous and subject to loss by mutational degeneration. Under this scenario, a coordinated shift in the entire regulatory mechanism might be achieved by multiple loci, as the initiating event will have been experienced simultaneously by each of the relevant regulatory regions owing to their shared reliance on the first TF.

These kinds of observations have profound implications for how we study biology, the obvious concern being that the molecular details deciphered for the regulatory pathway in one model system need not be relevant to that operating in other species. Yet, most molecular, cellular, and developmental biologists eschew intraspecific variation, often concentrating on just a single strain of a single model species, sometimes for decades. The resultant exquisite, painstaking research has led to remarkable advances in our understanding of the details of subcellular mechanisms, but what is the generality of such findings?

Because virtually every complex trait in every species exhibits significant genetic variation (Lynch and Walsh 1998), it is likely that many textbook examples of regulatory pathways derived from single clones or inbred lines will turn out to be quite unrepresentative of the operational features of related phylogenetic lineages. Among those with interests in multicellular organisms, these kinds of observations have motivated interest in the process of 'developmental system drift', whereby seemingly similar morphological structures in closely related species are achieved by substantially different regulatory mechanisms (Johnson and Porter 2000, 2001, 2007; Weiss and Fullerton 2000; True and Haag 2001; Force et al. 2005; Haag and Molla 2005; Ruvinsky and Ruvkun 2005; Tsong et al. 2006; Lynch 2007b; Pavlicev and Wagner 2012; Sommer 2012; Metzger and Wittkopp 2019). Notably, even in the relatively simple bacterium *E. coli*, when a TF is duplicated and one copy is then engineered to have a non-orthologous regulatory region, there are no notable changes in organismal fitness (Isalan et al. 2008), suggesting a high degree of evolutionary flexibility of regulatory systems.

Network topology

As with all other genes, TF expression is often regulated by other TFs, whose control may ultimately be dictated by signal-transduction pathways induced by internal or external chemical signals (Chapter 22). Combined with the fact that TFs can operate as either enhancers or repressors, this opens up the possibility of multiple architectures for gene regulatory networks. For example, the joint operation of just two genes can be governed by six different topologies (Figure 21.10, above). A common form of network called the feed-forward loop involves three genes (two TFs, with one regulating the second, and both regulating a third target gene), but even this still has eight possible topologies not including self-regulatory loops, and expanding to 64 possibilities if the latter are included (Figure 21.10, below). Such loops are said to be coherent if the direct effect of the first TF is the same as its indirect effect through the second TF; otherwise, the loop is said to be incoherent.

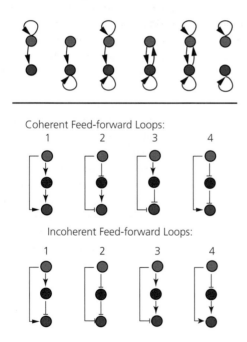

Coherent Feed-forward Loops:

Incoherent Feed-forward Loops:

Figure 21.10 Examples of possible topologies for simple gene networks. Arrows denote activation (including self-activation for loops), and blunt ends denote repression. **Above:** Two-gene model, with purple and green denoting alternative genes, one or both of which might be a TF. Note that examples of repression are not included here, which would further magnify the number of possible topologies. **Below:** Three-gene model, with purple and red denoting TFs, and green denoting the regulated gene. Examples of self-regulation are not included here, which again would further increase the number of topologies.

Many regulatory pathways are much more baroque in form than those just noted (Wilkins 2002, 2005; Lynch 2007b). For example, it is common for linear pathways to consist of a series of genes whose products are essential to the activation/deactivation of the next downstream member, with only the expression of the final component in the series being the ultimate determinant of the phenotype. For example, the product of gene D may be necessary to turn on gene C, whose product is necessary to turn on gene B, whose product finally turns on gene A. Pathways involving only inhibitory steps also exist, and these lead to an alternating series of high and low expression, depending on the state of the first gene in the pathway. For example, gene D may generate a product that inhibits the expression of gene C, whose silence allows gene B to be turned on, which inhibits the expression of gene A. It is often unclear that such complexity has any advantages over simpler two-gene pathways or even self-regulation.

The mechanisms by which genetic networks become established evolutionarily are far from clear. Many physicists, engineers, computer scientists, and cell and developmental biologists are convinced that biological networks are endowed with features that confer emergent properties that ostensibly could only be products of natural selection (Gerhart and Kirschner 1997; Milo et al. 2002; Shen-Orr et al. 2002; Alon 2003, 2007; Barabási and Oltvai 2004; Babu et al. 2006; Balaji et al. 2006; Davidson 2006; Tagkopoulos et al. 2008; Burda et al. 2011; Hong et al. 2018; Zitnik et al. 2019). Indeed, five popular concepts in biology today – redundancy, robustness, modularity, complexity, and evolvability – invoke a vision of the cell as an electronic circuit, designed by and for adaptation. However, the physical and genetic mechanisms giving give rise to genome architectural features are logically distinct from the adaptive processes utilizing such features as evolutionary resources (Lynch 2007b). Theoretical investigations of network evolution have only rarely examined these matters in the context of well-established evolutionary principles.

Qualitative observations suggest that the complexity of regulatory networks increases from prokaryotes to unicellular eukaryotes to multicellular eukaryotes, with simple autoregulatory loops being more common and multi-component loops less common in microbes (Thieffry et al. 1998; Lee et al. 2002; Wuchty and Almaas 2005; Sellerio et al. 2009). However, it is an open question as to whether complex pathway architectures are a necessary prerequisite for the evolution of complex phenotypes or whether the genome architectures of multicellular species are simply more conducive to the emergence of network connections owing to the elevated power of random genetic drift. The possibility that network-topology evolution is driven by the kinds of nonadaptive processes that generate changes in network participants (noted in the preceding section) clearly merits consideration.

The following arguments illustrate the ease with which commonly observed features of genetic pathways can emerge without any direct selection for such properties. In principle, pathway augmentation may be driven entirely by processes of duplication, degeneration, and random genetic drift. Consider the series of events in Figure 21.11. Initially, a single gene A carries out some function in a constitutive fashion, but in a series of steps, it becomes completely reliant on an upstream activation factor B. A scenario like this could unfold in the following way. First, gene A becomes sensitive

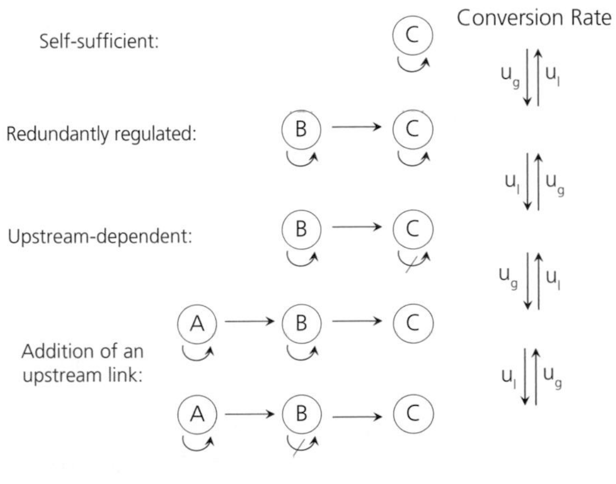

to activation by gene B, either because gene A has acquired a *cis*-modification that causes activation by B, or because some transcription factor B acquires a mutation that causes it to serve as a *trans*-activator of A. At this point, gene A has redundant activation pathways, and is therefore vulnerable to loss of one of them. Should a degenerative mutation cause gene A to lose the ability to self-regulate, gene B will have been established as an essential activator. This process can be repeated anew as gene B acquires sensitivity to a further upstream TF and loses the ability to constitutively express.

The probability of establishment of these types of changes is expected to depend on the effective population size (N_e). This is because a redundantly regulated allele has a weak mutational advantage equal to the rate of loss of a regulatory site (u_l) – one such mutation will result in the nonfunctionalization of either a self-regulated or an upstream-dependent allele, but will leave the function of a redundantly regulated allele unaltered. If $N_e \ll 1/u_l$, such an advantage will be impervious to selection, and the population will evolve to an allelic state that simply depends on the relative rates of gain and loss of regulatory sites by mutational processes (u_g and u_l in Figure 20.10), eventually leading to the establishment of an obligatory pathway. In contrast, if $N_e \gg 1/u_l$, the accumulation of upstream-dependent alleles will be inhibited by their weak mutational burden, as well as by the additional energetic burden (imposed by the expense of an additional pathway component). Although these arguments demonstrate that small population size provides a permissive environment for the emergence of complex genetic networks, without any direct selection for complexity, this does not mean that such alterations cannot occur in very large populations. However, if such changes are to occur in a large N_e context, they must have substantial enough additional advantages to offset the mutational and energetic burden of gene-structural complexity.

As discussed in detail in previous chapters, these simple arguments show how the relative power of the nonadaptive forces of evolution – genetic drift, mutation, and recombination – define the trajectories open to evolutionary exploitation. Although the incorporation of more technical details is needed,

Figure 21.11 Flow diagram for the series of evolutionary events leading to pathway expansions and contractions. Only activating interactions are shown, and for simplicity, gain and loss rates (u_g and u_l) are assumed to be the same for all links.

previous conclusions on the adaptive basis for the evolution of network topologies that rely on models devoid of population-genetic details should be interpreted with caution (Wagner 2005). Failure to reject a neutral hypothesis is not equivalent to ruling out selection as a governing force. However, the demonstration that the emergence of redundantly regulated genetic pathways is a function of population size and patterns of mutational bias raises doubts about the justification for the search for universal adaptive explanations for the evolution of genetic redundancy and robustness. For similar reasons, the conclusion that convergent evolution of network architectures in distantly related microbes provides compelling evidence for 'optimal design' (Conant and Wagner 2003) also appears to be questionable.

As a more explicit example of the issues, one of the most common pathways in bacteria, the coherent feed-forward loop (Figure 21.11) is, in fact, not particularly stable across phylogenetic lineages (Tsoy et al. 2012), and the case has been made that the relative utilization of alternative topologies in bacteria may largely be a consequence of random patterns of mutational loss and gain of substitutable links (Cordero and Hogeweg 2006; Lynch 2007b; Solé and Valverde 2008; Ruths and Nakhleh 2013). Finally, the ultimate output of a regulatory pathway is dictated not just by its topological form, but by the numerous dynamical properties (e.g., kinetic coefficients) and abundances of its participants, which together can override any supposed effects of topological structure (Ingram et al. 2006), leading to still more degrees of freedom for pathway rewiring. Although greatly simplified for presentational purposes, the verbal arguments presented above provide the seeds for the development of more biologically realistic models for the origins of pathway complexity, which will be necessary in future attempts to infer/reject the adaptive significance of such properties.

Summary

- A defining feature of all gene expression in all organisms is the production of RNA products from DNA templates, activated by various proteins called transcription factors (TFs) and carried out by RNA polymerases.

- Low rates of transcript production combined with high rates of degradation typically result in average steady-state mean numbers of transcripts per cell on the order of 20 or fewer, with bursty transcription generally resulting in a high level of noise in the number of transcripts per cell.

- As multiple proteins are commonly produced per mRNA and have elevated half-lives relative to the latter, the number of proteins per expressed gene typically vastly exceeds the number of mRNAs within a cell and the noise level is reduced accordingly.

- Transcription factors link to their binding sites (TFBSs) in ways that can be described by a simple biophysical model, which demonstrates that little is gained in terms of affinity for binding motifs longer than ~ 12 nucleotides. TF searches for target genes is facilitated by diffusion over the DNA with periodic jumps from one chromosomal location to another, and the search space in eukaryotes is reduced further by confining the genome to the nucleus.

- Despite these facilitating features, the time for a single TF to locate a specific TFBS can be on the order of minutes to hours. Thus, a basic cost of life's mechanism of gene expression is the necessity of producing substantially more copies of TFs than numbers of genes served in order to ensure a high probability of TFBS binding.

- Although the activation of gene expression by TFs must date to LUCA, there is a remarkable void of obvious TF orthologs between bacteria, archaea, and eukaryotes. In contrast, within eukaryotes, many of the TFs known to be associated with complex development in animals and land plants are also present in basal unicellular lineages.

- Changes in gene expression evolve through gene-specific modifications of TFBS motifs and/or shifts in the binding affinities and expression patterns of TFs. Nonetheless, many cases are known in which TFBS motifs wander among phylogenetic lineages, while continuity in gene expression is maintained. Such variation is particularly likely when a TF regulates only a few genes, as the TF and its binding sites are relatively susceptible to a coevolutionary dance so long as their mutual compatibility is maintained. In contrast, a TF with a large number of client genes can become frozen in time, as a slight improvement in the binding affinity to one gene may disrupt that for many others.

- TF systems provide the substrate for the development of a mechanistic evolutionary model directly linking genotype (number of nucleotides in a TFBS matching the optimal TF motif) to phenotype (expression level of the client gene) to fitness. This model predicts the existence of substantial variation of binding-site matching under a wide variety of conditions, especially when population sizes are relatively small.

- Transcription factors can operate as either activators or repressors of expression of client genes. The mode of regulation exploited by individual genes is generally the one that keeps the TF at the highest level of utilization, e.g., activation when the client gene is used frequently, and repression when client gene demands are low.

- A common feature of regulatory-pathway evolution is stasis of performance in the face of substantial regulatory rewiring in different phylogenetic lineages, in some cases to the point of using entirely different TFs to carry out the same tasks. Such cases of regulatory-system drift provide compelling examples of effectively neutral evolution at the subcellular level.

- A multitude of regulatory-pathway topologies exists among genes and organisms. Those in multicellular species are particularly elongated in structure, suggesting a syndrome of overdesign. In addition, the most common topologies in bacteria are often explainable with models invoking random gains and losses of links. Both kinds of observations raise questions about the common assertion that regulatory pathways are optimally designed to minimize expression noise and to maximize robustness and the capacity for future evolvability.

Foundations 21.1 Numbers of transcripts per cell

The rate of protein production depends on the number of mRNA molecules per cell, which in turn is a function of the rate of production of new transcripts and their subsequent loss by degradative processes. We first consider the situation for a constitutively expressed gene, with a constant rate of production of new transcripts k_m, and a rate of decay per transcript δ_m. With constant rates, regardless of the starting conditions, the stochastic probability distribution of the number of mRNA molecules per cell, $p(n_m)$, and eventually reach a steady state. At this point, the flux rate between the $n_m = 0$ and $n_m = 1$ states must be equal in both directions,

$$k_m p(0) = \delta_m p(1), \qquad (21.1.1a)$$

so

$$p(1) = p(0)(k_m/\delta_m). \qquad (21.1.1b)$$

Similarly, the flux rates in and out of class $n_m = 1$ must be equal, so

$$(k_m + \delta_m)p(1) = k_m p(0) + 2\delta_m p(2). \qquad (21.1.2a)$$

After subtracting Equation 21.1.1a and rearranging,

$$p(2) = p(0)(k_m/\delta_m)^2/2. \qquad (21.1.2b)$$

This approach generalizes to

$$p(n_m) = p(0)(k_m/\delta_m)^{n_m}/n_m!, \qquad (21.1.3)$$

where $n_m! = n_m(n_m - 1)(n_m - 2)\cdots 1$ is the factorial function.

To complete the solution, we require an expression for $p(0)$. Because the sum of the entire probability distribution, $p(n_m)$, is constrained to equal 1.0, it follows that $p(0)$ must equal a constant that ensures such equality. Noting that an exponential function can be written as the series expansion,

$$e^x = \sum_{n_m=0}^{\infty} \frac{x^{n_m}}{n_m!}, \qquad (21.1.4a)$$

which rearranges to

$$1 = e^{-x} \sum_{n_m=0}^{\infty} \frac{x^{n_m}}{n_m!}, \qquad (21.1.4b)$$

inspection of Equation 21.1.3, and substituting $x = k_m/\delta_m$, implies

$$p(0) = e^{-k_m/\delta_m}, \qquad (21.1.5a)$$

and more generally,

$$p(n_m) = (k_m/\delta_m)^{n_m} e^{-k_m/\delta_m}/n_m!. \qquad (21.1.5b)$$

This is the well-known Poisson distribution, which is a function of a single parameter (in this case k_m/δ_m), which in turn is equal to both the mean and the variance of the member of mRNA molecules per cell. Thus, under a model of constitutive gene expression, the mean number of transcripts per cell is simply equal to the ratio of the rates of production and elimination, k_m/δ_m.

Under more complex scenarios of gene regulation, the distribution of the number of transcripts per cell deviates from the Poisson, and needs to be evaluated by more complex methods (Thattai and van Oudenaarden 2001; Phillips et al. 2012). A solution for the two-state model in which the gene is turned on with some probability P_{on} was derived by Peccoud and Ycart (1995) and has the respective mean and variance

$$\mu_m = \frac{P_{on}k_m}{\delta_m} \tag{21.1.6a}$$

$$\sigma_m^2 = \mu_m \left(1 + \frac{(1 - P_{on})k_m}{k_{on} + k_{off} + \delta_m} \right), \tag{21.1.6b}$$

where k_{on} and k_{off} are, respectively, the rates of transition of cells from the off to on states, and vice versa. The complete distribution, worked out by Raj et al. (2006) and Shahrezaei and Swain (2008), is given by

$$p(n_m) = p^*(n_m) \frac{\delta(k'_{on} + n_m) \cdot \delta(k'_{on} + k'_{off})}{\delta(k'_{on} + k'_{off} + n_m)\delta(k'_{on})}$$
$$\cdot {}_1F_1[k'_{off}, (k'_{on} + k'_{off} + n_m); k_m/\delta_m], \tag{21.1.7}$$

where $p^*(n_m)$ is the Poisson distribution defined in Equation 21.1.5b, $k'_{on} = k_{on}/\delta_m$, and $k'_{off} = k_{off}/\delta_m$. Here, $\delta(\cdots)$ is the gamma function, and ${}_1F_1[\cdots]$ is the confluent hypergeometric function of the first kind, both of which can be approximated using expressions in Abramowitz and Stegun (1972).

Although this model is agnostic with respect to the mechanisms turning a gene on and off, it does assume that the switching events are completely random (i.e., have probabilities that do not depend on the length of stay in a previous state). Under this assumption,

$$P_{on} = \frac{k_{on}}{k_{on} + k_{off}}. \tag{21.1.8}$$

Alternative models that allow for the on/off rates being dependent of the state of the DNA, and/or influenced by the presence of cooperative factors, chromatin remodelling, etc., can be found in Phillips et al. (2012), Hammar et al. (2014), Corrigan et al. (2016), Sevier et al. (2016), and Skinner et al. (2016).

In Foundations 21.2, a more mechanistic description of P_{on} is provided in terms of TF binding. The central point here is that, owing to the population of cells being heterogeneous with respect to the on and off states, there is a greater dispersion in mRNA number per cell when $P_{on} < 1$ relative to the case of constitutive gene expression (as can be seen from the degree to which the variance of n_m exceeds the mean).

Foundations 21.2 Occupancy probability for a TFBS

Because gene activation requires that relevant TFBSs be occupied by their cognate TFs, an understanding of the mechanics of gene expression requires some basic theory for the probability that a particular TFBS is appropriately bound. This, in turn, requires information on the degree to which individual TF molecules are transiently tied to alternative substrates within the cell. Here we consider one particular target TFBS within a genome containing N_{ot} additional off-target but legitimate binding sites for the TF of interest, e.g., belonging to other client genes. In addition, we must account for the possibility of erroneous binding to illegitimate sites in the genome. Although such nonspecific binding is expected to be weak on a per-site basis, because each nucleotide site can serve as an initiation site for binding, the number of such sites is enormous, being close to the total number of bases in the genome (G).

Letting N_{tf} be the number of cognate TF molecules in the cell, we assume that $N_{ot} \ll N_{tf} \ll G$. The first inequality follows from the fact that a full repertoire of gene expression is extremely unlikely unless the number of TF molecules substantially exceeds the number of genes requiring their services. The second inequality follows from the sheer magnitude of genome sizes (generally, 10^6 to 10^{10} bp).

To compute the probability that a particular TFBS is bound by a cognate TF, we utilize a standard approach from statistical mechanics, evaluating the relative likelihoods of all possible ways in which N_{tf} TF molecules can be distributed within a cell (Bintu et al. 2005; Phillips et al. 2012). Here, we assume that essentially all such molecules are situated along the chromosome, either specifically bound to true cognate sites or non-specifically bound to random genomic regions, although this assumption need not literally be true as long as we appropriately account for all off-site sequestration. Ultimately, we require a measure scaling with the total probability that a TF is bound to the site of interest, Z_{on}, and another measure scaling with the probability that all N_{tf} TF molecules are engaged elsewhere on the genome, Z_{off}. The sum, $(Z_{on} + Z_{off})$, is known as the partition function, and it follows that the probability of a particular TFBS being occupied is simply

$$P_{on} = \frac{Z_{on}}{Z_{on} + Z_{off}} = \frac{1}{1 + (Z_{off}/Z_{on})}. \quad (21.2.1)$$

The first step to evaluating the two components of the partition function is to enumerate the full set of relevant configurations of the N_{tf} molecules within the cell, weighting each set of states by its multiplicity, that is, the number of ways in which a particular type of configuration can be distributed over the genome. Consider, for example, the situation in which the target TFBS is unoccupied. In this case, all N_{tf} TF molecules might be non-specifically bound, with none on off-target sites; here, there are $G!/[(G - N_{tf})!N_{tf}!]$ distinct ways in which the TFs can be distributed over the G non-specific sites (where $x! = x(x - 1)(x - 2)\cdots 1$ is the factorial product). Alternatively, $N_{tf} - 1$ TF molecules might be non-specifically bound, with one on an off-target site; there would then be $G!/[(G - N_{tf} - 1)!(N_{tf} - 1)!]$ distinct ways in which the TFs can be distributed over non-specific sites, and N_{ot} possible locations for the one off-target TFBS, yielding a total multiplicity of $N_{ot}G!/[(G - N_{tf} - 1)!(N_{tf} - 1)!]$. This general enumeration strategy must be extended to the opposite extreme in which all off-target sites are occupied, in each case following the general procedure for determining the distinct number of ways in which x TFs can be distributed over y sites. The same strategy for quantifying multiplicity of configurations applies to the situation in which the target TFBS is occupied, except in this case only $(N_{tf} - 1)$ TF molecules are distributed elsewhere.

Each of these multiplicities represents the potential for a particular configuration of TF locations within a cell. However, after such enumeration, all of the alternative states must be further weighted by their physical likelihoods dictated by the overall binding energy of each configuration. Here, we denote the binding energies of the TF to the target, off-target, and non-specific sites as E_t, E_{ot}, and E_{ns}, respectively. For example, for each configuration in which all TFs reside on non-specific binding sites, the total weight is $e^{-N_{tf}E_{ns}/(K_BT)}$. If one off-target site is occupied along with $(N_{tf} - 1)$ non-specific sites, the weight becomes $e^{-[E_{ot}+(N_{tf}-1)E_{ns})]/(K_BT)} = e^{-[(E_{ot}-E_{ns})+N_{tf}E_{ns}]/(K_BT)}$. If the target site is occupied, along with one off-target site and $(N_{tf} - 2)$ non-specific sites, the weight becomes $e^{-[E_t+E_{ot}+(N_{tf}-2)E_{ns}]/(K_BT)} = e^{-[(E_t-E_{ns})+(E_{ot}-E_{ns})+N_{tf}E_{ns}]/(K_BT)}$, etc. In these expressions, K_BT is the Boltzmann constant multiplied by the temperature (in degrees Kelvin), the standard measure of background thermal energy (Chapter 7). With both K_BT and the binding energies measured in the same units (usually kcal/mol), the weights are dimensionless. Because

the binding energies are negative, with stronger binding denoted by more negative E, the weights increase with the magnitude of binding strength to cognate sites relative to background expectations.

With this substantial amount of bookkeeping in place, we are now in a position to write down full expressions for each of the two components of the partition function. In each case, this is done by summing over all possible configurations the products of the multiplicity and the energetic weight of each configuration. In the following, we use the abbreviation $\beta = 1/(K_BT)$, and let $\Delta E_t = E_t - E_{ns}$ and $\Delta E_{ot} = E_{ot} - E_{ns}$ denote the differences in binding energies of target and off-target sites from background levels. Summing up, some rather complex-looking expressions arise,

$$Z_{off} \simeq \frac{G!N_{ot}!e^{-\beta N_{tf}E_{ns}}}{(G - N_{tf})!(N_{tf} - N_{ot})!N_{tf}^{N_{ot}}} \cdot$$
$$\sum_{i=0}^{N_{ot}} \frac{e^{-i\beta N_{tf}\Delta E_{ot}}}{(N_{ot} - i)!i!(G/N_{tf})^i} \quad (21.2.2a)$$

$$Z_{on} \simeq \frac{G!N_{ot}!e^{-\beta N_{tf}E_{ns}}e^{-\beta\Delta E_t}}{(G - N_{tf} + 1)!(N_{tf} - N_{ot} - 1)!N_{tf}^{N_{ot}}} \cdot$$
$$\sum_{i=0}^{N_{ot}} \frac{e^{-i\beta N_{tf}\Delta E_{ot}}}{(N_{ot} - i)!i!(G/N_{tf})^i} \quad (21.2.2b)$$

Noting, however, that the summations to the right of Equations 21.2.2a,b are identical, and that several of the components on the left are identical or very similar as well, substitution into Equation 21.2.1 leads to great simplification,

$$P_{on} = \frac{1}{1 + [G/(N_{tf} - N_{ot})]e^{\beta\Delta E_t}}. \quad (21.2.3)$$

In a succinct fashion, this expression reveals how the magnitude of gene expression is dictated by basic cellular features. First, the probability that a TFBS is occupied depends on the absolute difference in binding strengths between the target and non-specific sites; as E_t becomes more negative (implying stronger binding), $P_{on} \to 1$. Second, the probability of binding at the site declines with increasing concentration of non-specific sites (G) relative to the number of TF molecules available to the site, ($N_{tf} - N_{ot}$). The first effect is a function of the degree of match between the binding motif of the site of interest and the optimal sequence of its cognate TF, whereas the second effect is determined by the size of the genome (G), the degree of expression of the TF (the number of molecules in the cell, N_{tf}), and the number of additional legitimate sites serviced by the TF (N_{ot}).

Foundations 21.3 TFBS localization

Gene regulation requires that TFs navigate from their point of production by ribosomes to the genomic location of their cognate TFBSs. Such encounters are established through semi-random diffusive molecular motions, i.e., without the involvement of any directed guidance from specific transport mechanisms such as motor proteins. Here we consider the approximate timescale on which encounters are likely to occur, primarily to show that the rapid equilibration assumed in the previous section is indeed likely. We start with a focus on prokaryotic cells, which offer the relative simplicity of a fairly homogeneous cytoplasm. The biophysical principles underlying the formulae to be used are described in Chapter 7.

TFs have an inherent tendency to bind non-specifically to DNA. Thus, because the translation of prokaryotic mRNAs is performed in the close vicinity of the chromosome, often co-transcriptionally, it is reasonable to assume that a newly arisen TF is almost immediately bound weakly to a non-specific genomic site. This raises the possibility that a TF could then simply engage in a one-dimensional diffusion process over the chromosome until randomly encountering its cognate TF. The time required for such an encounter can be roughly estimated by noting that after t time units the average distance of a particle from its starting point in a one-dimensional diffusion process (and ignoring any boundary conditions) is

$$\bar{d} = \sqrt{2D_1 t}, \tag{21.3.1}$$

with D_1 being the one-dimensional diffusion coefficient (with units equal to the squared distance per time).

A central problem with linear diffusion is its redundancy – with random movement to the right and left, any diffusive event has a 50% probability of returning the molecule to its location in the preceding step. The *average* location of a molecule always remains at its starting position, with the probability distribution simply broadening, equally to the left and right with time. Because in the absence of any directional bias to movement, the particle will always reside to the left and right of the starting point with equal probabilities, the quantity $\bar{d}$ is generally referred to as the root-mean-square distance.

Assuming that a TF initially resides at a random location on the genome with respect to its target TFBS, how long would it take to locate a specific target site by one-dimensional diffusion? With the initial random TF position being half a genome away from the site (with G being the genome size in bp), and the TFBS being potentially on either strand, the TF will have to interrogate an average total of $\sim G$ potential sites to find a specific target. Thus, we require the time solution to Equation 21.3.1 that yields $\bar{d} = G$. Several studies suggest an average $D_1 \simeq 0.5 \times 10^6$ bp^2/sec for a protein moving along the DNA in an *E. coli* cell (Wang et al. 2006; Elf et al. 2007; Marklund et al. 2013). Noting that the *E. coli* genome is $G \simeq 5 \times 10^6$ bp in length, substituting D_1 into Equation 21.3.1 and rearranging, we find that the average time for a single TF molecule to encounter a specific TFBS by one-dimensional diffusion is $\sim 2.5 \times 10^6$ sec (or ~ 29 days). With N_{tf} TF molecules searching simultaneously, the average search time would be $1/N_{tf}$ times the single-molecule expectation, but even with 1000 TF molecules per cell (higher than what is seen in this species), the average search time would be ~ 0.7 hours. As this is too long to account for the fact that *E. coli* cells are capable of dividing in < 0.5 hours, it is clear that linear scanning cannot account for known rates of transcription.

An alternative way in which the search process might be accomplished is a form of three-dimensional diffusion. In this case, we make use of an expression for the encounter rate per unit concentration,

$$k_e = 4\pi(D_{3n} + D_{3p})(r_n + r_p), \tag{21.3.2}$$

where D_{3n} and D_{3p} are, respectively, the diffusion coefficients for the nucleic acid (TFBS) and the protein (TF), and r_n and r_p are their effective radii (Foundations 18.2). This equation assumes that an effective encounter occurs when the centers of the TF and TFBS fall within total distance $r_n + r_p$ of each other. Because of its bulk, it is reasonable to assume that the DNA molecule is effectively immobile relative to the TF, so that $D_{3n} \simeq 0$. Experimental estimates for proteins in *E. coli* suggest that $D_{3p} \simeq 3.5$ μm^2/sec (Elowitz et al. 1999; Elf et al. 2007). Taking an average TFBS motif in this species to be 20 bp in length, and noting that the length of a nucleotide on a DNA molecule is $\simeq 0.34 \times 10^{-3}$ μm, the effective radius of a potential binding site is approximately $r_n = 0.5 \times 20 \times 0.34 \times 10^{-3} = 0.0034$ μm. The effective radii of proteins of the size of a TF are roughly in the range of $r_p = 0.002$ to 0.010 μm (Wasyl et al. 1971; Erickson 2009), and we use an average of 0.006 μm. Substitution of these estimates into Equation 21.3.2 yields an estimated encounter rate of 0.4 μm^3/sec per unit concentration.

To obtain an estimate of the actual encounter rate, this specific rate must be multiplied by the products of the concentrations of the TFBS and TF within the cell, and we must also compute the number of times the TF must jump from the DNA to a new location prior to encountering its proper target. The volume of an *E. coli* cell is $\simeq 1\ \mu m^3$, and so with one TF molecule in search of $\sim 10^7$ non-specific binding sites (summed over both sides of the genome), the rate of encounter with any site on the DNA is $4 \times 10^6/\text{sec}$. The average time for a jump between chromosomal locations is the reciprocal of this quantity, 2.5×10^{-7} sec. Elf et al. (2007) estimate that once on the DNA, a TF spends ~ 0.0026 sec diffusing over ~ 100 bp, so essentially all of the search time is spent directly interrogating the DNA, rather than jumping from spot to spot. Thus, because approximately 10^5 100-bp scans are required to cover the entire genome, the estimated time to locate a site is $10^5 \times 0.0026 = 260$ sec. With N_{tf} molecules in the cell, the search time would be reduced to $260/N_{tf}$. A few prokaryotic species have cell volumes as small as $0.1\ \mu m^3$, which would reduce the search time further by a factor of 10, and few have volumes exceeding $100\ \mu m^3$, which would increase the time hundredfold.

How might these results extend to transcription in eukaryotes? First, because eukaryotic TFBSs are about half the length of those of prokaryotes, the encounter rate will be reduced by a factor of 0.5 on the basis of target size. Second, the average rate of diffusion in the nucleoplasm of mammalian cells is on the order of $D_{3p} \simeq 18\ \mu m^2/\text{sec}$ for proteins (Kühn et al. 2011), which will speed things up by a factor of $18/4 = 4.5$. Third, nuclear volumes in eukaryotic cells are typically larger than the volumes of entire prokaryotic cells, generally in the range of 100 to $10^4\ \mu m^3$ (Chapter 15). However, the concentration of DNA within nuclei appears to be higher than that within prokaryotic cells – averaging 57×10^6 bp/μm^3 in root-tip cells of land plants (Fujimoto et al. 2005), and 189×10^6 bp/μm^3 in the blood cells of amphibians (Cavalier-Smith 1982), which is $\sim 25\times$ the concentration in an *E. coli* cell. Taken together, these results suggest that, once within the nucleus, a TF will encounter DNA at a rate on the order of $0.5 \times 4.5 \times 25 = 56$ times faster than the rate calculated above for *E. coli*. Estimates for the one-dimensional diffusion parameters do not appear to be available for eukaryotes. However, assuming that they are roughly the same as in *E. coli*, because eukaryotic haploid genome sizes are generally in the range of 10 to 3000 million bp in length, we can anticipate search times on the order of 2–$600\times$ greater than that for *E.*

coli. On the other hand, a substantial fraction of eukaryotic chromosomes is spooled around histones, which will serve to reduce the time needed to search for an exposed TFBS.

Although fairly crude, these estimates clearly indicate that, given the architecture of cells, specific motor proteins are not required to guide TFs to their final destinations. All of the given calculations ignore the electrostatic interactions between proteins and nucleic acids, which by increasing the effective radii of interacting particles, would further speed up the localization process (Riggs et al. 1970; Halford 2009). Moreover, initial encounters are expected to be considerably sped up in prokaryotes where the TF is often encoded in a genomic location close to its target genes, ensuring that a newly translated TF has a starting point close to its final destination (Kolesov et al. 2007).

The preceding calculations for eukaryotes ignore the additional problem of a cytoplasmically translated TF finding its way to the nucleus. An estimate of this time can be obtained by referring back to Equation 21.3.2, which defines the expected encounter rate between two diffusing particles. Recalling the range of nuclear volumes cited and assuming a spherical shape, the radii of nuclei commonly fall in the range of $r_n = 3$ to $13\ \mu m$ (for nuclear volumes in the range of 100 to $10^4\ \mu m^3$, respectively). As this is far larger than the size of proteins, r_p, the latter can be ignored. From the standpoint of diffusion, it can be assumed that the position of the nucleus is relatively fixed, which implies $D_{3n} \simeq 0$, and we again let $D_{3p} \simeq 18\ \mu m^2/\text{sec}$. It then follows that the encounter rate falls in the approximate range of $k_e = 680$ to $2950\ \mu m^2/\text{sec}$. The concentration of a single particle is the reciprocal of the cell volume, and the reciprocal of the product of this and k_e provides an estimate of the mean encounter time. Eukaryotic cell volumes are on the order of 100 to $10^6\ \mu m^3$ (Chapter 8), and if we assume that the latter are $\sim 100\times$ the nuclear volume, we obtain time search estimates in the range of 2.5 to 5.5 min for a TF molecule randomly placed in the cytoplasm. These analyses ignore the additional time to locate and transport through a nuclear pore. Thus, adding in the search time within the nucleus, the total time for an individual eukaryotic TF to locate a specific TFBS is expected to be on the order of 10 minutes to several hours.

Foundations 21.4 Evolutionary dispersion of TFBS matching profiles

For any TF, given its specific binding domain, there will also be a specific TFBS sequence on the DNA that maximizes the strength of binding. However,

owing to the recurrent introduction of mutations, variation will inevitably arise among the TFBS sequences harbored by different genes. Selection will prevent extreme TFBS degeneration, but there is little to be gained above a high level of binding strength (Foundations 21.3). Thus, we can expect the levels of TF–TFBS matching to wander within the boundaries dictated by these extremes. Such variation will be manifest among the TFBS sequences associated with different genes within species as well as among orthologous genes across species. Here, we outline a simple model to predict the evolutionary dispersion of such sequences as a function of mutation pressure and the efficiency of selection.

We start with the assumption that all binding sites with the same number of matches (m) are equivalent with respect to binding probability, regardless of the position of the mismatches. In addition, for simplicity we assume that each of the four nucleotides mutates to each of the three other states at the same rate μ. Under these conditions, with a TF recognition motif of length ℓ nucleotides, there are ℓ genotypic classes to consider, each consisting of multiple subclasses with equal expected probabilities under selection–mutation balance. For example, class $m = \ell - 1$ consists of 3ℓ types, as the single mismatch can reside in sites 1 to ℓ and there are three mismatching nucleotide types per site. More generally, the multiplicity within each class can be determined simply from the binomial coefficient $3^n \ell! / [(\ell - m)! m!]$. This reduces a complex problem involving many classes to a more manageable level.

We further assume a population that usually resides in a near pure state, with a short enough timescale assumed that stochastic changes involve one-step transitions to adjacent states (Figure 21.12). Denoting the probability that a TFBS resides in class m at time t as $P(m, t)$, where m denotes the number of matches, the time-dependent behavior of the system is described by

$$\frac{\partial P(m, t)}{t} = N\mu \cdot \{(3(m + 1)\phi_{m+1,m}P(m + 1, t)$$
$$- [(\ell - m)\phi_{m,m+1} + (3m\phi_{m,m-1})]P(m, t)$$
$$+ (\ell - m + 1)\phi_{m-1,m}P(m - 1, t)\}. \tag{21.4.1}$$

The front term $N\mu$ denotes the rate of influx of new mutations, whereas all remaining terms denote the probabilities of fixation of various changes conditional on origin by mutation. The first term is dropped when $m = \ell$, and the last is dropped when $m = 0$. Here, we assume a haploid population of N individuals (for a diploid population, $2N$ should be substituted for N throughout).

This dynamical equation consists of three terms, the first denoting the influx of probability from the next higher class, with $(m + 1)$ functional sites mutating to non-matching states at rate 3μ in each gene copy (the 3 accounting for mutation to three alternative nucleotide

types), and going on to become fixed in the population with probability $\phi_{m+1,m}$. The second term accounts for the efflux from class m to the next upper and lower classes ($m + 1$ and $m - 1$), again accounting for the number of possible mutations that cause such movement and their probabilities of fixation. The final term describes the influx from the next lower class, which has $\ell - m + 1$ mismatches, each back-mutating to a matching state at rate μ.

The fixation probabilities are provided by Kimura's (1962) diffusion equation for newly arisen mutations,

$$\phi_{x,y} = \frac{1 - e^{-2N_e s_{x,y}/N}}{1 - e^{-2N_e s_{x,y}}} \tag{21.4.2}$$

where N_e is the effective population size, $1/N$ is the initial frequency of a mutation (for a haploid population), and $s_{x,y}$ is the fractional selective advantage of allelic class y over x (Chapter 4).

Despite its apparent complexity, Equation 21.4.1 can be solved in a relatively transparent way, which we clarify by starting with the assumption of neutrality, i.e., $s_{x,y} = 0$ for all (x, y). In this case, $\phi_{x,x+1} = \phi_{x,x-1} = 1/N$ for all x, and N cancels out in Equation 21.4.1. The entire array of TFBS states can be represented as a diagram with connecting arrows denoting the flux rates between adjacent classes (Figure 21.11). Because the rate of flux to matches declines and the rate of flux to mismatches increases as m increases, such a system must eventually reach an equilibrium, at which point for each class the net flux from above equals that from below. This condition is known as detailed balance.

For example, denoting the equilibrium solutions with a tilde, detailed balance requires that $3\ell\mu\widetilde{P}(\ell) = \mu\widetilde{P}(\ell - 1)$, that is, the flux from class ℓ to $(\ell - 1)$ matches must equal the reciprocal flux. This tells us that the probability mass in class $(\ell - 1)$ is 3ℓ times that in the perfectly matching class ℓ, that is, $\widetilde{P}(\ell - 1)/\widetilde{P}(\ell) = 3\ell$. More generally, for a linear model of this nature, the full solution for each class can be obtained by simply multiplying all of the coefficients on the arrows pointing up to the class with the product of all of the coefficients pointing down (Lynch 2013). For the case of neutrality, this greatly simplifies to

$$\widetilde{P}(m) = C3^{\ell-m} \binom{\ell}{m}, \tag{21.4.3}$$

where $C = 1/\sum_{i=0}^{\ell} 3^i \binom{\ell}{i}$ is a normalization constant that ensures a total probability mass of 1.0. There are two notable features of this solution. First, the equilibrium probabilities are completely independent of the mutation rate. Second, the term $3^{\ell-m} \binom{\ell}{m}$ is equivalent to the number of unique ways in which a sequence of length ℓ can harbor m matches.

Extending this approach to include selection is conceptually straightforward. The coefficient on each arrow in Figure 21.11 simply needs to be multiplied by the fixation probability between adjacent classes. For example, for the arrows connecting classes ℓ and $(\ell - 1)$, the coefficients become $3\ell\mu\phi_{\ell,\ell-1}$ and $\mu\phi_{\ell-1,\ell}$. The equilibrium probabilities are then again obtained using the rule noted – multiplying together all of the coefficients leading up to and down to each class. Here, two useful results lead to great simplification: 1) $\phi_{m-1,m}/\phi_{m,m-1} = e^{2N_e s_{m-1,m}}$; and 2) $s_{m-1,m} = W_m - W_{m-1}$, where W_m is the fitness of alleles with m matches in their TFBS. Using these equalities, Equation 21.4.3 generalizes to

$$\widetilde{P}(m) = C3^{\ell-m} \binom{\ell}{m} e^{2N_e W(m)}, \qquad (21.4.4)$$

where C is again a normalization constant (equal to the reciprocal of the sum of the terms to the right of C for all m). Equation 21.4.4 shows that with selection the equilibrium probability distribution of alternative binding states is equivalent to a simple modification of the neutral expectation, with each neutral genotypic probability being weighted exponentially by the product of its fitness and the effective population size (which influences the efficiency of selection). Further elaborations of this model can be found in Lynch and Hagner (2015) and Tuğrul et al. (2015).

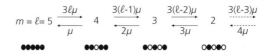

Figure 21.12 Flow diagram for the alternative states of a binding site of length $\ell = 5$, with the dot diagrams below simply illustrating one specific type within each category of numbers of matches (m, denoted by black balls). The transition rates are given on the arrows for the case of neutrality, where the probability of fixation is equal to the mutation rate per site, 3μ in the case of single-site losses (arrows to the right) because each appropriate nucleotide can mutate to three others, and μ in the case of site improvement (arrows to the left) because each mismatch can only mutate to the appropriate state in one way. Each coefficient further contains the number of relevant sites.

Literature Cited

Abramowitz, M., and I. A. Stegun (eds.) 1972. Handbook of Mathematical Functions. Dover Publications, Inc., New York, NY.

Alon, U. 2003. Biological networks: The tinkerer as an engineer. Science 301: 1866–1867.

Alon, U. 2007. Simplicity in biology. Nature 446: 497.

Aravind, L., V. Anantharaman, S. Balaji, M. M. Babu, and L. M. Iyer. 2005. The many faces of the helix-turn-helix domain: Transcription regulation and beyond. FEMS Microbiol. Rev. 29: 231–262.

Arnheim, N. 1986. The evolution of transcriptional control signals: Coevolution of ribosomal gene promoter sequences and transcription factors, pp. 37–51. In S. Karlin and E. Nevo (eds.) Evolutionary Processes and Theory. Academic Press, Inc., Orlando, FL.

Askew, C., A. Sellam, E. Epp, H. Hogues, A. Mullick, A. Nantel, and M. Whiteway. 2009. Transcriptional regulation of carbohydrate metabolism in the human pathogen *Candida albicans*. PLoS Pathog. 5: e1000612.

Babu, M. M., N. M. Luscombe, L. Aravind, M. Gerstein, and S. A. Teichmann. 2004. Structure and evolution of transcriptional regulatory networks. Curr. Opin. Struct. Biol. 14: 283–291.

Babu, M. M., S. A. Teichmann, and L. Aravind. 2006. Evolutionary dynamics of prokaryotic transcriptional regulatory networks. J. Mol. Biol. 358: 614–633.

Badis, G., M. F. Berger, A. A. Philippakis, S. Talukder, A. R. Gehrke, S. A. Jaeger, E. T. Chan, G. Metzler, A. Vedenko, X. Chen, et al. 2009. Diversity and complexity in DNA recognition by transcription factors. Science 324: 1720–1723.

Baker, C. R., L. N. Booth, T. R. Sorrells, and A. D. Johnson. 2012. Protein modularity, cooperative binding, and hybrid regulatory states underlie transcriptional network diversification. Cell 151: 80–95.

Baker, C. R., B. B. Tuch, and A. D. Johnson. 2011. Extensive DNA-binding specificity divergence of a conserved transcription regulator. Proc. Natl. Acad. Sci. USA 108: 7493–7498.

Balaji, S., L. M. Iyer, L. Aravind, and M. M. Babu. 2006. Uncovering a hidden distributed architecture behind scale-free transcriptional regulatory networks. J. Mol. Biol. 360: 204–212.

Balhoff, J. P., and G. A. Wray. 2005. Evolutionary analysis of the well characterized Endo16 promoter reveals substantial variation within functional sites. Proc. Natl. Acad. Sci. USA 102: 8591–8596.

Barabási, A. L., and Z. N. Oltvai. 2004. Network biology: Understanding the cell's functional organization. Nat. Rev. Genet. 5: 101–113.

Barrière, A., K. L. Gordon, and I. Ruvinsky. 2011. Distinct functional constraints partition sequence conservation in a *cis*-regulatory element. PLoS Genet. 7: e1002095.

Barrière, A., K. L. Gordon, and I. Ruvinsky. 2012. Coevolution within and between regulatory loci can preserve promoter function despite evolutionary rate acceleration. PLoS Genet. 8: e1002961.

Berg, J., S. Willmann, and M. Lässig. 2004. Adaptive evolution of transcription factor binding sites. BMC Evol. Biol. 4: 42.

Berg, O. G., and P. H. von Hippel. 1987. Selection of DNA binding sites by regulatory proteins. Statistical-mechanical theory and application to operators and promoters. J. Mol. Biol. 193: 723–750.

Berger, M. F., and M. L. Bulyk. 2009. Universal protein-binding microarrays for the comprehensive characterization of the DNA-binding specificities of transcription factors. Nat. Protoc. 4: 393–411.

Berggrun, A., and R. T. Sauer. 2001. Contributions of distinct quaternary contacts to cooperative operator binding by Mnt repressor. Proc. Natl. Acad. Sci. USA 98: 2301–2305.

Bernstein, J. A., A. B. Khodursky, P. H. Lin, S. Lin-Chao, and S. N. Cohen. 2002. Global analysis of mRNA decay and abundance in *Escherichia coli* at single-gene resolution using two-color fluorescent DNA microarrays. Proc. Natl. Acad. Sci. USA 99: 9697–9702.

Bintu, L., N. E. Buchler, H. G. Garcia, U. Gerland, T. Hwa, J. Kondev, T. Kuhlman, and R. Phillips. 2005. Transcriptional regulation by the numbers: Applications. Curr. Opin. Genet. Dev. 15: 125–135.

Borneman, A. R., T. A. Gianoulis, Z. D. Zhang, H. Yu, J. Rozowsky, M. R. Seringhaus, L. Y. Wang, M. Gerstein, and M. Snyder. 2007. Divergence of transcription factor binding sites across related yeast species. Science 317: 815–819.

Britton, C. S., T. R. Sorrells, and A. D. Johnson. 2020. Protein-coding changes preceded *cis*-regulatory gains in a newly evolved transcription circuit. Science 367: 96–100.

Brodsky, S., T. Jana, K. Mittelman, M. Chapal, D. K. Kumar, M. Carmi, and N. Barkai. 2020. Intrinsically disordered regions direct transcription factor *in vivo* binding specificity. Mol. Cell 79: 459–471.e4.

Burda, Z., A. Krzywicki, O. C. Martin, and M. Zagorski. 2011. Motifs emerge from function in model gene regulatory networks. Proc. Natl. Acad. Sci. USA 108: 17263–17268.

Carey, M. F., C. L. Peterson, and S. T. Smale. 2012. Experimental strategies for cloning or identifying genes encoding DNA-binding proteins. Cold Spring Harb. Protoc. 2012: 183–192.

Carroll, S. B., J. K. Grenier, and S. D. Weatherbee. 2001. From DNA to Diversity. Blackwell Science, Malden, MA.

Cavalier-Smith, T. 1982. Skeletal DNA and the evolution of genome size. Annu. Rev. Biophys. Bioeng. 11: 273–302.

Charoensawan, V., S. C. Janga, M. L. Bulyk, M. M. Babu, and S. A. Teichmann. 2012. DNA sequence preferences of transcriptional activators correlate more strongly than repressors with nucleosomes. Mol. Cell 47: 183–192.

Charoensawan, V., D. Wilson, and S. A. Teichmann. 2010. Genomic repertoires of DNA-binding transcription factors across the tree of life. Nucleic Acids Res. 38: 7364–7377.

Chong, S., C. Chen, H. Ge, and X. S. Xie. 2014. Mechanism of transcriptional bursting in bacteria. Cell 158: 314–326.

Cisneros, A. F., I. Gagnon-Arsenault, A. K. Dubé, P. C. Després, P. Kumar, K. Lafontaine, J. N. Pelletier, and C. R. Landry. 2023. Epistasis between promoter activity and coding mutations shapes gene evolvability. Sci. Adv. 9: eadd9109.

Conant, G. C., and A. Wagner. 2003. Convergent evolution of gene circuits. Nat. Genet. 34: 264–266.

Cordero, O. X., and P. Hogeweg. 2006. Feed-forward loop circuits as a side effect of genome evolution. Mol. Biol. Evol. 23: 1931–1936.

Corrigan, A. M., E. Tunnacliffe, D. Cannon, and J. R. Chubb. 2016. A continuum model of transcriptional bursting. eLife 5: e13051.

Cramer, P. 2019. Organization and regulation of gene transcription. Nature 573: 45–54.

Crocker, J., Y. Tamori, and A. Erives. 2008. Evolution acts on enhancer organization to fine-tune gradient threshold readouts. PLoS Biol. 6: e263.

Danko, C. G., N. Hah, X. Luo, A. L. Martins, L. Core, J. T. Lis, A. Siepel, and W. L. Kraus. 2013. Signaling pathways differentially affect RNA polymerase II initiation, pausing, and elongation rate in cells. Mol. Cell 50: 212–222.

Darzacq, X., Y. Shav-Tal, V. de Turris, Y. Brody, S. M. Shenoy, R. D. Phair, and R. H. Singer. 2007. *In vivo* dynamics of RNA polymerase II transcription. Nat. Struct. Mol. Biol. 14: 796–806.

Davidson, E. H. 2001. Genomic Regulatory Systems. Academic Press, San Diego, CA.

Davidson, E. H. 2006. The Regulatory Genome: Gene Regulatory Networks in Development and Evolution. Academic Press, New York, NY.

de Mendoza, A., A. Sebé-Pedrós, M. S. Šestak, M. Matejcic, G. Torruella, T. Domazet-Loso, and I. Ruiz-Trillo. 2013. Transcription factor evolution in eukaryotes and the assembly of the regulatory toolkit in multicellular lineages. Proc. Natl. Acad. Sci. USA 110: E4858–E4866.

Dermitzakis, E. T., and A. G. Clark. 2002. Evolution of transcription factor binding sites in Mammalian gene regulatory regions: Conservation and turnover. Mol. Biol. Evol. 19: 1114–1121.

Ding, C., D. W. Chan, W. Liu, M. Liu, D. Li, L. Song, C. Li, J. Jin, A. Malovannaya, S. Y. Jung, et al. 2013. Proteome-wide profiling of activated transcription factors with a concatenated tandem array of transcription factor response elements. Proc. Natl. Acad. Sci. USA 110: 6771–6776.

Djordjevic, M., A. M. Sengupta, and B. I. Shraiman. 2003. A biophysical approach to transcription factor binding site discovery. Genome Res. 13: 2381–2390.

Doniger, S. W., and J. C. Fay. 2007. Frequent gain and loss of functional transcription factor binding sites. PLoS Comput. Biol. 3: e99.

Dori-Bachash, M., E. Shema, and I. Tirosh. 2011. Coupled evolution of transcription and mRNA degradation. PLoS Biol. 9: e1001106.

Dowell, R. D. 2010. Transcription factor binding variation in the evolution of gene regulation. Trends Genet. 26: 468–475.

Duveau, F., A. Hodgins-Davis, B. P. Metzger, B. Yang, S. Tryban, E. A. Walker, T. Lybrook, and P. J. Wittkopp. 2018. Fitness effects of altering gene expression noise in *Saccharomyces cerevisiae*. eLife 7: e37272.

Elf, J., G. W. Li, and X. S. Xie. 2007. Probing transcription factor dynamics at the single-molecule level in a living cell. Science 316: 1191–1194.

Elowitz, M. B., M. G. Surette, P. E. Wolf, J. B. Stock, and S. Leibler. 1999. Protein mobility in the cytoplasm of *Escherichia coli*. J. Bacteriol. 181: 197–203.

Erickson, H. P. 2009. Size and shape of protein molecules at the nanometer level determined by sedimentation, gel filtration, and electron microscopy. Biol. Proceed. Online 11: 32–51.

Fields, D. S., Y. He, A. Y. Al-Uzri, and G. D. Stormo. 1997. Quantitative specificity of the Mnt repressor. J. Mol. Biol. 271: 178–194.

Fisher, S., E. A. Grice, R. M. Vinton, S. L. Bessling, and A. S. McCallion. 2006. Conservation of RET regulatory function from human to zebrafish without sequence similarity. Science 312: 276–279.

Force, A., W. Cresko, F. B. Pickett, S. Proulx, C. Amemiya, and M. Lynch. 2005. The origin of gene subfunctions and modular gene regulation. Genetics 170: 433–446.

Franco-Zorrilla, J. M., I. López-Vidriero, J. L. Carrasco, M. Godoy, P. Vera, and R. Solano. 2014. DNA-binding specificities of plant transcription factors and their potential to define target genes. Proc. Natl. Acad. Sci. USA 111: 2367–2372.

Fraser, H. B., A. E. Hirsh, G. Giaever, J. Kumm, and M. B. Eisen. 2004. Noise minimization in eukaryotic gene expression. PLoS Biol. 2: e137.

Fujimoto, S., M. Ito, S. Matsunaga, and K. Fukui. 2005. An upper limit of the ratio of DNA volume to nuclear volume exists in plants. Genes Genet. Syst. 80: 345–350.

Furey, T. S. 2012. ChIP-seq and beyond: New and improved methodologies to detect and characterize protein–DNA interactions. Nat. Rev. Genet. 13: 840–852.

Garfield, D., R. Haygood, W. J. Nielsen, and G. A. Wray. 2012. Population genetics of cis-regulatory sequences that operate during embryonic development in the sea urchin *Strongylocentrotus purpuratus*. Evol. Dev. 14: 152–167.

Gerhart, J., and M. Kirschner. 1997. Cells, Embryos and Evolution. Blackwell Science, Malden, MA.

Gerland, U., and T. Hwa. 2002. On the selection and evolution of regulatory DNA motifs. J. Mol. Evol. 55: 386–400.

Gerland, U., and T. Hwa. 2009. Evolutionary selection between alternative modes of gene regulation. Proc. Natl. Acad. Sci USA 106: 8841–8846.

Ghaemmaghami, S., W. K. Huh, K. Bower, R. W. Howson, A. Belle, N. Dephoure, E. K. O'Shea, and J. S. Weissman. 2003. Global analysis of protein expression in yeast. Nature 425: 737–741.

Golding, I., and E. C. Cox. 2004. RNA dynamics in live *Escherichia coli* cells. Proc. Natl. Acad. Sci. USA 101: 11310–11305.

Golding, I., J. Paulsson, S. M. Zawilski, and E. C. Cox. 2005. Real-time kinetics of gene activity in individual bacteria. Cell 123: 1025–1036.

Gotea, V., A. Visel, J. M. Westlund, M. A. Nobrega, L. A. Pennacchio, and I. Ovcharenko. 2010. Homotypic clusters of transcription factor binding sites are a key component of human promoters and enhancers. Genome Res. 20: 565–577.

Gowers, D. M., G. G. Wilson, and S. E. Halford. 2005. Measurement of the contributions of 1D and 3D pathways to the translocation of a protein along DNA. Proc. Natl. Acad. Sci. USA 102: 15883–15888.

Gruber, J. D., K. Vogel, G. Kalay, and P. J. Wittkopp. 2012. Contrasting properties of gene-specific regulatory, coding, and copy number mutations in *Saccharomyces cerevisiae*: Frequency, effects, and dominance. PLoS Genet. 8: e1002497.

Gunasekera, A., Y. W. Ebright, and R. H. Ebright. 1992. DNA sequence determinants for binding of the *Escherichia coli* catabolite gene activator protein. J. Biol. Chem. 267: 14713–14720.

Haag, E. S., and M. N. Molla. 2005. Compensatory evolution of interacting gene products through multifunctional intermediates. Evolution 59: 1620–1632.

Haberle, V., and A. Stark. 2018. Eukaryotic core promoters and the functional basis of transcription initiation. Nat. Rev. Mol. Cell Biol. 19: 621–637.

Hafner, A., and A. Boettiger. 2023. The spatial organization of transcriptional control. Nat. Rev. Genet. 24: 53–68.

Haimovich, G., D. A. Medina, S. Z. Causse, M. Garber, G. Millán-Zambrano, O. Barkai, S. Chávez, J. E. Pérez-Ortín, X. Darzacq, and M. Choder. 2013. Gene expression is circular: Factors for mRNA degradation also foster mRNA synthesis. Cell 153: 1000–1011.

Haldane, A., M. Manhart, and A. V. Morozov. 2014. Biophysical fitness landscapes for transcription factor binding sites. PLoS Comput Biol. 10: e1003683.

Halford, S. E. 2009. An end to 40 years of mistakes in DNA-protein association kinetics? Biochem. Soc. Trans. 37: 343–348.

Halford, S. E., and J. F. Marko. 2004. How do site-specific DNA-binding proteins find their targets? Nucleic Acids Res. 32: 3040–3052.

Hammar, P., M. Walldén, D. Fange, F. Persson, O. Baltekin, G. Ullman, P. Leroy, and J. Elf. 2014. Direct measurement of transcription factor dissociation excludes a simple operator occupancy model for gene regulation. Nature Genet. 46: 405–408.

Hare, E. E., B. K. Peterson, V. N. Iyer, R. Meier, and M. B. Eisen. 2008. Sepsid even-skipped enhancers are functionally conserved in *Drosophila* despite lack of sequence conservation. PLoS Genet. 4: e1000106.

He, B. Z., A. K. Holloway, S. J. Maerkl, and M. Kreitman. 2011. Does positive selection drive transcription factor binding site turnover? A test with *Drosophila cis*-regulatory modules. PLoS Genet. 7: e1002053.

He, Q., A. F. Bardet, B. Patton, J. Purvis, J. Johnston, A. Paulson, M. Gogol, A. Stark, and J. Zeitlinger. 2011. High conservation of transcription factor binding and evidence for combinatorial regulation across six *Drosophila* species. Nat. Genet. 43: 414–420.

Heinz, S., C. E. Romanoski, C. Benner, K. A. Allison, M. U. Kaikkonen, L. D. Orozco, and C. K. Glass. 2013. Effect of natural genetic variation on enhancer selection and function. Nature 503: 487–492.

Hill, M. S., P. Vande Zande, and P. J. Wittkopp. 2021. Molecular and evolutionary processes generating variation in gene expression. Nat. Rev. Genet. 22: 203–215.

Hogues, H., H. Lavoie, A. Sellam, M. Mangos, T. Roemer, E. Purisima, A. Nantel, and M. Whiteway. 2008. Transcription factor substitution during the evolution of fungal ribosome regulation. Mol. Cell 29: 552–562.

Hong, J., N. Brandt, F. Abdul-Rahman, A. Yang, T. Hughes, and D. Gresham. 2018. An incoherent feedforward loop facilitates adaptive tuning of gene expression. eLife 7: e32323.

Hsia, C. C., and W. McGinnis. 2003. Evolution of transcription factor function. Curr. Opin. Genet. Dev. 13: 199–206.

Ingram, P. J., M. P. Stumpf, and J. Stark. 2006. Network motifs: Structure does not determine function. BMC Genomics 7: 108.

Isalan, M., C. Lemerle, K. Michalodimitrakis, C. Horn, P. Beltrao, E. Raineri, M. Garriga-Canut, and L. Serrano. 2008. Evolvability and hierarchy in rewired bacterial gene networks. Nature 452: 840–845.

Ishihama, Y., T. Schmidt, J. Rappsilber, M. Mann, F. U. Hartl, M. J. Kerner, and D. Frishman. 2008. Protein abundance profiling of the *Escherichia coli* cytosol. BMC Genomics 9: 102.

Iyer, L. M., V. Anantharaman, M. Y. Wolf, and L. Aravind. 2008. Comparative genomics of transcription factors and chromatin proteins in parasitic protists and other eukaryotes. Int. J. Parasitol. 38: 1–31.

Johnson, N. A., and A. H. Porter. 2000. Rapid speciation via parallel, directional selection on regulatory genetic pathways. J. Theor. Biol. 205: 527–542.

Johnson, N. A., and A. H. Porter. 2001. Toward a new synthesis: Population genetics and evolutionary developmental biology. Genetica 112/113: 45–58.

Johnson, N. A., and A. H. Porter. 2007. Evolution of branched regulatory genetic pathways: Directional selection on pleiotropic loci accelerates developmental system drift. Genetica 129: 57–70.

Jolma, A., J. Yan, T. Whitington, J. Toivonen, K. R. Nitta, P. Rastas, E. Morgunova, M. Enge, M. Taipale, G. Wei, et al. 2013. DNA-binding specificities of human transcription factors. Cell 152: 327–339.

Jolma, A., Y. Yin, K. R. Nitta, K. Dave, A. Popov, M. Taipale, M. Enge, T. Kivioja, E. Morgunova, and J. Taipale. 2015. DNA-dependent formation of transcription factor pairs alters their binding specificity. Nature 527: 384–388.

Kimura, M. 1962. On the probability of fixation of mutant genes in a population. Genetics 47: 713–719.

Kinney, J. B., A. Murugan, C. G. Callan, Jr., and E. C. Cox. 2010. Using deep sequencing to characterize the biophysical mechanism of a transcriptional regulatory sequence. Proc. Natl. Acad. Sci. USA 107: 9158–9163.

Kolesov, G., Z. Wunderlich, O. N. Laikova, M. S. Gelfand, and L. A. Mirny. 2007. How gene order is influenced by the biophysics of transcription regulation. Proc. Natl. Acad. Sci. USA 104: 13948–13953.

Kolomeisky, A. B. 2011. Physics of protein-DNA interactions: Mechanisms of facilitated target search. Phys. Chem. Chem. Phys. 13: 2088–2095.

Kühn, T., T. O. Ihalainen, J. Hyväluoma, N. Dross, S. F. Willman, J. Langowski, M. Vihinen-Ranta, and J. Timonen. 2011. Protein diffusion in mammalian cell cytoplasm. PLoS One 6: e22962.

Lässig, M. 2007. From biophysics to evolutionary genetics: Statistical aspects of gene regulation. BMC Bioinformatics 8 (Suppl. 6): S7.

Lavoie, H., H. Hogues, J. Mallick, A. Sellam, A. Nantel, and M. Whiteway. 2010. Evolutionary tinkering with conserved components of a transcriptional regulatory network. PLoS Biol. 8: e1000329.

Le, D. D., T. C. Shimko, A. K. Aditham, A. M. Keys, S. A. Longwell, Y. Orenstein, and P. M. Fordyce. 2018. Comprehensive, high-resolution binding energy landscapes reveal context dependencies of transcription factor binding. Proc. Natl. Acad. Sci. USA 115: E3702–E3711.

Lee, T. I., N. J. Rinaldi, F. Robert, D. T. Odom, Z. Bar-Joseph, G. K. Gerber, N. M. Hannett, C. T. Harbison, C. M. Thompson, I. Simon, et al. 2002. Transcriptional regulatory networks in *Saccharomyces cerevisiae*. Science 298: 799–804.

Lestas, I., G. Vinnicombe, and J. Paulsson. 2010. Fundamental limits on the suppression of molecular fluctuations. Nature 467: 174–178.

Levo, M., and E. Segal. 2014. In pursuit of design principles of regulatory sequences. Nat. Rev. Genet. 15: 453–468.

Levy, S. F., N. Ziv, and M. L. Siegal. 2012. Bet hedging in yeast by heterogeneous, age-correlated expression of a stress protectant. PLoS Biol. 10: e1001325.

Li, G.-W., D. Burkhardt, C. Gross, and J. S. Weissman. 2014. Quantifying absolute protein synthesis rates reveals principles underlying allocation of cellular resources. Cell 157: 624–635.

Li, G. W., and X. S. Xie. 2011. Central dogma at the single-molecule level in living cells. Nature 475: 308–315.

Liu, J., H. Martin-Yken, F. Bigey, S. Dequin, J. M. Franois, and J. P. Capp. 2015. Natural yeast promoter variants reveal epistasis in the generation of transcriptional-mediated noise and its potential benefit in stressful conditions. Genome Biol. Evol. 7: 969–94.

Lozada-Chávez, I., S. C. Janga, and J. Collado-Vides. 2006. Bacterial regulatory networks are extremely flexible in evolution. Nucleic Acids Res. 34: 3434–3445.

Lu, P., C. Vogel, R. Wang, X. Yao, and E. M. Marcotte. 2007. Absolute protein expression profiling estimates the relative contributions of transcriptional and translational regulation. Nat. Biotechnol. 25: 117–124.

Ludwig, M. Z., R. K. Manu, K. P. White, and M. Kreitman. 2011. Consequences of eukaryotic enhancer architecture for gene expression dynamics, development, and fitness. PLoS Genet. 7: e1002364.

Luscombe, N. M., and J. M. Thornton. 2002. Protein–DNA interactions: Amino acid conservation and the effects of mutations on binding specificity. J. Mol. Biol. 320: 991–1009.

Lusk, R. W., and M. B. Eisen. 2010. Evolutionary mirages: Selection on binding site composition creates the illusion of conserved grammars in *Drosophila* enhancers. PLoS Genet. 6: e1000829.

Lynch, M. 2007a. The Origins of Genome Architecture. Sinauer Associates, Inc. Sunderland, MA.

Lynch, M. 2007b. The evolution of genetic networks by nonadaptive processes. Nat. Rev. Genet. 8: 803–813.

Lynch, M. 2013. Evolutionary diversification of the multimeric states of proteins. Proc. Natl. Acad. Sci. USA 110: E2821–E2828.

Lynch, M., and K. Hagner. 2015. Evolutionary meandering of intermolecular interactions along the drift barrier. Proc. Natl. Acad. Sci. USA 112: E30–8

Lynch, M., and J. B. Walsh. 1998. Genetics and Analysis of Quantitative Traits. Sinauer Associates, Inc., Sunderland, MA.

Lynch, V. J., and G. P. Wagner. 2008. Resurrecting the role of transcription factor change in developmental evolution. Evolution 62: 2131–2154.

Malmström, J., M. Beck, A. Schmidt, V. Lange, E. W. Deutsch, and R. Aebersold. 2009. Proteome-wide cellular protein concentrations of the human pathogen *Leptospira interrogans*. Nature 460: 762–765.

Mariño-Ramírez, L., I. K. Jordan, and D. Landsman. 2006. Multiple independent evolutionary solutions to core histone gene regulation. Genome Biol. 7: R122.

Marinov, G. K., B. A. Williams, K. McCue, G. P. Schroth, J. Gertz, R. M. Myers, and B. J. Wold. 2014. From single-cell to cell-pool transcriptomes: Stochasticity in gene expression and RNA splicing. Genome Res. 24: 496–510.

Marklund, E. G., A. Mahmutovic, O. G. Berg, P. Hammar, D. van der Spoel, D. Fange, and J. Elf. 2013. Transcription-factor binding and sliding on DNA studied using micro- and macroscopic models. Proc. Natl. Acad. Sci. USA 110: 19796–19801.

Martchenko, M., A. Levitin, H. Hogues, A. Nantel, and M. Whiteway. 2007. Transcriptional rewiring of fungal galactose-metabolism circuitry. Curr. Biol. 17: 1007–1013.

Martin-Perez, M., and J. Villén. 2017. Determinants and regulation of protein turnover in yeast. Cell Syst. 5: 283–294.e5.

Martínez-Antonio, A., and J. Collado-Vides. 2003. Identifying global regulators in transcriptional regulatory networks in bacteria. Curr. Opin. Microbiol. 6: 482–489.

Matsumoto, T., K. Mineta, N. Osada, and H. Araki. 2015. An individual-based diploid model predicts limited conditions under which stochastic gene expression becomes advantageous. Front. Genet. 6: 336.

Metzger, B. P., F. Duveau, D. C. Yuan, S. Tryban, B. Yang, and P. J. Wittkopp. 2016. Contrasting frequencies and effects of *cis*- and *trans*-regulatory mutations affecting gene expression. Mol. Biol. Evol. 33: 1131–1146.

Metzger, B. P. H., and P. J. Wittkopp. 2019. Compensatory *trans*-regulatory alleles minimizing variation in *TDH3* expression are common within *Saccharomyces cerevisiae*. Evol. Lett. 3: 448–461.

Miguel, A., F. Montón, T. Li, F. Gómez-Herreros, S. Chávez, P. Alepuz, and J. E. Pérez-Ortín. 2013. External conditions inversely change the RNA polymerase II elongation rate and density in yeast. Biochim. Biophys. Acta 1829: 1248–1255.

Milo, R., S. Shen-Orr, S. Itzkovitz, N. Kashtan, D. Chklovskii, and U. Alon. 2002. Network motifs: Simple building blocks of complex networks. Science 298: 824–827.

Mineta, K., T. Matsumoto, N. Osada, and H. Araki. 2015. Population genetics of non-genetic traits: Evolutionary roles of stochasticity in gene expression. Gene 562: 16–21.

Morgunova, E., Y. Yin, P. K. Das, A. Jolma, F. Zhu, A. Popov, Y. Xu, L. Nilsson, and J. Taipale. 2018. Two distinct DNA sequences recognized by transcription factors represent enthalpy and entropy optima. eLife 7: e32963.

Moses, A. M., D. A. Pollard, D. A. Nix, V. N. Iyer, X. Y. Li, M. D. Biggin, and M. B. Eisen. 2006. Large-scale turnover of functional transcription factor binding sites in *Drosophila*. PLoS Comput. Biol. 2: e130.

Mustonen, V., J. Kinney, C. G. Callan, Jr., and M. Lässig. 2008. Energy-dependent fitness: A quantitative model for the evolution of yeast transcription factor binding sites. Proc. Natl. Acad. Sci USA 105: 12376–12381.

Mustonen, V., and M. Lässig. 2005. Evolutionary population genetics of promoters: Predicting binding sites and functional phylogenies. Proc. Natl. Acad. Sci. USA 102: 15936–15941.

Nakagawa, S., S. S. Gisselbrecht, J. M. Rogers, D. L. Hartl, and M. L. Bulyk. 2013. DNA-binding specificity changes in the evolution of forkhead transcription factors. Proc. Natl. Acad. Sci. USA 110: 12349–12354.

Newman, J. R., S. Ghaemmaghami, J. Ihmels, D. K. Breslow, M. Noble, J. L. DeRisi, and J. S. Weissman. 2006. Single-cell proteomic analysis of *S. cerevisiae* reveals the architecture of biological noise. Nature 441: 840–846.

Nitta, K. R., A. Jolma, Y. Yin, E. Morgunova, T. Kivioja, J. Akhtar, K. Hens, J. Toivonen, B. Deplancke, E. E. Furlong, and J. Taipale. 2015. Conservation of transcription factor binding specificities across 600 million years of bilateria evolution. eLife 4: e04837.

Nocedal, I., E. Mancera, and A. D. Johnson. 2017. Gene regulatory network plasticity predates a switch in function of a conserved transcription regulator. eLife 6: e23250.

Normanno, D., M. Dahan, and X. Darzacq. 2012. Intranuclear mobility and target search mechanisms of transcription factors: A single-molecule perspective on gene expression. Biochim. Biophys. Acta 1819: 482–493.

Nourmohammad, A., and M. Lässig. 2011. Formation of regulatory modules by local sequence duplication. PLoS Comput. Biol. 7: e1002167.

Nutiu, R., R. C. Friedman, S. Luo, I. Khrebtukova, D. Silva, R. Li, L. Zhang, G. P. Schroth, and C. B. Burge. 2011. Direct measurement of DNA affinity landscapes on a high-throughput sequencing instrument. Nat. Biotechnol. 29: 659–664.

Oda, M., K. Furukawa, K. Ogata, A. Sarai, and H. Nakamura. 1998. Thermodynamics of specific and nonspecific DNA binding by the c-Myb DNA-binding domain. J. Mol. Biol. 276: 571–590.

Oda-Ishii, I., V. Bertrand, I. Matsuo, P. Lemaire, and H. Saiga. 2005. Making very similar embryos with divergent genomes: Conservation of regulatory mechanisms of Otx between the ascidians *Halocynthia roretzi* and *Ciona intestinalis*. Development 132: 1663–1674.

Omagari, K., H. Yoshimura, M. Takano, D. Hao, M. Ohmori, A. Sarai, and A. Suyama. 2004. Systematic single base-pair substitution analysis of DNA binding by the cAMP receptor protein in cyanobacterium *Synechocystis* sp. PCC 6803. FEBS Lett. 563: 55–58.

Paris, M., T. Kaplan, X. Y. Li, J. E. Villalta, S. E. Lott, and M. B. Eisen. 2013. Extensive divergence of transcription factor binding in *Drosophila* embryos with highly conserved gene expression. PLoS Genet. 9: e1003748.

Pavlicev, M., and G. P. Wagner. 2012. A model of developmental evolution: Selection, pleiotropy and compensation. Trends Ecol. Evol. 27: 316–322.

Peccoud, J., and B. Ycart. 1995. Markovian modeling of gene-product synthesis. Theor. Popul. Biol. 48: 222–234.

Perez, J. C., and E. A. Groisman. 2009a. Evolution of transcriptional regulatory circuits in bacteria. Cell 138: 233–244.

Perez, J. C., and E. A. Groisman. 2009b. Transcription factor function and promoter architecture govern the evolution of bacterial regulons. Proc. Natl. Acad. Sci. USA 106: 4319–4324.

Phillips, R., J. Kondev, J. Theriot, and H. Garcia. 2012. Physical Biology of the Cell, 2nd edn. Garland Science, New York, NY.

Pougach, K., A. Voet, F. A. Kondrashov, K. Voordeckers, J. F. Christiaens, B. Baying, V. Benes, R. Sakai, J. Aerts, B.

Zhu, et al. 2014. Duplication of a promiscuous transcription factor drives the emergence of a new regulatory network. Nat. Commun. 5: 4868.

Price, M. N., P. S. Dehal, and A. P. Arkin. 2007. Orthologous transcription factors in bacteria have different functions and regulate different genes. PLoS Comput. Biol. 3: 1739–1750.

Proshkin, S., A. R. Rahmouni, A. Mironov, and E. Nudler. 2010. Cooperation between translating ribosomes and RNA polymerase in transcription elongation. Science 328: 504–508.

Raj, A., C. S. Peskin, D. Tranchina, D. Y. Vargas, and S. Tyagi. 2006. Stochastic mRNA synthesis in mammalian cells. PLoS Biol. 4: e309.

Rajewsky, N., N. D. Socci, M. Zapotocky, and E. D. Siggia. 2002. The evolution of DNA regulatory regions for proteo-gamma bacteria by interspecies comparisons. Genome Res. 12: 298–308.

Reece-Hoyes, J. S., C. Pons, A. Diallo, A. Mori, S. Shrestha, S. Kadreppa, J. Nelson, S. Diprima, A. Dricot, B. R. Lajoie, et al. 2013. Extensive rewiring and complex evolutionary dynamics in a *C. elegans* multiparameter transcription factor network. Mol. Cell 51: 116–127.

Richter, D. J., P. Fozouni, M. B. Eisen, and N. King. 2018. Gene family innovation, conservation and loss on the animal stem lineage. eLife 7: e34226.

Riechmann, J. L., J. Heard, G. Martin, L. Reuber, C. Jiang, J. Keddie, L. Adam, O. Pineda, O. J. Ratcliffe, R. R. Samaha, et al. 2000. *Arabidopsis* transcription factors: Genome-wide comparative analysis among eukaryotes. Science 290: 2105–2110.

Riggs, A. D., S. Bourgeois, and M. Cohn. 1970. The lac repressor-operator interaction. 3. Kinetic studies. J. Mol. Biol. 53: 401–417.

Robison, K., A. M. McGuire, and G. M. Church. 1998. A comprehensive library of DNA-binding site matrices for 55 proteins applied to the complete *Escherichia coli* K-12 genome. J. Mol. Biol. 284: 241–254.

Romano, L. A., and G. A. Wray. 2003. Conservation of Endo16 expression in sea urchins despite evolutionary divergence in both *cis* and *trans*-acting components of transcriptional regulation. Development 130: 4187–4199.

Ruths, T., and L. Nakhleh. 2013. Neutral forces acting on intragenomic variability shape the *Escherichia coli* regulatory network topology. Proc. Natl. Acad. Sci. USA 110: 7754–7759.

Ruvinsky, I., and G. Ruvkun. 2003. Functional tests of enhancer conservation between distantly related species. Development 130: 5133–5142.

Sanchez, A., and I. Golding. 2013. Genetic determinants and cellular constraints in noisy gene expression. Science 342: 1188–1193.

Sarai, A., and Y. Takeda. 1989. Lambda repressor recognizes the approximately 2-fold symmetric half-operator sequences asymmetrically. Proc. Natl. Acad. Sci. USA 86: 6513–6517.

Sarda, S., and S. Hannenhalli. 2015. High-throughput identification of *cis*-regulatory rewiring events in yeast. Mol. Biol. Evol. 32: 3047–3063.

Savageau, M. A. 1974. Genetic regulatory mechanisms and the ecological niche of *Escherichia coli*. Proc. Natl. Acad. Sci. USA 71: 2453–2455.

Savageau, M. A. 1977. Design of molecular control mechanisms and the demand for gene expression. Proc. Natl. Acad. Sci. USA 74: 5647–5651.

Savageau, M. A. 1998. Demand theory of gene regulation. I. Quantitative development of the theory. Genetics 149: 1665–1676.

Sayou, C., M. Monniaux, M. H. Nanao, E. Moyroud, S. F. Brockington, E. Thévenon, H. Chahtane, N. Warthmann, M. Melkonian, Y. Zhang, et al. 2014. A promiscuous intermediate underlies the evolution of LEAFY DNA binding specificity. Science 343: 645–648.

Schildbach, J. F., A. W. Karzai, B. E. Raumann, and R. T. Sauer. 1999. Origins of DNA-binding specificity: Role of protein contacts with the DNA backbone. Proc. Natl. Acad. Sci. USA 96: 811–817.

Schmidt, D., M. D. Wilson, B. Ballester, P. C. Schwalie, G. D. Brown, A. Marshall, C. Kutter, S. Watt, C. P. Martinez-Jimenez, S. Mackay, et al. 2010. Five-vertebrate ChIP-seq reveals the evolutionary dynamics of transcription factor binding. Science 328: 1036–1040.

Schwanhäusser, B., D. Busse, N. Li, G. Dittmar, J. Schuchhardt, J. Wolf, W. Chen, and M. Selbach. 2011. Global quantification of mammalian gene expression control. Nature 473: 337–342.

Sellerio, A. L., B. Bassetti, H. Isambert, and M. Cosentino Lagomarsino. 2009. A comparative evolutionary study of transcription networks. The global role of feedback and hierarchical structures. Mol. Biosyst. 5: 170–179.

Sengupta, A. M., M. Djordjevic, and B. I. Shraiman. 2002. Specificity and robustness in transcription control networks. Proc. Natl. Acad. Sci. USA 99: 2072–2077.

Sepúlveda, L. A., H. Xu, J. Zhang, M. Wang, and I. Golding. 2016. Measurement of gene regulation in individual cells reveals rapid switching between promoter states. Science 351: 1218–1222.

Sevier, S. A., D. A. Kessler, and H. Levine. 2016. Mechanical bounds to transcriptional noise. Proc. Natl. Acad. Sci. USA 113: 13983–13988.

Shahrezaei, V., and P. S. Swain. 2008. Analytical distributions for stochastic gene expression. Proc. Natl. Acad. Sci. USA 105: 17256–17261.

Shen-Orr, S. S., R. Milo, S. Mangan, and U. Alon. 2002. Network motifs in the transcriptional regulation network of *Escherichia coli*. Nat. Genet. 31: 64–68.

Shinar, G., E. Dekel, T. Tlusty, and U. Alon. 2006. Rules for biological regulation based on error minimization. Proc. Natl. Acad. Sci. USA 103: 3999–4004.

Shultzaberger, R. K., D. S. Malashock, J. F. Kirsch, and M. B. Eisen. 2010. The fitness landscapes of *cis*-acting binding sites in different promoter and environmental contexts. PLoS Genet. 6: e1001042.

Skinner, S. O., H. Xu, S. Nagarkar-Jaiswal, P. R. Freire, T. P. Zwaka, and I. Golding. 2016. Single-cell analysis of transcription kinetics across the cell cycle. eLife 5: e12175.

Smith, R. P., L. Taher, R. P. Patwardhan, M. J. Kim, F. Inoue, J. Shendure, I. Ovcharenko, and N. Ahituv. 2013. Massively parallel decoding of mammalian regulatory sequences supports a flexible organizational model. Nat. Genet. 45: 1021–1028.

So, L. H., A. Ghosh, C. Zong, L. A. Sepúlveda, R. Segev, and I. Golding. 2011. General properties of transcriptional time series in *Escherichia coli*. Nat. Genet. 43: 554–60.

Solé, R. V., and S. Valverde. 2008. Spontaneous emergence of modularity in cellular networks. J. R. Soc. Interface 5: 129–133.

Sommer R. J. 2012. Evolution of regulatory networks: Nematode vulva induction as an example of developmental systems drift. Adv. Exp. Med. Biol. 751: 79–91.

Sorrells, T. R., L. N. Booth, B. B. Tuch, and A. D. Johnson. 2015. Intersecting transcription networks constrain gene regulatory evolution. Nature 523: 361–365.

Staller, M. V. 2022. Transcription factors perform a 2-step search of the nucleus. Genetics 222: iyac111.

Stefflova, K., D. Thybert, M. D. Wilson, I. Streeter, J. Aleksic, P. Karagianni, A. Brazma, D. J. Adams, I. Talianidis, J. C. Marioni, et al. 2013. Cooperativity and rapid evolution of cobound transcription factors in closely related mammals. Cell 154: 530–540.

Stewart, A. J., S. Hannenhalli, and J. B. Plotkin. 2012. Why transcription factor binding sites are ten nucleotides long. Genetics 192: 973–985.

Sun, M., B. Schwalb, D. Schulz, N. Pirkl, S. Etzold, L. Larivière, K. C. Maier, M. Seizl, A. Tresch, and P. Cramer. 2012. Comparative dynamic transcriptome analysis (cDTA) reveals mutual feedback between mRNA synthesis and degradation. Genome Res. 22: 1350–1359.

Tagkopoulos, I., Y. C. Liu, and S. Tavazoie. 2008. Predictive behavior within microbial genetic networks. Science 320: 1313–1317.

Takeda, Y., A. Sarai, and V. M. Rivera. 1989. Analysis of the sequence-specific interactions between Cro repressor and operator DNA by systematic base substitution experiments. Proc. Natl. Acad. Sci. USA 86: 439–443.

Tănase-Nicola, S., and P. R. ten Wolde. 2008. Regulatory control and the costs and benefits of biochemical noise. PLoS Comput. Biol. 4: e1000125.

Tanay, A., A. Regev, and R. Shamir. 2005. Conservation and evolvability in regulatory networks: The evolution of ribosomal regulation in yeast. Proc. Natl. Acad. Sci. USA 102: 7203–7208.

Taniguchi, Y., P. J. Choi, G. W. Li, H. Chen, M. Babu, J. Hearn, A. Emili, and X. S. Xie. 2010. Quantifying *E. coli* proteome and transcriptome with single-molecule sensitivity in single cells. Science 329: 533–538.

Thattai, M., and A. van Oudenaarden. 2001. Intrinsic noise in gene regulatory networks. Proc. Natl. Acad. Sci. USA 98: 8614–8619.

Thieffry, D., A. M. Huerta, E. Perez-Rueda, and J. Collado-Vides. 1998. From specific gene regulation to genomic networks: A global analysis of transcriptional regulation in *Escherichia coli*. Bioessays 20: 433–440.

Thurman, R. E., E. Rynes, R. Humbert, J. Vierstra, M. T. Maurano, E. Haugen, N. C. Sheffield, A. B. Stergachis, H. Wang, B. Vernot, et al. 2012. The accessible chromatin landscape of the human genome. Nature 489: 75–82.

True, J. R., and E. S. Haag. 2001. Developmental system drift and flexibility in evolutionary trajectories. Evol. Dev. 3: 109–119.

Tsong, A. E., B. B. Tuch, H. Li, and A. D. Johnson. 2006. Evolution of alternative transcriptional circuits with identical logic. Nature 443: 415–420.

Tsoy, O. V., M. A. Pyatnitskiy, M. D. Kazanov, and M. S. Gelfand. 2012. Evolution of transcriptional regulation in closely related bacteria. BMC Evol. Biol. 12: 200.

Tuch, B. B., H. Li, and A. D. Johnson. 2008. Evolution of eukaryotic transcription circuits. Science 319: 1797–1799.

Tuğrul, M., T. Paixão, N. H. Barton, and G. Tkačik. 2015. Dynamics of transcription factor binding site evolution. PLoS Genet. 11: e1005639.

van Nimwegen, E. 2003. Scaling laws in the functional content of genomes. Trends Genet. 19: 479–484.

von Hippel, P. H., and O. G. Berg. 1986. On the specificity of DNA-protein interactions. Proc. Natl. Acad. Sci. USA 83: 1608–1612.

von Hippel, P. H., and O. G. Berg. 1989. Facilitated target location in biological systems. J. Biol. Chem. 264: 675–678.

Wagner, A. 2005. Robustness and Evolvability in Living Systems. Princeton University Press, Princeton, NJ.

Wagner, G. P., and V. J. Lynch. 2008. The gene regulatory logic of transcription factor evolution. Trends Ecol. Evol. 23: 377–385.

Wang, X., H. Gao, Y. Shen, G. M. Weinstock, J. Zhou, and T. Palzkill. 2008. A high-throughput percentage-of-binding strategy to measure binding energies in DNA-protein interactions: Application to genome-scale site discovery. Nucleic Acids Res. 36: 4863–4871.

Wang, Y., C. L. Liu, J. D. Storey, R. J. Tibshirani, D. Herschlag, and P. O. Brown. 2002. Precision and functional specificity in mRNA decay. Proc. Natl. Acad. Sci. USA 99: 5860–5865.

Wang, Y. M., R. H. Austin, and E. C. Cox. 2006. Single molecule measurements of repressor protein 1D diffusion on DNA. Phys. Rev. Lett. 97: 048302.

Wang, Z., and J. Zhang. 2011. Impact of gene expression noise on organismal fitness and the efficacy of natural selection. Proc. Natl. Acad. Sci. USA 108: E67–E76.

Wasyl, Z., E. Luchter, and W. Bielanski, Jr. 1971. Determination of the effective radius of protein molecules by thin-layer gel filtration. Biochim. Biophys. Acta 243: 11–18.

Weiss, K. M., and S. M. Fullerton. 2000. Phenogenetic drift and the evolution of genotype–phenotype relationships. Theor. Popul. Biol. 57: 187–195.

Wilkins, A. S. 2002. The Evolution of Developmental Pathways. Sinauer Associates, Inc., Sunderland, MA.

Wilkins, A. S. 2005. Recasting developmental evolution in terms of genetic pathway and network evolution and the implications for comparative biology. Brain Res. Bull. 66: 495–509.

Wilson, M. D., N. L. Barbosa-Morais, D. Schmidt, C. M. Conboy, L. Vanes, V. L. Tybulewicz, E. M. Fisher, S. Tavaré, and D. T. Odom. 2008. Species-specific transcription in mice carrying human chromosome 21. Science 322: 434–438.

Wolf, L., O. K. Silander, and E. van Nimwegen. 2015. Expression noise facilitates the evolution of gene regulation. eLife 4: e05856.

Wolfe, K. H., and D. C. Shields. 1997. Molecular evidence for an ancient duplication of the entire yeast genome. Nature 387: 708–713.

Wray, G. A. 2007. The evolutionary significance of cis-regulatory mutations. Nat. Rev. Genet. 8: 206–216.

Wuchty, S., and E. Almaas. 2005. Evolutionary cores of domain co-occurrence networks. BMC Evol. Biol. 5: 24.

Yan, J., M. Enge, T. Whitington, K. Dave, J. Liu, I. Sur, B. Schmierer, A. Jolma, T. Kivioja, M. Taipale, and J. Taipale. 2013. Transcription factor binding in human cells occurs in dense clusters formed around cohesin anchor sites. Cell 154: 801–813.

Yang, L., T. Zhou, I. Dror, A. Mathelier, W. W. Wasserman, R. Gordân, and R. Rohs. 2014. TFBSshape: A motif database for DNA shape features of transcription factor binding sites. Nucleic Acids Res. 42 (Database issue): D148–D155.

Yokoyama, K. D., and D. D. Pollock. 2012. SP transcription factor paralogs and DNA-binding sites coevolve and adaptively converge in mammals and birds. Genome Biol. Evol. 4: 1102–1117.

Yokoyama, K. D., J. L. Thorne, and G. A. Wray. 2011. Coordinated genome-wide modifications within proximal promoter cis-regulatory elements during vertebrate evolution. Genome Biol. Evol. 3: 66–74.

Yuh, C. H., H. Bolouri, and E. H. Davidson. 1998. Genomic cis-regulatory logic: Experimental and computational analysis of a sea urchin gene. Science 279: 1896–1902.

Zenklusen, D., D. R. Larson, and R. H. Singer. 2008. Single-RNA counting reveals alternative modes of gene expression in yeast. Nat. Struct. Mol. Biol. 15: 1263–1271.

Zheng, W., T. A. Gianoulis, K. J. Karczewski, H. Zhao, and M. Snyder. 2011. Regulatory variation within and between species. Annu. Rev. Genomics Hum. Genet. 12: 327–346.

Zhou, H.-X. 2011. Rapid search for specific sites on DNA through conformational switch of nonspecifically bound proteins. Proc. Natl. Acad. Sci. USA 108: 8651–8656.

Zitnik, M., R. Sosič, M. W. Feldman, and J. Leskovec. 2019. Evolution of resilience in protein interactomes across the tree of life. Proc. Natl. Acad. Sci. USA 116: 4426–4433.

CHAPTER 22

Environmental Sensing

In order to survive, reproduce, and physiologically adjust in appropriate ways at the correct times, nearly all species constantly monitor their internal and external environments. Assessment of extracellular conditions usually involves trans-membrane proteins, with an external domain serving as an environmental sensor, and an internal domain transmitting signals to messenger proteins that further elicit appropriate cellular responses. These signal-transduction (ST) pathways may involve multiple steps, but information exchange almost always involves a series of chemical and/or physical changes in the pathway participants. ST systems are central to the nervous systems of metazoans, but for unicellular species, they are the nervous system.

Unravelling the features of both single molecules and ensembles of them is key to understanding the function and evolutionary properties of communication systems at the cellular level. At the single-molecule level, ST processes are digital in the sense that each molecular participant exhibits a finite number of effectively discrete phenotypes, e.g., active versus inactive conformational states. At the whole-cell level, such information will be distributed over all of the relevant ST molecules in the cell, providing a more graded and accurate assessment of environmental states.

The three main features of all ST pathways are sensitivity, accuracy, and speed. First, external chemical concentrations are often in the μM range (Chapter 18), so cells have to make decisions based on encounters with small numbers of molecules. Second, each ST pathway is devoted to a specific environmental stimulus (or small set of them), the menu of which is very broad, including organic metabolites, inorganic nutrients, markers of pathogens, atmospheric gases, osmolarity,

and antibiotics. With up to several dozen pathways operating simultaneously within the confines of single cells, avoidance of crosstalk among pathways is critical to maintaining coherent cellular responses. Third, the chemical systems involved must generate responses on appropriate time scales. Too slow a response can leave a cell in a compromised physiological state, but too rapid a response can be a major energy drain on the bewildered cell.

Many of the central players in signal transduction are enzymes, so understanding the operation of such systems requires an appreciation of the basic features of enzyme kinetics. But this is not enough. As reviewed in Foundations 19.1, the kinetics of simple one-enzyme systems are such that there is generally a smooth, hyperbolic relationship between substrate concentration and enzyme output. In contrast, ST pathways are generally constructed in ways that can yield much sharper responses of the whole system to external ligand concentrations. In extreme cases, this amplification of external signals can lead to near switch-like behavior in phenotypic outputs.

ST systems invoke many evolutionary questions that remain to be answered in a convincing fashion. First, as the accurate transmission of environmental information along chains of pathway molecules is key to signal transduction, the recurring theme of the evolution of cohesive molecular languages (Chapters 15, 19, and 21) again becomes central. Second, because the recording and erasing of information is energy demanding, questions emerge about the critical threshold above which the gain in information is offset by the energetic cost of building and maintaining a communication system. Third, the innate capacity of many ST systems to generate populations of individuals with discrete alternative

Evolutionary Cell Biology. Michael Lynch, Oxford University Press. © Michael Lynch (2024). DOI: 10.1093/oso/9780192847287.003.0022

states (in the absence of genetic variation) raises questions as to whether such systems are exploited by natural selection to operate as bet-hedging strategies, as opposed to being inadvertent by-products of the structure of ST networks.

In the following pages, these issues will be explored mainly from the standpoint of bacterial ST systems, which owing to their simplicity have been studied in much more detail than the more complicated ST networks typically operating in eukaryotes. A broad overview of the biology of ST systems, more focused on eukaryotes, is given by Lim et al. (2015). Drawing from many examples, Wan and Jékely (2021) have argued that eukaryotes owe their success to the evolution of more diverse and complex sensory systems, implying that these are also more refined in terms of speed and accuracy. However, as noted multiple times in previous chapters, success in the eye of the eukaryotic beholder is often a false caricature of the actual situation. Increased complexity need not mean increased functionality, and it remains unclear if prokaryotic sensory systems are less efficient in terms of time and/or energy, let alone less accurate than those of eukaryotes.

Bacterial Signal Transduction Systems

Relative to the complex ST systems of eukaryotes (below), those of bacteria typically have simple enough structures that their operational features can be dissected in detail. The simplest type of bacterial ST mechanism is the so-called one-component system, which consists of just a single protein (usually a cytosolic transcription factor) with an input domain serving as a signal receptor (sensor) and an output domain transmitting information to a receiver (Figure 22.1). Almost always, the incoming signal is a small ligand molecule that allosterically modifies the protein in such a way as to activate the response domain, which then induces transcription in one or more downstream target genes (Ulrich et al. 2005).

The second most common mechanism of signal transduction in bacteria is the two-component system (Stock et al. 2000; Capra and Laub 2012), the operation of which always involves post-translational modification of protein participants.

The first component in such systems, the signal receptor, is generally a histidine kinase (HK) embedded in the cell membrane. Kinases are enzymes that catalyse the transfer of a terminal phosphate from ATP to an amino acid on a target protein. The extracellular domain of a HK receives environmental information, usually in the form of a small ligand that induces autophosphorylation (addition of a phosphoryl group, PO_3^{-2}) of a specific histidine residue on the internal domain (Figure 22.1). The phosphoryl group is then transferred to a specific aspartate residue on the second (intracellular) component, known as the response regulator (RR). This transfer elicits a conformational change in the RR that in turn induces a specific cellular response, usually with the RR operating as a transcription factor. Almost all HK and RR proteins in such systems have homodimeric structures.

There are thus six key determinants of specificity in a two-component system: the ligand-binding, phosphotransfer, and dimerization domains of the HK, and the HK-contact, DNA-binding, and dimerization domains of the RR. Moreover, most HK proteins are bifunctional in that when not phosphorylated, they operate as phosphatases on their cognate RR proteins. The ratio of kinase to phosphatase activity dictates the output of the pathway. In *E. coli*, an average HK protein is present as 10 to 100 molecules per cell, whereas the cognate response regulators are generally 10 to 100× more abundant.

More complex phosphorelay systems exist in some bacteria, the simplest of which consists of three components (Figure 22.1). Here, as with two-component systems, a membrane-bound HK molecule first autophosphorylates at an internal histidine residue after receiving an appropriate extracellular signal. But in this case, the phosphoryl group is then transferred to a secondary receiver domain, often on the same molecule (usually an arginine residue). Given the presence of both a transfer and acceptor domain in the same molecule, such a protein is often referred to as a hybrid kinase. A separate protein, again a cytosolic RR protein, transfers the phosphoryl group to the final acceptor. Variants on this type of pathway are known, including chains of transfers from a histidine to an arginine to another histidine, and so on. The *Bacillus subtilis* sporulation-control system is an example of

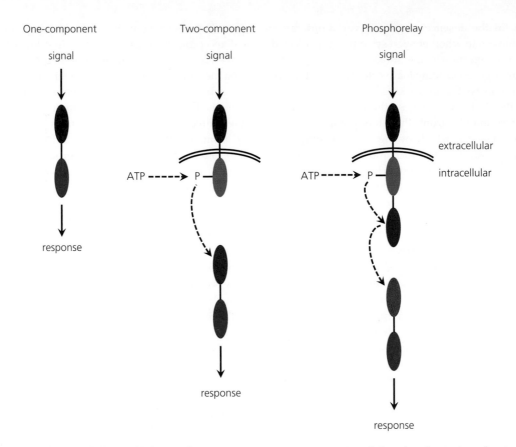

Figure 22.1 Idealized schematics for bacterial one-component, two-component, and phosphorelay systems for signal transduction. Different colored ovals denote protein domains (contained within the same protein if connected by a black line). Double lines denote cellular membranes. For the second two systems, the covalently attached phosphate (P) group is derived from ATP after receipt of an external signal. Phosphorelays can have more complex structures than the one illustrated, with the first P transfer sometimes being to a separate protein, and with multiple players involved in longer chains of reaction. The ultimate response generally involves the operation of the final activated molecule in the pathway as a transcription factor, which binds the regulatory DNA of a target gene.

a chain involving a series of four proteins (Sonen-shein 2000). Note that because all of these systems involve the production of mRNAs that must be translated, there can be a significant delay between the receipt of a signal and the ultimate cellular response.

Origin and diversification

The numbers of both one- and two-component systems scale with roughly the square of genome size in different bacterial species (Ulrich et al. 2005; Alm et al. 2006; Capra and Laub 2012). One-component systems are about 7× more abundant than two-component systems, with most bacterial species

containing dozens (in a few cases, hundreds) of such systems in total (Figure 22.2). The HK and RR genes for most two-component systems reside in the same operons, and hence are coexpressed. Nevertheless, many cases are known in which single members of an operon are duplicated, and as a consequence on average, orphan HK and RR genes are nearly as abundant as those in operons (Burger and van Nimwegen 2008). In addition, although they are a minority, many-to-one and one-to-many HK–RR systems exist (Goulian 2010), with the numbers of HK proteins in a genome usually being up to twice the number of RR proteins (Figure 22.2). This implies that multiple signals are often transmitted through the same response regulator.

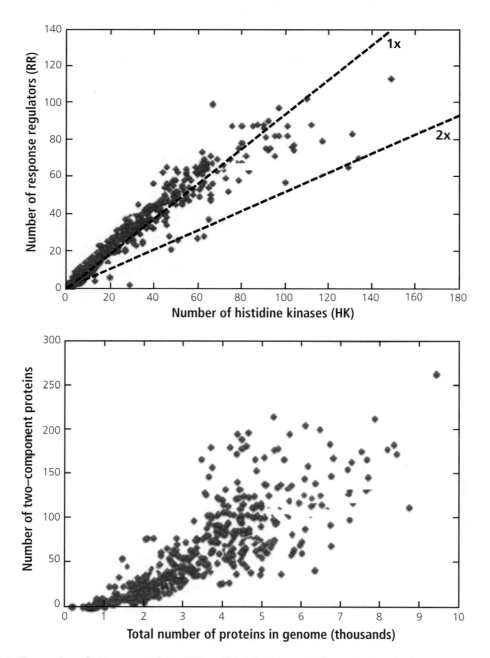

Figure 22.2 The number of response-regulator (RR) and histidine-kinase (HK) proteins involved in two-component systems in a wide range of bacterial species. **Above:** Scaling of the joint numbers of RRs and HKs within genomes. Diagonal lines denote positions of 1:1 and 1:2 ratios. **Below:** The number of proteins associated with two-component systems as a function of the total numbers of proteins encoded within genomes. From Capra and Laub (2012).

Given the very precise mode of operation of all such systems – initial phosphorylation of a histidine on one protein, followed by phospho-transfer to an aspartate on another, it is likely that this was the ancestral state of a primordial two-component system from which most others were subsequently derived by duplication and divergence. Why histidine and arginine became the chosen amino acids remains unclear. Why phosphorylation was settled upon as the key form of post-translational modification is also unclear, although the two negatively charged oxygens associated with each phosphate group provide opportunities for the modification of protein structure by binding with positively charged residues. Such structural changes can then be linked to alterations in protein function in a binary fashion.

The fact that the vast majority of both one- and two-component systems acquire environmental information via small-molecule binding, and elicit a final output via transcriptional regulation motivates the suggestion that one-component systems provide the evolutionary seeds of two-component systems (Ulrich et al. 2005). However, any such transition requires several modifications: minimally, the insertion of a histidine-kinase domain and a receptor domain, the physical separation of these two domains into two separate proteins, and the acquisition of a trans-membrane domain by the HK. Moreover, there is no compelling reason to rule out the alternative possibility that one-component systems are often derived and simplified versions of two-component systems.

Coevolutionary integration of components

The multiplicity of ST systems within cells and their operation by pairwise communication raise many questions about their collective evolutionary properties. One of the central issues concerns the mechanisms by which individual systems avoid miscommunication with non-cognate systems. Each HK generally communicates with a specific RR, although errors must occur. Indeed, some systems exhibit a low degree of crosstalk when examined *in vitro* (Yamamoto et al. 2005), and crosstalk can be greatly enhanced if cognate partners are eliminated *in vivo*, owing to the release from competitive binding (Siryaporn and Goulian 2008). Nonetheless, pathway insulation is in part an ingrained consequence of the biochemical nature of bacterial two-component pathways, in particular the editing-like properties of the HK molecules. When the proper signal for a particular pathway is lacking, its HK remains unphosphorylated and acts primarily as a phosphatase for the cognate RR, thereby tending

to erase inadvertent phosphorylation of an RR by non-cognate HKs.

Mutual HK–RR recognition is generally a function of coevolutionary changes accumulated on just a small number of amino-acid residues at phosphotransfer domains, typically < 6 sites on each protein (Li et al. 2003; Laub and Goulian 2007; Weigt et al. 2009; Capra et al. 2012b). As a consequence, the requirements for specificity rewiring are not high. For example, alterations of just three amino-acid residues of the *E. coli* HK EnvZ (involved in osmoregulation) are sufficient to both eliminate its ability to recognize its cognate RR and to confer full specificity toward another non-cognate RR (Skerker et al. 2008; Capra et al. 2010; Nocedal and Laub 2022). Just single amino-acid changes can lead to mutual recognition of cognate and non-cognate RRs, showing that under the right situations, a HK can gain an ability to recognize a novel RR without relinquishing its initial partner. Indeed, Siryaporn et al. (2010) found that single amino-acid substitutions in an *E. coli* sensor kinase called CpxA (involved in envelope-stress response) can cause the efficiency of signalling to a non-cognate RR to even exceed that of the latter's cognate HK.

In a broader attempt to understand the degree of recognition-motif degeneracy, Podgornaia and Laub (2015) made constructs of all possible $20^4 =$ 160,000 amino-acid motifs for the four key recognition residues in *E. coli* protein kinase PhoQ, a signal receptor for external magnesium concentration. From this pool of variants, 1659 were found to be functional. Extending this analysis even further to include the recognition motif on the cognate response regulator PhoP, McClune et al. (2019) found 58 unique PhoQ–PhoP motif combinations that yielded fully functional systems that were also completely insulated from the native system.

Given that many two-component systems operate in an essentially one-to-one manner, these kinds of observations also suggest the capacity for substantial neutral coevolutionary drift between interacting motifs, even in the face of selection for conserved function. Recalling the theory formally evaluated in Chapter 21, such systems drift in coupled regulatory vocabularies is expected in simple pairwise interactions, so long as the maintenance of a strong degree of mutual interaction is retained (Lynch and Hagner 2015). In principle, this could then lead to the evolution of incompatibilities among mixtures of orthologous HKs and RRs from different taxa, in the absence of any within-species functional changes.

The few experimental attempts to shed light on this matter have yielded mixed results. On the one hand, divergence of the recognition motifs of the components of the bacterial PhoR–PhoB system (involved in phosphate regulation) between members of the Alphaproteobacteria and Gammaproteobacteria is sufficient to nearly completely prevent crosstalk (Capra et al. 2012b). On the other hand, two studies of other systems have shown that the HK gene from a different bacterial phylum can complement the loss of the orthologous *E. coli* gene (Tabatabai and Forst 1995; Ballal et al. 2002). Likewise, the conserved ability to phosphorylate orthologous substrate proteins from distantly related species has been noted for a different class of bacterial signalling proteins, the tyrosine kinases, in this case despite the lack of obvious sequence homology (Shi et al. 2014). These kinds of observations are not necessarily incompatible with a hypothesis of neutral systems drift, although they do highlight uncertainties in the degree to which the evolution of sequence motifs in the individual participants are mutually constrained.

Emergence of new pathways

Although these types of observations make clear that two-component systems can be rewired with only a small number of changes, the challenges in establishing an entirely new ST system are numerous. A common idea is that the evolution of a novel system initiates with duplication of both members of the pair, a scenario made plausible in bacteria by the frequent joint occurrence of cognate HK and RR genes in the same operon. The high degree of congruence between the phylogenetic trees of HK and RR genes conjoined within operons supports this view (Koretke et al. 2000). Linkage within operons ensures that both members of an interacting pair will be coexpressed from their time of origin, an essential ingredient for coevolutionary reinforcement.

However, ultimate preservation by neofunctionalization requires the integration of a new signal input and/or output into a system in a way that avoids crosstalk with the ancestral system (Figure 22.3). This, in turn, requires divergence in the HK–RR communication language used within the duplicated pairs (Capra et al. 2012b). Evidence suggests that the amino-acid sequences within HK–RR interface regions evolve at high rates at least in the early stages of post-duplication divergence (Rowland and Deeds 2014), and dramatic changes in recognition motifs are known to have accumulated among duplicated systems within the Gammaproteobacteria (Capra et al. 2012b).

While these observations are potentially consistent with diversifying selection, modifications at the HK–RR phosphotransfer interface are not sufficient for the insulation of two pathways. There is also a need for changes in the dimerization interfaces of both the HK and RR molecules to prevent heterodimerization between the diverging copies. Consistent with more general observations on the relative simplicity of binding interfaces in multimeric enzymes (Chapter 13), experimental work suggests that changes involving fewer than four amino-acid residues in dimerization interfaces can suffice to establish a new homodimerization group (Ashenberg et al. 2011).

Capra and Laub (2012) and Rowland and Deeds (2014) have suggested that all of these crosstalk interactions must be removed before new input/output functions are acquired, arguing that in large bacterial populations even mutations with very mildly deleterious crosstalk effects would be immediately removed by selection. If this is indeed the case, then a transition to a novel signalling pathway would require an early order of events that is essentially neutral with little to no impact on the overall performance of the ancestral system.

A key difficulty with this hypothesis is the series of steps that must be accomplished – two losses of heterodimerization potential, one for the HK and one for the RR; loss of cross-phosphotransfer potential; and the emergence of at least one new HK–RR phosphotransfer interaction. During the period of time in which this series of events is achieved, both systems must also avoid non-functionalizing

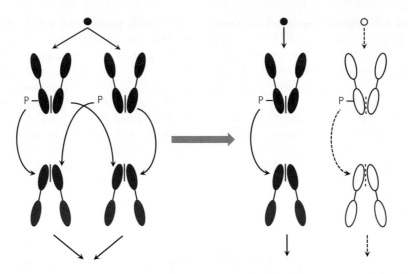

Figure 22.3 A general schematic for what might be necessary for the preservation and long-term functional divergence of a duplicated HK–RR pair. For the particular scenario noted, one system diverges and the other remains the same, although it is not necessary, and perhaps not even likely, that only one pair would undergo functionally significant evolutionary changes. If the two systems are to evolve to be completely insulated from each other: 1) new dimerization domains (vertical lines separating adjacent ovals) need to emerge for both the HK and RR proteins; and 2) a new feature of the phosphoryl-transfer mechanism (large curved arrows) needs to be incorporated. Open and closed black dots denote two types of signalling ligands in the extracellular environment. The red and green ovals denote domains of the ancestral membrane-bound HK proteins, red being the sensory domain, and green being the autophosphorylation domain. Purple and blue ovals denote domains of the ancestral intracellular RR proteins, purple denoting the phosphotransfer domain, and blue denoting the DNA binding (or other output) domain. Right panel: Open versus closed ovals and solid versus dashed lines denote functional changes, with the system on the far right no longer capable of crosstalk with the system on the left.

mutations (which would remove the entire system from selection). It may help that most bacterial populations are so large that most first-step variants (as well as double mutants) are always maintained by recurrent mutation. These might then provide the staging grounds for the emergence of downstream mutations, which can then proceed to fixation without any bottleneck in population fitness (Chapter 5). However, the population-genetic conditions necessary for the origin of an insulated, coevolving pair of proteins need to be worked out formally to resolve these numerous open questions (Foundations 15.1 provides a starting point).

Finally, a plausible case can be made that more complex phosphorelay systems (Figure 22.1) arise by fusion of the HK and RR components of two-component systems to produce a hybrid kinase. In principle, this may involve nothing more than deletion of an intergenic region within an operon,

provided the open-reading frames of both components remain intact (Zhang and Shi 2005; Cock and Whitworth 2007). Such a starting point would facilitate the evolution of a new signalling system, as the HK and RR are open to communication from the onset. In addition, their enforced proximity within the same molecule would reduce the necessity of high affinity between the pair, thereby enhancing the likelihood of evolutionary motif divergence. Consistent with this idea, empirical work has shown that when the kinase domain of a phosphorelay system is disconnected from its receiver domain, the level of crosstalk increases substantially (Wegener-Feldbrügge and Søgaard-Andersen 2009; Capra et al. 2012a). Once novel communication motifs with little potential for crosstalk are established, such a phosphorelay system might then revert back to the structure of a two-component system by a fission event.

Interconvertible Proteins and Ultrasensitivity

Despite their relatively simple modular structures, ST pathways exhibit an array of unusual features at the biochemical and cellular level. Central to understanding such behavior is the concept of an interconvertible protein whose active versus inactive states are defined by the presence/absence of post-translational modifications. As already discussed, the most common case by far is the phosphorylation/dephosphorylation cycle, in which a specific ATP-dependent kinase attaches a phosphoryl group to a particular amino-acid residue on the interconvertible protein, and a specialized phosphatase is responsible for the reverse reaction. In the simplest bacterial systems, the same enzyme is often used for both the addition and removal of the modification, but in eukaryotes different enzymes are generally deployed in each transformation. The joint activities of kinases and phosphatases, along with the concentration of the intermediate protein (here viewed as a response regulator), determine the fractional activity of the latter, and this ultimately dictates the cellular response.

The following discussion denotes three proteins as F (forwards converter, e.g., a kinase), R (reverse converter, e.g., a phosphatase), and I (interconvertible protein) (Figure 22.4). Although such systems have discrete on/off states at the single-molecule level, this is not the case for the entire ensemble of molecules at the cellular level. Instead, the proportional levels of alternative forms of I (active I_a, and inactive I_i) can fall over an essentially continuous range of 0.0 to 1.0, depending on the activities of the converter enzymes (F and R). In one range of parameter space, the forward (kinase) reaction will dominate, and the majority of I will exist in its active form, whereas for other parameter values, the reverse (phosphatase) reaction will dominate, rendering the average molecule of I inactive. However, the real novelty of such systems is their capacity to generate switch-like behavior, from nearly completely off to nearly completely on with just a small change in signal concentration.

The degree to which the fractional activation of I depends on the concentration of the external signal (in the form of a ligand S_F) dictates the magnitude of the overall cellular response. Recall that with standard two-parameter Michaelis–Menten enzyme kinetics, there is a fairly gradual response of the system output to the substrate concentration, describable with just two parameters (Chapter 19). In contrast, with a triad of interacting proteins, more than 10 parameters, including the kinetic coefficients of enzymes F and R, determine the partitioning of the total concentration of I into its active and inactive forms (Figure 22.4; Foundations 22.1). As a consequence, the level of I activation can exhibit a far richer array of behaviors than possible with basic enzyme kinetics, even though both enzymes F and R behave as Michaelis–Menten enzymes with I as their substrate.

Consider the situation in which the active versus inactive form of converter enzyme F depends on whether it is bound to its ligand S_F (Figure 22.4), and recall that with Michaelis–Menten enzymes, the rate of a reaction is hyperbolically related to the substrate concentration, $[S_F]$, as described with Equation 19.1.4 (throughout brackets denote concentration). With increasing $[S_F]$, enzyme F is expected to be increasingly converted to its active form F_a. If, however, F_a feeds into a loop involving the interconvertible enzyme I, the fraction of active enzyme, $I^* = [I_a]/[I_T]$, can reach much higher levels than the fraction of active F at low levels of the input substrate S_F (Figure 22.5, upper panel). In other words, the signal from the external ligand can be substantially amplified. This is because the total concentration of I constitutes a closed system, enabling F_a to cumulatively convert I_i to I_a.

The degree of amplification also depends on the kinetic parameters of the reverse converter, which provides the only route back to I_i. As the forward conversion reaction increasingly overwhelms the reverse reaction, the phosphorylation reaction dominates, and $[I_a] \rightarrow [I_T]$. On the other hand, as the kinetic efficiency or amount of enzyme R increases, the phosphatase reaction increasingly dominates, and the system can converge to situations in which I can never attain a fully active state, even at the highest concentrations of the external ligand (Figure 22.5, upper panel).

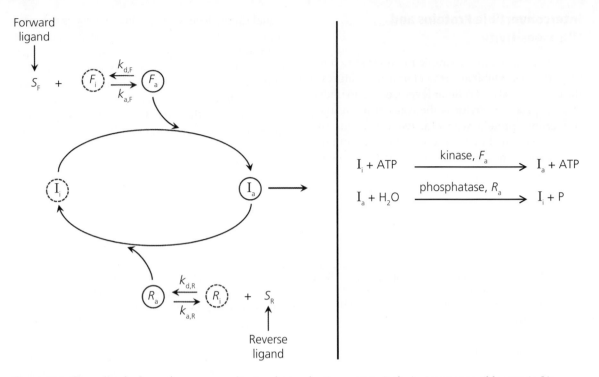

Figure 22.4 Generalized scheme for a monocyclic signal-transduction system. **Left:** An interconvertible protein I is transformed between active (a) and inactive (i) states by forward- and reverse-acting (F and R, respectively) enzymes, which themselves have active and inactive forms dependent on whether they are bound to their respective ligands, S_F and S_R. The k coefficients denote the association (a) and dissociation (d) constants between the ligands and the converting enzymes. **Right:** The reaction equations involving the interconvertible enzymes, with P denoting inorganic phosphate.

A key assumption underlying the preceding results is that the total concentration of I is substantially below the half-saturation constants for the forward and reverse reactions (Stadtman and Chock 1977), such that the active forms of enzymes F and R are not saturated by their substrate. In this case, a simple expression can be obtained for the fraction of activated intermediate enzyme,

$$I^* = \frac{[I_a]}{[I_T]} = \frac{\kappa_F[F_a]}{\kappa_R[R_a] + \kappa_F[F_a]}, \qquad (22.1)$$

where κ_x is the kinetic efficiency of enzyme x operating on substrate I (from Equations 22.1.7a,b). In this non-saturating case, the forward and reverse rates of conversion of I are both linearly related to their substrate concentrations, and I* is independent of the total concentration of intermediate enzyme, $[I_T]$, in the system. Furthermore, because the amounts of active converter enzymes, $[F_a]$

and $[R_a]$, are Michaelis–Menten functions of their ligand concentrations (Equations 22.2.3a,b), I* is also a conventional hyperbolic function, in this case of $\kappa_F[F_a]$.

Goldbeter and Koshland (1981) found that with increasing concentration of I in the system (so that the responses of F_a and R_a to their substrate concentrations are no longer linear), the response of I* to ligand concentrations is no longer hyperbolic or independent of $[I_T]$. Rather, the steepness of the activation response to ligand concentrations elevates dramatically with increasing $[I_T]$, in the extreme becoming an effectively stepwise process (Figure 22.5, lower panel). This sharp response, often referred to as zero-order ultrasensitivity, arises because high levels of I allow the converter enzymes to operate at maximum capacity, thereby sharpening their responses near the threshold between the kinase and phosphatase

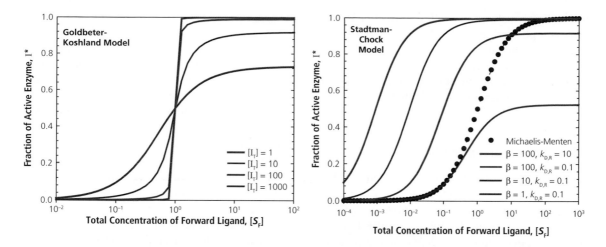

Figure 22.5 Response of the active fraction of an interconvertible protein (I^*) to the concentration of ligand for the forward converting enzyme, [S_F]. Both plots are derived using Equation 22.1.8, with the upper plot denoting the limiting behavior when the concentration of I is well below the half-saturation constants of the converting enzymes, Equation 22.1.6. **Left:** Results are given for the situation in which the concentration of the ligand for the reverse enzyme, [S_R], is set equal to 1.0, with increasing levels of [S_F] (according to Equation 22.1.6, the results depend only on the ratio of these two concentrations). The dissociation constant for the forward enzyme and its ligand (equivalent to its half-saturation constant) is $k_{D,F} = 1.0$, with results given for different values of $k_{D,R}$ (for the reverse enzyme). The parameter β is the ratio of kinetic potentials of the forward- and reverse-converter enzymes, as defined in Foundations 22.2. As β and $k_{D,R}$ increase, autophosphorylation increasingly dominates and the response curves shift to the left. The black dotted line denotes the situation that would be expected if the response followed the Michaelis–Menten enzyme kinetics of the forward converter and its ligand. **Right:** Results are given for increasing total concentrations of the interconvertible enzyme. Both of the ligand concentrations and all of the catalytic coefficients and half-saturation constants are arbitrarily set to 1.0.

domains on the scale of ligand concentrations. At low levels of the external ligand, the reverse (phosphatase) reaction dominates, and with a high level of I pushes the rate of a conversion to I_i to the maximum. At high levels of the external ligand, the forward reaction overwhelms the reverse reaction, pushing I_a to the maximum level.

To summarize, the use of an interconvertible-enzyme system can alter both the sensitivity and the amplitude of response of a signalling system to the input ligand concentration. Most notably, for sufficiently high concentrations of the central enzyme I, near switch-like behavior of the population of active I molecules arises. Thus, should a selective scenario exist in which switch-like behavior is advantageous, mutational fine tuning of the kinetic parameters of the enzymes underlying the ST system, combined with the maintenance of a sufficiently high level of

I, can provide an evolutionary path towards such behavior.

This being said it is an open question as to whether the switch-like behavior implied by the mathematics of these kinds of systems commonly occurs in cells, let alone is promoted by selection. In fact, zero-order ultrasensitivity has not yet been directly demonstrated in *in vivo* ST systems (Blüthgen 2006). Moreover, as pointed out by Ortega et al. (2002) and Xu and Gunawardena (2012), such an extreme response requires that the phosphorylation/dephosphorylation reactions are intrinsically irreversible, such that a predominating kinase can literally drive I to the point at which all molecules are in the active state. With most enzymes, a high product concentration drives reactions in their reverse direction, leading to a steady-state situation in which both active and inactive molecules coexist within a cell. The situation is even more complicated in bacterial two-component systems, where the same enzyme often serves both the F and R functions, rendering the maintenance of high substrate concentrations for both enzymatic functions (necessary for ultrasensitivity) difficult to achieve.

The cost of signal transduction

Acquisition, processing, and propagation of information requires energy, whether via a computer (Landauer 1988) or by the nervous system of a metazoan (Mehta and Schwab 2012; Niven 2016; Kempes et al. 2017; Levy and Calvert 2021). Quiescent nerve tissue consumes energy just to maintain a steady response capacity, and the same is true for the interconvertible enzymes at the heart of ST systems. In addition to the structural costs, there is the matter that the relay signal (the relative concentrations of active and inactive I molecules) must be constantly adjusted by the simultaneous running of two pathways (phosphorylation and dephosphorylation) in opposing directions in order to convey information on the cellular environment.

For this reason, the dynamics of the proteins underlying ST systems are often referred to as push-and-pull or futile cycles – even when the system settles into a steady-state in a constant environment, active and inactive I molecules are continuously being interconverted, leading to a cyclical flux. However, the word futile is a bit of a misnomer here, in that phosphorylation/dephosphorylation cycles are the price a cell must pay to keep a ST system in a constant state of readiness.

Because each phosphorylation event requires hydrolysis of an ATP molecule, a rough idea of the cost of maintaining a particular level of system activity can be obtained by noting that at steady-state, the reciprocal rates of activation/inactivation of I molecules must be completely balanced and equal to the rate of ATP consumption. The cost of running an ST system can then be obtained from measures of *in vivo* rates of ATP consumption by the pathway kinases. For example, Shacter et al. (1984) noted that the flux through hepatic (liver) pyruvate kinase is $V_{ATP} = 20$ to $200 \, \mu M/min$, depending on the level of activation of the intermediate protein.

Assuming a cell volume of $\sim 5000 \, \mu m^3 = 5 \times 10^{-12}$ liters, a cell-division time of 24 hours, and converting moles to number of molecules using Avogadro's number, the total rate of ATP consumption per cell by this kinase is then of order 10^{10} to 10^{12} molecules/cell division.

As usual, a quantitative understanding of what this energetic cost means to the cell requires information on the total cellular energy budget. From Chapter 8, for a cell of this size and cell-cycle length, total cell maintenance costs are $\sim 2 \times 10^{13}$ ATP hydrolyses/cell cycle, implying that a single mammalian ST system can demand up to 5% of a cell's basal-maintenance energy budget. From a knowledge of the concentration of the kinase and phosphatase in this system and their molecular sizes (~ 2600 and ~ 1430 amino acids, respectively), the construction cost of the entire system can be shown to be $\sim 2 \times 10^{11}$ ATPs (not including the cost of the interconvertible enzyme). Thus, in this particular example, the costs of running and building the system are of the same order of magnitude.

Estimates for other mammalian kinases derived in Goldbeter and Koshland (1987) are in this rough range as well. Studies with yeast, ciliates, and zebrafish embryos have shown that such expenditures in cell signalling can be sufficient enough to be discerned as heat-dissipation oscillations during the eukaryotic cell cycle (Poole et al. 1973; Lloyd et al. 1978; Rodenfels et al. 2019).

Similarities and differences in eukaryotic systems

Eukaryotic signal-transduction systems are generally much more complex than those in bacteria. Although kinases and phosphatases are still broadly utilized, different amino acids typically serve as the sites of phosphorylation – usually serine, threonine, and tyrosine, with the latter largely confined to metazoans. Moreover, whereas bacterial histidine kinases produce phosphoramidates by phosphorylating side-chain nitrogen atoms, eukaryotes phosphorylate oxygen atoms on serine, threonine, and tyrosine residues, creating phosphoesters, which are much more energetically stable. The reasons for this shift in target residues is unclear, but one possibility is that enhanced stability is essential for

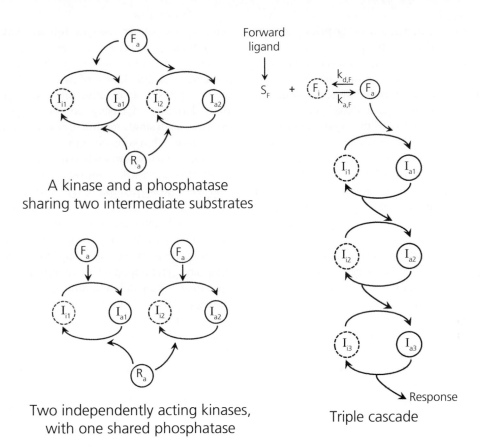

A kinase and a phosphatase
sharing two intermediate substrates

Two independently acting kinases,
with one shared phosphatase

Triple cascade

Figure 22.6 Three variants on the structure of signal-transduction pathways commonly found in eukaryotes, with notation as in Figure 22.4.

activated molecules that have to travel larger distances in eukaryotic cells. Some two-component systems involving autophosphorylating histidine kinases are known in plants, fungi, and slime molds, but they are absent from a number of eukaryotic lineages, notably metazoans (Loomis et al. 1997; Schaller et al. 2011).

Given that eukaryotes are derived from archaea, it is of interest to know the nature of ST systems in the latter, but there is an unfortunate void of knowledge here. It is known that many archaea are entirely lacking histidine kinases, and that perhaps all have serine/threonine kinases (Makarova et al. 2017). Most unusual is the apparent exploitation of KaiC-like ATPases in a wide variety of signalling pathways across the archaeal phylogeny. Recalling from Chapter 18 that KaiC is at the heart of

the phosphorylation/dephosphorylation cycle that forms the circadian clock in cyanobacteria, the latter may have been acquired by horizontal transfer from archaea.

The numbers of kinases in eukaryotes scale with roughly the square of the total number of proteins, similar to what is seen in bacteria, and substantial lineage-specific expansions of particular families have arisen (Anantharaman et al. 2007). However, in eukaryotes, each of the interacting enzymes typically engages with multiple substrate proteins (Figure 22.6), often more than a dozen, leading to complex networks quite unlike the well-insulated systems of bacteria. Hence, the elegant molecular dissections that have been accomplished for bacterial ST systems are a rarity for eukaryotes. Nonetheless, for the few simple eukaryotic systems that have been investigated, many of the principles outlined above for bacteria with respect to motif evolution still apply. For example, a transcription factor involved in the response to amino-acid starvation in the yeast *S. cerevisiae* (Gcn4) is targeted for degradation by a specialized kinase (Pho85), but whereas the same system operates in the yeast *Candida albicans*, the cross-species components are incapable of molecular recognition (Gildor et al. 2005). As discussed for bacteria, this kind of coevolutionary wandering of motif language in the face of conserved function is consistent with the operation of mutually constrained systems drift.

There can, however, be limits to the degree to which such wandering extends. For example, Zarrinpar et al. (2003) found that a kinase (Pbs2) involved in the osmoregulation pathway in *S. cerevisiae* interacts specifically with one particular membrane-bound sensor protein, despite the presence in this species of 26 other related sensor proteins with rather similar recognition sequences. However, when orthologs of these off-target proteins from other distantly related species were presented to Pbs2, strong cross-species recognition often occurred. The implication is that there has been significant negative selection within yeast to avoid off-target interactions, with the suppressing motifs diverging among lineages.

Unlike the situation in bacterial ST systems where just a single amino-acid residue is typically modified on the intermediate protein to elicit a response, eukaryotic ST proteins commonly have multiple phosphosites (Chapter 14). Because the activity of the modified enzyme can require a complete set of phosphorylated sites, and the sequential ordering of marks may follow a rigid recipe, this introduces novel twists to the types of models discussed in the preceding section. Contrary to the switch-like behavior found with simple systems with single-site modifications, the use of multiple phosphorylation sites for activation leads to a more graded response. Although there is still a threshold level of substrate below which the system is inactive, there can be a simple Michaelis–Menten-like response above the threshold (Gunawardena 2005). Thus, although the reason for the use of multiple phosphosites in eukaryotic ST-pathway enzymes remains unclear,

refined enhancement of switch-like behavior does not seem to be a viable hypothesis.

Limited attention has been given to the consequences of the kinds of shared utilization of kinases/phosphatases illustrated in Figure 22.6 (left). However, focusing on the simple case of enzymes with single modifiable sites, Rowland et al. (2012) found that coupled systems can often behave in a transitive fashion, such that if one intermediate substrate saturates the controlling enzymes in a way that leads to switch-like ultrasensitivity, all other connected intermediate enzymes will behave in the same way. In effect, saturation by one intermediate substrate alters the controlling enzymes (F and R) to fully active states, governing the entire system. This can even happen when all multiple intermediate substrates are below saturation levels, provided the aggregate is sufficient for saturation. Whether these kinds of collective effects are evolved mechanisms for coordinated ultrasensitivity in eukaryotic cells or simple inadvertent consequences of complex networks remains unclear.

One of the most pronounced differences between eukaryotic and bacterial ST systems is the extended length of the former, with a relay of three kinases being particularly common (Figure 22.6, right). For example, MAP (mitogen-activated protein) kinase kinase kinases phosphorylate MAP kinase kinases, which in turn phosphorylate MAP kinases that finally transmit information to a response regulator. Many variants of the MAP kinase family are deployed in wide variety of eukaryotic cellular processes, and their efficiency of operation and degree of insulation is enhanced by the use of scaffold proteins that physically link all three layers of each pathway into single complexes.

How and why ST pathways with extra steps evolve remains unclear. Armbruster et al. (2014) argue that additional steps enable pathways to integrate out the effects of environmental noise on the elicited response, but again, this leaves unanswered the question as to why bacteria would not take advantage of such possibilities, especially given the likelihood that small bacterial cells may be more subject to stochastic effects than their larger eukaryotic counterparts. An alternative possibility is that, as with gene-regulatory systems (Chapter 21), more complex systems passively emerge in lineages experiencing elevated levels of random genetic drift.

Finally, eukaryotes harbor another broad class of intracellular messaging systems, the so-called G proteins, used in a wide range of cellular activities, including import/export through nuclear pores and vesicle transport (Chapter 15). These proteins operate in a quite different way than those noted earlier, through binding of GTP rather than via phosphorylation of amino acids. Nonetheless, the kinetics of the overall systems follow the same general plan as interconvertible proteins. G proteins have alternative on/off states driven by opposing enzymes responsible for GTP addition and removal, which in turn lead to conformational changes in the substrate G protein: writers called guanine nucleotide exchange factors (GEFs) add GTP to the G protein, putting it in the active state, whereas erasers called GTPase-activator proteins (GAPs) hydrolyse the GTP to GDP. Most GEFs are membrane-bound G-protein-coupled receptors (GPCRs) that become activated upon binding an appropriate ligand, presenting a still further analogy with the types of systems noted above. A whole realm of theoretical investigation, involving all of the issues noted in Chapter 21 for gene expression, awaits exploration here (Kapp et al. 2012).

Chemotaxis

The most ornate ST systems in bacteria drive chemotaxis, which directs motility towards particular chemo-attractants (or away from repellents). Although such systems have a kinase and a response regulator at their core, the output is modulated by up to nine other participants (Figure 22.7). The system operates like a 'bacterial eye'. Rather than being membrane-bound, the histidine kinase, CheA, is located at the base of a sensory complex, connected to a special set of chemoreceptors (called methyl-accepting chemotaxis proteins or MCPs) residing at the cell surface. Rather than functioning as a transcription factor, the response regulator, CheY, interacts directly with the base of the flagellum, influencing the direction of flagellar rotation and thereby the swimming behavior. This allows for a much more rapid response than with

conventional two-component systems that regulate gene expression.

The MCPs typically sit in one or two large hexagonal arrays at the cell surface, organized in a honeycomb-like form of thousands of receptor dimers, which serve as relays to CheA (Briegel et al. 2009). In *E. coli*, there are five forms of MCPs, with various sensitivities to different ligands such as sugars or amino acids. As the different forms are mixed together within the arrays, this allows the cell to simultaneously process complex information about the environment. Cooperative interactions between adjacent elements sharpen the response (Mello et al. 2004). Absent from eukaryotes, this type of sense organ appears to have been horizontally transferred to some archaeal lineages (Briegel et al. 2015), even though the archaeal flagellum evolved independently (Chapter 16).

Although the complex kinetic and dynamical features of the *E. coli* chemotaxis system have been worked out in considerable detail (Barkai and Leibler 1997; Keymer et al. 2006; Mello and Tu 2007; Bitbol and Wingreen 2015; Colin and Sourjik 2017), these technicalities are not covered here. The key point is that bacterial chemotactic responses are achieved by comparing a current ligand concentration with that in the recent past, reflected in part by the level of receptor methylation. Specialized methyltransferases and methylesterases regulate the methylation level of four key residues on the MCPs, providing them with a capacity for adjusting the level of sensitivity over a broad range of ligand concentrations. High levels of methylation result from high levels of ligand concentration, but by altering ligand affinity, this resensitizes the system to higher levels of chemoattractant that would otherwise be saturating. Thus, the system operates similarly to the way in which a vertebrate eye adjusts to different levels of light, enabling the cell to maintain a constant sensitivity to changes in ligand concentration regardless of the absolute ligand concentration.

Transmission of information on the external ligand concentration to the modified response regulator is accomplished through mechanisms similar to those noted above. In this case, CheA and then CheY become phosphorylated/dephosphorylated in the absence/presence of ligand binding, and

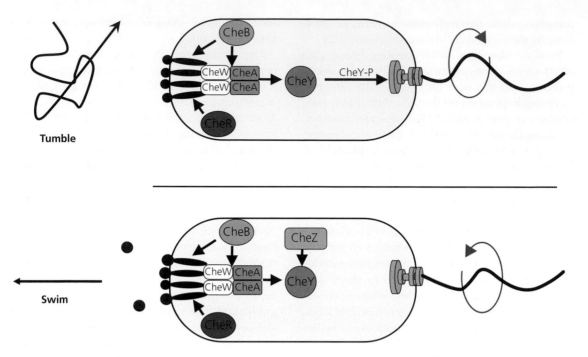

Figure 22.7 Idealized schematic of the chemotaxis pathway in *E. coli*. The histidine kinase CheA is linked to the external sensors by another protein called CheW. In the absence of ligand binding, CheA becomes phosphorylated, thereby phosphorylating CheY, which binds to the base of the flagellum, inducing clockwise rotation and random tumbling. In the presence of ligand binding (red dots) and dephosphorylation of CheY, the flagellum rotates in a counter-clockwise fashion, causing the cell to propel forward in a directed manner. Proteins not specifically mentioned in the text: CheZ is a phosphatase that acts on CheY-P; CheR and CheB are methylases and demethylases that operate on the sensor proteins (MCPs, methyl-accepting chemotaxis proteins; dark circles and ovals on the left) and modify their sensitivity to external ligand concentrations.

the resultant switch dictates whether the flagellum rotates in a clockwise versus counter-clockwise fashion. When phosphorylated, CheY becomes bound to the flagellar base, which causes tumbling and a change in swimming direction, whereas CheY disconnects when dephosphorylated by phosphatase CheZ, inducing the swimming that propels the cell forward. This sort of guided behavior then leads to a biased random walk in the direction of a perceived chemical gradient.

As noted above, the maintenance of the information utilized in ST systems requires an energy expenditure, and in this rapidly responding CheA/CheY system, the costs are quite high. For an *E. coli* cell, the opposing processes of phosphorylation and dephosphorylation consume $\sim 5 \times 10^6$ ATPs per hour (Lan et al. 2012; Govern and ten

Wolde 2014), which is about 10% of the cost of swimming in this species (Chapter 16). Additional ATPs must also be consumed in the opposing methylation and demethylation reactions operating on the MCPs, although the numbers are unclear.

Approximately 50% of bacterial species have a chemotaxis system, although the architectural features vary widely, with some having additional proteins outside of the core CheA and CheY (Wuichet and Zhulin 2010; Abedrabbo et al. 2017). Whereas *E. coli* has five types of receptor proteins, the number ranges from 1 to 30 in other bacteria (Wadhams et al. 2005). For example, the purple photosynthetic bacterium *Rhodobacter sphaeroides* has nine different receptor proteins, as well as four versions of CheA and six of CheY, which engage in non-random crosstalk, presumably broadening the capacity for environmental differentiation (Porter and Armitage 2002, 2004).

A clear example of the evolutionary rewiring of such systems is revealed by the contrast between the *E. coli* network and that in the soil bacterium *Bacillus subtilis*. Whereas interaction of phosphorylated CheY with the flagellar motor induces clockwise rotation and tumbling in *E. coli*, it induces counterclockwise rotation and directional swimming in *B. subtilis*; and whereas CheZ dephosphorylates CheY-P in *E. coli*, this function is carried out by a flagellar motor protein in *B. subtilis* (Szurmant et al. 2004; Yang et al. 2015).

Chemotaxis enables organisms to move up resource gradients, thereby leading to an elevation of cell growth and clonal expansion. However, the mechanism of accrual of any such advantage may be more nuanced. In an otherwise homogeneous environment, populations of organisms at the edge of their range will indirectly generate a gradient of a chemo-attractant through their own activities, thereby causing continued migration (Adler 1966). Indeed, Cremer et al. (2019) found that even under conditions in which resources are non-limiting to growth, *E. coli* still migrate towards a chemo-attractant of no nutritional value. Such range expansion increases the overall population growth rate, as the migrating cells at the leading edge extend the distribution over a greater area, while the laggards utilize the still plentiful nutrients in the space left behind. Clones modified to be insensitive to chemo-gradients experience lower overall growth rates because more cells experience local resource limitation. Thus, although the usual view is that chemotaxis has evolved as a mechanism for moving towards more immediately beneficial conditions, these observations suggest a role for simply expanding into unoccupied locations even under nutrient-replete conditions.

It is, however, unclear whether the kinds of spatial structure that can arise on a completely unoccupied solid surface, as employed in these experiments, generalize to other settings. Notably, when *E. coli* is grown in a well-mixed liquid environment (which prevents the development of chemical gradients), cells increase their investment in motility when grown in nutrient-poor conditions, consistent with the idea that such a shift is a searching mechanism for more nutrient-rich situations (Ni et al. 2020). In addition, although laboratory populations of *E. coli* exhibit chemotaxis towards amino acids that serve as more nutritional resources, such a correlation does not exist for *B. subtilis* (Yang et al. 2015). Such mixed results may exist because utility in a laboratory setting need not reflect the conditions under which differential chemosensitivity evolved in nature. For example, whereas amino acids are commonly used as nutrients in the intestinal bacterium *E. coli*, this may not be the case in the soil bacterium *B. subtilis*.

Much less attention has been given to the mechanisms of chemotaxis in eukaryotes, outside of issues related to cell migration and signalling in metazoan development. Unlike small bacteria, whose arrayed chemoreceptors monitor nutrients in a temporal manner, larger eukaryotic cells populate their entire surfaces with receptors and are able to sense spatial concentration gradients of <5% from the front to the rear of the cell (van Haastert and Postma 2007).

The best eukaryotic example comes from the slime mold *Dictyostelium discoideum*. During times of nutrient scarcity, the amoeboid cells of *Dictyostelium* aggregate into multicellular slugs, with recruitment being induced by waves of cyclic AMP emanating from the aggregation center about every six minutes. Surrounding cells respond by producing pseudopods in the direction of the front of the plume, and continue to do so even after the wavefront has passed such that the cells are confronted with a downward gradient. Skoge et al. (2014) found that the solution to this 'back of the wave' problem involves memory-like processes associated with positive feedback. As a wave approaches, the front of the cell is sensitized, and the message to move forward persists for several minutes owing to the slow decay of the positive-feedback mechanism. The kinetics of the response system appear to have coevolved with the signalling system, as exposure to waves with periodicities exceeding six minutes leads to reversals in migratory behavior.

Accuracy of environmental assessment

To improve fitness, the environmental sensing mechanisms of cells must provide accurate information on the concentrations of relevant ligands in the surrounding medium. However, the capacity to make environmental assessments resides in the degree to which signal receptors on cell surfaces bind to their ligands, which is an inherently stochastic process, owing to fluctuations in molecular arrival times and binding success at individual receptors. This raises significant questions about the conditions under which chemoreception can actually convey accurate information about environmental conditions. These problems were first analysed by Berg and Purcell (1977), who viewed the expected degree of occupancy of a receptor as a counting mechanism for assessing ligand concentrations (Foundations 22.2).

Considering the features of a single receptor molecule, the basis for their primary result starts with the assumption that the long-term average probability of occupancy of a receptor is described by a function of Michaelis–Menten form $p = c_0/(K_D + c_0)$, where c_0 is the environmental concentration of the ligand, and K_D is the dissociation constant, equivalent to the concentration at which the receptor has a 50% probability of being bound (Chapter 19). Rearrangement of this expression shows how p provides information about c_0. For low ligand concentrations, $c_0 \ll K_D$, the relationship between p and c_0 is essentially linear ($p \simeq c_0/K_D$), but with increasing c_0, the expected degree of occupancy approaches saturation (i.e., $p \simeq 1.0$). Thus, to be most effective in transmitting information, i.e., to maximize the response of p, a receptor should have a dissociation constant larger than the typical environmental concentration.

Owing to the transient nature of binding, individual receptors have binary states at any particular time (bound or unbound). Thus, accurate assessment of information via the degree of occupancy requires a long enough time for the averaging of repeated instances of binding and unbinding. Assuming the cell continuously monitors the environment for a time period T, a measure of the error in inference of the true environmental concentration (c_0) is provided by the coefficient of variation

(CV, ratio of the standard deviation of the inferred concentration to c_0)

$$\frac{\sigma_c}{c_0} = \sqrt{\frac{1}{2Dr_s c_0 (1-p)T}}, \tag{22.2}$$

where D is the diffusion coefficient for the ligand, r_s is the radius of the receptor at the cell surface (the target size), and p is the function of c_0 defined above (Berg and Purcell 1977).

Full derivation of this expression is given in Foundations 22.2, but its final structure meets intuitive expectations. The quantity $4Dr_s c_0$ is the rate at which a ligand particle diffuses to a receptor, and a longer T means that the receptor can integrate information over a longer series of bound and unbound states. Thus, the level of noise in assessment scales negatively with the encounter rate and time, but positively with the degree of occupancy p. As $p \to \infty$, the surface receptor becomes saturated, providing little information on the environmental state, i.e., $\sigma_c/c_0 \to \infty$. As noted in Foundations 22.2, given typical estimates of D, even for quite low ligand concentrations, a measurement duration of several seconds can be sufficient to reduce the CV to near 0.01, equivalent to $\sim 99\%$ accuracy in the estimation of the true environmental concentration.

The statistical relationship conveyed by Equation 22.2 is just one of many possible criteria for evaluating the accuracy of an environmental-monitoring mechanism. Arguing that only the length of unbound periods provides information on the environmental concentration of ligand, Endres and Wingreen (2009) showed that if the cell were instead somehow able to sense the duration of unbound intervals and use this alone as an estimate of c_0, the uncertainty in Equation 22.2 would be reduced by a factor of $1/\sqrt{2}$. Another alternative arises if the 'counter' resides within the cellular interior, in which case the result in Equation 22.2 would need to be multiplied by 1.6 (Berg and Purcell 1977); the noise is elevated in this case because ligand molecules transiently trapped within the cell (as opposed to being released into the environment) can be recounted, effectively reducing the number of independent evaluations per unit time. Finally, whereas the preceding calculations assume that the cell is evaluating a constant environment, gradient sensing (i.e., monitoring the rate of change of ligand concentration, as in swimming up a gradient, or contrasting the inferred concentration at two ends of a stationary cell) might be employed. Endres and Wingreen (2008) found that gradient sensing at the cell surface yields a measure identical to Equation 22.2, whereas monitoring inside the cell leads to a 2.9× increase in the noise level, again emphasizing the advantages of monitoring at the cell surface.

Notably, all of these refinements only change Equation 22.2 by a constant multiplier. However, these measures also ignore the biochemical aspects of binding to receptors, under the assumption that the whole sensing process is essentially diffusion limited; the necessary modification for including the former is described in Foundations 22.2. In addition, note that the measure of noise outlined here does not necessarily translate linearly to that expected after transmission to the downstream response regulator, i.e., to the resultant behavioral change, an issue taken up by Mehta and Schwab (2012). Although uncertainty remains as to how cells actually count, these varied formulations illustrate an array of potential mechanisms that may be just as accurate as human decision-making processes.

Questions remain as to the optimal spatial configuration of collections of sensor molecules (Iyengar and Rao 2014). Spatial clustering reduces the sampling error in the vicinity of the array, whereas the spreading of receptors across the cell surface improves average sensing in spatially variable environments. Even in an environment that is spatially uniform on the scale of cell length, sensor aggregation can enhance information transfer into the cellular interior if cooperative interactions exist among adjacent receptor molecules, as appears to be the case for bacterial sensor arrays (Briegel et al. 2009). Although information on the key microanatomical features remains to be determined, with n effectively independent receptors, the denominator of Equation 22.3 would just be multiplied by $\sqrt{n}$. With cooperation among receptors, the denominator needs to be multiplied by $\sqrt{n^x}$ with $x > 1$.

There remain many unanswered evolutionary questions in this area. Relative to the situation with the nervous systems of metazoans, how much of the energy budgets of single-celled organisms is devoted to environmental monitoring and decision making? What allocation of resources to environmental sensors optimally balances the costs of production of such molecules and the advantages accrued? Confronted with increasing levels of environmental variation, at what point does sensory overload and the energetic cost of running a chemoreception system offset the advantages of environmental tracking? What role does the timescale of environmental fluctuations (e.g., within vs. between generations) play in these processes? Evolutionary theory relevant to these questions can be found in Lynch and Gabriel (1987), Lan et al. (2012), and Govern and ten Wolde (2014), but to be of full use, this work will need to be integrated with the known cell biological features of chemosensory systems.

Phenotypic Bimodality and Bet-Hedging

ST systems provide one means for physiological acclimation within the lifespan of an individual cell. An alternative mechanism for dealing with environmental stochasticity is to generate phenotypic diversity independent of current environmental information. Here, the focus is not on genetic polymorphism, but on the production of variable offspring by individual genotypes. This second option provides a potential genotypic advantage in that a segment of a clonal population may be immediately poised to deal with an environmental shift, but this comes at the cost of being maladapted at other times. Moreover, the possibility exists that all phenotypes will be suboptimal on some occasions. If phenotypic diversification among clonal progeny is to be promoted by natural selection, it must increase the long-term genotypic growth rate relative to that in other clones.

Although phenotypic distributions of many biological traits are continuous in form (Lynch and Walsh 1998), striking cases of discrete bimodal states are known in microbes, e.g., dispersing versus sedentary states, vegetative reproduction versus spore formation, and activation versus silencing of metabolic pathways. In some situations, different lineages of genetically identical cells can become trapped in alternative states for indefinite periods, even after the initiating environmental signal has dissipated. Such a condition is known as hysteresis.

There are a number of ways by which such phenotypic switching can be modulated by the types of ST-pathway architecture noted earlier. Consider the case of ultrasensitivity illustrated in Figure 22.5. Should cells straddle the threshold point, either because of stochastic internal cellular variation (e.g., molecular inheritance and/or transcriptional noise) or external environmental variation in ligand concentrations, individuals would receive entirely different messages for downstream phenotypic modification. In other words, adding noise to the otherwise deterministic model will lead to some level of phenotypic switching. In this case, the duration of dwell times in alternative states would depend on the magnitude of fluctuations and the degree to which they are sustained.

Sustained bimodality can arise when there is positive feedback between the activated intermediate substrate and its activation enzyme, as shown graphically in Figure 22.8. In the absence of feedback, the rate of activation of the intermediate substrate declines smoothly to 0.0 as the fraction of active intermediate enzyme (I^*) approaches 1.0. In this case, with the rate of deactivation increasing smoothly with increasing I^*, the intersection between these opposing functions implies a single stable steady-state point for I^* (Figure 22.8a). When I^* exceeds this point, the rate of deactivation exceeds that of activation, and I^* declines, and vice versa, if the starting point is below the equilibrium steady-state.

If, however, there is positive feedback between I^* and the activating enzyme, the form of the activation function can be altered in such a way that there are up to three alternative equilibria for I^* (Figure 22.8b). In this case, when I^* is sufficiently low, increasing I^* further accelerates its own production by positive feedback, but eventually the rate of activation must decline (as in the case of no feedback, as a natural consequence of the reduction in inactivated substrate). If the inflection in the activation function leads to three intersections with the

deactivation function, the intermediate equilibrium will be unstable, with deviations in either direction resulting in movement towards the flanking equilibria, both of which are stable. Thus, depending on their starting states, cells will gravitate towards one or the other alternative stable states, and remain there until sufficiently large fluctuations in the internal and/or external environments shift I^* into an alternative basin of attraction. Such a system is said to exhibit bistability.

Assuming that the different equilibrium levels of I^* are sufficient to lead to altered downstream patterns of gene expression, bistability of an underlying ST system provides a basis for eliciting discrete differences in phenotypic states among otherwise genetically identical cells. The relative frequencies of alternative states will depend on the underlying enzyme kinetics of the system and the magnitude and frequency of fluctuations in the governing parameters.

Bistability can arise by a number of other mechanisms, including those involving inhibition, provided the number of inhibitory steps is even. If, for example, I_a inhibits the reverse enzyme, while I_i inhibits the forward enzyme (Figure 22.4), a situation can arise in which one of the modifying enzymes, but not both simultaneously, can be common. In addition, many bacterial ST systems exhibit autoregulation, with the response regulator (RR) activating the transcription of the operon containing it (Gao and Stock 2013). This too can generate bistability – when the RR level is high, RR gene expression remains high because of the positive feedback loop; but when the RR level is low, the concentration remains at the basal level of expression (Ferrell 2002; Igoshin et al. 2008; Hermsen et al. 2011; Ram and Goulian 2013).

Adaptive fine-tuning versus inadvertent by-products of pathway structure

Taken together, these theoretical results indicate that without any direct selection at all, the basic structure of ST systems endows genotypically uniform populations of cells with a capacity to develop bistable phenotypic polymorphism. This is reflected in a number of dramatic dimorphisms in cell morphology and/or behavior. However, less visible,

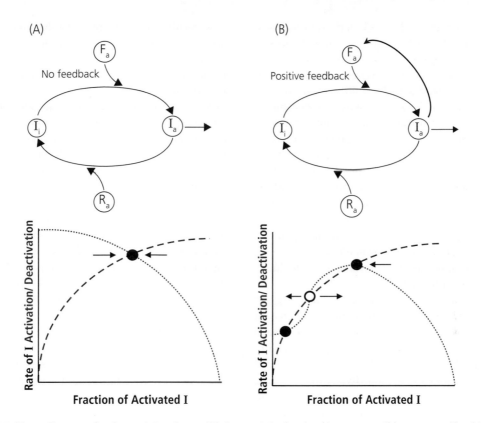

Figure 22.8 Phase diagrams for determining the equilibrium activity levels of interconvertible enzymes (I) subject to activation and inactivation cycles by enzymes F and R, respectively. The lines depict how the two rates change with increasing fraction of activated I. All points of intersection denote equilibria, but only the solid points are stable, as in these cases, deviations in both directions result in differences in activation and deactivation rates that return the system to the point. The fraction of activated I increases when the rate of activation (blue) exceeds the rate of deactivation, and vice versa, under the opposite condition. **A:** With a system with no feedback, there is a single stable equilibrium. **B:** When activated enzyme feeds back positively to the forward enzyme, the dynamics are altered in such a way that there can be as many as three equilibria, with the central one being unstable (as deviations in either direction move the system to one of the alternative stable equilibria). Depending on the elevation and angularity of the activation curve, there might be only a single equilibrium in this case.

molecular-level shifts may also commonly arise from the structural underpinnings of ST systems.

This being said, while being conducive to bistability, feedback-containing networks do not guarantee it (Cherry and Adler 2000; Angeli et al. 2004; Nichol et al. 2016). The conditions for existence of dual equilibria and the relative sizes of the basins of attraction for the alternative stable points (when they exist) are sensitive to the underlying kinetic parameters of the pathway

constituents. For example, changes in the elevation of the activation/deactivation response curve in Figure 22.8b can lead to there being just a single intersection, implying monostability. This then suggests the feasibility of the evolutionary fine-tuning of underlying features of ST systems via the selection of appropriate mutations to favor phenotypic switching versus uniformity. Indeed, because bistability may often be deleterious, with one or both cellular states far from optimal, selection may often operate to move the key kinetic parameters of ST systems to levels that minimize the chances of phase shifting (Hermsen et al. 2011).

A striking example of the types of processes that govern the dynamics of bistability is known for the bacterium *Bacillus subtilis*, which stochastically switches between a motile single-celled state and a non-motile chained state (Norman et al. 2013). In this system, switches to alternative states are governed by a double-negative feedback loop. One protein confers the motile, colonizing state, and the other promotes biofilm formation, but each protein suppresses the expression of the other. The molecular details can be found in the original paper. The focus here is simply on the statistical properties of the phase-shifting processes, which are quite different in the two directions.

The *B. subtilis* motile state is memory-less, in the sense that once initiated, there is a constant probability of switching to the chained state at each subsequent cell division. Letting this probability be p, the probability of switching after the first cell division is p, after the second division is $(1 - p)p$, and after the nth division is $(1 - p)^n p$. This is an exponential (roughly L-shaped) distribution, with the mean (and standard deviation) of the number of generations to switching both equal to $1/p$, in this case ~81 cell divisions. Mechanistically, the stochastic switching appears to be due to rare random fluctuations of the concentrations of proteins underlying the double-negative feedback loop, allowing the previously silenced proteins to escape suppression. Notably, although the distribution of switching times suggests randomness at the population level, the behavior of closely related cells (e.g., sister cells) is correlated, owing to shared effects inherited from the maternal cell (Kaufmann et al. 2007; see Chapter 9).

The distribution of switching times for *B. subtilis* chains is quite different, being approximately normal (bell-shaped), with a mean of ~ 8 cell divisions, implying a tight degree of regulation. However, the underlying molecular mechanism for chain termination is again fairly simple – upon chain initiation, there is a substantial pulse of production of material involved in cell adhesion, which is then diluted over subsequent divisions until a minimum threshold is reached. Taken together,

these results illustrate remarkably simple molecular paths to dramatic developmental changes – stochastic molecular switches influencing alternative master regulators for flagellar production and cell adhesion.

One of the key evolutionary questions about the dynamics of switching behavior concerns the relative longevities of the alternative phase states and the degree to which natural selection molds them in relationship to the scale of temporal variation in environmental conditions. Not surprisingly, a considerable amount of attention has been given to the idea that bistability is an adaptively evolved 'bet-hedging' strategy enabling individual genotypes to maximize their long-term fitness without resorting to potentially costly mechanisms for short-term physiological acclimation (Kussell Leibler 2005; Smits et al. 2006; Veening et al. 2008; Wei et al. 2014; Norman et al. 2015). For reasons of tractability, most of the theoretical work has focused on simple systems with just two discrete environments, each lasting for a time period in excess of cell-generation lengths, and with two discrete phenotypes, each better adapted to an alternative environment.

In a temporally variable environment, the genotype with the highest long-term exponential growth rate will be favored, and for the simple two-state/two-environment model, the optimal average random phenotypic switching time is equal to the average of the periods between shifts between the two environmental states, provided there are equally large (but opposite in sign) selective pressures on the two alternative phenotypes in the alternative environments (Thattai and van Oudenaarden 2004; Kussell et al. 2005; Salathé et al. 2009; Gaál et al. 2010). On the other hand, if the selection differential between the two environments is sufficiently large, stochastic switching can be disfavored, as the monomorphic genotype favored in the environment with large effects can overwhelm the smaller, short-lived disadvantage in the opposite environment.

Some aspects of this model have been explored experimentally with a system involving two alternative growth phenotypes in the yeast *S. cerevisiae* (Acar et al. 2008). Here, strains were engineered to switch between two physiological states that yielded different growth rates in two alternative environments. The strain that rapidly shifted from one state to the other experienced an early advantage whenever the environment shifted, as it quickly produced progeny adapted in the changed environment. However, as the duration between environmental shifts was lengthened, the slow-shifting variant gained an advantage, owing to its reduced production of the maladapted type.

Although somewhat contrived, these results show that the long-term advantage of phenotypic switching depends on the temporal dynamics of both environmental change and phenotypic response. However, it is prudent to consider that most environmental variables are continuous in nature, and can vary on both within- and between-generation timescales, as well as across spatial scales, and that these kinds of alternative scenarios can lead to rather different expectations (Lynch and Gabriel 1987). Thus, without direct empirical evidence, there is no justification for assuming that all instances of phenotypic polymorphisms reflect adaptive fine tuning.

Finally, in all of the preceding discussion, it has been assumed that phenotypic switching is an intrinsic feature of a cellular network, occurring without regard to the current environmental state. However, using environmental cues, organisms can, in principle, reinforce various phenotypic outcomes. This might happen, for example, through appropriate epigenetic modifications such as DNA methylation and/or histone modifications if these somehow encouraged individuals expressing a particular phenotype to produce offspring with an elevated frequency of the same phenotype (Xue and Leibler 2016). Provided that the individuals with inappropriate phenotypes are removed by selection, such a system could then lead to a form of transgenerational acclimation that superficially appears like learning or the inheritance of acquired characteristics. Such transient shifts in mean phenotype without any underlying genetic change will revert back to an alternative phenotype distribution upon environmental change (see Foundations 9.5).

Summary

- Unicellular organisms respond to external environmental stimuli through the use of signal transduction (ST) pathways that relay information from the cell surface to intracellular effectors, such as transcription factors. In most species, dozens to hundreds of such systems are specialized to different environmental indicators. In bacteria, information relay systems often involve just two proteins: an external sensor at the cell surface and an internal response regulator.

- ST systems are modular in nature, being based on several small motifs that specify proper communication between sensor and regulator proteins to the exclusion of members of other parallel pathways. The vast majority of such systems operate via additions and removals of phosphoryl groups on the participating proteins.

- Owing to the simplicity of these communication systems, rewiring of ST pathways is readily accomplished by changes in just a few key amino acids. Although this opens up opportunities for the establishment of novel signalling pathways following gene duplication, it also promotes the neutral drift of the recognition vocabulary in the absence of selection for altered functions.

- Despite the fact that the individual proteins driving ST systems operate as conventional Michaelis–Menten enzymes, the pathways through which they operate are often structured in such a way as to potentially generate very steep responses to external ligand concentrations. In some cases, the response approaches switch-like behavior wherein the downstream target is essentially 100% on or off when the ligand concentration is above versus below the threshold value.

- The continuous operation of opposing phosphorylation/dephosphorylation reactions at the heart of ST systems imposes a substantial energetic cost of processing and transmitting information, even in a constant environment.

- Owing to fluctuations in the arrival and binding of external signals to cell-surface receptors, environmental sensing is also an inherently noisy process. Noise buffering is facilitated by increasing the numbers of receptors and setting the binding/unbinding kinetics to levels that allow the cell to repeatedly make independent measures of the degree of receptor binding at rates that exceed the internal cellular response.

- Eukaryotic ST systems tend to be much more complicated than those in bacteria, commonly with expansions to cascades of multiple intermediate steps, use of multiple phosphosites per communicating molecule, and kinases and phosphatases cross-talking with multiple interacting partners. It remains unclear whether this complexity enhances the speed, efficiency, or accuracy of environmental assessment, and whether differences between prokaryotic versus eukaryotic ST systems have been driven by adaptive forces.

- Chemotaxis provides a rapid mechanism for adjusting the direction of bacterial motility in response to environmental gradients. These systems, which vary in structure among species, often have simple built-in feedback mechanisms for adjusting sensitivity to the prevailing environmental state, much like the visual systems of metazoans adjust to different light levels.

- The structure of ST systems is such that the addition of positive feedback loops (or pairs of negative feedbacks) can give rise to bistable responses to external ligand concentrations. By this means, genetically uniform populations can generate dimorphic populations of cells, potentially enhancing long-term genotypic fitness in environments presenting certain levels of variation, but also yielding maladaptive responses in inappropriate environments.

Foundations 22.1 Behavior of a monocycling system

A key issue with respect to an interconvertible enzyme (I) is the degree of activity expected under various conditions. The total concentration of the enzyme $[I_T]$ partitions into subsets of active and inactive molecules, $[I_a]$ and $[I_i]$, to a degree that depends on the relative concentrations of the active forms of converter enzymes (denoted F and R, respectively, for forwards and reverse reactions). The fractions of active versus inactive converter enzymes depend in turn on the concentrations of their ligands and their affinities for them. Because the two converter enzymes push the interconvertible enzyme in opposite directions, the relative concentrations of their active forms dictate the level of activity of enzyme I.

Here, we consider the steady-state situation in which the concentrations of both converter enzymes and their ligands are kept constant by ambient cellular conditions. Initially, we further assume that the fractions of both converter enzymes tied up with the interconvertible enzyme are negligible, which requires that the latter not be at a saturating level. Under these conditions (first-order rate kinetics), the active fractions of both converting enzymes will reach steady-state levels independent of the amount of enzyme I, and determined only by the rates of association and dissociation with their ligands. Using the terms defined in Figure 22.4, for the forward enzyme, a steady state requires that the rate of production of the active (a) enzyme from inactive (i) enzyme equals the flux in the opposite direction (resulting from the deactivation of F_a),

$$k_{a,F}[F_i][S_F] = k_{d,F}[F_a]. \qquad (22.1.1)$$

Noting that the total concentration of forward enzyme in the system is

$$[F_T] = [F_i] + [F_a], \qquad (22.1.2)$$

solving these two equations leads to the steady-state concentration of the active forward enzyme

$$[F_a] = \frac{[F_T][S_F]}{k_{D,F} + [S_F]}, \qquad (22.1.3a)$$

where $k_{D,F} = k_{d,F}/k_{a,F}$ is the dissociation constant of enzyme F. Likewise, the equilibrium concentration of the active form of the reverse enzyme is

$$[R_a] = \frac{[R_T][S_R]}{k_{D,R} + [S_R]}. \qquad (22.1.3b)$$

Provided the concentrations of the converter enzymes are at steady state, the alternative forms of the central enzyme I will also attain steady state. This occurs when the rate of production of active from inactive I equals the rate in the opposite direction. Using the familiar Michaelis–Menten formulations (Chapter 19), these forward and reverse reaction rates can be written as

$$V_F = \frac{k_{cat,F}[F_a][I_i]}{k_{S,F} + [I_i]}, \qquad (22.1.4a)$$

and

$$V_R = \frac{k_{cat,R}[R_a][I_a]}{k_{S,R} + [I_a]}. \qquad (22.1.4b)$$

Letting the total concentration of interconvertible enzyme in the system to be

$$[I_T] = [I_i] + [I_a], \qquad (22.1.5a)$$

the quantity of interest is the fraction of molecules that are in the active state,

$$I^* = [I_a]/[I_T]. \qquad (22.1.5b)$$

The general solution can be obtained by setting Equations 22.1.4a,b equal to each other, letting $[I_i] = [I_T] - [I_a]$, and solving for the level of $[I_a]$ that satisfies the equality.

The full solution is quite complicated, but as pointed out by Stadtman and Chock (1977) and Shacter-Noiman et al. (1983), provided the total amount of enzyme I in the system is small relative to the half-saturation constants in Equations 22.1.4a,b (non-saturating conditions), the concentrations of I in the denominators of these equations can be ignored, and this leads to an expression of the form

$$I^* = \frac{\beta C}{1 + \beta C}, \qquad (22.1.6)$$

where

$$\beta = \frac{\kappa_F[F_T]}{\kappa_R[R_T]} \qquad (22.1.7a)$$

is the ratio of kinetic potentials of the forward and reverse converter enzymes, with $\kappa_x = k_{cat,x}/k_{S,x}$ being the specificity constant of enzyme x (see Foundations 19.1), and

$$C = \frac{[S_F](k_{D,R} + [S_R])}{[S_R](k_{D,F} + [S_F])} \qquad (22.1.7b)$$

is the ratio of degrees of saturation of the input reactions (see Equations 22.1.3a,b). Although Equation 22.1.6 has a simple hyperbolic form, the underlying function is quite complex, as it actually depends on 10 different parameters

(two each of the k_{cat}, k_S, and k_D terms, and the concentrations of the two converter enzymes and their input ligands). Equation 22.3 in the main text gives an expression equivalent to Equation 22.1.6 in terms of the active concentrations of the forward and reverse enzymes.

Several significant points are revealed by Equation 22.1.6. First, for this case of low overall concentration of I, I^* is independent of the total concentration $[I_T]$. Second, as in the case of simple Michaelis–Menten kinetics, I^* is a hyperbolic function, in this case of C. Although the latter is itself a complex function, inspection shows a hyperbolic relationship with either ligand concentration. Third, although $I^* \to 1$ as $(\beta C) \to \infty$, because the relative concentrations of active enzymes are limited by the properties of the system (the total enzyme concentrations, total ligand concentrations, and the dissociation constants), there is an upper bound to βC. Thus, the maximum fractional activity of the central enzyme is generally <1.0 (Figure 22.5, upper panel).

Finally, a more general expression allowing for any concentration of I was derived by Goldbeter and Koshland (1981). In this case, the solution does depend on $[I_T]$, bringing the total number of relevant parameters to eleven, but can be written as a function of three composite parameters,

$$I^* = \frac{(\alpha-1)-(k_F^*+\alpha k_R^*)+\sqrt{[(\alpha-1)-(k_F^*+\alpha k_R^*)]^2+4\alpha(\alpha-1)k_R^*}}{2\alpha}, \qquad (22.1.8)$$

where $\alpha = (k_{cat,F}[F_a])/(k_{cat,R}[R_a])$, $k_F^* = k_{S,F}/[I_T]$, $k_R^* = k_{S,R}/[I_T]$, and $[F_a]$ and $[R_a]$ are defined by Equations 22.1.3a,b. Contrary to the limiting situation in which $[I_T]$ is low, the relationship of I^* is no longer a simple hyperbola, as discussed in the main text (Figure 22.5, lower panel).

Additional types of systems, including those with inhibitor interactions and with linked (multicyclic) cycles, are explored in Chock and Stadtman (1977), Stadtman and Chock (1977), and Goldbeter and Koshland (1984). Not surprisingly, these exhibit even richer behavior than those noted. For the case of bacterial two-component systems, where the kinase often has a dual function as the phosphatase (the reverse converter enzyme in the above scheme), expressions similar in form to Equation 22.1.8 have been developed by Batchelor and Goulian (2003) and Rowland and Deeds (2014). An excellent overview of all of these models, and the logic underlying them, is provided by Qian (2007). Malaguti and ten Wolde (2021) extend things to time-varying signal concentrations.

Foundations 22.2 Accuracy of environmental sensing

A successful sensing system requires that the receptors be capable of assaying the current environmental state accurately enough to transmit a reliable signal to the downstream responders to elicit appropriate changes in cell behavior. Here, we consider the degree to which a single molecular receptor at the cell surface can assess the concentration of a ligand in the surrounding environment. The initial assumption is that the fractional time during which the receptor is bound to the external ligand (p) provides the best information that the cell can utilize for environmental prediction. However, this is by no means the only possible approach to the problem.

At any single point in time, the receptor is either occupied or not, so a single snap-shot assessment provides little information. Over time, however, as ligand molecules become unbound, the receptor can make repeated assays of the environment, so that the average occupancy during a particular period becomes an estimate of p. As a measure of accuracy, we utilize the coefficient of variation (CV), which equals the ratio of the standard deviation (σ_x) to the mean (μ_x) of repeated measures of a variable. Because it is easier to work with measures of variance, which is the square of the standard deviation, the derivations to follow are based on the squared coefficient of variation, σ_x^2/μ_x^2.

It is assumed here that on the timescale of environmental assessment, the cell resides in a homogeneous environment with constant ligand concentration (c_0), so we are obtaining a pure measurement of environmental sensitivity based on the properties of the receptor molecule and its ligand molecule. If the environment is variable within the time-frame of environmental assessment, the variance of ligand concentration would need to be incorporated into the measure of noise presented below. The following derivations are based on the first study of the problem by Berg and Purcell (1977) and subsequent refinements by Kaizu et al. (2014). Some uncertainties about the precise nature of the final formulation are addressed in an excellent overview by Aquino et al. (2016).

To describe the temporal behavior of receptor occupancy, consider the stochastic differential equation

$$\frac{d\Gamma_t}{dt} = k_{on}c_0(1 - \Gamma_t) - k_{off}\Gamma_t + \epsilon_t, \tag{22.2.1}$$

where Γ_t denotes the occupancy (0 or 1) of a single receptor at time point t, k_{on} is the rate of ligand binding to an unbound receptor (per unit of external concentration, c_0), k_{off} is the rate of dissociation of a ligand molecule from a bound receptor, and ε_t is a stochastic variable with mean zero. By setting the derivative to zero and solving, the equilibrium probability of occupancy is found to be

$$\overline{\Gamma} = p = \frac{k_{on}c_0}{k_{on}c_0 + k_{off}} = \frac{c_0}{c_0 + k_D}, \tag{22.2.2}$$

where $k_D = k_{off}/k_{on}$.

Although the cell perceives the environment through the act of ligand binding, the ultimate goal of environmental sensing is to obtain an estimate of the ligand concentration (c) that closely approximates the true concentration (c_0). This requires an estimate of the variance among sample estimates of c_0 inferred from the cell's readout Γ. To obtain this, we start with a general rule from statistics that the variance of a dependent variable is approximately equal to the variance of a causal variable times the squared derivative of the first with respect to the second (Lynch and Walsh 1998, Appendix A), which in this case implies,

$$\sigma_\Gamma^2 = (\partial p/\partial c)^2 \cdot \sigma_c^2. \tag{22.2.3}$$

Rearranging and dividing by c_0^2 yields our desired measure of accuracy, the squared coefficient of variation of inferred concentration,

$$\frac{\sigma_c^2}{c_0^2} = \frac{1}{c_0^2} \cdot \frac{\sigma_\Gamma^2}{(\partial p/\partial c)^2}. \tag{22.2.4}$$

From Equation 22.2.2, the partial derivative evaluated at c_0 is

$$\frac{\partial p}{\partial c} = \frac{k_D}{(c_0 + k_D)^2}, \tag{22.2.5}$$

and substitution into Equation 22.2.4, after some rearrangement, leads to

$$\frac{\sigma_c^2}{c_0^2} = \frac{c_0^2}{p^4 k_D^2} \cdot \sigma_\Gamma^2. \tag{22.2.6}$$

The final step requires an expression for the variance in the mean occupancy σ_Γ^2 over some period of time T of continuous assessment, as this ultimately determines the degree of accuracy of overall environmental assessment. This is complicated by the fact that the realized Γ at any one particular time is not independent of that in adjacent time periods, owing to the time spans between ligand binding and release. Taking these autocorrelations into consideration, Berg and Purcell (1977) showed that

$$\sigma_\Gamma^2 = \frac{2p(1 - p)^2}{Tk_{off}}. \tag{22.2.7a}$$

Noting from Equation 22.2.2 that

$$k_{on}c_0(1 - p) = k_{off}p, \tag{22.2.7a}$$

Equation 22.2.7a can be equivalently written as

$$\sigma_{\Gamma}^2 = \frac{2p^2(1-p)}{Tk_{on}}c_0. \qquad (22.2.7b)$$

Finally, substituting Equation 22.2.7b into 22.2.6, again with some downstream rearrangement, leads to a remarkably simple expression

$$\frac{\sigma_c^2}{c_0^2} = \frac{2}{pTk_{off}} = \frac{2}{(1-p)Tk_{on}c_0}. \qquad (22.2.8)$$

The accuracy of assessment increases (i.e., σ_c^2/c_0^2 decreases) with increasing time over which the cell integrates environmental information, and also with increasing k_{off}. The latter feature arises because the inverse of k_{off} is equal to the average release time of ligands, which means that higher k_{off} allows the receptor to make more evaluations of the environment.

There are at least two other ways to express the accuracy. First, the average time between consecutive ligand-binding events is equal to the sum of the mean times for the length of binding to an occupied receptor and that of the time for an unoccupied receptor to accept another ligand, each of which is the reciprocal of the respective rate,

$$\tau_b = \frac{1}{k_{off}} + \frac{1}{k_{on}c_0}. \qquad (22.2.9)$$

Noting that the mean number of expected bindings in interval T is $\overline{N} = T/\tau_b$, substitution of $T = \overline{N}\tau_b$ and Equation 22.2.2 into 22.2.8 leads to

$$\frac{\sigma_c^2}{c_0^2} = \frac{2}{\overline{N}}, \qquad (22.2.10)$$

showing that the squared CV of the cell's estimate of c_0 is inversely proportional to the expected number of molecules bound during the assessment period (which itself is a function of time and ligand concentration).

Second, k_{on} is the inverse of the mean time to binding of an unoccupied receptor (per unit ligand concentration), which in turn is equal to the sum of expected times for particles to diffuse to the receptor (k_e) and of binding upon contact (k_+),

$$k_{on} = \left(\frac{1}{k_e} + \frac{1}{k_+}\right)^{-1} = \frac{k_e k_+}{k_e + k_+}. \qquad (22.2.11)$$

Substituting Equation 22.2.11 into 22.2.8 yields

$$\frac{\sigma_c^2}{c_0^2} = \frac{2}{Tc_0(1-p)}\left(\frac{1}{k_e} + \frac{1}{k_+}\right). \qquad (22.2.12)$$

Assuming that the receptor binding site can be approximated as disc of radius r_s on the cell surface, the encounter rate (per unit concentration) by diffusion is

$$k_e = 4Dr_s, \qquad (22.2.13)$$

where D is the diffusion constant for the ligand. Berg and Purcell (1977) assumed the case of diffusion limitation, such that $k_e \ll k_+$, which reduces Equation 22.2.12 to

$$\frac{\sigma_c^2}{c_0^2} = \frac{1}{2Dr_s c_0(1-p)T}. \qquad (22.2.14)$$

To gain more quantitative insight into the accuracy of monitoring as inferred by Equation 22.2.14, let $D = 10^{-5}$ cm^2/sec, which closely approximates true values for single amino acids (with cations and anions having values only $\sim 2\times$ higher; Chapter 7). Estimates of r_s for chemoreceptors are sparse, but can be inferred to be on the order of 2 nm (= 2×10^{-7} cm) given that the total area of receptor arrays in a wide range of bacteria implies an area/receptor of ~ 100 nm^2 with most of the array space being empty (Briegel et al. 2009). Supposing the dissociation constant $k_D = k_{off}/k_{on}$ is such that $p \simeq 0.5$ (Equation 22.2.2), and a ligand concentration of $c_0 = 1\,\mu$M $= 6 \times 10^{14}$ molecules/cm^3, the coefficient of variation of measurement then becomes

$$\frac{\sigma_c}{c_0} \simeq \sqrt{\frac{1}{1000T}},$$

where the units of T are in seconds. Thus, monitoring a constant environment for just 10 seconds is sufficient to reduce the level of estimation noise to 0.01 (i.e., a standard deviation of the inferred concentration just 1% of the true value). Assuming the same p, with a thousandfold lower concentration ($c_0 = 1$ nM), the level of uncertainty will be increased by a factor of $\sqrt{1000} \simeq 32$, and to achieve a level of accuracy of 0.01, T has to be thousandfold higher.

A broad overview of the biophysical constraints associated with other modes of environmental sensing (e.g., mechanoreception, vision, and hearing, all of which are exploited by metazoans) is given in Martens et al. (2015), who demonstrate that intrinsic limits associated with the scale of environmental noise and signal transmission restrict the utility of mechanosensing in an open-water environment to eukaryote-sized cells.

Literature Cited

Abedrabbo, S., J. Castellon, K. D. Collins, K. S. Johnson, and K. M. Ottemann. 2017. Cooperation of two distinct coupling proteins creates chemosensory network connections. Proc. Natl. Acad. Sci. USA 114: 2970–2975.

Acar, M., J. T Mettetal, and A. van Oudenaarden. 2008. Stochastic switching as a survival strategy in fluctuating environments. Nat. Genet. 40: 471–475.

Adler, J. 1966. Chemotaxis in bacteria. Science 153: 708–716.

Alm, E., K. Huang, and A. Arkin. 2006. The evolution of two-component systems in bacteria reveals different strategies for niche adaptation. PLoS Comput. Biol. 2: e143.

Anantharaman, V., L. M. Iyer, and L. Aravind. 2007. Comparative genomics of protists: New insights into the evolution of eukaryotic signal transduction and gene regulation. Ann. Rev. Microbiol. 61: 453–475.

Angeli, D., J. E. Ferrell, Jr., and E. D. Sontag. 2004. Detection of multistability, bifurcations, and hysteresis in a large class of biological positive-feedback systems. Proc. Natl. Acad. Sci. USA 101: 1822–1827.

Aquino, G., N. S. Wingreen, and R. G. Endres. 2016. Know the single-receptor sensing limit? Think again. J. Stat. Phys. 162: 1353–1364.

Armbruster, D., J. Nagy, and J. Young. 2014. Three level signal transduction cascades lead to reliably timed switches. J. Theor. Biol. 361: 69–80.

Ashenberg, O., K. Rozen-Gagnon, M. T. Laub, and A. E. Keating. 2011. Determinants of homodimerization specificity in histidine kinases. J. Mol. Biol. 413: 222–235.

Ballal, A., R. Heermann, K. Jung, M. Gassel, K. Apte, and K. Altendorf. 2002. A chimeric *Anabaena/Escherichia coli* KdpD protein (Anacoli KdpD) functionally interacts with *E. coli* KdpE and activates kdp expression in *E. coli*. Arch. Microbiol. 178: 141–148.

Barkai, N., and S. Leibler. 1997. Robustness in simple biochemical networks. Nature 387: 913–917.

Batchelor, E., and M. Goulian. 2003. Robustness and the cycle of phosphorylation and dephosphorylation in a two-component regulatory system. Proc. Natl. Acad. Sci. USA 100: 691–696.

Berg, H. C., and E. M. Purcell. 1977. Physics of chemoreception. Biophys. J. 20: 193–219.

Bitbol, A. F., and N. S. Wingreen. 2015. Fundamental constraints on the abundances of chemotaxis proteins. Biophys. J. 108: 1293–1305.

Blüthgen, N. 2006. Sequestration shapes the response of signal transduction cascades. IUBMB Life 58: 659–663.

Briegel, A., D. R. Ortega, E. I. Tocheva, K. Wuichet, Z. Li, S. Chen, A. Müller, C. V. Iancu, G. E. Murphy, M. J. Dobro, et al. 2009. Universal architecture of bacterial chemoreceptor arrays. Proc. Natl. Acad. Sci. USA 106: 17181–17186.

Briegel, A., D. R. Ortega, A. N. Huang, C. M. Oikonomou, R. P. Gunsalus, and G. J. Jensen. 2015. Structural conservation of chemotaxis machinery across Archaea and Bacteria. Environ. Microbiol. Rep. 7: 414–419.

Burger, L., and E. van Nimwegen. 2008. Accurate prediction of protein–protein interactions from sequence alignments using a Bayesian method. Mol. Syst. Biol. 4: 165.

Capra, E. J., and M. T. Laub. 2012. Evolution of two-component signal transduction systems. Annu. Rev. Microbiol. 66: 325–347.

Capra, E. J., B. S. Perchuk, O. Ashenberg, C. A. Seid, H. R. Snow, J. M. Skerker, and M. T. Laub. 2012a. Spatial tethering of kinases to their substrates relaxes evolutionary constraints on specificity. Mol. Microbiol. 86: 1393–1403.

Capra, E. J., B. S. Perchuk, E. A. Lubin, O. Ashenberg, J. M. Skerker, and M. T. Laub. 2010. Systematic dissection and trajectory-scanning mutagenesis of the molecular interface that ensures specificity of two-component signaling pathways. PLoS Genet. 6: e1001220.

Capra, E. J., B. S. Perchuk, J. M. Skerker, and M. T. Laub. 2012b. Adaptive mutations that prevent crosstalk enable the expansion of paralogous signaling protein families. Cell 150: 222–232.

Cherry, J. L., and F. R. Adler. 2000. How to make a biological switch. J. Theor. Biol. 203: 117–133.

Chock, P. B., and E. R. Stadtman. 1977. Superiority of interconvertible enzyme cascades in metabolite regulation: Analysis of multicyclic systems. Proc. Natl. Acad. Sci. USA 74: 2766–2770.

Cock, P. J., and D. E. Whitworth. 2007. Evolution of prokaryotic two-component system signaling pathways: Gene fusions and fissions. Mol. Biol. Evol. 24: 2355–2357.

Colin, R., and V. Sourjik. 2017. Emergent properties of bacterial chemotaxis pathway. Curr. Opin. Microbiol. 39: 24–33.

Cremer, J., T. Honda, Y. Tang, J. Wong-Ng, M. Vergassola, and T. Hwa. 2019. Chemotaxis as a navigation strategy to boost range expansion. Nature 575: 658–663.

Endres, R. G., and N. S. Wingreen. 2008. Accuracy of direct gradient sensing by single cells. Proc. Natl. Acad. Sci. USA 105: 15749–15754.

Endres, R. G., and N. S. Wingreen. 2009. Maximum likelihood and the single receptor. Phys. Rev. Lett. 103: 158101.

Ferrell, J. E., Jr. 2002. Self-perpetuating states in signal transduction: Positive feedback, double-negative feedback and bistability. Curr. Opin. Cell Biol. 14: 140–148.

Gaál, B., J. W. Pitchford, and A. J. Wood. 2010. Exact results for the evolution of stochastic switching in variable asymmetric environments. Genetics 184: 1113–1119.

Gao, R., and A. M. Stock. 2013. Evolutionary tuning of protein expression levels of a positively autoregulated two-component system. PLoS Genet. 9: e1003927.

Gildor, T., R. Shemer, A. Atir-Lande, and D. Kornitzer. 2005. Coevolution of cyclin Pcl5 and its substrate Gcn4. Eukaryot. Cell 4: 310–318.

Goldbeter, A., and D. E. Koshland, Jr. 1981. An amplified sensitivity arising from covalent modification in biological systems. Proc. Natl. Acad. Sci. USA 78: 6840–6844.

Goldbeter, A., and D. E. Koshland, Jr. 1984. Ultrasensitivity in biochemical systems controlled by covalent modification: Interplay between zero-order and multistep effects. J. Biol. Chem. 259: 14441–14447.

Goldbeter, A., and D. E. Koshland, Jr. 1987. Energy expenditure in the control of biochemical systems by covalent modification. J. Biol. Chem. 262: 4460–4471.

Goulian, M. 2010. Two-component signaling circuit structure and properties. Curr. Opin. Microbiol. 13: 184–189.

Govern, C. C., and P. R. ten Wolde. 2014. Optimal resource allocation in cellular sensing systems. Proc. Natl. Acad. Sci. USA 111: 17486–17491.

Gunawardena, J. 2005. Multisite protein phosphorylation makes a good threshold but can be a poor switch. Proc. Natl. Acad. Sci. USA 102: 14617–14622.

Hermsen, R., D. W. Erickson, and T. Hwa. 2011. Speed, sensitivity, and bistability in auto-activating signaling circuits. PLoS Comput. Biol. 7: e1002265.

Igoshin, O. A., R. Alves, and M. A. Savageau. 2008. Hysteretic and graded responses in bacterial two-component signal transduction. Mol. Microbiol. 68: 1196–1215.

Iyengar, G., and M. Rao. 2014. A cellular solution to an information-processing problem. Proc. Natl. Acad. Sci. USA 111: 12402–1247.

Kaizu, K., W. de Ronde, J. Paijmans, K. Takahashi, F. Tostevin, and P. R. ten Wolde. 2014. The Berg–Purcell limit revisited. Biophys. J. 106: 976–985.

Kapp, G. T., S. Liu, A. Stein, D. T. Wong, A. Reményi, B. J. Yeh, J. S. Fraser, J. Taunton, W. A. Lim, and T. Kortemme. 2012. Control of protein signaling using a computationally designed GTPase/GEF orthogonal pair. Proc. Natl. Acad. Sci. USA 109: 5277–5282.

Kaufmann, B. B., Q. Yang, J. T. Mettetal, and A. van Oudenaarden. 2007. Heritable stochastic switching revealed by single-cell genealogy. PLoS Biol. 5: e239.

Kempes, C. P., D. Wolpert, Z. Cohen, and J. Pérez-Mercader. 2017. The thermodynamic efficiency of computations made in cells across the range of life. Philos. Trans. A Math. Phys. Eng. Sci. 375: 20160343.

Keymer, J. E., R. G. Endres, M. Skoge, Y. Meir, and N. S. Wingreen. 2006. Chemosensing in *Escherichia coli*: Two

regimes of two-state receptors. Proc. Natl. Acad. Sci. USA 103: 1786–1791.

Koretke, K. K., A. N. Lupas, P. V. Warren, M. Rosenberg, and J. R. Brown. 2000. Evolution of two-component signal transduction. Mol. Biol. Evol. 17: 1956–1970.

Kussell, E., R. Kishony, N. Q. Balaban, and S. Leibler. 2005. Bacterial persistence: A model of survival in changing environments. Genetics 169: 1807–1814.

Kussell, E., and S. Leibler. 2005. Phenotypic diversity, population growth, and information in fluctuating environments. Science 309: 2075–2078.

Lan, G., P. Sartori, S. Neumann, V. Sourjik, and Y. Tu. 2012. The energy–speed–accuracy tradeoff in sensory adaptation. Nat. Physics 8: 422–428.

Landauer, R. 1988. Dissipation and noise immunity in computation and communication. Nature 335: 779–784.

Laub, M. T., and M. Goulian. 2007. Specificity in two-component signal transduction pathways. Annu. Rev. Genet. 41: 121–145.

Levy, W. B., and V. G. Calvert. 2021. Communication consumes 35 times more energy than computation in the human cortex, but both costs are needed to predict synapse number. Proc. Natl. Acad. Sci. USA 118: e2008173118.

Li, L., E. I. Shakhnovich, and L. A. Mirny. 2003. Amino acids determining enzyme-substrate specificity in prokaryotic and eukaryotic protein kinases. Proc. Natl. Acad. Sci. USA 100: 4463–4468.

Lim, W., B. Mayer, and T. Pawson. 2014. Cell Signaling. Garland Science, New York, NY.

Lloyd, D., C. A. Phillips, and M. Statham. 1978. Oscillations of respiration, adenine nucleotide levels and heat evolution in synchronous cultures of *Tetrahymena pyriformis* ST prepared by continuous-flow selection. Microbiology 106: 19–26.

Loomis, W. F., G. Shaulsky, and N. Wang. 1997. Histidine kinases in signal transduction pathways of eukaryotes. J. Cell Sci. 110: 1141–1145.

Lynch, M., and W. Gabriel. 1987. Environmental tolerance. Amer. Natur. 129: 283–303.

Lynch, M., and K. Hagner. 2015. Evolutionary meandering of intermolecular interactions along the drift barrier. Proc. Natl. Acad. Sci. USA 112: E30–E38.

Lynch, M., and J. B. Walsh. 1998. Genetics and Analysis of Quantitative Traits. Sinauer Associates, Inc., Sunderland, MA.

Makarova, K. S., M. Y. Galperin, and E. V. Koonin. 2017. Proposed role for KaiC-like ATPases as major signal transduction hubs in Archaea. mBio 8: e01959-17.

Malaguti, G., and P. R. ten Wolde. 2021. Theory for the optimal detection of time-varying signals in cellular sensing systems. eLife 10: e62574.

Martens, E. A., N. Wadhwa, N. S. Jacobsen, C. Lindemann, K. H. Andersen, and A. Visser. 2015. Size structures sensory hierarchy in ocean life. Proc. Biol. Sci. 282: 20151346.

McClune, C. J., A. Alvarez-Buylla, C. A. Voigt, and M. T. Laub. 2019. Engineering orthogonal signalling pathways reveals the sparse occupancy of sequence space. Nature 574: 702–706.

Mehta, P., and D. J. Schwab. 2012. Energetic costs of cellular computation. Proc. Natl. Acad. Sci. USA 109: 17978–17982.

Mello, B. A., L. Shaw, and Y. Tu. 2004. Effects of receptor interaction in bacterial chemotaxis. Biophys. J. 87: 1578–1595.

Mello, B. A., and Y. Tu. 2007. Effects of adaptation in maintaining high sensitivity over a wide range of backgrounds for *Escherichia coli* chemotaxis. Biophys. J. 92: 2329–2337.

Ni, B., R. Colin, H. Link, R. G. Endres, and V. Sourjik. 2020. Growth-rate dependent resource investment in bacterial motile behavior quantitatively follows potential benefit of chemotaxis. Proc. Natl. Acad. Sci. USA 117: 595–601.

Nichol, D., M. Robertson-Tessi, P. Jeavons, and A. R. Anderson. 2016. Stochasticity in the genotype–phenotype map: Implications for the robustness and persistence of bet-hedging. Genetics 204: 1523–1539.

Niven, J. E. 2016. Neuronal energy consumption: Biophysics, efficiency and evolution. Curr. Opin. Neurobiol. 41: 129–135.

Nocedal, I., and M. T. Laub. 2022. Ancestral reconstruction of duplicated signaling proteins reveals the evolution of signaling specificity. eLife 11: e77346.

Norman, T. M., N. D. Lord, J. Paulsson, and R. Losick. 2013. Memory and modularity in cell-fate decision making. Nature 503: 481–486.

Norman, T. M., N. D. Lord, J. Paulsson, and R. Losick. 2015. Stochastic switching of cell fate in microbes. Annu. Rev. Microbiol. 69: 381–403.

Ortega, F., L. Acerenza, H. V. Westerhoff, F. Mas, and M. Cascante. 2002. Product dependence and bifunctionality compromise the ultrasensitivity of signal transduction cascades. Proc. Natl. Acad. Sci. USA 99: 1170–1175.

Podgornaia, A. I., and M. T. Laub. 2015. Pervasive degeneracy and epistasis in a protein–protein interface. Science 347: 673–677.

Poole, R. K., D. Lloyd, and R. B. Kemp. 1973. Respiratory oscillations and heat evolution in synchronously dividing cultures of the fission yeast *Schizosaccharomyces pombe* 972h. Microbiol. 77: 209–220.

Porter, S. L., and J. P. Armitage. 2002. Phosphotransfer in *Rhodobacter sphaeroides* chemotaxis. J. Mol. Biol. 324: 35–45.

Porter, S. L., and J. P. Armitage. 2004. Chemotaxis in *Rhodobacter sphaeroides* requires an atypical histidine protein kinase. J. Biol. Chem. 279: 54573–54580.

Qian, H. 2007. Phosphorylation energy hypothesis: Open chemical systems and their biological functions. Annu. Rev. Phys. Chem. 58: 113–142.

Ram, S., and M. Goulian. 2013. The architecture of a prototypical bacterial signaling circuit enables a single point mutation to confer novel network properties. PLoS Genet. 9: e1003706.

Rodenfels, J., K. M. Neugebauer, and J. Howard. 2019. Heat oscillations driven by the embryonic cell cycle reveal the energetic costs of signaling. Dev. Cell 48: 646–658.

Rowland, M. A., and E. J. Deeds. 2014. Crosstalk and the evolution of specificity in two-component signaling. Proc. Natl. Acad. Sci. USA 111: 5550–5555.

Rowland, M. A., W. Fontana, and E. J. Deeds. 2012. Crosstalk and competition in signaling networks. Biophys. J. 103: 2389–2398.

Salathé, M., J. Van Cleve, and M. W. Feldman. 2009. Evolution of stochastic switching rates in asymmetric fitness landscapes. Genetics 182: 1159–1164.

Schaller, G. E., S. H. Shiu, and J. P. Armitage. 2011. Two-component systems and their co-option for eukaryotic signal transduction. Curr. Biol. 21: R320–R330.

Shacter, E., P. B. Chock, and E. R. Stadtman. 1984. Energy consumption in a cyclic phosphorylation/dephosphorylation cascade. J. Biol. Chem. 259: 12260–12264.

Shacter-Noiman, E., P. B. Chock, and E. R. Stadtman. 1983. Protein phosphorylation as a regulatory device. Philos. Trans. R. Soc. Lond. B Biol. Sci. 302: 157–166.

Shi, L., N. Pigeonneau, V. Ravikumar, P. Dobrinic, B. Macek, D. Franjevic, M.-F. Noirot-Gros, and I. Mijakovic. 2014. Cross-phosphorylation of bacterial serine/threonine and tyrosine protein kinases on key regulatory residues. Front. Microbiol. 5: 495.

Siryaporn, A., and M. Goulian. 2008. Cross-talk suppression between the CpxA-CpxR and EnvZ-OmpR two-component systems in *E. coli*. Mol. Microbiol. 70: 494–506.

Siryaporn, A., B. S. Perchuk, M. T. Laub, and M. Goulian. 2010. Evolving a robust signal transduction pathway from weak cross-talk. Mol. Syst. Biol. 6: 452.

Skerker, J. M., B. S. Perchuk, A. Siryaporn, E. A. Lubin, O. Ashenberg, M. Goulian, and M. T. Laub. 2008. Rewiring the specificity of two-component signal transduction systems. Cell 133: 1043–1054.

Skoge, M., H. Yue, M. Erickstad, A. Bae, H. Levine, A. Groisman, W. F. Loomis, and W. J. Rappel. 2014. Cellular memory in eukaryotic chemotaxis. Proc. Natl. Acad. Sci. USA 111: 14448–144453.

Smits, W. K., O. P. Kuipers, and J. W. Veening. 2006. Phenotypic variation in bacteria: The role of feedback regulation. Nat. Rev. Microbiol. 4: 259–271.

Sonenshein, A. L. 2000. Control of sporulation initiation in *Bacillus subtilis*. Curr. Opin. Microbiol. 3: 561–566.

Stadtman, E. R., and P. B. Chock. 1977. Superiority of interconvertible enzyme cascades in metabolic regulation: Analysis of monocyclic systems. Proc. Natl. Acad. Sci. USA 74: 2761–2765.

Stock, A. M., V. L. Robinson, and P. N. Goudreau. 2000. Two-component signal transduction. Annu. Rev. Biochem. 69: 183–215.

Szurmant, H., T. J. Muff, and G. W. Ordal. 2004. *Bacillus subtilis* CheC and FliY are members of a novel class of CheY-P-hydrolyzing proteins in the chemotactic signal transduction cascade. J. Biol. Chem. 279: 21787–21792.

Tabatabai, N., and S. Forst. 1995. Molecular analysis of the two-component genes, *ompR* and *envZ*, in the symbiotic bacterium *Xenorhabdus nematophilus*. Mol. Microbiol. 17: 643–652.

Thattai, M., and A. van Oudenaarden. 2004. Stochastic gene expression in fluctuating environments. Genetics 167: 523–530.

Ulrich, L. E., E. V. Koonin, and I. B. Zhulin. 2005. One-component systems dominate signal transduction in prokaryotes. Trends Microbiol. 13: 52–56.

van Haastert, P. J., and M. Postma. 2007. Biased random walk by stochastic fluctuations of chemoattractant–receptor interactions at the lower limit of detection. Biophys. J. 93: 1787–1796.

Veening, J. W., W. K. Smits, and O. P. Kuipers. 2008. Bistability, epigenetics, and bet-hedging in bacteria. Annu. Rev. Microbiol. 62: 193–210.

Wadhams, G. H., A. C. Martin, A. V. Warren, and J. P. Armitage. 2005. Requirements for chemotaxis protein localization in *Rhodobacter sphaeroides*. Mol. Microbiol. 58: 895–902.

Wan, K. Y., G. Jékely. 2021. Origins of eukaryotic excitability. Philos. Trans. R. Soc. Lond. B Biol. Sci. 376: 20190758.

Wegener-Feldbrügge, S., and L. Søgaard-Andersen. 2009. The atypical hybrid histidine protein kinase RodK in *Myxococcus xanthus*: spatial proximity supersedes kinetic preference in phosphotransfer reactions. J. Bacteriol. 191: 1765–1776.

Wei, K., M. Moinat, T. R. Maarleveld, and F. J. Bruggeman. 2014. Stochastic simulation of prokaryotic two-component signalling indicates stochasticity-induced active-state locking and growth-rate dependent bistability. Mol. Biosyst. 10: 2338–2346.

Weigt, M., R. A. White, H. Szurmant, J. A. Hoch, and T. Hwa. 2009. Identification of direct residue contacts in protein–protein interaction by message passing. Proc. Natl. Acad. Sci. USA 106: 67–72.

Wuichet, K., B. J. Cantwell, and I. B. Zhulin. 2010. Evolution and phyletic distribution of two-component signal transduction systems. Curr. Opin. Microbiol. 13: 219–225.

Xu, Y., and J. Gunawardena. 2012. Realistic enzymology for post-translational modification: Zero-order ultrasensitivity revisited. J. Theor. Biol. 311: 139–152.

Xue, B., and S. Leibler. 2016. Evolutionary learning of adaptation to varying environments through a transgenerational feedback. Proc. Natl. Acad. Sci. USA 113: 11266–11271.

Yamamoto, K., K. Hirao, T. Oshima, H. Aiba, R. Utsumi, and A. Ishihama. 2005. Functional characterization *in vitro* of all two-component signal transduction systems from *Escherichia coli*. J. Biol. Chem. 280: 1448–1456.

Yang, Y., A. Pollard, C. Höfler, G. Poschet, M. Wirtz, R. Hell, and V. Sourjik. 2015. Relation between chemotaxis and consumption of amino acids in bacteria. Mol. Microbiol. 96: 1272–1282.

Zarrinpar, A., S. H. Park, and W. A. Lim. 2003. Optimization of specificity in a cellular protein interaction network by negative selection. Nature 426: 676–680.

Zhang, W., and L. Shi. 2005. Distribution and evolution of multiple-step phosphorelay in prokaryotes: Lateral domain recruitment involved in the formation of hybrid-type histidine kinases. Microbiology 151: 2159–2173.

PART 7

Organismal Complexity

CHAPTER 23

Endosymbiosis

Eukaryotic biology is replete with cases in which cells are inhabited by descendants of other species, often bacterial in origin, and frequently with the latter having relinquished a capacity for free living. The most celebrated examples involve the mitochondrion and the plastid (called a chloroplast in photosynthesizing tissues). Almost all eukaryotes contain a mitochondrion or modified version of one, ultimately derived from an alphaproteobacterium, and nearly all eukaryotic photosynthesis involves chloroplasts with ultimate cyanobacterial ancestry. Although the bulk of the proteomes within these organelles is encoded in the nuclear genome, mitochondria and chloroplasts contain diminutive remnant genomes, and it is the sequence information within these that confirms their bacterial roots. Speculation that other eukaryotic features, including the mitotic apparatus and the eukaryotic flagellum, owe their origins to endosymbiosis (Sagan 1967; Margulis 1970), has garnered no support. There are, however, numerous other examples of lineage-specific endosymbionts in eukaryotes, ranging from bacteria residing within specialized organs in sap-feeding insects and tube worms to those within single-celled ciliates and amoebozoans.

Various words have been used to describe such intracellular occupants, e.g., endosymbionts, endocytobionts, and proto-organelles. However, the degree to which the interactants derive a benefit varies and is often unclear. Endosymbiosis implies the living of one type of cell within another, but the interaction may be jointly favorable (mutualism), beneficial for one member to the detriment of the other (parasitism), or essentially neutral (commensalism). Moreover, depending on the environmental context, the same consortium may switch from one of these conditions to another.

One operational distinction between an organelle and an endosymbiont is that the former relies on protein import from the host cell for at least some cell functions other than nutrition (Cavalier-Smith and Lee 1985). Mitochondria and plastids meet this criterion, as both are generally locked into obligate mutualisms with their host cells (with a few exceptions). But even here, there can be significant uncertainty regarding costs and benefits depending on one's point of reference (Keeling and McCutcheon 2017; McCutcheon et al. 2019). For example, although today's mitochondria cannot survive without their host cells, and vice versa, this does not necessarily mean that the interdependency started from a mutually beneficial situation.

The primary focus of this chapter is on the origins of mitochondria and chloroplasts and the remodelling of their functional operations following integration into host cellular environments. Special attention is paid to bioenergetic consequences, as there is debate regarding the benefits that host cells derive from endosymbiosis.

The comparative biology of organelles provides an ideal platform for integrating evolutionary theory with cell biology for two reasons. First, the population-genetic environment of organelles is often dramatically different from that experienced by genes residing within the nucleus. Unlike the latter, the former are typically inherited uniparentally and without recombination, and often exhibit dramatically altered mutation rates, sometimes elevated and other times reduced. Second, most of the protein-coding genes in organelle genomes produce products that coassemble with nuclear-encoded subunits. This raises unique issues with respect to intermolecular coevolution between participants

Evolutionary Cell Biology. Michael Lynch, Oxford University Press. © Michael Lynch (2024). DOI: 10.1093/oso/9780192847287.003.0023

residing in the same cell but experiencing different population-genetic constraints.

Mitochondria

One of the grandest events in the history of the biosphere was the emergence of the mitochondrion, which ultimately became associated with the entire domain of eukaryotic life. Often referred to as the 'powerhouse of the cell', the mitochondrion is the location of ATP production by oxidative phosphorylation, driven by the electron-transport chain. Via the citric-acid cycle (also known as the tricarboxylic-acid or Krebs cycle), the mitochondrion also fuels pathways for amino-acid and lipid biosynthesis, and is the site of synthesis of iron-sulfur clusters that are incorporated into numerous biomolecules, including those involved in the electron-transport chain.

Although a few eukaryotes harbor 'mitochondrion-related organelles', phylogenetic evidence indicates that all such variants are descendent from the same stock as the mitochondrion itself (Figure 23.1). For example, some anaerobic ciliates and parabasalids (e.g., *Trichomonas*) independently evolved mitochondrial derivatives called hydrogenosomes, which are incapable of oxidative phosphorylation, and instead regenerate ATP from ADP by substrate-level phosphorylation and generate molecular hydrogen as a by-product of the conversion of pyruvate to acetyl-CoA (Lewis et al. 2020). At least three other biochemically modified forms of mitochondria are known across various unicellular lineages, including the mitosomes of diplomonads (e.g., *Giardia*) and microsporidians (parasitic fungi), again independently evolved, which do not generate ATP at all but retain the ancestral trait of synthesizing iron-sulfur clusters (Zimorski et al. 2019). Also, a few eukaryotes are completely devoid of any form of mitochondrion, having lost them secondarily.

Origins

The early debate as to whether mitochondria were derived from endosymbiotic bacteria or instead

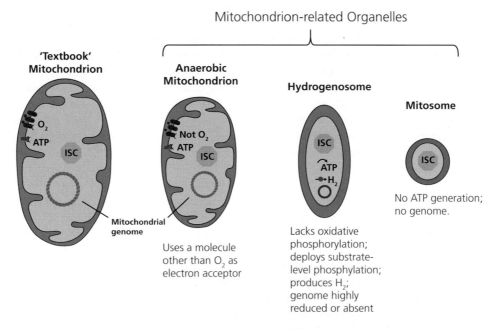

Mitochondrion-related Organelles

'Textbook' Mitochondrion

O_2
ATP
ISC

Mitochondrial genome

Anaerobic Mitochondrion

Not O_2
ATP
ISC

Uses a molecule other than O_2 as electron acceptor

Hydrogenosome

ISC
ATP
H_2

Lacks oxidative phosphorylation; deploys substrate-level phosphylation; produces H_2; genome highly reduced or absent

Mitosome

ISC

No ATP generation; no genome.

Figure 23.1 Idealized variants of classical mitochondria, and a few of their key modifications, including the presence/absence of a genome (pink rings), an electron-transport chain (red membrane-bound complexes), and ability to produce ATP. All three variants are sites of production of iron-sulfur clusters (ISCs, yellow). The dark surrounding lines denote the double membrane, and the internal invaginations are called cristae. From Burki (2016).

somehow arose endogenously was resolved with the realization that mitochondria contain their own genomes (generally circular as in bacteria), as this provided the gold standard for the determination of phylogenetic relationships by DNA-sequence comparisons. Multiple analyses of this sort point to a single origin of the mitochondrion from an alphaproteobacterium, but this leaves many questions unanswered (Archibald 2015; López-Garcia et al. 2017; Martin et al. 2017). From what specific lineage of the large alphaproteobacterial group did the mitochondrion emerge, and what might this tell us about the nature of the initial colonizer? From what microbial lineage was the host cell derived – bacterial, archaeal, or eukaryotic, and is the eukaryotic nucleus a descendant of that cell? Did the mitochondrion evolve after the establishment of the many other eukaryotic-specific attributes, or did it come first, with its presence somehow facilitating the origin of the latter? What, if anything, did the original host cell gain from the presence of its colonist, and vice versa?

There are three central challenges to achieving definitive answers to these questions. First, all of today's mitochondrial genomes contain <100 protein-coding genes (Figure 23.2), greatly reducing any remaining phylogenetic signal. Second, the long time span between the establishment of the mitochondrion and the most recent common ancestor of today's eukaryotes (LECA) blurs the signal from the few genes that remain. Third, although the large number of genes from the primordial mitochondrion transferred to the nuclear genome on the branch to LECA expands the range of informative sequence, background nuclear host-cell acquisitions of genes from other bacteria further cloud the issue.

Although numerous lineages of the Alphaproteobacteria phylum have been evaluated by genome sequencing, the picture remains murky as to which particular branch of the group gave rise to the mitochondrion (Rochette et al. 2014; Gray 2015; Martijn et al. 2018; Muñoz-Gómez et al. 2019). The initial view was that the base of the mitochondrial lineage resides near the order Rickettsiales (Andersson et al. 1998; Emelyanov 2001). As all members of this and closely related groups (e.g., *Rickettsia*, *Wolbachia*, *Anaplasma*, and *Orientia*) are intracellular parasites of eukaryotic cells, this raises the

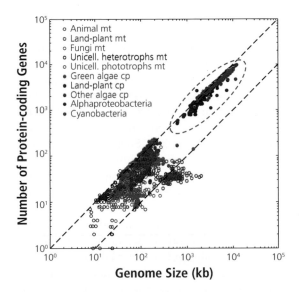

Figure 23.2 The number of protein-coding genes and genome sizes of fully sequenced organelle genomes and of the bacterial lineages containing the putative ancestors of the mitochondrion (Alphaproteobacteria) and of the chloroplast (Cyanobacteria). Data are from the NCBI Genomes Database (October 2020); mt and cp denote mitochondria and chloroplasts, respectively. Upper and lower diagonal dashed lines denote scalings of 1 and 0.1 genes per kilobase of genomic sequence, the first of which is closely adhered to by bacterial species (upper right).

possibility that the primordial mitochondrion was an energy parasite (Amiri et al. 2003; Andersson et al. 2003; Sassera et al. 2011; Wang and Wu 2014).

If this hypothesis is correct, the genome contents of the Rickettsiales and related species imply that the founder mitochondrion may have initially harbored ~1200 genes, contained an ATP/ADP antiporter that enabled ATP import from the host cell, and had a flagellum. In terms of metabolism, it was likely capable of driving a citric-acid cycle, had an electron-transport chain allowing for oxidative phosphorylation, and carried out ribosomal biogenesis and fatty-acid synthesis. A limited capacity of amino-acid biosynthesis would have been accommodated by the presence of transporters for acquiring amino acids from the host cell. One analysis suggesting that the mitochondrion emerged from an unknown clade that is sister to the entire Alphaproteobacteria leaves the nature of the ancestral state more ambiguous (Muñoz-Gómez et al. 2022).

An alternative hypothesis, based primarily on morphological observations, is that the mitochondrion arose from an anaerobic autotroph (Martin and Muller 1998; Cavalier-Smith 2006; Muñoz-Gómez et al. 2015, 2017). A hallmark feature of mitochondria is their internal network of invaginated membranes (cristae) upon which the electron-transport system complexes and ATP synthase reside (Figure 23.1). Such structures appear to be homologous to intracellular membranes used in bioenergetic transactions by members of a large alphaproteobacterial clade containing anaerobic photosynthesizers (purple non-sulfur bacteria), methanotrophs, and nitrite-oxidizing bacteria. The argument for inheritance of such features (as opposed to independent establishment after the origin of mitochondria) is strengthened by the observation that mitochondrial cristae junctions are organized by a protein orthologous to the one used for similar functions in these bacteria. Under this view, the primordial mitochondrion would have been capable of free living (Martijn et al. 2018; Muñoz-Gómez et al. 2019), although it may still have derived more benefits from its host cell than it provided in return, i.e., it may have been a facultative parasite.

Adding to the uncertainties about mitochondrial ancestry are an enormous number of hypotheses for the mechanism of mitochondrial establishment and the nature of the host cell. A historical compendium of proposed hypotheses assembled by Martin et al. (2015) outlines how these ideas vary in terms of the postulated timing relative to the origin of other eukaryote-specific traits, the types of metabolic transactions between host and endosymbiont, and the downstream evolutionary consequences for eukaryotes. Only a few of the more frequently invoked narratives are presented here.

One idea, called the hydrogen hypothesis, is that a methanogenic host cell consumed fuel (waste products) provided by an associated hydrogen-producing bacterium, which eventually became integrated as the primordial mitochondrion (initially, a hydrogenosome) (Martin and Muller 1998). In this and related models, the host cell is generally assumed to be an anaerobic member of the archaea (Rivera and Lake 1992; Vellai et al. 1998), an idea that is consistent with the emerging consensus

that eukaryotes are derived archaea (Chapter 3). An alternative idea, the oxygen-scavenging hypothesis, proposes that the mitochondrion arose as a mechanism to remove toxic oxygen from an anaerobic host cell (Sagan 1967; Andersson et al. 2003). One thing is clear. Assuming the host cell was indeed a member of the archaea, its metabolic machinery must have been largely displaced by that derived from the mitochondrion (or perhaps other bacteria by horizontal gene transfer), as most genes associated with metabolism in today's eukaryotes are bacterial derived (Chapter 3).

A final matter of concern is the morphological nature of the primordial host cell. Under the mitochondria-early view, the hydrogen hypothesis being one example, the initial host was prokaryotic in form, lacking internal membranes. The strongest variant of this argument is that eukaryogenesis was impossible without the presumed energetic boost provided by the endosymbiont (Lane and Martin 2010; Raval et al. 2022). This then raises the question as to how host-cell invasion could occur in the absence of phagocytotic engulfment, which is thought to require a well-developed cytoskeleton and vesicle system.

There are examples of bacteria living inside of other bacteria. These include *Bdellovibrio*, which burrows into host cells (Davidov and Jurkevitch 2009; Martin et al. 2017), and a betaproteobacterium *Tremblaya* that serves as an endosymbiont in insects and is itself inhabited by a gammaproteobacterium *Moranella* (von Dohlen et al. 2001; Husnik et al. 2013; Husnik and McCutcheon 2016). However, *Bdellovibrio* is a predator, and the cell envelope of *Tremblaya* has been so heavily modified that its ancestral bacterial nature is no longer apparent. Although the cyanobacterium *Pleurocapsa* has been reported to be occupied by other bacteria (Wujek 1979), the latter may simply be internal reproductive structures called baeocytes, a normal part of the *Pleurocapsa* life cycle. Thus, there are as yet no definitive examples of bacteria stably coexisting in free-living cells with bacterial membranes.

In contrast, under the mitochondria-late view, many of the early stages in eukaryogenesis, including the origin of a nucleus and internal membranes, are assumed to have been present prior to mitochondrial entry (de Duve 2007; Cavalier-Smith 2009). This view was initially championed by Cavalier-Smith (1987) at a time when basal-branching lineages of amitochondriate eukaryotes were thought to exist, a position that is no longer tenable. However, comparative analyses focused on the estimated times of origin of various classes of eukaryotic genes support the idea that the mitochondrion arose subsequent to the establishment of a number of eukaryotic-specific features (including aspects of internal membrane systems), potentially derived from genes acquired from lineages outside of the Alphaproteobacteria (Pittis and Gabaldón 2016; Vosseberg et al. 2021). Although interpretation of the genomic data bearing on this issue has been debated (Esposti 2016; Martin et al. 2017; Gabaldón 2018; Tria et al. 2021), this is true for the majority of hypotheses on the roots of the mitochondrion.

Most discussion of the origin of the mitochondrion starts (and often ends) with imagined benefits gained by the host cell, for example, the oxygen-scavenging and energy-production hypotheses. However, the eukaryotic consortium consists of two participants, and evolutionary stability of a mutualism usually demands that both partners acquire more resources than would be possible by living alone. How might this condition have been achieved, particularly given that one member of the original consortium may have been a parasite, rather than a benevolent partner? Despite its disadvantages to the host, such a system would have been rendered stable if: 1) the host lost a key function that was complemented by the presence of the endosymbiont; and 2) the emerging mitochondrion relocated just a single self-essential gene to the host genome, cementing its dependence on the host.

This scenario almost certainly played out in the lineage leading to LECA, as all mitochondria have forfeited nearly all genes for biosynthesis, replication, and maintenance to the nuclear genome, and hence depend entirely on their host cells for these essential metabolic functions. Likewise, the host cell abandoned key metabolic functions, such as membrane bioenergetics and iron-sulfur cluster biosynthesis, to the endosymbiont, thereby establishing reciprocal interdependence. Such an outcome represents a

grand example of the preservation of two ancestral components by complementary degenerative mutations (Force et al. 1999). Notably, the process of subfunctionalization (Chapter 6) is most likely to proceed in relatively small populations because the end state is slightly deleterious (both mutationally and bioenergetically) owing to the additional investment required to carry out individual tasks (Lynch et al. 2001). Thus, a plausible scenario is that the full eukaryotic cell plan emerged at least in part by initially nonadaptive processes made possible by a very strong and prolonged population bottleneck (Lynch 2007; Koonin 2015).

This type of functional partitioning appears to be the rule in endosymbiont evolution. The many origins of bacterial endosymbiosis in various lineages of sap-feeding insects (e.g., aphids, mealybugs, whiteflies, and psyllids) provide a case in point (McCutcheon et al. 2019). One of the more dramatic examples of genome reduction concerns the mealybug endosymbiont *Tremblaya*, whose genome is just 139 kb in length and contains only 120 protein-coding genes, whereas its own inhabitant *Moranella* has a somewhat larger genome containing 406 protein-coding genes (Husnik and McCutcheon 2016). Remarkably, *Tremblaya* has a number of metabolic pathways that appear to be assembled piecemeal, with some components imported from *Moranella* and others from nuclear-encoded host genes derived from horizontal transfers from still other bacteria (e.g., Bublitz et al. 2019). Such metabolic-pathway mosaicisms are consistent with the principles of intergenomic subfunctionalization just noted, resonate well with the theory of remodelling of metabolic pathways by non-orthologous gene replacement (Chapter 19), and provide further evidence in support of the concept of serial endosymbiosis (below).

Energetic boost or burden

The preceding section highlights numerous uncertainties regarding even the most basic features of mitochondrial evolution. We have a vague understanding of the phylogenetic roots of the mitochondrion, but not so refined as to be certain of the metabolic nature of the original foundress. Likewise, most aspects of the host-cell's biology remain unclear. However, these voids in our knowledge need not constrain our understanding of the evolutionary biology of modern mitochondria. In particular, as touched upon in Chapter 17, observations of today's eukaryotes raise considerable doubts about the enduring belief among some cell biologists that mitochondria endow eukaryotes with exceptional bioenergetic capacity.

Dating to Sagan (1967), this idea has been pushed most intensively by Lane and Martin (Lane 2006; Lane and Martin 2010; Martin 2017; Martin et al. 2017). They argue that the primordial mitochondrion-bearing host was not only promoted by positive selection on the basis of its enhanced energetic capabilities, but that this boost was essential for the downstream emergence of essentially all eukaryotic features, including increased cell volume and genome size, capacity for phagocytosis, and many other elaborations of morphological and behavioral complexity. However, diverse lines of evidence are inconsistent with this hypothesis (Lynch and Marinov 2015, 2017; Hampl et al. 2018; Schavemaker and Muñoz-Gómez 2022).

First, as discussed in several prior chapters (7, 8, 9, and 17), direct measures of metabolic rates and growth potential provide no evidence that eukaryotic cells are superior relative to prokaryotes. For all well-studied eukaryotic groups, maximum specific-growth rates decline with cell size and are less than those for similar-sized prokaryotes. Moreover, in contrast to the negative scaling in eukaryotes, growth rates of bacterial species increase with cell size, contrary to the hypothesis that prokaryotic bioenergetics is surface-area limited.

Second, the idea that complex internal structures cannot be sustained in the absence of mitochondria is contradicted by the presence of intracytoplasmic membranes in several Alphaproteobacterial species (noted earlier) and in other bacterial lineages discussed in Chapters 3 and 15. It has been suggested that phagotrophy (and by extension, complex internal cell structure) by a mitochondrion-free archaeal species would be selectively disadvantageous relative to a higher energetic yield achieved by the absorption of small dissolved metabolites (Martin et al. 2017). However, this argument assumes that the latter resources are available in unlimited supply, an unlikely scenario. Natural selection operates on features in the context of realized environments, and as there are very few settings in which resources are unlimited, there will always be a premium on moving into new ecological niches that minimize competition for prevailing resources. Bamboo and eucalyptus are not particularly nutritious, but pandas and koalas have found unique ways to exploit them. Moreover, contrary to the supposed impossibility of phagocytosis without mitochondria, members of the bacterial planctomycete group do ingest and digest bacterial and eukaryotic cells (Shiratori et al. 2019).

Third, eukaryotic species with reduced mitochondria or none at all still have elaborate internal and external complexities. The most extreme case is the oxymonad *Monocercomonoides exilis*, an excavate that lives in the guts of chinchillas. The former not only retains the standard internal cellular structure of eukaryotes, but uses four energy-consuming flagella for motility (Karnkowska et al. 2016, 2019). *M. exilis* consumes bacteria and is not a parasite, although it has no citric-acid cycle and instead makes ATP by glycolysis. *Henneguya salminicola*, a multicellular animal (related to hydras), which parasitizes salmon, also appears to be completely free of oxidative phosphorylation, although it does retain a genome-free mitochondrion-related organelle of unknown function (Yahalomi et al. 2020). Many other parasitic eukaryotes with highly modified mitochondria, such as *Giardia* and *Trichomonas*, generate their ATP by substrate-level phosphorylation. Thus, membrane-bound mitochondrial energetics is not a requirement for the maintenance of the complex morphological features of eukaryotic cells.

Fourth, the expansion of genome size in eukaryotes, thought by some to be essential to eukaryogenesis (Lane and Martin 2010), is readily explained by the increased power of random genetic drift in such lineages relative to prokaryotes (Lynch 2007; Chapters 17 and 24). Although genome sizes increase by factors of 10^2 to 10^3 from prokaryotes to unicellular eukaryotes to multicellular species, the vast majority of this expansion is a consequence of the proliferation of noncoding DNA, in particular the colonization of introns and mobile-element insertions, rather than an increase in gene number and functional diversification.

Finally, aside from these direct lines of evidence against the quantum boost in energetic capacity engendered by the mitochondrion, a more funda-mental problem is the basic premise that an increase in 'energy availability per gene' drives evolutionary diversification and a natural progression towards complexity (Lane and Martin 2010; Raval et al. 2022). Energy is a requirement for life, but no convincing argument has been offered as to why increased access to energy should promote pheno-typic evolutionary divergence by either adaptive or nonadaptive processes. Long-term rates of evolu-tion are a function of the rate of introduction of vari-ation by mutation, but whereas the mutation rate increases with organism size, this is not a function of energy, but of an increase in the power of ran-dom genetic drift and the resultant inability to retain high replication fidelity (Chapter 4). Recombination rates decline with increases in organism size, but again this has nothing to do with energy, but with the growth of chromosome sizes by nonadaptive processes (Chapter 4). Increased rates of adaptive evolution require increases in directional selection pressure, and there is no obvious reason why organ-isms with greater energetic capacity would burden themselves by inhabiting environments imposing stronger selection pressures.

To sum up, the idea that more energy allows evolution the freedom to do more tinkering and diversification (Martin et al. 2017; Lane 2020), with apparently no harmful side effects, remains to be explained in terms of known evolutionary mech-anisms. There is no evidence of a relentless push by natural selection towards complexity, and given that simpler structures are energetically less expen-sive, it is these that should be promoted by selec-tion. If there is a causal connection between the establishment of the mitochondrion and the radi-ation of eukaryotes, it does not appear to involve a revolution in bioenergetic potential. An alter-native is the passive establishment of new repro-ductively isolated lineages by subfunctionaliza-tion of the genomes of hosts and endosymbionts (Chapter 3).

Functional remodelling

Although almost all of the genes in the primor-dial mitochondrial genome were lost prior to LECA, either by outright deletion or transfer to the nuclear genome, mitochondria generally have proteomes consisting of ∼1000 nuclear-encoded proteins. In yeast (Karlberg et al. 2000) and the ciliate *Tetrahy-mena* (Smith et al. 2007), about half of the nuclear-encoded proteins appearing in mitochondria have bacterial affinities. Only a small fraction of these are clearly alphaproteobacterial in origin (Gray 2015; Ku et al. 2015), and may or may not have originated with the primordial mitochondrion. Such observa-tions are consistent with the idea that the road from FECA to LECA experienced serial endosymbiosis, with successive cycles of endosymbiont coloniza-tion, extinction, and transfer of genetic material to the nucleus (Pittis and Gabaldón 2016), a well-documented feature of insect endosymbionts (Hus-nik and McCutcheon 2016). Under this view, prior bacterial inhabitants may have contributed to host-cell modifications that in turn paved the way for the later arrival of the mitochondrion.

Likewise, energetic issues aside, the emergence of the mitochondrion would have altered the con-text in which all other eukaryotic cellular features evolved downstream. For example, numerous mitochondrially derived genes residing in the nucleus address their products to organelles other than the mitochondrion, and some ancestral host genes evolved new functions in the mitochon-drial proteome (Sloan et al. 2018). The key point is that the establishment of the mitochondrion was followed by the emergence of a diversity of cellular functions with no precedent in the host or endosymbiont, and that much of this remodelling occurred pre-LECA (Huynen et al. 2013; Ku et al. 2015; Raval et al. 2022). One of the more astounding such alterations is the independent origin of editing of transcript sequences in multiple lineages, often by remarkably complex mechanisms (Foundations 23.1). Here, we briefly describe four additional functional modifications that followed the genesis of the mitochondrion.

First, some of the key morphological innova-tions of mitochondria involve the mechanisms by which proteins from nuclear-encoded genes are imported. Such orchestrations involve the evolution of organelle-localization signals on the front ends of the traveling proteins and of novel molecular-recognition systems for them on organelle mem-branes. For example, the sorting and assembly machinery (SAM) complex, which resides in the outer mitochondrial membrane plays a central role in incorporating other outer-membrane proteins, all of which are nuclear encoded. SAM appears to be related to a similar outer membrane protein in bacteria called Omp85, and therefore may have resided in the original endosymbiont. However, for proteins that function within the central chamber of the mitochondrion, the situation is more challenging, as two mitochondrial membranes must be traversed. The TOM and TIM (translocases of the outer and inner membranes) import hundreds of cytoplasmic proteins marked with specific N-terminal localization signals, which are cleaved upon translocation. As in the case of SAM, the TIM and TOM proteins are nuclear encoded, but their origins remain unclear and are not obviously prokaryotic (Dolezal et al. 2006).

Second, critical to the maintenance of a stable endosymbiosis are mechanisms for preventing uncontrolled organelle proliferation while reliably promoting organelle division and inheritance at appropriate times. Within the lifespan of a host cell, mitochondria can undergo multiple rounds of fission and fusion. Mitochondrial division in a number of eukaryotic lineages relies on an inner constriction produced by the GTPase protein FtsZ as in most bacteria (Chapter 10; Leger et al. 2015). However, other eukaryotes appear to have independently evolved a mitochondrial division mechanism that also requires dynamin-related proteins (Friedman and Nunnari 2014; Leger et al. 2015), which pinch from the outside while FtsZ pulls from the inside. Phylogenetic analysis suggests that the ancestral mitochondrial dynamins had dual functions of mitochondrial division and vesicle scission (Purkanti and Thattai 2015), raising the possibility that internal vesicles preceded the origin of the mitochondrion. On at least three occasions (ancestors to alveolates, green algae, and opisthokonts), the dynamin gene was duplicated and then subfunctionalized into the two separate functions, and in each case, FtsZ was lost.

Third, a number of novel relationships have evolved between mitochondria and endogenously evolved organelles. For example, contact sites between the endoplasmic reticulum and mitochondria serve to coordinate mitochondrial division and play important roles in lipid and ion exchange (Friedman et al. 2011; Wideman et al. 2013; Lewis et al. 2016; Murley and Nunnari 2016; Prinz et al. 2020).

Fourth, while mitochondria are critical to the growth and maintenance of nearly all eukaryotic cells, in a wide range of species they also play a central role in targeting certain cells for death. In particular, the loss of mitochondrial membrane potential often elicits a cascade of molecular events resulting in either homeostatic rebalancing, or if beyond recovery, signaling a death sentence to the cell (Galluzzi et al. 2012). Eukaryotic cells also have an internal mechanism called mitophagy that enables the selective recognition and removal of mitochondria containing defective proteins (Youle 2019). Such a mechanism also occurs during inheritance, at least in metazoans, where maternal gametes induce the elimination of paternal mitochondrial DNA transmitted by fused sperm (Satoh and Kuroiwa 1991; Zhou et al. 2016).

All of the molecules participating in these signalling cascades for cell death are nuclear encoded in today's eukaryotes. However, the capacity for inducing cell death might have been carried by a selfish ancestral mitochondrion that killed cells not containing it, ensuring its permanent residence (Kobayashi 1998). As discussed at the close of this chapter, bacteria with exactly these properties exist in *Paramecium* (Preer et al. 1971), further demonstrating the potential for some endosymbionts to promote their own existence without providing any advantages to their host cells.

The Extreme Population-Genetic Environments of Mitochondria

As emphasized in previous chapters, the population-genetic environment (defined by the power of mutation, recombination, and random genetic drift) is a critical determinant of the ways in which genotypes respond to imposed selective pressures. The matter is of particular interest with respect to the evolution of mitochondria, which exhibit dramatic shifts in population-genetic features relative to both their extracellular ancestors and their adopted hosts. The historical consequences of such shifts are reflected in a wide array of changes in genomic architecture (Lynch et al. 2006; Lynch 2007), and extend to multiple aspects of organelle transcriptome and proteome integrity. To provide the setting for discussing these issues, we first consider the three major population-genetic alterations experienced by organelle genomes: mutation-rate modifications, reductions in population size, and loss of recombinational activity, all of which alter the efficiency of natural selection in fundamental ways (Chapter 4).

Mutation rates

Despite the fact that the replication and repair of organelle genomes are almost exclusively carried out by nuclear-encoded gene products, the mutation rates of organelle genomes often diverge substantially from those in the nucleus. Two molecular factors may contribute to unusual patterns of organelle-genome mutation rates. First, as sites of metabolic activity, organelles generate high levels of free-oxygen radicals, which encourage DNA damage via the deamination of cytosine to uracil and the oxidative modification of guanine to 8-oxoG. If not repaired, these two types of premutations respectively cause C:G → T:A transitions and C:G → A:T transversions, and likely are responsible for the near universal A/T bias in organelle genomes (Lynch 2007). Second, in contrast to nuclear DNA, which replicates only once per cell division, organellar DNA can replicate continuously, even within non-dividing cells, magnifying the opportunities for replication errors per cell cycle.

The realized mutation rate is also a function of the accuracy of the replication machinery and the ability of repair enzymes to correct pre-replication damage. Unfortunately, information on these matters in mitochondria is largely derived from studies of budding yeast and human cells, a small fraction of overall biodiversity (Bohr et al. 2002; Kang and Hamasaki 2002; Mason and Lightowlers 2003). Even here, there is a mix of observations. Base-misincorporation rates of organelle DNA polymerases may be lower than those for polymerases deployed in the nucleus (Kunkel and Alexander 1986; Johnson and Johnson 2001), but proofreading may be less accurate in the mitochondrion (Anderson et al. 2020). Base-excision repair may often replace damaged bases with incorrect nucleotides in mitochondria (Phadnis et al. 2006; Stein and

Sia 2017), and nucleotide excision repair may be entirely absent. In addition, the lack of strand-specificity in mismatch repair implies that correct and incorrect bases are equally likely to be altered (Mason et al. 2003).

The central question concerns the net effect of this diversity of factors on overall mitochondrial mutation rates, the most reliable estimates of which derive from mutation-accumulation experiments (Chapter 4). For metazoans, these rates are extraordinarily high. In the nematode *C. elegans*, the mitochondrial base substitution mutation rate is 9.7×10^{-8}/nucleotide site/generation (Denver et al. 2000), $\sim 70\times$ the nuclear rate for this species (Denver et al. 2004, 2009). In the fly *Drosophila melanogaster*, the mitochondrial rate of 4.4×10^{-8} (Haag-Liautard et al. 2008) is $\sim 9\times$ that for the nuclear genome. For two species of the microcrustacean *Daphnia*, the average mitochondrial mutation rates is 5.2×10^{-7}/site/generation (Xu et al. 2012; Ho et al. 2020), $68\times$ that in the nuclear genome. Using pedigree data, the average mutation rate estimate for humans, 3.6×10^{-5}, is $\sim 2700\times$ the nuclear rate (Howell 1996; Santos et al. 2005). Unfortunately, the only data of this sort outside of metazoans are for the diatom *Phaeodactylum tricornutum*, where the mitochondrial rate of 1.1×10^{-9} is just $2.3\times$ the nuclear rate (Krasovec et al. 2019), and for the yeast *S. cerevisiae*, where the mitochondrial rate is $\sim 10\times$ that in the nuclear genome (Liu and Zhang 2019).

All other attempts to estimate organelle mutation rates have relied upon the enumeration of substitutions at silent sites in pairs of species with geologically based divergence-time estimates (see Chapter 4). These estimates are not fully concordant with the direct observations noted earlier. However, the data do uphold the interpretation that, relative to rates in the nuclear genome, mitochondrial mutation rates are inflated to a greater extent in metazoans than in unicellular species.

Averaging over a wide range of vertebrates and invertebrates, the ratio of indirect mitochondrial to nuclear rates falls mostly in the range of 2 to 20, whereas the average for a range of unicellular species is $\simeq 1.5$ (Lynch 2007; Popescu and Lee 2007; Smith et al. 2014; Smith 2015). Aside from the inaccuracy of fossil-based estimates of divergence times, one concern here is that phylogenetically based mutation-rate estimates can be biased by factors such as selection (Stewart et al. 2008).

In striking contrast to all of these results, mitochondrial mutation rates in land plants are typically ~5% of nuclear rates, and at least 100× lower than in metazoan mitochondria (Wolfe et al. 1987; Lynch 2007). Although the mechanisms responsible for the extraordinary mutational quiescence of land-plant mitochondria are unclear, they are not invariant features, as some plant species are known to have mitochondrial silent-site substitution rates up to 5000× greater than the usual background rate (Palmer et al. 2000; Cho et al. 2004; Richardson et al. 2013).

From the standpoint of evolutionary theory, explaining this diversity of organelle mutation rates is a challenge. As discussed in Chapter 4, the mutation rate is expected to be driven down by selection to the lowest level compatible with the power of random genetic drift, owing to the association of mutator alleles with the linked deleterious mutations that they create. As the replicative and DNA repair machinery associated with mitochondria is nuclear encoded, in a sexually reproducing species a mutator allele will be quickly dissociated from its instigated damage, minimizing the power of linked selection to modulate the mutation rate. The elevated mutation rates in animal and some land-plant mitochondria are consistent with this argument. In addition, the reduced deviation between mutation rates in organelle and nuclear genomes of unicellular eukaryotes may be explained by the fact that such organisms reproduce in a predominantly clonal fashion, ensuring a long-term one-to-one relationship between nuclear and organelle genomes.

Less clear is the mechanism by which land-plant mitochondria are able to maintain some of the lowest mutation rates known in eukaryotes, while presumably experiencing random segregation between organelles and nuclear autosomes, a small effective population size (relative to unicellular species), and an ~100-fold in reduction in the gene number target size for mutations relative to the nuclear genome. One potential explanation for this apparent puzzle is that key plant enzymes involved in nuclear replication and/or repair are also utilized in the organelles, which would result in the organelle mutation rates being a by-product of

selection on the nuclear rate. Wu et al. (2020) found that a mismatch-repair pathway shared between mitochondrial and plastid genomes in *Arabidopsis* reduces the organelle mutation rate 10 to 100×, but this is no greater than the efficacy of such systems in nuclear genomes, and so taken alone cannot fully explain the further reduced rate in plant organelle genomes. Reconciliation of such an elevation in genome stability with conventional theory would be possible if land plants acquired a mechanism for substantial enough improvement of replication fidelity or DNA repair to make a quantum leap beyond the typical location of the drift barrier (Chapter 20), but no such land-plant-specific mechanisms have been revealed.

Modes of inheritance

Unlike nuclear genomes, all organellar genomes are replicated ameiotically (as in bacteria). In most species, they are also inherited uniparentally, usually through the mother in multicellular species. This raises significant questions about the genetic effective population sizes of genes within organelles relative to those in the nucleus. The simplest view is that with uniparental inheritance the mitochondrial effective population size (N_e) would be one-quarter that for nuclear genes in a diploid species, as there is one of the former for each of the four alleles of a nuclear-encoded locus in a mating pair (Palumbi et al. 2001). However, as emphasized in Chapter 4, when population sizes are even moderately large, a major determinant of N_e is the influence of background selection operating on linked genes. If there is essentially no recombination among organelle genomes, this effect could be quite pronounced.

Might such effects be compensated by the presence of multiple organelles, each containing multiple genome copies, within individual hosts cells? Although hundreds to thousands of mitochondrial-genome copies may exist in some growing cells, strong transmission bottlenecks typically occur during progeny production. The issue has been addressed on several occasions through the serendipitous discovery of heteroplasmic females (carrying two distinct mitochondrial types). Letting p and $(1 - p)$ denote the frequencies of two haplotypes in a mother, assuming random assortment, the variance in frequencies among progeny follows from simple binomial sampling, $p(1 - p)/n_o$, where n_o (the effective number of mitochondrial genomes per individual) can be ascertained from the degree of dispersion of the haplotype frequencies among progeny. (This formula is identical in form to that for genetic drift of two autosomal alleles within a population; Chapter 4). Using this approach, or a close variant of it, the effective number of mitochondrial genomes per female transmission is estimated to be ~ 2–10 in mammals (Ashley et al. 1989; Jenuth et al. 1996; Marchington et al. 1997), 30–300 in insects (Solignac et al. 1983; Rand and Harrison 1986; Haag-Liautard et al. 2008), and four in the plant *Arabidopsis* (Broz et al. 2022). Less is known on the matter for unicellular organisms, although one might surmise that for small-celled species, n_o would be reduced even further, as the number of mitochondria/cell approaches just one in the smallest cells. Consistent with this view, $n_o \simeq 5$ in the slime mold *Physarum* (Meland et al. 1991) and in the yeast *S. cerevisiae* (Birky et al. 1978).

Although the mechanisms responsible for transmission bottlenecks are unclear (candidates include direct organelle destruction, differential replication, and random partitioning of cytoplasm; Burt and Trivers 2006), if inheritance is primarily uniparental, rapid sorting of variants bears on the issue of organelle N_e in two significant ways. First, all genomic copies within an individual will coalesce genealogically to a single ancestral molecule in just a few generations (Birky et al. 1983). Second, even though organelle genomes are physically capable of recombination (Thyagarajan et al. 1996; Kazak et al. 2012), the opportunities for generating novel recombinant genotypes will be restricted, as this requires the participation of two molecules differing at a minimum of two nucleotide sites, an unlikely mutational scenario with rapid within-individual sorting.

Thus, genetically effective recombination between organelles will generally require biparental transmission to bring divergent genomes into contact. The degree to which such situations arise has been debated considerably (Eyre-Walker and Awadalla 2001; McVean 2001; Piganeau et al. 2004), although most of the discussion has revolved around multicellular species. Sperm mitochondria are generally targeted for destruction upon delivery, but the process is not perfect, and low levels of biparental inheritance have been revealed in nematodes (Lunt and Hyman 1997), *Daphnia* (Ye et al. 2022) and humans (Luo et al. 2018). Mitochondrial inheritance in fungi is often biparental, and recombination does occur (Wilkie and Thomas 1973; Silliker et al. 1996; MacAlpine et al. 1998; Saville et al. 1998; Ling et al. 2000; Anderson et al. 2001), but little is known on the matter in other unicellular species.

The presence of genomically autonomous mitochondria within host cells raises significant 'levels of selection' issues. Most notably, within-host selection has the potential for the promotion of mutant organelle genomes with a replicative advantage despite the disadvantage to the host cells (Havird et al. 2019). The most dramatic examples of such expansions involve mitochondrial deletion mutants, which, despite having lost key genes, proceed through replication more rapidly than ancestral molecules (Clark et al. 2012; Jasmin and Zeyl 2014; Phillips et al. 2015). In principle, these same processes may elicit counter-adaptations on the part of the host species to prevent its own loss of fitness due to the competing interests of the endosymbiont. Uniparental inheritance and selective mitophagy are potential ameliorating factors, as both minimize the chances for competition between divergent mitochondrial genomes.

Although the fixation of deletion mutants for essential genes is ordinarily not possible, when combined with processes involving gene transfer to the nucleus, the proliferative advantage of mitochondrial genomes of reduced size may have cascading effects. Once a gene transfer to the nucleus has been integrated to the extent that its products are efficiently fed back to the mitochondrion, deletions of the mitochondrial gene will be free to advance. This perhaps explains why inter-genomic transfer between mitochondrial and nuclear genomes has been essentially unidirectional, and provides an extension to Doolittle's (1998) suggestion that sheer mutational pressure created a ratchet-like mechanism that ultimately ensured the relocation of organelle genes to the nucleus.

Finally, uniparental inheritance further alters the selective environment for the endosymbiont by favoring features in the latter that enhance the fitness in the transmitting sex, while severing the selective connection with the non-transmitting sex (Cosmides and Tooby 1981; Frank and Hurst 1996; Gemmell et al. 2004). In the case of maternal inheritance, for example, a mitochondrial genome with beneficial female effects can be promoted through mothers even if it has severe negative effects on male offspring, as males with superior mitochondria do not pass them on to offspring (assuming no paternal leakage). The presumed outcomes of such a process, sometimes referred to as the 'mother's curse', are commonly observed in land plants with cytoplasmic male sterility. Of course, such a situation will also select for mutations in nuclear genes that suppress the male fitness-reducing effects of female-driven mutations, setting up a sort of coevolutionary arms race between the sexes. Not surprisingly, mutations that restore male fertility are commonly found in plants (Fujii and Toriyama 2009; Gaborieau et al. 2016; Yamauchi et al. 2019).

Muller's ratchet

The magnitude of reduction of mitochondrial N_e is a key determinant of the evolutionary limitations of mitochondrial genomes, as mutations with selective effects smaller than the inverse of N_e are essentially immune to selection (Chapter 4). The combination of low (to no) recombination and uniparental inheritance is of particular relevance to mitochondria, but given the multitude of additional effects that define N_e, resolving the totality of the issues requires empirical analysis. Recall from Chapter 4 that the usual approach here is to estimate standing levels of genetic variation at neutral genomic sites (generally third positions in redundant codons), setting this equal to the drift-mutation equilibrium expectation $2N_e u$ for haploids, and factoring out the mutation rate u to obtain N_e. Our interest here is in the depression of N_e in mitochondrial relative to nuclear genomes in the same species.

Unfortunately, there are very few species with the diversity and mutation-rate data necessary for such a computation. Consider, however, the situation in humans where the ratio of silent-site diversity in mitochondrial versus nuclear genomes $\simeq 5.5$ (Lynch et al. 2006). Letting $2N_{gn}u_n$ be the expected nucleotide diversity per silent site in the nuclear genome at mutation-drift equilibrium, with N_{gn} and u_n being, respectively, the effective number of chromosomes and mutation rate per nuclear genomic site, and $2N_{gm}u_m$ be the similar expression for the mitochondrion, then 5.5 provides an estimate of $(N_{gm}u_m)/(N_{gn}u_n)$. If, as suggested, $u_m/u_n \simeq 2700$, this implies a ratio of effective sizes of just $N_{gm}/N_{gn} \simeq 0.002$. For arthropods, the average ratio of diversities is close to 1.0 (Lynch et al. 2006), but as noted, the ratio of mutation rates is $\simeq 40$, suggesting $N_{gm}/N_{gn} \simeq 0.025$. On the other hand, the mean ratio of diversities for the few unicellular species with available data is $\simeq 0.5$ (Lynch et al. 2006), and using a mutation rate ratio of 1.5 (from earlier) implies $N_{gm}/N_{gn} \simeq 0.1$. Thus, based on the few systems for which data are available, the power of drift operating on mitochondrial genes can be as much as 500× that in nuclear genes in some metazoans, with the situation being much less extreme in unicellular species.

The inability to shed mutations by recombination has inspired the idea that organelle genomes are uniquely susceptible to mutational degradation by a process known as Muller's ratchet (Muller 1964; Felsenstein 1974; Figure 23.3). In the absence of recombination, parent molecules cannot produce offspring molecules with a reduced number of deleterious mutations, except in the rare case of back or compensatory mutations. Thus, in an asexual population, when by chance the best class of individuals produces either no surviving offspring or only offspring with at least one new deleterious mutation, a nearly irreversible decline in fitness is experienced. Each generation, there is an appreciable chance of such an event because recurrent mutation pressure generally reduces the best-fit class to just a small fraction of the total population (Haigh 1978). Moreover, each time the currently best-fit class is lost from the population, the previously second-best class becomes subject to the same stochastic process, eventually suffering an identical fate. If this process proceeds to the point at which the mutation load is so high that the average individual cannot replace itself, the population size must begin to decline.

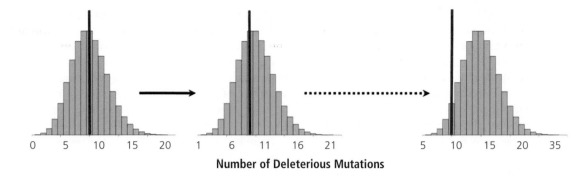

Number of Deleterious Mutations

Figure 23.3 Muller's ratchet, as illustrated by the stochastic, but progressive, movement of the distribution of the numbers of deleterious mutations among members of the population. The red vertical line denotes the initial mean number, whereas the leftmost numbers on the x axes denote the class of individuals with the minimum number of mutations.

This further enhances the magnitude of random genetic drift, promoting increasingly higher rates of deleterious-mutation accumulation and ultimately culminates in population extinction by mutational meltdown (Lynch and Gabriel 1990; Lynch et al. 1993, 1995a,b).

Aside from recombination, which enables pairs of parental genomes to produce progeny molecules with reduced numbers of deleterious mutations (Maynard Smith 1978; Charlesworth et al. 1993; Lynch et al. 1995a,b), the only remedy to this problem is back (or compensatory) mutation (Wagner and Gabriel 1990; Poon and Otto 2000; Goyal et al. 2012). However, even then, one expects the long-term mean phenotype to deviate from the optimum to a degree that depends on the power of random genetic drift (Lynch 2020). It has been argued that random segregation of mitochondrial variants at the intracellular level can reduce the rate of the ratchet by creating more variation upon which natural selection can act (Takahata and Slatkin 1984; Bergstrom and Pritchard 1998; Radzvilavicius et al. 2017; Edwards et al. 2021), but this alone cannot prevent the ultimate accumulation of deleterious mutations.

Given that LECA dates to > 1.5 billion years ago, it is clear that mitochondrial mutational meltdowns have been avoided on this timescale in a large number of eukaryotic lineages. Nonetheless, it remains possible that individual lineages have succumbed

to such decay and that others may still be predisposed to do so (Takahata and Slatkin 1984; Hastings 1992; Gabriel et al. 1993; Reboud and Zeyl 1994; Loewe 2006). Moreover, as outlined in the next section, organelle genomes have commonly gone down paths of degradative evolution in ways never seen in nuclear genomes.

Organelle Genome Degradation

The central point of the preceding discussion is that the peculiar population-genetic environment of the mitochondrion, combined with the asymmetry of interests of organelle and nuclear genomes, have played a key role in the remodelling of eukaryotic cell functions. The reach goes far beyond cellular energetic performance, mutation rates, and uniparental inheritance. Here, we further explore how the resultant reduction in mitochondrial N_e and the associated decline in the efficiency of selection against deleterious mutations have had cascading effects across key genes in both the mitochondrial and nuclear genomes. Despite the centrality of the mitochondrion for eukaryotic cell integrity, mitochondria have likely been compromised by mildly deleterious mutation accumulation since their inception.

As mentioned above, the most obvious manifestation of mitochondrial decay is the massive reductive evolution in genome size that occurred on the road from FECA to LECA (Figure 23.2). Most mitochondrial genomes are < 100 kb in length, with the exception of those of land plants, which can reach 1 Mb. They are typically > 10× smaller than their partner nuclear genomes, retaining < 100 of the ∼ 1000 protein-coding genes likely harbored in the ancestral mitochondrion. A number of genes in the primordial mitochondrion were simply lost, presumably because the fitness advantages were reduced to the point of effective neutrality, and perhaps in some cases owing to disadvantages in the host-cell environment. Most of the few protein-coding genes remaining in mitochondrial genomes are involved in energy metabolism. Nearly all genes involved in mitochondrial DNA replication and repair, transcription, and translation reside in the nucleus.

The presence of recent fragments of mitochondrial DNA in the nuclear genomes of most eukaryotes highlight the ample opportunities that exist for such transfer even today (Rho et al. 2009; Hazkani-Covo et al. 2010). However, physical relocation need not lead to functional transfer, which requires some form of positive selection for gene relocation. As noted, selection on host cells to resist the expansion of rogue mitochondria with deletion mutations is one potential source of such selection. The need to escape from the consequences of Muller's ratchet provides another long-term advantage for transfer to a recombining nuclear genome. Kelly (2021) also makes the case for a short-term advantage – a substantial reduction in the energetic cost of organelle genes residing in nuclear genomes. The large number of organelle genomes typically present in cells greatly magnifies the DNA-level cost relative to that of a nuclear gene present in just one (haploid) to two (diploid) copies per cell.

Several indirect lines of evidence are consistent with the idea that the few genes retained in organelle genomes are vulnerable to deleterious-mutation accumulation. For example, in a wide variety of animals and land plants, within-population surveys of nucleotide sequence variation in organelles consistently reveal that ratios of nonsynonymous (amino-acid replacement) to synonymous (silent-site) polymorphisms are 2–10× greater than those for divergence between closely related species (Ballard and Kreitman 1994;

Nachman et al. 1994, 1996; Rand et al. 1994; Templeton 1996; Hasegawa et al. 1998; Wise et al. 1998; Fry 1999; Städler and Delph 2002; James et al. 2016). These patterns are dramatically different than those seen in nuclear genomes, where there is often an excess of replacement substitution at the level of divergence, which is a presumed reflection of fixation of adaptive mutations (Chapter 4).

There is need to extend this sort of work to unicellular eukaryotes, but the most reasonable interpretation of the existing patterns is that significant numbers of deleterious mitochondrial mutations are able to expand to high enough frequencies to be observed in population surveys but not so high as to go to fixation. If this is correct, because mildly deleterious mutations have a wide range of selective effects (Chapter 5), it also follows that some deleterious mutations with very mild individual effects must be vulnerable to advancing all the way to fixation in organelle genomes with small N_e.

Animal mitochondrial tRNAs

More direct insight into the matter of mutation accumulation can be obtained by examining the evolutionary fates of parallel sets of mitochondrial and nuclear genes with identical general functions in their respective cellular environments. tRNA and rRNA genes provide an ideal venue for such analysis, as many organelle genomes contain full sets of both, with parallel nuclear-encoded sets producing RNAs that operate in the cytosol. In particular, the extreme conservation of the primary, secondary, and tertiary structure of tRNAs (Kimura 1983; Söll and RajBhandary 1995) across the entire Tree of Life of prokaryotes and eukaryotic nuclear genes testifies to the power of natural selection at maintaining the optimal molecular architecture of these genes. Transfer RNAs have a standard cloverleaf secondary structure, with their ∼ 70 bases being contained mostly in three loops and four stems. Thirteen of these base positions are essentially invariant across tRNAs for all amino acids in all prokaryotes and all nuclear genomes (Lynch 1996, 1997; Lynch and Blanchard 1998).

Given this extraordinary degree of phylogenetic constancy, the same features must also have been present in the tRNAs within the primordial mitochondrial genome. Thus, any deviations from the canonical tRNA architecture are likely to reflect a reduction in the efficiency of selection imposed by mitochondrial population-genetic environments. The evidence for such a shift is compelling.

First, contrary to the situation in the nuclear genome, there are no invariant sites in organelle tRNAs, and every region of such molecules evolves at a higher rate than the homologous region in nuclear tRNAs. This is not simply a consequence of elevated mitochondrial mutation rates, as the ratio of the observed substitution rate to the neutral expectation is elevated several-fold in mitochondrial tRNAs of all phylogenetic groups. Second, animal mitochondrial tRNAs exhibit a wide array of structural deviations from the canonical form of the prokaryotic/nuclear tRNAs, including losses of entire arms in some cases (Wolstenholme 1992). Third, for animals, plants, and fungi, the average binding strengths of mitochondrial tRNA stems are 40–90% of those in nuclear tRNAs, largely due to the higher incidence of A:U versus G:C bonds (two vs. three hydrogen bonds) in the former (Lynch 1996, 1997).

Although compensatory mutations in tRNA molecules, such as the restoration of Watson–Crick base pairs in stem positions, may eventually mitigate some negative effects of single-base changes (Steinberg and Cedergren 1994; Steinberg et al. 1994; Watanabe et al. 1994; Wolstenholme et al. 1994; Kern and Kondrashov 2004), experimental evidence suggests that the structural modifications noted above compromise the efficiency of protein synthesis. For example, bovine mitochondrial tRNAs have unusually low rates of amino-acid loading (Kumazawa et al. 1989, 1991; Hanada et al. 2001). Thus, although the bizarre architectures of animal mitochondrial tRNAs have been accompanied by dramatic changes in the recognition mechanisms used by their nuclear-encoded cognate tRNA amino-acyl synthetases (Kuhle et al. 2020), the compensating effects are less than perfect.

Could the increased width of the selective sieve for organelle-encoded genes be a simple consequence of the relaxation of selection in organelles (i.e., smaller selection coefficients), rather than an outcome of a reduction in the efficiency of selection owing to a smaller effective population size

(Brown et al. 1982; Kumazawa and Nishida 1993)? Analyses laid out in Lynch (2007) and Popadin et al. (2013) suggest that the absolute strength of selection (i.e., the selection coefficient) against deleterious mutations in the mitochondrion is equivalent, if not higher, than in the nuclear genome.

Coevolutionary drive and compensatory mutations

The conclusion that mitochondrial genomes harbor a reduced ability to purge deleterious mutations motivates the additional idea that such mutations secondarily drive the fixation of compensatory mutations (Rand et al. 2004). In principle, such fitness-restoring mutations may arise in the organelle genes themselves (Oliveira et al. 2008; Meer et al. 2010; James et al. 2016), although most attention has been given to alterations in key interacting nuclear-encoded genes. Three types of nuclear genes with intimate connections with organelle partners are of particular interest: 1) the sets of tRNA amino-acyl synthetases (noted earlier), each of which attaches a specific amino acid to its cognate tRNA, either in the mitochondrion or in the cytosol; 2) the ribosomal protein-coding genes designated for cytosolic versus mitochondrial ribosomes; and 3) the nuclear-encoded components of the complexes in the mitochondrial oxidative phosphorylation (OXPHOS) pathway.

As noted in Chapter 6, relative to the bacterial ancestral state, mitochondrial ribosomes have experienced a dramatic increase in the number of protein subunits, all encoded in the nuclear genome. Early in eukaryotic evolution, on the order of 75 new subunits were added to the mitochondrial ribosome, and this was then followed by multiple lineage-specific gains (and losses in some cases), with patchy additions suggesting recruitment to ameliorate pre-existing structural instabilities (van der Sluis et al. 2015; Petrov et al. 2019). Thus, as in the case of tRNAs, the diversification of the structural features of mitochondrial ribosomes contrasts dramatically with the high degree of phylogenetic stasis in cytosolic ribosomes. In addition, as with tRNAs, many of the rRNA stem pairs that are G:C in bacterial ribosomal RNAs are A:U in mitochondrial rRNAs, resulting in a loss of ~ 260 hydrogen bonds and hence, a substantial reduction in stability (van der Sluis et al. 2015).

These kinds of observations on structural modifications generalize to the mitochondrial OXPHOS complexes. Despite the overall mass migration of organelle genes to the nuclear genome, several large mitochondrial complexes involved in the electron-transport chain retain a few mitochondrially encoded subunits. However, as with ribosomal proteins, along the path from FECA to LECA, the individual OXPHOS complexes acquired multiple novel protein subunits (in addition to the nuclear transferred units), nearly tripling the numbers of components relative to the ancestral state (Hirst 2011; Huynen et al. 2013; van der Sluis et al. 2015). For example, mitochondrial complex I has 40–64 protein subunits in most eukaryotic lineages, whereas the orthologous structure in bacteria has only 14 (Gabaldón et al. 2005; Cardol 2011; Huynen and Elurbe 2022). As with ribosomes, an early (pre-LECA) phase of expansion was followed by smaller numbers of lineage-specific gains and losses, such that not all eukaryotic lineages have the same sets of supernumerary proteins. Although there is evidence that such proteins play important roles in maintaining enzyme stability (Angerer et al. 2011; Stroud et al. 2016), this is expected for an evolved compensatory modification and need not imply improved overall enzyme performance. Indeed, there is no evidence that the simpler bacterial complexes are less stable or inferior in any way (Hirst 2011).

Finally, several studies at the amino-acid sequence level have shown that lineages with rapidly evolving mitochondrial-encoded proteins show parallel elevations in rates of evolution of nuclear-encoded subunits (Osada and Akashi 2012; Zhang and Broughton 2013; Sloan et al. 2014; Adrion et al. 2016; Havird et al. 2017). For example, studies in animals and yeast indicate that, even after accounting for mutation-rate differences, mitochondrial ribosomal-protein sequences evolve >10× more rapidly than those for cytoplasmic ribosomes, despite both being encoded in the nuclear genome (Pietromonaco et al. 1986; Barreto and Burton 2013; Barreto et al. 2018). Notably, components of OXPHOS complexes that are fully encoded in the nuclear genome do not exhibit such elevated rates of evolution (Havird et al. 2015). Although it has been suggested that elevated rates of amino-acid sequence evolution in nuclear-encoded mitochondrial versus cytosolic proteins may simply be due to lower expression levels of the former (and hence, potentially weaker purifying selection against deleterious mutations), this idea does not have wide support (Osada and Akashi 2012; Barreto and Burton 2013; Barreto et al. 2018). This general set of observations is consistent with theory suggesting that a gene in a high mutation/low effective population-size environment can drive accelerated rates of fixation of compensatory mutations in interacting partners encoded in a more mutationally quiescent/high N_e environment (Lynch 2023).

Plastid Evolution

Long after the establishment of the mitochondrion, on the order of 1.0 BYA, another endosymbiotic event forever changed the eukaryotic world: the colonization of a lineage that would go on to form the base of the Archaeplastida, which subsequently diversified into the red algae, green plants (including green algae), and glaucophytes (a basal group of unicellular algae) (Parfrey et al. 2011; Keeling 2013; Eme et al. 2014). Giving rise to the chloroplast, this brought photosynthesis into the eukaryotic domain.

As with the mitochondrion, the search for the historical roots of this event is made possible by the presence of a genome within the plastid. Although the plastid has clear affinities with cyanobacteria, this large phylogenetic group contains lineages with diverse properties (including multicellular forms, and those capable of fixing nitrogen). From comparative genomics, an emerging consensus is that the closest living relative of the plastid is *Gloeomargarita lithophora*, a non-nitrogen-fixing species (Ponce-Toledo et al. 2017; Sánchez-Baracaldo et al. 2017). The fact that this species and its relatives, as well as the basal glaucophytes, are restricted to freshwater environments (Price et al. 2012), suggests the possibility that photosynthetic eukaryotes arose in a terrestrial freshwater environment, with colonization of the oceans occurring secondarily.

Unlike the universal spread of the mitochondrion across the entire eukaryotic phylogeny by simple vertical inheritance, eukaryotic photosynthesis acquired a punctate phylogenetic distribution by horizontal transfer among a small number of lineages (Keeling 2013). Secondary plastids have arisen on multiple occasions as heterotrophic eukaryotes from one lineage engulfed photosynthetic species from another and then retained them in a permanent endosymbiotic state. For example, the basal ancestors of euglenoids and chlorarachniophytic algae independently acquired photosynthesis via the capture and domestication of green algae. Morphological support for such transfer derives from the presence of four membranes surrounding secondary plastids: two from the primary plastid, a third from the plasma membrane of the donating eukaryotic cell, and a fourth putatively from the phagosomal membrane of the host cell. Thus, whereas primary plastids float freely in the cytoplasm, secondary plastids are topologically integrated into the endomembrane system.

Secondary plastids have also arisen on multiple occasions via engulfment of members of the red-algal lineage. Although uncertainty remains as to how many independent events have occurred, such transfers have led to the spread of photosynthesis across a wide array of eukaryotic groups, among them the stramenopiles (including diatoms), cryptomonads, haptophytes, and dinoflagellates (Keeling 2013). A few cases are even known in which dinoflagellate species absorbed another cell containing secondary plastids, endowing them with a tertiary plastid.

Cases of secondary and tertiary transfer must initiate with an endosymbiont containing three genomes – plastid, mitochondrion, and nucleus – but in all known cases, the mitochondrion has been lost. In a few cases, however, a remnant of the nuclear genome has been retained in the form of a nucleomorph. These include the cryptophytes, chlorarachniophytes, and some dinoflagellates (Sarai et al. 2020). As a result of endosymbiotic gene transfer, the nuclear genomes of the host cells in these lineages are substantially chimeric.

Aside from the primary plastid colonization at the base of the Archaeplastida and its secondary and tertiary spreads, still more introductions of photosynthesis into eukaryotes have occurred by independent primary events. For example, about 200 MYA, a freshwater amoeba called *Paulinella chromatophora* became colonized by a cyanobacterium related to *Synechococcus* (Nowack and Grossman 2012; Nowack 2014; Singer et al. 2017). Given that about 30 genes of cyanobacterial origin have been relocated to the nuclear genome and produce products addressed back to the endosymbiont, the latter can be regarded as a legitimate organelle. An even more recent establishment, probably < 20 MYA, involves the colonization of the diatom *Rhopalodia* by a nitrogen fixer related to the cyanobacterium *Cyanothece*, which is no longer photosynthetic in its host (which itself is photosynthetic via a secondary plastid) (Nakayama et al. 2011, 2014).

Examples also exist in which cyanobacteria are less fully integrated with their eukaryotic partners. For example, *Atelocyanobacterium thalassa*, a cyanobacterium lacking both oxygen-producing photosystem II and the citric-acid cycle, associates extracellularly with a marine prymnesiophyte, to which it provides fixed nitrogen while gaining fixed carbon in return (Thompson et al. 2012). A dinoflagellate called *Ornithocercus* carries in an extracellular chamber a load of cyanobacteria, distantly related to marine *Prochlorococcus/Synechococcus*, apparently periodically digesting them in a farming-like process (Nakayama et al. 2019). In both of these cases, the cyanobacterial genome is highly reduced in size, consistent with genome reduction following a long-term association. Finally, kleptoplasty, wherein a heterotrophic consumer ingests a photosynthetic prey item and then retains its chloroplasts, is found in a number of lineages, dinoflagellates in particular (Hehenberger et al. 2019).

Conventional chloroplasts, derived from the ancestral archaeplastid event, provide dramatic examples of parallel evolution of the types of remodelling observed in mitochondria. Just a few of these are mentioned here. First, as in mitochondria, chloroplast division typically proceeds with a cyanobacterial-derived FtsZ protein operating on the inside and a eukaryotic dynamin protein on the outside of the membrane (Miyagishima et al. 2014). However, glaucophyte chloroplasts, which branched off from the Archaeplastida prior to the red and green algae, are unusual in having a peptidoglycan layer between the inner and outer membranes and divide using only FtsZ. Thus, the integration of dynamin into chloroplast division apparently occurred after the loss of the peptidoglycan layer.

Second, nuclear-encoded chloroplast genes require the presence of terminal targeting sequences for localization to the plastid, with the situation being even more extreme in the case of secondary plastids, which must carry dual targeting messages, one to the external endosymbiont membrane and the other to its internal organelle membrane. Independent of mitochondrial TIM and TOM, plastid TIC and TOC evolved as inner- and outer-membrane chloroplast channels tethered together to control protein import from the cytoplasm (Chen et al. 2018).

Third, although there are unique features in individual lineages, for the most part genome evolution has proceeded down parallel pathways in mitochondria and plastids (Lynch et al. 2006; Smith and Keeling 2015), including substantial genome size reduction (Figure 23.2), the gravitation towards A:T richness, and the emergence of uniparental inheritance. Plastid genomes are highly diminished relative to those of free-living cyanobacteria, generally containing just 30–230 protein-coding genes. Many of the original genes were transferred to the nucleus, some of which have taken on entirely novel functions (Martin et al. 2002). Some land plants have multi-chromosomal plastid genomes, but the same is true for mitochondrial genomes in a number of metazoan lineages (Lavrov and Pett 2016).

One of the most unique features of plastid evolution concerns the degree to which the mutation rate has been reduced – commonly 5–10× lower than that of the nuclear genome in land plants (Gaut et al. 1996), although less extreme than in plant mitochondria and less diminished in algal lineages (Smith and Keeling 2012; Ness et al. 2015). Despite the lower mutation rates than in metazoan mitochondria, the plastid proteome has evolved in a number of ways that suggest an influence from deleterious-mutation accumulation, similar to that observed in mitochondria. For example,

new protein subunits have been recruited to plastid ribosomes, although not as extensively as in the case of mitochondria (Yamaguchi et al. 2000; Sharma et al. 2007).

As with plant mitochondria, it remains a mystery as to how selection has been able to promote the extremely low mutation rates of plastid genomes. However, a few land-plant lineages have evolved dramatic increases in plastid mutation rates, and these exhibit the kinds of alterations in protein-sequence evolution noted for mitochondria, including enhanced rates of amino-acid substitutions in nuclear-encoded subunits of plastid molecular complexes (Sloan et al. 2014; Zhang et al. 2015; Rockenbach et al. 2016; Weng et al. 2016).

Addiction to Endosymbionts

Given their roles in energy production, it is common to think of mitochondria and other endosymbionts as having been driven to fixation by adaptive mechanisms, most notably by magnifying the growth potential of the host species. However, as this view ignores the apparent adaptive losses incurred by endosymbionts, it is not entirely consistent with the general postulates of evolution by natural selection. A more fundamental issue, emphasized earlier, is whether the host species achieves a net advantage even in the long run.

To be sure, well-embedded endosymbiotic systems are essential to their host species – once locked in by processes of reciprocal subfunctionalization, reversion is no longer possible. However, although such relationships are often viewed as cooperative mutualisms, it need not follow that the total productivity of the pair (or even just the host cell) exceeds the pre-mutualism condition. For this reason, McCutcheon et al. (2019) advocate the use of the label 'host-beneficial endosymbiont' to describe the internal inhabitant.

Further insights into these issues have been gained through attempts to establish microbial mutualisms in experimental systems. Although a symbiosis may start out with a recipient simply benefiting from a waste product of a donor, the recipient can then respond evolutionarily to feed the donor so as to create more by-product, with the donor then becoming more enslaved by the original recipient and provisioning it even more (Harcombe et al. 2018). Thus, the establishment of a stable mutualism incurs bi-directional costs, which can only be revealed in the early stages of establishment when the participants can still be grown alone. Once established, functional interdependence enhances the likelihood of loss of genes essential to independent living but no longer required (Hillesland et al. 2014).

In this sense, mutualisms represent a sort of reciprocal addiction. In some cases, this may enable the system to survive in a novel nutritional environment, e.g., with species A providing B with a critical supply of carbon, and B being an essential source of nitrogen for A (Hom and Murray 2014; Fritts et al. 2020), but it remains unclear whether evolved mutualisms can ever outcompete virgin host cells in ancestral environments. Of course, to the extent that endosymbioses initiate as host–pathogen systems, an idea with considerable support (Sachs et al. 2011, 2013), then the eventual taming of the pathogen by the host cell can be viewed as beneficial. Again, however, this need not mean that the host has become better off than in the complete absence of the pathogen.

We close with a brief overview of some of the more dramatic examples of addiction of host cells to externally acquired agents: the toxin-antitoxin systems in bacteria. Many bacteria produce toxins that are excreted into the environment, and in doing so eliminate susceptible competing species or other innocent bystanders. However, thousands of bacterial systems are known in which a toxin is released intracellularly along with an antitoxin molecule, which prevents autotoxicity. Nearly all bacterial species harbor one or more such systems, frequently carried as linked toxin–antitoxin (TA) cassettes on extrachromosomal DNAs called plasmids (Hayes and Van Melderen 2011; Goeders and Van Melderen 2014). Generally, TA phylogenies are incongruent with host-cell phylogenies, implying horizontal transfer of the plasmids. Moreover, the toxicity mechanisms associated with TA cassettes are highly diverse, ranging from transcription/translation inhibition to transcript destruction to interference with membranes or cell division, and the antitoxins can be either proteins or RNAs, implying the independent evolution of such systems.

TA systems are exquisitely constructed to ensure self-proliferation. As the antitoxin is less stable than the toxin, unless the plasmid is retained after host-cell division, death will rapidly ensue as the toxin is freed from inhibition. In this sense, the host becomes addicted to the TA-carrying plasmid. Over evolutionary time, many TA systems become incorporated into bacterial chromosomes, in effect serving as partial cure to addiction, allowing the loss of the plasmid, although not fully relieving the host from carrying the TA cassette.

The key point here is that plasmid-born TA systems provide a compelling example of highly successful parasites that ensure their own selfish proliferation by selectively eliminating non-cooperating host cells. In principle, such systems can be stable for a long time, provided a subpopulation of plasmid-free host cells does not emerge (e.g., by fortuitous deletion of the toxin gene). Moreover, a TA system can become an essentially permanent fixture if the toxin and/or antitoxin provides enough additional side benefit to the host to offset the occasional loss by post-segregation killing (Rankin et al. 2012). In some cases, such secondary benefits involve alterations of the host-cell response to additional stresses (Van Melderen 2010; Yamaguchi et al. 2011).

Finally, although the mechanisms are not fully understood, TA-like systems of a different nature seem to exist in ciliated protozoans. Numerous *Paramecium* strains are carriers of so-called killer bacteria that, when released into the environment, are lethal to naive (bacteria-free) host cells (Görtz and Fokin 2009; Schrallhammer and Schweikert 2009). These bacteria are phylogenetically diverse, distributed over multiple proteobacterial lineages, and exhibit a wide array of killing mechanisms, including paralysis, osmotic imbalance, mate killing, and out-of-control spinning. Thus, as in the case of bacterial TA systems, there appear to be multiple independent origins of killer-bacterial systems in ciliates.

Under certain physiological conditions, killer bacteria produce a huge ribbon-like inclusion called an R body, which unwinds into a spear-like structure at low pH, and appears to carry defective phage particles that might be the carriers of toxins (Pond et al. 1989). The four small proteins that polymerize to form R bodies are often carried on plasmids, but in some cases, they have moved to the bacterial chromosome (Jeblick and Kusch 2005), reminiscent of the fates of TA systems. Moreover, genomic surveys suggest that R bodies are widely distributed across the bacterial phylogeny, although their functions are generally unknown (Raymann et al. 2013). It remains unclear how the *Paramecium* carriers of killer bacteria acquire an immunity to toxicity. However, it is known that, if the host cells are grown at maximum rates, the bacterial cells are gradually lost owing to their inability to proliferate as fast as the *Paramecium*, i.e., the host cells can be progressively weaned from their addiction. Upon loss of the bacteria, resistance to the toxin is lost (Grosser et al. 2018), as expected if the endocytobiont is a carrier of a toxin–antitoxin system.

Summary

- The eukaryotic phylogeny is replete with endosymbiotic mutualisms, the most prominent being the mitochondrion, derived from an alphaproteobacterium that colonized the stem eukaryote, and the younger plastid, derived from a cyanobacterium. The basal mitochondrion led to the establishment of internal-membrane bioenergetics throughout all eukaryotes. The plastid, which brought photosynthesis into the eukaryotic domain, acquired a more punctate phylogenetic distribution by spreading horizontally into pre-existing heterotrophic lineages.

- It remains unclear whether the establishment of the mitochondrion preceded the emergence of other eukaryotic-specific traits such as internal membranes and cytoskeletons, but numerous observations from comparative genomics suggest the opposite order of events.

- It is commonly argued that the establishment of mitochondrial membrane bioenergetics led to a massive increase in the bioenergetic and evolutionary capacity of eukaryotes relative to prokaryotes, and that this boost was an essential pre-requisite for all things related to eukaryogenesis. However, multiple lines of evidence are inconsistent with this view, including direct bioenergetic measurements, the bi-directional costs and conflicts expected when mutualisms evolve, and the accumulation of mildly deleterious mutations in non-recombining genomes.

- These caveats aside, substantial intracellular remodelling did follow the establishment of the mitochondrion. The vast majority of surviving genes of mitochondrial origin was relocated to the nuclear genome; mechanisms for targeting the nuclear-encoded products of these genes back to the mitochondria evolved; host mechanisms for regulating mitochondrial proliferation emerged; and mitochondria became progressively intertwined in entirely new functions, such as targeted cell death. This diverse set of observations demonstrates the power of the intracellular environment at directing the course of evolution.

- Relative to the situation for nuclear genes, organelle genes often experience a substantial reduction in effective population size, an absence of recombination, and altered mutation rates. Taken together, these conditions have arguably led to the accumulation of excess deleterious mutations in organelle genomes, generating as coevolutionary side effects compensatory mutations in interacting nuclear-encoded genes. As a result, the structural features of mitochondrial proteins and RNAs have sometimes been modified in dramatic ways never seen in nuclear encoded genes or in prokaryotic genomes.

- Beyond the mitochondrion and plastid, there are many additional examples of intracellular inhabitants within unicellular eukaryotes. The advantages of such consortia to the host cell are generally unknown, and in many cases may not exist, with the endocytobiont simply ensuring its own existence by use of a toxin–antitoxin mechanism that serves as an addictive leash on the host.

Foundations 23.1 Messenger-RNA editing

A long history of research in molecular biology inspires confidence that coding-sequence information at the DNA level provides a reliable prediction of protein sequences. However, some organelles use post-transcriptional editing to modify mRNA, tRNA, and/or rRNA sequences. The most spectacular display of editing occurs in the mitochondrial genes of kinetoplastids (e.g., *Trypanosoma*), where insertion and deletion of Us (uridines) affects about 90% of all codons (Simpson et al. 2000; Horton and Landweber 2002). With editing at such a massive scale, the underlying genomic sequences for genes are literally nonsensical. Editing in such species relies on the baroque structure of the kinetoplastid mitochondrial genome – a vast network of intertwined molecules, including several 20–40 kb maxicircles carrying the cryptic gene sequences, and thousands of 0.5–3.0 kb minicircles carrying guide RNA templates for the addition/removal of Us in the immature maxicircle-derived mRNAs (Koslowsky et al. 1992).

Although no other mitochondrial lineage engages in editing as extensively as kinetoplastids, some slime molds insert nucleotides every 25–40 nucleotides in mRNAs, rRNAs, and tRNAs (Horton and Landweber 2000; Cheng et al. 2001; Byrne and Gott 2004), and 2–4% of the amino-acid replacement sites in dinoflagellate mitochondrial genomes are edited (Lin et al. 2002; Zhang and Lin 2005). Multiple animals, rhizopod amoebae, and basal fungi use editing to restore base mismatches in the stems of mitochondrial tRNAs (Janke and Pääbo 1993; Lonergan and Gray 1993; Yokobori and Pääbo 1995; Tomita et al. 1996; Paquin et al. 1997; Lavrov et al. 2000; Laforest et al. 2004).

This sporadic distribution of diverse forms of editing strongly suggests that such processes have evolved independently in different lineages and, with very few minor exceptions, always confined to organelle genomes. To date, no compelling explanation has been promoted as to how such substantial investments in editing may have arisen by adaptive mechanisms. However, a plausible case has been made that such seemingly superfluous systems can in some cases be inadvertently promoted by effectively neutral processes (Stoltzfus 1999; Gray 2012). Given one of the central themes of the book – that neutral evolutionary mechanisms can drive modifications at the cellular level, further exploration of the matter of editing is warranted. Here, the focus is primarily on the organelle genomes of plants, where the phenomenon has been studied most intensely.

Messenger-RNA editing is used extensively in land-plant organelles. For example, in *Arabidopsis* mitochondria, 441 editing sites are present in coding regions along with smaller numbers in introns and intergenic DNA,

nearly all of them changing C to U (Giegé and Brennicke 1999). Similar levels of C→U mRNA editing are found in the mitochondria of other land plants, including some liverworts (Malek et al. 1996; Freyer et al. 1997). Although editing is less extensive in land-plant plastids, there are commonly 25–30 editing sites per genome in angiosperms (Tsudzuki et al. 2001), and up to several hundred sites in ferns and hornworts (Kugita et al. 2003; Wolf et al. 2004). These observations, combined with the absence of mRNA editing in the organelles of green algae (Rüdinger et al. 2012), suggest a dramatic expansion of organelle editing with the origin of multicellular plants (Hiesel et al. 1994).

The vast majority of mRNA editing in land-plant organelles occurs at amino-acid replacement (rather than silent) sites, often ensuring the preservation of amino acids that are highly conserved across distantly related species (Maier et al. 1996; Tsudzuki et al. 2001). Although this type of observation motivates the idea that editing provides a genomic buffer against the accumulation of deleterious mutations (Cavalier-Smith 1997; Horton and Landweber 2002; Smith 2006), several observations raise doubts about this adaptive interpretation. First and most notably, if the buffering hypothesis is correct, we would expect editing to be most common in genomes with high mutation rates, which is exactly the opposite of the pattern actually seen.

Second, the buffering hypothesis ignores the complexities of the editing process, which necessarily relies on the sequence stability of both the recognition sites in the organelle and the significant investment in the editing apparatus itself. The *cis*-recognition sites for editing factors span at least 23 bp (Choury et al. 2004; Miyamoto et al. 2004), and as further discussed below, the *trans*-acting factors minimally involve a large family of nuclear-encoded proteins, each devoted to just one or two specific editing sites. To be promoted by positive selection, any advantage to editing a particular site would have to exceed the prices paid.

Third, editing in plant organelles is quite noisy, resulting in the production of a heterogeneous pool of transcripts, some being incompletely edited and/or containing erroneous editorial changes (Phreaner et al. 1996; Inada et al. 2004; Guo et al. 2015). In many cases, completely edited transcripts are the exception, rather than the rule (Schuster et al. 1990).

A fourth significant challenge to the hypothesis that mRNA editing is maintained by selection derives from the following observation. Recall that the vast majority of mRNA editing in plant organelles involves conversions of C to U. Because C→T mutations at these sites eliminate the need for editing, such mutations are expected to accumulate at the neutral rate under the buffering hypothesis, as an allele with an encoded T should be selectively equivalent to one that simply acquires a C→U replacement by editing. However, the rate of conversion of C→U editing sites to non-edited Ts is 4× greater than the neutral expectation (Shields and Wolfe 1997; Tsudzuki et al. 2001; Fujii and Small 2011). Thus, if anything, editing sites in land-plant organelles are at least mildly deleterious.

As just noted, the disadvantages associated with editing a site include the inaccuracies of editing and the energetic burden of maintaining the editing apparatus. An additional issue is the excess degenerative mutation rate for alleles bearing editing sites (Lynch 2007). The intrinsic mutational disadvantage of an editing site will be approximately equal to the total mutation rate over the nucleotide sites reserved for editing-site recognition, i.e., $\sim 23\times$ the mutation rate per site. Thus, the reduced level of editing in land-plant plastids versus mitochondria is consistent with this mutational-hazard hypothesis, as the former have higher mutation rates than the later. The dramatic reduction in the incidence of editing in the organelle genomes of plants that have experienced massive increases in the mutation rate (Palmer et al. 2000; Parkinson et al. 2005; Fan et al. 2019) is also consistent with the hypothesis. In contrast, the mitochondrion of the tulip tree, which has the one of the most mutationally quiescent genomes known, is heavily edited (Richardson et al. 2013).

The preceding arguments lead to a reasonably satisfying hypothesis for the phylogenetic distribution of editing, but substantial questions remain as to how such processes initially become established. To account for such a system, a mechanism must exist for the establishment of dozens to hundreds of nuclear-encoded editing factors, each specialized to recognize a small number of (perhaps even single) organelle sites. Although gene duplication can plausibly allow the expansion and specialization of a well-operating system, its initial establishment requires the existence of factors with latent editing potential prior to the origin of mRNA editing, presumably as a byproduct of some other essential cellular function (Covello and Gray 1993). In addition, the emergence of site-specific refinements by natural selection is difficult with nuclear-encoded factors, which would have to remain in tight linkage disequilibrium with their serviced organelle sites while both are en route to fixation. Uniparental inheritance of organelles facilitates such associations, as half of the gametes of the transmitting parentwill contain the appropriate nuclear-cytoplasmic combination, but full dissociation becomes almost certain beyond two generations of sexual reproduction.

Although the precise mechanism of C→U mRNA editing remains unclear, the process intimately involves one of the largest nuclear-encoded gene families in land plants: the PPR (pentatricopeptide) proteins. Containing up to 600 genes per land-plant genome, many of the members of this family are targeted to specific mRNAs within organelles, where they likely recruit another enzyme to complete the editing step (Schmitz-Linneweber and Small 2008; Fujii and Small 2011). Some PPR genes have functions unassociated with editing (e.g., RNA folding and translation), but one subclass in particular, the DYW-domain-containing PPR genes, is largely restricted to land plants that carry out editing. Species such as *Marchantia* that have lost editing have also lost the DYW family, and others with small number of editing sites have greatly diminished numbers of PPR genes, e.g., the moss *Physcomitrella* has only 10 organellar editing sites and just a single DYW PPR gene. Unicellular lineages of green algae without editing lack members of this family, whereas the amoeboid protist *Naegleria*, which has C→U editing, also has DYW PPRs (Fritz-Laylin et al. 2010).

In summary, the origin of mRNA editing is one of more enigmatic aspects of genome evolution. There is no evidence that such processes have originated to buffer mutational damage. In addition, the hypothesis that editing promotes the generation of adaptive variation at the RNA level (Tillich et al. 2006) is entirely without support, as is the idea that editing arises as a resolution of a nucleo-cytoplasmic conflict (Castandet and Araya 2011). Whereas it is difficult to reject the hypothesis that mRNA editing in organelles has arisen by nearly neutral processes, the mechanisms by which editing factors acquire their apparent site-specificity remain unclear. A really creative selfish editor might inflict the organelle genomic change necessary to ensure its own survival, but no such element is known to exist. It is, however, intriguing that some proteins involved in nuclear-mRNA editing are capable of inducing site-specific mutations (Smith 2006; Iyer et al. 2011).

Literature Cited

Adrion, J. R., P. S. White, and K. L. Montooth. 2016. The roles of compensatory evolution and constraint in aminoacyl tRNA synthetase evolution. Mol. Biol. Evol. 33: 152–161.

Amiri, H., O. Karlberg, and S. G. Andersson. 2003. Deep origin of plastid/parasite ATP/ADP translocases. J. Mol. Evol. 56: 137–150.

Anderson, A. P., X. Luo, W. Russell, and Y. W. Yin. 2020. Oxidative damage diminishes mitochondrial

DNA polymerase replication fidelity. Nucleic Acids Res. 48: 817–829.

Anderson, J. B., C. Wickens, M. Khan, L. E. Cowen, N. Federspiel, T. Jones, and L. M. Kohn. 2001. Infrequent genetic exchange and recombination in the mitochondrial genome of *Candida albicans*. J. Bacteriol. 183: 865–872.

Andersson S. G, O. Karlberg, B. Canback, and C. G. Kurland. 2003. On the origin of mitochondria: A genomics perspective. Phil. Trans. Roy. Soc. Lond. B Biol. Sci. 358: 165–177.

Andersson, S. G., A. Zomorodipour, J. O. Andersson, T. Sicheritz-Ponten, U. C. Alsmark, R. M. Podowski, A. K. Naslund, A. S. Eriksson, H. H. Winkler, and C. G. Kurland. 1998. The genome sequence of *Rickettsia prowazekii* and the origin of mitochondria. Nature 396: 133–140.

Angerer, H., K. Zwicker, Z. Wumaier, L. Sokolova, H. Heide, M. Steger, S. Kaiser, E. Nübel, B. Brutschy, M. Radermacher, et al. 2011. A scaffold of accessory subunits links the peripheral arm and the distal proton-pumping module of mitochondrial complex I. Biochem. J. 437: 279–288.

Archibald, J. 2015. One Plus One Equals One: Symbiosis and the Evolution of Complex Life. Oxford University Press, Oxford, UK.

Ashley, M. V., P. J. Laipis, and W. W. Hauswirth. 1989. Rapid segregation of heteroplasmic bovine mitochondria. Nucleic Acids Res. 17: 7325–7331.

Ballard, J. W., and M. Kreitman. 1994. Unraveling selection in the mitochondrial genome of *Drosophila*. Genetics 138: 757–772.

Barreto, F. S., and R. S. Burton. 2013. Evidence for compensatory evolution of ribosomal proteins in response to rapid divergence of mitochondrial rRNA. Mol. Biol. Evol. 30: 310–314.

Barreto, F. S., E. T. Watson, T. G. Lima, C. S. Willett, S. Edmands, W. Li, and R. S. Burton. 2018. Genomic signatures of mitonuclear coevolution across populations of *Tigriopus californicus*. Nat. Ecol. Evol. 2: 1250–1257.

Bergstrom, C. T., and J. Pritchard. 1998. Germline bottlenecks and the evolutionary maintenance of mitochondrial genomes. Genetics 149: 2135–2146.

Birky, C. W. Jr., C. A. Demko, P. S. Perlman, and R. Strausberg. 1978. Uniparental inheritance of mitochondrial genes in yeast: Dependence on input bias of mitochondrial DNA and preliminary investigations of the mechanism. Genetics 89: 615–651.

Birky, C. W. Jr., T. Maruyama, and P. Fuerst. 1983. An approach to population and evolutionary genetic theory for genes in mitochondria and chloroplasts, and some results. Genetics 103: 513–527.

Bohr, V. A., T. Stevnsner, and N. C. de Souza-Pinto. 2002. Mitochondrial DNA repair of oxidative damage in mammalian cells. Gene 286: 127–134.

Brown, W. M., E. M. Prager, A. Wang, and A. C. Wilson. 1982. Mitochondrial DNA sequences of primates: Tempo and mode of evolution. J. Mol. Evol. 18: 225–239.

Broz, A. K., A. Keene, M. Fernandes Gyorfy, M. Hodous, I. G. Johnston, and D. B. Sloan. 2022. Sorting of mitochondrial and plastid heteroplasmy in *Arabidopsis* is extremely rapid and depends on MSH1 activity. Proc. Natl. Acad. Sci. USA 119: e2206973119.

Bublitz, D. C., G. L. Chadwick, J. S. Magyar, K. M. Sandoz, D. M. Brooks, S. Mesnage, M. S. Ladinsky, A. I. Garber, P. J. Bjorkman, V. J. Orphan, and J. P. McCutcheon. 2019. Peptidoglycan production by an insect-bacterial mosaic. Cell 179: 703–712.

Burki, F. 2016. Mitochondrial evolution: Going, going, gone. Curr. Biol. 26: R410–R412.

Burt, A., and R. Trivers. 2006. Genes in Conflict. Harvard University Press, Cambridge, MA.

Byrne, E. M., and J. M. Gott. 2004. Unexpectedly complex editing patterns at dinucleotide insertion sites in *Physarum* mitochondria. Mol. Cell. Biol. 24: 7821–7828.

Cardol, P. 2011. Mitochondrial NADH: Ubiquinone oxidoreductase (complex I) in eukaryotes: A highly conserved subunit composition highlighted by mining of protein databases. Biochim. Biophys. Acta 1807: 1390–1397.

Castandet, B., and A. Araya. 2011. RNA editing in plant organelles. Why make it easy? Biochemistry (Mosc.) 76: 924–931.

Cavalier-Smith, T. 1987. The origin of eukaryotic and archaebacterial cells. Ann. N. Y. Acad. Sci. 503: 17–54.

Cavalier-Smith, T. 1997. Cell and genome coevolution: Facultative anaerobiosis, glycosomes and kinetoplastan RNA editing. Trends Genet. 13: 6–9.

Cavalier-Smith, T. 2006. Origin of mitochondria by intracellular enslavement of a photosynthetic purple bacterium. Proc. Roy. Soc. Lond. B 273: 1943–1952.

Cavalier-Smith, T. 2009. Predation and eukaryote cell origins: A coevolutionary perspective. Int. J. Biochem. Cell Biol. 41: 307–322.

Cavalier-Smith, T., and J. J. Lee. 1985. Protozoa as hosts for endosymbioses and the conversion of symbionts into organelles. J. Euk. Microbiol. 32: 371–558.

Charlesworth, B., M. T. Morgan, and D. Charlesworth. 1993. The effect of deleterious mutations on neutral molecular variation. Genetics 134: 1289–303.

Chen, Y. L., L. J. Chen, C. C. Chu, P. K. Huang, J. R. Wen, and H. M. Li. 2018. TIC236 links the outer and inner membrane translocons of the chloroplast. Nature 564: 125–129.

Cheng, Y. W., L. M. Visomirski-Robic, and J. M. Gott. 2001. Non-templated addition of nucleotides to the 3′ end of nascent RNA during RNA editing in *Physarum*. EMBO J. 20: 1405–1414.

Cho, Y., J. P. Mower, Y. L. Qiu, and J. D. Palmer. 2004. Mitochondrial substitution rates are extraordinarily elevated and variable in a genus of flowering plants. Proc. Natl. Acad. Sci. USA 101: 17741–17746.

Choury, D., J. C. Farre, X. Jordana, and A. Araya. 2004. Different patterns in the recognition of editing sites in plant mitochondria. Nucleic Acids Res. 32: 6397–6406.

Clark, K. A., D. K. Howe, K. Gafner, D. Kusuma, S. Ping, S. Estes, and D. R. Denver. 2012. Selfish little circles: Transmission bias and evolution of large deletion-bearing mitochondrial DNA in *Caenorhabditis briggsae* nematodes. PLoS One 7: e41433.

Cosmides, L. M., and J. Tooby. 1981. Cytoplasmic inheritance and intragenomic conflict. J. Theor. Biol. 89: 83–129.

Covello, P. S., and M. W. Gray. 1993. On the evolution of RNA editing. Trends Genet. 9: 265–268.

Davidov, Y., and E. Jurkevitch. 2009. Predation between prokaryotes and the origin of eukaryotes. Bioessays 31: 748–757.

de Duve, C. 2007. The origin of eukaryotes: A reappraisal. Nat. Rev. Genet. 8: 395–403.

Denver, D. D., P. C. Dolan, L. J. Wilhelm, W. Sung, J. I. Lucas-Lledó, D. K. Howe, S. C. Lewis, K. Okamoto, M. Lynch, W. K. Thomas, et al. 2009. A genome-wide view of *Caenorhabditis elegans* base-substitution mutation processes. Proc. Natl. Acad. Sci. USA 106: 16310–16314.

Denver, D. R., K. Morris, M. Lynch, and W. K. Thomas. 2004. High mutation rate and predominance of insertions in the *Caenorhabditis elegans* nuclear genome. Nature 430: 679–682.

Denver, D. R., K. Morris, M. Lynch, L. L. Vassilieva, and W. K. Thomas. 2000. High direct estimate of the mutation rate in the mitochondrial genome of *Caenorhabditis elegans*. Science 289: 2342–2344.

Dolezal, P., V. Likic, J. Tachezy, and T. Lithgow. 2006. Evolution of the molecular machines for protein import into mitochondria. Science 313: 314–318.

Doolittle, W. F. 1998. You are what you eat: A gene transfer ratchet could account for bacterial genes in eukaryotic nuclear genomes. Trends Genet. 14: 307–311.

Edwards, D. M., E. C. Royrvik, J. M. Chustecki, K. Giannakis, R. C. Glastad, A. L. Radzvilavicius, and I. G. Johnston. 2021. Avoiding organelle mutational meltdown across eukaryotes with or without a germline bottleneck. PLoS Biol. 19: e3001153.

Eme, L., S. C. Sharpe, M. W. Brown, and A. J. Roger. 2014. On the age of eukaryotes: Evaluating evidence from fossils and molecular clocks. Cold Spring Harb. Perspect. Biol. 6: a016139.

Emelyanov, V. V. 2001. Evolutionary relationship of Rickettsiae and mitochondria. FEBS Lett. 501: 11–18.

Esposti, M. D. 2016. Late mitochondrial acquisition, really? Genome Biol. Evol. 8: 2031–2035.

Eyre-Walker, A., and P. Awadalla. 2001. Does human mtDNA recombine? J. Mol. Evol. 53: 430–435.

Fan, W., W. Guo, L. Funk, J. P. Mower, and A. Zhu. 2019. Complete loss of RNA editing from the plastid genome and most highly expressed mitochondrial genes of *Welwitschia mirabilis*. Sci. China Life Sci. 62: 498–506.

Felsenstein, J. 1974. The evolutionary advantage of recombination. Genetics 78: 737–756.

Force, A., M. Lynch, B. Pickett, A. Amores, Y.-L. Yan, and J. Postlethwait. 1999. Preservation of duplicate genes by complementary, degenerative mutations. Genetics 151: 1531–1545.

Frank, S. A., and L. D. Hurst. 1996. Mitochondria and male disease. Nature 383: 224.

Freyer, R., M. C. Kiefer-Meyer, and H. Kossel. 1997. Occurrence of plastid RNA editing in all major lineages of land plants. Proc. Natl. Acad. Sci. USA 94: 6285–6290.

Friedman, J. R., L. L. Lackner, M. West, J. R. DiBenedetto, J. Nunnari, and G. K. Voeltz. 2011. ER tubules mark sites of mitochondrial division. Science 334: 358–362.

Friedman, J. R., and J. Nunnari. 2014. Mitochondrial form and function. Nature 505: 335–343.

Fritts, R. K., J. T. Bird, M. G. Behringer, A. Lipzen, J. Martin, M. Lynch, and J. B. McKinlay. 2020. Enhanced nutrient uptake is sufficient to drive emergent cross-feeding between bacteria in a synthetic community. ISME J. 14: 2816–2828.

Fritz-Laylin, L. K., S. E. Prochnik, M. L. Ginger, J. B. Dacks, M. L. Carpenter, M. C. Field, A. Kuo, A. Paredez, J. Chapman, J. Pham, et al. 2010. The genome of *Naegleria gruberi* illuminates early eukaryotic versatility. Cell 140: 631–642.

Fry, A. J. 1999. Mildly deleterious mutations in avian mitochondrial DNA: Evidence from neutrality tests. Evolution 53: 1617–1620.

Fujii, S., and I. Small. 2011. The evolution of RNA editing and pentatricopeptide repeat genes. New Phytol. 191: 37–47.

Fujii, S., and K. Toriyama. 2009. Suppressed expression of retrograde-regulated male sterility restores pollen fertility in cytoplasmic male sterile rice plants. Proc. Natl. Acad. Sci. USA 106: 9513–9518.

Gabaldón, T. 2018. Relative timing of mitochondrial endosymbiosis and the 'pre-mitochondrial symbioses' hypothesis. IUBMB Life 70: 1188–1196.

Gabaldón, T., D. Rainey, and M. A. Huynen. 2005. Tracing the evolution of a large protein complex in the eukaryotes, NADH: Ubiquinone oxidoreductase (Complex I). J. Mol. Biol. 348: 857–870.

Gaborieau, L., G. G. Brown, and H. Mireau. 2016. The propensity of pentatricopeptide repeat genes to evolve into restorers of cytoplasmic male sterility. Front. Plant Sci. 7: 1816.

Gabriel, W., M. Lynch, and R. Bürger. 1993. Muller's ratchet and mutational meltdowns. Evolution 47: 1744–1757.

Galluzzi, L., O. Kepp, and G. Kroemer. 2012. Mitochondria: Master regulators of danger signalling. Nat. Rev. Mol. Cell Biol. 13: 780–788.

Gaut, B. S., B. R. Morton, B. C. McCaig, and M. T. Clegg. 1996. Substitution rate comparisons between grasses and palms: Synonymous rate differences at the nuclear gene Adh parallel rate differences at the plastid gene rbcL. Proc. Natl. Acad. Sci. USA 93: 10274–10279.

Gemmell, N. J., V. J. Metcalf, and F. W. Allendorf. 2004. Mother's curse: The effect of mtDNA on individual fitness and population viability. Trends Ecol. Evol. 19: 238–244.

Giegé, P., and A. Brennicke. 1999. RNA editing in *Arabidopsis* mitochondria effects 441 C to U changes in ORFs. Proc. Natl. Acad. Sci. USA 96: 15324–15329.

Goeders, N., and L. Van Melderen. 2014. Toxin-antitoxin systems as multilevel interaction systems. Toxins (Basel) 6: 304–324.

Görtz, H.-D., and S. I. Fokin. 2009. Diversity of endosymbiotic bacteria in *Paramecium*, pp. 131–160. In M. Fujishima (ed.) Endosymbionts in *Paramecium*. Springer-Verlag, Berlin, Germany.

Goyal, S., D. J. Balick, E. R. Jerison, R. A. Neher, B. I. Shraiman, and M. M. Desai. 2012. Dynamic mutation-selection balance as an evolutionary attractor. Genetics 191: 1309–1319.

Gray, M. W. 2012. Mitochondrial evolution. Cold Spring Harb. Perspect. Biol. 4: a011403.

Gray, M. W. 2015. Mosaic nature of the mitochondrial proteome: Implications for the origin and evolution of mitochondria. Proc. Natl. Acad. Sci. USA 112: 10133–10138.

Grosser, K., P. Ramasamy, A. D. Amirabad, M. H. Schulz, G. Gasparoni, M. Simon, and M. Schrallhammer. 2018. More than the 'killer trait': Infection with the bacterial endosymbiont *Caedibacter taeniospiralis* causes transcriptomic modulation in *Paramecium* host. Genome Biol. Evol. 10: 646–656.

Guo, W., F. Grewe, and J. P. Mower. 2015. Variable frequency of plastid RNA editing among ferns and repeated loss of uridine-to-cytidine editing from vascular plants. PLoS One 10: e0117075.

Haag-Liautard, C., N. Coffey, D. Houle, M. Lynch, B. Charlesworth, and P. D. Keightley. 2008. Direct estimation of the mitochondrial DNA mutation rate in *D. melanogaster*. PLoS Biology 6: 1706–1714.

Haigh, J. 1978. The accumulation of deleterious genes in a population – Muller's ratchet. Theor. Pop. Biol. 14: 251–267.

Hampl, V., Čepička I, and M. Eliaš. 2019. Was the mitochondrion necessary to start eukaryogenesis? Trends Microbiol. 27: 96–104.

Hanada, T., T. Suzuki, T. Yokogawa, C. Takemoto-Hori, M. Sprinzl, and K. Watanabe. 2001. Translation ability of mitochondrial tRNAsSer with unusual secondary structures in an *in vitro* translation system of bovine mitochondria. Genes Cells 6: 1019–1030.

Harcombe, W. R., J. M. Chacón, E. M. Adamowicz, L. M. Chubiz, and C. J. Marx. 2018. Evolution of bidirectional costly mutualism from byproduct consumption. Proc. Natl. Acad. Sci. USA 115: 12000–12004.

Hasegawa, M., Y. Cao, and Z. Yang. 1998. Preponderance of slightly deleterious polymorphism in mitochondrial DNA: Nonsynonymous/synonymous rate ratio is much higher within species than between species. Mol. Biol. Evol. 15: 1499–1505.

Hastings, I. M. 1992. Population genetic aspects of deleterious cytoplasmic genomes and their effect on the evolution of sexual reproduction. Genet. Res. 59: 215–225.

Havird, J. C., E. S. Forsythe, A. M. Williams, J. H. Werren, D. K. Dowling, and D. B. Sloan. 2019. Selfish mitonuclear conflict. Curr. Biol. 29: R496–R511.

Havird, J. C., P. Trapp, C. M. Miller, I. Bazos, and D. B. Sloan. 2017. Causes and consequences of rapidly evolving mtDNA in a plant lineage. Genome Biol. Evol. 9: 323–336.

Havird, J. C., N. S. Whitehill, C. D. Snow, and D. B. Sloan. 2015. Conservative and compensatory evolution in oxidative phosphorylation complexes of angiosperms with highly divergent rates of mitochondrial genome evolution. Evolution 69: 3069–3081.

Hayes, F., and L. Van Melderen. 2011. Toxins-antitoxins: Diversity, evolution and function. Crit. Rev. Biochem. Mol. Biol. 46: 386–408.

Hazkani-Covo, E., R. M. Zeller, and W. Martin. 2010. Molecular poltergeists: Mitochondrial DNA copies (numts) in sequenced nuclear genomes. PLoS Genet. 6: e1000834.

Hehenberger, E., R. J. Gast, and P. J. Keeling. 2019. A kleptoplastidic dinoflagellate and the tipping point between transient and fully integrated plastid endosymbiosis. Proc. Natl. Acad. Sci. USA 116: 17934–17942.

Hiesel, R., B. Combettes, and A. Brennicke. 1994. Evidence for RNA editing in mitochondria of all major groups of land plants except the Bryophyta. Proc. Natl. Acad. Sci. USA 91: 629–633.

Hillesland, K. L., S. Lim, J. H. J. Flowers, S. Turkarslan, N. Pinel, G. M. Zane, N. Elliott, Y. Qin, L. Wu, N. S.

Baliga, et al. 2014. Erosion of functional independence early in the evolution of a microbial mutualism. Proc. Natl. Acad. Sci. USA 111: 14822–14827.

Hirst, J. 2011. Why does mitochondrial complex I have so many subunits? Biochem. J. 437: e1—3.

Ho, E. K. H., F. Macrae, L. C. Latta, P. McIlroy, D. Ebert, P. D. Fields, M. J. Benner, and S. Schaack. 2020. High and highly variable spontaneous mutation rates in *Daphnia*. Mol. Biol. Evol. 36: 1942–1954.

Hom, E. F., and A. W. Murray. 2014. Niche engineering demonstrates a latent capacity for fungal–algal mutualism. Science 345: 94–98.

Horton, T. L., and L. F. Landweber. 2000. Evolution of four types of RNA editing in myxomycetes. RNA 6: 1339–1346.

Horton, T. L., and L. F. Landweber. 2002. Rewriting the information in DNA: RNA editing in kinetoplastids and myxomycetes. Curr. Opin. Microbiol. 5: 620–626.

Howell, N. 1996. Mutational analysis of the human mitochondrial genome branches into the realm of bacterial genetics. Am. J. Hum. Genet. 59: 749–755.

Husnik, F., N. Nikoh, R. Koga, L. Ross, R. P. Duncan, M. Fujie, M. Tanaka, N. Satoh, D. Bachtrog, A. C. Wilson, et al. 2013. Horizontal gene transfer from diverse bacteria to an insect genome enables a tripartite nested mealybug symbiosis. Cell 153: 1567–1578.

Husnik, F., and J. P. McCutcheon. 2016. Repeated replacement of an intrabacterial symbiont in the tripartite nested mealybug symbiosis. Proc. Natl. Acad. Sci. USA 113: E5416–E5424.

Huynen, M. A., I. Duarte, and R. Szklarczyk. 2013. Loss, replacement and gain of proteins at the origin of the mitochondria. Biochim. Biophys. Acta 1827: 224–231.

Huynen, M. A., and D. M. Elurbe. 2022. Mitochondrial complex complexification. Science 376: 794–795.

Inada, M., T. Sasaki, M. Yukawa, T. Tsudzuki, and M. Sugiura. 2004. A systematic search for RNA editing sites in pea chloroplasts: An editing event causes diversification from the evolutionarily conserved amino acid sequence. Plant Cell Physiol. 45: 1615–1622.

Iyer, L. M., D. Zhang, I. B. Rogozin, and L. Aravind. 2011. Evolution of the deaminase fold and multiple origins of eukaryotic editing and mutagenic nucleic acid deaminases from bacterial toxin systems. Nucleic Acids Res. 39: 9473–9497.

James, J. E, G. Piganeau, and A. Eyre-Walker. 2016. The rate of adaptive evolution in animal mitochondria. Mol. Ecol. 25: 67–78.

Janke, A., and S. Pääbo. 1993. Editing of a tRNA anticodon in marsupial mitochondria changes its codon recognition. Nucleic Acids Res. 21: 1523–1525.

Jasmin, J. N., and C. Zeyl. 2014. Rapid evolution of cheating mitochondrial genomes in small yeast populations. Evolution 68: 269–275.

Jeblick, J., and J. Kusch. 2005. Sequence, transcription activity, and evolutionary origin of the R-body coding plasmid pKAP298 from the intracellular parasitic bacterium *Caedibacter taeniospiralis*. J. Mol. Evol. 60: 164–173.

Jenuth, J. P., A. C. Peterson, K. Fu, and E. A. Shoubridge. 1996. Random genetic drift in the female germline explains the rapid segregation of mammalian mitochondrial DNA. Nat. Genet. 14: 146–151.

Johnson, A. A., and K. A. Johnson. 2001. Fidelity of nucleotide incorporation by human mitochondrial DNA polymerase. J. Biol. Chem. 276: 38090–38096.

Kang, D., and N. Hamasaki. 2002. Maintenance of mitochondrial DNA integrity: Repair and degradation. Curr. Genet. 41: 311–322.

Karlberg, O., B. Canbäck, C. G. Kurland, and S. G. Andersson. 2000. The dual origin of the yeast mitochondrial proteome. Yeast 17: 170–187.

Karnkowska, A., S. C. Treitli, O. Brzoň, L. Novák, V. Vacek, P. Soukal, L. D. Barlow, E. K. Herman, S. V. Pipaliya, T. Pánek, et al. 2019. The oxymonad genome displays canonical eukaryotic complexity in the absence of a mitochondrion. Mol. Biol. Evol. 36: 2292–2312.

Karnkowska, A., V. Vacek, Z. Zubáčová, S. C. Treitli, R. Petrželková, L. Eme, L. Novák, V. Žárský, L. D. Barlow, E. K. Herman, et al. 2016. A Eukaryote without a Mitochondrial Organelle. Curr. Biol. 26: 1274–1284.

Kazak, L., A. Reyes, and I. J. Holt. 2012. Minimizing the damage: Repair pathways keep mitochondrial DNA intact. Nat. Rev. Mol. Cell Biol. 13: 659–671.

Keeling, P. J. 2013. The number, speed, and impact of plastid endosymbioses in eukaryotic evolution. Annu. Rev. Plant Biol. 64: 583–607.

Keeling, P. J., and J. P. McCutcheon. 2017. Endosymbiosis: The feeling is not mutual. J. Theor. Biol. 434: 75–79.

Kelly, S. 2021. The economics of endosymbiotic gene transfer and the evolution of organellar genomes. Genome Biol. 22: 345.

Kern, A. D., and F. A. Kondrashov. 2004. Mechanisms and convergence of compensatory evolution in mammalian mitochondrial tRNAs. Nat. Genet. 36: 1207–1212.

Kimura, M. 1983. The Neutral Theory of Molecular Evolution. Cambridge University Press, Cambridge, U.K.

Kobayashi, I. 1998. Selfishness and death: Raison d'être of restriction, recombination and mitochondria. Trends Genet. 14: 368–374.

Koonin, E. V. 2015. Origin of eukaryotes from within archaea, archaeal eukaryome and bursts of gene gain:

Eukaryogenesis just made easier? Philos. Trans. R. Soc. Lond. B Biol. Sci. 370: 20140333.

Koslowsky, D. J., H. U. Goringer, T. H. Morales, and K. Stuart. 1992. *In vitro* guide RNA/mRNA chimaera formation in *Trypanosoma brucei* RNA editing. Nature 356: 807–809.

Krasovec, M., S. Sanchez-Brosseau, and G. Piganeau. 2019. First estimation of the spontaneous mutation rate in diatoms. Genome Biol. Evol. 11: 1829–1837.

Ku, C., S. Nelson-Sathi, M. Roettger, F. L. Sousa, P. J. Lockhart, D. Bryant, E. Hazkani-Covo, J. O. McInerney, G. Landan, and W. F. Martin. 2015. Endosymbiotic origin and differential loss of eukaryotic genes. Nature 524: 427–432.

Kugita, M., Y. Yamamoto, T. Fujikawa, T. Matsumoto, and K. Yoshinaga. 2003. RNA editing in hornwort chloroplasts makes more than half the genes functional. Nucleic Acids Res. 31: 2417–2423.

Kuhle, B., J. Chihade, and P. Schimmel. 2020. Relaxed sequence constraints favor mutational freedom in idiosyncratic metazoan mitochondrial tRNAs. Nat. Commun. 11: 969.

Kumazawa, Y., and M. Nishida. 1993. Sequence evolution of mitochondrial tRNA genes and deep-branch animal phylogenetics. J. Mol. Evol. 37: 380–398.

Kumazawa, Y., C. J. Schwartzbach, H. X. Liao, K. Mizumoto, Y. Kaziro, K. Miura, K. Watanabe, and L. L. Spremulli. 1991. Interactions of bovine mitochondrial phenylalanyl-tRNA with ribosomes and elongation factors from mitochondria and bacteria. Biochim. Biophys. Acta 1090: 167–172.

Kumazawa, Y., T. Yokogawa, E. Hasegawa, K. Miura, and K. Watanabe. 1989. The aminoacylation of structurally variant phenylalanine tRNAs from mitochondria and various nonmitochondrial sources by bovine mitochondrial phenylalanyl-tRNA synthetase. J. Biol. Chem. 264: 13005–13011.

Kunkel, T. A., and P. S. Alexander. 1986. The base substitution fidelity of eucaryotic DNA polymerases. Mispairing frequencies, site preferences, insertion preferences, and base substitution by dislocation. J. Biol. Chem. 261: 160–166.

Laforest, M. J., C. E. Bullerwell, L. Forget, and B. F. Lang. 2004. Origin, evolution, and mechanism of 5′ tRNA editing in chytridiomycete fungi. RNA 10: 1191–1199.

Lane, N. 2006. Power, Sex, Suicide: Mitochondria and the Meaning of Life. Oxford University Press, Oxford, UK.

Lane N. 2020. How energy flow shapes cell evolution. Curr. Biol. 30: R471–R476.

Lane, N., and W. Martin. 2010. The energetics of genome complexity. Nature 467: 929–934.

Lavrov, D. V., W. M. Brown, and J. L. Boore. 2000. A novel type of RNA editing occurs in the mitochondrial tRNAs

of the centipede *Lithobius forficatus*. Proc. Natl. Acad. Sci. USA 97: 13738–13742.

Lavrov, D. V., and W. Pett. 2016. Animal mitochondrial DNA as we do not know it: mt-genome organization and evolution in nonbilaterian lineages. Genome Biol. Evol. 8: 2896–2913.

Leger, M. M., M. Petrů, V. Zárský, L. Eme, Č. Vlček, T. Harding, B. F. Lang, M. Eliáš, P. Doležal, and A. J. Roger. 2015. An ancestral bacterial division system is widespread in eukaryotic mitochondria. Proc. Natl. Acad. Sci. USA 112: 10239–10246.

Lewis, S. C., L. F. Uchiyama, and J. Nunnari. 2016. ER-mitochondria contacts couple mtDNA synthesis with mitochondrial division in human cells. Science 353: aaf5549.

Lewis, W. H., A. E. Lind, K. M. Sendra, H. Onsbring, T. A. Williams, G. F. Esteban, R. P. Hirt, T. J. G. Ettema, and T. M. Embley. 2020. Convergent evolution of hydrogenosomes from mitochondria by gene transfer and loss. Mol. Biol. Evol. 37: 524–539.

Lin, S., H. Zhang, D. F. Spencer, J. E. Norman, and M. W. Gray. 2002. Widespread and extensive editing of mitochondrial mRNAS in dinoflagellates. J. Mol. Biol. 320: 727–739.

Ling, F., H. Morioka, E. Ohtsuka, and T. Shibata. 2000. A role for MHR1, a gene required for mitochondrial genetic recombination, in the repair of damage spontaneously introduced in yeast mtDNA. Nucleic Acids Res. 28: 4956–4963.

Liu, H., and J. Zhang. 2019. Yeast spontaneous mutation rate and spectrum vary with environment. Curr. Biol. 29: 1584–1591.

Loewe, L. 2006. Quantifying the genomic decay paradox due to Muller's ratchet in human mitochondrial DNA. Genet. Res. 87: 133–159.

Lonergan, K. M., and M. W. Gray. 1993. Editing of transfer RNAs in *Acanthamoeba castellanii* mitochondria. Science 259: 812–816.

López-Garcia, P., L. Eme, and D. Moreira. 2017. Symbiosis in eukaryotic evolution. J. Theor. Biol. 434: 20–33.

Lunt, D. H., and B. C. Hyman. 1997. Animal mitochondrial DNA recombination. Nature 387: 247.

Luo, S., C. A. Valencia, J. Zhang, N. C. Lee, J. Slone, B. Gui, X. Wang, Z. Li, S. Dell, J. Brown, et al. 2018. Biparental inheritance of mitochondrial DNA in humans. Proc. Natl. Acad. Sci. USA 115: 13039–13044.

Lynch, M. 1996. Mutation accumulation in transfer RNAs: Molecular evidence for Muller's ratchet in mitochondrial genomes. Mol. Biol. Evol. 14: 914–925.

Lynch, M. 1997. Mutation accumulation in nuclear, organelle, and prokaryotic transfer RNA genes. Mol. Biol. Evol. 13: 209–220.

Lynch, M. 2007. The Origins of Genome Architecture. Sinauer Associates, Inc., Sunderland, MA.

Lynch, M. 2020. The evolutionary scaling of cellular traits imposed by the drift barrier. Proc. Natl. Acad. Sci. USA 117: 10435–10444.

Lynch, M. 2023. Mutation pressure, drift, and the pace of molecular coevolution. Proc. Natl. Acad. Sci. U S A 120: e2306741120.

Lynch, M., and J. L. Blanchard. 1998. Deleterious mutation accumulation in organelle genomes. Genetica 102–103: 29–39.

Lynch, R. Bürger, D. Butcher, and W. Gabriel. 1993. Mutational meltdowns in asexual populations. J. Heredity 84: 339–344.

Lynch, M., J. Conery, and R. Bürger. 1995a. Mutational meltdowns in sexual populations. Evolution 49: 1067–1080.

Lynch, M., J. Conery, and R. Bürger. 1995b. Mutation accumulation and the extinction of small populations. Amer. Natur. 146: 489–518.

Lynch, M., and W. Gabriel. 1990. Mutation load and the survival of small populations. Evolution 44: 1725–1737.

Lynch, M., B. Koskella, and S. Schaack. 2006. Mutation pressure and the evolution of organelle genomic architecture. Science 311: 1727–1730.

Lynch, M., and G. K. Marinov. 2015. The bioenergetic costs of a gene. Proc. Natl. Acad. Sci. USA 112: 15690–15695.

Lynch, M., and G. K. Marinov. 2017. Membranes, energetics, and evolution across the prokaryote–eukaryote divide. eLife 6: e20437.

Lynch, M., M. O'Hely, B. Walsh, and A. Force. 2001. The probability of preservation of a newly arisen gene duplicate. Genetics 159: 1789–1804.

MacAlpine, D. M., P. S. Perlman, and R. A. Butow. 1998. The high mobility group protein Abf2p influences the level of yeast mitochondrial DNA recombination intermediates *in vivo*. Proc. Natl. Acad. Sci. USA 95: 6739–6743.

Maier, R. M., P. Zeltz, H. Kossel, G. Bonnard, J. M. Gualberto, and J. M. Grienenberger. 1996. RNA editing in plant mitochondria and chloroplasts. Plant Mol. Biol. 32: 343–365.

Malek, O., K. Lattig, R. Hiesel, A. Brennicke, and V. Knoop. 1996. RNA editing in bryophytes and a molecular phylogeny of land plants. EMBO J. 15: 1403–1411.

Marchington, D. R., G. M. Hartshorne, D. Barlow, and J. Poulton. 1997. Homopolymeric tract heteroplasmy in mtDNA from tissues and single oocytes: Support for a genetic bottleneck. Amer. J. Hum. Genet. 60: 408–416.

Margulis, L. 1970. Origin of Eukaryotic Cells: Evidence and Research Implications for a Theory of the Origin and Evolution of Microbial, Plant, and Animal Cells on the Precambrian Earth. Yale University Press, New Haven, CT.

Martijn, J., J. Vosseberg, L. Guy, P. Offre, and T. J. G. Ettema. 2018. Deep mitochondrial origin outside the sampled alphaproteobacteria. Nature 557: 101–105.

Martin, W. F. 2017. Physiology, anaerobes, and the origin of mitosing cells 50 years on. J. Theor. Biol. 434: 2–10.

Martin, W. F., S. Garg, and V. Zimorski. 2015. Endosymbiotic theories for eukaryote origin. Philos. Trans. R. Soc. Lond. B Biol. Sci. 370: 20140330.

Martin, W. F., and M. Muller. 1998. The hydrogen hypothesis for the first eukaryote. Nature 392: 37–41.

Martin, W., T. Rujan, E. Richly, A. Hansen, S. Cornelsen, T. Lins, D. Leister, B. Stoebe, M. Hasegawa, and D. Penny. 2002. Evolutionary analysis of *Arabidopsis*, cyanobacterial, and chloroplast genomes reveals plastid phylogeny and thousands of cyanobacterial genes in the nucleus. Proc. Natl. Acad. Sci. USA 99: 12246–12251.

Martin, W. F., A. G. M. Tielens, M. Mentel, S. G. Garg, and S. B. Gould. 2017. The physiology of phagocytosis in the context of mitochondrial origin. Microbiol. Mol. Biol. Rev. 81: e00008–17.

Mason, P. A., and R. N. Lightowlers. 2003. Why do mammalian mitochondria possess a mismatch repair activity? FEBS Lett. 554: 6–9.

Mason, P. A., E. C. Matheson, A. G. Hall, and R. N. Lightowlers. 2003. Mismatch repair activity in mammalian mitochondria. Nucleic Acids Res. 31: 1052–1058.

Maynard Smith, J. M. 1978. The Evolution of Sex. Cambridge University Press, Cambridge, UK.

McCutcheon, J. P., B. M. Boyd, and C. Dale. 2019. The life of an insect endosymbiont from the cradle to the grave. Curr. Biol. 29: R485–R495.

McVean, G. A. 2001. What do patterns of genetic variability reveal about mitochondrial recombination? Heredity 87: 613–620.

Meer, M. V., A. S. Kondrashov, Y. Artzy-Randrup, and F. A. Kondrashov. 2010. Compensatory evolution in mitochondrial tRNAs navigates valleys of low fitness. Nature 464: 279–282.

Meland, S., S. Johansen, T. Johansen, K. Haugli, and F. Haugli. 1991. Rapid disappearance of one parental mitochondrial genotype after isogamous mating in the myxomycete *Physarum polycephalum*. Curr. Genet. 19: 55–59.

Miyagishima, S. Y., M. Nakamura, A. Uzuka, and A. Era. 2014. FtsZ-less prokaryotic cell division as well as FtsZ- and dynamin-less chloroplast and non-photosynthetic plastid division. Front. Plant Sci. 5: 459.

Miyamoto, T., J. Obokata, and M. Sugiura. 2004. A site-specific factor interacts directly with its cognate RNA editing site in chloroplast transcripts. Proc. Natl. Acad. Sci. USA 101: 48–52.

Muller, H. J. 1964. The relation of recombination to mutational advance. Mutat. Res. 106: 2–9.

Muñoz-Gómez, S. A., S. Hess, G. Burger, B. F. Lang, E. Susko, C. H. Slamovits, and A. J. Roger. 2019. An updated phylogeny of the *Alphaproteobacteria* reveals that the parasitic *Rickettsiales* and *Holosporales* have independent origins. eLife 8: e42535.

Muñoz-Gómez, S. A., C. H. Slamovits, J. B. Dacks, K. A. Baier, K. D. Spencer, and J. G. Wideman. 2015. Ancient homology of the mitochondrial contact site and cristae organizing system points to an endosymbiotic origin of mitochondrial cristae. Curr. Biol. 25: 1489–1495.

Muñoz-Gómez, S. A., E. Susko, K. Williamson, L. Eme, C. H. Slamovits, D. Moreira, P. López-Garcia, and A. J. Roger. 2022. Site-and-branch-heterogeneous analyses of an expanded dataset favour mitochondria as sister to known Alphaproteobacteria. Nat. Ecol. Evol. 6: 253–262.

Muñoz-Gómez, S. A., J. G. Wideman, A. J. Roger, and C. H. Slamovits. 2017. The origin of mitochondrial cristae from Alphaproteobacteria. Mol. Biol. Evol. 34: 943–956.

Murley, A., and J. Nunnari. 2016. The emerging network of mitochondria–organelle contacts. Mol. Cell. 61: 648–653.

Nachman, M. W., S. N. Boye, and C. F. Aquadro. 1994. Nonneutral evolution at the mitochondrial NADH dehydrogenase subunit 3 gene in mice. Proc. Natl. Acad. Sci. USA 91: 6364–6368.

Nachman, M. W., W. M. Brown, M. Stoneking, and C. F. Aquadro. 1996. Nonneutral mitochondrial DNA variation in humans and chimpanzees. Genetics 142: 953–963.

Nakayama, T., Y. Ikegami, T. Nakayama, K. Ishida, Y. Inagaki, and I. Inouye. 2011. Spheroid bodies in rhopalodiacean diatoms were derived from a single endosymbiotic cyanobacterium. J. Plant Res. 124: 93–97.

Nakayama, T., R. Kamikawa, G. Tanifuji, Y. Kashiyama, N. Ohkouchi, J. M. Archibald, and Y. Inagaki. 2014. Complete genome of a nonphotosynthetic cyanobacterium in a diatom reveals recent adaptations to an intracellular lifestyle. Proc. Natl. Acad. Sci. USA 111: 11407–11412.

Nakayama, T., M. Nomura, Y. Takano, G. Tanifuji, K. Shiba, K. Inaba, Y. Inagaki, and M. Kawata. 2019. Single-cell genomics unveiled a cryptic cyanobacterial lineage with a worldwide distribution hidden by a dinoflagellate host. Proc. Natl. Acad. Sci. USA 116: 15973–15978.

Ness, R. W., A. D. Morgan, R. B. Vasanthakrishnan, N. Colegrave, and P. D. Keightley. 2015. Extensive *de novo* mutation rate variation between individuals and across the genome of *Chlamydomonas reinhardtii*. Genome Res. 25: 1739–1749.

Nowack, E. C. M. 2014. *Paulinella chromatophora* – rethinking the transition from endosymbiont to organelle. Acta Soc. Bot. Pol. 83: 387–397.

Nowack, E. C., and A. R. Grossman. 2012. Trafficking of protein into the recently established photosynthetic organelles of *Paulinella chromatophora*. Proc. Natl. Acad. Sci. USA 109: 5340–5345.

Oliveira, D. C., R. Raychoudhury, D. V. Lavrov, and J. H. Werren. 2008. Rapidly evolving mitochondrial genome and directional selection in mitochondrial genes in the parasitic wasp *Nasonia* (hymenoptera: pteromalidae). Mol. Biol. Evol. 25: 2167–2180.

Osada, N., and H. Akashi. 2012. Mitochondrial-nuclear interactions and accelerated compensatory evolution: Evidence from the primate cytochrome C oxidase complex. Mol. Biol. Evol. 29: 337–346.

Palmer, J. D., K. L. Adams, Y. Cho, C. L. Parkinson, Y. L. Qiu, and K. Song. 2000. Dynamic evolution of plant mitochondrial genomes: Mobile genes and introns and highly variable mutation rates. Proc. Natl. Acad. Sci. USA 97: 6960–6966.

Palumbi, S. R., F. Cipriano, and M. P. Hare. 2001. Predicting nuclear gene coalescence from mitochondrial data: The three-times rule. Evolution 55: 859–868.

Paquin, B., M. J. Laforest, L. Forget, I. Roewer, Z. Wang, J. Longcore, and B. F. Lang. 1997. The fungal mitochondrial genome project: Evolution of fungal mitochondrial genomes and their gene expression. Curr. Genet. 31: 380–395.

Parfrey, L. W., D. J. Lahr, A. H. Knoll, and L. A. Katz. 2011. Estimating the timing of early eukaryotic diversification with multigene molecular clocks. Proc. Natl. Acad. Sci. USA 108: 13624–13629.

Parkinson, C. L., J. P. Mower, Y. L. Qiu, A. J. Shirk, K. Song, N. D. Young, C. W. DePamphilis, and J. D. Palmer. 2005. Multiple major increases and decreases in mitochondrial substitution rates in the plant family Geraniaceae. BMC Evol. Biol. 5: 73.

Petrov, A. S., E. C. Wood, C. R. Bernier, A. M. Norris, A. Brown, and A. Amunts. 2019. Structural patching fosters divergence of mitochondrial ribosomes. Mol. Biol. Evol. 36: 207–219.

Phadnis, N., R. Mehta, N. Meednu, and E. A. Sia. 2006. Ntg1p, the base excision repair protein, generates mutagenic intermediates in yeast mitochondrial DNA. DNA Repair 5: 829–839.

Phillips, W. S., A. L. Coleman-Hulbert, E. S. Weiss, D. K. Howe, S. Ping, R. I. Wernick, S. Estes, and D. R. Denver. 2015. Selfish mitochondrial DNA proliferates and diversifies in small, but not large, experimental populations of *Caenorhabditis briggsae*. Genome Biol. Evol. 7: 2023–2037.

Phreaner, C. G., M. A. Williams, and R. M. Mulligan. 1996. Incomplete editing of rps12 transcripts results in the synthesis of polymorphic polypeptides in plant mitochondria. Plant Cell 8: 107–117.

Pietromonaco, S. F., R. A. Hessler, and T. W. O'Brien. 1986. Evolution of proteins in mammalian cytoplasmic and mitochondrial ribosomes. J. Mol. Evol. 24: 110–117.

Piganeau, G., M. Gardner, and A. Eyre-Walker. 2004. A broad survey of recombination in animal mitochondria. Mol. Biol. Evol. 21: 2319–2325.

Pittis, A. A., and T. Gabaldón. 2016. Late acquisition of mitochondria by a host with chimaeric prokaryotic ancestry. Nature 531: 101–104.

Pond, F. R., I. Gibson, J. Lalucat, and R. L. Quackenbush. 1989. R-body-producing bacteria. Microbiol. Rev. 53: 25–67.

Poon, A., and S. P. Otto. 2000. Compensating for our load of mutations: Freezing the meltdown of small populations. Evolution 54: 1467–1479.

Ponce-Toledo, R. I., P. Deschamps, P. López-García, Y. Zivanovic, K. Benzerara, and D. Moreira. 2017. An early-branching freshwater cyanobacterium at the origin of plastids. Curr. Biol. 27: 386–391.

Popadin, K. Y., S. I. Nikolaev, T. Junier, M. Baranova, and S. E. Antonarakis. 2013. Purifying selection in mammalian mitochondrial protein-coding genes is highly effective and congruent with evolution of nuclear genes. Mol. Biol. Evol. 30: 347–355.

Popescu, C. E., and R. W. Lee. 2007. Mitochondrial genome sequence evolution in *Chlamydomonas*. Genetics 175: 819–826.

Preer, J. R., Jr., L. B. Preer, B. Rudman, and A. Jurand. 1971. Isolation and composition of bacteriophage-like particles from kappa of killer Paramecia. Mol. Gen. Genet. 111: 202–208.

Price, D. C., C. X. Chan, H. S. Yoon, E. C. Yang, H. Qiu, A. P. Weber, R. Schwacke, J. Gross, N. A. Blouin, C. Lane, et al. 2012. *Cyanophora paradoxa* genome elucidates origin of photosynthesis in algae and plants. Science 335: 843–847.

Prinz, W. A., A. Toulmay, and T. Balla. 2020. The functional universe of membrane contact sites. Nat. Rev. Mol. Cell Biol. 21: 7–24.

Purkanti, R., and M. Thattai. 2015. Ancient dynamin segments capture early stages of host-mitochondrial integration. Proc. Natl. Acad. Sci. USA 112: 2800–2805.

Radzvilavicius, A. L., H. Kokko, and J. R. Christie. 2017. Mitigating mitochondrial genome erosion without recombination. Genetics 207: 1079–1088.

Rand, D. M., M. Dorfsman, and L. M. Kann. 1994. Neutral and non-neutral evolution of *Drosophila* mitochondrial DNA. Genetics 138: 741–756.

Rand, D. M., R. A. Haney, and A. J. Fry. 2004. Cytonuclear coevolution: The genomics of cooperation. Trends Ecol. Evol. 19: 645–653.

Rand, D. M., and R. G. Harrison. 1986. Mitochondrial DNA transmission genetics in crickets. Genetics 114: 955–970.

Rankin, D. J., L. A. Turner, J. A. Heinemann, and S. P. Brown. 2012. The coevolution of toxin and antitoxin genes drives the dynamics of bacterial addiction complexes and intragenomic conflict. Proc. Biol. Sci. 279: 3706–3715.

Raval, P. K., S. G. Garg, and S. B. Gould. 2022. Endosymbiotic selective pressure at the origin of eukaryotic cell biology. eLife 11: e81033.

Raymann, K., L. M. Bobay, T. G. Doak, M. Lynch, and S. Gribaldo. 2013. A genomic survey of Reb homologs suggests widespread occurrence of R-bodies in proteobacteria. G3 (Bethesda) 3: 505–516.

Reboud, X., and C. Zeyl 1994. Organelle inheritance in plants. Heredity 72: 132–140.

Rho, M., M. Zhou, X. Gao, S. Kim, H. Tang, and M. Lynch. 2009. Parallel mammalian genome contractions following the KT boundary. Genome Biol. Evol. 1: 2–12.

Richardson, A. O., D. W. Rice, G. J. Young, A. J. Alverson, and J. D. Palmer. 2013. The 'fossilized' mitochondrial genome of *Liriodendron tulipifera*: Ancestral gene content and order, ancestral editing sites, and extraordinarily low mutation rate. BMC Biol. 11: 29.

Rivera, M. C., and J. A. Lake. 1992. Evidence that eukaryotes and eocyte prokaryotes are immediate relatives. Science 257: 74–76.

Rochette, N. C., C. Brochier-Armanet, and M. Gouy. 2014. Phylogenomic test of the hypotheses for the evolutionary origin of eukaryotes. Mol. Biol. Evol. 31: 832–845.

Rockenbach, K., J. C. Havird, J. G. Monroe, D. A. Triant, D. R. Taylor, and D. B. Sloan. 2016. Positive selection in rapidly evolving plastid–nuclear enzyme complexes. Genetics 204: 1507–1522.

Rüdinger, M., U. Volkmar, H. Lenz, M. Groth-Malonek, and V. Knoop. 2012. Nuclear DYW-type PPR gene families diversify with increasing RNA editing frequencies in liverwort and moss mitochondria. J. Mol. Evol. 74: 37–51.

Sachs, J. L., R. G. Skophammer, and J. U. Regus. 2011. Evolutionary transitions in bacterial symbiosis. Proc. Natl. Acad. Sci. USA 108 (Suppl. 2): 10800–10807.

Sachs, J. L., R. G. Skophammer, N. Bansal, and J. E. Stajich. 2013. Evolutionary origins and diversification of proteobacterial mutualists. Proc. Biol. Sci. 281: 20132146.

Sagan, L. 1967. On the origin of mitosing cells. J. Theor. Biol. 14: 255–274.

Sánchez-Baracaldo, P., J. A. Raven, D. Pisani, and A. H. Knoll. 2017. Early photosynthetic eukaryotes inhabited low-salinity habitats. Proc. Natl. Acad. Sci. USA 114: E7737–E7745.

Santos, C., R. Montiel, B. Sierra, C. Bettencourt, E. Fernandez, L. Alvarez, M. Lima, A. Abade, and M. P. Aluja. 2005. Understanding differences between phylogenetic and pedigree-derived mtDNA mutation rate: A model using families from the Azores Islands (Portugal). Mol. Biol. Evol. 22: 1490–1505.

Sarai C., G. Tanifuji, T. Nakayama, R. Kamikawa, K. Takahashi, E. Yazaki, E. Matsuo, H. Miyashita, K. I. Ishida, M. Iwataki, et al. 2020. Dinoflagellates with relic endosymbiont nuclei as models for elucidating organellogenesis. Proc. Natl. Acad. Sci. USA 117: 5364–5375.

Sassera, D., N. Lo, S. Epis, G. D'Auria, M. Montagna, F. Comandatore, D. Horner, J. Peretó, A. M. Luciano, F. Franciosi, et al. 2011. Phylogenomic evidence for the presence of a flagellum and cbb(3) oxidase in the free-living mitochondrial ancestor. Mol. Biol. Evol. 28: 3285–3296.

Satoh, M., and T. Kuroiwa. 1991. Organization of multiple nucleoids and DNA molecules in mitochondria of a human cell. Exp. Cell Res. 196: 137–140.

Saville, B. J., Y. Kohli, and J. B. Anderson. 1998. mtDNA recombination in a natural population. Proc. Natl. Acad. Sci. USA 95: 1331–1335.

Schavemaker, P. E., and S. A. Muñoz-Gómez. 2022. The role of mitochondrial energetics in the origin and diversification of eukaryotes. Nat. Ecol. Evol. 6: 1307–1317.

Schmitz-Linneweber, C., and I. Small. 2008. Pentatricopeptide repeat proteins: A socket set for organelle gene expression. Trends Plant Sci. 13: 663–670.

Schrallhammer, M., and M. Schweikert. 2009. The killer effect of Paramecium and its causative agents, pp. 227–246. In M. Fujishima (ed.) Endosymbionts in Paramecium. Springer-Verlag, Berlin, Germany.

Schuster, W., B. Wissinger, M. Unseld, and A. Brennicke. 1990. Transcripts of the NADH-dehydrogenase subunit 3 gene are differentially edited in Oenothera mitochondria. EMBO J. 9: 263–269.

Sharma, M. R., D. N. Wilson, P. P. Datta, C. Barat, F. Schluenzen, P. Fucini, and R. K. Agrawal. 2007. Cryo-EM study of the spinach chloroplast ribosome reveals the structural and functional roles of plastid-specific ribosomal proteins. Proc. Natl. Acad. Sci. USA 104: 19315–19320.

Shields, D. C., and K. H. Wolfe. 1997. Accelerated evolution of sites undergoing mRNA editing in plant mitochondria and chloroplasts. Mol. Biol. Evol. 14: 344–349.

Shiratori, T., S. Suzuki, Y. Kakizawa, and K. I. Ishida. 2019. Phagocytosis-like cell engulfment by a planctomycete bacterium. Nat. Commun. 10: 5529.

Silliker, M. E., M. R. Liotta, and D. J. Cummings. 1996. Elimination of mitochondrial mutations by sexual reproduction: Two Podospora anserina mitochondrial mutants yield only wild-type progeny when mated. Curr. Genet. 30: 318–324.

Simpson, L., O. H. Thiemann, N. J. Savill, J. D. Alfonzo, and D. A. Maslov. 2000. Evolution of RNA editing in trypanosome mitochondria. Proc. Natl. Acad. Sci. USA 97: 6986–6993.

Singer, A., G. Poschmann, C. Mühlich, C. Valadez-Cano, S. Hänsch, V. Hüren, S. A. Rensing, K. Stühler, and E. C. M. Nowack. 2017. Massive protein import into the early-evolutionary-stage photosynthetic organelle of the Amoeba paulinella chromatophora. Curr. Biol. 27: 2763–2773.e5.

Sloan, D. B., D. A. Triant, M. Wu, and D. R. Taylor. 2014. Cytonuclear interactions and relaxed selection accelerate sequence evolution in organelle ribosomes. Mol. Biol. Evol. 31: 673–682.

Sloan, D. B., J. M. Warren, A. M. Williams, Z. Wu, S. E. Abdel-Ghany, A. J. Chicco, and J. C. Havird. 2018. Cytonuclear integration and co-evolution. Nat. Rev. Genet. 19: 635–648.

Smith, D. G., R. M. Gawryluk, D. F. Spencer, R. E. Pearlman, K. W. Siu, and M. W. Gray. 2007. Exploring the mitochondrial proteome of the ciliate protozoon Tetrahymena thermophila: Direct analysis by tandem mass spectrometry. J. Mol. Biol. 374: 837–863.

Smith, D. R. 2015. Mutation rates in plastid genomes: They are lower than you might think. Genome Biol. Evol. 7: 1227–1234.

Smith, D. R., C. J. Jackson, and A. Reyes-Prieto. 2014. Nucleotide substitution analyses of the glaucophyte Cyanophora suggest an ancestrally lower mutation rate in plastid vs. mitochondrial DNA for the Archaeplastida. Mol. Phylogenet. Evol. 79: 380–384.

Smith, D. R., and P. J. Keeling. 2012. Twenty-fold difference in evolutionary rates between the mitochondrial and plastid genomes of species with secondary red plastids. J. Eukaryot. Microbiol. 59: 181–184.

Smith, D. R., and P. J. Keeling. 2015. Mitochondrial and plastid genome architecture: Reoccurring themes, but significant differences at the extremes. Proc. Natl. Acad. Sci. USA 112: 10177–10184.

Smith, H. C. 2006. Editing informational content of expressed DNA sequences and their transcripts, pp. 248–265. In L. H. Caporale (ed.) The Implicit Genome. Oxford University Press, Oxford, UK.

Solignac, M., M. Monnerot, and J. C. Mounolou. 1983. Mitochondrial DNA heteroplasmy in *Drosophila mauritiana*. Proc. Natl. Acad. Sci. USA 80: 6942–6946.

Söll, D., and U. L. RajBhandary. 1995. tRNA: Structure, Biosynthesis, and Function. ASM Press, Washington, DC.

Städler, T., and L. F. Delph. 2002. Ancient mitochondrial haplotypes and evidence for intragenic recombination in a gynodioecious plant. Proc. Natl. Acad. Sci. USA 99: 11730–11735.

Stein, A., and E. A. Sia. 2017. Mitochondrial DNA repair and damage tolerance. Front. Biosci. 22: 920–943.

Steinberg, S., and R. Cedergren. 1994. Structural compensation in atypical mitochondrial tRNAs. Nat. Struct. Biol. 1: 507–510.

Steinberg, S., D. Gautheret, and R. Cedergren. 1994. Fitting the structurally diverse animal mitochondrial tRNAs(Ser) to common three-dimensional constraints. J. Mol. Biol. 236: 982–989.

Stewart, J. B., C. Freyer, J. L. Elson, A. Wredenberg, Z. Cansu, A. Trifunovic, and N. G. Larsson. 2008. Strong purifying selection in transmission of mammalian mitochondrial DNA. PLoS Biol. 6: e10.

Stoltzfus, A. 1999. On the possibility of constructive neutral evolution. J. Mol. Evol. 49: 169–181.

Stroud, D. A., E. E. Surgenor, L. E. Formosa, B. Reljic, A. E. Frazier, M. G. Dibley, L. D. Osellame, T. Stait, T. H. Beilharz, D. R. Thorburn, et al. 2016. Accessory subunits are integral for assembly and function of human mitochondrial complex I. Nature 538: 123–126.

Takahata, N., and M. Slatkin. 1984. Mitochondrial gene flow. Proc. Natl. Acad. Sci. USA 81: 1764–1767.

Templeton, A. R. 1996. Gene lineages and human evolution. Science 272: 1363–1364.

Thompson, A. W., R. A. Foster, A. Krupke, B. J. Carter, N. Musat, D. Vaulot, M. M. Kuypers, and J. P. Zehr. 2012. Unicellular cyanobacterium symbiotic with a single-celled eukaryotic alga. Science 337: 1546–1550.

Thyagarajan, B., R. A. Padua, and C. Campbell. 1996. Mammalian mitochondria possess homologous DNA recombination activity. J. Biol. Chem. 271: 27536–27543.

Tillich, M., P. Lewahrk, B. R. Morton, and U. G. Maier. 2006. The evolution of chloroplast RNA editing. Mol. Biol. Evol. 23: 1912–1921.

Tomita, K., T. Ueda, and K. Watanabe. 1996. RNA editing in the acceptor stem of squid mitochondrial tRNA(Tyr). Nucleic Acids Res. 24: 4987–4991.

Tria, F. D. K., J. Brueckner, J. Skejo, J. C. Xavier, N. Kapust, M. Knopp, J. L. E. Wimmer, F. S. P. Nagies, V. Zimorski, S. B. Gould, et al. 2021. Gene duplications trace mitochondria to the onset of eukaryote complexity. Genome Biol. Evol. 13: evab055.

Tsudzuki, T., T. Wakasugi, and M. Sugiura. 2001. Comparative analysis of RNA editing sites in higher plant chloroplasts. J. Mol. Evol. 53: 327–332.

van der Sluis, E. O., H. Bauerschmitt, T. Becker, T. Mielke, J. Frauenfeld, O. Berninghausen, W. Neupert, J. M. Herrmann, and R. E. Beckmann. 2015. Parallel structural evolution of mitochondrial ribosomes and OXPHOS complexes. Genome Biol. Evol. 7: 1235–1251.

Van Melderen, L. 2010. Toxin–antitoxin systems: Why so many, what for? Curr. Opin. Microbiol. 13: 781–785.

Vellai, T., K. Takács, and G. Vida. 1998. A new aspect to the origin and evolution of eukaryotes. J. Mol. Evol. 46: 499–507.

von Dohlen, C. D., S. Kohler, S. T. Alsop, and W. R. McManus. 2001. Mealybug beta-proteobacterial endosymbionts contain gamma-proteobacterial symbionts. Nature 412: 433–436.

Vosseberg, J., J. J. E. van Hooff, M. Marcet-Houben, A. van Vlimmeren, L. M. van Wijk, T. Gabaldón, and B. Snel. 2021. Timing the origin of eukaryotic cellular complexity with ancient duplications. Nat. Ecol. Evol. 5: 92–100.

Wagner, G. P., and W. Gabriel. 1990. Quantitative variation in finite parthenogenetic populations: What stops Muller's ratchet in the absence of recombination? Evolution 44: 715–731.

Wang, Z., and M. Wu. 2014. Phylogenomic reconstruction indicates mitochondrial ancestor was an energy parasite. PLoS One 9: e110685.

Watanabe, Y., H. Tsurui, T. Ueda, R. Furushima, S. Takamiya, K. Kita, K. Nishikawa, and K. Watanabe. 1994. Primary and higher order structures of nematode (*Ascaris suum*) mitochondrial tRNAs lacking either the T or D stem. J. Biol. Chem. 269: 22902–22906.

Weng, M. L., T. A. Ruhlman, and R. K. Jansen. 2016. Plastid-nuclear interaction and accelerated coevolution in plastid ribosomal genes in Geraniaceae. Genome Biol. Evol. 8: 1824–1838.

Wideman, J. G., R. M. Gawryluk, M. W. Gray, and J. B. Dacks. 2013. The ancient and widespread nature of the ER–mitochondria encounter structure. Mol. Biol. Evol. 30: 2044–2049.

Wilkie, D., and D. Y. Thomas. 1973. Mitochondrial genetic analysis by zygote cell lineages in *Saccharomyces cerevisiae*. Genetics 73: 367–377.

Wise, C. A., M. Sraml, and S. Easteal. 1998. Departure from neutrality at the mitochondrial NADH dehydrogenase subunit 2 gene in humans, but not in chimpanzees. Genetics 148: 409–421.

Wolf, P. G., C. A. Rowe, and M. Hasebe. 2004. High levels of RNA editing in a vascular plant chloroplast genome: Analysis of transcripts from the fern *Adiantum capillus-veneris*. Gene 339: 89–97.

Wolfe, K. H., W. H. Li, and P. M. Sharp. 1987. Rates of nucleotide substitution vary greatly among plant mitochondrial, chloroplast, and nuclear DNAs. Proc. Natl. Acad. Sci. USA 84: 9054–9058.

Wolstenholme, D. R. 1992. Animal mitochondrial DNA: Structure and evolution, pp. 173–216. In D. R. Wolstenholme and K. W. Jeon (eds.) Mitochondrial Genomes. Academic Press, New York, NY.

Wolstenholme, D. R., R. Okimoto, and J. L. Macfarlane. 1994. Nucleotide correlations that suggest tertiary interactions in the TV-replacement loop-containing mitochondrial tRNAs of the nematodes, *Caenorhabditis elegans* and *Ascaris suum*. Nucleic Acids Res. 22: 4300–4306.

Wu, Z., G. Waneka, A. K. Broz, C. R. King, and D. B. Sloan. 2020. *MSH1* is required for maintenance of the low mutation rates in plant mitochondrial and plastid genomes. Proc. Natl. Acad. Sci. USA 117: 16448–16455.

Wujek, D. E., and B. C. Asmund. 1978. *Mallomonas cyathellata* sp. nov. and *Mallomonas cyathellata* var. *kenyana* var. nov. (Chrysophyceae) studied by means of scanning and transmission electron microscopy. Phycologia 18: 115–119.

Xu, S., S. Schaack, A. Seyfert, E. Choi, M. Lynch, and M. E. Cristescu. 2012. High mutation rates in the mitochondrial genomes of *Daphnia pulex*. Mol. Biol. Evol. 29: 763–769.

Yahalomi, D., S. D. Atkinson, M. Neuhof, E. S. Chang, H. Philippe, P. Cartwright, J. L. Bartholomew, and D. Huchon. 2020. A cnidarian parasite of salmon (Myxozoa: Henneguya) lacks a mitochondrial genome. Proc. Natl. Acad. Sci. USA 117: 5358–5363.

Yamaguchi, K., K. von Knoblauch, and A. R. Subramanian. 2000. The plastid ribosomal proteins. Identification of all the proteins in the 30 S subunit of an organelle ribosome (chloroplast). J. Biol. Chem. 275: 28455–28465.

Yamaguchi, Y., J. H. Park, and M. Inouye. 2011. Toxin–antitoxin systems in bacteria and archaea. Annu. Rev. Genet. 45: 61–79.

Yamauchi, A., T. Yamagishi, R. Booton, A. Telschow, and G. Kudo. 2019. Theory of coevolution of cytoplasmic male-sterility, nuclear restorer and selfing. J. Theor. Biol. 477: 96–107.

Ye, Z., C. Zhao, R. T. Raborn, M. Lin, W. Wei, Y. Hao, and M. Lynch. 2022. Genetic diversity, heteroplasmy, and recombination in mitochondrial genomes in *Daphnia pulex*, *Daphnia pulicaria*, and *Daphnia obtusa*. Mol. Biol. Evol. 39: msac059.

Yokobori, S, and S. Pääbo. 1995. Transfer RNA editing in land snail mitochondria. Proc. Natl. Acad. Sci. USA 92: 10432–10435.

Youle, R. J. 2019. Mitochondria – striking a balance between host and endosymbiont. Science 365: eaaw9855.

Zhang, F., and R. E. Broughton. 2013. Mitochondrial–nuclear interactions: Compensatory evolution or variable functional constraint among vertebrate oxidative phosphorylation genes? Genome Biol. Evol. 5: 1781–1791.

Zhang, H., and S. Lin. 2005. Mitochondrial cytochrome b mRNA editing in dinoflagellates: Possible ecological and evolutionary associations? J. Eukaryot. Microbiol. 52: 538–545.

Zhang, J., T. A. Ruhlman, J. Sabir, J. C. Blazier, and R. K. Jansen. 2015. Coordinated rates of evolution between interacting plastid and nuclear genes in Geraniaceae. Plant Cell 27: 563–573.

Zhou, Q., H. Li, H. Li, A. Nakagawa, J. L. Lin, E. S. Lee, B. L. Harry, R. R. Skeen-Gaar, Y. Suehiro, D. William, et al. 2016. Mitochondrial endonuclease G mediates breakdown of paternal mitochondria upon fertilization. Science 353: 394–399.

Zimorski, V., M. Mentel, A. G. M. Tielens, and W. F. Martin. 2019. Energy metabolism in anaerobic eukaryotes and Earth's late oxygenation. Free Radic. Biol. Med. 140: 279–294.

CHAPTER 24

Origins of Organismal Complexity

In the antecedent to this book, a case was made for the emergence of genome complexity as a passive response to mutational bias and the limited reach of natural selection in lineages experiencing elevated levels of genetic drift (Lynch 2007). This left open the question as to whether differences at the next highest level of organization – the cell – might also be a product of effectively neutral processes. One might anticipate that the higher the level of organization and the closer to outward phenotypic expression, the more likely it is that natural selection will be fully responsible for patterns of variation. With this reasoning, some would even argue that none of the internal molecular and cellular details matter. However, multiple observations summarized in the preceding chapters make clear that this is not the case.

What follows is a synthesis of results from comparative cell biology and evolutionary theory relevant to the matter of the emergence and diversification of cellular and organismal complexity. Evolutionary biology is more than the application of phylogenetic comparative analyses and the concoction of adaptive hypotheses for observed historical patterns. Understanding the mechanisms by which evolution proceeds requires an appreciation of the functional constraints on cells imposed by the historical roots of biology's invariants (e.g., the dependence on DNA-based genomes, RNA-based transcriptomes, protein- and lipid-based infrastructures, and the reliance on ATP as a supplier of energy). As the biochemical, bioenergetic, and biophysical properties of life's cellular components ultimately define the limited ways by which mutational processes can introduce variation into populations, a mechanistic understanding of evolution is inconceivable without information on its material

basis. Likewise, conditional on the material endowment of life, the imposition of key population-genetic features dictates the paths of possible evolutionary trajectories, defining the types of changes that are open versus closed to exploitation by natural selection and mutational biases.

At the time of Wallace and Darwin, essentially nothing was known about genetics or the molecular constituents of cells, removing all inhibitions to thinking that natural selection is essentially all powerful. Things might have been different had biology's great phase of molecular and cell emphasis come earlier. We now know that the need to expand evolutionary biology beyond a pan-adaptationist framework is deeply rooted in biological reality. Mutations are particulate in nature, with effects that are typically small and deleterious, and the transmission of alleles across generations is a stochastic process, with the role of chance in evolution varying by several orders of magnitude across phylogenetic lineages. Every genomic nucleotide site in every species is subject to mutational change and to the vagaries of genetic drift. However, genomes are also finite in size, and these matters, along with issues such as recombination rates and non-additive gene interactions, dictate the kinds of pathways that are open to evolutionary exploration.

Most notably, the capacity of natural selection to use genetic fuel to drive the expansion of biological diversity has significant bounds dictated by intrinsic and extrinsic factors influencing effective population sizes. Not all mutations are available for adaptive evolutionary utilization, as the effective size of a population dictates the granularities of mutational effects that are visible to the eyes of natural selection. In small populations, adaptive change can only proceed by use of beneficial mutations

Evolutionary Cell Biology. Michael Lynch, Oxford University Press. © Michael Lynch (2024). DOI: 10.1093/oso/9780192847287.003.0024

of relatively large effects, and mildly deleterious mutations cannot be purged. In contrast, in large populations, evolution is much more finely tuned, as its arsenal expands to mutations with very small effects. Thus, even in the ideal case in which mutation and selection pressures operate in identical ways, phylogenetic lineages are expected to vary in their response to selection, simply as a consequence of variation in the power of random genetic drift. Drift does not modify the production of existing variation, but through history defines the genomic architectures from which mutational changes can be built.

Given various assumed selection coefficients, mutation and recombination rates, and aspects of population demography, evolutionary genetics theory has been very good at defining how evolution is expected to proceed in a generic way, usually in terms of changes in allele frequencies without any explicit reference to phenotypes. The next step in the development of a more mature science of evolutionary biology is to elucidate from first principles the connections between molecular/cell-level features and their selective consequences – the so-called genotype–phenotype–fitness mapping. The critical importance of cell biology to evolutionary biology resides in the exacting details by which the functional links between genotypes and phenotypes can be defined. This being said, although more than one physicist has claimed that all of biology is physics, this final chapter highlights why biophysics, biochemistry, and cell biology alone are unlikely to ever be sufficient to understand evolutionary processes.

Deconstructing the Great Chain of Being

It is commonly asserted that eukaryotes are superior life forms relative to prokaryotes, that animals represent the pinnacle of the evolutionary process within eukaryotes, and that vertebrates occupy a higher rung on the imagined ladder of ascendancy than invertebrates, with humans highest of all. One extreme view is that prokaryotes are condemned to a pathetic and perpetual fate of simplicity in form and function owing only to the absence of a mitochondrion (Lane and Martin 2010).

There is, however, no justification for this view of life. As enticing as the idea might be, there is no evidence that the ultimate goal of natural selection is to build bigger, bonier, and more complex organisms. Humans are better at being human than are bacteria, and bacteria are better at carrying out microbial functions than vertebrates, but one cannot directly compare the genetic fitnesses of reproductively isolated organisms out of context with their ecologies. Evolutionary fitness is a function of the relative transmission rate of genomic material within a cohesive population of compatible individuals. Nonetheless, even where performances can be compared in an absolute sense, simpler organisms generally come out on top.

For example, heterotrophic bacteria, particularly large-celled species, are capable of assimilating biomass at substantially higher rates than morphologically more complex unicellular eukaryotes of comparable size (Chapter 8). The highest rate of cell division in the smallest unicellular eukaryotes is nearly an order of magnitude less than the highest rate in bacteria, and the eukaryotic rate progressively declines with increasing cell size. In multicellular eukaryotes, the maximum mass-specific growth rate (at any life stage) continues to decline with increasing size at maturity, such that the highest growth rates achievable by animals and land plants are 10–$100\times$ below those for unicellular species (Figure 24.1). This ability of prokaryotes to produce daughter cells at higher rates than eukaryotes is accomplished without subsidizing a vast intracellular network of specialized organelles. It is also done with fewer and smaller proteins, which on average appear to have higher catalytic efficiencies and higher accuracy of substrate utilization (Chapters 4 and 12). Thus, by all accounts, given identical metabolic tasks, prokaryotes are just as good as (if not better than) their eukaryotic counterparts. The same appears to be true for motility and environmental sensing when properly scaled against organismal size (Chapters 16 and 22).

As in the case of reduced genome-replication fidelity in organisms of increasing size (Chapter 4), this gradient of declining maximum growth rates in eukaryotes may be a consequence of the accumulation of mildly deleterious growth-reducing mutations in small-N_e species (Lynch 2020; Lynch

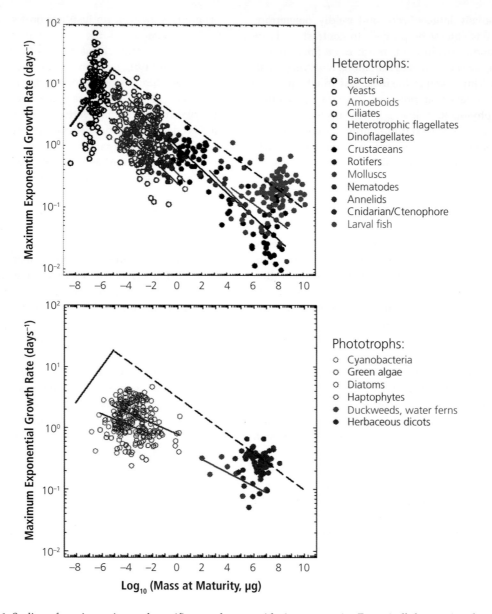

Figure 24.1 Scaling of maximum interval-specific growth rates with size at maturity. For unicellular species, the growth rates are equivalent to cell-division rates, whereas for multicellular species, they are the maximum rates among all age classes during individual development. All data are standardized to 20°C. The upper dashed line is the approximate position of the upper limit to evolved maximum growth rates (not including larval fishes, with maternal provisioning); solid lines are fitted regressions for individual phylogenetic groups. Note that this figure is an extension of Figure 8.5 to multicellular species. From Lynch et al. (2022).

et al. 2022). Particularly noteworthy in Figure 24.1 are the approximately −0.20 and −0.10 power-law scaling relationships with size for eukaryotic heterotrophs and autotrophs. As noted in Chapter 8, there has been an ongoing debate as to whether mass-specific metabolic rate scales with the −0.25 or −0.33 power, but the observations for maximum growth rate are inconsistent with both hypotheses.

One might imagine that biophysical constraints would set the baseline pattern, with deleterious-mutation accumulation then increasingly diminishing the capacity of larger organisms (with smaller N_e) relative to the biophysics barrier. However, this would lead to a steeper slope than the biophysical expectations, that is, slopes < -0.25, contrary to what is actually seen.

If the drift-barrier hypothesis is the correct explanation for the patterns observed in Figure 24.1, the maintenance of a constant power-law relationship across a wide range of body sizes demands a particular distribution of growth-altering effects of mutations – for each proportional increase in organism size, there must be a constant proportional increase in the total fixed growth-reducing mutational load. As organism size increases, N_e declines (Figure 4.3), remarkably also with the -0.2 power of organism mass, and this subjects windows of more deleterious mutational effects to fixation by drift. However, to keep the increment in load constant with increasing organism size, a specific relationship is required: the number of genomic sites with a particular effect must be inversely proportional to the effect size, as the product of the two is the total load.

Although direct estimation of very small mutational effects is currently intractable, observational limits are not grounds for dismissing a hypothesis. Numerous areas of science, including bioenergetics and genetics, have advanced significantly by building around theoretical constructs that required decades of work by thousands of investigators and enormous funding to achieve validation with direct observations. Not convinced? Consider the multiple billion-dollar projects currently funding team projects in particle physics and astronomy.

Genome complexity and organismal complexity

Three decades of whole-genome sequencing have revealed clear phylogenetic patterns of genome structure and organization (Lynch 2007). Rarely exceeding 10 Mb in length, prokaryotic genomes typically contain 1000–8000 protein-coding genes, and these generally comprise >95% of the entire genome, with slightly less than 1 kb of genomic space allocated per gene (Figure 24.2). Prokaryotic gene regulation is often simplified to the point that

multiple genes are coordinately expressed from the same operons. Transposons and retrotransposons are nearly entirely absent from most prokaryotic genomes.

In contrast, eukaryotic genomes are rarely smaller than 8 Mb, with unicellular species commonly having genomes in the range of 20 to 100 Mb, and metazoan and land-plant genomes often expanding beyond 1 Gb. This vast expansion of genome size is largely a consequence of the increase of intergenic DNA caused by the proliferation of multiple classes of mobile-genetic elements (essentially parasitic DNAs) and by the colonization of protein-coding genes by spliceosomal introns. The latter are intergenic DNAs that must be spliced out of transcripts to achieve productive messenger RNAs, and are completely absent from prokaryotic genomes.

Much less pronounced is the expansion of gene number in eukaryotes, which except in extreme cases of recent polyploidy is typically $< 50,000$, and quite often $< 20,000$ (Figure 24.2). The smallest eukaryotic genomes are quite similar in nature to those in prokaryotes, being devoid of mobile elements and generally highly depauperate of intronic

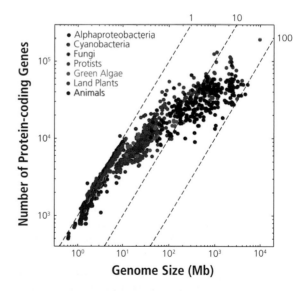

Figure 24.2 The scaling of the number of protein-coding genes per genome versus total genome size (in millions of bases) across the Tree of Life. The numbers on the dashed isoclines denote the average kb of genomic DNA per gene for points located on the lines. Data obtained from the NCBI genome summaries.

DNA. In contrast, in the enormously bloated genomes of animals and land plants, < 2% of DNA is associated with protein coding, and in the extreme cases, neighboring genes are separated by ~100 or more kb.

The origins of these massive changes in gene and genome organization across the Tree of Life have little to do with adaptive evolutionary differentiation (Lynch 2007). For example, mobile-genetic elements are a burden to host cells for two reasons: 1) each such element typically spans a few hundred to a few thousand base pairs, thereby imposing an energetic cost in terms of excess DNA synthesis; and 2) random insertions of offspring elements into novel genomic sites often have negative consequences for host-cell fitness as a consequence of gene disruption. Consistent with this view, despite having no innate immunity to mobile-element invasion, the genomes of microbial eukaryotes with relatively large effective population sizes are generally nearly devoid of such elements (as in prokaryotes). Such insertions are simply kept rare in these species by purifying natural selection. Likewise, introns, which can exceed 10 kb in length in multicellular species, are an energetic burden at both the DNA and RNA levels, and also impose upon their host genes an elevated rate of mutation to defective alleles owing to the need to maintain specific splice-site recognition sequences.

The fact that the smallest eukaryotic genomes contain fewer genes than bacterial genomes in cells of comparable size (Figure 24.2) makes clear that an expanded gene set is not a requirement for the development of a eukaryotic cell plan. The data also uphold the hypothesis that most of the changes in eukaryotic genome organization are simple by-products of the shift in the population-genetic environment permissive to the expansion of non-coding DNA. Recall from Chapter 17 that, based on the energetic costs of DNA synthesis, species with small cells and large effective population sizes are able to selectively purge insertions of non-functional DNA as small as a few base pairs. In contrast, in large eukaryotic species, natural selection is essentially blind to the relative energetic costs of insertions as large as several kb.

A final point of caution in interpreting the features of contemporary biology in terms of adaptive establishment is that the current function of a cellular trait may have little to do with its origin. For example, introns enable multicellular species to engage in tissue-specific alternative splicing of precursor messenger RNAs, thereby increasing proteome diversity without adding more genes. However, introns were present in large numbers in LECA (Rogozin et al. 2012; Irimia and Roy 2014), well before the emergence of multicellularity. Similar arguments can be made with respect to the increase of intergenic DNA as a substrate for the development of novel mechanisms of gene regulation (Lynch 2007). Thus, although an expansion of genome size in the ancestral eukaryote was likely promoted by mutation pressure despite the intrinsic disadvantages, this set the stage for later adaptive exploitation. These observations again provide compelling support for the strong role of historical contingencies with nonadaptive roots in guiding the future paths of cellular evolution.

A shake-up of genomic organization in the ancestral eukaryote

Although part of the expansion of eukaryotic gene number was a consequence of mitochondrial-to-nuclear genome transfer (Chapter 23), other additions came from duplications within the nuclear genome itself (Lynch 2007; Vosseberg et al. 2021). Incremental single-gene duplication is an ongoing process in all genomes, with rates per gene often of the same magnitude as base-substitution mutation rates per nucleotide site, i.e., in the range of 10^{-9} to 10^{-8}/gene copy/generation. However, such processes are typically balanced in the long run by an approximately equal rate of gene loss, and there is no evidence of an inexorable climb in gene number in any domain of life. There are, nonetheless, occasional episodes of massive genome expansion with more permanent effects. For example, many whole-genome duplication events have been recorded in lineages of plants, animals, yeasts, ciliates, and many other eukaryotes, although these are unknown within prokaryotes.

Of special note is a significant period of genome expansion on the road from FECA to LECA that led to the addition of several thousand genes. Given the low probability of preservation of duplicate genes (Chapter 6), the initial burst in gene number likely exceeded 10,000, with some eukaryotic lineages then losing more duplicates than others. One of the most pronounced lines of evidence for a basal expansion of eukaryotic gene number draws from the increased complexity of multimeric proteins in this lineage. Virtually every aspect of eukaryotic cell biology reveals the use of heteromeric protein complexes (with the component parts encoded by different genes) whose orthologs in prokaryotes are frequently homomeric (with all subunits encoded by the same genetic locus). A large fraction of these changes occurred in the stem eukaryote, with the added subcomponents arising by gene duplication. Already discussed in detail in prior chapters, just a few examples are summarized here.

First, as outlined in Chapter 10, many of the homomeric proteins involved in prokaryotic maintenance of chromosome integrity and replication have eukaryotic orthologs (involved in mitosis and meiosis) that assemble into heteromers consisting of subunits encoded by duplicated genes. The existence of numerous duplications in archaeal orthologs, not present in bacteria, implies that the nuclear genome of the ancestral eukaryote (likely archaeal in origin) may have been endowed with such features from the outset (Makarova and Koonin 2013). However, the dimeric form of histones in archaea transitioned to a hetero-octomeric form in eukaryotes (Henneman et al. 2018). Second, numerous proteins involved in RNA processing exhibit a similar syndrome. Consider, for example, the Sm family of proteins, which are involved in the processing of single-stranded RNAs and comprise the core of the eukaryotic spliceosome that removes introns from transcripts. Whereas bacterial Sm proteins form six- or seven-subunit homomeric rings, they are fully heteromeric in eukaryotes, with each subunit encoded by a different genetic locus (Scofield and Lynch 2008). Third, many of the key molecular machines involved in protein surveillance and processing, including chaperones, the proteasome, and the exosome, obtained their heteromeric structures in pre-LECA eukaryotes (Chapter 14). Fourth, the guardian of the nuclear environment, the nuclear-pore complex, consists of a layered series of duplications, and this is also true

of nearly every aspect of the eukaryotic vesicle-transport system (Chapter 15). Finally, the α and β subunits of structural tubulin filaments emerged prior to LECA, as did the δ and ε subunits deployed in the eukaryotic flagellum (Chapter 16).

Although many more examples could be given, these observations suffice to reveal that there was a substantial amount of gene duplication in the stem eukaryote, with an especially significant amount of duplicate preservation associated with the structural features of protein complexes. Given the frequency with which whole-genome duplications occur in modern-day lineages of eukaryotes, the possibility that one or more of such events precipitated this expansion in gene number in the ancestral eukaryote cannot be ruled out (Makarova et al. 2005; Zhou et al. 2010). Unfortunately, a formal evaluation of the matter by phylogenetic analysis is made difficult by secondary chromosome rearrangements and removals of large numbers of duplicates over the vast reach of time since the origin of eukaryotes.

Whatever the mechanism – one or more whole-genome duplications or cumulative duplications of smaller chromosomal regions – Marakova et al. (2005) suggest that the basal increase in eukaryotic gene number may have been precipitated by a cataclysmic event inducing a sudden and prolonged reduction in population size. Part of the rationale for this argument is the fact that the preservation of duplicate genes by subfunctionalization (as opposed to neofunctionalization) is facilitated in populations of small size. Recall from Chapter 6 that subfunctionalization is a process by which duplicate genes are preserved by the complementary loss of key subfunctions. Unless there is a resolution of an adaptive conflict, no fitness gain results from such a process. There are weak bioenergetic and mutational costs of relying on duplicated genes relative to single-copy genes with the same functions, but such costs will be invisible to the eyes of purifying selection in populations of sufficiently small size.

Was this expansion of the complexity of key molecular machines of any relevance to the establishment of the altered eukaryotic cell plan or to the subsequent diversification of eukaryotes into morphologically diverse descendent lineages? Evidence that increases in multimeric complexity endow organisms with superior molecular performance or with a dramatic shift in function or diversity of functions has been elusive (Chapter 13). On the other hand, as outlined in Chapter 6, the differential loss of duplicate genes in parallel lineages can passively lead to genetic-map changes that manifest as post-reproductive isolating mechanisms between sister taxa, thereby establishing novel and independent lineages. This again illustrates the point that cellular modifications arising entirely by nonadaptive mechanisms may have had a foundational role in channelling downstream evolutionary pathways, not just by the creation of novel molecular functions and structures, but by expanding the potential for novel lineage proliferation.

Multicellularity

Under the common belief that organismal complexity represents adaptive progress, the evolution of multicellularity is often viewed as the ultimate goal of natural selection. In this view, single-celled lineages (the vast majority of the Tree of Life) are condemned to a perpetual state of simplicity by the lack of one or more critical ingredients, such as the mitochondrion, rather than by a lack of selective incentive. For this reason, the evolution of multicellularity is often touted as a 'major transition' in the history of life (Maynard Smith and Szathmáry 1995). The icons of multicellularity are metazoans and land plants, made not just of multiple cells but multiple cell types. It should be remembered, however, that complex multicellularity with multiple cell types has arisen many times across the eukaryotic Tree of Life (Bonner 2001; Grosberg and Strathmann 2007; Ruiz-Trillo and Nedelcu 2015; Sebé-Pedrós et al. 2017), including on multiple occasions in fungi, red and green algae, slime molds, ciliates, and even in bacteria. Moreover, this broad phylogenetic distribution still understates the ease with which multicellularity can evolve.

Under appropriate laboratory settings, it is not difficult to coax multicellularity out of cultures of unicellular organisms on short timescales (Boraas et al. 1998; Hammerschmidt et al. 2014; Fisher et al. 2016). This is most dramatically illustrated by an experiment that self-selected for aggregations of *S. cerevisiae* cells, so-called snowflake yeast colonies,

that rapidly settle in test-tube environments (Ratcliff and Travisano 2014; Ratcliff et al. 2015). In parallel experiments, the initial transition to snowflake form results from a mutation in a single gene preventing mother–daughter cell separation. As discussed below, a challenge for the stable establishment of multicellularity is the emergence of cheater genotypes that harvest the benefits of group living without paying the price incurred by other group members (e.g., the release of group-beneficial products). However, in this particular system, the potential for genetic conflict is thwarted, as selection favors large colonies produced by fragmentation, eliminating the opportunities for transmission of cheaters that fail to engage in aggregation.

In situations like this, as long as the ecological conditions selecting for multicellularity remain, secondary genomic changes that refine such a state will be further promoted, provided the advantages of group living outweigh the interests of the selfish individual. However, it helps if the group members are close relatives. With clonally reproducing snowflake yeast, a crude form of cellular differentiation emerges, including apoptosis (suicide) of central cells, which, although disadvantageous to the individual cell, is essential to offspring colony production. In effect, the colony (rather than the individual cell) becomes the unit of selection. That is, the colony becomes the individual.

Constructing a more contrived system, Wahl and Murray (2016) developed another form of multicellular yeast with somatic differentiation. In this case, a construct was engineered to enable the recurrent production of two cell types, one from the other. Upon chemical induction, a faster-growing 'germ cell' gave rise to a slower growing 'somatic' cell type that secreted invertase into the medium. This enzyme then hydrolyzed non-utilizable sucrose into fructose and glucose molecules essential for the growth of the germ cells. Multicellularity was maintained in this system because cheating cells (which don't invest in invertase secretion) rapidly segregate into cheating-only colonies, with reduced reproductive rates.

These two examples clearly show that the key issue with respect to the evolution of complex multicellularity is not the presence of an intrinsic barrier to initial emergence. Rather, multicellularity is typically held back by the limited ecological opportunities that encourage such body plans, by the investment costs imposed by larger somas, and by the presence of internal threats that can thwart persistence after initial establishment. Before addressing these matters in more detail, a brief consideration of bacteria will make clear that multicellularity is by no means restricted to the eukaryotic domain.

Multicellularity and cooperativity in bacteria

Although they have not gone to the extremes of metazoans and land plants, prokaryotes are not immune to evolving physical or behavioral consortia of mutually beneficial cells with specialized functions (Claessen et al. 2014). In effect, the individual cells of such collectives are similar to those in asexually propagating, multicellular eukaryotes, in that cell fitness is a function of the emergent properties of the group. Cell-cell communication is typically involved, and in some cases, subsets of cells terminally differentiate to the point of relinquishing the capacity to reproduce. This is reminiscent of the altruistic behaviors observed in social animals, where the sacrifice of individual fitness is advantageous provided sufficient benefits are accrued by close relatives, e.g., warning calls in prairie dogs (Hamilton 1964a,b; Diggle et al. 2007). The one major difference between multicellularity in eukaryotes and bacteria is that in the former case, cells are usually aggregated from the outset (slime molds being exceptions), whereas in bacteria individual cells often come together to form a unit (cyanobacteria and actinomycetes being exceptions). Just a few examples are given here.

Numerous species of filamentous cyanobacteria are capable of producing specialized cells with anaerobic interiors for nitrogen fixation (heterocysts) as well as specialized spore cells (akinetes). The soil bacterium *Bacillus subtilis* undergoes periodic switching between motile unicellular and sessile filamentous states, with one transition being essentially random and the other involving a timer (Chapter 22). Multicellular magnetotactic bacteria have no known unicellular stage, and are linked together via structural filaments contained within a central acellular chamber (Keim et al. 2004;

Shapiro et al. 2011). Mycobacteria harbor a family of excretion-system loci that coordinate intercellular communication and a form of sexual reproduction known as conjugation (Gray et al. 2016). A cave-dwelling species, *Jeongeupia sacculi*, initially grows into well-organized sheets that then differentiate into mounds exceeding 1 mm in diameter, finally developing into closed volcano-like structures that explosively extrude single-celled propagules (Mizuno et al. 2022)

Particularly striking are the widespread quorum-sensing systems that bacteria use to induce intraspecific neighbors to cooperate in the production of biofilms, planktonic aggregates, or swarming motility (Ng and Bassler 2009). Such systems generally involve simple positive-feedback loops, wherein above a certain cell density, cells begin to release a chemical pheromone (a low molecular-weight chemical called an autoinducer). Pheromones are costly, and of little use at low population densities, but when the external autoinducer concentration exceeds a threshold level, cognate receptors bind them, triggering a signal-transduction cascade that coordinately turns on a group of response genes, including the gene that produces the pheromone.

The signalling molecules used in quorum sensing range widely among species. There is clear coevolution between these and their receptors (Wellington Miranda et al. 2021), although the extent to which this is driven by selection to avoid cross-species messaging or a simple consequence of drift in bivariate communication systems (Chapters 21 and 22) remains unclear. In some systems, the pheromone molecules are small enough to simply diffuse through the cell membrane and are picked up by cytoplasmic factors (e.g., lactones in *Pseudomonas*; Ng and Bassler 2009). In other cases, larger message molecules (e.g., modified peptides in Gram-positive bacteria) are bound by transporters located on the cell surface, which then transfer them to kinases used to phosphorylate response regulator proteins (Chapter 22). Some species have multiple quorum-sensing systems, providing a basis for complex environmental-sensing mechanisms and combinatorial communication (Cornforth et al. 2014; Feng et al. 2015; Jemielita et al. 2018), behavioral attributes that are typically assumed to be restricted to metazoans.

Bacteria even have memories, in the sense that prior exposure to an environmental challenge can enhance the response at a later point in time, even extending into the next generation (Chapter 22). This occurs when the lifespans of key proteins exceed the time needed for cell division (Mitchell et al. 2009; Lambert and Kussell 2014; Mathis and Ackermann 2016). The benefits of collective behavior are not always clear, but include presumed advantages associated with biofilm formation, such as aggregative food-degradation and/or tissue invasion.

Given that signalling proteins are bioenergetically expensive, such aggregative systems are vulnerable to invasion by mutant cheaters that relinquish the costs of signalling while still profiting from the public goods produced by the remaining group members, a classical weak-link in social systems. A powerful defense against such invasions is made possible by kin recognition, wherein individuals can discriminate close from distant relatives, thereby providing opportunities for selectively dispensing benefits to individuals likely harboring similar genetic constitutions, including the underlying genes for helping (Diggle et al. 2007).

Such a recognition system is known to operate in the social soil bacterium *Myxococcus xanthus*, a striking example of prokaryotic multicellularity. In this species, individual cells aggregate into swarming masses that assemble into fruiting bodies, within which only a fraction of cells produces spores. Such aggregations are genetically highly homogeneous, with kin detection and restricted cell adhesion being governed by a simple two-locus system sensitive to even single amino-acid changes (Cao and Wall 2017). Dozens to hundreds of mutually exclusive recognition groups, each operating in effect as a unit of selection, can coexist in single soil samples (Vos and Velicer 2009). Strikingly, kin discrimination in these myxobacteria involves a two-tiered system. Following the first coarse-grained recognition step, specific toxins are injected that kill recipients that do not carry the appropriate immunity genes, thereby virtually assuring that members of individual aggregates are clonemates (Vassallo and Wall 2019).

The deployment of kin-recognition groups extends beyond *Myxococcus*. For example, when grown on agar substrates, the soil bacterium *B. subtilis* forms distinct boundaries between incompatibility groups (Stefanic et al. 2015). Eukaryotic slime molds (*Dictyostelium*) are social amoebae that form stalks and fruiting bodies after solitary cells aggregate into kin-recognition groups with a simple genetic basis (Benabentos et al. 2009; Hirose et al. 2011). With collective benefits being dispensed only to close kin, these kinds of systems are organized in ways that are not much different than multicellular eukaryotes where all cells within individuals are clone-mates. However, there is a bit of double-edged sword here – too little discrimination at the time of aggregation will lead to the loss of kin-group fitness, but too much discrimination will lead to aggregates that are physically too small to reap overall group-level benefits (Márquez-Zacarías et al. 2021).

How does a single species come to consist of multiple kin-recognition groups, as opposed to operating as a single species-wide unit? One possibility, demonstrated by experiments with *B. subtilis* (Pollak et al. 2016), is that the coexistence of incompatibility groups is facilitated by a form of frequency-dependent selection. In a mixed population, rare strains with one particular sensing mechanism are unable to activate their own quorum-sensing system because the concentration of their messenger molecule is too low, but such strains can still gain advantages elicited by common strains with activated quorum-sensing systems. However, as initially rare strains increase in abundance, and their own systems become active, they then become subject to exploitation by other rare strains. In principle, the entire system facilitates coexistence, with the fitnesses of strains increasing as their relative frequencies decrease, thereby bounding them away from extinction.

Finally, it is worth noting that just as the individual cells of multicellular eukaryotes exhibit division of labor, there are many examples of bacterial consortia in which different members provide complementary resources for metabolic cross-feeding. For example, Rosenthal et al. (2018) found that genetically uniform laboratory populations of *Bacillus subtilis* can subdivide into two subpopulations of metabolically differentiated cells, one producing harmful acetate, and another converting the latter into a benign storage molecule. Likewise, simple test-tube populations of *E. coli* often develop spatial structure and cross-feeding in ways that promote coexistence of different evolved clades (Good et al. 2017; Behringer et al. 2018). In an engineered example, Pande et al. (2014) constructed two strains of *E. coli* with complementary amino-acid requirements – each lacked the ability to synthesize one amino-acid but released the other into the environment. Over a short time period, the strains evolved a division of metabolic labor that enhanced the productivity of the overall system, as the cost of producing an excess amino-acid for the partner was less than the advantage gained from being provisioned by the other.

All of these observations on bacteria highlight the importance of a key issue in evolutionary biology: the unit of selection. In principle, selection can operate at any level, discriminating among individuals within populations, among groups of individuals based on their emergent properties, or even among species on very long time scales. Moreover, a trait that is beneficial at one level need not be beneficial at another. Altruistic behavior is a case in point, as in this case the individual incurs a fitness cost while other members of the population experience a fitness gain. Generally, when conflicts like this arise, we expect selection at the individual level to prevail for the simple reason that the turnover rate of individuals exceeds that of groups (Williams 1966), but the distinction between individual and group selection becomes blurred when the interacting members of the population are close relatives.

What is an individual from an evolutionary perspective? In almost all organisms, individuality is clearly demarcated in a physical sense, e.g., by cell membranes in the case of unicellular species, and by distinct soma in the case of multicellulars. However, things are not so clear at the genetic level, as some individuals are more closely related than others, the extreme cases being monozygotic twins or members of an asexually propagating lineage. This matters because behavior that is harmful to an individual but sufficiently beneficial to close relatives can be promoted by kin selection (Wilson 1975). For example, a suicidal behavior that sufficiently elevates the fitness of multiple clone-mates (who also carry the genes for such behavior) will be advanced

by selection. Given the clonal nature of microbial aggregates, it is then clear that the unit of selection at the genetic level extends beyond individual cells to the local kin group, not greatly different from the situation in multicellular eukaryotes in which the soma is derived by clonal expansion. Thus, many microbes are multicellular in an evolutionary sense, even in the absence of structural connections between cells. The key point is that eukaryogenesis was not a prerequisite for admission into the multicellular world.

The costs of multicellularity

Certain forms of multicellularity can open up access to novel resource pools not physically possible for single cells, e.g., the ability of animals to overwhelm smaller prey items and land plants to experience the full spectrum of ambient sunlight. Multicellularity can also provide added benefits via improvements in extracellular metabolism, protection against physical environmental challenges, stress resistance, and avoiding predation, although increased size may invite the attention of still larger consumers (Tong et al. 2022). However, to be promoted by natural selection, transitions to novel ecological niches require that the benefits accrued outweigh the costs, at least in the early stages of establishment. For example, a mutation that expands dietary breadth needs to do so in a way that magnifies the net rate of acquisition of resources beyond what was possible in the ancestral state, and needs to do so for a long enough time period to move to fixation (with the number of required generations being on the order of the effective population size in the simplest of cases; Chapter 4).

With their structural support systems, animals and land plants were predisposed to proliferate across uncolonized land masses, provided the advantages outweighed the costs of cellular cooperation. However, once enough multicellular lineages had evolved diverse and refined mechanisms for occupying most niches afforded by large size, there would be no continuing evolutionary incentive for the mass movement of microbes towards multicellularity, any more than established multicellular species would be expected to undergo regressive evolution to unicellularity.

The often-stated beauty and wondrous nature of land plants and animals may have artistic appeal, but belies the underlying costs of multicellularity. The prices to be paid are not simply trade-offs involving allocations to different functions, but include more fundamental consequences of the shift in the population-genetic environment. Once propelled into a new ecological niche, the reduction in N_e incurred by larger, multicellular organisms would have had downstream effects, including subjection of the genome and associated cellular features to drift and mutational pressures towards undesirable states.

First, as discussed above, whereas large complex organisms may have access to novel resources, they also ultimately succumb to a reduction in bioenergetic capacity (Figure 24.1). Specifically, maximum growth rates decline with organism size, such that even with an unlimited food supply, large eukaryotes are unable to convert resources into biomass at the rates of smaller organisms. Thus, the long-term benefits of multicellularity do not reside in a greater capacity for assimilating biomass essential for growth and reproduction, at least not at high resource levels. Are there energetic advantages at low resource levels? Given the increased maintenance costs of larger organisms per unit time and their longer generation times (Chapter 8), this too seems unlikely.

Second, the idea that selection relentlessly promotes increased complexity to confer robustness in the face of environmental and mutational changes also comes up short, as it ignores both the physical costs of complexity and the constraints on evolutionary processes. As noted above, more complex gene structures magnify both the bioenergetic costs and the mutational vulnerability of alleles. There is no evidence that genes endowed with introns, proteins of greater length, or expansions of mobile-element families were promoted via advantages conferred upon their host genomes. Quite the contrary – the initial establishment of all of these genomic features appears to have been made possible only by the reduction in the efficiency of negative selection.

Third, as outlined in Chapter 20, while added layers of intracellular surveillance, such as DNA polymerase proof-readers, constitute an increase in complexity by any definition of the term and superficially lead to the impression of an elevation in robustness, this too is an illusion. Although the initial establishment of a new layer of protection may be promoted by positive selection, in the long run, the overall advantage is expected to dissipate as a consequence of the further accumulation of mildly deleterious mutations to the multiple-component system. This then leads to a more complex, seemingly more robust system, but with no better capacity than the originally simpler version and now with added costs of energetic investment and mutational vulnerability. A rough analogy from human behavior is Parkinson's (1958) law, which states that the work required to complete a task expands as the time necessary for completion increases.

Fourth, multicellularity imposes the necessity of costly mechanisms for the suppression of renegade cells. As described above for species with simple aggregative groups, the constant threat from emerging cheater cells is arguably one of the greatest challenges to the maintenance of multicellularity (Frank 1995; Michod 1999). This is also true for organisms with complex development, particularly for metazoans. A recurrent problem for species with somatic tissues is the emergence of cells that proliferate at the expense of the overall organism, cancer being one of the most obvious manifestations of this kind of problem. Indeed, because somatic mutations are not inherited across generations, selection for genomic-repair mechanisms appears to be substantially reduced relative to the situation in the germline, with mutation rates in somatic cells typically being elevated at least tenfold (Lynch 2010; Blokzijl et al. 2016; Abascal et al. 2021; Ueda et la. 2022). As intrasomatic selection progressively selects for clonal variants that expand at the expense of their neighboring cells, somatic mutations may also inevitably lead to senescence (Nelson and Masel 2017; Rozhok and DeGregori 2019).

The claim here is not that complex multicellularity emerged in the face of short-term disadvantages, but rather that once set in motion, the transition to such body plans creates unavoidable downstream consequences. During the evolutionary diversification of eukaryotes, short-term opportunities guided particular lineages down trajectories leading to larger body plans with structural support only possible with multicellularity. However, once established, such body plans not only resulted in lineages with reduced bioenergetic capacity, but also altered the population-genetic environment in ways that encourage the establishment of genomic and subcellular features with added expenses. In particular, multicellularity sets up a scenario by which effectively neutral (but absolutely deleterious) changes accumulate through time by mutation pressure, perhaps in some cases even being driven by positive selection for mechanisms that enhance competitive ability while reducing individual productivity (e.g., physical dominance of larger over smaller individuals). Combined with the partitioning of the full repertoire of functions essential in unicellular organisms to specialized cell types (discussed in the following section), complex multicellularity then becomes an evolutionary trap in the sense that reversion to unicellularity becomes nearly impossible.

This being said, additional considerations remain to be explored from the standpoint of population-genetic mechanisms. In particular, in the initial stages of a transition, mutations to multicellularity would need to proliferate in a background population of unicellularity. Assuming a life cycle with a sexual phase, this leaves many questions unanswered with respect to genetic compatibility of unicellular and multicellular variants. These are key issues given that, owing to initial rarity in their early phase of establishment, all mutations in sexual populations face the challenge of the fitness consequences of mixing with foreign types, effectively being tested only on heterotypic backgrounds. Thus, long-term phases of clonal propagation, haploidy, and/or local inbreeding would seem to facilitate the emergence of multicellularity. If so, how is inbreeding depression and mutational meltdown avoided in the initially small isolates of multicellulars? Are multicellular variants immediately isolated from their unicellular ancestors? Can multicellularity drive itself through the unicellular subpopulation by contagious spread through outcrossing?

The emergence of cell type specialization

As noted above, the average land plant and animal genome harbors two- to threefold more genes than in their unicellular relatives. However, little of this increase seems to be related to necessities associated with multicellularity. Consider, for example, that in the early days of genome sequencing, many thought that the human genome would contain >100,000 genes, whereas the final count is only ~20,000, about the same as in a typical nematode. The subsequent observation that numerous genes in mammalian genomes are absent from *Drosophila* and *Caenorhabditis* genomes led to the idea that such genes play key roles in the development of vertebrates. However, as a broader phylogenetic survey of genome contents began to emerge, it became clear that these apparent 'vertebrate-specific' genes were actually cases of gene loss in invertebrate lineages.

Similarly, many of the genes originally thought to have been unique to metazoans have since been found in basal unicellular lineages (e.g., choanoflagellates and ichthyosporeans), and in some cases even more deeply (Ocaña-Pallarès et al. 2022). For example, the integrin proteins used for cell–cell adhesion in metazoans are present in the apusozoan lineage (basal to both animals and fungi) but absent from fungi and choanoflagellates (basal lineages outside of metazoans) (Sebé-Pedrós et al. 2010). Likewise, many proteins initially thought to be uniquely involved in cell signalling, immune response, and development in metazoans were subsequently found to be present in choanoflagellates (King et al. 2008; Richter et al. 2018). While there are many gene gains on the branch subtending the metazoan lineage, the number of gene losses appears to be just as great, and included among them are the genes for the biosynthesis of nine amino-acids (Richter et al. 2018). Thus, the evolution of complex multicellularity did not involve a major influx of new genes. Rather, many of the genes deployed in the unique features of complex multicellular organisms are modified descendants of those used in related functions in unicellular ancestors. Notably, the genesis of a number of novel cell functions in multicellular species seems to involve the co-option and repurposing of pre-existing stress-response pathways (Love and Wagner 2022).

Under the assumption that division of labor leads to the whole being more than the sum of its parts, it has been suggested that cell-type specialization is critical to the evolution of complex multicellularity (Maynard Smith and Szathmáry 1995; Michod 1999). However, although metazoans and land plants exhibit dozens to hundreds of cell types, the increase in complexity of cellular functions is relatively small. Moreover, although some metazoan cell types have evolved new sets of tasks (e.g., nerve and bone cells), most cell types in multicellular species have simply lost a range of features found in ancestral unicellular species. A plausible explanation for such outcomes is that, whereas single-celled organisms must be capable of multitasking with respect to nutrient acquisition, avoidance of predators, dealing with unfavorable environments, etc., multicellularity offers the possibility of subfunctionalization at the level of cell-type specialization. Indeed, drawing from such patterns of partitioning, some have suggested that cell types can be classified based on shared and divergent patterns of gene expression (Arendt 2008; Arendt et al. 2016; Kishi and Parker 2021). The key point is that, as with many aspects of genome and cell biology, the emergence of multicellularity relies much more heavily on the subdivision of ancestral functions than on neofunctionalization.

A common way in which cell specialization emerges from single-celled life appears to be the conversion of a prior temporal pattern of cell life-cycle differentiation into a developmentally regulated spatial pattern (Mikhailov et al. 2009; Sebé-Pedrós et al. 2017). Examples of pre-existing phenotypic variants include bimodal phases in bacteria (Chapter 22) and alterations between asexual and sexual generations in unicellular eukaryotes. The expected scenario here is one in which cell differentiation based on external environmental cues is progressively eliminated and replaced by internal signalling mechanisms based on cell–cell communication and developmental regulation. This potential for partitioning pre-existing gene functions combined with regulatory rewiring clarifies why the evolution of multicellularity does not require a massive investment in new genomic real estate.

Just as the loss of complementary functions by subfunctionalization leads to the preservation of duplicate genes, division of labor among cell types plays a key role in ensuring the stability of complex body plans. That is, the establishment of cell-type specific features that enhance overall organismal performance but come at an expense to free-living cells (e.g., functional partitioning) reduces the likelihood of reversion to unicellularity even if this were to be beneficial. In this sense, the division of labor in multicellular organisms, whether intrinsically beneficial or not, has a ratchet-like effect on the stability of multicellularity – as cell types become more and more specialized, the likelihood of reversion of any single cell type to a form containing all ancestral-cell features necessary for independent living becomes less and less likely (Libby and Ratcliff 2014; Cooper and West 2018).

Separation of germline and somatic cells represents an extreme form of division of labor, as the former become increasingly specialized for the single function of propagation, releasing the latter from any special requirements associated with meiosis and sexual reproduction of offspring. However, to see that such an extreme form of partitioning does not depend on the prior establishment of multicellularity, one need only look to the ciliated protozoa (Chapter 10). These highly diverse and globally distributed protists are binucleate, with the diploid micronucleus serving as a transcriptionally silent germline, which recombines during sexual reproduction. The macronucleus, a highly polyploid and edited version of the micronucleus, serves as the locale of somatic gene expression, and is transmitted without recombination during asexual cell division, but after numerous cell divisions and following sexual reproduction, the old macronucleus is disposed of and replaced by a processed copy of the new micronucleus.

Studies on the Volvocales, an order of green algae that includes the unicellular *Chlamydomonas*, provide further insight into many of the above points. The order includes species with two, four, eight celled body plans, etc., extending up to large spherical forms known as *Volvox*, containing thousands of cells. Notably, the increase in colony size is accompanied by a substantial expansion in non-coding DNA in both the mitochondrial and plastid genomes, consistent with a shift in the population-genetic environment given that these organelles are uninvolved in cell-type differentiation (Smith et al. 2013). Although it has been argued that collective aggregates are more efficient at responding to weak chemotactic signals (Colizzi et al. 2020), and that flagellar stirring of boundary layers can magnify advective transport beyond the limits of diffusion (Solari et al. 2006), this leaves unexplained the predominance of the simplest body plans. Thus, the advantages of multicellularity versus unicellularity in the group remain unclear.

According to the phylogeny, the first steps in the transition to multicellularity in the Volvocales involve the evolution of colonies of undifferentiated cells, with division of labor arising secondarily (Kirk 2005; Herron et al. 2009; Herron 2016; Featherston et al. 2018; Matt and Umen 2018). Notably, the genus *Volvox* is polyphyletic, as the volvocine form has actually evolved on several independent occasions, and these changes are accompanied by significant reductions in gene number (Lindsey et al. 2021; Jiménez-Marín and Olson 2022). The most extreme form is *Volvox carteri*, which harbors 16 large nonmobile germ cells embedded in an extracellular matrix surrounded by a single-layered sphere of ~2000 terminally differentiated flagellated cells. The germ cells exhibit the greatest breadth of gene expression, relative to the more specialized transcriptomes of the somatic cells, a pattern similar to that seen in pluripotent stem cells in metazoans (Matt and Umen 2018).

Finally, we return to the common assertion that the establishment of the mitochondrion was central to the emergence of multicellularity (e.g., Bendich 2010; Lane and Martin 2010; Medini et al. 2020). The reasoning behind this speculation is diverse, ranging from the supposed bioenergetic advantages of mitochondria (shown in preceding chapters to be incorrect), to the sequestration of germ cells from mutagenic by-products of organelle respiration, to the involvement of mitochondria in various aspects of developmental control. All of these arguments ignore the existence of multicellular prokaryotes and fail to make the distinction between mechanisms of origin of organismal features and their secondary, downstream modifications. As just one example, cellular apoptosis (directed cell death) is often viewed as a unique innovation that emerged after the evolution of multicellularity. The process is triggered by events associated with the mitochondrion, and can be essential to development and cancer suppression. However, the apoptotic machinery appears to predate not only the evolution of complex multicellularity but even the origin of eukaryotes, as it is present in multiple lineages of unicellular eukaryotes as well as in bacteria (Koonin and Aravind 2002; Klim et al. 2018).

Closing Comments

We get excited by things that we see day to day. As a consequence, the majority of current research in evolutionary biology has focused on things like butterflies, vertebrates, and flowering plants. The bulk of the work done in the remaining areas of biology is concentrated on a few model organisms, such as yeast and *E. coli*, often pursued under the guise of better understanding of human biology and promoting biomedical applications.

From work on metazoans and land plants, we have learned a lot about agents of natural and sexual selection, at least for multicellular species. What remains to be understood are the molecular/cellular mechanisms underlying phenotypic divergence and the degree to which these vary among phylogenetic lineages. Although the issues are common to all organisms, unicellular organisms provide a logical starting point for such investigation. The root of the Tree of Life as well as most of the branches is unicellular, and each cell in all of today's organisms is a product of a continuous cell lineage tracing back to the beginning of biological time.

We also gravitate towards simple explanations for biodiversification. However, although it is relatively easy to understand the superficial features of the process of natural selection and to concoct adaptive stories as to how biodiversity arose, evolution is not a simple matter of natural selection pushing around mean phenotypes. Rather, the paths open to exploitation by selection are governed by the nonadaptive processes of mutation, recombination, and random genetic drift, all of which have been universal genetic forces since the origin of life. Although we do not know their relative strengths at the earliest stages in evolution, they vary by orders of magnitude among today's phylogenetic lineages, often in ways that scale with organism size, and there is no reason to think that this has not been the case for the past three billion or so years.

Thus, whereas the technical field of population genetics is often conveniently viewed as being marginal with respect to questions concerning deep phylogenetic divergence, it is actually front and centre. Likewise, while the details of molecular and cellular biology are often viewed as largely irrelevant to the ways in which populations respond to natural selection, it is precisely here that structural and functional modifications must be made to yield new phenotypes, so these details also matter.

Formally demonstrating that selection operates in an effectively deterministic manner, with the underlying details being immaterial, would be a great achievement for evolutionary biology. Indeed, such a demonstration would effectively confine the future of the field to the monotonous cataloging of selective forces operating on different organisms. However, the preceding pages challenge this view. Evolutionary processes are inherently stochastic, with the potential paths open to evolutionary exploitation depending on historical contingencies and the granularity of mutational effects relative to the power of random genetic drift. This makes the development of evolutionary theory more challenging than desirable for those confined to a pure Darwinian mode of thinking, but a desire for simplicity is no substitute for the scientific goal of describing reality.

Although no serious scientist any longer argues against the central role played by natural selection in evolutionary change, it is now clear that organisms exposed to identical selection pressures will respond in qualitatively different ways depending on their population and cellular environments, in some cases being completely impervious to selection and largely driven by mutation pressure. Once the subject of a long debate confined to molecular evolution, effectively neutral processes now appear to have played a key role in the evolution of genome architecture, cellular features, and by extension, whole-organism biology. If this view is correct, it offers a unifying and mechanistic approach to thinking about the history and dynamics of evolutionary change across the Tree of Life and across levels of organization, liberating us from the century-old tradition of assuming that all of biodiversity reflects an optimization process dictated by a supreme designer called natural selection.

Summary

- Were life given the chance to start anew, cells bounded by external envelopes and harboring polymeric genomes would likely evolve. However, the underlying elemental features of life might be entirely different from the particular hand dealt by the early Earth. With different baseline genomic features and ecological settings, population-genetic environments might differ radically as well. It is interesting to contemplate how such altered beginnings might modify downstream paths of biodiversification. However, a big enough problem for now is explaining the emergence and maintenance of organismal simplicity versus complexity across this particular planet's biosphere.

- Contrary to popular belief, evolution is not on an inexorable path to build more complex organisms. There is no evidence for an intrinsic advantage of complexity at the molecular, cellular, or organismal levels, and empirical observations show that simple bacteria often carry out the most basic functions of life more efficiently than do the more complex cells of eukaryotes.

- The emergence of eukaryotes from prokaryotic ancestors was accompanied by a substantial increase in genome size, particularly in multicellular lineages. However, the vast majority of this expansion is a consequence of the colonization by non-coding and non-functional DNA. Rather than being driven by positive selection, such changes appear to have arisen as passive by-products of a reduction in effective population size, which diminishes the ability of natural selection to oppose insertions of excess DNA.

- There is no evidence that an increase in gene number played a primary role in the establishment of the eukaryotic cell plan. Instead, it appears that the reorganization of the ancestral eukaryotic genome, possibly spawned in part by one or two whole-genome duplication events followed by gene loss and subfunctionalization, led to the emergence of a new form of genomic architecture, which in turn opened up new paths for evolutionary change. One of the most striking sets of such changes involved transitions of homomeric molecular complexes in prokaryotes to heteromeric forms in eukaryotes.

- Complex multicellularity, involving large numbers of cell types, as embodied in land plants and metazoans, is restricted to eukaryotes. With so few instances of such evolution across the Tree of Life, there is no statistical basis for arguing that such evolution was dependent on the prior emergence of the eukaryotic cell plan.

- Bacterial species exhibit a wide range of features that enable collectives of individuals to operate in ways that are more than the sum of their parts, including division of labor among different cell types. Such traits include morphological differentiation at the cellular level, quorum-sensing mechanisms that promote coordinated behavioral responses to changes in cell density, and cross-feeding of complementary resources. Many of these mutual benefits are specifically dispensed to close relatives (kin-recognition groups), an essential behavior for thwarting the emergence of cheater cells.

- Ecological opportunity played a central role in the emergence of land plants and animals, but such establishment induced downstream costs, some of which were inevitable consequences of a reduction in long-term effective population sizes. These include a reduced ability to assimilate biomass, a substantial increase in the energetic cost and mutational vulnerability of genomes, gratuitous investment in overly complex cellular features, and the constant threat of mutant somatic cheater cells.

- Arguments about the causes and consequences of multicellularity may profit from a broader consideration of the unicellular branches on the Tree of Life. For example, although multicellular organisms often have sequestered germlines, multicellularity is not a prerequisite for this condition, as all ciliates have separate germline and somatic nuclei. Likewise, although mitochondria are involved in apoptotic cell death in multicellular species, apoptosis is not a unique feature of multicellular organisms, as its antecedents can be found in unicellular lineages.

- The emergence of multicellularity relies much less on the evolution of novel gene functions than on the partitioning and/or loss of ancestral cell functions leading to the division of labor among cell types. As multicellular organisms become more complex, this leads to a situation in which reversion to unicellularity becomes effectively impossible. As a consequence, multicellular lineages become terminal branches in the Tree of Life, whereas unicellular lineages retain the capacity to become multicellular should the ecological opportunity emerge.

- A void left by the complete extinction of all animals and land plants would likely be entered quickly by multicellular descendants of unicellular species. Although the specific phenotypes to emerge under such a scenario would likely deviate from the situation today, the same syndrome of genomic changes would be expected to unfold.

Literature Cited

Abascal, F., L. M. R. Harvey, E. Mitchell, A. R. J. Lawson, S. V. Lensing, P. Ellis, A. J. C. Russell, R. E. Alcantara, A. Baez-Ortega, Y. Wang, et al. 2021. Somatic mutation landscapes at single-molecule resolution. Nature 593: 405–410.

Arendt, D. 2008. The evolution of cell types in animals: Emerging principles from molecular studies. Nat. Rev. Genet. 9: 868–882.

Arendt, D., J. M. Musser, C. V. H. Baker, A. Bergman, C. Cepko, D. H. Erwi, M. Pavlicev, G. Schlosser, S. Widder, M. D. Laubichler, and G. P. Wagner. 2016. The origin and evolution of cell types. Nat. Rev. Genet. 17: 744–757.

Behringer, M. G., B. I. Choi, S. F. Miller, T. G. Doak, J. A. Karty, W. Guo, and M. Lynch. 2018. *Escherichia coli* cultures maintain stable subpopulation structure during long-term evolution. Proc. Natl. Acad. Sci. USA 115: E4642–E4650.

Benabentos, R., S. Hirose, R. Sucgang, T. Curk, M. Katoh, E. A. Ostrowski, J. E. Strassmann, D. C. Queller, B. Zupan, G. Shaulsky, et al. 2009. Polymorphic members of the lag gene family mediate kin discrimination in *Dictyostelium*. Curr. Biol. 19: 567–572.

Bendich, A. J. 2010. Mitochondrial DNA, chloroplast DNA and the origins of development in eukaryotic organisms. Biol. Direct 5: 42.

Blokzijl, F., J. de Ligt, M. Jager, V. Sasselli, S. Roerink, N. Sasaki, M. Huch, S. Boymans, E. Kuijk, P. Prins, et al.

2016. Tissue-specific mutation accumulation in human adult stem cells during life. Nature 538: 260–264.

Bonner, J. T. 2001. First Signals: The Evolution of Multicellular Development. Princeton University Press, Princeton, NJ.

Boraas, M. E., D. B. Seale, and J. E. Boxhorn. 1998. Phagotrophy by a flagellate selects for colonial prey: A possible origin of multicellularity. Evol. Ecol. 12: 153–164.

Cao, P., and D. Wall. 2017. Self-identity reprogrammed by a single residue switch in a cell surface receptor of a social bacterium. Proc. Natl. Acad. Sci. USA 114: 3732–3737.

Claessen, D, D. E. Rozen, O. P. Kuipers, L. Søgaard-Andersen, and G. P. van Wezel. 2014. Bacterial solutions to multicellularity: A tale of biofilms, filaments and fruiting bodies. Nat. Rev. Microbiol. 12: 115–124.

Colizzi, E. S., R. M. Vroomans, and R. M. Merks. 2020. Evolution of multicellularity by collective integration of spatial information. eLife 9: e56349.

Cooper, G. A., and S. A. West. 2018. Division of labour and the evolution of extreme specialization. Nat. Ecol. Evol. 2: 1161–1167.

Cornforth, D. M., R. Popat, L. McNally, J. Gurney, T. C. Scott-Phillips, A. Ivens, S. P. Diggle, and S. P. Brown. 2014. Combinatorial quorum sensing allows bacteria to resolve their social and physical environment. Proc. Natl. Acad. Sci. USA 111: 4280–4284.

Diggle, S. P., A. S. Griffin, G. S. Campbell, and S. A. West. 2007. Cooperation and conflict in quorum-sensing bacterial populations. Nature 450: 411–414.

Featherston, J., Y. Arakaki, E. R. Hanschen, P. J. Ferris, R. E. Michod, B. J. S. C. Olson, H. Nozaki, and P. M. Durand. 2018. The 4-celled *Tetrabaena socialis* nuclear genome reveals the essential components for genetic control of cell number at the origin of multicellularity in the volvocine lineage. Mol. Biol. Evol. 35: 855–870.

Feng, L., S. T. Rutherford, K. Papenfort, J. D. Bagert, J. C. van Kessel, D. A. Tirrell, N. S. Wingreen, and B. L. Bassler. 2015. A qrr noncoding RNA deploys four different regulatory mechanisms to optimize quorum-sensing dynamics. Cell 160: 228–240.

Fisher, R. M., T. Bell, and S. A. West. 2016. Multicellular group formation in response to predators in the alga *Chlorella vulgaris*. J. Evol. Biol. 29: 551–559.

Frank, S. A. 1995. Mutual policing and repression of competition in the evolution of cooperative groups. Nature 377: 520–522.

Good, B. H., M. J. McDonald, J. E. Barrick, R. E. Lenski, and M. M. Desai. 2017. The dynamics of molecular evolution over 60,000 generations. Nature 551: 45–50.

Gray, T. A., R. R. Clark, N. Boucher, P. Lapierre, C. Smith, and K. M. Derbyshire. 2016. Intercellular communication and conjugation are mediated by ESX secretion systems in mycobacteria. Science 354: 347–350.

Grosberg, R. K., and R. R. Strathmann. 2007. The evolution of multicellularity: A minor major transition? Ann. Rev. Ecol. Evol. Syst. 38: 621–654.

Hamilton, W. D. 1964a. The genetical evolution of social behavior. I. J. Theor. Biol. 7: 1–16.

Hamilton, W. D. 1964b. The genetical evolution of social behavior. II. J. Theor. Biol. 7: 17–52.

Hammerschmidt, K., C. J. Rose, B. Kerr, and P. B. Rainey. 2014. Life cycles, fitness decoupling and the evolution of multicellularity. Nature 515: 75–79.

Henneman, B., C. van Emmerik, H. van Ingen, and R. T. Dame. 2018. Structure and function of archaeal histones. PLoS Genet. 14: e1007582.

Herron, M. D, J. D. Hackett, F. O. Aylward, and R. E. Michod. 2009. Triassic origin and early radiation of multicellular volvocine algae. Proc. Natl. Acad. Sci. USA 106: 3254–3258.

Herron, M. D. 2016. Origins of multicellular complexity: *Volvox* and the volvocine algae. Mol. Ecol. 25: 1213–1223.

Hirose, S., R. Benabentos, H. I. Ho, A. Kuspa, and G. Shaulsky. 2011. Self-recognition in social amoebae is mediated by allelic pairs of tiger genes. Science 333: 467–70.

Irimia, M., and S. W. Roy. 2014. Origin of spliceosomal introns and alternative splicing. Cold Spring Harb. Perspect. Biol. 6: a016071.

Jemielita, M., N. S. Wingreen, and B. L. Bassler. 2018. Quorum sensing controls *Vibrio cholerae* multicellular aggregate formation. eLife 7: e42057.

Jiménez-Marín B., and B. J. S. C. Olson. 2022. The curious case of multicellularity in the volvocine algae. Front. Genet. 13: 787665.

Keim, C. N., F. Abreu, U. Lins, H. Lins de Barros, and M. Farina. 2004. Cell organization and ultrastructure of a magnetotactic multicellular organism. J. Struct. Biol. 145: 254–262.

King, N., M. J. Westbrook, S. L. Young, A. Kuo, M. Abedin, J. Chapman, S. Fairclough, U. Hellsten, Y. Isogai, I. Letunic, et al. 2008. The genome of the choanoflagellate *Monosiga brevicollis* and the origin of metazoans. Nature 451: 783–788.

Kirk, D. L. 2005. A twelve-step program for evolving multicellularity and a division of labor. Bioessays 27: 299–310.

Kishi, Y., and J. Parker. 2021. Cell type innovation at the tips of the animal tree. Curr. Opin. Genet. Dev. 69: 112–121.

Klim, J., A. Gładki, R. Kucharczyk, U. Zielenkiewicz, and S. Kaczanowski. 2018. Ancestral state reconstruction of the apoptosis machinery in the common ancestor of eukaryotes. G3 (Bethesda) 8: 2121–2134.

Koonin, E. V., and L. Aravind. 2002. Origin and evolution of eukaryotic apoptosis: The bacterial connection. Cell Death Differ. 9: 394–404.

Lambert, G., and E. Kussell. 2014. Memory and fitness optimization of bacteria under fluctuating environments. PLoS Genet. 10: e1004556.

Lane, N., and W. Martin. 2010. The energetics of genome complexity. Nature 467: 929–934.

Libby, E., and W. C. Ratcliff. 2014. Ratcheting the evolution of multicellularity. Science 346: 426–427.

Lindsey, C. R., F. Rosenzweig, and M. D. Herron. 2021. Phylotranscriptomics points to multiple independent origins of multicellularity and cellular differentiation in the volvocine algae. BMC Biol. 19: 182.

Love, A. C., and G. P. Wagner. 2022. Co-option of stress mechanisms in the origin of evolutionary novelties. Evolution 76: 394–413.

Lynch, M. 2007. The Origins of Genome Architecture. Sinauer Associates, Inc., Sunderland, MA.

Lynch, M. 2010. Evolution of the mutation rate. Trends Genet. 26: 345–352.

Lynch, M. 2020. The evolutionary scaling of cellular traits imposed by the drift barrier. Proc. Natl. Acad. Sci. USA 117: 10435–10444.

Lynch, M., B. Trickovic, and C. P. Kempes. 2022. Evolutionary scaling of maximum growth rate with organism size. Sci. Reports 12: 22586.

Makarova, K. S., and E. V. Koonin. 2013. Archaeology of eukaryotic DNA replication. Cold Spring Harb. Perspect. Biol. 5: a012963.

Makarova, K. S., Y. I. Wolf, S. L. Mekhedov, B. G. Mirkin, and E. V. Koonin. 2005. Ancestral paralogs and pseudoparalogs and their role in the emergence of the eukaryotic cell. Nucleic Acids Res. 33: 4626–4638.

Márquez-Zacarías, P., P. L. Conlin, K. Tong, J. T. Pentz, and W. C. Ratcliff. 2021. Why have aggregative multicellular organisms stayed simple? Curr. Genet. 67: 871–876.

Mathis, R., and M. Ackermann. 2016. Response of single bacterial cells to stress gives rise to complex history dependence at the population level. Proc. Natl. Acad. Sci. USA 113: 4224–4229.

Matt, G. Y., and J. G. Umen. 2018. Cell-type transcriptomes of the multicellular green alga *Volvox carteri* yield insights into the evolutionary origins of germ and somatic differentiation programs. G3 (Bethesda) 8: 531–550.

Maynard Smith, J., and E. Szathmáry. 1995. The Major Transitions in Evolution. Oxford University Press, Oxford, UK.

Medini, H., T. Cohen, and D. Mishmar. 2020. Mitochondria are fundamental for the emergence of metazoans: On metabolism, genomic regulation, and the birth of complex organisms. Annu. Rev. Genet. 54: 151–166.

Michod, R. E. 1999. Darwinian Dynamics: Evolutionary Transitions in Fitness and Individuality. Princeton University Press, Princeton, NJ.

Mikhailov, K. V., A. V. Konstantinova, M. A. Nikitin, P. V. Troshin, L. Y. Rusin, V. A. Lyubetsky, Y. V. Panchin, A. P. Mylnikov, L. L. Moroz, S. Kumar, et al. 2009. The origin of Metazoa: A transition from temporal to spatial cell differentiation. Bioessays 31: 758–768.

Mitchell, A., G. H. Romano, B. Groisman, A. Yona, E. Dekel, M. Kupiec, O. Dahan, and Y. Pilpel. 2009. Adaptive prediction of environmental changes by microorganisms. Nature 460: 220–224.

Mizuno, K., M. Maree, T. Nagamura, A. Koga, S. Hirayama, S. Furukawa, K. Tanaka, and K. Morikawa. 2022. Novel multicellular prokaryote discovered next to an underground stream. eLife 11: e71920.

Nelson, P., and J. Masel. 2017. Intercellular competition and the inevitability of multicellular aging. Proc. Natl. Acad. Sci. USA 114: 12982–12987.

Ng, W. L., and B. L. Bassler. 2009. Bacterial quorum-sensing network architectures. Annu. Rev. Genet. 43: 197–222.

Ocaña-Pallarès, E., T. A. Williams, D. López-Escardó, A. S. Arroyo, J. S. Pathmanathan, E. Bapteste, D. V. Tikhonenkov, P. J. Keeling, G. J. Szöllõsi, and I. Ruiz-Trillo. 2022. Divergent genomic trajectories predate the origin of animals and fungi. Nature 609: 747–753.

Pande, S., H. Merker, K. Bohl, M. Reichelt, S. Schuster, L. F. de Figueiredo, C. Kaleta, and C. Kost. 2014. Fitness and stability of obligate cross-feeding interactions that emerge upon gene loss in bacteria. ISME J. 8: 953–962.

Parkinson, C. N. 1958. Parkinson's Law, or The Pursuit of Progress. John Murray, London, UK.

Pollak, S., S. Omer-Bendori, E. Even-Tov, V. Lipsman, T. Bareia, I. Ben-Zion, and A. Eldar. 2016. Facultative cheating supports the coexistence of diverse quorum-sensing alleles. Proc. Natl. Acad. Sci. USA 113: 2152–2157.

Ratcliff, W. C., J. D. Fankhauser, D. W. Rogers, D. Greig, and M. Travisano. 2015. Origins of multicellular evolvability in snowflake yeast. Nat. Commun. 6: 6102.

Ratcliff, W. C., and M. Travisano. 2014. Experimental evolution of multicellular complexity in *Saccharomyces cerevisiae*. BioScience 64: 383–393.

Richter, D. J., P. Fozouni, M. B. Eisen, and N. King. 2018. Gene family innovation, conservation and loss on the animal stem lineage. eLife 7: e34226.

Rogozin, I. B., L. Carmel, M. Csuros, and E. V. Koonin. 2012. Origin and evolution of spliceosomal introns. Biol. Direct. 7: 11.

Rosenthal, A. Z., Y. Qi, S. Hormoz, J. Park, S. H. Li, and M. B. Elowitz. 2018. Metabolic interactions between dynamic bacterial subpopulations. eLife 7: e33099.

Rozhok, A., and J. DeGregori. 2019. Somatic maintenance impacts the evolution of mutation rate. BMC Evol. Biol. 19: 172.

Ruiz-Trillo, I., and A. M. Nedelcu. (eds.) 2015. Evolutionary Transitions to Multicellular Life: Principles and Mechanisms. Springer-Verlag, New York, NY.

Scofield, D. G., and M. Lynch. 2008. Evolutionary diversification of the Sm family of RNA-associated proteins. Mol. Biol. Evol. 25: 2255–2267.

Sebé-Pedrós, A., B. M. Degnan, and I. Ruiz-Trillo. 2017. The origin of Metazoa: A unicellular perspective. Nat. Rev. Genet. 18: 498–512.

Sebé-Pedrós, A., A. J. Roger, F. B. Lang, N. King, and I. Ruiz-Trillo. 2010. Ancient origin of the integrin-mediated adhesion and signaling machinery. Proc. Natl. Acad. Sci. USA 107: 10142–10147.

Shapiro, O. H., R. Hatzenpichler, D. H. Buckley, S. H. Zinder, and V. J. Orphan. 2011. Multicellular photomagnetotactic bacteria. Environ. Microbiol. Rep. 3: 233–238.

Smith, D. R., T. Hamaji, B. J. Olson, P. M. Durand, P. Ferris, R. E. Michod, J. Featherston, H. Nozaki, and P. J. Keeling. 2013. Organelle genome complexity scales positively with organism size in volvocine green algae. Mol. Biol. Evol. 30: 793–797.

Solari, C. A., S. Ganguly, J. O. Kessler, R. E. Michod, and R. E. Goldstein. 2006. Multicellularity and the functional interdependence of motility and molecular transport. Proc. Natl. Acad. Sci. USA 103: 1353–1358.

Stefanic, P., B. Kraigher, N. A. Lyons, R. Kolter, and I. Mandic-Mulec. 2015. Kin discrimination between sympatric *Bacillus subtilis* isolates. Proc. Natl. Acad. Sci. USA 112: 14042–14047.

Tong, K., G. O. Bozdag, and W. C. Ratcliff. 2022. Selective drivers of simple multicellularity. Curr. Opin. Microbiol. 67: 102141.

Ueda, S., S. Yamashita, M. Nakajima, T. Kumamoto, C. Ogawa, Y. Y. Liu, H. Yamada, E. Kubo, N. Hattori, H. Takeshima, et al. 2022. A quantification method of somatic mutations in normal tissues and their accumulation in pediatric patients with chemotherapy. Proc. Natl. Acad. Sci. USA 119: e2123241119.

Vassallo, C. N., and D. Wall. 2019. Self-identity barcodes encoded by six expansive polymorphic toxin families discriminate kin in myxobacteria. Proc. Natl. Acad. Sci. USA 116: 24808–24818.

Vos, M., and G. J. Velicer. 2009. Social conflict in centimeter-and global-scale populations of the bacterium *Myxococcus xanthus*. Curr. Biol. 19: 1763–1767.

Vosseberg, J., J. J. E. van Hooff, M. Marcet-Houben, A. van Vlimmeren, L. M. van Wijk, T. Gabaldón, and B. Snel. 2021. Timing the origin of eukaryotic cellular complexity with ancient duplications. Nat. Ecol. Evol. 5: 92–100.

Wahl, M. E., and A. W. Murray. 2016. Multicellularity makes somatic differentiation evolutionarily stable. Proc. Natl. Acad. Sci. USA 113: 8362–8367.

Wellington Miranda, S., Q. Cong, A. L. Schaefer, E. K. MacLeod, A. Zimenko, D. Baker, and E. P. Greenberg. 2021. A covariation analysis reveals elements of selectivity in quorum sensing systems. eLife 10: e69169.

Williams, G. C. 1966. Adaptation and Natural Selection. Princeton Univ. Press, Princeton, NJ.

Wilson, E. O. 1975. Sociobiology: The New Synthesis. Harvard University Press, Cambridge, MA.

Zhou, X., Z. Lin, and H. Ma. 2010. Phylogenetic detection of numerous gene duplications shared by animals, fungi and plants. Genome Biol. 11: R38.

Index